Lecture Notes in Computer Science 6667

Commenced Publication in 1973

Measuring Geodesic Distances via the Uniformization Theorem

Yonathan Aflalo[1] and Ron Kimmel[2]

[1] Faculty of Electrical Engineering
[2] Faculty of Computer Science
Technion University, Haifa 3200, Israel

Abstract. According to the Uniformization Theorem any surface can be conformally mapped into a flat domain, that is, a domain with zero Gaussian curvature. The *conformal factor* indicates the local scaling introduced by such a mapping. This process could be used to compute geometric quantities in a simplified flat domain. For example, the computation of geodesic distances on a curved surface can be mapped into solving an eikonal equation in a plane weighted by the conformal factor. Solving an eikonal equation on the weighted plane can then be done with regular sampling of the domain using, for example, the *fast marching method*. The connection between the conformal factor on the plane and the surface geometry can be justified analytically. Still, in order to construct consistent numerical solvers that exploit this relation one needs to prove that the conformal factor is bounded.

In this paper we provide theoretical bounds over the conformal factor and introduce optimization formulations that control its behavior. It is demonstrated that without such a control the numerical results are unboundedly inaccurate. Putting all ingredients in the right order, we introduce a method for computing geodesic distances on a two dimensional manifold by using the fast marching algorithm on a weighed flat domain.

1 Introduction

Consistent and efficient distance computation on various domains is a key component in many important applications. Several papers tackle the problem of geodesic distance computation on triangulated surfaces. The celebrated *fast marching method* [7,9] enabled the solution in isotropic inhomogeneous domains that are regularly sampled. It was later generalized [3] through a geometric interpretation of the numerical update step, that enabled consistent and efficient computation of distances in anisotropic domains. So far, the fast marching method was implemented on manifolds given as either a triangulated mesh, a parametrized surface [10,8], or implicitly defined in a narrow band numerically sampled with a regular grid [5]. Traditionally, the *fast marching method* is executed on the manifold itself where some parametrization is provided. In these cases, usually there is some processing involved in order to overcome the iregularity of the numerical sampling. This is the case for the unfolding initialization

A.M. Bruckstein et al. (Eds.): SSVM 2011, LNCS 6667, pp. 471–482, 2012.
© Springer-Verlag Berlin Heidelberg 2012

Alfred M. Bruckstein
Bart M. ter Haar Romeny
Alexander M. Bronstein
Michael M. Bronstein (Eds.)

Scale Space and Variational Methods in Computer Vision

Third International Conference, SSVM 2011
Ein-Gedi, Israel, May 29 – June 2, 2011
Revised Selected Papers

 Springer

Volume Editors

Alfred M. Bruckstein
Technion – Israel Institute of Technology, Computer Science Department
718 Taub Building Technion, Haifa 32000, Israel
E-mail: freddy@cs.technion.ac.il

Bart M. ter Haar Romeny
Eindhoven University of Technology, Department of Biomedical Engineering
Biomedical Image Analysis and Interpretation
Den Dolech 2 – WH 2.101, 5600 MB Eindhoven, The Netherlands
E-mail: b.m.terhaarromeny@tue.nl

Alexander M. Bronstein
Tel Aviv University, Faculty of Engineering, School of Electrical Engineering
Ramat Aviv 69978, Israel
E-mail: bron@eng.tau.ac.il

Michael M. Bronstein
Università della Svizzera Italiana
Institute of Computational Science, Faculty of Informatics SI-109
Via Giuseppe Buffi 13, 6904 Lugano, Switzerland
E-mail: michael.bronstein@usi.ch

ISSN 0302-9743   e-ISSN 1611-3349
ISBN 978-3-642-24784-2   ISBN 978-3-642-24785-9 (eBook)
DOI 10.1007/978-3-642-24785-9
Springer Heidelberg Dordrecht London New York

Library of Congress Control Number: 2011939303

CR Subject Classification (1998): I.4, I.5, I.3.5, I.2.10, G.1.2, F.2.2

LNCS Sublibrary: SL 6 – Image Processing, Computer Vision, Pattern Recognition,
and Graphics

Typesetting: Camera-ready by author, data conversion by Scientific Publishing Services, Chennai, India

Printed on acid-free paper

Springer is part of Springer Science+Business Media (www.springer.com)

Preface

The International Conference on Scale Space and Variational Methods in Computer Vision (SSVM 2011, www.ssvm2011.org) was the third instance of the conference born in 2007 as the joint edition of the Scale-Space Conferences (since 1997, Utrecht) and the Workshop on Variational, Geometric, and Level set Methods (VLSM) that first took place in Vancouver in 2001. Previous editions in Ischia, Italy (2007), and Voss, Norway (2009), were very successful, materializing the hope of the first SSVM organizers, Professors Sgallari, Murli and Paragios, that the conference would 'become a reference in the domain.'

This year, SSVM was held in Kibbutz Ein-Gedi, Israel—a unique place on the shores of the Dead Sea, the global minimum on earth. Despite its small size, Israel plays an important role on the worldwide scientific arena, and in particular in the fields of computer vision and image processing.

Following the tradition of the previous SSVM conferences, we invited outstanding scientists to give keynote presentations. This year, it was our pleasure to welcome Haim Brezis (Rutgers University, USA; Technion, Israel; and University of Paris VI, France), Remco Duits, (Eindhoven University, The Netherlands), Stèphane Mallat (École Polytechnique, France), and Joachim Weickert (Saarland University, Germany). Additionally, we had six review lectures on topics of broad interest, given by experts in the field: Philip Rosenau (Tel Aviv University, Israel), Jing Yuan (University of Western Ontario, Canada), Patrizio Frosini (University of Bologna, Italy), Radu Horaud (INRIA, France), Gérard Medioni (University of Southern California, USA), and Elisabetta Carlini (La Sapienza, Italy).

From the submitted papers, 24 were selected to be presented orally and 44 as posters. Over 100 people attended the conference, representing countries from all over the world, including Austria, China, France, Germany, Hong Kong, Israel, Italy, Japan, Korea, The Netherlands, Norway, Singapore, Slovakia, Switzerland, Turkey, and the USA. We would like to thank the authors for their contributions, the members of the Program Committee for their dedication and timely reviews, and to Yana Katz and Boris Princ for local arrangements and organization without which this conference would not have been possible. Finally, our special thanks to the Technion Department of Computer Science, HP Laboratories Israel, Haifa, Rafael Ltd., Israel, BBK Technologies Ltd., Israel, and the European Community's FP7 ERC/FIRST programs for their generous sponsorship.

May–June 2011

Alfred M. Bruckstein
Bart ter Haar Romeny
Alexander M. Bronstein
Michael M. Bronstein

Organization

General Chairs

Alfred M. Bruckstein   Technion, Haifa Israel
Bart ter Haar Romeny   Eindhoven University of Technology,
            The Netherlands

Local Chairs

Alex Bronstein   Tel Aviv University, Israel
Michael Bronstein   University of Lugano, Switzerland

Scientific and Program Committee

Luis Alvares   Universidad de Las Palmas de Gran Canaria,
            Spain
Thomas Brox   University of Freiburg, Germany
Vicent Caselles   Universitat Pompeu Fabra, Spain
Raymond Chan   Chinese University of Hong Kong, SAR China
Laurent Cohen   CEREMADE, France
Daniel Cremers   Technical University of Munich, Germany
Françoise Dibos   Université Paris 13, France
Remco Duits   Eindhoven University, The Netherlands
Michael Elad   Technion, Israel
Michael Felsberg   Linkopings Universitet, Sweden
Luc Florack   Eindhoven University of Technology,
            The Netherlands
Lewis Griffin   University College London, UK
Atsushi Imiya   Chiba University, Japan
Ron Kimmel   Technion, Israel
Nahum Kiryati   Tel Aviv University, Israel
Arjan Kuijper   Fraunhofer Institute, Germany
Antonio Leitao   Federal University of Santa Catarina, Brazil
Riccardo March   CNR, Italy
Antonio Marquina   Universidad de Valencia, Spain
Gerard Medioni   USC, USA
Étienne Mémin   IRISA, France
Jan Modersitzki   McMaster University, Canada
Jean-Michel Morel   ENS Cachan, France
Mads Nielsen   IT University of Copenhagen, Denmark
Mila Nikolova   École Normale Supérieur Cachan, France
Stanley Osher   UCLA, USA

Table of Contents

Part I: Denoising and Enhancement (O1 and O5)

Fiber Enhancement in Diffusion-Weighted MRI 1
  Remco Duits, Tom C.J. Dela Haije, Arpan Ghosh, Eric J. Creusen,
  Anna Vilanova, and Bart ter Haar Romeny

Numerical Schemes for Linear and Non-linear Enhancement of
DW-MRI 14
  Eric J. Creusen, Remco Duits, and Tom C.J. Dela Haije

Optimising Spatial and Tonal Data for Homogeneous Diffusion
Inpainting 26
  Markus Mainberger, Sebastian Hoffmann, Joachim Weickert,
  Ching Hoo Tang, Daniel Johannsen, Frank Neumann, and
  Benjamin Doerr

Nonlocal Surface Fairing 38
  Serena Morigi, Marco Rucci, and Fiorella Sgallari

Nonlocal Filters for Removing Multiplicative Noise 50
  Tanja Teuber and Annika Lang

Volumetric Nonlinear Anisotropic Diffusion on GPUs 62
  Andreas Schwarzkopf, Thomas Kalbe, Chandrajit Bajaj,
  Arjan Kuijper, and Michael Goesele

A Statistical Multiresolution Strategy for Image Reconstruction 74
  Klaus Frick and Philipp Marnitz

A Variational Approach for Exact Histogram Specification 86
  Raymond Chan, Mila Nikolova, and You-Wei Wen

Joint ToF Image Denoising and Registration with a CT Surface in
Radiation Therapy 98
  Sebastian Bauer, Benjamin Berkels, Joachim Hornegger, and
  Martin Rumpf

Either Fit to Data Entries or Locally to Prior: The Minimizers
of Objectives with Nonsmooth Nonconvex Data Fidelity and
Regularization 110
  Mila Nikolova

A Study on Convex Optimization Approaches to Image Fusion 122
  Jing Yuan, Juan Shi, Xue-Cheng Tai, and Yuri Boykov

Efficient Beltrami Flow in Patch-Space 134
    Aaron Wetzler and Ron Kimmel

A Fast Augmented Lagrangian Method for Euler's Elastica Model...... 144
    Yuping Duan, Yu Wang, Xue-Cheng Tai, and Jooyoung Hahn

Deblurring Space-Variant Blur by Adding Noisy Image 157
    Iftach Klapp, Nir Sochen, and David Mendlovic

Fast Algorithms for *p*-elastica Energy with the Application to Image
Inpainting and Curve Reconstruction 169
    Jooyoung Hahn, Ginmo J. Chung, Yu Wang, and Xue-Cheng Tai

The Beltrami-Mumford-Shah Functional 183
    Nir Sochen and Leah Bar

An Adaptive Norm Algorithm for Image Restoration 194
    Daniele Bertaccini, Raymond H. Chan, Serena Morigi, and
    Fiorella Sgallari

Variational Image Denoising with Adaptive Constraint Sets 206
    Frank Lenzen, Florian Becker, Jan Lellmann, Stefania Petra, and
    Christoph Schnörr

Simultaneous Denoising and Illumination Correction via Local
Data-Fidelity and Nonlocal Regularization 218
    Jun Liu, Xue-cheng Tai, Haiyang Huang, and Zhongdan Huan

Anisotropic Non-Local Means with Spatially Adaptive Patch Shapes ... 231
    Charles-Alban Deledalle, Vincent Duval, and Joseph Salmon

Part II: Segmentation (O2)

Entropy-Sale Profiles for Texture Segmentation 243
    Byung-Woo Hong, Kangyu Ni, and Stefano Soatto

Non-local Active Contours 255
    Miyoun Jung, Gabriel Peyré, and Laurent D. Cohen

From a Modified Ambrosio-Tortorelli to a Randomized Part Hierarchy
Tree.. 267
    Sibel Tari and Murat Genctav

A Continuous Max-Flow Approach to Minimal Partitions with Label
Cost Prior 279
    Jing Yuan, Egil Bae, Yuri Boykov, and Xue-Cheng Tai

Robust Edge Detection Using Mumford-Shah Model and Binary Level
Set Method 291
    Li-Lian Wang, Yu-Ying Shi, and Xue-Cheng Tai

Bifurcation of Segment Edge Curves in Scale Space.................... 302
    Tomoya Sakai, Haruhiko Nishiguchi, Hayato Itoh, and Atsushi Imiya

Efficient Minimization of the Non-local Potts Model 314
    Manuel Werlberger, Markus Unger, Thomas Pock, and Horst Bischof

Sulci Detection in Photos of the Human Cortex Based on Learned
Discriminative Dictionaries.. 326
    *Benjamin Berkels, Marc Kotowski, Martin Rumpf, and
    Carlo Schaller*

An Efficient and Effective Tool for Image Segmentation, Total
Variations and Regularization 338
    Dorit S. Hochbaum

Supervised Scale-Invariant Segmentation (and Detection) 350
    Yan Li, David M.J. Tax, and Marco Loog

A Geodesic Voting Shape Prior to Constrain the Level Set Evolution
for the Segmentation of Tubular Trees 362
    Youssef Rouchdy and Laurent D. Cohen

Amoeba Active Contours 374
    Martin Welk

A Hybrid Scheme for Contour Detection and Completion Based on
Topological Gradient and Fast Marching Algorithms - Application to
Inpainting and Segmentation 386
    Y. Ahipo, D. Auroux, L.D. Cohen, and M. Masmoudi

A Segmentation Quality Measure Based on Rich Descriptors and
Classification Methods.. 398
    David Peles and Michael Lindenbaum

Framelet-Based Algorithm for Segmentation of Tubular Structures 411
    *Xiaohao Cai, Raymond H. Chan, Serena Morigi, and
    Fiorella Sgallari*

Weakly Convex Coupling Continuous Cuts and Shape Priors 423
    Bernhard Schmitzer and Christoph Schnörr

Part III: Image Representation and Invariants (O3)

Wasserstein Barycenter and Its Application to Texture Mixing......... 435
    Julien Rabin, Gabriel Peyré, Julie Delon, and Marc Bernot

Theoretical Foundations of Gaussian Convolution by Extended Box
Filtering 447
    *Pascal Gwosdek, Sven Grewenig, Andrés Bruhn, and
    Joachim Weickert*

From High Definition Image to Low Space Optimization 459
    Micha Feigin, Dan Feldman, and Nir Sochen

Measuring Geodesic Distances via the Uniformization Theorem 471
    Yonathan Aflalo and Ron Kimmel

Polyakov Action on (ρ, G)-Equivariant Functions Application to Color
Image Regularization 483
    Thomas Batard and Nir Sochen

Curvature Minimization for Surface Reconstruction with Features 495
    Juan Shi, Min Wan, Xue-Cheng Tai, and Desheng Wang

Should We Search for a Global Minimizer of Least Squares Regularized
with an ℓ_0 Penalty to Get the Exact Solution of an under Determined
Linear System? 508
    Mila Nikolova

Weak Statistical Constraints for Variational Stereo Imaging of Oceanic
Waves 520
    *Guillermo Gallego, Anthony Yezzi, Francesco Fedele, and
    Alvise Benetazzo*

Novel Schemes for Hyperbolic PDEs Using Osmosis Filters from Visual
Computing 532
    Kai Hagenburg, Michael Breuß, Joachim Weickert, and Oliver Vogel

Fast PDE-Based Image Analysis in Your Pocket 544
    *Andreas Luxenburger, Henning Zimmer, Pascal Gwosdek, and
    Joachim Weickert*

A Sampling Theorem for a 2D Surface 556
    Deokwoo Lee and Hamid Krim

Quadrature Nodes Meet Stippling Dots 568
    Manuel Gräf, Daniel Potts, and Gabriele Steidl

Part IV: Shape Analysis (O4)

Discrete Minimum Distortion Correspondence Problems for Non-rigid
Shape Matching 580
    *Chaohui Wang, Michael M. Bronstein,
    Alexander M. Bronstein, and Nikos Paragios*

A Correspondence-Less Approach to Matching of Deformable Shapes ... 592
    *Jonathan Pokrass, Alexander M. Bronstein, and
    Michael M. Bronstein*

Hierarchical Matching of Non-rigid Shapes 604
Dan Raviv, Anastasia Dubrovina, and Ron Kimmel

Photometric Heat Kernel Signatures................................ 616
Artiom Kovnatsky, Michael M. Bronstein,
Alexander M. Bronstein, and Ron Kimmel

Human Activity Modeling as Brownian Motion on Shape Manifold 628
Sheng Yi, Hamid Krim, and Larry K. Norris

3D Curve Evolution Algorithm with Tangential Redistribution for
a Fully Automatic Finding of an Ideal Camera Path in Virtual
Colonoscopy 640
Karol Mikula and Jozef Urbán

Distance Images and Intermediate-Level Vision 653
Pavel Dimitrov, Matthew Lawlor, and Steven W. Zucker

Shape Palindromes: Analysis of Intrinsic Symmetries in 2D Articulated
Shapes 665
Amit Hooda, Michael M. Bronstein, Alexander M. Bronstein, and
Radu P. Horaud

Kernel Bundle EPDiff: Evolution Equations for Multi-scale
Diffeomorphic Image Registration 677
Stefan Sommer, François Lauze, Mads Nielsen, and Xavier Pennec

Deformable Shape Retrieval by Learning Diffusion Kernels 689
Yonathan Aflalo, Alexander M. Bronstein,
Michael M. Bronstein, and Ron Kimmel

Part V: Optical Flow (O6)

Stochastic Models for Local Optical Flow Estimation 701
Thomas Corpetti and Étienne Mémin

Optic Flow Scale Space 713
Oliver Demetz, Joachim Weickert, Andrés Bruhn, and
Henning Zimmer

Group-Valued Regularization Framework for Motion Segmentation of
Dynamic Non-rigid Shapes 725
Guy Rosman, Michael M. Bronstein, Alexander M. Bronstein,
Alon Wolf, and Ron Kimmel

Wavelet-Based Fluid Motion Estimation 737
Pierre Dérian, Patrick Héas, Cédric Herzet, and Étienne Mémin

Multiscale Weighted Ensemble Kalman Filter for Fluid Flow
Estimation   749
   Sai Gorthi, Sébastien Beyou, Thomas Corpetti, and Étienne Mémin

Over-Prameterized Optical Flow Using a Stereoscopic Constraint   761
   Guy Rosman, Shachar Shem-Tov, David Bitton, Tal Nir,
   Gilad Adiv, Ron Kimmel, Arie Feuer, and Alfred M. Bruckstein

Robust Optic-Flow Estimation with Bayesian Inference of Model and
Hyper-parameters   773
   P. Héas, C. Herzet, and Étienne Mémin

Regularization of Positive Definite Matrix Fields Based on
Multiplicative Calculus   786
   Luc Florack

Author Index   797

Fiber Enhancement in Diffusion-Weighted MRI

Remco Duits[1,2], Tom C.J. Dela Haije[2], Arpan Ghosh[1], Eric Creusen[1,2],
Anna Vilanova[2], and Bart ter Haar Romeny[2]

Eindhoven University of Technology, The Netherlands,
[1] Department of Mathematics and Computer Science
[2] Department of Biomedical Engineering
{R.Duits,A.Ghosh,E.J.Creusen,B.M.terhaarRomeny,A.Vilanova}@tue.nl,
T.C.J.Dela.Haije@student.tue.nl

Abstract. Diffusion-Weighted MRI (DW-MRI) measures local water diffusion in biological tissue, which reflects the underlying fiber structure. In order to enhance the fiber structure in the DW-MRI data we consider both (convection-)diffusions and Hamilton-Jacobi equations (erosions) on the space $\mathbb{R}^3 \rtimes S^2$ of 3D-positions and orientations, embedded as a quotient in the group $SE(3)$ of 3D-rigid body movements. These left-invariant evolutions are expressed in the frame of left-invariant vector fields on $SE(3)$, which serves as a moving frame of reference attached to fiber fragments. The linear (convection-)diffusions are solved by a convolution with the corresponding Green's function, whereas the Hamilton-Jacobi equations are solved by a morphological convolution with the corresponding Green's function. Furthermore, we combine dilation and diffusion in pseudo-linear scale spaces on $\mathbb{R}^3 \rtimes S^2$. All methods are tested on DTI-images of the brain. These experiments indicate that our techniques are useful to deal with both the problem of limited angular resolution of DTI and the problem of spurious, non-aligned crossings in HARDI.

Keywords: DTI, HARDI, DW-MRI, sub-Riemannian geometry, scale spaces, Lie groups, Hamilton-Jacobi equations, erosion.

1   Introduction

Diffusion-Weighted Magnetic Resonance Imaging (DW-MRI) involves magnetic resonance techniques for non-invasively measuring local water diffusion in tissue. Local water diffusion profiles reflect underlying biological fiber structure. For instance in the brain, diffusion is less constrained parallel to nerve fibers than perpendicular to them.

The diffusion of water molecules in tissue over time t is described by a transition density function p_t, cf. [2]. Diffusion Tensor Imaging (DTI), introduced by Basser et al. [3], assumes that p_t can be described for each position $\mathbf{y} \in \mathbb{R}^3$ by an anisotropic Gaussian. If $\{Y_t\}$ denotes the stochastic process describing the movement of water-molecules in $\mathbb{R}^3$, then one has

$$p_t(Y_t = \mathbf{y}' \mid Y_0 = \mathbf{y}) = (4\pi t)^{-\frac{3}{2}} |\det(D(\mathbf{y}))|^{-\frac{1}{2}} e^{-\frac{(\mathbf{y}'-\mathbf{y})^T (D(\mathbf{y}))^{-1}(\mathbf{y}'-\mathbf{y})}{4t}} ,$$

A.M. Bruckstein et al. (Eds.): SSVM 2011, LNCS 6667, pp. 1–13, 2012.
© Springer-Verlag Berlin Heidelberg 2012

where D is a tensor field of positive definite symmetric tensors on $\mathbb{R}^3$ estimated from the MRI data. In a DTI-visualization one plots the surfaces

$$\mathbf{y} + \{\mathbf{v} \in \mathbb{R}^3 \mid \mathbf{v}^T D^{-1}(\mathbf{y})\mathbf{v} = \mu^2\}, \tag{1}$$

where $\mu > 0$ is fixed and $\mathbf{y} \in \Omega$ with Ω some compact subset of $\mathbb{R}^3$. From now on we refer to these surfaces as DTI-glyphs.

The drawback of this anisotropic Gaussian function approximation is the limited angular resolution of the corresponding probability density $U : \mathbb{R}^3 \times S^2 \to \mathbb{R}^+$ on positions and orientations

$$U(\mathbf{y}, \mathbf{n}) = \frac{3}{4\pi \int_\Omega \text{trace}\{D(\mathbf{y}')\}d\mathbf{y}'}\, \mathbf{n}^T D(\mathbf{y})\mathbf{n}, \quad \mathbf{y} \in \mathbb{R}^3, \mathbf{n} \in S^2. \tag{2}$$

Thereby unprocessed DTI is not capable of representing crossing fibers [2].

High Angular Resolution Diffusion Imaging (HARDI) is another recent DW-MRI technique for imaging water diffusion processes in fibrous tissues. HARDI provides for each position in $\mathbb{R}^3$ and for each orientation in S^2 an MRI signal attenuation profile, which can be related to the local diffusivity of water molecules in the corresponding direction. As a result, HARDI images are distributions $(\mathbf{y}, \mathbf{n}) \mapsto U(\mathbf{y}, \mathbf{n})$ over positions and orientations. HARDI is not restricted to functions on S^2 induced by a quadratic form and is thus capable of reflecting crossing information. See Fig. 1, where a HARDI data set is depicted using glyph visualization as defined below. In HARDI modeling the Fourier transform of the estimated transition densities is typically considered at a fixed characteristic radius (generally known as the *b-value*), cf. [8].

Definition 1. *A glyph of a distribution $U : \mathbb{R}^3 \times S^2 \to \mathbb{R}^+$ on positions and orientations is a surface $\mathcal{S}_\mu(U)(\mathbf{y}) = \{\mathbf{y} + \mu\, U(\mathbf{y}, \mathbf{n})\, \mathbf{n} \mid \mathbf{n} \in S^2\} \subset \mathbb{R}^3$ for some $\mathbf{y} \in \mathbb{R}^3$, and some suitably chosen $\mu > 0$. A glyph visualization of the distribution $U : \mathbb{R}^3 \times S^2 \to \mathbb{R}^+$ is a visualization of a field $\mathbf{y} \mapsto \mathcal{S}_\mu(U)(\mathbf{y})$ of glyphs.*

For the purpose of detecting and visualizing biological fibers, DTI and HARDI data should be enhanced by fiber propagation models such that fiber junctions are more visible and high frequency noise and non-aligned glyphs are reduced. Promising research has been done on constructing diffusion/regularization processes on the 2-sphere defined at each spatial locus separately [8,13] as an essential pre-processing step for robust fiber tracking. In these approaches position- and orientation space are decoupled, and diffusion is only performed over the angular part, disregarding spatial context. Consequently, these methods tend to fail precisely at the interesting locations where fibres cross or bifurcate.

In contrast to previous work on enhancement of DW-MRI [8,13,15,5], we consider both the spatial and the orientational part to be included in the *domain*, so a HARDI dataset is considered as a function $U : \mathbb{R}^3 \times S^2 \to \mathbb{R}^+$. Furthermore, we explicitly employ the proper underlying group structure, that arises by embedding the coupled space of positions and orientations

$$\mathbb{R}^3 \rtimes S^2 := SE(3)/(\{0\} \times SO(2))$$

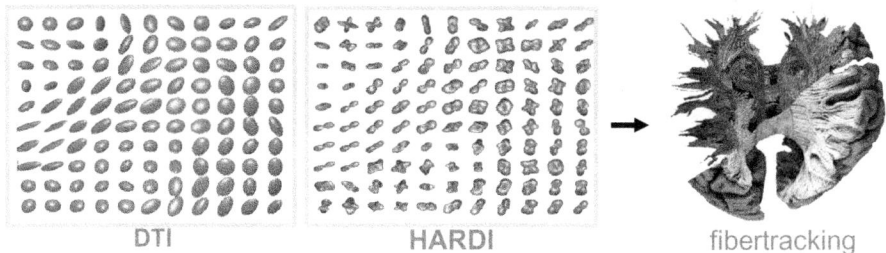

Fig. 1. This figure shows glyph visualizations of HARDI and DTI-images of a 2D-slice in the brain where neural fibers in the corona radiata cross with neural fibers in the corpus callosum. Here DTI and HARDI are visualized differently; HARDI is visualized according to Def. 1, whereas DTI is visualized using Eq. (1).

as the partition of left cosets into the group $SE(3) = \mathbb{R}^3 \rtimes SO(3)$ of 3D-rigid motions. The group product on $SE(3)$ is given by

$$(\mathbf{x}, R)(\mathbf{x}', R') = (\mathbf{x} + R\mathbf{x}', RR'),$$

for all positions $\mathbf{x}, \mathbf{x}' \in \mathbb{R}^3$ and rotations $R, R' \in SO(3)$. Throughout this article we use the following identification between the DW-MRI image $(\mathbf{y}, \mathbf{n}) \to U(\mathbf{y}, \mathbf{n})$ and functions $\tilde{U} : SE(3) \to \mathbb{R}$ given by

$$\tilde{U}(\mathbf{y}, R) = U(\mathbf{y}, R\mathbf{e}_z) \text{ with } \mathbf{e}_z = (0, 0, 1)^T. \tag{3}$$

The general advantage of our approach on $SE(3)$ is that we can enhance the original HARDI/DTI data using orientational and spatial neighborhood information simultaneously. This can create crossings in DTI data and allows a reduction of scanning directions in areas where the random walks that underly (hypo-elliptic) diffusion [11, ch:4.2] on $\mathbb{R}^3 \rtimes S^2$ yield reasonable fiber extrapolations, cf. [11,19,18] and see Fig. 2. HARDI already produces more detailed information about complex-fiber structures. Application of the same (hypo-elliptic) diffusion on HARDI then removes spurious crossings, see Fig. 3 and [19]. Here we will address the following issues that arise from our previous work [18,11,19]:

– Can we replace the grey-scale transformations [18,11,19] by Hamilton-Jacobi equations (erosions) on $\mathbb{R}^3 \rtimes S^2$ to visually sharpen the fibers in the data?
– Can we find the viscosity solutions of these Hamilton-Jacobi equations?
– Can we find analytic approximations for the viscosity solutions of these left-invariant Hamilton-Jacobi equations on $\mathbb{R}^3 \rtimes S^2$, similar to the analytic approximations of the linear left-invariant diffusions, cf. [11, ch:6.2]?
– Can we combine left-invariant diffusions and left-invariant dilations in a pseudo-linear scale space on $\mathbb{R}^3 \rtimes S^2$, generalizing [14] to DW-MRI images?

To address these issues, we introduce besides linear scale spaces, morphological and pseudo-linear scale spaces, all defined on $(\mathbb{R}^3 \rtimes S^2) \times \mathbb{R}^+$:

$$(\mathbf{y}, \mathbf{n}, t) \mapsto W(\mathbf{y}, \mathbf{n}, t) \text{ for all } \mathbf{y} \in \mathbb{R}^3, \mathbf{n} \in S^2, t > 0,$$

where the input DW-MRI image serves as initial condition $W(\mathbf{y}, \mathbf{n}, 0) = U(\mathbf{y}, \mathbf{n})$.

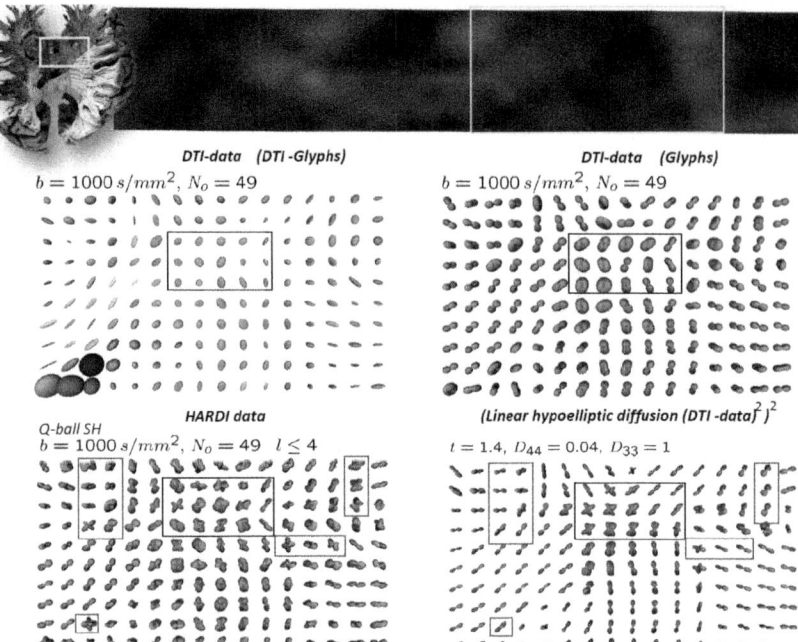

Fig. 2. DTI and HARDI data containing fibers of the corpus callosum and the corona radiata in a human brain, with b-value $1000s/mm^2$ on voxels of $(2mm)^3$, cf. [18]. We visualize a 10×16-slice of interest (162 samples on S^2 using icosahedron tessellations) from $104 \times 104 \times 10 \times (162 \times 3)$ datasets. Top row: region of interest with fractional anisotropy intensities with colorcoded DTI-principal directions. Middle row, DTI data U visualized according to Eq.(1) resp. Def. 1. Bottom row: HARDI data (Q-ball with $l \leq 4$, [8]) of the same region, hypo-elliptically diffused DTI data $(\mathbf{y}, \mathbf{n}) \mapsto W(\mathbf{y}, \mathbf{n}, t)$, Eq. (9). We applied min-max normalization of $W(\mathbf{y}, \cdot, t)$ for all positions $\mathbf{y}$.

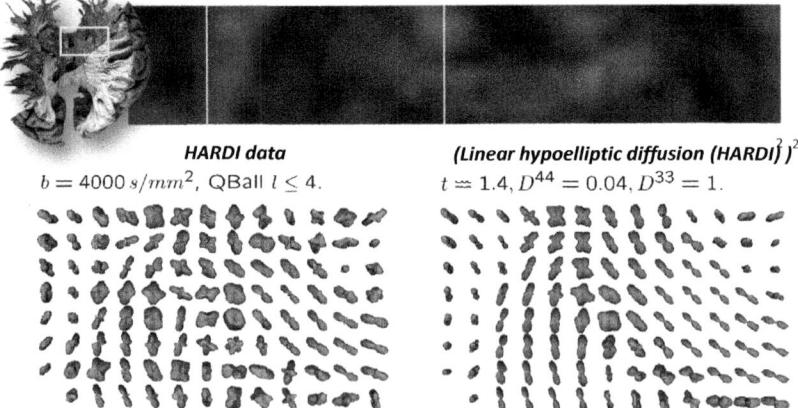

Fig. 3. Same settings as Fig:2, except for a different b-value and region of interest. The (hypo-elliptic) diffusion, Eq. (9), is applied to the HARDI dataset.

To get a preview of how these evolutions perform on the same neural DTI dataset (different slice) considered in [18], see Fig. 4, where we used

$$\mathcal{V}(U)(\mathbf{y},\mathbf{n}) = \left(\frac{U(\mathbf{y},\mathbf{n}) - U_{min}(\mathbf{y})}{U_{max}(\mathbf{y}) - U_{min}(\mathbf{y})}\right)^2, \text{ with } U_{\substack{min\\max}}(\mathbf{y}) = \min_{\substack{\max}}\{U(\mathbf{y},\mathbf{n}) \mid \mathbf{n} \in S^2\}. \quad (4)$$

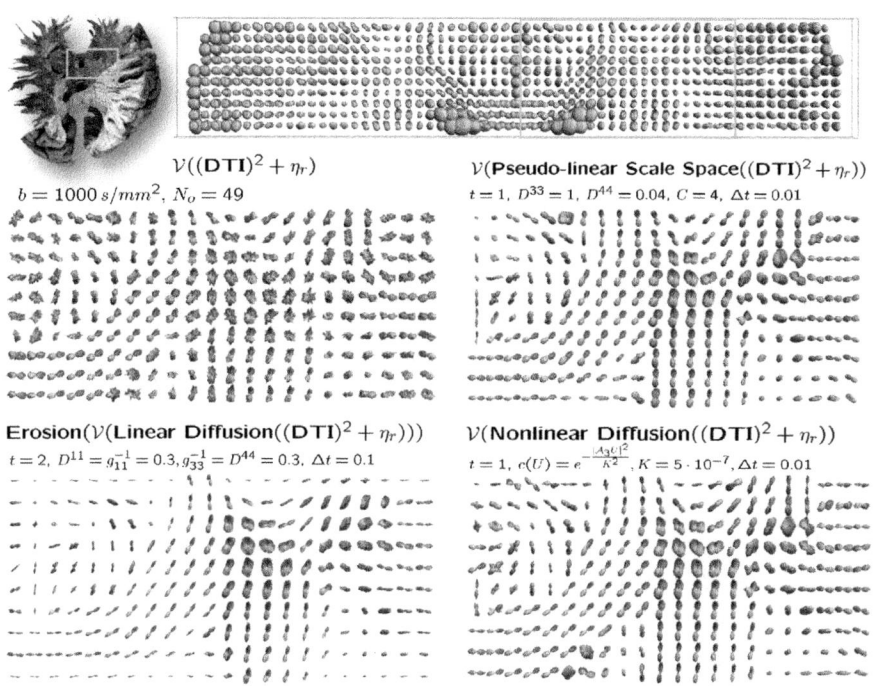

Fig. 4. DTI data of corpus callosum and corona radiata fibers in a human brain with b-value $1000s/mm^2$ on voxels of $(2mm)^3$. Top row: DTI-visualization according to Eq. (1). The yellow box contains $13 \times 22 \times 10$ glyphs with 162 orientations of the input DTI-data depicted in the left image of the middle row. This input-DTI image U is visualized using Eq. (2) and Rician noise η_r [11, Eq. 90] with $\sigma = 10^{-4}$ has been included. Operator $\mathcal{V}$ is defined in Eq. (4). Middle row, right: output of pseudo-linear scale space, Eq. (12). Bottom row, left: output erosion, Eq. (11) using the diffused DTI-data set as input, Eq. (9) with $(D^{44} = 0.04, D^{33} = 1, t = 1)$, right: output of non-linear diffusions with adaptive scalar diffusivity explained in our companion work [7]. All evolutions are implemented by finite difference schemes, [9], with step size Δt.

1.1   Motivation for Morphological Scale Spaces on $\mathbb{R}^3 \rtimes S^2$

Typically, if linear diffusions are directly applied to DTI the fibers visible in DTI are propagated in too many directions. Therefore we combined these diffusions with monotonic transformations in the codomain $\mathbb{R}^+$, such as squaring input and output cf. [11,19,18]. Visually, this produces anatomically plausible results, cf. Fig. 2

and Fig. 3, but does not allow large global variations in the data. This is often problematic around ventricle areas in the brain, where the glyphs are usually larger than those along the fibers as can be seen in the top row of Fig. 4. In order to achieve a better way of sharpening the data where global maxima do not dominate the sharpening of the data, cf. Fig. 5, we propose morphological scale spaces on $\mathbb{R}^3 \rtimes S^2$ where transport takes place orthogonal to the fibers, both spatially and spherically, see Fig. 7. The result of such an erosion after application of a linear diffusion is depicted down left in Fig. 4, where the diffusion has created crossings in the fibers and where the erosion has visually sharpened the fibers.

Fig. 5. From left to right. Noisy artificial dataset, output diffused dataset (thresholded), squared output diffused dataset as in [18,11,19], $\mathbb{R}^3 \rtimes S^2$-eroded output, Eq. (11), diffused dataset, Eq. (9).

2   A Moving Frame of Reference for Scale Spaces on $\mathbb{R}^3 \rtimes S^2$

Evolutions on DW-MRI must commute with rotations and translations. Therefore our evolutions on DW-MRI and the underlying metric-tensor are expressed in a local frame of reference attached to fiber fragments. This frame of reference $\{\mathcal{A}_1, \ldots, \mathcal{A}_6\}$ consists of 6 left-invariant vector fields on $SE(3)$ given by

$$\mathcal{A}_i \tilde{U}(\mathbf{y}, R) = \lim_{h \downarrow 0} \frac{\tilde{U}((\mathbf{y}, R) e^{hA_i}) - \tilde{U}((\mathbf{y}, R) e^{-hA_i})}{2h} \tag{5}$$

where $\{A_1, \ldots, A_6\}$ is the basis for the Lie-algebra at the unity element and $T_e(SE(3)) \ni A \mapsto e^A \in SE(3)$ is the exponential map in $SE(3)$. For more explicit, non-trivial, analytical formulas of the exponential map and corresponding left-invariant vector fields (5) we refer to [11, ch:3.3,Eq. 23–25,ch:5.1 Eq. 54]. However, these technical formulas are only needed for analytic approximation of Green's functions, see [11, ch:6]. In practice one uses finite difference approximations [11, ch:7], where spherical interpolation in between higher order tessellation of the icosahedron can be done by means of the discrete spherical harmonic transform [11, ch:7.1] or by triangular interpolation [7]. For an intuitive preview of this moving frame of reference attached to points in $\mathbb{R}^3 \rtimes S^2 = (SE(3)/(\{0\} \times SO(2)))$ we refer to Fig. 7.

The associated left-invariant dual frame $\{d\mathcal{A}^1, \ldots, d\mathcal{A}^6\}$ is determined by

$$\langle d\mathcal{A}^i, \mathcal{A}_j \rangle := d\mathcal{A}^i(\mathcal{A}_j) = \delta^i_j, i, j = 1, \ldots, 6, \tag{6}$$

where $\delta_j^i = 1$ if $i = j$ and zero else. Then all possible left-invariant metric tensors on $SE(3)$ are given by $\mathbf{G}_{(\mathbf{y},R_\mathbf{n})} = \sum\limits_{i,j=1}^{6} g_{ij} \, d\mathcal{A}^i\big|_{(\mathbf{y},R_\mathbf{n})} \otimes d\mathcal{A}^j\big|_{(\mathbf{y},R_\mathbf{n})}$ with $g_{ij} \in \mathbb{C}$ and where $\mathbf{y} \in \mathbb{R}^3$, $\mathbf{n} \in S^2$, and where $R_\mathbf{n} \in SO(3)$ is *any* rotation that maps $\mathbf{e}_z$ onto the normal $\mathbf{n} \in S^2$, i.e.

$$R_\mathbf{n}\mathbf{e}_z = \mathbf{n}. \tag{7}$$

Necessary and sufficient conditions on g_{ij} to induce a well-defined left-invariant metric on $\mathbb{R}^3 \rtimes S^2$ are derived in [9, App.E]. It turns out that the matrix $[g_{ij}]$ must be constant and diagonal $g_{ij} = \frac{1}{D^{ii}}\delta_{ij}$, $i,j = 1\ldots,6$ with $D^{ii} \in \mathbb{R}^+ \cup \infty$, with $D^{11} = D^{22}$, $D^{44} = D^{55}$, $D^{66} = 0$. The metric is thereby parameterized by the values of D^{11}, D^{33} and D^{44}, and we write the metric as a tensor product of left-invariant co-vectors:

$$\mathbf{G} = \frac{1}{D^{11}}(d\mathcal{A}^1 \otimes d\mathcal{A}^1 + d\mathcal{A}^2 \otimes d\mathcal{A}^2) + \frac{1}{D^{33}}(d\mathcal{A}^3 \otimes d\mathcal{A}^3) + \frac{1}{D^{44}}(d\mathcal{A}^4 \otimes d\mathcal{A}^4 + d\mathcal{A}^5 \otimes d\mathcal{A}^5)$$

The metric tensor on the quotient $\mathbb{R}^3 \rtimes S^2 = (SE(3)/(\{0\} \times SO(2)))$ now reads

$$\mathbf{G}_{(\mathbf{y},\mathbf{n})}\left(\sum_{i=1}^{5} c^i \mathcal{A}_i\big|_{(\mathbf{y},\mathbf{n})}, \sum_{j=1}^{5} d^j \mathcal{A}_i\big|_{(\mathbf{y},\mathbf{n})}\right) = \frac{c^1 d^1 + c^2 d^2}{D^{11}} + \frac{c^3 d^3}{D^{33}} + \frac{c^4 d^4 + c^5 d^5}{D^{44}}, \tag{8}$$

where vector fields are described by the differential operators on $C^1(\mathbb{R}^3 \times S^2)$:

$$(\mathcal{A}_j\big|_{(\mathbf{y},\mathbf{n})} U)(\mathbf{y},\mathbf{n}) = \lim_{h\to 0} \frac{U(\mathbf{y}+hR_\mathbf{n}\mathbf{e}_j,\mathbf{n}) - U(\mathbf{y}-hR_\mathbf{n}\mathbf{e}_j,\mathbf{n})}{2h},$$
$$(\mathcal{A}_{3+j}\big|_{(\mathbf{y},\mathbf{n})} U)(\mathbf{y},\mathbf{n}) = \lim_{h\to 0} \frac{U(\mathbf{y},(R_\mathbf{n}R_{\mathbf{e}_j,h})\mathbf{e}_z) - U(\mathbf{y},(R_\mathbf{n}R_{\mathbf{e}_j,-h})\mathbf{e}_z)}{2h}, \quad j = 1,2,3,$$

where $R_{\mathbf{e}_j,h}$ denotes the counter-clockwise rotation around axis $\mathbf{e}_j$ by angle h, with $\mathbf{e}_1 = (1,0,0)^T$, $\mathbf{e}_2 = (0,1,0)^T$, $\mathbf{e}_3 = (0,0,1)^T$. The induced metric is well-defined on the quotient $\mathbb{R}^3 \rtimes S^2$ since the choice of $R_\mathbf{n}$, as defined in Eq.(7), does not matter as the metric tensor is isotropic in the planes depicted in Fig. 7. In the remainder of this article we sometimes use short notation $\mathcal{A}_i$ for $\mathcal{A}_i\big|_{(\mathbf{y},\mathbf{n})}$.

3 The Evolution Equations for Scale Spaces on DW-MRI

The spherical and the spatial Laplacian can be expressed in the left-invariant vector fields as $\Delta_{S^2} = (\mathcal{A}_4)^2 + (\mathcal{A}_5)^2$ and $\Delta_{\mathbb{R}^3} = (\mathcal{A}_1)^2 + (\mathcal{A}_2)^2 + (\mathcal{A}_3)^2$. These Laplacians generate diffusion over S^2 and $\mathbb{R}^3$ separately and are thereby likely to destroy the fiber structure in DW-MRI, [11]. Therefore we introduce the following evolutions (with time $t > 0$) for respectively, linear contour enhancement[1]:

$$\begin{cases} \frac{\partial W}{\partial t}(\mathbf{y},\mathbf{n},t) = ((D^{33}(\mathcal{A}_3)^2 + D^{44}\,\Delta_{S^2})\,W)(\mathbf{y},\mathbf{n},t)\,, \\ W(\mathbf{y},\mathbf{n},0) = U(\mathbf{y},\mathbf{n})\,, \end{cases} \tag{9}$$

---

[1] Eq. (9) boils down to hypo-elliptic diffusion and corresponds to Brownian motion on $\mathbb{R}^3 \rtimes S^2$ [11, ch:4.2], generalizing some of the results in [17,10,4] to 3D.

for linear contour completion[2]:

$$\begin{cases} \frac{\partial W}{\partial t}(\mathbf{y}, \mathbf{n}, t) = ((-\mathcal{A}_3 + D^{44}\Delta_{S^2})\, W)(\mathbf{y}, \mathbf{n}, t)\,, \\ W(\mathbf{y}, \mathbf{n}, 0) = U(\mathbf{y}, \mathbf{n})\,, \end{cases} \qquad (10)$$

and for morphological scale spaces:

$$\begin{cases} \frac{\partial W}{\partial t}(\mathbf{y}, \mathbf{n}, t) = \pm \frac{1}{2\eta} \left(\mathbf{G}_{(\mathbf{y}, \mathbf{n})}^{-1} (\ dW(\cdot, \cdot, t)|_{\mathbf{y}, \mathbf{n}}, dW(\cdot, \cdot, t)|_{\mathbf{y}, \mathbf{n}}) \right)^{\eta} \\ \quad = \pm \frac{1}{2\eta} \left(D^{11} \left(|\mathcal{A}_1 W(\mathbf{y}, \mathbf{n}, t)|^2 + |\mathcal{A}_2 W(\mathbf{y}, \mathbf{n}, t)|^2 \right) + \\ \qquad D^{44} \left(|\mathcal{A}_4 W(\mathbf{y}, \mathbf{n}, t)|^2 + |\mathcal{A}_5 W(\mathbf{y}, \mathbf{n}, t)|^2 \right) \right)^{\eta}, \\ W(\mathbf{y}, \mathbf{n}, 0) = U(\mathbf{y}, \mathbf{n}), \end{cases} \qquad (11)$$

with $\eta \in [\frac{1}{2}, 1]$, cf. Fig. 6. Finally, for pseudo-linear scale spaces:

$$\begin{cases} \frac{\partial W}{\partial t}(\mathbf{y}, \mathbf{n}, t) = ((D^{33}(\mathcal{A}_3)^2 + D^{44}\Delta_{S^2})W)(\mathbf{y}, \mathbf{n}, t) + \\ C\left(D^{33} |\mathcal{A}_3 W(\mathbf{y}, \mathbf{n}, t)|^2 + D^{44} \left(|\mathcal{A}_4 W(\mathbf{y}, \mathbf{n}, t)|^2 + |\mathcal{A}_5 W(\mathbf{y}, \mathbf{n}, t)|^2 \right) \right), \\ W(\mathbf{y}, \mathbf{n}, 0) = U(\mathbf{y}, \mathbf{n}), \end{cases} \qquad (12)$$

where $C > 0$ balances between infinitesimal dilation and diffusion. These evo-

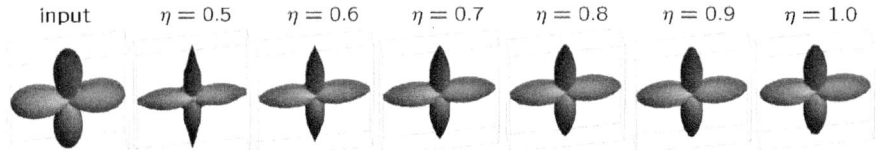

Fig. 6. The effect of $\eta \in [\frac{1}{2}, 1]$ on angular erosion Eq. (11), $D^{44} = 0.4$, $D^{11} = 0$ and $t = 0.4$. Left: original glyph, right eroded glyphs (normalized) for $\eta = 0.5, \ldots, 1.0$.

lutions are either solved by (morphological) convolution with the corresponding Green's function or by finite difference schemes. To get an intuition on the underlying geometrical ideas behind these evolutions see Fig. 7.

4   Solving the Evolutions by Convolution on $\mathbb{R}^3 \rtimes S^2$

Operators on DW-MRI data must commute with rotations and translations. This means they must be left-invariant, i.e. they must commute with $\mathfrak{L}_g$ for all $g = (\mathbf{x}, R) \in SE(3)$, where

$$(\mathfrak{L}_g U)(\mathbf{y}, \mathbf{n}) = U(g^{-1} \cdot (\mathbf{y}, \mathbf{n})) = U(R^{-1}(\mathbf{y} - \mathbf{x}), R^{-1}\mathbf{n}),$$

for all $U \in \mathbb{L}_2(\mathbb{R}^3 \rtimes S^2)$, $(\mathbf{y}, \mathbf{n}) \in \mathbb{R}^3 \rtimes S^2$. According to the theorem below, all reasonable linear, left-invariant operators on DW-MRI are $\mathbb{R}^3 \rtimes S^2$-convolutions.

---

[2] Eq. (10) boils down to hypo-elliptic convection-diffusion, direction process on $\mathbb{R}^3 \rtimes S^2$ [11, ch:4.2], generalizing [16,12].

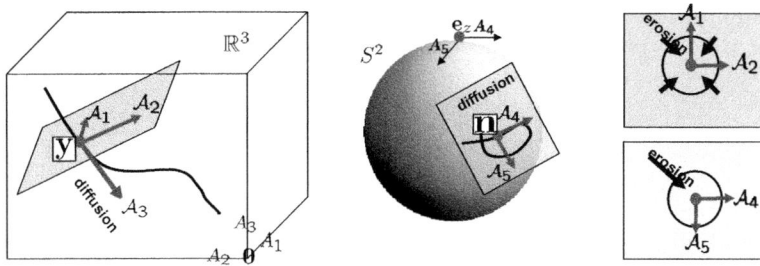

Fig. 7. A curve $[0,1] \ni s \mapsto \gamma(s) = (\mathbf{x}(s), \mathbf{n}(s)) \to \mathbb{R}^3 \rtimes S^2$ consists of a spatial part $s \mapsto \mathbf{x}(s)$ (left) and an angular part $s \mapsto \mathbf{n}(s)$ (right). Along this curve we have the moving frame of reference $\{\mathcal{A}_i|_{\tilde{\gamma}(s)}\}_{i=1}^5$ with $\tilde{\gamma}(s) = (\mathbf{x}(s), R_{\mathbf{n}(s)})$ where $R_{\mathbf{n}(s)} \in SO(3)$ is *any* rotation such that $R_{\mathbf{n}(s)}\mathbf{e}_z = \mathbf{n}(s) \in S^2$. Here $\mathcal{A}_i$, with $A_i = \mathcal{A}_i|_{(\mathbf{0},I)}$ denote the left-invariant vector fields in $SE(3)$, Eq. (5). To ensure that the diffusions and erosions do not depend on the choice $R_{\mathbf{n}(s)} \in SO(3)$, Eq. (7), these left-invariant evolution equations must be isotropic in the tangent planes $\mathrm{span}\{\mathcal{A}_1, \mathcal{A}_2\}$ and $\mathrm{span}\{\mathcal{A}_4, \mathcal{A}_5\}$. Diffusion/convection primarily takes place along $\mathcal{A}_3$ in space and (outward) in the plane $\mathrm{span}\{\mathcal{A}_4, \mathcal{A}_5\}$ tangent to S^2. Erosion takes place both inward in the tangent plane $\mathrm{span}\{\mathcal{A}_1, \mathcal{A}_2\}$ in space and inward in the plane $\mathrm{span}\{\mathcal{A}_4, \mathcal{A}_5\}$.

Theorem 1. *Let $\mathcal{K}$ be a bounded operator from $\mathbb{L}_2(\mathbb{R}^3 \rtimes S^2)$ into $\mathbb{L}_\infty(\mathbb{R}^3 \rtimes S^2)$. Then there exists an integrable kernel $k : (\mathbb{R}^3 \rtimes S^2) \times (\mathbb{R}^3 \rtimes S^2) \to \mathbb{C}$ such that $\|\mathcal{K}\|^2 = \sup\limits_{(\mathbf{y},\mathbf{n}) \in \mathbb{R}^3 \rtimes S^2} \int_{\mathbb{R}^3 \rtimes S^2} |k(\mathbf{y},\mathbf{n}; \mathbf{y}',\mathbf{n}')|^2 \mathrm{d}\mathbf{y}'\mathrm{d}\sigma(\mathbf{n}') < \infty$, and we have*

$$(\mathcal{K}U)(\mathbf{y},\mathbf{n}) = \int_{\mathbb{R}^3 \rtimes S^2} k(\mathbf{y},\mathbf{n}; \mathbf{y}',\mathbf{n}')U(\mathbf{y}',\mathbf{n}')\mathrm{d}\mathbf{y}'\mathrm{d}\sigma(\mathbf{n}') \ ,$$

for almost every $(\mathbf{y},\mathbf{n}) \in \mathbb{R}^3 \rtimes S^2$ and all $U \in \mathbb{L}_2(\mathbb{R}^3 \rtimes S^2)$. Now $\mathcal{K}_k := \mathcal{K}$ is left-invariant iff k is left-invariant, meaning

$$\forall_{g \in SE(3)}\forall_{\mathbf{y},\mathbf{y}' \in \mathbb{R}^3}\forall_{\mathbf{n},\mathbf{n}' \in S^2} : k(g \cdot (\mathbf{y},\mathbf{n}); g \cdot (\mathbf{y}',\mathbf{n}')) = k(\mathbf{y},\mathbf{n}; \mathbf{y}',\mathbf{n}').$$

Then to each positive left-invariant kernel $k : \mathbb{R}^3 \rtimes S^2 \times \mathbb{R}^3 \rtimes S^2 \to \mathbb{R}^+$ with $\int_{\mathbb{R}^3} \int_{S^2} k(\mathbf{0}, \mathbf{e}_z ; \mathbf{y}, \mathbf{n})\mathrm{d}\sigma(\mathbf{n})\mathrm{d}\mathbf{y} = 1$ we associate a unique probability density $p : \mathbb{R}^3 \rtimes S^2 \to \mathbb{R}^+$ by means of $p(\mathbf{y},\mathbf{n}) = k(\mathbf{y},\mathbf{n}; \mathbf{0},\mathbf{e}_z)$. The convolution now reads

$$\mathcal{K}_k U(\mathbf{y},\mathbf{n}) = (p *_{\mathbb{R}^3 \rtimes S^2} U)(\mathbf{y},\mathbf{n}) = \int_{\mathbb{R}^3}\int_{S^2} p(R_{\mathbf{n}'}^T(\mathbf{y}-\mathbf{y}'), R_{\mathbf{n}'}^T\mathbf{n})\, U(\mathbf{y}',\mathbf{n}')\mathrm{d}\sigma(\mathbf{n}')\mathrm{d}\mathbf{y}',$$

where σ is the surface measure on S^2 and where $R_{\mathbf{n}'} \in SO(3)$ s.t. $\mathbf{n}' = R_{\mathbf{n}'}\mathbf{e}_z$.

For a proof see [11]. Consequently, the linear scale spaces (9) and (10) are solved by $\mathbb{R}^3 \rtimes S^2$ convolution with the corresponding Green's functions! Next we extend the ideas in [6,1] and replace the $(+, \cdot)$-algebra by the $(\max, +)$-algebra to solve the morphological scale spaces (11) by dilation and erosion on $\mathbb{R}^3 \rtimes S^2$ given by

$$
\begin{aligned}
(k^- \oplus_{\mathbb{R}^3 \rtimes S^2} U)(\mathbf{y}, \mathbf{n}) &= \sup_{(\mathbf{y}', \mathbf{n}') \in \mathbb{R}^3 \rtimes S^2} \left[k^- (R_{\mathbf{n}'}^T (\mathbf{y} - \mathbf{y}'), R_{\mathbf{n}'}^T \mathbf{n}) + U(\mathbf{y}', \mathbf{n}') \right], \\
(k^+ \ominus_{\mathbb{R}^3 \rtimes S^2} U)(\mathbf{y}, \mathbf{n}) &= \inf_{(\mathbf{y}', \mathbf{n}') \in \mathbb{R}^3 \rtimes S^2} \left[k^+ (R_{\mathbf{n}'}^T (\mathbf{y} - \mathbf{y}'), R_{\mathbf{n}'}^T \mathbf{n}) + U(\mathbf{y}', \mathbf{n}') \right].
\end{aligned}
\tag{13}
$$

where dilation kernels k^- are negative and erosion kernels k^+ are positive.

Definition 2. *A viscosity solution of Eq. (11) is a bounded and continuous weak solution $W : (\mathbb{R}^3 \rtimes S^2) \times \mathbb{R}^+ \to \mathbb{R}$ of (11) such that*

1. *for any smooth function $V : (\mathbb{R}^3 \rtimes S^2) \times \mathbb{R}^+ \to \mathbb{R}$ s.t. $W - V$ attains a local maximum at $(\mathbf{y}_0, \mathbf{n}_0, t_0)$ one has $\frac{\partial V}{\partial t}(\mathbf{y}_0, \mathbf{n}_0, t_0) \mp (H(\mathrm{d}V(\cdot, \cdot, t)))(\mathbf{y}_0, \mathbf{n}_0) \le 0$.*
2. *for any smooth function $V : (\mathbb{R}^3 \rtimes S^2) \times \mathbb{R}^+ \to \mathbb{R}$ s.t. $W - V$ attains a local minimum at $(\mathbf{y}_0, \mathbf{n}_0, t_0)$ one has $\frac{\partial V}{\partial t}(\mathbf{y}_0, \mathbf{n}_0, t_0) \mp (H(\mathrm{d}V(\cdot, \cdot, t)))(\mathbf{y}_0, \mathbf{n}_0) \ge 0$.*

with Hamiltonian $H(\mathrm{d}V(\cdot, \cdot, t)) = \frac{1}{2\eta} \left(\mathbf{G}^{-1}(\mathrm{d}V(\cdot, \cdot, t), \mathrm{d}V(\cdot, \cdot, t)) \right)^\eta$ and with gradient $\mathrm{d}V(\mathbf{y}, \mathbf{n}, t) = \sum_{i=1}^5 \mathcal{A}_i V(\cdot, t)|_{(\mathbf{y}, \mathbf{n})} \ \mathrm{d}\mathcal{A}^i|_{(\mathbf{y}, \mathbf{n})}$.

Theorem 2. *The unique viscosity solutions of the Hamilton-Jacobi-Bellman equations on $\mathbb{R}^3 \rtimes S^2$, Eq. (11), are resp. given by (+ case) left-invariant erosion*

$$
W(\mathbf{y}, \mathbf{n}, t) = (k_t^{D^{11}, D^{44}, \eta, +} \ominus_{\mathbb{R}^3 \rtimes S^2} U)(\mathbf{y}, \mathbf{n})
\tag{14}
$$

and $(-$ case) left-invariant dilation $W(\mathbf{y}, \mathbf{n}, t) = (k_t^{D^{11}, D^{44}, \eta, -} \oplus_{\mathbb{R}^3 \rtimes S^2} U)(\mathbf{y}, \mathbf{n})$ where $k_t^{D^{11}, D^{44}, \eta, -} = -k_t^{D^{11}, D^{44}, \eta, +}$ and where

$$
k_t^{D^{11}, D^{44}, \eta, +}(\mathbf{y}, \mathbf{n}) := \inf_{\substack{\gamma = (\mathbf{x}(\cdot), R(\cdot)) \in C^\infty((0, t), SE(3)), \\ \gamma(0) = (0, I = R_{e_z}), \gamma(t) = (\mathbf{y}, R_\mathbf{n}), \\ \langle \mathrm{d}\mathcal{A}^3|_\gamma, \dot{\gamma} \rangle = \langle \mathrm{d}\mathcal{A}^6|_\gamma, \dot{\gamma} \rangle = 0}} \int_0^t \overline{\mathcal{L}}_\eta(\gamma(p), \dot{\gamma}(p)) \left(\frac{\mathrm{d}p}{\mathrm{d}s} \right)^{\frac{1}{2\eta - 1}} \mathrm{d}p,
\tag{15}
$$

with spatial arclength $s > 0$ (of $\mathbf{x}(\cdot)$) and with Lagrangian

$$
\begin{aligned}
\overline{\mathcal{L}}_\eta(\gamma(p), \dot{\gamma}(p)) &:= \frac{2\eta - 1}{2\eta} \left(\frac{1}{D^{11}}((\dot{\gamma}^1(p))^2 + (\dot{\gamma}^2(p))^2) + \frac{1}{D^{44}}((\dot{\gamma}^4(p))^2 + (\dot{\gamma}^5(p))^2) \right)^{\frac{\eta}{2\eta - 1}} \\
&= \frac{2\eta - 1}{2\eta}
\end{aligned}
$$

with $\dot{\gamma}^i(p) = \langle \mathrm{d}\mathcal{A}^i|_{\gamma(p)}, \dot{\gamma}(p) \rangle$ and with $\mathbb{R}^3 \rtimes S^2$- "erosion arclength" p given by

$$
p(\tau) = \int_0^\tau \sqrt{\mathbf{G}_{\gamma(\tilde{\tau})}(\dot{\gamma}(\tilde{\tau}), \dot{\gamma}(\tilde{\tau}))} \, \mathrm{d}\tilde{\tau} = \int_0^\tau \sqrt{\sum_{i \in \{1,2,4,5\}} \frac{|\langle \mathrm{d}\mathcal{A}^i|_{\gamma(\tilde{\tau})}, \dot{\gamma}(\tilde{\tau}) \rangle|^2}{D^{ii}}} \, \mathrm{d}\tilde{\tau}.
\tag{16}
$$

For proof see our technical report [9, App.B]. The Lagrangian in Theorem 2 relates to the Hamiltonian in Def. 2 by Fenchel transform [1] on the Lie algebra of left-invariant vector fields on $SE(3)$, for all $\frac{1}{2} \le \eta \le 1$, cf. [9, App.B,ch:8.3].

A *sub-Riemannian manifold* is a Riemannian manifold with the extra constraint that certain subspaces of the tangent space are prohibited. For example, curves in $(SE(3), d\mathcal{A}^1, d\mathcal{A}^2, d\mathcal{A}^6)$ are curves $\tilde{\gamma} : [0,1] \to SE(3)$ such that

$$\langle d\mathcal{A}^1 |_{\tilde{\gamma}(s)}, \dot{\tilde{\gamma}}(s) \rangle = \langle d\mathcal{A}^2 |_{\tilde{\gamma}(s)}, \dot{\tilde{\gamma}}(s) \rangle = \langle d\mathcal{A}^6 |_{\tilde{\gamma}(s)}, \dot{\tilde{\gamma}}(s) \rangle = 0, \qquad (17)$$

for all $s \in [0,1]$. Curves satisfying (17) are called *horizontal curves* in $SE(3)$ and we depicted such a curve in Fig. 7.

In [11, ch:6.2] we have analytically approximated the Green's functions of contour completion, Eq. (10) and contour enhancement, Eq. (9) that take place on the sub-Riemannian manifold $(SE(3), d\mathcal{A}^1, d\mathcal{A}^2, d\mathcal{A}^6)$. These Green's functions coincide with the diffusion kernels for respectively the direction process and Brownian motion om $\mathbb{R}^3 \times S^2$ in probability theory, [9, ch:8]. Moreover, in [9, App. A, B, C] we applied similar techniques to approximate the dilation/erosion kernels that describe the growth of balls in the sub-Riemannian manifold $(SE(3), d\mathcal{A}^3, d\mathcal{A}^6)$, cf. Theorem 2. Again there exists a connection with probability theory as these erosion kernels coincide with transition-cost densities of Bellman-processes defined on $(SE(3), d\mathcal{A}^3, d\mathcal{A}^6)$, see [9, ch:8.3].

The next theorem provides some of the *approximations* for the Green's functions, cf. [11, ch:6.2], [9, App.B]:

Theorem 3. *Let $1 \geq \eta > \frac{1}{2}$, $D^{11} > 0, D^{33} > 0, D^{44} > 0$. Then for the morphological erosion (+) and dilation kernel (-) on $\mathbb{R}^3 \rtimes S^2$ one can use the following approximation*

$$k_t^{D^{11}, D^{44}, \pm}(\mathbf{y}, \mathbf{n}) \approx \frac{(2\eta-1)(c^{-2\eta}t)^{-\frac{1}{2\eta-1}}}{\pm 2\eta} \left(\left(\frac{|c^1|^2 + |c^2|^2}{D^{11}} + \frac{|c^4|^2 + |c^5|^2}{D^{44}} \right)^2 + \frac{|c^3|^2}{D^{11} D^{44}} \right)^{\frac{\eta}{2(2\eta-1)}}$$

for $t > 0$ small, where $\tilde{\mathbf{n}}(\tilde{\beta}, \tilde{\gamma})) = (\sin \tilde{\beta}, -\cos \tilde{\beta} \sin \tilde{\gamma}, \cos \tilde{\beta} \cos \tilde{\gamma})^T$, $c > 0$, with $\tilde{\beta} \in (-\frac{\pi}{2}, \frac{\pi}{2})$, $\tilde{\gamma} \in (-\frac{\pi}{2}, \frac{\pi}{2})$. For the Green's functions of Eq. (9), the heat kernels, we have the approximation

$$p_t^{D^{33}, D^{44}}(\mathbf{y}, \mathbf{n}) \approx \frac{1}{16\pi^2 (D^{33})^2 (D^{44})^2 t^4} e^{-\frac{\sqrt{\frac{|c^1|^2+|c^2|^2}{D^{33}D^{44}} + \frac{|c^6|^2}{D^{44}} + \left(\frac{(c^3)^2}{D^{33}} + \frac{|c^4|^2+|c^5|^2}{D^{44}} \right)^2}}{4t}}.$$

In both cases the functions $c^i := c^i(\mathbf{y}, \tilde{\alpha} = 0, \tilde{\beta}, \tilde{\gamma})$ are given by

$$\mathbf{c}^{(1)} := (c^1, c^2, c^3)^T = \mathbf{y} - \frac{1}{2} \mathbf{c}^{(2)} \times \mathbf{y} + \tilde{q}^{-2}(1 - (\frac{\tilde{q}}{2}) \cot(\frac{\tilde{q}}{2})) \, \mathbf{c}^{(2)} \times (\mathbf{c}^{(2)} \times \mathbf{y}),$$
$$\mathbf{c}^{(2)} := (c^4, c^5, c^6)^T = \frac{\tilde{q}}{\sin(\tilde{q})} \left(\sin \tilde{\gamma} \cos^2(\frac{\tilde{\beta}}{2}), \, \sin \tilde{\beta} \cos^2(\frac{\tilde{\gamma}}{2}), \, \frac{1}{2} \sin \tilde{\gamma} \sin \tilde{\beta} \right)^T$$
with $\tilde{q} = \arcsin \sqrt{\cos^4(\tilde{\gamma}/2) \sin^2(\tilde{\beta}) + \cos^2(\tilde{\beta}/2) \sin^2(\tilde{\gamma})}$.

For $\eta = \frac{1}{2}$ we obtain the erosion kernel approximation (take $\eta \downarrow \frac{1}{2}$ in Theorem 3):

$$k_t^{D^{11}, D^{44}, \frac{1}{2}, -}(\mathbf{y}, \mathbf{n}) \approx \begin{cases} \infty & \text{if } \sqrt{\left(\frac{|c^1|^2+|c^2|^2}{D^{11}} + \frac{|c^4|^2+|c^5|^2}{D^{44}} \right)^2 + \frac{|c^3|^2}{D^{11}D^{44}}} \geq t^2, \\ 0 & \text{else.} \end{cases} \qquad (18)$$

The solutions of the pseudo-linear spaces, Eq. (12), are given by

$$W(\mathbf{y}, \mathbf{n}, t) = \chi_C^{-1}((e^{t(D^{33}(\mathcal{A}_3)^2 + D^{44}\Delta_{S^2})} \circ \chi_C \circ U)(\mathbf{y}, \mathbf{n})),$$

i.e. a linear hypo-elliptic diffusion conjugated with the grey-value transformation $\chi_C(I) = \frac{e^{CI}-1}{e^C-1}$ if $C \neq 0$ and $\chi_C(I) = I$ if $C = 0$, $I \in \mathbb{R}^+$, cf. [9, ch], [14].

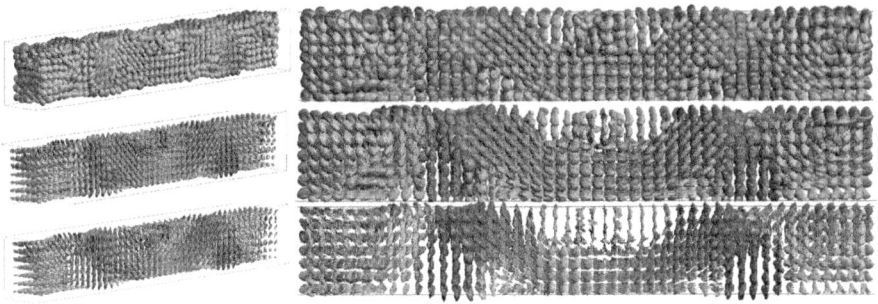

Fig. 8. 1st row: Input DTI-data. 2nd row: Output squared linear diffusion on squared data-set. 3rd row: Output erosion applied to the diffused dataset in the 2nd row.

5 Conclusion

We have developed crossing preserving, rotation- and translation covariant scale spaces on DW-MRI. The underlying evolutions are convection-diffusion equations and Hamilton-Jacobi-Bellman equations of respectively stochastic and cost processes cf. [9], on the space of positions and orientations $\mathbb{R}^3 \rtimes S^2$. These scale spaces are expressed in a moving frame of reference allowing (hypo-elliptic) diffusion along fibers and erosion orthogonal to fibers. They extrapolate complex fiber-structures (crossings) from DTI, while reducing non-aligned crossings in HARDI. They can be implemented by finite difference methods [7] (e.g. Fig. 4 and Fig. 8), or by convolutions with analytic kernels (e.g. Fig. 1 and 2).

References

1. Akian, M., Quadrat, J., Viot, M.: Bellman processes. Lecture Notes in Control and Information Science 199, 302–311 (1994)
2. Alexander, D.C., Barker, G.J., Arridge, S.R.: Detection and modeling of non-gaussian apparent diffusion coefficient profiles in human brain data. Magnetic Rosonance in Medicine 48, 331–340 (2002)
3. Basser, P.J., Mattiello, J., Lebihan, D.: MR diffusion tensor spectroscopy and imaging. Biophysical Journal 66, 259–267 (1994)
4. Boscain, U., Duplaix, J., Gauthier, J.P., Rossi, F.: Anthropomorphic image reconstruction via hypoelliptic diffusion. (accepted for publication in JMIV, to appear)
5. Burgeth, B., Pizarro, L., Didas, S., Weickert, J.: 3d-coherence-enhancing diffusion filtering for matrix fields. In: Florack, Duits, Jongbloed, van Lieshout, Davies (eds.) Locally Adaptive Filters in Signal and Image Processing (to appear, 2011)

6. Burgeth, B., Weickert, J.: An explanation for the logarithmic connection between linear and morphological systems. In: Griffin, L.D., Lillholm, M. (eds.) Scale-Space 2003. LNCS, vol. 2695, pp. 325–339. Springer, Heidelberg (2003)
7. Creusen, E.J., Duits, R., Dela Haije, T.C.J.: Numerical schemes for linear and non-linear enhancement of DW-MRI. In: Bruckstein, A.M., et al. (eds.) SSVM 2011. LNCS, vol. 6667, pp. 14–25. Springer, Heidelberg (2011)
8. Descoteaux, M.: High Angular Resolution Diffusion MRI: From Local Estimation to Segmentation and Tractography. PhD thesis, Universite de Nice (2008)
9. Duits, R., Creusen, E.J., Ghosh, A., Dela Haije, T.C.J.: Diffusion, convection and erosion on SE(3)/(0×SO(2)) and their application to the enhancement of crossing fibers. Published on Arxiv, nr. arXiv:1103.0656v4 also available as CASA-report (2011), `http://arxiv.org/abs/1103.0656v4`, `http://www.win.tue.nl/analysis/reports/rana11-22.pdf`
10. Duits, R., Franken, E.M.: Left-invariant parabolic evolutions on SE(2) and contour enhancement via invertible orientation scores, part I: Linear left-invariant diffusion equations on SE(2). Quarterly of Appl. Math., A.M.S. 68, 255–292 (2010)
11. Duits, R., Franken, E.M.: Left-invariant diffusions on the space of positions and orientations and their application to crossing-preserving smoothing of HARDI images. International Journal of Computer Vision, IJCV 92(3), 231–264 (2011), `http://www.springerlink.com/content/511j713042064t35/`
12. Duits, R., van Almsick, M.A.: The explicit solutions of linear left-invariant second order stochastic evolution equations on the 2d-Euclidean motion group (April 2008)
13. Florack, L.: Codomain scale space and regularization for high angular resolution diffusion imaging. In: CVPR Workshop on Tensors in Image Processing and Computer Vision, Anchorage, Alaska, The United States, vol. 20 (June 2008)
14. Florack, L.M.J., Maas, R., Niessen, W.J.: Pseudo-linear scale-space theory. International Journal of Computer Vision 31(2/3), 247–259 (1999)
15. Gur, Y., Sochen, N.: Regularizing Flows over Lie groups. Journal of Mathematical Imaging and Vision 33(2), 195–208 (2009)
16. Mumford, D.: Elastica and computer vision. In: Algebraic Geometry and Its Applications, pp. 491–506. Springer, Heidelberg (1994)
17. Petitot, J.: Neurogéomètrie de la vision–Modèles mathématiques et physiques des architectures fonctionelles. Les Éditions de l'École Polytechnique (2008)
18. Prckovska, V., Rodrigues, P., Duits, R., Vilanova, A., ter Haar Romeny, B.M.: Extrapolating fiber crossings from DTI data. Can we infer similar fiber crossings as in HARDI? In: CDMRI 2010 Proc. MICCAI Workshop Computational Diffusion MRI, China (2010)
19. Rodrigues, P., Duits, R., Vilanova, A., ter Haar Romeny, B.M.: Accelerated Diffusion Operators for Enhancing DW-MRI. In: Eurographics Workshop on Visual Computing for Biology and Medicine, Leipzig, Germany, pp. 49–56 (2010)

Numerical Schemes for Linear and Non-linear Enhancement of DW-MRI

Eric J. Creusen[1,2], Remco Duits[1,2], and Tom C.J. Dela Haije[2]

[1] Eindhoven University of Technology, The Netherlands
Department of Mathematics and Computer Science
[2] Department of Biomedical Engineering
{E.J.Creusen,R.Duits}@tue.nl, T.C.J.Dela.Haije@student.tue.nl

Abstract. We consider left-invariant diffusion processes on DTI data by embedding the data into the space $\mathbb{R}^3 \rtimes S^2$ of 3D positions and orientations. We then define and solve the diffusion equation in a moving frame of reference defined using left-invariant derivatives. The diffusion process is made adaptive to the data in order to do Perona-Malik-like edge preserving smoothing, which is necessary to handle fiber structures near regions of large isotropic diffusion such as the ventricles of the brain. The corresponding partial differential systems are solved using finite difference stencils. We include experiments both on synthetic data and on DTI-images of the brain.

Keywords: DTI, DW-MRI, scale spaces, Lie groups, adaptive diffusion, Perona-Malik diffusion.

1 Introduction

Diffusion-Weighted Magnetic Resonance Imaging (DW-MRI) are MRI techniques for non-invasively measuring local water diffusion inside tissue. The water diffusion profiles of the imaged area allow inference of the underlying tissue structure. For instance in the brain, diffusion is less constrained parallel to nerve fibers than perpendicular to them, and so the water diffusion gives information about the fiber structures present. This allows for the extraction of clinical information concerning biological fiber structures from DW-MRI scans.

The diffusion of water molecules in tissue over some time interval t can be described by a diffusion propagator which is the probability density function $\mathbf{y} \mapsto p_t(Y_t = \mathbf{y} \,|\, Y_0 = \mathbf{y}_0)$ of finding a water molecule at time $t \geq 0$ and at position $\mathbf{y}$ given that it started at $\mathbf{y}_0$ on $t = 0$. Here the family of random variables $(Y_t)_{t \geq 0}$ describes the distribution of water molecules over time. The function p_t can be directly related to MRI signal attenuation of diffusion weighted image sequences and so can be estimated given enough measurements. The exact methods to do this are described by e.g. Alexander [1].

Diffusion Tensor Imaging(DTI), introduced by Basser et al. [2] assumes that p_t can be described for each voxel by an anisotropic Gaussian function, i.e.

$$p_t(Y_t = \mathbf{y} \,|\, Y_0 = \mathbf{y}_0) = \frac{1}{\sqrt{(4\pi t)^3 \det(D(\mathbf{y}_0))}} \exp\left(\frac{-(\mathbf{y} - \mathbf{y}_0)^T D(\mathbf{y}_0)^{-1} (\mathbf{y} - \mathbf{y}_0)}{4t} \right),$$

A.M. Bruckstein et al. (Eds.): SSVM 2011, LNCS 6667, pp. 14–25, 2012.

where D is a tensor field of 3×3 positive definite symmetric tensors that each describe the local Gaussian diffusion process. The tensors contain 6 parameters for each voxel, which means the tensor field requires at least 6 DW-MRI images.

The drawback of approximating p_t with an anisotropic Gaussian function is that it is only able to estimate one preferred direction per voxel. However, if more complex structures such as crossing, kissing or diverging fibers are present the Gaussian assumption fails, as was demonstrated by Alexander et al. [1]. In practice though, large areas of the brain can be approximated well with DTI tensors and in the regions where complex fiber structures are present the diffusion profile can be inferred by taking contextual information into consideration [12].

Since DTI tensors cannot contain information regarding crossings the DTI data needs to be represented in a form that does allow crossing fiber structures. A representation that suits these demands can be obtained by viewing a DTI image as a probability function on positions and orientation: $U : \mathbb{R}^3 \times S^2 \to \mathbb{R}^+$, where $S^2 = \{\mathbf{x} \in \mathbb{R}^3 \mid ||\mathbf{x}|| = 1\}$ denotes the 2-sphere and where for each position $\mathbf{y}$ and orientation $\mathbf{n}$, $U(\mathbf{y}, \mathbf{n})$ gives the probability density that a water molecule starts at $\mathbf{y}$ and travels in direction $\mathbf{n}$. Using the Gaussian assumption this distribution is given by

$$U(\mathbf{y}, \mathbf{n}) = \frac{3}{4\pi \int_\Omega \operatorname{trace}\{D(\mathbf{y'})\}d\mathbf{y'}} \mathbf{n}^T D(\mathbf{y})\mathbf{n}, \; \mathbf{y} \in \mathbb{R}^3, \mathbf{n} \in S^2.$$

Such functions on position and orientation are then visualized by the surfaces $S_\mu(U)(\mathbf{x}) = \{\mathbf{x} + \mu\, U(\mathbf{x}, \mathbf{n})\, \mathbf{n} \mid \mathbf{n} \in S^2\} \subset \mathbb{R}^3$, which are called glyphs. A figure is generated by visualizing all these surfaces for varying $\mathbf{x}$ and with a suitable value for $\mu > 0$ that determines the size of the glyphs. Note that for DTI data a different visualization based on ellipsoids is commonly used, which isn't suitable to visualize crossing fibers.

To reduce noise and to infer information about fiber crossings contextual information can be used. This enhancement is useful both for visualization purposes and as a preprocessing step for other algorithms such as fiber tracking algorithms, which may have difficulty in noisy or incoherent regions. This enhancement, done through linear and nonlinear adaptive diffusion processes, is the main focus of this paper. Special attention is given to the implementation of these algorithms through the use of finite difference schemes.

1.1   The Euclidean Motion Group $SE(3)$

A function on $\mathbb{R}^3 \times S^2$ can also be seen as a function embedded in the *Euclidean motion group* $SE(3) = \mathbb{R}^3 \rtimes SO(3)$, where $SO(3)$ represents the (noncommutative) group of 3D rotations defined as a matrix group by

$$SO(3) = \{\mathbf{R} \mid \mathbf{R} \in \mathbb{R}^{3\times3}, \mathbf{R}^T = \mathbf{R}^{-1}, \det(\mathbf{R}) = 1\}.$$

Expressed in Euler angles, this becomes

$$\mathbf{R}_{(\alpha,\beta,\gamma)} = \mathbf{R}_\gamma^{\mathbf{e}_x} \mathbf{R}_\beta^{\mathbf{e}_y} \mathbf{R}_\alpha^{\mathbf{e}_z}, \tag{1}$$

where $\mathbf{e}_1 = \mathbf{e}_x$, $\mathbf{e}_2 = \mathbf{e}_y$ and $\mathbf{e}_3 = \mathbf{e}_z$ are the unit vectors in the coordinate axes and $\mathbf{R}_\alpha^{\mathbf{e}_i}$ denotes a counterclockwise rotation of α around vector $\mathbf{e}_i$. Here, an Euler angle parametrization is used that has a discontinuity at $\mathbf{n} = (\pm 1, 0, 0)$, so that the tangent space of $SE(3)$ is well defined at the unity element $(\mathbf{0}, \mathbf{I})$.

For $g, g' \in SE(3)$ the group product and inverse are given by

$$gg' = (\mathbf{x}, \mathbf{R})(\mathbf{x}', \mathbf{R}') = (\mathbf{x} + \mathbf{R} \cdot \mathbf{x}', \mathbf{R}\mathbf{R}')$$
$$g^{-1} = (\mathbf{x}, \mathbf{R})^{-1} = (-\mathbf{R}^{-1}\mathbf{x}, \mathbf{R}^{-1}).$$

To get correspondence between $SO(3)$ and S^2, we introduce equivalence classes on $SO(3)$. Two group elements $g, h \in SO(3)$ are equivalent if $g^{-1}h = \mathbf{R}_\alpha^{\mathbf{e}_z}$ for some angle $\alpha \in [0, 2\pi)$. This equivalence relation induces sections of equivalent group members, called the *left cosets* of $SO(3)$. If we associate $SO(2)$ with rotations around the z-axis, then formally we can use this equivalence to write $S^2 \equiv SO(3)/SO(2)$ to denote these left cosets.

If we extend this equivalence relation to $SE(3)$, i.e. $g, h \in SE(3)$, g is equivalent to h if $g^{-1}h = (\mathbf{0}, \mathbf{R}_\alpha^{\mathbf{e}_z})$, we obtain the left coset of $SE(3)$ which equals the space of positions and orientations. To stress that this space has been embedded in $SE(3)$ and to stress the induced (quotient)group structure we write the space of positions and orientations as $\mathbb{R}^3 \rtimes S^2 := (\mathbb{R}^3 \rtimes SO(3))/(\{\mathbf{0}\} \times SO(2))$.

Now, we can express any function on position and orientation $U : \mathbb{R}^3 \rtimes S^2 \to \mathbb{R}$ with an equivalent function on $SE(3)$ $\tilde{U} : \mathbb{R}^3 \rtimes SO(3) \to \mathbb{R}$ by solving for $(\mathbf{x}, \mathbf{n}) = (\mathbf{x}, \mathbf{R}_\mathbf{n}\mathbf{e}_z)$, where $\mathbf{R}_\mathbf{n}$ is *any* rotation matrix that maps $\mathbf{e}_z$ to $\mathbf{n}$.

Every group element from $SE(3)$ can be associated with a representation, which is nothing else than an action that translates and rotates a function. The left- and right-regular representations on $\mathbb{L}_2(SE(3))$ are given by

$$(\mathcal{L}_g \circ U)(h) = U(g^{-1}h), \quad g, h \in SE(3), \ U \in \mathbb{L}_2(SE(3))$$
$$(\mathcal{R}_g \circ U)(h) = U(hg). \tag{2}$$

Duits and Franken [8,9] demonstrated that every reasonable linear operation on functions on $SE(3)$ must be left-invariant by showing that the orientation marginal $\int_{S^2} U(\mathbf{y}, \mathbf{n}) d\sigma(\mathbf{n})$ commutes with rotations and translations only under such operations, which explains our choice for left-invariant processes in this paper. Formally an operator $\Phi : \mathbb{L}_2(SE(3)) \to \mathbb{L}_2(SE(3))$ is left invariant iff

$$\forall g \in SE(3) : (\mathcal{L}_g \circ \Phi \circ U) = (\Phi \circ \mathcal{L}_g \circ U).$$

The right regular representation is left-invariant and can be used to generate left-invariant derivatives, as is shown in the next section. It should be noted that because of the non commutative structure of $SE(3)$ the left regular representation itself is *not* left-invariant.

1.2 Left-Invariant Derivatives

By viewing functions on $\mathbb{R}^3 \rtimes S^2$ as probability density functions of oriented particles, it becomes a natural idea to describe these particles in a moving coordinate system. This is done by attaching a coordinate system to each point

$(\mathbf{x}, \mathbf{R}) \in SE(3)$ such that one of the spatial axes points in the direction of $\mathbf{n} = \mathbf{R_n}\mathbf{e}_z$. In this section we will introduce diffusion equations for these oriented particles, and for these processes this coordinate system is the natural choice to easily differentiate between motion forward, sideways and rotations. We can obtain such a coordinate system by starting at the identity element $(\mathbf{0}, \mathbf{I})$ of $SE(3)$, which corresponds to $(\mathbf{0}, \mathbf{e}_z)$ and attaching a suitable coordinate system using Euler angles. We express a basis of tangent vectors at the unity element by

$$A_1 = \partial_x, A_2 = \partial_y, A_3 = \partial_z, A_4 = \partial_\gamma, A_5 = \partial_\beta, A_6 = \partial_\alpha,$$

where we use the coordinate system in the parametrization of $SE(3)$:$(\mathbf{x}, R) = (x, y, z, R_{(\alpha,\beta,\gamma)})$ (see Eq. (1)). Here A_i can be viewed both as tangent vectors and as local differential operators.

We construct a moving frame of reference attached to fibers in the space $\mathbb{R}^3 \rtimes S^2$ by using the derivative of the right-regular representation $\mathcal{R}$:

$$\mathcal{A}_i|_g U = (d\mathcal{R}(A_i)U)(g) = \lim_{t\downarrow 0} \frac{U(g\, e^{tA_i}) - U(g)}{t}, \quad i = 1, 2, 3, 4, 5, 6, \quad (3)$$

where $\mathcal{R}$ is defined by Eq. (2) and e^{tA_i} is the exponential map in $SE(3)$ [8], which can be seen as the group element obtained by traveling distance t in the A_i direction from the identity element. We note that $\mathcal{A}_6 U(\mathbf{y}, \mathbf{n}) = 0$, because $U(\mathbf{y}, \mathbf{n})$ is constant within equivalence classes and that $\mathcal{A}_1, \mathcal{A}_2, \mathcal{A}_4$ and $\mathcal{A}_5$ are defined on $SE(3)$ and not on $\mathbb{R}^3 \rtimes S^2$. We therefor use combinations of these operators that *are* well-defined on $\mathbb{R}^3 \rtimes S^2$ in the diffusion generator (see section 1.3).

Analytical formulas for these left-invariant derivatives, expressed in charts of Euler angles can be found in [8] where they are used to analytically approximate Green's functions of convection diffusion processes. Here, we focus on the numerical aspects, and instead give only the finite difference schemes (section 2) since they do not suffer from the discontinuities of the Euler angle parametrization.

1.3   Convection-Diffusion Processes

The left-invariant derivatives given in the previous section can be used to write the equations for diffusion processes on $SE(3)$ [8], which can remove noise while preserving complex structures such as crossings and junctions[12].

The general convection-diffusion equation with diffusion matrix $\mathbf{D}$ and convection parameters $\mathbf{a}$ is given by:

$$\begin{cases} \partial_t W(\mathbf{y}, \mathbf{n}, t) = Q^{\mathbf{D},\mathbf{a}}(\mathcal{A}_1, \mathcal{A}_2, \ldots, \mathcal{A}_5)W(\mathbf{y}, \mathbf{n}, t) \\ W(\mathbf{y}, \mathbf{n}, 0) = U(\mathbf{y}, \mathbf{n}) \end{cases} \quad (4)$$

where the convection-diffusion generator $Q^{\mathbf{D},\mathbf{a}}$ is given by

$$Q^{\mathbf{D},\mathbf{a}}(\mathcal{A}_1, \mathcal{A}_2, \ldots, \mathcal{A}_5) = \sum_{i=1}^{5} \left(-a_i \mathcal{A}_i + \sum_{j=1}^{5} \mathcal{A}_i D_{ij} \mathcal{A}_j \right) \quad (5)$$

and a_i are convection parameters and D_{ij} diffusion coefficients. In this paper, $a_i = 0$ for all $i = 1, \ldots, 6$ because only pure diffusion processes are studied, but other processes, like contour completion [4], can be obtained by also including convection terms. In the linear case, a_i and D_{ij} are chosen constant and the solution to these evolution equations can be obtained by an $SE(3)$-convolution of the initial data with the process's Green's function [8,9,5] or by using finite difference methods. $SE(3)$-convolutions (see [8]) are generally computationally more expensive than finite difference stencils and they can not handle adaptive schemes, so finite difference schemes are used exclusively in this paper.

2   Finite Difference Schemes for $\mathbb{R}^3 \rtimes S^2$ Diffusion

To approximate the required left-invariant derivatives of the evolution equations of Eq. (4), we use finite difference approximations [8] of Eq. (3). These derivatives are approximated in the usual way, with the (conceptually) small difference that the steps are taken in the $\mathcal{A}_i$ direction rather than the $\mathbf{e}_i$ direction. The forward finite difference approximation of the left-invariant derivatives are given by

$$
\begin{aligned}
\mathcal{A}_1^f U(\mathbf{y}, \mathbf{n}) &= \tfrac{U(\mathbf{y}+h\,R_\mathbf{n}\mathbf{e}_x\,,\,\mathbf{n})-U(\mathbf{y}\,,\,\mathbf{n})}{h} \,, & \mathcal{A}_4^f U(\mathbf{y}, \mathbf{n}) &= \tfrac{U(\mathbf{y}\,,\,R_\mathbf{n}\,R_{h_a}^{\mathbf{e}_x}\,\mathbf{e}_z)-U(\mathbf{y}\,,\,\mathbf{n})}{h_a} \,, \\
\mathcal{A}_2^f U(\mathbf{y}, \mathbf{n}) &= \tfrac{U(\mathbf{y}+h\,R_\mathbf{n}\mathbf{e}_y\,,\,\mathbf{n})-U(\mathbf{y}\,,\,\mathbf{n})}{h} \,, & & \\
\mathcal{A}_3^f U(\mathbf{y}, \mathbf{n}) &= \tfrac{U(\mathbf{y}+h\,R_\mathbf{n}\mathbf{e}_z\,,\,\mathbf{n})-U(\mathbf{y}\,,\,\mathbf{n})}{h} \,, & \mathcal{A}_5^f U(\mathbf{y}, \mathbf{n}) &= \tfrac{U(\mathbf{y}\,,\,R_\mathbf{n}\,R_{h_a}^{\mathbf{e}_y}\,\mathbf{e}_z)-U(\mathbf{y}\,,\,\mathbf{n})}{h_a} \,,
\end{aligned}
\tag{6}
$$

where h is the spatial stepsize and h_a the angular step size in radians.

Analogously, the backward and central finite difference approximations can be obtained. For example:

$$
\mathcal{A}_3^b U(\mathbf{y}, \mathbf{n}) = \tfrac{U(\mathbf{y}, \mathbf{n})-U(\mathbf{y}-h\,R_\mathbf{n}\mathbf{e}_z\,,\,\mathbf{n})}{h} \,, \quad \mathcal{A}_4^b U(\mathbf{y}, \mathbf{n}) = \tfrac{U(\mathbf{y}, \mathbf{n})-U(\mathbf{y}\,,\,R_\mathbf{n}\,R_{-h_a}^{\mathbf{e}_x}\,\mathbf{e}_z)}{h_a} \,,
$$

and

$$
\begin{aligned}
\mathcal{A}_3^c U(\mathbf{y}, \mathbf{n}) &= \frac{U(\mathbf{y} + h\,R_\mathbf{n}\mathbf{e}_z, \mathbf{n}) - U(\mathbf{y} - h\,R_\mathbf{n}\mathbf{e}_z, \mathbf{n})}{2h} \,, \\
\mathcal{A}_4^c U(\mathbf{y}, \mathbf{n}) &= \frac{U(\mathbf{y}, R_\mathbf{n} R_{h_a}^{\mathbf{e}_x}\mathbf{e}_z) - U(\mathbf{y}, R_\mathbf{n} R_{-h_a}^{\mathbf{e}_x}\mathbf{e}_z)}{2h_a} \,.
\end{aligned}
\tag{7}
$$

We take second order centered finite differences by applying the discrete operators in the righthand side of Eq. (7) twice (where we replaced $2h \mapsto h$), e.g. we have for $p = 1, 2, 3$:

$$
\begin{aligned}
((\mathcal{A}_p^c)^2 U)(\mathbf{y}, \mathbf{n}) &= \frac{U(\mathbf{y} + hR_\mathbf{n}\mathbf{e}_p, \mathbf{n}) - 2\,U(\mathbf{y}, \mathbf{n}) + U(\mathbf{y} - hR_\mathbf{n}\mathbf{e}_p, \mathbf{n})}{h^2} \,, \\
((\mathcal{A}_{p+3}^c)^2 U)(\mathbf{y}, \mathbf{n}) &= \frac{U(\mathbf{y}, R_\mathbf{n}\,R_{\mathbf{e}_p, h_a}\,\mathbf{e}_z) - 2\,U(\mathbf{y}, \mathbf{n}) + U(\mathbf{y}, R_\mathbf{n}\,R_{\mathbf{e}_p, -h_a}\,\mathbf{e}_z)}{h_a^2} \,.
\end{aligned}
$$

2.1   Efficient Computation of Left-invariant Derivatives

So far, all approximations assume $U : \mathbb{R}^3 \rtimes S^2 \to \mathbb{R}^+$ to be continuously differentiable. In practice we have discretized functions $U[i, j, k, \mathbf{n}_l]$, where i,j and

k enumerate the discrete spatial grid and $\mathbf{n}_l$ is an orientation from a tessellation of the sphere enumerated by $l \leq N_o$. The tessellation used in this paper is obtained by taking an icosahedron and regularly subdividing each face into 16 triangles before projecting the vertices back to the sphere. Every vertex of this shape becomes a sampling orientation and thus $N_o = 162$.

Because of this sampling, interpolation is necessary to approximate the (left-invariant) derivatives. Spatially, any regular 3D interpolation scheme such as linear interpolation or spline interpolation can be used. Since the approximations in Eq. (6) are only first order accurate, we use linear interpolation.

Since the three spatial derivatives only require neighboring samples with the same $\mathbf{n}$, they can be efficiently computed through a regular $\mathbb{R}^3$ convolution or correlation for each orientation separately.

$$A_p U[i, j, k, \mathbf{n}_l] \approx (\mathbf{M}_l^p \star U[\cdot, \cdot, \cdot, \mathbf{n}_l]) \, [i, j, k] \quad p = 1, 2, 3,$$

where $\star$ denotes the discrete spatial correlation and $\mathbf{M}_l^p$ can be obtained by linear interpolation from the finite difference stencils Eq. (6). For example, in the case of forward finite difference stencils

$$\mathbf{M}_l^p[i, j, k] = \frac{1}{2h}(w_{l,p}^1[i, j, k] - w_{l,p}^2[i, j, k]),$$

with $w_{l,p}^1[i, j, k] = N_{\mathbf{y}_l^p}[i, j, k]$ and $w_{l,p}^2[i, j, k] = \delta_{i0}\delta_{j0}\delta_{k0}$,

where $\mathbf{y}_l^p = h R_{\mathbf{n}_l} e_p$ and $N_{\mathbf{y}_l^p}[i, j, k]$ is the interpolation matrix required to interpolate point $\mathbf{y}_l^p$. Assuming cubic voxels of size 1, $|\mathbf{y}_l^p| < 1$, and using linear interpolation $N_{\mathbf{y}_l^p}[i, j, k]$ is given by

$$N_{\mathbf{y}}[i, j, k] = \prod_{m=1}^{3} v_{(\mathbf{y}_{p,l})^m}[\mathbf{z}^m] \ , \ \mathbf{z} = (i, j, k) \ \ i, j, k \in \{-1, 0, 1\},$$

$$v_a[b] = \begin{cases} 1 - |a| & \text{if } b = 0 \\ H(ab)|a| & \text{if } b \in \{-1, 1\} \\ 0 & \text{otherwise.} \end{cases}$$

with heaviside function $H(u)$ and where $(\mathbf{y}_{p,l})^m$ and $\mathbf{z}^m$ denote the m-th component of vectors $\mathbf{y}_{p,l}$ and $\mathbf{z}$.

For angular interpolation, either linear interpolation or spherical harmonics can be used. Spherical harmonics were used in previous work by Franken et al. [10]. In terms of stability in the diffusion process, both perform equally well (see section 2.2, Theorem 1), but Franken had to add an angular diffusion term t_{reg} which in practice was very sensitive: when set too large the data becomes too isotropic (destroying fiber structures) and when set too small the algorithm becomes unstable. As we will show next, linear interpolation is also computationally cheaper.

The angular derivatives only require samples of neighbors with the same $\mathbf{y}$ and can therefor be computed by a matrix multiplication for each point $\mathbf{y}$:

$$\mathcal{A}_{p+3}^f U[i, j, k, \mathbf{n}_l] \approx \frac{1}{h}\sum_{l'=1}^{N_o} \mathbf{M}_{l\,l'}^p U[i, j, k, \mathbf{n}_{l'}]) - \frac{1}{h}U[i, j, k, \mathbf{n}_l], \tag{8}$$

where $\mathbf{M}_{ll'}^{p}$ is the interpolation matrix to interpolate $\mathbf{n}_{p,l} = R_{\mathbf{n}_l} R_{\mathbf{e}_p,h_a} \mathbf{e}_3$ and is given by

$$
M_{ll'} = \begin{cases} 1 - \sum_{\mathbf{n}_j \in A_{p,l}} (\mathbf{n}_{p,l} - \mathbf{n}_{l'}) \cdot (\mathbf{n}_j - \mathbf{n}_{l'}) & \text{if } \mathbf{n}_{l'} \in A_{p,l} \\ 0 & \text{otherwise,} \end{cases}
$$

where $A_{p,l}$ is the triangle that contains point $\mathbf{n}_{p,l}$. $A_{p,l}$ is sparse due to the linear interpolation which enables Eq. (8) to be computed cheaply. If $M_{ll'}^{p}$ is created using spherical harmonics then it becomes a full matrix and thus Eq. (8) is much more expensive to calculate.

2.2   Numerical Contour Enhancement

The *contour enhancement* process on $\mathbb{R}^3 \rtimes S^2$ can be obtained from Eq. (4) by setting $a_i = 0$ (no convection), $D_{33} \geq 0$, $D_{44} = D_{55} \geq 0$ and other diffusion coefficients D_{ij} are set to zero. These settings yield the following evolution equation

$$
\begin{cases} \partial_t W(\mathbf{y}, \mathbf{n}, t) = \left(D_{33}(\mathcal{A}_3)^2 + D_{44}((\mathcal{A}_4)^2 + (\mathcal{A}_5)^2) \right) W(\mathbf{y}, \mathbf{n}, t) \\ W(\mathbf{y}, \mathbf{n}, 0) = U(\mathbf{y}, \mathbf{n}). \end{cases} \tag{9}
$$

This process can intuitively be understood as a discription of the Brownian motion of oriented particles both in space (diffusion in direction $\mathbf{n}$) and angular (changing direction) [8]. The simulation of this PDE is done by taking standard centered second order finite differences according to Eq. (2), and using a forward Euler scheme for the time discretization:

$$
\begin{cases} W(\mathbf{y}, \mathbf{n}, t+\Delta t) = W(\mathbf{y}, \mathbf{n}, t) + \Delta t \left(D_{33}(\mathcal{A}_3^c)^2 + D_{44}((\mathcal{A}_4^c)^2 + (\mathcal{A}_5^c)^2) \right) W(\mathbf{y}, \mathbf{n}, t) \\ W(\mathbf{y}, \mathbf{n}, 0) = U(\mathbf{y}, \mathbf{n}). \end{cases}
$$

Of these parameters, D_{44} and simulation time t are most important. D_{33} may be set to 1, as changing D_{33} is equivalent to scaling D_{44} and t, while Δt needs only be sufficiently small for the algorithm to remain stable and accurate.

Theorem 1. *The stability bound for the Euler forward finite difference scheme of the evolution described by Eq. (9) using the interpolation described in section 2.1 is given by*

$$
\Delta t \leq \frac{1}{\frac{(4D_{11}+2D_{33})}{h^2} + \frac{4D_{44}}{h_a^2}} \tag{10}
$$

when using linear interpolation for the angular derivatives and by

$$
\begin{aligned} \Delta t &\leq \frac{1}{\frac{4D_{11}+2D_{33}}{h^2} + D_{44}\frac{L(L+1)}{2e^{t_{reg}L(L+1)}}} & \text{if } t_{reg}.L(L+1) \leq 1 \\[2mm] \Delta t &\leq \frac{1}{\frac{4D_{11}+2D_{33}}{h^2} + D_{44}\frac{1}{2et_{reg}}} & \text{if } t_{reg}.L(L+1) > 1 \end{aligned} \tag{11}
$$

when using spherical harmonics.

for proof see [4, Section 2.2.1] *and* [7, Appendix B].

In terms of stability, both algorithms can be made equally stable because both have regularizing parameters (h_a for linear interpolation and t_{reg} for spherical harmonics). There are, however other reasons to prefer linear interpolation over spherical harmonics (see section 2.1).

3 Perona-Malik Diffusion on $\mathbb{R}^3 \rtimes S^2$

Linear contour enhancement has the disadvantage that it performs diffusion across areas where the gradient is very large. In particular, the neural tracts of the brain are sometimes located near the ventricles of the brain. These ventricles are structures that contain cerebrospinal fluid which shows up in DTI as unrestricted, isotropic glyphs much larger in magnitude than the restricted, anisotropic glyphs of the neural tracts. It is undesirable that these large isotropic diffusion profiles start to interfere with the oriented structures of the neural tracts when we apply a diffusion scheme, because they are likely to destroy fiber structures. A Perona-Malik [11] type scheme for diffusion can separate these two regions, and apply the diffusion within the neural tracts and within the ventricles, but prevents transport from one to the other.

Our approach is similar to recent work by Burgeth et al. [3] who used adaptive, edge preserving diffusion on the DTI tensor components separately. The difference is that here the diffusion considers both positions and orientations in the domain and therefor separates two crossing fibers in the domain so that it is better equipped to handle crossing structures.

We test the algorithm on a synthetic test image consisting of two crossing fibers consisting of oriented glyphs surrounded by isotropic spheres, (see Fig. 1) in which linear diffusion destroys the fiber structure, whereas nonlinear adaptive diffusion both preserves the fiber structures and denoises the entire dataset.

Mutual influence of the anisotropic regions (fibers) and isotropic regions (ventricles) is avoided by replacing the constant diffusivity D_{33} in 5 by

$$\mathcal{A}_3 D_{33} \mathcal{A}_3 \mapsto \mathcal{A}_3 \circ D_{33} e^{-\frac{(\mathcal{A}_3 W(\cdot, t))^2}{K^2}} \circ \mathcal{A}_3, \tag{12}$$

where for $K \to \infty$, linear contour enhancement is obtained. The idea is to set a soft threshold (determined by K) on the amount of diffusion in $\mathcal{A}_3$ direction. Within homogeneous regions one expects $|\mathcal{A}_3 W(\mathbf{y}, \mathbf{n}, t)|$ to be small, whereas in the transition areas between ventricles and white matter where one needs to block the diffusion process, one expects a large $|\mathcal{A}_3 W(\mathbf{y}, \mathbf{n}, t)|$.

To implement this, we propose the following discretization scheme

$$\mathcal{A}_3(\tilde{D}_{33}\mathcal{A}_3)W(\mathbf{y}, \mathbf{n}, t) \approx \frac{\tilde{D}_{33}(\mathbf{y}+\frac{1}{2}\mathbf{h}, \mathbf{n})\mathcal{A}_3 W(\mathbf{y}+\frac{1}{2}\mathbf{h}, \mathbf{n}, t)}{h} - \frac{\tilde{D}_{33}(\mathbf{y}-\frac{1}{2}\mathbf{h}, \mathbf{n})\mathcal{A}_3 W(\mathbf{y}-\frac{1}{2}\mathbf{h}, \mathbf{n}, t)}{h}$$

$$\mathcal{A}_3 W(\mathbf{y}+\frac{1}{2}\mathbf{h}, \mathbf{n}, t) \approx \frac{W(\mathbf{y}+\mathbf{h}, \mathbf{n}, t) - W(\mathbf{y}, \mathbf{n}, t)}{h} = \mathcal{A}_3^f W(\mathbf{y}, \mathbf{n}, t)$$

$$\mathcal{A}_3 W(\mathbf{y}-\frac{1}{2}\mathbf{h}, \mathbf{n}, t) \approx \frac{W(\mathbf{y}, \mathbf{n}, t) - W(\mathbf{y}-\mathbf{h}, \mathbf{n}, t)}{h} = \mathcal{A}_3^b W(\mathbf{y}, \mathbf{n}, t)$$

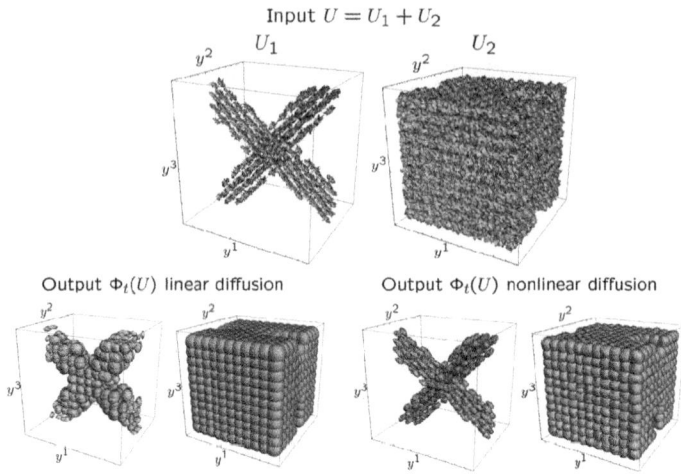

Fig. 1. Adaptive Perona-Malik diffusion based on the data. Top row: Artificial $15 \times 15 \times 15 \times 162$ input data that is the sum of a noisy fiber part and a noisy isotropic part. For the sake of visualization, we depict these parts separately. Bottom row left: Output of linear diffusion with $t = 1$, $D_{33} = 1$, $D_{44} = 0.04$ and $\Delta t = 0.01$. Bottom right: Output of Perona-Malik adaptive diffusion with $D_{33} = 1$, $D_{44} = 0.015$, $K = 0.05$, $\Delta t = 0.01$, $t = 1$

combining these three equations leads to

$$\mathcal{A}_3(\tilde{D}_{33}\mathcal{A}_3)W(\mathbf{y},\mathbf{n},t) \approx \frac{\tilde{D}_{33}(\mathbf{y}+\frac{1}{2}\mathbf{h},\mathbf{n})\mathcal{A}_3^f W(\mathbf{y},\mathbf{n},t) - \tilde{D}_{33}(\mathbf{y}-\frac{1}{2}\mathbf{h},\mathbf{n})\mathcal{A}_3^b W(\mathbf{y},\mathbf{n},t)}{h}$$

where for notational convenience $\mathbf{h} = h\,R_{\mathbf{n}}\mathbf{e}_z$ and $\tilde{D}_{33} = D_{33}e^{-\frac{(\max(|\mathcal{A}_3^f W|,|\mathcal{A}_3^b W|))^2}{K^2}}$. The $\tilde{D}_{33}$ terms can easily be calculated with linear interpolation. Combined with the finite difference operators of section 2, this give the full discretization scheme.

The discretization scheme for $\tilde{D}_{33}$ uses $\max(|\mathcal{A}_3^f W|,|\mathcal{A}_3^b W|)$ because forward and backwards finite difference schemes individually induce shifts near discontinuities while central finite difference schemes sometimes allow diffusion across region boundaries. This happens when fiber voxels have more than one isotropic neighbor, then $\mathcal{A}_3^c W$ may be close to zero because the stencil does not depend on the center point. Because of the spatial discretization and because every direction $\mathbf{n}$ is considered, this is very likely to occur in almost all geometries.

4   Enhancement of DTI of the Human Brain

To test the algorithm on real data, a DTI brain scan was acquired from a healthy volunteer with 132 gradient directions and a b-value of $1000s/mm^2$. Linear contour enhancement (Eq. (9)) as well as Perona-Malik adaptive diffusion (Eq. (12)) was performed on it, as can be seen in Fig. 2.

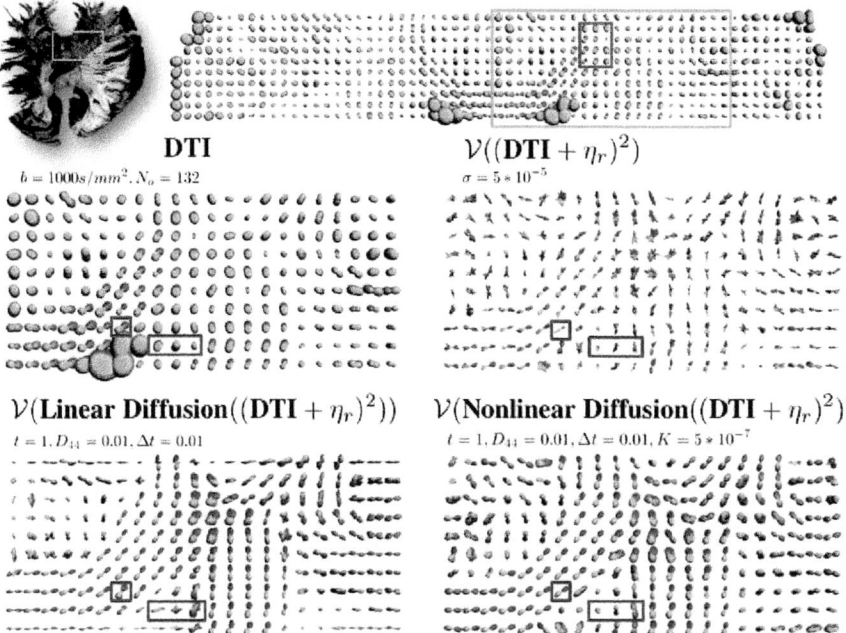

Fig. 2. DTI data of the corpus callosum and corona radiata fibers in a human brain with b-value $1000s/mm^2$ and 132 gradient directions on voxels of $(2mm)^3$. Top row: A coronal slice of the original data with a region of interest in the yellow square. The region in the blue square is shown for multiple values of K in Fig. 3. Middle row: The unprocessed region of interest (left) and with added Rician noise($\sigma = 5 \cdot 10^{-5}$, see [4] for definition) and sharpening according to Eq.(13) (right). Bottom row: The result of linear contour enhancement (left) and Perona-Malik diffusion (right). Marked in red are areas in which the ventricles have induced crossing structures in the linear diffusion process.

Prčkovska et al. [12] showed that DTI combined with enhancement techniques can extrapolate crossing information from contextual information. It is interesting to see if such a method can be improved with a Perona-Malik type scheme, especially since the ventricles may make such methods unreliable in those areas.

Since visualization of larger datasets is difficult, only coronal slices through the center of the brain are depicted, where the ventricles are visible as large, isotropic spheres. Because of the relative isotropy of real data, sharpening techniques have to be employed. Squaring the input data is the simplest way to do this (and is used here), but other techniques such as $\mathbb{R}^3 \rtimes S^2$-erosions are also an option [6]. For visualization, a min-max-normalization and another sharpening step are used, given by operator

$$\mathcal{V}(U)(\mathbf{y}, \mathbf{n}) = \left(\frac{U(\mathbf{y}, \mathbf{n}) - U_{min}(\mathbf{y})}{U_{max}(\mathbf{y}) - U_{min}(\mathbf{y})} \right)^2, \text{ with } U_{\substack{min \\ max}}(\mathbf{y}) = \min_{max}\{U(\mathbf{y}, \mathbf{n}) \mid \mathbf{n} \in S^2\}. \quad (13)$$

From Fig. 2 it can be seen that the Perona-Malik method performs better than linear contour enhancement. The first effect is visible on the boundary of the data. Linear contour enhancement diffuses signal outside of the boundaries of the image (because of zero padding boundary conditions for the calculation of derivatives), which causes artifacts visible as horizontal structures near the top and bottom edges. The same zero padding ensures a large derivative at these places so Perona-Malik does not suffer from this problem.

The second effect is visible around the ventricles (marked by red in Fig. 2). Linear diffusion shows some crossing structures directly to the right of the ventricles, while the surrounding glyphs do not suggest there should be any crossings there. It also affects the fibers of the corpus callosum to the top left of the ventricles by bending them a bit upwards and away from the ventricles.

Figure 3 shows the effect of parameter K on the diffusion profile of the area of fiber crossings where it can be seen that setting K too small leads to a shortcoming of the algorithm to correctly infer crossing information while setting it too large leads to the same result as linear diffusion.

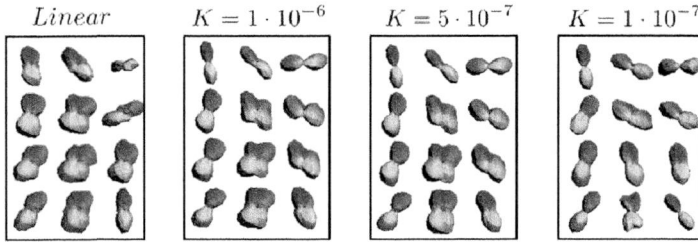

Fig. 3. Area with fiber crossings of the corpus callosum and corona radiata for different values of K. All images created with $D_{33} = 1$, $D_{44} = 0.01$, $t = 1$ and $\Delta t = 0.01$

5   Conclusion

We have developed an edge-preserving, adaptive Perona-Malik smoothing process using finite difference schemes that can be used to remove high frequency noise and extrapolate fiber crossing information from DTI data by embedding the DTI data into a function on the space of positions and orientations $\mathbb{R}^3 \rtimes S^2$. Our experiments have shown that the adaptive diffusion process performs better than linear processes in areas with large isotropic diffusion (such as the ventricles of the brain) since at these areas adaptive diffusivity strongly reduces the interference between isotropic glyphs in the ventricles and anisotropic glyphs of the fibers.

References

1. Alexander, D.C., Barker, G.J., Arridge, S.R.: Detection and modeling of non gaussian apparent diffusion coefficient profiles in human brain data. Magnetic Resonance in Medicine 48, 331–340 (2002)
2. Basser, P.J., Mattiello, J., Lebihan, D.: MR diffusion tensor spectroscopy and imaging. Biophysical Journal 66, 259–267 (1994)

3. Burgeth, B., Pizarro, L., Didas, S., Weickert, J.: Coherence-Enhancing Diffusion for Matrix Fields. In: Proc. of Locally Adaptive Filtering in Signal and Image (to appear)
4. Creusen, E.J.: Numerical schemes for linear and non-linear enhancement of HARDI data. Master's thesis, Technische Universiteit Eindhoven, the Netherlands (2010)
5. Duits, R., Creusen, E., Ghosh, A., Dela Haije, T.: Diffusion, convection and erosion on $\mathbb{R}^3 \rtimes S^2$ and their application to the enhancement of crossing fibers. Technical report, TU/e, CASA, The Netherlands, Eindhoven, CASA-report nr.6 (January 2011), http://www.win.tue.nl/casa/research/casareports/2011.html
6. Duits, R., Dela Haije, T., Ghosh, A., Creusen, E.J., Vilanova, A., ter Haar Romeny, B.: Fiber enhancement in DW-MRI. In: Bruckstein, A.M., ter Haar Romeny, B.M., Bronstein, A.M., Bronstein, M.M. (eds.) SSVM 2011. LNCS, vol. 6667, pp. 1–13. Springer, Heidelberg (2011)
7. Duits, R., Franken, E.: Left-invariant diffusions on the space of positions and orientations and their application to crossing-preserving smoothing of HARDI images. CASA-report 18, Department of Mathematics and Computer Science, Technische Universiteit Eindhoven (May 2009),
http://www.win.tue.nl/casa/research/casareports/2009.html
8. Duits, R., Franken, E.: Left-invariant diffusions on the space of positions and orientations and their application to crossing-preserving smoothing of HARDI images. International Journal of Computer Vision 40 (2010)
9. Franken, E.: Enhancement of crossing elongated structures in images. PhD thesis, Eindhoven University of Technology (2008)
10. Franken, E.M., Duits, R., ter Haar Romeny, B.M.: Diffusion on the 3D Euclidean motion group for enhancement of HARDI data. In: Tai, X.-C., Mørken, K., Lysaker, M., Lie, K.-A. (eds.) SSVM 2009. LNCS, vol. 5567, pp. 820–831. Springer, Heidelberg (2009)
11. Perona, P., Malik, J.: Scale-space and edge detection using anisotropic diffusion. IEEE Trans. Pattern Anal. Mach. Intell. 12(7) (1990)
12. Prčkovska, V., Rodrigues, P., Duits, R., Vilanova, A., ter Haar Romeny, B.M.: Extrapolating fiber crossings from DTI data. Can we infer similar fiber crossings as in HARDI? In: CDMRI 2010 Proc. MICCAI Workshop Computational Diffusion MRI, China (2010)

Optimising Spatial and Tonal Data
for Homogeneous Diffusion Inpainting

Markus Mainberger[1], Sebastian Hoffmann[1], Joachim Weickert[1], Ching Hoo Tang[2],
Daniel Johannsen[3], Frank Neumann[4], and Benjamin Doerr[2]

[1] Mathematical Image Analysis Group,
Faculty of Mathematics and Computer Science, Campus E1.1
Saarland University, 66041 Saarbrücken, Germany
{mainberger,hoffmann,weickert}@mia.uni-saarland.de
[2] Department 1: Algorithms and Complexity
Max Planck Institute for Informatics, Campus E1.4
66123 Saarbrücken, Germany
{doerr,chtang}@mpi-inf.mpg.de
[3] School of Mathematical Sciences,
Tel Aviv University, Ramat Aviv, Tel Aviv 69978, Israel
johannse@tau.ac.il
[4] School of Computer Science, Innova21 Building
University of Adelaide, Adelaide, SA 5005, Australia
frank@cs.adelaide.edu.au

Abstract. Finding optimal inpainting data plays a key role in the field of image compression with partial differential equations (PDEs). In this paper, we optimise the spatial as well as the tonal data such that an image can be reconstructed with minimised error by means of discrete homogeneous diffusion inpainting. To optimise the spatial distribution of the inpainting data, we apply a probabilistic data sparsification followed by a nonlocal pixel exchange. Afterwards we optimise the grey values in these inpainting points in an exact way using a least squares approach. The resulting method allows almost perfect reconstructions with only 5% of all pixels. This demonstrates that a thorough data optimisation can compensate for most deficiencies of a suboptimal PDE interpolant.

Keywords: image compression, partial differential equations (PDEs), inpainting, optimisation, homogeneous diffusion.

1   Introduction

Research on PDE-based data compression suffers from poverty, but enjoys liberty [1, 2, 8, 18]: Unlike in pure inpainting research [14, 3], one has an extremely tight pixel budget for reconstructing some given image. However, one is free to choose where and how one spends this budget.

Let us explain the problem of PDE-based image compression in more detail. The basic idea is to reconstruct some given image by inpainting from a sparse set of pixels with a suitable partial differential equation (PDE). There is an evident tradeoff between the number of pixels to store and the achievable reconstruction quality. Even if the

A.M. Bruckstein et al. (Eds.): SSVM 2011, LNCS 6667, pp. 26–37, 2012.

Fig. 1. Reconstruction of the test image *trui* using only 5% of all pixels and homogeneous diffusion inpainting. **(a)** Original image. **(b)** Unoptimised data (randomly selected from original image). **(c)** Optimised tonal and spatial data

number of pixels and the PDE are already specified, we still have many degrees of freedom: On one hand we can place the pixels wherever we want. On the other hand we can freely choose the grey value (or colour value) in each selected pixel.

The goal of the present paper is to optimise this spatial and tonal data selection. In order to show the real potential behind this approach, we choose an extremely simple PDE that has a bad reputation for inpainting tasks: We interpolate with the steady state of a homogeneous diffusion process, i.e. we solve the Laplace equation. Figure 1 illustrates the huge potential that one can exploit with this optimisation. Even with homogeneous diffusion and a pixel density of only 5%, astonishing results can be achieved. One should note that we did not optimise our algorithm with respect to its runtime, as we regard it as a proof-of-concept only. Thus, our methods can require several hours to days to process typical images. However, we are confident that this runtime can be significantly reduced, and are going to address this issue in our ongoing research.

Organisation of the paper. Section 2 gives a brief introduction to homogeneous diffusion inpainting. In Section 3 we present two approaches that are applied sequentially to optimise the pixel locations: a probabilistic sparsification method, followed by a nonlocal pixel exchange. Afterwards, in Section 4, we show how the results can be improved further by an exact optimisation of the tonal data. Finally, we summarise and conclude our paper in Section 5.

Related work. The most similar work to our paper is a recent publication by Belhachmi et al. [2] where a continuous analysis on spatially optimal data selection for homogeneous diffusion interpolation is presented. Their framework is based on the theory of shape optimisation and suggests to choose a pixel density that is an increasing function of the modulus of the image Laplacian. In order to make this result applicable to the practically relevant discrete setting, dithering techniques must be applied that can introduce additional errors. In the experiments we compare our results with the ones from [2]. It should be mentioned that in [2] no tonal optimisation is performed.

There is a long tradition to restore image data by homogeneous diffusion inpainting from edges [5,7,9,13,20] or specific feature points in Gaussian scale space [10,11,12]. Although such features can be perceptually relevant, one cannot expect that they are optimal w.r.t. some error norm.

In order to come up with data-adaptive point distributions, some publications use subdivision strategies in connection with anisotropic diffusion [8, 18]. They offer the advantage that the resulting tree structures allow an inexpensive coding of the selected pixels, but they severely constrain the set of admissible point distributions. For a more sensitive interpolant such as homogeneous diffusion, this restriction is too prohibitive.

The holographic image representation presented in [4] maps the image into a sequence of sample pixels, such that any partition of this sequence allows for a reconstruction of the whole image with similar quality. This requires the samples in each portion to be equally optimal. On the contrary, our goal is to reduce the image to only one set of optimal samples.

From the Green function of the Laplace operator it follows that homogeneous diffusion inpainting involves radial basis functions. These functions are popular for scattered data interpolation, and some of them have also been used for inpainting corrupted images [6, 19]. However, such problems usually do not allow to optimise the location and the grey values of the inpainting data set.

2   Image Inpainting with Homogeneous Diffusion

Continuous formulation. Let $f(x)$ be a continuous grey value image, where $x = (x, y)^\top$ denotes the location within a rectangular image domain $\Omega \subset \mathbb{R} \times \mathbb{R}$. Furthermore, let $\Omega_K \subset \Omega$ be a subset of the image domain, denoting *known* data. A reconstruction $u(x)$ by means of homogeneous diffusion inpainting can be obtained by keeping known data and using them as Dirichlet boundary conditions, while solving the Laplace equation on the set of *unknown* data $\Omega \setminus \Omega_K$:

$$
\begin{aligned}
u(x) &= f(x) \quad && \text{for } x \in \Omega_K , \\
\Delta u(x) &= 0 \quad && \text{for } x \in \Omega \setminus \Omega_K ,
\end{aligned}
\tag{1}
$$

with homogeneous (reflecting) Neumann boundary conditions across the image boundary $\partial \Omega$. These two equations can be combined to a single equation

$$
c(x)(u(x) - f(x)) - (1 - c(x))\Delta u(x) = 0 ,
\tag{2}
$$

by using a confidence function $c(x)$ which specifies whether a point is known or not:

$$
c(x) = \begin{cases} 1 & \text{for } x \in \Omega_K , \\ 0 & \text{for } x \in \Omega \setminus \Omega_K . \end{cases}
\tag{3}
$$

Discrete formulation. To apply the homogeneous inpainting process to a digital image, we need a discrete formulation of Equation 2. The discrete version of a continuous image f is represented as a one-dimensional vector $\boldsymbol{f} = (f_1, \ldots, f_N)^\top = (f_i)_{i \in J}$, where $J = \{1, \ldots, N\}$ denotes the set of all pixel indices. Analogously, $\boldsymbol{u}$ describes the solution vector and $\boldsymbol{c}$ the binary pixel mask that indicates whether a pixel is known or not. The set K contains the pixel indices i of known pixels, i.e. for which $c_i = 1$. The Laplacian Δu is discretised by means of finite differences [15]. Then the discrete formulation reads

$$
\boldsymbol{C}(\boldsymbol{u} - \boldsymbol{f}) - (\boldsymbol{I} - \boldsymbol{C})\boldsymbol{A}\boldsymbol{u} = \boldsymbol{0} ,
\tag{4}
$$

where I is the identity matrix, $C := \mathrm{diag}(c)$ is a diagonal matrix having the components of c as diagonal entries, and A is a symmetric $N \times N$ matrix, describing the discrete Laplace operator Δ with homogeneous Neumann boundary conditions. Its entries are given by

$$
a_{i,j} = \begin{cases}
\dfrac{1}{h_\ell^2} & (j \in \mathcal{N}_\ell(i)) \\
-\displaystyle\sum_{\ell\in\{x,y\}}\sum_{j\in\mathcal{N}_\ell(i)} \dfrac{1}{h_\ell^2} & (j = i) \\
0 & (\text{else}) ,
\end{cases}
\tag{5}
$$

where $\mathcal{N}_\ell(i)$ are the neighbours of pixel i in ℓ-direction.

Reformulating Equation 4 yields a linear system of equations:

$$
\underbrace{(C - (I - C)A)}_{=:M}\, u = C f .
\tag{6}
$$

This linear system of equations has a unique solution and can be solved efficiently by using bidirectional multigrid methods [13].

3   Optimising Spatial Data

Now that we know how an image can be reconstructed by means of homogeneous diffusion inpainting, let us optimise the spatial data. This means we are looking for a pixel mask that selects for example only 5% of all pixels and that minimises the reconstruction error.

The good news on the pixel selection is that it is a discrete problem and thus it is finite and a global optimum exits. The bad news is that selecting the best 5% pixels of a 256×256 image offers already $\binom{65536}{3277} \approx 1.72 \cdot 10^{5648}$ possible solutions.

We overcome this problem by introducing two optimisation approaches. The first one is the *probabilistic sparsification*, which step by step removes pixels until the desired amount of pixels is left. Since this method can be trapped in local minima, we apply in a second step a method which we call *nonlocal pixel exchange*. It takes the mask that was created by the probabilistic sparsification and tries to improve the result by globally exchanging mask pixels with non-mask pixels.

3.1   Probabilistic Sparsification

Given an fixed discrete image f, let $r(c, f)$ be a function which computes the solution u of the discrete homogeneous inpainting process described by Equation 6, depending on a mask c. Our goal is to obtain a pixel mask c, marking only a predefined fraction d of all pixels J such that the *mean squared error (MSE)*

$$
\mathrm{MSE}(u) = \frac{1}{|J|}\sum_{i\in J}(f_i - u_i)^2
\tag{7}
$$

Input:	Original image f, fraction p of mask pixels used for candidate set, fraction q of candidate pixels that are finally removed, desired pixel density d.	
Output:	Pixel mask c, s.t. $\sum_{i \in J} c_i = d \cdot |J|$.	
Initialisation:	$c := (1, \ldots, 1)^\top$, thus $K = J$.	

While $|K| > d \cdot |J|$ do

1. Choose randomly $p \cdot |K|$ pixel indices from K into a candidate set T.
2. For all $i \in T$ reassign $c_i := 0$.
3. Compute $u := r(c, f)$.
4. For all $i \in T$ compute the local error $e_i = |u_i - f_i|$.
5. For all i of the $(1 - q) \cdot |T|$ largest values of $\{e_i | i \in T\}$, reassign $c_i := 1$.
6. Update K and clear T.

Fig. 2. Probabilistic sparsification

is minimised. To obtain a suitable, approximatively optimal pixel mask, we suggest a method that we call *probabilistic sparsification*.

In each iteration, we first randomly remove a fraction of mask pixels, inpaint, compute the error in each removed pixel, and put a subset of the removed pixels with largest error back into the mask again. Thus, pixels which are supposed to be least significant are step by step removed until the desired fraction of pixels remains. The algorithm is given in detail in Figure 2.

Note that our algorithm removes $p \cdot q \cdot |K|$ pixels in each step. Thus, in the k-th iteration, there are $(1 - pq)^k \cdot |J|$ mask pixels left, since K is initially J. In total, we need $\log_{(1-pq)} d$ many steps to obtain the desired fraction d. Hence, the larger p and q are chosen, the faster the algorithm converges. On the other hand, it is then more likely that significant pixels are removed. Since we aim for an optimal pixel mask, we suggest to choose small values, such as $p = 0.02$ and $q = 0.02$.

3.2 Nonlocal Pixel Exchange

The previously presented method has the disadvantage that once a pixel is removed from the mask, it will never be put back again. Moreover, by selecting pixels randomly, it might be possible that we also remove some significant pixels. To this end, we add a post-optimisation step, called *nonlocal pixel exchange*.

In each iteration, we choose randomly a fixed amount of non-mask pixels into a candidate set. A subset of those which exhibit the largest inpainting error, are exchanged with randomly chosen mask pixels. If the inpainting error for the new mask does not decrease, we reset the mask to its previous configuration. Thus, we allow mask pixels to move globally as long as the reconstruction result improves. The details of the algorithm are given in Figure 3.

For our experiments, we exchange only one pixel per iteration ($n = 1$) and keep the candidate set small by choosing $m = 10$. Interestingly, the results cannot be improved by larger candidate sets. Restricting it to this size adds some moderate amount of randomness such that we are not trapped in the next local minimum.

Input: Original image $\boldsymbol{f}$, (pre-optimised) pixel mask $\boldsymbol{c}$,
              size m of candidate set and number n of mask pixels
              exchanged per iteration.
Output: Post-optimised pixel mask $\boldsymbol{c}$.
Initialisation: $\boldsymbol{u} := r(\boldsymbol{c}, \boldsymbol{f})$ and $\boldsymbol{c}^{\text{new}} := \boldsymbol{c}$.

Repeat:

1. Choose randomly $m \leq |K|$ pixel indices from $J \setminus K$ into a candidate set T and compute for all $i \in T$ the local error $e_i = |u_i - f_i|$.
2. Choose randomly $n \leq |T|$ pixel indices i from K and reassign $c_i^{\text{new}} := 0$.
3. For all i of the n largest values of $\{e_i | i \in T\}$, reassign $c_i^{\text{new}} := 1$.
4. Compute $\boldsymbol{u}^{\text{new}} := r(\boldsymbol{c}^{\text{new}}, \boldsymbol{f})$.
5. If $\text{MSE}(\boldsymbol{u}) > \text{MSE}(\boldsymbol{u}^{\text{new}})$
      $\boldsymbol{u} := \boldsymbol{u}^{\text{new}}$ and $\boldsymbol{c} := \boldsymbol{c}^{\text{new}}$.
      Update K.
   else
      Reset $\boldsymbol{c}^{\text{new}} := \boldsymbol{c}$.
6. Clear T.

Fig. 3. Nonlocal pixel exchange

3.3   Results

Let us now evaluate the capabilities of the probabilistic sparsification and the nonlocal pixel exchange. To this end, we consider the test image *trui*, which is depicted in Figure 1(a). We apply the probabilistic sparsification to select only 5% of all pixels. For comparison, we choose the same amount of pixels randomly. In addition, we compare our method with an inpainting mask which relies on the analytic approach of Belhachmi et al. [2]: We first compute the Laplace magnitude $|\Delta f_\sigma|$ of the Gaussian presmoothed original image, using a standard deviation σ. Then we rescale the obtained data and apply electrostatic halftoning [17] such that we obtain a dithered version which contains only 5% of all pixels. We decided to favour the electrostatic halftoning over simpler dithering approaches, since it has proven to be the state-of-the-art method for discretising a continuous distribution function. In the following, we say "analytic approach" when referring to this method and choose the standard deviation σ of the Gaussian presmoothing such that the MSE is minimal.

The resulting masks as well as the corresponding reconstruction results are depicted in Figure 6(a)–(c) and (e)–(g). As expected, the random mask gives poor quality reconstructions. Comparing the reconstructed images of the analytic approach and the probabilistic sparsification, we observe that the latter has a lower reconstruction error. This shows that we cannot immediately deduce an optimal pixel set from the optimal continuous theory.

If we now additionally apply the nonlocal pixel exchange to the mask that was obtained by the probabilistic sparsification, we also get a visually more pleasant result (see Figure 6(d) and (h)). The MSE is decreased to 23.21, which is much better than the MSE of the analytic approach that is 49.47.

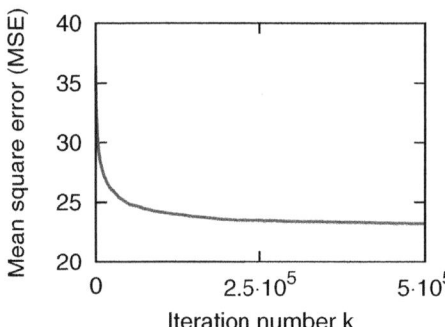

Fig. 4. Convergence behaviour of the nonlocal pixel exchange ($m = 10$, $n = 1$, 500,000 iterations) applied to the mask obtained by the probabilistic sparsification (see Figure 6(c)) with 5% of all pixels

The plot in Figure 4 shows that the nonlocal pixel exchange achieves the most significant improvement during the first 50,000 iterations. On the other hand, it illustrates that after 500,000 iterations the real optimum is still not reached, even though we are probably rather close to it.

4   Optimising Tonal Data

So far we explained how to obtain approximatively optimal positions for a predefined amount of pixels. This is considered as spatial data optimisation. However, it is also possible to optimise the data with respect to the tonal data (i.e. the co-domain).

For inpainting, we usually use the original grey values of the input image. Now we allow arbitrary grey values and thus accept to introduce some error at the positions of mask pixels in favour of a lower overall reconstruction error.

4.1   Grey Value Optimisation

Let us start by stressing that the homogeneous inpainting function $r(c, f)$ is a linear function with respect to the grey values f. This allows us to formulate a least squares approach, with which we can compute the optimal grey values for a given mask exactly.

For a given mask c and given data f the solution $u = r(c, f)$ of the discrete homogeneous inpainting process (6) is given by

$$r(c, f) := M^{-1} C f . \tag{8}$$

Since M only depends on c it follows directly that r is a linear function in f.

Least squares approximation. Our goal is to find grey values g such that $\text{MSE}(r(c, g))$ becomes minimal for a fixed mask c. To this end, we suggest the following minimisation approach:

$$\operatorname*{argmin}_{\alpha} \| f - r(c, f + \alpha) \|^2 , \tag{9}$$

such that $g = f + \alpha$. Let e_i denote the vector with a 1 in the i-th coordinate and zeros elsewhere. Then we call $r(c, e_i)$ the *inpainting echo* of the i-th pixel. By linearity and $\alpha = \sum_{i \in J} \alpha_i e_i$ it follows that

$$r(c, f + \alpha) = r(c, f) + r(c, \alpha) = r(c, f) + \sum_{i \in J} \alpha_i r(c, e_i) . \tag{10}$$

Since the $r(c, e_i)$ is 0 if $c_i = 0$ (i.e. $i \in J \setminus K$), we get

$$r(c, f + \alpha) = r(c, f) + \sum_{i \in K} \alpha_i r(c, e_i) . \tag{11}$$

For our minimisation problem (9), this means that α_i can be chosen arbitrarily if $i \in J \setminus K$. Thus, for the sake of simplicity, we set $\alpha_i = 0$ for $i \in J \setminus K$. The remaining α_i with $i \in K$ can be obtained by considering the least squares problem:

$$\underset{\alpha_K}{\operatorname{argmin}} \|U \alpha_K - b\|^2 , \tag{12}$$

where $b = r(c, f) - f$ is a vector of size $|J|$, $\alpha_K = (\alpha_i)_{i \in K}$ is a vector of size $|K|$, and U is a $|J| \times |K|$ matrix which contains the vectors $r(c, e_i)$, $i \in K$ as columns.

Its solution is given by solving the normal equations:

$$U^{\top} U \alpha_K = U^{\top} b . \tag{13}$$

Let us first prove that the matrix $U^{\top} U$ is invertible: Since U contains the vectors $r(c, e_i)$, $i \in K$ as columns, it is sufficient to show that the vectors $r(c, e_i)$ with $i \in K$ are linearly independent. It holds that

$$r(c, e_i) = M^{-1} C e_i \overset{i \in K}{=} M^{-1} e_i . \tag{14}$$

Hence, $r(c, e_i)$ is the i-th column of M^{-1} and since M^{-1} exists [13], the vectors $r(c, e_i)$ have to be linearly independent. Thus, $U^{\top} U$ is invertible.

Iterative approach. The linear system given by Equation 13 can be solved exactly by using standard methods such as an LU-decomposition. Since this is rather slow, we suggest the following iterative solver.

Let us for a moment consider the simplified optimisation problem, where a vector g and the inpainting result $u = r(c, g)$ are initially given. We want to optimise only the i-th grey value and keep the remaining grey values fixed:

$$\underset{\alpha}{\operatorname{argmin}} \|f - r(c, g + \alpha e_i)\|^2 . \tag{15}$$

Then the solution is

$$\alpha = \frac{r(c, e_i)^{\top} (f - u)}{r(c, e_i)^{\top} r(c, e_i)} . \tag{16}$$

The optimised grey value can be computed as $g_i := g_i + \alpha$. Moreover, provided we have precomputed all inpainting echos $r(c, e_i)$, $i \in K$, we can not only efficiently compute

Input: Original image f, attenuation factor ω.
Output: Optimised grey values g.
Initialisation: $u := r(c, f)$ and $g := f$.

Do

    For all $i \in K$ (randomly chosen):
      1. Get the inpainting echo $u_i := r(c, e_i)$.
      2. Compute the correction term $\alpha := \dfrac{u_i^\top (f - u)}{u_i^\top u_i}$.
      3. Set $u_{\mathrm{old}} := u$.
      4. Update the reconstruction $u := u + \omega \cdot \alpha \cdot u_i$.
        and the grey value $g_i := g_i + \omega \cdot \alpha$.

while $|\mathrm{MSE}(u) - \mathrm{MSE}(u_{\mathrm{old}})| > \varepsilon$.

Fig. 5. Grey value optimisation

α, but also the inpainting result for the updated image g. To this end, we exploit again the linearity of r:

$$r(c, g) := r(c, g + \alpha e_i) = r(c, g) + \alpha \cdot r(c, e_i) = u + \alpha \cdot r(c, e_i) . \qquad (17)$$

If we apply this optimisation for each $i \in K$ iteratively, and update the grey values in each step directly, we obtain an algorithm which corresponds to the Gauss-Seidel method [16] for the previously presented linear system of equations (see Equation 13).

Optimising one grey value at a time means that this grey value might be shifted extremely in order to reduce the inpainting error in its neighbourhood. However, there could be mask points nearby which are not optimised yet and a combined optimisation would lead to a smaller shift for each of them. Thus, to prevent over- and undershoots we suggest to introduce an attenuation factor ω. This can be seen as a variant of the so-called *successive over-relaxation method (SOR)* [16] with under-relaxation instead of over-relaxation. Figure 5 summarises our iterative algorithm.

We terminate our algorithm when the qualitative improvement from one to the next iteration step decreases to a value smaller then $\varepsilon = 0.001$. Moreover, note that we choose the indices $i \in K$ randomly in each run. This allows more stable results, since we do not rely on a specific pixel ordering for each run and thus the approximation error is better distributed over the whole image.

4.2 Results

We apply the presented grey value optimisation to the test image *trui* and the mask obtained by our spatial optimisation method (see Figure 6(d)). However, we actually can use the grey value optimisation to optimise the grey values for any fixed mask. Thus, for the sake of comparison, we also consider the random mask, the mask obtained by the analytic approach, and the mask created with the probabilistic sparsification (see Figure 6(a),(b) and (c)). The reconstruction results are depicted in the last row of Figure 6.

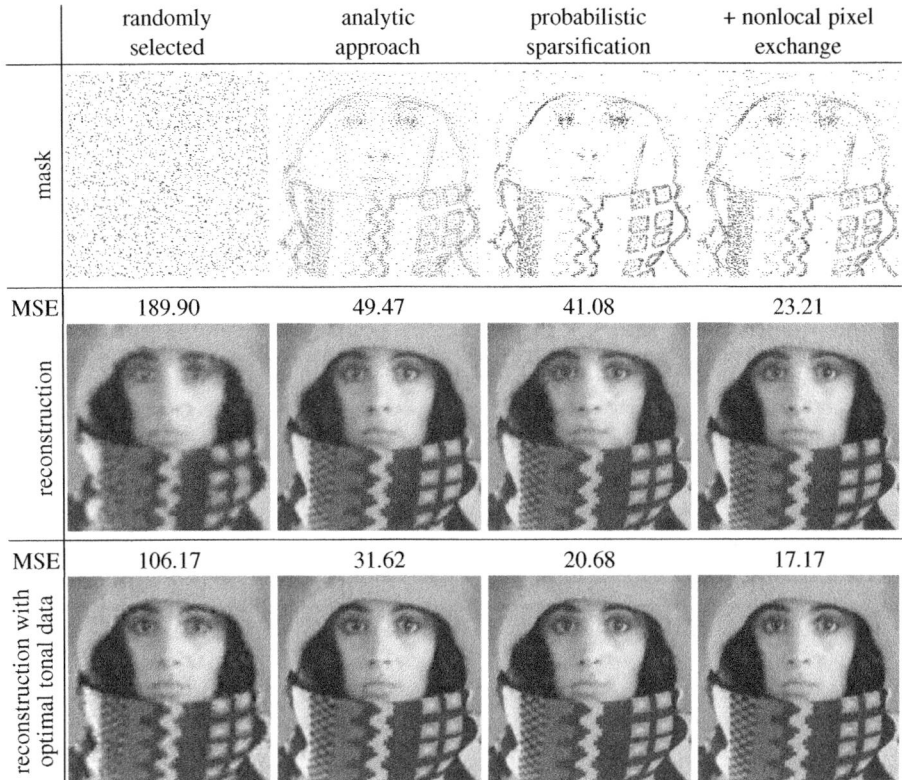

	randomly selected	analytic approach	probabilistic sparsification	+ nonlocal pixel exchange
MSE	189.90	49.47	41.08	23.21
MSE	106.17	31.62	20.68	17.17

Fig. 6. Evaluation of different inpainting data using 5% of all pixels. **Top row:** Different masks obtained by **(a)** random selection, **(b)** analytic approach ($\sigma = 1.44$), **(c)** probabilistic sparsification ($p = 0.02$, $q = 0.02$, $d = 0.05$), **(d)** nonlocal pixel exchange ($m = 10$, $n = 1$, 500,000 iterations) applied to (c). **Middle row: (e-h)** Reconstructions with homogeneous diffusion inpainting using the masks (a-d). **Bottom row: (i-l)** As middle row but using optimal tonal data

For all examples the MSE has decreased. However, we observe that the worse the spatial data are selected, the larger the improvement that can be achieved by the grey value optimisation. The explanation for this behaviour is simple. Both our spatial optimisation method as well as the analytic approach select the pixels depending on the grey values of the original image. Thus, the spatial data are optimised by incorporating these tonal data. If we choose random spatial data, it is more likely that we can compensate bad locations by adapting the grey values.

Besides this observation, the smallest MSE of only 17.17 is obtained by the combination of the presented spatial and tonal optimisation methods. To further evaluate this combined approach, we apply it to two other test images. The results are depicted in Figure 7. Moreover, Table 1 gives a comparison with the analytic approach and the

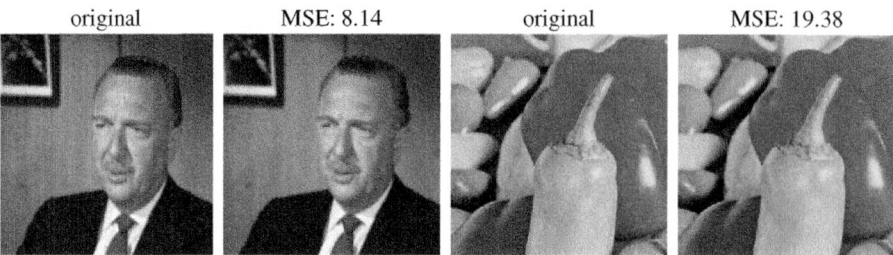

| original | MSE: 8.14 | original | MSE: 19.38 |

Fig. 7. Reconstruction results with 5% of all pixels, spatially and tonally optimised, for the test images *walter* and *peppers256*

Table 1. Comparison of the reconstruction error (MSE) with 5% of all pixels for different test images and different inpainting data

	unoptimised (randomly selected)	analytic approach ($\|\Delta f_\sigma\|$ dithered)	spatially and tonally optimised
trui	198.90	49.47 ($\sigma = 1.44$)	17.17
walter	183.37	24.59 ($\sigma = 1.37$)	8.14
peppers256	179.22	49.71 ($\sigma = 1.15$)	19.38

results obtained with a random pixel mask. In all cases, we obtain by far the best reconstruction results with our new approach. This confirms that it is a suitable method for the selection of optimal inpainting data.

5   Conclusion

While many researchers have tried to find highly sophisticated PDEs for inpainting problems with given data, we have investigated the opposite way: finding optimal data for a given PDE. We have shown that even for the simplest inpainting PDE, namely homogeneous diffusion, one can obtain reconstructions of astonishing quality using only 5% of all pixels. However, this requires to optimise the data carefully in the domain and the co-domain.

Since we are able to reduce the amount of data needed for high quality reconstructions drastically, our ongoing research addresses the problem how these data can be encoded efficiently. This includes appropriate adaptations of the grey value optimisation to quantised data. Moreover, we are interested in applying our optimisation framework also to nonlinear inpainting methods. As a result, we might obtain similar qualitative reconstructions with even less data, allowing further cuts in our pixel budget.

Acknowledgements. Our research is partly funded by the Deutsche Forschungsgemeinschaft (DFG) through a Gottfried Wilhelm Leibniz Prize. This is gratefully acknowledged. We also thank Pascal Gwosdek and Christian Schmaltz for providing the electrostatic halftoning images for us.

References

1. Bae, E., Weickert, J.: Partial differential equations for interpolation and compression of surfaces. In: Dæhlen, M., Floater, M., Lyche, T., Merrien, J.-L., Mørken, K., Schumaker, L.L. (eds.) MMCS 2008. LNCS, vol. 5862, pp. 1–14. Springer, Heidelberg (2010)
2. Belhachmi, Z., Bucur, D., Burgeth, B., Weickert, J.: How to choose interpolation data in images. SIAM Journal on Applied Mathematics 70(1), 333–352 (2009)
3. Bertalmío, M., Sapiro, G., Caselles, V., Ballester, C.: Image inpainting. In: Proc. SIGGRAPH 2000, New Orleans, LI, pp. 417–424 (July 2000)
4. Bruckstein, A.M., Holt, R.J., Netravali, A.N.: Holographic representations of images. IEEE Transactions on Image Processing 7(11), 1583–1597 (1998)
5. Carlsson, S.: Sketch based coding of grey level images. Signal Processing 15, 57–83 (1988)
6. Di Blasi, G., Francomano, E., Tortorici, A., Toscano, E.: A smoothed particle image reconstruction method. Calcolo 48(1), 61–74 (2011), http://dx.doi.org/10.1007/s10092-010-0028-3
7. Elder, J.H.: Are edges incomplete? International Journal of Computer Vision 34(2/3), 97–122 (1999)
8. Galić, I., Weickert, J., Welk, M., Bruhn, A., Belyaev, A., Seidel, H.P.: Image compression with anisotropic diffusion. Journal of Mathematical Imaging and Vision 31(2–3), 255–269 (2008)
9. Hummel, R., Moniot, R.: Reconstructions from zero-crossings in scale space. IEEE Transactions on Acoustics, Speech, and Signal Processing 37, 2111–2130 (1989)
10. Johansen, P., Skelboe, S., Grue, K., Andersen, J.D.: Representing signals by their toppoints in scale space. In: Proc. Eighth International Conference on Pattern Recognition, Paris, France, pp. 215–217 (October 1986)
11. Kanters, F.M.W., Lillholm, M., Duits, R., Janssen, B.J.P., Platel, B., Florack, L.M.J., ter Haar Romeny, B.M.: On image reconstruction from multiscale top points. In: Kimmel, R., Sochen, N.A., Weickert, J. (eds.) Scale-Space 2005. LNCS, vol. 3459, pp. 431–442. Springer, Heidelberg (2005)
12. Lillholm, M., Nielsen, M., Griffin, L.D.: Feature-based image analysis. International Journal of Computer Vision 52(2/3), 73–95 (2003)
13. Mainberger, M., Bruhn, A., Weickert, J., Forchhammer, S.: Edge-based image compression of cartoon-like images with homogeneous diffusion. Pattern Recognition 44(9), 1859–1873 (2011)
14. Masnou, S., Morel, J.M.: Level lines based disocclusion. In: Proc.1998 IEEE International Conference on Image Processing, Chicago, IL, vol. 3, pp. 259–263 (October 1998)
15. Morton, K.W., Mayers, L.M.: Numerical Solution of Partial Differential Equations, 2nd edn. Cambridge University Press, Cambridge (2005)
16. Saad, Y.: Iterative Methods for Sparse Linear Systems, 2nd edn. SIAM, Philadelphia (2003)
17. Schmaltz, C., Gwosdek, P., Bruhn, A., Weickert, J.: Electrostatic halftoning. Computer Graphics Forum 29(8), 2313–2327 (2010)
18. Schmaltz, C., Weickert, J., Bruhn, A.: Beating the quality of JPEG 2000 with anisotropic diffusion. In: Denzler, J., Notni, G., Süße, H. (eds.) DAGM 2009. LNCS, vol. 5748, pp. 452–461. Springer, Heidelberg (2009)
19. Uhlir, K., Skala, V.: Reconstruction of damaged images using radial basis functions. In: Proc.13th European Signal Processing Conference (EUSIPCO), Antalya, Turkey, pp. 160–163 (September 2005)
20. Zeevi, Y., Rotem, D.: Image reconstruction from zero-crossings. IEEE Transactions on Acoustics, Speech, and Signal Processing 34, 1269–1277 (1986)

Nonlocal Surface Fairing

Serena Morigi, Marco Rucci, and Fiorella Sgallari

Department of Mathematics-CIRAM, University of Bologna, Bologna, Italy
{morigi,rucci,sgallari}@dm.unibo.it

Abstract. We propose a new variational model for surface fairing. We extend nonlocal smoothing techniques for image regularization to surface smoothing or fairing, with surfaces represented by triangular meshes. Our method is able to smooth the surfaces and preserve features due to geometric similarities using a mean curvature based local geometric descriptor. We present an efficient two step approach that first smoothes the mean curvature normal map, and then corrects the surface to fit the smoothed normal field. This leads to a fast implementation of a feature preserving fourth order geometric flow. We demonstrate the efficacy of the model with several surface fairing examples.

1 Introduction

A surface smoothing method, in the following named fairing, removes undesirable noise and uneven edges from discrete surfaces. The fairing problem arises mainly when creating high-fidelity computer graphics objects using imperfectly-measured data from the real world, captured for example from 3D laser scanner devices. Fairing can be applied either before or after generating the mesh from sampled data. The advantage of denoising a mesh rather than a point-cloud, is that the connectivity information implicitly defines the surface topology and can be exploited as a means for fast access to neighboring samples.

The goal is to remove noise from a surface while keeping features, e.g. sharp edges, corners and ridges. There are essentially two ways to represent surfaces. Explicit representations, such as meshes, are most commonly used by the computer graphics community, while implicit representations, usually based on level set functions, are mostly considered by partial differential equations (PDEs) community. Explicit surface representations offer a easier way to discretize differential operators, but topological changes are harder to handle. However, for smoothing processing driven by PDE models, the evolution is sufficiently slow to avoid both topological modifications and triangle flips in triangular meshes.

In this paper we will focus on explicitly represented surfaces, that is we process a triangular mesh M which represents a piecewise-linear approximation of a smooth surface $\mathcal{M}$. We assume that the surface $\mathcal{M}$ is a two-dimensional manifold embedded in $\mathbb{R}^3$, and we denote by (Ω, X) a chart of $\mathcal{M}$, where $\Omega \subset \mathbb{R}^2$ is an open reference domain and X is the corresponding coordinate map, that is the parametrization of $\mathcal{M}$ at a given point. Let M be defined by a set T of triangles $T_i, i = 1, \ldots, N_t$, that cover M, and a set X of vertices $X_i, i = 1, \ldots, N_v$, where $X_i \in \mathbb{R}^3$ is the ith vertex.

A.M. Bruckstein et al. (Eds.): SSVM 2011, LNCS 6667, pp. 38–49, 2012.
© Springer-Verlag Berlin Heidelberg 2012

The most common surface degradation model, when the observed data $X^0 \in \mathbb{R}^{n^3}$ are corrupted by a random variation of the vector field, is

$$X = X^0 + \vec{E}, \tag{1}$$

where $\vec{E} \in \mathbb{R}^{n^3}$ accounts for the vector perturbations.

We present an original two step approach which implements nonlocal surface diffusion flow on surfaces represented by meshes. First, we smooth the mean curvature normal map of a surface, and next we manipulate the surface to fit the processed smoothed curvature normal vector field. We show we can efficiently implement geometric fourth-order flow by solving a set of second order PDEs discretized on the mesh M. Inspired by [7] we integrate a nonlocal approach into this framework driven by a mean curvature based local geometric descriptor. We look for patches that have similarities in order to reduce noise while preserving the surface details.

Variational and PDE-based surface denoising models have had great success in the past ten years. Several authors presented isotropic/anisotropic denoising of surfaces applying image processing methodology based on linear/nonlinear diffusion equations [1][11][12][13][14].

In [15] the authors propose a point cloud nonlocal denoising using the signed distance function as local surface descriptor in a point-wise process. Similar descriptors are used in the nonlocal denoising method proposed in [16] where instead of the moving least square representation, the authors used local radial basis functions. In [7] a nonlocal diffusion process is derived as steepest descent of a nonlocal quadratic functional of weighted differences. This formulation is an excellent framework for nonlocal variational image denoise, Bregman iterations, and segmentation. A nonlocal heat equation for denoising surfaces has been introduced in [8], where the signed distance function is used to define the similarity weights, and the PDE evolution is solved using a level set formulation on an implicitly defined surface.

Let us briefly introduce a key ingredient in our variational denoising formulation. The Laplace-Beltrami on $\mathcal{M}$, $\Delta_{\mathcal{M}}$, is a local operator acting on a smooth function η. When the mesh M approximates the manifold $\mathcal{M}$, the discretization of $\Delta_{\mathcal{M}}(\eta)$ on M, (denoted by L) evaluated at the vertex X_i, is given by

$$L\eta(X_i) = \sum_{j \in N(i)} w_{ij}(\eta(X_j) - \eta(X_i)), \tag{2}$$

where $N(i)$ is the set of 1-ring neighbor vertices of vertex X_i, and the weights w_{ij} are positive numbers and satisfy the normalization condition $\sum_{j \in N(i)} w_{ij} = 1$. The weights w_{ij} are "local", thus the summation in (2) is "local". Different geometric discretizations of the Laplacian can be obtained for different choices of the weights in (2), the most common, introduced by Meyer et al. in [3], is

$$w_{ij} = (\cot \alpha_{ij} + \cot \beta_{ij}), \tag{3}$$

where α_{ij} and β_{ij} are the two angles opposite to the edge in the two triangles sharing the edge (X_j, X_i).

The paper is organized as follows. We briefly describe the nonlocal image denoising algorithm in Section 2. The variational nonlocal approach is described in Section 3, and the proposed algorithm together with its numerical aspects are discussed in Section 4. Numerical examples and comments are provided in Section 5. Section 6 contains concluding remarks.

2   Nonlocal Means Image Denoising

Nonlocal denoising is an algorithm for image denoising introduced in [9]. The algorithm aims to denoise a gray-scale image I, defined over a rectangular bounded domain Ω, by replacing each pixel with a weighted mean of the neighborhoods. The new value of the image pixel is

$$NL[I](x) = \int_\Omega W(x,y)I(y)dy, \tag{4}$$

where the convolution kernel $W(x,y)$ is given by

$$\begin{aligned} W(x,y) &= \tfrac{1}{C(x)}e^{-D(I(x),I(y))/c} \\ D(I(x),I(y)) &= \|I(x) - I(y)\|_2^2, \quad y \in N(x) \end{aligned} \tag{5}$$

with a normalization factor $C(x) = \int_\Omega W(x,y)$, $N(x)$ represents a neighborhood of x, and c is a filtering parameter which is related to the noise level. The similarity between pixels is measured by the similarity kernel D in (5) and depends on the similarity of gray-level intensities in the neighborhood of x and y, that is, the algorithm not only compares the (color) value at a single pixel but the geometrical configuration in a whole neighborhood. The algorithm gives excellent results in image denoising (see [10],[9]). For a more detailed analysis on the NL-means algorithm see [9].

3   The Nonlocal Variational Fairing

We propose a variational formulation in order to derive our nonlocal approach to surface fairing.

For a surface parameterization X of $\mathcal{M}$ on a domain Ω, and a given vector field $f \in \mathbb{R}^{n^3}$, we consider the minimization of the following functional

$$Inf_X \int_\Omega |\nabla_W X|^2 + \frac{\lambda}{2}(X - f)^2 d\omega, \tag{6}$$

where $\lambda > 0$ is a regularization parameter and ∇_W is a weighted gradient operator. The corresponding Euler-Lagrange descent flow can be written as

$$\frac{\partial X}{\partial t} = \int_\Omega (X(y) - X(x))W(x,y)d\omega + \lambda(f - X), \tag{7}$$

with $x, y \in \Omega$, (see [8] for a similar definition). Here $W(x,y)$ is the weight function, which satisfies $W(x,y) \geq 0$, and is symmetric $W(x,y) = W(y,x)$.

For image processing the weight function can be defined as in (5). The spatial discretization of (7) on the mesh M, is

$$\frac{\partial X_i}{\partial t} = \sum_{j \in N(i)} W_{ij}(X_j - X_i) + \lambda(f_i - X_i), \tag{8}$$

where X_i denotes the value of X at the ith vertex, $i = 1, \ldots, N_v$, and $N(i)$ is the set of 1-ring neighbor vertices of the ith vertex.

Let $f(x) := (f^1, f^2, f^3)(x)$ be a vector field on $\mathcal{M}$, $W(x,y)$ is the same for all vector components. Let $X(x) := (X^1, X^2, X^3)(x)$ be the coordinate function vector on M, where X^1 is the scalar function that defines the first coordinate of point $x \in \mathcal{M}$, and analogously for the second and the third coordinate scalar functions. Then the regularizing formulation (8) for each vector component $X^k, k = 1, 2, 3$, is

$$\frac{\partial X_i^k}{\partial t} = \sum_{j \in N(i)} W_{ij}(X_j^k - X_i^k) + \lambda(f_i^k - X_i^k), \tag{9}$$

by initializing, e.g., each component k of X as $X^k|_{t=0} = f^k$.

If we let $W_{ij} = w_{ij}$, with w_{ij} defined by (3), then the regularized PDEs (9) can be interpreted as the spatial discetization on M of the well know *mean curvature flow* (MCF)

$$\frac{\partial X}{\partial t} = \triangle_\mathcal{M} X + \lambda(X^0 - X), \quad X|_{t=0} = X^0, \tag{10}$$

with initial surface X^0. The first term in (10) is the *regularization* term, while the second one is the *fidelity* term.

The mean curvature flow is known to have a strong regularization effect, because it is the gradient flow for the area functional. In a discrete setting, the mean curvature flow moves every vertex in the normal direction with the speed equal to a discrete approximation of the mean curvature at the vertex. It is also well known that the mean curvature flow performs well in smoothing (fairing) but produces uneven distribution of vertices.

In [1] the authors present finite element schemes for MCF on triangulated surfaces. analogously, in [2], implicit and explicit discretizations using cotangent discretization are considered. Unfortunately, MCF not only decreases the geometric noise due to unprecise measurements, but also smoothes out geometric features such as edges and corners of the surfaces.

In view of (2) we propose the following nonlocal operator that we define the *weighted Laplace-Beltrami operator* on M,

$$L_w X_i = \sum_{j \in N(i)} (X_j - X_i) W_{ij} w_{ij}, \tag{11}$$

where w_{ij} is defined as in (2), while W_{ij} depends on a similarity measure between ith and jth vertex. A proposal of similarity weight functions in surface processing is discussed in Section 4.

By initializing, e.g., with $X|_{t=0} = X^0$, and using the nonlocal operator (11), then (8) can be rewritten as

$$\frac{\partial X_i}{\partial t} = L_w X_j + \lambda(X_i^0 - X_i). \tag{12}$$

In Section 4, we apply the nonlocal variational approach to mesh fairing and develop a new mesh smoothing method which solves a fourth order surface diffusion equation on $\mathcal{M}$.

4   Non Local Surface Diffusion Flow (NL-SDF)

Let H be the mean curvature function on the mesh M, defined as the sum of the two principal curvatures $H(X) = k_1 + k_2$, and $\overrightarrow{H}(X) = H(X)\overrightarrow{N}(X)$ be the mean curvature normal vector field.

Replacing X with $\overrightarrow{H}$ in (9), and considering a uniform discretization of the time interval $[0, T], T > 0$, with a temporal time step dt, then (12) can be fully discretized using a variety of explicit or implicit time integration schemes. In our computational method, we used the forward Euler scheme which yields a first order scheme in time. Therefore, applying an implicit scheme to (12), without the fidelity term, we get the iterative scheme

$$(I - dt L_w)\overrightarrow{H}_i^{n+1} = \overrightarrow{H}_i^n, \quad \overrightarrow{H}|_{t=0} = \overrightarrow{H}^0, \tag{13}$$

where L_w is computed as given in (11), with initial condition $\overrightarrow{H}^0$ determined from X^0.

The number of time iterations n is chosen by the user; from our experimental work we tuned up $n_{MAX} \leq 20$ (see the numerical Section 5 for more details).

For image processing the weight function is defined by image features and represents the similarity between two pixels, based on features in their neighborhood, see [9]. Working with surfaces, the way of choosing weight W_{ij} in (11) should characterize the similarities between two local surface patches. We propose to use the mean curvature values. Therefore, according to (5), we define the weights as follows

$$W_{ij} = \frac{1}{\sum_{j \in N(i)} W_{ij}} e^{-D(X_i, X_j)/\sigma},$$
$$D(X_i, X_j) = \|\overrightarrow{H}(X_i) - \overrightarrow{H}(X_j)\|_2^2, \quad j \in N(i). \tag{14}$$

The parameter σ controls how much the similarities of two patches are penalized. Larger σ gives results with sharper features. By using (14) we get a good measurement of similarity, which penalizes the contribution in (11) of the vertices with different curvature features and is rotationally invariant.

We propose a two-step strategy which first smoothes the normal vectors allowing the mean curvature normals to diffuse on M, then the second step refits the parameterization X according to a given mean curvature distribution.

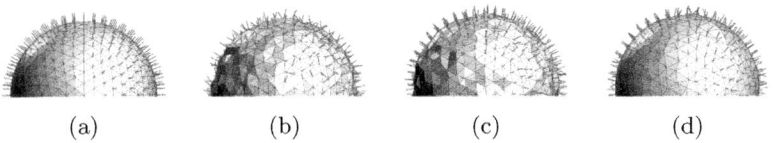

(a)   (b)   (c)   (d)

Fig. 1. (a) The noise-free sphere mesh; (b) the perturbed sphere; (c) the smoothed mean curvature vector field obtained by step 1; (d) reconstructed sphere by step 2.

The normal vector smoothing (13) is "nonlocal". By this we mean that a "nonlocal" operator is used which includes weights that penalize the similarity between patches.

The nonlocal approach is described by the following algorithm, where in step 1 we solve (13) by a sequence of linear systems, the smoothed mean curvature normal vector field is then plugged into the constrained least square problem in step 2, which is solved by a LSQR iterative method. Here L is defined by (2) and L_w as in (11).

Non Local SDF Algorithm
Given an initial position vector X^0,
STEP 1: SOLVE FOR H:
For each $n = 1, \cdots, n_{MAX}$
 $(I - dtL_w)\overrightarrow{H}^{n+1} = \overrightarrow{H}^n$
end for
STEP 2: PLUG IN $\overrightarrow{H}$ AND SOLVE FOR X:
 $min_X \|LX - \overrightarrow{H}\|_2^2 + \lambda \|X_0 - X\|_2^2$

Fig. 1 shows how the two step NL-SDF algorithm works. A noise-free sphere mesh together with the associated mean curvature normal field is shown in (Fig. 1 (a)). The mesh is perturbed by a randomly chosen noise vector field. The perturbed sphere is illustrated in Fig. 1(b). The smoothed mean curvature vector field obtained by applying 10 iterations of step 1 is shown in Fig. 1 (c), while the recovered sphere resulting from applying step 2 using the smoothed normal vector field, is shown in Fig. 1 (d).

We shall assume a certain level of connectivity in the mesh such that there will not be any disjoint regions where no information is exchanged between them throughout the evolution. Thus we assume that M consists of only one connected mesh. The matrix L has $rank(L) = N_v - k$, where k is the number of connected components of M, and it is positive semi-definite. Since we imposed that M is connected, that is $k = 1$, then L has a zero eigenvalue with multiplicity 1. The linear system derived from solving step 2 is uniquely solvable by fixing a vertex to have an assigned value.

When the perturbation on the initial mesh affects only the magnitude of the normal field, that is $\overrightarrow{E}$ in (1) are in the normal directions, we can replace $\overrightarrow{H}$ with H in step 1 and step 2, thus processing the mean curvature scalar field instead of the mean curvature normal vector field.

In the following we theoretically justify the NL-SDF algorithm, which approaches to the solution of a fourth-order PDE representing a nonlocal surface diffusion flow on $\mathcal{M}$.

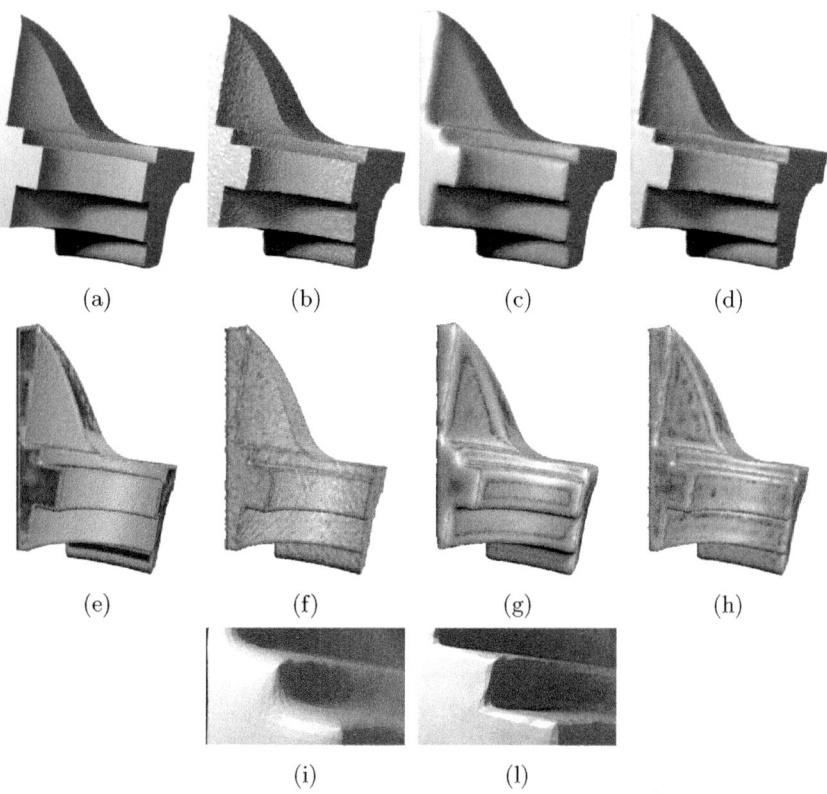

Fig. 2. fandisk mesh: (a) Noise-free mesh and its curvature map (e); (b) noisy mesh and its curvature map (f); (c) restored mesh by the two step SDF and its curvature map (g); (d) restored mesh by NL-SDF algorithm and its curvature map (h); (i) and (l) zoomed details from (c) and (d), respectively.

Let us suppose that the weight functions $W(x, y)$ are defined as in (14), and $\lambda = 0$. Then the sequence $\{X^{(n)}\}$, generated by the NL-SDF algorithm is convergent to the solution X^* of the fourth order **Non Local Surface Diffusion Flow (NL-SDF)** on $\mathcal{M}$

$$\frac{\partial X}{\partial t} = \triangle_{w\mathcal{M}} H(X), \quad X(0) = X_0, \tag{15}$$

where $\triangle_{w\mathcal{M}}$ is a nonlocal Laplace Beltrami operator, and M is the piecewise linear representation of $\mathcal{M}$.

We factorize (15) into a set of two nested second order PDEs

$$1. \frac{\partial \vec{H}}{\partial t} = \triangle_{w\mathcal{M}} \vec{H}(X), \quad \vec{H}(0) = \vec{H}_0,$$
$$2. \triangle_{\mathcal{M}} X = \vec{H}, \tag{16}$$

implemented in step 1 and step 2 of the NL-SDF algorithm, respectively.

The nonlinear parabolic PDE (16).1 can be interpreted as a diffusion flow for the vectors H_i. The unknown mean curvature vectors at the vertices are determined by an implicit scheme which leads to a nonsingular linear system $(I - dtL_w)$, with the matrix L_w defined as in (11) that discretizes $\triangle_{w\mathcal{M}}$. The computed mean curvature normals $\vec{H}_i, i = 1, ..., N_v$ are then used to move each vertex ith, according to the well known relation [3]

$$\vec{H} = H(X)\vec{N}(X) = -\triangle_{\mathcal{M}} X, \tag{17}$$

where $\vec{N}(X)$ is the unit outward normal of the surface at point X. The discretization L in (2) with weights defined by (3) is shown to be convergent to the Laplace-Beltrami operator $\triangle_{\mathcal{M}}$ applied to $f \in C^2(\mathcal{M})$ except for special cases, see [17]. This justifies the use of L to discretize the second step.

On the other hands, considering the similarity weights $W_{ij} = 1, \forall i, j$, and $\lambda = 0$, then the NL-SDF algorithm approaches to the solution of the Surface Diffusion Flow (SDF): $\frac{\partial X}{\partial t} = \triangle_{\mathcal{M}} H(X)$.

Moreover, if $\mathcal{M}(t)$ is a closed surface then the volume of the bounded domain computed by both NL-SDF and SDF is preserved.

In [4] the two step method is applied to solve the elliptic fourth order PDE $\triangle_M H = 0$. A pioneer approach to the two-step denoising procedure with a fourth order model is introduced in [18]. A level set formulation of a two step geometric denoising via normal maps is also presented in [12].

5   Numerical Results

The results of the proposed algorithm are demonstrated applying perturbations to the meshes shown in Fig. 2(a), Fig.3(a) and Fig.4(a). The meshes present different characteristics in terms of details, "sharpness", and level of refinement, as summarized in Table 1.

Table 1. Data for the meshes used in the examples

Mesh	Faces	Vertices	Volume	c_1	c_2
fandisk	51784	25894	0.234024	-0.6	-0.6
oilpump	82176	41090	0.184494	-0.8	-0.8
igea	268686	134345	0.376882	0.8	0.8

The meshes are corrupted by adding a perturbation vector $\overrightarrow{E_i}$ for each vertex i of the mesh according to (1). We let $\overrightarrow{E_i}$ be a weighted sum of the normal vector $\overrightarrow{N(X_i)}$, and a random-direction unitary vector $\overrightarrow{v}$,

$$\overrightarrow{E_i} = \frac{c_1 \overrightarrow{N(X_i)} + c_2 \overrightarrow{v}}{\bar{e}}, \quad c_1, c_2 \in [-1, 1], \tag{18}$$

where $\bar{e}$ is a scaling factor determined by the edge length average of the mesh, and c_1 and c_2 are assigned scalar parameters that control the maximum length of the corresponding vectors.

Table 2. Data for the examples shown in Fig. 2, 3 and 4

Mesh	Algorithm	dt	n_{MAX}	$\Delta V(\%)$	σ
fandisk	MCF	0.013	10	8.11×10^{-4}	-
fandisk	SDF	0.139	10	0.89×10^{-4}	-
fandisk	NL-SDF	0.139	10	0.92×10^{-4}	0.6
oilpump	MCF	0.013	10	5.93×10^{-4}	-
oilpump	SDF	0.077	10	0.32×10^{-4}	-
oilpump	NL-SDF	0.022	10	0.32×10^{-4}	0.5
igea	MCF	0.146	20	3.90×10^{-4}	-
igea	SDF	0.141	20	0.02×10^{-4}	-
igea	NL-SDF	0.141	20	0.02×10^{-4}	0.6

The amount of noise added to the meshes is then controlled by parameters c_1 and c_2, whose values are reported in Table 1. The perturbed versions of the meshes in the examples are shown in Fig. 2(b), Fig.3(b) and Fig.4(b).

Table 2 summarizes the experiments illustrated in this section. We compared the performance of the proposed NL-SDF method with MCF and SDF algorithms. The parameter λ for the fidelity term in step 2 of the algorithm is set to be 0.5. In Table 2 for each mesh (first column), the algorithm applied is shown in the second column, the corresponding time step used (dt) is provided in the third column, while the number of iteration steps (n_{MAX}) is in the fourth column. The differences in volume are labeled by $\Delta V\%$, and σ is the parameter in the weight functions (14).

The three models compared NL-SDF, MCF and SDF are all discretized by implicit schemes to avoid stability conditions on the time step. The time step dt for the iterative process is automatically chosen using the formula

$$dt = \frac{10\bar{e}}{\max_i(\|\ell_i X\|)},$$

where the denominator represents the maximum norm value of the displacement vectors, and ℓ_i is the ith row of L_w. This choice allows for producing a good quality denoised mesh, using about 5 to 20 iterations, independently on the mesh

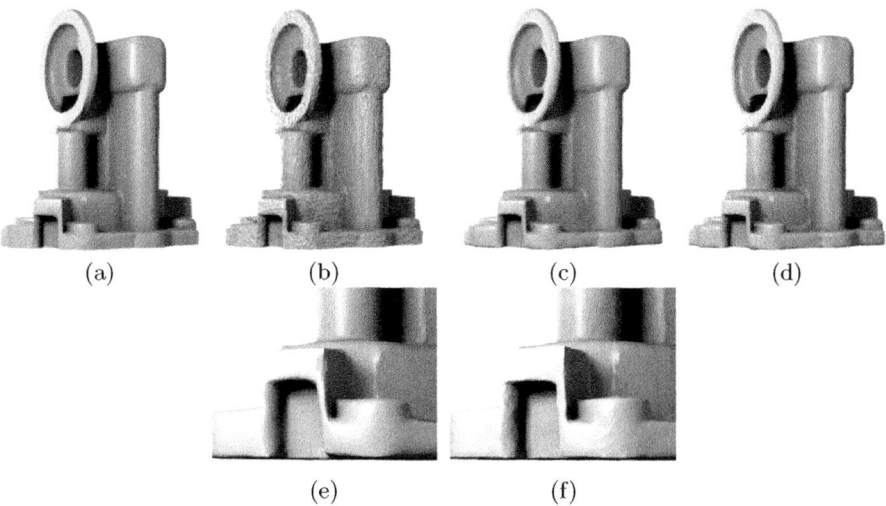

Fig. 3. olipump mesh: (a) noise-free mesh; (b) noisy mesh; (c) restored mesh by the SDF algorithm; (d) restored mesh by NL-SDF algorithm; (e) and (f) zoomed details from (c) and (d), respectively.

characteristics or the Laplacian weights in (2). In Fig.2 and Fig.3 we compare the recovered fandisk and oilpump meshes by applying algorithms SDF and NL-SDF. In Fig.2, second row, by false colors we represented the value of the norm of the mean curvature vector associated to each vertex of the corresponding mesh in the first row. In Fig. 4 we compare the recovered igea meshes by applying algorithms MCF and NL-SDF. The example shown in Fig. 4 demonstrates that the proposed method can produce better results even on more naturally smooth meshes. From a visual inspection of Fig.2 and Fig.3, we can observe that, while the SDF and MCF algorithms well accomplish the task of denoising the surface, they fail in distinguishing the edges and sharp corners from the noise. The NL-SDF algorithm clearly enhances sharp features of the object while removing the noise in the flat areas. The overhead of computational effort for NL-SDF with respect to SDF, is negligible and it consists in computing the weights W_{ij} in (14). The superiority of the NL-SDF method can be better appreciated in the more detailed and sharp areas of the mesh, where the features are reconstructed preserving the sharpness of the original noise-free mesh as shown in Fig.2(i), (l), and Fig.3(e) and (f).

In Table 2 we labeled by $\Delta V(\%)$ the difference between the volume of the noise-free mesh (see Table 1, column marked by $Volume$), and the volume of the restored mesh. The NL-SDF algorithm ensures that the volume of the mesh is preserved after each smoothing iteration.

6 Conclusions

In this paper we present a novel two step algorithm that solves a nonlocal surface diffusion flow PDE. The proposed similarity functions which measure the distance between two patches are based on the mean curvature values. This allows for a fairing method which is able to remove spurious oscillations while preserving and even restoring sharp features. Numerical experiments seem to confirm that our algorithm is promising. We plan to extend the variational framework to general weighted operators.

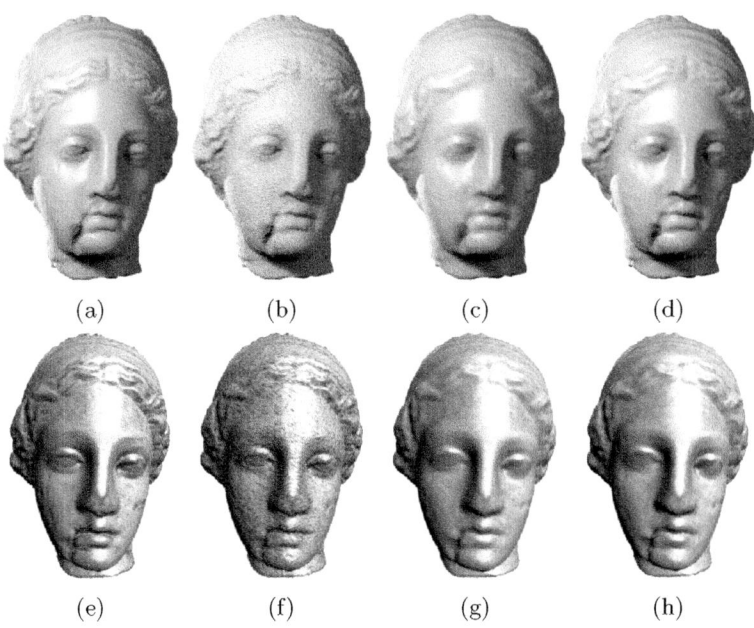

(a)   (b)   (c)   (d)

(e)   (f)   (g)   (h)

Fig. 4. igea mesh: (a) Noise-free mesh and its curvature map (e); (b) noisy mesh and its curvature map (f); (c) the MCF smoothing and its curvature map (g); (d) restored mesh by NL-SDF algorithm and its curvature map (h).

Acknowledgments. This work has been supported by MIUR-Prin 2008, $ex60\%$ project by University of Bologna "Funds for selected research topics" and by GNCS-INDAM.

References

1. Clarenz, U., Diewald, U., Dziuk, G., Rumpf, M., Rusu, R.: A finite element method for surface restoration with smooth boundary conditions. Computer Aided Geometric Design 21(5), 427–445 (2004)
2. Desbrun, M., Meyer, M., Schroeder, P., Barr, A.: Implicit fairing of Irregular meshes using diffusion and curvature flow. In: Computer Graphics (SIGGRAPH 1999 Proceedings), pp. 317–324 (1999)
3. Meyer, M., Desbrun, M., Schroeder, P., Barr, A.: Discrete Differential Geometry Operators for Triangulated 2-Manifolds. In: Proc. VisMath 2002, Berlin-Dahlem, Germany, pp. 237–242 (2002)

4. Schneider, R., Kobbelt, L.: Geometric fairing of irregular meshes for free-form surface design. Computer Aided Geometric Design 18(4), 359–379 (2001)
5. Tasdizen, T., Whitaker, R., Burchard, P., Osher, S.: Geometric surface processing via normal maps. ACM Transactions on Graphics (TOG) 22/4, 1012–1033 (2003)
6. Buades, A., Coll, B., Morel, J.: A Non-Local Algorithm for Image Denoising. In: Proceedings of the 2005 IEEE Computer Society Conference on Computer Vision and Pattern Recognition (CVPR 2005), vol. 2, pp. 60–65. IEEE Computer Society, Washington, DC (2005)
7. Gilboa, G., Osher, S.: Nonlocal linear image regularization and supervised segmentation. Multiscale Modeling and Simulation 6(2), 595–630 (2007)
8. Dong, B., Ye, J., Osher, S., Dinov, I.: Level Set Based Nonlocal Surface Restoration. Multiscale Modeling and Simulation 7(2), 589–598 (2008)
9. Buades, A., Coll, B., Morel, J.M.: A review of image denoising algorithms, with a new one. Multiscale Modeling and Simulation (SIAM Interdisciplinary Journal) 4(2), 490–530 (2005)
10. Jung, M., Bresson, X., Vese, L.: Nonlocal Mumford-Shah Regularizers for Color Image Restoration. IEEE Trans. Image Process (2010)
11. Xu, G., Pan, Q., Bajaj, C.L.: Discrete surface modelling using partial differential equations. Computer Aided Geometric Design 23(2), 125–145 (2006)
12. Tasdizen, T., Whitaker, R., Burchard, P., Osher, S.: Geometric surface processing via normal maps. ACM Transactions on Graphics 22(4), 1012–1033 (2003)
13. Ohtake, Y., Belyaeva, A., Bogaevski, I.: Mesh regularization and adaptive smoothing. Computer-Aided Design 33/11, 789–800 (2001)
14. Morigi, S.: Geometric Surface Evolution with Tangential Contribution. Journal of Computational and Applied Mathematics 233, 1277–1287 (2010)
15. Deschaud, J.E., Goulette, F.: Point cloud non local denoising using local surface descriptor similarity. In: Paparoditis, N., Pierrot-Deseilligny, M., Mallet, C., Tournaire, O. (eds.) IAPRS, vol. XXXVIII, Part 3A - Saint-Mandé, France (2010)
16. Yoshizawa, S., Belyaev, A., Seidel, H.P.: Smoothing by Example: Mesh Denoising by Averaging with Similarity-based Weights. In: Proc. IEEE International Conference on Shape Modeling and Applications (SMI), Matsushima, Japan, June 14-16, pp. 38–44 (2006)
17. Xu, G.: Convergent Discrete Laplace-Beltrami Operators over Triangular Surfaces. In: Proceedings of the Geometric Modeling and Processing 2004, GMP 2004 (2004)
18. Lysaker, M., Osher, S., Tai, X.C.: Noise Removal Using Smoothed Normals and Surface Fitting. IEEE Transaction on Image Processing 13(10), 1345–1457 (2004)

Nonlocal Filters for Removing
Multiplicative Noise

Tanja Teuber[1] and Annika Lang[2]

[1] Department of Mathematics, University of Kaiserslautern, Germany
`tteuber@mathematik.uni-kl.de`
[2] Seminar for Applied Mathematics, ETH Zurich, Switzerland
`annika.lang@sam.math.ethz.ch`

Abstract. In this paper, we propose nonlocal filters for removing multiplicative noise in images. The considered filters are deduced in a weighted maximum likelihood estimation framework and the occurring weights are defined by a new similarity measure for comparing data corrupted by multiplicative noise. For the deduction of this measure we analyze a probabilistic measure recently proposed for general noise models by Deledalle et al. and study its properties in the presence of additive and multiplicative noise. Since it turns out to have unfavorable properties facing multiplicative noise we propose a new similarity measure consisting of a density specially chosen for this type of noise. The properties of our new measure are examined theoretically as well as by numerical experiments. Afterwards, it is applied to define the weights of our nonlocal filters and different adaptations are proposed to further improve the results. Throughout the paper, our findings are exemplified for multiplicative Gamma noise. Finally, restoration results are presented to demonstrate the good properties of our new filters.

1 Introduction

In 2005, Buades et al. introduced the well-known *nonlocal (NL) means filter* [3]. For the restoration this filter uses information gained by comparing various image regions, so-called patches, with each other. In detail, for a discrete image $f \in \mathbb{R}^{m,n}$, $N = mn$ with pixels f_i, $i = 1, \ldots, N$, the restored pixels are set to be

$$\widetilde{u}_i = \frac{1}{C_i} \sum_{j=1}^{N} w_{NL}(i,j) f_j \qquad \text{with} \quad C_i := \sum_{j=1}^{N} w_{NL}(i,j). \tag{1}$$

If the image patches with centers f_i, f_j are given by f_{i+I}, resp. f_{j+I} for I denoting an appropriate index set, then the weights are given by

$$w_{NL}(i,j) = \exp\left(-\frac{1}{h} \sum_{k \in I} g_k |f_{i+k} - f_{j+k}|^2\right).$$

Here, $h > 0$ controls the amount of filtering. The vector $g = (g_k)_{k \in I}$ represents usually a sampled two dimensional Gaussian kernel with mean zero and standard deviation a, which steers the influence of neighboring pixels on the weight.

A.M. Bruckstein et al. (Eds.): SSVM 2011, LNCS 6667, pp. 50–61, 2012.

This filter has been extensively studied in the past five years and further improved in various directions. An overview is for example given in [4]. One improvement was that several authors proposed different approaches to adapt the NL means filter to noise statistics. Kervrann et al. proposed the so-called *Bayesian NL means filter* [10], which was applied for the removal of speckle noise in ultrasound images in [5]. For Rician noise an approach was presented in [16]. Another relative of the original NL means filter in a probabilistic framework was proposed by Deledalle et al. in [6]. Their approach involved a new noise dependent similarity measure for the patch comparison and was demonstrated to perform well for images corrupted by additive Gaussian noise, noise following a Nakagami-Rayleigh distribution as well as Poisson noise studied in [7].

The aim of this paper is to present nonlocal filters for removing multiplicative noise. To exemplify our results we concentrate on multiplicative Gamma noise. Note that all missing proofs and further examples including different types of noise can be found in [15]. In Section 2 we start by defining our filters by maximum likelihood estimation. For the weight definition we propose a new similarity measure specially designed for comparing data corrupted by multiplicative noise. To obtain this measure we analyze the similarity measure of [6] in the framework of conditional densities in Section 3 and study its properties facing additive and multiplicative noise. Since it turns out to be well suited for additive noise, but to have unfavorable properties for multiplicative noise, we deduce our new measure by logarithmically transformed random variables in Section 4. The advantages of our measure are shown theoretically and by numerical experiments. In Section 5, we consider variants of the weight definition, which further improve the results. Finally, the very good performance of our novel nonlocal filters is demonstrated for images corrupted by multiplicative Gamma noise in Section 6.

2 Nonlocal Filters for Multiplicative Noise

As proposed in [6,12], we will deduce our nonlocal filters by weighted maximum likelihood estimation. Throughout this paper, all random variables are supposed to be continuous and defined on a fixed probability space $(\Omega, \mathcal{F}, P)$. Moreover, for a random variable X and a constant $c \in \mathbb{R}$ we denote by p_{cX} the density of the random variable cX. For $x \in \mathbb{R}$ with $p_X(x) > 0$, the *conditional density* of Y given $X = x$ is defined by $p_{Y|X}(\cdot \,|\, x) := \frac{p_{Y,X}(\cdot, x)}{p_X(x)}$, see, e.g., [9, p. 104]. Now, assume that for $i = 1, \ldots, N$ all noisy image pixels f_i are realizations of independent random variables F_i and the corresponding initial noise free pixels u_i are realizations of independent and identically distributed (i.i.d.) random variables U_i. Moreover, suppose that all f_i are corrupted by the same noise model with equal parameters. Then, we define our restored pixels by

$$\widetilde{u}_i := \operatorname*{argmax}_t \sum_{j=1}^N w(i,j) \ln p_{F_j|U_j}(f_j \,|\, t) \quad \text{s.t. } p_{U_1}(t) = \cdots = p_{U_N}(t) > 0, \quad (2)$$

where $w(i,j) \in [0,1]$ is ideally one if $u_i = u_j$ and zero otherwise. If $w = w_{NL}$, we obtain for additive Gaussian noise and positive p_{U_i} that $\widetilde{u}_i$ is given by (1) as outlined in [6]. For the case of *multiplicative Gamma noise*, we assume that

$$F_i = U_i V_i, \qquad \text{with } p_{U_i}(t) = 0 \quad \forall\, t < 0,\ i = 1, \dots, N, \tag{3}$$

where all V_i are continuous random variables with density

$$p_{V_i}(v) = \frac{L^L}{\Gamma(L)} v^{L-1} \exp(-Lv)\, 1_{\mathbb{R}_{\geq 0}}(v), \quad L \geq 1 \tag{4}$$

and Γ denotes the Gamma function. Besides, all U_i, V_i are considered pairwise independent. Then, for $j = 1, \dots, N$ and any $f_j, t > 0$ with $p_{U_j}(t) > 0$ we have

$$p_{F_j \mid U_j}(f_j \mid t) = \frac{1}{|t|}\, p_{V_j}\left(\frac{f_j}{t}\right) = \frac{L^L}{\Gamma(L)} \frac{f_j^{L-1}}{t^L} \exp\left(-L\,\frac{f_j}{t}\right). \tag{5}$$

For $f_j > 0$, $j = 1, \dots, N$, this implies

$$\tilde{u}_i = \operatorname*{argmax}_{\substack{t > 0 \\ p_{U_i}(t) > 0}} \sum_{j=1}^{N} w(i,j)\, \ln p_{F_j \mid U_j}(f_j \mid t) = \operatorname*{argmin}_{\substack{t > 0 \\ p_{U_i}(t) > 0}} \sum_{j=1}^{N} w(i,j)\left(\ln(t) + \frac{f_j}{t}\right).$$

Similarly, $H(f, u) := \sum_{i=1}^{N} \ln(u_i) + \frac{f_i}{u_i}$ has been deduced as a data fidelity term for a variational approach to remove multiplicative Gamma noise in [1]. If $p_{U_i}(t) > 0$ for $t > 0$ or p_{U_i} is simply unknown, we omit the restriction $p_{U_i}(t) > 0$ and obtain for $f_j > 0$, $j = 1, \dots, N$, by the first order optimality condition that

$$\tilde{u}_i = \frac{1}{C_i} \sum_{j=1}^{N} w(i,j) f_j \qquad \text{with } C_i := \sum_{j=1}^{N} w(i,j). \tag{6}$$

Hence, we get for multiplicative Gamma noise an ordinary weighted average filter like the original NL means filter in (1). Next, we would like to define the weights similarly to w_{NL}, but incorporate the statistics of the noise. By

$$w_{NL}(i,j) = \prod_{k \in I} s_{NL}(f_{i+k}, f_{j+k})^{\frac{g_k}{h}} \quad \text{with } s_{NL}(x,y) := \exp(-|x - y|^2) \tag{7}$$

we see that $w_{NL}(i,j)$ can be written as the product of all $s_{NL}(f_{i+k}, f_{j+k})^{\frac{g_k}{h}}$, where f_{i+k}, f_{j+k} are pairs of pixels of two fix image patches. The function $s_{NL} : \mathbb{R} \times \mathbb{R} \to (0, 1]$ acts as a similarity measure, where $s_{NL}(f_{i+k}, f_{j+k})$ should be close to 1 if $u_{i+k} = u_{j+k}$ and close to 0 if not. Facing additive Gaussian noise, s_{NL} is known to perform well, but it can be far from optimal for other types of noise. Hence, the challenge is now to find a suitable similarity measure for our noise model.

3   The Similarity Measure of Deledalle et al.

To measure whether $u_1 = u_2$ by noisy observations f_1, f_2, Deledalle, Denis and Tupin suggest in [6] to use a so-called 'similarity probability' denoted by

$p(\theta_1 = \theta_2 | f_1, f_2)$. In their paper, θ_i is a parameter depending deterministically on u_i and we set $\theta_i = u_i$ for $i = 1, 2$. Since in general it is not clear what the probability or even conditional density of $U_1 = U_2$ given $F_1 = f_1$, $F_2 = f_2$ is, see e.g. [9, p. 111], we start by rewriting the 'similarity probability' as a conditional density: By definition we have for $p_{F_i}(f_i) > 0$, $i = 1, 2$, that

$$p(u_1 = u_2 | f_1, f_2) := \int_S p_{U_1|F_1}(u \mid f_1) \, p_{U_2|F_2}(u \mid f_2) \, du \qquad (8)$$

and set $S := \mathrm{supp}(p_{U_i})$. Applying the definition of the conditional density and Jacobi's Transformation Formula, see e.g., [13, p. 135f], we obtain that

$$p(u_1 = u_2 | f_1, f_2) = p_{U_1-U_2|(F_1,F_2)}(0 \mid f_1, f_2). \qquad (9)$$

Besides, we have

$$p_{U_1-U_2|(F_1,F_2)}(0 \mid f_1, f_2) = \frac{\int_S p_{U_1}(u) \, p_{U_2}(u) \, p_{F_1|U_1}(f_1 \mid u) \, p_{F_2|U_2}(f_2 \mid u) \, du}{p_{F_1}(f_1) \, p_{F_2}(f_2)}. \qquad (10)$$

Since normally p_{U_i} is unknown, Deledalle et al. propose to neglect this density and p_{F_i}, $i = 1, 2$, on the right hand side and to consider only

$$s_{DDT}(f_1, f_2) := \int_S p_{F_1|U_1}(f_1 \mid u) \, p_{F_2|U_2}(f_2 \mid u) \, du. \qquad (11)$$

This measure is very close to the one investigated for block matching in [11]. To study its properties we start by considering data corrupted by additive noise.

3.1   Properties in the Presence of Additive Noise

For $i = 1, 2$ let the random variables V_i be i.d.d. and follow some noise distribution. Moreover, let f_i be corrupted by *additive noise*, i.e. $f_i := u_i + v_i$ and

$$F_i := U_i + V_i, \qquad i = 1, 2.$$

Here, v_i is a realization of V_i and all U_i, V_i, $i = 1, 2$, are considered to be pairwise independent. In this case, we can show that s_{DDT} has the following properties:

Proposition 1. *For our additive noise model with $S = \mathrm{supp}(p_{U_i}) = \mathbb{R}$ we have*

$$s_{DDT}(f_1, f_2) = p_{V_1-V_2}(f_1-f_2) = p_{F_1-F_2|U_1-U_2}(f_1-f_2|0), \quad f_1, f_2 \in \mathbb{R}. \quad (12)$$

Moreover, s_{DDT} is symmetric and has the following properties:

i) $s_{DDT}(f, f) = const$ for all $f \in \mathbb{R}$,
ii) $0 \leq s_{DDT}(f_1, f_2) \leq s_{DDT}(f, f) = p_{V_1-V_2}(0)$ for all $f_1, f_2, f \in \mathbb{R}$.

For the proof of this and the following propositions see [15]. The last property implies that $s_{DDT}(f_1, f_2)$ is maximal whenever $f_1 = f_2$ and that it is bounded so that it can be scaled to the interval $[0, 1]$, i.e. the range of s_{NL}. For the special case that V_i, $i = 1, 2$, are normally distributed with standard deviation σ, it follows that

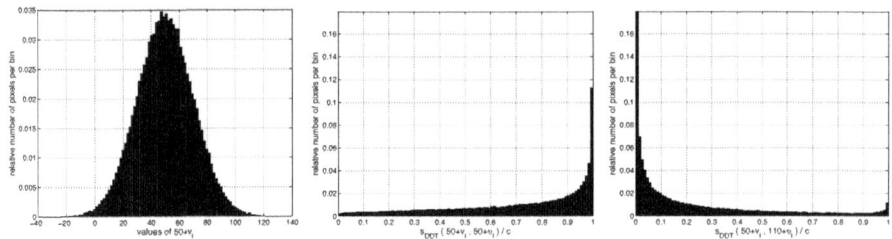

Fig. 1. *Left*: Histogram of a constant image of gray value 50 corrupted by additive Gaussian noise with $\sigma = 20$. *Middle*: Histogram of $(s_{DDT}(f_i, \widetilde{f}_i)/c)_{i=1}^N$, where f, $\widetilde{f}$ are images with gray value distributions as on the left. *Right*: Same as in the middle, but now $\widetilde{f}$ represents a constant image of gray value 110 corrupted by noise.

$$s_{DDT}(f_1, f_2) = \frac{1}{2\sqrt{\pi}\sigma} \exp\left(-\frac{|f_1 - f_2|^2}{4\sigma^2}\right) = c\left(s_{NL}(f_1, f_2)\right)^{\frac{1}{4\sigma^2}}$$

with $c := \max_{x,y \in \mathbb{R}} s_{DDT}(x, y) = \frac{1}{2\sqrt{\pi}\sigma}$. The behavior of s_{DDT} for additive Gaussian noise is illustrated in Fig. 1. In the middle, the distribution of the values $s_{DDT}(f_i, \widetilde{f}_i)/c$ is depicted if both images f, $\widetilde{f}$ are corrupted versions of the same constant image. As expected, most values are close to 1, i.e. s_{DDT}/c detected that the corresponding noisy pixels belong to the same noise free pixel. Only a few values are close to zero, where the measure did not recognize that also these noisy pixels have the same initial gray value. On the right, where the initial gray values have been different, most values $s_{DDT}(f_i, \widetilde{f}_i)/c$ are close to zero and only few pixels are falsely detected to correspond to the same noise free pixel.

3.2 Properties in the Presence of Multiplicative Noise

Next, we want to investigate the case of multiplicative noise. We suppose that the random variables V_i, $i = 1, 2$, are i.i.d., pairwise independent with both U_i and $p_{V_i}(x) = 0$ for $x < 0$. Besides, we assume that F_i follows the multiplicative noise model (3) so that $F_i > 0$ almost surely for $i = 1, 2$. For this setting, we obtain the following properties of s_{DDT}:

Proposition 2. *For our multiplicative noise model with $S = \mathrm{supp}(p_{U_i}) = \mathbb{R}_{\geq 0}$ and $f_1, f_2 > 0$ it holds that*

$$s_{DDT}(f_1, f_2) = \int_0^\infty \frac{1}{u^2} p_{V_1}\left(\frac{f_1}{u}\right) p_{V_2}\left(\frac{f_2}{u}\right) du = p_{f_2 V_1 - f_1 V_2}(0). \tag{13}$$

Besides, s_{DDT} is symmetric and has the following properties:

i) $s_{DDT}(f, f) = \frac{1}{f} p_{V_1 - V_2}(0)$ *for all $f = f_1 = f_2 > 0$,*
ii) s_{DDT} *is not bounded from above.*

These properties stand in sharp contrast to the additive case. The first property implies that s_{DDT} always considers small values $f = f_1 = f_2$ more likely to

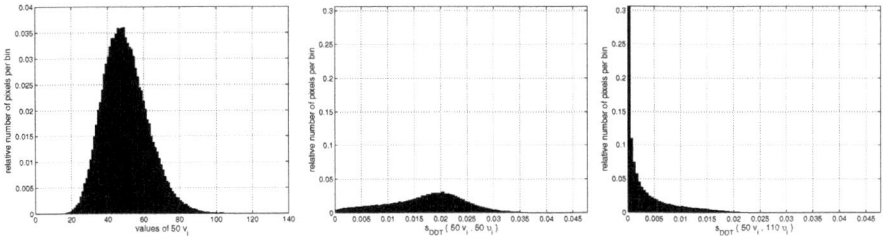

Fig. 2. *Left*: Histogram of a constant image with gray value 50 corrupted by multiplicative Gamma noise with $L = 16$. *Middle*: Histogram of $(s_{DDT}(f_i, \tilde{f}_i))_{i=1}^N$, where f, $\tilde{f}$ have gray value distributions as on the left. *Right*: Same as in the middle, but now $\tilde{f}$ represents a constant image of gray value 110 corrupted by noise.

have the same initial gray value than bigger ones. Besides, the unboundedness is problematic with regard to the weight definition of our nonlocal filters, since a single pixel could get an arbitrarily large weight and dominate all others.

For multiplicative Gamma noise we obtain for $f_1, f_2 > 0$ and $S = \mathbb{R}_{\geq 0}$ that

$$s_{DDT}(f_1, f_2) = L \frac{\Gamma(2L-1)}{\Gamma(L)^2} \frac{(f_1 f_2)^{L-1}}{(f_1 + f_2)^{2L-1}} \propto \frac{1}{f_1 + f_2} \left(2 + \frac{f_1}{f_2} + \frac{f_2}{f_1} \right)^{1-L}.$$

One may expect that for fixed f_1, s_{DDT} is maximal if $f_2 = f_1$. However, for $L > 1$ and a given value f_1 it is maximal for $f_2 = \frac{L-1}{L} f_1$. This is again in sharp contrast to the properties of s_{DDT} in the additive case. For $L = 1$ we have $s_{DDT}(f_1, f_2) = \frac{1}{f_1 + f_2}$. Thus, $s_{DDT}(f_1, f_2)$ is large whenever f_1, f_2 are small. Further properties of this measure are illustrated for $L = 16$ in Fig. 2. In contrast to Fig. 1 (middle), the peak of the histogram at Fig. 2 (middle) is no longer at the largest obtained value of the measure, but at some intermediate value. This is not desirable with respect to the weight definition of a nonlocal filter, since for a large number of pixels it would not definitely determine whether the true pixels have been the same or not. Hence, s_{DDT} does not seem to be optimal for multiplicative noise.

4   A New Similarity Measure for Multiplicative Noise

To deduce a different measure for our multiplicative noise model, we consider the transformed random variables $\tilde{F}_i = \ln(F_i)$, $\tilde{U}_i = \ln(U_i)$, $\tilde{V}_i = \ln(V_i)$, where

$$\tilde{F}_i = \ln(F_i) = \ln(U_i V_i) = \tilde{U}_i + \tilde{V}_i, \qquad i = 1, 2.$$

The new random variables $\tilde{F}_i$ follow an additive noise model now and the supports of $p_{\tilde{U}_i}$, $p_{\tilde{V}_i}$ may be the whole of $\mathbb{R}$. By computing (9) for these new random variables we can show the following:

Lemma 1. *For $f_1, f_2 > 0$ with $p_{F_i}(f_i) > 0$ and $\tilde{S} = \text{supp}(p_{\tilde{U}_i})$ it holds that*

$$p_{\tilde{U}_1 - \tilde{U}_2 | (\tilde{F}_1, \tilde{F}_2)}(0 \mid \ln(f_1), \ln(f_2)) = p_{\frac{U_1}{U_2} | (F_1, F_2)}(1 \mid f_1, f_2). \tag{14}$$

Compared to (9), we have replaced $U_1 - U_2 = 0$ by $U_1/U_2 = 1$ now. Next, we use (10) for the transformed variables and omit $p_{\tilde{U}_i}$, $p_{\tilde{F}_i}$, $i = 1, 2$. Supposing that $\tilde{S} = \mathbb{R}$, i.e. $S = \mathbb{R}_{\geq 0}$, and using (12) for the right hand side, we thus obtain

$$\int_{\tilde{S}} p_{\tilde{F}_1|\tilde{U}_1}(\ln(f_1)\,|\,t)\, p_{\tilde{F}_2|\tilde{U}_2}(\ln(f_2)\,|\,t)\,dt \;=\; p_{\tilde{V}_1 - \tilde{V}_2}(\ln(f_1) - \ln(f_2)).$$

Defining our new similarity measure by

$$s(f_1, f_2) := p_{\tilde{V}_1 - \tilde{V}_2}(\ln(f_1) - \ln(f_2)) = p_{\tilde{F}_1 - \tilde{F}_2|\tilde{U}_1 - \tilde{U}_2}(\ln(f_1) - \ln(f_2)\,|\,0), \quad (15)$$

it has the following properties similar to s_{DDT} for $S = \mathbb{R}$ in the additive case:

Proposition 3. *For our multiplicative noise model and $f_1, f_2 > 0$ it holds that*

$$s(f_1, f_2) = p_{\frac{f_2}{f_1}\frac{V_1}{V_2}}(1) = \frac{f_1}{f_2}\, p_{\frac{F_1}{F_2}|\frac{U_1}{U_2}}\left(\frac{f_1}{f_2}\,\Big|\,1\right) = \int_0^\infty \frac{f_1 f_2}{u^3}\, p_{V_1}\left(\frac{f_1}{u}\right) p_{V_2}\left(\frac{f_2}{u}\right) du. \tag{16}$$

Moreover, $s(\cdot, \cdot)$ is symmetric and has the following properties:

i) $s(f, f) = const$ for all $f > 0$,
ii) $0 \leq s(f_1, f_2) \leq s(f, f) = p_{\frac{V_1}{V_2}}(1)$ for all $f_1, f_2, f > 0$.

Note that (16) differs from (13) only by the factor $\frac{f_1 f_2}{u}$ within the integral. Regarding (12) and (14), our similarity measure is not exactly $p_{\frac{F_1}{F_2}|\frac{U_1}{U_2}}\left(\frac{f_1}{f_2}\,|\,1\right)$, but a scaled version of it. For multiplicative Gamma noise we have

$$s(f_1, f_2) = \frac{\Gamma(2L)}{\Gamma(L)^2}\frac{(f_1 f_2)^L}{(f_1 + f_2)^{2L}} = \frac{\Gamma(2L)}{\Gamma(L)^2}\left(2 + \frac{f_1}{f_2} + \frac{f_2}{f_1}\right)^{-L}, \qquad f_1, f_2 > 0,$$

with a maximum of $c = p_{\frac{V_1}{V_2}}(1) = \frac{\Gamma(2L)}{\Gamma(L)^2}\frac{1}{4^L}$. Fig. 3 shows that for multiplicative Gamma noise we obtain by $s(\cdot, \cdot)/c$ similar histograms as initially for additive Gaussian noise in Fig. 1. Hence, a similar good performance can be expected if applied for nonlocal filtering.

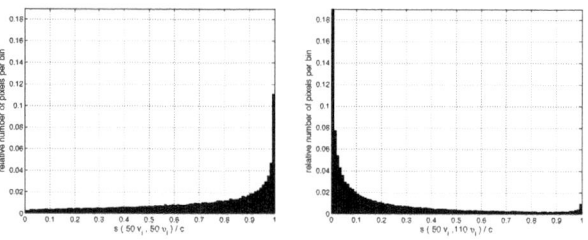

Fig. 3. *Left*: Histogram of $(s(f_i, \tilde{f}_i)/c)_{i=1}^N$, where f, $\tilde{f}$ are both constant images of gray value 50 corrupted by multiplicative Gamma noise with $L = 16$. *Right*: Same as on the left, but now $\tilde{f}$ represents a constant image of gray value 110 corrupted by noise.

5   Weight Definition of Our Nonlocal Filters

For random variables U_i, V_i, F_i, $i = 1, \ldots, N$, fulfilling the multiplicative noise model in Subsection 3.2 with unknown distribution p_{U_i}, the weights can now be defined similarly to (7) by

$$w(i,j) = \prod_{k \in I} \left(\frac{1}{c} \, s(f_{i+k}, f_{j+k}) \right)^{\frac{g_k}{h}} = \prod_{k \in I} \left(p_{\frac{f_{j+k}}{f_{i+k}} \frac{V_{i+k}}{V_{j+k}}}(1) \, / \, p_{\frac{V_{i+k}}{V_{j+k}}}(1) \right)^{\frac{g_k}{h}}. \quad (17)$$

As before, $h > 0$ and $g = (g_k)_{k \in I}$ represents a sampled two dimensional Gaussian kernel with mean zero and standard deviation a, which we normalize such that $\sum_{k \in I} g_k = 1$. Besides, the index set I is set to be a squared grid of size $l \times l$ centered at 0 using reflecting boundary conditions for f.

Fig. 4 (top) shows the histograms of the weights (17) for different constant patches corrupted by multiplicative Gamma noise. As visible here, multiplying the values of the similarity measure over a whole patch significantly changes the histograms compared to Fig. 3. Now, the weights of the left histogram are all larger than on the right. Unfortunately, the histogram on the left is no longer maximal at 1. Even worse, weights close to 1 have never been assigned.

To overcome this drawback we propose an additional adaptation of the weights inspired by the implementation of the NL means filter described at [2]. Here, we use that for random variables X, Y and a continuous function b, where $\mathbb{E}(b(Y))$ exists, the *conditional expectation* of $b(Y)$ given $X = x$ is

$$\mathbb{E}(b(Y)|X = x) := \int_{-\infty}^{\infty} b(y) \, p_{Y|X}(y|x) \, dy \qquad \forall \, x \text{ with } p_X(x) > 0,$$

see, e.g., [13, p. 168]. In detail, for two sets of random variables $F_{i+k} = U_{i+k}V_{i+k}$, $F_{j+k} = U_{j+k}V_{j+k}$, $k \in I$, we set

$$b_k \left(\frac{f_{i+k}}{f_{j+k}} \right) := \left(\frac{1}{c} \, p_{\frac{f_{j+k}}{f_{i+k}} \frac{V_{i+k}}{V_{j+k}}}(1) \right)^{\frac{g_k}{h}} = \left(\frac{1}{c} \, s(f_{i+k}, f_{j+k}) \right)^{\frac{g_k}{h}}$$

and compute for disjoint index sets $i + I$, $j + I$ the conditional expectation

$$\mu := \mathbb{E} \left(\prod_{k \in I} b_k \left(\frac{F_{i+k}}{F_{j+k}} \right) \, \Big| \, \left(\frac{U_{i+k}}{U_{j+k}} = 1 \right)_{k \in I} \right) = \prod_{k \in I} \mathbb{E} \left(b_k \left(\frac{F_{i+k}}{F_{j+k}} \right) \, \Big| \, \frac{U_{i+k}}{U_{j+k}} = 1 \right).$$

Since $w(i,j)$ is a realization of $\prod_{k \in I} b_k \left(\frac{F_{i+k}}{F_{j+k}} \right)$, the variable μ denotes the value we can expect for $w(i,j)$ if the (non-overlapping) image patches f_{i+I}, f_{j+I} have been generated from the same noise free patch. We can show that

$$\mu = \prod_{k \in I} \int_0^{\infty} b_k(t) \, p_{\frac{V_{i+k}}{V_{j+k}}}(t) \, dt.$$

using properties of the conditional expectation. For *multiplicative Gamma noise* we obtain by technical computations that

$$\mu = \prod_{k \in I} 4^{Lg_k/h} \frac{\Gamma(2L) \, \Gamma(L(1 + \frac{g_k}{h}))^2}{\Gamma(L)^2 \, \Gamma(2L(1 + \frac{g_k}{h}))}.$$

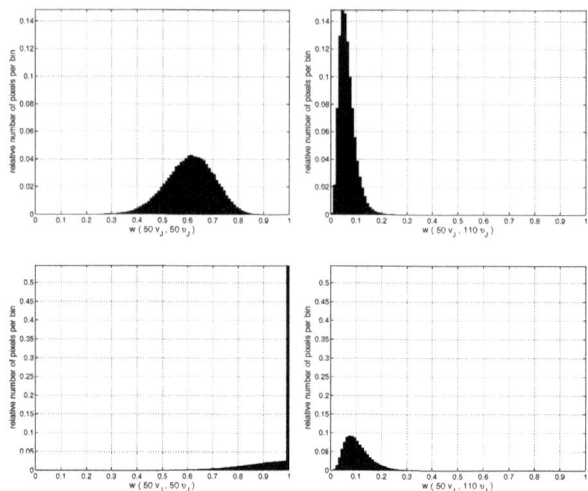

Fig. 4. Histograms of the weights (17) (top) and (18) (bottom) used to compare N different image patches f_I, $\widetilde{f}_I$ ($l = 5$, $a = 1.5$, $h = 1$, $q = 0$). *Left*: Both f_I, $\widetilde{f}_I$ are image patches of gray value 50 corrupted by multiplicative Gamma noise with $L = 16$. *Right*: Same as on the left, but now $\widetilde{f}_I$ is of gray value 110 and corrupted by noise.

Now, we set for $q \in [0, 1)$

$$w_{\mu,q}(i,j) := \begin{cases} 1 & \text{if } w(i,j) \geq \mu, \\ \frac{w(i,j)}{\mu} & \text{if } q\mu \leq w(i,j) < \mu, \\ 0 & \text{otherwise} \end{cases} \qquad \forall\, i,j \in \{1,\dots,N\} \qquad (18)$$

and use these weights in our nonlocal filters deduced from (2). Here, μ is used as an approximation of the true expectation value for all overlapping image patches.

The effect of this additional adaptation compared to (17) is visualized in Fig. 4 (bottom). The histogram for the image patches generated from the same noise free patch has now a significant peak at 1. By setting, e.g., $q = 0.5$ we can additionally achieve that all weights of the right histogram obtain an optimal weight of 0 without effecting the weights of the left histogram.

As usually done, we finally restrict the number of patches being compared to a so-called *similarity window*. Thus, we set all weights $w(i,j)$, $w_{\mu,q}(i,j)$ automatically to zero if pixel j is outside of a squared image region of size $\omega \times \omega$ centered at pixel i. This reduces the computational costs as well as the risk of falsely assigning nonzero weights to a large number of patches.

Updating the Similarity Neighborhoods

In [6] Deledalle et al. suggest to refine the weights of their nonlocal filters iteratively using the former result $u^{(r-1)}$. To obtain $u^{(r)}$, the filter is again applied to the initial noisy image using the new weights. The idea for this updating scheme was taken from [12]. In the following, we apply a variant of this updating strategy, where we perform only one updating step. For this second step we use within the similarity windows for $i, j = 1, \ldots, N$, $i \neq j$ the weights

$$\widetilde{w}_{i,j}(u^{(1)}) = \exp\left(-\frac{1}{d} \sum_{k \in \widetilde{I}} \widetilde{g}_k \, K_{\text{sym}} \left(p_{F_{i+k}|U_{i+k}}(\cdot\,|u^{(1)}_{i+k}), p_{F_{j+k}|U_{j+k}}(\cdot\,|u^{(1)}_{j+k}) \right) \right)$$

and set $\widetilde{w}_{i,i}(u^{(1)}) = \max_j \widetilde{w}_{i,j}(u^{(1)})$. Here, $d > 0$ and $\widetilde{g} = (\widetilde{g}_k)_{k \in \widetilde{I}}$ is again a sampled two dimensional Gaussian kernel with mean zero, but with standard deviation $\widetilde{a}$. As before, $\widetilde{g}$ is normalized such that $\sum_{k \in \widetilde{I}} \widetilde{g}_k = 1$. Moreover, $\widetilde{I} = \widetilde{l} \times \widetilde{l}$ may vary from I. Usually, we choose $\widetilde{a} < a$ and $\widetilde{l} < l$. Furthermore,

$$K_{\text{sym}}(p_X, p_Y) := \int_{-\infty}^{\infty} (p_X(t) - p_Y(t)) \ln\left(\frac{p_X(t)}{p_Y(t)} \right) \, dt$$

denotes the *symmetric Kullback-Leibler divergence* of p_X, p_Y. If we assume that $p_{U_i}(x) > 0$ for all $x \geq 0$, we can show using (5) that the sought symmetric Kullback-Leibler divergence for multiplicative Gamma noise is given by

$$K_{\text{sym}}\left(p_{F_i|U_i}(\cdot\,|u^{(1)}_i), p_{F_j|U_j}(\cdot\,|u^{(1)}_j) \right) = L \frac{(u^{(1)}_i - u^{(1)}_j)^2}{u^{(1)}_i u^{(1)}_j} \qquad \text{for } u^{(1)}_i, u^{(1)}_j > 0.$$

6 Numerical Results

Finally, we present two examples demonstrating the good performance of our novel nonlocal filters for images corrupted by multiplicative Gamma noise. The implementation was done with MATLAB and the parameters were chosen to obtain the best visual results. Note that all images, especially the noisy one, are displayed in the gray scale of the original image to have a consistent coloring for each example. To this purpose, all image values outside of the range of the original image are projected on this range.

For our first example we use the same test image as in [14, Fig. 6]. Obviously, our reconstructions in Fig. 5 (bottom middle and right) are superior to the result by the I-divergence - TV method at top left. Moreover, the difference of applying (6) with weights $w(i, j)$ or $w_{\mu,q}(i, j)$ is illustrated. By using $w_{\mu,q}(i, j)$ instead of $w(i, j)$ more noise has been removed, especially in the background. Moreover, an appropriate value q helps to improve the contrast, e.g., visible at the camera, and leads to sharper edges and contours. By the final updating step used for Fig. 5 (bottom right) we further improved the contrast and small amounts of possibly remained noise are finally removed.

Our second example in Fig. 6 shows our result for the noisy image in [8, Fig. 8]. For a better comparison we included its peak signal to noise ratio (PSNR) and mean absolute-deviation error (MAE) as, e.g, defined in [8]. Obviously, our result is superior or at least competitive to the results obtained by various methods in [8, Fig. 8]. There, the best result was obtained by the proposed hybrid multiplicative noise removal method, which combines variational and sparsity-based shrinkage methods involving curvelets and TV regularization.

Fig. 5. *Top*: Original image with values in $[0, 255]$ (left), corrupted version by multiplicative Gamma noise with $L = 4$ (middle) and restored image by the I-divergence - TV model as presented in [14]. *Bottom*: Results by our new nonlocal filter (6) using just (17) with $l = 7$, $\omega = 29$, $a = 1.5$, $h = 1$ (left), using (18) with $q = 0.35$ (middle) and after an additional updating step with $\widetilde{l} = 3$, $\widetilde{a} = 0.5$, $d = 0.25$ (right).

Fig. 6. *Left*: Original image of the French city of Nîmes (512×512) with values in $[1, 256]$, which has been corrupted by multiplicative Gamma noise with $L = 4$ in [8, Fig. 8]. *Right*: Restoration result by our nonlocal filter (6) applied to the noisy image using (18) and an additional updating step with $l = 7$, $\omega = 29$, $a = 2$, $h = 0.5$, $q = 0.7$, $\widetilde{l} = 5$, $\widetilde{a} = 1$, $d = 0.1$ (PSNR = 26.01, MAE = 8.60).

Acknowledgment. The authors would like to thank the authors of [8], in particular Mila Nikolova, for kindly providing the initial data used in Fig. 6.

References

1. Aubert, G., Aujol, J.-F.: A variational approach to removing multiplicative noise. SIAM Journal on Applied Mathematics 68(4), 925–946 (2008)
2. Buades, A., Coll, B., Morel, J.-M.: Online demo: Non-local means denoising, http://www.ipol.im/pub/algo/bcm_non_local_means_denoising
3. Buades, A., Coll, B., Morel, J.-M.: A non-local algorithm for image denoising. In: IEEE Conf. on CVPR, vol. 2, pp. 60–65 (2005)
4. Buades, A., Coll, B., Morel, J.-M.: Image denoising methods. A new nonlocal principle. SIAM Review 52(1), 113–147 (2010)
5. Coupé, P., Hellier, P., Kervrann, C., Barillot, C.: Nonlocal means-based speckle filtering for ultrasound images. IEEE Trans. Image Process. 18(10), 2221–2229 (2009)
6. Deledalle, C.-A., Denis, L., Tupin, F.: Iterative weighted maximum likelihood denoising with probabilistic patch-based weights. IEEE Trans. Image Process. 18(12), 2661–2672 (2009)
7. Deledalle, C.-A., Tupin, F., Denis, L.: Poisson NL means: Unsupervised non local means for Poisson noise. In: Proceedings of IEEE International Conference on Image Processing, pp. 801–804 (2010)
8. Durand, S., Fadili, J., Nikolova, M.: Multiplicative noise removal using L1 fidelity on frame coefficients. J. Math. Imaging Vision 36(3), 201–226 (2010)
9. Grimmett, G.R., Stirzaker, D.R.: Probability and random processes, 3rd edn. Oxford University Press, Oxford (2001)
10. Kervrann, C., Boulanger, J., Coupé, P.: Bayesian non-local means filter, image redundancy and adaptive dictionaries for noise removal. In: Sgallari, F., Murli, A., Paragios, N. (eds.) SSVM 2007. LNCS, vol. 4485, pp. 520–532. Springer, Heidelberg (2007)
11. Matsushita, Y., Lin, S.: A probabilistic intensity similarity measure based on noise distributions. In: IEEE Conf. Computer Vision and Pattern Recognit. (2007)
12. Polzehl, J., Spokoiny, V.: Propagation-separation approach for local likelihood estimation. Probability Theory and Related Fields 135(3), 335–362 (2006)
13. Rohatgi, V.K.: An Introduction to Probability Theory and Mathematical Statistics. John Wiley & Sons, Inc., Chichester (1976)
14. Steidl, G., Teuber, T.: Removing multiplicative noise by Douglas-Rachford splitting methods. J. Math. Imaging Vision 36(2), 168–184 (2010)
15. Teuber, T., Lang, A.: A new similarity measure for nonlocal filtering in the presence of multiplicative noise. University of Kaiserslautern (preprint, 2011)
16. Wiest-Daesslé, N., Prima, S., Coupé, P., Morrissey, S.P., Barillot, C.: Rician noise removal by non-local means filtering for low signal-to-noise ratio MRI: Applications to DT-MRI. In: Metaxas, D., Axel, L., Fichtinger, G., Székely, G. (eds.) MICCAI 2008, Part II. LNCS, vol. 5242, pp. 171–179. Springer, Heidelberg (2008)

Volumetric Nonlinear Anisotropic Diffusion on GPUs

Andreas Schwarzkopf[1], Thomas Kalbe[1], Chandrajit Bajaj[2],
Arjan Kuijper[1,3], and Michael Goesele[1]

[1] Technische Universität Darmstadt, Germany
[2] ICES-CVC University of Texas at Austin, USA
[3] Fraunhofer IGD, Darmstadt, Germany

Abstract. We present an efficient implementation of volumetric nonlinear anisotropic image diffusion on modern programmable graphics processing units (GPUs). We avoid the computational bottleneck of a time consuming eigenvalue decomposition in $\mathbb{R}^3$. Instead, we use a projection of the Hessian matrix along the surface normal onto the tangent plane of the local isodensity surface and solve for the remaining two tangent space eigenvectors. We derive closed formulas to achieve this resulting in efficient GPU code. We show that our most complex volumetric nonlinear anisotropic diffusion gains a speed up of more than 600 compared to a CPU solution.

1 Introduction and Motivation

Diffusion equations smooth out noise effectively and provide a scale space representation [1–5] of the image, when time is considered as a natural, continuous scale space parameter. They are well known in the field of image processing and have been subject to many enhancements during the last decades.

These equations are widely used for 2D images processing, see e.g. [3, 6] for an introduction. Recent publications apply this diffusion for smoothing of normal maps [7], and fairing of surfaces and functions on surfaces and meshes [8, 9], to mention only some possibilities. Anisotropic diffusion of whole volume images or general meshes [10, 11] and smoothing vector valued volume images [12] are also common tasks arising in medical applications.

As this diffusion requires to solve second order partial differential equations (PDEs) numerically for a rapidly increasing amount of discretized data, it is a perfect application for modern graphic cards, which can easily handle large data sets. The current GPU SIMD architecture allows to solve each iteration in a few milliseconds due to massively parallel processing. This holds for equations that lead to an efficient parallelization. However, this is at least difficult for most interesting, non-linear, PDEs due to the local structure in each voxel that determines in which direction smoothing can be performed.

1.1 Contribution

As main contribution we show how one can obtain the so-called local structure frame for volumetric data sets easily. This leads to nearly unconditional code, performing

A.M. Bruckstein et al. (Eds.): SSVM 2011, LNCS 6667, pp. 62–73, 2012.

extremely well on GPUs. For this purpose, we build on a technique described by Hadwiger et al. [13]. Anisotropic nonlinear diffusion on symmetric multiprocessor (SMP) clusters for volumetric data was discussed in [14], showing a maximum speedup of 20 on SMP clusters with up to 30 processors. In our GPU approach, we do not need to slice the volume and distribute it across a platform, as all shading processors of modern GPUs can access the same memory. We therefore achieve speedups of up to 640 on a comparably cheap GPU.

Volumetric anisotropic diffusion on GPUs using shader programs in the standard graphics pipeline has been discussed in, e.g, the works by Jeong et al. [15], Zhao [16] or Beyer et al. [17]. In contrast to this, our approach is based on NVidia's *CUDA* which is better suited for GPGPU (general purpose GPU) algorithms like volumetric diffusion. Intermediate values, such as the Hessian, are recomputed on-the-fly in each iteration and are stored temporarily in per-thread local memory. Therefore, larger data sets can be processed on the GPU. Further, to our knowledge, we are the first to compute the local structure frame in 2D in the context of volumetric diffusion. This significantly simplifies the algorithm while the results of the diffusion are very good, see Fig. 1, right, and 4.

2   Prerequisites

The linear homogeneous diffusion equation removes noise from images by solving the heat equation, a second order parabolic PDE. Initial and boundary conditions are required to find a particular solution. A general diffusion equation can then be defined as follows:

$$\frac{\partial}{\partial t}\Phi(\boldsymbol{x},t) = \text{div}\left(D\nabla\Phi(\boldsymbol{x},t)\right) \text{ for } \boldsymbol{x} \in \Omega, t > 0, \tag{1}$$

$$\Phi(\boldsymbol{x},0) = \Phi_0(\boldsymbol{x}) \text{ for } \boldsymbol{x} \in \Omega, \tag{2}$$

$$\frac{\partial}{\partial \boldsymbol{n}}\Phi(\boldsymbol{x},0) = 0 \text{ for } \boldsymbol{x} \in \partial\Omega. \tag{3}$$

Here, Φ denotes the noisy image function defined on a region Ω of the Euclidean space. D is a function, which determines the diffusion speed through the medium. D is constant (usually 1 or $1/2$) in the linear case. The initial condition (2) initializes the function at time $t = 0$ with the original noisy image Φ_0. The boundary values are defined in (3) by their derivative in normal direction $\boldsymbol{n}$ to the border of the considered volume: Since the directional derivative is assumed to be 0, no flow through the boundary $\partial\Omega$ is induced.

Solving the heat equation for $D = 1$ at time $t = 1/2\,\sigma^2$ equates to convolving the image function with a Gaussian of size σ (see [1, 3, 18]). Thus, this diffusion equation has exactly the same smoothing characteristics as the well known Gaussian filter. In particular boundaries blur out fast and therefore edge information gets lost quickly, see Fig. 1, left.

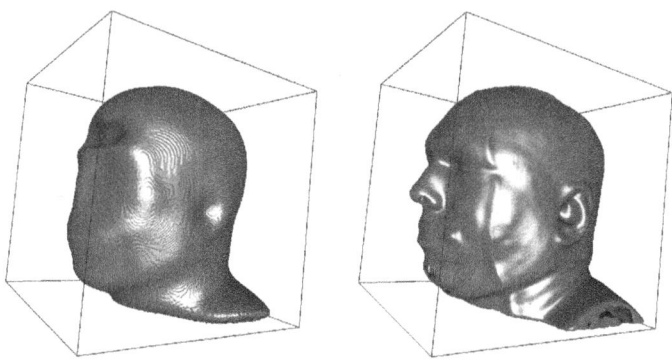

Fig. 1. 100 iterations of diffusion with a time step $\Delta t = 0.05$. *Left*: Homogeneous diffusion. The frontal sinus destroys the surface structure of the forehead and small scaled details (e.g. lips, nose, ears) are lost. *Right*: Nonlinear anisotropic diffusion (here: edge enhancing diffusion) steered by a diffusion tensor based on local structure preserves fine scaled features.

2.1   Inhomogeneous Diffusion

In the context of image processing, the heat equation was modified significantly by Perona and Malik [19] by replacing D in Eq. (1) with an edge detector, a monotonically decreasing non-negative real function g, which attenuates the induced flow close to edges and therefore effectively prevents edges from being washed out:

$$\frac{\partial}{\partial t}\Phi = \mathrm{div}\left(g(|\nabla\Phi|)\nabla\Phi\right), \text{ for } \boldsymbol{x} \in \Omega, t > 0. \tag{4}$$

Perona and Malik proposed the diffusivity functions $g(\nabla\Phi) = e^{-(|\nabla\Phi|/\lambda)^2}$ and $g(\nabla\Phi) = \frac{1}{1+(|\nabla\Phi|/\lambda)^2}$. They designated this diffusion anisotropic, but it is only locally adapting and still isotropic, as it is steered by a scalar diffusion coefficient. Weickert calls this locally adapting diffusion *inhomogeneous* [3]. Inhomogeneous diffusion is able to preserve edges over a long period of time, but its smoothing capabilities close to edges are rather poor.

2.2   Nonlinear Anisotropic Diffusion

Weickert introduced a new nonlinear anisotropic diffusion, using a tensor for D in Eq. (1). This allows for anisotropic adjustment of the diffusion flow [3, 6, 20]. The diffusion tensor D aligns the diffusion flow along the surface structure and its exact definition is mainly dependent on the desired results of the smoothing process.

Edge enhancing diffusion (EED) attenuates diffusion flow normal to the edge or surface but promotes flow along the edge or parallel to the surface, see Fig. 1, right. Furthermore, *coherence enhancing diffusion* (CED) tries to steer diffusion along line-like structures and is able to reconnect interrupted lines [6]. In a hybrid approach, joining EED and CED to locally adapting diffusion, one is able to enhance edges, smooth out noise and to connect broken lines.

In all cases the definition of a useful diffusion tensor involves the construction of a local structure frame: One needs to find a transformation which aligns the coordinate system orthogonally to the surface of the submanifold. Let V be such a coordinate transformation, aligning the third axis normal to the surface, then we can define the EED tensor D in three dimensions as follows:

$$D = VD^*V^T = V \begin{pmatrix} 1 & 0 & 0 \\ 0 & 1 & 0 \\ 0 & 0 & g(|\nabla \Phi|) \end{pmatrix} V^T. \tag{5}$$

Matrix D transformed to the new basis V is a diagonal matrix D^*, as the diffusion flow is aligned perfectly along the principal directions of the surface structure. The crucial point when designing anisotropic diffusion is the efficient construction of this frame V. The traditional way is to obtain this basis by analyzing the structure tensor, defined as the outer product of the gradient $\nabla \Phi$ with itself. In the next section, we will present a method to find such a frame by efficiently analyzing the Hessian, which holds structural information, as it describes the change of the surface normal.

3   Surface Structure and the Hessian

When defining a diffusion tensor, it is utterly important to find a basis V whose axes are aligned exactly along the principal curvature directions of the surface. Theoretically, this was also possible by eigen-decomposition of the structure tensor in 3D space. As the structure tensor is a real symmetric matrix, the eigenvalues are real and the eigenvectors are existent. But the characteristic polynomial of a 3×3 matrix has degree 3 and therefore it is rather time consuming to solve for the roots.

On the other hand, the eigenvalues of a 2×2 matrix are computed easily by evaluating only a few closed formulas. Especially for machine code executed on modern GPUs, this is of advantage as the single execution paths are not divergent (not branching) and parallel execution on the hardware is achieved ideally. In the following, we will show how to obtain the structure frame V by evaluating closed formulas only.

3.1   Tangent Space Projection of the Hessian

The following considerations are aimed at finding a basis transformation $V : \mathbb{R}^3 \mapsto \mathbb{R}^3$ with as few computations as possible and which will describe a coordinate system normal to the tangent plane of the isosurface at a given point. Despite that, the remaining two basis vectors of V spanning the tangent plane should be aligned with the orthogonal principal curvature directions.

Assuming that the inner region of a volume consists of higher density volumes, we define the *surface normal* by the gradient $\nabla \Phi = \begin{bmatrix} \frac{\partial \Phi}{\partial x} & \frac{\partial \Phi}{\partial y} & \frac{\partial \Phi}{\partial z} \end{bmatrix}^T$ as $n = -\nabla \Phi / |\nabla \Phi|$. Since $\nabla \Phi$ points towards the direction of the greatest density ascent inside the volume, n lies inside the linear span of the gradient. Therefore it points into the direction the surface moves when the iso value is increased. We can choose $n(x)$ to be the first vector of our frame V.

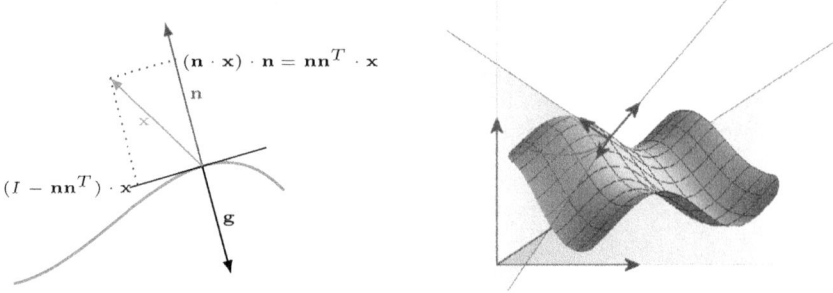

Fig. 2. *Left:* Projecting a point x into the span of the normal n and its complement. *Right:* The local surface frame.

The *curvature* of a surface is defined as the ratio between change of surface normal and change of position, which is described by the gradient of n: If we move in an infinitesimal close area around the point x, the normal will change according to the surface.

In [13] and [21] one finds methods on how to characterize the curvature of a surface based on gradient informations and, moreover, how to obtain the principal curvature directions. The derivative of the normal field ∇n^T at some point x contains curvature information of the surface. Note that ∇n^T is a 3×3 matrix. According to Kindlmann et al. [21] it holds that

$$\nabla n^T = -\frac{1}{|\nabla \Phi|}(I - nn^T)H. \tag{6}$$

Here, I denotes the 3×3 identity matrix, and $H = \nabla(\nabla \Phi)^T$ is the Hessian containing all combinations of partial second order derivatives of the image Φ.

While the gradient describes the amount of change of Φ, the Hessian describes the amount of change of the gradient, that is the amount of change of the surface normal in an infinitesimal close region to a given point x. This amount of change of the gradient can be decomposed into two components, namely the changes along the gradient direction and changes in the tangent space. Only the latter is required for isosurface curvature computation.

To perform the *tangent space projection*, we proceed as follows: We may omit the scaling factor $|\nabla \Phi|^{-1}$ in Eq. (6) and concentrate on the remaining term. It is easy to see that $(nn^T)x = (nx)n$, and the operator (nn^T) projects any point $x \in \Omega$ onto the linear span of the normal. Therefore we are able to define a linear map

$$P = (I - nn^T) = \left(I - \frac{\nabla \Phi (\nabla \Phi)^T}{|\nabla \Phi|^2}\right). \tag{7}$$

which projects any given point x into the complement of the linear span of n, which is the iso surface (see Fig. 2).

Projection P extracts the gradients change of direction from the Hessian inside the tangent space. By using P we define the *shape operator*

$$S = P^T \frac{H}{|\nabla \Phi|} P. \tag{8}$$

Since S is symmetric, solving the characteristic polynomial gives us three real roots and associated orthogonal eigenvectors. Still, the computational overhead of a full eigendecomposition of a 3×3 matrix is rather high, especially as one eigenvector, the normal n, is already known. The remaining eigenvectors in the tangent plane are the principle curvature directions with corresponding eigenvalues $\lambda_{1,2}$ which amount to the principle curvatures.

According to Hadwiger et al. [13], we can solve for the eigenvalues directly in 2D tangent space without explicitly computing S. The transformation of S into any arbitrary orthogonal basis (u, v) of the tangent space is defined as

$$S' = \begin{pmatrix} s_{11} & s_{12} \\ s_{12} & s_{22} \end{pmatrix} = (u, v)^T \frac{H}{|\nabla \Phi|} (u, v). \tag{9}$$

For finding an arbitrary orthogonal basis u, v in the tangent plane, we may proceed as follows: We choose the canonical unit vector $e_1 = (1, 0, 0)^T$ assuming that $e_1 \nparallel n$ holds and compute the cross product $u = e_1 \times n$. In case $u = 0$ we compute the cross product again, now using the second unit vector, $e_2 = (0, 1, 0)^T$. u is now normal to n and therefore it must be part of the tangent plane. We finish the new basis by adding $v = u \times n$.

By using Eq. (9) we are now able to compute the eigenvalues $\lambda_{1,2}$ of S' by solving the characteristic polynomial

$$\det(S' - \lambda I) = \begin{vmatrix} s_{11} - \lambda & s_{12} \\ s_{12} & s_{22} - \lambda \end{vmatrix} = 0$$

$$\Rightarrow \lambda_{1,2} = \frac{\text{trace}(S')}{2} \pm \sqrt{\frac{\text{trace}(S')^2}{4} - \det(S')}. \tag{10}$$

From the eigenvalues $\lambda_{1,2}$ we compute the corresponding eigenvectors. The appropriate formula in [13] is incorrect and can be found in the correct formulation in [22, p. 96]. The eigenvectors $w^*_{1,2}$ are computed with reference to the basis (u, v) at first, and afterwards they are transformed back into 3D space:

$$w_1^* = \begin{pmatrix} w_{1u}^* \\ w_{1v}^* \end{pmatrix} = \begin{cases} \begin{pmatrix} \lambda_1 - s_{22} \\ s_{12} \end{pmatrix}, & \text{for } s_{12} \neq 0 \\ \begin{pmatrix} 1 \\ 0 \end{pmatrix}, & \text{for } s_{12} = 0 \end{cases}$$

$$w_2^* = \begin{pmatrix} w_{2u}^* \\ w_{2v}^* \end{pmatrix} = \begin{cases} \begin{pmatrix} \lambda_2 - s_{22} \\ s_{12} \end{pmatrix}, & \text{for } s_{12} \neq 0 \\ \begin{pmatrix} 0 \\ 1 \end{pmatrix}, & \text{for } s_{12} = 0 \end{cases} \tag{11}$$

The transformation of the 2D eigenvectors into object space is accomplished by extending the tangent space basis with n to 3D and a retransformation into the original orientation by means of $V = \{u, v, n\}$:

$$
w_i = \begin{pmatrix} u_x & v_x & n_x \\ u_y & v_y & n_y \\ u_z & v_z & n_z \end{pmatrix} \begin{pmatrix} w_{iu}^* \\ w_{iv}^* \\ 0 \end{pmatrix} = \begin{pmatrix} u_x w_{iu}^* + v_x w_{iv}^* \\ u_y w_{iu}^* + v_y w_{iv}^* \\ u_z w_{iu}^* + v_z w_{iv}^* \end{pmatrix}. \tag{12}
$$

3.2 Algorithm: Retrieving the Diffusion Tensor

Building on the results of the previous sections, we now depict a compact and easy to implement algorithm to define the anisotropic diffusion tensor. The core of our algorithm is a 2×2 eigen-decomposition of the Hessian projected into the tangent space of the iso surface, computed with simple, closed formulas. The algorithm can be outlined as follows:

1. Calculate the gradient $\nabla\Phi$ and the normal of the isosurface $n = -\nabla\Phi/|\nabla\Phi|$
2. Build the Hessian $H = \nabla(\nabla\Phi)^T$ (see Sect. 4)
3. Complete n with any arbitrary u and v to an orthonormal basis, whose u, v plane is tangential to the isosurface
4. Using Eq. (9), project H into the tangent plane to obtain the 2×2 matrix S'
5. Using Eq. (10) we can calculate the eigenvalues $\lambda_{1,2}$
6. Now, using Eq. (11), we obtain the corresponding eigenvectors $w_{1,2}^*$ with respect to the (u, v) basis
7. W.l.o.g. we might – if this was necessary for the definition of our diffusion tensor – reorder the eigenvalues and eigenvectors: $\lambda_1 < \lambda_2$
8. Transform the 2D eigenvectors back to object space using Eq. (12), receiving the 3D eigenvectors w_1, w_2
9. Set $V = (w_1, w_2, n)$ and $V^{-1} = V^T$
10. Define $D = V \cdot \mathrm{diag}(1, 1, g(\nabla\Phi)) \cdot V^{-1}$

The last step defines the EED diffusion tensor, smoothing along the isosurface and attenuating the diffusion flow normal to the edge. It is further possible to define other diffusion tensors with different properties upon the frame V.

4 Implementation

Our prototype was implemented using C for CUDA which allows for high-parallel computations on NVidia GPUs. As divergent program execution – arising from conditional code which leads to branching – and sequential calculations of the GPU multiprocessors could eliminate speed advantages it is important to find a reduction of the dimension: As one surface frame basis vector – the normal – is known, we can search for the remaining ones in the hyper plane.

The volume data was stored as a 3D texture on the GPU. Coalesced memory access is not possible when processing volumetric data, so the best speedup was achieved by using cached texture memory. Clamping the textures automatically keeps track of all the border values: As gradients at the borders equal zero, no flow will be induced and we avoid conditional code which would slow down the program.

Besides, we can access values between the grid centers and request the hardware to do trilinear interpolation. This gives us a slight speedup for discretizing the Hessian as explained below. Finally, and most important, texture memory is cached, which compensates for the uncoalesced access and results in faster access to neighboring voxels. Performing multiple iterations can be achieved by synchronizing all threads and copying the output back to the texture memory. Copying memory within the device is a fast solution to circumvent the read only issue of texture memory. Each CUDA thread processes one voxel at a time. The code was straight forward developed from the discretized formulation of the anisotropic diffusion.

4.1 Discretization

In the following $\widehat{\Phi}$ denotes the discretized image function $\Phi : \Omega \subset \mathbb{R}^3 \mapsto \mathbb{R}$. Each grid point is associated with a value $\widehat{\Phi}_{x,y,z}$. Neighboring voxels are labeled $\widehat{\Phi}_{x_-} = \widehat{\Phi}_{x-1,y,z}$, likewise $\widehat{\Phi}_{x_+}, \widehat{\Phi}_{y_-}, \widehat{\Phi}_{y_+}, \widehat{\Phi}_{z_-}$ and $\widehat{\Phi}_{z_+}$.

The isotropic grid structure provides a natural spatial discretizing scheme for central differences. For temporal discretization we use forward differences. For nonlinear anisotropic diffusion we obtain the following discretization:

$$\widehat{\Phi}(t + \Delta t) \approx \widehat{\Phi}(t) + \Delta t \cdot \mathrm{div}(D(\nabla \Phi)\nabla \Phi)$$

$$= \widehat{\Phi}(t) + \Delta t \cdot \left(\frac{\partial}{\partial x} \left(d_{11} \frac{\partial \widehat{\Phi}}{\partial x} + d_{12} \frac{\partial \widehat{\Phi}}{\partial y} + d_{13} \frac{\partial \widehat{\Phi}}{\partial z} \right) + \right.$$

$$\frac{\partial}{\partial y} \left(d_{12} \frac{\partial \widehat{\Phi}}{\partial x} + d_{22} \frac{\partial \widehat{\Phi}}{\partial y} + d_{23} \frac{\partial \widehat{\Phi}}{\partial z} \right) +$$

$$\left. \frac{\partial}{\partial z} \left(d_{13} \frac{\partial \widehat{\Phi}}{\partial x} + d_{23} \frac{\partial \widehat{\Phi}}{\partial y} + d_{33} \frac{\partial \widehat{\Phi}}{\partial z} \right) \right) \tag{13}$$

The entries d_{ij} represent the components of the diffusion tensor D. We discretize Eq (13) over an isotropic grid with central differences.

The diffusion tensor of EED as defined in step 10 of the algorithm outlined in Sect. 3.2 yields a symmetric matrix D consisting of the eigenvectors $\boldsymbol{w_1}$ and $\boldsymbol{w_2}$ of the Hessian projected along $\boldsymbol{n}$ to the tangent plane of the isosurface:

$$D = \begin{pmatrix} d_{11} & d_{12} & d_{13} \\ d_{12} & d_{22} & d_{23} \\ d_{13} & d_{23} & d_{33} \end{pmatrix}, \tag{14}$$

with $d_{11} = w_{1x}w_{1x} + w_{2x}w_{2x} + n_x n_x g(\nabla \Phi)$ and accordingly for the remaining cases. The eigenvalues from Eq. (10) and eigenvectors from Eq. (11) are obtained directly in tangent space and are transformed back to object space, see Eq. (12).

The step size Δt needs to be small enough in order to guarantee numerical stability. Following [14], $\Delta t < 0.5/N_d$, with N_d the dimension of the problem, i.e. 3. We used the conservative value $\Delta t = 0.05$.

5   Results

The results of our GPU implementation as well as a CPU implementation for a volume with size 512^3 voxels are given in Table 1. Unless specified otherwise, the timings were taken on a system consisting of an Intel Xeon E5430 CPU (2.66 GHz) and a NVidia GeForce GTX 480 with 480 shader cores. The CPU variants were ported from the GPU code in a straightforward way without any further optimizations.

The table shows the timings, separated into transfer times of the volume to the GPU (obviously not applicable to CPU) and the times needed for one iteration of the code. The speedups are given for one iteration alone. Since diffusion equations typically need many iterations we neglect the transfer times and only take the iteration timings into account. We achieve a speedup of about 170 to 320 for the homogeneous and inhomogeneous diffusion. For the nonlinear anisotropic diffusion (EED) the results are even better: Five iterations require 10 minutes CPU time compared to 1 second on the GPU. As other anisotropic diffusion equations have even higher arithmetic intensity one can expect them to be even faster compared to their CPU variants.

The runtimes of single iterations are also visualized in Fig. 3. As the filters are independent of the data, execution times are proportional to volume sizes and particularly the fraction of CPU to GPU times is constant, so one can easily extrapolate to other data sizes. Modern GPUs provide up to 4 GB memory, so it is possible to process data sets with up to 800^3 voxels on the GPU.

Table 1. Timings and speedups of one iteration step of the discretized PDE, Eq. (13), with a data set of size 512^3 voxels (CPU: Intel Xeon E5430; GPU: NVIDIA GTX 480)

		Time ([ms])		Speedup
		MemCpy	Iteration	Iteration
Linear	cpu	0	1 917	1
homogeneous	gpu	256	11	173
Nonlinear inhomo-	cpu	0	33 499	1
geneous (Perona-Malik)	gpu	281	105	320
Nonlinear	cpu	0	118 717	1
anisotropic (EED)	gpu	276	185	641

Overall it is important to acquire high-quality first and second order derivatives for the gradient as well as the Hessian. Especially at the beginning of the diffusion process significant noise components may disturb the discrete computation of the derivatives enormously. To initialize the diffusion process optimally, usually some sort of (pre-) filtering is applied to the image. for instance using a box, a Gaussian, or a median filter. Bajaj et al. propose to use bilateral filtering, since it removes noise while preserving edge or curvature information, which is important for constructing the diffusion tensor [18]. The bilateral filter can be seen as an expansion to Gaussian filtering by applying an additional edge term [23–25]. In our GPU implementation we achieve a speedup

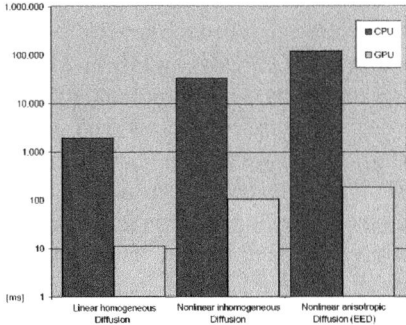

Fig. 3. Runtimes for one iteration of the presented kernels from table 1 on a logarithmic scale (CPU: Intel Xeon E5430; GPU: NVIDIA GTX 480)

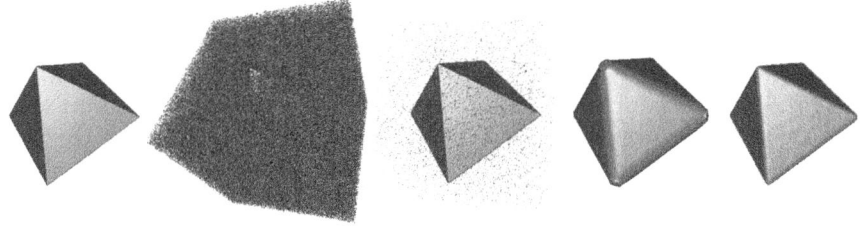

Fig. 4. Restoring the functional data $\Phi(x, y, z) = |x| + |y| + |z|$ (256^3 voxels). *From Left to Right*: Ground truth, Gaussian noise added ($\approx 20\%$ voxels affected, variance 30%), after bilateral prefiltering, after 100 iterations ($\Delta t = 0.05$) of homogeneous and edge enhancing diffusion, respectively, of the prefiltered data set.

of 800 for the bilateral filter. Simpler filters also benefit from massively parallelization, albeit less due to their simplicity. The box filter has a speed up factor of 43, the Gaussian filter one of 151, and the median filter is 76 times faster. All filters took around 1 to 2 seconds on the 512^3 data set.

In Fig. 4 an example of a 3D model endowed with a significant amount of noise is given. Here applying the bilateral filter before the diffusion definitely makes sense.

6   Conclusions

Using the example of EED, we have shown that volumetric nonlinear anisotropic diffusion can be mapped *efficiently* onto the GPU. As efficient GPU code should avoid branching if possible, we derived *closed formulas for the 3D eigenvalue analysis* of the shape operator that allows for reducing the problem from 3D object space onto 2D tangent space: We have presented closed formulas for creating a structure frame along the three principal curvature directions. Building on that, we defined the diffusion tensor for nonlinear anisotropic diffusion and achieved over 600 times the speed compared to a conventional CPU solution. Among the different possible pre-filters for very noisy images we have seen that the bilateral filter is a promising candidate for being processed on the GPU, achieving 830 times the speed of our CPU solution.

As the technical development of GPUs is rapidly progressing and available memory expands, increasingly larger data sets can be processed directly on the GPU. Apart from scalar volume data, one could also process vector data sets on the GPU, as they arise, for example in DW-MRI (diffusion-weighted magnetic resonance imaging). A starting point for this could be 3D-RGBA-textures, representing 4D vectors. Another question concerns automatic parameter detection. Presumably it was necessary to construct and analyze complete or statistically representative image and gradient histograms, which could be done directly on the GPU. Also, one could examine how CED or hybrid diffusion performs on GPUs, as the arithmetic intensity is higher. Building on successful (pre-)filtering and the diffusion process one could try to deal with segmentation as well, in order to present a seamless GPU solution.

References

1. Koenderink, J.J.: The structure of images. Biological Cybernetics 50, 363–370 (1984)
2. Lindeberg, T.: Scale-Space Theory in Computer Vision. The Kluwer International Series in Engineering and Computer Science. Kluwer Academic Publishers, Dordrecht (1994)
3. Weickert, J.: A review of nonlinear diffusion filtering. In: ter Haar Romeny, B.M., Florack, L.M.J., Viergever, M.A. (eds.) Scale-Space 1997. LNCS, vol. 1252, pp. 1–28. Springer, Heidelberg (1997)
4. Lindeberg, T.: Generalized Gaussian scale-space axiomatics comprising linear scale-space, affine scale-space and spatio-temporal scale-space. Journal of Mathematical Imaging and Vision, 1–46 (2010)
5. Kuijper, A.: Geometrical PDEs based on second order derivatives of gauge coordinates in image processing. Image and Vision Computing 27(8), 1023–1034 (2009)
6. Weickert, J.: Coherence enhancing diffusion filtering. International Journal of Computer Vision 31, 111–127 (1999)
7. Tasdizen, T., Whitaker, R., Burchard, P., Osher, S.: Geometric surface smoothing via anisotropic diffusion of normals. In: Proc. VIS 2002, pp. 125–132 (2002)
8. Bajaj, C.L., Xu. G.: Adaptive surfaces fairing by geometric diffusion. In: Symp. CAGD, pp. 731–737 (2001)
9. Bajaj, C.L., Xu, G.: Anisotropic diffusion of surfaces and functions on surfaces. ACM Trans. Graph. 22(1), 4–32 (2003)
10. Lipnikov, K., Shashkov, M., Svyatskiy, D., Vassilevski, Y.: Monotone finite volume schemes for diffusion equations on unstructured triangular and shape-regular polygonal meshes. Journal of Computational Physics 227(1), 492–512 (2007)
11. Agelas, L., Masson, R.: Convergence of the finite volume MPFA O scheme for heterogeneous anisotropic diffusion problems on general meshes. Comptes Rendus Mathematique 346(17-18), 1007–1012 (2008)
12. Zhang, X., Chen, W., Qian, L., Ye, H.: Affine invariant non-linear anisotropic diffusion smoothing strategy for vector-valued images. Imaging Science Journal 58(3), 119–124 (2010)
13. Hadwiger, M., Sigg, C., Scharsach, H., Bühler, K., Gross, M.H.: Real-time ray-casting and advanced shading of discrete isosurfaces. Comp. Graph. Forum 24(3), 303–312 (2005)

14. Tabik, S., Garzon, E., Garcia, I., Fernandez, J.: Implementation of anisotropic nonlinear diffusion for filtering 3D images in structural biology on SMP clusters. In: Proc. Int. Conf. Parallel Computing: Current & Future Issues of High-End Computing, ParCo., vol. 33, pp. 727–734 (2005)
15. Interactive 3D seismic fault detection on the graphics hardware. In: Proc. Volume Graphics (2006)
16. Zhao, Y.: Lattice Boltzman based PDE solver on the GPU. The Visual Computer 24(5), 323–333 (2008)
17. Beyer, J., Langer, C., Fritz, L., Hadwiger, M., Wolfsberger, S., Bühler, K.: Interactive diffusion-based smoothing and segmentation of volumetric datasets on graphics hardware. Methods Inf. Med. 46(3), 270–274
18. Bajaj, C.L., Wu, Q., Xu, G.: Level set based volumetric anisotropic diffusion for 3D image denoising. ICES TR03-10, UTexas, Austin USA (2003)
19. Perona, P., Malik, J.: Scale-space and edge detection using anisotropic diffusion. IEEE Trans. Pattern Analysis and Machine Intelligence 12, 629–639 (1990)
20. Weickert, J.: Anisotropic Diffusion in Image Processing. B.G. Teubne, Stuttgart (1998)
21. Kindlmann, G., Whitaker, R., Tasdizen, T., Mller, T.: Curvature-based transfer functions for direct volume rendering: Methods and applications. In: Proc. IEEE Vis., pp. 513–520 (2003)
22. Sigg, C.: Representation and Rendering of Implicit Surfaces. PhD thesis, ETH Zurich (2006)
23. Tomasi, C., Manduchi, R.: Bilateral filtering for gray and color images. In: Proc. IEEE Int. Conf. on Computer Vision 1998, pp. 839–846 (1998)
24. Durand, F., Dorsey, J.: Fast bilateral filtering for the display of high-dynamic-range images. ACM Trans. Graph. 21(3), 257–266 (2002)
25. Paris, S., Kornprobst, P., Tumblin, J., Durand, F.: A gentle introduction to bilateral filtering and its applications. In: SIGGRAPH Course (2007)

A Statistical Multiresolution Strategy for Image Reconstruction

Klaus Frick and Philipp Marnitz

Institute for Mathematical Stochastics
University of Göttingen
Goldschmidtstrasse 7, 37077 Göttingen
{frick,marnitz}@math.uni-goettingen.de
http://www.stochastik.math.uni-goettingen.de

Abstract. In this paper we present a fully data-driven and locally-adaptive method for image reconstruction that is based on the concept of *statistical multiresolution estimation* as introduced in [1]. It constitutes a statistical regularization technique that uses a ℓ_∞-type distance measure as data fidelity combined with a convex cost functional. The resulting convex optimization problem is approached by a combination of an inexact augmented Lagrangian method and Dykstra's projection algorithm.

Keywords: statistical multiresolution, extreme-value statistics, total-variation regularization, statistical inverse problems, statistical imaging.

1 Introduction

In this paper we are concerned with the reconstruction of an unknown gray-valued image $u^0 \in L^2(\Omega)$ with $\Omega = [0,1]^2$ given the data

$$Y_{ij} = (Ku^0)_{ij} + \varepsilon_{ij}, \quad 1 \leq i \leq m, 1 \leq j \leq n. \tag{1}$$

We assume that ε_{ij} are independent and identically distributed Gaussian random variables with $\mathbf{E}(\varepsilon_{11}) = 0$ and $\mathbf{E}(\varepsilon_{11}^2) = \sigma^2 > 0$ and that $K : L^2(\Omega) \to \mathbb{R}^{m \times n}$ is a linear and bounded operator. K is assumed to model image acquisition and sampling at the same time, i.e. $(Ku)_{ij}$ is assumed to be a sample at the pixel $(i/m, j/n)$ of a smoothed version of u.

Numerous methods for reconstructing the image u^0 from the data Y in the recent literature are covered by a common variational idea: an estimator $\hat{u}$ of u^0 is computed as the solution of the optimization problem

$$J(u) \to \inf \quad \text{s.t.} \quad \sup_{S \in \mathcal{S}} c_S^{-1} \sum_{(i,j) \in S} |(Ku)_{ij} - Y_{ij}|^2 \leq 1, \tag{2}$$

where $J : L^2(\Omega) \to \overline{\mathbb{R}}$ is a convex and lower-semicontinuous regularization functional. Moreover $\mathcal{S}$ denotes a system of subsets of the grid $G = \{1, \ldots, m\} \times$

A.M. Bruckstein et al. (Eds.): SSVM 2011, LNCS 6667, pp. 74–85, 2012.

$\{1, \ldots, n\}$ and $\{c_S \; : \; S \in \mathcal{S}\}$ is a set of positive regularization parameters that govern the trade-off between data-fit and regularity. Solutions of (2) are a special case of *statistical multiresolution estimators (SMRE)* as studied in [1]. In this context the statistic $T : \mathbb{R}^{m \times n} \to \mathbb{R}$ defined by

$$T(v) = \sup_{S \in \mathcal{S}} c_S^{-1} \sum_{(i,j) \in S} |v_{ij}|^2, \quad v \in \mathbb{R}^{m \times n} \tag{3}$$

is referred to as *multiresolution (MR) statistic*. Summarizing, we find the estimator $\hat{u}$ of u^0 such that $J(u)$ is minimal und the condition $T(Ku - Y) \le 1$.

The most popular instance of (2) is obtained by choosing $S = \{G\}$. Then, the MR-statistic coincides with the *quadratic fidelity* and problem (2) can be rewritten into

$$\hat{u}(\lambda) \in \underset{u \in L^2(\Omega)}{\operatorname{argmin}} \frac{\lambda}{2} \sum_{(i,j) \in G} |(Ku)_{ij} - Y_{ij}|^2 + J(u) \tag{4}$$

for a suitable multiplier $\lambda > 0$. In the seminal work [2], for example, the authors proposed the *total variation semi-norm*

$$J(u) = \begin{cases} |\mathrm{D}u| \, (\Omega) & \text{if } u \in \mathrm{BV}(\Omega) \\ +\infty & \text{else} \end{cases} \tag{5}$$

as penalization functional which has been a widely used model in imaging ever since. Here, $|\mathrm{D}u| \, (\Omega)$ denotes the total variation of the (measure-valued) gradient of u which coincides with $\int_{\Omega} |\nabla u|$ if u is smooth. Numerous efficient solution methods for (2) [3–5] and various modifications have been suggested so far (cf. [6–9] to name but a few).

However, the quadratic fidelity has an essential drawback: the information in the residual is incorporated *globally*, that is each pixel value $(Ku)_{ij} - Y_{ij}$ contributes equally to the statistic T *independent of its spatial position*. In practical situations this is clearly undesirable: images usually contain features of different scales and modality, i.e. constant and smooth portions as well as oscillating patterns both of different sizes. A solution $\hat{u}$ of (2) with a global fidelity T is hence likely to exhibit under- and oversmoothed regions at the same time.

Recently, also non-trivial choices of $\mathcal{S}$ that result in *locally adaptive* fidelity measures were considered. In [10] $\mathcal{S}$ is chosen to consist of a partition of G which is obtained beforehand by a Mumford-Shah segmentation. In [11, 12], a subset $S \subset G$ is fixed and afterwards $\mathcal{S}$ is defined as the collection of all translates of S. Both approaches allow for an approximate solution of (2) by means of an analogon of (4) with locally varying regularization parameter, i.e.

$$\hat{u} \in \underset{u \in L^2(\Omega)}{\operatorname{argmin}} \frac{1}{2} \sum_{(i,j) \in G} \lambda_{ij} |(Ku)_{ij} - Y_{ij}|^2 + J(u). \tag{6}$$

In this work we amend this paradigm and present a numerical framework that is capable of directly solving (2) without any restrictions to $\mathcal{S}$. To this end we extend the algorithmic ideas in [1] and propose a combination of an inexact augmented Lagrangian method [9, 13] with Dykstra's projection algorithm [14]. We also propose a novel a priori parameter choice rule for the constants c_S that allows for a statistical interpretation of the latter. We illustrate the capability of our approach by numerical examples, focusing on total variation regularization.

In the following we denote by $|S|$ the cardinality of $S \in \mathcal{S}$. We often refer to $|S|$ as the *scale* of S. We assume that $m, n \in \mathbb{N}$ are fixed and denote by $\langle \cdot, \cdot \rangle$ and $\|\cdot\|$ the Euclidean inner-product and norm on $\mathbb{R}^{m \times n}$ and by $\|u\|_{L^2}$ the L^2-norm of u.

2   Statistical Multiresolution Estimation

In this section we review sufficient conditions that guarantee existence of SMRE, that is of a solution of (2). Moreover we propose a statistically sound parameter choice model for the constants c_S and discuss how to choose the system $\mathcal{S}$.

2.1   Existence of SMRE

For the time being, let $\{c_S : S \in \mathcal{S}\}$ be a set of positive real numbers. We rewrite (2) to an equality constrained problem by introducing the slack variable $v \in \mathbb{R}^{m \times n}$. To be more precise, we aim for the solution of

$$J(u) + H(v) \rightarrow \inf \quad \text{s.t.} \quad Ku + v = Y \tag{7}$$

where H denotes the indicator function on the feasible set $\mathcal{C}$ of (2), i.e.

$$\mathcal{C} = \{v \in \mathbb{R}^{m \times n} : T(v) \le 1\} \quad \text{and} \quad H(v) = \begin{cases} 0 & \text{if } v \in \mathcal{C} \\ \infty & \text{else} \end{cases}. \tag{8}$$

Problems of type (7) are studied e.g. in [15, Chap. III]. There, Lagrangian multiplier methods are employed to solve (7). Recall the definition of the *augmented Lagrangian* of (7):

$$L_\lambda(u, v; p) = \frac{1}{2\lambda} \|Ku + v - Y\|^2 + J(u) + H(v) - \langle p, Ku + v - Y \rangle, \quad \lambda > 0. \tag{9}$$

Here $p \in \mathbb{R}^{m \times n}$ denotes the Lagrange multiplier for the linear constraint in (7). It is well known that existence of a saddle point of L_λ follows from certain constraint qualifications of the MR-statistic T. One typical example is given in Proposition 1 (see [1, Thm. 2.1] for a proof).

Proposition 1. *Assume that*

1. *there exists $\bar{u} \in L^2(\Omega)$ such that $J(\bar{u}) < \infty$ and $T(K\bar{u} - Y) < 1$ and that*
2. *for all $c \in \mathbb{R}$, the following sets are bounded:*

$$\left\{ u \in L^2(\Omega) : \sup_{S \in \mathcal{S}} \sum_{(i,j) \in S} |(Ku)_{ij} - Y_{ij}|^2 + J(u) \le c \right\}. \tag{10}$$

Then, there exist $\hat{u} \in L^2(\Omega)$ and $\hat{v}, \hat{p} \in \mathbb{R}^{m \times n}$ such that

$$L_\lambda(\hat{u}, \hat{v}; p) \leq L_\lambda(\hat{u}, \hat{v}; \hat{p}) \leq L_\lambda(u, v; \hat{p}), \quad \forall \left(u \in L^2(\Omega), v, p \in \mathbb{R}^{m \times n}\right).$$

Remark 1. 1. If $\hat{u} \in L^2(\Omega)$ and $\hat{v}, \hat{p} \in \mathbb{R}^{m \times n}$ are as in Proposition 1, then $\hat{u}$ and $\hat{v}$ solve (7) and hence $\hat{u}$ is an SMR estimator.
  2. Assumption 1) in Proposition 1 is called *Slater's constraint qualification*. It is for instance satisfied if the set $\left\{Ku \ : \ u \in L^2(\Omega) \text{ and } J(u) < \infty\right\}$ is dense in $\mathbb{R}^{m \times n}$.
  3. If J is chosen as the total variation semi-norm (5), then a sufficient condition for assumption (10) will be that there exists $(i, j) \in G$ such that $(K\mathbf{1})_{ij} \neq 0$, where $\mathbf{1} \in L^2(\Omega)$ is the constant 1-function. This is immediate from Poincaré's inequality for functions in $\mathrm{BV}(\Omega)$ (cf. [16, Thm.5.11.1]).

2.2 An a Priori Parameter Selection Method

The choice of the *regularization parameters* c_S in (2) is of utmost importance for they determine the trade-off between smoothing and data-fit. We propose an a priori parameter choice method that is based on quantile values of extremes of transformed χ^2 distributions.

To this end, observe that for $S \in \mathcal{S}$ the random variable

$$t_S(\varepsilon) = \sigma^{-2} \sum_{(i,j) \in S} \varepsilon_{ij}^2$$

is χ^2-distributed with $|S|$ degrees of freedom (d.o.f.). We first aim for transforming $t_S(\varepsilon)$ to normality. It was shown in [17] that the *fourth root transform* $\sqrt[4]{t_S(\varepsilon)}$ is approximately normal with mean and variance

$$\mu_S = \sqrt[4]{|S| - 0.5} \quad \text{and} \quad \sigma_S^2 = \left(8\sqrt{|S|}\right)^{-1},$$

respectively. The fourth root transform outperforms other power transforms in the sense that the Kullback-Leibler distance to the normal distribution is minimized, see [17]. In particular, the approximation works well for small d.o.f.

Next, we consider the extreme value statistic

$$\sup_{S \in \mathcal{S}} \frac{\sqrt[4]{t_S(\varepsilon)} - \mu_S}{\sigma_S}. \tag{11}$$

We note that due to the transformation of the random variable $t_S(\varepsilon)$ to normality each scale contributes equally to the supremum in (11). Hence a parameter choice strategy based on the statistic (11) - like the one suggested in Proposition 2 below - is likely to balance the different scales occurring in $\mathcal{S}$.

It is important to note that the random variable $t_S(\varepsilon)$ and $t_{S'}(\varepsilon)$ are independent if and only if $S \cap S' = \emptyset$. As we do not assume that $\mathcal{S}$ consists of pairwise disjoint sets, (11) constitutes an extreme value statistic of *dependent* random variables. Except for special cases, little is known about the distribution of such statistics as a consequence of which the empirical distribution of (11) is considered in practice.

Proposition 2. *For $\alpha \in (0,1)$ and $S \in \mathcal{S}$ let q_α be the α-quantile of the statistic (11) and set $c_S = (q_\alpha \sigma_S + \mu_S)^4$. Then we get for each solution of (2):*

$$\mathbb{P}(J(\hat{u}) \leq J(u^0)) \geq \alpha. \tag{12}$$

Proof. From (1) and monotonicity of the fourth root transform it follows that

$$\mathbb{P}\left(T(Ku^0 - Y) \leq 1\right) = \mathbb{P}\left(t_S(\varepsilon) \leq c_S \; \forall S \in \mathcal{S}\right)$$

$$= \mathbb{P}\left(\sqrt[4]{t_S(\varepsilon)} \leq q_\alpha \sigma_S + \mu_S \; \forall S \in \mathcal{S}\right)$$

$$= \mathbb{P}\left(\sup_{S \in \mathcal{S}} \frac{\sqrt[4]{t_S(\varepsilon)} - \mu_S}{\sigma_S} \leq q_\alpha\right) = \alpha.$$

In other words, the constants c_S are chosen such that the true signal u^0 satisfies the constraints with probability α. By the fact that $\hat{u}$ is a solution of (2) it follows that $\mathbb{P}(T(Ku^0 - Y) \leq 1) \leq \mathbb{P}(J(\hat{u}) \leq J(u^0))$.

Remark 2. By the rule $c_S = (q_\alpha \sigma_S + \mu_S)^4$ in Proposition 2 the problem of selecting the *set* of regularization parameters c_S is reduced to the question on how to choose the *single* value $\alpha \in (0,1)$. The probability α plays the role of a regularization parameter and allows for a precise statistical interpretation: it constitutes a lower bound on the probability that the SMRE $\hat{u}$ is more regular than the true object u^0.

2.3 On the Choice of $\mathcal{S}$

In the previous section we addressed the question on how to select the regularization parameters $\{c_S\}_{S \in \mathcal{S}}$ for a given system of subsets $\mathcal{S}$ of the grid G. We will now comment on the choice of $\mathcal{S}$.

On the one hand, $\mathcal{S}$ should be chosen rich enough to resolve local features of the image sufficiently well. On the other hand, it is desirable to keep the cardinality of $\mathcal{S}$ small such that the optimization problem in (2) remains solvable within reasonable time. We suggest two different choices of $\mathcal{S}$, namely the set $\mathcal{S}_0$ of *all discrete squares in G* and the set $\mathcal{S}_2$ of *dyadic partitions of G*. The latter is obtained by recursively splitting the grid into four equal subsets until the lowest level of single pixels is reached. For the case $m = n$ it can be formally defined as

$$\mathcal{S}_2 = \bigcup_{l=1}^{\lfloor \log_2(n) \rfloor} \left\{ \left\{ k2^l, \ldots, (k+1)2^l \right\}^2 \; : \; k = 0, \ldots, 2^{\lfloor \log_2(n) \rfloor} \right\}.$$

Obviously, $\mathcal{S}_0$ contains much more elements than $\mathcal{S}_2$ and is hence likely to achieve a higher resolution. We indicate this behaviour in Figure 1.

Here, a solution $\bar{u}$ of (4) of a natural image from perturbed data is depicted (first row). Since this reconstruction method does not adapt the amount of regularization to the local image features, the reconstruction exhibits both over-

Fig. 1. True signal u^0, data Y with $\sigma = 0.1$ and solution of (4) $\bar{u}$ with $\lambda = 0.75$ (upper row). Oversmoothed regions identified on the scales $|S| = 4, 8$ and 16 (from left to right) for the system $\mathcal{S}_0$ (middle row) and $\mathcal{S}_2$ (lower row).

and undersmoothed regions. The oversmoothed regions can be identified via the MR-statistic T in (3) by marking those sets S in $\mathcal{S}$ for which

$$c_S^{-1} \sum_{(i,j)\in S} |Y_{ij} - (K\bar{u})_{ij}|^2 > 1.$$

The union of these sets for the systems $\mathcal{S}_0$ (second row) and $\mathcal{S}_2$ (third row) are highlighted in Figure 1 where we examine the scales $|S| = 4, 8, 16$ (from left to right). The parameters c_S are chosen according to Section 2.2 with $\alpha = 0.9$.

3 Algorithmic Methodology

In what follows, we present an algorithmic approach to the numerical computation of SMRE in practice that extends the methodology in [1]. We use an *inexact Uzawa-type algorithm* which decomposes the original problem into a series of subproblems which are substantially easier to solve.

3.1 Inexact Uzawa Algorithm

In order to compute the desired saddle point of the augmented Lagrangian function L_λ in (9), we use a modified version of the Uzawa-Algorithm (see e.g. [15,

Chap. III]). Starting with some initial $p_0 \in \mathbb{R}^{m \times n}$, the original algorithm consists in iteratively computing

1. $(u_k, v_k) \in \operatorname{argmin}_{u \in L^2(\Omega), v \in \mathbb{R}^{m \times n}} L_\lambda(u, v; p_{k-1})$
2. $p_k = p_{k-1} - \lambda(Ku_k + v_k - Y)$.

Item 1. amounts to an implicit minimization step w.r.t. to the variabels u and v whereas 2. constitutes an explicit maximization step for the Lagrange multiplier p. The algorithm is usually stopped once the constraint in (7) is fulfilled up to a certain tolerance (e.g. with respect to the L^2-norm as described in Algorithm 1).

Rather than applying this algorithm in a straightforward manner, however, we carry out two modifications. Firstly, we add in the k-th step the following additional term to L_λ:

$$\frac{1}{2}\left(M \|u - u_{k\ 1}\|_{L^2}^2 - \|K(u - u_{k-1})\|^2\right)^2 + \frac{\beta}{2}\|v - v_{k-1}\|^2. \tag{13}$$

Here M is chosen such that $M \geq \|K\|^2$ and $\beta \geq 0$. By adding (13) to L_λ the distance to the previous iterate is additionally penalized. As a result, we won't have to evaluate K repeatedly within an iterative minimization scheme, but only once at u_{k-1} as we will see when our algorithmic methodology will be addressed in the following subsection. Secondly, we perform successive minimization w.r.t. u and v instead of minimizing simultaneously. The resulting two subproblems can be tackled much more efficiently than the original problem. For details, we again refer to the next subsection .

After some rearrangements of the terms in L_λ and (13) and by exploiting the fact that H is the indicator function of the convex set $\mathcal{C}$, the modified Uzawa algorithm with successive minimization can be summarized as in Algorithm 1. In practice, Algorithm 1 is very stable and straightforward to implement, provided that efficient methods to solve (14) and (15) are at hand. However, a sound convergence analysis for Algorithm 1 in the present general setting is not available so far (see e.g. [18] for the linear case and [1, Thm. 2.2] for the case when the additional term in (13) is skipped).

3.2   Subproblems

Closer inspection of Algorithm 1 reveals that the original problem - computing a saddle point of L_λ - has been replaced by an iterative series of subproblems (14) and (15). We will now examine these two subproblems and propose methods that are suited to solve them. Here we proceed as in [1].

We focus on (15) first. Note that the problem given there amounts to computing the L_2-projection of $v_k := Y + \alpha p_{k-1} - Ku_{k-1}$ onto the feasible region $\mathcal{C}$ as defined in (8). Due to the supremum taken in the definition (3) of the statistic T, we can decompose $\mathcal{C}$ into $\mathcal{C} = \bigcap_{S \in \mathcal{S}} \mathcal{C}_S$ where

$$\mathcal{C}_S = \left\{ v \in \mathbb{R}^{m \times n} : c_S^{-1} \sum_{(i,j) \in S} |v_{ij}|^2 \leq 1 \right\}, \tag{17}$$

Algorithm 1. Inexact Uzawa Algorithm

Require: $Y \in \mathbb{R}^{m \times n}$ (data), $\lambda > 0$ (step size), $\tau \geq 0$ (tolerance).
Ensure: $(u[\tau], v[\tau])$ is an approximate solution of (7) computed in $k[\tau]$ iteration steps.

$u_0 \leftarrow \mathbf{0}_{L^2}$ and $v_0 = p_0 \leftarrow 0$.
$r \leftarrow \|Ku_0 + v_0 - Y\|$ and $k \leftarrow 0$.
while $r > \tau$ **do**
$\quad k \leftarrow k + 1$.
$\quad$ Minimize $L_\lambda(\cdot, v_{k-1}; p_{k-1}) + \frac{1}{2}\left(M \|\cdot - u_{k-1}\|_{L^2}^2 - \|K(\cdot - u_{k-1})\|^2\right)$:

$$u_k \leftarrow \underset{u \in L^2(\Omega)}{\arg\min} \frac{1}{2} \|u - (u_{k-1} - K^*(Ku_{k-1} + v_{k-1} - (Y + \lambda p_{k-1})))\|_{L^2}^2 + \frac{\lambda}{M} J(u).$$

(14)

$\quad$ Minimize $L_\lambda(u_k, \cdot; p_{k-1}) + \frac{\beta}{2}\|\cdot - v_{k-1}\|^2$:

$$v_k \leftarrow \underset{\mathcal{C}}{\text{proj}}\left(\frac{Y + \lambda p_{k-1} + \beta v_{k-1} - Ku_k}{1 + \beta}\right).$$

(15)

$\quad$ Update dual variable:

$$p_k \leftarrow p_{k-1} - \lambda^{-1}(Ku_k + v_k - Y).$$

(16)

$\quad r \leftarrow \max(\|Ku_k + v_k - Y\|, \|K(u_k - u_{k-1})\|)$.
end while
$u[\tau] \leftarrow u_k$ and $v[\tau] \leftarrow v_k$ and $k[\tau] \leftarrow k$.

---

i.e. each $\mathcal{C}_S$ refers to the feasible region that would result if $\mathcal{S}$ contained S only. Note that all $\mathcal{C}_S$ are closed and convex sets. If we fix a $\mathcal{C}_S$ and consider some $v \notin \mathcal{C}_S$, the projection from v onto $\mathcal{C}_S$ can be stated explicitly as

$$(P_{\mathcal{C}_S}(v))_{i,j} = \begin{cases} v_{i,j} & \text{if } (i,j) \notin S \\ v_{i,j}(1 + \sqrt{c_S / \sum_{(k,l) \in S} |v_{k,l}|^2}) & \text{if } (i,j) \in S. \end{cases}$$

(18)

This insight leads us to the conclusion that any method which computes the projection onto the intersection of closed and convex sets by projecting on the individual sets only would be feasible to solve (15). As it turns out, Dykstra's Algorithm [14] works exactly in this way and is hence our method of choice to solve (15). For a detailed statement of the algorithm and how the total number of sets that enter it may be decreased to speed up runtimes, see [1, Sec. 2.3].

We now turn our attention to (14). In contrast to the standard version of the Uzawa algorithm as stated in [15], this second subproblem in Algorithm 1 does not involve the inversion of the operator K, at least as long as a suitable constant M is chosen in (13). For this reason, (14) here simply amounts to solving an unconstrained denoising problem with a least-squares data-fit. Numerous methods for a wide range of different choices of J are available in order to cope with this problem. If J is chosen as the total variation seminorm, for example, the methods introduced in [3–5] will be suited (we will use the one in [3]).

4   Numerical Results

We conclude this paper by demonstrating the performance of SMRE as computed
by our methodology introduced in Section 3. We will show SMRE computed
for the *denoising* problem in Paragraph 4.1 as well as for *deconvolution* and
inpainting problems in Paragraph 4.2. When it comes down to computation, we
think of an image u as an $m \times n$ array of pixels rather than an element in $L^2(\Omega)$.
Accordingly, the operator K is realized as a $mn \times mn$ matrix.

4.1   Denoising

In this paragraph we consider data Y given by (1) when K is the identity matrix
and u^0 is the test image in Figure 1 ($m = 341$ and $n = 512$). We compute SMRE
based on the systems S_0 and S_2 as introduced in Paragraph 2.3 where we fixed
$\alpha = 0.9$. To this end we utilize Algorithm 1 with $M = 1$ and $\beta = 0$, i.e. the
standard Uzawa Algorithm.

   We compare our estimators to the global estimators $\hat{u}(\lambda)$ ($\lambda > 0$) as defined
in (4). We choose $\lambda = \lambda_2$ and $\lambda = \lambda_B$ such that the mean squared distance
and the mean symmetric Bregman distance to the true signal u^0 is minimized,
respectively. To be more precise, we set

$$\lambda_2 = \mathbf{E}\left(\operatorname*{argmin}_{\lambda>0} \left\| u^0 - \hat{u}(\lambda) \right\|^2\right) \quad \text{and} \quad \lambda_B = \mathbf{E}\left(\operatorname*{argmin}_{\lambda>0} D_J^{\mathrm{sym}}(u^0, \hat{u}(\lambda))\right), \quad (19)$$

where the symmetric Bregman distance for J as in (5) reads as

$$D_J^{\mathrm{sym}}(u, v) = \int_\Omega \left(\frac{\nabla u}{|\nabla u|} - \frac{\nabla v}{|\nabla v|}\right) \cdot (\nabla u - \nabla v) \, \mathrm{d}x.$$

Since the parameters λ_2 and λ_B are not accessible in practice as u^0 is unknown,
we refer to $\hat{u}(\lambda_2)$ and $\hat{u}(\lambda_B)$ as L^2- and *Bregman-oracle*, respectively. In addition,
we compare our approach to the *spatially adaptive TV (SA-TV)* method as
introduced in [11]. The SA-TV algorithm approximates solutions of (2) for the
case where $\mathcal{S}$ constitutes the set of all translates of a fixed window $S \subset G$ by
computing a solution of (6) with a suitable spatially dependent regularization
parameter λ. Starting from a (constant) initial parameter $\lambda = \lambda_0$ the SA-TV
algorithm iteratively adjusts λ by increasing it in regions which were poorly
reconstructed before according to the MR statistic.

   For our numerical comparisons, we used the SA-TV-Algorithm as formulated
in [11], considering square windows with side lengths 5 and 9, respectively. All
parameters involved in the algorithm were chosen as suggested in [11]. As a
breaking condition, we used the discrepancy principle which ended the reconstruction process after exactly three iteration steps in all of our experiments.

   The reconstructions are displayed in Figure 2. By visual inspection, we find
that the oracles are globally under- (L^2) and over-regularized (Bregman),

Fig. 2. Upper row: L^2 - and Bregman oracles. Middle row: SA-TV reconstruction with window size 5 and 9. Lower row: SMRE w.r.t. $\mathcal{S}_2$ and $\mathcal{S}_0$ with $\alpha = 0.9$.

respectively. While the scalar parameter λ was chosen optimally w.r.t. the different distance measures, it still cannot cope with the spatially varying smoothness of the true object u^0.

In contrast, SMRE and SA-TV reconstructions exhibit the desired locally adaptive behaviour. Still the SMRE as formulated in this paper has the advantage that multiple scales are taken into account *at once*, while SA-TV only adapts the parameter on a single given scale. As a result, SA-TV reconstructions are of varying quality for finer and coarser features of the object, while the SMRE is capable of reconstructing such features equally well.

4.2 Deconvolution and Inpainting

We finally investigate the performance of our approach if the operator K in (1) is non-trivial. To be exact, we consider *inpainting* and *deconvolution* problems. For the first we consider an inpainting domain that occludes 15% of the image with noise level $\sigma = 0.1$ (upper left panel in Figure 3) and for the latter a Gaussian convolution kernel with variance 2 and noise level $\sigma = 0.02$ (lower left panel in Figure 3). For all experiments we use the dyadic system $\mathcal{S}_2$ and $\alpha = 0.9$.

Fig. 3. Inpainting (upper row): data Y with $\sigma = 0.1$ (left) and SMRE (right). Deconvolution (lower row): data Y with $\sigma = 0.02$ (left) and SMRE (right).

Note that in both cases we have $K = K^*$ and $\|K\| = 1$; we therefore set $M = 1.01$ and $\beta = M - 1$ in (14) and (15), respectively. We use $\tau = 10^{-3}$ as breaking tolerance which results in both cases in $k[\tau] \sim 30$ iterations in Algorithm 1 and a total computation time of less than 4 min. The results are depicted in the upper right and lower right images of Figure 3, respectively.

Again, the results indicate that a reasonable trade-off between data fit and smoothing is found by the proposed a priori parameter choice rule and that the amount of smoothing is adapted according to the image features.

5   Conclusion

In this paper we showed how statistical multiresolution estimators, that is solutions of (2), can be employed for image reconstruction. We stressed that our method, combined with an a priori parameter selection rule, locally adapts the amount of regularization according to the image geometry. For the solution of the optimization problem (2) we suggested an inexact Uzawa algorithm. The performance of our method was illustrated for standard problems in imaging.

Acknowledgments. K.F. is supported by the DFG-SNF Research Group FOR916 *Statistical Regularization* (Z-Project). P.M is supported by the BMBF project 03MUPAH6 *INVERS* and by the SFB755 *Photonic Imaging on the Nanoscale*.

References

1. Frick, K., Marnitz, P., Munk, A.: Statistical multiresolution estimation in imaging: Fundamental concepts and algorithmic approach (2011),
   http://arxiv.org/abs/1101.4373v1
2. Rudin, L.I., Osher, S., Fatemi, E.: Nonlinear total variation based noise removal algorithms. Phys. D 60, 259–268 (1992)
3. Dobson, D.C., Vogel, C.R.: Convergence of an iterative method for total variation denoising. SIAM J. Numer. Anal. 34, 1779–1791 (1997)
4. Chambolle, A.: An algorithm for total variation minimization and applications. J. Math. Imaging Vision 20, 89–97 (2004)
5. Hintermüller, M., Kunisch, K.: Total bounded variation regularization as a bilaterally constrained optimization problem. SIAM J. Appl. Math. 64, 1311–1333 (2004)
6. Chambolle, A., Lions, P.L.: Image recovery via total variation minimization and related problmes. Numer. Math. 76, 167–188 (1997)
7. Osher, S., Burger, M., Goldfarb, D., Xu, J., Yin, W.: An iterative regularization method for total variation-based image restoration. Multiscale Model. Simul. 4, 460–489 (2005) (electronic)
8. Frick, K., Scherzer, O.: Convex inverse scale spaces. In: Sgallari, F., Murli, A., Paragios, N. (eds.) SSVM 2007. LNCS, vol. 4485, pp. 313–325. Springer, Heidelberg (2007)
9. Zhang, X., Burger, M., Bresson, X., Osher, S.: Bregmanized nonlocal regularization for deconvolution and sparse reconstruction. SIAM Journal on Imaging Sciences 3, 253–276 (2010)
10. Bertalmio, M., Caselles, V., Rougé, B., Solé, A.: TV based image restoration with local constraints. J. Sci. Comput. 19, 95–122 (2003)
11. Dong, Y., Hintermüller, M., Rincon-Camacho, M.: Automated regularization parameter selection in a multi-scale total variation model for image restoration. Technical report, Inst. of Mathematics and Scientific Computing, IFB Report 22 (2008)
12. Dong, Y., Hintermüller, M.: Multi-scale total variation with automated regularization parameter selection for color image restoration. In: Tai, X.C., Mørken, K., Lysaker, M., Lie, K.A. (eds.) SSVM 2009. LNCS, vol. 5567, pp. 271–281. Springer, Heidelberg (2009)
13. Elman, H.C., Golub, G.H.: Inexact and preconditioned Uzawa algorithms for saddle point problems. SIAM Journal on Numerical Analysis 31, 1645–1661 (1994)
14. Boyle, J.P., Dykstra, R.L.: A method for finding projections onto the intersection of convex sets in Hilbert spaces. In: Advances in Order Restricted Statistical Inference (Iowa City, Iowa, 1985). LNS, vol. 37, pp. 28–47. Springer, Berlin (1986)
15. Fortin, M., Glowinski, R.: Augmented Lagrangian methods. Studies in Mathematics and its Applications, vol. 15. North-Holland Publishing Co., Amsterdam (1983); Applications to the numerical solution of boundary value problems
16. Ziemer, W.P.: Weakly differentiable functions. Springer, New York (1989)
17. Hawkins, D.M., Wixley, R.: A note on the transformation of chi-squared variables to normality. Amer.Statist. 40, 296–298 (1986)
18. Zulehner, W.: Analysis of iterative methods for saddle point problems: a unified approach. Math. Comput. 71, 479–505 (2002)

A Variational Approach
for Exact Histogram Specification[*]

Raymond Chan[1], Mila Nikolova[2], and You-Wei Wen[1,3]

[1] Department of Mathematics, The Chinese University of Hong Kong,
Shatin, NT, Hong Kong
[2] Centre de Mathématiques et de Leurs Applications, ENS de Cachan, 61 av. du
Président Wilson, 94235 Cachan Cedex, France
[3] Department of Mathematics, Kunming University of Science and Technology,
Yunnan, P.R. China
rchan@math.cuhk.edu.hk, nikolova@cmla.ens-cachan.fr, wenyouwei@gmail.com

Abstract. We focus on exact histogram specification when the input
image is quantified. The goal is to transform this input image into an
output image whose histogram is exactly the same as a prescribed one. In
order to match the prescribed histogram, pixels with the same intensity
level in the input image will have to be assigned to different intensity
levels in the output image. An approach to classify pixels with the same
intensity value is to construct a strict ordering on all pixel values by
using auxiliary attributes. Local average intensities and wavelet coefficients have been used by the past as the second attribute. However, these
methods cannot enable strict-ordering without degrading the image. In
this paper, we propose a variational approach to establish an image preserving strict-ordering of the pixel values. We show that strict-ordering
is achieved with probability one. Our method is image preserving in the
sense that it reduces the quantization noise in the input quantified image. Numerical results show that our method gives better quality images
than the preexisting methods.

Keywords: Exact histogram specification, strict-ordering, variational
methods, restoration from quantization noise, smooth nonlinear optimization, convex minimization.

1 Introduction

Image histogram processing is the act of altering each individual pixel of an
image by modifying its dynamic range in order to improve the contrast of the
whole image. It is an important image processing task with many real-world
applications, such as contrast enhancement, segmentation, watermarking, among
many others.

In histogram processing, image intensity level is viewed as a random variable
characterized by its probability density function. The histogram of an image

---

[*] The research was supported in part by HKRGC Grant CUHK400510 and DAG
Grant 2060408.

A.M. Bruckstein et al. (Eds.): SSVM 2011, LNCS 6667, pp. 86–97, 2012.

shows the empirical distribution of the intensity levels of its pixels. One of the basic histogram processing problem is histogram equalization [10,18]. It aims to find a transformation so that the output image has a uniform histogram. In the continuous setting the random variable defined by the cumulative distribution function of the intensity levels is uniformly distributed in $[0, 1]$, and hence such a function can always be found. More generally, we may want to yield an output image with pre-specified histogram shapes. This problem is called *histogram specification* or *histogram matching*. The prescribed histogram can be given according to various needs. For example, it can be the histogram of another image, a modified version of the original histogram [19], or a "weighted" histogram of two histograms [6,7].

Numerous methods have been proposed to modify the histogram of an input image. The simplest method is histogram linear stretching [13]. Histogram clipping method [19] limits the maximum number of pixels for each intensity level to a given constant and the clipped pixels are then uniformly distributed among the other intensity levels where the numbers of pixels are less than the clip limit. Several other methods were proposed to preserve the mean brightness of the input image [3,12,23]. In [20], Sapiro and Caselles proposed histogram modification via image evolution equations. Arici *et al.* proposed a general framework for histogram modification [1].

The principle behind histogram specification methods is straightforward for real-valued (analog) images: the histogram of the input image and the prescribed histogram should be equalized to uniform distribution first, say by T_i and T_t respectively. Then the output image can be obtained from the composite transformation $T_t^{-1} \circ T_i$. Since the images are real-valued, T_i and T_t are one-to-one functions, and hence $T_t^{-1} \circ T_i$ is well-defined. The principle fails, however, for quantized (digital) images, which is the case of all digital video systems. The reason is that for quantized images, the intensity levels of all pixels take a limited number of discrete values. Therefore their cumulative density functions are staircase functions rather than strictly increasing functions like those for the real-valued images. Indeed, there are groups of pixels with the same intensity value. Some pixels in such a group will have to be mapped to pixels with different intensity values to match the prescribed histogram. This task cannot be achieved without the use of some auxiliary information on pixel values.

Methods to obtain strict ordering for a quantized image were proposed in [4,5,22]. Once all pixels are strictly ordered, the prescribed intensity values are assigned exactly according to the specified histogram. Coltuc *et al.* considered to use the average intensities of neighboring pixels as the auxiliary attribute [5]. Considering two pixels with the same intensity value, the mean values over the neighborhoods centered on each pixel are compared to order these two pixels. If the mean values are still the same, then they choose larger neighborhoods and continue in the same way until all pixels are ordered. Wan and Shi argued that the local mean approach fails to sharpen the edges of the output image [22]. They proposed to order the pixels according to the absolute values of its wavelet coefficients. The wavelet-based approach tends to amplify the noise since a noise in a smooth region may be mistaken as an edge and hence is sharpened. Post-processing approach or iterative methods can be applied to suppress the amplified noises [2]. We emphasize that both the local mean approach and the wavelet-based approach cannot realize strict ordering without degrading the input quantized image. This is a major drawback.

In this paper, we propose a variational method that enables us to strictly order the pixel values of a quantified image by restoring it from the quantization noise. We prove that the pixels of the restored image can be totally-ordered with probability equal to one. Our experimental results show that the proposed method is very efficient and produces images of better quality than both the local mean method [5] and the wavelet-based method [22].

The outline of the paper is as follows. In Section 2, we present the proposed method. In Section 3, numerical examples are given to demonstrate the effectiveness of the proposed model. Concluding remarks are given in Section 4.

2   Variational Approach for Exact Histogram Specification

In this section, we introduce the definition of strict-ordering and then we propose our variational approach for exact histogram specification. First, let us present the problem of exact histogram specification.

Consider an M-by-N input quantized image u whose pixel values live in the set $\mathcal{P} = \{p_1, \cdots, p_L\}$. We assume, without loss of generality, that p_i are in increasing order. For 8-bit images, $\mathcal{P} = \{0, \cdots, 255\}$. Let the grid of u be denoted by

$$\Omega := \{\mathbf{x} : \mathbf{x} = (i, j),\ 1 \le i \le M, 1 \le j \le N\}.$$

The intensity of u at the pixel $\mathbf{x}$ is given by $u_{\mathbf{x}}$. Define

$$\Omega_k := \{\mathbf{x} \in \Omega : u_{\mathbf{x}} = p_k\}, \quad k = 1, 2, \cdots, L.$$

The associated histogram of u is the L-tuple $(|\Omega_1|, |\Omega_2|, \ldots, |\Omega_L|)$, where $|\cdot|$ denotes the cardinality of the set. The problem of exact histogram specification that we consider can be stated as follows: given the input image u, obtained from an original real-valued (analog) image u_o by quantization, and a pre-specified histogram $\boldsymbol{h} = (h_1, h_2, \ldots, h_L)$, find an output image v such that its histogram is $\boldsymbol{h}$ and for any $\mathbf{x}, \mathbf{y} \in \Omega$, we have $v_{\mathbf{x}} \le v_{\mathbf{y}}$ if $u_{o,\mathbf{x}} \le u_{o,\mathbf{y}}$.

2.1   Sorting Algorithms

Since $MN \gg L$ generally, there are many pixels that share the same intensity value. In order to order strictly the pixels with the same intensity, auxiliary information must be used. Combining the auxiliary information, we can create a K-vector defined as $(u_{\mathbf{x}}, \kappa_{\mathbf{x}}^1, \ldots, \kappa_{\mathbf{x}}^{K-1})$ for $\mathbf{x} \in \Omega$, where $\kappa_{\mathbf{x}}^i \in \mathbb{R}$ is the i-th auxiliary information of the pixel $\mathbf{x}$. Our approach to determine the auxiliary information will be outlined later.

Now we can define an ascending ordering "$\prec$" for pixels in Ω based on such K-tuples. To facilitate the discussions, let $\kappa_{\mathbf{x}}^0 := u_{\mathbf{x}}$. For any two pixels $\mathbf{x}$ and $\mathbf{y}$ in Ω, we say that $\mathbf{x} \prec \mathbf{y}$ if for some $0 \leq \ell \leq K - 1$

$$\kappa_{\mathbf{x}}^j = \kappa_{\mathbf{y}}^j \text{ for all } 0 \leq j \leq \ell - 1 \text{ and } \kappa_{\mathbf{x}}^\ell < \kappa_{\mathbf{y}}^\ell. \tag{1}$$

For good choices of auxiliary information and K sufficiently large, one can in principle sort all pixels $\mathbf{x}$ in Ω according to the ordering $\prec$. That is, we can order the pixels $\mathbf{x}$ in Ω in such a way that $\mathbf{x}_1 \prec \mathbf{x}_2 \prec \ldots \prec \mathbf{x}_{NM}$.

Once such a strict-ordering is obtained, matching the input histogram to the prescribed one is straightforward. This can be done by dividing the ordered list from left to right into L groups. Starting from $\mathbf{x}_1$ on the list, the first h_1 pixels belong to the first group, and are assigned the intensity of p_1. The next h_2 pixels belong to the second group and are assigned the intensity of p_2, and so on until all pixels are assigned to their new intensities.

Several ideas have been proposed for the auxiliary information. Coltuc *et al.* proposed to use the local average intensities of a pixel's neighborhood as auxiliary information [5]. For pixels having the same intensity, if the average intensities of their neighborhoods are the same, then a larger neighborhood will be chosen to compute the average intensity. This procedure is repeated until all pixels are ordered. The author claimed that $K = 6$ is appropriate for any application. Wan and Shi proposed to order the pixels according to the absolute values of the wavelet coefficients of the whole image [22]. Here we propose a variational approach to obtain pertinent auxiliary information.

2.2 A Variational Approach

Let the input (digital) image u be obtained from an original real-valued (analog) image u_o by quantization. Since the pixels of u_o have a continuous range, they can be totally-ordered with probability one. The input image u contains quantization noise. The most natural way to define the ordering for the pixels of u is to restore the original real-valued image u_o using u and a good prior knowledge. Such a restoration can efficiently be done using a detail preserving variational method as the one we are proposing here.

For any $\mathbf{x} \in \Omega$, let $\mathcal{N}_{\mathbf{x}} \subset \Omega$ be the set of neighboring pixels of $\mathbf{x}$. In our experiment, we choose $\mathcal{N}_{\mathbf{x}}$ to be the four neighboring pixels of $\mathbf{x}$ in the vertical and horizontal directions. Now we order the pixels by minimizing f in the cost functional $\mathcal{J} : \mathbb{R}^{M \times N} \times \mathbb{R}^{M \times N} \to \mathbb{R}$ given below

$$\mathcal{J}(f, u) = \sum_{\mathbf{x} \in \Omega} \left(\psi\big(f(\mathbf{x}) - u(\mathbf{x})\big) + \beta \sum_{\mathbf{y} \in \mathcal{N}_{\mathbf{x}}} \phi\big(f(\mathbf{x}) - f(\mathbf{y})\big) \right). \tag{2}$$

Here $\beta > 0$ is the regularization parameter and

H1 $\phi : \mathbb{R} \mapsto \mathbb{R}$ *and* $\psi : \mathbb{R} \mapsto \mathbb{R}$ *are even functions in* $\mathcal{C}^s$ *with* $s \geq 2$, *such that* $\phi''(t) > 0$ *and* $\psi''(t) > 0$, $\forall t \in \mathbb{R}$.

For instance we can choose

$$\psi(t) = \sqrt{t^2 + \alpha_1} \quad \text{and} \quad \phi(t) = \sqrt{t^2 + \alpha_2} \,, \quad \alpha_1 > 0, \ \alpha_2 > 0 \qquad (3)$$

which are $\mathcal{C}^\infty$ and analytic. The minimizer of $\mathcal{J}$ in (2) is denoted by $\hat{f}$.

We know that the quantization noise is bounded, $\|u_o - u\|_\infty \leq 0.5$. This constraint should not be used explicitly however because many pixels may then be stuck on the box constraint which will make strict ordering impossible. Instead, the constraint can be satisfied in a relaxed way by using a slightly smoothed ℓ_1 data-fidelity term like ψ in (3) for $\alpha_1 \gtrsim 0$ and $\beta \gtrsim 0$ in (2). By choosing $\beta \gtrsim 0$, data-fidelity is enhanced. If $\psi(t) = |t|$, some data entries would be kept intact [15] and since data-fidelity is enhanced we would find $\hat{f} = u$. But taking ψ as in (3) for $\alpha_1 \gtrsim 0$ entails that $\hat{\hat{f}} \neq u$. A prior holding for large classes of natural images is that they are almost nowhere constant (see [11]) and that they involve edges and fine structures. Nowhere constant implies that ϕ must be smooth at the origin [16]. For edges and fine structures, ϕ must be affine or nonconvex away from the origin. Since pixels must change no more than $|0.5|$ for an image range equal to 255, the best choice is a convex ϕ of the form (3) for $\alpha_2 \gtrsim 0$. Below we show that the pixels of $\hat{f}$ can be ordered with probability one.

Definition 1. *A function $\mathcal{F} : O \mapsto \mathbb{R}^{M \times N}$, where O is an open domain in $\mathbb{R}^{M \times N}$, is said to be a minimizer function relevant to $\mathcal{J}(\cdot, O)$ if for every $u \in O$, the point $\hat{f} = \mathcal{F}(u)$ is a strict local minimizer of $\mathcal{J}(\cdot, u)$.*

For any $u \in \mathbb{R}^{M \times N}$, the functional $\mathcal{J}(\cdot, u)$ in (2), satisfying H1, is strictly convex and coercive, hence for any u and $\beta > 0$, it has a unique minimizer. What is more, one can show that $\mathcal{J}$ has a unique minimizer function $\mathcal{F} : \mathbb{R}^{M \times N} \mapsto \mathbb{R}^{M \times N}$ which is $\mathcal{C}^{s-1}$ continuous, see [14].

We denote by $\mathbb{L}^{M \times N}$ the Lebesgue measure on $M \times N$ subsets of matrices using the isomorphism between $M \times N$ real matrices and MN-length real vectors. Our main theoretical results, proven in [14], are summarized below. The components of the minimizer function $\mathcal{F}$ are denoted by $\mathcal{F}_\mathbf{x}, \mathbf{x} \in \Omega$.

Theorem 1. *Let $\mathcal{J}$ in (2) satisfy H1. For its minimizer function $\mathcal{F} : \mathbb{R}^{M \times N} \mapsto \mathbb{R}^{M \times N}$, define the sets $\mathcal{Q}$ and $\mathcal{R}$ as follows:*

$$\mathcal{Q} = \{u \in \mathbb{R}^{M \times N} \ : \ \mathcal{F}_\mathbf{x}(u) = \mathcal{F}_\mathbf{y}(u) \,, (\mathbf{x}, \mathbf{y}) \in \Omega \times \Omega, \ \mathbf{x} \neq \mathbf{y}\} \,, \qquad (4)$$

$$\mathcal{R} = \{u \in \mathbb{R}^{M \times N} \ : \ \mathcal{F}_\mathbf{x}(u) = u_\mathbf{y} \,, (\mathbf{x}, \mathbf{y}) \in \Omega \times \Omega, \ \mathbf{x} \neq \mathbf{y}\} \,. \qquad (5)$$

The sets $\mathcal{Q}$ and $\mathcal{R}$ are closed, and satisfy $\mathbb{L}^{M \times N}(\mathcal{Q}) = 0$ and $\mathbb{L}^{M \times N}(\mathcal{R}) = 0$.

The set $\mathcal{Q}$ in (4) contains all possible $u \in \mathbb{R}^{M \times N}$ such that the minimizer $\hat{f} = \mathcal{F}(u)$ might have two equal entries, $\mathcal{F}_\mathbf{x}(u) = \mathcal{F}_\mathbf{y}(u)$ for some $\mathbf{x} \neq \mathbf{y}$ belonging to Ω. The set $\mathcal{R}$ in (5) contains all possible $u \in \mathbb{R}^p$ such that the minimizer $\hat{f} = \mathcal{F}(u)$ might contain some quantized entries, $\mathcal{F}_\mathbf{x}(u) = u_\mathbf{y}$ for some $\mathbf{x}, \mathbf{y} \in \Omega$.

Even though $\mathcal{Q}$ is not empty, since $\mathcal{Q}$ is closed and of null Lebesgue measure, the chance that real-world quantized images u live in it is *null*. Thus, $\mathcal{F}_\mathbf{x}(u) \neq$

$\mathcal{F}_{\mathbf{y}}(u)$, for $\mathbf{x} \neq \mathbf{y}$, is a *generic* property of the minimizers $\mathcal{F}$ of $\mathcal{J}$, as given in (2) and satisfying H1. *For any real-world quantized image u, the entries of the minimizer $\hat{f} = \mathcal{F}(u)$ can be classified with probability one.* In the numerous experiments we have done, we never found natural quantized images belonging to $\mathcal{Q}$ nor to $\mathcal{R}$, i.e. in all cases we could perfectly order the pixels of $\hat{f}$.

There are many methods to compute the minimizer $\hat{f}$ of $\mathcal{J}(\cdot, u)$ in (2) [8, 9, 17, 21]. We applied fixed point iteration method [21] to find $\hat{f}$. Once we have find the minimizer $\hat{f}$, we establish the ordering of the pixels based on the 2-tuple $(u_{\mathbf{x}}, \hat{f}_{\mathbf{x}})$ to produce the quantized output image v.

3 Experimental Results

The performance of the proposed method for exact histogram specification was evaluated using extended numerical experiments. Some of them are presented below. We compare our method with the local mean (LM) algorithm [5] for $K = 6$ as recommended by the authors and with the wavelet-based algorithm (WA) in [22]. For our method, we set α_i, $i = 1, 2$, in (3) to 0.01, and β in (2) to 0.1. We stop the iteration when the relative difference between the iterant is less than 10^{-8}.

In order to measure the results quantitatively, we start out with a given true quantized image w with histogram $\boldsymbol{h}_w$; then we degrade it to obtain an input quantized image u. By applying the three methods on u with prescribed histogram $\boldsymbol{h}_w$, we obtain an output image v which is in fact a restored version of w. We use peak-signal-to-noise-ratio to measure the quality of the output image v with respect to w. It is defined as $\text{PSNR} = 20 \log_{10}(255NM/\|v - w\|_2)$. We tried two sets of degradation to obtain the input image u.

3.1 Contrast Compression

In our first set of degradation, the true quantized images w are chosen to be the 256-by-256 8-bit images of "Cameraman", "Lenna" and "Peppers". The input image u is obtained from w by the degradation: $u = \text{round}(\rho \cdot w)$, where $\rho < 1$ is a constant. This situation arises when a picture is taken with insufficient exposure time, or when we want to compress the image by reducing the number of intensity levels. For example, a 7-bit image can be obtained from an 8-bit image by using $\rho = 0.5$. The input images u for $\rho = 0.3$ are shown in the first row of Figure 1. In the tests, we used LM, WA and our method to obtain the output images v having a prescribed histogram $\boldsymbol{h}_w$.

The comparisons of LM, WA and our algorithm are shown in Table 1. We see from the PSNR values that our method outperforms LM and WA in all cases. In order to save space, we just show the output images v by our method, see the second row of Figure 1. The difference images between the true image w and the output image v are shown in Figure 2. We can discern more features in the first row and the second row than in the third row. It demonstrates that our algorithm yields the best restoration.

Fig. 1. First row: the input images. Second row: the output images by our method.

Table 1. The PSNR (dB) between the true image w and the output image v

	Cameraman			Lenna			Peppers		
ρ	LM	WA	Ours	LM	WA	Ours	LM	WA	Ours
0.8	55.97	55.86	56.07	55.64	55.50	55.73	55.98	55.68	56.05
0.7	54.15	54.07	54.33	53.93	53.77	53.98	54.17	53.82	54.24
0.6	52.96	52.84	53.09	52.69	52.50	52.74	52.93	52.64	53.02
0.5	51.93	51.84	52.06	51.67	51.51	51.74	51.97	51.66	52.04
0.4	49.24	49.12	49.45	49.01	48.72	49.15	49.40	48.90	49.58
0.3	47.16	46.98	47.42	46.74	46.32	47.00	47.19	47.42	47.50
0.2	44.07	33.87	44.46	43.70	43.14	44.07	44.39	43.57	44.94
0.1	38.75	38.55	39.36	38.72	37.83	39.38	39.44	38.36	40.38

One important indicator for a good exact histogram specification algorithm is to see if it can establish a strict ordering for all the pixels. If a sorting method yields two pixels sharing the same value we call them a pair-pixel, and consider that as a failure of the method. Table 2 shows the numbers of pair-pixels produced by the three methods. We find that LM and WA have a high number of pair-pixels while our method can give a total ordering of all pixels for all three images. Incidentally, for the "Cameraman" image, when $\rho = 0.1$ there are 13,859 pair-pixels for WA. Compared with the image size, which has 65,532 pixels, the ordering failure rate is about 21%.

3.2 Histogram Equalization Inversion

The second set of degradation is done as follows. Given the true quantized image w with histogram $\boldsymbol{h}_w$, we apply each individual method to get the pixel ordering of w. Then we use the ordering to match w to an image with uniform histogram. The resulting image is used as the input image u of our experiment. Given u

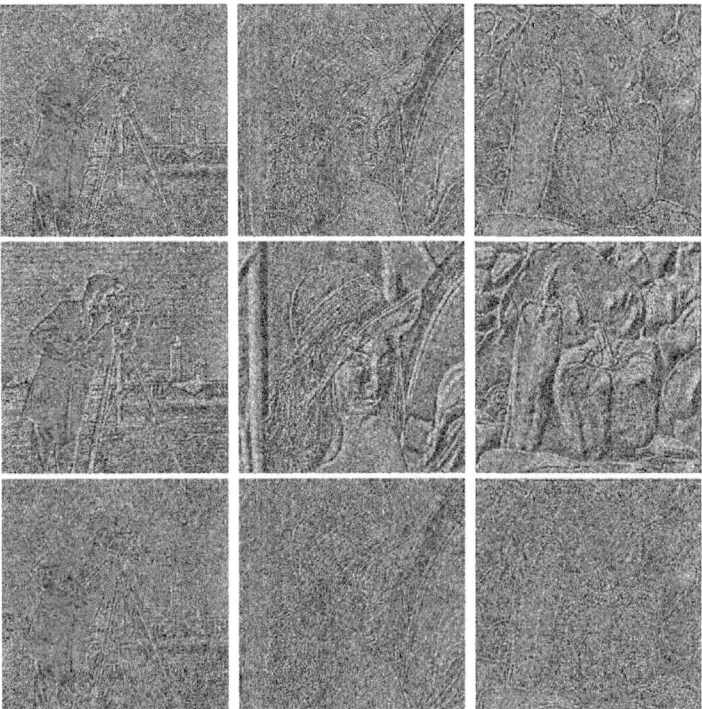

Fig. 2. The difference images between the true quantized image w and the output image v. First row: LM method. Second row: WA method. Third row: our method.

Table 2. The numbers of pair-pixels from the three methods

ρ	Cameraman			Lenna			Peppers		
	LM	WA	Ours	LM	WA	Ours	LM	WA	Ours
0.8	3	154	0	0	7	0	0	13	0
0.7	90	377	0	0	8	0	0	15	0
0.6	88	437	0	0	15	0	0	35	0
0.5	76	587	0	0	26	0	1	38	0
0.4	344	1,267	0	1	66	0	1	145	0
0.3	829	2,293	0	1	177	0	20	403	0
0.2	2,146	4,529	0	36	803	0	109	1,205	0
0.1	6,517	13,859	0	1,493	5,499	0	3,211	7,230	0

and the prescribed histogram $\boldsymbol{h}_w$, we apply each individual method to obtain the output image v. If the ordering among the pixels is preserved by the method, we should have $v = w$ exactly.

For this experiment, we tried the three images in Section 3.1 together with 15 real 768-by-512 8-bit images available at http://r0k.us/graphics/kodak/. Color images are converted to the gray-scale images first. Table 3 shows the PSNR of

Table 3. The PSNR (dB) between the true image w and output images v

Image	LM	WA	Ours	Image	LM	WA	Ours
Cameraman	48.25	48.44	48.79	Kadim07	43.74	43.83	48.09
Lenna	51.24	51.75	51.50	Kadim08	48.33	48.55	50.77
Peppers	51.99	52.66	52.14	Kadim09	44.85	44.94	48.71
Kadim01	41.77	41.81	43.36	Kadim10	44.74	44.85	47.29
Kadim02	43.32	43.38	45.12	Kadim11	45.26	45.35	46.63
Kadim03	44.69	44.76	47.95	Kadim12	40.66	40.70	45.64
Kadim04	45.92	45.99	46.86	Kadim13	47.42	47.58	50.39
Kadim05	49.41	49.71	49.81	Kadim14	45.76	45.86	47.19
Kadim06	44.88	44.95	48.80	Kadim15	49.00	49.23	49.71

Fig. 3. The difference image between w and v by LM (first row), WA (second row) and our method (third row). Our method yields fewest features in the difference images.

Fig. 4. Top-left corner: the given true quantized image w. The difference image between w and the output image v by LM method (top-right), WA method (bottom-left) and our method (bottom-right). Our method yields fewest features in the difference images.

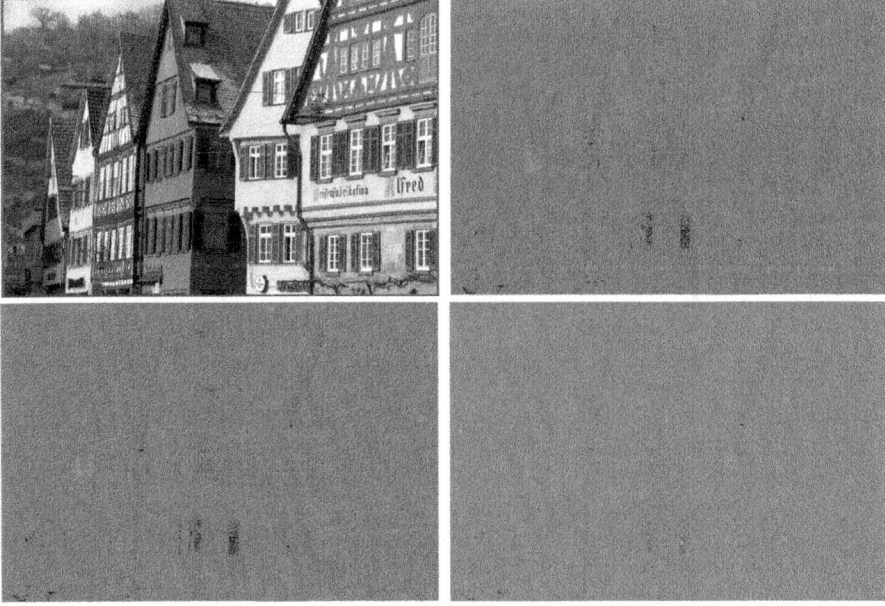

Fig. 5. Top-left corner: the given true image w. The difference image between w and the output image v by LM method (top-right), WA method (bottom-left) and our method (bottom-right). Our method yields fewest features in the difference images.

the results by the three methods. Figures 3–5 give the difference images between w and v on "Cameraman", "Lenna", "Peppers" and two of the 15 images. We notice from Table 3 that WA method yields better PSNR than LM method in all images, but worse than our method in all cases except for the "Lenna" and "Peppers" images. Though WA method yields better PSNR than our method in those two, from Figure 3, we can discern more features in the difference images by WA method than by our method. This indicates that our method is more accurate.

4   Conclusions

In this paper, we propose a variational approach for exact histogram specification. Since the energy we minimize is smooth, its minimizers enable us to strictly order all the pixels in the image. Noticing also that our method reduces the quantification noise, the obtained results outperform the preexisting methods.

References

1. Arici, T., Dikbas, S., Altunbasak, Y.: A histogram modification framework and its application for image contrast enhancement. IEEE Transactions on Image Processing 18(9), 1921–1935 (2009)
2. Avanaki, A.N.: Exact global histogram specification optimized for structural similarity. Optical Review 16(6), 613–621 (2009)
3. Chen, S., Ramli, A.R.: Contrast enhancement using recursive mean-separate histogram equalization for scalable brightness preservation. IEEE Transactions on Consumer Electronics 49(4), 1301–1309 (2003)
4. Coltuc, D., Bolon, P.: An inverse problem: histogram equalization. In: Signal Processing IX, Theories and Applications, EUSIPCO 1998, vol. 2, pp. 861–864 (1998)
5. Coltuc, D., Bolon, P., Chassery, J.-M.: Exact histogram specification. IEEE Transactions on Image Processing 15(5), 1143–1152 (2006)
6. Cox, I.J., Roy, S., Hingorani, S.L.: Dynamic histogram warping of image pairs for constant image brightness. In: Proceedings of International Conference on Image Processing, vol. 2, pp. 366–369. IEEE, Los Alamitos (2002)
7. Delon, J.: Midway image equalization. Journal of Mathematical Imaging and Vision 21(2), 119–134 (2004)
8. Geman, D., Reynolds, G.: Constrained restoration and the recovery of discontinuities. IEEE Trans. Pattern Anal. Mach. Intell. 14(3), 367–383 (1992)
9. Geman, D., Yang, C.: Nonlinear image recovery with half-quadratic regularization. IEEE Trans. Image Process. 4(7), 932–946 (1995)
10. Gonzalez, R.C., Woods, R.E.: Digital image processing. Publishing House of Electronics Industry, Beijing (2005)
11. Gousseau, Y., Morel, J.-M.: Are natural images of bounded variation? SIAM Journal on Mathematical Analysis 33(3), 634–648 (2001)
12. Kim, Y.: Contrast enhancement using brightness preserving bi-histogram equalization. IEEE Transactions on Consumer Electronics 43(1), 1–8 (1997)
13. Narasimhan, S., Nayar, S.: Contrast restoration of weather degraded images. IEEE Trans. Pattern Anal. Mach. Intell. 25, 713–724 (2003)

14. Nikolova, M.: On the minimizers of smooth detail preserving regularized objectives. CMLA-Report 2011, ENS, Cachan (2011)
15. Nikolova, M.: A variational approach to remove outliers and impulse noise. Journal of Mathematical Imaging and Vision 20(1–2), 99–120 (2004)
16. Nikolova, M.: Weakly constrained minimization. Application to the estimation of images and signals involving constant regions. Journal of Mathematical Imaging and Vision 21(2), 155–175 (2004)
17. Nikolova, M., Ng, M.: Analysis of half-quadratic minimization methods for signal and image recovery. SIAM J. Sci. Comput. 27, 937–966 (2005)
18. Pratt, W.K.: Digital Image Processing, 2nd edn. Wiley, New York (1991)
19. Reza, A.M.: Realization of the contrast limited adaptive histogram equalization (CLAHE) for real-time image enhancement. The Journal of VLSI Signal Processing 38(1), 35–44 (2004)
20. Sapiro, G., Caselles, V.: Histogram modification via partial differential equations. In: Proceedings of International Conference on Image Processing, vol. 3, pp. 632–635. IEEE, Los Alamitos (2002)
21. Vogel, C., Oman, M.: Iterative method for total variation denoising. SIAM J. Sci. Comput. 17, 227–238 (1996)
22. Wan, Y., Shi, D.: Joint exact histogram specification and image enhancement through the wavelet transform. IEEE Transactions on Image Processing 16(9), 2245–2250 (2007)
23. Wang, C., Ye, Z.: Brightness preserving histogram equalization with maximum entropy: a variational perspective. IEEE Transactions on Consumer Electronics 51(4), 1326–1334 (2005)

Joint ToF Image Denoising and Registration with a CT Surface in Radiation Therapy

Sebastian Bauer[1], Benjamin Berkels[3],
Joachim Hornegger[1,2], and Martin Rumpf[4]

[1] Pattern Recognition Lab, Dept. of Computer Science
[2] Erlangen Graduate School in Advanced Optical Technologies (SAOT)
Friedrich-Alexander-Universität Erlangen-Nürnberg,
Martensstr. 3, 91058 Erlangen, Germany
{sebastian.bauer,joachim.hornegger}@informatik.uni-erlangen.de
[3] Interdisciplinary Mathematics Inst., University of South Carolina
Columbia, SC 29208, USA
berkels@mailbox.sc.edu
[4] Inst. for Numerical Simulation, Rheinische Friedrich-Wilhelms-Universität Bonn,
Endenicher Allee 60, 53115 Bonn, Germany
martin.rumpf@ins.uni-bonn.de

Abstract. The management of intra-fractional respiratory motion is becoming increasingly important in radiation therapy. Based on in advance acquired accurate 3D CT data and intra-fractionally recorded noisy time-of-flight (ToF) range data an improved treatment can be achieved. In this paper, a variational approach for the joint registration of the thorax surface extracted from a CT and a ToF image and the denoising of the ToF image is proposed. This enables a robust intra-fractional full torso surface acquisition and deformation tracking to cope with variations in patient pose and respiratory motion. Thereby, the aim is to improve radiotherapy for patients with thoracic, abdominal and pelvic tumors. The approach combines a Huber norm type regularization of the ToF data and a geometrically consistent treatment of the shape mismatch. The algorithm is tested and validated on synthetic and real ToF/CT data and then evaluated on real ToF data and 4D CT phantom experiments.

1 Introduction

In this paper, we propose a variational framework that simultaneously solves denoising of time-of-flight (ToF) range data and its registration to a surface extracted from computed tomography (CT) data. Thereby, we underline the benefits of such a joint variational approach. As a case study we show its potential for improvements in radiation therapy planning and treatment. Our algorithm is tested on synthetic and real ToF/CT data using a rigid torso phantom with real ToF data and a 4D CT phantom. We show that the method is capable to cope both with deformations caused by a variation in the patient positioning and by the respiratory motion.

A.M. Bruckstein et al. (Eds.): SSVM 2011, LNCS 6667, pp. 98–109, 2012.
© Springer-Verlag Berlin Heidelberg 2012

Compensation of Respiratory Motion as a Challenge in Radiation Therapy. The management of respiratory motion in diagnostic imaging, interventional imaging and therapeutic applications is an evolving field with many current and future issues still to be adequately addressed. In particular, effects due to organ and tumor motion attract considerable attention in radiation oncology [1]. Technologies that allow an increased dose to the tumor while sparing healthy tissue will improve the balance between complication and cure. Besides a typical patient setup error of 3-5 mm (1 standard deviation) with thoracic radiotherapy [2], a fundamental source of error and uncertainties in radiation therapy is caused by respiratory motion during delivery. Thus, real-time tumor-tracking methods based on the proper identification of thorax deformations due to breathing will significantly improve the radiation therapy. Recently, it has been demonstrated that respiratory motion can be effectively monitored using real-time 3D surface imaging [3]. Schaller et al. [4] presented a time-of-flight respiratory motion detection system that estimates at the ToF frame rate of 25 fps two 1D-signals for the thorax and abdomen movement, respectively. Fayad et al. [5] proposed to use ToF as surrogate to develop a respiration model using PCA. In [6,7] a patient specific respiration model for use in radiotherapy has been investigated.

Time-of-Flight Imaging. ToF imaging directly acquires 3D metric surface information with a single sensor based on the phase shift ρ between an actively emitted and the reflected optical signal [8]. Based on ρ, the radial distance (range) r from the sensor element to the object can be computed as $r = \frac{c\rho}{4\pi f_{mod}}$ where f_{mod} denotes the modulation frequency and c the speed of light. The technology has recently been proposed for diagnostic, interventional and therapeutic medical applications such as patient positioning [9] and respiratory motion detection [4]. However, due to physical limitations of the sensor, depth data from ToF cameras are subject to high temporal noise and exhibit systematic errors. Temporal noise is usually reduced by temporal averaging and can be further smoothed by employing edge preserving filters [10].

Joint Variational Methods in Imaging. Given a pre-fractionally acquired CT image and an intra-fractionally recorded sequence of ToF images of a torso we set up a variational approach, which combines the two highly intertwined tasks of denoising the ToF image and registration of the ToF surface of the thorax with the corresponding surface extracted from the CT data. Indeed, tackling each task would benefit significantly from prior knowledge of the solution of the other tasks. Joint variational methods have proven to be powerful approaches in imaging. E.g. already in 2001 Yezzi, Zöllei and Kapur [11] and Unal et al. [12] have combined segmentation and registration and Feron and Mohammad-Djafari [13] proposed a Bayesian approach for the joint segmentation and fusion of images. Droske and Rumpf proposed in [14] a variational scheme for morphological image denoising and registration based on nonlinear elastic functionals. Recently, in [15] Buades et al. proposed sharpening methods for images, based on joint denoising and matching of images taken as an image burst.

The paper is organized as follows. In Section 2, we introduce the model for joint registration and denoising, including the functional definitions and variational formulations, while Section 3 covers the numerical implementations. In Section 4, we study the parameter setting of the method and show experimental results. Eventually, we draw a conclusion in Section 5.

2   A Joint Registration and Denoising Approach

In this section, we will describe the underlying geometric configuration, derive the variational model and prove the existence of minimizers.

Geometric Configuration

Let us assume that we have already extracted a reliable surface $\mathcal{G}_{\text{CT}} \subset \mathbb{R}^3$ from the given CT image. Now, given the ToF camera parameters, we denote by $\mathcal{G}_r$ the corresponding (unknown) noise free surface geometry uniquely described by the range data (ToF) r. Indeed, for each point ξ on the image plane Ω a range value $r(\xi)$ describes a position vector $X_r(\xi) \in \mathbb{R}^3$ with

$$X_r(\xi) = r(\xi)\gamma(\xi),$$

where the transformation $\gamma : \Omega \to S^2$; $\gamma(\xi) = \left(|\xi|^2 + d_f^2\right)^{-\frac{1}{2}} (\xi_1, \xi_2, d_f)$ is based on the pinhole camera model with d_f denoting the focal length. Now, the pre-fractionally acquired surface $\mathcal{G}_{\text{CT}}$ differs from the intra-fractionally found surface $\mathcal{G}_r$ (cf. Fig. 1).

In our application scenario, the shape of $\mathcal{G}_r$ depends on the actual positioning of the patient on the therapy table and the current state of the respiratory motion at the acquisition time of the ToF image. Hence, we consider a deformation ϕ matching $\mathcal{G}_r$ and $\mathcal{G}_{\text{CT}}$ in the sense that $\phi(\mathcal{G}_r) \subset \mathcal{G}_{\text{CT}}$ and that this deformation can best be represented by a displacement u defined on the parameter domain Ω with

$$\phi(X_r(\xi)) = X_r(\xi) + u(\xi).$$

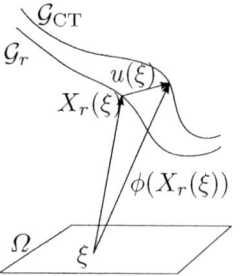

Fig. 1. A geometric sketch of the registration configuration

To quantify the closeness of $\phi(\mathcal{G}_r)$ to $\mathcal{G}_{\text{CT}}$ we represent $\mathcal{G}_{\text{CT}}$ by the corresponding signed distance function d with $d(x) := \pm\text{dist}(x, \mathcal{G}_{\text{CT}})$, where the sign is positive outside the object domain bounded by $\mathcal{G}_{\text{CT}}$ and negative inside. In particular $d = 0$ on $\mathcal{G}_{\text{CT}}$. Furthermore, $|\nabla d| = 1$ and $\nabla d(x)$ is the outward pointing normal on $\mathcal{G}_{\text{CT}}$. Based on this signed distance map we can define the projection $P(x) := x - d(x)\nabla d(x)$ of a point x in a neighborhood of $\mathcal{G}_{\text{CT}}$ onto the closest point on $\mathcal{G}_{\text{CT}}$. Thus $|P(\phi(x)) - \phi(x)|$ is a quantitative pointwise measure for the closeness of $\phi(x)$ to $\mathcal{G}_{\text{CT}}$ for $x \in \mathcal{G}_r$.

Variational Formulation

Now, we are in the position to develop a suitable variational framework which allows us to cope with significantly noisy range data r_0 from the ToF camera and to simultaneously restore a reliable range function r^* and extract a suitable matching displacement u^* as a minimizer of a functional

$$\mathcal{E}[u, r] := \mathcal{E}_{\text{fid}}[r] + \kappa \mathcal{E}_{r,\text{reg}}[r] + \lambda \mathcal{E}_{\text{match}}[u, r] + \mu \mathcal{E}_{u,\text{reg}}[u]$$

consisting of a fidelity energy $\mathcal{E}_{\text{fid}}$ for the range function r given the input range function r_0, a suitable variational prior $\mathcal{E}_{r,\text{reg}}$ for the estimated range function, a matching functional $\mathcal{E}_{\text{match}}$ depending on both the range data r and the displacement u, and finally a prior $\mathcal{E}_{u,\text{reg}}$ for the displacement. Here, κ, λ, μ are positive constants which weight the contributions of the different energies.

Fidelity Energy for the Range Function. We confine here to a simple least square type functional enforcing closeness of the restored range function r to the given input data r_0 and define

$$\mathcal{E}_{\text{fid}}[r] := \int_{\Omega} |r - r_0|^2 \, d\xi \,.$$

Let us remark that nowadays ToF devices deliver together with a dense sequence of range data frames an indicator of the reliability of the output separately for each pixel. This allows to get rid of true outliers. Denoting by $r_0^i(\xi)$ the range value at a position $\xi \in \Omega$ at time t^i and by $\chi^i(\xi)$ the corresponding reliability indicator ($\chi^i(\xi) = 1$ if $r_0^i(\xi)$ is reliable and 0 else) we actually consider time averaged input data and define at a particular time t^j the input range function r_0 of our method as $r_0^j(\xi) = \left(\sum_{i=j-m}^{i=j} \chi^i(\xi) \right)^{-1} \sum_{i=j-m}^{i=j} \chi^i(\xi) r_0^i(\xi)$ for a fixed m (in our application $m = 4$). In fact, in our model we take into account this L^2-fidelity term instead of a in general more robust L^1-functional since in the application considered here large outliers are already eliminated by this time averaging using the reliability indicator of the ToF device.

Prior for the Range Function. Range images of the thorax taken from above are characterized by steep gradients in particular at the boundary of the projected thorax surface and by pronounced contour lines. To preserve these features properly a *TV*-type regularization prior for the range function is decisive. On the other hand, we would like to avoid the well-known staircasing artifacts of a standard *TV* regularization. Hence, we take into account a pseudo Huber norm $|y|_\delta = \sqrt{|y|^2 + \delta^2}$ for $y \in \mathbb{R}^2$ and a suitably fixed regularization parameter $\delta > 0$ and define

$$\mathcal{E}_{r,\text{reg}}[r] := \int_{\Omega} |\nabla r|_\delta \, d\xi \,.$$

Decreasing this energy comes along with a strong smoothing in flat regions which avoids staircasing and at the same time preserves large gradient magnitudes at contour lines or boundaries.

Matching Energy. The purpose of the matching functional is to ensure that $\phi(\mathcal{G}_r) \approx \mathcal{G}_{CT}$ with $\phi(x) = x + u(x)$. Thus, we pick up the pointwise measure $|P(\phi(x)) - \phi(x)|$ of the mismatch at a position $x \in \mathcal{G}_r$ and obtain a first ansatz for the functional

$$\mathcal{E}_{\text{match}}[u, r] := \int_{\mathcal{G}_r} |P(\phi(x)) - \phi(x)|^2 \, da = \int_{\Omega} d(\phi(X_r(\xi)))^2 \sqrt{\det DX_r(\xi)^T DX_r(\xi)} \, d\xi.$$

Here, we have used that $|\nabla d| = 1$ and thus

$$|P(\phi(x)) - \phi(x)| = |d(\phi(x))\nabla d(\phi(x))| = |d(\phi(x))|.$$

The area weight $\sqrt{\det DX_r(\xi)^T DX_r(\xi)}$ with $DX_r(\xi) = Dr(\xi) \otimes \gamma(\xi) + r(\xi)D\gamma(\xi)$ involves first derivatives of r, which can be regarded as a further first order prior for the range function. We experimented with this at first glance geometrically appealing approach, but observed a strong bias between this local weight for the quality of the matching and the actual matching term $d(\phi(X_r(\xi)))^2$ leading to less accurate matching results in particular in regions of steep gradients in $r(\cdot)$ corresponding to edges or the boundary contour of $\mathcal{G}_r$.

Thus, we considered the functional

$$\mathcal{E}_{\text{match}}[u, r] := \int_{\Omega} d(\phi(X_r(\xi)))^2 \, d\xi = \int_{\Omega} d(r(\xi)\gamma(\xi) + u(\xi))^2 \, d\xi.$$

This functional directly combines the range map r and the displacement u and together with the corresponding prior functions both for r and u substantiates the joined optimization approach of our method. In fact, an insufficient and possibly noisy range function r prevents a regular and suitable matching displacement and vice versa.

Prior for the Displacement. Finally, we have to take into account a regularizing prior for the displacement $u : \Omega \to \mathbb{R}^3$. Here, we consider

$$\mathcal{E}_{u,\text{reg}}[u] := \int_{\Omega} |Du(\xi)|^2 \, d\xi$$

with $|A|^2 := \text{tr}(A^T A)$, which leads to satisfying results in our applications with a moderate rigid body motion component in the underlying deformation. Let us mention that a generalized model, which strictly incorporates rigid body motion invariance will depend on the Cauchy Green strain tensor of the deformation $\phi \circ X_r$ and thus again combines gradients of the range function r and the displacement u in a functional of the type $\int_{\Omega} W(D(\phi \circ X_r)^T(\xi)D(\phi \circ X_r)(\xi)) \, d\xi$ with $D(\phi \circ X_r)(\xi) = (DX_r(\xi) + Du(\xi))$ for some energy density function W.

Joint Functional. All in all, we obtain the following joint functional

$$\mathcal{E}[u, r] = \int_{\Omega} |r - r_0|^2 + \kappa |\nabla r|_\delta + \lambda d(r(\xi)\gamma(\xi) + u(\xi))^2 + \mu |Du(\xi)|^2 \, d\xi$$

and can postulate the following result concerning the existence of an optimal range map and a corresponding optimal deformation.

Theorem 1 (Existence of Minimizers). *Let Ω be a bounded domain, $\mathcal{G}_{CT} \neq \emptyset$ and bounded, and $r_0 \in L^2(\Omega)$. Then there exists a minimizer (u^*, r^*) of $\mathcal{E}[u, r]$ on $(H^{1,2}(\Omega))^3 \times BV(\Omega)$.*

Proof. At first we observe that on a minimizing sequence the range functions are uniformly bounded in $BV(\Omega)$ because of the uniform boundedness of $\mathcal{E}_{fid}$ and $\mathcal{E}_{r,reg}$. From $\mathcal{G}_{CT} \neq \emptyset$ we deduce that $d(\cdot)$ is Lipschitz continuous. Furthermore, $d(\cdot)$ has linear growth outside a sufficiently large ball due to the boundedness of $\mathcal{G}_{CT}$. From this and the fact that the range maps are already uniformly bounded in $L^2(\Omega)$ we obtain that the displacements are uniformly bounded in $(L^2(\Omega))^3$. Taking into account the uniform bound on the displacement prior $\mathcal{E}_{u,reg}$ we finally get that the displacements are uniformly bounded in $(H^{1,2}(\Omega))^3$. Hence, we can extract a subsequence for which the range functions converge weak-$*$ in $BV(\Omega)$ and the displacements converge weakly in $(H^{1,2}(\Omega))^3$. Finally, $\mathcal{E}_{fid}$ and $\mathcal{E}_{match}$ are continuous in r and u, $\mathcal{E}_{r,reg}$ is weakly lower semicontinuous on $BV(\Omega)$, and $\mathcal{E}_{u,reg}$ is convex in the Jacobian of the displacement. Thus, by the usual arguments of the direct method in the calculus of variations one verifies the existence of a minimizing range function r^* and a minimizing deformation u^*.

3 Numerical Minimization Algorithm

We consider a gradient descent method for the numerical minimization of the energy functional $\mathcal{E}[\cdot, \cdot]$, which requires the computation of the first variations with respect to the range function and the displacement, respectively. The variations of $\mathcal{E}[u, r]$ in u and r are given as

$$\partial_u \mathcal{E}[u, r](\psi) = \int_\Omega 2\lambda d(r\gamma + u)(\nabla d)(r\gamma + u) \cdot \psi + 2\mu Du : D\psi \, d\xi \,,$$

$$\partial_r \mathcal{E}[u, r](\vartheta) = \int_\Omega 2(r - r_0)\vartheta + \kappa \frac{\nabla r \cdot \nabla \vartheta}{\sqrt{|\nabla r|^2 + \delta^2}} + 2\lambda d(r\gamma + u)(\nabla d)(r\gamma + u) \cdot \gamma \vartheta \, d\xi$$

where $\vartheta : \Omega \to \mathbb{R}$ is a scalar test function and $\psi : \Omega \to \mathbb{R}^3$ is vector-valued test displacement. Furthermore, $A : B = \mathrm{tr}(A^T B)$.

For the spatial discretization a piecewise bilinear Finite Element approximation on a uniform rectangular mesh covering the image domain Ω is applied. The distance function d is precomputed using a fast marching method [16] and stored on grid nodes. In the assembly of the functional gradient we use a Gauss quadrature scheme of order 3. The total energy $\mathcal{E}$ is highly non-linear due to the involved nonlinear distance function d and the pseudo Huber norm $|\cdot|_\delta$. We take a multiscale gradient descent approach [17], solving a sequence of joint matching and denoising problems from coarse to fine scales. On each scale a non-linear conjugate gradient method is applied on the space of discrete range maps and discrete deformations. As initial data for the range function r we take into account the raw (time averaged) range data r_0, respectively. The displacement is initialized with the zero mapping. The gradient descent is performed with respect to a regularizing metric

$$g((\delta_r, \delta_u), (\delta_r, \delta_u)) = \int_{\Omega} |\delta_r|^2 + \frac{\sigma^2}{2}|\nabla\delta_r|^2 + |\delta_u|^2 + \frac{\sigma^2}{2}|D\delta_u|^2$$

where δ_r and δ_u are increments in the range function r and the displacement u, respectively. Furthermore, σ corresponds to a Gaussian type filter width acting on the descent directions. As time step control the Armijo rule is taken into account [18]. We stop iterating as soon as the energy decay is sufficiently small.

4   Validation and Application of the Model

To validate our model we have investigated the validation on a real CT and synthetic ToF data (rigid torso phantom), on synthetic CT and ToF data (NCAT respiration phantom), and finally the application to a real CT and real ToF data (rigid torso phantom).

Underlying Data. CT data was acquired on a Siemens SOMATOM Sensation 64 for a male torso phantom at a resolution of $512 \times 512 \times 346$ voxels with a spacing of $0.95 \times 0.95 \times 2.50\,\mathrm{mm}^3$. The surface $\mathcal{G}_{\mathrm{CT}}$ with an approximate diameter of 33 cm is extracted from this data set using a thresholding based region growing segmentation, a marching cube algorithm on the resulting binary segmentation mask followed by a Laplacian mesh smoothing. ToF frame sequences were acquired using a CamCube 3.0 ToF camera from PMD Technologies GmbH[1] with a resolution of 200×200 pixels, a frame rate of 40 Hz, a modulation frequency of 20 MHz, an infrared wavelength centered at 870 nm, an integration time of 750 μs, and a lens with $40° \times 40°$ field of view. This frame rate renders a temporal averaging over 5 frames as acceptable. At the clinical working distance of 1-1.5 m, the noise level of the range measurements is $\sigma^2 \approx 40\,\mathrm{mm}^2$. In addition we have used the NCAT: 4D NURBS-based CArdiac-Torso phantom [19] and generated (artificial) CT data for 16 states within one respiration cycle. For each state, the phantom surface mesh is extracted with the segmentation and mesh generation pipeline sketched above (voxel spacing (x,y,z): $3.125 \times 3.125 \times 3.125\,\mathrm{mm}^3$ and overall resolution of $256 \times 256 \times 191$ voxels). The length of the underlying respiratory cycle is 5 s with an extent of diaphragm motion of 20 mm, an extent of the AP chest expansion of 12 mm (respiration start phase: full exhale, full inhale: 0.4). We generated a typical RT treatment scene by adding a treatment table plane. The synthetic data generation follows the proposal in [20] but has been simplified: Instead of simulating the photon mixing device we directly operate on simulated distance values based on the z-buffer representation of a 3D scene. We then approximate the temporal noise on a per-pixel basis by adding an individual offset drawn from a standard normal with $\sigma^2 = 40\,\mathrm{mm}^2$. This

---

[1] http://www.pmdtec.com/

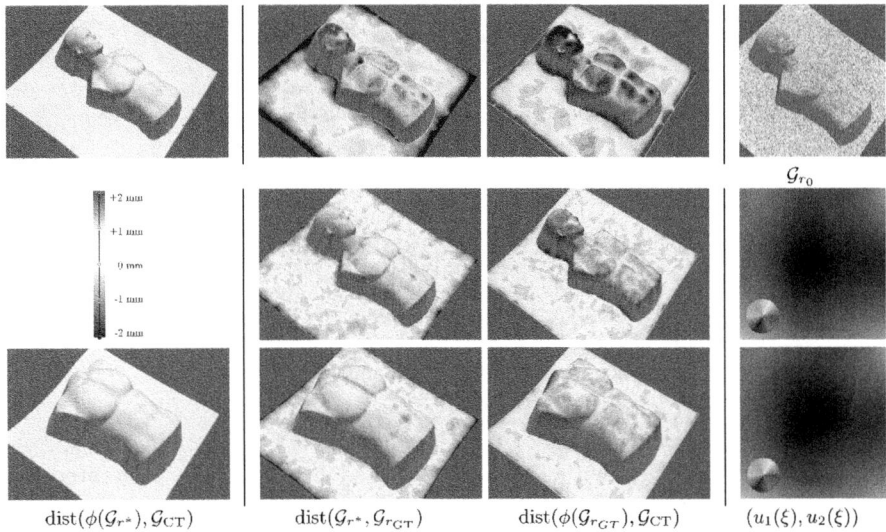

$$\mathrm{dist}(\phi(\mathcal{G}_{r^*}),\mathcal{G}_{\mathrm{CT}}) \qquad \mathrm{dist}(\mathcal{G}_{r^*},\mathcal{G}_{r_{GT}}) \qquad \mathrm{dist}(\phi(\mathcal{G}_{r_{GT}}),\mathcal{G}_{\mathrm{CT}}) \qquad (u_1(\xi),u_2(\xi))$$

Fig. 2. Validation of the model on a male phantom $\mathcal{G}_{\mathrm{CT}}$ (top left and bottom left). The first two lines correspond to results for the full torso incl. head, whereas the third line refers to results for the thorax and abdomen part of the phantom. As quantitative measure of the denoising and registration results we show the distance $\mathrm{dist}(\mathcal{G}_{r^*},\mathcal{G}_{r_{GT}})$ on $\mathcal{G}_{r^*}$ (middle left) and the distance $\mathrm{dist}(\phi(\mathcal{G}_{r_{GT}}),\mathcal{G}_{\mathrm{CT}})$ on $\phi(\mathcal{G}_{r_{GT}})$ (middle right) color coded from $-2\,\mathrm{mm}$ to $+2\,\mathrm{mm}$ using the color bar on the left. Results in the first row correspond to raw, non time averaged range data, whereas in the second and third row a time averaging with $m = 4$ is taken into account. Furthermore, $\mathcal{G}_{r_0}$ for time averaged range data r_0 is shown (top right) and a color coding of the resulting in plane displacement is rendered below for the full torso incl. head (second row) and the sole torso (third row) in case of the time averaged range data (angle and length of the vector $(u_1(\xi),u_2(\xi))$ are encoded as color and brightness, respectively).

variance is motivated by observations on real ToF data at the clinical working distance of about 1–1.5 m. As rather large synthetic deformation we have taken into account $u_{1/2}(x) = \alpha(\pm x_1(x_2 - 1/2) + (1 - x_1)(x_1 - 1/2))$ and $u_3(x) = 0$ with a comparably large deformation scale parameter $\alpha = 0.1$.

Algorithmic Validation Setup. The workflow of the preparatory phase of our validation experiments is as follows: At first we load the torso mesh (real CT phantom or NCAT). Next, we generate a ground truth range image r_{GT}. Then, we generate a synthetic ToF image by adding Gaussian noise with a particular standard derivation σ: $r_{\mathrm{noisy}} = r_{\mathrm{GT}} + \mathrm{noise}_{\sigma^2}$. Furthermore, we deform the phantom torso by the synthetic deformation (in the 2D table plane) to generate a planning CT surface $\mathcal{G}_{\mathrm{CT}}^{\mathrm{phantom}}$. Finally, we generate the discrete signed distance function from the triangular planning CT surface on a 3D mesh of grid resolution 257^3.

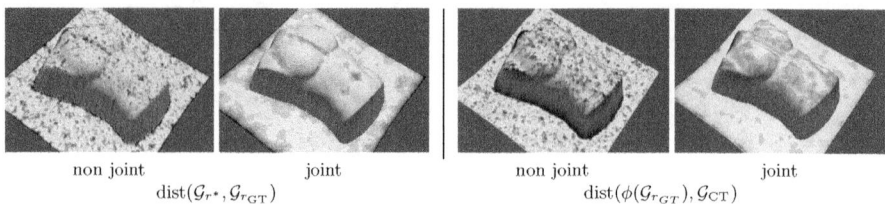

non joint   joint   non joint   joint

$\text{dist}(\mathcal{G}_{r^*}, \mathcal{G}_{r_{\mathrm{GT}}})$ $\text{dist}(\phi(\mathcal{G}_{r_{GT}}), \mathcal{G}_{\mathrm{CT}})$

Fig. 3. Comparison of denoising and subsequent registration to the proposed joint approach for time averaged range data r_0. The two left images show $\text{dist}(\mathcal{G}_{r^*}, \mathcal{G}_{r_{\mathrm{GT}}})$ on $\mathcal{G}_{r^*}$ for the non joint (left) and joint approach (right). The two right images depict the distance measure $\text{dist}(\phi(\mathcal{G}_{r_{GT}}), \mathcal{G}_{\mathrm{CT}})$ on $\phi(\mathcal{G}_{r_{GT}})$ for the non joint (left) and joint approach (right). The color coding is the same as in Fig. 2.

Validation Results for the Real CT and Synthetic Range Data. In Fig. 2, results of our algorithm are shown for a phantom torso and artificially generated range data. We compare the case of unfiltered range data with a suitable set of model parameters ($\kappa = 0.0004$, $\lambda = 10000$, $\mu = 0.004$) to the case of time averaged range data with an adapted set of parameters ($\kappa = 0.0001$, $\lambda = 2500$, $\mu = 0.001$). In addition, we evaluate the benefits of the joint approach in comparison to an algorithm, where one first denoises r_0 and then computes a matching of $\mathcal{G}_r$ and $\mathcal{G}_{\mathrm{CT}}$. Fig. 3 shows that the joint approach is superior to the subsequent denoising and registration approach. Obviously, incorporating prior knowledge about the target shape $\mathcal{G}_{\mathrm{CT}}$ helps substantially in the denoising process. On the other hand, proper denoising also renders the registration problem more robust.

Furthermore, we study the impact of different denoising models in Fig. 4, where the proposed regularization using the pseudo Huber norm is compared to a simple quadratic regularization energy $\kappa \int_\Omega |\nabla r|^2$ and an egde preserving TV regularization of r. The oversmoothing effect of the quadratic model and the staircasing artifacts of the TV model are clearly visible. Here, time averaged ToF data has been investigated and $\kappa = 0.0001$.

Application Benchmark for a 4D CT Respiration Phantom. In Fig. 5, we consider the joint denoising and registration of the synthetic ToF data ($\sigma^2 = 40mm^2$, time averaging over 5 frames) based on the 4D CT respiration phantom with

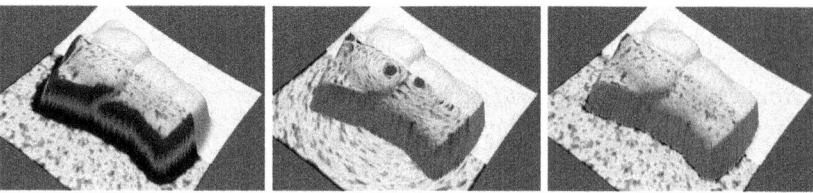

Fig. 4. An experimental evaluation of different denoising models is performed. From left to right the distance $\text{dist}(\mathcal{G}_{r^*}, \mathcal{G}_{r_{\mathrm{GT}}})$ is color coded on $\mathcal{G}_{r^*}$ for a quadratic regularization, a TV regularization, and the proposed regularization via the pseudo Huber norm of ∇r. The color coding is the same as in Fig. 2.

phase 3   phase 5   phase 7   phase 9

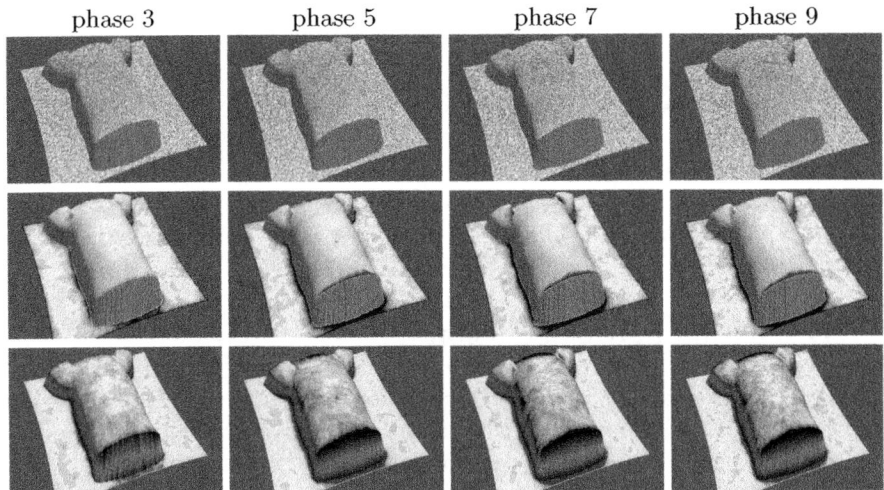

Fig. 5. Four different phases of a respiration cycle $\mathcal{G}_{r_0}$ for time averaged range data r_0 are depicted (first row). The distance $\mathrm{dist}(\mathcal{G}_{r^*}, \mathcal{G}_{r_{\mathrm{GT}}})$ on $\mathcal{G}_{r^*}$ (second row) and the distance $\mathrm{dist}(\phi(\mathcal{G}_{r_{GT}}), \mathcal{G}_{\mathrm{CT}})$ on $\phi(\mathcal{G}_{r_{GT}})$ (third row) are color-coded as in Fig. 2.

16 phases. Thereby, the phantom volume at full expiration is considered as the CT geometry $\mathcal{G}_{\mathrm{CT}}$ (phase 1 out of 16). To speed up the algorithm we now take into account the estimated deformation field and the denoised range data from the previous phase as initial data for our algorithm on the next phase. Table 1 compares this to an initialization of r with r_0 and u with the zero displacement. We observe a reduction of the required gradient descent steps by a factor $\frac{1}{3}$ without any change of the resulting minimal energy. Here, the model parameters are $\kappa = 0.0001$, $\lambda = 2500$, $\mu = 0.001$.

Application to Real CT and Real ToF Data. Finally, we study the performance of our algorithm on real CT and real ToF data based on the rigid torso phantom in Fig. 6. Here, we apply a time averaging of the range data over 5 frames and use the parameters $\kappa = 0.0001$, $\lambda = 2500$, and $\mu = 0.001$. We observe that even topological artifacts (systematical errors of the ToF data due to intensity related distance errors) can be removed and we obtain satisfying denoising and matching results using the proposed joint denoising and registration approach.

Table 1. The number of non-linear CG steps are reported for different phases of a respiration cycle for our method with and without initialization based on the previously processed respiration phase.

respiration phases	phase 3		phase 5		phase 7		phase 9	
	# it	$\mathcal{E}[u^*, r^*]$	# it	$\mathcal{E}[u^*, r^*]$	# it	$\mathcal{E}[u^*, r^*]$	# it	$\mathcal{E}[u^*, r^*]$
No initial.	1743	1.778	1602	1.818	1781	1.832	1812	1.832
incremental initial.	662	1.778	418	1.818	538	1.832	470	1.832

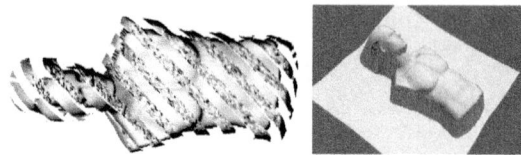

Fig. 6. On the left $\mathcal{G}_{r_0}$ for time averaged real ToF data and the underlying CT phantom $\mathcal{G}_{CT}$ are rendered in a single image using alternating slices. On the right the distance $\mathrm{dist}(\phi(\mathcal{G}_{r^*}), \mathcal{G}_{CT})$ on $\phi(\mathcal{G}_{r^*})$ is again color-coded as in Fig. 2.

5   Discussion and Conclusion

We have proposed a joint variational model for the denoising of ToF range data and the simultaneous matching with a surface extracted from CT data. The approach turned out to be of strong potential for the application in radiation therapy, where respiratory motion has to be compensated to improve therapy planning and treatment. The joint approach is capable of significantly reducing systematic errors from ToF imaging and the obtained quantitative results are within the intended tolerance margins. Based on this approach in a next step a reliable 3D extension of the matching displacement onto the whole geometric model can be computed, which would then finally allow an adaptive steering of the beam in the radiation therapy.

Acknowledgments. S. Bauer gratefully acknowledges the support by the European Union (Europäischer Fonds für regionale Entwicklung) and the Bayerisches Staatsministerium für Wissenschaft, Forschung und Kunst, in the context of the R&D program IUK Bayern under Grant No. IUK338/001.

References

1. Keall, P.J., Mageras, G.S., Balter, J.M., Emery, R.S., Forster, K.M., Jiang, S.B., Kapatoes, J.M., Low, D.A., Murphy, M.J., Murray, B.R., Ramsey, C.R., Herk, M.B.V., Vedam, S.S., Wong, J.W., Yorke, E.: The management of respiratory motion in radiation oncology, report of AAPM task group 76. Med. Phys. 33(10), 3874–3900 (2006)
2. Essapen, S., Knowles, C., Norman, A., Tait, D.: Accuracy of set-up of thoracic radiotherapy: prospective analysis of 24 patients treated with radiotherapy for lung cancer. Br. J. Radiol. 75(890), 162–169 (2002)
3. Johnson, U., Landau, D., Lindgren-Turner, J., Smith, N., Meir, I., Howe, R., Rodgers, H., Davit, S., Deehan, C.: Real time 3D surface imaging for the analysis of respiratory motion during radiotherapy. International Journal of Radiation Oncology Biology Physics 60(supplement 1), 603–604 (2004)
4. Schaller, C., Penne, J., Hornegger, J.: Time-of-Flight Sensor for Respiratory Motion Gating. Medical Physics 35(7), 3090–3093 (2008)
5. Fayad, H., Pan, T., Roux, C., Le Rest, C., Pradier, O., Clement, J., Visvikis, D.: A patient specific respiratory model based on 4D CT data and a time of flight camera (TOF). In: Proceedings of IEEE NSS/MIC, pp. 2594–2598 (2009)

6. Fayad, H., Pan, T., Roux, C., Le Rest, C., Pradier, O., Visvikis, D.: A 2D-spline patient specific model for use in radiation therapy. In: Proceedings of IEEE ISBI, pp. 590–593 (2009)
7. McClelland, J., Blackall, J., Tarte, S., Chandler, A., Hughes, S., Ahmad, S., Landau, D., Hawkes, D.: A continuous 4D motion model from multiple respiratory cycles for use in lung radiotherapy. Medical Physics 33(9), 3348–3358 (2006)
8. Kolb, A., Barth, E., Koch, R., Larsen, R.: Time-of-flight sensors in computer graphics. In: Proceedings of Eurographics, pp. 119–134 (2009)
9. Schaller, C., Adelt, A., Penne, J., Hornegger, J.: Time-of-flight sensor for patient positioning. In: Samei, E., Hsieh, J. (eds.) Proceedings of SPIE Medical Imaging, vol. 7258, p. 726110 (2009)
10. Lindner, M., Schiller, I., Kolb, A., Koch, R.: Time-of-flight sensor calibration for accurate range sensing. Computer Vision and Image Understanding 114(12), 1318–1328 (2010); Special issue on Time-of-Flight Camera Based Computer Vision
11. Kapur, T., Yezzi, L., Zöllei, L.: A variational framework for joint segmentation and registration. In: Proceedings of IEEE Workshop on Mathematical Methods in Biomedical Image Analysis, pp. 44–51 (2001)
12. Unal, G., Slabaugh, G., Yezzi, A., Tyan, J.: Joint segmentation and non-rigid registration without shape priors. Technical Report SCR-04-TR-7495, Siemens Corporate Research (2004)
13. Féron, O., Mohammad-Djafari, A.: Image fusion and unsupervised joint segmentation using a HMM and MCMC algorithms. J. of Electronic Imaging 15(02), 023014 (2004)
14. Droske, M., Rumpf, M.: Multi scale joint segmentation and registration of image morphology. IEEE Transaction on Pattern Recognition and Machine Intelligence 29(12), 2181–2194 (2007)
15. Buades, T., Lou, Y., Morel, J., Tang, Z.: A note on multi-image denoising. In: Proceedings of International Workshop on Local and Non-Local Approximation in Image Processing, pp. 1–15 (2009)
16. Russo, G., Smereka, P.: A remark on computing distance functions. Journal of Computational Physics 163, 51–67 (2000)
17. Álvarez, L., Weickert, J., Sánchez, J.: A scale-space approach to nonlocal optical flow calculations. In: Nielsen, M., Johansen, P., Fogh Olsen, O., Weickert, J. (eds.) Scale-Space 1999. LNCS, vol. 1682, pp. 235–246. Springer, Heidelberg (1999)
18. Armijo, L.: Minimization of functions having Lipschitz continuous first partial derivatives. Pacific Journal of Mathematics 16(1), 1–3 (1966)
19. Segars, W., Mori, S., Chen, G., Tsui, B.: Modeling respiratory motion variations in the 4D NCAT phantom. In: Proceedings of IEEE NSS/MIC, vol. 4, pp. 2677–2679 (2007)
20. Keller, M., Orthmann, J., Kolb, A., Peters, V.: A simulation framework for time-of-flight sensors. In: Proceedings of ISSCS, pp. 1–4 (2007)

Either Fit to Data Entries or Locally to Prior: The Minimizers of Objectives with Nonsmooth Nonconvex Data Fidelity and Regularization

Mila Nikolova

CMLA CNRS ENS Cachan UniverSud
61 av. du Prsident Wilson, 94235 Cachan Cedex, France
nikolova@cmla.ens-cachan.fr
http://www.cmla.ens-cachan.fr/~nikolova/

Abstract. We investigate coercive objective functions composed of a data-fidelity term and a regularization term. Both of these terms are non differentiable and non convex, at least one of them being strictly non convex. The regularization term is defined on a class of linear operators including finite differences. Their minimizers exhibit amazing properties. Each minimizer is the exact solution of an (overdetermined) linear system composed partly of linear operators from the data term, partly of linear operators involved in the regularization term. This is a strong property that is useful when we know that some of the data entries are faithful and the linear operators in the regularization term provide a correct modeling of the sought-after image or signal. It can be used to tune numerical schemes as well. Beacon applications include super resolution, restoration using frame representations, inpainting, morphologic component analysis, and so on. Various examples illustrate the theory and show the interest of this new class of objectives.

Keywords: Image processing, Inverse problems, Non-smooth analysis, Non-convex analysis, Regularization, Signal processing, Variational methods.

1 Introduction

We consider general linear problems where observed data $v[i]$, $1 \leqslant i \leqslant q$, are related to an object of interest $u \in \mathbb{R}^p$ according to

$$v[i] = \langle a_i, u \rangle \text{ with perturbations}, \quad 1 \leqslant i \leqslant q .$$

The object u can be a signal or an $n \times m$ image rearranged into a p-length vector. The family of linear operators $\{a_i \in \mathbb{R}^p, \ 1 \leqslant i \leqslant q\}$ can be any. For instance, it can describe direct observation, optical blurring, sub-sampling, missing data problems, a Radon or a Fourier transform (e.g. in computational tomography), and so on [6], [4], [1]. Following a regularization approach, see e.g. [10], [3], [5],

A.M. Bruckstein et al. (Eds.): SSVM 2011, LNCS 6667, pp. 110–121, 2012.

[2], [9], given data $v \in \mathbb{R}^q$, the sought-after solution $\hat{u}$ is defined as a minimizer of an objective $\mathfrak{F}(\cdot, v) : \mathbb{R}^p \mapsto \mathbb{R}$ of the form

$$\mathfrak{F}(u, v) = \sum_I \psi(\langle a_i, u \rangle - v[i]) + \beta \sum_{j \in J} \varphi(\langle g_j, u \rangle), \quad \beta > 0 \tag{1}$$

$$I = \{1, \cdots, q\} \quad \text{and} \quad J = \{1, \cdots, r\}. \tag{2}$$

The linear operators $\{g_j \in \mathbb{R}^p, \ j \in J\}$ can be any. In practice they produce finite differences of various orders, or discrete Laplacian operators. Let $\{d_j \in \mathbb{R}^p, \ j \in J\}$ denote one of these difference operators; another case of interest is when $g_j = (W^*)^\top d_j$ where W^* is the synthesis operator of a tight frame transform W. To avoid trivialities, it is assumed that

$$a_i \neq 0, \ \forall i \in I \quad \text{and} \quad g_j \neq 0, \ \forall j \in J .$$

Let us denote by $A \in \mathbb{R}^{q \times p}$ and $G \in \mathbb{R}^{r \times p}$ the matrices whose rows are all $a_i^\top$ and all $g_i^\top$, respectively:

$$A = [a_1, \cdots, a_q]^\top \quad \text{and} \quad G = [g_1, \cdots, g_r]^\top ,$$

where the superscript $^\top$ stands for transposed. We assume that

H1 $\ker A \cap \ker G = \{0\}$.

We adopt the classical notation

$$\mathbb{R}_+ = \{t \in \mathbb{R} : t \geqslant 0\} \quad \text{and} \quad \mathbb{R}_+^* = \{t \in \mathbb{R} : t > 0\} .$$

We investigate the case when *both* $\psi : \mathbb{R} \to \mathbb{R}_+$ *and* $\varphi : \mathbb{R} \to \mathbb{R}_+$ *are even nondifferentiable at zero and concave on* $\mathbb{R}_+$, *where at least one of them is strictly concave on* $\mathbb{R}_+$. Thus ψ and φ share some features. The precise assumptions on these functions are presented jointly.

H2 *For $f = \psi$ and $f = \varphi$, we have*

1. *$f : \mathbb{R} \to \mathbb{R}_+$ is even, C^2 on $\mathbb{R} \setminus \{0\}$ and $f(t) > f(0) = 0$ if $|t| \neq 0$;*
2. *$f'(0^+) > 0$ and $f'(t) > 0$ on $\mathbb{R}_+^*$;*
3. *f'' is increasing on $\mathbb{R}_+^*$, $f''(t) \leqslant 0, \ \forall t > 0$ and $\lim_{t \searrow 0} f''(t)$ is well defined.*

H3 *At least one of the functions $f = \psi$ or $f = \varphi$ satisfy*
   f is strictly concave on $\mathbb{R}_+$: $f''(t) < 0, \ \forall t > 0$ and $\lim_{t \searrow 0} f''(t) < 0$.

Several examples of functions f are shown in Table 1 and plotted in Fig. 1.

1.1 Motivation

An illustration of a minimizer of $\mathfrak{F}(\cdot, v)$ in (1) for $A = \mathrm{I}$, and (ψ, φ) satisfying H2 and H3, is given in Fig. 2. One observes that restored samples either fit data samples exactly or form constant patches.

Table 1. Functions $f\big|_{\mathbb{R}_+} : \mathbb{R}_+ \to \mathbb{R}_+$ satisfying H2. All functions except (f6) satisfy H3 as well. The functions (f3), (f4), (f5) and (f6) are coercive.

	(f1)	(f2)	(f3)	(f4)	(f5)	(f6)
$f\big|_{\mathbb{R}_+}$	$\dfrac{\alpha t}{\alpha t + 1}$	$1 - \alpha^t$	$\ln(\alpha t + 1)$	$(t + \varepsilon)^\alpha - \varepsilon^\alpha$	t^α	t
	$\alpha > 0$	$0 < \alpha < 1$	$\alpha > 0$	$0 < \alpha < 1, \varepsilon > 0$	$0 < \alpha < 1$	$\cdot$
$f'\big|_{\mathbb{R}_+^*}$	$\dfrac{\alpha}{(\alpha t+1)^2}$	$-\alpha^t \ln \alpha$	$\dfrac{\alpha}{\alpha t+1}$	$\alpha(t + \varepsilon)^{\alpha-1}$	$\alpha t^{\alpha-1}$	1
$f'(0^+)$	α	$-\ln \alpha$	α	$\alpha \varepsilon^{\alpha-1}$	$+\infty$	1
$f''\big|_{\mathbb{R}_+^*}$	$\dfrac{-2\alpha^2}{(\alpha t+1)^3}$	$-\alpha^t(\ln \alpha)^2$	$\dfrac{-\alpha^2}{(\alpha t+1)^2}$	$\alpha(\alpha - 1)(t + \varepsilon)^{\alpha-2}$	$\alpha(\alpha - 1)t^{\alpha-2}$	0
$\lim\limits_{t \searrow 0} f''(t)$	$-2\alpha^2$	$-(\ln \alpha)^2$	$-\alpha^2$	$\alpha(\alpha - 1)\varepsilon^{\alpha-2}$	$-\infty$	0

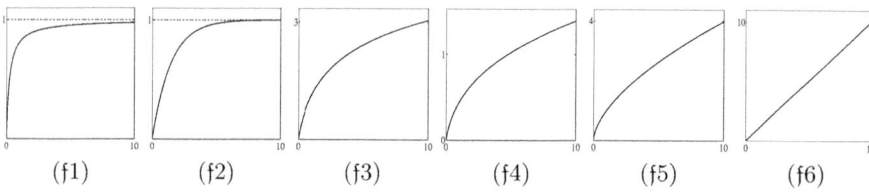

(f1)   (f2)   (f3)   (f4)   (f5)   (f6)

Fig. 1. Plots of the PFs $f\big|_{\mathbb{R}_+}$ given in Table 1

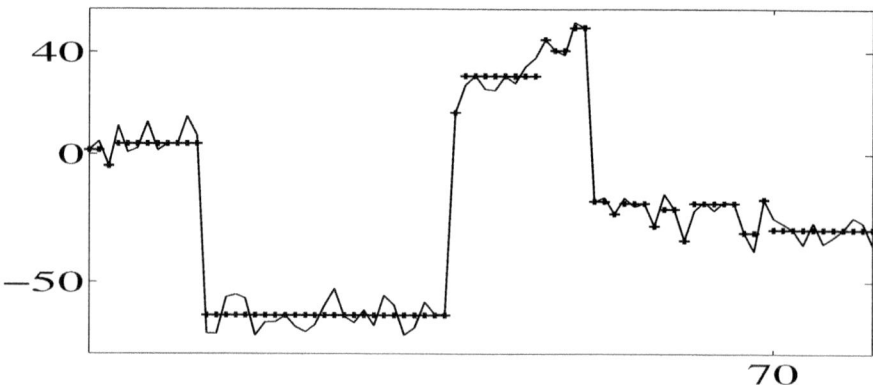

Fig. 2. $\mathfrak{F}(u, v) = \sum_{i=1}^{p} \psi(u[i] - v[i]) + \beta \sum_{i=1}^{p-1} \varphi(u[i + 1] - u[i])$ for $\psi(t) = |t|^{0.7}$ and $\varphi(t) = \dfrac{\alpha |t|}{\alpha |t|+1}$. Note that H1 is satisfied and that (ψ, φ) satisfy H2 and H3. Data v are plotted with "—", each sample of the minimizer $\hat{u}$ is marked with "+".

Example 1. This example is quite illuminating. Given $v \in \mathbb{R} \setminus \{0\}$, consider $\mathfrak{F}(\cdot, v) : \mathbb{R} \mapsto \mathbb{R}$ for $A = I$, and a pair of functions (ψ, φ) satisfying H2 and H3:

$$\mathfrak{F}(u, v) = \psi(u - v) + \beta\varphi(u) \ , \forall u \in \mathbb{R}, \tag{3}$$

$$F(u, v) = \mathfrak{F}(u, v), \quad \forall u \in \mathbb{R} \setminus \{0, v\}. \tag{4}$$

Note that F is the restriction of $\mathfrak{F}$ on $\mathbb{R} \setminus \{0, v\}$.

The differential of order j of a function f with respect to its k-th argument is denoted by $D_k^j f$. Since $\mathfrak{F}$ is coercive, it does admit minimizers. Let $\hat{u}$ be a minimizer of $\mathfrak{F}(\cdot, v)$. The necessary conditions for $\mathfrak{F}$ to have a (local) minimum at $\hat{u} \neq 0$ and $\hat{u} \neq v$, or equivalently, for F to have a (local) minimum at $\hat{u}$, namely $D_1 F(\hat{u}, v) = 0$ and $D_1^2 F(\hat{u}, v) \geqslant 0$, do not hold. Indeed, by H3, the second derivatives on $\mathbb{R} \setminus \{0, v\}$ of ψ and φ are non positive and at least one of them is negative. So

$$D_1^2 F(u, v) = \psi''(u - v) + \beta\varphi''(u) < 0 \quad \forall u \in \mathbb{R} \setminus \{0, v\} \ .$$

Hence there is no minimizer such that $\hat{u} \neq 0$ and $\hat{u} \neq v$. In this way, $F(\cdot, v)$ in (4) does not have minimizers. It follows that any minimizer of $\mathfrak{F}(\cdot, v)$ in (3) satisfies

$$\hat{u} \in \{0, v\}.$$

Example 2. Given $v \in \mathbb{R}$, consider $\mathfrak{F}(\cdot, v) : \mathbb{R}^2 \mapsto \mathbb{R}$ as given below:

$$\mathfrak{F}(u, v) = \psi\big(u[1] + u[2] - v\big) + \beta\big(\varphi(u[1]) + \varphi(u[2])\big), \quad 0 < \beta < 1 \ .$$

Let $\psi = \varphi$ satisfy H2 and H3. Then $\mathfrak{F}(\cdot, v)$ has two strict global minimizers

$$\hat{u}_1 = [v, \ 0]^\top \quad \text{and} \quad \hat{u}_2 = [0, \ v]^\top$$

yielding $\mathfrak{F}(\hat{u}_1, v) = \mathfrak{F}(\hat{u}_2, v) = \beta\varphi(v) < \varphi(v) = \psi(v) = \mathfrak{F}(0, v)$. When ψ and φ are nonsmooth and strictly nonconvex on $\mathbb{R}_+$, we have two strict global (sparse) minimizers.

If $\psi(t) = \varphi(t) = |t|$, then $\mathfrak{F}(\cdot, v)$ is convex and reaches its minimum for

$$\hat{u}_t = (1 - t) [v, \ 0]^\top + t [0, \ v]^\top, \quad 0 \leqslant t \leqslant 1 \ .$$

This yields $\mathfrak{F}(\hat{u}_t, v) = \beta|v|, \quad 0 \leqslant t \leqslant 1$. The minimum is hence *nonstrict*.

1.2 Notations

Given a $K \times p$ matrix B with rows $b_i^\top$, $1 \leqslant i \leqslant K$, a K-length vector w and a strictly increasing subsequence $\varpi \subset \{1, \cdots, K\}$, say $\varpi = (\varpi[1], \cdots, \varpi[n])$ with $\varpi[1] < \cdots < \varpi[n]$, where $n = \sharp\varpi$, we *systematically* denote

$$B_\varpi = [b_{\varpi[1]}, \cdots, b_{\varpi[n]}]^\top \quad \text{and} \quad w_\varpi[i] = w\big[\varpi[i]\big], \ 1 \leqslant i \leqslant n \ . \tag{5}$$

We write $B_\varpi^\top$ for the transposed of B_ϖ. The range of B_ϖ reads $\mathcal{R}(B_\varpi)$. We denote by $\mathbb{1}$ a column vector of whatever length appropriate to the context composed of ones. If necessary, $\mathbb{1}_K$ specifies that the vector is of length K. The canonical basis of $\mathbb{R}^K$ is denoted $\{e_i, \ i \in \{1, \cdots, K\}\}$.

1.3 Outline of the Paper

Existence and strictness of local minimizers are shown in section 2. Section 3 reveals that a strict (local) minimizer is the unique solution of a linear system. Stability of minimizers is studied in section 4. Section 5 focuses on the case when ψ and φ are coercive and strictly nonconvex on $\mathbb{R}_+$. The numerical examples in section 6 confirm the theoretical results. All proofs can be found in [7].

2 Preliminaries

2.1 The Objective $\mathfrak{F}$ Is Not Too Bad

Even though nonconvex and nonsmooth, $\mathfrak{F}(\cdot, v)$ does have minimizers. A general strong sufficient condition is evoked below.

Lemma 1. *Let (ψ, φ) satisfy H2 and H3. Let one of the following assumptions hold true:*

(a) $\operatorname{rank}(A) = p$ and ψ is coercive, i.e. $\lim_{t \to \infty} \psi(t) = \infty$;

(b) H1 holds, and ψ and φ are coercive.

Then $\forall v \in \mathbb{R}^q$ and $\forall \beta > 0$, the function $\mathfrak{F}(., v)$ in (1) does admit a minimum.

We should emphasize that Lemma 1 gives only strong sufficient conditions for the existence of a minimizer. They are not necessary, as illustrated by the example given below.

Example 3. Consider $\mathfrak{F}$ of the form (1) for $p = 3$ and $q = 2$ where

$$A = \begin{bmatrix} 1 & 0 & 0 \\ 0 & 0 & 1 \end{bmatrix}, \quad v = \begin{bmatrix} 1 \\ 3 \end{bmatrix}, \quad \begin{matrix} g_1 = [1\ -1\ 0]^\top, \\ g_2 = [0\ 1\ -1]^\top, \end{matrix} \quad \psi(t) = |t|, \quad \varphi(t) = \frac{\alpha|t|}{\alpha|t| + 1}. \quad (6)$$

Assumptions H1, H2 and H3 are satisfied. The objective $\mathfrak{F}$ reads

$$\mathfrak{F}(u, v) = \big|u[1] - v[1]\big| + \big|u[3] - v[2]\big| + \beta\Big(\varphi(u[1] - u[2]) + \varphi(u[2] - u[3])\Big).$$

Clearly, $\mathfrak{F}(., v)$ does not meet the conditions of Lemma 1 since $\operatorname{rank}(A) = 2 < p = 3$ and φ is not coercive. Nevertheless, one computes that for $\alpha = 1$ and $\beta = 2$ the global minimizer of $\mathfrak{F}(., v)$ reads

$$\hat{u} = [1\ \ 1\ \ 3]^\top. \quad (7)$$

Below we show that if a (local) minimizer $\hat{u}$ fits exactly some data entries, the relevant rows of A are almost surely linearly independent.

Lemma 2. *For $\nu \subseteq \{1, \cdots, \operatorname{rank} A\}$ such that $\operatorname{rank} A_\nu < \sharp\nu$, consider the subset $V_\nu \overset{\text{def}}{=} \{w \in \mathbb{R}^{\sharp\nu} : w \in \mathcal{R}(A_\nu)\}$. We have*

(i) $V_\nu \subsetneqq \mathbb{R}^{\sharp\nu}$ is closed and $\mathbb{L}^{\sharp\nu}(V_\nu) = 0$.

(ii) Given $v \in \mathbb{R}^q$ such that $v_\nu \in \mathbb{R}^{\sharp\nu} \setminus V_\nu$, let $\hat{u}$ be a (local) minimizer of $u \mapsto \mathfrak{F}(u,v)$ satisfying $\{i \in I : \langle a_i, \hat{u} \rangle = v[i]\} = \nu$. Then $\operatorname{rank} A_\nu = \sharp\nu$.

Given $v \in \mathbb{R}^q$, let $\hat{u}$ be a (local) minimizer of $u \mapsto \mathfrak{F}(u,v)$. With each such $\hat{u}$ we systematically associate the following subsets:

$$\nu = \{i \in I : \langle a_i, \hat{u} \rangle = v[i]\} \quad \text{and} \quad \nu^c = I \setminus \nu = \{i \in I : \langle a_i, \hat{u} \rangle \neq v[i]\} , \qquad (8)$$

$$\sigma = \{i \in J : \langle g_i, \hat{u} \rangle = 0\} \quad \text{and} \quad \sigma^c = J \setminus \sigma = \{i \in J : \langle g_i, \hat{u} \rangle \neq 0\} . \qquad (9)$$

In the case of *Example 3*, we have $\nu = \{1,2\} = I$ and $\sigma = \{1\}$, so $\nu^c = \varnothing$ and $\sigma^c = \{2\}$.

For $(u,v) \in \mathbb{R}^p \times \mathbb{R}^q$, denote

$$\psi_i(u) = \psi\big(\langle a_i, u \rangle - v[i]\big), \quad \forall\, i \in I, \qquad (10)$$
$$\varphi_i(u) = \varphi\big(\langle g_i, u \rangle\big), \qquad\qquad \forall\, i \in J. \qquad (11)$$

Since (ψ, φ) are $\mathcal{C}^2$ on $\mathbb{R} \setminus \{0\}$, one can expect that ψ_i and φ_j in (10)-(11) are locally $\mathcal{C}^2$ provided that $i \notin \nu$ and $j \notin \sigma$.

Lemma 3. *Given $v \in \mathbb{R}^q$, let $\mathfrak{F}(\cdot, v)$ reach a (local) minimum at $\hat{u}$. Let H2 and H3 hold. Put*

$$\rho = \min \left\{ \min_{i \in \nu^c} \frac{|\langle a_i, \hat{u} \rangle - v[i]|}{\|a_i\|_2}, \; \min_{j \in \sigma^c} \frac{|\langle g_j, \hat{u} \rangle|}{\|g_j\|_2} \right\}.$$

We have $\rho > 0$. Let $u \in B(\hat{u}, \rho) \overset{\text{def}}{=} \{w \in \mathbb{R}^p : \|w - \hat{u}\|_2 < \rho\}$ then

$$i \in \nu^c \Rightarrow \psi_i(u) \in \mathcal{C}^2\big(B(\hat{u}, \rho)\big) , \qquad (12)$$
$$j \in \sigma^c \Rightarrow \varphi_i(u) \in \mathcal{C}^2\big(B(\hat{u}, \rho)\big) . \qquad (13)$$

2.2 (Local) Minimizers Are Strict

A local minimizer $\hat{u}$ is *strict* if there is a neighborhood $\mathcal{O} \subset \mathbb{R}^N$, containing $\hat{u}$, such that $\mathfrak{F}(\hat{u}, v) < \mathfrak{F}(w, v)$ for any $w \in \mathcal{O}$. Such a minimizer is isolated.

With a (local) minimizer $\hat{u}$ of $\mathfrak{F}(\cdot, v)$ we associate the manifolds given below:

$$\mathcal{K}_{\hat{u}} = \{w \in \mathbb{R}^p : A_\nu w = v_\nu \text{ and } G_\sigma w = 0\} , \qquad (14)$$
$$\mathrm{K}_{\hat{u}} = \{w \in \mathbb{R}^p : A_\nu w = 0 \text{ and } G_\sigma w = 0\} , \qquad (15)$$

where ν and σ are defined in (8)-(9). Since

$$\hat{u} \in \mathcal{K}_{\hat{u}},$$

we are guaranteed that $\mathcal{K}_{\hat{u}}$ is nonempty. Note that $\mathrm{K}_{\hat{u}}$ is the vector subspace tangent to $\mathcal{K}_{\hat{u}}$. Equivalently, for any $w \in \mathrm{K}_{\hat{u}}$ we have $\hat{u} + w \in \mathcal{K}_{\hat{u}}$: thus $\mathrm{K}_{\hat{u}}$ contains directions in which the (local) minimizer $\hat{u}$ might be nonstrict.

Lemma 4. *Consider $\mathfrak{F}$ of the form (1). Let (ψ, φ) satisfy H2. For $v \in \mathbb{R}^q$, let $\hat{u}$ be a (local) minimizer of $u \mapsto \mathfrak{F}(u, v)$. The subsets ν and σ read according to (8) and (9), respectively. The vector subspace $\mathrm{K}_{\hat{u}}$ is defined in (15) and we suppose that*

$$\dim(\mathrm{K}_{\hat{u}}) \geqslant 1 \; .$$

(i) If ψ satisfies H3 and $\operatorname{rank} A_\nu < \operatorname{rank} A$, then $\exists\, w \in \mathrm{K}_{\hat{u}}$ such that $Aw \neq 0$.
(ii) If φ satisfies H3 and $\operatorname{rank} G_\sigma < \operatorname{rank} G$, then $\exists\, w \in \mathrm{K}_{\hat{u}}$ such that $Gw \neq 0$.
(iii) If ψ and ϕ satisfy H3 and we have $\operatorname{rank} A_\nu < \operatorname{rank} A$ or $\operatorname{rank} G_\sigma < \operatorname{rank} G$, then

$$\exists\, w \in \mathrm{K}_{\hat{u}} \quad \text{such that} \quad \Big[Aw \neq 0 \quad \text{or} \quad Gw \neq 0 \Big] \; .$$

Given $v \in \mathbb{R}^q$, we consider the function given below

$$F(\cdot, v) : \mathcal{K}_{\hat{u}} \mapsto \mathbb{R}$$
$$F(u, v) = \sum_{i \in \nu^c} \psi(\langle a_i, u \rangle - v[i]) + \beta \sum_{j \in \sigma^c} \varphi(\langle g_j, u \rangle) \tag{16}$$

where $\mathcal{K}_{\hat{u}}$ is defined in (14). Obviously, $F(\cdot, v)$ is the restriction of $\mathfrak{F}(\cdot, v)$ on $\mathcal{K}_{\hat{u}}$. One can remind the function F in Example 1. For any $w \in \mathrm{K}_{\hat{u}}$ we have

$$\langle D_1^2 F(\hat{u}, v) w, w \rangle = \sum_{i \in \nu^c} \psi''(\langle a_i, \hat{u} \rangle - v[i]) \langle a_i, w \rangle^2 + \beta \sum_{i \in \sigma^c} \varphi''(\langle g_j, \hat{u} \rangle) \langle g_j, w \rangle^2 \; .$$

Lemma 5. *Let $\mathfrak{F}$ be such that (ψ, φ) satisfy H2. For $v \in \mathbb{R}^q$, let $\hat{u}$ be a (local) minimizer of $\mathfrak{F}(\cdot, v)$. Suppose that the vector subspace $\mathrm{K}_{\hat{u}}$ in (15) satisfies*

$$\dim(\mathrm{K}_{\hat{u}}) \geqslant 1 \; .$$

Assume also that one of the following conditions is met:

1. ψ satisfies assumption H3 and $\operatorname{rank} A_\nu < \operatorname{rank} A$;
2. φ satisfies assumption H3 and $\operatorname{rank} G_\sigma < \operatorname{rank} G$;
3. ψ and ϕ satisfy H3, and we have $\operatorname{rank} A_\nu < \operatorname{rank} A$ or $\operatorname{rank} G_\sigma < \operatorname{rank} G$.

Then there exists $w \in \mathrm{K}_{\hat{u}}$ such that $\langle D_1^2 F(\hat{u}, v) w, w \rangle < 0$.

In general, this lema states quite an unusual result: the restriction of $\mathfrak{F}(\cdot, v)$ on $\mathcal{K}_{\hat{u}}$, namely $F(\cdot, v)$, does not have minimizers. The reader is invited to remind the restricted function F in Example 1 since it does not have minimizers neither.

Next we show that the (local) minimizers of $\mathfrak{F}(\cdot, v)$ are strict in general.

Theorem 1. *Consider $\mathfrak{F}$ of the form (1). Let (ψ, φ) satisfy H2. For $v \in \mathbb{R}^q$, let $\hat{u}$ be a (local) minimizer of $u \mapsto \mathfrak{F}(u, v)$. The subsets ν and σ are defined according to (8) and (9), respectively, and the vector subspace $\mathrm{K}_{\hat{u}}$ is defined in (15). Assume also that one of the conditions 1, 2 or 3 in Lemma 5 is met. Then*

$$\mathcal{K}_{\hat{u}} = \{\hat{u}\} \quad \text{and} \quad \mathrm{K}_{\hat{u}} = \{0\} \; , \tag{17}$$

so $\mathfrak{F}(\cdot, v)$ reaches a strict (local) minimum at $\hat{u}$.

Example 4. Let us consider again Example 3, p. 114. From the ingredients of $\mathfrak{F}$ given in (6), the minimizer in (7) and the definition of $\mathcal{K}_{\hat{u}}$ in (14), on finds

$$\begin{aligned}
\mathcal{K}_{\hat{u}} &= \{w \in \mathbb{R}^3 : \langle a_1, w \rangle = v[1], \ \langle a_2, w \rangle = v[2], \ \langle g_1, w \rangle = 0\} \\
&= \{w \in \mathbb{R}^3 : w[1] = v[1], \ w[3] = v[2], \ w[1] - w[2] = 0\} \\
&= \{w \in \mathbb{R}^3 : w[1] = v[1], \ w[3] = v[2], \ w[2] = w[1]\} \\
&= \{w \in \mathbb{R}^3 : w[1] = w[2] = v[1], \ w[3] = v[2]\} \\
&= \{w \in \mathbb{R}^3 : w[1] = 1, \ w[2] = 1, \ w[3] = 3\} \ = \ \{\hat{u}\}.
\end{aligned}$$

Then $\mathrm{K}_{\hat{u}} = \{0\}$.

Let us list the cases when we cannot guarantee that the minimum is strict.

1. $\mathrm{rank}\, A_\nu = \mathrm{rank}\, A$ and ψ meets H3 but φ does not.
   Given the fact that A_ν is defined according to (5), the condition given above means that all a_i, $i \in \nu^c$ are linear combinations of $\{a_i, \forall i \in \nu\}$. Then we have the equivalence $\big[A_\nu w = 0 \ \Leftrightarrow \ Aw = 0 \big]$. In the first instance, this situation occurs when

$$A\hat{u} = v \ .$$

   In case we wish to change some data equations (e.g. if there is some noise), such a minimizer does not do the job. Otherwise, $\mathrm{rank}\, A_\nu = \mathrm{rank}\, A < q$ means that we have reached the maximum among all data entries that can be fitted exactly as far as in general $v \notin \mathcal{R}(A_\nu) = \mathcal{R}(A)$ whose dimension is strictly smaller than the dimension of the data space.
2. $\mathrm{rank}\, G_\sigma = \mathrm{rank}\, G$ and φ meets H3 but ψ does not.
   A similar reasoning than above shows that

$$G_\sigma w = 0 \quad \Leftrightarrow \quad Gw = 0 \ .$$

For instance, if $\{g_j, j \in J\}$ are first-order differences, $G\hat{u} = 0$ means that $\hat{u}$ is constant, i.e. $\hat{u} = c\mathbb{1}$ for any $c \in \mathbb{R} \setminus \{0\}$. Such an $\hat{u}$ is certainly not a meaningful solution.

We conclude that all these cases, excluded from Theorem 1, are quite pathological.

3 Either Fidelity or Prior

3.1 Strict Minimizers Solve Exactly Linear Systems

In spite of the high nonlinearity of the minimization problem, it is shown below that every strict (local) minimizer of $\mathfrak{F}(\cdot, v)$ is the unique solution of a linear system composed out of some elements of $\{a_i, i \in I\}$ and of $\{g_j, j \in J\}$.

Theorem 2. *Let (ψ, φ) satisfy H2 and H3. For $\hat{u}$ a (local) minimizer of $\mathfrak{F}(\cdot, v)$, we posit the definitions of ν and σ in (8)-(9) and the one of $\mathrm{K}_{\hat{u}}$ in (15). Assume* also that one *of the conditions 1, 2 or 3 in Lemma 5 is met. Then* $\hat{u}$ *is the* unique *solution of the* linear system of equations *given below:*

$$\begin{aligned} \langle a_i, \hat{u} \rangle &= v[i] \quad \forall i \in \nu \ , \\ \langle g_j, \hat{u} \rangle &= 0 \quad \forall j \in \sigma \ . \end{aligned} \tag{18}$$

Let $H_{\nu,\sigma} \in \mathbb{R}^{p \times (\sharp\nu + \sharp\sigma)}$ *read*

$$H_{\nu,\sigma} = \begin{bmatrix} A_\nu^\top & G_\sigma^\top \end{bmatrix}^\top \ . \tag{19}$$

We have $\operatorname{rank} H_{\nu,\sigma} = p$. *Let* $v_{\nu,\sigma} \in \mathbb{R}^{\sharp\nu + \sharp\sigma}$ *have its first subvector equal to* v_ν *and its second* $\sharp\sigma$-*length subvector composed of zeros:* $v_{\nu,\sigma} = \begin{bmatrix} v_\nu^\top, (0 \ \mathbb{1}_{\sharp\sigma})^\top \end{bmatrix}^\top$. *Then*

$$\hat{u} = (H_{\nu,\sigma}^\top H_{\nu,\sigma})^{-1} H_{\nu,\sigma}^\top v_{\nu,\sigma} \ . \tag{20}$$

Example 5. Let $r = p$ and $a_i = y_i = e_i$ for $i = 1, \cdots, p$. Then $\mathfrak{F}$ reads

$$\mathfrak{F}(u, v) = \sum_{i=1}^{p} \Big(\psi\big(u[i] - v[i]\big) + \beta\varphi(u[i]) \Big) \ .$$

According to Theorem 2, we have

$$\text{either} \quad \hat{u}[i] = v[i] \quad \text{or} \quad \hat{u}[i] = 0, \quad \forall i \in \{1, \cdots, p\} \ .$$

Next consider that g_i are as in Fig. 2, i.e.

$$g_i[j] = \begin{cases} -1 & \text{if } j = i \\ 1 & \text{if } j = i + 1 \\ 0 & \text{if } j \notin \{i, i+1\} \end{cases} \quad \text{for} \quad i \in \{1, \cdots, p-1\} \ .$$

Now

$$\mathfrak{F}(u, v) = \sum_{i=1}^{p} \psi\big(u[i] - v[i]\big) + \beta \sum_{i=1}^{p-1} \varphi\big(u[i+1] - u[i]\big) \ .$$

By Theorem 2 we find that

$$\hat{u}[i] = v[i] \quad \text{or} \quad \hat{u}[i] = \hat{u}[i+1], \quad \forall\{i, i+1\} \in I \times I \ .$$

In words, the (local) minimizer is composed partly of constant patches, partly of pixels that fit data samples exactly, as seen in Fig. 2.

On the role of the regularization parameter $\beta > 0$. Theorem 2 and in particular the expression for a (local) minimizer $\hat{u}$ given in (20) does not make an explicit reference to the regularization parameter β. Usually $\mathfrak{F}(\cdot, v)$ has numerous (local) minimizers. According to the same theorem, each one of them is strict and is the unique solution of a linear system of the form (18). Any other such (local) minimizer $\hat{u}'$ corresponds to different subsets $\nu' \subset I$ and $\sigma' \subset J$ and in general, $\mathfrak{F}(\hat{u}, v) \neq \mathfrak{F}(\hat{u}', v)$. As far as a minimizer is determined by the subsets $\nu' \subset I$ and $\sigma' \subset J$, the selection of different local minimizers, including the global minimizer, is controlled by β.

4   Local Stability of Strict Minimizers

Here we study how local minimizers do behave under variations of the data.

Definition 1. *Let $\mathfrak{F} : \mathbb{R}^p \times \mathbb{R}^q \to \mathbb{R}$ and $\mathcal{O} \subseteq \mathbb{R}^q$ be open. We say that $\mathcal{U} : \mathcal{O} \to \mathbb{R}^p$ is a (local) minimizer function for the family of functions $\mathfrak{F}(\cdot, \mathcal{O}) = \{\mathfrak{F}(\cdot, v) : v \in \mathcal{O}\}$ if for any $v \in \mathcal{O}$, the function $\mathfrak{F}(\cdot, v)$ reaches a strict (local) minimum at $\mathcal{U}(v)$.*

Theorem 3. *Let (ψ, φ) satisfy H2 and H3. For $v \in \mathbb{R}^q \setminus \{0\}$, let $\hat{u}$ be a (local) minimizer of $u \mapsto \mathfrak{F}(u, v)$. We posit the definitions of ν and σ as given in (8)-(9), and of $K_{\hat{u}}$ in (15). Assume also that one of the conditions 1, 2 or 3 in Lemma 5 is met. Then there exists $\varrho > 0$ and a (local) minimizer function $\mathcal{U}$*

$$\|v' - v\|_2 < \varrho \quad \Rightarrow \quad \hat{u}' = \mathcal{U}(v') \tag{21}$$
$$\mathcal{U}(v') = (H_{\nu,\sigma}^\top H_{\nu,\sigma})^{-1} H_{\nu,\sigma}^\top v'_{\nu,\sigma} \tag{22}$$

where $H_{\nu,\sigma}$ is defined according to (19).

Note that the (local) minimizer function $\mathcal{U}$ is linear with respect to data v. The *global* minimizer function is *piecewise linear* with respect to data v.

5   A Special Case

Here we address a particular class of functions (ψ, φ), as given in H4 below.

H4 *Assume the following:*

- ψ *and* φ *satisfy H2 and H3 ;*
- ψ *and* φ *are coercive ;*
- $\psi'(0^+) = +\infty$ *and* $\varphi'(0^+) = +\infty$.

Popular examples are ℓ_p "norms" for $0 < p < 1$, see ($\mathfrak{f}4$) in Table 1.

Corollary 1. *Theorems 2 and 3 holds true only under assumption 3.*

It appears that each collection of a_i's and g_j's of rank p corresponds to a (local) minimizer of $\mathfrak{F}(\cdot, v)$. The result can be seen as the inverse of Theorem 2.

Theorem 4. *Given $v \in \mathbb{R}^q$, let $\nu \subset I$ and $\sigma \subset J$ be such that the system of linear equations given below does admit a unique solution $\hat{u}$:*

$$\begin{aligned} \langle a_i, \hat{u} \rangle &= v[i] \quad \forall i \in \nu , \\ \langle g_j, \hat{u} \rangle &= 0 \qquad \forall j \in \sigma . \end{aligned} \tag{23}$$

Then for any $\beta > 0$, $\hat{u}$ is a strict (local) minimizer of an objective $\mathfrak{F}(\cdot, v)$ of the form (1) where (ψ, φ) satisfy H4.

6   Numerical Examples

Here we consider a toy missing data recovery problem using $\mathfrak{F} : \mathbb{R}^p \times \mathbb{R}^q \to \mathbb{R}$

$$\mathfrak{F}(u, v) = \sum_{i \in I} \psi(\langle a_i, u \rangle - v[i]) + \beta \sum_{i=1}^{p-2} \varphi(u[i+2] - 2u[i+1] + u[i]) \quad (24)$$

where $p = 80$ for $\psi(t) = |t|^{0.7}$ and $\varphi(t) = \frac{\alpha |t|}{\alpha |t| + 1}$. Here $\langle g_i, \hat{u} \rangle = 0$ means that

$$\hat{u}[i+2] - 2\hat{u}[i+1] + \hat{u}[i] = 0 , \quad (25)$$

i.e. that three consecutive pixels form a piece of line. The original is shown in Figs. 3 and 4(c) with a dashed line. It contains large polynomial, nearly affine parts.

In the first experiment in Fig 3 we have $a_i = e_i$, $i \in I$ for $q = \sharp I = 25$. Thus $\langle a_i, u \rangle - v[i] = u[i] - v[i]$ in (24). Data samples are plotted with diamonds. These few data samples are largely enough to interpolate all missing parts by affine pieces. The minimizer is strict because φ meets H3.

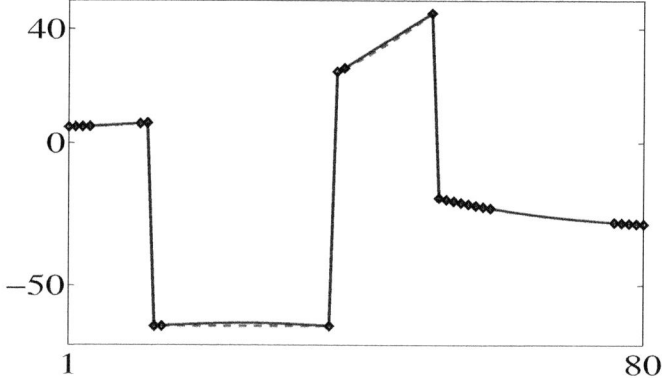

Fig. 3. Data v in $\diamond$, minimizer $\hat{u}$ in thick line, original in dashed line. Results correspond to $\alpha = 4$ and $\beta = 15$.

In the second experiment in Fig. 4 the same original is considered. Ten data samples ($q = \sharp I = 10$) are produced using randomly generated $\{a_i, i \in I\}$. The 10-length data vector v is shown in Fig. 4(a). Yet again, all polynomial parts are interpolated via affine pieces satisfying (25). It is likely that the obtained minimizers $\hat{u}$ yield just a local minimum of $\mathfrak{F}(\cdot, v)$. All data equations are satisfied exactly. Missing parts are fitted using the 2nd order differences in (24). The minimizer is strict because φ meets H3.

The numerical experiments corroborate the theoretical results presented above.

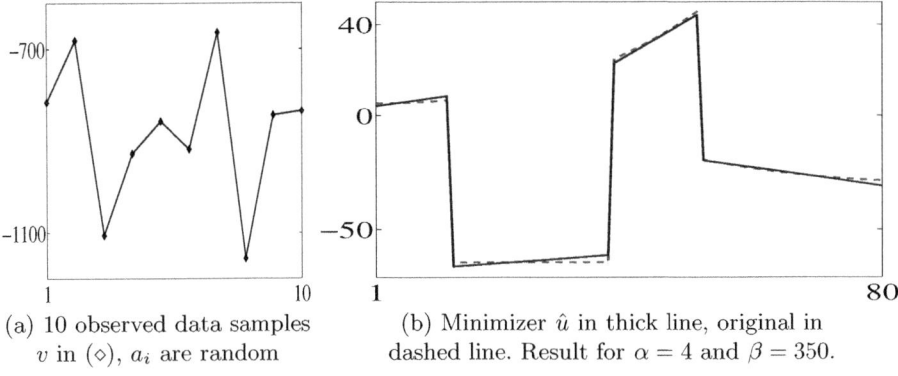

(a) 10 observed data samples v in ($\diamond$), a_i are random

(b) Minimizer $\hat{u}$ in thick line, original in dashed line. Result for $\alpha = 4$ and $\beta = 350$.

Fig. 4. Restoration from 10 random observations

7 Concluding Notes

We show that if ψ and φ are nonconvex and nonsmooth at zero, and at least one of them is strictly nonconvex on $\mathbb{R}_+$, (local) minimizers are generally strict and are given as the unique solution of a linear system composed of linear operators coming from the data term and from the regularization term. This result provides a flexible tool to check if an algorithm minimizing $\mathfrak{F}(\cdot, v)$ has found a strict local minimum.

References

1. Scherzer, O. (ed.): Handbook of Mathematical Methods in Imaging, 1st edn. Springer, Heidelberg (2011)
2. Aubert, G., Kornprobst, P.: Mathematical problems in image processing, 2nd edn. Springer, Berlin (2006)
3. Besag, J.E.: Digital image processing: Towards Bayesian image analysis. Journal of Applied Statistics 16(3), 395–407 (1989)
4. Bovik, A.C.: Handbook of image and video processing. Academic Press, New York (2000)
5. Demoment, G.: Image reconstruction and restoration: Overview of common estimation structure and problems. IEEE Transactions on Acoustics Speech and Signal Processing ASSP-37(12), 2024–2036 (1989)
6. Gonzalez, R., Woods, R.: Digital Image Processing. Addison-Wesley, Reading (1993)
7. Nikolova, M.: Properties of the minimizers of energies with nonsmooth nonconvex data fidelity and regularization. Report (2011)
8. Rockafellar, R.T., Wets, J.B.: Variational analysis. Springer, New York (1997)
9. Scherzer, O., Grasmair, M., Grossauer, H., Haltmeier, M., Lenzen, F.: Variational problems in imaging. Springer, New York (2009)
10. Tikhonov, A., Arsenin, V.: Solutions of Ill-Posed Problems, Winston, Washington DC (1977)

A Study on Convex Optimization Approaches to Image Fusion

Jing Yuan[1], Juan Shi[2], Xue-Cheng Tai[2,3], and Yuri Boykov[1]

[1] Computer Science Department, University of Western Ontario
London, Ontario, Canada N6A 5B7
cn.yuanjing@gmail.com, yuri@csd.uwo.ca
[2] Division of Mathematical Sciences, School of Phys. and Math. Sci.
Nanyang Technological University, Singapore
shij0004@e.ntu.edu.sg
[3] Department of Mathematics, University of Bergen, Norway
tai@mi.uib.no

Abstract. Image fusion is an imaging technique to visualize information from multiple images in one single image, which is widely used in remote sensing, medical imaging etc. In this work, we study two variational approaches to image fusion which are closely related to the standard TV-L_2 and TV-L_1 image approximation methods. We investigate their convex optimization models under the perspective of primal and dual and propose the associated new image decompositions. In addition, we consider the TV-L_1 based image fusion approach and study the problem of fusing two discrete-constrained images $f_1(x) \in \mathcal{L}_1$ and $f_2(x) \in \mathcal{L}_2$, where $\mathcal{L}_1$ and $\mathcal{L}_2$ are the sets of linearly-ordered discrete values. We prove that the TV-L_1 based image fusion actually gives rise to an exact convex relaxation to the corresponding nonconvex image fusion given the discrete-valued constraint $u(x) \in \mathcal{L}_1 \cup \mathcal{L}_2$. This extends the results for the global optimization of the discrete-constrained TV-L_1 image approximation [7,30] to the case of image fusion. The proposed dual models also lead to new fast and reliable algorithms in numerics, based on modern convex optimization techniques. Experiments of medical imaging, remote sensing and multi-focusing visibly show the qualitive differences between the two studied variational models of image fusion.

1   Introduction

Imaging fusion technologies have been developed to be an effective way to show different imagery information from various sources in one single image, which is especially interesting in many areas, e.g. remote sensing [26,10], medical imaging [24,27] and synthesis of multi-focused images [16,25]. More specifically, given two or more information data which are from different sources and properly aligned, image fusion integrates all such data into one visualized image, mostly with higher spatial or spectral resolution. For example, given two images, which may capture the same scene but with different focuses (see the left two images of Fig. 1), fusing these two images clearly gives a better visual result (see the

A.M. Bruckstein et al. (Eds.): SSVM 2011, LNCS 6667, pp. 122–133, 2012.

right two fused images of Fig. 1). In remote sensing and satellite imaging, the fused image, which is merged by multispectral data, effectively conveys more information [26,23]. In medical imaging, while both the Magnetic Resonance (MR) and Computed Tomography (CT) imagery provide standard diagnostic tools other than fluoroscopy and ultrasound techniques, it is well-known that a CT scan will adequately highlight the bone structure details while soft tissue information is not clearly visible; on the other hand, a T2 weighted MR scan produces significantly better details for images of these tissues. In this respect, it is highly desirable to have a combined view of CT and MR images, which illustrates significant details both from both CT and MR inputs and assists clinical diagnoses.

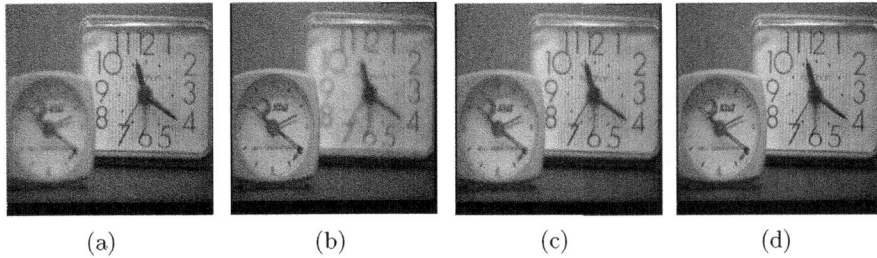

(a)   (b)   (c)   (d)

Fig. 1. Multi-focus image fusion: (a) and (b) give two images exposed with different focuses; (c) and (d) are the fused image computed by the proposed methods (1) and (3) in this work.

Parallel to recent developments of image processing, many pixelwise image fusion methods have been proposed to tackle such problem of combining multiple images or informative data, e.g. the wavelet or contourlet based approaches [19,18,27], high-pass filtering method [1,23] etc. In this paper, we concentrate on the variational approaches to image fusion, which were explored in [20,25,15]. Energy minimization and variational methods have been developed to be a standard way to effectively and reliably handle many practical topics of image processing and computer vision both in mathematics and numerics. Successful applications include image denoising [22,8,30], image decomposition [2,17] and image segmentation [9,8,29,28] etc. In this regard, variational image fusion methods [25,15,20] provide such an elegant approach for the tradeoff between redundant imagery information and image priors.

Contributions: In this work, we study the variational models to the integration of images with gray-scales, which were proposed or partially investigated by [25,15]. We propose the convex optimization approach to the studied variational problems under the novel duality-based perspective, and consider the exactness of the reduced convex relaxation model to the nonconvex TV-L1 based image fusion with the pixelwise constraint of discrete values. Our contributions can be summarized as follows:

We consider the two convex optimization models of image fusion based on standard techniques of TV-L_2 and TV-L_1 image approximation. We propose their equivalent convex formulations under the new perspective of primal and dual. We show the studied image fusion models actually result in two new image decompositions of the weighted input image, with helps of the proposed dual forumations. In addition, we focus on the TV-L_1 based image fusion method and prove it gives an exact convex relaxation model to the corresponding image fusion problem constrained by a linearly-ordered discrete-value set to each pixel, i.e. it solves such nonconvex image fusion problem globally and exactly. This properly extends the convex relaxation models of TV-L1 image approximation, proposed by Chan et al [7] and Yuan et al [30], to TV-L1 based image fusion applications. Clearly, direct and global solvers to such discrete-constrained image fusion, especially over a large number of linearly-ordered discrete values for instance medical imaging, definitely result in a high load of memory and computation, e.g. graph-cuts method [5,14] and the continuous min-cut method [3]. To this end, the convex relaxation approach proposed in this work leads to a much easier and more efficient approach to the given discrete-constrained optimization problem. We also derive fast and reliable algorithms to the studied two image fusion methods through their proposed dual formulations, which properly avoids nonsmoothness of the energy functions and leads to simple and efficient numerical implementations.

2   Convex Optimization Models

Given two input images $f_1(x)$ and $f_2(x)$, a total-variation based method for image fusion was proposed by Wang et al [25] such that

$$\min_{u \in BV(\Omega)} \frac{1}{2} \int_{\Omega} w_1 \, (u - f_1)^2 \, dx + \frac{1}{2} \int_{\Omega} w_2 \, (u - f_2)^2 \, dx + \alpha \int_{\Omega} |\nabla u| \, dx \quad (1)$$

where the functions $\omega_1(x)$ and $\omega_2(x)$ are the pixelwise weight functions such that

$$\omega_1(x) + \omega_2(x) = 1, \quad \omega_{1,2}(x) \geq 0; \quad \forall x \in \Omega. \quad (2)$$

In this work, we extend (1) to the convex optimization model with the L_1-normed data fidelity term:

$$\min_{u} \int_{\Omega} w_1 \, |u - f_1| \, dx + \int_{\Omega} w_2 \, |u - f_2| \, dx + \alpha \int_{\Omega} |\nabla u| \, dx. \quad (3)$$

Similar formulation as (3) was also studied in [15] where the weight functions are given constant.

Clearly, both models (1) and (3) formulate the integration of two input images as the problem of convex optimization which can be generalized as

$$\min_{u} \int_{\Omega} w_1 D_1(f_1 - u) \, dx + \int_{\Omega} w_2 D_2(f_2 - u) \, dx + \alpha \int_{\Omega} |\nabla u| \, dx \quad (4)$$

where $D_1(\cdot)$ and $D_2(\cdot)$ are positive convex functions. In this work, we call (4), along with (1) and (3), the *primal model*.

In the following parts, we investigate (4) under the perspective of primal and dual and build up its connections to variational image decomposition.

2.1 Equivalent Convex Formulations

Let $D_1^*(q)$ and $D_2^*(q)$ be the respective conjugate of the convex function $D_1(v)$ and $D_2(v)$ such that

$$D_1(v) = \max_{q_1} \{vq_1 - D_1^*(q_1)\} , \quad D_2(v) = \max_{q_2} \{vq_2 - D_2^*(q_2)\} . \quad (5)$$

For the model (1) where the functions D_1 and D_2 are in quadratic forms, i.e. $D_1(v) = D_2(v) = \frac{1}{2}v^2$, we have

$$D_1^*(q) = D_2^*(q) = \frac{1}{2}q^2 . \quad (6)$$

For the problem (3) where both D_1 and D_2 are absolute functions, i.e. $D_1(v) = D_2(v) = |v|$, we have

$$D_1^*(q) = D_2^*(q) = I_\delta(q \in [-1,1]) \quad (7)$$

where $I_\delta(q \in [-1,1])$ is the characteristic function of the convex set $q \in [-1,1]$.

We also recall that the dual formulation of the total-variation function [13]

$$\alpha \int_\Omega |\nabla u| \, dx = \max_{p \in C_\alpha} \int_\Omega u \operatorname{div} p \, dx \quad (8)$$

where C_α is a convex set defined by

$$C_\alpha := \{p \,|\, p \in C_c^1(\Omega, \mathbb{R}^2) , \ |p(x)| \le \alpha , \ \forall x \in \Omega \} . \quad (9)$$

By simple computation, in view of (5) and (8), the primal formulation (4) can be rewritten as

$$\min_u \max_{q_1, q_2} \max_{p \in C_\alpha} \int_\Omega w_1 \, (q_1 f_1 - D_1^*(q_1)) \, dx + \int_\Omega w_2 \, (q_2 f_2 - D_2^*(q_2)) \, dx \quad (10)$$
$$+ \langle \operatorname{div} p - (w_1 q_1 + w_2 q_2), u \rangle .$$

In this paper, we call (10) the equivalent *primal-dual model* of (4).

Observe that u is unconstrained and the convex formulation (10) suffices the minimax theorem [11,12] for our cases (1) and (3) in this study, the min and max operators of (10) are interchangeable. The minimization of (10) over u, therefore, leads to the linear equality

$$w_1 q_1 + w_2 q_2 = \operatorname{div} p , \quad (11)$$

and the corresponding constrained maximization problem as follows

$$\max_{q_1, q_2} \max_{p \in C_\alpha} \int_\Omega w_1 \, (q_1 f_1 - D_1^*(q_1)) \, dx + \int_\Omega w_2 \, (q_2 f_2 - D_2^*(q_2)) \, dx \quad (12)$$
$$\text{s.t.} \quad w_1 q_1 + w_2 q_2 = \operatorname{div} p .$$

Similarly, we call (12) the equivalent *dual model* of (4).

2.2   Variational Image Decompositions

With helps of the conjugates (5), we will see that the optimum of the variational image fusion (4) actually proposes a decomposition of the weighted input image $f(x) := w_1(x)f_1(x) + w_2(x)f_2(x)$, $x \in \Omega$. More specifically, we have

Proposition 1. *Given the optimum (q_1^*, q_2^*, p^*, u^*) of the equivalent primal-dual model (10), (q_1^*, q_2^*, p^*, u^*) just gives rise to the decomposition of the weighted input image $(w_1 f_1 + w_2 f_2)(x)$, $x \in \Omega$, such that*

$$f := w_1 f_1 + w_2 f_2 = u^* + v^* \tag{13}$$

where

$$v^* = w_1 v_1^* + w_2 v_2^*, \quad v_1^* \in \partial D_1(q_1^*), \ v_2^* \in \partial D_2(q_2^*).$$

Proof. Observe the conjugate formulations (5), we have

$$f_1 - u^* = v_1^* \in \partial D_1(q_1^*), \quad f_2 - u^* = v_2^* \in \partial D_2(q_2^*).$$

Recall that $w_1(x) + w_2(x) = 1$ for $\forall x \in \Omega$, then we have

$$w_1 v_1^* + w_2 v_2^* = w_1(f_1 - u^*) + w_2(f_2 - u^*) = (w_1 f_1 + w_2 f_2) - u^*.$$

Then (13) simply follows.

Consider the conjugates (6) and Prop. 1, the L_2-norm based image fusion problem (1) results in the following image decomposition:

Corollary 1. *Given the optimum (q_1^*, q_2^*, p^*, u^*) of the equivalent primal-dual model (10) corresponding to (1), (q_1^*, q_2^*, p^*, u^*) just gives rise to the decomposition of the weighted input image $(w_1 f_1 + w_2 f_2)(x)$, $x \in \Omega$, such that*

$$f := w_1 f_1 + w_2 f_2 = u^* + \operatorname{div} p^*. \tag{14}$$

Proof. In view of (6), we have $f_1 - u^* = q_1^*$ and $f_2 - u^* = q_2^*$. It follows

$$f := w_1 f_1 + w_2 f_2 = (w_1 q_1 + w_2 q_2) + u^*.$$

In view of the linear equality constraint (11), i.e. $w_1 q_1^* + w_2 q_2^* = \operatorname{div} p^*$, then we prove Coro. 1.

Moreover, we also have

Corollary 2. *The image fusion problem (1) is equivalent to*

$$\min_{p \in C_\alpha} \ \|(w_1 f_1 + w_2 f_2) - \operatorname{div} p\|^2, \tag{15}$$

i.e. the projection of the weighted input image $(w_1 f_1 + w_2 f_2)(x)$, $x \in \Omega$, to the convex set $\operatorname{div} C_\alpha$.

Proof directly follows from the image decomposition model of Coro. 1 and (6). Clearly, similar results of image decomposition and projections as Coro. 1 and Coro. 2 were proposed in [6], which showed that TV-L_2 image denoising amounts to image decomposition (14) and projection (15) of the single input image $f(x)$ instead of the weighted image $w_1 f_1 + w_2 f_2$.

For the image fusion problem (3), it leads to image decomposition as follows:

Corollary 3. *Given the optimum (q_1^*, q_2^*, p^*, u^*) of the equivalent primal-dual model (10) which is equivalent to (3), (q_1^*, q_2^*, p^*, u^*) just gives rise to the decomposition of the weighted input image $(w_1 f_1 + w_2 f_2)(x)$, $x \in \Omega$, such that*

$$f := w_1 f_1 + w_2 f_2 = u^* + v^* \tag{16}$$

where

$$v^* = w_1 v_1^* + w_2 v_2^*, \quad v_1^* \in \partial I_S(q_1^*), \ v_2^* \in \partial I_S(q_2^*),$$

I_S is the characteristic function of the set $S = \{q \,|\, q(x) \in [-1, 1], \ \forall x \in \Omega \}$.

Its proof directly follows by the conjugates (7) and Prop. 1.

3   Global and Exact Optimization

Now we focus on the TV-L_1 based approach (3) and consider the respective discrete-valued optimization problem

$$\min_{u(x) \in \mathcal{L}} \int_\Omega w_1 \,|u - f_1|\, dx + \int_\Omega w_2 \,|u - f_2|\, dx + \alpha \int_\Omega |\nabla u|\, dx \tag{17}$$

where we assume the two input images $f_1(x)$ and $f_2(x)$ take discrete values which are linearly ordered such that

$$f_i(x) \in \left(\mathcal{L}_i := \{l_1^i, \ldots, l_{n_i}^i\}\right), \quad l_1^i < l_2^i < \ldots < l_{n_i}^i; \quad i = 1, 2 \tag{18}$$

and $\mathcal{L} = \mathcal{L}_1 \cup \mathcal{L}_2$ is the combination set of $\mathcal{L}_1$ and $\mathcal{L}_2$. We also assume

$$\mathcal{L} = \{l_1, \ldots, l_n\}, \quad l_1 < l_2 < \ldots < l_n, \tag{19}$$

includes n discrete values.

In this section, we show that the TV-L_1 based image fusion problem (3) amounts to an exact convex relaxation model of the integer-constrained optimization problem (17), i.e. the optimum of the convex optimization problem (3) results in the global and exact integer-valued optimum of (17). A similar result was recently proposed by [30], where the authors proved that the convex TV-L_1 image approximation does give global and exact optima to the corresponding discrete-constrained TV-L_1 approximation. We directly state our result as the following proposition and omit the proof, due to the limit space. Its detailed proof can be derived by the same way as [30].

Proposition 4. *Given the optimum $u^*(x)$ to (3) and the set of discrete values $\mathcal{L} = \{l_1, \ldots, l_n\}$, $l_1 < \ldots < l_n$, which is the combination of two sets (18) of discrete image values given in $f_1(x)$ and $f_2(x)$, then for any given $n-1$ values γ_i, $i = 1, \ldots, n-1$, such that*

$$l_1 < \gamma_1 < l_2 < \ldots < \gamma_{n-1} < l_n, \tag{20}$$

we define the image function $u^\gamma(x)$ by the $n-1$ upper level sets of $u^*(x)$:

$$u^\gamma(x) = l_1 + \sum_{i=1}^{n-1} (l_{i+1} - l_i) U^{\gamma_i}(x). \tag{21}$$

$u^\gamma(x) \in \{l_1, \ldots, l_n\}$ and $u^\gamma(x)$ gives an exact and global optimum of (17).

4   Duality Based Algorithms

In this section, we propose fast numerical algorithms to image fusion problems (1) and (3) through their respective dual formulations.

By Coro. 2, we observe that the image fusion problem (1) corresponds to the projection of the image $w_1 f_1 + w_2 f_2$ to the convex set $\mathrm{div}\, C_\alpha$. It directly leads to the same duality-based algorithm as [6] proposed by Chambolle. We list its iterative projected-gradient descent steps for computing the dual variable p as follows:

$$p^{i+1} = \mathbf{Proj}_{C_\alpha}\left(p^i + \tau \nabla\big((w_1 f_1 + w_2 f_2) - \mathrm{div}\, p^i\big)\right),$$

where $\tau > 0$ gives the step-size at each iteration.

With helps of (7) and (10), the TV-L_1 based image fusion problem (3) can be equally written as the following primal-dual formulation:

$$\min_u \max_{q_1, q_2} \max_{p \in C_\alpha} \int_\Omega q_1 f_1 \, dx + \int_\Omega q_2 f_2 \, dx + \langle \mathrm{div}\, p - (q_1 + q_2), u \rangle \tag{22}$$

$$\text{s.t. } q_1(x) \in [-w_1(x), w_1(x)], \quad q_2(x) \in [-w_2(x), w_2(x)]. \tag{23}$$

Also in view of (12), its equivalent dual model can be formulated as

$$\max_{q_1, q_2} \max_{p \in C_\alpha} \int_\Omega q_1 f_1 \, dx + \int_\Omega q_2 f_2 \, dx \tag{24}$$

$$\text{s.t. } q_1(x) \in [-w_1(x), w_1(x)], \quad q_2(x) \in [-w_2(x), w_2(x)]$$

$$q_1 + q_2 = \mathrm{div}\, p. \tag{25}$$

We see that the image $u(x)$ in the primal-dual formulation (22), which is what we wish to obtain, just works as the multiplier function to the linear equality constraint (25) of the dual model (24). In addition, the energy function of (22) gives the corresponding Lagrangian function to the dual formulation (24). Through these observations, we define its augmented Lagrangian function as

$$L_c(q_1, q_2, p, u) = \langle q_1, f_1 \rangle + \langle q_2, f_2 \rangle + \langle \mathrm{div}\, p - (q_1 + q_2), u \rangle - \frac{c}{2} \|\mathrm{div}\, p - (q_1 + q_2)\|^2$$

where $c > 0$.

In this work, we apply the classical augmented Lagrangian algorithm [21,4] through its augmented Lagrangian function $L_c(q_1, q_2, p, u)$, which includes the following steps at k-th iteration:

1. Optimize q_1^{k+1} by fixing q_2^k, p^k and u^k, which gives

$$q_1^{k+1} := \arg\max_{|q_1(x)| \le w_1(x)} \langle q_1, f_1 \rangle - \frac{c}{2} \left\| q_1 - (\operatorname{div} p^k - q_2^k - u^k/c) \right\|^2.$$

It can be computed by the following step in a close form:

$$q_1^{k+1} = \mathbf{Proj}_{|q_1(x)| \le w_1(x)} (f_1/c + (\operatorname{div} p^k - q_2^k(x) - u^k/c)); \qquad (26)$$

2. Optimize q_2^{k+1} by fixing q_1^{k+1}, p^k and u^k, which gives

$$q_2^{k+1} := \arg\max_{|q_2(x)| \le w_2(x)} \langle q_2, f_2 \rangle - \frac{c}{2} \left\| q_2 - (\operatorname{div} p^k - q_1^{k+1} - u^k/c) \right\|^2.$$

It can be computed by the following step in a close form:

$$q_2^{k+1} = \mathbf{Proj}_{|q_2(x)| \le w_2(x)} (f_2/c + (\operatorname{div} p^k - q_1^{k+1}(x) - u^k/c)); \qquad (27)$$

3. Optimize p^{k+1} by fixing q_1^{k+1}, q_2^{k+1} and u^k, which gives

$$p^{k+1} := \arg\min_{p \in C_\alpha} \left\| \operatorname{div} p - (q_1^{k+1} + q_2^{k+1} + u^k/c) \right\|^2. \qquad (28)$$

It is the projection of $(q_1^{k+1} + q_2^{k+1} + u^k/c)$ to the convex set $\operatorname{div} C_\alpha$.

4. Update u^{k+1} by

$$u^{k+1} = u^k + c(q_1^{k+1} + q_2^{k+1} - \operatorname{div} p^{k+1}); \qquad (29)$$

and let $k = k + 1$, repeat until convergence.

The algorithm gives a splitting optimization framework over each dual variables q_1, q_2 and p respectively, by exploring projection to their corresponding convex set. To this end, we call it the *multiplier-based algorithm to TV-L_1 image fusion*. It explores three simple sub-steps: (26), (27) and (28) at each iteration, which properly avoids tackling the nonsmooth terms in (3) in a direct way. The substeps of (26) and (27) are easy and cheap to compute. For the projection substep (28), we can use one or a few steps of the iterative projected-gradient decent algorithm to approximately solve (28) as follows:

$$p^{i+1} = \mathbf{Proj}_{C_\alpha} \left(p^i + \tau \nabla \{ \operatorname{div} p^i - ((q_1^{i+1} + q_2^{i+1}) + u^i/c) \} \right). \qquad (30)$$

Interestingly, our experiments show that just one single step of the above iteration (30), with a proper step-size τ, is needed to make the algorithm converge! This implements the algorithm in a very fast way, mostly superlinear.

5   Experiments

In this section, we first fuse two binary images to show the fundamental differences between (1) and (3). Then experiments for both medical imaging and remote sensing are given for comparisons. Experiment results may vary with different choices of $w_1(x)$ and $w_2(x)$, but this is not the focus of this paper.

5.1   Fusing Binary Images

Given two binary images (see the two images on the leftside of Fig. 2), i.e. $f_{1,2}(x) \in \{0,1\}$, we computed the fused image by both two approaches: (1) and (3), where the weighted functions $w_1(x)$ and $w_2(x)$ are computed based image edges. For the TV-L_2 based method (1), we set $\alpha = 3$ and its fused result $u(x)$ is shown by the 3rd image of Fig. 2. For the TV-L_1 based method (3), we set $\alpha = 1$ and its fused result $u(x)$ is shown by the last image of Fig. 2. Clearly, the TV-L_1 based method gives the binary optimum which takes the value either 0 or 1 nearly everywhere. This is in contrast to the resultby (1).

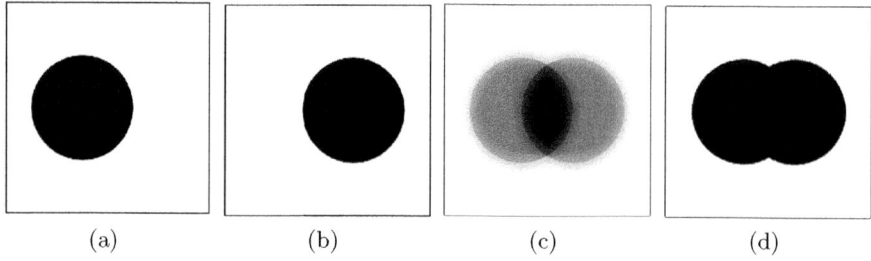

(a)   (b)   (c)   (d)

Fig. 2. Fusing binary images: (a) and (b) give the two input binary image; (c) and (d) show the results computed by the TV-L_2 and TV-L_1 based methods respectively.

5.2   Applications to Medical Imaging and Remote Sensing

Besides the fusion experiment of multi-focused images (shown in Fig. 1), we also made fusion experiments for medical imaging and remote-sensing images. Both experiments are computed by a Ubuntu desktop with AMD Athalon 64 X2 5600. The TV-L_1 based method (3) takes couple of seconds. Except one additional step of (26) and (27), its algorithmic scheme has the same complexities as the fast TV-L_1 method proposed in [30]. All the images are adjusted into the same grayscale range for comparisons. Fig. 3 shows the fusion experiment of medical imaging, which integrates the images from CT and MRI (see Fig. 3). The TV-L_1 based method performs visually better than the TV-L_2 based method in preserving high-contrast and details (see the enlarged image patches for comparisons). Fig. 4 shows the image fusion experiment of remote sensing, where two images from different spectral channels are fused by the studied two methods respectively. Detailed comparison of the enlarged patches (see the images at 2nd row of Fig. 4) clearly indicates better visual result by the TV-L_1 based method.

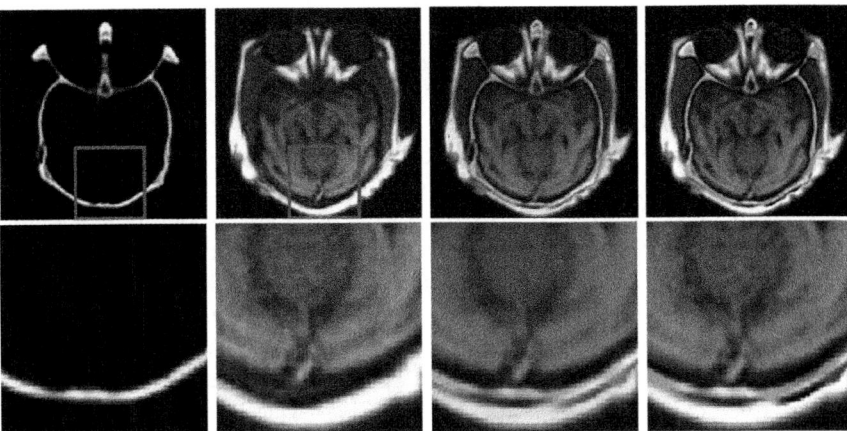

Fig. 3. Fusing medical images. 1st row: the left two images show two input images of spine discs by CT and MRI respectively; the right two images show the fused images computed by (1) and (3) respectively. **2nd row:** the left two images show the zoomed image patches cropped by the red lines on the same position of CT and MRI images respectively; the right two images show the fused results at the patched area computed by (1) and (3) respectively.

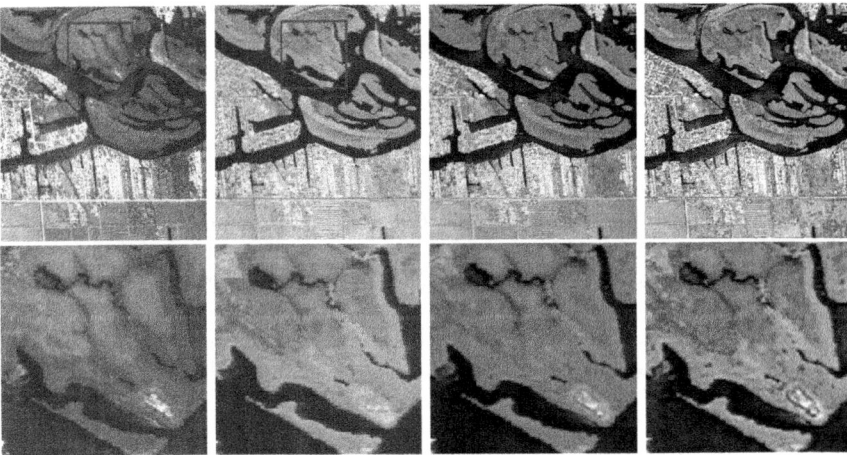

Fig. 4. Fusing images from two spectral bands. At 1st row: the left two images show the input images of remote sensing images from two different spectral channels; the right two images show the fused images computed by (1) and (3) respectively. **At 2nd row:** the left two images show the zoomed image patches croped by the red lines on the same position of the input images respectively; the right two images show the fused results at the patched area computed by (1) and (3) respectively.

6   Conclusion and Acknowledgements

In this work, we study two variational approaches to image fusion, which are related to TV-L_2 and TV-L_1 image approximation, under the new perspective in terms of primal and dual and show both result new image decompositions. We focus on the TV-L_1 based image fusion approach and consider fusing two discrete-valued images. In this regard, we prove that the TV-L_1 based image fusion actually gives the exact convex relaxation to its corresponding image fusion subject to the specified discrete-valued constraint. This extends recent developments for global optimization of the discrete-constrained TV-L_1 image approximation [7,30] to the case of image fusion. The proposed dual models lead to fast and reliable algorithmic schemes based on the standard convex optimization.

The research has been supported by MOE (Ministry of Education) Tier II project T207N2202, IDM project NRF2007IDM-IDM002-010; the Norwegian Research Council (eVita project 166075); Canadian NSERC discovery grant 298299-2007 RGPIN and accelerator grant for exceptional new opportunities 349757-2007 RGPAS. The authors especially thank Brandon Miles for his kind helps and discussions.

References

1. Aiazzi, B., Alparone, L., Baronti, S., Garzelli, A.: Context-driven fusion of high spatial and spectral resolution images based on oversampled multiresolution analysis. IEEE Geoscience and Remote Sensing 40(10), 2300–2312 (2002)
2. Aujol, J.-F., Gilboa, G., Chan, T.F., Osher, S.: Structure-texture image decomposition - modeling, algorithms, and parameter selection. International Journal of Computer Vision 67(1), 111–136 (2006)
3. Bae, E., Yuan, J., Tai, X.C., Boykov, Y.: A fast continuous max-flow approach to non-convex multilabeling problems. Technical Report CAM10-62, UCLA (2010)
4. Bertsekas, D.P.: Constrained optimization and Lagrange multiplier methods. Academic Press Inc., New York (1982)
5. Boykov, Y., Veksler, O., Zabih, R.: Fast approximate energy minimization via graph cuts. IEEE PAMI 23(11), 1222–1239 (2001)
6. Chambolle, A.: An algorithm for total variation minimization and applications. Journal of Mathematical Imaging and Vision 20(1-2), 89–97 (2004)
7. Chan, T.F., Esedoğlu, S.: Aspects of total variation regularized L^1 function approximation. SIAM J. Appl. Math. 65(5), 1817–1837 (2005) (electronic)
8. Chan, T.F., Esedoglu, S., Nikolova, M.: Algorithms for finding global minimizers of image segmentation and denoising models. SIAM J. Appl. Math. 66(5), 1632–1648 (2006)
9. Chan, T.F., Vese, L.A.: Active contours without edges. IEEE Transactions on Image Processing 10(2), 266–277 (2001)
10. Das, A., Revathy, K.: A comparative analysis of image fusion techniques for remote sensed images. In: World Congress on Engineering, pp. 639–644 (2007)
11. Ekeland, I., Téman, R.: Convex analysis and variational problems. Society for Industrial and Applied Mathematics, Philadelphia (1999)
12. Fan, K.: Minimax theorems. Proc. Nat. Acad. Sci. U.S.A. 39, 42–47 (1953)

13. Giusti, E.: Minimal surfaces and functions of bounded variation. Australian National University, Canberra (1977)
14. Ishikawa, H.: Exact optimization for markov random fields with convex priors. IEEE PAMI 25, 1333–1336 (2003)
15. Kluckner, S., Pock, T., Bischof, H.: Exploiting redundancy for aerial image fusion using convex optimization. In: Goesele, M., Roth, S., Kuijper, A., Schiele, B., Schindler, K. (eds.) DAGM 2010. LNCS, vol. 6376, pp. 303–312. Springer, Heidelberg (2010)
16. Li, H., Manjunath, B.S., Mitra, S.K.: Multisensor image fusion using the wavelet transform. Graphical Models and Image Processing 57(3), 235–245 (1995)
17. Meyer, Y.: Oscillating patterns in image processing and nonlinear evolution equations. University Lecture Series, vol. 22. American Mathematical Society, Providence (2001); The fifteenth Dean Jacqueline B. Lewis memorial lectures
18. Nez, J., Otazu, X., Fors, O., Prades, A., Pal'a, V., Arbiol, R.: Multiresolution-based image fusion with additive wavelet decomposition. IEEE Trans. On Geoscience And Remote Sensing 37(3), 1204–1211 (1999)
19. Pajares, G., de la Cruz, J.M.: A wavelet-based image fusion tutorial. Pattern Recognition 37(9), 1855–1872 (2004)
20. Piella, G.: Image fusion for enhanced visualization: A variational approach. International Journal of Computer Vision 83(1), 1–11 (2009)
21. Rockafellar, R.T.: Augmented Lagrangians and applications of the proximal point algorithm in convex programming. Math. of Oper. Res. 1, 97–116 (1976)
22. Rudin, L., Osher, S., Fatemi, E.: Nonlinear total variation based noise removal algorithms. Physica D 60(1-4), 259–268 (1992)
23. Schowengerdt, R.A.: Remote Sensing: Models and Methods for Image Processing, 3rd edn. Elsevier, Amsterdam (2007)
24. Sohn, M.-J., Lee, D.-J., Yoon, S.W., Lee, H.R., Hwang, Y.J.: The effective application of segmental image fusion in spinal radiosurgery for improved targeting of spinal tumours. Acta Neurochir 151, 231–238 (2009)
25. Wang, W.-W., Shui, P.-L., Feng, X.-C.: Variational models for fusion and denoising of multifocus images. IEEE Signal Processing Letters 15, 65–68 (2008)
26. Wang, Z., Ziou, D., Armenakis, C., Li, D., Li, Q.: A comparative analysis of image fusion methods. IEEE Geo. and Res. 43(6), 1391–1402
27. Yang, L., Guo, B.L., Ni, W.: Multimodality medical image fusion based on multiscale geometric analysis of contourlet transform. Neurocomputing 72, 203–211 (2008)
28. Yuan, J., Bae, E., Tai, X.-C., Boykov, Y.: A continuous max-flow approach to potts model. In: Daniilidis, K., Maragos, P., Paragios, N. (eds.) ECCV 2010. LNCS, vol. 6316, pp. 379–392. Springer, Heidelberg (2010)
29. Yuan, J., Bae, E., Tai, X.-C.: A study on continuous max-flow and min-cut approaches. In: CVPR 2010, pp. 2217–2224 (2010)
30. Yuan, J., Shi, J., Tai, X.-C.: A convex and exact approach to discrete constrained tv-l1 image approximation. Technical Report CAM-10-51, UCLA (2010)

Efficient Beltrami Flow in Patch-Space

Aaron Wetzler and Ron Kimmel

Department of Computer Science,
Technion, Israel

Abstract. The Beltrami framework treats images as two dimensional manifolds embedded in a joint features-space domain. This way, a color image is considered to be a two dimensional surface embedded in a hybrid special-spectral five dimensional $\{x, y, R, G, B\}$ space. Image selective smoothing, often referred to as a denoising filter, amounts to the process of area minimization of the image surface by mean curvature flow. One interesting variant of the Beltrami framework is treating local neighboring pixels as the feature-space. A distance is defined by the amount of deformation a local patch undergoes while traversing its support in the spatial domain. The question we try to tackle in this note is how to perform patch based denoising accurately, and efficiently. As a motivation we demonstrate the performance of the Beltrami filter in patch-space, and provide useful implementation considerations that allow for parameter tuning and efficient implementation on hand-held devices like smart phones.

Keywords: Beltrami flow, patch-space, denoising.

1 Introduction

Following the success of the Non Local Means denoising method as introduced by Buades et al. in [2] much attention has been devoted to developing various types of patch based denoising techniques. A patch, in terms of an image, is generally considered to be a square region of pixels of fixed size centered at the coordinates of an image pixel. Peyrè in [6] studies patch based manifolds while a more specific analysis of a generalized patch based denoising framework is done by Tschumperlè and Brun in [12]. They show that the NL means [2] and Bilateral [11] filters are isotropic versions of their patch based diffusion framework by choosing a specific patch size and metric. In much the same way Sochen et al. present the Beltrami framework and show in [8] how choices of different metrics can be used to produce filtering methods like the anisotropic diffusion process of Perona and Malik [5] as an example. Anisotropic diffusion was also shown by Barash in [1] to have a strong connection to the Bilateral filter through the adaptive smoothing filter and Elad in [4] demonstrated its connection to other classical filtering techniques.

In [7] Maragos and Roussos, explore a generalization of the Beltrami flow using weighted patches. We will use a similar formulation while setting the weights

A.M. Bruckstein et al. (Eds.): SSVM 2011, LNCS 6667, pp. 134–143, 2012.

of each neighboring pixel to be one. In this context, the Beltrami framework provides a general and natural substrate for diffusion based image manipulation and naturally extends to higher dimensions. We will show how it can be applied to an image manifold in patch-space with better visual results as well as the overall PSNR compared to strictly local-differential techniques. We will discuss numerical considerations and demonstrate how the use of an integral image eliminates the algorithm's dependency on the patch size allowing for good performance on a modern smartphone.

Fig. 1. Examples of Beltrami patch denoising for color images. From top to bottom, left to right a) Noisy F16, $\sigma = 20$ b) Denoised image, $PSNR = 31.51dB$ c) Noisy Lena, $\sigma = 30$ d) Denoised image, $PSNR = 29.54dB$ e) Noisy Mandrill, $\sigma = 50$ f) Denoised image, $PSNR = 21.43dB$.

2   The Beltrami Framework

We consider an image to be a $2D$ Riemannian manifold embedded in $D = d + 2$ dimensional space where $d = 1$ for grayscale images and $d = 3$ for color images. We can thus write the map $X : \Sigma \to M$ where X is the mapping of the image manifold into the embedding space feature manifold M. For a grayscale mapping we can write

$$X(\sigma_1, \sigma_2) = (x(\sigma_1, \sigma_2), y(\sigma_1, \sigma_2), z(\sigma_1, \sigma_2)). \qquad (1)$$

If we further specify that $\sigma_1 = x$, $\sigma_2 = y$ and I is the image intensity map, then from (1) we have the graph of I given by

$$X(x, y) = (x, y, I(x, y)). \qquad (2)$$

Both Σ and M are Riemannian manifolds and hence are equipped with metrics G and H respectively which enable measurement of lengths over each manifold.

We require the lengths as measured on each manifold to be the same. Thus we can write that

$$ds^2 = (dx \; dy \; dI) \, H \begin{pmatrix} dx \\ dy \\ dI \end{pmatrix} = (dx \; dy) \, G \begin{pmatrix} dx \\ dy \end{pmatrix}. \tag{3}$$

We can equate these and write the result compactly using Einstein notation where repeated upper and lower indices are summed over

$$g_{uv} = h_{ij} \partial_u X^i \partial_v X^j \quad u, v = 1..2 \quad i, j = 1..3 \tag{4}$$

Here the meaning of $\partial_{u,v}$ is just the partial derivative with respect to x or y. For the simple case in (4) where $H = (h_{ij})$ is the identity matrix we use the chain rule $dI = I_x dx + I_y dy$ and determine that for (2) the induced metric tensor $G = (g_{uv})$ is

$$G = \begin{pmatrix} 1 + I_x^2 & I_x I_y \\ I_x I_y & 1 + I_y^2 \end{pmatrix}. \tag{5}$$

Having a metric enables us to define a measure on the manifold which, for a Euclidean embedding in M, turns out to be the area of the surface as measured by the local coordinates in Σ

$$S[X, G] = \iint \sqrt{g} dx dy = A = \iint \sqrt{1 + I_x^2 + I_y^2} dx dy. \tag{6}$$

Here $g = \text{div}(G)$. There is a more general version of the above measure called the Polyakov action which can be useful for non-Euclidean embeddings and details of its application to the Beltrami framework can be found in [8]. We now minimize the functional in (6) using the methods of variational calculus with the resulting Euler-Lagrange relation given by

$$-\frac{d}{dx} \left(\frac{I_x}{\sqrt{g}} \right) - \frac{d}{dy} \left(\frac{I_y}{\sqrt{g}} \right) = -\text{div} \left(\sqrt{g} G^{-1} \nabla I \right) = 0. \tag{7}$$

We excersize freedom of paramaterization and multiply by $g^{-1/2}$ which allows (7) to be compactly written as $\Delta_g I = 0$ where Δ_g is the second order differential operator of Beltrami. We now formulate a geometric flow of the manifold

$$I_t = \Delta_g I, \tag{8}$$

which creates a scale space via the generalization of the Laplace operator onto Riemannian manifolds. The discretized version of (8) allows us to perform iterative traversal through this scale space on a computer and produces a very effective technique for denoising grayscale images when using the metric in (5).

3   Operating in Patch-Space

A patch is a window centered at a given pixel. We therefore define the mapping $P : \Sigma \to \mathbb{R}^{nw^2+2}$ in the form

$$P(x, y) = \left(x, y, \left\{ I^k(x + iw, y + jw) \right\} \right) \quad i, j = -w, .., w, \quad k = 1, .., n. \tag{9}$$

Here $w \in \mathbb{N}$ is known as the window size or patch size, and n is the number of channels in the image. For example, a single channel image where $n = 1$ and $w = 5$ produces patches of size 11×11 centered about each pixel in the image I. We can see that the above definition reduces to the grayscale embedding (2) for $w = 0$ and $n = 1$ as described in the previous section. From here on we will denote $\{I^k (x + iw, y + jw)\} \; i, j = -w, .., w \quad k = 1, .., n$, as $I^k_{i,j}$. Note that I^k is simply the kth color channel. We wish to derive the induced metric tensor G for this new embedding. For that goal we first consider the arclength measurement in the embedding space which we assume to be Euclidean and therefore

$$ds^2 = \langle dP, dP \rangle_H = dx^2 + dy^2 + \sum_{i,j,k} \left(dI^k_{i,j} \right)^2. \tag{10}$$

In reality, the coordinates x and y do not possess the same physical measure as the intensity values of the image so we need to introduce a scaling factor into the patch-space metric given by

$$h_{ij} = \begin{cases} \delta_{ij} & i, j \leqslant 2 \\ \beta^2 \delta_{ij} & \text{otherwise} \end{cases}, \tag{11}$$

where δ_{ij} is the Kronecker delta. Following the same procedure as before and using the chain rule $dI^k_{i,j} = I^k_{i,j\,x} dx + I^k_{i,j\,y} dy$ we pullback the metric from the embedding to determine that the new induced metric tensor for the 2D image manifold embedded into patch-space is given by

$$G = \begin{pmatrix} 1 + \beta^2 \sum_{i,j,k} I^{k\,2}_{i,j\,x} & \beta^2 \sum_{i,j,k} I^k_{i,j\,x} I^k_{i,j\,y} \\ \beta^2 \sum_{i,j,k} I^k_{i,j\,x} I^k_{i,j\,y} & 1 + \beta^2 \sum_{i,j,k} I^{k\,2}_{i,j\,y} \end{pmatrix}. \tag{12}$$

This metric combined with (8) gives the Beltrami flow in patch-space as

$$I_t = \Delta_g I = \frac{1}{\sqrt{g}} \text{div} \left(\sqrt{g} G^{-1} \nabla I \right). \tag{13}$$

4   Implementation and Results

We use the flow given by (13) to progress through the scale space on image manifolds embedded into patch-space for both grayscale and color images. The algorithm was tested on a desktop PC and the color version was efficiently implemented on an iPhone 4 smartphone. To measure the success we visually inspected the results as well as measured the standard Peak Signal to Noise Ratio for images: $PSNR = 10\log_{10} \left(255^2 / E \left[(I_{est} - I)^2 \right] \right)$ where I_{est} is the estimation of the denoised version of I.

4.1   Parameter Optimization

Given an image with additive Gaussian white noise and standard deviation σ we need to find a set of parameters that produces the best $PSNR$ value. The normal approach is to fix β and change the number of iterations which allows traversal of the scale space. The obvious disadvantage is that more iterations mean longer execution times. One efficient alternative which has been used here is to fix the number of iterations and vary β. This has the effect of artificially moving through the scale space by causing a change of the distances on the embedded image manifold. The output therefore depends on the window size, the number of iterations of the update, and the parameter β. The time complexity of the algorithm is $O(KN^2W^2)$ where K is the number of iterations, N is the width of an image (assuming it is square) and $W = 2w + 1$ for a patch size w. We

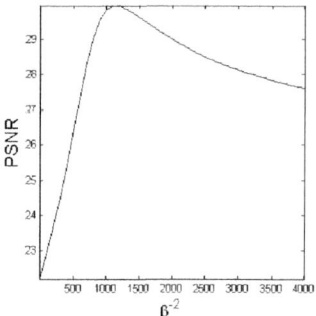

Fig. 2. Example of $PSNR$ as a function of β^{-2} with a typical global maximum

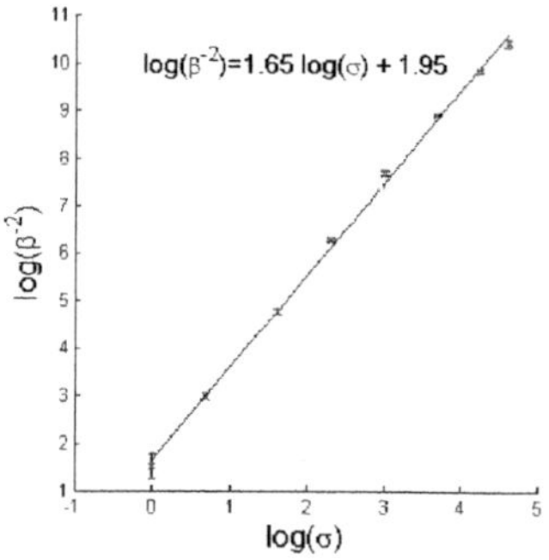

Fig. 3. Loglinear relationship between σ and β^{-2}. Error bars indicate one standard deviation from the mean over a set of different images.

fixed the variables depending on whether an image was grayscale or color. To optimize for β we ran a non-linear optimization program with the $PSNR$ as the target function for a particular image. With the variables held constant except for β, the $PSNR$ function was found to always have a global maximum over the search region. An example function is shown in Fig. 2. The analytical relationship between β and σ is non-trivial however we determined experimentally that the optimal value of β is approximately related to σ by a linear model in log space for values of σ up to 100^1 by the simple relation

$$\log(\beta^{-2}) = a\log(\sigma) + b. \tag{14}$$

Fig. 3 shows this relationship graphically. The graph was obtained by running the optimization program for a set of different images and then fitting the model in (14). It was found that the value of β that globally maximized the $PSNR$ of any representative image produced $PSNR$ values very close to maximum in other images corrupted by Gaussian noise with the same σ. The error bars in Fig. 3 show that there is almost negligible deviation from the mean for the optimal

---

[1] Pixel intensity values range from 0 to 255.

values of β for a given σ for different images. This fact is critical and illustrates that a and b obtained from the log-linear model only need to be calculated once for a predetermined window size, color type and iteration count. They can then be used to generate a β for any given σ for any image. Alternatively, a densely populated look-up table can be generated to relate the two for even greater accuracy.

4.2   Reducing Time Complexity

The weights of nearby pixels are unitary in our method. We take advantage of this property and eliminate the W^2 component by using an integral image to calculate the sums in (12) for each color channel yielding running time complexity of $O(KN^2)$. This allows for patch size independence in performance which is especially important for a practical implementation on a mobile device.

For low values of K and images of size 256×256 the iPhone implementation performs denoising in real time. A patch size of 5×5 ($w = 2$) produced the best results for grayscale images with negligible PSNR differences for the various iterations as shown in Table 1. The same behavior occurs for color images except

Table 1. $PSNR$ results from denoising of the Cameraman image corrupted with AGWN for $\sigma = 20$ using optimized β. Values are in dB.

	$w = 0$	$w = 1$	$w = 2$	$w = 3$	$w = 4$
10 iterations	28.04	29.11	**29.21**	29.09	28.96
50 iterations	27.58	29.04	**29.37**	29.27	29.15
100 iterations	27.35	28.94	**29.36**	29.28	29.16
150 iterations	27.22	28.88	**29.35**	29.28	29.16

that the optimal patch size appears to be 7×7 ($w = 3$). The $PSNR$ alone is not enough as can be seen in Fig. 4. The denoising properties of the Beltrami flow are reasonable for a small number of iterations, however higher quality visual results require more iterations. It was found that grayscale images are best denoised by $K = 150$ iterations and $w = 2$, whereas color images require only $K = 10$ iterations at a window size of $w = 3$.

Table 2. Run times in seconds for Patch Beltrami color denoising on an iPhone 4

	$N = 256$	$N = 512$
$K = 1$	0.14	0.54
$K = 5$	0.65	2.60
$K = 10$	1.27	5.13
$K = 20$	2.55	10.37
$K = 50$	6.42	25.45

Fig. 4. From left to right a) Noisy image at $\sigma = 20$ b) Denoised with 10 iterations, $PSNR = 29.21$ c) Denoised with 150 iterations $PSNR = 29.35$ d) Original image

Running times for different iteration counts of the iPhone implementation for color images are shown in Table 2. Depending on an application's speed requirements, K can be further reduced down to $K = 1$ with a gradual decrease in output quality as shown in Fig. 5, where it is seen that after $K = 10$ there is virtually no improvement. For each iteration the update of a pixel is independent of the update of any other pixel, so the process is highly parallelizable, however this characteristic has not been exploited in the current implementation.

Table 3. PSNR comparison between Regular Beltrami, Patch Beltrami and NL means for some standard grayscale images. All values are in dB.

	CMan	Lena	Barbara	House	
Regular Beltrami	36.97	36.90	35.85	36.92	
Patch Beltrami	**37.70**	**38.11**	**37.01**	**38.16**	$\sigma = 5$
NL means	33.91	37.55	36.06	38.05	
Regular Beltrami	27.22	29.03	26.20	28.98	
Patch Beltrami	29.35	**32.05**	28.71	**32.13**	$\sigma = 20$
NL means	**29.37**	31.56	**29.86**	31.97	
Regular Beltrami	20.55	23.00	20.95	22.63	
Patch Beltrami	23.83	**27.13**	23.52	**26.88**	$\sigma = 50$
NL means	**23.93**	26.46	**24.26**	26.09	

Using the optimally chosen parameters, the denoising process can now be used automatically with the only input parameter being σ as is the norm for denoising images. The experiments in Table 3 reveal that apart from causing a significant improvement over the original application of the Beltrami flow, the new patch-based metric in fact produces results comparable to or even better than the Non-Local means method [2]. Furthermore, the results are within about $2dB$ of the state of the art, such as the block matching algorithms of Dabov et al. [3].

4.3 Residual Noise

Another way to compare the effectiveness of a denoising process is by evaluating the residual as noise as introduced by Baudes et al. in [2]. Here, we look at

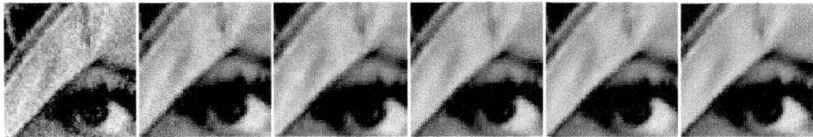

Fig. 5. Optimized denoising for different values of K a) Noisy image, $\sigma = 20$, $PSNR = 22.25$ b) $PSNR = 28.93$, $K = 1$ c) $PSNR = 29.91$, $K = 2$ d) $PSNR = 31.15$, $K = 5$ e) $PSNR = 31.45$, $K = 10$ f) $PSNR = 31.55$, $K = 40$

the differences between the estimated output image and the noisy input image. Ideally, the resulting difference image should also appear as Gaussian white noise. Fig. 6 shows that patch based Beltrami flow produces significantly less structure in the difference image compared to the original pixel based version.

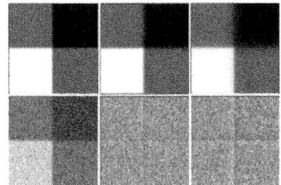

Fig. 6. Method noise. From top to bottom, left to right a) Original image b) Noisy image with $\sigma = 20$ c) Beltrami patch denoising. 150 iterations, $w = 2$ d) Method noise for Beltrami patch denoising e) Regular Beltrami denoising f) Method noise for regular Beltrami denoising.

4.4 Non-gaussian Denoising

In addition to filtering Gaussian noise, the Beltrami flow has other desirable properties. The process tends to align colors along boundaries which lends itself to solving the problem of antialiasing images with jagged, unmatched edges. Another fundamental characteristic of the method is the traversal of a scale space which flattens out smooth, weakly textured objects. An example of the effect of applying the Beltrami patch filter in both types of examples is shown in Fig. 7. It is interesting to note that a state of the art denoising method, BM3D [3], copes very poorly with these two situations because it is optimized for Gaussian noise removal alone.

5 Conclusions

We have shown that the extension of the original Beltrami filter with a more general metric produces significantly better results than the original Beltrami filter. The number of iterations required for denoising color images is $K \simeq 10$ resulting in a relatively fast algorithm of time complexity $O(KN^2)$ permitting an

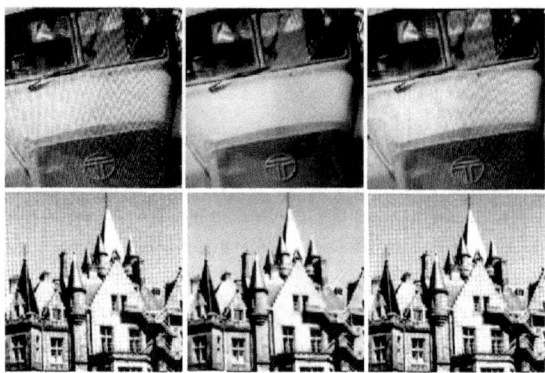

Fig. 7. Removal of aliasing and block textures. From left to right, top to bottom a) Photograph of truck with aliasing. b) Beltrami patch denoising, $\sigma = 20$ c) CBM3D denoising, $\sigma = 20$ d) Photograph of castle with weak block textures e) Beltrami patch denoising, $\sigma = 20$ f) BM3D denoising, $\sigma = 20$.

efficient implementation on a modern smartphone. We have also experimentally determined the relationship between the image intensities and their coordinates as posed in [8]. The proposed method produces $PSNR$ values close to state of the art techniques such as BM3D [3]. In addition to Gaussian denoising, the process accurately removes weak textures and aliasing while preserving the fine structure of the edges in images that other methods are not capable of dealing with.

6   Future Work

Modern smartphones have powerful graphics processing units as well as accelerated vector engine hardware. Neither of these features were utilized for the current application and further work is required to enable the method to work at optimal speed. Although this note has focused mainly on the denoising property of the Beltrami operator it would seem reasonable to further study other applications of the operator in patch-space such as inverse diffusion and other processes which control the eigen-values of the local diffusion operator. Many of these techniques have already been developed for the original Beltrami flow such as the FAB diffusion method as described by Gilboa et al. [9] and therefore it would be prudent to extend their application to patch-space. The same can be said for the short time Beltrami kernel as described by Spira et al. in [10] where it is approximated by finding local geodesic distances on the manifold via the fast marching method and the local metric tensor. The analysis and implementation of the same procedure would be a fruitful direction for future research in patch-space based flows.

Acknowledgments. This research was supported by European Community's FP7- ERC program, grant agreement no. 267414.

References

1. Barash, D.: A fundamental relationship between bilateral filtering, adaptive smoothing and the nonlinear diffusion equation. IEEE Transactions on Pattern Analysis and Machine Intelligence 24(6), 844–847 (2002)
2. Buades, A., Coll, B., Morel, J.M.: A non-local algorithm for image denoising. In: CVPR, pp. 60–65 (2005)
3. Dabov, K., Foi, A., Katkovnik, V., Egiazarian, K.: Image denoising by sparse 3-d transform-domain collaborative filtering. IEEE Transactions on Image Processing, 2080–2095 (2007)
4. Elad, M.: On the origin of the bilateral filter and ways to improve it. IEEE Transactions on Image Processing 11(10), 1141–1151 (2002)
5. Perona, P., Malik, J.: Scale-space and edge detection using anisotropic diffusion. IEEE Trans. Pattern Anal. Mach. Intell., 629–639 (1990)
6. Peyre, G.: Manifold models for signals and images. Computer Vision and Image Understanding 113, 249–260 (2009)
7. Roussos, A., Maragos, P.: Tensor-based image diffusions derived from generalizations of the total variation and Beltrami functionals. In: ICIP (September 2010)
8. Sochen, N., Kimmel, R., Malladi, R.: A general framework for low level vision. IEEE Trans. on Image Processing, 310–318 (1998)
9. Sochen, N.A., Gilboa, G., Zeevi, Y.Y.: Color image enhancement by a forward-and-backward adaptive Beltrami flow. In: Sommer, G., Zeevi, Y.Y. (eds.) AFPAC 2000. LNCS, vol. 1888, pp. 319–328. Springer, Heidelberg (2000)
10. Spira, A., Kimmel, R., Sochen, N.A.: Efficient Beltrami flow using a short time kernel. In: Griffin, L.D., Lillholm, M. (eds.) Scale-Space 2003. LNCS, vol. 2695, pp. 511–522. Springer, Heidelberg (2003)
11. Tomasi, C., Manduchi, R.: Bilateral filtering for gray and color images. In: Proc. IEEE ICCV, pp. 836–846 (1998)
12. Tschumperlé, D., Brun, L.: Non-local image smoothing by applying anisotropic diffusion pde's in the space of patches. In: ICIP, pp. 2957–2960 (2009)

A Fast Augmented Lagrangian Method for Euler's Elastica Model

Yuping Duan[1], Yu Wang[2], Xue-Cheng Tai[1,3], and Jooyoung Hahn[4]

[1] Division of Mathematical Sciences, School of Physical and Mathematical Sciences,
Nanyang Technological University, Singapore
duan0010@e.ntu.edu.sg
[2] Computer Science Department, Technion, Haifa 32000, Israel
yuwang@cs.technion.ac.il
[3] Department of Mathematics, University of Bergen,
Johannes Brunsgate 12, 5007 Bergen, Norway
tai@math.uib.no
[4] Institute for Mathematics and Scientific Computing, University of Graz, Austria
joo.hahn@uni-graz.at

Abstract. In this paper, a fast algorithm for Euler's elastica functional is proposed, in which the Euler's elastica functional is reformulated as a constrained minimization problem. Combining the augmented
Lagrangian method and operator splitting techniques, the resulting
saddle-point problem is solved by a serial of sub-problems. To tackle
the nonlinear constraints arising in the model, a novel fixed-point-based
approach is proposed so that all the sub-problems either are linear problems or have closed form solutions. Numerical examples are provided to
demonstrate the performance of the proposed method.

1 Introduction

Suppose that the observed image u_0 is the original image u perturbed by additive
gaussian noise η

$$u_0 = u + \eta.$$

Our task is to recover the image u from the noisy image u_0. The image denoising
problems are often solved by the variational methods using the total variation
minimization. Total variation was first introduced into image denoising in [13]
and played an important role since then. However, the total variation based
models always converge to a piecewise constant solution, which will cause the
staircasing effect in the results. To counteract the disadvantages, some high
order models are developed during recent years in [5,6,9,10]. The Euler's elastica
model is one of the important high order models, which is defined based on the
curvature of the image. It was introduced into computer vision in [11] and has
been successfully applied in various applications such as image inpainting and
denoising etc. [1,3,4,6,10]. We first define the curvature of a level curve Γ as a
function of u in the following way

$$\kappa(u) = \nabla \cdot \left(\frac{\nabla u}{|\nabla u|} \right). \tag{1}$$

A.M. Bruckstein et al. (Eds.): SSVM 2011, LNCS 6667, pp. 144–156, 2012.

We can express Euler's elastica of the curve Γ using the curvature as the following energy functional

$$E(\Gamma) = \int_{\Gamma} \left(a + b \cdot |\kappa|^{\beta}\right) ds, \tag{2}$$

where a, b are two tuning parameters and s is the arc length. In the functional (2), the first term minimizes the total length and the second term minimizes the power of the total curvature in the image. The power β can be set to either $\beta = 1$ as in [10], or $\beta = 2$ as in [6]. In this work, we set $\beta = 2$, but the techniques developed below can be extended to the case $\beta = 1$ without too many efforts. Therefore, the following minimization problem for image denoising is concerned

$$\min_{u} \int_{\Omega} \left(a + b\left(\nabla \cdot \frac{\nabla u}{|\nabla u|}\right)^2\right) |\nabla u| + \frac{\mu}{2} \int_{\Omega} (u - u_0)^2. \tag{3}$$

The traditional ways to solve Euler's elastica functional (3) are usually complex and time consuming due to the high nonlinearity of the partial differential equations (PDE) as in [6,10]. In [2,7], graph-cuts are applied to the high order models and Euler's elastica. Recently, the augmented Lagrangian method [12,15] has been successfully used to solve the Euler's elastica model in [14]. The authors provide us a fast algorithm by solving the sub-problems emerging from the augmented Lagrangian functional using FFT. In their approach, one quadratic penalty term in the augmented Lagrangian functional is relaxed to the first order and a frozen coefficient method with FFT is used to solve the coupled PDEs with variable coefficient. The high dependence of the FFT limits the applications of the algorithm in Tai-Hahn-Chung formulation, especially for the surface problems; see [8].

In this work, we propose a different augmented Lagrangian formulation for Euler's elastica model (3) and use the operator splitting technique to solve the corresponding saddle-point problem as well. Instead of disposing of the quadratic term in the sub-problem related to normal vector, we apply a fixed-point method to find a closed form solution. Furthermore, we introduce a new variable into the constrained problem to avoid to solve the PDE with variable coefficient. All other linear problems are solved efficiently by the iterative solver. Therefore, the proposed method is computationally economic in terms of both memory requests and computation costs.

2   Augmented Lagrangian Method for Euler's Elastica Model

In this section, we propose an augmented Lagrangian formulation for the Euler's elastica energy (3). First, we introduce two extra variables $\mathbf{p}$ and $\mathbf{n}$ to cast the functional (3) into a constrained minimization problem as follows

$$\min_{u,\mathbf{p},\mathbf{n}} \int_{\Omega} \left(a + b(\nabla \cdot \mathbf{n})^2\right) |\mathbf{p}| + \frac{\mu}{2} \int_{\Omega} (u - u_0)^2,$$

$$\text{s.t.} \quad \mathbf{p} = \nabla u; \quad \mathbf{n} = \frac{\mathbf{p}}{|\mathbf{p}|}. \tag{4}$$

The second constraint in (4) is equivalent to $\mathbf{p} = |\mathbf{p}|\mathbf{n}$. Therefore, the corresponding augmented Lagrangian functional for the constrained optimization problem (4) is defined as follows

$$\mathcal{L}(u, \mathbf{p}, \mathbf{n}; \lambda_1, \lambda_2) = \int_\Omega \left(a + b(\nabla \cdot \mathbf{n})^2 \right)|\mathbf{p}| + \frac{\mu}{2} \int_\Omega (u - u_0)^2 \tag{5}$$

$$+ \int_\Omega \lambda_1 \cdot (\mathbf{p} - |\mathbf{p}|\mathbf{n}) + \frac{r_1}{2} \int_\Omega \|\mathbf{p} - |\mathbf{p}|\mathbf{n}\|^2 + \int_\Omega \lambda_2 \cdot (\mathbf{p} - \nabla u) + \frac{r_2}{2} \int_\Omega \|\mathbf{p} - \nabla u\|^2,$$

where λ_1 and λ_2 are the Lagrange multipliers, r_1 and r_2 are positive penalty parameters. We aim to seek a saddle-point of the augmented Lagrangian functional (5), named as

$$\text{Find} \quad (u^*, \mathbf{p}^*, \mathbf{n}^*; \lambda_1^*, \lambda_2^*), \tag{6}$$

$$\text{s.t.} \mathcal{L}(u^*, \mathbf{p}^*, \mathbf{n}^*; \lambda_1, \lambda_2) \leq \mathcal{L}(u^*, \mathbf{p}^*, \mathbf{n}^*; \lambda_1^*, \lambda_2^*)$$

$$\leq \mathcal{L}(u, \mathbf{p}, \mathbf{n}; \lambda_1^*, \lambda_2^*), \quad \forall (u, \mathbf{p}, \mathbf{n}; \lambda_1, \lambda_2).$$

In the forthcoming sections, we first review the existing algorithm [14] for solving the minimization problem (3) in the augmented Lagrangian formulations in Section 3. Then, based on the augmented Lagrangian functional (5) a novel algorithm is proposed in Section 4, in which the advantages of the proposed algorithm compared to the existing augmented Lagrangian algorithm is discussed. In Section 5, the numerical solution of each sub-problem emerging from the augmented Lagrangian functional is discussed separately. Numerical results are given in Section 6 to demonstrate the efficiency of our proposed method.

3   The Existing Algorithm

In this section, we give a brief review of the augmented Lagrangian method applied to Euler's elastica in the Tai-Hahn-Chung formulation referred to [14]. It is difficult to solve the augmented Lagrangian functional (5) efficiently because of the non-differentiable quadratic term involving $|\mathbf{p}|$. Therefore, the authors introduce one more variable into the constrained problem (4), which is defined as

$$\mathbf{m} = \mathbf{n} \quad \text{and} \quad |\mathbf{m}| \leq 1.$$

By the constraint $|\mathbf{m}| \leq 1$, there exists the relationship $|\mathbf{p}| - \mathbf{m} \cdot \mathbf{p} \geq 0$, a.e. in Ω. Therefore, the quadratic penalty term $\int_\Omega (|\mathbf{p}| - \mathbf{m} \cdot \mathbf{p})^2$ in (5) can be relaxed to a first order penalty term. The augmented Lagrangian functional (5) can be reformulated as follows

$$\mathcal{L}(u, \mathbf{p}, \mathbf{n}, \mathbf{m}; \lambda_1, \lambda_2, \lambda_3) = \int_\Omega \left(a + b(\nabla \cdot \mathbf{n})^2 \right)|\mathbf{p}| + \frac{\mu}{2} \int_\Omega (u - u_0)^2 \tag{7}$$

$$+ \int_\Omega \lambda_1(|\mathbf{p}| - \mathbf{m} \cdot \mathbf{p}) + r_1 \int_\Omega (|\mathbf{p}| - \mathbf{m} \cdot \mathbf{p}) + \int_\Omega \lambda_2 \cdot (\mathbf{p} - \nabla u)$$

$$+ \frac{r_2}{2} \int_\Omega \|\mathbf{p} - \nabla u\|^2 + \int_\Omega \lambda_3 \cdot (\mathbf{n} - \mathbf{m}) + \frac{r_3}{2} \int_\Omega \|\mathbf{n} - \mathbf{m}\|^2 + \delta_\mathcal{R}(\mathbf{m}),$$

where $\delta_\mathcal{R}$ is the indicator function defined on the set $\mathcal{R}$; see [14].

In [14], the energy functional (7) is decomposed into the following sub-problems

$$\epsilon_1(u) = \frac{\mu}{2} \int_\Omega (u - u_0)^2 - \int_\Omega \boldsymbol{\lambda}_2 \cdot \nabla u + \frac{r_2}{2} \int_\Omega \|\mathbf{p} - \nabla u\|^2, \tag{8}$$

$$\epsilon_2(\mathbf{p}) = \int_\Omega \left(a + b(\nabla \cdot n)^2 \right) |\mathbf{p}| + \int_\Omega \lambda_1 (|\mathbf{p}| - \mathbf{m} \cdot \mathbf{p}) \tag{9}$$

$$+ r_1 \int_\Omega (|\mathbf{p}| - \mathbf{m} \cdot \mathbf{p}) + \int_\Omega \boldsymbol{\lambda}_2 \cdot \mathbf{p} + \frac{r_2}{2} \int_\Omega \|\mathbf{p} - \nabla u\|^2,$$

$$\epsilon_3(\mathbf{n}) = \int_\Omega b|\mathbf{p}|(\nabla \cdot \mathbf{n})^2 + \int_\Omega \boldsymbol{\lambda}_3 \cdot \mathbf{n} + \frac{r_3}{2} \int_\Omega \|\mathbf{n} - \mathbf{m}\|^2, \tag{10}$$

$$\epsilon_4(\mathbf{m}) = -(\lambda_1 + r_1) \int_\Omega \mathbf{p} \cdot \mathbf{m} - \int_\Omega \boldsymbol{\lambda}_3 \cdot \mathbf{m} + \frac{r_3}{2} \int_\Omega \|\mathbf{n} - \mathbf{m}\|^2 + \delta_{\mathcal{R}}(\mathbf{m}). \tag{11}$$

The above sub-problems are solved alternatively in one iteration of the algorithm. For the u-sub problem, since its Euler-Lagrange equation is a linear PDE, it is solved efficiently by the FFT. There are closed form solutions for the $\mathbf{p}$-sub and $\mathbf{m}$-sub problems referred to [14]. The most difficult and time-consuming part is to solve the Euler-Lagrange equation of the $\mathbf{n}$-sub problem, which is

$$-2\nabla(b|\mathbf{p}|\nabla \cdot \mathbf{n}) + r_3(n - m) + \lambda_3 = 0. \tag{12}$$

The coefficient of $\nabla \cdot \mathbf{n}$ is a variable in (12), which makes it difficult to handle. Aimed to use the FFT, a frozen coefficient method is applied to the coupled PDEs in [14]. However, there are some drawbacks coming with this method. First, the method needs the inner iterations. Second, to solve the coupled PDEs (12), people have to apply the FFT twice in one iteration. Supposed that the inner iteration of the $\mathbf{n}$-sub problem is once, together with the once FFT used in the u-sub problem, three times of FFT are involved in one outer iteration of the algorithm in [14]. Both the above two points will increase the computational costs. Besides, there exist cases that the domain or boundary condition assigned to the minimization problem (3) are not applicable to FFT. Therefore, we consider to find a better way to solve the augmented Lagrangian functional(5).

4   The Proposed Algorithm

In this section, we propose a more efficient algorithm to solve the augmented Lagrangian functional (5). Unlike the algorithm in [14], we keep the two constraints in the optimization problem (4) and introduce one more variable, which is defined as follows

$$h = \nabla \cdot \mathbf{n}.$$

We use the variable h to remove the variable coefficient in the $\mathbf{n}$-sub problem (10) in the Tai-Hahn-Chung formulation [14]. Moreover, in our algorithm the quadratic penalty term instead of first order penalty term is used in **p**-sub problem. To be precise, first, the Euler's elastica model (3) is reformulated as the following constrained optimization problem

$$\min_{u,\mathbf{p},\mathbf{n},h} \int_\Omega (a + bh^2)|\mathbf{p}| + \frac{\mu}{2}\int_\Omega (u - u_0)^2, \tag{13}$$
$$\text{s.t.} \quad \mathbf{p} = \nabla u; \quad \mathbf{p} = |\mathbf{p}|\mathbf{n}; \quad h = \nabla \cdot \mathbf{n}.$$

And, by the augmented Lagrangian method [15], the constrained problem (13) is changed into an unconstrained minimization problem as follows

$$\mathcal{L}(u,\mathbf{p},\mathbf{n},h;\boldsymbol{\lambda}_1,\boldsymbol{\lambda}_2,\lambda_3) = \int_\Omega (a + bh^2)|\mathbf{p}| + \frac{\mu}{2}\int_\Omega (u - u_0)^2 + \int_\Omega \boldsymbol{\lambda}_1 \cdot (\mathbf{p} - |\mathbf{p}|\mathbf{n})$$
$$+ \frac{r_1}{2}\int_\Omega \|\mathbf{p} - |\mathbf{p}|\mathbf{n}\|^2 + \int_\Omega \boldsymbol{\lambda}_2 \cdot (\mathbf{p} - \nabla u) + \frac{r_2}{2}\int_\Omega \|\mathbf{p} - \nabla u\|^2$$
$$+ \int_\Omega \lambda_3 (h - \nabla \cdot \mathbf{n}) + \frac{r_3}{2}\int_\Omega (h - \nabla \cdot \mathbf{n})^2. \tag{14}$$

To handle the difficulty caused by the quadratic penalty term $\int_\Omega \|\mathbf{p} - |\mathbf{p}|\mathbf{n}\|^2$, a novel fix-point-based technique is used, and details are postponed to Section 5. We apply an iterative algorithm to solve the saddle-point problem corresponding to the augmented Lagrangian functional (14); see Algorithm 1.

Algorithm 1. Augmented Lagrangian Method for the Euler's elastica model

---

1. Initialization: u^0, $\mathbf{p}^0$, $\mathbf{n}^0$, h^0 and $\boldsymbol{\lambda}_1^0$, $\boldsymbol{\lambda}_2^0$, λ_3^0.
2. For $k = 0, 1, 2\ldots$, compute $(u^k, \mathbf{p}^k, \mathbf{n}^k, h^k)$ as an approximate minimizer of the augmented Lagrangian functional with the Lagrange multiplier $\boldsymbol{\lambda}_1^k$, $\boldsymbol{\lambda}_2^k$ and λ_3^k, i.e.,

$$(u^k, \mathbf{p}^k, \mathbf{n}^k, h^k) \approx \arg\min \mathcal{L}(u, \mathbf{p}, \mathbf{n}, h; \boldsymbol{\lambda}_1^k, \boldsymbol{\lambda}_2^k, \lambda_3^k). \tag{15}$$

3. Update the Lagrange multipliers:

$$\boldsymbol{\lambda}_1^{k+1} = \boldsymbol{\lambda}_1^k + r_1(\mathbf{p}^k - |\mathbf{p}^k|\mathbf{n}^k), \tag{16}$$
$$\boldsymbol{\lambda}_2^{k+1} = \boldsymbol{\lambda}_2^k + r_2(\mathbf{p}^k - \nabla u^k), \tag{17}$$
$$\lambda_3^{k+1} = \lambda_3^k + r_3(h^k - \nabla \cdot \mathbf{n}^k). \tag{18}$$

Since the variables u, $\mathbf{p}$, $\mathbf{n}$, h are coupled together in the minimization problem (15), it is difficult to solve all the variables simultaneously. Therefore, we use the operator splitting method to separate the problem (15) into sub-problems. The four sub-problems are given as follows

$$\mathcal{E}_1(u) = \frac{\mu}{2}\int_\Omega (u - u_0)^2 - \int_\Omega \boldsymbol{\lambda}_2 \cdot \nabla u + \frac{r_2}{2}\int_\Omega \|\mathbf{p} - \nabla u\|^2, \tag{19}$$

$$\mathcal{E}_2(\mathbf{p}) = \int_\Omega (a + bh^2)|\mathbf{p}| + \int_\Omega \boldsymbol{\lambda}_1 \cdot (\mathbf{p} - |\mathbf{p}|\mathbf{n}) + \frac{r_1}{2} \int_\Omega \||\mathbf{p} - |\mathbf{p}|\mathbf{n}\|^2 \qquad (20)$$

$$+ \int_\Omega \boldsymbol{\lambda}_2 \cdot \mathbf{p} + \frac{r_2}{2} \int_\Omega \|\mathbf{p} - \nabla u\|^2,$$

$$\mathcal{E}_3(\mathbf{n}) = -\int_\Omega \boldsymbol{\lambda}_1 \cdot |\mathbf{p}|\mathbf{n} + \frac{r_1}{2} \int_\Omega \|\mathbf{p} - |\mathbf{p}|\mathbf{n}\|^2 - \int_\Omega \lambda_3 \nabla \cdot \mathbf{n} + \frac{r_3}{2} \int_\Omega (h - \nabla \cdot \mathbf{n})^2, \qquad (21)$$

$$\mathcal{E}_4(h) = \int_\Omega b|\mathbf{p}|h^2 + \int_\Omega \lambda_3 h + \frac{r_3}{2} \int_\Omega (h - \nabla \cdot \mathbf{n})^2. \qquad (22)$$

Sub-problems (19) to (22) can be efficiently solved. We will discuss the specific solution to each sub-problem in the forthcoming section.

5   Numerical Solutions for Subproblems

In this section, we explain how to find the minimizer of each sub-problem. We use a staggered grid system as in Fig. 1 to solve the energy functional minimization (19) to (22) and update the Lagrange multipliers from (16) to (18).

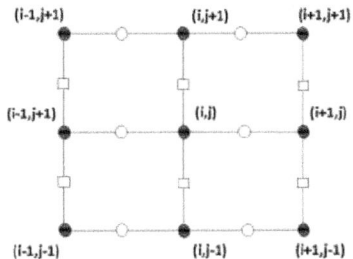

Fig. 1. Grid definition. The rule of indexing variables in the augmented Lagrangian functional (5): u, h and λ_3 are defined on ●-nodes. The first and second component of $\mathbf{p}$, $\mathbf{n}$, $\boldsymbol{\lambda}_1$ and $\boldsymbol{\lambda}_2$ are defined on ○-nodes and □-nodes, respectively.

5.1   Notations

We first give some basic notations at the beginning. An image is regarded as a function

$$u : \{1, \ldots, M\} \times \{1, \ldots, N\},$$

where $M, N \geq 2$. We denote the Euclidean space $\mathbb{R}^{M \times N}$ as V and define another inner product vector space: $Q = V \times V$.

For a given $(i, j) \in [1, M] \times [1, N]$, we see that

$$u \in V, \quad u(i,j) \in \mathbb{R} \quad \text{and} \quad \mathbf{p} \in Q, \quad \mathbf{p}(i,j) = (p_1(i,j), p_2(i,j)) \in \mathbb{R}^2,$$

which is equipped with the standard Euclidean inner products as follows

$$(u, v)_V = \sum_{i,j} u(i,j)v(i,j) \quad \text{and} \quad (\mathbf{p}, \mathbf{q})_Q = (p_1, q_1)_V + (p_2, q_2)_V.$$

We will use the discrete backward and forward differential operators for $u \in V$, which are defined with periodic boundary condition as follows

$$\partial_x^- u(i,j) = \begin{cases} u(i,j) - u(i-1,j) & 1 < i \le M, \\ u(1,j) - u(M,j) & i = 1. \end{cases}$$

$$\partial_y^- u(i,j) = \begin{cases} u(i,j) - u(i,j-1) & 1 < j \le N, \\ u(i,1) - u(i,N) & j = 1. \end{cases}$$

$$\partial_x^+ u(i,j) = \begin{cases} u(i+1,j) - u(i,j) & 1 \le i < M, \\ u(1,j) - u(M,j) & i = M. \end{cases}$$

$$\partial_y^+ u(i,j) = \begin{cases} u(i,j+1) - u(i,j) & 1 \le j < N, \\ u(i,1) - u(i,N) & j = N. \end{cases}$$

5.2   Sub-problems

In the following, for each sub-problem, we denote the fixed Lagrange multipliers in the previous $(k-1)^{th}$ iteration as $\lambda_1 = \lambda_1^k$, $\lambda_2 = \lambda_2^k$ and $\lambda_3 = \lambda_3^k$.

u-sub Problem. For the u-sub problem, the optimality condition of (19) gives a linear equation of u, which is

$$(\mu - r_2 \Delta)u = \mu u_0 - \nabla \cdot \lambda_2 - r_2 \nabla \cdot \mathbf{p}. \tag{23}$$

Therefore, the discretized form of the equation (23) is

$$\left(\mu - r_2(\partial_x^- \partial_x^+ + \partial_y^- \partial_y^+) \right) u = g,$$

where

$$g = \mu u_0 - (\partial_x^- \lambda_{21} + \partial_y^- \lambda_{22}) - r_2(\partial_x^- p_1 + \partial_y^- p_2).$$

The PDE problem (23) can be solved efficiently by a wide range of linear iterative methods, such as Jacobi method, Gauss-Seidel method. In this work, we choose to use one sweep of the Gauss-Seidel method, which is enough to approximate the solution.

p-sub Problem. For the p-sub problem, it is difficult to solve the Euler-Lagrange equation of (20) due to the non-differentiability element $|\mathbf{p}|$ in the quadratic term. To avoid this situation, we consider to apply a fixed-point formulation to the constraint $\mathbf{p} = |\mathbf{p}|\mathbf{n}$ in the kth iteration, which gives

$$\mathbf{p} = |\mathbf{p}^{k-1}|\mathbf{n}.$$

To get rid of the nonlinearity and non-differentiability term, we use $\mathbf{p}-|\mathbf{p}^{k-1}|\mathbf{n}$ to replace $\mathbf{p}-|\mathbf{p}|\mathbf{n}$ in the quadratic penalty term in (20). Therefore, we reformulate the energy functional of $\mathbf{p}$-sub problem $\mathcal{E}_2$ as follows

$$\mathcal{E}_2(\mathbf{p}) = \int_\Omega (a + bh^2 - \boldsymbol{\lambda}_1 \cdot \mathbf{n})|\mathbf{p}| + \frac{r_1 + r_2}{2} \int_\Omega \Big\|\mathbf{p} - \frac{r_1|\mathbf{p}^{k-1}|\mathbf{n} + r_2\nabla u - \boldsymbol{\lambda}_1 - \boldsymbol{\lambda}_2}{r_1 + r_2}\Big\|^2.$$

For the simplicity, let

$$c = a + bh^2 - \boldsymbol{\lambda}_1 \cdot \mathbf{n} \qquad \text{and} \qquad \mathbf{q} = \frac{r_1|\mathbf{p}^{k-1}|\mathbf{n} + r_2\nabla u - \boldsymbol{\lambda}_1 - \boldsymbol{\lambda}_2}{r_1 + r_2}.$$

Here, c can be either positive or negative. For each case, there is the closed form solution for solving $\mathbf{p}$-sub problem. If c is positive, we have the following closed form solution for $\mathbf{p}$

$$\mathbf{p}(i, j) = \max\Big\{0, \ 1 - \frac{c}{(r_1 + r_2)|\mathbf{q}(i, j)|}\Big\}\mathbf{q}(i, j).$$

And if c is negative, the solution for $\mathbf{p}$ is

$$\mathbf{p}(i, j) = \Big(1 - \frac{c}{(r_1 + r_2)|\mathbf{q}(i, j)|}\Big)\mathbf{q}(i, j),$$

which belongs to the case when c is positive. Therefore, for each case the closed form solution for $\mathbf{p}$ in (20) can be summarized as follows

$$\mathbf{p}(i, j) = \max\Big\{0, \ 1 - \frac{c}{(r_1 + r_2)|\mathbf{q}(i, j)|}\Big\}\mathbf{q}(i, j). \tag{24}$$

n-sub Problem. For the **n**-sub problem, the Euler-Lagrange equation for the energy (21) is the following linear coupled PDEs

$$-r_3\nabla(\nabla \cdot \mathbf{n}) + r_1|\mathbf{p}|^2\mathbf{n} = r_1\mathbf{p}|\mathbf{p}| + \boldsymbol{\lambda}_1|\mathbf{p}| - r_3\nabla h - \nabla\lambda_3.$$

Pay attention that the operator $\nabla(\nabla\cdot)$ is singular. Since it is possible that $|\mathbf{p}| = 0$, we add a quadratic penalty term to (21) to avoid the singularity. Therefore, we rewrite $\mathcal{E}_3(\mathbf{n})$ as follows

$$\mathcal{E}_3(\mathbf{n}) = -\int_\Omega \boldsymbol{\lambda}_1 \cdot |\mathbf{p}|\mathbf{n} + \frac{r_1}{2}\int_\Omega \||\mathbf{p} - |\mathbf{p}|\mathbf{n}\|^2 - \int_\Omega \lambda_3\nabla \cdot \mathbf{n} \tag{25}$$
$$+ \frac{r_3}{2}\int_\Omega (h - \nabla \cdot \mathbf{n})^2 + \frac{\gamma}{2}\int_\Omega \|\mathbf{n} - \mathbf{n}^{k-1}\|^2,$$

where γ is a positive constant. In the experiments, γ could be chosen to be a very small number.

We have the following optimality condition for the **n**-sub problem (25) by its Euler-Lagrange equation

$$\Big(\gamma + r_1|\mathbf{p}|^2 - r_3\nabla(\nabla\cdot)\Big)\mathbf{n} = \gamma\mathbf{n}^{k-1} + r_1\mathbf{p}|\mathbf{p}| + \boldsymbol{\lambda}_1|\mathbf{p}| - r_3\nabla h - \nabla\lambda_3. \tag{26}$$

The equation (26) is coupled PDEs of the variable $\mathbf{n} = (n_1, n_2)$. When we compute the component n_1, we use the n_2 in previous iteration

$$(\gamma + r_1|\mathbf{p}|^2 - r_3\partial_x^2)n_1 = \gamma n_1^{k-1} + r_4\partial_x\partial_y n_2 + r_1 p_1|\mathbf{p}| + \lambda_{11}|\mathbf{p}| - r_3\partial_x h - \partial_x\lambda_3,$$

and vice versa, when solve n_2, we use the n_1 in previous iteration

$$(\gamma + r_1|\mathbf{p}|^2 - r_3\partial_y^2)n_2 = \gamma n_2^{k-1} + r_4\partial_y\partial_x n_1 + r_1 p_2|\mathbf{p}| + \lambda_{12}|\mathbf{p}| - r_3\partial_y h - \partial_y\lambda_3.$$

Similarly to the u-sub problem, the above linear PDEs can be efficiently solved by the linear iterative methods. We still apply the one sweep Gauss-Seidel iteration to solve the $\mathbf{n}$-sub problem in our work. Compared to the frozen coefficient FFT method in [14], the proposed method for the $\mathbf{n}$-sub problem is easy to implement and solves the PDEs with low computational cost.

h-sub Problem. For the h-sub problem, we have the Euler-Lagrange equation of the functional (22) as follows

$$(2b|\mathbf{p}| + r_3)h = r_3\nabla \cdot \mathbf{n} - \lambda_3. \tag{27}$$

We can obtain a closed form solution for h by solving the first-order equation (27). The minimizer of $\mathcal{E}_4(h)$ is solved as follows

$$h = \frac{r_3\nabla \cdot \mathbf{n} - \lambda_3}{2b|\mathbf{p}| + r_3}.$$

Based on the discussion of each sub problem, we can use Algorithm 2 to solve the minimization problem (15). We set $N = 1$ for all experiments in Section 6.

Algorithm 2. Alternating minimization method to solve the sub problems of Eqn. (15)

---

1. Initialization: $u^{k+1,0} = u^k$, $\mathbf{p}^{k+1,0} = \mathbf{p}^k$, $\mathbf{n}^{k+1,0} = \mathbf{n}^k$ and $h^{k+1,0} = h^k$.
2. For $n = 0, 1, \ldots, N$, compute $u^{k+1,n+1}$ from Eqn. (23), $\mathbf{p}^{k+1,n+1}$ from Eqn. (24), $\mathbf{n}^{k+1,n+1}$ from Eqn. (26) and $h^{k+1,n+1}$ from Eqn. (27);
3. $u^{k+1} = u^{k+1,N}$, $\mathbf{p}^{k+1} = \mathbf{p}^{k+1,N}$, $\mathbf{n}^{k+1} = \mathbf{n}^{k+1,N}$ and $h^{k+1} = h^{k+1,N}$.

---

6   Numerical Examples

In this section, we consider the applications of the proposed Algorithm 1 for Euler's elastica to image denoising problems. All experiments are implemented in C++ language on a 2.4 GHz CPU and 4GB memory.

During the iterations, we define the relative error of the solution $\{u^k|\ k = 1, 2, \ldots\}$ as

$$\frac{||u^k - u^{k-1}||_{L^1}}{||u^{k-1}||_{L^1}}, \tag{28}$$

and we stop the iteration when the relative error is less than the given error tolerance. We define the numerical energy of the Euler's elastica model as follows

$$E^k = \int_\Omega \left(a + b(h^k)^2\right)|\mathbf{p}^k| + \frac{\mu}{2}\int_\Omega (u^k - u_0)^2. \tag{29}$$

During the iterations, we monitor the relative residuals of Lagrange multipliers defined by

$$(R_1^k, R_2^k, R_3^k) = \left(\frac{||\mathbf{p}^k - |\mathbf{p}^k|\mathbf{n}^k||_{L^1}}{|\Omega|}, \frac{||\mathbf{p}^k - \nabla u^k||_{L^1}}{|\Omega|}, \frac{||h - \nabla \cdot \mathbf{n}||_{L^1}}{|\Omega|}\right). \tag{30}$$

We also monitor the relative errors of the Lagrange multipliers:

$$(L_1^k, L_2^k, L_3^k) = \left(\frac{||\boldsymbol{\lambda}_1^k - \boldsymbol{\lambda}_1^{k-1}||_{L^1}}{||\boldsymbol{\lambda}_1^{k-1}||_{L^1}}, \frac{||\boldsymbol{\lambda}_2^k - \boldsymbol{\lambda}_2^{k-1}||_{L^1}}{||\boldsymbol{\lambda}_2^{k-1}||_{L^1}}, \frac{||\lambda_3^k - \lambda_3^{k-1}||_{L^1}}{||\lambda_3^{k-1}||_{L^1}}\right). \tag{31}$$

We choose three synthetic and two real images from [14] to test the proposed method. Therefore, we can compare the results from our algorithm with the results in [14]. The Gaussian white noise with mean zero and the standard deviation 10 is added to the test images. For all the experiments, we fix $\gamma = 0.01$. We display the results of the synthetic images in Fig. 2 while we show the results of real images in Fig. 3. In Table 1, we provide the size of the image, SNR, number of iterations and computational time for the test images shown in Fig. 2 and Fig. 3.

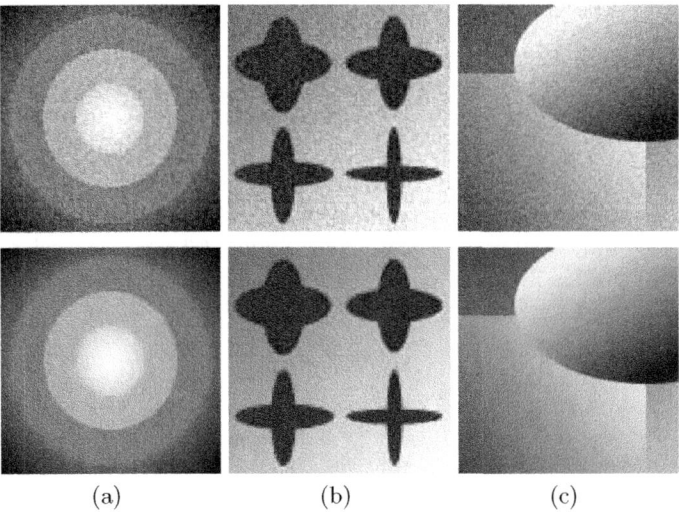

(a)   (b)   (c)

Fig. 2. Euler's elastica based image denoising. We set $a = 1$, $b = 10$, $\mu = 1$, $r_1 = 0.03$, $r_2 = 10$ and $r_3 = 10000$. The tolerance is $6 \cdot 10^{-4}$ for (a) and $5 \cdot 10^{-4}$ for (b) and (c).

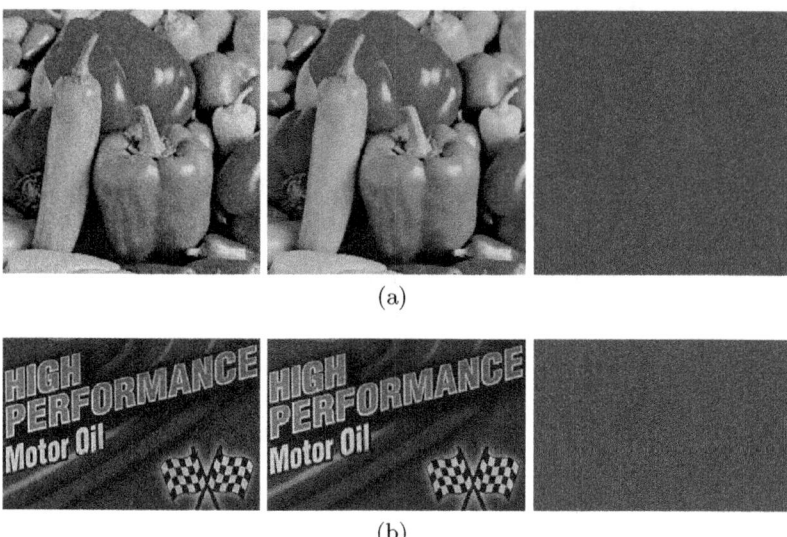

(a)

(b)

Fig. 3. Euler's elastica based image denoising. From left to right: noisy image, denoised image and difference between the denoised image and noisy image. We set $a = 1$, $b = 2$, $\mu = 1$, $r_1 = 0.02$, $r_2 = 10$ and $r_3 = 5000$. The tolerance is $1 \cdot 10^{-3}$ for (a) and $5 \cdot 10^{-4}$ for (b).

Table 1. Numerical Results of Euler's elastica model

images	size	SNR	# of iteration	time (sec)
Fig.2(a)	100×100	22.47	101	2.87
Fig.2(b)	100×100	25.35	109	3.11
Fig.2(c)	128×128	24.05	99	4.62
Fig.3(a)	256×256	18.23	84	15.21
Fig.3(b)	332×216	17.35	125	24.75

To test the performance of the proposed algorithm, we track the relative error in u (28), the numerical energy (29), the residuals (30) and the relative error in Lagrange multipliers (31) of the images in Fig. 3. We display the plots in Fig. 4.

7   Conclusion and Future Work

In this work, we propose a simple and efficient augmented Lagrangian approach for Euler's elastica. The operator splitting method is adopted to solve the saddle-point problem arising in the augmented Lagrangian formulation. We artfully apply the fixed-point method to one sub-problem to get a closed form solution. All the sub problems in our method can be solved efficiently by fast iterative methods and closed form solution. As the numerical results of image denoising problems demonstrate, our method yields good results in terms of computational time. We

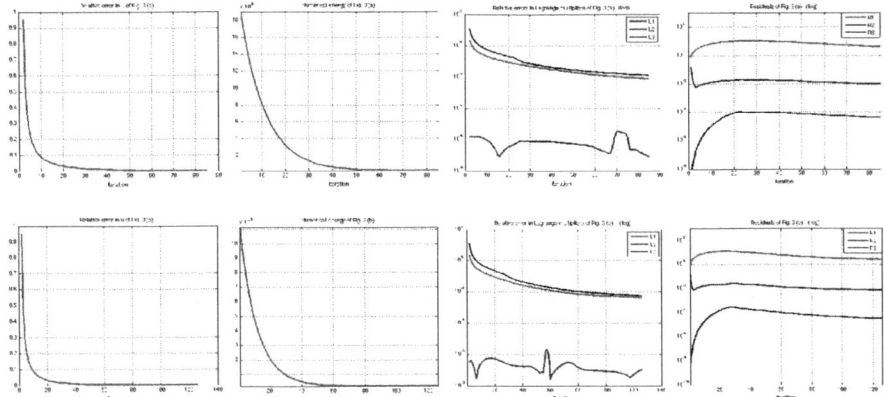

Fig. 4. Plots of (28), (29), (30) and (31) values versus iteration numbers for examples shown in Fig. 3 (a) (Row One) and Fig. 3 (b) (Row Two)

would like to mention that the proposed algorithm can be easily applied to image inpainting and zooming problems as well. Since numerical experiments suggest the convergence of our algorithm, we hope to do some convergence analysis in future.

Acknowledgement. The research has been supported by MOE (Ministry of Education) Tier II project T207N2202 and IDM project NRF2007IDM-IDM002-010. The second author is also supported in part at the Technion by a fellowship of the Israel Council for Higher Education.

References

1. Ambrosio, L., Masnou, S.: A direct variational approach to a problem arising in image reconstruction. Interfaces and Free Boundaries 5(1), 63–82 (2003)
2. Bae, E., Shi, J., Tai, X.C.: Graph Cuts for Curvature based Image Denoising. UCLA CAM Report 10-28, Department of Mathematics, UCLA, Los Angeles, CA (2010)
3. Ballester, C., Bertalmio, M., Caselles, V., Sapiro, G., Verdera, J.: Filling-in by joint interpolation of vector fields and gray levels. IEEE Transactions on Image Processing 10(8), 1200–1211 (2002)
4. Ballester, C., Caselles, V., Verdera, J.: Disocclusion by joint interpolation of vector fields and gray levels. Multiscale Modeling and Simulation 2, 80–123 (2004)
5. Chan, T., Marquina, A., Mulet, P.: High-order total variation-based image restoration. SIAM Journal on Scientific Computing 22(2), 503–516 (2000)
6. Chan, T.F., Kang, S.H., Shen, J.: Euler's elastica and curvature-based inpainting. SIAM Journal on Applied Mathematics, 564–592 (2002)
7. Komodakis, N., Paragios, N.: Beyond pairwise energies: Efficient optimization for higher-order MRFs. In: IEEE Computer Society Conference on Computer Vision and Pattern Recognition, CVPR (2009)

8. Lai, R., Chan, T.F.: A Framework for Intrinsic Image Processing on Surfaces. UCLA CAM Report 10–25, Department of Mathematics, UCLA, Los Angeles, CA (2010)
9. Lysaker, M., Lundervold, A., Tai, X.C.: Noise removal using fourth-order partial differential equation with applications to medical magnetic resonance images in space and time. IEEE Transactions on Image Processing 12(12), 1579 (2003)
10. Masnou, S., Morel, J.M.: Level lines based disocclusion. In: In Proc. IEEE Int. Conf. on Image Processing, pp. 259–263 (2002)
11. Mumford, D.: Elastica and computer vision. Algebraic Geometry and its Applications (1994)
12. Rockafellar, R.T.: Augmented Lagrangians and applications of the proximal point algorithm in convex programming. Mathematics of Operations Research 1(2), 97–116 (1976)
13. Rudin, L.I., Osher, S., Fatemi, E.: Nonlinear total variation based noise removal algorithms. Physica D: Nonlinear Phenomena 60(1-4), 259–268 (1992)
14. Tai, X.C., Hanh, J., Chung, G.J.: A fast algorithm for Euler's elastica model using augmented Lagrangian method. UCLA CAM Report 10-47, Department of Mathematics, UCLA, Los Angeles, CA (2010)
15. Tai, X.C., Wu, C.: Augmented Lagrangian method, dual methods and split Bregman iteration for ROF model. In: Tai, X.-C., Mørken, K., Lysaker, M., Lie, K.-A. (eds.) SSVM 2009. LNCS, vol. 5567, pp. 502–513. Springer, Heidelberg (2009)

Deblurring Space-Variant Blur
by Adding Noisy Image

Iftach Klapp[1], Nir Sochen[2], and David Mendlovic[1]

[1] Department of Physical Electronics,
[2] Department of Mathematics,
Tel-Aviv University
Tel-Aviv 69978
Israel

Abstract. Imaging restoration is an essential step in hybrid optical and image processing system which relays on poor optics. The poor optics makes the blur ill-conditioned and turns the deblurring process difficult and unstable. Recently the idea of *parallel optics* (PO) was introduced. In the parallel optics setup the optical system is composed of a main system and an auxiliary system. The auxiliary system is designed to improve the stability of the deblurring process by improving the condition number of the blurring operator. In this paper we show that in one such system the post processing acts as a noise filter hence allows to work with noisy data in the auxiliary channel. Using the singular value decomposition we derive analytical limit for the difference in SNR requirements of the auxiliary channel relative to that of the main channel. The gap between the SNR requirements of the two systems is analyzed theoretically and proved to be as large as 27.68 [db]. Image restoration comparison on simulations is performed between a blurred/noisy pair with average SNR gap of 20 [db] and a system without an auxiliary system. The average Mean Square Error Improvement Factor (MSEIF) achieved by the blurred/noisy pair, was 13.9 [db] higher than the system without a noisy auxiliary system.

Keywords: deblurring, parallel optics, blurred and noisy image pair.

1 Introduction

Traditional optical design should typically include few optical elements to create a required sharp image [9]. Due to its optical and mechanical complexity this optical design is expensive in cost and volume. The advances in digital cameras technology allow us the alternative design of a hybrid optical and image processing imaging system, which is based on simple and low cost optics. The strategy in such a design is that the poor optical performance, which are associate with the low cost optics, is compensated by sophisticated image processing algorithms. The process is done in a serial way (Fig. 1).

In [1] authors extended the optical system Depth Of Field (EDOF) on the expanse by a Optical Phase Mask (OPM), the resulted poor but depth insensitive

A.M. Bruckstein et al. (Eds.): SSVM 2011, LNCS 6667, pp. 157–168, 2012.

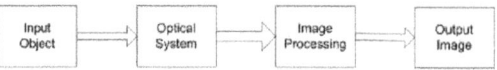

Fig. 1. Block diagram for a hybrid system

response was deblured. Similar approach was used for reducing zoom lens complexity and dimensions [3]. OPM for EDOF was implemented in cellular cameras
[3], Other Hybrid EDOF approach was suggested in [4] where the optical system
chromatic aberration was used to gain one or more sharp channel (color), where
its high frequency information used to enhance the other two channel. In [5] by
using OPM poor space variant response was improved and followed by image
processing. Another hybrid system is the compound eye optics where series of
low resolution images which captured by single imaging system used for reconstructing single high resolution image by supper resolution methods [6,7]. Other
system is the light field camera where by placing a system similar to the CMO in
the optical system image plane, one can recognize ray direction and reconstruct
sharp image of objects in various distances [8].

In order to have a bright image the optics should have high Numerical Aperture (NA). However, in low cost optics this demand is necessarily accompanied
by Space Variant (SV) blur. This blur is modeled in the continuum as a linear
operator $H(x, x')$ acting on the image $I(x)$ via

$$I^{img}(x) = \int H(x, x')I^{obj}(x')dx' \tag{1}$$

where $x, x' \in R$. In the discrete setup we arrange the image in row stack such
that an image with L pixels is a L dimensional vector and the operator is a
$L \times L$ matrix. In this notation the relation between the object and the image is
described as:

$$I^{img} = H \cdot I^{obj} \quad . \tag{2}$$

The columns of H are the main system PSFs for each field point. The Ill-
conditioning of the blur operator is manifesting itself in the condition number of
the matrix H. The condition number $\kappa_2(H)$ for the L_2 norm is bounded from
below by the ratio of the largest to smallest singular values of H

$$\kappa_2(H) = \frac{\sigma_1}{\sigma_n} \quad . \tag{3}$$

Realizing the importance of improving the matrix condition for the hybrid, simple optics based, imaging system we are searching for an optical design that
improvs the matrix condition number. The optical design that realizes this idea
is achieved by adding an auxiliary optical system (O) [10] as shown in Fig. 2.

The two systems are imaging the same object and the output is superimposed
optically. The result is an effective new blur

$$H_1 = H + O \quad . \tag{4}$$

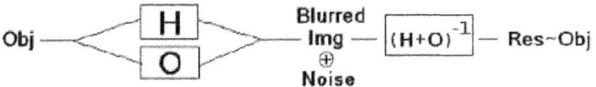

Fig. 2. The parallel optics process

The imaging process of the parallel optics system is:

$$I^{img} = H_1 \cdot I^{obj} = H \cdot I^{obj} + O \cdot I^{obj} \tag{5}$$

where the last stage in Fig. 2 depicts the deblurring process.

This paper deals with the design of the auxiliary system O and its properties. In particular we will see how the auxiliary operator can be derived, under few assumptions, from a functional. We will also demonstrate how the operator that is derived in an analytical form can be realized via simple optical means. Finally we will analyze the noise requirement of the auxiliary system in comparison to the same requirement of the main system. Our analysis shows that we can start from general mathematical requirement for the desired features of the auxiliary system, derive the desired operator from a functional and realize this operator with a simple optical design. Moreover, we find that the auxiliary system can add considerable noise while improving the final result!

This paper is organized as follows: In section 2 we discuss a functional whose minimization yields the regularization operator we need. We project this operator unto the space spanned by simple translation operators in section 3. Noise analysis in section 4 shows that the auxiliary system can be much noisier than in the usual setup. Section 5 presents the results of numerical simulation that we performed and we summarize and conclude in section 6.

2   Regularization

The condition number of a matrix quantifies its distance to a singular matrix. A regularization of the system means that an operator is added such that the combined operator/matrix are further away from the singular matrix. In practice this means that small and zero singular values of H are modified. This can be analyzed via the singular value decomposition (SVD) of the main system:

$$H = USV^t \tag{6}$$

We impose that the auxiliary system is of the form [10]:

$$O = U\left(\Delta S\right)V^t \tag{7}$$

Where ΔS is strong at least where the S singular values are weak. From the arbitrary choice of ΔS we suggest to assume that the case of that the non zero instance on ΔS diagonal equal to one.

The desired operator O can be derived from a functional. We start by writing a general regularization

$$S\left(I^{obj}\right) = \frac{1}{2}\int\left(I^{img}(x) - \int H(x,x')I^{obj}(x')dx'\right)^2 dx + \frac{\mu}{2}\int (PI^{obj}(x))^2 dx \quad.$$
(8)

The operator P can be differential one or integral one. After discretization and transforming to a row stack representation the functional reads

$$S_D\left(I^{obj}\right) = \frac{1}{2}\left\|I^{img} - H \cdot I^{obj}\right\|_F^2 + \frac{\mu}{2}\left\|P \cdot I^{obj}\right\|_F^2$$
(9)

Minimizing S_D with respect to I^{obj} yields the following

$$H^t\left(I^{img} - H \cdot I^{obj}\right) + \mu P^t P \cdot I^{obj} = 0$$
(10)

Rearranging

$$H^t I^{img} = \left(H^t H + \mu P^t P\right) \cdot I^{obj} \quad.$$
(11)

From Eq. 5 we find

$$H^t I^{img} = H^t\left(H + O\right) \cdot I^{obj} \quad.$$
(12)

Direct comparison leads to the following characterization of P

$$\frac{1}{\mu}H^t O = P^t P \quad.$$
(13)

Using the SVD and the fact that U and V are unitary matrices we find

$$H^t O = V S U^t U\left(\Delta S\right)V^t = V S\left(\Delta S\right)V^t$$
(14)

from which we conclude

$$P = \sqrt{\frac{S\Delta S}{\mu}}V^t \quad.$$
(15)

Thus, by defining that ΔS values to be 1 where S values are weak we force a solution that minimize the error function by minimizing the influence of the weakest singular values of the main system H i.e. improve the matrix condition.

3   Translation Based Operator Space

The required auxiliary system is determined mathematically, however optical implementation of such a system is not obvious. In this section we investigate the possibility to approximate the desired operator with simple optical design building blocks. In particular we will look into realization of operators via superposition of lenses with various transparency and linear phase shifts. In mathematical terms it boils down to superposition of translation operators with different weights. we present the duality between the basic optical operation and matrix representation of auxiliary optics with pixel confined PSF. This correspondence is elaborated in the next subsection. The required hardware is illustrated in figure 3. We add to the main system, an auxiliary optics with the same FOV magnification with pixel and confined PSF such that $I^{aux} = I^{obj} + n$ where n is the noise in the auxiliary system. The constructed regularization O acts on I^{aux}. In practice, however, the regularization operator is applied as a post-processing operation on the noisy image *before* it is added to the main system's blurred image.

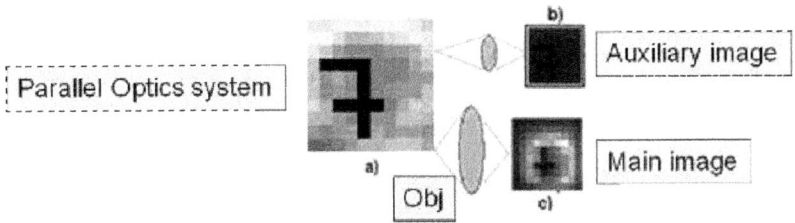

Fig. 3. Optical hardware for implementing the Trajectories methods

3.1   Summary of the Trajectory Methods

Phase shifts are simple to achieve in optics. Moreover, phase shifts in the frequency space are equivalent to translations in the spatial space. We will build in this subsection an approximation scheme for operators via superposition of these elementary optical operations.

Translations may be realized as operators acting on images. These operators are represented as matrices after discretization of the problem. These matrices have simple structure. The translation operators are defined via

$$\left(T_{kl} \cdot I^{obj}\right)_{ij} = \sum_{rs} (T_{kl})_{ij,rs}\, I_{rs}^{obj} = I_{i-k,j-l}^{obj} \tag{16}$$

We use as before the row stack notation with only one index standing for the translation. For conventional reasons we denote the translation operator in this representation O_l. This is a $L \times L$ matrix. One key feature of these operators is their orthogonality. Indeed

$$\frac{1}{L}Tr(O_k^t O_l) = \delta_{kl} \quad . \tag{17}$$

The operators O_l span thus a large subspace in the space of all linear operators. The question of completeness is out of scope in this paper and is differed to future publication. Empirically one can check that it is a good scheme of approximation. These operators are also called "trajectories" by the authors of [12] Indeed, In optics it is clear that the field of view allows many translations, hence we can define a series of matrices which can be optically realized, and can serve as a partial base to decompose the required (O) matrix.

$$O \approx \sum w_l O_l \tag{18}$$

where w_l is the weight associated with the translation O_l. The weights can be realized optically as the transparency of the lens that add the l'th phase shift. In figure 4 we show an example for 3X2 field of view (FOV) [11]. We show that each Trajectories matrix is associated with an image. The FOV is fixed and determine by the main system, designate by the black frame. From linearity we use the Trajectories matrices weights for decompose the image of (O) by the images of the Trajectories matrices. The decomposition can be stated as a minimization problem

$$e(w_1, \ldots, w_n) = \left\| O - \sum_{l=1}^{n} w_l O_l \right\|_F^2 \qquad (19)$$

the condition for extremum

$$\frac{\partial c}{\partial w_k} = -2 \left(O^t - \sum w_l O_l^t \right) O_k = 0 \qquad (20)$$

Taking the trace and normalizing we finally find

$$\frac{1}{L} Tr \left(O^t O_k \right) = \sum w_l \frac{1}{L} Tr \left(O_l^t O_k \right) = \sum w_l \delta_{lk} = w_k \qquad (21)$$

In figure 5 we present the over all process in block schema, the auxiliary optics matrix representation and image is post process to manipulate the required target auxiliary system matrix O [11].

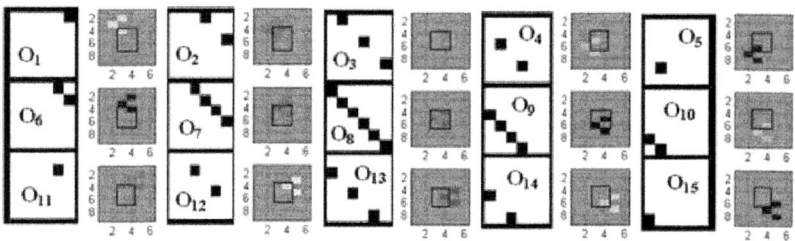

Fig. 4. Trajectories matr3ces and their related chessboard object image

4   Noise Contributors in Image Restoration

In the context of this work it is important to emphasize that the additive noise in the auxiliary optics, is added before post processing and, thus, the post processing has an influence on it. As mention above, typically this filtering causes that the SNR requirements from the auxiliary optics can be lower than that of the main system. In the next subsection we present an ideal limit for the gap between the SNR requirements of the two systems.

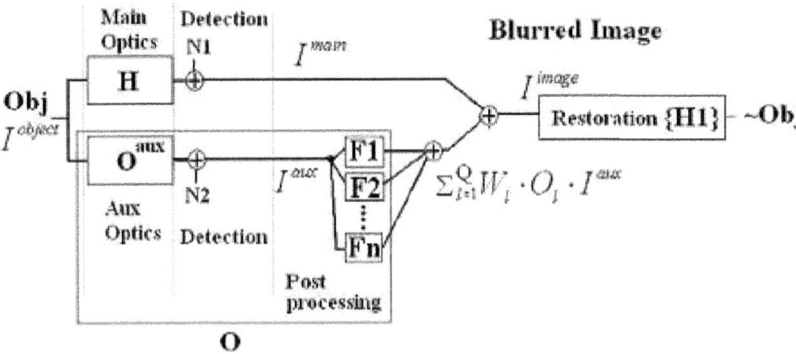

Fig. 5. Trajectories method schema for PO system implementation, the auxiliary system is composed of auxiliary optics imaging and a series of shifted and weighted operations which are done by post processing

4.1   The Relative Immunity to Noise of the Auxiliary System

In this section show that the special structure of the Parallel Optics allow to work with auxiliary system which inherently suffer from SNR which is much lower of the main system. The image of an auxiliary optics with pixel confined response is an darken replica of the object as seen in figure 3

$$I^{aux} = I^{obj}/AR \tag{22}$$

AR is the illumination ratio between the main system to the auxiliary system

$$AR = \left(\frac{NA_{main}}{NA_{aux}}\right)^2 \tag{23}$$

In imaging systems the signal depends on the radiant optical power (Popt [Watt]) Detectors responsivity $\Re$ ([Watt/A]). The optical power is proportional to the square or the system numerical aperture (NA) [13]:

$$P_{opt} \propto (NA)^2 \tag{24}$$

The additive noise of the imaging system have many sources, for simplicity we assume that it is governing by the electronics noise which is the case in low illumination. In addition we determine the signal as the peak to peak level relative to zero mean level [14] the simplified SNR expression is [13]:

$$SNR = \frac{(\Re P_{opt})^2}{\sigma_n^2} \simeq \frac{I_s^2}{\sigma_n^2} \tag{25}$$

Assuming that the reduction in the image contrast is governing by the NA:

$$I_s = \Re P_{opt} \propto (NA)^2 \tag{26}$$

the ratio between the auxiliary system signal to the main system signal can be determine by the NA ratio:

$$\frac{I_s^{aux}}{I_s^{main}} = \left(\frac{NA_{aux}}{NA_{main}}\right)^2 = \frac{1}{AR} \tag{27}$$

AR is the square of the NA ratio Eq. (23). Implementing this ratio in the SNR, under the assumption that both the auxiliary optics and the main system are subjecting to the same noise level such that:

$$\sigma_{n-aux} = \sigma_{n-main} \tag{28}$$

The relation between the main system SNR to the auxiliary lens SNR in [db] is:

$$SNR_{aux}^a = 20log\left(\frac{I_s^{aux}}{\sigma_{n-aux}}\right) = 20log\left(\frac{I_s^{main}}{\sigma_{n-main}AR}\right) = SNR_{main} - 20log\,(AR) \tag{29}$$

Equation (23) is reflecting a conservative maximal SNR gap between the main and the auxiliary SNR which point on the fact that in order to have pixel confined response from simple auxiliary optics we pay in SNR. This may suggest that the auxiliary system is more vulnerable to noise than the main system. However, bellow we show that in the special case of the Trajectories system, in terms of contribution to restoration noise the auxiliary system is generally more tolerant to the working SNR level. As was mentioned above, in the special implementation by the Trajectories method, based on the pixel confined auxiliary optics response, the required PO response is synthesized digitally with no additional optical effort.

Indeed, with mild assumptions and using the singular value decomposition the inputs noise ratio in output equilibrium is:

$$\frac{\sigma_{n-aux}^2}{\alpha_{n-main}^2} = \frac{\left\lVert\frac{S+\Delta S}{(S+\Delta S)^2+\alpha I}\right\rVert_F^2}{AR^2 \cdot \left\lVert\frac{(S+\Delta S)\Delta S}{(S+\Delta S)^2+\alpha I}\right\rVert_F^2} \tag{30}$$

where α is the Tikhonov regularization parameter. Assigning the standard deviation in the auxiliary SNR according to (27) the gap in SNR requirement between the main system and the auxiliary optics is found to be:

$$SNR_{aux}^b = 20log\left(\frac{I_s^{aux}}{\sigma_{n-aux}}\right) = SNR_{main} - 20log\left(\sqrt{\frac{\left\lVert\frac{(S+\Delta S)}{(S+\Delta S)^2+\alpha I}\right\rVert_F^2}{\left\lVert\frac{(S+\Delta S)\Delta S}{(S+\Delta S)^2+\alpha I}\right\rVert_F^2}}\right) \tag{31}$$

A full derivation of this limit will be presented in [15]. The multiplication by ΔS narrows the bandwidth thus:

$$\left\lVert\frac{S+\Delta S}{(S+\Delta S)^2+\alpha I}\right\rVert_F^2 \geq \left\lVert\frac{(S+\Delta S)\,\Delta S}{(S+\Delta S)^2+\alpha I}\right\rVert_F^2 \tag{32}$$

Hence for the same noise contribution to restoration we obtain

$$SNR_{aux}^b < SNR_{main} \tag{33}$$

In the system architecture, AR value is pre-determine by the required PSF of the auxiliary optics and ΔS is pre determine by the requirements for matrix condition improvements. AR tends to be larger than 1 thus to lower the auxiliary system SNR in a given electronics noise (27). In the other hand the matrix ΔS tend to be narrow band than the matrix S and, thus, to reduce the relative contribution of the auxiliary system detection noise to the restoration corruption. This in turns entails that the auxiliary system tends to be more tolerant in SNR requirements than the main system (39). For the general case of parallel optics the additive noise is added to the auxiliary system such as in figure 2, no filtering done over the noise of the auxiliary system thus in equation (37) $\Delta S = I_d$, where I_d is the identity matrix hence there is no reduction in the noise contribution of the auxiliary system to the restoration.

5   Numerical Examples

In this section we illustrate numerically the theoretical results. A study case was taken from previous investigated Trajectories system [11]. The main system is describe by 100X100 space variant PSF matrix (H), the matrix is ill-conditioned, with condition number of $\kappa = 87640$. The image pixel's size is 11.3?m. the image size is 10X10, the image distance is 0.69 mm, the lens diameter is 0.4 mm and the FOV is -/+ 4.67 deg. The auxiliary optics (O_{aux}), is a simple double convex lens with 0.16 mm diameter. The auxiliary optics PSF is almost space invariant with 90 of the energy confined in the central pixel and 98% of the energy is confined to 5X5 pixel area. We mould like to emphasize that this almost pixel confined response achieved on the expense of the low NA. In simulations we use the 5X5 section as the auxiliary system fixed PSF, Performing the Trajectories method we produce parallel optics system with condition number of $\kappa = 5.75$, for more details see [11]. For investigating the average restoration quality as function of the systems SNR, in each main SNR and auxiliary SNR pair we performed an average over 300 image restoration realizations. The restoration quality measure is taken as the Mean Square Error Improvement Factor (MSEIF) [16]:

$$MSEIF = 20log\left(\frac{\left\|I^{img} - I^{obj}\right\|_2}{\left\|I^{res} - I^{obj}\right\|_2}\right) \tag{34}$$

In this function the restoration error is in the denominator and the blur error is in the nominator. Both are measured relative to the ideal object (I^{obj}). When MSIEF< 0 [db], the restored image (I^{res}) is worse than the blur study case image (I^{img}), therefore there is no use in restoration. Since the object range is limited to 0 - 255 gray levels, it is reasonable to assume that imaging gray level values below 0 and above 255 are out of the instrument range and can be limited to 0 and 255 respectively.

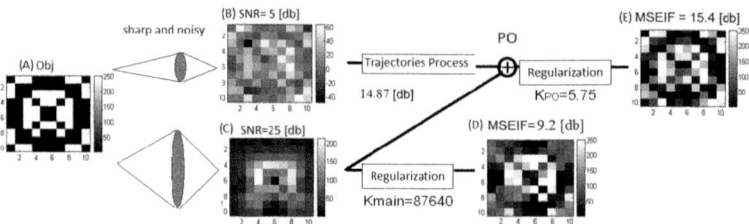

Fig. 6. Target. (A) The object, (B) The auxiliary optics image, (C) The blurred main system image, (D) The main system restoration by regularization, (E) The Trajectories restoration by regularization.

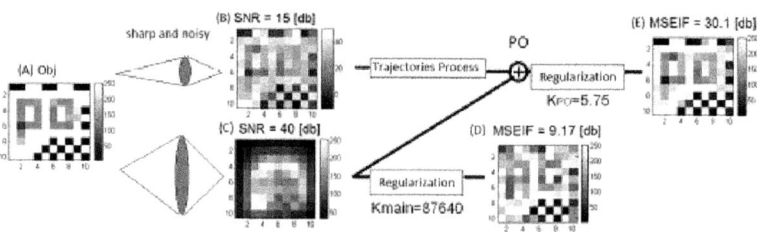

Fig. 7. PO. (A) The object, (B) The auxiliary optics image, (C) The blurred main system image, (D) The main system restoration by regularization, (E) The Trajectories restoration by regularization.

Calculating AR for this study case the auxiliary optics SNR is 15.9 [db] lower then of the main system SNR. However in the aspect of noise contribution to the restoration, the auxiliary optics SNR can be 27.68 [db] lower then the main system SNR before it equally contribute to restoration error. Hence, the working SNR of the auxiliary system can be 11.78 [db] lower than traditionally assumed. In order to demonstrate the system's performance we calculate the restoration quality over ensemble of three different objects, "PO", "Chessboard", and "Target". Simulation was carried out for four SNR levels 25, 35, 45 and 55 [db] when AWGN was added to each image. The above limits suggest that the auxiliary system is more tolerance to the working SNR then the main system. To verify that, we calculate the MSEIF performance, once when the main SNR is 20 [db] higher than the auxiliary SNR and once the opposite, results show that in all cases the MSEIF score was higher when the low SNR was of the auxiliary system with average gap was 5.4 [db]. Looking at the MSEIF scores, in Fig. 8, we see that the auxiliary filtering allow us working in very low SNR values 5-15 [db] and still score remarkable average MSEIF values of 15.38 to 33.21 [db] respectively, which allow working with blurred/noisy pair for improving matrix condition by the trajectories method. In figures 6, 7 we present two visual example of the noisy/blurred pair image restoration: "Target", and "PO" objects respectively. Also shown the restoration of the main system without the auxiliary system. Although the auxiliary optics image is corrupted by the noise, restoration results

Ensemble Δ<MSEIF> [db]	Ensemble <MSEIF> [db]	"PO" MSEIF [db]	"Chessboard" MSEIF [db]	"Target" MSEIF [db]	SNR$_{aux}$ [db]	SNR$_{main}$ [db]
7.37	15.38	11.64	16.6	17.9	5	25
	8.01	3.84	9.26	10.93	25	5
7.71	33.21	31.4	31.12	37.11	15	35
	25.5	19.16	26.94	30.4	35	15
5.14	46.65	46.98	43.8	49.18	25	45
	41.51	38.32	40.58	45.63	45	25
1.3	50.65	52.94	46	53	35	55
	49.35	50.31	45.51	52.23	55	35

Fig. 8. Columns 1+2 are the SNR state Columns 3-5 are the MSEIF average results of PO restoration. It was done by averaging MSEIF score of 300 restorations where in each restoration a different noise pattern was randomly produced. Column 6 is the average value of columns 3-5 Column 7 is the gap between the results in the two competitive SNR states (ex SNRmain=25 and SNRaux=5 Vs SNRmain=5 and SNRaux=25).

show both visually and in MSEIF score the advantage of adding the Trajectories over the use of the "main system alone. The average MSEIF achieved by the blurred/noisy pair, was 13.9 [db] higher than the system without the noisy auxiliary system.

6   Summary and Conclusions

In this work we showed that the specific parallel optics configuration which use an auxiliary channel based on post processing of auxiliary optics, the post processing which come after the detection have a secondary effect of noise reduction in this channel, hence its SNR requirements are significantly lower with compare to those of the main channel. An analytical limits, for SNR gap between the main and the auxiliary channels was developed in two cases. First assuming equal noise we investigate the gap dependence in numerical aperture ratio of the channels. Second the SNR gap dependency in post processing, assuming equal contribution to restoration level. The derivation was demonstrate on ill space variant study case where image restoration done by Tikhonov regularization. First by switching the high and low SNR between the main and the auxiliary channels, we test the claim, that the auxiliary channel is more tolerant to SNR level then the main channel. The restoration results confirm that it is the case, the average MSEIF value was 5.4 [db] to the favor of lower SNR in the auxiliary channel. Following that we showed that the low SNR requirement of the auxiliary channel is allowing blurred/noisy pair uses for parallel optics implementation. Comparing the main system image restoration quality to that of the parallel optics to noisy auxiliary system with SNR gap of 20 [db]. The average Mean Square Error Improvement Factor (MSEIF) achieved by the blurred/noisy pair, was 13.9 [db] higher of the system without the noisy auxiliary system.

References

1. Dowski, E.R., Chathey, T.: Extended depth of field through wave-front coding. App. Opt. 34(11), 1859 (1995)
2. Demenikov, M., Findlay, E., Harvey, A.R.: Miniaturization of zoom lenses with a single moving element. Opt. Express 17(8), 6118 (2009)
3. Shabtay, G., Goldenberg, E., Dery, E.: Imaging system with improved image quality and associated methods US patent 2009/122150 A1
4. Guichard, F., et al.: Extended depth-of-field using sharpness transport across color channels. In: Proc. SPIE 7250, 72500N (2009)
5. Muyo, G., et al.: Infrared imaging with a wavefront-coded singlet lens. Opt. Express 17(23), 21118 (2009)
6. Nitta, K., et al.: Image reconstruction for thin observation module by bound optics by using the iterative backprojection method. App. Opt. 45(13), 2893–2900 (2006)
7. Duparr, J.W., et al.: Ultra thin camera based on artificial apposition compound eyes, http://www.suss-microoptics.com/downloads/Publications/MOC-04.pdf
8. Ng, R., et al.: Light field photography with hand held Plenoptic camera, Stanford Tech Report CTSR 2005-02
9. Kidjer, M.J.: Principles of lens design. In: SPIE Proc., CR4, pp. 30–52 (1992)
10. Klapp, I., Mendlovic, D.: Improvement of matrix condition of Hybrid, space variant optics by the means of Parallel Optics design. Opt. Express 17, 11673–11689 (2009)
11. Klapp, I., Mendlovic, D.: Trajectories in Parallel Optics (submitted to JOSA)
12. Klapp, I., Mendlovic, D.: Optical Design for Improving Matrix Condition, SRS, OSA (2009) paper STuA7
13. Kopeika, N.S.: A system Engineering approach to imaging, pp. 517–520 (SPIE, 1998)
14. Jain, A.K.: Fundamentals of digital image processing, p. 59. Prentice Hall, Englewood Cliffs (1989)
15. Klapp, I., Mendlovic, D.: Blurred/noisy image pairs in Parallel Optics (to be submitted)
16. Bertero, M., Boccacci, P.: Introduction to inverse problems in imaging (IOP, 1998). Waterman, M.S.: Identification of Common Molecular Subsequences

Fast Algorithms for p-elastica Energy with the Application to Image Inpainting and Curve Reconstruction*

Jooyoung Hahn[1,**], Ginmo J. Chung[2], Yu Wang[2], and Xue-Cheng Tai[2,3]

[1] Institute for Mathematics and Scientific Computing, University of Graz, Austria
[2] Division of Mathematical Sciences, School of Physical Mathematical Sciences,
Nanyang Technological University, Singapore
[3] Mathematics Institute, University of Bergen, Norway
joo.hahn@uni-graz.at

Abstract. In this paper, we propose fast and efficient algorithms for p-elastica energy ($p = 1$ or 2). Inspired by the recent algorithm for Euler's elastica models in [16], the algorithm is extended to solve the problem related to p-elastica energy based on augmented Lagrangian method. The proposed algorithms are as efficient as the previous method in terms of low computational cost per iteration. We provide an algorithm which replaces fast Fourier transform (FFT) by a cheap arithmetic operation at each grid point. Numerical tests on image inpainting are provided to demonstrate the efficiency of the proposed algorithms. We also show examples of using the proposed algorithms in curve reconstruction from unorganized data set.

Keywords: p-elastica energy, Augmented Lagrangian method, Euler's elasitca, Image inpainting, Curve reconstruction, Unorganized data set.

1 Introduction

The curvature of the curve has been extensively used in minimization problems in image processing and computer vision. D. Mumford, M. Nitzberg, and T. Shiota [14] introduced segmentation with depth to find a continuation curve γ which minimizes Euler's elastica energy ($p = 2$):

$$\mathcal{E}(\gamma) = \int_{\gamma} (a + b|\kappa|^p) \, ds, \tag{1}$$

---

* The research is supported by MOE (Ministry of Education) Tier II project T207N2202 and IDM project NRF2007IDMIDM002-010. In addition, the support from SUG 20/07 is also gratefully acknowledged.

** This author is currently at Institute of Mathematics and Scientific Computing in University of Graz, Austria. He has been supported by the Austrian Science Fund (FWF) under the START-Program Y305 "Interfaces and Free Boundaries" and the SFB "Mathematical Optimization and Its Applications in Biomedical Sciences" since November 2010.

A.M. Bruckstein et al. (Eds.): SSVM 2011, LNCS 6667, pp. 169–182, 2012.

where κ is the curvature of the curve in $\mathbf{R}^2$, a and b are positive constants, and $p \geq 1$. In [2], Euler's elastica problem is reduced to solve a set of algebraic equations in Jacobi's functions. Semicontinuity and relaxation properties of (1) were presented in [6]. Following the work [14], Masnou and Morel [13] proposed a variational formulation in the geometrical recovery of the missing parts from a given image $u_0 : \mathcal{D} \setminus \tilde{\mathcal{D}} \subset \mathbf{R}^2 \to \mathbf{R}$, where $\mathcal{D} \supset \tilde{\mathcal{D}}$. In [13], an energy functional to complete of all level lines of u_0 is written by using change of variable and the coarea formula from (1):

$$\int_{\mathcal{D}} \left(a + b \left| \nabla \cdot \frac{\nabla u}{|\nabla u|} \right|^p \right) |\nabla u|, \tag{2}$$

where $p \geq 1$. Note that the standard Lebesgue measure in $\mathbf{R}^2$ is omitted in the rest of paper. The authors in [8] solved the minimization problem (2) with $p = 2$ by using the Euler-Lagrange equation and the gradient descent method. In [9], they showed that the curvature term is essential to achieve a connectivity principle. The properties of variational model and the existence of minimizing functional (2) are investigated by Ambrosio and Masnou [1]. The authors in [4] proposed an energy functional minimization with two arguments, $\mathbf{n}$ which represents the normalized image gradient and a gray image (real-valued function) u defined on $\mathcal{D}$:

$$\min_{\mathbf{n}, u} \left(\int_{\mathcal{D}} |\nabla \cdot \mathbf{n}|^p \left(c_1 + c_2 |\nabla k * u| \right) + \zeta \int_{\mathcal{D}} \left(|\nabla u| - \mathbf{n} \cdot \nabla u \right) \right), \tag{3}$$
$$|\mathbf{n}| \leq 1, \quad \|u\| \leq \|u_0\|_{L^\infty(\mathcal{D} \setminus \tilde{\mathcal{D}})},$$

where c_1 and c_2 are positive constants, k denotes a Gaussian kernel; see [4] for boundary conditions and the detail admissible sets. Note that the constraint term $|\nabla u| - \mathbf{n} \cdot \nabla u$ in (3) is crucially used in [16]. The existence of minimizers of a relaxed variant of (3) is proved in [5].

Efficient numerical algorithms for energy minimization related to the curvature are studied very recently. The authors [15] used a linear programming relaxation and the discrete elastica [7] to minimize energy functionals for image segmentation and inpainting with curvature regularity. The algorithm in [15] is independent of initialization and computes the global minimum. An improved fast algorithm to the elastica model in [15] is introduced in [10]. The authors in [3] proposed an efficient algorithm based on graph cuts for minimizing the Euler's elastica model for image denoising and inpainting. In [16], new variables and several constraint conditions are introduced to change the Euler's elastica model into a constraint minimization and then augmented Lagrangian method (ALM) is used to obtain a stationary point.

In this paper, we present fast and efficient algorithms for p-elastica energy:

$$\int_{\Omega} \left(a + b \left| \nabla \cdot \frac{\nabla u}{|\nabla u|} \right|^p \right) |\nabla u| + \frac{\eta}{q} \int_{\Gamma} |u - u_0|^q, \tag{4}$$

where $p \geq 1$, $q \geq 1$, Ω is the domain of image u, and $\Gamma \subsetneq \Omega$ is the domain of a given image u_0. The minimization of the functional (4) interpolates the values u_0 on the boundary $\partial \Gamma$ into the inpainting domain $\Omega \setminus \Gamma$. Inspired by the recent algorithm in [16], we extend the algorithm to minimize the p-elastica energy (4). The proposed algorithms use less memory and lower computational cost per iteration than [16]. Numerical tests on image inpainting are provided to demonstrate the efficiency of the proposed algorithms. Moreover, we present a model and numerical examples for curve reconstruction from unorganized data set which has the same regularity term in (4) with $p = 1$.

2   Review of ALM for Euler's Elastica Model

In this section, the augmented Lagrangian method for Euler's elastica model [16] is briefly introduced and we discuss properties of the algorithm and possible improvements in terms of computational cost. When $p = 2$ in (4), the authors [16] proposed several new variables to change the energy minimization of (4) into the constraint minimization problem:

$$\min_{v,u,\mathbf{m},\mathbf{p},\mathbf{n}} \int_\Omega \left(a + b(\nabla \cdot \mathbf{n})^2 \right) |\mathbf{p}| + \frac{\eta}{q} \int_\Gamma |v - u_0|^q \tag{5}$$
$$\text{with} \quad v = u, \quad \mathbf{p} = \nabla u, \quad \mathbf{n} = \mathbf{m}, \quad |\mathbf{p}| = \mathbf{m} \cdot \mathbf{p}, \quad |\mathbf{m}| \leq 1.$$

Note that the variable $\mathbf{m}$ plays an important role to avoid nonuniqueness of a solution in the Euler-Lagrange equation for $\mathbf{n}$-subproblem; see more details in [16]. In order to solve the constraint optimization problem (5), the following augmented Lagrangian functional is used:

$$\mathcal{L}(v, u, \mathbf{m}, \mathbf{p}, \mathbf{n}; \lambda_1, \lambda_2, \lambda_3, \lambda_4) = \int_\Omega \left(a + b(\nabla \cdot \mathbf{n})^2 \right) |\mathbf{p}| + \frac{\eta}{q} \int_\Gamma |v - u_0|^q$$
$$+ r_1 \int_\Omega (|\mathbf{p}| - \mathbf{m} \cdot \mathbf{p}) + \int_\Omega \lambda_1 (|\mathbf{p}| - \mathbf{m} \cdot \mathbf{p}) + \frac{r_2}{2} \int_\Omega |\mathbf{p} - \nabla u|^2$$
$$+ \int_\Omega \lambda_2 \cdot (\mathbf{p} - \nabla u) + \frac{r_3}{2} \int_\Omega (v - u)^2 + \int_\Omega \lambda_3 (v - u) \tag{6}$$
$$+ \frac{r_4}{2} \int_\Omega |\mathbf{n} - \mathbf{m}|^2 + \int_\Omega \lambda_4 \cdot (\mathbf{n} - \mathbf{m}) + \delta_\mathcal{R}(\mathbf{m}),$$

where λ_1, λ_2, λ_3, and λ_4 are Lagrange multipliers, r_1, r_2, r_3, and r_4 are positive penalty parameters, and an indicator function $\delta_\mathcal{R}(\cdot)$ on $\mathcal{R} = \{\mathbf{m} \in \mathbf{L}^2(\Omega) \mid |\mathbf{m}| \leq 1$ a.e. in $\Omega\}$ is defined by

$$\delta_\mathcal{R}(\mathbf{m}) = \begin{cases} 0 & \mathbf{m} \in \mathcal{R}, \\ +\infty & \text{otherwise.} \end{cases}$$

Note that the constraint $|\mathbf{m}| \leq 1$ is imposed by the indicator function and then we have $|\mathbf{p}| - \mathbf{m} \cdot \mathbf{p} \geq 0$, a.e. in Ω. That is, it is not necessarily to use L^2 penalization for the term multiplied by r_1 which causes nonlinearity in the $\mathbf{p}$-subproblem.

An iterative algorithm is suggested to find a stationary point of (6). Lagrange multipliers λ_1^0, $\boldsymbol{\lambda}_2^0$, λ_3^0, and $\boldsymbol{\lambda}_4^0$ and the variables v^0, u^0, $\mathbf{m}^0$, $\mathbf{p}^0$, and $\mathbf{n}^0$ are initialized to zero. For $k \geq 0$, an approximate minimizer

$$\left(v^{k+1}, u^{k+1}, \mathbf{m}^{k+1}, \mathbf{p}^{k+1}, \mathbf{n}^{k+1}\right) \simeq \underset{v,u,\mathbf{m},\mathbf{p},\mathbf{n}}{\arg\min} \ \mathcal{L}(v, u, \mathbf{m}, \mathbf{p}, \mathbf{n}; \lambda_1^k, \boldsymbol{\lambda}_2^k, \lambda_3^k, \boldsymbol{\lambda}_4^k)$$

is obtained by alternatingly solving the subproblems. Let $\tilde{v}^0 = v^k$, $\tilde{u}^0 = u^k$, $\tilde{\mathbf{m}}^0 = \mathbf{m}^k$, $\tilde{\mathbf{p}}^0 = \mathbf{p}^k$, and $\tilde{\mathbf{n}}^0 = \mathbf{n}^k$. For $l = 0, \cdots, L-1$, minimizers $\tilde{v}^{l+1}$, $\tilde{u}^{l+1}$, $\tilde{\mathbf{m}}^{l+1}$, $\tilde{\mathbf{p}}^{l+1}$, and $\tilde{\mathbf{n}}^{l+1}$ are approximately obtained by alternatingly minimizing the following energy functionals:

$$\mathcal{E}_1(v) = \frac{\eta}{q} \int_\Gamma |v - u_0|^q + \int_\Omega \frac{r_3}{2} \left(v - \tilde{u}^l\right)^2 + \lambda_3^k v, \tag{7}$$

$$\mathcal{E}_2(u) = \int_\Omega \frac{r_2}{2} \left|\tilde{\mathbf{p}}^l - \nabla u\right|^2 - \boldsymbol{\lambda}_2^k \cdot \nabla u + \frac{r_3}{2} \left(\tilde{v}^{l+1} - u\right)^2 + \lambda_3^k(-u), \tag{8}$$

$$\mathcal{E}_3(\mathbf{m}) = \delta_{\mathcal{R}}(\mathbf{m}) + \int_\Omega \frac{r_4}{2} \left|\tilde{\mathbf{n}}^l - \mathbf{m}\right|^2 - \boldsymbol{\lambda}_4^k \cdot \mathbf{m} - (r_1 + \lambda_1^k)\mathbf{m} \cdot \tilde{\mathbf{p}}^l, \tag{9}$$

$$\mathcal{E}_4(\mathbf{p}) = \int_\Omega \left(a + b \left(\nabla \cdot \tilde{\mathbf{n}}^l\right)^2\right) |\mathbf{p}| + (r_1 + \lambda_1^k) \left(|\mathbf{p}| - \tilde{\mathbf{m}}^{l+1} \cdot \mathbf{p}\right)$$

$$+ \int_\Omega \frac{r_2}{2} \left|\mathbf{p} - \nabla \tilde{u}^{l+1}\right|^2 + \boldsymbol{\lambda}_2^k \cdot \mathbf{p}, \tag{10}$$

$$\mathcal{E}_5(\mathbf{n}) = \int_\Omega b(\nabla \cdot \mathbf{n})^2 \left|\tilde{\mathbf{p}}^{l+1}\right| + \frac{r_4}{2} \left|\mathbf{n} - \tilde{\mathbf{m}}^{l+1}\right|^2 + \boldsymbol{\lambda}_4^k \cdot \mathbf{n}. \tag{11}$$

After L iterations, variables at $(k+1)^{\text{th}}$ step are updated:

$$\left(v^{k+1}, u^{k+1}, \mathbf{m}^{k+1}, \mathbf{p}^{k+1}, \mathbf{n}^{k+1}\right) = \left(\tilde{v}^L, \tilde{u}^L, \tilde{\mathbf{m}}^L, \tilde{\mathbf{p}}^L, \tilde{\mathbf{n}}^L\right).$$

A large number of iteration L may be necessary to find the minimizers of the above functional $\mathcal{L}(v, u, \mathbf{m}, \mathbf{p}, \mathbf{n}; \lambda_1^k, \boldsymbol{\lambda}_2^k, \lambda_3^k, \boldsymbol{\lambda}_4^k)$. However, it is empirically enough to use $L = 1$ according to recent literatures [17,19]. Lagrange multipliers λ_1^{k+1}, $\boldsymbol{\lambda}_2^{k+1}$, λ_3^{k+1}, and $\boldsymbol{\lambda}_4^{k+1}$ are updated by the standard method in augmented Lagrangian method; see details in [16]. For $q = 1$ or 2, subproblems to minimize $\mathcal{E}_1(v)$, $\mathcal{E}_3(\mathbf{m})$, and $\mathcal{E}_4(\mathbf{p})$ can be solved by closed form formulas which only take arithmetic operations at each grid point. Subproblems to minimize $\mathcal{E}_2(u)$ and $\mathcal{E}_5(\mathbf{n})$ need to solve a linear partial differential equation (PDE) and a coupled PDE with variable coefficients, respectively.

The most difficult and time-consuming process in the algorithm [16] is to solve the Euler-Lagrange equation of (11):

$$-2\nabla \left(b \left|\tilde{\mathbf{p}}^{l+1}\right| \nabla \cdot \mathbf{n}\right) + r_4 \left(\mathbf{n} - \tilde{\mathbf{m}}^{l+1}\right) + \boldsymbol{\lambda}_4^k = 0. \tag{12}$$

A frozen coefficient method with FFT is suggested to solve (12) and it needs an inner iteration. Since the equation is a coupled PDE, FFT is used twice per each inner iteration; see details in [16]. Since FFT is also used to solve the Euler-Lagrange equation of (8), the algorithm needs FFT more than three times for each outer iteration k. Therefore, if an algorithm uses FFT three times per outer iteration, it will be optimal. Such an optimality can be achieved as long as the variable coefficient $b|\mathbf{p}^{l+1}|$ in (12) is eliminated.

When the boundary condition is directly imposed without using the fidelity term $\frac{\eta}{q}\int_{\Gamma}|u-u_0|^q$ in (5) and the irregular inpainting domain is assigned, it is obvious that FFT cannot be used to solve the Euler-Lagrange equations of (8) and (11). Moreover, there are also some applications in [12] which we cannot use FFT in the augmented Lagrangian method [17]. In the case of u-subproblem (8), even though FFT is not used, it does not make any difficulties to find the minimizer because the Euler-Lagrange equation of (8) is a linear PDE:

$$-r_2\triangle u + r_3 u = r_3 \tilde{v}^{l+1} + \lambda_3^k - r_2 \nabla \cdot \tilde{\mathbf{p}}^l - \nabla \cdot \boldsymbol{\lambda}_2^k.$$

We can use linear iterative methods for symmetric positive definite matrix. However, it is not straightforward to solve the equation (12) in this manner. Considering some applications defined on a two dimensional surface in $\mathbf{R}^3$, it is necessary to develop an algorithm to efficiently minimize p-elastica model (4) without using FFT.

Note that one may use the gradient descent method to find a minimizer of (11). If the explicit method is applied, a small time step should be used because of large variation of $|\tilde{\mathbf{p}}^{l+1}|$ in the domain and then it makes a slow convergence. Since the coefficient $b|\tilde{\mathbf{p}}^{l+1}|$ in (12) varies in the domain and the equations are coupled, the implicit method is difficult to be applied.

3   Proposed Algorithms

In this section, we propose two algorithms. The first method is designed to eliminate the variable coefficients in (12) and then an optimal number of using FFT is achieved for each outer iteration. The second method replaces FFT procedure in [16] for minimizing (8) and (17) into a very simple updating scheme and it is memory efficient comparing with the first method. Both algorithms are as fast as the algorithm in Section 2.

3.1   Method 1

In order to extend the algorithm [16] to solve the p-elastica problem ($p=1$ or 2) and remove the variable coefficient in (12), we simply introduce a new variable

$$g = \nabla \cdot \mathbf{n}. \tag{13}$$

That is, we use the augmented Lagrangian functional for p-elastica problem with the additional positive penalty parameter r_5 and the Lagrange multiplier λ_5:

$$\mathcal{L}^1\left(v, u, \mathbf{m}, \mathbf{p}, g, \mathbf{n}; \lambda_1, \boldsymbol{\lambda}_2, \lambda_3, \boldsymbol{\lambda}_4, \lambda_5\right) = \int_\Omega \left(a + b|g|^p\right)|\mathbf{p}| + \frac{\eta}{q}\int_\Gamma |v - u_0|^q$$

$$+ r_1 \int_\Omega \left(|\mathbf{p}| - \mathbf{m}\cdot\mathbf{p}\right) + \int_\Omega \lambda_1(|\mathbf{p}| - \mathbf{m}\cdot\mathbf{p}) + \frac{r_2}{2}\int_\Omega |\mathbf{p} - \nabla u|^2 + \delta_{\mathcal{R}}(\mathbf{m})$$

$$+ \int_\Omega \boldsymbol{\lambda}_2 \cdot (\mathbf{p} - \nabla u) + \frac{r_3}{2}\int_\Omega (v - u)^2 + \int_\Omega \lambda_3(v - u) + \frac{r_4}{2}\int_\Omega |\mathbf{n} - \mathbf{m}|^2 \tag{14}$$

$$+ \int_\Omega \boldsymbol{\lambda}_4 \cdot (\mathbf{n} - \mathbf{m}) + \frac{r_5}{2}\int_\Omega (\nabla\cdot\mathbf{n} - g)^2 + \int_\Omega \lambda_5(\nabla\cdot\mathbf{n} - g).$$

We use the same iterative algorithm for (6) to find a stationary point of (14). After all variables and Lagrange multipliers are initialized to zero, an approximate minimizer for $k \geq 0$

$$\left(v^{k+1}, u^{k+1}, \mathbf{m}^{k+1}, \mathbf{p}^{k+1}, g^{k+1}, \mathbf{n}^{k+1}\right)$$

$$\simeq \underset{v,u,\mathbf{m},\mathbf{p},g,\mathbf{n}}{\arg\min}\ \mathcal{L}^1(v, u, \mathbf{m}, \mathbf{p}, g, \mathbf{n}; \lambda_1^k, \boldsymbol{\lambda}_2^k, \lambda_3^k, \boldsymbol{\lambda}_4^k, \lambda_5^k)$$

is obtained by alternatingly solving the subproblems. Letting $\tilde{v}^0 = v^k$, $\tilde{u}^0 = u^k$, $\tilde{\mathbf{m}}^0 = \mathbf{m}^k$, $\tilde{\mathbf{p}}^0 = \mathbf{p}^k$, $\tilde{g}^0 = g^k$, and $\tilde{\mathbf{n}}^0 = \mathbf{n}^k$, for $l = 0, \cdots, L-1$, we find minimizers $\tilde{v}^{l+1}$, $\tilde{u}^{l+1}$, $\tilde{\mathbf{m}}^{l+1}$, $\tilde{\mathbf{p}}^{l+1}$, $\tilde{g}^{l+1}$, and $\tilde{\mathbf{n}}^{l+1}$ of the following energy functionals:

$$\mathcal{E}_1^1(v) = \mathcal{E}_1(v), \quad \mathcal{E}_2^1(u) = \mathcal{E}_2(u), \quad \mathcal{E}_3^1(\mathbf{m}) = \mathcal{E}_3(\mathbf{m}),$$

$$\mathcal{E}_4^1(\mathbf{p}) = \int_\Omega \left(a + b\left|\tilde{g}^l\right|^p\right)|\mathbf{p}| + (r_1 + \lambda_1^k)\left(|\mathbf{p}| - \tilde{\mathbf{m}}^{l+1}\cdot\mathbf{p}\right)$$

$$+ \int_\Omega \frac{r_2}{2}\left|\mathbf{p} - \nabla\tilde{u}^{l+1}\right|^2 + \boldsymbol{\lambda}_2^k \cdot \mathbf{p}, \tag{15}$$

$$\mathcal{E}_5^1(g) = \int_\Omega b\left|\tilde{\mathbf{p}}^{l+1}\right||g|^p + \frac{r_5}{2}\left(\nabla\cdot\tilde{\mathbf{n}}^l - g\right)^2 + \lambda_5^k(-g), \tag{16}$$

$$\mathcal{E}_6^1(\mathbf{n}) = \int_\Omega \frac{r_4}{2}\left|\mathbf{n} - \tilde{\mathbf{m}}^{l+1}\right|^2 + \boldsymbol{\lambda}_4^k \cdot \mathbf{n} + \frac{r_5}{2}\left(\nabla\cdot\mathbf{n} - \tilde{g}^{l+1}\right)^2 + \lambda_5^k\nabla\cdot\mathbf{n}. \tag{17}$$

After L iterations, we update

$$\left(v^{k+1}, u^{k+1}, \mathbf{m}^{k+1}, \mathbf{p}^{k+1}, g^{k+1}, \mathbf{n}^{k+1}\right) = \left(\tilde{v}^L, \tilde{u}^L, \tilde{\mathbf{m}}^L, \tilde{\mathbf{p}}^L, \tilde{g}^L, \tilde{\mathbf{n}}^L\right).$$

In practice, $L = 1$ is used to make a consistent and fair comparison with the algorithm in Section 2. For p and $q = 1$ or 2, there are closed form formulas to find minimizers of $\mathcal{E}_1^1(v)$, $\mathcal{E}_3^1(\mathbf{m})$, $\mathcal{E}_4^1(\mathbf{p})$, and $\mathcal{E}_5^1(g)$ and it takes simply arithmetic operations at each grid point. We use FFT to solve the Euler's Lagrange equation of $\mathcal{E}_2^1(u)$.

Now, the Euler-Lagrange equation of $\mathcal{E}_6^1(\mathbf{n})$ is a linear coupled PDE:

$$-r_5\nabla\left(\nabla\cdot\mathbf{n}\right) + r_4\mathbf{n} = r_4\tilde{\mathbf{m}}^{l+1} - \boldsymbol{\lambda}_4^k - r_5\nabla\tilde{g}^{l+1} + \nabla\lambda_5^k. \tag{18}$$

The variable coefficient in (12) is removed and solution $\tilde{\mathbf{n}}^{l+1}$ of PDE are unique with a suitable boundary condition and the positive penalty parameter r_4. In the discrete frequency domain, the coupled PDE (18) yields a 2 by 2 system of equation for each frequency and the determinant of the coefficient matrix is not zero if $r_4 > 0$. Note that the operator $\nabla(\nabla\cdot)$ is singular and then the coupled PDE becomes unstable if $r_4 \simeq 0$. Unlike the iterative method for solving (12) in [16], the coupled PDE (18) can be directly solved with just two FFT algorithms to obtain a minimizer of $\mathcal{E}_6^1(\mathbf{n})$.

One may use the gradient descent method to find the minimizer of $\mathcal{E}_6^1(\mathbf{n})$ (17). However, it introduces another variable for time step which should be properly chosen depending on r_4 and r_5.

In numerical examples of Method 1, we use FFT for subproblems $\mathcal{E}_2^1(u)$ and $\mathcal{E}_6^1(\mathbf{n})$ to make a fair comparison with the algorithm in Section 2.

3.2 Method 2

In this subsection, we propose an algorithm which is more effective than previous algorithms in terms of using memory and is as fast as the Method 1 in subsection 3.1 and the algorithm in Section 2. Note that the algorithms for (6) and (14) need to use 14 and 16 arrays, respectively, which have the same size as a given image u_0. The easiest technique to reduce memory usage is to eliminate unnecessary variables. Even though all variables in (14) play an important role in separating nonlinear properties and dissolving higher order derivatives in p-elastica model (4), the variable $\mathbf{m}$ may not be very crucial because $|\mathbf{n}| \leq 1$ can be achieved by a brute force method. Therefore, we simply propose the following augmented Lagrangian functional for p-elastica problem:

$$
\begin{aligned}
\mathcal{L}^2\left(v, u, \mathbf{p}, g, \mathbf{n}; \mu_1, \boldsymbol{\mu}_2, \mu_3, \mu_4\right) &= \int_\Omega (a + b|g|^p)\,|\mathbf{p}| + \frac{\eta}{q}\int_\Gamma |v - u_0|^q \\
&+ c_1\int_\Omega (|\mathbf{p}| - \mathbf{n}\cdot\mathbf{p}) + \int_\Omega \mu_1(|\mathbf{p}| - \mathbf{n}\cdot\mathbf{p}) + \frac{c_2}{2}\int_\Omega |\mathbf{p} - \nabla u|^2 \\
&+ \int_\Omega \boldsymbol{\mu}_2\cdot(\mathbf{p} - \nabla u) + \frac{c_3}{2}\int_\Omega (v - u)^2 + \int_\Omega \mu_3(v - u) \\
&+ \frac{c_4}{2}\int_\Omega (\nabla\cdot\mathbf{n} - g)^2 + \int_\Omega \mu_4(\nabla\cdot\mathbf{n} - g), \quad \text{with} \quad |\mathbf{n}| \leq 1,
\end{aligned}
\tag{19}
$$

where μ_1, $\boldsymbol{\mu}_2$, μ_3, and μ_4 are Lagrange multipliers, c_1, c_2, c_3, and c_4 are positive penalty parameters. Note that the algorithm for $\mathcal{L}^2$ needs 12 arrays which have the same size as a given image u_0.

We use the same iterative algorithm for (6) to find a stationary point of (19). After all variables and Lagrange multipliers are initialized to zero, for $k \geq 0$, an approximate minimizer

$$
\left(v^{k+1}, u^{k+1}, \mathbf{p}^{k+1}, g^{k+1}, \mathbf{n}^{k+1}\right) \simeq \underset{v,u,\mathbf{m},\mathbf{p},g,\mathbf{n}}{\arg\min}\ \mathcal{L}^2(v, u, \mathbf{m}, \mathbf{p}, g, \mathbf{n}; \mu_1^k, \boldsymbol{\mu}_2^k, \mu_3^k, \mu_4^k)
$$

is obtained by alternatingly solving the subproblems. Letting $\tilde{v}^0 = v^k$, $\tilde{u}^0 = u^k$, $\tilde{\mathbf{p}}^0 = \mathbf{p}^k$, $\tilde{g}^0 = g^k$, and $\tilde{\mathbf{n}}^0 = \mathbf{n}^k$, for $l = 0, \cdots, L-1$, we find minimizers $\tilde{v}^{l+1}$, $\tilde{u}^{l+1}$, $\tilde{\mathbf{p}}^{l+1}$, $\tilde{g}^{l+1}$, and $\tilde{\mathbf{n}}^{l+1}$ of the following energy functionals:

$$\mathcal{E}_1^2(v) = \frac{\eta}{q} \int_\Gamma |v - u_0|^q + \int_\Omega \frac{c_3}{2} \left(v - \tilde{u}^l\right)^2 + \mu_3^k v, \tag{20}$$

$$\mathcal{E}_2^2(u) = \int_\Omega \frac{c_2}{2} \left|\tilde{\mathbf{p}}^l - \nabla u\right|^2 - \boldsymbol{\mu}_2^k \cdot \nabla u + \frac{c_3}{2} \left(\tilde{v}^{l+1} - u\right)^2 + \mu_3^k(-u), \tag{21}$$

$$\mathcal{E}_3^2(\mathbf{p}) = \int_\Omega \left(a + b\left|\tilde{g}^l\right|^p\right) |\mathbf{p}| + (c_1 + \mu_1^k) \left(|\mathbf{p}| - \tilde{\mathbf{n}}^{l+1} \cdot \mathbf{p}\right),$$

$$+ \int_\Omega \frac{c_2}{2} \left|\mathbf{p} - \nabla \tilde{u}^{l+1}\right|^2 + \boldsymbol{\mu}_2^k \cdot \mathbf{p}, \tag{22}$$

$$\mathcal{E}_4^2(g) = \int_\Omega b\left|\tilde{\mathbf{p}}^{l+1}\right| |g|^p + \frac{c_4}{2} \left(\nabla \cdot \tilde{\mathbf{n}}^l - g\right)^2 + \mu_4^k(-g), \tag{23}$$

$$\mathcal{E}_5^2(\mathbf{n}) = \int_\Omega \frac{c_4}{2} \left(\nabla \cdot \mathbf{n} - \tilde{g}^{l+1}\right)^2 + \mu_4^k \nabla \cdot \mathbf{n} - (c_1 + \mu_1) \mathbf{n} \cdot \tilde{\mathbf{p}}^{l+1}. \tag{24}$$

After L iterations, we update

$$\left(v^{k+1}, u^{k+1}, \mathbf{p}^{k+1}, g^{k+1}, \mathbf{n}^{k+1}\right) = \left(\tilde{v}^L, \tilde{u}^L, \tilde{\mathbf{p}}^L, \tilde{g}^L, \tilde{\mathbf{n}}^L\right).$$

Similar to the Method 1, we observe that $L > 1$ does not make quite different numerical results from using $L = 1$.

Now, one may easily notice that we have a huge problem in the $\mathbf{n}$-subproblem for minimizing the functional (24) whose the Euler-Lagrange equation is

$$-c_4 \nabla \left(\nabla \cdot \mathbf{n} - \tilde{g}^{l+1}\right) - (c_1 + \mu_1) \tilde{\mathbf{p}}^{l+1} - \nabla \mu_4^k = 0. \tag{25}$$

Obviously, the solution of PDE is not unique because of the operator $\nabla (\nabla \cdot)$. Comparing with (12) and (18), such a problem is caused by the lack of linear term in the coupled PDE (25), which has been provided by the new variable $\mathbf{m}$ in the augmented Lagrangian functionals in (6) and (14). However, we simply generate a linear term by the linearization of L^2 penalization for $g = \nabla \cdot \mathbf{n}$, inspired by the linearized proximal alternating minimization algorithm in [20]. That is, the energy functional (24) can be approximated by linearization of $(\nabla \cdot \mathbf{n} - g)^2$ at $\tilde{\mathbf{n}}^l$:

$$\mathcal{E}_5^2(\mathbf{n}) = \int_\Omega \mu_4^k \nabla \cdot \mathbf{n} - (c_1 + \mu_1) \mathbf{n} \cdot \tilde{\mathbf{p}}^{l+1} + \frac{c_4}{2} \left(\nabla \cdot \mathbf{n} - \tilde{g}^{l+1}\right)^2$$

$$\simeq \int_\Omega \mu_4^k \nabla \cdot \mathbf{n} - (c_1 + \mu_1) \mathbf{n} \cdot \tilde{\mathbf{p}}^{l+1}$$

$$+ \int_\Omega \frac{c_4}{2} \left(\left(\nabla \cdot \tilde{\mathbf{n}}^l - \tilde{g}^{l+1}\right)^2 - 2\nabla \left(\nabla \cdot \tilde{\mathbf{n}}^l - \tilde{g}^{l+1}\right) \cdot \left(\mathbf{n} - \tilde{\mathbf{n}}^l\right) + \delta \left|\mathbf{n} - \tilde{\mathbf{n}}^l\right|^2\right),$$

where δ is a constant. Therefore, we approximately obtain a minimizer $\tilde{\mathbf{n}}^{l+1}$ of $\mathcal{E}_5^2(\mathbf{n})$ by a cheap arithmetic operation at each grid point:

$$\tilde{\mathbf{n}}^{l+1} = \tilde{\mathbf{n}}^l + \frac{1}{c_4 \delta} \left((c_1 + \mu_1) \tilde{\mathbf{p}}^{l+1} + c_4 \nabla \left(\nabla \cdot \tilde{\mathbf{n}}^l - \tilde{g}^{l+1}\right) + \nabla \mu_4\right). \tag{26}$$

Table 1. Computational costs are presented for Fig. 1: number of outer iteration / computational time (sec)

	Fig. 1-(a)	Fig. 1-(b)	Fig. 1-(c)
size	80×80	100×100	300×235
Algorithm in [16]	449/6.57	307/7.75	329/80.64
Method 1	177/2.23	323/6.90	430/79.97
Method 2	187/1.31	445/4.95	383/32.38

More interestingly, the closed form formula (26) is also obtained in a different way. The linear term can be added by an explicit time discretization of the gradient decent method. That is, if we use the gradient descent method for approximately finding a minimizer of (24), we have

$$\frac{\partial \mathbf{n}}{\partial \tau} = (c_1 + \mu_1)\, \tilde{\mathbf{p}}^{l+1} + c_4 \nabla \left(\nabla \cdot \mathbf{n} - \tilde{g}^{l+1} \right) + \nabla \mu_4. \tag{27}$$

Then, explicit Euler scheme gives the same formula as (26) with the time step $\tau = \frac{1}{c_4\delta}$.

For p and $q = 1$ or 2, there are closed form formulas to find minimizers of $\mathcal{E}_1^2(v)$, $\mathcal{E}_4^1(\mathbf{p})$, and $\mathcal{E}_5^1(g)$ and it takes simply arithmetic operations at each grid point. In the proposed Method 2, we use the GS method for u-subproblem of minimizing $\mathcal{E}_2^2(u)$. Considering a method for low computational cost, one sweep of GS iteration is practically enough for approximately solving the equation. For the $\mathbf{n}$-subproblem $\mathcal{E}_5^2(\mathbf{n})$ (24), we use a simple and cost effective formula (26).

4   Numerical Results

We demonstrate numerical examples using the proposed algorithms in image inpainting and curve reconstruction from unorganized points set. We use the staggered grid system to obtain finite difference discretization of our models; see more details in [16]. The test system is a Intel(R) Core(TM) i7 CPU Q720 1.6GHz with 4GB RAM.

4.1   Image Inpainting

Numerical tests on image inpainting are provided to demonstrate the efficiency of the proposed algorithms. In Fig. 1, we choose the same examples shown in [16]. The inpainting results from Method 1 and Method 2 numerically show that the curvature term works to connect the level curves of image on a large inpainting domain.

In Table 1, the improved computational speed is demonstrated. In order to show efficiency of our algorithms, we use smaller (or same) value of stoping criterion (relative residuals) than [16]. Even though the number of outer iteration is

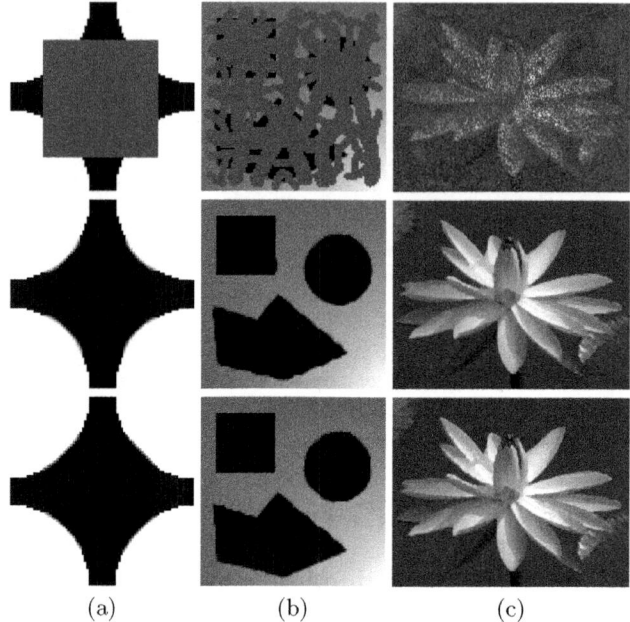

(a)   (b)   (c)

Fig. 1. The red regions in the first row indicate the inpainting domain. The images in the second and third row are image inpainting results from Method 1 and Method 2, respectively. For all results obtained by Method 1, we use $a = 1$ and $\eta = 10^3$. The remaining parameters are $b = 50$, $r_1 = 1$, $r_2 = r_3 = 20$, $r_4 = 10^2$, $r_5 = 1$ in (a), $b = 50$, $r_1 = 1$, $r_2 = 10^2$, $r_3 = 50$, $r_4 = 5 \cdot 10^2$, $r_5 = 10$ in (b), and $b = 30$, $r_1 = 2$, $r_2 = 6 \cdot 10^2$, $r_3 = 10^2$, $r_4 = 10^3$, $r_5 = 10$ in (c). For all results obtained by Method 2, we use $a = 1$, $\eta = 10^3$, and $\delta = 1$. The remaining parameters are $b = 10$, $c_1 = 1$, $c_2 = 5$, $c_3 = 10$, $c_4 = 10^2$ in (a), $b = 50$, $c_1 = 1$, $c_2 = 50$, $c_3 = 10$, and $c_4 = 10^3$ in (b), and $b = 30$, $c_1 = 2$, $c_2 = 4 \cdot 10^2$, $c_3 = 10^2$, and $c_4 = 10^3$ in (c).

larger than the algorithm in [16], the computational time in Method 2 is reduced because we use a very cheap arithmetic operation at each grid point. Method 1 usually may have a similar computational cost to the algorithm in [16] because the number of inner iteration in the frozen coefficient method for solving (12) is empirically less than 5 in an early stage of outer iteration. Moreover, the number of inner iteration tends to be reduced as long as the outer iteration is increased. Since our results are obtained by smaller (or same) relative residual error bound than [16] and they are converged faster than the previous method in Section 2, the proposed algorithms improve the computational cost.

In Fig. 2, we also show the graphs (log scales on xy-axis) of residuals, relative errors in Lagrange multipliers, relative error in u, and energy for Method 2 of the example in Fig. 1-(a); see more details in [16]. The profile of graphs are very similar to the results from Method 1 and [16]. The proposed algorithms are numerically verified that they are practically faster than the previous algorithm in [16].

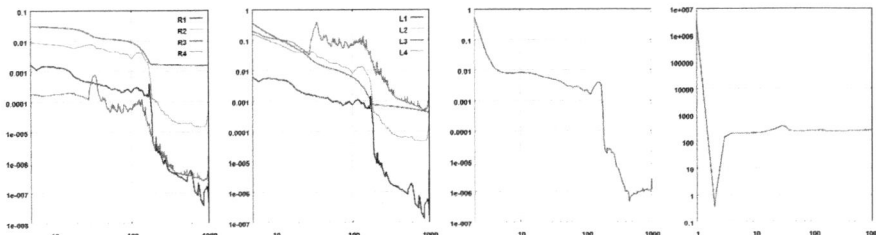

Fig. 2. From the left, the log scale plots of residuals, relative errors in Lagrange multipliers, relative error in u, and energy on y-axis versus iteration on x-axis for Method 2 of the example in Fig. 1-(a). Note that graphs from Method 1 have almost same profiles.

4.2   Curve Reconstruction

To address a reconstruction problem, the following model

$$\int_{\Omega} \left(a\psi + b \left| \nabla \cdot \frac{\nabla u}{|\nabla u|} \right| \right) |\nabla u| + \frac{1}{2} \int_{\Gamma} \eta u((c_1 - u_0)^2 - (c_2 - u_0)^2) \quad \text{s.t.} \quad 0 \le u \le 1$$

is minimized by the proposed methods, where c_1 and c_2 are positive constants and ψ is the unsigned distance function induced from the unorganized points set. As in [18], u_0 is an initial guess obtained with region-growing methods, and the fidelity parameter is a function rather than a constant, which suggests that to what extent the initial guess is faithful. Specifically, we use

$$\eta(x) = \begin{cases} c_\eta & \psi(x) > 5h, \\ 0 & \psi(x) \le 5h, \end{cases}$$

for each point x in the domain Ω, where h is the mesh size and c_η is a constant. For simplicity, we use the domain $\Omega = [-1, 1] \times [-1, 1]$ and we update c_1 and c_2 for every 100 iterations in our implementation as [11]:

$$c_1 - \frac{1}{\mathcal{A}(\mathcal{R})} \int_{\mathcal{R}} u \quad \text{and} \quad c_2 = \frac{1}{\mathcal{A}(\Omega \setminus \mathcal{R})} \int_{\Omega \setminus \mathcal{R}} u,$$

where $\mathcal{R} \equiv \{x \in \Omega : u(x) \ge 0.5\}$ and $\mathcal{A}(\cdot)$ measures the area of a set.

To impose the constraint on u, a projection operator in [11] is carried on a new variable v in the augmented Lagrangian formulations in Sections 2 and 3. For example, in Method 1, the v-subproblem is solved as follows:

$$\begin{cases} \tilde{v} = \arg\min_{\tilde{v}} \tilde{\mathcal{E}}_1(v) = \arg\min_{\tilde{v}} \int_{\Omega} \frac{\eta}{2} \tilde{v} f + \int_{\Omega} \frac{r_3}{2} \left(\tilde{v} - \tilde{u}^l \right)^2 + \lambda_3^k \tilde{v}, \\ v = \max\{\min\{\tilde{v}, 0\}, 1\}. \end{cases}$$

Since u converges to v, the constraint is therefore imposed on u correspondingly. The same projection operation is used in Method 2 as well. In all experiments, it is observed that u converges to a function between 0 and 1. Figure 3 presents a reconstruction example.

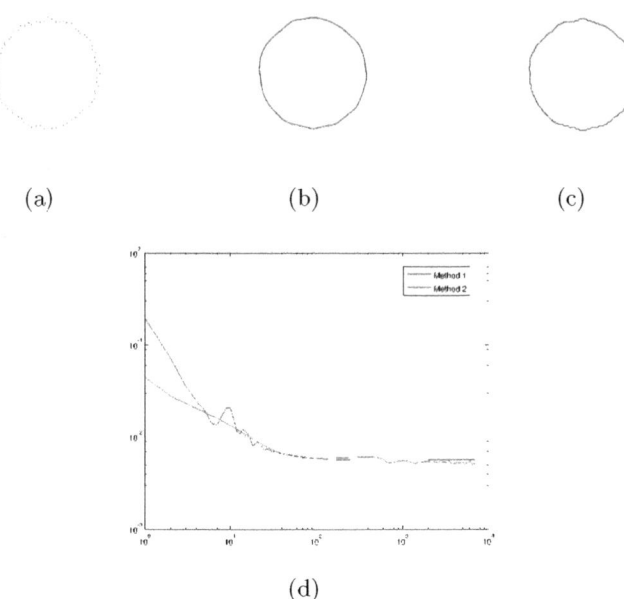

(a)   (b)   (c)

(d)

Fig. 3. (a) Noisy points set sampled from a circle; (b) The result produced by Method 1; (c) The result produced by Method 2; (d) Error (28) vs. iteration times (loglog). In this example, we use $a = 1$, $b = 10^4$, $t = 2 \cdot 10^4$, $r_1 = 1$, $r_2 = 0.1$, $r_3 = 2 \cdot 10^2$, $r_4 = 2$, and $r_5 = 10^2$ in Method 1. The parameters for Method 2 are $a = 0.5$, $b = 10^4$, $t = 2 \cdot 10^4$, $c_1 = 1$, $c_2 = 0.1$, $c_3 = 1.5 \cdot 10^2$, $c_4 = 30$, and $\delta = 1$.

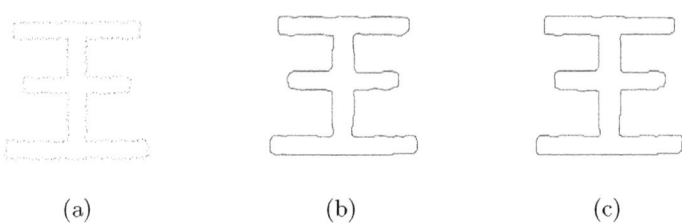

(a)   (b)   (c)

Fig. 4. Results from different b in Method 2: (a) Noisy point set; (b) Reconstructed curve with $b = 1$; (c)Reconstructed curve with $b = 10$. The remaining parameters are selected as $a = 1$, $t = 2 \cdot 10^4$, $c_1 = 1$, $c_2 = 0.1$, $c_3 = 2 \cdot 10^2$, $c_4 = 2$, and $\delta = 1$.

Fig. 3-(a) shows the points set with 5% noise sampled from a unit circle. From the noisy points set, Fig. 3-(b) illustrates the reconstructed circle produced by Method 1 and Fig. 3-(c) is the result with Method 2. It can be seen that Method 1 yields smoother result because it solves (18) completely. Fig. 3(d) gives the error against iteration times. Here, the error is the measurement of comparison between the reconstructed curves and the exact unit circle as follows:

$$E_u = \frac{|\mathbf{1}_{exact} - \mathbf{1}_u|_{L^1}}{|\mathbf{1}_{exact}|_{L^1}}, \tag{28}$$

where $\mathbf{1}_u$ and $\mathbf{1}_{exact}$ are the indicator functions, which take value 1 or 0 for each point in the domain. $\mathbf{1}_{exact}$ is an indicator function of the circle centered at origin and with radius 0.3

$$\mathbf{1}_{exact}(x) = \begin{cases} 1 & \|x\| \leq 0.3, \\ 0 & \|x\| > 0.3, \end{cases}$$

where $|x|$ is the Euclidean length of the point x and

$$\mathbf{1}_u(x) = \begin{cases} 1 & u(x) > 0.5, \\ 0 & u(x) \leq 0.5. \end{cases}$$

Although the convergent rates of Methods 1 and 2 are almost similar, Method 2 is much faster than Method 1 for each iteration.

Fig. 4 gives another reconstruction example by Method 2. Fig. 4-(a) shows noisy points set sampled from a chinese character. Figs. 4-(b) and 2-(c) illustrate reconstructed curves with different parameters. We numerically observe that a better result is obtained by increasing parameter b.

5   Conclusion

We proposed two algorithms to efficiently solve the p-elastica model in image inpainting and curve reconstruction from unorganized point set. Inspired by the recent work [16], we used augmented Lagrangian method and extend the algorithm in [16]. The first algorithm eliminates an inner iterative steps in [16] and the second algorithm replaces FFT into a very cheap arithmetic operation. From the numerical results, the efficiency of the algorithms are demonstrated. In the future, we would like to extend the model in subsection 4.2 into 3D to reconstruct a surface which minimizes its mean curvature. Moreover, the Euler's elastica model on the surface is a possible extension of using the proposed Method 2.

References

1. Ambrosio, L., Masnou, S.: A direct variational approach to a problem arising in image reconstruction. Interfaces Free Bound 5(1), 63–81 (2003)
2. Ardentov, A., Sachkov, Y.L.: Solution to Euler's elastica problem. Automation and Remote Control 70, 633–643 (2009)
3. Bae, E., Shi, J., Tai, X.C.: Graph cuts for curvature based image. IEEE Trans. Image Process (to appear)

4. Ballester, C., Bertalmio, M., Caselles, V., Sapiro, G., Verdera, J.: Filling-in by joint interpolation of vector fields and gray levels. IEEE Trans. Image Processing 10(8), 1200–1211 (2001)
5. Ballester, C., Caselles, V., Verdera, J.: Disocclusion by joint interpolation of vector fields and gray levels. Multiscale Model. Simul. 2(1), 80–123 (2003)
6. Bellettini, G., Dal Maso, G., Paolini, M.: Semicontinuity and relaxation properties of a curvature depending functional in 2d. Ann. Scuola Norm. Sup. Pisa Cl. Sci (4) 20(2), 247–297 (1993)
7. Bruckstein, A., Netravali, A., Richardson, T.: Epi-convergence of discrete elastica. Appl. Anal. 79, 137–171 (2001)
8. Chan, T.F., Kang, S.H., Shen, J.: Euler's elastica and curvature based inpaintings. SIAM J. Appl. Math. 63(2), 564–594 (2002)
9. Chan, T.F., Shen, J.: Nontexture inpainting by curvature driven diffusion (CDD). J. Visul Comm. Image Rep. 12, 436–449 (2001)
10. El-Zehiry, N., Grady, L.: Fast global optimization of curvature. In: IEEE Conference on CVPR, pp. 3257–3264 (2010)
11. Goldstein, T., Bresson, X., Osher, S.: Geometric applications of the split bregman method: Segmentation and surface reconstruction. Journal of Scientific Computing 45, 272–293 (2010)
12. Lai, R., Chan, T.F.: A framework for intrinsic image processing on surfaces. Tech. rep., UCLA CAM Report 10-25 (2010)
13. Masnou, S., Morel, J.M.: Level lines based disocclusion. In: Proc. IEEE Int. Conf. on Image Processing, Chicago, IL, pp. 259–263 (1998)
14. Nitzberg, M., Mumford, D., Shiota, T.: Filtering, Segmentation and Depth, vol. 662. Springer, Berlin (1993)
15. Schoenemann, T., Kahl, F., Cremers, D.: Curvature regularity for region-based image segmentation and inpainting: A linear programming relaxation. In: IEEE Int. Conf. Comp, Vision, Kyoto, Japan (2009)
16. Tai, X.C., Hahn, J., Chung, G.J.: A fast algorithm for Euler's elastica model using augmented Lagrangian method. SIAM J. Img. Sci. 4, 313–344
17. Tai, X.-C., Wu, C.: Augmented Lagrangian method, dual methods and split Bregman iteration for ROF model. In: Tai, X.-C., Mørken, K., Lysaker, M., Lie, K.-A. (eds.) SSVM 2009. LNCS, vol. 5567, pp. 502–513. Springer, Heidelberg (2009)
18. Wan, M., Wang, Y., Wang, D.: Variational surface reconstruction based on delaunay triangulation and graph cut. International Journal for Numerical Methods in Engineering 85, 206–229 (2011)
19. Wu, C., Tai, X.C.: Augmented Lagrangian method, dual methods, and split Bregman iteration for ROF, vectorial TV, and high order models. SIAM J. Imaging Sci. 3(3), 300–339 (2010)
20. Yun, S., Woo, H.: Linearized proximal alternating minimization algorithm for motion deblurring by nonlocal regularization. In: Pattern Recognition (to appear in, 2011)

The Beltrami-Mumford-Shah Functional

Nir Sochen and Leah Bar

Department of Mathematics,
Tel-Aviv University
Tel-Aviv, 69978
Israel

Abstract. We present in this paper a unifying generalization of the Mumford-Shah functional, in the Ambrosio-Totorelli set up, and the Beltrami framework. The generalization of the Ambrosio-Tortorelli is in using a diffusion tensor as an indicator of the edge set instead of a function. The generalization of the Beltrami framework is in adding a penalty term on the metric such that it is defined dynamically from minimization of the functional.

We show that we are able, in this way, to have the benefits of true anisotropic diffusion together with a dynamically tuned metric/diffusion tensor. The functional is naturally defined in terms of the vielbein-the metric's square root. Preliminary results show improvement on both the Beltrami flow and the Mumford-Shah flow.

Keywords: Inhomogeneous diffusion, anisotropic diffusion, Mumford-Shah functional, Ambrosio-Tortorelli functional, Beltrami framework.

1 Introduction

The seminal work of Mumford and Shah[6] was a breakthrough, both conceptually and technically. From conceptual standpoint it reveled the need to unify the de-noising and edge detection problems into one problem where the two tasks are simultaneously solved. Technically it introduced a continuous functional to give the conceptual understanding a mathematical language and used calculus of variations to derive partial differential equations for the solution. It was interpreted later, via the relation to statistical inference ideas, as the prior on images that favors piecewise smooth functions over other possible functions.

In the early nineties the concept of inhomogeneous diffusion, coined "anisotropic diffusion" by Perona and Malik [7], gained popularity. With this method (which was discovered earlier in an independent manner in mathematical physics by Rosenau [8]) it was possible to construct a controlled non-linear filtering that reduces noise on one hand and conserves the sharpness of the image on the other. The relation between "anisotropic diffusion" and the variational approach became clearer after the "Total Variation" (TV) functional was introduced by Rudin, Osher and Fatemi [9] and was later generalized by Faugeras and Deriche in the Φ-formalism [5]. In these inhomogeneous diffusion methods

A.M. Bruckstein et al. (Eds.): SSVM 2011, LNCS 6667, pp. 183–193, 2012.

the form of the local diffusion coefficient is predefined in advance. Usually the diffusion coefficient is given as a known function of the amplitude of the local gradient. One can show that, similarly to the Mumford-Shah approach, all these methods treat images as functions and impose piecewise smoothness as a prior.

Inhomogeneous diffusion was linked to the Mumford-Shah functional by the seminal work of Ambrosio and Tortorelli [1]. In that approach the set of discontinuities in the image is represented by an auxiliary, soft indicator, function. This function serves as a local diffusion coefficient in the Euler-Lagrange or gradient descent equations for the image. The difference from other inhomogeneous diffusion methods is in the fact that this diffusion coefficient is determined *dynamically* by the minimization of the functional. This dynamic choice of discontinuities position and magnitude enhances the performance of de-noising/de-blurring algorithms [3,4].

Towards the end of the nineties another distinction, and consequently, an advancement was achieved. Weickert in the "Coherence-Enhancing diffusion" (CD) [11] and Sochen et al. in the "Beltrami flow" (BF) [10] introduced true anisotropic diffusion where the local diffusion function was replaced with a full rank diffusion tensor. The latter was linked in the Beltrami framework to a Riemannian metric. In this approach the image is not a function any more but a Riemannian manifold and the anisotropic diffusion is a consequence of diffusion of the image on an image-induced non-flat manifold. Both in the CD and in the BF the diffusion tensor's form is given in advance either as a variant of the structure tensor in CD or as the induced metric in the BF.

It is the aim of this paper to generalize the Beltrami framework and the Mumford-Shah approach by extending the respective functionals to a unifying one. The starting point is the seminal work of Ambrosio and Tortorelli [1]. We extend their approach that treat the image and it set of discontinuities as two different *dynamical* variables that should be optimized by the same functional. We present in this work a functional where the (color) image and the diffusion tensor are treated both as dynamical variables. We interpret the diffusion tensor as a metric and end up with an extension of the Beltrami framework.

The paper is organized as follows: We review inhomogeneous diffusion methods and its derivation as a minimization of a functional in Section 1. The Mumford-Shah functional and the Ambrosio-Tortorelli approach are presented in Section 2. In that section we will also point out to the relation of the minimization of the Ambrosio-Tortorelli's functional to inhomogeneous diffusion. Section 3 presents anisotropic diffusion via the Coherent diffusion and the Beltrami frameworks. Generalizing the anisotropic diffusion in an "Ambrosio-Tortorelli like" functional is presented In section 4. The generalization of the Mumford-Shah functional is find to generalize the Polyakov action of the Beltrami framework at the same time. preliminary results are shown in Section 5 and we summarize and conclude in Section 6.

2   Inhomogeneous Diffusion

2.1   Isotropic Diffusion

Inhomogeneous diffusion started with the work of Perona and Malik [7]. In order to better situated this formalism we first discuss isotropic diffusion. In the isotropic case the filtering of the image is done via the solution of the isotropic diffusion equation

$$u_t = c\Delta u = c\,\mathrm{div}(\nabla u) = \mathrm{div}(c\nabla u)$$
$$u(t = 0) = u_0 \quad .$$

where Δu is the Laplacian of u, div is the divergent, ∇u is the gradient and c is a constant. This equation is called in image processing context linear scale-space because of its relation to convolution with a Gaussian with time dependent variance. The relation to linear filtering is done via the Green function (kernel) of this Partial Differential Equation (PDE):

$$u(x, y, t) = \int G(x - x', y - y';\ t)u_0(x', y')dx'dy'$$

$$G(x, y;\ t) = \frac{1}{4\pi t}e^{-\frac{x^2+y^2}{4t}} \quad .$$

The relation of isotropic diffusion to the calculus of variations is given by the functional

$$S[u] = \int ||\nabla u(x, y)||_2^2 dx dy$$

and its gradient descent

$$u_t = \frac{\partial u}{\partial t} = -\frac{\delta S}{\delta u}$$

2.2   Inhomogeneous Diffusion

The idea of Perona and Malik was to use isotropic-like filtering far from the edges of the image and to reduce the smoothing near the edges in order to preserve the sharpness of the image. This was achieved in the PDE formulation via a local diffusion function

$$u_t = \mathrm{div}(c(x, y)\nabla u)$$
$$u(t = 0) = u_0 \quad .$$

The function $c(x, y)$ is the local diffusion coefficient and is usually taken as a monotonically decreasing function of $||\nabla u||_2$.

   The relation to minimization of a functional was nicely formulated by the Φ-formalism of Deriche and Faugeras [5]. The functional they proposed is

$$S_\Phi[u] = \int \Phi(||\nabla u(x, y)||_2)\,dx dy \quad ,$$

and the gradient descent equation is

$$u_t = div\left(\frac{\Phi'(||\nabla u(x,y)||_2)}{||\nabla u(x,y)||_2}\nabla u\right)$$

$$u(t=0) = u_0 \quad .$$

The relation to Perona-Malik is given by $c(s) = \Phi'(s)/s$.

2.3   TV and MAP

Another approach that links functional minimization and inhomogeneous diffusion is Total Variation. In the original paper the functional is given by

$$S_{TV}[u] = \int\left[\frac{1}{2}\left(h*u(x,y) - u_0(x,y)\right)^2 + \lambda||\nabla u(x,y)||_2\right]dxdy \quad .$$

where h is a blur kernel and $*$ denotes convolution. The first term is called fidelity term and the second term is referred to as the smoothing term. The smoothing term is of the form of the Φ-formalism with $\Phi(s) = s$. The gradient descent equation reads

$$u_t = \lambda div\left(\frac{1}{||\nabla u(x,y)||_2}\nabla u\right) - \bar{h}*(h*u(x,y) - u_0(x,y))$$

$$u(t=0) = u_0 \quad , \quad \bar{h} = h(-x,-y) \quad .$$

This equation is related to the Maximum A-posteriori Probability (MAP) method of statistical inference. Indeed by the Bayes rule the conditional probability of u given u_0 denoted by $P(u|u_0)$ is given by $P(u|u_0,h) \propto P(u_0|u,h)P(u)$ where $P(u)$ is the prior on the space of images. The relation to TV is given by

$$P(u_0|u,h) \propto \exp\{-\frac{1}{2}\int\left(h*u(x,y) - u_0(x,y)\right)^2 dxdy\}$$

$$P(u) \propto \exp\{-\lambda\int ||\nabla u(x,y)||_2 dxdy\}$$

The a-posteriori probability function is proportional to $\exp\{-S_{TV}\}$ and the MAP approximation is given by

$$\hat{u} = \arg\max_u P(u|u_0,h) = \arg\min_u S_{TV}[u] \quad .$$

We will refer to the relations between PDEs, functionals, filters and statistical inference in all the following analysis.

3   The Mumford-Shah Functional

The Mumford-Shah functional aims to simultaneously solve the problems of denoising and edge detection. For this end the functional is formulated as

$$S_{MS}[u,K] = \frac{1}{2}\int_\Omega (u-u_0)^2\,dxdy + \lambda\int_{\Omega/K}||\nabla u||_2^2 dxdy + \alpha\,(\text{length of } K) \quad ,$$

where the first term is the fidelity term. The second term dictates smoothing far from the set K of image discontinuities. The set K is assumed to be a set of continues curves and the last term is a penalty on the total length of these curves. The minimization and analysis of this functional are not simple since the set of discontinuities intervenes in the boundary of the integration. This is a free boundaries problem which is notoriously difficult. One of the best ways to deal with this problem is via the Γ-convergence technique, which was proposed by Ambrosio and Tortorelli.

3.1   The Ambrosio-Tortorelli Functional

In this approach one constructs a series (or a one-parametric family) of functionals that converge to a functional such that the limit of the series minimizers converges to the minimizer of the limit functional. The functionals in the series are easier to analyze. Ambrosio and Tortorelli suggested the following family of ϵ dependent functionals

$$S_{AT}^{\epsilon}[u, v] = \int_{\Omega} \left[\frac{1}{2} \left(u - u_0 \right)^2 + \frac{\lambda}{2} v^2 ||\nabla u||_2^2 + \alpha \left(\epsilon ||\nabla v||_2^2 + \frac{(v-1)^2}{4\epsilon} \right) \right] dx dy$$

which Γ-converges to the Mumford-Shah functional when $\epsilon \to 0$. Here v is an auxiliary function that encodes the images discontinuity set: It approaches one in smooth regions and approach zero near an edge. The gradient descent for the image u leads to an inhomogeneous diffusion

$$u_t = \lambda \mathrm{div} \left(v^2 \nabla u \right) - \left(u - u_0 \right)$$
$$u(t = 0) = u_0 \quad .$$

where the edge function v^2 plays the role of local diffusion coefficient.

The great difference from the TV and the Φ-formalism is in the way this diffusion coefficient is determined. In the latter methods the form of the diffusion coefficient is predefined. Here, in the Ambrosio-Tortorelli approach, this coefficient is a *dynamical variable* that is found by minimizing of the functional! The gradient descent equations read

$$v_t = \alpha \epsilon \Delta v - \left(\lambda ||\nabla u||_2^2 v + \frac{\alpha(v-1)}{2\epsilon} \right)$$
$$v(t = 0) = 1 \quad .$$

The advantage of dynamic determination of the edge set or equivalently the diffusion coefficient was shown in [3] for the case of de-blurring. The Ambrosio Tortorelli functional was shown to perform better than the TV and the Beltrami framework. The latter is the subject of the next section.

4   The Beltrami Framework

In this framework a two-dimensional (multi-channel) image is considered to be an imbedding of a surface in a higher dimensional manifold, or in more general terms an image is a section of the spatial-feature trivial bundle. The section is endowed with a metric and is, thus, a Riemannian manifold. The functional over the space of sections is the Polyakov action:

$$S_B[u, G] = \int \sum_{rs} (\nabla u^r)^T G^{-1} \nabla u^s H_{rs}(u) \sqrt{\det G} \, dx dy$$

where G is the metric of the image manifold, $H_{r,s}$ are entries of the metric of the spatial-feature space and the indices are for the different spatial and channel/feature, e.g. colors, of the image. One important parameter of the Beltrami framework is the ratio between spatial distance and feature distance. This ratio, termed here β is needed to measure distances in the combined spatial-feature space.

The inner product on the manifold is

$$< f, h >_G = \int f(x, y) h(x, y) \sqrt{\det G} \, dx dy$$

The EL equations of the functional with respect to this inner product necessitate division by $\sqrt{\det G}$. Assuming that H is the identity matrix the gradient descent equation for the image read

$$u_t^r = \frac{1}{\sqrt{\det G}} \operatorname{div} \left(\sqrt{\det G} G^{-1} \nabla u^r \right) \quad .$$

Note that G^{-1} plays the role of the diffusion tensor and leads to a true anisotropic diffusion flow. The metric is determined by minimizing the functional. The analytic solution is the induced metric.

5   The Beltrami-Mumford-Shah Functional

The main idea of this paper is to generalize the Mumford-Shah functional. Indeed, one can rewrite the MS functional via the AT approach as follows

$$S_{AT}^\epsilon[u, v] = \int_\Omega \left[\frac{1}{2}(u - u_0)^2 + \frac{\lambda}{2}(\nabla u)^T \begin{pmatrix} v^2 & 0 \\ 0 & v^2 \end{pmatrix} \nabla u + \alpha \left(\epsilon ||\nabla v||_2^2 + \frac{(v-1)^2}{4\epsilon} \right) \right] dx dy$$

This is a suggestive form that can be easily generalized to

$$S_{AT}^\epsilon[u, V] = \int_\Omega \left[\frac{1}{2}(u - u_0)^2 + \frac{\lambda}{2}(\nabla u)^T V^T V \nabla u + \alpha \left(\epsilon ||\nabla V||_F^2 + \frac{||V - Id||_F^2}{4\epsilon} \right) \right] dx dy \,,$$

where

$$V = \begin{pmatrix} v_{11} & v_{12} \\ v_{12} & v_{22} \end{pmatrix} \quad , \quad ||V||_F^2 = v_{11}^2 + 2v_{12}^2 + v_{22}^2 \quad .$$

and $||\nabla V||_F^2 = ||V_x||_F^2 + ||V_y||_F^2$. This turns the dynamic diffusion coefficient v into a *dynamic diffusion tensor* V!

The next observation is that we can write

$$G^{-1} = V^T V$$

and reinterpret the functional in a new way: Let u and u_0 be functions on a Riemannian manifold. We demand that the two functions be similar in the L_2 norm *on the manifold*. The metric is G and V^{-1} is the vielbein i.e. the symmetric square root of the metric. The fidelity and smoothness terms should be written on the manifold. The penalty term regards the metric (or the vielbein) only. It enforces it to be close to the identity matrix in smooth regions and drive the metric to a singular matrix that aligns along the discontinuity near an strong edge. The penalty regularizes the metric as well. The new functional, thus, read

$$S_{BMS}^\epsilon[u,V] = \int \left[\frac{1}{2}(u - u_0)^2 + \lambda \, (\nabla u)^T V^T V \nabla u \right] \frac{dxdy}{\det V}$$

$$+ \alpha \int \left(\epsilon ||\nabla V||_F^2 + \frac{||V - Id||_F^2}{4\epsilon} \right) dxdy \quad .$$

This functional generalizes the Mumford-Shah functional from a scalar diffusion coefficient to a tensor one going from inhomogeneous smoothing to a true anisotropic one. It also generalizes the Beltrami framework since the metric, that serves here as the diffusion tensor, is not predefined but is a dynamical variable that is fixed along the flow by the functional.

For multi-channel image, e.g. color image one may write

$$S_{BMS}^\epsilon[u,V] = \int \left[\frac{1}{2}\sum_r (u^r - u_0^r)^2 + \frac{\lambda}{2}\sum_{r,s}(\nabla u^r)^T V^T V \nabla u^s H_{r,s}(u) \right] \frac{dxdy}{\det V}$$

$$+ \alpha \int \left(\epsilon ||\nabla V||_F^2 + \frac{||V - Id||_F^2}{4\epsilon} \right) dxdy$$

where $H_{r,s}$ is the metric in the feature space e.g. color space.

The minimization equations, assuming that $H_{ab} = \delta_{ab}$, are

$$u_t^a = \lambda(\det V)\mathrm{div}\left(\frac{1}{\det V} V^T V \nabla u^a \right) - (u^r - u_0^r)$$

$$(V_{ij})_t = (V^{-1})_{ij}\sum_a (u^a - u_0^a) + \alpha\epsilon\Delta V_{ij} - \frac{\alpha(V - Id)_{ij}}{2\epsilon} - \frac{\lambda}{2}\sum_r (\nabla u^r)^T W_{ij}\nabla u^r$$

where

$$W_{11} = \begin{pmatrix} v_{11} & v_{12} \\ v_{12} & 0 \end{pmatrix} \quad , \quad W_{22} = \begin{pmatrix} 0 & v_{12} \\ v_{12} & v_{22} \end{pmatrix} \quad , \quad W_{12} = \begin{pmatrix} v_{12} & 2v_{11} \\ 2v_{22} & v_{12} \end{pmatrix} \quad ,$$

6 Results

6.1 Numerical Implementation

Let
$$\partial_x^f u := u(x+1, y) - u(x, y)$$
and
$$\partial_y^f u := u(x, y+1) - u(x, y)$$
be the forward finite difference approximation of $\partial_x(u)$ and $\partial_y(u)$ respectively. Similarly, backward derivatives are defined as
$$\partial_x^b u := u(x, y) - u(x-1, y)$$
and
$$\partial_y^b u := u(x, y) - u(x, y-1).$$
The forward gradient is therefore
$$\nabla^f(u) := (\partial_x^f, \partial_y^f)^T(u),$$
and the the backward gradient is given by
$$\nabla^b(u) := (\partial_x^b, \partial_y^b)^T(u).$$

Numerical scheme of the functional derivatives takes the form:

$$\delta\mathcal{F}_{,v_{11}} = -\frac{v_{22}\sum_c(u^c - u_0^c)^2}{2(v_{11}v_{22} - v_{12}^2)^2} + \lambda\left[v_{11}(\partial_x^f u)^2 + 2v_{12}\partial_x^f\partial_y^f u\right] + \frac{\alpha}{2\epsilon}(v_{11} - 1) - 2\alpha\epsilon\nabla^b\cdot\nabla^f v_{11}.$$

$$\delta\mathcal{F}_{,v_{22}} = -\frac{v_{11}\sum_c(u^c - u_0^c)^2}{2(v_{11}v_{22} - v_{12}^2)^2} + \lambda\left[v_{22}(\partial_y^f u)^2 + 2v_{12}\partial_x^f\partial_y^f u\right] + \frac{\alpha}{2\epsilon}(v_{22} - 1) - 2\alpha\epsilon\nabla^b\cdot\nabla^f v_{22}.$$

$$\delta\mathcal{F}_{,v_{12}} = -\frac{v_{12}\sum_c(u^c - u_0^c)^2}{(v_{11}v_{22} - v_{12}^2)^2}$$
$$+ \lambda\left[v_{12}(\partial_x^f u)^2 + v_{12}(\partial_y^f u)^2 + 2(v_{11} + v_{22})\partial_x^f u\partial_y^f u\right] + \frac{\alpha}{\epsilon}v_{12} - 4\alpha\epsilon\nabla^b\cdot\nabla^f v_{12}.$$

$$\delta\mathcal{F}_{,u^r} = (u^r - u_0^r) - \lambda\det V\nabla^b\left(\frac{1}{\det V}V^T V\nabla^f u^c\right)$$

Optimization was carried out using the alternate minimization technique using the line search strategy. Descent direction was computed as gradient descent, and step size was calculated by Armijo rule [2]. The algorithm stops whenever all variables have reached convergence tolerance ε. The algorithm of Tensor-MS method is given below.

(a) Original image

(b) Beltrami-Mumford-Shah flow. PSNR= 30.21

(c) Mumford-Shah flow. PSNR= 29.56

(d) Beltrami flow. PSNR= 29.13

Algorithm Energy Descent(u_0)

- Initialize $u^0 = u_0, v_{11}^0 = 1, v_{22}^0 = 1, v_{12}^0 = 1, k = 0$
- Do
  1. $\tau^{v_{11}} = \text{ArmijoStep}(\mathcal{F}, v_{11}^k)$
  2. $v_{11}^{k+1} = v_{11}^k - \tau^{v_{11}} \delta\mathcal{F}_{,v_{11}}(v_{11}^k, v_{22}^k, v_{12}^k, u^k)$
  3. $\tau^{v_{22}} = \text{ArmijoStep}(\mathcal{F}, v_{22}^k)$
  4. $v_{22}^{k+1} = v_{22}^k - \tau^{v_{22}} \delta\mathcal{F}_{,v_{22}}(v_{11}^{k+1}, v_{22}^k, v_{12}^k, u^k)$
  5. $\tau^{v_{12}} = \text{ArmijoStep}(\mathcal{F}, v_{12}^k)$
  6. $v_{12}^{k+1} = v_{12}^k - \tau^{v_{12}} \delta\mathcal{F}_{,v_{12}}(v_{11}^{k+1}, v_{22}^{k+1}, v_{12}^k, u^k)$
  7. $\tau^u = \text{ArmijoStep}(\mathcal{F}, u^k)$
  8. $u^{k+1} = v_{12}^k - \tau^u \delta\mathcal{F}_{,u}(v_{11}^{k+1}, v_{22}^{k+1}, v_{12}^{k+1}, u^k)$
- while $\|v_{11}^{k+1} - v_{11}^k\|, \|v_{22}^{k+1} - v_{22}^k\|, \|v_{12}^{k+1} - v_{12}^k\|, \|u^{k+1} - u^k\| \geq \varepsilon$

Beltrami and MS methods are similarly implemented using the corresponding derivatives. Parameter set for two images are given in the following table, where in all cases tolerance was set to $\varepsilon = 10^{-3}$.

	Tensor MS			MS			Beltrami	
	α	λ	ϵ	α	λ	ϵ	λ	β
Ballet	0.1	0.5	0.01	0.1	0.6	0.01	0.65	1.0
Lenna	0.1	0.5	0.1	0.1	0.6	0.01	0.68	1.0

6.2   Results

(e) Original image

(f) Beltrami-Mumford-Shah flow. PSNR= 32.57

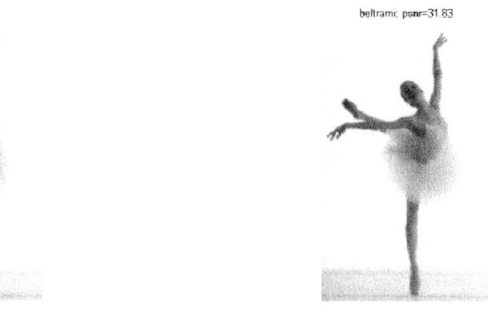

(g) Mumford-Shah flow. PSNR= 31.99   (h) Beltrami flow. PSNR= 31.83

7   Summary and Conclusions

We present in this paper a unifying generalization of the Mumford-Shah functional and the Beltrami framework. We show that we are able, in this way, to have the benefits of true anisotropic diffusion together with a dynamically tuned metric/diffusion tensor. The functional is naturally defined in terms of the vielbein-the metric's square root. preliminary results show improvement on both the Beltrami flow and the Mumford-Shah flow.

References

1. Ambrosio, L., Tortorelli, V.M.: Approximation of functionals depending on jumps by elliptic functionals via Γ-convergence. Comm. Pure Appl. Math. 43, 999–1036 (1990)
2. Armijo, L.: Minimization of functions having Lipschitz continuous first partial derivatives. Pacific J. Math. 16(1), 13 (1966)
3. Bar, L., Sochen, N., Kiryati, N.: Image Deblurring in the Presence of Impulsive Noise. International Journal of Computer Vision 70(3), 279–298 (2006)
4. Bar, L., Sochen, N., Kiryati, N.: Semi-Blind Image Restoration via Mumford-Shah Regularization. IEEE Trans. on Image Processing 15(2), 483–493 (2006)
5. Deriche, R., Faugeras, O.: Les EDP en Traitement des Images et Vision par Ordinateur. Traitement du Signal 13(6) (1996)
6. Mumford, D., Shah, J.: Optimal Approximations by Piecewise Smooth Functions and Associated Variational Problems. Comm. on Pure and Applied Math. XLII(5), 577–684 (1989)
7. Perona, P., Malik, J.: Scale-space and edge detection using anisotropic diffusion. IEEE Transactions on Pattern Analysis and Machine Intelligence 12(7), 629–639 (1990)
8. Rosenau, P.: Free energy functionals at the high-gradient limit. Phys. Rev. A (Rapid Communications) 41(4), 2227–2230 (1990)
9. Rudin, L., Osher, S., Fatemi, E.: Nonlinear Total Variation based noise removal algorithms. Physica D 60, 259–268 (1992)
10. Sochen, N., Kimmel, R., Malladi, R.: A general framework for low level vision. IEEE Transactions on Image Processing 7, 310–318 (1998)
11. Weickert, J.: Coherence-Enhancing Diffusion Filtering. International Journal of Computer Vision 31(2/3), 111–127 (1999)

An Adaptive Norm Algorithm
for Image Restoration

Daniele Bertaccini[1], Raymond H. Chan[2], Serena Morigi[3], and Fiorella Sgallari[3]

[1] Department of Mathematics, University of Roma "Tor Vergata", Rome, Italy
bertaccini@mat.uniroma2.it
[2] Department of Mathematics, The Chinese University of Hong Kong,
Hong Kong, P.R. China
rchan@math.cuhk.edu.hk
[3] Department of Mathematics-CIRAM, University of Bologna, Bologna, Italy
{morigi,sgallari}@dm.unibo.it

Abstract. We propose an adaptive norm strategy designed for the restoration of images contaminated by blur and noise. Standard Tikhonov regularization can give good results with Gaussian noise and smooth images, but can over-smooth the output. On the other hand, L_1-TV (Total Variation) regularization has superior performance with some non-Gaussian noise and controls both the size of jumps and the geometry of the object boundaries in the image but smooth parts of the recovered images can be blocky. According to a coherence map of the image which is obtained by a threshold structure tensor, and can detect smooth regions and edges in the image, we apply L_2-norm or L_1-norm regularization to different parts of the image. The solution of the resulting minimization problem is obtained by a fast algorithm based on the half-quadratic technique recently proposed in [2] for L_1-TV regularization. Some numerical results show the effectiveness of our adaptive norm image restoration strategy.

1 Introduction

The recent increase in the widespread use of digital imaging technologies in consumer (e.g., digital camera and video) and other markets (e.g., medicine imaging) has brought with it a simultaneous demand for image denoising and deblurring.

The most common image degradation model, where the observed data $f \in \mathbb{R}^{n^2}$ are related to the underlying $n \times n$ image rearranged into a vector $u \in \mathbb{R}^{n^2}$, is

$$f = Bu + e, \tag{1}$$

where $e \in \mathbb{R}^{n^2}$ accounts for the perturbations and B is a $n^2 \times n^2$ matrix representing the optical blurring.

The computation of a useful approximation of u can be accomplished by replacing the linear system of equations (1) by a nearby system, whose solution is less sensitive to the noise e. This replacement is commonly referred to as regularization.

A.M. Bruckstein et al. (Eds.): SSVM 2011, LNCS 6667, pp. 194–205, 2012.
© Springer-Verlag Berlin Heidelberg 2012

The standard Tikhonov regularized solution of the inverse problem for two dimensional image restoration of the observed image f, is the minimum of the functional

$$J(u) = \frac{1}{p}\|Bu - f\|_p^p + \frac{\mu}{q}\|Au\|_q^q, \tag{2}$$

for $p = 2, q = 2$, where A is a regularization operator, and μ is the regularization parameter that controls the trade-off between data fitting term and the regularization term. The use of the Euclidean norm in (2) yields a least squares problem to which many efficient algorithms exist [15,16]. However, the result is only optimal when noise in the image f is white Gaussian noise, (e.g. no outliers) and the solution is smooth, i.e., without discontinuities.

For the regularization term, there has been a growing interest in using the L_1 norm ($q = 1$). The minimization problem (2) with $p = 2, q = 1$, and $A = \nabla$ (the gradient operator), becomes convex but non-smooth and it is denoted by L_2-TV regularization. While a number of algorithms [9,10], have been proposed to solve this optimization problem, it remains a computationally expensive task that can be prohibitively costly for large problems and for operators without a fast implicit implementation or a sparse explicit matrix representation. Recently, the L_1-TV functional, corresponding to the choice $p = 1, q = 1$, $A = \nabla$ in (2) [2,12,13,11], has attracted attention due to a number of advantages, including superior performance with non-Gaussian noise such as impulse noise. The solutions are very stable with respect to outliers and moreover TV controls both the size of jumps and the geometry of the object boundaries in the image.

The main goal of this work is to adaptively consider a suitable norm ($q = 1$ or $q = 2$) according to the determined image structures (smooth regions or edges). Although, the same presented framework can be considered on p, that is on the data fitting term. The L_2-norm regularization well restores corrupted images with wide smooth regions but it oversmoothes the resulting images. On the other hands, the L_1-TV regularization has been successfully applied to restore images because of its good property in preserving edges but in general, the resulting images are blocky. Driven by a suitable map of the structures of the image, we can apply the appropriate norm to selected parts of the image domain.

To achieve this aim, we introduce a measure of the coherence in the image by mean of a threshold structure tensor [8] which provides a coherence map of the image. Following the coherence map we use L_2-norm for pixels in smooth regions and L_1-TV for pixels along edges and corners.

In Fig. 1(a) a simple test image is shown with a white square in a black background corrupted by Gaussian blur and Gaussian noise. The restored image obtained by solving (2) with $p = 2, q = 2$ is shown in Fig. 1(b), The restored images with $p = 1, q = 1$ is shown in Fig. 1(c), while the proposed adaptive approach is shown in Fig. 1(d). Comparing the images in Fig. 1(b),(c),(d) it is clear how an adaptive choice can lead to denoised homogenous regions without blocky effects. In fact it takes advantage of the L_2 approach in the homogeneous regions while keeping the edges thanks to the L_1-TV method.

The paper is organized as follows. We briefly describe the half-quadratic algorithm for L_1-TV image restoration in Section 2. The proposed model and its

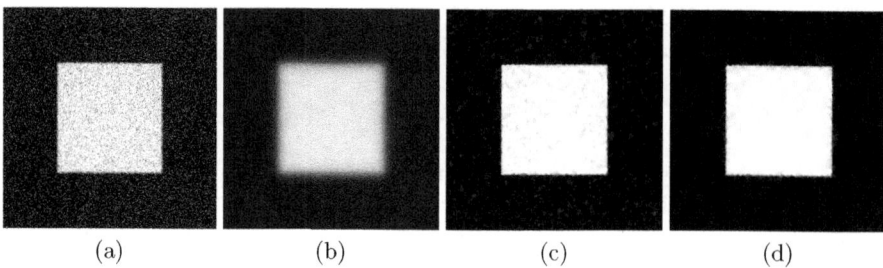

$$\text{(a)} \qquad\qquad \text{(b)} \qquad\qquad \text{(c)} \qquad\qquad \text{(d)}$$

Fig. 1. (a) corrupted image by white Gaussian noise with band $= 5$, sigma $= 3$ and noise level $\nu = 0.05$; (b) restoration by (2) with $p = 2$, $q = 2$, $(SNR = 9.62)$; (c) restoration by L_1-TV (2) with $p = 1$, $q = 1$, $(SNR = 20.30)$; (d) restoration by our proposal (adaptive-norm model (11) with $p = 1$), $(SNR = 20.93)$, $\mu_1 = 0.5$, $\mu_2 = 90$.

numerical aspects are discussed in Section 3. Numerical examples and comments are provided in Section 4. Section 5 contains concluding remarks.

2 Description of the HQ-Algorithm for L_1-TV Regularization

Let us briefly summarize a recently proposed algorithm [2] that minimizes in a fast and accurate way (2) with $p = q = 1$, that is, with the notation of [2],

$$\min_{u} \left\{ \sum_{i=1}^{n^2} |B_i\, u - f_i|_\gamma + \mu |\nabla u_i|_\beta \right\}, \tag{3}$$

where B_i is the ith row of the discrete data fidelity operator B; and β and γ are both small positive regularization parameters which prevents the denominator from vanishing in numerical implementations. The specification of β and γ involves trade-offs between the quality of edges restored and the speed in converging. Precisely, the smaller β and γ are, the higher quality of the restoration on the edges will be. We used the notation

$$|\nabla u_i|_\beta = \left((\nabla_x u_i)^2 + (\nabla_y u_i)^2 + \beta\right)^{1/2},$$

$$|B_i u - f_i|_\gamma = \left((B_i u - f_i)^2 + \gamma\right)^{1/2},$$

where ∇_x, ∇_y are the first order finite difference operators in the horizontal and vertical directions, respectively.

The proposed idea is based on an iterative reweighting of a *half-quadratic* algorithm (HQA) for L_1-TV image restoration. Half-quadratic regularization, was introduced in [4,14], and is based on the following expression for the modulus of a real, nonzero number x:

$$|x| = \min_{v > 0} \left\{ v\, x^2 + \frac{1}{4\,v} \right\}, \tag{4}$$

whose minimum is at $v = \frac{1}{2|x|}$, and the function in the curly bracket in (4) is quadratic in x but not in v; hence the name *half-quadratic*.

By using (4), the minimum of the function in (3) can be found by applying an alternate minimization procedure to minimize the operator $\mathcal{L}$, i.e.,

$$\min_{u,\, v>0,\, w>0} \mathcal{L}(u, v, w),$$

where

$$\mathcal{L}(u, v, w) = \sum_{i=1}^{n^2} \left[\mu \left(v_i |\nabla u_i|_\beta^2 + \frac{1}{4\,v_i} \right) + w_i |B_i\, u - f_i|_\gamma^2 + \frac{1}{4w_i} \right] \qquad (5)$$

With the notation in [2], we need to perform in sequence the three iterative minimizations, for each iteration step k, that is

$$v^{(k+1)} = \arg\min_{v>0} \mathcal{L}(u^{(k)}, v, w^{(k)}),$$

with explicit solution

$$v_i^{(k+1)} = \left(2|\nabla u_i^{(k)}|_\beta \right)^{-1} \qquad (6)$$

and

$$w^{(k+1)} = \arg\min_{w>0} \mathcal{L}(u^{(k)}, v^{(k+1)}, w),$$

with explicit solution

$$w_i^{(k+1)} = \left(2|B_i u^{(k)} - f_i|_\gamma \right)^{-1} \qquad (7)$$

and

$$u^{(k+1)} = \arg\min_{u} \mathcal{L}(u, v^{(k+1)}, w^{(k+1)}), \qquad (8)$$

whose solution u can be found by imposing that:

$$\nabla_u \left(\mathcal{L}(u, v^{(k+1)}, w^{(k+1)}) \right) = 0. \qquad (9)$$

This leads to the sequence of linear systems for updating $u^{(k+1)}$:

$$\left[\mu\, A^T\, \widehat{D}_\beta(u^{(k)})\, A + B^T\, D_\gamma(u^{(k)})\, B \right] u^{(k+1)} = B^T\, D_\gamma(u^{(k)})\, f, \qquad (10)$$

where $A \in \mathbb{R}^{2n^2 \times n^2}$ is the matrix discretizing the gradient operator $[\nabla_x^T;\ \nabla_y^T]$ with, e.g., first order finite differences (this is the choice in our experiments), $\widehat{D}_\beta(u^{(k)}) := diag(D_\beta(u^{(k)}), D_\beta(u^{(k)}))$, and the weight component matrices $D_\beta(u^{(k)})$, $D_\gamma(u^{(k)})$ are diagonal matrices whose ith entries are given by

$$\left(D_\beta(u^{(k)}) \right)_i = 2v_i^{(k+1)} = \frac{1}{|\nabla u_i^{(k)}|_\beta},$$

$$\left(D_\gamma(u^{(k)})\right)_i = 2w_i^{(k+1)} = \frac{1}{|B_i\,u^{(k)} - f_i|_\gamma}, i = 1, ..., n^2.$$

In order to get an approximate L_1-TV restoration, given an initial image f and an initial guess for the recovered image $u^{(0)}$, there is just the need to apply an iterative linear solver to (10) like conjugate gradients method for $k = 0, 1, \ldots, k_{\max}$.

We note that other reweighted least squares approachs can be considered such as the one in [6] where an inexact Newton strategy is used to solve the system (10).

3   The Adaptive Norm Algorithm (ANA)

The L_1-TV restoration algorithm HQA, developed by [2] and summarized in Section 2, works very well especially in presence of salt-and-pepper noise and near edges and corners. On the other hand, with different types of image perturbations, like the white Gaussian noise, and in the presence of smooth, homogeneous regions and weak edges it can provide a less accurate restoration and can give artifacts like, e.g., a blocky effect in smooth regions. A *selective reweighted of the half-quadratic approach* is one of the possible natural ways to overcome these well-known issues of the L_1-TV restoration. With the word *selective* we mean "using different norm for different pixels" of the image. To achieve this aim, one could work with a norm continuously changing from 1 to 2, but this would lead to the solution of a PDE derived from the variational problem similar to (2). A preliminary step in this direction has been proposed by [18]. In contrast, we choose not to change the norm continuously from 1 to 2, but, using a suitable *coherence map*, we classify the pixels in the image as pixels belonging to homogeneous region or pixels belonging to edges or corners, and we associate them with norm L_2 or L_1, respectively. Details on the construction of the coherence map C are given in Section 3.1.

Driven by the coherence map, we use the L_2 norm for smooth and homogeneous regions and the L_1 norm near edges and corners. Let C be a diagonal matrix with the ith entry $(C)_i = 1$ if the ith pixel belongs to a homogeneous region identified by the coherence map, while $(C)_i = 0$ near edges. Let $\bar{C} = I - C$, with I the identity matrix. In view of this, we propose to modify the functional in (2) to the following functional

$$\Phi(u) = \|Bu - f\|_1^1 + \mu_1\|CAu\|_1^1 + \mu_2\|\bar{C}Lu\|_2^2, \tag{11}$$

where μ_1, μ_2 are regularization parameters, L is a regularization operator, such as for example $L = A$ or, e.g., the discrete Laplacian, and A is defined as in (10). This new functional also caters for different regularization operators. Moreover, the adaptive norm strategy can also be applied to the data fidelity term $Bu - f$ in (11).

The minimization of the functional (11) can be obtained in a way similar to what is done for the half quadratic L_1-TV and, in particular, by solving the following linear systems for updating $u^{(k+1)}$

$$\left[(\mu_1 A^T C \widehat{D}_\beta(u^{(k)})CA + \mu_2 L^T \bar{C} L + B^T D_\gamma(u^{(k)})B\right] u^{(k+1)} =$$
$$= B^T D_\gamma(u^{(k)})f \qquad (12)$$

where the weight component matrix $\widehat{D}_\beta(u^{(k)})$, incorporates the selective L_1/L_2 reweighted by the diagonal matrix C. We initialized the iterative process by setting $u^{(0)} = f$, and the coherence map is computed at each iteration step k. In the following we will name our algorithm Adaptive Norm Algorithm (ANA).

In order to accelerate the solution of (12) the strategy used for sequences of linear systems proposed in [1] can be used. However we found that stopping the conjugate gradient solver after a few iterations already gives good results; see Section 4 for numerical examples.

The model (11) allows the use of the techniques in [2] to prove the convergence of sequence $\{u^{(k)}\}$ to a minimum of $\Phi(u)$. An analysis of the convergence of the sequence $\{u^{(k)}\}$ generated by the proposed adaptive norm strategy can be based on the analysis of the convergence of half-quadratic algorithm in [2].

3.1 The Coherence Matrix Construction

In order to detect if a pixel of the given image belongs to an edge or a homogeneous region, we need a strategy that is able to mark each pixel with a score that we normalize in the range $[0, 1]$.

Coherence enhancing image smoothing has been introduced in [8] and successfully applied in image filtering by anisotropic diffusion. This type of nonlinear diffusion includes the construction of a diffusion tensor which is built as follows. Given an image u, and its Gaussian-smoothed version u_σ, a regularized shape descriptor is provided by

$$S_\delta(\nabla u_\sigma) := (K_\delta * (\nabla u_\sigma \otimes \nabla u_\sigma)) \qquad (13)$$

where K_δ is a Gaussian kernel with $\delta \geq 0$. The matrix S_δ is symmetric positive semi-definite and its eigenvalues $\lambda_1 \geq \lambda_2$ integrate the variation of the gray values within a neighborhood of size $O(\delta)$. They describe the average contrast in the corresponding eigendirections $\mathbf{v}_1$ and $\mathbf{v}_2$. The orientation of the eigenvector $\mathbf{v}_2$, corresponding to the smaller eigenvalue, represents the direction of lowest fluctuations, the so-called coherence orientation. In this way, constant areas are characterized by $\lambda_1 = \lambda_2 = 0$, while straight edges give $\lambda_1 \gg \lambda_2 = 0$.

The normalized coherence value which measures the anisotropic structures within a window of scale δ is thus defined as

$$c = \frac{(\lambda_1 - \lambda_2)^2}{\max\{(\lambda_1 - \lambda_2)^2\}}, \quad c \in [0, 1]. \qquad (14)$$

Thus, for anisotropic structures, c approaches 1, while it tends to zero for isotropic structures. Let c_i be the coherence value obtained by computing (14) for the ith pixel in the vectorized image u. We use a "selective" threshold parameter τ (typically $0 \ll \tau < 1$) to construct the diagonal matrix C, with $(C)_i = 1$

when the ith pixel belongs to a homogeneous region, that is when $c_i < \tau$, while $(C)_i = 0$ near edges. This aim to partition the image into homogeneous and non-homogeneous regions, different partitioning will be further investigated.

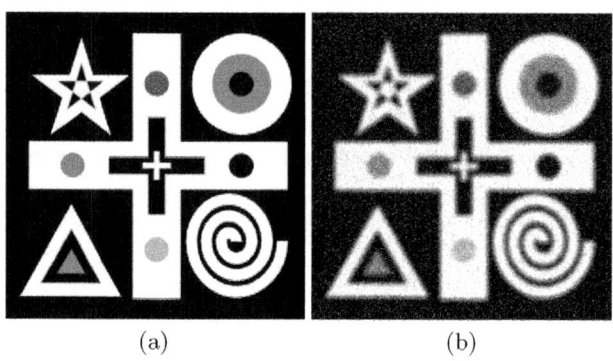

(a)   (b)

Fig. 2. Example 1:(a) Blur- and noise-free 320×320 image; (b) corrupted image by symmetric Gaussian blur with band $= 7$ and sigma $= 5$, noise level $\nu = 0.02$

4   Experiments and Results

Let $u \in \mathbb{R}^{n^2}$ represent a blur- and noise-free image. We generate an associated blurred and noise-free image $\hat{f}$ by multiplying u by a block Toeplitz matrix $B \in \mathbb{R}^{n^2 \times n^2}$ with Toeplitz blocks. The matrix B represents a symmetric Gaussian blurring operator and has two parameters band and sigma. The former specifies the half-bandwidth of the Toeplitz blocks and the latter the variance of the Gaussian point spread function. The larger the sigma is, the more the blurring will be. A blur- and noise-contaminated image $f \in \mathbb{R}^{n^2}$ is obtained by adding an error vector $e \in \mathbb{R}^{n^2}$ to $\hat{f}$.

Thus,

$$f = Bu + e.$$

The corrupted image $f \in \mathbb{R}^{n^2}$ is assumed to be available and we would like to determine the blur- and noise-free image u. In our experiments, e has normally distributed entries with mean zero, scaled to yield a desired noise-level

$$\nu = \frac{\|e\|}{\|u\|}.$$

In all the examples we take the parameters $\beta = 10^{-3}$ and $\gamma = 10^{-6}$ in (3) and we consider periodic boundary conditions for the difference matrix A. Equation (12) is solved by the conjugate gradient method where we stopped when the Euclidean norm of the relative error between successive approximations is less than $5 \cdot 10^{-5}$. The solver is very fast and we do not need to accelerate the solution of (12) by preconditioning strategies.

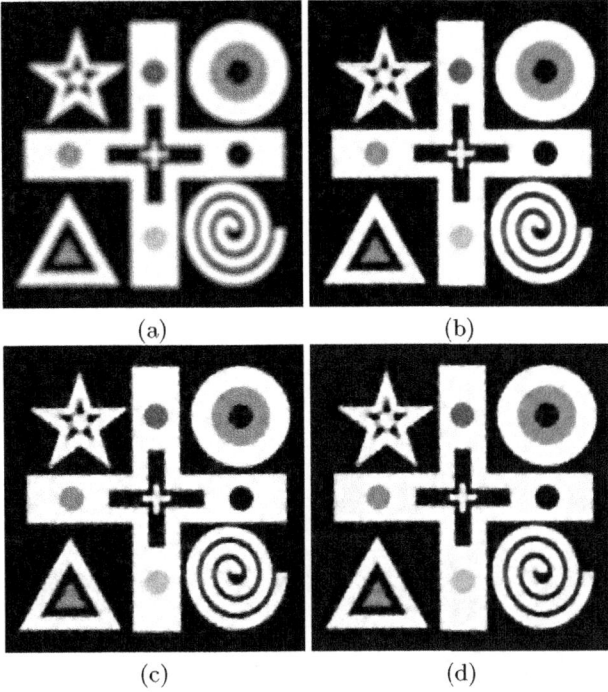

Fig. 3. Example 1: (a) restoration by (2) with $p = 1$, $q = 2$, $\mu = 10$ ($SNR = 10.68$); (b) restoration by L_1-TV (2) with $p = 1$, $q = 1$, $\mu = 0.5$ ($SNR = 15.72$) (c) restoration by ANA with $p = 1$, $\mu_1 = 0.2$, $\mu_2 = 10$ ($SNR = 16.76$); (d) restoration by ANA with adaptivity also for the fidelity term, $\mu_1 = 0.2$, $\mu_2 = 10$ ($SNR = 16.47$)

The displayed restored images provide a qualitative comparison of the performance of the proposed adaptive norm algorithm. A quantitative comparison is given by the Signal-to-Noise Ratio (SNR),

$$\text{SNR} := 10 \log_{10} \frac{\|u - E(u)\|_2^2}{\|\hat{u} - u\|_2^2} \text{ dB}, \tag{15}$$

where u denotes the blur- and noise-free image, $\hat{u}$ the restored image and $E(u)$ is the mean grey-level value of the original image.

The choice of the parameters μ_1 and μ_2 in (11) clearly affect the quality of the restored image, in our experimentation we have empirically chosen μ_1 in the range $[0, 1]$, and μ_2 in the range $[10, 100]$, but further investigations will be planned. In the literature there are several regularization parameter selection methods for Tikhonov regularization problems ($p = 2, q = 2$), e.g. the discrepancy principle, the L-curve and the Generalized Cross-Validation (GCV) methods [7]. Recently in [17] a generalization of GCV for the case $p = 2, q = 1$ has been proposed.

Table 1. Example 2: Results for restorations of image corrupted by Gaussian blur corresponding to different band,sigma values, and noise-levels ν. $\mu_1 = 0.2$, $\mu_2 = 5$ (first 8 rows) $\mu_1 = 0.5$, $\mu_2 = 10$ (last 4 rows).

band	sigma	ν	SNR(L_1-TV)	SNR(ANA)
7	5	0.01	22.09	22.90
7	5	0.02	20.08	20.95
7	5	0.05	16.74	17.61
7	5	0.1	13.69	15.20
5	3	0.01	23.63	24.53
5	3	0.02	21.07	22.16
5	3	0.05	18.02	18.76
5	3	0.1	15.10	15.68
3	1	0.01	26.35	26.78
3	1	0.02	23.20	23.98
3	1	0.05	18.69	19.38
3	1	0.1	14.62	15.35

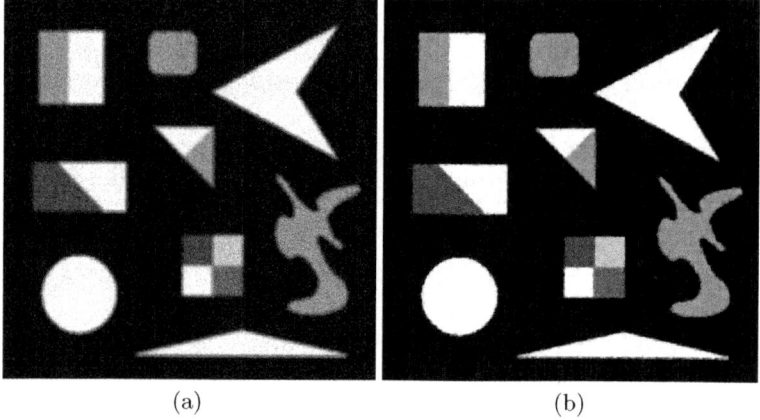

(a)   (b)

Fig. 4. Example 2: (a) corrupted 512×512 image by symmetric Gaussian blur with band = 7 and sigma = 5, noise level $\nu = 0.02$; (b) restoration by ANA with $p = 1$, $\mu_1 = 0.2$, $\mu_2 = 1$ ($SNR = 20.95$).

Example 1. In this example the image in Fig. 2(a) is corrupted by Gaussian noise, characterized by noise level $\nu = 0.02$, and symmetric Gaussian blur with band = 7 and sigma = 5. The corrupted image is shown in Fig. 2(b). The restorations obtained by applying the three approaches (L_2-TV, L_1-TV, ANA) are shown in Figure 3. In Fig. 3 (d) the reconstructed image is obtained by solving (11) with the adaptivity also in the fidelity term. In all the algorithms we have considered $k_{max} = 40$ outer steps in (12) but also less outer steps give satisfactory results. Our ANA gives the best $SNR = 16.76$. From a visual inspection of Fig. 3(c),(d), we observe that $white - homogemous$ regions are clearly better restored.

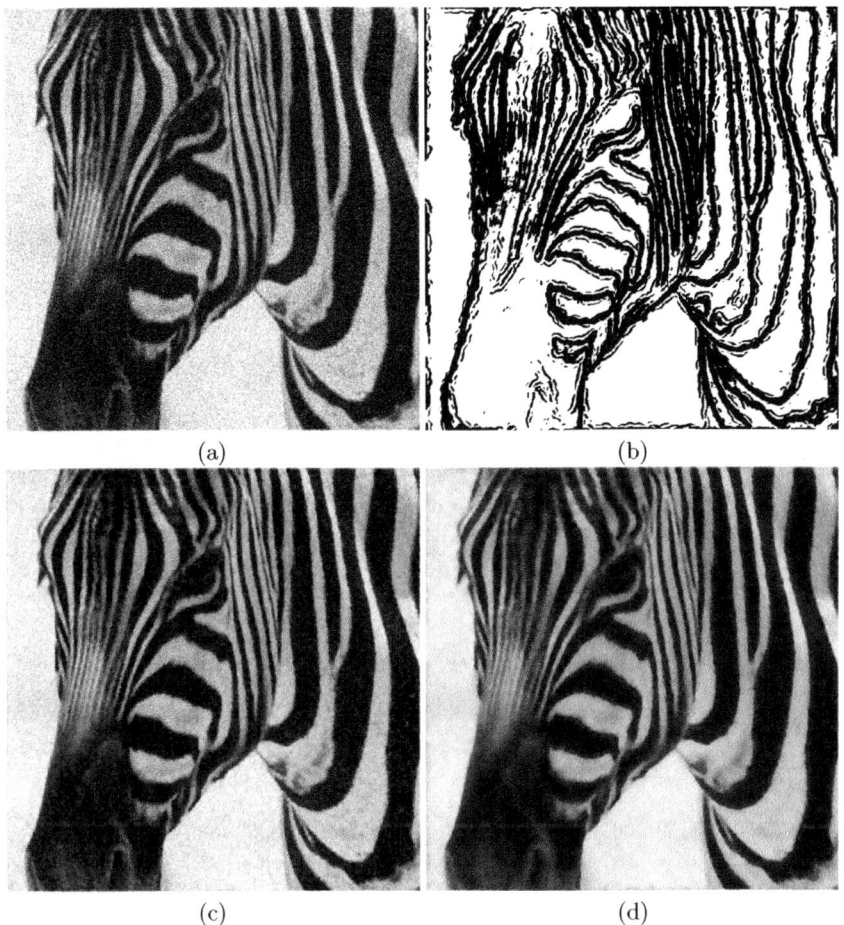

(a)   (b)

(c)   (d)

Fig. 5. Example 3: (a) corrupted image ($SNR = 9.43$); (b) coherence map; (c) restoration by L_1-TV (2) with $p = 1$, $q = 1$, $\mu = 0.5$ ($SNR = 17.15$) (d) restoration by ANA with $p = 1$, $\mu_1 = 0.5$, $\mu_2 = 80$ ($SNR = 17.47$)

Example 2. In this example a 512×512 image is contaminated by different noise levels and incremental Gaussian blur. In Fig. 4(a) the corrupted image by Gaussian noise, characterized by noise level $\nu = 0.02$, and symmetric Gaussian blur with band $= 7$ and sigma $= 5$ is shown, while Fig. 4(b) shows the image restored by ANA using $k_{max} = 30$ outer steps. In Table 1, algorithms L_1-TV and ANA are compared and their SNR values are reported in the fourth and fifth columns, respectively. Table 1 and other additional numerical experiments, indicate that the performance of our method is better for images with quite large homogenous regions, for medium blur and for quite high noise levels.

Example 3. In this example we test our approach on a photographic image of size 800×800 corrupted by symmetric Gaussian blur with band $= 5$ and sigma $= 3$ and noise level $\nu = 0.02$, shown in Fig. 5(a). The results of applying $k_{max} = 10$ outer steps in (12) are shown in Fig. 5(c), (d) for algorithm L_1-TV and ANA, respectively. We can appreciate the good quality results we get with just a few steps, that demonstrate the efficiency of the algorithm. In Fig. 5(b) the used coherence map is illustrated.

5   Conclusions

In this paper we propose a fast algorithm that allow L_1 or L_2-norm regularization for different image areas according to the image structures (e.g., smooth regions or edges). Numerical experiments seem to confirm that our algorithm is promising. We plan to extend this framework to general L_1-regularized problems and to consider other choices for p and q in model (2).

Acknowledgments. This work has been partially supported by MIUR-Prin 2008, $ex60\%$ project by University of Bologna "Funds for selected research topics" and by GNCS-INDAM.

References

1. Bertaccini, D., Sgallari, F.: Updating preconditioners for nonlinear deblurring and denoising image restoration. Applied Numerical Mathematics 60, 994–1006 (2010)
2. Chan, R.H., Liang, H.X.: A fast and efficient half-quadratic algorithm for TV-L1 Image restoration, CHKU research report 370 (submitted, 2010), ftp://ftp.math.cuhk.edu.hk/report/2010-03.ps.Z
3. Chan, T., Mulet, P.: On the convergence of the lagged diffusivity fixed point method in total variation image restoration. SIAM J. Numer. Anal. 36, 354–367 (1999)
4. Geman, D., Yang, C.: Nonlinear image recovery with half-quadratic regularization and FFTs. IEEE Trans. Image Proc. 4, 932–946 (1995)
5. Jacobson, M., Fessler, J.: An expanded theoretical treatment of iteration-dependent majorize-minimize algorithms. IEEE Trans. Image Proc. 16, 2411–2422 (2007)
6. Rodriguez, P., Wohlberg, B.: Efficient minimization method for a generalized total variation functional. IEEE Transactions on Image Processing 18, 322–332 (2009)
7. Hansen, P.: Rank-Deficient and Discrete Ill-Posed Problems. SIAM, Philadelphia (1998)

8. Weickert, J., Scharr, H.: A scheme for coherence enhancing diffusion filtering with optimized rotation invariance. J. of Visual Communication and Image Representation 13, 103–118 (2002)
9. Vogel, C., Oman, M.: Iterative methods for total variation denoising. SIAM J. Sci. Comp. 17(1-4), 227–238 (1996)
10. Chambolle, A.: An algorithm for total variation minimization and applications. J. of Math. Imaging and Vision 20, 89–97 (2004)
11. Nikolova, M.: A variational approach to remove outliers and impulse noise. J. of Math. Imaging and Vision 20, 99–120 (2004)
12. Nikolova, M.: Minimizers of cost-functions involving nonsmooth datafidelity terms application to the processing of outliers. SIAM J. Numerical Analysis 40, 965–994 (2002)
13. Chan, T.F., Esedoglu, S.: Aspects of total variation regularized L1 function approximation. SIAM J. Appl. Math. 65, 1817–1837 (2005)
14. Nikolova, M., Chan, R.: The equivalence of half-quadratic minimization and the gradient linearization iteration. IEEE Trans. Image Proc. 16, 1623–1627 (2007)
15. Reichel, L., Sgallari, F., Ye, Q.: Tikhonov regularization based on generalized Krylov subspace methods. Appl. Numer. Math. (2010), doi:10.1016/j.apnum.2010.10.002
16. Morigi, S., Reichel, L., Sgallari, F.: An interior-point method for large constrained discrete ill-posed problems. J. Comput. Appl. Math. 233, 1288–1297 (2010)
17. Liao, H., Li, F., Ng, M.: On Selection of Regularization Parameter in Total Variation Image Restoration. Journal of the Optical Society of America A 26, 2311–2320 (2009)
18. Chen, Q., Montesinos, P., Sun, Q.S., Heng, P.A., Xia, D.S.: Adaptive total variation denoising based on difference curvature. Image and Vision Computing 28(3), 298–306

Variational Image Denoising
with Adaptive Constraint Sets

Frank Lenzen[1,2], Florian Becker[1], Jan Lellmann[1],
Stefania Petra[1], and Christoph Schnörr[1]

[1] HCI & IPA, Heidelberg University,
Speyerer Str. 6, 69115 Heidelberg, Germany
frank.lenzen@iwr.uni-heidelberg.de
{becker,lellmann,petra,schnoerr}@math.uni-heidelberg.de
http://hci.iwr.uni-heidelberg.de,
http://ipa.iwr.uni-heidelberg.de
[2] Intel Visual Computing Institute, Saarland University,
Campus E2 1, 66123 Saarbrücken, Germany
http://www.intel-vci.uni-saarland.de

Abstract. We propose a generalization of the total variation (TV) minimization method proposed by Rudin, Osher and Fatemi. This generalization allows for adaptive regularization, which depends on the minimizer
itself. Existence theory is provided in the framework of quasi-variational
inequalities. We demonstrate the usability of our approach by considering
applications for image and movie denoising.

Keywords: solution dependent adaptivity, quasi-variational inequalities, spatio-temporal TV, anisotropic TV, image denoising.

1 Introduction

One of the most widely used methods for image denoising is total variation (TV)
minimization. The TV method proposed by Rudin, Osher and Fatemi in [13]
(ROF) consists in minimizing the functional

$$\frac{1}{2}\|u - f\|^2 + \alpha \operatorname{TV}(u), \tag{1}$$

w.r.t. u over $\mathrm{BV}(\Omega), \Omega \subset \mathbb{R}^d$ for given noisy data f. Here $\operatorname{TV}(u)$ is the total variation semi-norm and α is a regularization parameter. We consider the
formulation of the regularization term $\alpha \operatorname{TV}(u)$ based on constraint sets:

$$\alpha \operatorname{TV}(u) = \sigma_{\mathcal{C}}(u), \quad \mathcal{C} = \operatorname{div} \mathcal{D}, \quad \mathcal{D} = \{p \in C_c^\infty(\Omega; \mathbb{R}^d) \colon \|p(x)\| \le \alpha\}, \tag{2}$$

where $\sigma_{\mathcal{C}}$ is the support function of the set $\mathcal{C}$ and div is applied elementwise.

In this paper, we generalize the ROF functional (1) by introducing the dependency $\mathcal{C} = \mathcal{C}(u)$. This allows for variants of the TV method, where the set
$\mathcal{C}(u)$ locally adapts to the image content *depending on the solution u itself.*

A.M. Bruckstein et al. (Eds.): SSVM 2011, LNCS 6667, pp. 206–217, 2012.

In the literature, *adaptive* TV methods have been proposed e.g. in [6,9], with locally varying regularization parameter, and [1,14,15,10], where anisotropic regularization is steered by local structures. Except for [10], these methods gather the required local information either in a preprocessing or as an additional unknown of the variational problem, not depending on the minimizer itself. The variational framework presented below differs from [10]; possible connections will be explored in future work. Another kind of denoising methods are non-local methods, cf. e.g. [11,8,3]. Although these methods can be applied in an iterated fashion, a dependency of the regularization on the minimizer is not modeled explicitly.

Our paper is organized as follows. In Sect. 2, the proposed generalization of the TV minimization functional (1) is described. The mathematical framework is presented in terms of variational inequalities. Considering a sequence of convex variational inequalities, we provide an existence result for fixed points (see Sect. 3), using only general assumptions on the convex set $\mathcal{C}(u)$. In particular, non-local information can be used in the definition of $\mathcal{C}(u)$. Moreover, we provide a first basic algorithm. The usability of this novel concept is supported by applications for image and movie denoising. In particular, we generalize the approach of anisotropic TV with double orientations, proposed by Steidl & Teuber [15] (Sect. 4), and present an anisotropic spatio-temporal TV method for denoising image sequences (Sect. 5). In Sect. 6 we provide experimental results. Concluding remarks are given in Sect. 7.

2   Problem

We begin with the primal TV denoising approach (1). Inserting (2) in (1) yields

$$\min_{u \in BV(\Omega)} \left\{ \frac{1}{2} \|u - f\|^2 + \sigma_{\mathcal{C}}(u) \right\}. \tag{3}$$

We follow [4] to derive the corresponding dual problem. With the fact that $v \in \partial \sigma_{\mathcal{C}}(u) \Leftrightarrow u \in \partial(\sigma_{\mathcal{C}})^*(v) = \partial(\delta_{\mathcal{C}})(v)$ for the subdifferentials of the support function $\sigma_{\mathcal{C}}$ and the indicator function $\delta_{\mathcal{C}}$, where * denotes the Legendre-Fenchel transform, we find $\partial \sigma_{\mathcal{C}}(\overline{u}) = \{v \in L^2(\Omega) \colon \langle \overline{u}, v - u \rangle \geq 0, \ \forall u \in \mathcal{C}\}$. Thus the optimality condition for $\overline{u}$ minimizing (3) reads

$$f - \overline{u} \in \partial \sigma_{\mathcal{C}}(\overline{u}) \quad \Leftrightarrow \quad \langle \overline{u}, f - \overline{u} - u \rangle \geq 0, \quad \forall u \in \mathcal{C}.$$

Using the additive decomposition $f = \overline{u} + \overline{v}$, we find $\overline{v} = \Pi_{\overline{\mathcal{C}}}(f)$, where $\Pi_{\overline{\mathcal{C}}}$ denotes the projection onto the closure $\overline{\mathcal{C}}$ of $\mathcal{C}$. Finally, we end up with the dual problem

$$\inf_{p \in \mathcal{D}} F(p), \qquad F(p) := \frac{1}{2} \|f - \operatorname{div} p\|^2. \tag{4}$$

In the following, we study generalized adaptive denoising approaches that take into account dependencies of the primal and dual constraint sets $\mathcal{C}(u)$ and $\mathcal{D}(p)$, respectively, on the solutions themselves. To this end, let

$$\mathcal{C}(u) := \operatorname{div}\{p \in C_c^\infty(\Omega, \mathbb{R}^d) : p(x) \in \tilde{D}(x, u)\}, \tag{5}$$

where $\tilde{D}(x,u) : \Omega \times \mathrm{BV}(\Omega) \rightrightarrows \mathbb{R}^d$ is a set-valued mapping. In view of the dual problem (4), we define $D(x,p) := \tilde{D}(x, f - \operatorname{div} p)$ and

$$\mathcal{D}(p) := \{\tilde{p} \in C_c^\infty(\Omega, \mathbb{R}^d) : \tilde{p}(x) \in D(x,p)\}. \tag{6}$$

3   Approach

3.1   A Quasi-Variational Inequality

We approximate the space $C_c^\infty(\Omega, \mathbb{R}^d)$ by a finite-dimensional space $\mathbb{R}^{nd}$. Then the set-valued mapping $\mathcal{D}$ in (4) takes the form

$$\mathcal{D} : \mathbb{R}^{nd} \rightrightarrows \mathbb{R}^{nd}, \qquad p \mapsto \{\tilde{p} \in \mathbb{R}^{nd} : \tilde{p}_i \in D_i(p), i = 1, \ldots, n\}. \tag{7}$$

Here, $D_i(p) : \mathbb{R}^d \rightrightarrows \mathbb{R}^d, i = 1, \ldots, n$ is the discrete analogue of $D(x,p)$. Note that in the finite dimensional setting $\mathcal{D}(p)$ is compact. In order to show existence of a solution, if the constraint sets (5) and (6) vary, in analogy to [2, Prop. 4.7.1], we formulate our approach as a generalization of the variational inequality corresponding to the dual problem (4): find $\bar{p} \in \mathcal{D}(\bar{p})$ such that

$$\langle \nabla F(\bar{p}), p - \bar{p} \rangle \geq 0, \qquad \forall p \in \mathcal{D}(\bar{p}). \tag{8}$$

Notice the dependency of the dual constraint set on $\bar{p}$, that significantly generalizes the dual TV minimization problem.

3.2   Existence of Solutions

The existence of a solution to (8) can be shown under the following assumption:

Assumption 1. $D_i(p) : \mathbb{R}^d \rightrightarrows \mathbb{R}^d, i = 1, \ldots, n$ *have the following properties:*

1. *For fixed p the set $D_i(p)$ is a closed convex subset of $\mathbb{R}^d$.*
2. *There exists $c > 0$, such that for all i, p: $\{0\} \subset D_i(p) \subset \overline{B_c(0)}$, where $\overline{B_c(0)}$ is the closed unit ball. In particular, $D_i(p)$ is non-empty.*
3. *The projection $\Pi_{D_i(p)}(q)$ of q onto $D_i(p)$ for a fixed q is continuous w.r.t. p.*

Proposition 1. *Let $F := \frac{1}{2}\|f - \operatorname{div} p\|^2$ and $\mathcal{D}$ be defined as in (7), such that $D_i(p), i = 1, \ldots, n$ satisfy Assumption 1. Then the problem*

$$\text{find } \bar{p} \in \mathbb{R}^{nd} \text{ such that} \qquad \langle \nabla F(\bar{p}), p - \bar{p} \rangle \geq 0, \quad \forall p \in \mathcal{D}(\bar{p}) \tag{9}$$

has a solution.

The proof of Proposition 1 utilizes the following theorem and lemma.

Theorem 2. *(cf. Theorem 5.2 in [5]) Let $G \colon \mathbb{R}^m \to \mathbb{R}^m$ be a point-valued and $\mathcal{D} \colon \mathbb{R}^m \rightrightarrows \mathbb{R}^m$ be a set-valued mapping. Suppose that there exists a nonempty compact convex set P such that*

1. $\mathcal{D}(P) = \cup_{p \in P} \mathcal{D}(p) \subseteq P$;
2. $\mathcal{D}$ takes nonempty closed convex sets as values;
3. $\mathcal{D}$ is continuous, that is $\mathcal{D}(p^k) \to \mathcal{D}(\overline{p})$ whenever $p^k \to \overline{p}$, or in view of (2), denoting the projection onto $\mathcal{D}$ by $\Pi_{\mathcal{D}}(p)$, equivalently $\Pi_{\mathcal{D}(p^k)}(p) \to \Pi_{\mathcal{D}(\overline{p})}(p)$ for all p

Then there exists $\overline{p} \in \mathbb{R}^{nd}$ such that $\langle G(\overline{p}), p - \overline{p} \rangle \geq 0$ for all $p \in \mathcal{D}(\overline{p})$.

Lemma 1. *Let $\mathcal{D}$ be defined as in (7). Assume that for every $i = 1, \dots, n$ and $q \in \mathbb{R}^d$ the projection $\Pi_{D_i(p)}(q)$ is continuous w.r.t. p. Then $\Pi_{\mathcal{D}(p)}(q)$ is continuous for fixed $q \in \mathbb{R}^{nd}$.*

Proof. $\Pi_{\mathcal{D}}(q)$ can be written as $\Pi_{\mathcal{D}(p)}(q) = (\Pi_{D_1(p)}(q_1), \dots, \Pi_{D_n(p)}(q_n))^\top$. Thus each component of $\Pi_{\mathcal{D}(p)}$ is continuous, from which the continuity of $\Pi_{\mathcal{D}(p)}$ follows immediately. $\qquad\square$

Proof of Proposition 1: We apply Thm. 2. Conditions (1) and (2) follow from Assumption 1, that, in turn, has to be verified later, see Prop. 2, 3 and 4. Lemma 1 shows that also condition (3) holds. $\qquad\square$

3.3   Algorithm

We propose an algorithm for solving (9). Let us first consider the case where $\mathcal{D}$ does not dependend on the dual variable p. The problem then can be solved by a projected gradient method:

$$p^{k+1} = \Pi_{\mathcal{D}}\big(p^k - \tau \nabla F(p^k)\big), \qquad 0 < \tau < 2/L,$$

where L denotes the Lipschitz-constant of ∇F. In order to adapt to the dependency of $\mathcal{D}$ on p, we propose to use

$$p^{k+1} = p^k - \frac{1}{\lambda}\Big(p^k - \Pi_{\mathcal{D}^k}\big(p^k - \tau \nabla F(p^k)\big)\Big), \quad \mathcal{D}^k := \mathcal{D}(p^k),$$

with sufficiently large $\lambda \in (0, 1)$. In practice, two nested iterations, one outer iteration for updating $\mathcal{D}$ and one inner iteration for updating p, are used. Providing convergence results will be part of our future work. However, our experiments show that this iteration converges for λ sufficiently large.

4   Adaptive Anisotropic TV Minimization for Image Denoising

In order to improve the image quality of TV methods for denoising, Steidl & Teuber [15], proposed an anisotropic TV method based on two independent orientations. In this section, we demonstrate how this approach can be modified in order to fit into the ansatz presented above. As a consequence, Prop. 1 provides a theoretical underpinning. Before discussing the approach in [15] (Sect. 4.2), we describe the required modifications by means of a simpler model (Sect. 4.1).

4.1 Anisotropic TV with a Single Direction

Consider $u \in L^2(\Omega), \Omega \subset \mathbb{R}^2$. The aim is to define a convex set $\mathcal{D}(p)$ (cf. (6)) satisfying Assumption 1 in order to derive an anisotropic TV measure.

In our approach, we are interested in a *local* description of the set $\mathcal{D}$. To this end, we define $D(x, p)$ for any $x \in \Omega$, based on local edge information obtained from $u = f - \operatorname{div} p$. To be precise, for an edge being present at a location x the set $D(x, p)$ will be defined as a square with one side parallel to the edge.

In order to detect edges, we utilize the structure tensor $J(x, u)$ defined as follows: Let

$$J_0(x, u) := \nabla u_\sigma(x) \nabla u_\sigma(x)^\top, \tag{10}$$

where $u_\sigma := K_\sigma * u$ is a smoothed version of u, obtained by convolution with a Gaussian kernel K_σ with standard deviation $\sigma > 0$. The structure tensor $J(x, u)$ is given as

$$J(x, u) := K_\rho * J_0(x, u), \tag{11}$$

with $\rho > 0$. (Here the convolution is applied componentwise). Moreover, let $v_i(x, u)$ and $\lambda_i(x, u), i = 1, 2$ be the eigenvectors and eigenvalues of J, respectively. We assume w.l.o.g. that the eigenvalues of $J(x, u)$ are ordered, $\lambda_1(x, u) \geq \lambda_2(x, u) \geq 0$, with corresponding eigenvectors $v_1(x, u)$ and $v_2(x, u)$. For simplicity of notation, we omit the dependency of D, J, v_i and λ_i on x in the following.

Consider for a moment some arbitrary $r \in \mathbb{R}^2, \|r\| = 1$. We define the square $\mathcal{S}(r)$ with sides parallel to r and $r^\perp$ as

$$\mathcal{S}(r) := \{p \in \mathbb{R}^2 : |r^\top p| \leq \alpha, |(r^\perp)^\top p| \leq \alpha\}. \tag{12}$$

We would like to set $D(p) = \mathcal{S}(r(f - \operatorname{div} p))$ with $r(u) = v_1(u)$. But then the projection $\Pi_{D(p)}$ would not depend continuously on p, since the eigenvector $v_1(u)$ in general does not depend continuously on the entries of $J(u)$.

On the other hand, for $\mathcal{S}(r)$ as defined in (12), the mapping $r \to \Pi_{\mathcal{S}(r)}$ is continuous, as the following lemma shows. Moreover, $u = f - \operatorname{div} p$ depends continuously on p. Thus, asserting the continuity of $r(u)$ is sufficient to guarantee the continuity of $\Pi_{D(p)}$.

Lemma 2. *Let $\mathcal{S}(r)$ be defined as in (12). Then $\Pi_{\mathcal{S}(r)}(q)$ depends continuously on r for fixed but arbitrary q.*

Proof. For $q \in \mathcal{S}(r)$, we have $\Pi_{\mathcal{S}(r)}(q) = q$. For $q \notin \mathcal{S}(r)$ the projection onto $\mathcal{S}(r)$ can be calculated as follows: Let $j^* := \operatorname{argmin}_{j=1,\dots,4} \|q - \Pi_j(q)\|$, where Π_j is the projection on the j-th side of the square. Then $\Pi_{\mathcal{S}(r)}(q) = \Pi_{j^*}(q)$.

Each of the projections Π_j is a composition of the orthogonal projection onto a line and a projection from the line onto a line segment. Only the projection onto the line depends on the parameter r. Since the orthogonal projection Π onto a line $\{a + tb \mid t \in \mathbb{R}\}, \|b\| = 1$, which is given by $\Pi(q) = a + \langle q - a, b \rangle b$, depends continuously on a, b, the continuity of $\Pi_{\mathcal{S}(r)}(q)$ w.r.t. r follows. Obviously, the transition between the cases $q \in \mathcal{S}(r)$ and $q \notin \mathcal{S}(r)$ is continuous. $\square$

In the following, we describe the construction of a vector $r(u)$ depending continuously on u, such that $r(u) = v_1(u)$, if $\lambda_1(u) \gg \lambda_2(u)$.

Note that the eigenvectors of $J(u) \in \mathbb{R}^{2 \times 2}$ depend continuously on u, as long as the eigenvalues $\lambda_1(u)$ and $\lambda_2(u)$ differ (cf. Theorem 3 in [12]). We define

$$\text{coh}(u) := \lambda_1(u) - \lambda_2(u) \geq 0.$$

Note that $\text{coh}(u)$ depends continuously on u, since the eigenvalues depend continuously on $J(u)$ (cf. e.g. Theorem of Wielandt-Hoffman in [16]), and $J(u)$, which is a composition of convolution and differentiation, is continuous w.r.t. u.

Now let $g : \mathbb{R}_0^+ \to [0,1]$ be a continuous and increasing function, such that $g(0) = 0$ and $\lim_{x \to \infty} g(x) = 1$. Moreover, let $I(p,q,t) : S^1 \times S^1 \times [0,1] \to S^1$ be a continuous interpolation from p to q on the unit sphere S^1 with the properties, that $I(p,q,1) = p$, $I(p,q,0) = q$ and $\|I(p,q,t) - q\| \leq C\|t\|$ for some $C > 0$. (For example, a steady rotation of vector p onto q suffices.) We set

$$D(p) := \mathcal{S}(r(f - \text{div}\, p)), \tag{13}$$

where $r(u) := I(v_1(u), (1,0)^\top, g(\text{coh}(u)))$. The set $D(p)$ satisfies Assumption 1:

Proposition 2. *Let $D(p)$ be defined as in (13). Then*

1. $D(p)$ *is closed, convex and satisfies* $\{0\} \subset D(p) \subset \overline{B_{\sqrt{2}\alpha}(0)}$.
2. *For fixed* $q \in \mathbb{R}^2$, $u \to \Pi_{D(p)}(q)$ *is continuous.*

The proof of Prop. (2) utilizes the following lemma:

Lemma 3. *Let q be fixed. Then $r(u) = I(v_1(u), q, g(\text{coh}(u)))$ depends continuously on u.*

Proof. We distinguish between the cases $\text{coh}(u) > 0$, and $\text{coh}(u) = 0$. In the first case, $v_1(u)$ is an eigenvector to an isolated eigenvalue and thus depends continuously on $J(u)$, see [12]. Moreover $\text{coh}(u)$ depends continuously on $J(u)$ (cf. [16]). Since $J(u)$ is a composition of convolutions and differentiation, it depends continuously on u; thus $\text{coh}(u)$ and $v_1(u)$ are continuous. The continuity of $r(u)$ at $u, \text{coh}(u) > 0$ then follows from the continuity of I and g.

In the second case, $\text{coh}(u) = 0$, we find from the continuity of $\text{coh}(u)$ that $\text{coh}(u^k) \to 0$ for every sequence u^k converging to u. Then the continuity of $r(u)$ follows from

$$\|r(u^k) - r(u)\| = \|I(u^k, q, g(\text{coh}(u^k))) - I(u, q, g(\text{coh}(u)))\|$$
$$= \|I(u^k, q, g(\text{coh}(u^k))) - q\| \leq Cg(\text{coh}(u^k)) \to 0,$$

using the properties of the interpolation I. □

Proof of Prop. 2: (i) The set $D(p)$ is a closed square with center 0 and sides of length $2\alpha > 0$. (ii) Lemma 3 provides the continuity of $r(u)$. Moreover, $u = f - \text{div}\, p$ depends continuously on p. Together with Lemma 2 the continuity of $D(p)$ follows. □

4.2   Anisotropic TV with Double Directions

Steidl & Teuber [15] proposed an anisotropic TV method based on the estimation of two orientations $r_1, r_2 : \Omega \to \mathbb{R}^2$. They consider the variational problem:

$$\min_u \frac{1}{2}\|u - f\|^2 + \alpha \left(|r_1^\top \nabla u| + |r_2^\top \nabla u| \right), \tag{14}$$

where two models for obtaining r from the data f are proposed. As an alternative to (14), they propose to use infimal convolution. In the dual formulation of (14) the set $D = D(f)$ is a parallelogram with sides $r_i, i = 1, 2$:

$$\mathcal{P}(r_1, r_2) := \{p \in \mathbb{R}^2 : |r_1^\top p| \le \alpha, \ |r_2^\top p| \le \alpha\}. \tag{15}$$

In our considerations, we concentrate on the 'occlusion model' described in [15]. Moreover, we consider r_i depending on the unknown $u := f - \operatorname{div} p$ and, by introducing slight changes of the original approach, guarantee the applicability of the theoretical results of Sect. 3. The orientations r_i are obtained as follows.

Let $\nu(u) := \left((\partial_x u_\sigma)^2, \partial_x u_\sigma \partial_y u_\sigma, (\partial_y u_\sigma)^2 \right)^\top$, where u_σ is defined as in Sect. 4.1. For the occlusion model, the following structure tensor is utilized:

$$J_0(u) := \nu(u)\nu^\top(u), \qquad J(u) := K_\rho * J_0(u),$$

where the convolution is applied componentwise.

Now let $\lambda_1(u) \ge \lambda_2(u) \ge \lambda_3(u) \ge 0$ denote the eigenvalues of $J(u)$, and $v_1(u)$, $v_2(u)$ and $v_3(u)$ the corresponding eigenvectors.

Analogously to the previous section, in view of the continuity of $v_i(u)$ we have to deal with non-isolated eigenvalues. To this end, we define

$$\operatorname{coh}_1(u) := \lambda_1(u) - \lambda_2(u), \quad \operatorname{coh}_2(u) := \lambda_2(u) - \lambda_3(u).$$

In order to define $r_1(u), r_2(u)$, we consider the following cases:

Case 1 & 2 – corners ($\operatorname{coh}_2(u) > 0$): Steidl & Teuber distinguish between the cases $v_{3,1} \ne 0$ and $v_{3,1} = 0$ ($v_{3,1}$ being the first entry of v_3). In the case $v_{3,1} \ne 0$, they propose to use the unit vectors $r_1^1(u) \parallel (v_{3,1}(u), y_1(u))^\top$ and $r_2^1(u) \parallel (v_{3,1}(u), y_2(u))^\top$, where $y_1(u), y_2(u)$ are the solutions of the quadratic equation $y^2 + v_{3,2}(u)\, y + v_{3,1}(u)\, v_{3,3}(u) = 0$. Otherwise, the unit vectors $r_1^2(u) \parallel (v_{3,2}(u), v_{3,3}(u))^\top$ and $r_2^2(u) \parallel (-v_{3,3}(u), v_{3,2}(u))^\top$ can be used.

Case 3 – edges ($\operatorname{coh}_2(u) \approx 0$, $\operatorname{coh}_1(u) > 0$): Since we can only guarantee that eigenvalue $\lambda_1(u)$ is isolated, we determine r_1, r_2 depending on the eigenvector $v_1(u)$. Along straight edges, the eigenvector v_1 is parallel to the normal of the edge. Therefore v_1 and $v_1^\perp$ are suitable for defining the orientation for anisotropic TV at edges. We set $r_1^3(u) \parallel (v_{1,1}(u), v_{1,2}(u))^\top$ and $r_2^3(u) \parallel (-v_{1,2}(u), v_{1,1}(u))^\top$.

Case 4 – homogeneous regions ($\operatorname{coh}_1(u) \approx \operatorname{coh}_2(u) \approx 0$): We use the default orientations $r_1^4(u) := (1, 0)^\top$ and $r_2^4(u) := (0, 1)^\top$.

In general, $r_1(u), r_2(u)$ have to be continuous interpolations between the above cases. For $i = 1, 2$ let

$$r_i(u) = I\Big(I\left(r_i^1(u), r_i^2(u), g(|v_{3,1}(u)|) \right), I\left(r_i^3(u), r_i^4(u), g(\operatorname{coh}_1(u)) \right), g(\operatorname{coh}_2(u)) \Big) \tag{16}$$

using g and I as defined in the previous section.

Proposition 3. *Let $D(p) = \mathcal{P}(r_1(f - \operatorname{div} p), r_2(f - \operatorname{div} p))$ with $\mathcal{P}(r_1, r_2)$ being the parallelogram defined in (15) and $r_i(u)$, $i = 1, 2$ defined as in (16).*

1. *$D(p)$ is closed, convex and satisfies $\{0\} \subset D(p) \subset \overline{B_{2\alpha}(0)}$.*
2. *$\Pi_{D(p)}(q)$ for fixed q depends continuously on p.*

In particular, $D(p)$ satisfies Assumption 1.

Proof. The first claim follows from the fact that $D(p)$ is a closed parallelogram with sides of length α. For the continuity of $\Pi_{D(p)}(q)$, we observe that the vectors $r_i^j(u)$, $i = 1, 2$, $j = 1, \ldots, 4$ are defined in a way that they depend continuously on $u = f - \operatorname{div} p$. The continuity of $r_i(u)$, $i = 1, 2$ is guaranteed by smooth interpolation (c.f. proof of Proposition 3). The proof of the continuity of $\mathcal{P}(r_1, r_2)$ is analogous to the proof of Lemma 2. □

5   Anisotropic Spatio-temporal TV Minimization

In the following we describe a spatio-temporal TV minimization approach. We interpret time as third coordinate, thus $u, f : \Omega \subset \mathbb{R}^3 \to \mathbb{R}$.

To obtain directional information, we utilize the three-dimensional structure tensor $J_\rho(u)$ defined analogously to (10) and (11). Let $\lambda_1(u) \geq \lambda_2(u) \geq \lambda_3(u) > 0$ denote the eigenvalues and $v_1(u), v_2(u), v_3(u)$ the eigenvectors of $J_\rho(u)$.

Let us assume that a two-dimensional surface is present in $u_\sigma(x)$. Then $\lambda_1(u) \gg \lambda_2(u)$ and $v_1(u)$ approximates the normal to this surface. The idea is to penalize variations mainly in directions tangential to the surface. To this end we set

$$D(p) := \mathcal{E}(v_1(f - \operatorname{div} p), \alpha, \beta),$$

where $\mathcal{E}(r, \alpha, \beta) := \{q \in \mathbb{R}^3 : |r^\top q|^2/\beta^2 + \|q - rr^\top q\|^2/\alpha^2 \leq 1\}$, $0 < \beta \ll \alpha$.

In homogeneous regions, where a unique orientation r can not be estimated, we choose $D(p) := B_\alpha(0)$. A continuous transition between both cases is obtained by defining

$$D(p) := \mathcal{E}(r(f - \operatorname{div} p), \tilde{\alpha}(f - \operatorname{div} p), \beta), \tag{17}$$

where

$$\operatorname{coh}_1(u) := \lambda_1(u) - \lambda_2(u) \gg 0,$$
$$r(u) := I\left(v_1(u), (0, 0, 1)^\top, g(\operatorname{coh}_1(u))\right),$$
$$\tilde{\alpha}(u) := g(\operatorname{coh}_1(u))\alpha + (1 - g(\operatorname{coh}_1(u)))\beta.$$

In order to remove speckles and similar kinds of distortions, an adaptation of (17) is required. This is due to the fact that at speckles, $v_1(u)$ is in direction of $(0, 0, 1)^\top$. Using (17) with the above $\tilde{\alpha}$ then would lead to a penalization of ∇u mainly in spatial directions, which is not suitable for removing distortions of medium/large scale in spatial directions. Instead we propose to use (17) with

$$\tilde{\alpha}(u) = g(\operatorname{coh}_1(u))g(\phi(u))\alpha + (1 - g(\operatorname{coh}_1(u))g(\phi(u)))\beta, \tag{18}$$

where $\phi(u)$ is the angle between $v_1(u)$ and $(0, 0, 1)^\top$. The above modification leads to stronger smoothing of surfaces parallel to the x_1, x_2-axes.

Fig. 1. 2D anisotropic TV filtering of artificial test image. Left: noisy test images, middle: filtering with the standard ROF model, right: anisotropic filtering with double directions. All images are scaled with respect to the intensity range of the original test image. Undesirable smoothing effects are considerably reduced on the right.

Fig. 2. 2D anisotropic filtering of real-world test image. Left: noisy test images, middle: result of standard ROF minimization, right: result of anisotropic TV minimization with double directions. Undesirable smoothing effects are considerably reduced on the right, see Fig. 3 for detailed views.

Proposition 4. *The set $D(p)$ defined in* (17) *with the above definitions of $\tilde{\alpha}(u)$ satisfies Assumption 1.*

Proof. The set $D(p)$ is a closed ellipsoid and therefore is convex. Its half-axes are bounded by $\max\{\alpha, \beta\}$, thus $0 \subset D(p) \subset B_{\max\{\alpha,\beta\}}(0)$. The projection onto the ellipsoid $\mathcal{E}(r, \tilde{\alpha}, \beta)$ can be expressed as a continuous function of r, $\tilde{\alpha}$, β and one distinct root of a rational function, see [7]. In a surrounding of this root, the function depends continuously on the half-axes. Thus the root depends continuously on r, $\tilde{\alpha}$ and β. r and $\tilde{\alpha}$ depend continuously on $u = f - \operatorname{div} p$, as $\operatorname{coh}_1(u)$ and $\phi(u)$ do. Moreover, u depends continuously on p. $\qquad\square$

6   Experiments

6.1   Anisotropic TV Minimization with Double Directions

We present experimental results for the anisotropic TV model with $D(p)$ as defined in (15) and r_1, r_2 as defined in (16). We compare this method with standard ROF minimization, using the same regularization parameter α. We consider two different test images, both with artificial noise.

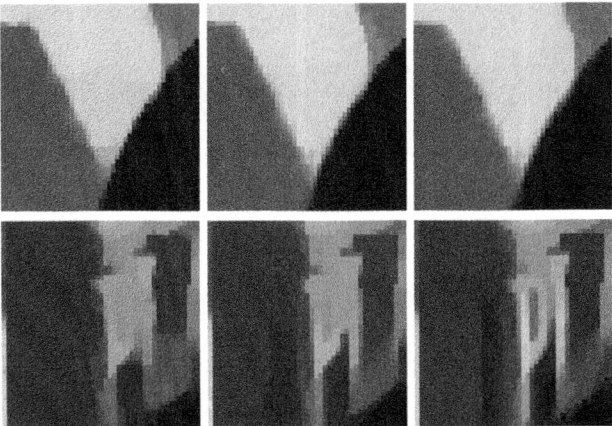

Fig. 3. Zoom into two regions of the filtered images shown in Fig. 2. Left: standard ROF, middle: anisotropic TV minimization with $D = D(f)$, right: adaptive anisotropic TV minimization with $D = D(p)$. It can be observed that adaptivity of the TV regularization improves with increasing number of iterations.

For the first test image (cf. Fig. 1, left) we use $\alpha = 0.6$ and 10 outer iteration steps. The results of the standard and anisotropic TV model are shown in Fig. 1, middle and right, respectively. A comparison shows, that anisotropic TV minimization better reconstructs corners of parallelogram and produces less smoothing at corners (as already demonstrated in [15]).

The second test image is a real world image with artificial noise, cf. Fig. 2, left. The result of standard ROF and anisotropic TV minimization for $\alpha = 0.4$ and 10 outer iteration steps is depicted in Fig. 2, middle and right, respectively.

In order to highlight differences, we zoom into two regions of the image: Fig. 3 shows the results for the standard ROF model (left), the result of applying anisotropic TV minimization with double directions, where the constraint set depends only on the data f, i.e. $D = D(f)$ (middle), and the result of anisotropic TV with the constraint set depending on the solution, $D = D(p)$ (right). It can be observed that anisotropic filtering leads to an improved and more regular reconstruction of edges and less stair-casing. If the constraint sets depend on the solution itself, an adaption to local structures can be observed during the iterations, see Fig. 3, bottom right. Here, the reconstruction of the characters improves when using fully adaptive constraint sets.

6.2 Adaptive Motion-Based TV Minimization for Image Sequences

In our example for spatio-temporal TV minimization, we consider an image sequence taken with a time-of-flight (ToF) camera, see Fig. 4 (4 frames out of the whole sequence). ToF cameras provide a depth map of the captured scene. The noise and speckles, which can be observed in the original data, are introduced by the camera system.

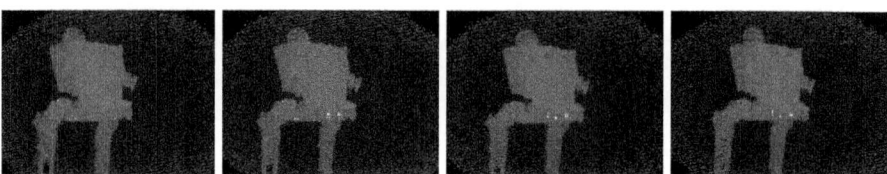

Fig. 4. Four exemplarily selected frames of a sequence of depth maps taken with a time-of-flight camera

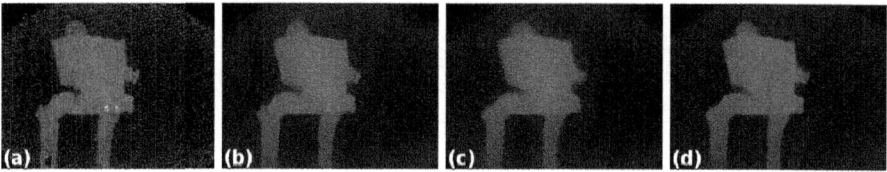

Fig. 5. (a) one of the original frames with real noise. (b) frame filtered with standard 2D ROF. (c) frame filtered with standard 3D ROF. (d) frame filtered with proposed adaptive TV minimization. Only the spatio-temporal methods are able to remove both noise and speckles. Anisotropic TV keeps the result sharper than isotropic 3D TV minimization.

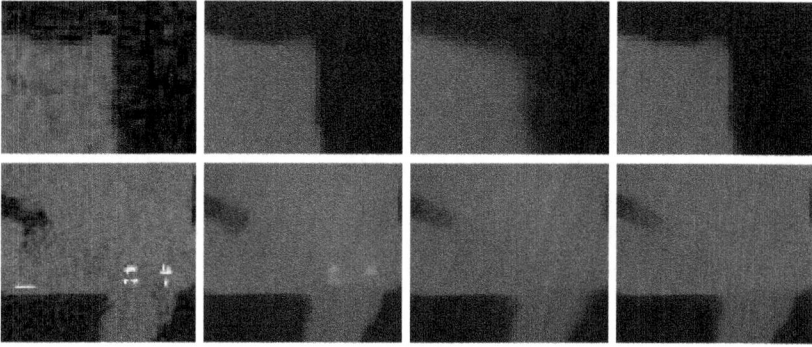

Fig. 6. Zoom into two regions of the depth map shown in Fig. 5. First column: original data, second column: result of standard 2D ROF filtering, third column: 3D ROF filtering, fourth column: proposed adaptive TV minimization. Only the spatio-temporal methods are able to remove both noise and speckles. Anisotropic TV keeps the result sharper than 3D ROF minimization.

For filtering, we propose to use spatio-temporal anisotropic TV with $D(p)$ as defined in (17) and $\tilde{\alpha}$ defined as in (18). As parameters, we chose $\alpha = 0.3$, $\beta = 0.001$ and 10 steps for the outer iteration. The result for one specific frame is depicted in Fig. 5, right. We compare this method with standard 2D ROF (Fig. 5, second left) and 3D ROF in the spatio-temporal domain (Fig. 5, second right), using the same parameter $\alpha = 0.3$. Additionally, we zoom into two image regions, see Fig. 6. We observe that standard 2D ROF filtering provides a good noise removal with preserving edges, but is not able to remove the speckles. 3D ROF filtering removes both noise and speckles, but introduces some blurring of edges, which is due to the stair-casing effect in 3D. The proposed adaptive anisotropic TV comprises both the advantages of the 2D and 3D isotropic model: it removes noise and speckles, while edges in each individual frame are kept sharp.

7   Conclusion

In this work we have presented a general approach for adaptive total variation. Existence results as well as a first basic algorithm have been provided. Several applications demonstrate the usability of our concept. As future work, we will support our framework with convergence results and investigate efficient numerical solvers.

References

1. Berkels, B., Burger, M., Droske, M., Nemitz, O., Rumpf, M.: Cartoon extraction based on anisotropic image classification. In: Vision, Modeling, and Visualization Proceedings, pp. 293–300 (2006)
2. Bertsekas, D.P., Nedic, A., Ozdaglar, A.E.: Convex Analysis and Optimization (2003)
3. Buades, A., Coll, B., Morel, J.M.: A review of image denoising algorithms, with a new one. Multiscale Model. Simul. 4(2), 490–530 (2005)
4. Chambolle, A.: An algorithm for total variation minimization and applications. J. Math. Imaging Vision 20(1–2), 89–97 (2004)
5. Chan, D., Pang, T.S.: The generalized quasi-variational inequality problem. Math. Operat. Res. 7(2), 211–222 (1982)
6. Dong, Y., Hintermüller, M.: Multi-scale total variation with automated regularization parameter selection for color image restoration. In: Tai, X.-C., Mørken, K., Lysaker, M., Lie, K.-A. (eds.) SSVM 2009. LNCS, vol. 5567, pp. 271–281. Springer, Heidelberg (2009)
7. Eberly, D.: Distance from a point to an ellipse in 2D. Technical report (2002)
8. Gilboa, G., Osher, S.: Nonlocal operators with applications to image processing. Multiscale Model. Simul. 7(3), 1005–1028 (2008)
9. Grasmair, M.: Locally adaptive total variation regularization. In: Tai, X.-C., Mørken, K., Lysaker, M., Lie, K.-A. (eds.) SSVM 2009. LNCS, vol. 5567, pp. 331–342. Springer, Heidelberg (2009)
10. Grasmair, M., Lenzen, F.: Anisotropic Total Variation Filtering. Appl. Math. Optim. 62(3), 323–339 (2010)
11. Kindermann, S., Osher, S., Jones, P.W.: Deblurring and denoising of images by nonlocal functionals. Multiscale Model. Simul. 4(4), 1091–1115 (2005) (electronic)
12. Rellich, F.: Störungstheorie der Spektralzerlegung, I. Math. Ann (1936)
13. Rudin, L.I., Osher, S., Fatemi, E.: Nonlinear total variation based noise removal algorithms. Phys. D 60(1–4), 259–268 (1992)
14. Setzer, S., Steidl, G., Teuber, T.: Restoration of images with rotated shapes. Numerical Algorithms 48, 49–66 (2008)
15. Steidl, G., Teuber, T.: Anisotropic smoothing using double orientations. In: Tai, X.-C., Mørken, K., Lysaker, M., Lie, K.-A. (eds.) SSVM 2009. LNCS, vol. 5567, pp. 477–489. Springer, Heidelberg (2009)
16. Wilkinson, J.H.: The algebraic eigenvalue problem. In: Numerical Mathematics and Scientific Computation. Oxford University Press, Oxford (1988)

Simultaneous Denoising and Illumination Correction via Local Data-Fidelity and Nonlocal Regularization

Jun Liu[1], Xue-cheng Tai[2], Haiyang Huang[1], and Zhongdan Huan[1]

[1] School of Mathematical Sciences
Laboratory of Mathematics and Complex Systems,
Beijing Normal University, Beijing 100875, P.R. China
[2] School of Physical and Mathematical Sciences,
Nanyang Technological University, Singapore and
Department of Mathematics, University of Bergen, Norway

Abstract. In this paper, we provide a new model for simultaneous denoising and illumination correction. A variational framework based on local maximum likelihood estimation (MLE) and a nonlocal regularization is proposed and studied. The proposed minimization problem can be efficiently solved by the augmented Lagrangian method coupled with a maximum expectation step. Experimental results show that our model can provide more homogeneous denoisng results compared to some earlier variational method. In addition, the new method also produces good results under both Gaussian and non-Gaussian noise such as Gaussian mixture, impulse noise and their mixtures.

1 Introduction

Image denoising is a fundamental technique of image processing. A large number of denoising methods have been proposed. It is common to assume that the noise is additive, i.e.

$$f(x) = u(x) + n(x),$$

where $f, u, n : \Omega \subseteq \mathbb{R}^2 \mapsto \mathbb{R}$ are the observed noisy image, true image and noise, respectively. Image denoising is to recover u for any given f and a priori knowledge of n. Variational method is one of the most efficient methods. It has now grown as a popular and widely used tool in image processing. Since the ROF model was proposed in [1], many variants based on total variation (TV) had been designed for different denoising tasks due to its good edges-preserving properties. Extending ROF, the authors in [2, 3, 4, 5] have used L_1 norm or its linear combinations as the fidelity term to removing impulse noise. In order to better preserve some small structures such as textures, an efficient method called nonlocal mean was discussed in [6]. Motivated by the nonlocal mean and the graph theory, the nonlocal TV variational framework base on nonlocal operators was proposed in [7]. In [8], it was extended to nonlocal Mumford-Shah regularizers for image restoration. However, all these methods do not consider

A.M. Bruckstein et al. (Eds.): SSVM 2011, LNCS 6667, pp. 218–230, 2012.

the varying illumination in the images. Moreover, it is also hard to treat cases that the intensity values are inhomogeneous.

Illumination correction or bias field correction is very important for real images. The artifacts caused by smooth, spatially varying illumination, although not usually a problem for visual inspection, can dramatically impede automated processing of the images. A widely accepted bias model, such as in MRI data, is the multiplicative bias field, which assumes that the observed signal f is equal to an uncorrupted signal u scaled by some bias β, i.e. $f = \beta u$. Then the application of a logarithmic transformation to the intensities allows the artifact to be modeled as an additive bias field $\ln f = \ln u + \ln \beta$. There are some works based on this logarithmic additive model for image segmentation such as [9,10,11] etc..

Motivated by modeling the illumination bias with a multiplicative field in segmentation problem, in this paper, we propose an unified model for denoising and correcting illumination simultaneously with different types of noise include Gaussian noise, impulse noise and their mixtures. Our model is built on MLE and nonlocal regularization. To be different from the traditional regularized MLE, we construct a novel block-based adaptive data-fidelity term to handle inhomogeneous illumination and the noise. Besides, our approach do not need any additional constraints such as regularization on the bias function β to keep it smooth. Anther superiority of this model is that it can work well under different types noise like Gaussian, impulse noise, Gaussian noise plus impulse noise. The new model can be efficiently optimized by an extended augmented Lagrangian method (ALM) for nonlocal regularization according to the recently proposed ALM framework [13,14] together with a maximum expectation process. These algorithms extend the Split-Bregman method of [17].

The rest of the paper is organized as follows: Section 2 gives our proposed model. Section 3 contains the optimization algorithms, while numerical experiments are presented in Section 4

2   The Proposed Model

2.1   Some Model Assumptions

In this paper, we consider the noise model with illumination bias

$$f(x) = \beta(x)(u(x) + n(x)), \tag{1}$$

where f is an observed noisy image, u stands for the ground truth image, n represents noise and β is a illumination bias function. In order to get a suitable denoising cost functional, we have the following assumptions:

- A1: the noise $n(x)$ at each location x is a realization of a random variable ξ with Gaussian mixture probability density function (PDF) $\sum_{k=1}^{K} \gamma_k p_k(z; c_k, \sigma_k^2)$. Here $p_k(z; c_k, \sigma_k^2)$ is the 1-D Gaussian PDF parameterized by mean c_k and variance σ_k^2, and γ_k is the mixture ratio which satisfies $\sum_{k=1}^{K} \gamma_k = 1$. This is an extended Gaussian noise model.

– A2: the bias function $\beta(x) > 0$, and β is smoothly varying. Motivated by [12], we use the following method to describe the smoothness of β: In a small neighborhood O_x centered at x, β satisfies $\beta(y) \approx \beta(x)$ when $y \in O_x$.

Now, we suppose the intensity value of the observed pixels at location x, namely $f(x)$, is a realization of a random variable η, then according to assumption A1 and model (1), we have

Proposition 1. *The PDF of η has the expression*

$$p_\eta(z) = \sum_{k=1}^{K} \frac{\gamma_k}{\sqrt{2\pi}\sigma_k \beta(x)} \exp\left(-\frac{[z - c_k \beta(x) - u(x)\beta(x)]^2}{2\sigma_k^2 \beta^2(x)}\right).$$

In the next, we shall use this PDF to construct a local data-fidelity in terms of MLE and some model assumptions.

2.2   The Local Fidelity Term

Let us construct a new local data term according to the pixel density function in the section. Let $\Theta = \{\gamma_1, \cdots, \gamma_K, c_1, \cdots, c_K, \sigma_1^2, \cdots, \sigma_K^2, \beta\}$ is a parameter set. By independence assumption of $f(x)$, the PDF expression in the proposition 1 and a likelihood process, one can get the continuous functional in a neighborhood O_y centered at y

$$L_y(u, \Theta; f) = \int_{O_y} \ln \sum_{k=1}^{K} \frac{\gamma_k}{\sqrt{2\pi}\sigma_k \beta(x)} \exp\left(-\frac{[f(x) - c_k \beta(x) - u(x)\beta(x)]^2}{2\sigma_k^2 \beta^2(x)}\right) dx.$$

Note that $\beta(x)$ can be replaced by $\beta(y)$ when $x \in O_y$ in terms of assumption A2, thus L_y becomes

$$L_y(u, \Theta; f) = \int_{O_y} \ln \sum_{k=1}^{K} \frac{\gamma_k}{\sqrt{2\pi}\sigma_k \beta(y)} \exp\left(-\frac{[f(x) - c_k \beta(y) - u(x)\beta(y)]^2}{2\sigma_k^2 \beta^2(y)}\right) dx.$$

At this time, we get a local data fidelity term $D_y(u, \Theta) = -L_y(u, \Theta; f(x))$ in O_y. If we consider the different contributions to the fidelity D_y in terms of the distance from the neighborhood center, then we can assign some weights to different pixels. A common choice for this is the so-called Gaussian smoothness, and thus we get a cost functional

$$D_y(u, \Theta) = -\int_{O_y} G_\sigma(y - x) \ln \sum_{k=1}^{K} \frac{\gamma_k}{\sqrt{2\pi}\sigma_k \beta(y)} \exp\left(-\frac{[f(x) - c_k \beta(y) - u(x)\beta(y)]^2}{2\sigma_k^2 \beta^2(y)}\right) dx.$$

Here G_σ is a Gaussian kernel with a given standard deviation σ. Our objective is to recover all the degraded pixels. Thus we need to minimize all the local data fidelity. We shall use the cost functional

$$D(u, \Theta) = \int_\Omega D_y(u, \Theta) dy = -\int_\Omega \int_{O_y} G_\sigma(y - x) \ln \sum_{k=1}^{K} \frac{\gamma_k}{\sqrt{2\pi}\sigma_k \beta(y)} \exp\left(-\frac{[\frac{f(x)}{\beta(y)} - c_k - u(x)]^2}{2\sigma_k^2}\right) dx dy.$$

Using properties of the Gaussian kernel, $G_\sigma(y-x) \approx 0$ when $x \notin O_y$ by choosing an appropriate σ, the neighborhood O_y in the second integration can be dropped. However, this new data-fidelity term is not easy to minimize due to the log-sum function. We use the conclusion of the following proposition [15,16] to overcome this difficult.

Proposition 2. *For all $\alpha_k(x) > 0$, let $\Delta = \{\phi(x) = (\phi_1(x), \cdots, \phi_k(x)) : \sum_{k=1}^{K} \phi_k(x) = 1, \phi_k(x) > 0\}$, then*

$$-\ln \sum_{k=1}^{K} \alpha_k(x) \exp(-\psi_k(x)) = \min_{\phi(x) \in \Delta} \left\{ \sum_{k=1}^{K} (\psi_k(x) - \ln \alpha_k(x))\phi_k(x) + \sum_{k=1}^{K} \phi_k(x) \log \phi_k(x) \right\}.$$

By applying Proposition 2 with $\alpha_k(x) = \dfrac{\gamma_k}{\sqrt{2\pi}\sigma_k \beta(y)}$, $\psi_k(x) = \dfrac{[\frac{f(x)}{\beta(y)} - c_k - u(x)]^2}{2\sigma_k^2}$, $D(u, \Theta)$ becomes

$$D(u, \Theta) = \int_\Omega \int_\Omega G_\sigma(y-x) \min_{\phi(x) \in \Delta} \{ \sum_{k=1}^{K} [\frac{(\frac{f(x)}{\beta(y)} - c_k - u(x))^2}{2\sigma_k^2} - \ln \gamma_k + \ln(\sqrt{2\pi}\sigma_k \beta(y)) + \ln \phi_k(x)]\phi_k(x) \} dx dy.$$

Unlike the common methods to choose the negative log-likelihood as the data-fidelity term, we introduce a functional $E(u, \Theta, \phi)$ with an additional variable ϕ:

$$E(u, \Theta, \phi) =$$
$$\int_\Omega \int_\Omega G_\sigma(y-x) \sum_{k=1}^{K} \left[\frac{(\frac{f(x)}{\beta(y)} - c_k - u(x))^2}{2\sigma_k^2} - \ln \gamma_k + \ln(\sqrt{2\pi}\sigma_k \beta(y)) + \ln \phi_k(x) \right] \phi_k(x) dx dy,$$

and consider the the minimization problem

$$(u^*, \Theta^*, \phi^*) = \underset{u, \Theta, \phi \in \Delta}{\arg\min} \ E(u, \Theta, \phi) \tag{2}$$

to be solved by the following alternative minimization procedure:

$$\begin{cases} \phi^{\nu+1} = \underset{\phi \in \Delta}{\arg\min} \ E(u^\nu, \Theta^\nu, \phi), \\ (u^{\nu+1}, \Theta^{\nu+1}) = \underset{u, \Theta}{\arg\min} \ E(u, \Theta, \phi^{\nu+1}). \end{cases} \tag{3}$$

Actually, the above iteration scheme can be interpreted as the well-known expectation-maximization (EM) algorithm. The updating of ϕ and Θ corresponding to the E-step and M-step, respectively. One can also prove that

Proposition 3. *The sequence u^ν, Θ^ν produced by iteration scheme (3) satisfies $D(u^{\nu+1}, \Theta^{\nu+1}) \leqslant D(u^\nu, \Theta^\nu)$.*

Thus we can take $E(u, \Theta, \phi)$ as the data-fidelity term. Compared to the model that directly uses $D(u, \Theta)$, we get some close-form solutions for the sub-problems when optimizing $E(u, \Theta, \phi)$.

2.3   Nonlocal TV

Nonlocal regularization could preserve repeated structures and textures and at the same time remove noise. The nonlocal denoising method was first proposed by Buades *etc.* [6]. In [7], Gilboa and Osher defined a variational framework based nonlocal operators. Let us review some definitions and notations on nonlocal TV regularization. Let $\Omega \subset \mathbb{R}^2$, $H_1 = L^2(\Omega)$, $H_2 = L^2(\Omega \times \Omega)$ and $\omega(x, y) \in H_2$ be a nonnegative symmetric weight function. The nonlocal gradient operator $\nabla_\omega : H_1 \mapsto H_2$ is defined as the vector of all partial derivatives at x such that:

$$(\nabla_\omega \circ u)(x) \mapsto \nabla_\omega u(x, y) \triangleq (u(y) - u(x))\sqrt{\omega(x, y)}.$$

The inner product in H_1 and H_2 is defined as

$$< u, v >_{H_1} = \int_\Omega u(x)v(x)\mathrm{d}x, \quad < p, q >_{H_2} = \int_\Omega \int_\Omega p(x, y)q(x, y)\mathrm{d}y\mathrm{d}x.$$

Naturally, the isotropic L_1 and L_2 norms in H_2 is

$$||p||_1 = \int_\Omega \sqrt{\int_\Omega p(x, y)^2 \mathrm{d}y}\mathrm{d}x, \quad ||p||_2 = \sqrt{\int_\Omega \int_\Omega p(x, y)^2 \mathrm{d}y\mathrm{d}x}.$$

The nonlocal divergence operator $\mathrm{div}_\omega : H_2 \mapsto H_1$ is given by the standard adjoint relation

$$< \nabla_\omega u, p >_{H_2} = - < \mathrm{div}_\omega p, u >_{H_1},$$

which leads to

$$\mathrm{div}_\omega p(x) = \int_\Omega (p(x, y) - p(y, x))\sqrt{\omega(x, y)}\mathrm{d}y.$$

Thus the nonlocal Laplacian operator $\Delta_\omega : H_1 \mapsto H_1$ is given by

$$\Delta_\omega u(x) = \mathrm{div}_\omega \nabla_\omega u(x) = 2\int_\Omega (u(y) - u(x))\omega(x, y)\mathrm{d}y.$$

With these notations, the nonlocal TV functional

$$R_\omega(u) = ||\nabla_\omega u||_1 = \int_\Omega \sqrt{\int_\Omega (u(x) - u(y))^2 \omega(x, y)\mathrm{d}y}\mathrm{d}x.$$

In this paper, we shall use the following weighting function [6]:

$$\omega^f(x, y) = \exp\{-\frac{\int_\Omega G_a(z)(f(x + z) - f(y + z))^2 \mathrm{d}z}{2h^2}\}. \tag{4}$$

2.4   The Proposed Cost Functional

The data-fidelity term $E(u, \Theta, \phi)$ together with the nonlocal TV norm yield the following new cost functional for simultaneous denoising and illumination correction :

$$J(u, \Theta, \phi) = E(u, \Theta, \phi) + \mu ||\nabla_\omega u||_1,$$

where $\mu > 0$ is a regularization parameter.

We need to impose some constraint condition on the parameters

$$\Theta = \{\gamma_1, \cdots, \gamma_K, c_1, \cdots, c_K, \sigma_1^2, \cdots, \sigma_K^2, \beta\}$$

and ϕ. For γ_k, we require $\sum_{k=1}^{K} \gamma_k = 1$ since it represents the mixture ratio. The $\phi_k(x)$ is actually a probability distribution of the pixel $f(x)$ contaminated by the noise comes from the k-th Gaussian distribution with mean c_k and variance σ_k. Thus, the constraint $\phi \in \Delta$ can guarantee this.

3   Algorithm: Augmented Lagrangian Method and EM

Operator splitting is an efficient method to solve L_1 minimization. In recent years, many efficient algorithms based on operator splitting have appeared, such as split Bregman method [17], augmented Lagrangian method (ALM) [13, 14], Douglas-Rachford splitting [18] and so on. These methods are all equivalent under certain conditions. In [13, 14], the authors only considered the local L_1 regularization, here we extend the split Bregman method [17] following the framework of Tai and Wu [13, 14]. The nonlocal TV in our model can be efficiently optimized with ALM.

In order to apply augmented Lagrangian method, the original minimization problem

$$(u^*, \Theta^*, \phi^*) = \underset{u, \Theta, \phi \in \Delta}{\arg\min} \; J(u, \Theta, \phi)$$

is reformulated as a constraint optimization minimization problem:

$$(u^*, d^*, O^*, \phi^*) = \underset{u, d, \Theta, \phi \in \Delta}{\arg\min} \; E(u, \Theta, \phi) \mid \mu ||d||_1 \;\; \text{s.t.} \;\; d = \nabla_\omega u. \tag{5}$$

The augmented Lagrangian functional for this constrained minimization problem is:

$$\mathcal{L}(u, d, \Theta, \phi, \lambda) = E(u, \Theta, \phi) + \mu ||d||_1 + < \lambda, (d - \nabla_\omega u) >_{H_2} + \frac{r}{2} ||d - \nabla_\omega u||_2^2,$$

where the Lagrangian multiplier $\lambda(x, y) \in H_2$, and $r > 0$ is a penalty parameter. It can be shown that one of the saddle points $(\hat{u}, \hat{d}, \hat{\Theta}, \hat{\phi}, \hat{\lambda})$ of $\mathcal{L}(u, d, \Theta, \phi, \lambda)$ is a solution of (5). We can search a saddle point by the following alternative algorithm:

$$\begin{cases} (u^{\nu+1}, d^{\nu+1}, \Theta^{\nu+1}, \phi^{\nu+1}) = \underset{u, d, \Theta, \phi \in \Delta}{\arg\min} \; \mathcal{L}(u, d, \Theta, \phi, \lambda^\nu), \\ \lambda^{\nu+1} = \lambda^\nu + r(d^{\nu+1} - \nabla_\omega u^{\nu+1}). \end{cases} \tag{6}$$

First, let us derive the updating formulations for $u^{\nu+1}$ and $d^{\nu+1}$. Denote

$$H(u) = \int_\Omega \int_\Omega G_\sigma(y-x) \sum_{k=1}^{K} \frac{[\frac{f(x)}{\beta(y)^\nu} - c_k^\nu - u(x)]^2}{2(\sigma_k^2)^\nu} \phi_k^\nu(x) dx dy.$$

Ignoring the constant terms, the minimization problem for u and d can be rewritten as

$$(u^{\nu+1}, d^{\nu+1}) = \arg\min_{u,d} \ H(u) + \mu|d|_1 + \frac{r}{2}||d - \nabla_\omega u + \frac{\lambda^\nu}{r}||_2^2.$$

We define $b = -\frac{\lambda}{r}$, together with the second updating formula in (6), one get the following iterative scheme:

$$\begin{cases} (u^{\nu+1}, d^{\nu+1}) = \arg\min_{u,d} \ H(u) + \mu|d|_1 + \frac{r}{2}||d - \nabla_\omega u - b^\nu||_2^2, \\ b^{\nu+1} = b^\nu + \nabla_\omega u^{\nu+1} - d^{\nu+1}. \end{cases} \quad (7)$$

Note that (7) actually is the split Bregman iteration and b is the Bregman vector [19]. We shall use an alternative minimization for u and d. The Euler-Lagrange equation for u is:

$$(\sum_{k=1}^{K} \frac{\phi_k^\nu}{(\sigma_k^2)^\nu} - r\triangle_\omega)u = \sum_{k=1}^{K} \frac{\phi_k^\nu}{(\sigma_k^2)^\nu}\left[f(x)\int_\Omega G_\sigma(y-x)\frac{1}{\beta(y)^\nu}dy - c_k^\nu\right] + r\mathrm{div}_\omega(b^\nu - d^\nu),$$

$$(8)$$

The above equation is linear. Its approximate solution $u^{\nu+1}$ can be easily computed by a Gauss-Seidel process.

Once $u^{\nu+1}$ and b^ν is known, the minimizer $d^{\nu+1}$ is given by the following shrinkage operation:

$$d^{\nu+1} = \mathrm{shrink}(\nabla_\omega u^{\nu+1} + b^\nu, \frac{\mu}{r}) = \frac{\nabla_\omega u^{\nu+1} + b^\nu}{|\nabla_\omega u^{\nu+1} + b^\nu|}\max\{|\nabla_\omega u^{\nu+1} + b^\nu| - \frac{\mu}{r}, 0\}$$

$$(9)$$

For $\phi_k^{\nu+1}$ and $\Theta^{\nu+1}$, both of them have explicit solutions. To simplify the notations, we define

$$q_k^\nu(x) \triangleq \frac{\gamma_k^\nu}{\sqrt{(\sigma_k^2)^\nu}}\exp\left(-\frac{1}{2(\sigma_k^2)^\nu}\int_\Omega G_\sigma(y-x)\left(\frac{f(x)}{\beta^\nu(y)} - c_k^\nu - u^{\nu+1}(x)\right)^2 dy\right),$$

$$s^{\nu+1}(y) \triangleq \sum_{l=1}^{K}\int_\Omega G_\sigma(y-x)\frac{c_l^{\nu+1} + u^{\nu+1}(x)}{(\sigma_l^2)^{\nu+1}}f(x)\phi_l^{\nu+1}(x)dx,$$

$$t^{\nu+1}(y) \triangleq \sum_{l=1}^{K}\frac{1}{(\sigma_l^2)^{\nu+1}}\int_\Omega G_\sigma(y-x)f^2(x)\phi_l^{\nu+1}(x)dx.$$

Then the solutions for the E-step and M-step are given by:

$$
\begin{cases}
\phi_k^{\nu+1}(x) = \dfrac{q_k^\nu(x)}{\displaystyle\sum_{l=1}^{K} q_l^\nu(x)}, \\[4ex]
\gamma_k^{\nu+1} = \dfrac{\displaystyle\int_\Omega \phi_k^{\nu+1}(x)\mathrm{d}x}{\displaystyle\int_\Omega 1\,\mathrm{d}x}, \\[4ex]
c_k^{\nu+1} = \dfrac{\displaystyle\int_\Omega \phi_k^{\nu+1}(x)[f(x)\int_\Omega G_\sigma(y-x)\dfrac{1}{\beta^\nu(y)}\mathrm{d}y - u^{\nu+1}(x)]\mathrm{d}x}{\displaystyle\int_\Omega \phi_k^{\nu+1}(x)\mathrm{d}x}, \\[4ex]
(\sigma_k^2)^{\nu+1} = \dfrac{\displaystyle\int_\Omega \phi_k^{\nu+1}(x)\int_\Omega G_\sigma(y-x)\left(\dfrac{f(x)}{\beta^\nu(y)} - c_k^{\nu+1} - u^{\nu+1}(x)\right)^2 \mathrm{d}y\mathrm{d}x}{\displaystyle\int_\Omega \phi_k^{\nu+1}(x)\mathrm{d}x}, \\[4ex]
\beta^{\nu+1}(y) = \dfrac{-s^{\nu+1}(y) + \sqrt{(s^{\nu+1}(y))^2 + 4t^{\nu+1}(y)}}{2}.
\end{cases}
\tag{10}
$$

Our algorithm with weight ω updating can be summarized as in the following:
Algorithm 1 (ALM-EM algorithm). *Given K, Choosing $\Theta^0, u^0, \phi^0, b^0, d^0$, and the parameters μ, r. Let $\nu = 0$ and calculate the initial weight ω^f. Do:*
1. ALM step: updating $u^{\nu+1}, d^{\nu+1}$ and $b^{\nu+1}$ according to (8), (9) and the second equation in (7), respectively.
2. If $||u^{\nu+1} - u^\nu||_2^2 < 10^{-5}||u^\nu||_2^2$, end the algorithm; else go to the next step.
3. E-step: updating $\phi^{\nu+1}$ using the first equation of (10).
4. M-step: updating the parameter set $\Theta^{\nu+1}$ using in (10).
5. Updating weight: if $mod(\nu + 1, 5) == 1$, compute $\omega^{u^{\nu+1}/\beta^{\nu+1}}$ using (4). Set $\nu = \nu + 1$, and go to the ALM step.

4   Numerical Experiments

4.1   Parameters and Initial Values Selection

In this section, we give some guidelines and criterions on selection of the parameters and initial values. Here we suppose the observed image $f(x) \in [0, 1]$.

The parameter K is the number of the Gaussian PDF and it usually set to 2 or 3. Larger K can better models the true distribution of noise in some real applications, but the algorithm would be more time-consuming. In this paper, we set $K = 2$ for all the experiments.

The σ in the Gaussian kernel G_σ controls the smoothness of bias function β. Generally speaking, we need to choose a large value to keep the β smooth due to the fact that the illumination or intensity inhomogeneity in an images is often slow-varying. In our tests, we choose $\sigma = 10$.

The regularization parameter μ depends on the noise level. We find that μ in our model is not so sensitive to the noise level as in the nonlocal ROF model. This might be related to the fact that the introduction of the noise variance parameter σ_k^2, and σ_k^2 can adaptively balance the data-fidelity and the nonlocal TV terms together with μ. Experimental results show that $\mu \in [1, 15]$ can yield good results for different noise levels. In the experiments, unless otherwise specified, we set $\mu = 5$. In addition, we set penalty parameter $r = 200$.

The initial value $b^0 = d^0 = 0, \gamma_1^0 = \gamma_2^0 = \frac{1}{2}, c_1^0 = c_2^0 = 0, \sigma_1^2 = 0.1, \sigma_2^2 = 0.01, \phi_1 = 1, \phi_2 = 0$ are used. We can assume the desirable β to be around 1, and thus we set $\beta^0 = (G_\sigma * f + 1.5)/2$. Finally, we let $u^0 = G_\sigma * \frac{f}{\beta}$.

4.2   Experimental Results

We first mention that the proposed model will reduced to the nonlocal ROF model by setting control parameter $\beta = 1$ and others in Θ to be equal, i.e. $\gamma_1 = \gamma_2, \sigma_1^2 = \sigma_2^2$, and so on. Thus if the illumination of an image is very homogeneous and the noise obeys a single Gaussian distribution, our method produces similar results as the nonlocal ROF model.

The superiority of our model is that it can work well under inhomogeneous illumination even with noise mixing. We shall tests these out.

Fig. 1 shows the results of the nonlocal ROF model [7] and our model under Gaussian mixture noise. The original image is displayed in Fig.1(i), one can find that the illumination of the original image itself is not homogeneous and the intensity on the left side is slightly lighter than the one on the right. We add noise and get the observed image f as shown in Fig.1(a). Here, the image f is corrupted by two additive white Gaussian noise with standard deviation $\frac{75}{255}$ and $\frac{20}{255}$, respectively. The mixture ratio is about $1 : 3$. As can be seen from the Fig.1(b) and Fig.1(c), the denoising result provided by the proposed method is better than the nonlocal ROF model. Firstly, the intensity of the reconstructed image in Fig.1(c) is more homogeneous than the one in Fig.1(b). This is caused by the use of β in our model. It can correct the inhomogeneous illumination. Secondly, our method can better preserve details in the texture areas and simultaneously clean the noise in the flat areas by adaptively adjusting the data term and nonlocal TV term through the control parameters σ_k^2 and ϕ_k. We use PSNR $= 10 \log_{10} \frac{1}{\text{var}(f-\hat{f})}$ to evaluate the quality of the denoising images, where $f, \hat{f}$ are observed and reconstructed images, respectively. For the proposed model, obviously, we need to define $\hat{f} = \beta u$ and then calculate the PSNR to make comparisons with other methods. The PSNR values for nonlocal ROF and the proposed are 23.94 and 27.34, respectively. Some estimated functions and parameters in the proposed approach are illustrated in Fig.1(d)-1(g). For visualization, we normalized β in $[0,1]$ in Fig.1(d). The corrected noisy image can be found in Fig.1(e). We also calculate the variances $\sigma_f, \sigma_{\frac{f}{\beta}}$ of noisy image f and the corrected image $\frac{f}{\beta}$ respectively. We get $\sigma_f = 0.0576, \sigma_{\frac{f}{\beta}} = 0.0486$, which indicates the intensity in the latter image is more uniform. As mentioned earlier,

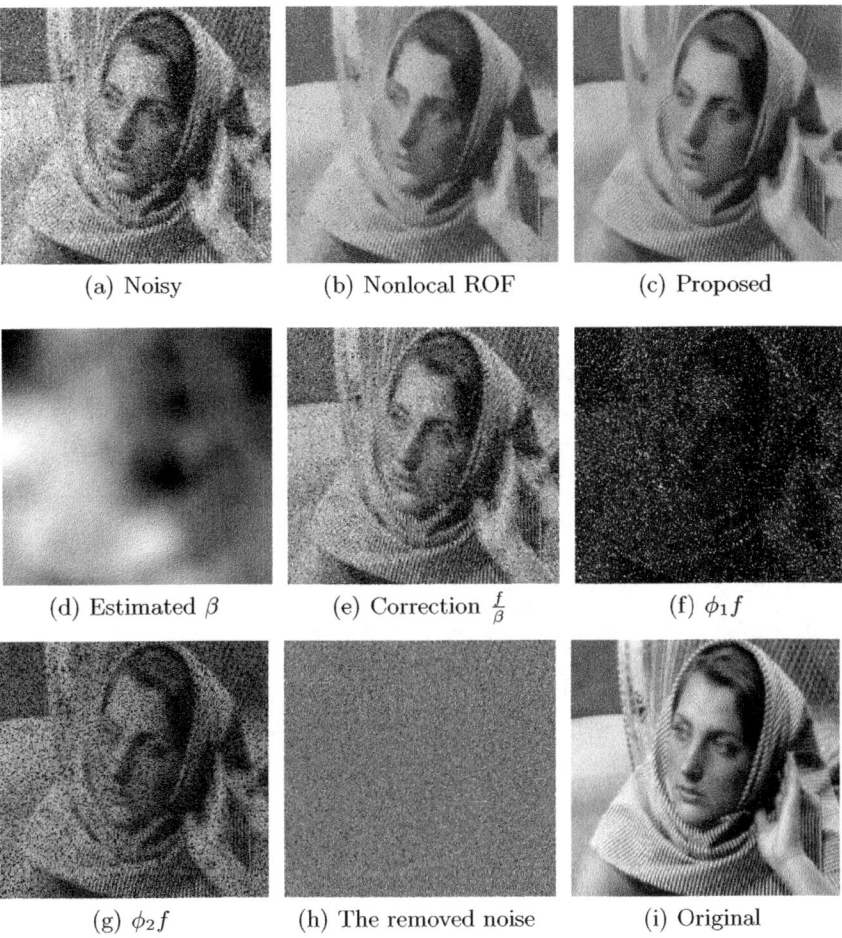

(a) Noisy   (b) Nonlocal ROF   (c) Proposed

(d) Estimated β (e) Correction $\frac{f}{\beta}$ (f) $\phi_1 f$

(g) $\phi_2 f$ (h) The removed noise   (i) Original

Fig. 1. A Comparison of nonlocal ROF [7] and the proposed model. (a)noisy image with PSNR= 17.01; (b) result with nonlocal ROF, PSNR= 23.94; (c) u, result with the proposed model, PSNR= 27.34 ; (d) the estimated bias function β; (e) the corrected image $\frac{f}{\beta}$; (f),(g) the estimated pixels with high level noise and low level noise, respectively; (h) the removed noise by the proposed method, i.e. $\frac{f}{\beta} - u$; (i) the original image.

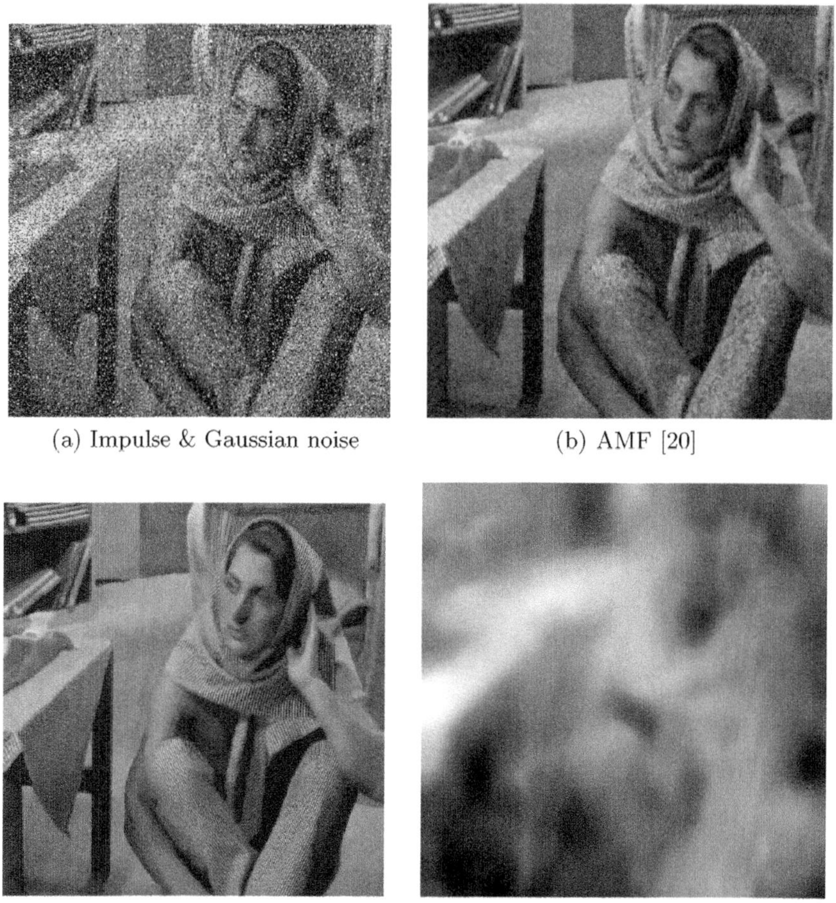

(a) Impulse & Gaussian noise

(b) AMF [20]

(c) Proposed

(d) Estimated β

Fig. 2. Results under impulse noise plus Gaussian noise. (a)noisy image; (b) denoising with adaptive median filter (AMF) in [20], PSNR=23.47; (c) denoising with the proposed model, PSNR= 27.16 ; (d) the estimated bias function β;

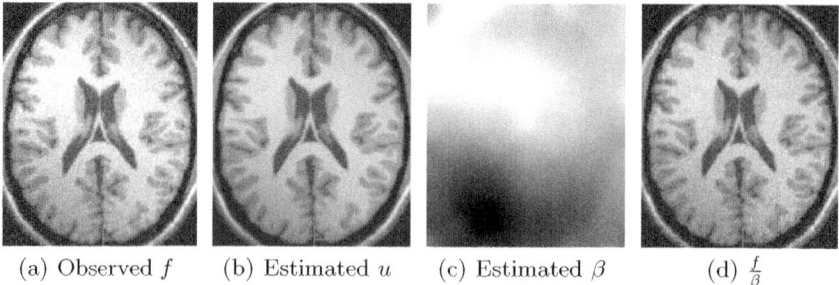

(a) Observed f (b) Estimated u (c) Estimated β (d) $\frac{f}{\beta}$

Fig. 3. Applying to MR image

our model can group the pixels into several clusters using different variances of the noise. In Fig. 1(f) and 1(g), the finally estimated partitions are displayed.

A denoising result with the proposed model under impulse noise plus Gaussian noise are given in Fig. 2. In this experiment, the image is contaminated by 25% salt-and-pepper noise together with Gaussian noise with standard deviation $\frac{15}{255}$. Here we take the common used adaptive median filter (AMF) [20] for comparison. It can be seen the AMF can clean impulse noise efficiently, but it fails in removing Gaussian noise and retaining the textures. Compared with the AMF, our method can give much better results. The denoised images and the estimated bias function provided by our method are displayed in the last two figures.

Fig. 3 shows result of applying the algorithm to MR images. In this experiment, we need to tune the regularization parameter $\mu = 15$ to get a smoothed image since the level of noise in the images is low. The denoised, corrected images and the estimated bias function β with the proposed algorithm are all illustrated in the last three figures. A benefit of the intensity correction is that the corrected images can be segmented easily with some center-based clustering methods such as Chan-Vese model, but it is very difficult to obtain a desirable segmentation result from the original data f.

5   Conclusion

We have presented an approach for simultaneous illumination correction and denoising. Numerical experiments demonstrated the method is very superior for mixed noise (e.g. impulse noise, Gaussian noise plus impulse noise etc.) compared to some earlier proposed nonlocal variational PDE based models. In addition, the non-uniform illumination function in the original data can be estimated and corrected by using the bias function. Our method can be extended to image segmentation, registration and some other computer vision problems.

References

1. Rudin, L., Osher, S., Fatemi, E.: Nonlinear total variation based noise removal algorithms. Physica D 60, 259–268 (1992)
2. Nikolova, M.: A variational approach to remove outliers and impulse noise. Journal of Mathematical Imaging and Vision 20, 99–120 (2004)
3. Bar, L., Sochen, N., Kiryati, N.: Image deblurring in the presence of impulsive noise. International Journal of Computer Vision 70, 279–298 (2006)
4. Cai, J., Chan, R., Nikolova, M.: Two-phase approach for deblurring images corrupted by impulse plus Gaussian noise. Inverse Problems and Imaging 2, 187–204 (2008)
5. Liu, J., Huang, H., Huan, Z., Zhang, H.: Adaptive variational method for restoring color images with high density impulse noise. International Journal of Computer Vision 90, 131–149 (2010)
6. Buades, A., Coll, B., Morel, J.M.: A review of image denoising algorithms, with a new one. Multiscale Modeling & Simulation 4(2), 490–530 (2005)

7. Gilboa, G., Osher, S.: Nonlocal operators with applications to image processing. Multiscale Modeling & Simulation 7(3), 1005–1028 (2008)
8. Jung, M., Vese, L.A.: Image restoration via nonlocal Mumford-Shah regularizers. UCLA C.A.M. Report 08-35 (2008)
9. Wells, W.M., Grimson, E.L., Kikinis, R., Jolesz, F.A.: Adaptive segmentation of MRI data. IEEE Transactions on Medical Imaging 15, 429–442 (1996)
10. Ahmed, M.N., Yamany, S.M., Mohamed, N., Farag, A.A., Moriarty, T.: A modified fuzzy c-means algorithm for bias field estimation and segmentation of MRI data. IEEE Transactions on Medical Imaging 21, 193–198 (2002)
11. Ashburner, J., Friston, K.J.: Unified segmentation. Neuroimage 26(3), 839–851 (2005)
12. Li, C., Huang, R., Ding, Z., Gatenby, C., Metaxas, D.N., Gore, J.C.: A variational level set approach to segmentation and bias correction of images with intensity inhomogeneity. In: Metaxas, D., Axel, L., Fichtinger, G., Székely, G. (eds.) MICCAI 2008, Part II. LNCS, vol. 5242, pp. 1083–1091. Springer, Heidelberg (2008)
13. Tai, X.C., Wu, C.L.: Augmented Lagrangian method, dual methods and split Bregman iteration for ROF model. UCLA C.A.M. Report 09-05 (2009)
14. Wu, C.L., Tai, X.C.: Augmented Lagrangian method, dual methods, and split bregman iteration for ROF, vectorial TV, and high order models. UCLA C.A.M. Report 09-76 (2009)
15. Rockafellar, R.T.: Convex analysis. Princeton University Press, Princeton (1970)
16. Bae, E., Yuan, J., Tai, X.C.: Global minimization for continuous multiphase partitioning problems using a dual approach. UCLA C.A.M. Report 09-75 (2009)
17. Goldstein, T., Osher, S.: The split Bregman method for L1 regularized problems. UCLA C.A.M. Report 08-29 (2008)
18. Setzer, M.: Split Bregman algorithm, Douglas-Rachford splitting and frame shrinkage. In: Tai, X.-C., Mørken, K., Lysaker, M., Lie, K.-A. (eds.) SSVM 2009. LNCS, vol. 5567, pp. 464–476. Springer, Heidelberg (2009)
19. Zhang, X., Burgery, M., Bresson, X., Osher, S.: Bregmanized nonlocal regularization for deconvolution and sparse reconstruction. UCLA C.A.M. Report 09-03 (2009)
20. Wang, H.H., Haddad, R.A.: Adaptive median filters: new algorithms and results. IEEE Transactions on Image Processing 4, 499–502 (1995)

Anisotropic Non-Local Means with Spatially Adaptive Patch Shapes

Charles-Alban Deledalle[1], Vincent Duval[1], and Joseph Salmon[2]

[1] Institut Telecom – Telecom ParisTech – CNRS LTCI
46, rue Barrault 75634 Paris cedex 13, France
perso.telecom-paristech.fr/~deledall/,
perso.telecom-paristech.fr/~vduval/
[2] Université Paris 7 – Diderot– LPMA – CNRS-UMR 7599
175 rue du Chevaleret 75013 Paris, France
www.math.jussieu.fr/~salmon/

Abstract. This paper is about extending the classical Non-Local Means (NLM) denoising algorithm using general shapes instead of square patches. The use of various shapes enables to adapt to the local geometry of the image while looking for pattern redundancies. A fast FFT-based algorithm is proposed to compute the NLM with arbitrary shapes. The local combination of the different shapes relies on Stein's Unbiased Risk Estimate (SURE). To improve the robustness of this local aggregation, we perform an anistropic diffusion of the risk estimate using a properly modified Perona-Malik equation. Experimental results show that this algorithm improves the NLM performance and it removes some visual artifacts usually observed with the NLM.

Keywords: Image denoising, non-local means, spatial adaptivity, aggregation, risk estimation, SURE.

1 Introduction

During the last decades, the problem of image denoising in the presence of additive white Gaussian noise has drawn a lot of efforts. A wide variety of strategies were proposed, from partial differential equations (PDE) to transform-domain methods (e.g., wavelets), approximation theory or stochastic analysis.

A major difficulty in image denoising is to handle efficiently regular parts while preventing edges from being blurred, thus one needs spatial adaptive methods to deal with images. In PDE-driven image processing, this is often achieved using anisotropic diffusion [1–3]. Spatial adaptivity can also be reached by considering adaptive neighborhood filters, as the Yaroslavsky [4] or Bilateral [5] filters, or by applying Lepski's method [6] (cf. [7, 8]). Though efficient at dealing with edges and smooth regions, such methods cannot proceed efficiently in textured regions.

To overcome this drawback, many authors have proposed to work with small sub-images, called patches, to take into account the redundancy in natural images, especially in textured parts. The interest of using patches lies in their

A.M. Bruckstein et al. (Eds.): SSVM 2011, LNCS 6667, pp. 231–242, 2012.

robustness to noise. The Non-Local Means algorithm (NLM) [9] and its variants [10, 11] are typical examples consisting in averaging similar pixels, measuring their similarity with patches. Dictionnary learning on patches achieves *state-of-the-art* performance for denoising [12–14]. The key point of this method is to get a good representation for each patch of the image by using ℓ_1 regularization or greedy algorithms. Another *state-of-the-art* method in denoising is BM3D [15]. It also relies on patches and combines classical filtering techniques, such as wavelet thresholding and Wiener's Filter, applied in the space of patches.

The NLM is quite efficient at dealing with smooth regions and textures. However, since it uses patches with fixed (square) shape and scale over the whole image, the performance is limited when dealing with edges with high contrast. Such edges can appear in natural images and in high dynamic range images (HDR) since these images present high contrasted features. They present few redundancies in term of patches, and their denoising versions suffer from a persistence of residual noise: this is called the *noise halo*. A way to overcome this drawback is to use locally chosen scales and orientations of shapes. As far as we know, few attempts have taken advantage of several patch sizes [13, 16] and only one handle variable shapes rather than squares ([17], to improve the BM3D algorithm).

In the NLM framework, spatial-adaptivity may be reached by locally selecting the parameters according to a local estimate of the risk [18]. This relies on Stein's Unbiased Risk Estimate (SURE) [19] which was first used with NLM to globally select the bandwidth [20]. SURE-based methods were widely used in image processing [21, 22] after their introduction for wavelet thresholding [23].

Our contributions — We investigate the potential benefit of replacing the simple square patches with more general shapes, in the classical NLM filter. We give in Section 2 a general overview of the NLM method. We propose in Section 3 a fast algorithm, Non-Local Means with Shape-Adaptive Patches (NLM-SAP), based on the FFT, which allows to compute the solution of the NLM for arbitrary shapes. In Section 4, we locally select or combine the shape-based estimates by measuring the performance of their associated denoisers with SURE. As in [18], one has to regularize SURE to make a local decision. Since the choice of shape is an anistropic decision, a specificity of our approach is that it uses an anisotropic diffusion scheme in the spirit of Perona and Malik [1]. In Section 5, we illustrate numerically, and above all visually, the gain in aggregating various shape-based estimates: using adaptive patch shapes in the context of NLM reduces the *noise halo* produced around edges.

2   An Overview of the NLM

We focus on the problem of denoising: an observed image $\mathbf{Y}$ is assumed to be a noisy version of an unobserved image $\mathbf{f}$ corrupted by a white Gaussian noise. Let $\Omega \subset \mathbb{Z}^2$ be the indexing set of the pixels. For any pixel $x \in \Omega$:

$$\mathbf{Y}(x) = \mathbf{f}(x) + \varepsilon(x), \tag{1}$$

where ε is a centered Gaussian random variable with known variance σ^2 and the noise components $\varepsilon(x)$ are independent. First, let us present the definition of the NLM as introduced in [9]. For each pixel the output of the procedure is a weighted average of the whole image. The weights used are selected using a "metric" which determines whether two pixels are similar or not. The core idea of the NLM is to create a metric governed by patches surrounding each pixel, regardless of their position, i.e., non-local in the image space. For a fixed (odd) width p, a patch P_x is a subimage of width p, centered around the pixel x, and the NLM estimator of $\mathbf{f}(x)$ is then:

$$\hat{\mathbf{f}}(x) = \frac{\sum_{x' \in \Omega} \omega(x, x') \mathbf{Y}(x')}{\sum_{x' \in \Omega} \omega(x, x')} , \text{ where } \omega(x, x') = \exp\left(-\frac{\|\mathsf{P}_x - \mathsf{P}_{x'}\|_{2,a}^2}{2h^2}\right), \quad (2)$$

where $h > 0$ is the bandwidth, $\|\cdot\|_{2,a}$ is a weighted Euclidean norm in $\mathbb{R}^{|\mathsf{P}|}$ ($|\mathsf{P}| = p^2$) using a Gaussian kernel, a controlling the concentration of the norm around the central pixel. The denominator is a normalizing factor ensuring the weights sum to one. Let us briefly recall the influence of each parameter.

The bandwidth h plays the same role as the bandwidth for kernel methods: the larger the bandwidth, the smoother the image. In [11], the authors set its value according to the quantile of a χ^2 distribution, due to the metric they consider to compare patches. We adapt this method for our more general shapes.

The search window size ℓ determines the pixels to be averaged in Eq. (2). The summation is restricted to an $\ell \times \ell$ search window W around the pixel of interest. This was proposed in [9] for computational acceleration. However, some authors have noticed that choosing locally the best search window [11] or using small ones [18, 24] could benefit to the NLM.

The patch size p is usually set globally (between 5 and 9). Choosing $p = 1$ would lead to a method close to the Bilateral Filter [5] or Yaroslavsky Filter [4].

3   From Patches to Shapes: Beyond the *Rare Patch Effect*

The NLM algorithm suffers from a *noise halo* around edges, due to an abrupt lack of redundancy of the image, sometimes referred to as the *rare patch effect*. It occurs because the NLM has large variance around edges. Several solutions have already been proposed to handle this drawback [16, 18, 25]. We extend the latter two approaches by considering general shapes instead of simple square patches. To deal with arbitrary shapes, we reformulate the way the distance between pixels is measured. We generalize the distance $\|\cdot\|_{2,a}$ used in Eq. (2) by:

$$d_{\mathbf{S}}^2(x, x') = \sum_{\tau \in \Omega} \mathbf{S}(\tau) \left(\mathbf{Y}(x + \tau) - \mathbf{Y}(x' + \tau)\right)^2, \quad (3)$$

where $\mathbf{S}$ encodes the shape we aim at. We can use several shapes, so we need to choose the collection of shapes and a way to take the most of each proposed one. We provide an efficient algorithm to compute the distances in Eq. (3). It relies on

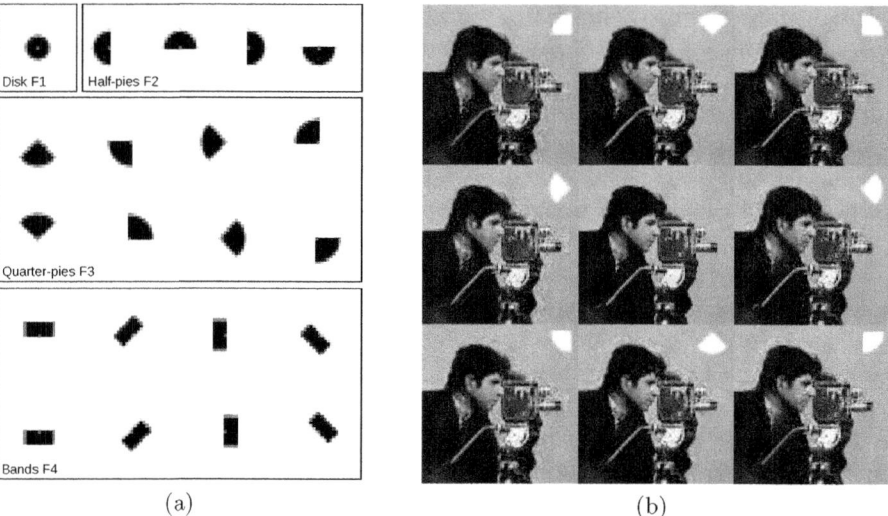

Fig. 1. (a) Examples of shapes with the "central" pixel shown in red. Shapes are grouped in four categories: F1. the *disk* family, F2. the *half-pies* family, F3. the *quarter-pies* family and F4. the *bands* family. (b) Eight denoised images obtained for different oriented pie slices. Each denoiser provides good performance in a specific target direction but suffers from *noise halos* in the other directions. The final aggregate (center) takes advantage of every oriented-denoiser to provide high quality restored edges.

the FFT and is independent of the shape S. We extend to general shapes, works initiated to speed up the NLM [26, 27] by computing the distances between patches with "Summed Area Tables" (also referred to as "Integral Images"). We modify the original algorithm by swapping the two loops: instead of considering all the shifts for each pixel, we consider all the pixels for each shift (see Fig. 2 for details). This reduces the computational cost from $O(|W| \cdot |\Omega| \cdot |\mathsf{P}|)$ to $O(|W| \cdot |\Omega| \cdot \log(|\Omega|))$, where $|W| = \ell^2$, $|\Omega|$ is the image size and $|\mathsf{P}| = p^2$.

The main purpose of this paper is to show that the use of different shapes allows to reduce the *rare patch effect*. Another alternative consists in properly handling overlapping square patches. Indeed, we get $|\mathsf{P}|$ estimates for each pixel. In [9, 11], those $|\mathsf{P}|$ estimates are uniformly averaged while a weighted average is performed in [16]. In our framework, these blockwise approaches are equivalent to combine $|\mathsf{P}|$ (possibly) decentered square shapes. Now, the challenge is to find shapes with enough similar candidates in the search window. We have considered new shapes: *disks*, *bands* and *pies* (see Fig 1).

4   Aggregation of Shape-Based Estimates

For any pixel x, we can build a collection of K pixel estimators $\hat{\mathbf{f}}_1(x), \cdots, \hat{\mathbf{f}}_K(x)$ based on different shapes, as estimates of their corresponding performance. We can now focus on different aggregation procedures.

Algorithm 2D-FFT NLM for an arbitrary shape

Inputs: noisy image $\mathbf{Y}$, 2D-FFT of the shape $\mathcal{F}(\boldsymbol{S})$
Parameters: search window W, bandwidth h
Output: estimated image $\hat{\mathbf{f}}$
Initialize accumulator images $\mathbf{A}$ and $\mathbf{B}$ to zero
for all shift vector δ in the search window W **do**

$\quad\quad$ Compute $\quad\quad\boldsymbol{\Delta}_\delta(x) := (\mathbf{Y}(x) - \mathbf{Y}(x + \delta))^2$ for all pixels x

$\quad$ Compute the 2D-FFT $\mathcal{F}(\boldsymbol{\Delta}_\delta)$
$\quad$ Perform the convolution of $\boldsymbol{\Delta}_\delta$ by the shape $\boldsymbol{S}$

$$d_{\boldsymbol{S}}^2(\cdot, \cdot + \delta) \leftarrow \left(\mathcal{F}^{-1}\left(\overline{\mathcal{F}(\boldsymbol{S})} \mathcal{F}(\boldsymbol{\Delta}_\delta) \right) \right)(\cdot)$$

$\quad$ **for all** pixels x in Ω **do**

$\quad\quad\quad$ Compute $\quad w(x, x + \delta) = \exp\left(-\dfrac{d_{\boldsymbol{S}}^2(x, x + \delta)}{2h^2} \right)$

$\quad\quad\quad$ Update the accumulators $\quad\quad \mathbf{A}(x) \leftarrow \mathbf{A}(x) + w(x, x + \delta)\mathbf{Y}(x + \delta)$
$\quad\quad\quad\quad\quad\quad\quad\quad\quad\quad\quad\quad\quad\quad\quad \mathbf{B}(x) \leftarrow \mathbf{B}(x) + w(x, x + \delta)$

$\quad$ **end for**
end for
Final (normalized) estimator $\quad \hat{\mathbf{f}}(x) = \frac{\mathbf{A}(x)}{\mathbf{B}(x)}$ for all pixel x

---

Fig. 2. NLM pseudo-code for an arbitrary patch shape $\boldsymbol{S}$. Pre-computations (2D-FFT) of distances between shapes from the noisy image and shapes from its shifted version leads to a complexity of $O(|W| \cdot |\Omega| \cdot \log |\Omega|)$, independent of the shape $\boldsymbol{S}$.

4.1 Classical Methods

Uniformly weighted aggregation (UWA). The idea to give the same weight to any shape-based estimator was already proposed for (possibly decentererd) square patches in [9, 11], leading to the pixel-estimate $\hat{\mathbf{f}}_{\text{UWA}}(x)$. With few shapes it is already an improvement in practice (see Table 2), but as the number of shapes increases, we can take into account irrelevant positions. Moreover, such a procedure still suffers from the *rare patch effect*.

Variance-based decision, Weighted Average (WAV). A possible way to limit the *noise halo* is to adapt WAV-reprojection [16] to general shapes. The idea, also proposed by Dabov *et al.* [15] in a different context, is to perform a weighted average of the estimates $\hat{\mathbf{f}}_1(x), \cdots, \hat{\mathbf{f}}_K(x)$, where each weight is chosen inversely proportional to the (estimated) variance of the corresponding estimator. However, this method tends to over-smooth edges and thin details since it does not consider the bias of each estimator.

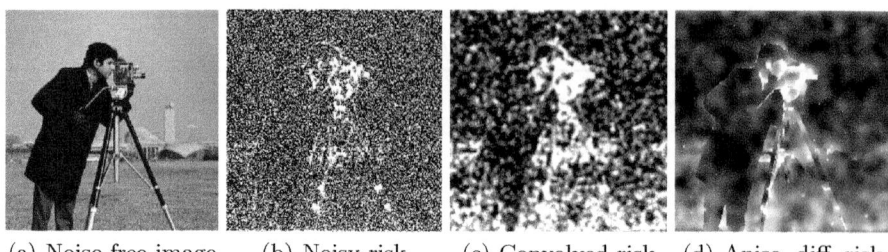

| (a) Noise-free image | (b) Noisy risk | (c) Convolved risk | (d) Aniso. diff. risk |

Fig. 3. Maps of the estimated risk associated with disk shape. From left to right, the noise-free image, the map of the risk without regularization, with convolution and with regularization based on anisotropic diffusion. Low risks are black, high ones are white.

4.2   SURE-Based Methods

In [20], a closed-form expression of SURE for the NLM allows to select the best bandwidth h for the whole image. Our approach is different and closer to the one in [18] (where SURE locally determines the parameter h and p), since we use SURE to locally combine the shape-based estimators. Stein's Lemma [19] still holds when considering shapes: for the pixel x and the k-th shape-based estimate

$$r_k(x) = (\hat{\mathbf{f}}_k(x) - \mathbf{Y}(x))^2 + 2\sigma^2 \frac{\partial \hat{\mathbf{f}}_k(x)}{\partial \varepsilon(x)} - \sigma^2, \qquad (4)$$

is an unbiased estimate of the risk. Thanks to Eq. (2), the derivative is:

$$\frac{\partial \hat{\mathbf{f}}_k(x)}{\partial \varepsilon(x)} = \left(1 + \sum_{x'} \mathbf{Y}(x') \frac{\partial \omega(x, x')}{\partial \varepsilon(x')} - \left(\frac{\sum_{x'} \mathbf{Y}(x') \omega(x, x')}{C_x}\right) \sum_{x''} \frac{\partial \omega(x, x'')}{\partial \varepsilon(x)}\right) / C_x.$$

where $C_x = \sum_{x'} \omega(x, x')$. Our shape-based norm defined in Eq. (3) leads to the following expression of the derivative of the weights $\omega(x, x')$:

$$\frac{\partial \omega(x, x')}{\partial \varepsilon(x')} = \frac{\mathbf{S}(0) \left[\mathbf{Y}(x) - \mathbf{Y}(x')\right] + \mathbf{S}(x - x') \left[\mathbf{Y}(x) - \mathbf{Y}(2x - x')\right]}{h^2}. \qquad (5)$$

where $\mathbf{S}$ encodes the shape of our k-th shape-based estimator. Combining the last equations leads to unbiased risk estimates $r_1(x), \cdots, r_K(x)$ for our K denoisers.

Minimizer of the risk estimates (MRE). A simple proposition is to select the shape that minimizes the local risk estimates we have at hand:

$$\hat{\mathbf{f}}_{\mathrm{MRE}}(x) := \hat{\mathbf{f}}_{k^*}(x) \quad \text{where} \quad k^* = \arg\min_k r_k(x). \qquad (6)$$

This rule is all the more relevant as the estimators are different. Selecting the locally optimal shape yields satisfying results, but combining some of the best performing estimators may improve the results.

(a) *Cameraman*   (b) *City*   (c) *Windmill*   (d) *Lake*

Fig. 4. Chosen 256×256 noise-free images for our experiments

Exponentially Weighted Aggregation (EWA). It might be better to combine several estimators rather than just selecting one. This happens if the best estimators are diversified enough or if the risk of the MRE was wrongly underestimated. Thus, we have used the statistical method of Exponentially Weighted Aggregation, studied for instance in [28] and adapted for patch-based denoising in [29]. It consists in aggregating the estimators by performing a weighted average, with higher weights for estimators with low risks:

$$\hat{\mathbf{f}}_{\text{EWA}}(x) := \sum_{k=1}^{K} \alpha_k \hat{\mathbf{f}}_k(x) , \quad \text{with} \quad \alpha_k = \frac{\exp(-\boldsymbol{r}_k(x)/T)}{\sum_{k'=1}^{K} \exp(-\boldsymbol{r}_{k'}(x)/T)} .$$

The temperature $T > 0$ is a smoothing parameter that controls the confidence attributed to the risk estimates. If $T \to \infty$, the EWA is simply the uniform aggregate $\hat{\mathbf{f}}_{\text{UWA}}$ defined before. Conversely, if $T \to 0$, then $\hat{\mathbf{f}}_{\text{EWA}} \to \hat{\mathbf{f}}_{\text{MRE}}$.

The problem of using SURE to take a local decision for each pixel x is difficult since this estimator has large oscillations (see Fig. 3), so that regularizing the risk maps $\boldsymbol{r}_1, \cdots, \boldsymbol{r}_K$ is required.

4.3 Regularizing the Risk Maps with Anisotropic Diffusion

To make the risk estimates more robust, it is necessary to regularize it. The convolution of the risk map is an efficient way to estimate the local risk in view of setting h since on both sides of an edge a large value of h should be used [18]. Here, the anisotropy of the shapes implies that on one side of an edge the risk may be low whereas it may be high on the other side.

Since convolutions diffuse the risks across the edge, the risk maps become blurred and their comparison becomes difficult. To diffuse the risks on each side of edges, we have adopted a heat equation with spatially and timely dependent coefficients (inspired by the Perona-Malik equation [1]).

More precisely, we let the risk maps $\boldsymbol{r}_1, \cdots, \boldsymbol{r}_K$ evolve according to:

$$\begin{cases} \dfrac{\partial \boldsymbol{r}_k}{\partial t}(x, t) = \text{div}\left(g(|\nabla u(x,t)|)\nabla \boldsymbol{r}_k(x,t)\right) , \\ \boldsymbol{r}_k(x, 0) \ = (\hat{\mathbf{f}}_k(x) - \mathbf{Y}(x))^2 + 2\sigma^2 \dfrac{\partial \hat{\mathbf{f}}_k(x)}{\partial \varepsilon(x)} - \sigma^2 , \end{cases} \quad (7)$$

Table 1. Gain in using anisotropic or mixture of isotropic and anisotropic shapes in terms of PSNR/SSIM. The studied patch shapes are the isotropic *disks*, the *half-pies*, the *quarter-pies*, the *bands* and some combination of them (see Fig. 1.a).

$\sigma = 20$		Cameraman	City	Windmill	Lake
Disk shapes	(F1)	29.45/0.832	28.16/0.885	30.97/0.904	28.68/0.863
Half-pie shapes	(F2)	29.43/0.832	28.08/0.886	30.97/0.906	28.60/0.863
Quarter-pie shapes	(F3)	29.31/0.831	27.87/0.883	30.95/0.909	28.49/0.862
band shapes	(F4)	29.46/0.832	28.05/0.885	31.05/0.906	28.61/0.862
Combination: F1, F2		**29.50/0.833**	**28.21/0.887**	31.11/0.907	**28.73/0.865**
Combination: F1, F2, F3, F4		**29.50/0.833**	28.20/0.887	**31.19/0.909**	28.72/0.865

where $g(x) = \exp(-x^2/\kappa^2)$, the parameter κ controls the anisotropy of the diffusion (the larger κ, the more isotropic the diffusion), and u is the smoothed noisy image which jointly evolves using the Perona-Malik equation:

$$\begin{cases} \dfrac{\partial u}{\partial t}(x,t) = \operatorname{div}\left(g(|\nabla u(x,t)|)\nabla u(x,t)\right), \\ u(x,0) = \mathbf{Y}(x). \end{cases} \tag{8}$$

Curiously, we have noticed that we obtain better risk maps by diffusing $\sqrt{r_k}$ instead of r_k itself. Figure 3 shows that this regularization procedure provides smooth risk maps, following edges of the underlying noise-free image, and finer than without regularization or with convolution.

5   Numerical and Visual Results

The corrupted images are obtained from 256×256 images: *cameraman*, *city*, *windmill* and *lake*[1] (Fig. 4). These images are interesting to study since they present highly contrasted edges for which the classical NLM suffers from the *rare patch effect*. In all the experiments, unless otherwise specified, the NLM-SAP is used with the following default parameters: the search window width $\ell = 11$ px, the shape family combines 15 shapes from families F1 and F2 (Fig. 1.a) with shape areas of 12.5, 25 and 50 px^2, we use EWA with $T = 0.02\sigma^2$ and anisotropic risk regularization with 50 iterations, time-step $dt = 1/8$ and $\kappa = 30$. The parameter h is adapted to the size of the shapes using the rule given by [11]. For the central pixel, we set its central weight as recommended in [24].

Table 1 gives numerical results for different families. The compared families are (see Fig. 1.a): the *disks*, the *half-pies*, the *quarter-pies* and the *bands* and combinations of these families. Our experiments show that suitable families should contain isotropic shapes, directional shapes and various scales of shapes. Increasing the number of shapes does not necessarily improve the quality.

Table 2 presents the numerical performance for the four aggregation procedures: UWA, WAV, MRE and EWA. MRE suffers from brutal transitions, since it selects only one shape per pixel, while EWA evolves in a smoother way due to the weighted combination of shapes for each pixel and provides best results.

---

[1] Images from L. Condat's database: http://www.greyc.ensicaen.fr/~lcondat

Table 2. Comparisons of different aggregation procedures in terms of PSNR/SSIM: UWA, WAV, MRE and EWA

$\sigma = 20$	Cameraman	City	Windmill	Lake
UWA	29.40/0.830	27.99/0.880	30.76/0.897	28.53/0.858
WAV	29.46/0.830	27.98/0.879	30.82/0.898	28.48/0.856
MRE	29.33/0.829	28.02/0.885	30.88/0.905	28.58/0.862
EWA	**29.50/0.833**	**28.21/0.887**	**31.11/0.907**	**28.73/0.865**

We have studied the influence of the regularization of the risk maps on the aggregation results. Three methodologies are compared: aggregation using the noisy risk maps (i.e., SURE maps), the convolved risk maps (using a disk kernel of radius 4) and the risk maps obtained by anisotropic diffusions (Fig. 3). The choices of the local sizes and orientations of the patch shapes are more relevant with the maps obtained by anisotropic diffusions, in terms of scale adaptivity, feature directions and spatial coherency (Fig. 5). Using anisotropic diffusion, the NLM-SAP acts as expected, selecting big sizes of shapes, even around edges, since the shape orientations have been chosen properly to reduce the *rare patch effect*.

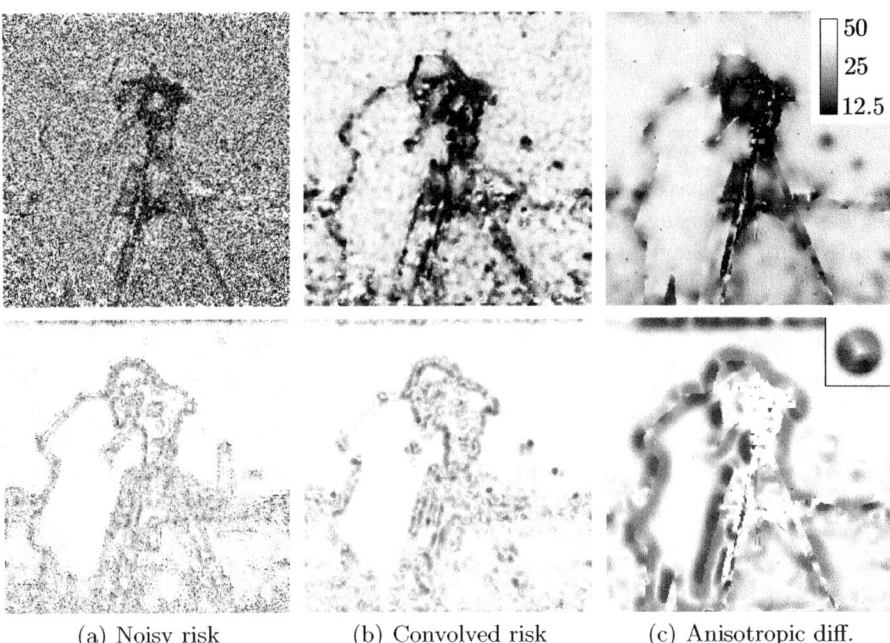

(a) Noisy risk   (b) Convolved risk   (c) Anisotropic diff.

Fig. 5. (top) Average areas and (bottom) average orientations of selected shapes for different risk maps. From left to right, results using the noisy risk maps, the convolved risk maps and the risk maps obtained by anisotropic diffusions. The average areas and the average orientations are represented using gray level colors.

Table 3. Comparisons of denoising approaches for various noise levels in terms of PSNR/SSIM: pixelwise NLM [9], blockwise NLM using UWA reprojection [9], blockwise NLM using WAV reprojection [16], pixelwise NL-means using SURE-based adaptive bandwidth selection [18], BM3D denoiser [15], and our proposed NLM-SAP.

	Cameraman	City	Windmill	Lake
	$\sigma = 5$			
NLM [9]	36.92/0.951	35.87/0.965	38.10/0.972	36.76/0.964
UWA Blockwise NLM [9]	36.99/0.953	35.94/0.966	38.18/0.973	36.77/0.963
WAV Blockwise NLM [16]	37.31/0.956	36.34/0.972	38.79/0.978	37.10/0.970
SURE adaptive NLM [18]	37.46/0.956	36.76/0.975	39.14/0.978	37.28/0.970
BM3D [15]	**38.17/0.962**	**37.48/0.978**	**39.91/0.983**	**38.15/0.977**
NLM-SAP	37.80/0.957	37.26/0.975	39.60/0.979	37.92/0.974
	$\sigma = 10$			
NLM [9]	32.46/0.905	31.11/0.932	33.62/0.945	32.07/0.926
UWA Blockwise NLM [9]	32.43/0.913	30.99/0.926	33.49/0.942	32.04/0.924
WAV Blockwise NLM [16]	32.84/0.922	31.48/0.941	34.07/0.953	32.37/0.936
SURE adaptive NLM [18]	33.11/0.918	32.11/0.948	34.78/0.954	32.61/0.935
BM3D [15]	**34.06/0.931**	**33.15/0.956**	**35.84/0.966**	**33.63/0.950**
NLM-SAP	33.44/0.914	32.84/0.950	35.28/0.955	33.27/0.940
	$\sigma = 20$			
NLM [9]	28.72/0.820	27.11/0.870	30.04/0.897	28.12/0.855
UWA Blockwise NLM [9]	28.88/0.830	27.02/0.868	29.92/0.890	28.14/0.860
WAV Blockwise NLM [16]	29.16/0.838	27.27/0.877	30.17/0.901	28.12/0.865
SURE adaptive NLM [18]	29.49/0.845	27.85/0.889	30.96/0.906	28.46/0.867
BM3D [15]	**30.35/0.871**	**29.07/0.912**	**32.07/0.936**	**29.38/0.895**
NLM-SAP	29.50/0.833	28.21/0.887	31.11/0.907	28.73/0.865

Comparisons have been performed with the classical (pixelwise) NLM [9], the blockwise NLM using UWA reprojection [9], the blockwise NLM using WAV reprojection [16], the pixelwise NL-means using SURE-based adaptive bandwidth selection [18], BM3D [15], and our proposed NLM-SAP approach. Table 3 shows that NLM-SAP outperforms all other NLM improvements. NLM-SAP brings a gain of PSNR of about 1 dB compared to the classical NLM. The BM3D approach leads to better numerical results than all NLM variants. While the presence of the *rare patch effect* is well illustrated by the *noise halos* for NLM, BM3D and NLM-SAP have reduced a lot this phenomenon. Our NLM-SAP provides smooth results with accurate details: the quality of the images we obtained challenges those by BM3D.

6   Conclusion

We have addressed the problem of the *rare patch effect* arising in the NLM and responsible of the *noise halos* around edges. Our method consists in substituting the square patches of fixed size by spatially adaptive shapes. A fast implementation based on the FFT has been proposed to handle arbitrary shapes. Several estimates are obtained by using different patch shapes, and we have extended

Fig. 6. Comparisons of the NLM [9], the BM3D [15] and the proposed NLM-SAP on images damaged by additive white Gaussian noise with standard deviation $\sigma = 20$

SURE-based approaches to aggregate them. The SURE-based risk maps require regularization, and diffusions can be satisfactorily used. Future work is to reduce computation time and treat other regularization strategies.

References

1. Perona, P., Malik, J.: Scale space and edge detection using anisotropic diffusion. IEEE Trans. Pattern Anal. Mach. Intell. 12, 629–639 (1990)
2. Alvarez, L., Guichard, F., Lions, P.L., Morel, J.M.: Axioms and fundamental equations of image processing. Arch. Rational Mech. Anal. 123(3), 199–257 (1993)
3. Weickert, J.: Anisotropic diffusion in image processing. European Consortium for Mathematics in Industry. B. G. Teubner, Stuttgart (1998)
4. Yaroslavsky, L.P.: Digital picture processing. Springer Series in Information Sciences, vol. 9. Springer, Berlin (1985)
5. Tomasi, C., Manduchi, R.: Bilateral filtering for gray and color images. In: ICCV, pp. 839–846 (1998)
6. Lepski, O.V., Mammen, E., Spokoiny, V.G.: Optimal spatial adaptation to inhomogeneous smoothness: an approach based on kernel estimates with variable bandwidth selectors. Ann. Statist. 25(3), 929–947 (1997)
7. Polzehl, J., Spokoiny, V.G.: Adaptive weights smoothing with applications to image restoration. J. R. Stat. Soc. Ser. B Stat. Methodol. 62(2), 335–354 (2000)

8. Katkovnik, V., Foi, A., Egiazarian, K.O., Astola, J.T.: Directional varying scale approximations for anisotropic signal processing. In: EUSIPCO, pp. 101–104 (2004)
9. Buades, A., Coll, B., Morel, J.M.: A review of image denoising algorithms, with a new one. Multiscale Model. Simul. 4(2), 490–530 (2005)
10. Awate, S.P., Whitaker, R.T.: Unsupervised, information-theoretic, adaptive image filtering for image restoration. IEEE Trans. Pattern Anal. Mach. Intell. 28(3), 364–376 (2006)
11. Kervrann, C., Boulanger, J.: Optimal spatial adaptation for patch-based image denoising. IEEE Trans. Image Process. 15(10), 2866–2878 (2006)
12. Aharon, M., Elad, M., Bruckstein, A.: K-SVD: An algorithm for designing over-complete dictionaries for sparse representation. IEEE Trans. Signal Process. 54(11), 4311–4322 (2006)
13. Mairal, J., Sapiro, G., Elad, M.: Learning multiscale sparse representations for image and video restoration. Multiscale Model. Simul. 7(1), 214–241 (2008)
14. Mairal, J., Bach, F., Ponce, J., Sapiro, G., Zisserman, A.: Non-local sparse models for image restoration. In: ICCV (2009)
15. Dabov, K., Foi, A., Katkovnik, V., Egiazarian, K.O.: Image denoising by sparse 3-D transform-domain collaborative filtering. IEEE Trans. Image Process. 16(8), 2080–2095 (2007)
16. Salmon, J., Strozecki, Y.: From patches to pixels in non-local methods: Weighted-Average reprojection. In: ICIP (2010)
17. Dabov, K., Foi, A., Katkovnik, V., Egiazarian, K.O.: BM3D image denoising with shape-adaptive principal component analysis. In: Proc. Workshop on Signal Processing with Adaptive Sparse Structured Representations, SPARS 2009 (2009)
18. Duval, V., Aujol, J.F., Gousseau, Y.: On the parameter choice for the non-local means. Technical Report hal-00468856, HAL (2010)
19. Stein, C.M.: Estimation of the mean of a multivariate normal distribution. Ann. Statist. 9(6), 1135–1151 (1981)
20. Van De Ville, D., Kocher, M.: SURE-based Non-Local Means. IEEE Signal Process. Lett. 16, 973–976 (2009)
21. Blu, T., Luisier, F.: The SURE-LET approach to image denoising. IEEE_J_IP 16(11), 2778–2786 (2007)
22. Ramani, S., Blu, T., Unser, M.: Monte-Carlo SURE: a black-box optimization of regularization parameters for general denoising algorithms. IEEE Trans. Image Process. 17(9), 1540–1554 (2008)
23. Donoho, D.L., Johnstone, I.M.: Adapting to unknown smoothness via wavelet shrinkage. J. Amer. Statist. Assoc. 90(432), 1200–1224 (1995)
24. Salmon, J.: On two parameters for denoising with Non-Local Means. IEEE Signal Process. Lett. 17, 269–272 (2010)
25. Zimmer, S., Didas, S., Weickert, J.: A rotationally invariant block matching strategy improving image denoising with non-local means. In: LNLA (2008)
26. Wang, J., Guo, Y.W., Ying, Y., Liu, Y.L., Peng, Q.S.: Fast non-local algorithm for image denoising. In: ICIP, pp. 1429–1432 (2006)
27. Darbon, J., Cunha, A., Chan, T.F., Osher, S., Jensen, G.J.: Fast nonlocal filtering applied to electron cryomicroscopy. In: ISBI, pp. 1331–1334 (2008)
28. Dalalyan, A.S., Tsybakov, A.B.: Aggregation by exponential weighting, sharp pac-bayesian bounds and sparsity. Mach. Learn. 72(1-2), 39–61 (2008)
29. Salmon, J., Le Pennec, E.: NL-Means and aggregation procedures. In: ICIP, pp. 2977–2980 (2009)

Entropy-Scale Profiles for Texture Segmentation

Byung-Woo Hong, Kangyu Ni*, and Stefano Soatto

School of Computer Science and Engineering, Chung-Ang University, Seoul, Korea
School of Mathematical and Statistical Sciences, Arizona State University, USA
Computer Science Department, University of California, Los Angeles, CA, USA

Abstract. We propose a variational approach to unsupervised texture segmentation that depends on very few parameters and is robust to imaging conditions. First, the uneven illumination in the observed image is removed by the proposed image decomposition model that approximates the illumination and well retains the textures and features in the image. Then, from the obtained intrinsic image, we introduce a new data, multiscale local entropy, which is the entropy of each location's neighborhood histogram with various scales. The proposed segmentation model uses multiscale local entropy as data. Together with a length penalizing term, minimizing the energy functional locates the contours so that the local entropy within each region is similar to one another. Since entropy is the only feature, there are very few parameters. Moreover, the segmentation model can be solved by a fast global minimization method. Experimental results on natural images show the proposed method is able to robustly segment various texture patterns with uneven illumination in the original images.

1 Introduction

One of the challenges of unsupervised texture segmentation is due to the difficulties to well define textures. There are many tools to analyze texture, from statistical models to filtering methods, to geometric approaches. There have been a large number of texture features: orientations, scales, frequencies, etc. Therefore, partitioning an image domain into several texture regions, or identify homogeneous regions in the sense of texture, without any given knowledge is very difficult. One of the earliest unsupervised segmentation model [1] approximates an image by a piecewise smooth image and a length penalizing term in an energy functional to locate the boundary of each region. This model satisfies many desired mathematical properties but is difficult to solve in practice. In [2], the one-dimensional contour/edge set is approximated by a two-dimensional smooth function, making the functional easier to solve. The model in [3] approximates an image by a piecewise constant image and furthermore incorporates the level set method with the variational model, which makes it easy to solve. However, these classical methods do not handle textures, especially when the average intensities of each texture region are similar.

---

* Kangyu Ni was supported by the US NSF-DMS grant #0652833.

A.M. Bruckstein et al. (Eds.): SSVM 2011, LNCS 6667, pp. 243–254, 2012.

There has been numerous works on texture segmentation. For instance, the authors in [4], [5], and [6] use Gabor transforms to represent texture features for segmentations and authors in [7] use wavelet transforms. These generally have a large set of texture features and therefore involve selecting a large set of parameters. Probability density function (PDF)/ histogram-based approaches, such as [8], [9], [10], [11], and [12], also involve some parameters associated with the assumptions on the histograms or selected texture features. Methods in [13], [14], [15], [16], and [17] use the entire PDFs or histograms without extracting predefined features for segmentation using histogram distances, such as χ^2 statistics, mutual information, Kullback-Leibler divergence, and Bhattacharyya distance. The data of the histogram is not limited to intensity. Any features and transforms of the image can be used. However, finding the solutions of these methods requires differentiating histograms with respect to the contour or region. Local histogram-based methods [18] do not require histograms to be differentiated and can employ a fast global minimization method. However, the scale of the local histogram windows is fixed and has to be chosen. In this paper, we introduce a new data, multiscale local entropy, which is the entropy of a local neighborhood's histogram with various scales. Therefore, the window size is unbiased.

Another challenge of unsupervised texture segmentation is due to the imaging conditions/nuisance factors in real images. Most of the above-mentioned segmentation models are not robust to imaging conditions, because these are not taken into account in the segmentation models. The proposed segmentation model in [19] simultaneously estimates the illumination and reflectance and segments the image using reflectance. This allows global smooth changes within a region due to uneven lighting and is therefore robust with respect to nuisance factors. However, this model approximates images by piecewise constant functions and therefore does not handle textures. Note also the Mumford-Shah segmentation model, even though is difficult to solve, also deals with smooth changes in the image. However, it also does not handle textures.

For robust texture segmentation, we add a pre-processing step that approximately decomposes an image into an illuminance component and a reflectance component. The image model is described in section 2.1. The proposed decomposition model is described in section 2.2. For segmentation, we only use the reflectance component. In section 2.3, we proposed a new data for texture segmentation, multiscale local entropy. The segmentation model is described in section 2.4. Finally, we show some experimental results in section 3 and conclude in section 4.

2   Methodology

2.1   Image Model

The image of a natural scene captured by a camera does not solely depend on the objects in the scene. The lighting condition, or illumination, also plays an important role. Therefore, for robust image segmentation, illumination should be taken into account. Let $I : [0, 1] \times [0, 1] \to [0, 1]$ be the observed image after normalization. One simple way to express the image with illumination is by the following multiplicative model:

$$I(x) = U(x)V(x),$$ (1)

where $x \in [0,1] \times [0,1]$, U is the illumination and V is the reflectance, or the intrinsic image structure. This model was formulated in [20] and was used for robust segmentation in [21] and [19]. Multiplicative noise model has been used for denoising and deblurring in [22] and segmentation in [23].

From this image model (1), we wish to find the reflectance component V and then use it for texture segmentation. To obtain V, we first take log of (1), which transforms the product model into the following sum:

$$\log I = \log U + \log V.$$ (2)

It is easier to decompose this additive expression than the multiplicative expression. In the next section, we provide a variational decomposition model for (2).

2.2   Image Decomposition

We take a variational PDE-based approach to decompose $\log I$. Let $f = \log I$, the decomposition is found by solving the following minimization problem:

$$\min_u \frac{1}{2} \int |\nabla u|^2 + \lambda \int |f - u|,$$ (3)

where λ is a parameter that controls the balance between the two penalty terms. The first term of the energy functional uses the ℓ_2 norm on the gradient of u because the illuminance component is approximately smooth. Note that more accurately the illumination is piecewise smooth, but the above approximation will suffice for the purpose of segmentation. The second term uses the ℓ_1 norm, rather than the ℓ_2 norm, on the residual, $f - u$, in order to better capture texture.

The solution of (3) can be found by using the gradient descent method:

$$\frac{du}{dt} = \triangle u + \lambda \frac{f - u}{|f - u|},$$ (4)

where the parameter λ can be chosen by methods, such as in [24].

Figure 1 demonstrates this image decomposition method using several images from the Yale Face Database. The images, in row (a) from left to right, have lighting from different directions: center, right, and left, respectively. Row (b) shows the respective illumination components U, and row (c) shows the respective reflectance components V. For all three decompositions, the parameter $\lambda = 0.0004$. These experiments show the robustness of extracting the nuisance factors from the intrinsic image structure using the method described here. Specifically, the illumination components desirably exclude the image structure, and the reflectance components show uniform lighting on the faces. In addition, note that this decomposition model does not take into account shadows as part

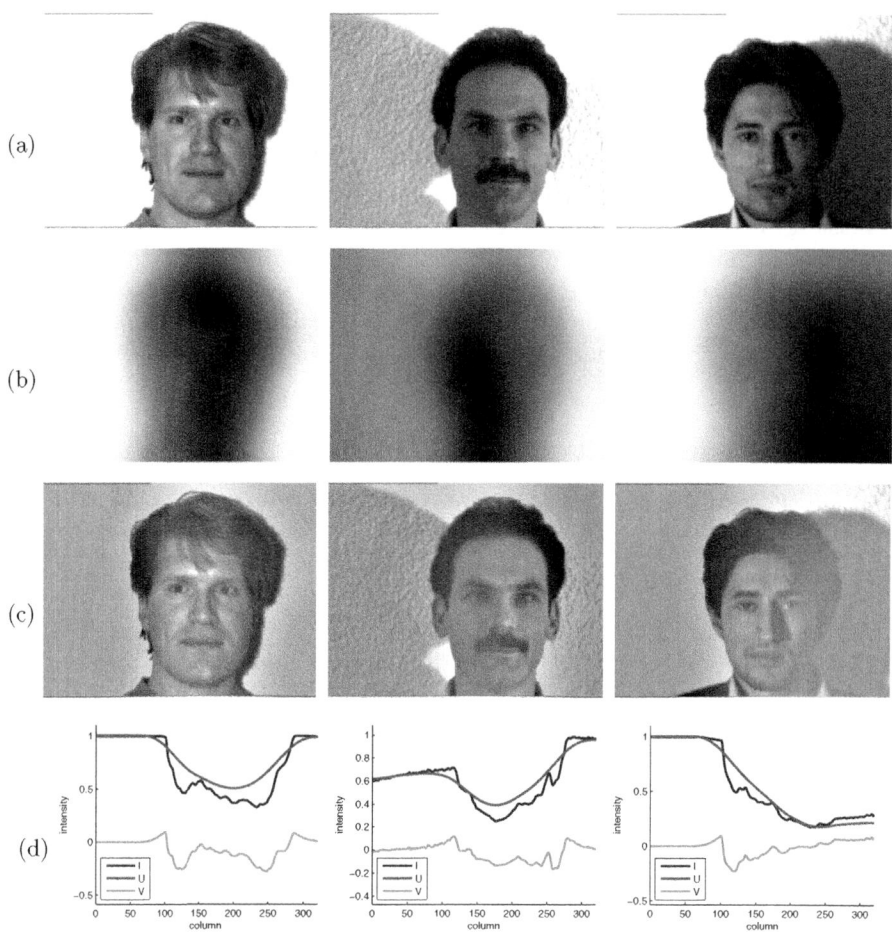

Fig. 1. Model (3) robustly decomposes images of faces with various lighting conditions into the illumination and intrinsic components. Row (a) shows original images I, row (b) shows illumination components U, row (c) shows reflectance components V, and row (d) shows vertical sum of intensity for each image

of illuminance and therefore is present in the reflectance component. Moreover, even though the areas with less lighting in the original images do not look as sharp in the reflectance image compared to the areas with more lighting originally, all areas possess similar levels of illuminance. This can also be seen in plots row (d), which shows the vertical sum of intensity for each original image, illumination, and reflectance components. For instance, since the lighting of the original image in the third row is from the left, we have high-left and low-right profile for the original image, high-left and low-right smooth profile for the illumination component, and more or less horizontal profile for the reflectance component.

2.3   Entropy Profile

In this section, we propose a descriptor that is calculated based on the reflectance component V and will be used for the proposed segmentation model described in the next section. First, let $h_{x,s}$ be the probability density function of image intensity on the square patch centered at location x with scale s. Note that in the discrete setting, the dimension of the patch is $(2s + 1) \times (2s + 1)$. Then, define $H_{x,s}$ as the entropy of $h_{x,s}$ by

$$H_{x,s} = -\int_0^1 h_{x,s}(y) \log(h_{x,s}(y))\, dy \, . \tag{5}$$

For a fixed location x, the entropy profile, $H_x(s)$, is a function of scale. In the following, we analyze the proposed entropy profile with a few examples..

Fig. 2 (a) is a synthetic image consisting of two textures with the same information (entropy) and different scales. The regions of each texture are indicated in (b). Four locations are selected in (c), and their respective entropy profiles are depicted in (e). Since both locations a and b are in the same texture region, their entropy profiles resemble each other. Similarly, the profiles of c and d resemble each other. In (f), the scale of entropy profile is adjusted by the logarithm, which is denoted by log-scale. Since entropy changes less as scale increases, entropy profiles in log-scale are more distinguishable. Figure (g) represents the median entropy profiles over all locations in each texture region, and (h) is the median entropy profiles in log-scale. The median entropy profiles are shown here because in the next section, the proposed segmentation model approximates the homogeneity of each region by using median entropy profile.

Fig. 3 illustrates with a synthetic image of two textures with the same scale and different information (entropy). Similarly, we see in (f) that the difference between profiles from different textures in log-scale is more prominent than without taking logarithms. Interestingly, the difference in entropy profiles in this case is in the vertical direction, instead of the horizontal direction in the previous example in Fig. 2. This is because the textures in Fig. 2 differ in scale and textures in Fig. 3 differ in information. However, if two textures have the same information and scale, the proposed entropy profile will not be able to distinguish them.

Fig. 4 shows a different perspective of entropy using a real image. Instead of looking at a entropy profile $H_x(s)$, which is a function of scale with a fixed location, each image is an entropy map $H_s(x)$, which is defined as the entropy of each location with a fixed scale. The scales are from 1 to 24, from left to right and top to bottom. Each row shares the same color bar at the end of the row, where dark red represents the highest value and dark blue represents the lowest value. The entropy maps change quickly when the scales are small, as shown in the first row, and do not change very much when the scales are large, as shown in the third and fourth rows. Therefore, for segmentation, we use log-scale for the scale in entropy maps.

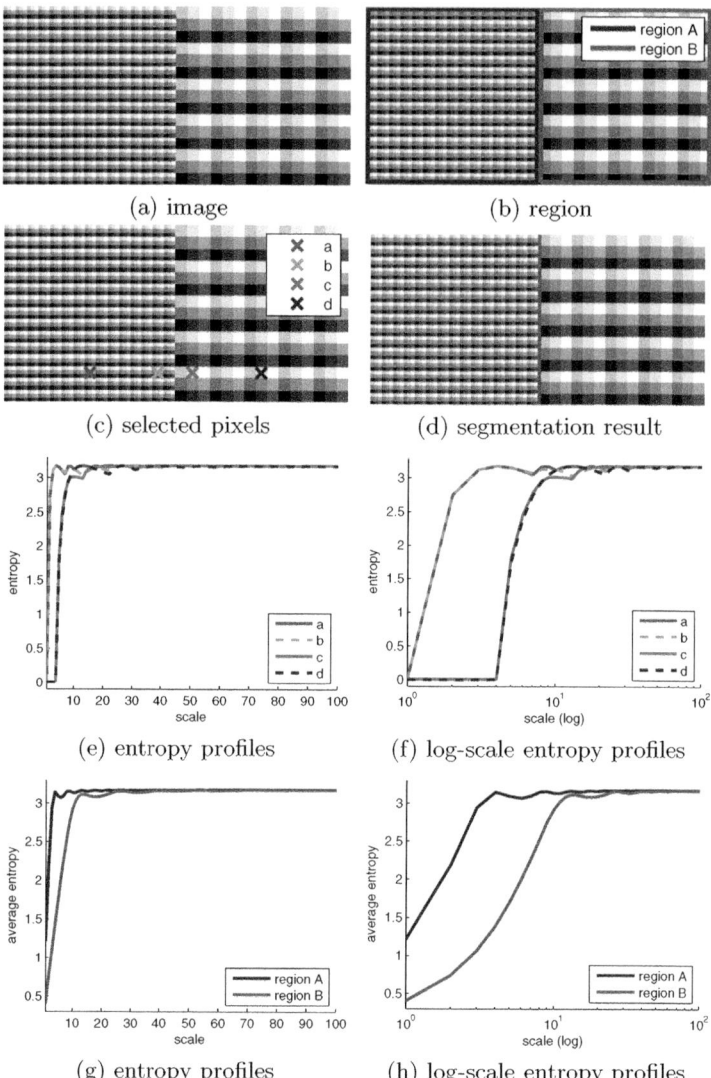

Fig. 2. The entropy profiles $H_x(s)$ of textures with same information and different scales are distinct

2.4   Texture Segmentation

The proposed texture segmentation model uses the entropy profile $H_x(s)$ of the reflectance component V in

$$\min_{u, H_1, H_2} \int |\nabla u(x)| \, dx + \lambda \int u(x) \, d(H_1, H_x) + [1 - u(x)] \, d(H_2, H_x) \, dx \,, \quad (6)$$

where $0 \leq u \leq 1$, H_1 and H_2 are unknown histograms, λ is a parameter, and the distance between two histograms is defined as

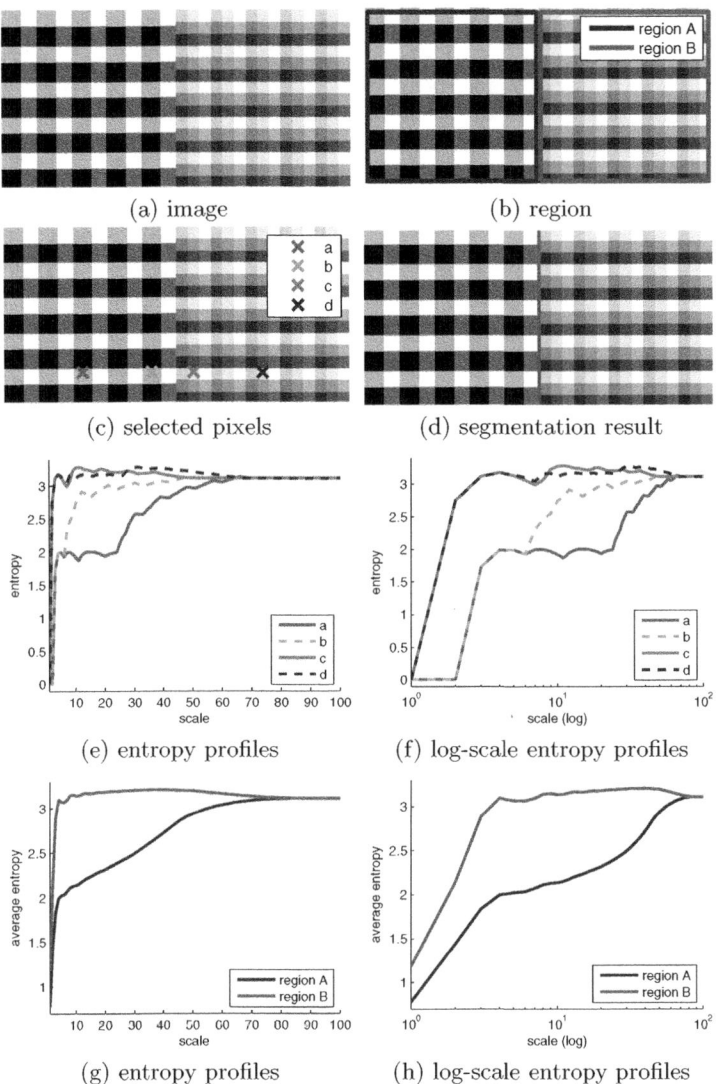

Fig. 3. The entropy profiles $H_x(s)$ of textures with same scale and different information are distinct

$$d(H_1, H_x) = \int |H_1(s) - H_x(s)| \, \log(s) \, ds, \tag{7}$$

which incorporates log-scale. The variable u represents the segmented regions. The set of u close to 1 is inside the contour and the set of u close to 0 is outside the contour. According to [25], minimizing this energy functional with respect to u is a convex problem. The data terms encourage finding contours so that the local entropy profiles are similar to one another within each region. The proposed segmentation model (6) resembles the local histogram based segmentation model

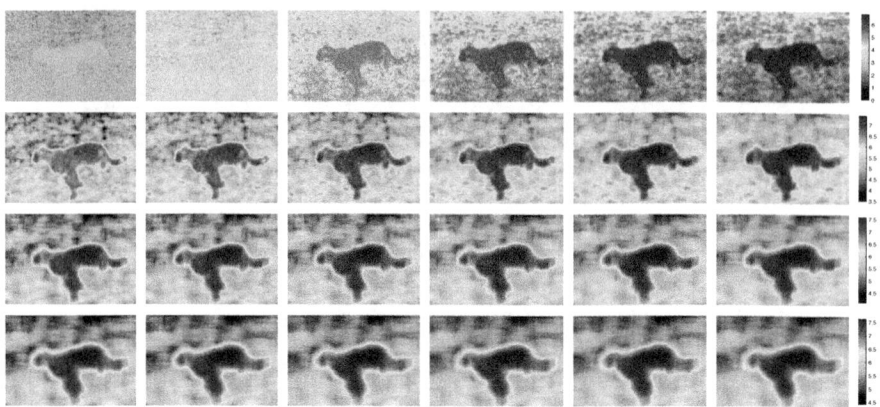

Fig. 4. Entropy maps $H_s(x)$ of the center image in Fig. 5 (a) with scale from 1 to 24, from left to right and top to bottom. The difference in maps becomes small when the scale increases.

with the Wasserstein distance in [18]. Nevertheless, it is in essence different, since entropy profile takes into account of various scales, rather than using a fixed-size window. As a result, this segmentation is more robust than local histogram-based methods, whose patch size needs to be close to the texture scale in the image.

To solve (6), we may follow the fast global minimization method described in [18] and [26]. Therefore, without repeating the derivations, the minimization is solved by repeating the following steps until convergence:

$$H_1(s) = \text{weighted (by } u(x)) \text{ median of } H_x(s) \tag{8}$$

$$H_2(s) = \text{weighted (by } 1 - u(x)) \text{ median of } H_x(s) \tag{9}$$

$$\overrightarrow{p}(x) = \frac{\overrightarrow{p}(x) + \delta t \, \nabla(\text{div}\,\overrightarrow{p}(x) - v(x)/\theta)}{1 + \delta t \, |(\text{div}\,\overrightarrow{p}(x) - v(x)/\theta)|} \tag{10}$$

$$u(x) = v(x) - \theta \, \text{div}\,\overrightarrow{p}(x) \tag{11}$$

$$v(x) = \max\{\min\{u(x) - \theta\lambda r_{x,H_1,H_2}, 1\}, 0\}, \tag{12}$$

where θ is a parameter, δt is a time-step that is $\leq \frac{1}{8}(\delta x)^2$, $\overrightarrow{p}(x) = (p_1(x), p_2(x))$, and

$$r_{x,H_1,H_2} = \int |H_1(s) - H_x(s)| - |H_2(s) - H_x(s)| \, ds.$$

The initializations can be arbitrary since this is a global minimization model. Therefore, one may initially choose an arbitrary contour and let $u = 1$ inside the contour and $u = 0$ outside the contour. Initializations for v and $\overrightarrow{p}$ can be done by setting $v = u$ and $\overrightarrow{p} = \overrightarrow{0}$.

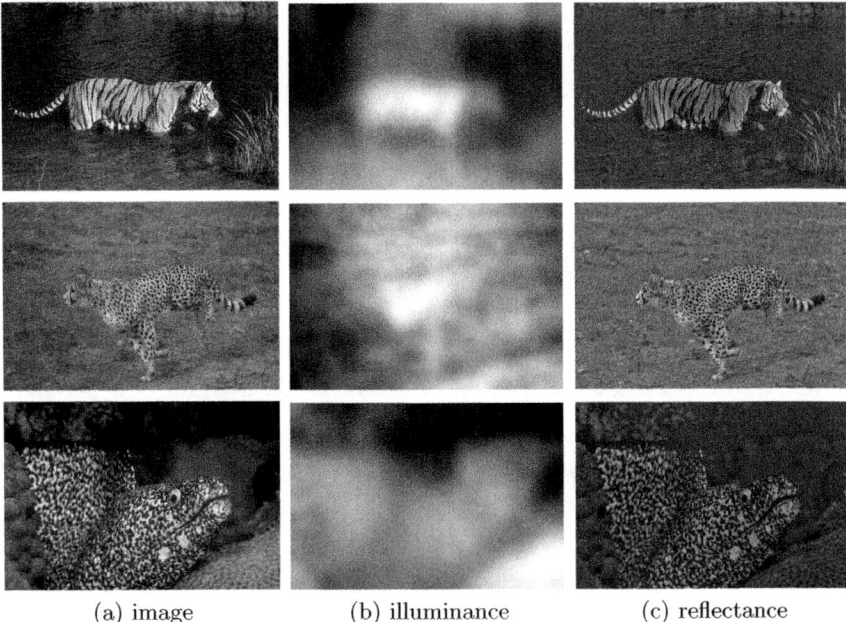

(a) image   (b) illuminance   (c) reflectance

Fig. 5. Model (3) robustly decomposes real images from Berkeley segmentation database. The reflectance components appear to have even lighting.

Fig. 2 (d) and fig. 3 (d) show the segmentation model (6) is able to accurately distinguish two textures, in which one pair of textures has the same information but different scales and the other pair has the same scale but different informations.

3   Experimental Results

Fig. 5 evaluates the proposed decomposition model with a few images from the Berkeley segmentation database, as shown in column (a). Their respective illuminance components U are shown in column (b), and the reflectance components V are shown in column (c). The illuminance appears to be faithfully extracted. As one can see, for instance, the front of the cheetah body is more illuminated than other areas in the original image. The left side of the background is less illuminated than other areas in the original image. Therefore, the reflectance in (c) desirably looks flat because the lighting is forced to be homogeneous. Similar observations can be made for the other two images.

Fig. 6 shows segmentation results using the proposed model (6) and other methods for the purpose of comparison. Row (b) shows segmentation results using the fast global minimization of active contour (GAC) in [26], which approximates an image by a piecewise constant function and therefore performs poorly for images with rich texture patterns. Rows (c) and (d) show segmentation

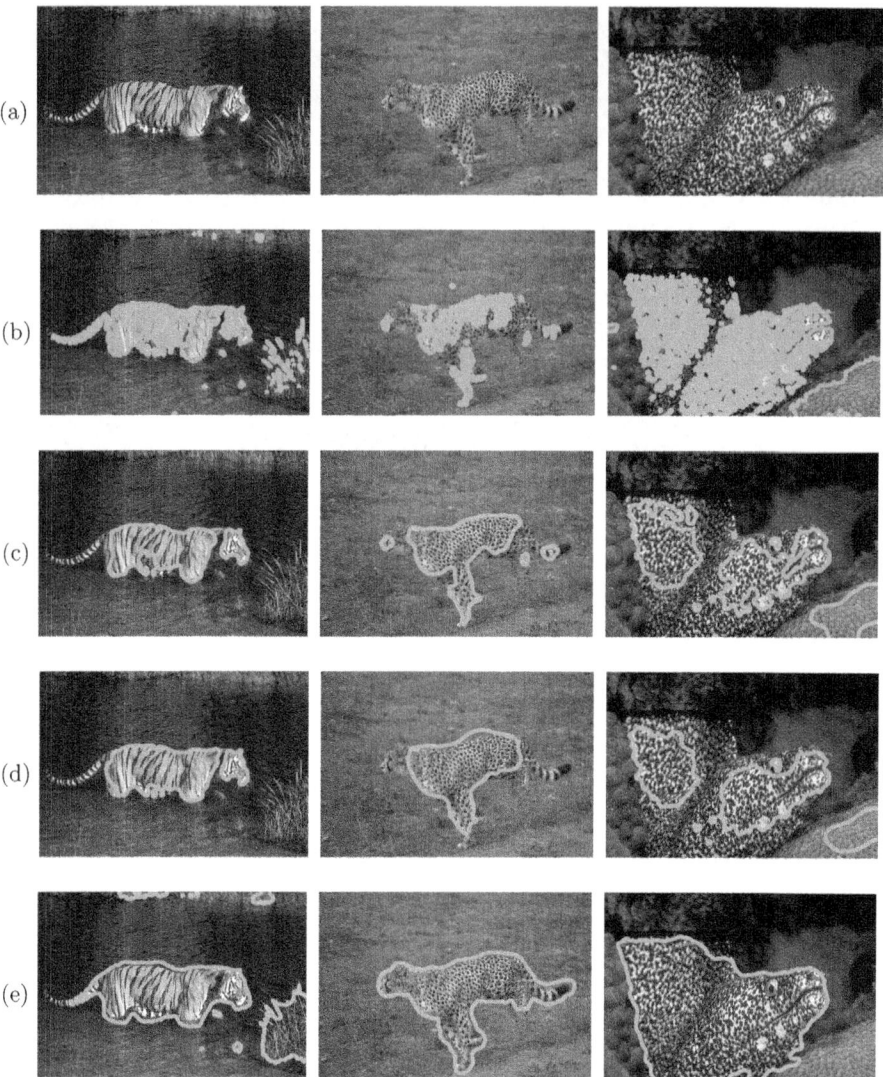

Fig. 6. (a) are the original images. (b) are the segmentation results by the fast global minimization of active contour (GAC) in [26]. (c) are the results by local histogram based segmentation using the Wasserstein distance (LHSWD) [18] with scale = 10. (d) are the results by LHSWD with scale = 30. (e) are the results by the proposed method, which is more robust to illumination than GAC and LHSWD and is able to segment the texture patterns more accurately.

results by the local histogram based segmentation method using the Wasserstein distance (LHSWD) in [18]. For the local histograms, the binning size is 100 and the scale sizes are 10 and 30 for (c) and (d), respectively. The parameters are $\theta = 0.001$ and $\lambda = 1$. Row (e) shows results of the proposed method with the same parameters θ and λ. The results are far better as one can see that the patterns of tiger, cheetah, and fish are more accurately segmented. We believe that this is due to two reasons. First, illuminance in an image plays an important role in segmentation, and it is to beneficial to even out the illuminance. Second, all scales of local histograms were taken into account, rather than using a fixed scale.

4   Conclusion

We propose a method for texture segmentation that is robust to imaging conditions using very few parameters. We propose a multiscale local entropy as a data descriptor and an image decomposition model for illumination removal. While it is possible to put the decomposition and segmentation models in one formulation, it is in practice difficult to solve. Therefore, the decomposition is done as a pre-processing step. The experimental results show that the proposed method is able to accurately segment natural images that contain texture patterns. In the future, we would like to analyze and extend the use of entropy profile.

Acknowledgments. This work was supported by the Korea Research Foundation Grant funded by the Korean Government (NRF-2010-220-D00078).

References

1. Mumford, D., Shah, J.: Optimal approximations by piecewise smooth functions and associated variational problems. Comm. Pure Appl. Math. 42(5), 577–685 (1989)
2. Ambrosio, L., Tortorelli, V.M.: Approximation of functionals depending on jumps by elliptic functionals via gamma convergence. Comm. on Pure and Applied Math. 43, 999–1036 (1990)
3. Chan, T., Vese, L.A.: Active contours without edges. IEEE Trans. on Image Processing 10(2), 266–277 (2001)
4. Jain, A.K., Farrakhonia, F.: Unsupervised texture segmentation using gabor filters. Pattern Recognition 23(12), 1167–1186 (1991)
5. Chan, T., Sandberg, B., Vese, L.: Active contours without edges for textured images. UCLA Comput. Appl. Math. Rep. 02-39 (2002)
6. Sagiv, C., Sochen, N.A., Zeevi, Y.Y.: Integrated active contours for texture segmentation. IEEE Transactions on Image Processing 15(6), 1633–1646 (2006)
7. Portilla, J., Simoncelli, E.P.: A parametric texture model based on joint statistics of complex wavelet coefficients. International Journal of Computer Vision 40(1)
8. Manjunath, B.S., Chellappa, R.: Unsupervised texture segmentation using markov random field models. IEEE Transactions on Pattern Analysis and Machine Intelligence 13(5), 478–482 (1991)
9. Zhu, S.C., Yuille, A.: Region competition: Unifying snakes, region growing, and bayes/mdl for multiband image segmentation. IEEE Trans. on Pattern Analysis and Machine Intelligence 18(9), 884–900 (1996)
10. Yezzi, J.A., Tsai, A., Willsky, A.: A statistical approach to snakes for bimodal and trimodal imagery. In: Int. Conf. on Computer Vision, pp. 898–903 (1999)

11. Paragios, N., Deriche, R.: Geodesic active regions and level set methods for supervised texture segmentation. International Journal of Computer Vision 46(3)
12. Rousson, M., Brox, T., Deriche, R.: Active unsupervised texture segmentation on a diffusion based feature space. In: Computer Vision and Pattern Recognition
13. Aubert, G., Barlaud, M., Faugeras, O., Jehan-Besson, S.: Image segmentation using active contours: Calculus of variations or shape gradients? SIAM Appl. Math. 1(2), 2128–2145 (2003)
14. Herbulot, A., Jehan-Besson, S., Duffner, S., Barlaud, M., Aubert, G.: Segmentation of vectorial image features using shape gradients and information measures. J. Math. Imaging and Vision 25(3), 365–386 (2006)
15. Kim, J., Fisher, J.W., Yezzi, A., Cetin, M., Willsky, A.S.: A nonparametric statistical method for image segmentation using information theory and curve evolution. IEEE Trans. on Image Processing 14, 1486–1502 (2005)
16. Awate, S.P., Tasdizen, T., Whitaker, R.T.: Unsupervised texture segmentation with nonparametric neighborhood statistics. In: Leonardis, A., Bischof, H., Pinz, A. (eds.) ECCV 2006. LNCS, vol. 3952, pp. 494–507. Springer, Heidelberg (2006)
17. Michailovich, O., Rathi, Y., Tannenbaum, A.: Image segmentation using active contours driven by the bhattacharya gradient flow. IEEE Trans. on Image Processing 16(11), 2787–2801 (2007)
18. Ni, K., Bresson, X., Chan, T., Esedoglu, S.: Local histogram based segmentation using the wasserstein distance. International Journal of Computer Vision 84(1), 97–111 (2009)
19. Li, C., Li, F., Kao, C.-Y., Xu, C.: IImage Segmentation with Simultaneous Illumination and Reflectance Estimation: An Energy Minimization Approach. In: Proc. of ICCV (2009)
20. Horn, B.K.P.: Robot Vision. MIT Press, Cambridge (1986)
21. Chen, T., Yin, W., Zhou, X.S., Comaniciu, D., Huang, T.S.: Total variation models for variable lighting face recognition. IEEE Trans. on Pattern Analysis and Machine Intelligence 28(9), 1519–1524 (2006)
22. Rudin, L., Lions, L., Osher, S.: Multiplicative denoising and deblurring: theory and algorithms, vol. 445. Springer, Heidelberg (2003)
23. Le, T.M., Vese, L.A.: Additive and Multiplicative piecewise-smooth segmentation models in a functional minimization approach. Contemporary Mathematics (2007)
24. Aujol, J.F., Gilboa, G., Chan, T., Osher, S.: Structure-texture image decomposition - modeling, algorithms, and parameter selection. International Journal of Computer Vision 67(1), 111–136 (2006)
25. Chan, T., Esedoglu, S., Nikolova, M.: Algorithms for finding global minimizers of image segmentation and denoising models. SIAM Journal on Applied Mathematics 66(5), 1632–1648 (2006)
26. Bresson, X., Esedoglu, S., Vandergheynst, P., Thiran, J.P., Osher, S.: Fast global minimization of the active contour/snake model. Journal of Mathematical Imaging and Vision 28(2), 151–167 (2007)

Non-local Active Contours

Miyoun Jung, Gabriel Peyré, and Laurent D. Cohen

Ceremade, UMR 7534 CNRS Université Paris-Dauphine, 75775 Paris, France
{jung,peyre,cohen}@ceremade.dauphine.fr

Abstract. This article introduces a new image segmentation method that makes use of non-local comparisons between pairs of patches of features. A non-local energy is defined by summing the interactions between pairs of patches inside and outside the segmented domain. A maximum radius of interaction can be adapted to fit the amount of variation of the features inside and outside the region to be segmented. This non-local energy is minimized using a level set approach. The corresponding curve evolution defines a non-local active contour that converges to a local minimum of our energy. In contrast to previous segmentation methods, this approach only requires a local homogeneity of the features inside and outside the region to be segmented. This does not impose a global homogeneity as required by region-based segmentation methods. This comparison principle is also less sensitive to initialization than edge-based approaches. We instantiate this novel framework using patches of intensity or color values as well as Gabor features. This allows us to segment regions with smoothly varying intensity or colors as well as complicated textures with a spatially varying local orientation.

1 Introduction

Image segmentation refers to the process of partitioning an image into several regions or locating objects and boundaries. This paper considers a variational minimization problem for segmentation, which aims to find a contour representing the boundary of objects, by minimizing an energy functional composed of a contour smoothing term and an attraction term that pulls the contour towards the object boundaries. The curve (locally) minimizing the energy functional, located at the object boundaries, is obtained by curve evolution or active contours: starting with a given initial curve and evolving it to the correct steady state, the object boundaries. Active contours have been represented either by explicit parametric representation [1] or by the implicit level set representation of [2]. The level set representation has widely been used because it allows automatic topology changes of the contour such as merging and breaking, and the computations are made on a fixed rectangular grid. Many existing active contour models segment an image according to edge information and/or region information.

Edge-based approaches. Edge-based active contour models use edge detection functions depending on the image gradient and evolve contours towards sharp gradients of pixel intensity. The first work was the snakes model by Kass et al.

A.M. Bruckstein et al. (Eds.): SSVM 2011, LNCS 6667, pp. 255–266, 2012.

[1]. Then, many edge-based active contour models such as balloon [3], geometric [4], [5], [6] models were proposed. In particular, Caselles et al. [6] proposed an intrinsic geometric model, geodesic active contours, where the curve evolution is handled by the level set method [2] proposed by Osher and Sethian. In this model, the evolving curve moves by mean curvature, but with an extra factor in the speed, by the stopping edge-function. Therefore, the curve stops on the edges, where the edge-function vanishes. Although these classical snakes or geometric active contour models are quite effective, they are usually not robust to noise because noise also has large gradients. These models need in addition to perform a-priori smoothing, to smooth out the noise. This can therefore produce a not very accurate location of edges.

Region-based approaches. Region-based active contour models incorporate region information so that image within each segmented region has a homogeneous characteristics, such as intensities and textures. A region-based energy for an active contour was proposed in [7]. This was a reduced form of the Mumford-Shah functional [8] where the image was approximated by a piecewise smooth function inside objects and a smooth background. Chan and Vese [9] proposed an active contours without edges model, which is also based on techniques of curve evolution and level set methods, but the gradient-based information is replaced by a criterion related to region homogeneity. This model approximates an image by a two-phase piecewise constant function. The active contours without edges model was also extended to vector valued images [10] and to texture segmentation [11].

Kimmel [12] proposed a hybrid model by incorporating a more general weighted arc-length in the active contours without edges model. Sagiv et al. [13] applied the integrated approach, by incorporating multi-channel approaches [10], [11], to the problem of texture segmentation.

In this article, we propose an active contour model with a novel energy functional using pairwise interaction of features inside and outside the object, which allows to only constrain the local homogeneity, in contrast to the Chan-Vese approach. The local homogeneity property allows our model to capture regions with features that vary spatially in a smooth way, as well as to segment several separated objects with different features.

Several region-based methods [14], [15], [16] have been proposed to address the segmentation of locally homogeneous images using piecewise smooth image models. We extend these methods by making use of patches. Our approach is also conceptually different, since it makes use of pairwise patches comparison, and thus does not require the estimation of a piecewise smooth parameter.

Image features. In this work, we consider different image features based on given images. The choice of features is difficult and critical to get an optimal segmentation result. For a scalar image, the gray-level value or intensity can be enough to characterize each pixel. If the image is composed of multiple channels (such as color images), then each pixel is described by a vector of intensities. For texture images, the pixel intensity value does not give pertinent information. A very popular class of texture features are the filter-based features of the given image.

For instance, Gabor filter has often been implemented in texture segmentation [17], [11], [13] because it can segment images having region differences in spatial frequency, density of elements, orientation and phase. In particular, Sandberg et. al [11] incorporated the multiple Gabor transforms, obtained by convolving the Gabor functions with the original textured image, with the vector valued active contours without edges algorithm [10]. A recent promising image feature to represent and process textures is the image intensity patch around the current pixel. The information on a close neighborhood around the current pixel is extracted and leads to semi-local information at each pixel. The patch idea as feature vector was first introduced for texture synthesis [18], [19], then for image denoising, illustrated in the following paragraph.

We incorporate the patch idea with selected image features: for instance, for texture images, we use Gabor transforms as image features, and we consider the non-local interaction between pairs of patches of the features.

Non-local image processing. Nonlocal methods in image processing have been explored in many papers because they are well adapted to texture denoising. Buades et al [20] proposed to compute the weight matrix with patch differences and denoise the image with a non-local averaging, which is the well-known non-local means filter. Kinderman et al. [21], Gilboa and Osher [22], and Peyré et al. [23] proposed non-local energy functionals, and these functionals were used to solve various image processing problems such as denoising, inpainting, super-resolution and compressive sensing. The idea of functionals on nonlocal-graphs, in a regularization process, has also been used for image segmentation in a semi-supervised [24], [25], [26] (an extension of the work of Shi and Malik [29]) or an unsupervised [27], [28] settings.

In our work, we use a non-local energy that enforces the similarity of features both located either inside or outside the object. Using a level set formulation, this defines an attraction term pulling the contour towards the object boundaries. This is contrast to the existing non-local based segmentation methods that use non-local energy terms only as regularization terms.

Contributions. This article introduces a novel non-local energy for image/texture segmentation. In contrast to existing energy, we use pairwise interaction of features, which allows to only constrain the local homogeneity. This local homogeneity is crucial to capture regions with smoothly spatially varying features, such as color gradient or oriented textures. This is also useful to segment several separated objects with different features.

2   Non-local Active Contours

The goal is to segment an image $f : [0,1]^2 \rightarrow \mathbb{R}^d$, where d is dimensionality of the feature space. For instance one might consider $d = 1$ for gray-valued images, $d = 3$ for color images. To segment a texture image, $f(x)$ is computed as a high dimensional vectors which is the output of a directional filter bank.

Since we aim at proposing a generic segmentation framework, we do not specify the exact nature of the features in this section. Sections 3.2 and 3.3 detail some typical examples of features spaces.

2.1   Pairwise Patch Interaction

To be able to be less sensitive to noise in the image, we consider patches of features around each pixel $x \in [0,1]^2$:

$$\forall t \in [-\tau/2, \tau/2]^2, \quad p_x(t) = f(x+t).$$

Patch-based processing of images has been used extensively for a very long time in stereo and image matching in general, and has been very popular since the introduction of the non-local means denoising method.

Similarly to non-local denoising, we consider the non local interaction between pairs of patches, measured using a weighted L^2 distance

$$d(p_x, p_y) = \int_t G_a(t) \|p_x(t) - p_y(t)\|^2 dt \quad \text{where} \quad G_a(t) = e^{-\frac{\|t\|^2}{2a^2}}.$$

The Gaussian weight is used to give more influence to the central pixel.

2.2   Pairwise Interaction Energy

In its simplest form, the segmentation problem corresponds to the computation of some region $\Omega \subset [0,1]^2$ that should capture the objects of interest. This is usually performed in some variational framework where Ω solves an optimization problem.

The local homogeneity of the region (and of its complementary) is measured by considering all possible pairwise patch interaction at a given scale $\sigma > 0$. This gives rise to the following pairwise interaction energy of a region

$$E(\Omega) = \iint_{\Omega \times \Omega} G_\sigma(x-y) d(p_x, p_y) dx dy + \iint_{\Omega^c \times \Omega^c} G_\sigma(x-y) d(p_x, p_y) dx dy. \quad (1)$$

where $\Omega^c = [0,1]^2 \backslash \Omega$ is the complementary of the region.

The parameter $\sigma > 0$ is important since it controls the scale of the local homogeneity one requires for the segmented object. If the region is made of a nearly constant pattern, one should use a large σ. In contrast, if the region exhibits fast feature variations, σ should be chosen smaller. For simplicity, we use the same scale for both inside and outside the region, but one could of course use two distinct parameters.

2.3   Non-local Active Contour Energy

In order to perform the segmentation, we use a level set framework [2] where one computes a function $\varphi : [0,1] \to \mathbb{R}$ so that $\Omega = \{x \backslash \varphi(x) > 0\}$.

The integration inside and outside the domain is carried over using a smoothed Heaviside function

$$H(x) = \frac{1}{2} + \frac{1}{\pi}\mathrm{atan}(x/\varepsilon).$$

The parameter ε should be chosen small enough to obtain a sharp region boundary, but not too small to avoid numerical instabilities. In the numerical examples, we use $\varepsilon = 1/n$ for a discretized image of $n \times n$ pixels.

The energy (1) on regions is turned into an energy on the level set function φ, enforcing the similarity of features located inside and outside Ω,

$$E(\varphi) = \iint \rho(H(\varphi(x)), H(\varphi(y)))G_\sigma(x - y)d(p_x, p_y)\mathrm{d}x\mathrm{d}y$$

where ρ is an indicator function such that $\rho(u, v) = 1$ if $u = v$, 0 otherwise. In practice, we used $\rho(u, v) = 1 - |u - v|$. The meaning of this term is a way to consider only pairs of points for which φ has the same sign. Note that other binary interaction function ρ could be used as well, such as $\rho(u, v) = uv + (1 - u)(1 - v)$ (when $u = H(\varphi(x))$, $v = H(\varphi(y))$) and $\rho(u, v) = 1 - |u - v|^2$.

To enforce the regularity of the extracted region, following previous works in active contours, we penalize the length of the boundary, which is computed as

$$L(\varphi) = \int \|\nabla H(\varphi(x))\|\mathrm{d}x = \int H'(\varphi(x))\|\nabla\varphi(x)\|\mathrm{d}x \qquad (2)$$

where $\nabla H(\varphi(x))$ is the gradient at point x of the function $H(\varphi)$.

Our non-local active contour method compute the segmentation as a stationary point of the energy

$$\min_{\varphi} E(\varphi) + \gamma L(\varphi)$$

where $\gamma > 0$ is a parameter that should be adapted to the expected regularity of the boundary of the region.

Using the gradient descent with an artificial time $t \geqslant 0$ leads to the evolution equation for φ:

$$\frac{\partial \varphi}{\partial t} = -\left(\nabla E(\varphi) + \gamma \nabla L(\varphi)\right), \qquad (3)$$

where the gradients are computes as

$$\nabla E(\varphi)(x) = \int (\partial_1 \rho)(H(\varphi(x)), H(\varphi(y)))G_\sigma(x - y)d(p_x, p_y)\mathrm{d}y\, H'(\varphi(x)),$$

$$\nabla L(\varphi)(x) = -\mathrm{div}\left(\frac{\nabla\varphi(x)}{\|\nabla\varphi(x)\|}\right)H'(\varphi(x)).$$

Numerical implementation details. The segmentation is applied to a discretized image f of $n\times n$ pixels. The length energy (2) is computed using a finite difference approximation of the gradient.

The algorithm consists of two steps: given an image f, the weight function $w(x, y) = G_\sigma(x - y)d(p_x, p_y)$ is constructed based on the selected features, and then the evolution equation (3) for φ is solved with an explicit scheme. Note that $H'(\varphi)$ is replaced by $\|\nabla\varphi\|$. To ensure the stability of the level set evolution

(3), one needs to re-initialize it from time to time. This corresponds to replacing φ by the signed distance function to the level set $\{x \setminus \varphi(x) = 0\}$.

The size of the windowing function $G_\sigma(x - y)$ depends on the initial curve: if the initial curve is far away from the object boundaries, then a large windowing function may be required. Here, 31×31 or 41×41 are used with a fixed $\sigma = 10$ for 100×100 or 200×200 images. The choice of the size of patch and the parameter a in G_a depends on the image features: for instance, for image features depending on intensity, 3×3 patch with $a = 0.5$ is used.

3 Experimental Results and Comparisons

This section presents experimental results with synthetic and real images.

3.1 Hybrid Region/Edge Based Active Contours

We compare our approach with both region-based and edge-based active contour segmentation methods. We compare our method with segmentations obtained by minimizing a hybrid energy of the form

$$\min_{\varphi,p} \alpha E_r(\varphi, p) + (1 - \alpha)E_c(\varphi) + \gamma L(\varphi) \tag{4}$$

where α weights the influence of the region term E_r and the edge term E_c:

$$E_r(\varphi, p) = \lambda_1 \int H(\varphi(x))d(p_x, p_1)\mathrm{d}x + \lambda_2 \int (1 - H(\varphi(x)))d(p_x, p_2)\mathrm{d}x,$$

$$E_c(\varphi) = \mu \int \|\nabla H(\varphi(x))\|g(x)\mathrm{d}x,$$

with positive parameters λ_1, λ_2, μ and a positive edge function g, and where p represents the expected constant value of the features inside and outside the object. In particular, we consider the geodesic active contour model ($\alpha = 0$, GAC model [6]) with adding balloon force term $\eta g(x)\|\nabla \varphi(x)\|$, the region-based model ($\alpha = 1$) of Chan and Vese [9], and the integrated region/edge based model ($\alpha = 1/2$) of Sagiv et al. [13], called IAC model:

$$\text{GAC:}\quad \frac{\partial\varphi}{\partial t} = \mu\|\nabla\varphi\|\mathrm{div}\left(g\frac{\varphi}{\|\varphi\|}\right) + \eta g\|\nabla\varphi\|,$$

$$\text{Chan-Vese:}\quad \frac{\partial\varphi}{\partial t} = \|\nabla\varphi\|\left\{-\lambda_1 d(p_x, p_1) + \lambda_2 d(p_x, p_2) + \gamma\mathrm{div}\left(\frac{\varphi}{\|\varphi\|}\right)\right\},$$

$$\text{IAC:}\quad \frac{\partial\varphi}{\partial t} = \frac{1}{2}\|\nabla\varphi\|\left\{\mu\mathrm{div}\left(g\frac{\varphi}{\|\varphi\|}\right) - \lambda_1 d(p_x, p_1) + \lambda_2 d(p_x, p_2)\right\},$$

where $\eta g\|\nabla\varphi\|$ is a balloon force term that helps to avoid poor local minima by forcing moving the curve forward/outward (depending on the sign of η). Note that, in practice, we use an edge function $g(x) = \frac{1}{\delta^2 + G_{b_1} * \|\nabla(G_{b_2} * f)(x)\|^p}$ with $\delta^2 = 0.1$ and $p, b_1, b_2 > 0$, and then we normalize it from 0 to 1. And we let $\lambda_1 = \lambda_2 = 1$.

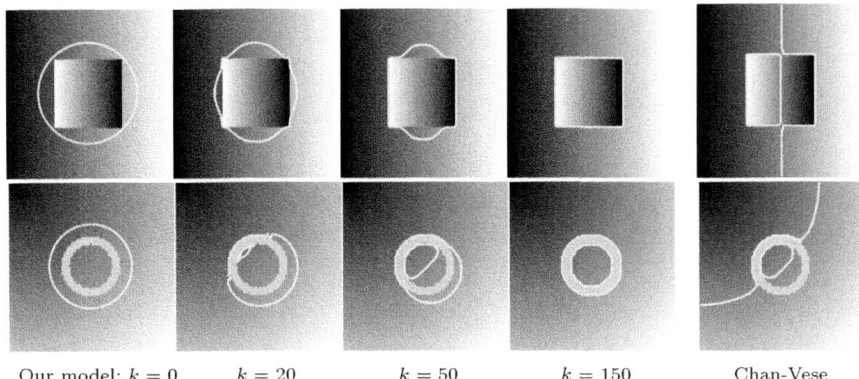

Our model: $k = 0$ $k = 20$ $k = 50$ $k = 150$ Chan-Vese

Fig. 1. Detection of object with spatially varying background or object, and comparison with Chan-Vese model ($\alpha = 1$) in (4). 100×100 image and 31×31 windowing function are used. k is the iteration number.

3.2   Gray-Level and Color Features

The simplest features $f(x)$ are the values of the image itself.

In the numerical examples, we use the edge function with $p = 1$, G_{b_1} with $b_1 = 0.5$ (or 1 for noisy image) and $G_{b_2} = 1$.

In Fig. 1 and 2, we test our method on several synthetic images with spatially varying background and/or object, or with several separated objects with different intensities. In all the examples, our model correctly detects the objects. This is due to the local homogeneity property of our model mentioned in Section 2.3, which is contrast to the two-phase Chan-Vese model requiring a global homogeneity in each region. The first example in Fig. 1 well demonstrates the effect of this property. The second example shows in addition the detection of interior contour. In the first example in Fig. 2, the bottom object has spatially varying intensities, and moreover the intensities of its left side are close to the ones of the background. Thus, Chan-Vese model (see Fig. 3) fails to segment this piecewise smooth object, regarding its left side as background, while our model captures the boundary with small gradients. Furthermore, Fig. 2 shows the detection of multiple separated objects with different intensities, unlike two phase Chan-Vese model. Lastly, we note that our model needs small number of iterations (around 150 iterations) to obtain final curves, even with an explicit scheme.

Fig. 3 presents the results of existing edge-based and/or region-based models, given in (4): $\alpha = 0$ with balloon force term $\eta g(x)\|\nabla\varphi(x)\|$ (geodesic model), $\alpha = 1/2$ (integrated active contour model), $\alpha = 1$ (region-based Chan-Vese model). For the IAC model, two final curves are shown with two different but close parameters μ ($\mu_1 > \mu_2$). Because μ is a balancing term between the region-based and edge-based energies, when $\mu > \mu_1$ (or $\mu < \mu_2$), the model tends to act like the geodesic snake model (or Chan-Vese model). Thus, with the given initial curves, all the models fail to detect the correct object boundaries. Note that, with good initial curves surrounding all the boundaries, IAC model was

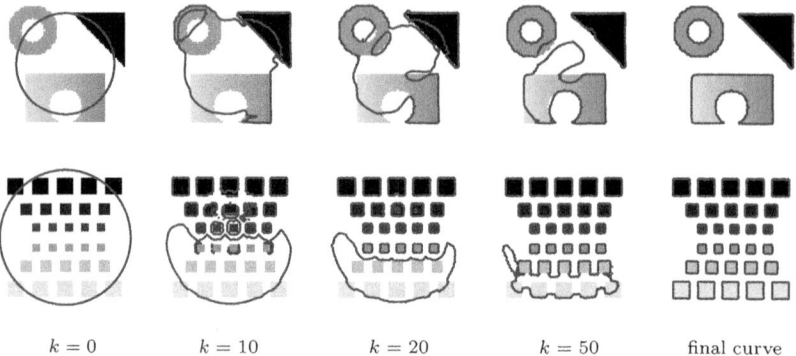

$k = 0$ $k = 10$ $k = 20$ $k = 50$ final curve

Fig. 2. Detection of objects with spatially varying object, or with several separated objects with different intensities, or with various shapes, using our model. k is the iteration number, and final curves are obtained at $k = 80$ (top) and $k = 120$ (bottom).

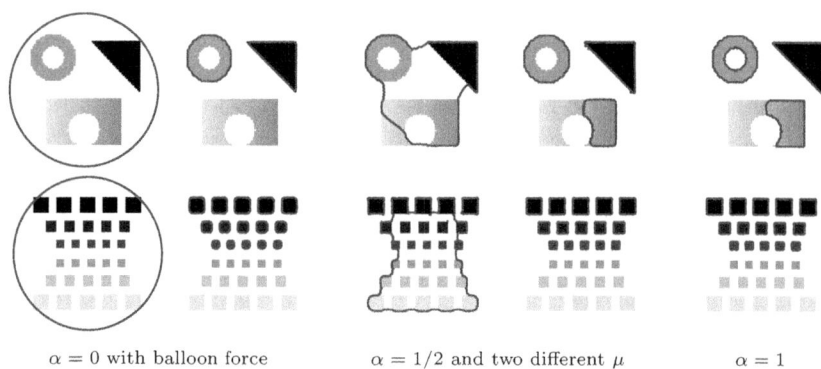

$\alpha = 0$ with balloon force $\alpha = 1/2$ and two different μ $\alpha = 1$

Fig. 3. Final curves of models given in (4): $\alpha = 0$ with balloon force term (GAC), $\alpha = 1/2$ (IAC) with two different but close parameters μ, $\alpha = 1$ (Chan-Vese). GAC: $\mu = 1$, $\eta = -0.3$. IAC: (top) $\mu = 3.6$ and 3.5, (bottom) $\mu = 1$ and 0.9. IAC and Chan-Vese models used initial curves given in Fig. 2.

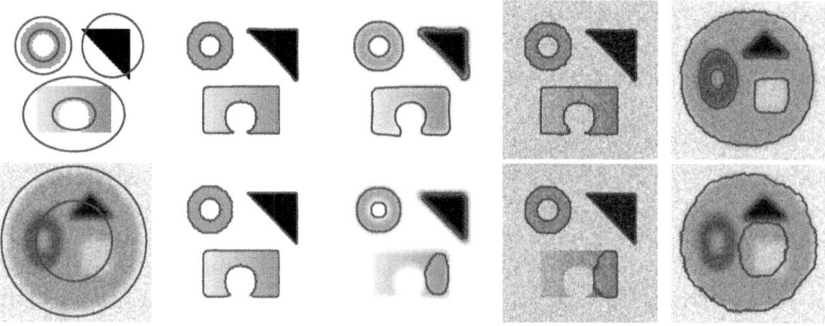

Fig. 4. Detection of objects from blurred and/or noisy images. 1st column: initial curves used, 2nd-5th columns: final curves of our model (Top) and IAC model (Bottom).

able to detect the boundaries, as shown in the 2nd column in Fig. 4, with large values of μ, while in our model one circle around objects as an initial curve was enough for segmentation. Thus, our model is less sensitive to the choice of initial curves than edge-based active contour models.

Fig. 4 presents how our model works on noisy images, and detection of objects with blurred boundaries. The 2nd-4th columns present a clear and clean image (2nd), given in Fig. 2, and a blurred and noisy version of it, respectively. However, IAC model fails to locate boundaries with small gradients in blurred or noisy images, even with good initial curves. These examples show that our model detects object boundaries with small gradients as well as that it is not sensitive to noise unlike edge-based models.

In Fig. 5, we test our method on real color images. We compare our model with the vector-valued Chan-Vese model [10] and IAC model. By using an initial curve near the boundary of object(s) and a small windowing function, our model could detect the boundary of non-homogeneous object(s). The segmentation result is fairly good, comparing with Chan-Vese model and IAC model that only capture part of object(s). On the other hand, these examples also show a limitation of our model: in order to detect the boundary of non-homogeneous objects, the initial curve needs to be located near the object boundary so that a small windowing function can be used.

3.3   Gabor Features

To segment a texture image, one can use the energy of the output of a dictionary multi-scale filter bank. Given an image f_0, one computes each $f(x) \in \mathbb{R}^d$ as the magnitude of d complex filters

$$\forall \ell \in \{0, \ldots, d-1\}, \quad f_\ell(x) = |f_0 \star h_\ell| \tag{5}$$

with $\forall x = (x_1, x_2) \in [0,1]^2$ and $h_\ell(x) = e^{\frac{2i\pi}{n}\eta_\ell(\cos(\theta_\ell)x_1 + \sin(\theta_\ell)x_2)} G_{s_\ell}(x)$.

The parameter $\eta_\ell > 0$ is the frequency of the filtering, $\theta_\ell \in [0, \pi)$ is the orientation and $s_\ell > 0$ is the spacial width of the filter. In the numerical examples, the parameters η_ℓ, θ_ℓ, s_ℓ are fine-tuned to obtain the best texture representation. Note that the energies (4) incorporating Gabor features and multi-channel approach have been used for texture segmentation in [11] (Gabor based multi-channel Chan-Vese model) and [13] (IAC model).

Fig. 6 presents our texture segmentation result and comparison with Gabor based Chan-Vese model [11]. In this case, we use $d = 8$ filters with $\eta_\ell \in \{2, 2.5, 3, 3.5\}$, $\theta_\ell \in \{0, \pi/2\}$, $s_\ell = 2$. Gabor based Chan-Vese model fails to detect the object on the top right side (even with $d = 64$ filters with $\eta_\ell \in \{2, 3, 4, 5\}$, $\theta_\ell \in \{0, \pi/4, \pi/2, 3\pi/4\}$, $s_\ell \in \{2, 2\sqrt{2}, 4, 4\sqrt{2}\}$) because the intensity values of that object in Gabor transforms are very small compared with the ones of the other objects and close to the one of the background. But, our model detects all the objects well due to the local homogeneity.

| initial curve | our model | Chan-Vese model | IAC model |

Fig. 5. Real color images. Final curves of our model, vector-valued Chan-Vese model [10], and integrated active contour model (IAC).

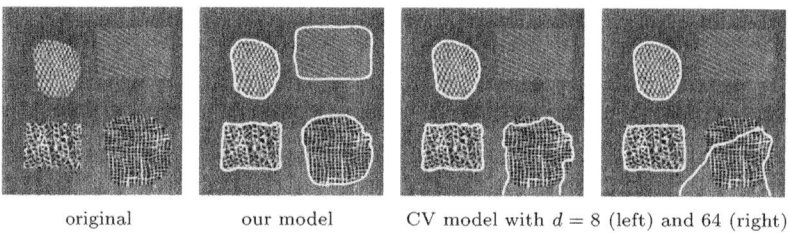

| original | our model | CV model with $d = 8$ (left) and 64 (right) |

Fig. 6. Texture segmentation with Gabor transforms. Comparison with Gabor based Chan-Vese model [11].

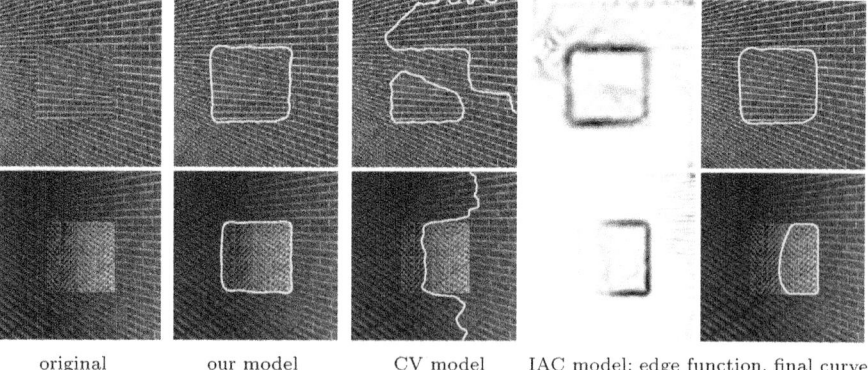

| original | our model | CV model | IAC model: edge function, final curve |

Fig. 7. Texture segmentation with Gabor transforms. Image composed of a background and an object with smoothly varying features. Comparison with Gabor based Chan-Vese model [11] and IAC model [13].

In Fig. 7, the images are composed of a background and an object with smoothly varying features. Here, we use $d = 4$ filters with $\eta_\ell \in \{0.7, 1.6\}$, $\theta_\ell = 0$, $s_\ell \in \{4, 4\sqrt{2}\}$ in the first example, and $d = 8$ filters with $\eta_\ell \in \{2, 3, 4, 5\}$, $\theta_\ell = 0$, $s_\ell \in \{4, 4\sqrt{2}\}$ in the second one. Due to a reason similar with the one in Fig. 1, Gabor based Chan-Vese model [11] fails to segment the actual object boundary, while our model detects it. For the IAC model, we use $p = 2$, $G_{b_1} = 1$, G_{b_2} with $b_2 = 3.75$ (top), 0.5 (bottom) for the edge function. IAC model detects the object in the first example but not in the second one, which depends on the edge function $g(x)$. However, our model could segment the object in both cases without any prior work on the edge function like the IAC model.

Conclusion

In this article, we have proposed a novel non-local energy for image/texture segmentation. We have compared our active contour model with state of the art. We have illustrated the superiority of our model over the existing region-based and/or edge-based active contour models. Due to the local homogeneity property, our segmentation model could detect regions with smoothly spatially varying features and segment several separated object with different features. Furthermore, our model is less sensitive to the choice of initial curves as well as to noise than edge-based active contour models.

Acknowledgments. This work was partially supported by Fondation Sciences Mathématiques de Paris and ANR grant MESANGE ANR-08-BLAN-0198.

References

1. Kass, M., Witkin, A., Terzopoulos, D.: Snakes: Active contour models. International Journal of Computer Vision 1, 321–331 (1988)
2. Osher, S., Sethian, J.: Fronts propagating with curvature-dependent speed: Algorithms based on Hamilton-Jacobi formulations. Journal of Computational Physics 79, 12–49 (1988)
3. Cohen, L.: On active contour models and balloons. CVGIP: Image Underst. 53, 211–218 (1991)
4. Caselles, V., Catté, F., Coll, T., Dibos, F.: A geometric model for active contours in image processing. Numerische Mathematik 66, 1–31 (1993)
5. Malladi, R., Sethian, J.A., Vemuri, B.C.: Shape modeling with front propagation: A level set approach. IEEE Trans. Patt. Anal. and Mach. Intell. 17, 158–175 (1995)
6. Caselles, V., Kimmel, R., Sapiro, G.: Geodesic active contours. International Journal of Computer Vision 22, 61–79 (1997)
7. Cohen, L.D.: Avoiding local minima for deformable curves in image analysis. In: Le Méhauté, A., Rabut, C., Schumaker, L.L. (eds.) Curves and Surfaces with Applications in CAGD. Vanderbilt University Press, Nashville (1997)
8. Mumford, D., Shah, J.: Optimal approximations by piecewise smooth functions and associated variational problems. Communications on Pure and Applied Mathematics XLII (1989)

9. Chan, T., Vese, L.: Active contours without edges. IEEE Trans. Image Proc. 10, 266–277 (2001)
10. Chan, T., Sandberg, B., Vese, L.: Active contours without edges for vector-valued images. J. Vis. Comm. Image Repr. 11, 130–141 (2000)
11. Sandberg, B., Chan, T., Vese, L.: A level-set and Gabor based active contour algorithm for segmenting textured images. UCLA CAM Report 02-39 (2002)
12. Kimmel, R.: Fast edge integration. In: Osher, S., Paragios, N. (eds.) Geometric Level Set Methods in Imaging, Vision, and Graphics. Springer, New York (2003)
13. Sagiv, C., Sochen, N.A., Zeevi, Y.Y.: Integrated active contours for texture segmentation. IEEE Trans. Image Proc. 15, 1633–1646 (2006)
14. Tsai, A., Yezzi, A., Willsky, A.S.: Curve evolution implementation of the mumford-shah functional for image segmentation, denoising, interpolation, and magnification. IEEE Trans. Image Proc. 10, 1169–1186 (2001)
15. Li, C., Kao, C., Gore, J., Ding, Z.: Implicit active contours driven by local binary fitting energy. In: Proceedings of the CVPR 2007, pp. 1–7 (2007)
16. Wang, X., Huang, D., Xu, H.: An efficient local chan–vese model for image segmentation. Pattern Recognition 43, 603–618 (2010)
17. Lee, T.S., Mumford, D., Yuille, A.: Texture segmentation by minimizing vector-valued energy functionals: The coupled-membrane model. In: Sandini, G. (ed.) ECCV 1992. LNCS, vol. 588, pp. 165–173. Springer, Heidelberg (1992)
18. Efros, A., Leung, T.: Texture synthesis by non-parametric sampling. In: IEEE International Conference on Computer Vision, vol. 2, pp. 10–33 (1999)
19. Efros, A., Freeman, W.T.: Image quilting for texture synthesis and transfer. In: Proceedings of SIGGRAPH, pp. 341–346 (2001)
20. Buades, A., Coll, B., Morel, J.M.: A review of image denoising algorithms, with a new one. SIAM Mul. Model. and Simul. 4, 490–530 (2005)
21. Kindermann, S., Osher, S., Jones, P.W.: Deblurring and denoising of images by nonlocal functionals. SIAM Mult. Model. and Simul. 4, 1091–1115 (2005)
22. Gilboa, G., Osher, S.: Nonlocal operators with applications to image processing. SIAM Multiscale Modeling and Simulation 7, 1005–1028 (2008)
23. Peyré, G., Bougleux, S., Cohen, L.: Non-local regularization of inverse problems. In: Forsyth, D., Torr, P., Zisserman, A. (eds.) ECCV 2008, Part III. LNCS, vol. 5304, pp. 57–68. Springer, Heidelberg (2008)
24. Gilboa, G., Osher, S.: Nonlocal linear image regularization and supervised segmentation. SIAM Mul. Model. and Simul. 6, 595–630 (2007)
25. Elmoataz, A., Lezoray, O., Bougleux, S.: Nonlocal discrete regularization on weighted graphs: a framework for image and manifold processing. IEEE Trans. Image Process 17, 1047–1060 (2008)
26. Houhou, N., Bresson, X., Szlam, A., Chan, T., Thiran, J.: Semi-supervised segmentation based on non-local continuous min-cut. In: Tai, X.-C., Mørken, K., Lysaker, M., Lie, K.-A. (eds.) SSVM 2009. LNCS, vol. 5567, pp. 112–123. Springer, Heidelberg (2009)
27. Bresson, X., Chan, T.: Non-local unsupervised variational image segmentation models. UCLA CAM Report 08-67 (2008)
28. Caldairou, B., Rousseau, F., Passat, N., Habas, P., Studholme, C., Heinrich, C.: A non-local fuzzy segmentation method: Application to brain mri. In: Jiang, X., Petkov, N. (eds.) CAIP 2009. LNCS, vol. 5702, pp. 606–613. Springer, Heidelberg (2009)
29. Shi, J., Malik, J.: Normlaized cuts and image segmentation. IEEE Trans. Patt. Anal. and Mach. Intell. 22, 888–905 (2002)

From a Modified Ambrosio-Tortorelli to a Randomized Part Hierarchy Tree

Sibel Tari* and Murat Genctav

Middle East Technical University, Department of Computer Engineering,
Ankara, TR-06531
stari@metu.edu.tr, muratgenctav@gmail.com

Abstract. We demonstrate the possibility of coding parts, features that are higher level than boundaries, using a modified AT field after augmenting the interaction term of the AT energy with a non-local term and weakening the separation into boundary/not-boundary phases. The iteratively extracted parts using the level curves with double point singularities are organized as a proper binary tree. Inconsistencies due to non-generic configurations for level curves as well as due to visual changes such as occlusion are successfully handled once the tree is endowed with a probabilistic structure. The work is a step in establishing the AT function as a bridge between low and high level visual processing.

Keywords: phase fields, non-local variational shape analysis.

1 Introduction

The phase field of Ambrosio and Tortorelli [1] (AT function) serving as a continuous indicator for the boundary/not-boundary state at every domain point has proven to be an indispensable tool in image and shape analysis. It is a minimizer of an energy composed of two competing terms: One term favors configurations that take values close to either 0 or 1 (separation into boundary/not-boundary phases) and the other term encourages local interaction in the domain by penalizing spatial inhomogeneity. A parameter controls the relative influence of these two terms, hence, the interaction. As this "interaction" parameter tends to 0, the separation term is strongly emphasized; consequently, the field tends to the characteristic function $1 - \chi_S$ of the boundary set S and the AT energy tends (following the Γ convergence framework [4]) to the boundary length.

In computer vision, the AT function first appeared as a technical device to apply gradient descent to the Mumford-Shah functional [15]. Over the years, it has been extended in numerous ways to address a rich variety of visual applications. Earlier works include Shah and colleagues [22,23,25,19], March and Dozio [13], Proesman, Pauwels and van Gool [21], Teboul *et al.* [29]. During the last couple of years we have witnessed an increasing number of promising works modifying or extending Ambrosio-Tortorelli/Mumford-Shah based models. Some examples

---

* Corresponding author.

A.M. Bruckstein et al. (Eds.): SSVM 2011, LNCS 6667, pp. 267–278, 2012.

are Bar, Sochen and Kiryati [3], Rumpf and colleagues [6,20], Erdem, Sancar-Yilmaz and Tari [7], Patz and Preusser [17], Jung and Vese [9]. These works together with many others collaboratively established the role of AT function in variational formulations that jointly involve region and boundary terms.

In the majority of the works, the AT function serves as an auxiliary variable to facilitate discontinuity-preserving smoothing and boundary detection. Relatedly, the interaction parameter is chosen sufficiently small to better localize boundaries. In contrast, Shah and Tari, starting with [27,28] in late 90's, have focused on the ability of the AT function in coding morphologic properties of shapes, regions construed by boundaries. Relatedly, they have weakened boundary/not-boundary separation either by choosing a large interaction parameter or by other means [2] and focused on the geometric properties of the level curves after constructing the AT function (reviewed in [24]) for shapes as

$$\arg\min_{v} \iint_{\Omega} \frac{1}{\rho} \underbrace{(v(\mathbf{x}) - \chi_{\Omega}(\mathbf{x}))^2}_{\text{boundary/interior separation}} + \rho \underbrace{|\nabla v(\mathbf{x})|^2}_{\text{local interaction}} \, dx \, dy$$

$$\text{with } v(\mathbf{x}) = 0 \text{ for } \mathbf{x} = (x, y) \in \partial\Omega \tag{1}$$

where $\Omega \in \mathbf{R}^2$ is a bounded open set with a boundary $\partial\Omega$ (denoting a shape); $\chi_{\Omega}(\mathbf{x})$ is the shape indicator function which attains 1 in Ω and 0 on $\partial\Omega$; ρ is the parameter. The first term forces strong boundary/interior separation while the second one forces smoothness.

The AT function of shape is related to a variety of morphological concepts. For instance, it is a weighted distance transform [11,12] with its level curves approximating curvature-dependent motion [16,10]. Thus, it enables extraction of local symmetries and skeletons directly from grayscale images; that is, it bridges image segmentation and shape description. The ability of level curves in coding morphological information is also exploited by Droske and Rumpf [20] to measure equivalence of two shapes in a registration problem.

In this paper, following Shah and Tari [27,28,2], we explore and extend the ability of an AT-like field in coding features that are at a higher level than boundaries. Whereas the previous works focus on local symmetry axes, we focus on shape's intuitive components as coded via upper and lower level sets. Our constructions are based on a new field obtained as the minimizer of a modified AT energy. We discuss the geometry of the level curves of the new minimizer and exploit it to extract a part hierarchy tree endowed with a probabilistic structure.

The considered modification involves an additive augmentation of the interaction term with a non-local term in a way that the upper and lower zero level sets of the minimizer yield disjoint domains [26] within which the minimizer is morphologically equivalent to the AT function. Following the pioneering work of Buades, Coll and Morel [5], UCLA group formulated interesting non-local variational formulations, including non-local versions of the Ambrosio-Tortorelli/Shah approximations of the Mumford-Shah functional [9,8] by replacing local image derivatives with non-local ones. This kind of modification is very different from our modification which modifies the phase field itself.

In this paper, we focus on shapes. Nevertheless, the long term goal of our work is to bridge low level processes such as segmentation and image registration with the high level process of shape abstraction. Integration of the presented developments to Mumford-Shah type models via coupled PDEs framework is a future work.

2   A Modified Energy and Its Minimizer

Let us consider

$$\iint_{\Omega} \frac{1}{\rho}(\omega(\mathbf{x}) - f(\mathbf{x}))^2 + \rho \left[|\nabla\omega(\mathbf{x})|^2 + (\mathbf{E}_{\mathbf{x}\in\Omega}\omega(\mathbf{x}))^2 \right] \, \mathrm{d}x \, \mathrm{d}y$$

$$\text{with } \omega(\mathbf{x}) = 0 \text{ for } \mathbf{x} = (x, y) \in \partial\Omega \tag{2}$$

where $\mathbf{E}_{\mathbf{x}\in\Omega}\omega(\mathbf{x})$ is the expectation of ω given by $\frac{1}{|\Omega|} \iint \omega(\mathbf{x}) \, \mathrm{d}x \, \mathrm{d}y$ and $f(\mathbf{x})$ is the distance transform. The new energy to be minimized is composed of three terms and obtained by modifying the AT energy in (1) in two aspects.

Firstly, the interaction term of (1) is additively augmented with $(\mathbf{E}_{\mathbf{x}\in\Omega}\omega(\mathbf{x}))^2$. This new term forces the minimizer to acquire a low average value with the average being computed over the entire domain. At a first glance, this seems to favor spatial homogeneity by forcing the minimizer to attain values close to zero. Yet, the minimum of $\iint (\mathbf{E}_{\mathbf{x}\in\Omega}\omega(\mathbf{x}))^2 \, \mathrm{d}x\mathrm{d}y$ is also reached when ω oscillates, that is, when it attains both negative and positive values adding up to 0. In this respect, the third term is a separation term partitioning Ω into subdomains of opposing signs. Due to the influence of the $|\nabla.|^2$ term which penalizes spatial inhomogeneities, locations of identical sign tend to form spatial groups. Obviously, the minimizer of $\iint_{\Omega} \left[|\nabla\omega(\mathbf{x})|^2 + (\mathbf{E}_{\mathbf{x}\in\Omega}\omega(\mathbf{x}))^2 \right] \mathrm{d}x\mathrm{d}y$ subject to homogeneous Dirichlet boundary condition is the flat function $\omega = 0$ unless accompanied by an external inhomogeneity.

Indeed, the purpose of the second modification is to influence spatial grouping of positive and negative values of ω in a particular way that the sign change separates the *gross* structure from the boundary detail. In particular, the upper zero level set $\{\Omega^+ = (x, y) \in \Omega : \omega(x, y) > 0\}$ covers central regions whereas the lower zero level set $\{\Omega_- = (x, y) \in \Omega : \omega(x, y) < 0\}$ covers peripheral regions containing limbs, protrusions and boundary texture or noise. Towards this end, the indicator $\chi_\Omega(\mathbf{x})$ is replaced by a weighted indicator that is a monotonically increasing function of the shortest distance to the boundary, namely, the distance transform. As before, the first term favors separation of the domain into phases; however, the phases are the level curves of the distance transform. Since, however, the level curves of the AT function in (1) are equivalent to the level curves of a smooth distance transform [28], this change merely scales ω without qualitatively affecting the geometry of its level curves. Nevertheless, when the terms considered together, the minimizer tends to have positive values at central locations and negative values at peripheral locations because the penalty incurred by assigning negative values to central locations with higher positive f values is higher than the penalty incurred by assigning negative values to locations with lower f values.

Similar to the AT function, the new minimizer is a compromise between inhomogeneity and homogeneity though the inhomogeneity is forced both externally (by f) and internally (by the third term); or it is the best approximation of an external inhomogeneity f subject to internal constraints.

The parameter ρ should be chosen large enough so that the attachment to the external inhomogeneity should not dominate over the tendency to interact. Indeed, in the absence of the third term, a good practice is to chose ρ at least on the order of the maximum thickness for the diffusive effect of $|\nabla.|^2$ to influence the entire shape ([2]; Fig. 1 in [28]). The same argument also holds here since the effect of the third term is to partition Ω into subdomains within which ω is morphologically similar to the AT function. Additionally, notice that the expression responsible for sign change, $\iint \omega(\mathbf{x})\,dx\,dy$, has been already normalized by $\frac{1}{|\Omega|}$. As such, ρ should be larger than $\sqrt{|\Omega|}$.

In Fig. 1 (a), an illustration for a 1-D case is given. ω is plotted for four different values of ρ ranging between $\sqrt{|\Omega|}$ and $0.5 * |\Omega|$. Naturally, ω gets flatter as ρ increases. (The flattening can be avoided by scaling either f or ω.) Nevertheless, the locations of the extrema and the zero crossings remain the same unless ρ is significantly smaller than $\sqrt{|\Omega|}$. Similarly in 2-D, the geometry of the level curves is stable as long as ρ is chosen suitably large. Illustrative level curves are depicted in Fig. 1 (b-c). Absolute values of ω separately normalized within regions of identical sign are used for convenience of color visualization. Zero level curves separate central and peripheral structures in the form of upper and lower zero level sets: $\{\Omega^+ = (x,y) \in \Omega : \omega(x,y) > 0\}$ and $\{\Omega_- = (x,y) \in \Omega : \omega(x,y) < 0\}$. The peripheral structure includes all the detail: limbs, protrusions, and boundary texture or noise. In contrast, the central structure is a very coarse blob-like form; it can even be thought as an interval estimate of the center whereas the centroid is the point estimate.

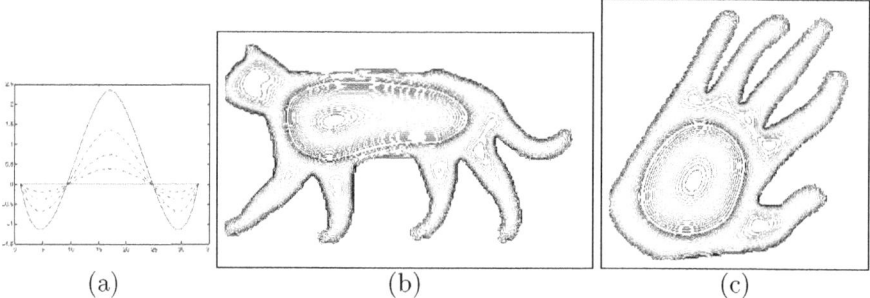

(a)   (b)   (c)

Fig. 1. (a) ω for an interval for varying values of ρ ranging between $\sqrt{|\Omega|}$ and $0.5 * |\Omega|$. (b) Illustrative level curves of ω.

Most commonly, Ω^+ is a simply connected set. Of course, it may also be either disconnected or multiply connected. For instance, it is disconnected for a dumbbell-like shape (two blobs of comparable radii combined through a thin neck) whereas it is multiply connected for an annulus formed by two concentric circles. Indeed, the annulus gets split into three concentric rings where the middle ring is the Ω^+. For quite a many shapes, however, Ω^+ is a simply connected set.

Firstly, shapes obtained by protruding a blob as well as shapes whose peripheral parts are smaller or thinner than their main parts always have a simply connected Ω^+. This is expected: When the width of a part is small, the highest value of f inside the part is small. That is, the local contribution to $(\omega - f)^2$ incurring due to negative values is less significant for such a part as compared to locations with higher positive values of f. Consequently, ω tends to attain negative values on narrow or small parts as well as on protrusions. Shapes with holes also have a simply connected Ω^+ as long as the holes are far from the center.

Secondly, even a dumbbell-like shape may have a simply connected Ω^+. This happens if the join area, namely, the neck is wide enough. Nevertheless, this does not cause any representational instability: Whereas the Ω^+ for a blob-like shape has a unique maximum located roughly at its centroid, the Ω^+ for a dumbbell-like shape has two local maxima indicating two bodies. Each body is captured by a connected component of an upper level set whose bounding curve passes through a saddle point $\boldsymbol{p}$, such that $\omega(\boldsymbol{p}) = s$, the $s - level$ curve has a double point singularity, i.e. it forms a cross. As such, the upper level set $\{\Omega^s = (x, y) \in \Omega^+ : \omega(x, y) > s\}$ yields two disjoint connected components capturing the two parts of the central structure.

In contrast to Ω^+, the peripheral structure Ω_- is often multiply connected. Indeed, its hole(s) are carved by Ω^+. It is also possible that Ω_- is disconnected. For instance, for an annulus, it is two concentric rings. Additionally, Ω_- may be disconnected when there are several elongated limbs organized around a rather small central body, e.g., a palm tree. Ω^+, being small, is tolerated to grow and reach to the most concave parts of the shape boundary creating a split of Ω_- by the zero-level curve. Similar to those in Ω^+, the level curves in Ω_- that are passing through saddle points provide further partitioning. The partitions are in the form of lower level sets $\{\Omega_s = (x, y) \in \Omega_- : \omega(x, y) < s\}$.

To sum up, within both Ω^+ and Ω_-, nested open sets (upper level sets inside Ω^+ and lower level sets inside Ω_-) characterize the domain. The level curves bounding the level sets are either closed curves or closed curves with crossing points. The ones with crossing points are of particular interest because the respective level set is partitioned at those points into two distinct connected components. A crossing of a level curve occurs at a saddle point of ω. Of course, each lower level set may contain other saddle points. Consequently, the partitioning is binary and iterative and determined by the order of saddle points.

It is not generically possible that a level curve has singular points of higher order because such singular points are unstable and may be removed by a slight change in ω. It is also highly unlikely that a connected component of an $s - level$ curve has two distinct crossing points. This issue is tackled in §3 via randomization.

3   Randomized Hierarchy Tree

Since the partitioning inside both the Ω^+ and Ω_- of a shape are iterative and binary, the parts can be organized starting from the second level in the form of a proper binary tree. Let the shape be the root node and its children be the upper and lower zero level sets, namely, the disjoint regions of the central and peripheral structures. Suppose the central and peripheral structures are respectively composed of N_c and N_p disjoint sets. Let us enumerate the nodes holding these sets as $11, 12, \cdots, 1N_c$ for the Ω^+ and as $21, 22, \cdots, 2N_p$ for the Ω_-. This is the second level of the tree and the first level of the partitioning. Of course, the root may have more than two children. Nevertheless, starting from the children of the root, each subtree is a proper binary tree because all the splits inside an Ω^+ or Ω_- occur at saddle points; that is, each connected component of the second level and its children either get split into two level sets or remain as they are. We call this hierarchical organization as the **Initial Part Tree**. A hypothetical initial part tree is illustrated in Fig. 2 (a). In a real example, the nodes hold application dependently selected attributes of the respective level sets.

Binary splits according to saddle points produce collections of parts which are at the leaf level consistent across visual changes. However, the hierarchical order and granularity of parts are not necessarily consistent. For instance, a weak saddle is easily removed when the shape is slightly smoothed. Likewise, certain non-generic configurations such as level curves with spatially distinct saddle point singularities or triple point singularities cannot occur; indeed, such configurations are easily replaced by one of the corresponding generic configurations which may differ for similar shapes. Furthermore, when a shape is occluded by another shape, added peripheral parts change the positions of some of the previous level sets in the hierarchy. Nevertheless, the relative values of ω at saddle points prompting consecutive splits are stable indicators of the organizational hierarchy. Of course, attempting to convert a saddle point value to a tree depth by discretization brings back the previous robustness issue.

Instead, we use the difference between the values of two successive saddle points as a measure of saliency for the partitioning prompted by the latter saddle point. Converting the saliency measure to a probability measure and considering probability measures for all nodes, we endow the initial part tree with a random structure from which possible re-organizations of the initial hierarchy tree are to be sampled. We call the new structure as the **Randomized Part Hierarchy Tree**. Below, we give the details of the randomization procedure. In contrast to the respective initial part tree, a random sample from a randomized part hierarchy tree is not necessarily a proper binary tree.

The randomization starts from level 3 nodes and propagates through their children. Recall that this is the first level of nodes that are created via saddle points. For each pair of siblings, there are two possible events: The pair of siblings either maintain their depth (no change in the local tree structure) or inherit the depth of their parent (change in the local tree structure). In the latter case, the node and its sibling replace their parent and become the children of their grandparent. The probabilities of the two events are derived from a quantity which we denote by $D\omega$. It is a property of a split meaning that the $D\omega$ values of a pair of siblings are identical. Specifically, it is the difference between the saddle point values of a node and its parent divided by the saddle point value of the node. Because the magnitude of the saddle point value of a node is always greater than that of its parent, $0 < D\omega \leq 1$; the equality is attained at level 3.

A small value of $D\omega$ implies that the consecutive saddle points are closer in value; that is, a slight change in their value changes their order hence the local tree structure. Equivalently a large value of $D\omega$ implies that the consecutive saddle points are well separated, therefore, the local structure is stable. We require that the probability p that a local structure change is necessary approaches 1 as $D\omega$ approaches to the smallest possible value which is 0. Equivalently, p should approach 0 as $D\omega$ approaches to its largest possible value. The function $e^{-4D\omega}$ is a good candidate for estimating p; there is less than 2% chance for reorganization since $e^{-4} = 0.018$ for the largest possible $D\omega$.

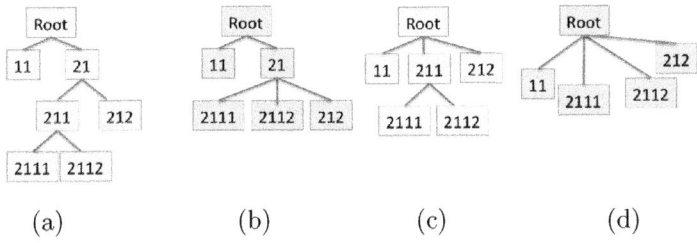

Fig. 2. (a) An initial part tree (a) and its possible re-organizations (b-d). A random sample should be in one of the four forms. See the text.

Let us consider the initial part tree in Fig. 2 (a). Assume that $D\omega = 0.301$ for nodes 211 and 212. With probability $(1 - p) = 0.7$, the local structure is preserved, whereas with probability $p = 0.3$ nodes 211 and 212 replace their parent and become children of their grandparent, the root. Assume that $D\omega = 0.128$ for 2111 and 2112. Then with probability $(1 - q) = 0.4$, the local structure is preserved, while with probability $q = 0.6$ nodes 2111 and 2112 replace their parent and become children of their grandparent which is either node 21 with $(1-p) = 0.7$ or the root with $p = 0.3$. Thus, there are four possible organizations: With probability $(1 - p)(1 - q) = 0.28$, the entire structure is preserved. With probability $(1 - p)q = 0.42$, the tree is re-organized as in (b). With probability $p(1 - q) = 0.12$, the tree is re-organized as in (c). With probability $pq = 0.18$, the tree is re-organized as in (d).

4   Experimental Results and Discussion

To evaluate the effectiveness of endowing the part hierarchy tree with a probabilistic structure, we consider a pairwise matching problem. It is formulated as finding a maximal clique in the joint association graph of the pair of trees to be matched, e.g. [18]. At each experiment, each of the two randomized part hierarchy trees is independently sampled several times and then all the sample pairs are matched. The trees in the sample pair that yields the highest matching score are called as the winning re-organizations.

Depending on the application, various properties related to the level sets stored at nodes can be used as node attributes. For instance, we extract parts enclosing each of the stored level sets and then use their area and the maximum ω values inside them as attributes. We remark that the maximum value of ω is related to the part width for the finest parts. Because boundaries of stored level sets pass through saddle points, an enclosing part is easily obtained as a morphologic watershed zone, whose seed is the respective level set [14]. Both to keep illustrations simple and resource requirements low, we require that each part to neighbor the central structure and each seed and part to have a certain size. Specifically, splits are performed only through the saddle points that reside on the boundaries of the watershed zones that are touching to the closure of the central structure and any split producing a seed that is less than 0.05% of the shape or a part that is less than 0.5% of the shape is ignored.

We present four illustrative examples. In each case, correct associations are found despite several order and granularity inconsistencies resulting from occluders, non-generic splits and weak saddles.

The first example is a matching between a human silhouette and its occluded version. The winning re-organizations for each of the two trees are shown in Fig. 3. At each node, the watershed (enclosing part) is depicted as dark gray and the respective level set (seed part) depicted as black is superimposed on the part; the neighboring part (the light gray) is also shown even though it is not used for the matching. Due to page limits, we cannot provide the initial trees and probability distributions, but the numbering of the nodes already reveals the structure of the initial binary tree.

Firstly, notice that the arms on the left and on the right reside at different levels in both of the initial trees as revealed by their respective five versus four digit node numbers. Ideally, the almost symmetric upper bodies (nodes 211) should contain two distinct saddle points p_1 and p_2 such that $\omega(p_1) = \omega(p_2) = s$; that is, two distinct saddle points on a single $s - level$ curve should simultaneously yield the three nodes: $21111, 2112$ (arms) and 21112 (head). However, certain configurations including this one are not generic; even the slightest perturbation imposes a strict order on the saddle points. Thus, firstly, the combination of the head and either one of the arms is separated from the other arm then the head-arm combination is partitioned. Nevertheless, in each case, the saddle point value separating the head and arm on the left combination from the arm on the right is very close to the saddle point value separating the head from the arm on the left; e.g., for the first shape, the respective saddle point values after

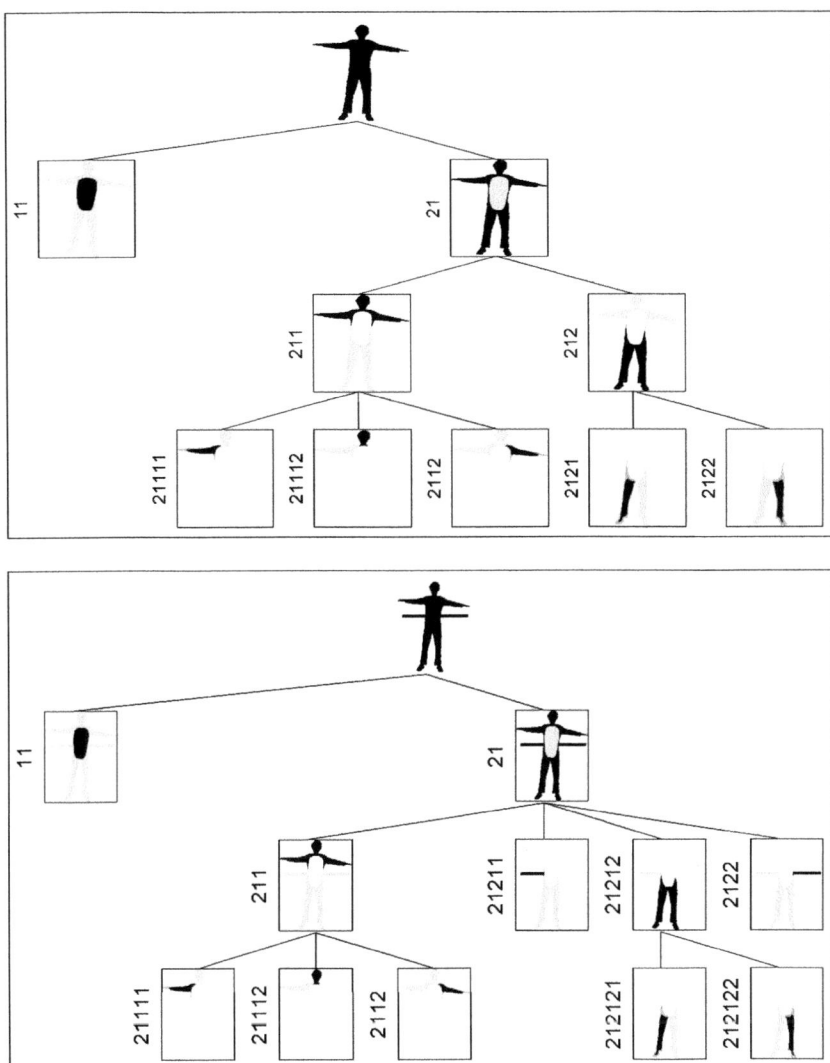

Fig. 3. The winning re-organizations for two shapes. The numbering of the nodes reveal their order in the initial binary tree. See the text.

normalization with respect to the global maximum of ω are -0.683 and -0.687 while the saddle point value separating the entire upper body from the entire lower body is -0.053. Clearly, the hierarchical order between the upper body and its children is more stronger than the hierarchical order among its children. We remark that even though the arm re-organization is not necessary for finding correct part correspondences since the structures of the upper bodies are already the same for the two initial trees, the left and right arms are brought to the same level. This is because the probabilities of retaining the initial binary local structures is very low due to the closeness of the consecutive saddle point values.

Secondly, notice that the legs of the occluded figure are at the sixth level whereas the legs of the un-occluded one are at the fourth level, as revealed by their node numbers. This is due to the influence of two additional parts (watershed regions) belonging to the occluder and poses a challenge for the tree matching. Nevertheless, the legs are brought to the same level as well as all of the corresponding parts and correct associations are found: 11 ⇔ 11 (central regions), 211 ⇔ 211 (upper bodies), 21111 ⇔ 21111 (arms on the left), 21112 ⇔ 21112 (heads), 2112 ⇔ 2112 (arms on the right), 212 ⇔ 21212 (lower bodies) 2121 ⇔ 212121 (legs on the left) 2122 ⇔ 212122 (legs on the right).

In the next three examples, due to limited space, only the matchings where at least one member of the matching pair is a leaf are depicted even though entire hierarchical structures are matched. Non-leaf nodes are circled. These examples also demonstrate the necessity of not restricting the correspondence search to leaf nodes. The matching between a cat and a horse in Fig. 4 illustrates a granularity inconsistency. Due to a weak saddle marked by the arrow in the left, the front legs of the horse are not further partitioned. Nevertheless, this inconsistency is resolved by matching the respective leaf node of the horse tree to a non-leaf node of the cat tree, the parent of the two nodes each holding a front leg of the cat.

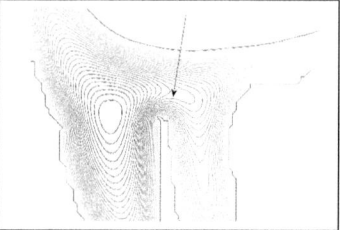

Fig. 4. A granularity inconsistency. Due to the weak saddle marked by the arrow in the left, the part of the horse corresponding to its front legs cannot be further partitioned. Nevertheless, it is correctly associated to a non-leaf node of the cat.

Fig. 5 (a) depicts the matching of the same horse to another horse. In addition to the previous granularity inconsistency, there are several order inconsistencies which are not noticeable at the leaf level presentation. For instance, the rear body of the first horse firstly splits into the fourth leg and tail versus the third leg, and then the fourth leg is separated from the tail. On the other hand, the rear body of the second horse after a spurious division gets splits into the rear legs versus the tail, and then the two legs are separated. Consequently, a two level difference between the third leg of the first horse (node 2121) and the third leg of the second horse (node 212221) is formed. Despite both granularity and order inconsistencies, all of the parts are correctly matched.

The final example (Fig. 5 (b)) involves several difficulties due to three weak saddles resulting with three unintuitive partitions for the second cat: Firstly,

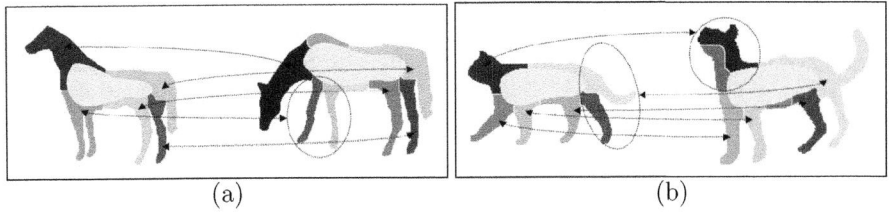

Fig. 5. Two more cases involving both level and granularity inconsistencies

its head is fragmented; secondly, its rear body goes through a spurious division causing an erroneous shift in the levels of its sub-parts; thirdly, its fourth leg and tail are not separated. Nevertheless, the selected clique contains all of the correct associations. The rear body and its parts for the second cat are properly lifted one level up; consequently, the correct associations of the parts of the rear bodies are found successfully. The head of the first cat matches to the parent of the two leaves holding two unintuitive parts of the head of the second cat. The two head fragments of the second cat as well as the fourth leg and the tail of the first cat are correctly excluded from the selected clique as there are no corresponding parts in the other tree.

Acknowledgements. ST is supported by the Alexander von Humboldt Foundation. She extends her gratitude to the members of Sci. Comp. Dept. of Tech. Universität München, in particular to Folkmar Bornemann, for providing a wonderful sabbatical stay in every respect. MG is funded as MS student via Tübitak project 108E015 to ST.

References

1. Ambrosio, L., Tortorelli, V.: On the approximation of functionals depending on jumps by elliptic functionals via Γ-convergence. Commun. Pure Appl. Math. 43(8), 999–1036 (1990)
2. Aslan, C., Tari, S.: An axis-based representation for recognition. In: ICCV, pp. 1339–1346 (2005)
3. Bar, L., Sochen, N., Kiryati, N.: Image deblurring in the presence of impulsive noise. Int. J. Comput. Vision 70(3), 279–298 (2006)
4. Braides, A.: Approximation of Free-discontinuity Problems. Lecture Notes in Mathematics, vol. 1694. Springer, Heidelberg (1998)
5. Buades, A., Coll, B., Morel, J.M.: A non-local algorithm for image denoising. In: CVPR, pp. 60–65 (2005)
6. Droske, M., Rumpf, M.: Multi scale joint segmentation and registration of image morphology. IEEE T-PAMI 29(12), 2181–2194 (2007)
7. Erdem, E., Sancar-Yilmaz, A., Tari, S.: Mumford-shah regularizer with spatial coherence. In: Sgallari, F., Murli, A., Paragios, N. (eds.) SSVM 2007. LNCS, vol. 4485, pp. 545–555. Springer, Heidelberg (2007)
8. Jung, M., Bresson, X., Chan, T.F., Vese, L.A.: Color image restoration using non-local mumford-shah regularizers. In: Cremers, D., Boykov, Y., Blake, A., Schmidt, F.R. (eds.) EMMCVPR 2009. LNCS, vol. 5681, pp. 373–387. Springer, Heidelberg (2009)

9. Jung, M., Vese, L.: Nonlocal variational image deblurring models in the presence of gaussian or impulse noise. In: Tai, X.-C., Mørken, K., Lysaker, M., Lie, K.-A. (eds.) SSVM 2009. LNCS, vol. 5567, pp. 401–412. Springer, Heidelberg (2009)
10. Kimia, B., Tannenbaum, A., Zucker, S.: Shapes, shocks, and deformations I: The components of two-dimensional shape and the reaction-diffusion space. Int. J. Comput. Vision 15(3), 189–224 (1995)
11. Kimmel, R., Kiryati, N., Bruckstein, A.: Sub-pixel distance maps and weighted distance transforms. J. of Math. Imag. and Vis. 6(2-3), 223–233 (1996)
12. Maragos, P., Butt, M.A.: Curve evolution, differential morphology and distance transforms as applied to multiscale and eikonal problems. Fundamentae Informatica 41, 91–129 (2000)
13. March, R., Dozio, M.: A variational method for the recovery of smooth boundaries. Image and Vision Computing 15(9), 705–712 (1997)
14. Meyer, F.: Topographic distance and watershed lines. Signal Processing 38, 113–125 (1994)
15. Mumford, D., Shah, J.: Optimal approximations by piecewise smooth functions and associated variational problems. Com. Pure App. Math. 42, 577–685 (1989)
16. Osher, S., Sethian, J.: Fronts propagating with curvat. dependent speed: algs. based on Hamilton-Jacobi formulations. J. Comp. Phys. 79, 12–49 (1988)
17. Pätz, T., Preusser, T.: Ambrosio-tortorelli segmentation of stochastic images. In: Daniilidis, K., Maragos, P., Paragios, N. (eds.) ECCV 2010. LNCS, vol. 6315, pp. 254–267. Springer, Heidelberg (2010)
18. Pelillo, M., Siddiqi, K., Zucker, S.W.: Matching hierarchical structures using association graphs. IEEE Trans. Pattern Anal. Mach. Intell. 21, 1105–1120 (1999)
19. Pien, H.H., Desai, M., Shah, J.: Segmentation of mr images using curve evolution and prior information. IJPRAI 11(8), 1233–1245 (1997)
20. Preußer, T., Droske, M., Garbe, C., Rumpf, M., Telea, A.: A phase field method for joint denoising, edge detection and motion estimation. SIAM Journal on Applied Mathematics 68(3), 599–618 (2007)
21. Proesman, M., Pauwels, E., van Gool, L.: Coupled geometry-driven diffusion equations for low-level vision. In: Romeny, B. (ed.) Geometry Driven Diffusion in Computer Vision. Kluwer, Dordrecht (1994)
22. Shah, J.: Segmentation by nonlinear diffusion. In: CVPR, pp. 202–207 (1991)
23. Shah, J.: A common framework for curve evolution, segmentation and anisotropic diffusion. In: CVPR, pp. 136–142 (1996)
24. Shah, J.: Skeletons and segmentation of shapes. Technical report, Northeastern University (2005), http://www.math.neu.edu/shah/publications.html
25. Shah, J., Pien, H.H., Gauch, J.: Recovery of shapes of surfaces with discontinuities by fusion of shading and range data within a variational framework. IEEE Trans. on Image Processing 5(8), 1243–1251 (1996)
26. Tari, S.: Hierarchical shape decomposition via level sets. In: Wilkinson, M.H.F., Roerdink, J.B.T.M. (eds.) ISMM 2009. LNCS, vol. 5720, pp. 215–225. Springer, Heidelberg (2009)
27. Tari, S., Shah, J., Pien, H.: A computationally efficient shape analysis via level sets. In: MMBIA, pp. 234–243 (1996)
28. Tari, S., Shah, J., Pien, H.: Extraction of shape skeletons from grayscale images. CVIU 66(2), 133–146 (1997)
29. Teboul, S., Blanc-Féraud, L., Aubert, G., Barlaud, M.: Variational approach for edge preserving regularization using coupled PDE's. IEEE Trans. Imag. Pr. 7, 387–397 (1998)

A Continuous Max-Flow Approach
to Minimal Partitions with Label Cost Prior

Jing Yuan[1], Egil Bae[2], Yuri Boykov[1], and Xue-Cheng Tai[2,3]

[1] Computer Science Department, University of Western Ontario
London, Ontario, Canada N6A 5B7
cn.yuanjing@gmail.com, yuri@csd.uwo.ca
[2] Department of Mathematics, University of Bergen
Bergen, Norway
{egil.bae,tai}@math.uib.no
[3] Division of Mathematical Sciences, School of Phys. and Math. Sci.
Nanyang Technological University, Singapore

Abstract. This paper investigates a convex relaxation approach for minimum description length (MDL) based image partitioning or labeling, which proposes an energy functional regularized by the spatial smoothness prior joint with a penalty for the total number of appearances or labels, the so-called *label cost prior*. As common in recent studies of convex relaxation approaches, the total-variation term is applied to encode the spatial regularity of partition boundaries and the auxiliary label cost term is penalized by the sum of convex infinity norms of the labeling functions. We study the proposed such convex MDL based image partition model under a novel continuous flow maximization perspective, where we show that the label cost prior amounts to a relaxation of the flow conservation condition which is crucial to study the classical duality of max-flow and min-cut! To the best of our knowledge, it is new to demonstrate such connections between the relaxation of flow conservation and the penalty of the total number of active appearances. In addition, we show that the proposed continuous max-flow formulation also leads to a fast and reliable max-flow based algorithm to address the challenging convex optimization problem, which significantly outperforms the previous approach by direct convex programming, in terms of speed, computation load and handling large-scale images. Its numerical scheme can by easily implemented and accelerated by the advanced computation framework, e.g. GPU.

1   Introduction

In this work, we study image labeling with the minimum description length principle (MDL) which naturally leads to regularities on both the spatial features, e.g. the minimum perimeter, and the total number of 'appearence' models. The MDL principle provides both an important concept of information theory and powerful tool to compress data, which states that 'the best hypothesis for a given set of data is the one that leads to the best compression of the data' (we refer to

A.M. Bruckstein et al. (Eds.): SSVM 2011, LNCS 6667, pp. 279–290, 2012.
© Springer-Verlag Berlin Heidelberg 2012

[22] for a detailed review). It naturally leads to use fewer symbols or models to describe the given data [14]. In fact, such requirement of model reduction have been considered in model fitting problems of computer vision for a long history, e.g. image segmentation [18,31,25], motion segmentation [20,24] etc.

In image segmentation or partitioning, it boils down to the penalization of the total number of appearence models or segments in addition to fitting data and regularities of segmentation boundaries. For the given n models/labels l_i, $i = 1 \ldots n$, Zhu and Yuille [31] proposed to partition images based on the minimization of the following energy function:

$$\min_{\Omega_i} \sum_{i=1}^{n} \left\{ \int_{\Omega_i} \rho(l_i, x)\, dx + \lambda \int_{\partial \Omega_i} ds \right\} + \gamma M , \tag{1}$$

where Ω_i, $i = 1, \ldots, n$, are homogeneous partitions corresponding to l_i, $M = \#\{1 \leq i \leq n \mid \Omega_i \neq \emptyset\}$ gives the number of nonempty partitions, i.e. the so-called label cost prior. The data fidelity function $\rho(l_i, x) = -\log P(I_x|l_i)$ is a negative log-likelihood for model l_i at pixel x. The second term in (1) describes the total perimeter of the partitions and favors spatially regular boundaries with minimum length. Zhu and Yuille applied a local searching method, namely region competition, to approximate the highly nonconvex optimization problem (1). Their method slowly converges to a local minimum. Such MDL principle was further developed in the evolution of level sets to assist merging, e.g. [17,6,1]. On the other hand, the label cost prior was also considered in the recent developments of graph cuts: Hoeim et al [16] introduced a technique of $\alpha-$expansion combined with MDL to the application of object recognition; Delong et al [9] independently developed another $\alpha-$expansion method which can efficiently optimize more general energy functions with incorporated label cost prior.

Recently, Yuan & Boykov [29] studied the MDL based image partitioning problem (1) in the spatially continuous setting such that

$$\min_{u_i(x) \in \{0,1\}} \sum_{i=1}^{n} \left\{ \int_{\Omega} u_i(x)\rho(l_i, x)\, dx + \lambda \int_{\Omega} |\nabla u_i|\, dx \right\} + \gamma M , \tag{2}$$

subject to $\sum_{i=1}^{n} u_i(x) = 1$, where $u_i(x) \in \{0, 1\}$, $i = 1 \ldots n$, is the indicator function of $\Omega_i \subset \Omega$. Here M is the total number of 'active' partitions and the total-variation terms encode the total perimeter of partitions. The authors [29] proposed a convex relaxation formulation of (2) as

$$\min_{u_i(x)} \sum_{i=1}^{n} \left\{ \int_{\Omega} u_i(x)\rho(l_i, x)\, dx + \lambda \int_{\Omega} |\nabla u_i(x)|\, dx \right\} + \gamma \sum_{i=1}^{n} \max_{x \in \Omega} u_i(x) \tag{3}$$

$$\text{s.t.} \sum_{i=1}^{n} u_i(x) = 1, \ u_i(x) \geq 0; \ \forall x \in \Omega \tag{4}$$

where the labeling functions $u_i(x)$, $i = 1 \ldots n$, are relaxed by the pixelwise simplex constraint (4) and the label cost term in (2) is encoded by the sum of convex infinity norms of $u_i(x)$ instead. (3) proposes the minimization of a convex energy function over a convex constraint. It was optimized globally by a direct convex programming based solver in [29], which is not feasible to handle large-scale image data and highly time-consuming.

Actually, the first two terms of (3) together with the pixelwise convex constraint (4) correspond to the convex relaxation formulation of the minimal partition model, i.e. Potts model,

$$\min_{u_i(x)} \sum_{i=1}^{n} \int_{\Omega} \left\{ u_i(x)\rho(l_i, x) + \lambda |\nabla u_i| \right\} dx, \quad \text{subject to (4)}. \tag{5}$$

(5) was actively studied during the last years, e.g. [7,19,2,3,27], and fast algorithms were developed at the same time, upon standard theories of convex optimization. It is well-known that the regularities of the partition boundaries, i.e. the second term of (3) helps to smooth out small-scale partitions, hence reduce the total number of 'appearences' implicitly. However, only considering such smoothness prior often fails to recover correct labeling results and often leads to either over-partition or over-smoothness (see Fig. 1). This is in contrast to the model (3) which explicitly couples the label cost prior. Its result possesses optimalities of both geometry and model simplicity. We show this by Fig. 1.

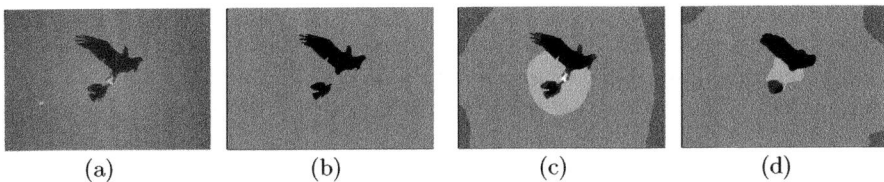

(a)   (b)   (c)   (d)

Fig. 1. (a) shows the given image. (b) shows the image partition result of (3) computed by the proposed method in this paper. It gives only two segments left along with properly smoothed boundaries! (c)-(d) show the partition results computed by the Potts model (5) without the label cost prior, which give the results either oversegmented (more labels) or oversmoothed. In this example we have used 11 evenly spaced labels.

Contributions: we focus on the convex relaxation model (3) of the MDL based image partition and propose a novel flow maximization perspective, i.e. the continuous max-flow formulation which is dual to (3). We show that the label cost prior in (3) just corresponds to a new flexible flow conservation constraint on the proposed continuous max-flow formulation, i.e. relaxation of flow conservation amounts to minimizing the number of 'active' labels! This is new to the best knowledge of the authors. It is in contrast to the crucial flow conservation condition of the classical max-flow models, where the flow excess given at each image node or pixel strictly vanishes, e.g. [27,26]. Moreover, we derive an efficient and reliable max-flow based algorithm which significantly outperforms the direct convex programming based method proposed by [29] in terms of speed, memory load and handling large-scale data. Compared to graph-cut based approaches

[16,9], our continuous max-flow approach comes with an elegant mathematical theory and is computed in the spatially continuous setting, which properly avoids metrication errors and can be easily implemented and accelerated on the advanced computation environment, e.g. GPU.

2   Previous Works

2.1   Convex Relaxation Approaches

Image labeling subject to the minimum perimeter, i.e. the Potts model, was intensively studied in both graph configuration [5] and spatially continuous settings [7,27] etc. Current studies [19,7,3,27,30] focus on computing the associated convex relaxation formulation (5) in the spatially continuous context, which avoid directly tackling the non-convex energies, as level-sets or active-contour method, and can be solved efficiently.

Let the convex set S denote the pixelwise simplex constraint (4) of $u(x) = (u_1(x), \ldots, u_n(x))^\mathsf{T}$. [30,19] proposed an optimization method which involves two substeps within each iteration: one explores the pointwise simplex constraint $u(x) \in S$ and the other tackles the total-variation term. In [7,23], Pock et al introduced a variant implementation of the constraint $u(x) \in S$, i.e. a tighter relaxation based on a multi-layered configuration, and gives a more complex constraint on the concerning dual variable p to avoid multiple counting. In contrast to the works of [30,19,7,23] which tried to compute the labeling functions $u(x)$ of (5) directly, Bae et al [3] proposed to solve (5) based on its equivalent dual formulation. The nonsmooth dual formulation can then be efficiently approximated by a smooth convex energy function.

Max-Flow and Flow Conservation: In the very recent studies of [26,28,27], Yuan et al proposed the new continuous max-flow model which regards (5) as its dual formulation in the spatially continuous setting. As the hard constraint of the proposed max-flow model, the flow conservation condition should be strictly satisfied.

For the Potts model (5), i.e. $n \geq 3$, the spatially continuous flow configurations are given as [27]: Let Ω_i, $i = 1 \ldots n$, be the n copies of the image domain Ω. For each $x \in \Omega$, the source flow $p_s(x)$ streams from the source s to the same position x of each Ω_i, $i = 1 \ldots n$, simultaneously. For each $x \in \Omega$, the sink flow $p_i(x)$ is directed from x of each Ω_i, $i = 1 \ldots n$, to the sink t. The spatial flow fields $q_i(x)$, $i = 1 \ldots n$, are defined within each Ω_i, $i = 1 \ldots n$.

The sink and spatial flow fields $p_i(x)$ and $q_i(x)$, $i = 1 \ldots n$, are constrained by the capacities such that

$$|q_i(x)| \leq C_i(x), \; p_i(x) \leq \rho(l_i, x); \; i = 1 \ldots n. \tag{6}$$

Especially, at each $x \in \Omega$, the source flow $p_s(x)$, the sink and spatial flows $p_i(x)$ and $q_i(x)$, $i = 1 \ldots n$, satisfy the exact flow conservation conditions:

$$\operatorname{div} q_i(x) - p_s(x) + p_i(x) = 0, \quad i = 1 \ldots n. \tag{7}$$

Likewise, the continuous max-flow problem is formulated as [27]:

$$\max_{p_s,p,q} \int_\Omega p_s(x) \, dx \tag{8}$$

subject to (6) and (7). [27] proved that (8) is dual to (5). Clearly, the three types of flow fields $p_s(x)$, $p_i(x)$ and $q_i(x)$, $i = 1\ldots n$, are connected by the flow conservation constraints (7). The labeling functions $u_i(x)$, $i = 1\ldots n$, just amount to the Lagrangian multipliers to the crucial flow-conservations [27].

Clearly, the flow conservation condition (7) plays the central role in the studies of the continuous max-flow model (8), and so for the theories of continuous max-flow and min-cut [26,28].

2.2 Convex Relaxed MDL Approach

Now we review the convex relaxation approach [29] to the challenging nonconvex problem (2): Given n labels $\{l_1,\ldots,l_n\}$, if the maximum of the labeling function $u_k(x) \in \{0,1\}$, $1 \le k \le n$, over the whole image domain Ω is 1, there must be some pixel $x \in \Omega$ which is labeled by l_k, i.e. the label l_k must present in the final image labeling result. Hence we can apply the sum of labeling functions' infinity norms $\sum_{i=1}^n \max_{x\in\Omega} u_i(x)$ to denote the total number M of 'active' models. Then, (2) can be equivalently reformulated by

$$\min_{u_i(x)\in\{0,1\}} \sum_{i=1}^n \left\{ \int_\Omega u_i(x)\rho(l_i,x) \, dx + \lambda \int_\Omega |\nabla u_i| \, dx \right\} + \gamma \sum_{i=1}^n \max_{x\in\Omega} u_i(x) \tag{9}$$

$$\text{s.t.} \quad \sum_{i=1}^n u_i(x) = 1, \quad \forall x \in \Omega.$$

Relax the binary constraint of the labeling functions $u_i(x) \in \{0,1\}$ together with $\sum_{i=1}^n u_i(x) = 1$ to the pointwise simplex constraint (4), i.e. $u(x) := (u_1(x),\ldots, u_n(x))^\mathsf{T} \in S$. The nonconvex optimization problem (9) can then be written as the continuous convex optimization problem (3), i.e.

$$\min_{u(x)\in S} \sum_{i=1}^n \left\{ \int_\Omega u_i(x)\rho(l_i,x) \, dx + \lambda \int_\Omega |\nabla u_i(x)| \, dx \right\} + \gamma \sum_{i=1}^n \max_{x\in\Omega} u_i(x) \tag{10}$$

where S denotes the pointwise simplex constraint (4). Obviously, the convex constrained convex optimization problem (10) can be solved globally. Its third term penalizes the infinity norm of each labeling function $u_i(x)$, $i = 1\ldots n$, which amounts to convex relaxation of the label cost prior.

3 Continuous Max-Flow Approach

In this section, we adopt the flow setting proposed in [27] and introduce a novel continuous max-flow formulation which is dual to the convex relaxed MDL-based labeling model (3) or (10). We show the label cost term is reduced to new flexible flow conservation constraints.

3.1 Continuous Max-Flow Formulation

In this section, we adopt the flow-maximization configurations and notations proposed in [27] and follow discussions in the above section.

By virtue of such continuous flow settings, the flow capacity constraints of flows $p_i(x)$ and $q_i(x)$, at $x \in \Omega$, are given in the same way as (6).

The flow conservation condition is formulated in a new flexible way:

$$\big(\mathrm{div}\, q_i - p_s + p_i \big)(x) \in R_i^{\gamma}, \quad R_i^{\gamma} := \{ r_i(x) \,|\, \int_{\Omega} |r_i(x)|\, dx \leq \gamma \}; \quad i=1\ldots n. \quad (11)$$

Note: The new flow conservation condition (11) proposes that at each $x \in \Omega$, the total in-coming flow is not balanced by the total out-going flow. However, the total absolute flow excesses associated with each label l_i, $i = 1\ldots n$, is controlled below γ as (11). This is in contrast to the exact flow conservation condition (7) in the classical max-flow theory, where the total in-coming flow should be strictly balanced by the total out-going flow.

We propose our new continuous max-flow model such that

$$\max_{p_s,p,q,r} \left\{ P(p_s, p, q) := \int_{\Omega} p_s(x)\, dx \right\} \quad (12)$$

subject to (6) and (11). In the following section, we study the equivalence between the proposed continuous max-flow formulation (12) and the convex relaxed MDL-based labeling model (3) or (10), especially for the case where $C(x) = \lambda$.

3.2 Equivalent Primal-Dual Model

We introduce the multiplier functions $u_i(x)$, $i = 1, \ldots, n$, to the new flow conservation condition (11). Therefore, we have its equivalent primal-dual model:

$$\max_{p_s,p,q,r} \min_{u} \left\{ E(p_s, p, q, r; u) := \int_{\Omega} p_s\, dx + \sum_{i=1}^{n} \int_{\Omega} u_i(\mathrm{div}\, q_i - p_s + p_i - r_i)\, dx \right\} \quad (13)$$

$$\text{s.t. } p_i(x) \leq \rho(l_i, x), \; |q_i(x)| \leq C_i(x); \quad \int_{\Omega} |r_i(x)|\, dx \leq \gamma; \; i = 1\ldots n$$

Rearranging the energy function $E(p_s, p, q, r; u)$ of (13), we have

$$E = \int_{\Omega} (1 - \sum_{i=1}^{n} u_i)\, p_s\, dx + \sum_{i=1}^{n} \left\{ \int_{\Omega} u_i\, p_i\, dx - \int_{\Omega} u_i\, r_i\, dx + \int_{\Omega} u_i\, \mathrm{div}\, q_i\, dx \right\}. \quad (14)$$

For the primal-dual model (13), the conditions of the minimax theorem [11] are all satisfied. That is, the constraints of flows are convex, and the energy function is linear to both the multiplier u and the flow functions p_s, p and q, hence convex l.s.c. for fixed u and concave u.s.c. for fixed p_s, p and q. This confirms the strong dualities of (13) and the existence of at least one saddle point [11,12]. It follows that the min and max operators of (13) can be interchanged:

$$\max_{p_s,p,q,r} \left\{ \min_{u} E(p_s, p, q, r; u) \right\} = \min_{u} \left\{ \max_{p_s,p,q,r} E(p_s, p, q, r; u) \right\}. \quad (15)$$

3.3   Equivalent Dual Model

Now we consider the optimization of (13) by switching the max-min order of the left hand side of (15), i.e. first maximizing $E(p_s, p, q, r; u)$ over the functions $p_s(x)$, $p(x)$, $q(x)$ and $r(x)$. Then we have

Proposition 1. *The maximization of* (13) *over the flow functions* $p_s(x)$, $p(x)$, $q(x)$ *and* $r(x)$ *amounts to the following dual model:*

$$\min_{u(x) \in S} D(u) := \sum_{i=1}^{n} \left\{ \int_{\Omega} u_i(x)\, \rho(l_i, x)\, dx + \int_{\Omega} C_i(x)\, |\nabla u_i|\, dx \right\} + \gamma \sum_{i=1}^{n} \max_{x \in \Omega} u_i(x) \tag{16}$$

which is equivalent to (3) *and* (10) *for the special case when* $C(x) = \lambda$.

To see Prop. 1, we follow the same analyzes as [26,28] , which gives

$$\max_{p_i(x) \leq \rho(l_i, x)} \int_{\Omega} u_i p_i\, dx = \int_{\Omega} u_i(x) \rho(l_i, x)\, dx \tag{17}$$

together with $u_i(x) \geq 0$, $i = 1 \ldots n$.

For the maximization of (14) over q_i and r_i, $i = 1 \ldots n$, it is well-known [13,15] that

$$\max_{|q_i(x)| \leq C_i(x)} \int_{\Omega} u_i \operatorname{div} q_i\, dx = \int_{\Omega} C_i(x)\, |\nabla u_i|\, dx\,, \tag{18}$$

and by the symmetry of the L_1-ball R_i^γ, we have

$$\max_{r_i(x) \in R_i^\gamma} - \int_{\Omega} u_i r_i\, dx = \gamma \max_{x \in \Omega} u_i(x)\,. \tag{19}$$

Moreover, observe the source flow $p_s(x)$ is unconstrained, then the maximization of (14) over p_s gives $1 - \sum_{i=1}^{n} u_i(x) = 0$, $\forall x \in \Omega$. Therefore, we have

Proposition 2. *The continuous max-flow model* (12), *the primal-dual model* (13) *and the dual model* (16) *are equivalent to each other.*

4   Fast Continuous Max-Flow Algorithm

Observe that the energy function of the primal-dual model (13) is nothing but the Lagrangian function of the proposed max-flow formulation (12) and the labeling functions $u_i(x)$, $i = 1 \ldots n$, give the corresponding multipliers to the introduced new flow conservation constraints (11). Observe this, we derive the new algorithm for (3) based on its equivalent continuous max-flow model (12).

We define the augmented Lagrangian function

$$L_c(p_s, p, q, r, u) := \int_{\Omega} p_s\, dx + \sum_{i=1}^{n} \langle u_i, \operatorname{div} q_i - p_s + p_i - r_i \rangle - \frac{c}{2} \sum_{i=1}^{n} \| \operatorname{div} q_i - p_s + p_i - r_i \|^2$$

where $c > 0$ and the auxiliary L_2 penalty term facilitates the vanishing of div $q_i(x) - p_s(x) + p_i(x) - r_i(x)$ at each $x \in \Omega$.

Now we construct our multiplier-based max-flow algorithm based on the augmented Lagrangian scheme [4]. Each k-th iteration includes the following steps:

- Maximize the energy $L_c(p_s, p, q, r, u)$ over the spatial flows $q_i(x)$, $i = 1 \ldots n$, by fixing other variables, which amounts to:

$$q_i^{k+1} := \arg \max_{\|q_i\|_\infty \leq \lambda} -\frac{c}{2} \left\| \mathrm{div}\, q_i - C^k(x) \right\|^2 , \tag{20}$$

where

$$C^k(x) = -p_i^k(x) + p_s^k(x) + r_i^k(x) + u_i^k(x)/c.$$

(20) can be approximated by a Chambolle-like projection-descent step [8].
- Maximize the energy $L_c(p_s, p, q, r, u)$ over the sink flows $p_i(x)$, $i = 1 \ldots n$, by fixing other variables, which corresponds to

$$p_i^{k+1} := \arg \max_{p_i(x) \leq \rho(\ell_i, x)} -\frac{c}{2} \left\| p_i - D^k \right\|^2 , \tag{21}$$

where

$$D^k(x) = -\mathrm{div}\, q_i^{k+1}(x) + p_s^k(x) + r_i^k(x) + u_i^k(x)/c.$$

(21) can be directly computed pointwise at each $x \in \Omega$.
- Maximize the energy $L_c(p_s, p, q, r, u)$ over $r_i(x)$, $i = 1 \ldots n$, by fixing other variables, which amounts to

$$r_i^{k+1} := \arg \max_{r_i(x) \in R_i^\gamma} -\frac{c}{2} \left\| r_i - F^k \right\|^2 \tag{22}$$

where

$$F^k(x) = \mathrm{div}\, q_i^{k+1}(x) - p_s^k(x) + p_i^k(x) - u_i^k(x)/c.$$

(22) can be addressed by the projection of $F^k(x)$ to the L_1-ball R_i^γ with the fast projection algorithm of linear $O(N)$ complexity [21,10].
- Optimize the energy $L_c(p_s, p, q, r, u)$ over the unconstrained source flow p_s and

$$p_s^{k+1} := \arg \max_{p_s} \int_\Omega p_s\, dx - \frac{c}{2} \sum_{i=1}^{n} \left\| p_s - G^k \right\|^2 \tag{23}$$

where

$$G^k(x) = p_i^{k+1}(x) + \mathrm{div}\, q_i^{k+1}(x) - r_i^{k+1}(x) - u_i^k(x)/c.$$

Finally, update the multiplier functions $u_i(x)$, $i = 1 \ldots n$, as follows

$$u_i^{k+1} = u_i^k - c\,(\mathrm{div}\, q_i^{k+1} - p_s^{k+1} + p_i^{k+1}). \tag{24}$$

Both (23) and (24) can be computed in a closed form.

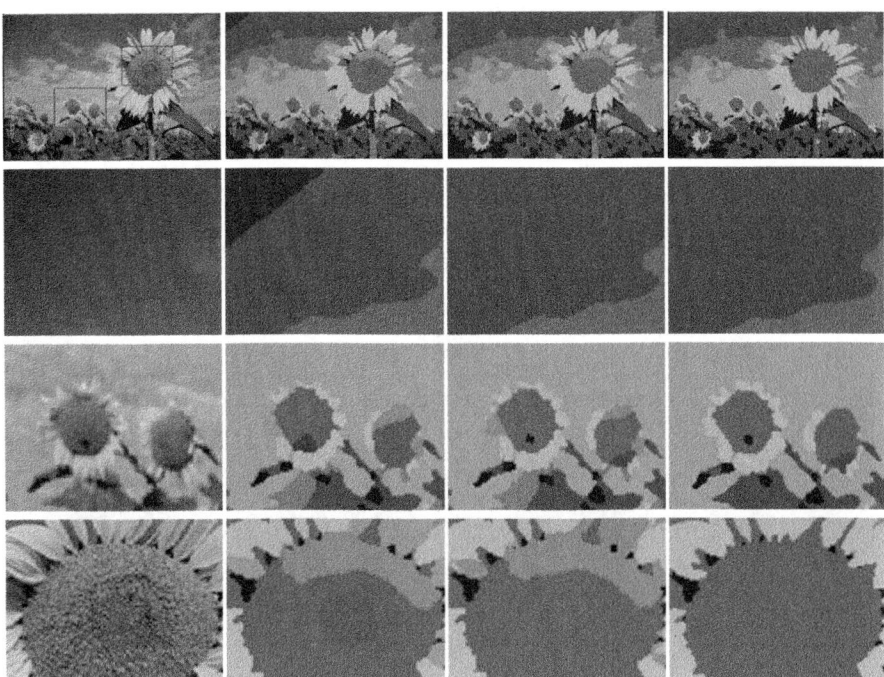

Fig. 2. Image segmentation with 15 labels. **From left to right:** input image (508 ×
336); labeling with Potts model; label cost model with $\gamma = 250$; label cost model with
$\gamma = 1000$. *Top:* Full image. *2nd - 4th row:* zoomed parts (red-line croped areas of the
input image). Visible differences can be clearly noticed in the zoomed images.

5   Numerical Experiments

Experiments demonstrate the advantages of the label cost model over Potts
model and the superior efficiency of the new max-flow algorithm over the previous SOCP method [29]. Gray scale image segmentation can be modeled as (1)
with the data term

$$\rho(l_i, x) = |f(x) - l_i|^p , \quad i = 1 \ldots n; \quad p = 1 \text{ or } 2$$

where $l_1, ..., l_n$ are predefined gray values, for instance the gaussian distribution
model of images. For colour image segmentation, the labels are instead colour
vectors $(l_1^j ... l_n^j)$, where $j \in \{r, g, b\}$. The data term is modeled as

$$\rho(\ell_i, x) = \sum_{j \in \{r,g,b\}} \left| f(x) - l_i^j \right|^p , \quad i = 1, \ldots, n; \quad p = 1 \text{ or } 2 .$$

In the experiments of Fig. 2 - 3, $\ell_1, ... \ell_n$ are chosen as evenly spaced gray values
in the interval between the smallest and largest gray value. For the color image,
$\ell_1, ..., \ell_n$ are evenly spaced color vectors. The results of Potts model are shown in

Fig. 3. Top: labeling by 11 labels. From left to right: input (481×321); Potts model $\lambda = 0.1$; MDL model $\lambda = 0.1, \gamma = 2000$; MDL model $\lambda = 0.1, \gamma = 3000$. **Middle:** labeling by 21 labels. From left to right: input (321×481); Potts model $\lambda = 0.05$; MDL model $\lambda = 0.05, \gamma = 3000$; MDL model $\lambda = 0.05, \gamma = 8000$, **Bottom:** labeling by 21 labels. From left to right: input (321×481); Potts model $\lambda = 0.15$; MDL model $\lambda = 0.15, \gamma = 2000$; MDL model $\lambda = 0.15, \gamma = 3000$.

the 2nd coloumns. It may produce more labels than desired. On the other hand, the label cost prior, 3rd and 4th coloumn, greatly greatly helps to generate less labels along with properly smoothed edges, such that the objects are more clearly distinguished. The label cost model allows to reduce the number of labels without simultaneously oversmoothing the partition boundaries, as Potts model does (see also Fig. 1).

The efficiency of the proposed max-flow algorithm is significantly superior to the SOCP implementation in [29]. Whereas [29] requires several hours to converge for even one small input image (150×150), the proposed max-flow algorithm converges around 2 minutes for a large image (about 500×500) (serial matlab implementation). The convergence is just a little slower than the max-flow algorithm [27] without label cost prior, due to the projections onto the L^1-ball (22). The algorithm [29] even fails to converge when the problem size is too big, due to the intense memory requirement. For instance, the problem in Fig. 3 bottom (21 labels) could not be handled by [29] for the Ubuntu desktop we used (Intel Xeon 3.06G, 16G Memory).

6 Conclusions

We studied a convex relaxed MDL based labeling model (3) in this work, and showed its effectiveness for image partitioning in minimizing both the total number of 'active' labels and the perimeter of partitions [29]. More specially, we proposed and investigated a novel continuous max-flow model which is dual to (3). We showed that the label cost prior introduced in (3) just corresponds to the new flexible constraint of flow conservation under the flow-maximization perspective. This is in contrast to the strict flow balance for the classical max-flow/min-cut theories. In numerics, the proposed continuous max-flow model naturally leads to a new fast max-flow based algorithm, which greatly outperforms the direct convex programming method proposed in [29] in terms of efficiency, computation load, implementation on GPUs, and handling large-scale image data.

Acknowledgements. This research has been supported by Canadian NSERC discovery grant 298299-2007 RGPIN and accelerator grant for exceptional new opportunities 349757-2007 RGPAS, the Norwegian Research Council (eVita project 166075), MOE (Ministry of Education) Tier II project T207N2202 and IDM project NRF2007IDMIDM002-010.

References

1. Ayed, I.B., Mitiche, A.: A region merging prior for variational level set image segmentation. IEEE Trans. Image Processing 17(12), 2301–2311 (2008)
2. Bae, E., Yuan, J., Tai, X.-C., Boykov, Y.: A fast continuous max-flow approach to non-convex multilabeling problems. Technical report CAM-10-62, UCLA (2010)
3. Bae, E., Yuan, J., Tai, X.-C.: Global minimization for continuous multiphase partitioning problems using a dual approach. International Journal of Computer Vision 92(1), 112–129 (2011)
4. Bertsekas, D.P.: Nonlinear Programming. Athena Scientific, Belmont (1999)
5. Boykov, Y., Kolmogorov, V.: An experimental comparison of min-cut/max-flow algorithms for energy minimization in vision. PAMI 26, 359–374 (2001)
6. Brox, T., Weickert, J.: Level set segmentation with multiple regions. IEEE Transactions on Image Processing 15(10), 3213–3218 (2006)
7. Chambolle, A., Cremers, D., Pock, T.: A convex approach for computing minimal partitions. Technical Report TR-2008-05, University of Bonn (2008)
8. Chambolle, A.: An algorithm for total variation minimization and applications. JMIV 20(1), 89–97 (2004)
9. Delong, A., Osokin, A., Isack, H., Boykov, Y.: Fast approximate energy minimization with label costs. In: CVPR (2010)
10. Duchi, J., Shalev-Shwartz, S., Singer, Y., Chandra, T.: Efficient projections onto the $_1$-ball for learning in high dimensions. In: ICML, pp. 272–279 (2008)
11. Ekeland, I., Téman, R.: Convex analysis and variational problems. Society for Industrial and Applied Mathematics, Philadelphia (1999)
12. Fan, K.: Minimax theorems. Proc. Nat. Acad. Sci. U.S.A. 39, 42–47 (1953)
13. Giusti, E.: Minimal surfaces and functions of bounded variation. Australian National University, Canberra (1977)

14. Gruenwald, P.D.: The Minimum Description Length Principle. MIT Press Books, vol. 1. The MIT Press, Cambridge (2007)
15. Hiriart-Urruty, J.-B., Lemaréchal, C.: Convex analysis and minimization algorithms I. Springer, Berlin (1993); Fundamentals
16. Hoiem, D., Rother, C., Winn, J.: 3D LayoutCRF for Multi-View Obect Class Recognition and Segmentation. In: CVPR (2007)
17. Kadir, T., Brady, M.: Unsupervised non-parametric region segmentation using level sets. In: ICCV, pp. 1267–1274 (2003)
18. Leclerc, Y.G.: Constructing simple stable descriptions for image partitioning. IJCV 3(1), 73–102 (1989)
19. Lellmann, J., Kappes, J., Yuan, J., Becker, F., Schnörr, C.: Convex multi-class image labeling by simplex-constrained total variation. Technical report, IWR, Uni. Heidelberg (November 2008)
20. Li, H.: Two-view motion segmentation from linear programming relaxation. In: CVPR (2007)
21. Liu, J., Ye, J.: Efficient euclidean projections in linear time. In: International Conference on Machine Learning (2009)
22. MacKay, D.J.C.: Information Theory, Inference, and Learning Algorithms. Cambridge University Press, Cambridge (2003)
23. Pock, T., Chambolle, A., Bischof, H., Cremers, D.: A convex relaxation approach for computing minimal partitions. In: CVPR, Miami, Florida (2009)
24. Vidal, R., Tron, R., Hartley, R.: Multiframe motion segmentation with missing data using powerfactorization and gpca. IJCV 79(1), 85–105 (2008)
25. Yang, A.Y., Wright, J., Ma, Y., Sastry, S.S.: Unsupervised segmentation of natural images via lossy data compression. Computer Vision and Image Understanding 110(2), 212–225 (2008)
26. Yuan, J., Bae, E., Tai, X.C.: A study on continuous max-flow and min-cut approaches. In: CVPR, USA, San Francisco (2010)
27. Yuan, J., Bae, E., Tai, X.-C., Boykov, Y.: A continuous max-flow approach to potts model. In: Daniilidis, K., Maragos, P., Paragios, N. (eds.) ECCV 2010. LNCS, vol. 6316, pp. 379–392. Springer, Heidelberg (2010)
28. Yuan, J., Bae, E., Tai, X.C., Boykov, Y.: A study on continuous max-flow and min-cut approaches. Technical Report CAM 10-61, UCLA (August 2010)
29. Yuan, J., Boykov, Y.: A continuous max-flow approach to image labelings with label cost prior. In: BMVC (2010)
30. Zach, C., Gallup, D., Frahm, J.-M., Niethammer, M.: Fast global labeling for real-time stereo using multiple plane sweeps. In: Vision, Modeling and Visualization Workshop, VMV (2008)
31. Zhu, S.C., Yuille, A.: Region competition: Unifying snakes, region growing, and bayes/mdl for multi-band image segmentation. PAMI 18, 884–900 (1996)

Robust Edge Detection Using Mumford-Shah Model and Binary Level Set Method

Li-Lian Wang[1], Yu-Ying Shi[2], and Xue-Cheng Tai[3]

[1] Division of Mathematical Sciences, School of Physical and Mathematical Sciences, Nanyang Technological University, 637371, Singapore
[2] Department of Mathematics and Physics, North China Electric Power University, Beijing, 102206 China
[3] Division of Mathematical Sciences, School of Physical and Mathematical Sciences, Nanyang Technological University, 637371, Singapore, and Department of Mathematics, University of Bergen, Bergen, Norway

Abstract. A new approximation of the Mumford-Shah model is proposed for edge detection, which could handle open-ended curves and closed curves as well. The essential idea is to treat the curves by narrow regions, and use a sharp interface technique to solve the approximate Mumford-Shah model. A fast algorithm based on the augmented Lagrangian method is developed. Numerical results show that the proposed model and method are very efficient and have the potential to be used for edge detections for real complicated images.

1 Introduction

Edge detection is one of the fundamental problems for image processing and computer vision [29], and its application also spans many other areas such as boundary extraction and solid-liquid interface detection [39,8] in material science and physics. In the context of image processing, edge detection is to find the boundaries of objects in a digital image. Many methods have been developed for this purpose. Classical approaches attempted to detect edges by discontinuities in image intensity values. Witkin [43] proposed the scale space framework to analyze images, where the extreme of the first gradient and zero-crossing of the second gradient are used to detect edges. However, the edges are usually disconnected, so Canny [11] proposed the so-called Canny edge detector to detect disconnected edges. The main principle is to classify a pixel into the edges, if the gradient magnitude of the pixel is larger than those of pixels at both sides in the image domain. The Canny's edge detector inspired many subsequential works, e.g., Susan [33] and Canny-Deriche [18].

Another important class of methods are based on nonlinear isotropic diffusive equations [13,1,30]. In general, such models are designed to prevent smoothing near the edges and to encourage diffusion in the homogeneous regions. However, they may break down when the gradient generated by noise is comparable to the targeted edges. Along this line, some reaction and diffusion equations [35,42],

A.M. Bruckstein et al. (Eds.): SSVM 2011, LNCS 6667, pp. 291–301, 2012.

edge flow [24], snake/active contour models [20,12,5] and universal gravity [34] are also proposed for edge detection.

It is obviously impossible to summarize all the methods and the interested readers are referred to [10,25,23,31] for a review of many other methods. Given the vast amount of existing algorithms, we feel compelled to provide sufficient justification for developing yet another method. Motivated by the success of piecewise constant level set methods for image segmentation [22] and the recent developed fast algorithms [6], we shall introduce a new edge detector based on a robust binary level set method for the Mumford-Shah model [27]. In contrast with image segmentation, which is to find a partition of the image domain, edge detection aims more at finding the discontinuities of the intensity function. Hence, we are mostly concerned with the determination of open curves. However, there is no a natural way to represent open curves, since there is no distinction of interior and exterior regions. Accordingly, the standard level set methods [28] can not be directly applied. Among very few approaches for dealing with open curves, the work of [32] for modeling spiral crystal growth used two level set functions to represent the codimension-two boundary of the open curve by $\{x : \phi(x) = 0 \text{ and } \psi(x) > 0\}$, where ϕ, ψ are two signed distance function. In [7], an open-ended curve was represented by the "centerline" of the level set function defined on the curve, and the motion of the curve was essentially driven by the evolution a small region surrounding the curve. Interestingly, a recent work of Leung and Zhao [21] proposed a grid based particle method to represent an open curve by the most relevant points in the neighborhood on the discrete grids. The authors also commented on the limitations of the methods in [32,7] for open curve evolution.

In this paper, we embed an open (or a closed) curve of interest in a narrow region (or band) with the curve being part of the (one-sided) boundary (see Figure 1 below for an illustration). From geometric point of view, such a region is formed by the parallel curve (also known as the offset curve) [38]. We define a binary piecewise constant level set function on the small region, and show that the total variation of the level set function gives a good approximation of the length of the curve. Moreover, we add intrinsic forces to enforce the level set function to converge to binary values, and this helps to enhance and sharpen the edges. Moreover, applying this notion to translate the Mumford-Shah model leads to fast and robust algorithms for edge detection. On the other hand, we simultaneously solve the edge set and the optimal piecewise smooth approximation of the given image in the Mumford-Shah model, so the proposed methods can be used for mutliphase piecewise smooth image segmentation (which is yet a challenging topic [40]).

The rest of the paper is organized as follows. In Section 2, we formulate the Mumford-Shah model based on an embedding of the curve in a narrow band and the binary piecewise constant level set method. In Section 3, we describe the fast algorithm based on the augmented Lagrangian method. Section 4 is devoted to the numerical experiments for detection of open curves and segmentation of multiphase piecewise smooth images. Some concluding remarks are given in Section 5.

2 Formulation of the Model

Given an image I on an open bounded domain $\Omega \in \mathbb{R}^2$, Mumford and Shah in their seminal paper [27] suggested minimizing the following functional to find a piecewise smooth approximation u of I and the edge set Γ:

$$E(u, \Gamma) = \frac{\alpha}{2} \int_{\Omega} (u - I)^2 d\boldsymbol{x} + \frac{\beta}{2} \int_{\Omega \backslash \Gamma} |\nabla u|^2 d\boldsymbol{x} + |\Gamma|, \qquad (1)$$

where $|\Gamma|$ is the length of the edges picked from all over the image domain, and α, β are positive parameters to balance three terms. They also conjectured that E has a minimizer and the edges (the discontinuity Γ) are the union of a finite set of $C^{1,1}$ embedded curves with three possible configurations [4]: (i) a *crap tip* (look like a half-line or a single arc ends without meeting others); (ii) a *triple junction* (three curves meeting at their endpoints with $2\pi/3$ angle between each pair); and (iii) *boundary points* (a curve meets $\partial\Omega$ perpendicularly). The Mumford-Shah model and conjecture have inspired deep mathematical investigation and extensive applications [4,17]. The reduced model (without the second term called the piecewise constant Mumford-Shah model), together with an appropriate level set implementation, becomes a fundamental tool for piecewise constant image segmentation (see, e.g., [15,22]). We are interested in detecting Γ by solving the full model (1). The first issue is to characterize the edge set. An important idea is to associate Γ with the jump Γ_u of the unknown u, and this leads to the remarkable approximation of (1) by Ambrosio and Tortorelli [2,3]:

$$E_\varepsilon(u, \Gamma) = \frac{\alpha}{2} \int_{\Omega} (u-I)^2 d\boldsymbol{x} + \frac{\beta}{2} \int_{\Omega} v^2 |\nabla u|^2 d\boldsymbol{x} + \int_{\Omega} \left(\varepsilon |\nabla v|^2 + \frac{1}{4\varepsilon} (v-1)^2 \right) d\boldsymbol{x}, \quad (2)$$

where v is an auxiliary variable such that $v \approx 0$ if $x \in \Gamma_u$, and $v \approx 1$ otherwise. A rigorous analysis (see [2,9]) shows that the last term converges to $|\Gamma|$ in Gamma-convergence sense [16]. The width of transition from $v = 0$ to 1 is about $O(\varepsilon)$.

In what follows, we take a different point of view to characterize Γ and approximate its length. Based on Mumford and Shah conjecture, we assume that the targeted edge set Γ consists of a finite union of simple (i.e., non-self-intersecting) curves with suitable regularity. Let $\boldsymbol{r}$ be a single curve in the edge set Γ with a definite parameterization: $\boldsymbol{r}(t) = (x(t), y(t)), t \in [0, 1]$. Without loss of generality, we assume that $\boldsymbol{r}(t)$ is regular (i.e., $|\boldsymbol{r}'(t)| \neq 0$ for all $t \in (0, 1)$), and it has finite length and curvature. Recall that the parallel or offset curve generated by $\boldsymbol{r}(t)$ is defined by (cf. [19,38]):

$$\boldsymbol{r}_d(t) = \boldsymbol{r}(t) + d\,\boldsymbol{n}(t), \quad \forall t \in [0, 1], \qquad (3)$$

where $\boldsymbol{n}(t)$ is the unit normal to $\boldsymbol{r}(t)$ at each point, and d is a preassigned signed distance. This defines a positive (exterior, $d > 0$) or negative (interior, $d < 0$) offset (see some examples in Figure 1). For clarity of presentation, we denote the total length of $\boldsymbol{r}$ and $\boldsymbol{r}_d$ by L and L_d, respectively. According to Lemma 3.1 in [19], the total length of $\boldsymbol{r}_d$ is

$$L_d = \int_0^1 |1 + \kappa d| \, |\boldsymbol{r}'| dt = \int_0^L |1 + \kappa d| \, ds = |L + d\, \Delta\theta|, \qquad (4)$$

where κ is the curvature and $\Delta\theta$ is the total angle of rotation of the normal n to r between $t = 0$ and $t = 1$, measured in the right-handed sense. Moreover, the area A_d between the generator r and the offset r_d is given by

$$A_d = \frac{1}{2}(L + L_d)|d|. \tag{5}$$

The interested reader may refer to [19] for detailed analysis.

Hereafter, we assume that $0 < d \ll 1$, and denote the corresponding (closed) narrow band (with area A_d) by R_d. Recall that the total variation of the indicator function $\mathbf{1}_{R_d}$ characterizes the perimeter of R_d, so we deduce from (3) that

$$TV(\mathbf{1}_{R_d}) = 2L + O(d), \quad 0 < d \ll 1, \tag{6}$$

where

$$TV(u) = \sup_{\boldsymbol{p} \in S} \int_\Omega u \operatorname{div} \boldsymbol{p}\, d\boldsymbol{x}, \quad S := \Big\{ \boldsymbol{p} \in C_c^1(\Omega; \mathbb{R}^2) : |\boldsymbol{p}| \le 1 \Big\}. \tag{7}$$

We introduce an auxiliary function ϕ to approximate the characteristic function:

$$\mathbf{1}_{R_d} \approx \frac{1 + \phi}{2}, \quad 1 - \mathbf{1}_{R_d} \approx \frac{1 - \phi}{2}. \tag{8}$$

In other words, $\phi \approx 1$ if $\boldsymbol{x} \in R_d$ and $\phi \approx -1$ otherwise. More importantly, we enforce the constraint $\phi^2 = 1$, which acts as an intrinsic force and enables to enhance and sharpen the edges.

After collecting all the necessary facts, we present the approximation of the Mumford-Shah model (1):

$$\min_{u, \phi^2 = 1} \Big\{ \frac{\alpha}{2} \int_\Omega (1 - \phi)^2 |\nabla u|^2 d\boldsymbol{x} + \frac{\beta}{2} \int_\Omega |u - I|^2 d\boldsymbol{x} + TV(\phi) \Big\}. \tag{9}$$

Compared with (2), we characterize and approximate the edge set and the length term in a different manner. The use of total variation regularization, together with the constraint, can be viewed as a sharp interface approach. It appears that the presence of TV-term and the constraint may increase the difficulty in resolving the model. To alleviate this concern, we next introduce a fast dual-type algorithm based on the Augmented Lagrangian method (ALM).

3   Description of the Algorithm

In this section, we introduce the algorithm to minimize the model (9). For clarity, we use boldface letters to denote vectors. As in [37], we handle the total variation term by introducing an auxiliary variable $\boldsymbol{q}$, and reformulate (9) as

$$\min_{\substack{u, \phi^2 = 1, \\ \boldsymbol{q} = \nabla\phi}} \Big\{ \frac{\alpha}{2} \int_\Omega (1 - \phi)^2 |\nabla u|^2 d\boldsymbol{x} + \frac{\beta}{2} \int_\Omega |u - I|^2 d\boldsymbol{x} + \int_\Omega |\boldsymbol{q}| d\boldsymbol{x} \Big\}. \tag{10}$$

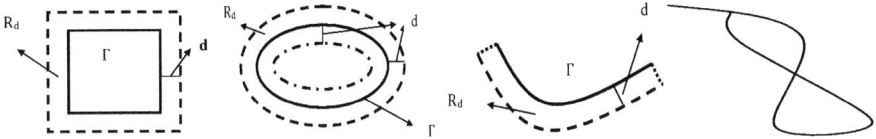

Fig. 1. The closed narrow region R_d formed by the curve $\Gamma : r(t)$ and its exterior parallel curve (dashed line) for closed curves and open curves (the dotted lines connected the corresponding starting points and end-points of the curve Γ and its exterior parallel curve, respectively), where the dot-dashed line is the interior parallel curve. The last one is an example of an intersected curve that can be split into simple open or closed curves.

The augmented Lagrangian formulation for this constrained problem takes the form

$$
\min_{u,\phi,\boldsymbol{q}} \max_{\boldsymbol{p},\lambda} \Big\{ \mathcal{L}(u,\phi,\boldsymbol{q};\boldsymbol{p},\lambda)
$$

$$
:= \frac{\alpha}{2}\int_\Omega (1-\phi)^2|\nabla u|^2 d\boldsymbol{x} + \frac{\beta}{2}\int_\Omega |u-I|^2 d\boldsymbol{x} + \int_\Omega |\boldsymbol{q}|d\boldsymbol{x} + (\boldsymbol{p},\boldsymbol{q}-\nabla\phi) \quad (11)
$$

$$
+ \frac{r}{2}\int_\Omega (\boldsymbol{q}-\nabla\phi)^2 d\boldsymbol{x} + \frac{1}{2}\int_\Omega \lambda\cdot(\phi^2-1)d\boldsymbol{x} + \frac{r_\phi}{2}\int_\Omega (\phi^2-1)^2 d\boldsymbol{x} \Big\},
$$

where $\boldsymbol{p}$ and λ are the Lagrange multipliers, and r and r_ϕ are positive constants.

Thus the minimizer problem (11) is to seek a saddle point of the augmented Lagrangian functional $\mathcal{L}(u,\phi,\boldsymbol{q};\boldsymbol{p},\lambda)$. But the problem (11) is not convex for both variables ϕ, u, which means a global minimizer may not be guaranteed.

A typical approach (see, e.g., [41,44]) is to split the problem (11) into several subproblems and minimize them consecutively.

It is clear that the optimality conditions for $\boldsymbol{p}$ and λ leads to the constraint: $\boldsymbol{q} = \nabla\phi$ and $\phi^2 = 1$, respectively. Therefore, we consider the following three subproblems:

- u-subproblem: Given $\phi, \boldsymbol{q}$,

$$
\min_u \Big\{ \frac{\alpha}{2}\int_\Omega (1-\phi)^2|\nabla u|^2 d\boldsymbol{x} + \frac{\beta}{2}\int_\Omega (u-I)^2 d\boldsymbol{x} \Big\}. \quad (12)
$$

- ϕ-subproblem: Given $u, \boldsymbol{q}, \boldsymbol{p}, \lambda$,

$$
\min_\phi \Big\{ \frac{\alpha}{2}\int_\Omega (1-\phi)^2|\nabla u|^2 d\boldsymbol{x} + (\boldsymbol{p},\boldsymbol{q}-\nabla\phi) + \frac{r}{2}\int_\Omega (\boldsymbol{q}-\nabla\phi)^2 d\boldsymbol{x}
$$

$$
+ \frac{1}{2}\int_\Omega \lambda\cdot(\phi^2-1)d\boldsymbol{x} + \frac{r_\phi}{2}\int_\Omega (\phi^2-1)^2 d\boldsymbol{x} \Big\}. \quad (13)
$$

- $\boldsymbol{q}$-subproblem: Given $u, \phi, \boldsymbol{p}$,

$$
\min_{\boldsymbol{q}} \Big\{ \int_\Omega |\boldsymbol{q}|d\boldsymbol{x} + (\boldsymbol{p},\boldsymbol{q}) + \frac{r}{2}\int_\Omega (\boldsymbol{q}-\nabla\phi)^2 d\boldsymbol{x} \Big\}. \quad (14)
$$

The optimality conditions for (12)-(14) yield

$$-\alpha \text{div}((1-\phi)^2 \nabla u) + \beta(u-I) = 0, \tag{15}$$

$$\alpha(\phi-1)|\nabla u|^2 + \text{div}\boldsymbol{p} + r\text{div}(\boldsymbol{q}-\nabla\phi) + \lambda\phi + 2r_\phi(\phi^2-1)\phi = 0, \tag{16}$$

$$\frac{\boldsymbol{q}}{|\boldsymbol{q}|} + \boldsymbol{p} + r(\boldsymbol{q}-\nabla\phi) = 0. \tag{17}$$

Since $\boldsymbol{q} = \nabla\phi$ and $\phi^2 = 1$, we find from (16) that

$$\alpha(\phi-1)|\nabla u|^2 + \text{div}\boldsymbol{p} + \lambda\phi = 0. \tag{18}$$

That is,

$$\left(\lambda + \alpha|\nabla u|^2\right)\phi = -\left(\text{div}\boldsymbol{p} - \alpha|\nabla u|^2\right). \tag{19}$$

Using the constraint $\phi^2 = 1$ to (19) again yields

$$(\text{div}\boldsymbol{p} - \alpha|\nabla u|^2)^2 = (\lambda + \alpha|\nabla u|^2)^2,$$

which implies the relation between λ and $\boldsymbol{p}$:

$$\lambda = -\text{div}\boldsymbol{p} \quad \text{or} \quad \lambda = \text{div}\boldsymbol{p} - 2\alpha|\nabla u|^2. \tag{20}$$

Therefore, we obtain from (19) that

$$\left|\text{div}\boldsymbol{p} - \alpha|\nabla u|^2\right|\phi = -\left(\text{div}\boldsymbol{p} - \alpha|\nabla u|^2\right). \tag{21}$$

Since $\boldsymbol{q} = \nabla\phi$, we derive from (17) that

$$\boldsymbol{p} = -\frac{\boldsymbol{q}}{|\boldsymbol{q}|} = -\frac{\nabla\phi}{|\nabla\phi|} \quad \Rightarrow \quad \nabla\phi + |\nabla\phi|\boldsymbol{p} = 0. \tag{22}$$

In summary, we need to solve the system for $(u, \phi, \boldsymbol{p})$:

$$\begin{cases} \left|\text{div}\boldsymbol{p} - \alpha|\nabla u|^2\right|\phi = -\left(\text{div}\boldsymbol{p} - \alpha|\nabla u|^2\right), \\ \nabla\phi + |\nabla\phi|\boldsymbol{p} = 0, \\ -\alpha\text{div}((1-\phi)^2\nabla u) + \beta(u-I) = 0. \end{cases} \tag{23}$$

Since the binary level set function ϕ is expected to satisfy $\phi^2 = 1$, we adopt a MBO-type projection (see [26,36]):

$$\phi = \mathcal{P}_B\left(\text{div}\boldsymbol{p} - \alpha|\nabla u|^2\right) \quad \text{with} \quad \mathcal{P}_B(t) := \begin{cases} 1, & \text{if } t \leq 0, \\ -1, & \text{if } t > 0. \end{cases} \tag{24}$$

The second equation in (23) can be solved in a very similar idea in [14]. For fixed ϕ, the last linear equation for u can be solved efficiently. We summarize the algorithm as follows.

Algorithm

1. Initialization: set $p^0 = 0$ and $u^0 = I$;
2. For $k = 0, 1, \cdots$,

   (i) Compute
   $$\phi^{k+1} = \mathcal{P}_B\big(\mathrm{div}p^k - \alpha|\nabla u^k|^2\big);$$

   (ii) Update p by the Chambolle's algorithm:
   $$p^{k+1} = \frac{p^k + \tau\nabla\phi^{k+1}}{1 + \tau|\nabla\phi^{k+1}|}; \tag{25}$$

   (iii) Update u by
   $$u^{k+1} = u^k + \tilde{\tau}\big\{\alpha\mathrm{div}((1 - \phi^{k+1})\nabla u^k) - \beta(u^k - I)\big\}; \tag{26}$$

3. End for till some stopping rule meets.

Some remarks are in order.

- Using the augmented Lagrangian formulation, we derive the simplified system (23), which does not depend on the parameters r, r_ϕ and λ. Hence, the algorithm might be more efficient than the algorithm, e.g., the Uzawa method, for the full model.
- The Lagrangian multiplier p turns out to be a dual variable, so the above algorithm is based on the primal-dual formulation with the complexity comparable to the fast algorithm in [14] for image denoising.
- An inner iteration can be applied to the equation for u, and it can be solved more efficiently other than (26). Indeed, a much deeper study can be conducted for (23), although we find the above algorithm works well.

4 Numerical Experiments

In this section, we provide numerical results to show the efficiency and robustness of the proposed model and algorithm. We also compare our method with Canny edge detector [11]. In the numerical tests, we take $\alpha = 10, \beta = 1, \tau = 0.12$ and $\tilde{\tau} = 10^{-4}$. The stopping rule is based on the maximum pixel-wise errors $\|\phi^{k+1} - \phi^k\|_\infty \leq 10^{-2}$ and $\|u^{k+1} - u^k\|_\infty \leq 10^{-4}$.

We first test two images configured with some typical open-ended and closed curves as boundaries in Figure 2. In particular, the testing image in Figure 2 (e) consists of objects with multiple constant intensities (i.e., multi-phases). We see that in all cases, the proposed method produces much sharper edges than the commonly used Canny detector. Indeed, the binary level set function ϕ converges to the expected values, that is, 1 in the very narrow region surrounding

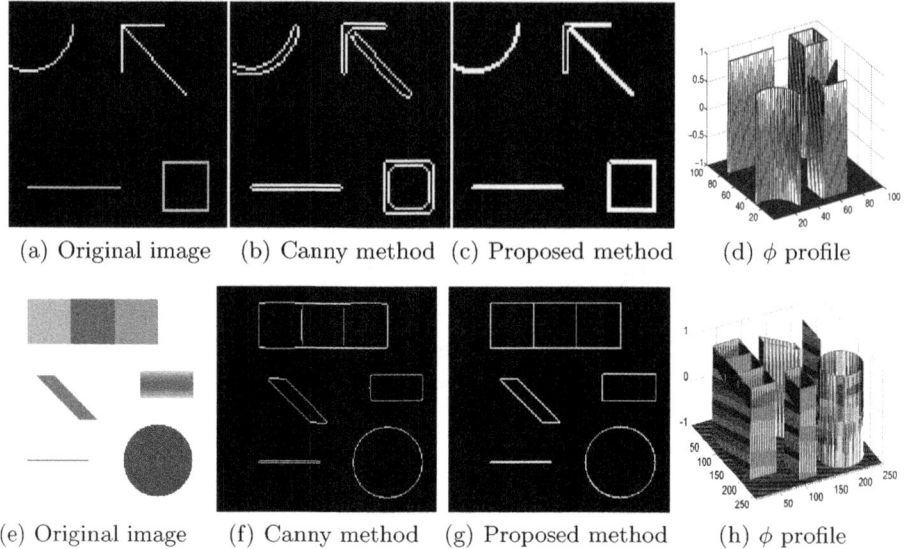

(a) Original image   (b) Canny method   (c) Proposed method   (d) ϕ profile

(e) Original image   (f) Canny method   (g) Proposed method   (h) ϕ profile

Fig. 2. Two tests of edge detection by Canny method and our proposed method

the curves, and -1 elsewhere. Under the stopping criterion, our proposed algorithm takes about 30 iterations to converge. In fact, satisfactory results can usually be obtained within 10 iterations. Moreover, the algorithm is robust for the initialization and parameters. The sharp interface model together with the fast algorithm could be a very promising tool for real image processing.

Next, we test our method for images with more features. In Figure 3, we show the results of our method and compare it with the results obtained by Canny edge detector, and the Ambrosio-Tortorelli method (2). In the comparison, we take the same parameters $\alpha = 10, \beta = 1$ as our proposed method, and the time step is 10^{-4} and the ϵ in (2) is 10^{-4}. The stopping rule is based on the maximum pixel-wise errors $\|v^{k+1} - v^k\|_\infty \leq 10^{-6}$ and $\|u^{k+1} - u^k\|_\infty \leq 10^{-6}$. Observe that our new algorithm is able to detect all the meaningful edges. More importantly, the smooth approximate solution u produces a very satisfactory recovery of the original image. Once again, the outcome of the proposed method is better than that of the Canny approach. We point out that our algorithm for the Cameraman image with size 256-by-256 takes about 5 seconds with about 200 iterations to converge, while that of the Ambrosio-Tortorelli method takes more than 20 seconds and 900 iterations.

5   Concluding Remarks

In this paper, we propose a new approach to approximate the Mumford-Shah model for edge detection. Some features of this work are highlighted below.

- An edge is viewed as a narrow region, and a binary level set method is applied to formulate the model. Therefore, compared with the approximation in

Fig. 3. Column 1: original images; Column 2: detected edges by the Canny method; Column 3: detected edges by the Ambrosio-Tortorelli method; Column 4: detected edges by our proposed method; Column 5: reconstructed image u by our proposed method.

Ambrosio and Tortorelli [2], the total variation regularization is adopted to approximate the length of edge set. In general, our method can be regarded as a sharp interface approach.

- A fast primal-dual algorithm based on the augmented Lagrangian method is proposed, which contains the minimal number of parameters and is naturally initialized. The computational cost is comparable to that of the efficient algorithm in [14].

Acknowledgments. The authors would like to thank the editor and the referee for their thoughtful comments. This research is supported by Singapore MOE Grant T207B2202, and Singapore NRF2007IDM-IDM002-010. The research of the second author is also partially supported by NSFC (NO. 10801049) and Foundation of North China Electric Power University. The second author is also grateful to the hospitality of the Mathematical Image and Vision Group at the Division of Mathematical Sciences in School of Physical and Mathematical Sciences of Nanyang Technological University during the visit.

References

1. Alvarez, L., Lions, P., Morel, J.: Image selective smoothing and edge detection by nonlinear diffusion ii. SIAM J. Numer. Anal. 29(3), 845–866 (1992)
2. Ambrosio, L., Tortorelli, V.: Approximation of functions depending on jumps by elliptic functions via gamma-convergence. Comm. Pure Appl. Math. 13, 999–1036 (1990)
3. Ambrosio, L., Tortorelli, V.: On the approximation of functionals depending on jumps by quadratic, elliptic functions. Boll. Un. Mat. Ital. 6-B, 105–123 (1992)
4. Aubert, G., Kornprobst, P.: Mathematical problems in image processing: partial differential equations and the calculus of variations. Springer-Verlag, New York Inc., Secaucus (2006)

5. Badshah, N., Chen, K.: Image selective segmentation under geometrical constraints using an active contour approach. Commun. Compu. Phys. 7(4), 759–778 (2010)
6. Bae, E., Tai, X.: Graph cut optimization for the piecewise constant level set method applied to multiphase image segmentation. In: Tai, X.-C., Mørken, K., Lysaker, M., Lie, K.-A. (eds.) SSVM 2009. LNCS, vol. 5567, pp. 1–13. Springer, Heidelberg (2009)
7. Basu, S., Mukherjee, D., Acton, S.: Implicit evolution of open ended curves. In: IEEE International Conference on Image Processing, vol. 1, pp. 261–264 (2007)
8. Berkels, B., Rätz, A., Rumpf, M., Voigt, A.: Extracting grain boundaries and macroscopic deformations from images on atomic scale. J. Sci. Comput. 35(1), 1–23 (2008)
9. Braides, A.: Approximation of free-discontinuity problems. Springer, Heidelberg (1998)
10. Brook, A., Kimmel, R., Sochen, N.: Variational restoration and edge detection for color images. J. Math. Imaging Vis. 18(3), 247–268 (2003)
11. Canny, J.: A computational approach to edge detection. IEEE Trans. Pattern. Anal. PAMI-8(6), 679–698 (1986)
12. Caselles, V., Kimmel, R., Sapiro, G.: Geodesic active contours. Int. J. Comput. Vis. 22(1), 61–79 (1997)
13. Catté, F., Lions, P., Morel, J., Coll, T.: Image selective smoothing and edge detection by nonlinear diffusion. SIAM J. Numer. Anal. 29(1), 182–193 (1992)
14. Chambolle, A.: An algorithm for total variation minimization and applications. J. Math. Imaging Vis. 20(1-2), 89–97 (2004)
15. Chan, T., Vese, L.: Active contours without edges. IEEE Trans. Image Process. 10(2), 266–277 (2001)
16. Dal Maso, G.: Introduction to Γ-convergence. Birkhauser, Basel (1993)
17. Dal Maso, G., Morel, J., Solimini, S.: A variation method in image segmentation-existence and approximation results. Acta Mathematica 168(1-2), 89–151 (1992)
18. Deriche, R.: Using canny's criteria to derive a recursively implemented optimal edge detector. Int. J. Comput. Vis. 1(2), 167–187 (1987)
19. Farouki, R., Neff, C.: Analytic properties of plane offset curves. Computer Aided Geometric Design 7(1-4), 83–99 (1990)
20. Kass, M., Witkin, A., Terzopoulos, D.: Snakes: Active contour models. Int. J. Comput. Vis. 1(4), 321–331 (1988)
21. Leung, S., Zhao, H.: A grid based particle method for evolution of open curves and surfaces. J. Comput. Phys. 228(20), 7706–7728 (2009)
22. Lie, J., Lysaker, M., Tai, X.: A binary level set model and some applications to Mumford-Shah image segmentation. IEEE Trans. Image Process. 15(5), 1171–1181 (2006)
23. Llanas, B., Lantaró, S.: Edge detection by adaptive splitting. J. Sci. Comput. 46(3), 486–518 (2011)
24. Ma, W., Manjunath, B.: Edgeflow: a technique for boundary detection and image segmentation. IEEE Trans. Image Process. 9(8), 1375–1388 (2000)
25. Meinhardt, E., Zacur, E., Frangi, A., Caselles, V.: 3D edge detection by selection of level surface patches. J. Math. Imaging Vis. 34(1), 1–16 (2009)
26. Merriman, B., Bence, J., Osher, S.: Motion of multiple functions: a level set approach. J. Comput. Phys. 112(2), 334–363 (1994)
27. Mumford, D., Shah, J.: Optimal approximations by piecewise smooth functions and associated variational problems. Comm. Pure Appl. Math 42(5), 577–685 (1989)

28. Osher, S., Sethian, J.: Fronts propagating with curvature dependent speed: Algorithms based on hamilton-jacobi formulations. J. Comput. Phys. 79(1), 12–49 (1988)
29. Paragios, N., Chen, Y., Faugeras, O.: Handbook of mathematical models in computer vision. Springer-Verlag New York Inc., Secaucus (2006)
30. Perona, P., Malik, J.: Scale-space and edge-detection using anisotropic diffusion. IEEE Trans. Pattern. Anal. 12(7), 629–639 (1990)
31. Pock, T., Cremers, D., Bischof, H., Chambolle, A.: An algorithm for minimizing the Mumford-Shah functional. In: 12th International Conference on Computer Vision, pp. 1133–1140. IEEE, Los Alamitos (2009)
32. Smereka, P.: Spiral crystal growth. Physica D: Nonlinear Phenomena 138(3-4), 282–301 (2000)
33. Smith, S.: Edge thinning used in the susan edge detector. Technical Report, TR95SMS5 (1995)
34. Sun, Y., Wu, P., Wei, G., Wang, G.: Evolution-operator-based single-step method for image processing. Int. J. Biomed. Imaging, 1–28 (2006)
35. Suzuki, Y., Takayama, T., Motoike, I., Asai, T.: A reaction-diffusion model performing stripe-and spot-image restoration and its lsi implementation. Electronics and Communications in Japan (Part III: Fundamental Electronic Science) 90(1), 20–29 (2007)
36. Tai, X., Christiansen, O., Lin, P., Skjælaaen, I.: Image segmentation using some piecewise constant level set methods with MBO type of projection. International Journal of Computer Vision 73(1), 61–76 (2007)
37. Tai, X.C., Wu, C.: Augmented lagrangian method, dual methods and split bregman iteration for ROF model. In: Tai, X.-C., Mørken, K., Lysaker, M., Lie, K.-A. (eds.) SSVM 2009. LNCS, vol. 5567, pp. 502–513. Springer, Heidelberg (2009)
38. Toponogov, V.: Differential geometry of curves and surfaces: a concise guide. Birkhauser, Basel (2006)
39. Upmanyu, M., Smith, R., Srolovitz, D.: Atomistic simulation of curvature driven grain boundary migration. Interface Sci. 6, 41–58 (1998)
40. Vese, L., Chan, T.: A multiphase level set framework for image segmentation using the mumford and shah model. Int. J. Comput. Vis. 50(3), 271–293 (2002)
41. Wang, Y., Yang, J., Yin, W., Zhang, Y.: A new alternating minimization algorithm for total variation image reconstruction. SIAM J. Imaging Sci. 1(3), 248–272 (2008)
42. Wei, G., Jia, Y.: Synchronization-based image edge detection. EPL (Europhysics Letters) 59(6), 814–819 (2002)
43. Witkin, A.P.: Scale-space filtering. In: Proc. 8th Int. Joint Conf. Art. Intell., Karlsruhe, Germany, pp. 1019–1022 (1983)
44. Wu, C., Zhang, J., Tai, X.: Augmented lagrangian method for total variation restoration with non-quadratic fidelity. In: UCLA, CAM09-82, pp. 1–26 (2009)

Bifurcation of Segment Edge Curves
in Scale Space

Tomoya Sakai[1], Haruhiko Nishiguchi[2], Hayato Itoh[3], and Atsushi Imiya[4]

[1] Department of Computer and Information Sciences, Nagasaki University, Japan
Bunkyou-cho 1-44, Nagsaki, Japan
[2] School of Science and Technology, Chiba University, Japan
[3] School of Advanced Integration Sciences, Chiba University, Japan
[4] Institute of Media and Information Technology, Chiba University, Japan
Yayoi-cho 1-33, Inage-ku, Chiba, Japan, 263-8522

Abstract. In this paper, we aim to develop a criterion to select scale parameters, which control pre-smoothing for edge detection. We first formalise the Canny edge detector which extracts the zeros of bilinear form of the first- and the second-order derivatives of image intensity. Then, we show the bifurcation property of the edge curves at the singular points in the linear scale space. Finally using the scale space hierarchy of the singular point, we derive a criterion to select scale parameters for edge detection.

1 Introduction

In this paper, we first show that segment edges detected by the Canny edge detector [1, 2] is a zero-crossing set defined by second-order differentials of an image [2–4] in the linear scale space [5–9]. Using this geometrical property of the segment edges, we second show a bifurcation property of edge segments in the linear scale space. This bifurcation property allows us to select appropriate scales for the accurate detection of the segment and the unification of segments extracted in various scales.

The segmentation of an image with high resolution inherently suffers from the oversegmentation problem, in which invalid segments are misinterpreted as pattern features of the image. In most cases, the oversegmentation due to randomness in the image, such as texture and noise, is suppressible by low-pass filtering or smoothing the image. However, smoothing operation reduces the image features as well. The segmentation of the smoothed image fails to extract valid segments related to the target objects in the image. This is the undersegmentation problem. Therefore, the selection of image resolution is crucial to avoid under- and oversegmentation.

A typical smoothing operation is the Gaussian convolution with an appropriate deviation [5, 6], which corresponds to the bandwidth of a low-pass filter. The Gaussian scale-space theory provides a mathematical framework on the multiresolution analysis of images with the Gaussian filter. A mathematical framework

A.M. Bruckstein et al. (Eds.): SSVM 2011, LNCS 6667, pp. 302–313, 2012.
© Springer-Verlag Berlin Heidelberg 2012

for the analysis of images smoothed by Gaussian filtering is the Gaussian scale-space theory [5, 6, 8, 9]. Since the Canny edge [1, 2] is a collection of singular curves in the linear scale space, we derive a method to select Gaussian smoothing operation using the scale space hierarchy of the singular point. This geometrical property also defines a hierarchical relations amon segments crossing the scales in the scale space.

2   Gaussian Scale Space

The scale space images are described as images blurred by Gaussian filtering [5, 6, 10]. The convolution of an N-dimensional grey-scale image $f(\boldsymbol{x})$, $\boldsymbol{x} \in \mathbb{R}^N$ and the isotropic Gaussian kernel with the deviation $\sigma = \sqrt{2\tau}$ derives a one-parameter family of non-negative functions,

$$f(\boldsymbol{x}, \tau) = G(\boldsymbol{x}, \sqrt{2\tau}) * f(\boldsymbol{x}), \tag{1}$$

where "$*$" expresses the N-dimensional convolution. The function $f(\boldsymbol{x}, \tau)$ is the solution of the partial differential equation

$$\frac{\partial}{\partial \tau} f(\boldsymbol{x}, \tau) = f(\boldsymbol{x}, \tau), \ f(\boldsymbol{x}, 0) = f(\boldsymbol{x}), \ \tau > 0. \tag{2}$$

The Gaussian scale space is a $(N + 1)$-dimensional space $(\boldsymbol{x}, \tau)$, in which the generalised image $f(\boldsymbol{x}, \tau)$ is defined. The generalised image satisfies the scale-space axioms: non-negativity, linearity, scaling invariance, translation invariance, rotation invariance, and non-enhancement of local extrema [5, 7, 8, 18]. As the geometric features of the scale-space image are reduced with increasing scale, the structure of the image is simplified.

Next, we briefly summarise singular points in the scale space.

Definition 1. *Stationary points S are defined as points where the spatial gradient vanishes, that is,*

$$S = \{\boldsymbol{x} \mid \nabla f(\boldsymbol{x}, \tau) = \boldsymbol{0}\}. \tag{3}$$

The structure of an image indicated by the critical curves in scale space has been investigated by various authors [10, 16, 19]. The bifurcation properties of image features in the scale space imply that the image structure across the scale is hierarchical.

Definition 2. *The annihilation point is a singular point where* $\det \boldsymbol{H}(\boldsymbol{x}, \tau) = 0$, *for the Hessian matrix $\boldsymbol{H}(\boldsymbol{x}, \tau)$ of $f(\boldsymbol{x}, \tau)$.*

The singular stationary points are also called the catastrophe points [16] in the scale-space theory.

Definition 3. *Catastrophe points T are the points where both the spatial gradient and the determinant of the Hessian matrix vanish, that is,*

$$T = \{(\boldsymbol{x}, \tau) \mid \nabla f(\boldsymbol{x}, \tau) = \boldsymbol{0}, \det \boldsymbol{H}(\boldsymbol{x}, \tau) = 0\}. \tag{4}$$

Every singular point generically has a zero eigenvalue, or a zero principal curvature since $\det \boldsymbol{H}(\boldsymbol{x}, \tau) = \prod \lambda_i(\tau) = 0$. The singular points are the points at which the stationary points meet in the scale space [15].

3   Segmentation Using Second-Order Derivatives

Enomoto *et al.* [3], and later Krueger *et al.* [2], define an edge manifold using the spatial gradient and the Hessian of the image.

Definition 4. *The edge manifold is defined as the set of points*

$$E = \{\boldsymbol{x} \mid \nabla f^\top \boldsymbol{H} \nabla f = 0\}. \tag{5}$$

E is called the edge surface and the edge line for a 3D image and a 2D image, respectively.

The critical points where $\nabla f = \boldsymbol{0}$ are in E of an image of arbitrary dimension. The edge line E of a two-dimensional image includes the points at which $\boldsymbol{H}$ is degenerated and ∇f is in the null space of $\boldsymbol{H}$. Such null space is equivalent to the subspace perpendicular to the edge direction. The edge line E also includes the points at which ∇f and $\boldsymbol{H}\nabla f$ are mutually perpendicular, if $\boldsymbol{H}$ is regular. Such regular points are found near the saddle points where the two eigenvalues of $\boldsymbol{H}$ are positive and negative.

One of the practical methods to extract the image segments is the Canny edge detection [1]. For each pre-determined deviation τ_i for $i = 1, 2, \ldots n$, edges of the image $f(\boldsymbol{x})$ are detected by the following procedure.

Image Smoothing The image $f(\boldsymbol{x})$ is smoothed by a Gaussian filter with the deviation τ_i to yield $f_i(\boldsymbol{x}) = G(\boldsymbol{x}, \tau_i) * f(\boldsymbol{x})$.
Differentiation Compute the gradient $\nabla f_i(\boldsymbol{x})$.
Edge Decision Assume the normal direction $\boldsymbol{n}$ to the edge to be the direction of the gradient, that is, $\boldsymbol{n} = \nabla f_i(\boldsymbol{x})$. Edge points are the points of inflection of $f_i(\boldsymbol{x})$ in the direction of $\boldsymbol{n}$.

$$E_i = \left\{ \boldsymbol{x} \mid \frac{\partial^2}{\partial n^2} f_i(\boldsymbol{x}) = 0 \right\}. \tag{6}$$

The smoothing and differentiation can be combined into a convolution of Gaussian derivatives. The detection of zero crossing in eq. (6) is implemented as hysteresis thresholding using a pair of thresholds. The ratio of two thresholds is regulated on the basis of the signal-to-noise ratio of the image[1].

Canny observed edges with small deviations τ_i [1]. From the viewpoint of the scale-space theory, the Canny edge E_i can be regarded as the scale-space version of the edge manifold denoted by eq. (5). Since the directional derivative in the direction of $\boldsymbol{n}$ is calculated as $\partial f_i / \partial n = \boldsymbol{n}^\top \nabla f_i$, equation (6) can be described as

$$\frac{\partial^2 f_i}{\partial n^2} = \boldsymbol{n}^\top \nabla (\boldsymbol{n}^\top \nabla f_i) = \boldsymbol{n}^\top \boldsymbol{H}_i \boldsymbol{n}. \tag{7}$$

Here, ∇f_i and $\boldsymbol{H}_i$ are the spatial gradient and the Hessian matrix of the Gaussian-smoothed image $f_i = f(\boldsymbol{x}, \tau_i)$, that is, the scale-space image. Therefore, we redefine the Canny edge in the scale-space fashion.

Definition 5. *The Canny edge in the scale space is defined as a one-parameter family of the edge manifold*

$$E(\tau) = \{x | \nabla f(x, \tau)^\top H(x, \tau) \nabla f(x, \tau) = 0\}. \tag{8}$$

Figure 1 shows segments extracted using the Canny operation for various scale parameters. Segments are white regions. Although, with small scale parameters, the operation extracts principal regions, with large scale parameters, the Canny operator extracts blurred regions. Therefore, for stable region extraction, the selection of the scale parameter is an essential task.

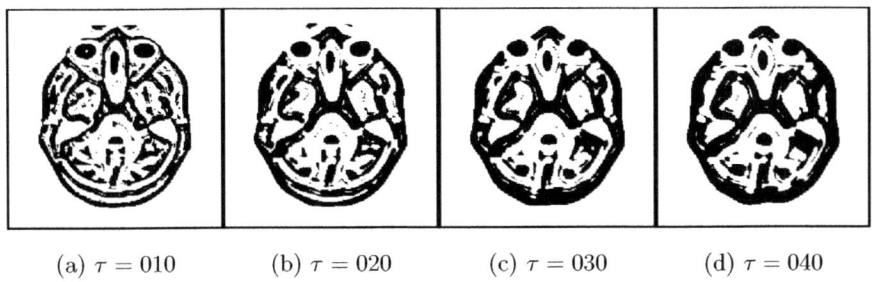

| (a) $\tau = 010$ | (b) $\tau = 020$ | (c) $\tau = 030$ | (d) $\tau = 040$ |

Fig. 1. Edge curves in the linear scale space. Segments are white regions. With small scale parameters, the operation extracts principal regions. However, with large scale parameters, the Canny operator extracts blurred regions.

4   Scale Space Hierarchy

The trajectory of the singular points is called the stationary curves[10] or the critical curves [15].

Definition 6. *Stationary curves are the trajectories of stationary points in scale space.*

Zhao *et al.* [10] showed that the stationary curves are solutions to the system of differential equations

$$H\frac{dx(\tau)}{d\tau} = -\nabla \Delta f(x(\tau), \tau). \tag{9}$$

Equation (9) indicates that the spatial velocity of the critical point with respect to scale becomes infinite at the annihilation point. Therefore, each top endpoint of the critical curves in scale space is a singular point, and does not have any connections by the critical curves to a higher scale, generically.

Using the principal axis coordinates of $H(x, \tau)$, eq. (9) is redescribed as

$$\frac{dp}{d\tau} = -\Lambda^{-1}\nabla_p \Delta f, \tag{10}$$

where $p(\tau) = V^\top x(\tau)$, and $\nabla_p = V^\top \nabla$ is the gradient operator in the principal axis coordinates. In the principal axis coordinates, the annihilation event [13, 14, 17], is modeled as

$$f(x, \tau) = x_1^3 + 6x_1\tau + \sum_{i=2}^{N} \gamma_i(x_i^2 + 2\tau), \tag{11}$$

where $\sum_{i=2}^{N} \gamma_i \neq 0$ and $\forall \gamma_i \neq 0$. τ is the scale parameter so that the annihilation event occurs at $\tau = 0$. For N-dimensional ($N > 1$) images, it is sufficient to consider the catastrophes in a two-dimensional case described as

$$f(x_1, x_2, \tau) = x_1^3 + 6x_1\tau + \gamma(x_2^2 + 2\tau). \tag{12}$$

This model of the scale-space image $f(x_1, x_2, \tau)$ has a local maximum and a saddle point if $\tau < 0$ and $\gamma < 0$. These two stationary points meet at the origin at $\tau = 0$. The parameterised stationary curves are obtained from eqs. (10) and (12) as

$$p(\tau) = (\pm\sqrt{-2\tau}, 0)^\top, \tag{13}$$

where the upper and lower signs correspond to the saddle curve and local maximum curve, respectively. The principal curvatures (λ_1, λ_2) are $(\sqrt{-2\tau}, 2\gamma)$ on the saddle curve and $(-\sqrt{-2\tau}, 2\gamma)$ on the local maximum curve. Therefore, the zero principal curvature direction at the annihilation scale $s = 0$ is in the x_1-axis. The pattern of the spatial gradient field [11, 12] clarifies the topological structure explicitly.

Definition 7. *The figure field F is defined as the negative of the vector field of the scale-space image.*

$$F = -\nabla f(x, \tau) \tag{14}$$

Definition 8. *The figure-flow curves are the directional flux curves of the figure field.*

The figure field can be considered as the current density flow of the image intensity with respect to scale, since the figure field satisfies the continuity equation.

Proposition 1. *The figure field F satisfies the equation of continuity*

$$\frac{\partial f}{\partial \tau} + \nabla^\top F = 0. \tag{15}$$

Equation (15) is directly obtained from eqs. (2) and (14).
Setting $\nabla_p = V^\top \nabla$, eqs. (12) and (14) imply that

$$F = -(3x_1^2 + 6\tau, 2\gamma x_2)^\top. \tag{16}$$

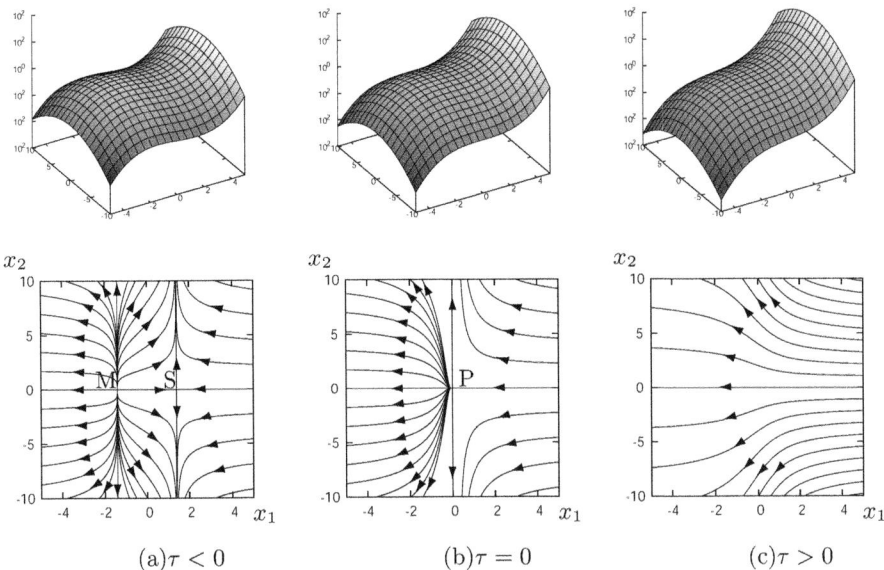

(a)$\tau < 0$ (b)$\tau = 0$ (c)$\tau > 0$

Fig. 2. Surface plot of $f(x_1, x_2)$ and corresponding figure flow curves (a) before, (b) at, and (c) after the Fold catastrophe event

The solution for $f(x_1, x_2, \tau)$ is

$$
x_2 = \begin{cases}
A \left| \dfrac{x_1 - \sqrt{-2\tau}}{x_1 + \sqrt{-2\tau}} \right|^{\frac{\gamma}{3\sqrt{-2\tau}}} & (\tau < 0) \\[2ex]
A \exp\left(-\dfrac{2\gamma}{3x_1} \right) & (\tau = 0) \\[2ex]
A \exp\left(\dfrac{2\gamma}{3\sqrt{2\tau}} \tan^{-1} \dfrac{x_1}{\sqrt{2\tau}} \right) & (\tau > 0) \, .
\end{cases}
\qquad (17)
$$

These relations show that the antidirectional figure-flow curve coincides with the zero principal curvature direction, the p_1-axis.

Definition 9. *A nongeneric figure-flow curve starts or ends at a singular point. An antidirectional figure-flow curve is defined as the figure-flow curve which ends at the annihilation point of the local maximum and saddle, or starts at the annihilation point of the local minimum and saddle.*

The motion of the stationary points described by eq. (9) implies the following geometrical properties of the annihilation point:

- The annihilation point is singular.
- The velocity of a stationary point is infinite at the annihilation point.
- The direction of the velocity of a stationary point is in that of the zero principal curvature at the annihilation point.

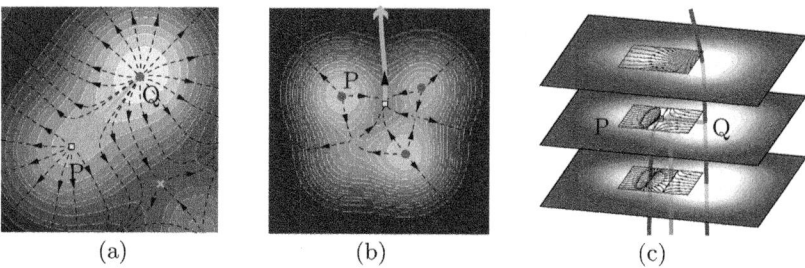

(a)	(b)	(c)

Fig. 3. Contour map of scale-space image and figure-flow curves at annihilation scale. (a) Annihilation of local maximum. The antidirectional figure-flow curve (solid line) links the annihilation point P and the local maximum Q. (b) Annihilation of local minimum. The antidirectional figure-flow curve to the boundary of the image indicates the connection to the point at infinity. (c)The connection by the stationary points and the antidirectional figure-flow curve define the hierarchical structure of the image.

Considering the infinite velocity at the annihilation point, the antidirectional figure-flow curve can be regarded as the continuation of the stationary curve at the annihilation scale. The antidirectional figure-flow curve connects the annihilation point and another local extremum. Since we can regard the figure field as the density flow of the image intensity with respect to scale, the image intensity of the annihilation point is provided only by the local maximum at the annihilation scale. Therefore, the local maximum is identified as the parent of the annihilation point. In the same manner as the local maximum, the antidirectional figure-flow curve identifies a local minimum as the parent of the annihilation point of a local minimum and a saddle point. Therefore, the connection of the local minima involves the local minimum at infinity. These topological properties of the singular points are shown in Fig. 3. Figure 3(a) shows the annihilation of the local maximum. The antidirectional figure-flow curve (solid line) links the annihilation point P and the local maximum Q. Figure 3(b) shows the annihilation of the local minimum. The antidirectional figure-flow curve to the boundary of the image indicates the connection to the point at infinity. Figure 3(c) shows that the connection by the stationary points and the antidirectional figure-flow curve define the hierarchical structure of the image.

5   Segment Edge Curve Bifurcation

We analyse topological property of the edge-lines when a saddle point merges to a local maximum or minimum point.

Setting $\gamma > 0$ and ϵ to be a small positive constant, in the neighbourhood of extrema, an image in the linear scale space is expressed as

$$f(x_1, x_2, \tau) = - \left\{ x_1^4 + 12x_1^2\tau + 12\tau^2 + \epsilon x_1 + \gamma(x_2^2 + 2\tau), \right\}. \tag{18}$$

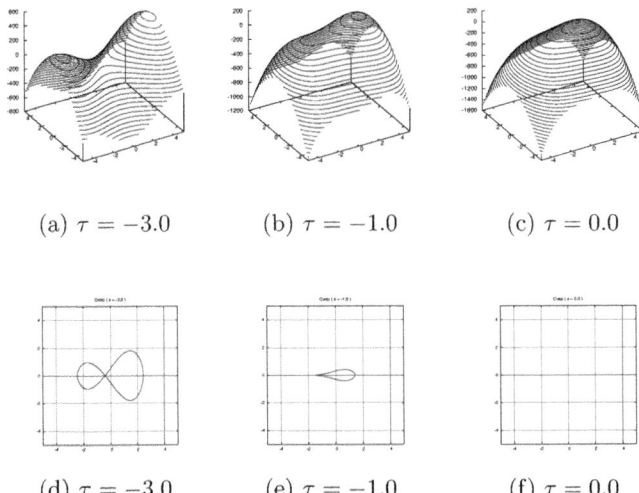

(a) $\tau = -3.0$ (b) $\tau = -1.0$ (c) $\tau = 0.0$

(d) $\tau = -3.0$ (e) $\tau = -1.0$ (f) $\tau = 0.0$

Fig. 4. The edge bifurcation in the neighbourhood of two local maximal points. (a) Before a pair of maximal points is merged, the edge curve is locally ∞-shaped. (b) A nonsimple curve becomes a simple loop. (c) Then, a closed curve disappears, since the image is smoothed by filtering.

If $\epsilon = 0$ and $\epsilon \neq 0$, the curve corresponds to the cusp catastrophe of two local maxima and a saddle and the fold catastrophe of a local maximum and a saddle. From eq. (18), since we have the relations

$$\nabla f = -(4x_1^2 + 24x_1\tau + \epsilon, 2\gamma x_2)^\top, \quad \boldsymbol{H}\nabla f = -(12(x_1^2 + 2\tau)(4x_1^3 + 12x_1\tau + \epsilon), 4\gamma^2 x_2)^\top \tag{19}$$

the edge curve is expressed as

$$12(x_1^2 + 2\tau)(4x_1^3 + 12x_1\tau + \epsilon) + 8\gamma^3 x_2^2 = 0 \tag{20}$$

For a negative value τ, eq. (20) yields a ∞-shape curve as shown in Fig. 4 (a). Then, for $\tau = 0$, since these two local maximal point is merged to a single local maximum as shown in Fig. 4(c), a local closed curve disappears.

For $\gamma < 0$, we have the relations

$$f(x_1, x_2, \tau) = x_1^4 + 12x_1^2\tau + 12\tau^2 + \epsilon x_1 + \gamma(x_2^2 + 2\tau), \tag{21}$$

and

$$\nabla f = (4x_1^2 + 24x_1\tau + \epsilon, 2\gamma x_2)^\top, \quad \boldsymbol{H}\nabla f = (12(x_1^2 + 2\tau)(4x_1^3 + 12x_1\tau + \epsilon), 4\gamma^2 x_2)^\top \tag{22}$$

the edge curve is expressed as

$$12(x_1^2 + 2\tau)(4x_1^3 + 12x_1\tau + \epsilon) + 8\gamma^3 x_2^2 = 0 \tag{23}$$

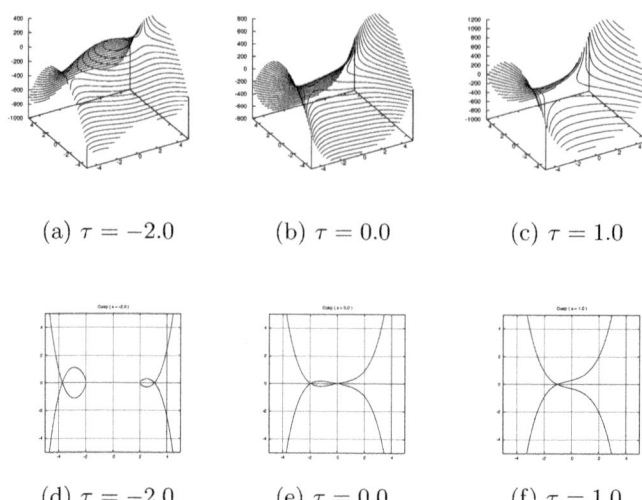

(a) $\tau = -2.0$ (b) $\tau = 0.0$ (c) $\tau = 1.0$

(d) $\tau = -2.0$ (e) $\tau = 0.0$ (f) $\tau = 1.0$

Fig. 5. Bifrucation of the local saddle point and a pair of local maximal points. (a) A pair of looped curves. (b) A pair of looped curves is merged. (c) A pair of simple curves contacts each other at a point.

In the scale space, eq. (23) changes from a pair of separate curves to a loop as shown in Fig. 5.

Figure 6 shows the transition of the edge curve in the scale space for

$$f(x, y) = \frac{3}{4} \exp\left(-\frac{(x+3)^2}{2^2} - \frac{(y-3)^2}{2^2}\right) + \exp\left(-\frac{(x-3)^2}{2^2} - \frac{(y+3)^2}{2}\right). \quad (24)$$

The curve configurations of Fig. 6, which is computed for a prob image, coincide with that of Figs. 4 and 5.

Using the stationary curves, for the point $det\boldsymbol{H}(\boldsymbol{x}, \tau) \neq 0$, Zhao [10] defined the stable points.

Definition 10. *For $s(\tau) = |\frac{d\boldsymbol{x}(\tau)}{d\tau}|$, the stable points V on the trajectory curves of singular points are*

$$V = V_0 \cup V_i,$$
$$V_0 = \{(\boldsymbol{x}, \tau) \mid s(\tau) = 0\} \quad (25)$$
$$V_i = \{(\boldsymbol{x}, \tau) \mid \textit{isolated points } s_\tau(\tau) = 0, \ s_{\tau\tau}(\tau) = 0\},$$

where $s_\tau(\tau) = \frac{ds(\tau)}{d\tau}$ and $s_{\tau\tau}(\tau) = \frac{d^2 s(\tau)}{d\tau^2}$.

Denoting a stable point on the stationary curves as $(\boldsymbol{x}_i, \tau_i)$, the region

$$\boldsymbol{R}(\boldsymbol{x}_i, \tau_i) = \{\boldsymbol{x} | |\boldsymbol{x} - \boldsymbol{x}_i| \leq \sqrt{2\tau_i}\} \quad (26)$$

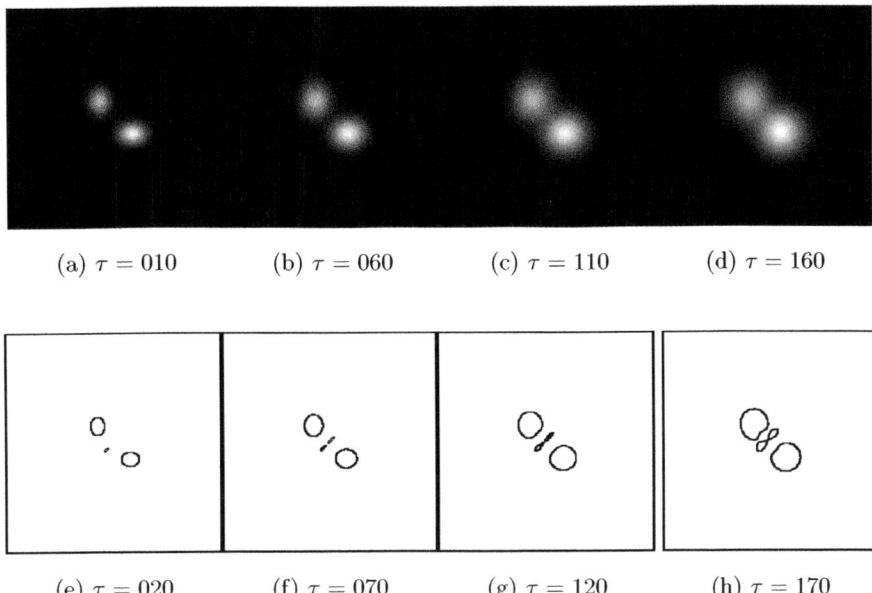

(a) $\tau = 010$ (b) $\tau = 060$ (c) $\tau = 110$ (d) $\tau = 160$

(e) $\tau = 020$ (f) $\tau = 070$ (g) $\tau = 120$ (h) $\tau = 170$

Fig. 6. Bifurcation of edge curves in the linear scale space. The bifurcation of the Canny edge for image $f(x,y) = \frac{3}{4}\exp\left(-\frac{(x+3)^2}{2^2} - \frac{(y-3)^2}{2^2}\right) + \exp\left(-\frac{(x-3)^2}{2^2} - \frac{(y+3)^2}{2}\right)$ is shown.

expresses a dominant part of $f(\boldsymbol{x}, \tau_i)$. Therefore, the stable points define dominant parts of an image in the scale space. We use this property of the stable points for the extraction of segments in the scale space.

Setting P to be the collection of the scale parameters for the saddles in the scale space, we order the elements of P as

$$0 < \tau_1 \leq \tau_2 \leq \cdots < \tau_n \leq \infty. \tag{27}$$

The scale parameter $\tau_i^* \in [\tau_i, \tau_{i+1}]$ does not produce any saddle points in the scale space. Furthermore, the parameters of stable points satisfy the condition $\tau \in [\tau_i, \tau_{i+1}]$. We introduce a selection criterion for the scale parameter of the Canny edge detector.

Proposition 2. *We select scales $\{\tau_i^*\}_{i=1}^n$ for the Canny edge detection from scales at stable points on stationary curves.*

Figure 7 shows the results of segmentation based on this criterion. In the figure, from top to bottom, images, stable points, and segments in the linear scale space are shown. The criterion derive a theoretical methodology for the selection of scale parameters for the Canny edge detector.

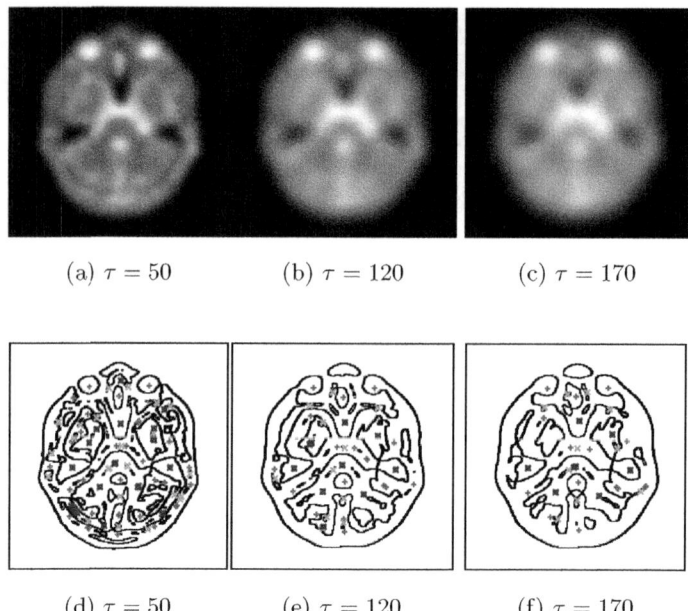

(a) $\tau = 50$ (b) $\tau = 120$ (c) $\tau = 170$

(d) $\tau = 50$ (e) $\tau = 120$ (f) $\tau = 170$

Fig. 7. Stationary points of stable state and segments. Top to bottom, images, stationary points, and segments in the linear scale space. Segment edges are extracted for τ^* which satisfies the condition $S(\boldsymbol{x}, \tau) = 0$.

6   Conclusions

We have first showed that segment edges detected by Canny edge detector is a zero-crossing set defined by second order differential of images in the linear scale space. Using this geometrical property of the segment edges, we have examined the bifurcation property of edge segments in the linear scale space. This bifurcation property allows the selection of scales for the accurate detection of segments and the unification of segments extracted in various scales.

This research was supported by "Computational anatomy for computer-aided diagnosis and therapy: Frontiers of medical image sciences" funded by Grant-in-Aid for Scientific Research on Innovative Areas, MEXT, Japan, Grants-in-Aid for Scientific Research founded by Japan Society of the Promotion of Sciences and Grant-in-Aid for Young Scientists (A), NEXT, Japan.

References

1. Canny, J.: A computational approach to edge detection. PAMI 8, 679–698 (1986)
2. Krueger, W.M., Phillips, K.: The geometry of differential operator with application to image processing. PAMI 11, 1252–1264 (1989)

3. Enomoto, H., Yonezaki, N., Watanabe, Y.: Application of structure lines to surface construction and 3-dimensional analysis. In: Fu, K.-S., Kunii, T.L. (eds.) Picture Engineering, pp. 106–137. Springer, Berlin (1982)

4. Najman, L., Schmitt, M.: Watershed of a continuous function. Signal Processing 38, 99–112 (1994)

5. Iijima, T.: Pattern Recognition, Corona-sha, Tokyo (1974) (in Japanese)

6. Witkin, A.P.: Scale space filtering. In: Proc. of 8th IJCAI, pp. 1019–1022 (1993)

7. Lindeberg, T.: Scale-Space Theory in Computer Vision. Kluwer, Boston (1994)

8. Koenderink, J.J.: The structure of images. Biological Cybernetics 50, 363–370 (1984)

9. ter Haar Romeny, B.M.: Front-End Vision and Multi-Scale Image Analysis Multiscale Computer Vision Theory and Applications, written in Mathematica. Springer, Berlin (2003)

10. Zhao, N.-Y., Iijima, T.: Theory on the method of determination of view-point and field of vision during observation and measurement of figure. IECE Japan, Trans. D. J68-D, 508–514 (1985) (in Japanese)

11. Zhao, N.-Y., Iijima, T.: A theory of feature extraction by the tree of stable viewpoints. IECE Japan, Trans. D. J68-D, 1125–1135 (1985) (in Japanese)

12. Sakai, T., Imiya, A.: Scale-space hierarchy of singularities. In: Fogh Olsen, O., Florack, L.M.J., Kuijper, A. (eds.) DSSCV 2005. LNCS, vol. 3753, pp. 181–192. Springer, Heidelberg (2005)

13. Damon, J.: Local Morse theory for solutions to the heat equation and Gaussian blurring. Journal of Differential Equations 115, 368–401 (1995)

14. Damon, J.: Generic properties of solutions to partial differential equations. Archive for Rational Mechanics and Analysis 140, 353–403 (1997)

15. Florack, L.M.J., Kuijper, A.: The topological structure of scale-space images. Journal of Mathematical Imaging and Vision 12, 65–79 (2000)

16. Kuijper, A., Florack, L.M.J., Viergever, M.A.: Scale space hierarchy. Journal of Mathematical Imaging and Vision 18, 169–189 (2003)

17. Kuijper, A.: The Deep Structure of Gaussian Scale Space Images, PhD thesis, Utrecht University (2002)

18. Duits, R., Florack, L.M.J., Graaf, J., ter Haar Romeny, B.: On the axioms of scale space theory. Journal of Mathematical Imaging and Vision 20, 267–298 (2004)

19. Griffin, L.D., Colchester, A.: Superficial and deep structure in linear diffusion scale space: Isophotes, critical points and separatrices. Image and Vision Computing 13, 543–557 (1995)

Efficient Minimization
of the Non-local Potts Model[⋆]

Manuel Werlberger, Markus Unger, Thomas Pock, and Horst Bischof

Institute for Computer Graphics and Vision, Graz University of Technology, Austria
{werlberger,unger,pock,bischof}@icg.tugraz.at
http://www.icg.tugraz.at

Abstract. The Potts model is a well established approach to solve different multi-label problems. The classical Potts prior penalizes the total interface length to obtain regular boundaries. Although the Potts prior works well for many problems, it does not preserve fine details of the boundaries. In recent years, non-local regularizers have been proposed to improve different variational models. The basic idea is to consider pixel interactions within a larger neighborhood. This can for example be used to incorporate low-level segmentation into the regularizer which leads to improved boundaries. In this work we study such an extension for the multi-label Potts model. Due to the increased model complexity, the main challenge is the development of an efficient minimization algorithm. We show that an accelerated first-order algorithm of Nesterov is well suited for this problem, due to its low memory requirements and its potential for massive parallelism. Our algorithm allows us to minimize the non-local Potts model with several hundred labels within a few minutes. This makes the non-local Potts model applicable for computer vision problems with many labels, such as multi-label image segmentation and stereo.

1 Introduction

The multiphase partitioning problem consists in finding a certain label for every pixel, tiling the image domain into multiple pairwise disjoint regions. Starting with the seminal work of Mumford and Shah [21] research on computing minimal partitions was ignited by typical Computer Vision problems such as segmentation, stereo or 3D reconstruction. In a discrete version, the Potts model [26], has been known much longer. It was originally invented to model phenomena in solid state mechanics in 1952 and generalizes the two-state model of Ising [18] (1925). The Potts model is a special case of the general multi-labeling problem, relying on a pairwise interaction term that does not assume any ordering of the labels. Minimizing the Potts energy is known to be NP-hard and in general cannot be solved exactly in reasonable time. For this reason various approximations have been proposed to convexify the optimization problem and hence to approximate its solution as effectively as possible.

[⋆] This work was supported by the BRIDGE project HD-VIP (no. 827544).

A.M. Bruckstein et al. (Eds.): SSVM 2011, LNCS 6667, pp. 314–325, 2012.

For the two label case, Chan and Vese [12] used the level set framework for optimization but do not yield any optimality. Later, Chan *et al.* [11] showed in a continuous setting that optimality for this problem can be achieved by solving this problem on a relaxed convex set. As the optimization task of the Potts model was originally formulated in a discrete setting, graph cut based approaches have often been used to solve such multi-label tasks. Most notable are move making algorithms of Boykov *et al.* [4] approximately minimizing the Potts model by solving a sequence of globally optimal binary segmentation problems. Although such sequential approaches often generate useful solutions, non of them is able to find a global minimizer. Ishikawa [17] showed that an exact solution can be computed in polynomial time for certain cases, namely when the labels are linearly ordered and the pairwise term is a convex function. Recently, it was shown by Pock *et al.* [25] that the same is true in the continuous case. Unfortunately, the constraint of having linearly ordered labels is not fulfilled in the segmentation task.

For solving the Potts model, several convex relaxations were proposed by *e.g.* Zach *et al.* [34], Lellmann *et al.* [19], Bae *et al.* [2] and Pock *et al.* [24], whereas the latter provides the tightest relaxation with respect to the original problem. The Potts formulation in a spatially continuous setting is given as the energy minimization problem

$$\min_{E_l} \left\{ \frac{1}{2} \sum_{l=1}^{K} Per\left(E_l; \Omega\right) + \sum_{l=1}^{K} \int_{E_l} f_l(x) \, dx \right\},$$

$$\text{s.t.} \quad \bigcup_{l=1}^{K} E_l = \Omega, \quad E_i \cap E_j = \emptyset \quad \forall i \neq j,$$

(1)

where the first term measures the interface length of the set E_l enforcing smooth label boundaries and the second term is the data term, a point-wise defined weighting function. Minimizing such an energy partitions the image domain $\Omega \subseteq \mathbb{R}^2$ into K pairwise disjoint regions E_l.

Rewriting the Potts model in terms of a convex total variation optimization problem (*cf.* [34,24,27]) yields the minimization task

$$\min_{u} \left\{ E(u) \right\} = \min_{u} \left\{ J(u) + \lambda \sum_{l=1}^{K} \int_{\Omega} u_l(x) f_l(x) \, dx \right\},$$

$$\text{s.t.} \quad u_l(x) \geq 0, \quad \sum_{l=1}^{K} u_l(x) = 1, \quad \forall x \in \Omega$$

(2)

with the labeling function $u = (u_1, \ldots, u_K) : \Omega \to [0,1]^K$ and the weighting function $f = (f_1, \ldots, f_K) : \Omega \to \mathbb{R}^K$. The regularizer $J(u)$ can for example be a simple total variation regularization

$$J(u) = \frac{1}{2} \sum_{l=1}^{K} \int_{\Omega} |\nabla u_l(x)| \, dx$$

(3)

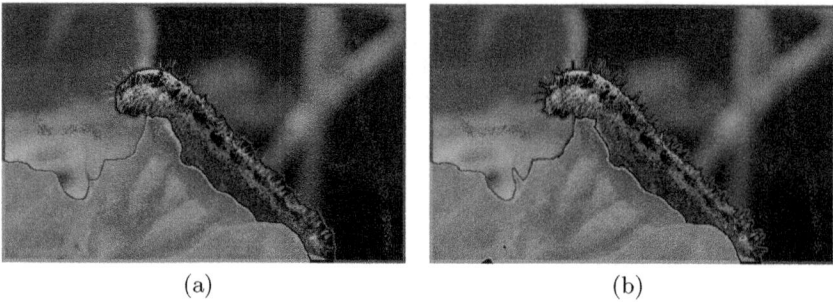

(a)   (b)

Fig. 1. An example of segmenting an image into 3 different regions. (a) shows the effect on minimizing the interface length and lose fine details like the tiny hairs (although an edge weighing is used) whereas the proposed method (b) is able to preserve those details.

where the minimization results in the perimeter as in (1). As a more sophisticated variant an anisotropic regularization like

$$J(u) = \frac{1}{2}\sum_{l=1}^{K}\int_{\Omega}\sqrt{\nabla u_l(x)^T D(x)\nabla u_l(x)}\,dx \qquad (4)$$

can be used. $D(x)$ denotes a symmetric tensor for weighting the total variation regularization. A simple variant of this tensor is the weighted total variation $\int_{\Omega} g(x)|\nabla u|dx$, studied by Bresson *et al.* in [7]. It can be obtained by setting

$$D(x) = \operatorname{diag}(g(x), g(x))$$

with an edge detector function $g(x)$, often defined as $g(x) = e^{-\alpha|\nabla I|^{\beta}}$, with the image gradient ∇I and some $\alpha, \beta > 0$ forcing the total variation regularization towards strong image edges and hence improving the labeling quality. On the other hand, a major drawback is its sensitivity to noise.

   In this paper we pursue a different approach to overcome this problem. We include a larger neighborhood in the regularizer $J(u)$ of the multiphase partitioning problem. Adapting the regularization towards local image structures enables the approach to obtain more accurate label boundaries without the dependence on an edge weighting function, which can be very sensitive to noise and strong texture. A result of this approach is depicted in Figure 1. It compares the result of the Potts model incorporating neighborhood relations to the edge-weighted variant on an image of the multi-label benchmark data set [27]. Especially fine structures are preserved and the segmentation results get enhanced towards the users expectations. A related approach was introduced in the field of unsupervised segmentation by Bresson *et al.* [6] where non-local variants of specialized energy functionals are presented by extending the Chan-Vese segmentation [12] and the Mumford-Shah segmentation [21]. While this paper concentrates on the two-label case, we study non-local generalizations of the multi-label Potts model. The main challenge hereby is the development of an efficient minimization algorithm. To achieve this we adopt the accelerated first-order algorithm of Nesterov [22]. Using this algorithm, we are able to compute the Potts model with several hundred labels.

The contribution of the paper is the definition of the non-local Potts model within a variational framework (Section 2) . The efficient minimization using Nesterov's algorithm (Section 3) makes the approach applicable to various Computer Vision Problem. In Section 4 some applications demonstrate the achieved improvements on multi-label segmentation and on disparity estimation of a stereo image pair. The shown examples provide insight into the possibilities of the non-local Potts model and we are convinced that the evident improvements can also be transferred to other Computer Vision problems. Finally, Section 5 concludes our work.

2 Non-local Potts Model

The main intention of the proposed approach is to enhance the labeling quality especially at the label boundaries. Therefore we exploit the affinity of neighboring pixels and steer the regularization towards coherent regions. In terms of image restoration such neighborhood relations have been introduced with *e.g.* the bilateral filtering [29] or non-local means [9], a generalization of the Yaroslavsky filter [32]. For image inpainting, patch-based methods for texture synthesis [14] are related to such non-local approaches and also in stereo applications, Yoon *et al.* [33] incorporated a so-called soft-segmentation to associate certain neighboring pixels for the regularization process. Consequently, the variational interpretation of these neighborhood filters leads to non-local total variation regularization [8,15]. Recently, the approach of non-local regularization in a variational framework was also introduced in the field of optical flow estimation [28,31]. To incorporate neighborhood relations directly into the objective function the non-local total variation regularizer is formulated as

$$J(u) = \sum_{l=1}^{K} \int_{\Omega} \int_{\mathcal{N}_x} w(x,y) \left| u_l(y) - u_l(x) \right| \, dy \, dx \ , \tag{5}$$

where the function $w(x,y)$ defines the support weights between the pixel x and its neighbors y. The neighborhood system $\mathcal{N}_x \subseteq \Omega$ contains all pixels y with a certain photometric and geometric vicinity around x. The support weight within $\mathcal{N}_x$ is defined in the sense of [33,31] using a low level segmentation combining an Euclidean distance in a color space $\Delta_c(x,y)$ (*e.g.* Lab, RGB or grayscale) and the spatial proximity $\Delta_p(x,y)$ as the Euclidean distance yielding

$$w(x,y) = e^{-\left(\frac{\Delta_c(x,y)}{\alpha} + \frac{\Delta_p(x,y)}{\beta} \right)} \ . \tag{6}$$

The parameters α and β weight the influence of color similarity and proximity. An exemplary neighboring patch of a specific pixel x is depicted in Figure 2.

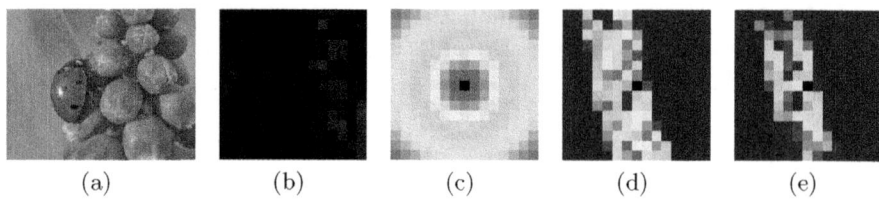

(a)   (b)   (c)   (d)   (e)

Fig. 2. Exemplar patch (b) of an image (a), the resultant proximity weighting (c), color similarities (d) and the final weighting (e) for the specific neighborhood. x is denoted as a dark red pixel in the center of (c-e), blue color means small weights (reduce regularization influence) and the increasing reddish color shows an increase in the weighting function (strengthen the regularizer).

Using the non-local TV regularizer (5) in the Potts energy (2) yields the energy minimization problem

$$\min_u \left\{ \sum_{l=1}^{K} \int_\Omega \int_{\mathcal{N}_x} w(x,y)\,|u_l(y) - u_l(x)|\ dy\ dx + \lambda \sum_{l=1}^{K} \int_\Omega u_l(x) f_l(x)\,dx \right\},$$
(7)
$$\text{s.t.}\quad u_l(x) \geq 0\ ,\quad \sum_{l=1}^{K} u_l(x) = 1\ .$$

3   Minimization

Let us first introduce the discrete setting. We consider a Cartesian grid G of size $M_x \times M_y$

$$G = \{(1,1) \leq (hx, hy) \leq (M_x, M_y)\}\ ,$$

with the pixel size h and (x,y) the discrete pixel location on the grid. For the ease of presentation we will enumerate the discrete pixel locations (x,y) with an index i, for example by scanning the image domain line by line. The discretized labeling function u is defined on the unit simplex

$$U = \left\{ u = (u_1, \ldots, u_K) \in [0,1]^{K \times M_x \times M_y} : \right.$$

$$\left. (u_l)_i \geq 0,\quad \sum_{l=1}^{K} (u_l)_i = 1,\quad i = 1 \ldots M_x \times M_y \right\}\ .$$
(8)

The non-negative discrete weighting function for the non-local regularization between discrete pixels i and j is defined as $w_{i,j} \geq 0$. $\mathcal{N}_i$ defines the set of neighbors for pixel i, where $N = |\mathcal{N}_i|$ is the number of pixels within the neighborhood. The weight matrix $(w_{i,j})$ is defined as

$$w_{i,j} = \begin{cases} e^{-\left(\frac{(\Delta_c)_{i,j}}{\alpha} + \frac{(\Delta_p)_{i,j}}{\beta} \right)} & \text{if}\quad j \in \mathcal{N}_i \\ 0 & \text{else}\ . \end{cases}$$

Now, we are ready to define the non-local gradient operator

$$(\nabla_w u_l)_{i,j} = w_{i,j} \left((u_l)_j - (u_l)_i \right) \ ,$$

which simply holds the weighted non-local pixel differences. Then, (7) can be rewritten in the discrete setting as the following minimization problem

$$\min_u \left\{ \sum_l \|\nabla_w u_l\|_{\ell_1} + \lambda \sum_l \langle u_l, f_l \rangle \right\} \ . \tag{9}$$

Minimizing (9) depicts a convex and non-smooth optimization problem. Solving it with off-the-shelf LP solvers or first-order primal-dual approaches [10] has the problem that each non-local link will demand for a dual variable. For a 512×512 image, 32 labels and a neighborhood size of 15×15 pixels this results in at least one billion dual variables. Hence, these approaches are not feasible for our purposes.

Instead we rely on an old first-order algorithm proposed by Nesterov [22] in 1983, which can be used to minimize a differentiable convex function of a convex set. Furthermore, Nesterov's algorithm comes along with an improved convergence rate. It can be shown that Nesterov's algorithm can approach the optimal function value with rate $O(1/n^2)$, where n is the number of iterations. This rate of convergence is still sublinear but improves the convergence rate of standard projected gradient schemes by one order of magnitude. Recently, there are several improved variants using Nesterov's algorithmic framework [3,30,1,13] and Nesterov himself proposed new algorithms [23]. The major benefit of Nesterov's first-order primal method is that the algorithm only depends on function values and gradient evaluations which removes the need of the large amount of dual variables. In order to apply Nesterov's algorithm for our problem (9) we have to find a differentiable approximation of the ℓ_1 norm. We do this by replacing any $|\cdot|$ function by Huber's function [16].

$$|q|_\varepsilon = \begin{cases} \frac{|q|^2}{2\varepsilon} & |q| \le \varepsilon \\ |q| - \frac{\varepsilon}{2} & \text{else} \ . \end{cases}$$

The function is quadratic for small values of ε and linear for the others.

Algorithm: Nesterov's algorithm for the non-local Potts model: We choose $u^0 = 0, \bar{u}^0 = 0, t^0 = 1$ and iterate for $n \ge 0$.

$$\begin{cases} u_l^{n+\frac{1}{2}} &= \bar{u}_l^n - \dfrac{1}{L} \left(\nabla_w^T \dfrac{\nabla_w \bar{u}_l^n}{\max\{\varepsilon, |\nabla_w \bar{u}_l^n|\}} + f_l \right) \ , \quad l = 1, \dots, K \\ u^{n+1} &= \Pi_U \left(u^{n+\frac{1}{2}} \right) \\ t^{n+1} &= \dfrac{1}{2} \left(1 + \sqrt{1 + 4(t^n)^2} \right) \\ \bar{u}_l^{n+1} &= u_l^{n+1} + \dfrac{t^n - 1}{t^{n+1}} \left(u_l^{n+1} - u_l^n \right) \ , \quad l = 1, \dots, K \end{cases} \tag{10}$$

Here, $L = \|\nabla_w\|$ is the norm of the non-local operator which we compute as $L = \frac{4N}{\varepsilon}$ and t^n a variable over-relaxation parameter. The projection Π_U is an orthogonal projection onto the unit simplex U. It is known that this projection is highly separable and it can be performed in a finite number of iterations. An exemplary method for computing such successive projections is given in [20].

Although Nesterov's algorithm allows to precompute the maximum number of iterations which are necessary to find an approximate solution in terms of the function values, we found it to be more practical to stop the iterations after the maximal change between two successive iterations is below some threshold. In Figure 3 we compare the convergence of the algorithm for different smoothing parameters ε for an unsupervised segmentation problem. Increasing the smoothing behavior of the Huber function improves the rate of convergence but worsens the approximation quality of the ℓ_1 norm which introduces inaccurate label boundaries. For all our experiments we set $\varepsilon = 0.01$. Observe that after 300 iterations the minimum energy is already attained. For this setup the algorithm runs with 120 iterations/second for an image with size $M_x \times M_y = 404 \times 320$, with $K = 10$ number of labels and a neighborhood size of 7×7 pixels.

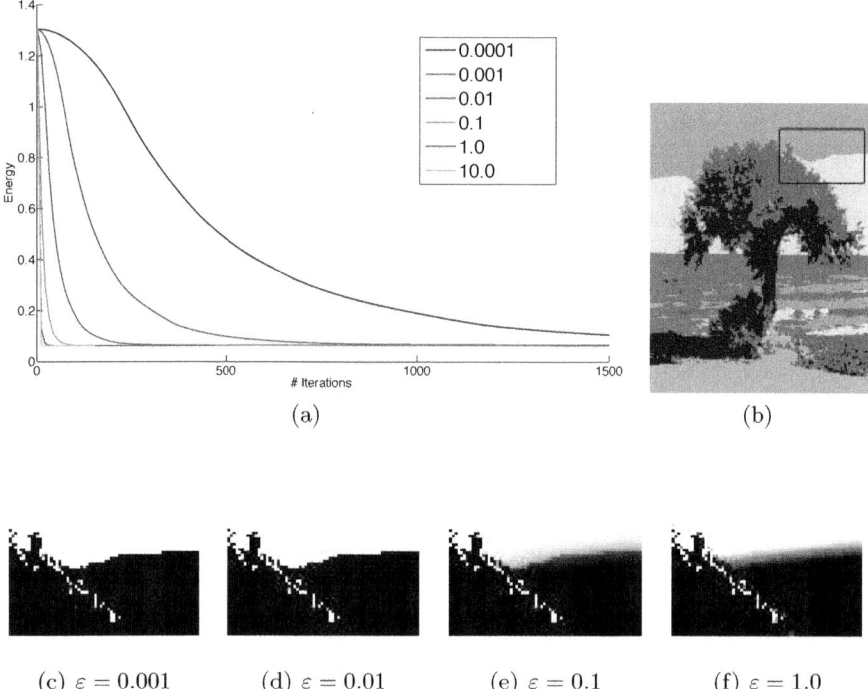

(a)   (b)

(c) $\varepsilon = 0.001$ (d) $\varepsilon = 0.01$ (e) $\varepsilon = 0.1$ (f) $\varepsilon = 1.0$

Fig. 3. Comparing convergence behavior of different smoothing parameters ε. (b) is the color-coded labeling result for $\varepsilon = 0.01$ with the marked crop region for (c)-(f) showing a single label u_l for varying ε demonstrating the smoothing effect on label borders for increasing smoothing factors.

4   Applications

4.1   Multi-label Segmentation

Image segmentation is one of the fundamental Computer Vision problems and therefore a vast amount of literature investigates this task. For a general overview on object segmentation we refer to [5]. Very recent work of Santner *et al.* [27] demonstrates the usage of (2) for interactive multi-label segmentation. There, the data term is modeled with different types of features. For a comparison of three different regularization terms, namely the total variation, the edge-weighted and the proposed non-local regularization, we compute a color histogram of scribbles drawn by the user and use this as a feature for the data term in the sense of [27].

Fig. 4. Comparing different regularization terms in terms of interactive multi-label segmentation (*cf.* [27]): TV regularization (first column), edge-weighted TV regularization (second column) and the proposed non-local variant (third column). The scribbles are the users input to mark the corresponding region.

In the first row of Figure 4 the improvement on fine details are visible over all three variants. The edge-weighted TV already yields reasonable accuracy when it comes to segmentation boundaries and with the non-local variant the borders are accurately segmenting the desired region including all fine details. The example in the second row of Figure 4 demonstrates the drawbacks on solely using edges to steer the regularization strength. Sometimes edges that do not coincide with label borders pull the label boundaries away from the desired objects. In Figure 7 we demonstrate especially the improvements on fine details, elongated regions and cavities between labels on segmentation results of the benchmark data set [27].

Next, we want to show the effects when the edge-weighting function of (4) is modified to obtain accurate label boundaries. In Figure 5 an unsupervised labeling routine splits the image domain into several piecewise constant regions.

The data term is solely based on RGB values that are clustered with a standard mean-shift algorithm. Tuning the edge-detector function towards accurate boundaries also introduces some clutter within label regions as a direct consequence of having strong edges within those areas. This weights the regularizer to obtain more and smaller labels and therefore introduces clutter. Using the proposed non-local regularization yields the same precise label borders but also gains a smoother result and keeps coherent regions together.

(a)   (b)   (c)

Fig. 5. Unsupervised segmentation splitting an input image (a) into $K = 10$ piecewise constant regions. The tuned edge-weighted regularization (b) achieves nice boundaries but exhibits more clutter within regions compared to the non-local regularization (c).

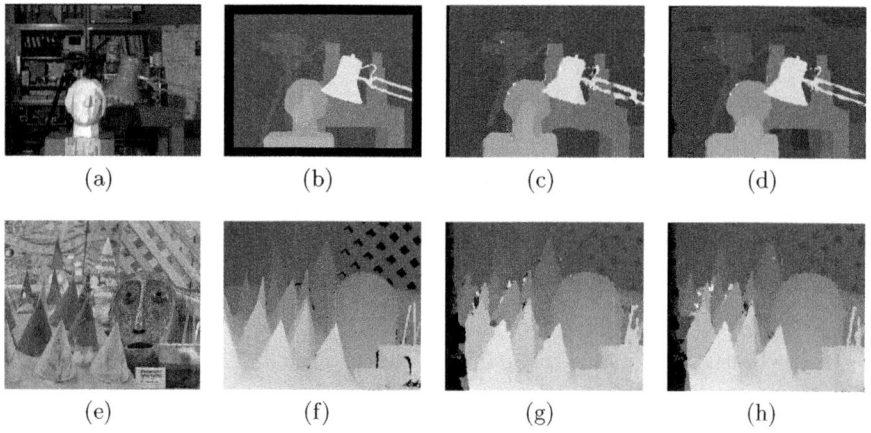

(a)   (b)   (c)   (d)

(e)   (f)   (g)   (h)

Fig. 6. Disparity estimation of an input image pair (a,e) from the Middlebury stereo data set (http://vision.middlebury.edu/stereo); (b,f) the ground truth disparities; disparity estimation with the Potts model (c,g) and the non-local Potts model (d,h);

4.2   Stereo

As we have already shown the improvements on the image segmentation problem we want to continue with a second Computer Vision problem. We use the Potts model for stereo estimation. The data term for the disparities are modeled using

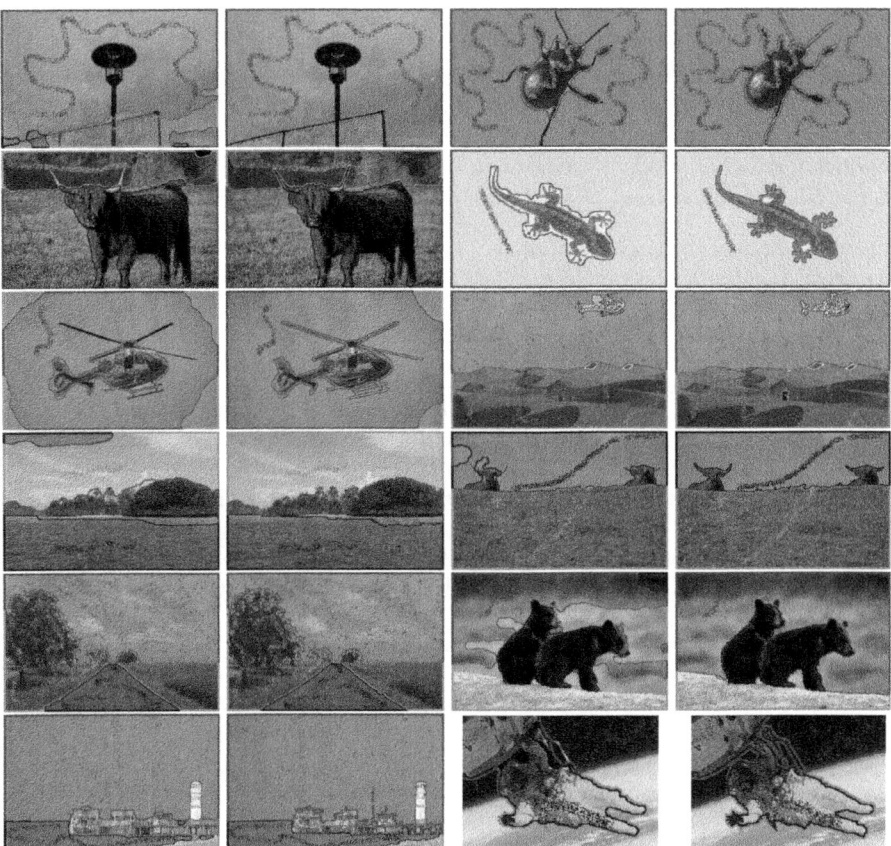

Fig. 7. Multi-Label Segmentation: Comparing results from the edge-weighted Potts model (left image of each pair) and the non-local Potts model (right image of each pair)

absolute differences on gray values. The labels correspond to distinct disparities. For the example in Figure 6 the benefits of the proposed method become apparent with more details and crisper label borders. For the Tsukuba image pair (*cf.* first row of Figure 6) the calculation for $M_x \times M_y \times K = 384 \times 288 \times 16$ takes 25 seconds for 500 iterations. For the Cones data set (*cf.* second row) with $M_x \times M_y \times K = 450 \times 375 \times 61$ the nonlocal Potts model takes 305 seconds to converge in 1000 iterations. We use a 15×15 neighborhood region for the non-local regularization in both examples.

5   Conclusion

Based on a variational formulation of the Potts model we showed how to incorporate neighborhood relations with a non-local total variation regularization term. Utilizing low-level image segmentation to steer the regularization towards local image structure enables the method to preserve fine details in the labeling process. The benefits are demonstrating on two typical Computer Vision applications and evident improvements are demonstrated by a comparison with a total variation regularization and its edge-weighted variant. The version of Nesterov's algorithm yields a memory-conscious algorithm and enables the usage of large neighborhoods and several labels with reasonable computational effort.

References

1. Aujol, J.F.: Some first-order algorithms for total variation based image restoration. J. Math. Imaging Vis. 34(3), 307–327 (2009)
2. Bae, E., Yuan, J., Tai, X.C.: Global minimization for continuous multiphase partitioning problems using a dual approach. International Journal of Computer Vision, 1–18 (2010), doi:10.1007/s11263-010-0406-y
3. Beck, A., Teboulle, M.: A fast iterative shrinkage-thresholding algorithm for linear inverse problems. SIAM J. Imaging Sci. 2(1), 183–202 (2009)
4. Boykov, Y., Veksler, O., Zabih, R.: Fast approximate energy minimization via graph cuts. IEEE Trans. Pattern Analysis and Machine Intelligence 23(11), 1222–1239 (2001)
5. Boykov, Y.Y., Lea, G.F.: Graph cuts and efficient N-D image segmentation. International Journal of Computer Vision 70(2), 109–131 (2006)
6. Bresson, X., Chan, T.F.: Non-local unsupervised variational image segmentation models. Tech. rep., UCLA CAM Report 08-67 (October 2008)
7. Bresson, X., Esedoglu, S., Vandergheynst, P., Thiran, J.P., Osher, S.J.: Fast global minimization of the active contour/snake model. J. Math. Imaging Vis. 28(2), 151–167 (2007)
8. Brox, T., Kleinschmidt, O., Cremers, D.: Efficient nonlocal means for denoising of textural patterns. IEEE Trans. on Image Processing 17(7), 1083–1092 (2008)
9. Buades, A., Coll, B., Morel, J.M.: Nonlocal image and movie denoising. International Journal of Computer Vision 76(2), 123–139 (2007)
10. Chambolle, A., Pock, T.: A first-order primal-dual algorithm for convex problems with applications to imaging. J. Math. Imaging Vis. (2010)
11. Chan, T., Esedoglu, S., Nikolova, M.: Algorithms for finding global minimizers of image segmentation and denoising models. SIAM Journal of Applied Mathematics 66(5), 1632–1648 (2006)
12. Chan, T., Vese, L.: Active contours without edges. IEEE Trans. Image Processing 10(2), 266–277 (2001)
13. Dahl, J., Hansen, P.C., Jensen, S.H., Jensen, T.L.: Algorithms and software for total variation image reconstruction via first-order methods. Numerical Algorithms 53(1), 67–92 (2010)
14. Efros, A.A., Leung, T.K.: Texture synthesis by non-parametric sampling. In: ICCV, pp. 1033–1038 (1999)
15. Gilboa, G., Osher, S.J.: Nonlocal operators with applications to image processing. Tech. rep., UCLA CAM Report 07-23 (July 2007)
16. Huber, P.J.: Robust Statistics. Wiley Series in Probability and Statistics (1981)
17. Ishikawa, H.: Exact optimization for markov random fields with convex priors. IEEE Trans. Pattern Analysis and Machine Intelligence 25(10), 1333–1336 (2003)

18. Ising, E.: Beitrag zur Theorie des Ferromagnetismus. Zeitschrift für Physik 23, 253–258 (1925)
19. Lellmann, J., Kappes, J., Yuan, J., Becker, F., Schnörr, C.: Convex Multi-class Image Labeling by Simplex-Constrained Total Variation. In: Tai, X.-C., Mørken, K., Lysaker, M., Lie, K.-A. (eds.) SSVM 2009. LNCS, vol. 5567, pp. 150–162. Springer, Heidelberg (2009)
20. Michelot, C.: A finite algorithm for finding the projection of a point onto the canonical simplex of $\mathbb{R}^n$. Journal of Optimization Theory and Applications 50, 195–200 (1986)
21. Mumford, D., Shah, J.: Optimal approximation by piecewise smooth functions and associated variational problems. Comm. Pure Appl. Math. 42, 577–685 (1989)
22. Nesterov, Y.: A method for solving the convex programming problem with convergence rate $O(1/k^2)$. Dokl. Akad. Nauk USSR 269(3), 543–547 (1983)
23. Nesterov, Y.: Smooth minimization of non-smooth functions. Math. Program. 103(1, ser. A), 127–152 (2005)
24. Pock, T., Chambolle, A., Cremers, D., Bischof, H.: A convex relaxation approach for computing minimal partitions. In: IEEE Computer Society Conference on Computer Vision and Pattern Recognition, CVPR (2009)
25. Pock, T., Cremers, D., Bischof, H., Chambolle, A.: Global solutions of variational models with convex regularization. SIAM Journal on Imaging Sciences 3(4), 1122–1145 (2010)
26. Potts, R.B.: Some generalized order-disorder transformations. Proc. Camb. Phil. Soc. 48, 106–109 (1952)
27. Santner, J., Pock, T., Bischof, H.: Interactive multi-label segmentation. In: Kimmel, R., Klette, R., Sugimoto, A. (eds.) ACCV 2010, Part I. LNCS, vol. 6492, pp. 397–410. Springer, Heidelberg (2011)
28. Sun, D., Roth, S., Black, M.J.: Secrets of optical flow estimation and their principles. In: IEEE Computer Society Conference on Computer Vision and Pattern Recognition (CVPR), San Francisco, CA, USA (June 2010)
29. Tomasi, C., Manduchi, R.: Bilateral filtering for gray and color images. In: ICCV, pp. 839–846 (1998)
30. Weiss, P., Blanc-Féraud, L., Aubert, G.: Efficient schemes for total variation minimization under constraints in image processing. SIAM J. Scientific Computing 31(3), 2047–2080 (2009)
31. Werlberger, M., Pock, T., Bischof, H.: Motion estimation with non-local total variation regularization. In: IEEE Computer Society Conference on Computer Vision and Pattern Recognition (CVPR), San Francisco, CA, USA (June 2010)
32. Yaroslavsky, L.P.: Digital Picture Processing — an Introduction. Springer, Berlin (1985)
33. Yoon, K.J., Kweon, I.S.: Adaptive support-weight approach for correspondence search. IEEE Trans. on Pattern Analysis and Machine Intelligence 28(4), 650–656 (2006)
34. Zach, C., Gallup, D., Frahm, J.M., Niethammer, M.: Fast global labeling for real-time stereo using multiple plane sweeps. In: Proceedings of the Vision, Modeling, and Visualization Conference, pp. 243–252 (2008)

Sulci Detection in Photos of the Human Cortex Based on Learned Discriminative Dictionaries

Benjamin Berkels[1], Marc Kotowski[2], Martin Rumpf[3], and Carlo Schaller[2]

[1] Interdisciplinary Mathematics Institute,
University of South Carolina, Columbia, SC 29208, USA
berkels@mailbox.sc.edu
[2] Hôpitaux Universitaires de Genève,
Rue Micheli-du-Crest 24, 1211 Genève, Switzerland
[3] Institut für Numerische Simulation,
Rheinische Friedrich-Wilhelms-Universität Bonn,
Endenicher Allee 60, 53115 Bonn, Germany
martin.rumpf@ins.uni-bonn.de

Abstract. The use of discriminative dictionaries is exploited for the segmentation of sulci in digital photos of the human cortex. Manual segmentation of the geometry of sulci by an experienced physician on training data is taken into account to build pairs of such dictionaries. It is demonstrated that this approach allows a robust segmentation of these brain structures on photos of the brain as long as the training data contains sufficiently similar images. Concerning the methodology an improved minimization algorithm for the underlying variational approach is presented taking into account recent advances in orthogonal matching pursuit. Furthermore, the method is stable since it ensures an energy decay in the dictionary update.

1 Introduction

In neurosurgery, a major challenge is the adaption of pre-surgery acquired brain images and cortex geometry to the intra-interventional brain configuration. Digital photos can be easily taken through the microscope and provide information on the currently observed brain shift. Sulci are the most prominent geometric characteristics visible on such photos. As illustrated by Figure 1, the detection of sulci in such images is a very challenging task. For instance, some of the sulci are covered by blood vessels while the very same blood vessels also cover part of the gyri. Therefore, pixelwise segmentation approaches based on the color values cannot be sufficient to handle this segmentation problem, not even when color distributions learned from images manually marked by an expert are used. In this paper, we use the concept of learned discriminative dictionaries to segment the geometry of sulci in 2D digital photos. Thereby, on a training data set an experienced physician marks the sulci geometry, which will then be used to built a suitable discriminative dictionary.

A.M. Bruckstein et al. (Eds.): SSVM 2011, LNCS 6667, pp. 326–337, 2012.

Nowadays, sparse signal representations based on overcomplete dictionaries are used for a wide range of signal and image processing tasks. The key assumption of these models is that finite dimensional signals can be well approximated by sparse linear combinations of so-called *atoms* or *atom signals*. Due to their finite dimensionality, the signals and the atoms are considered to be elements of $\mathbb{R}^N$. A set of atoms $d_1, \ldots, d_K$ is called *dictionary* and represented by the matrix $D \in \mathbb{R}^{N \times K}$ whose j-th column is the atom d_j.

There are two main variants of the sparse approximation problem, the *error-constrained* approaches and the *sparsity-constrained* approaches. Here, we are considering an approach of the latter type: For a given input signal y (in our application a patch from a digital photo of the brain) we ask for its best approximation under the constraint that at most $L \in \mathbb{N}$ atoms are used, i. e.

$$\min_{x \in \mathbb{R}^K} \|y - Dx\|^2 \text{ such that } \|x\|_0 \leq L,$$

where $\|\cdot\|_0$ denotes the l_0 "norm", i. e. the number of nonzero components.

One of the major challenges in the context of sparse representations is the design of suitable dictionaries. The sparse representation itself usually is just a means to an end and used to solve a certain task like, for instance, denoising or compression. Thus, the dictionary has to be tailored to the actual imaging task. In general, there are two distinct approaches to dictionary design: The simpler and more traditional route is to use a predefined dictionary generated by a transform like the short-time Fourier transform [2], the wavelet [12], curvelet [6] or contourlet transform [8], to name just a few. The more sophisticated approach is to learn the dictionary from the input data or some representative training data. A very popular and highly efficient representative of this kind is the K-SVD algorithm [1].

Like K-SVD, most of the existing dictionary learning algorithms aim at generating *reconstructive* dictionaries, i. e. dictionaries that are optimized to sparsely represent a certain class of input signals or images. In this paper, our goal is to detect sulci on the human cortex in digital photographs. Thus, we need to distinguish between different types of signals which gives rise to so-called *discriminative* dictionaries. These kind of dictionaries not only aim to give a suitable representation of a given type of signals, but are also optimized to be not as suitable for the reconstruction of a different given class of signals. Mairal et al. [10] introduced a variational approach to learning discriminative dictionaries for local image analysis and presented a multiscale extension applied to class-specific edge detection [11].

Zhao et al. [15] combine the discriminative dictionary model from [10] with additional pre- and post-processing stages to optimize the discriminative approach for text detection in images. Zhang and Li [14] propose a different route to discriminative dictionary learning: They extend the K-SVD algorithm to solve for a dictionary and a classifier simultaneously and claim that this kind of algorithm is less likely to get stuck in local minima than the one from [10]. Let us remark here, that their K-SVD extension still uses an alternating minimization scheme to solve a non-convex minimization problem. Hence, there is no guarantee that the global optimum is finally found.

The contributions of this paper are twofold: On the one hand, we introduce an improved minimization algorithm for the variational approach to discriminative dictionaries from [10]. This algorithm is more efficient because it incorporates recent advances in orthogonal matching pursuit made by Rubinstein et al. [13] and it is more stable since it ensures an energy decay in the dictionary update unlike the truncated Newton iteration used in [10,11]. On the other hand, we study the applicability of discriminative dictionaries to detect sulci on the intra-operative digital photographs of the human cortex. As we will see in this paper, manually marked images can indeed be used to learn a discriminative dictionary pair and thereby allow to detect sulci on images as long as the training data contains sufficiently similar brain images.

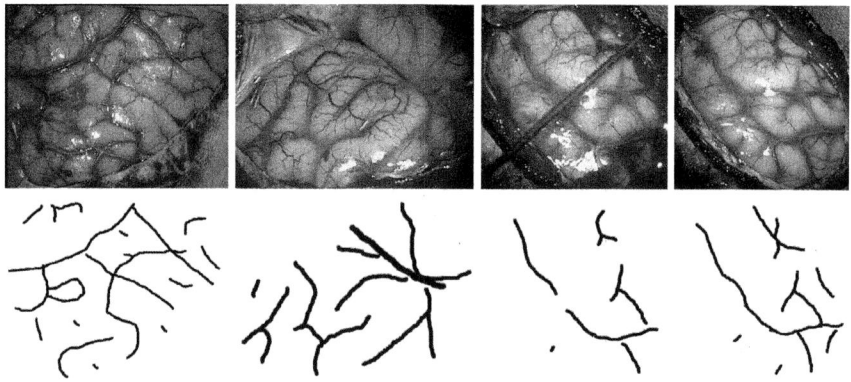

Fig. 1. Four typical digital photographs of the exposed human cortex (top row) and the sulci regions of the cortex manually marked by an expert (bottom row)

2   Learning Discriminative Dictionaries

Given M input patches $y_1, ..., y_M \in \mathbb{R}^N$, a reconstructive dictionary tailored to these patches can be learned with the minimization problem

$$\min_{X \in \mathbb{R}^{K \times M}, D \in \mathbb{R}^{N \times K}} \sum_{l=1}^{M} R(y_l, D, x_l) \text{ such that } \|x_l\|_0 \leq L \text{ for } 1 \leq l \leq M.$$

Here, x_l denotes the l-th column of X and $R(y, D, x) := \|y - Dx\|^2$ is the reconstruction error of a patch $y \in \mathbb{R}^N$ for a dictionary $D \in \mathbb{R}^{N \times K}$ and dictionary coefficients $x \in \mathbb{R}^K$. Well-known algorithms to tackle this minimization problem are the method of optimal directions (MOD) [9] or K-SVD [1].

Denoting the coefficients of the sparse best approximation of y using D by

$$x^*(y, D) := \underset{x \in \mathbb{R}^K, \|x\|_0 \leq L}{\text{argmin}} R(y, D, x),$$

the best approximation error is $\mathcal{R}(y, D) := R(y, D, x^*(y, D))$. Then the minimization problem for reconstructive dictionary learning is equivalent to

$$\min_{D \in \mathbb{R}^{N \times K}} \sum_{l=1}^{M} \mathcal{R}(y_l, D). \tag{1}$$

With this notation we can formulate the discriminative dictionary approach of Mairal et al. [10]. Since our application, the detection of sulci on the human cortex, only requires two labels, we here explicitly formulate only the two label case. The extension to multiple labels is straightforward and our algorithm can be easily adapted to more than two labels.

Given input patches $y_1, ..., y_{M_1+M_2}$ of two different classes P_1 and P_2, where $P_i := \{y_l : l \in S_i\}$, $S_1 = \{1, ..., M_1\}$ and $S_2 = \{M_1 + 1, ..., M_1 + M_2\}$, a pair of discriminative dictionaries can be found solving the minimization problem

$$\min_{D_1, D_2} \sum_{i=1}^{2} \frac{1}{M_i} \sum_{l \in S_i} \left[C_\lambda \left((-1)^{i+1} (\mathcal{R}(y_l, D_2) - \mathcal{R}(y_l, D_1)) \right) + \lambda \gamma \mathcal{R}(y_l, D_i) \right]. \tag{2}$$

Here, C_λ denotes the logistic loss function, i. e. $C_\lambda(s) = \ln(1 + \exp(-\lambda s))$, and λ, γ are nonnegative constant parameters. The last summand is already known from the reconstructive learning problem (1) and handles the reconstructive properties of our dictionary pair. The first summand is responsible for the discriminative properties of the dictionaries. For instance, for $i = 1$ and $l \in S_1$ we have

$$C_\lambda \left((-1)^2 (\mathcal{R}(y_l, D_2) - \mathcal{R}(y_l, D_1)) \right) \begin{cases} \approx 0 & \mathcal{R}(y_l, D_1) \ll \mathcal{R}(y_l, D_2) \\ \gg 0 & \mathcal{R}(y_l, D_1) \gg \mathcal{R}(y_l, D_2). \end{cases} \tag{3}$$

In other words, this logistic loss term is small, if and only if D_1 is more suitable to reconstruct P_1 (the signals from the first class) than D_2 is.

3   Minimization Algorithm

The discriminative minimization problem (2) is highly nonconvex and requires a carefully chosen numerical minimization strategy. Like [10], our minimization strategy is based on the K-SVD algorithm and consists of a *sparse coding stage* and a *codebook update stage*.

In the sparse coding stage, the sparse approximation coefficients are computed for all patches using the current estimates for both dictionaries, i.e. $x_l^i \approx x^*(y_l, D_i)$ for $i = 1, 2$ and $l = 1, ..., M_1 + M_2$, cf. Algorithm 1.1. Instead of using OMP for this as suggested in [10,11], we propose to use the Batch-OMP algorithm from [13]. This algorithm is based on the fact that the *same* dictionary is used to code a *large* set of signals. In particular, it exploits the fact that in the atom selection step of OMP neither the residual r nor the coefficients x need to be known, but only $D^T r$. As shown in [13], Batch-OMP is almost an order of magnitude faster than OMP when used on sufficiently many input signals.

In [11], a different way to speed up the algorithm from [10] is proposed: Also noting that the sparse coding stage is computationally expensive, they propose to update the dictionaries and the coefficients with fixed sparsity pattern in the codebook update stage by alternating till convergence instead of doing so only once to reduce the number of sparse coding steps. This idea is complementary to our proposal to speed up the algorithm and thus can be used in combination with it.

In the codebook update stage, the dictionaries and the coefficients are updated while keeping the obtained sparsity pattern fixed during the sparse coding stage. [10,11] propose to do this update with a truncated Newton method. "Truncated" here refers to the fact that this method neglects the second derivatives of C_λ. In our experiments with manually marked images of the human cortex, this algorithm unfortunately had numerically stability problems and didn't always produce sufficiently discriminative dictionary pairs. This is most likely because the truncated Newton method does not guarantee an energy decay of the target functional due to the lack of an appropriate step size control. Furthermore, C_λ is not approximately linear at 0, the transition between the nearly linear and the nearly constant part of C_λ which is the important transition region between the two cases outlined in (3). Therefore, neglecting of the second derivatives of C_λ is questionable.

To update a single entry of one of the dictionaries, we use a step size controlled gradient descent on the functional from (2) while freezing the coefficients, i. e. $\mathcal{R}(y_l, D_i)$ is approximated by $R(y_l, D_i, x_l^i)$ and thus we use the functional

$$E[D_1, D_2] = \sum_{i=1}^{2} \frac{1}{M_i} \sum_{l \in S_i} \left[C_\lambda \left((-1)^{i+1} (R(y_l, D_2, x_l^2) - R(y_l, D_1, x_l^1)) \right) \right.$$
$$\left. + \lambda \gamma R(y_l, D_i, x_l^i) \right]$$

and update d_j^1 by $d_j^1 - \tau \partial_{d_j^1} E[D_1, D_2]$ where τ is determined using the *Armijo rule* [3,5]. Note that the specific choice of the step size control is not important here, but it is important to use a step size control that guarantees an energy decay. Like K-SVD, we assume the dictionary entries to be normalized, i. e. $\|d_j^i\| = 1$, and therefore scale the dictionary entry accordingly after the gradient descent update. Using a straightforward calculation one obtains

$$\partial_{d_j^1} E[D_1, D_2] = 2 \sum_{i=1}^{2} \sum_{l \in S_i} w_l^i (x_l^1)_j \left(D_1 x_l^1 - y_l \right) ,$$

where

$$w_l^i = \frac{1}{M_i} \left((-1)^i C_\lambda' \left((-1)^{i+1} (R(y_l, D_2, x_l^2) - R(y_l, D_1, x_l^1)) \right) + \delta_{i1} \lambda \gamma \right).$$

Denoting the j-th entry of x_l^1 by $(x_l^1)_j$ and using

$$E_l^1[D, j] = \left(y_l - Dx_l^1 + (x_l^1)_j d_j\right)$$

as well as the indices of patches that use d_j^1, i.e. $\omega_j^1 := \{l : (x_l^1)_j \neq 0\}$, the variation can be expressed as

$$\partial_{d_j^1} E[D_1, D_2] = 2 \sum_{i=1}^{2} \sum_{l \in S_i \cap \omega_j^1} w_l^i (x_l^1)_j \left[(x_l^1)_j d_j^1 - E_l^1[D_1, j]\right].$$

Replacing the sum $\sum_{l \in S_i}$ by $\sum_{l \in S_i \cap \omega_j^1}$ is crucial to keep the computational cost for the codebook update stage within reasonable limits. The same replacement can be done in E when it needs to be evaluated for the Armijo rule. After updating a dictionary entry, we update the corresponding coefficients keeping the sparsity pattern. Similarly to the representation of $\partial_{d_j^1} E$, one now obtains

$$\partial_{(x_l^1)_j} E = \sum_{i=1}^{2} w_l^i \partial_{(x_l^1)_j} R(y_l, D_1, x_l^1) = \partial_{(x_l^1)_j} R(y_l, D_1, x_l^1) \sum_{i=1}^{2} w_l^i.$$

Therefore, $\partial_{(x_l^1)_j} E = 0$ holds when $\partial_{(x_l^1)_j} R(y_l, D_1, x_l^1) = 0$. Using

$$\partial_{(x_l^1)_j} R(y_l, D_1, x_l^1) = 2 \left((x_l^1)_j \left\|d_j^1\right\|^2 - E_l^1[D_1, j] \cdot d_j^1\right)$$

and $\left\|d_j^1\right\|^2 = 1$ leads to the update formula $(x_l^1)_j \leftarrow E_l^1[D_1, j] \cdot d_j^1$. The dictionary D_2 and its corresponding coefficients can be updated analogously.

Like [10], we use an ascending series for the parameter λ and a descending series for γ. Since our codebook update stage is guaranteed not to increase E, we do not need the sophisticated strategy to adaptively adjust the parameters used in [10]. Instead, in all our experiments, we simply used $\lambda = 100k$ and $\gamma = 1/k$ in the k-iteration of the algorithm. A sketch of this computational procedure is given in Algorithm 1.1.

4   Segmentation with Discriminative Dictionaries

For the detection of sulci in images of the human cortex, we assume to be given a number of human cortex images where the sulci were manually marked by a physician. These images are then separated into small patches of a user selectable patch size and the patches are divided into two sets, sulci and non-sulci patches depending on whether the central pixel of the patch belongs to the sulci region marked by the physician. Using these two sets of patches, a discriminative dictionary pair is learned using the Algorithm 1.1.

Using this dictionary pair, images can be segmented into sulci and non-sulci regions using a binary Mumford–Shah model where the reconstruction errors with the two dictionaries are used as the two indicator functions. A global minimizer of this model is calculated using the convex reformulation of the problem proposed in [4] and using [7, Algorithm 2] to efficiently calculate a minimizer of the convex functional.

Algorithm 1.1: General minimization strategy

> **given** input patches $y_1, ..., y_{M_1+M_2}$ of two different classes P_1 and P_2;
> initialize D_1 and D_2 with K-SVD from P_1 and P_2 respectively;
> initialize $k = 0$;
> **repeat**
> > $k \leftarrow k + 1$;
> > $\lambda = 100k, \gamma = 1/k$;
> > *Sparse coding stage*;
> > **for** $i = 1$ **to** 2 **do**
> > > **for** $l = 1$ **to** $M_1 + M_2$ **do**
> > > > Calculate $x_l^i \approx x^*(y_l, D_i)$ using Batch-OMP;
> > >
> > > **end**
> >
> > **end**
> > *Codebook update stage*;
> > **for** $i = 1$ **to** 2 **do**
> > > **for** $j = 1$ **to** K **do**
> > > > Calculate $\omega_j^i = \{ l : (x_l^i)_j \neq 0 \}$;
> > > > $d_j^i \leftarrow d_j^i - \tau \partial_{d_j^i} E$ determining τ using the Armijo rule;
> > > > $d_j^i \leftarrow d_j^i / \|d_j^i\|$;
> > > > **for** $l \in \omega_j^i$ **do**
> > > > > $(x_l^i)_j \leftarrow E_l^i[D_i, j] \cdot d_j^i$.;
> > > >
> > > > **end**
> > >
> > > **end**
> >
> > **end**
> **until** *convergence* ;

5 Results

As first experiment we verify the general applicability of our discriminative dictionary approach to detect sulci on intra-operative images of the human cortex. For this we learn a discriminative dictionary pair from a single manually marked image as described in the previous section. The first row of Figure 2 shows that in this optimal case, the segmentation almost perfectly matches the manual markings made by the expert. Here, we used a patch size of 13×13, and $K = 256$, $L = 4$ and $\gamma = 0.00002$ as parameters, where γ denotes the weighting of the regularity term in the Mumford–Shah model segmentation model.

In order to cut down the computational time given the fact that there are considerably more non-sulci than sulci input patches, we randomly selected 30% of the non-sulci patches instead of using all of them for the dictionary learning in a second experiment, cf. second row of Figure 2. Indeed, this only slightly reduces the accuracy of the segmentation. Henceforth, we only use 30% of the non-sulci patches to learn the dictionaries in the remaining experiments.

In the next experiment, we use three frames from an intra-operative video, all with sulci manually marked by a physician, to learn a discriminative dictionary pair. We use the same values for K, L and γ as in the previous experiments and

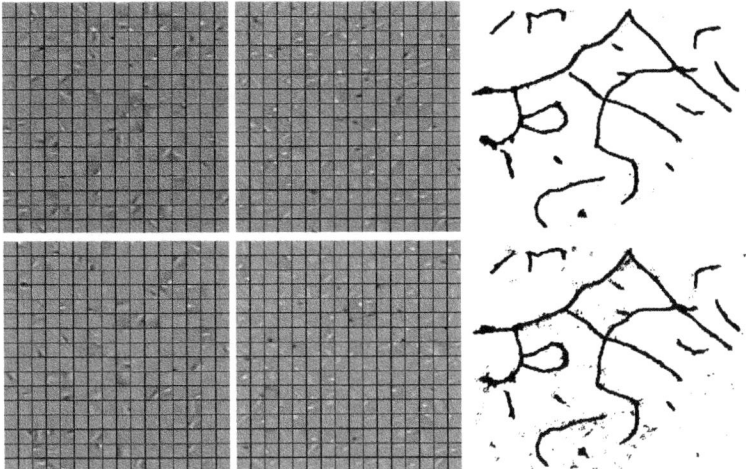

Fig. 2. Discriminative dictionary pairs learned from the manually marked image shown on the left of Figure 1 and segmentation of this image based on these dictionary pairs. In the top row all available non-sulci patches were used to learn the dictionaries, while in the bottom row only 30% of the non-sulci patches were used.

a patch size of 12×12 and 20×20. Figure 3 shows the resulting dictionaries while Figure 4 shows the segmentation obtained with these dictionaries. Segmentation on the frames already used in the learning phase is almost perfect. There are only a few minor artifacts compared to the manual segmentation performed by the physician. Although not surprising, this confirms that a discriminative dictionary pair has no problems encoding information from multiple input frames. It can also be seen from this figure that increasing the patch size from 12×12 to 20×20 slightly improves the results. The second row of Figure 4 is more interesting: It shows that the dictionaries can also be used to segment frames that were not used during the learning process. The dictionary based segmentation shows some artifacts away from the cortex region, but this is due to the fact that images used to learn the dictionaries were cropped to the cortex region and thus the dictionaries cannot contain information about these areas. This kind of effect can also be seen inside the cortex region: On the top right of the manual markings of the physician for this image is a small sulci that is not found in the dictionary based segmentation. This is just natural since the physician did not mark this sulci in the frames that were used for the learning, cf. top row of Figure 4.

In the final set of experiments, we use thirteen manually marked images, the three images already used in the previous experiment and ten from another intra-operative video to learn discriminative dictionaries. We use the same values for K, L and γ as in the previous experiments and a patch size of 13×13 and 19×19. Figure 5 shows the resulting dictionaries while Figure 6 shows the segmentation obtained with these dictionaries on multiple frames. The first three images were

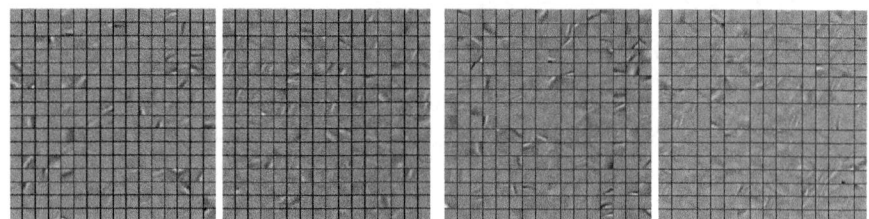

Fig. 3. Discriminative dictionary pairs with patch size of 12×12 (left pair) and 20×20 (right pair) learned from three frames of an intra-operative video

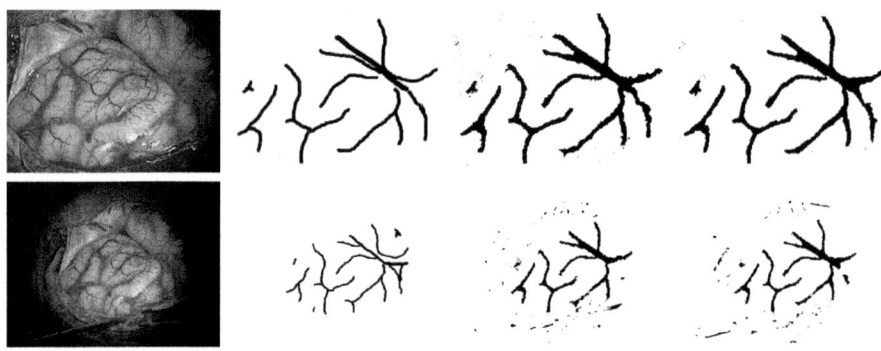

Fig. 4. Two cortex images (first column), manual segmentation of the sulci by an expert (second column) and segmentation obtained using the dictionaries from Figure 3 of patch size 12×12 (third column) and 20×20 (forth column). Note that the manual marking from the top row was used during the dictionary learning but the one from the bottom row was not.

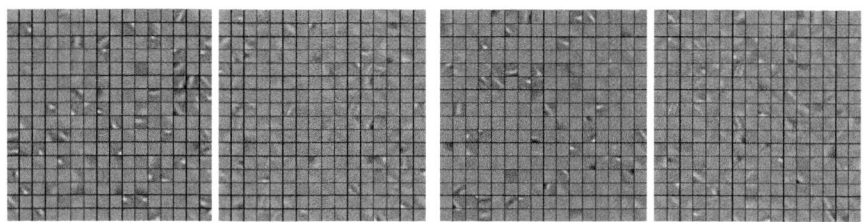

Fig. 5. Discriminative dictionary pairs with patch size of 13×13 (left pair) and 19×19 (right pair) learned from a total of thirteen different frames originating from two different intra-operative videos

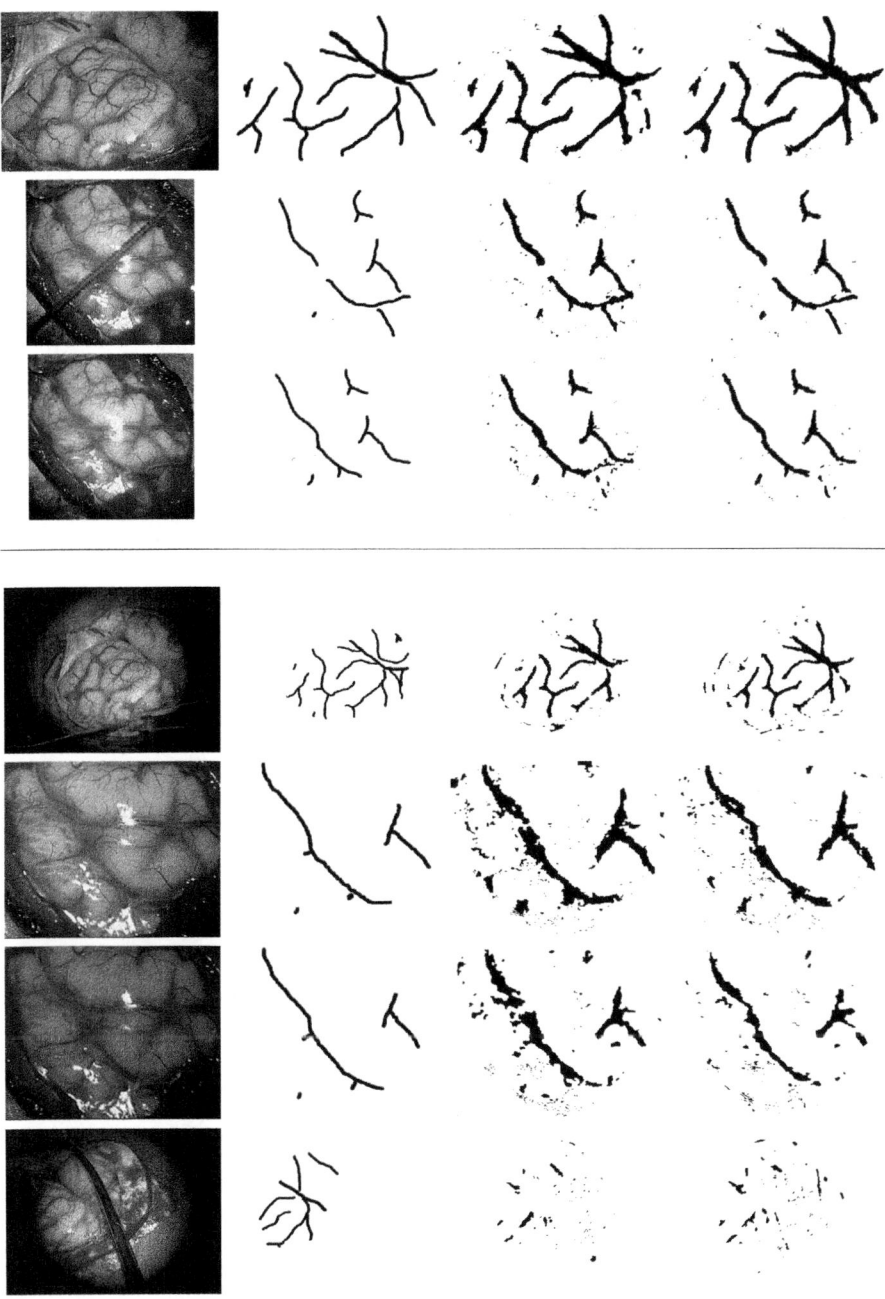

Fig. 6. Multiple cortex images (first column), manual segmentation of the sulci by an expert (second column) and segmentation obtained using the dictionaries from Figure 5 of patch size 13×13 (third column) and 19×19 (forth column). Note that the manual markings from the first three rows were used during the dictionary learning but the markings from the other rows were not

used in the dictionary learning process, so it comes as no surprise that the obtained segmentations are close to the manual markings. One observation here is remarkable though. In the image shown in the third row, the physician did not mark the sulci in the lower right part of the image even though he marked that sulci in other frames of the same video sequence, for instance in the frame shown in the second row. Nevertheless, the dictionary based segmentation is able to detect traces of these sulci because underlying information is encoded of more than just the marking of this single image and thus the method can average out conflicting markings.

The remaining four rows show results of the dictionary based segmentation on images that were not used while learning the dictionaries. While the segmentation understandably is not as good on these frames as on the frames used during the learning, the sulci structures are still clearly identified in the forth to sixth row. Here, it is also evident that increasing the patch size from 13×13 to 19×19 has a positive effect on the quality of the segmentation. With the larger patch size the width of the detected sulci is more accurate and there are less artifacts in the non-sulci regions. Finally, in the last row, the limits of the dictionary based segmentation approach become visible. The cortex region in this frame differs too much from the cortex regions in the learning frames and thus the sulci are not properly detected here.

6 Conclusion

We have studied the applicability of discriminative dictionaries to segment the geometry of sulci in intra-operative digital photographs of the human cortex. It turned out that human cortex images manually marked by an experienced physician can be used to learn discriminative dictionary pairs that allow for a robust segmentation of these brain structures on photos of the cortex as long as the training data contains sufficiently similar images.

Furthermore, we have presented an improved minimization strategy for the discriminative dictionary functional of Mairal et al. [10] that is more efficient by leveraging recent advances in orthogonal matching pursuit and more stable due to a new dictionary update step that ensures an energy decay.

References

1. Aharon, M., Elad, M., Bruckstein, A.: K-SVD: An algorithm for designing overcomplete dictionaries for sparse representation. IEEE Transactions on Signal Processing 54(11), 4311–4322 (2006)
2. Allen, J.B.: Short term spectral analysis, synthesis, and modification by discrete fourier transform. IEEE Transactions on Acoustics, Speech and Signal Processing ASSP-25(3), 235–238 (1977)
3. Armijo, L.: Minimization of functions having Lipschitz continuous first partial derivatives. Pacific Journal of Mathematics 16(1), 1–3 (1966)

4. Berkels, B.: An unconstrained multiphase thresholding approach for image segmentation. In: Tai, X.-C., Mørken, K., Lysaker, M., Lie, K.-A. (eds.) SSVM 2009. LNCS, vol. 5567, pp. 26–37. Springer, Heidelberg (2009)
5. Bertsekas, D.P.: Nonlinear Programming, 2nd edn. Athena Scientific, Belmont (1999)
6. Candès, E.J., Donoho, D.L.: Curvelets – a surprisingly effective nonadaptive representation for objects with edges. In: Schumaker, L.L., et al. (eds.) Curves and Surfaces. Vanderbilt University Press, Nashville (1999)
7. Chambolle, A., Pock, T.: A first-order primal-dual algorithm for convex problems with applications to imaging. Tech. Rep. 685, Ecole Polytechnique, Centre de Mathématiques appliquées, UMR CNRS 7641, 91128 Palaiseau Cedex (France) (May 2010)
8. Do, M.N., Vetterli, M.: The contourlet transform: an efficient directional multiresolution image representation. IEEE Transactions on Image Processing 14(12), 2091–2106 (2005)
9. Engan, K., Aase, S.O., Husøy, J.H.: Frame based signal compression using method of optimal directions (MOD). In: Proceedings of the IEEE International Symposium on Circuits and Systems (ISCAS 1999), vol. 4, pp. 1–4 (1999)
10. Mairal, J., Bach, F., Ponce, J., Sapiro, G., Zisserman, A.: Discriminative learned dictionaries for local image analysis. In: IEEE Computer Society Conference on Computer Vision and Pattern Recognition (CVPR), pp. 1–8 (2008)
11. Mairal, J., Leordeanu, M., Bach, F., Hebert, M., Ponce, J.: Discriminative sparse image models for class-specific edge detection and image interpretation. In: Forsyth, D., Torr, P., Zisserman, A. (eds.) ECCV 2008, Part III. LNCS, vol. 5304, pp. 43–56. Springer, Heidelberg (2008)
12. Mallat, S.: A wavelet tour of signal processing. Academic Press, London (1999)
13. Rubinstein, R., Zibulevsky, M., Elad, M.: Efficient implementation of the K-SVD algorithm using batch orthogonal matching pursuit. Tech. rep., CS Technion (April 2008)
14. Zhang, Q., Li, B.: Discriminative K-SVD for dictionary learning in face recognition. In: IEEE Conference on Computer Vision and Pattern Recognition (CVPR), pp. 2691–2698 (2010)
15. Zhao, M., Li, S., Kwok, J.: Text detection in images using sparse representation with discriminative dictionaries. Image and Vision Computing 28(12), 1590–1599 (2010)

An Efficient and Effective Tool
for Image Segmentation, Total Variations
and Regularization

Dorit S. Hochbaum[*]

Department of Industrial Engineering and Operations Research, University of
California, Berkeley
hochbaum@ieor.berkeley.edu

Abstract. One of the classical optimization models for image segmentation is the well known Markov Random Fields (MRF) model. MRF
formulates many total variation and other optimization criteria used in
image segmentation. In spite of the presence of MRF in the literature,
the dominant perception has been that the model is not effective for image segmentation. We show here that the reason for the non-effectiveness
is not due to the power of the model. Rather it is due to the lack of access to the optimal solution. Instead of solving optimally, heuristics have
been engaged. Those heuristic methods cannot guarantee the quality of
the solution nor the running time of the algorithm.

We describe here an implementation of a very efficient polynomial
time algorithm, which is provably fastest possible, delivering the optimal
solution to the MRF problem, Hochbaum (2001). It is demonstrated
here that many continuous models, common in image segmentation, have
a discrete analogs to various special cases of MRF. As such they are
solved optimally and efficiently, rather than with the use of continuous
techniques such as PDE methods that can only guarantee convergence
to a local minimum.

The MRF algorithm is enhanced here demonstrating that the set of
labels can be any discrete set. Other enhancements include dynamic features that permit adjustments to the input parameters and solves optimally for these changes with minimal computation time. Modifications
in the set of labels (colors), for instance, are executed instantaneously.
Several theoretical results on the properties of the algorithm are proved
here and are demonstrated for examples in the context of medical and
biological imaging.

1 Introduction

Partitioning and grouping of similar objects plays a fundamental role in image
segmentation and in clustering problems. In such problems the goals are to group
together similar objects, or pixels in the case of image processing. Given an input
image, the objective of image segmentation is to recognize the salient features

---

[*] Research supported in part by NSF award No. DMI-0620677 and CBET-0736232.

A.M. Bruckstein et al. (Eds.): SSVM 2011, LNCS 6667, pp. 338–349, 2012.

in the image. Each feature set is grouped together in one segment represented by some uniform color area.

A noisy or corrupted image is characterized by lacking uniform color areas, which are assumed to characterize a true image. Rather, in such image there are adjacent pixels of different color areas. To achieve higher degree of uniform color areas, it is reasonable to assign a penalty to neighboring pixels that have different colors associated with them. On the other hand, the purpose of the segmentation is to represent the "true" image. For that purpose the given assignment of colors in the input image is considered to be the "priors" on the colors of the pixels, and as such, the best estimate available on their true labels. Therefore, any change in those priors is assigned a penalty for deviating from the priors.

The Markov Random Fields problem for image segmentation is to assign colors to the pixels so that the total penalty is minimized. The penalty consists of two terms. One is the *separation* penalty, or *smoothing* term, and the second is the *deviation* penalty, or *fidelity* term. For this reason we refer to this penalty minimization problem also as the *separation-deviation* problem. This problem has been extensively studied over the past two decades, see e.g. [3], [5], [11], [12], [16], [17]. The problem formulation, described in full detail in Section 3 is

$$\text{(MRF)} \quad \min \quad \sum_{i \in V} G_i(x_i) + \sum_{i \in P} \sum_{j \in N(i)} F_{ij}(x_i - x_j)$$
$$\text{subject to} \quad x_i \in X \ \ \forall \ i \in P.$$

It is noted that the concept of "colors" associated with pixels can be replaced by any other scalar characterization of pixels or voxels, such as texture. We refer here to colors as a representation of such characterizations.

The complexity of MRF depends on the form of the penalty functions. A full classification of the problem's complexity is given in [15] showing that for convex penalty functions the problem is polynomially solvable, and for non-convex the problem is NP-hard. The cases when the deviation penalty functions are convex and the separation penalty functions are linear was shown by Hochbaum [15] to be solvable in polynomial time using a parametric cut procedure. Furthermore, it was shown that the complexity of the algorithm is the fastest possible. The case when both separation and deviation penalty functions are convex were also shown to be solvable very efficiently by [1,2]. For non-convex penalty functions the MRF problem is NP-hard.

Problems of total variations and regularization have been utilized in image analysis for the purpose of *denoising* an image. These employ continuous methodologies. Recent works that provide approximate methods for solving MRF utilize convex relaxations (e.g. Pock et al. [21]) along with primal-dual approaches, may not converge to an optimal solution, and the running time cannot be determined in advance. This is surprising, given that the exact discrete problem can be solved within guaranteed polynomial time complexity. Moreover, digital images are inherently discrete, and considering them as continuous causes loss of accuracy. The output of a continuous method must be mapped back to digital image information, entailing further loss of accuracy. We demonstrate that several classical continuous models are better represented with the MRF model and thus benefit from the algorithmic efficiency of solving it.

2   Relationship to Continuous Models

In the total variation method [19,23] the recorded image is represented by the function which maps each pixel to its label (color). It is assumed that u_0 can be decomposed as $u_0 = u + v$ where u contains homogeneous regions with sharp prominent edges, and v contains additional texture and noise. The goal of the total variation method is to find u by minimizing the functional

$$\int_\Omega |\nabla u| dx dy + \alpha ||u - u_0||.$$

This functional is define on the plane, where (x, y) designate the position of each pixel in the image.

Although not immediately apparent, there is a connection between this problem and the MRF problem: The term $|\nabla u|$ captures the difference between each pixel and its neighborhood. The neighborhood can be set to any desirable set – it is not restricted to the commonly used grid neighborhood. This gradient term is thus the *separation* term. The second term $\alpha ||u - u_0||$ is the deviation of the mapped function u from the recorded image u_0.

This total variation problem is solved by continuous techniques. One such method solves the associated Euler-Lagrange equation

$$u = u_0 + \frac{1}{2\alpha} \nabla \cdot \left(\frac{\nabla u}{||\nabla u||}\right).$$

In contrast to MRF, this method does not guarantee to deliver an optimal solution and its complexity is undetermined. For this problem MRF does deliver an optimal solution to this problem, and in polynomial time.

In a more general set-up, the total variation regularization problem (TVR) the image is represented as $s(x)$ – a given function define on an open subset Ω, and $f(x)$ is its *regularized* version, or for images, it is called the *denoising* of s. We define two real functions $\gamma : R \to [0, \infty)$ and $\beta : R \to [0, \infty)$ which assume the value 0 for the argument of 0,

$$F(f) = \int_\Omega \gamma(f(x) - s(x)) dx$$

In the denoising literature F is called a *fidelity* term since it measures deviation from $s()$ which could be a noisy grayscale image. In our terminology, the fidelity term is the *deviation*.

A second function is the total variations on f, $TV(f)$: The discrete form of the total variations function is represented as a function f on a grid of discrete values in Ω and associated with a defined *neighborhood* of each grid point. Let the set of neighboring pairs be denoted by E. Then the total variation of f is $\sum_{[i,j] \in E} \beta(f(i) - f(j))$ for a function β often selected as the absolute value function: $\beta(x) = \max\{0, x\}$. For a constant α the *total variation regularization* of $s()$ is the function f that minimizes the weighted combination of the total variations and fidelity of f:

$$\min TV(f) + \alpha F(f)$$

Rudin, Osher and Fatemi [23] have studied TVRs of F where $\gamma(y) = y^2$, and Chan and Esedoglu [7] studied $\gamma(y) = |y|$.

Since MRF is solved in polynomial time for convex γ and convex β, consequently, the problem of Chan and Esedoglu is a special case solved by parametric cut, and the problem of Rudin et al. is a special case solved by the quadratic convex dual of min cost network flow. Both cases are efficiently solvable and the MRF algorithm guarantees an optimal solution in polynomial time.

The MRF problem can also be used to represent certain classes of the Mumford-Shah problem, as well as several image analysis problems that are addressed with the eigenvector technique. The details of these mappings are to be described in the full version of this paper.

3   The Methodology

For the MRF model for the image segmentation problem the input is an image constituting of a set of pixels each with a given color and a neighborhood relation between pairs of pixels. The decision is to assign each pixel a color assignment, that may be different from the given color of the pixel, so that neighboring pixels will tend to have the same color assignment. The aim is to modify the given color values as little as possible while penalizing changes in color between neighboring pixels. The penalty function thus has two components: the deviation cost that accounts for modifying the color assignment of each pixel, and the separation cost that penalizes the extent of pairwise discontinuities in color assignment for each pair of neighboring pixels.

Formally, we are given a graph $G = (V, A)$, or an image which is a set of pixels V, with a real-valued intensity r_i for each pixel $i \in V$. The neighborhood of pixel i, which contains pixels adjacent to i, is denoted by $N(i)$. The set of pairs of nodes and their neighbors is denoted by A. So $A = \{(i,j)|j \in N(i)\}$. Note that for every pair of neighbors $\{i, j\}$ the graph G contains two arcs $(i,j).(j,i) \in A$. We wish to assign each pixel $i \in V$ an intensity x_i that belongs to a discrete finite set $X = \{i_1, i_2, \ldots, i_k\}$ so that the sum over all pixels of the *deviation* cost $G_i(\cdot)$ and the *separation* cost $F_{ij}(\cdot)$ is minimized. Note that the values of x_i do not have to be selected from the same set as the value of r_i, as shown here for the first time in Lemma 1. The deviation function depends on the deviation of the assigned color from the given intensity $G_i(x_i - r_i)$. The separation is a function of the difference in assigned intensities between adjacent pixels $F_{ij}(x_i - x_j)$. The problem is stated as follows.

$$\begin{aligned} \min \quad & \textstyle\sum_{i \in V} G_i(x_i) + \sum_{i \in V} \sum_{j \in N(i)} F_{ij}(x_i - x_j) \\ \text{subject to} \quad & x_i \in X \ \ \forall \ i \in V. \end{aligned}$$

We refer to the special case of the MRF problem with each variable x_i taking an integer value in an interval $[\ell_i, u_i]$ as the separation-deviation problem.

The separation-deviation problem was shown in [1,2,15] to be solvable in polynomial time when the functions $G_i(\cdot)$ and $F_{ij}(\cdot)$ are convex. Note that when those functions are not convex the problem is NP-hard, although when only the functions $G_i(\cdot)$ are nonlinear, and $F_{ij}(\cdot)$ are convex the problem is solved in pseudopolynomial time with run time that depends on the number of values of X, k, [1]. The important case we will focus on here is with $G_i(\cdot)$ convex and $F_{ij}(x_i - x_j)$ *bi-linear* forming a two piecewise linear function which is linear in the range $x_i \geq x_j$ and linear in the range $x_j \geq x_i$. For constants u_{ij}, u_{ji} the function is defined as:

$$F_{ij}(x_i - x_j) = \begin{cases} u_{ij} & \text{if } x_i > x_j \\ 0 & \text{if } x_i = x_j \\ u_{ji} & \text{if } x_i < x_j. \end{cases}$$

For convex functions $G_i(\cdot)$ and bi-linear functions $F_{ij}(\cdot)$, the formulation is equivalent to the following constrained optimization problem, referred to as (SD) (standing for Separation-Deviation):

$$\begin{aligned} \text{(SD) min} \quad & \sum_{j \in V} G_j(r_j, x_j) + \sum_{(i,j) \in A} F_{ij}(z_{ij}) \\ \text{subject to} \quad & x_i - x_j \leq z_{ij} \quad \text{for } (i,j) \in A \\ & u_j \geq x_j \geq \ell_j \quad j = 1, \ldots, n \\ & z_{ij} \geq 0 \quad (i,j) \in A. \end{aligned}$$

The complexity of this problem was shown in [15] to be $O(T(n,m) + n \log U)$ where $T(n,m)$ is the complexity of solving the minimum cut problem on a graph with n nodes and m arcs and U is the length of the interval for the color values – the number of possible labels – or as we show here, $|X|$. For the formulation above $U = \max_j \{u_j - \ell_j\}$. The second complexity term is required to find the minima of convex functions. In all our implementations the convex functions are piecewise linear (e.g. absolute value function) or quadratic. In those cases the second term vanishes and the complexity of the procedure is $T(n,m)$. The algorithm used solves the (SD) problem for any size of color set as a parametric minimum cut problem, in the complexity of a single minimum cut procedure. The algorithm used to solve the parametric minimum cut problem is the pseudoflow algorithm of [14], for which the software is available to download at [8]. The complexity of this algorithm was shown recently in [13] to be $T(n,m) = O(mn \log \frac{n^2}{m})$.

We show next that the algorithm solving (SD) extends to the MRF problem with $x_i \in X$ for any set of discrete values X. We first review the algorithm of [15] and then prove, in Lemma 1, that it extends to the MRF problem with an arbitrary discrete set of feasible values.

We define an s,t-graph $G_\alpha = (V_{st}, A_{st})$ from the adjacency graph of the image (V, A) where V is the set of pixels and A the set of adjacency arcs. For $\ell = \min_j \ell_j$ and $u = \max_j u_j$, we choose a parameter value $\alpha \in (\ell, u)$. For each arc (i,j) the arc capacity is u_{ij}.

We add to the set of nodes V a source s and sink t, $V_{st} = V \cup \{s, t\}$. Next let $G_i'(\alpha)$ be the subgradient of $G_i()$ at α, $G_i(\alpha) - G_i(\alpha - 1)$. Let the subgradient value of function $G_i(x)$ to be equal to M at values of $x > u_i$, and to $-M$ for values $x < \ell_i$, for M a suitably large value. With this extension the box constraints are uniform for all variables, $u \geq x_j \geq \ell$ we replace the weights of the nodes and set, for each node $v \in V$, by an arc adjacent to the source of capacity $c_{sv} = \max\{0, G_v'(\alpha)\}$, and an arc adjacent to the sink t of capacity

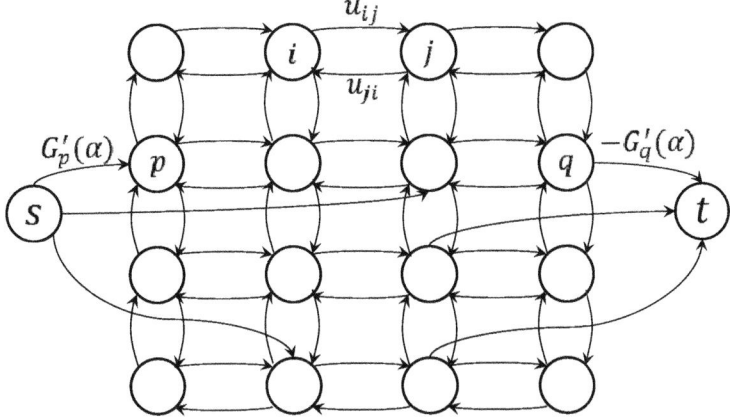

Fig. 1. The graph G_α

$c_{vt} = \max\{0, -G'_v(\alpha)\}$. Let the set of arcs of positive capacity adjacent to the source be denoted by A_s, and the set of arcs of positive capacity adjacent to the sink, A_t. The remainder of the arcs, for each arc (i, j), $j \in N(i)$ have capacities u_{ij}. Let the minimum cut $(\{s\} \cup S, \bar{S} \cup \{t\})$ in the graph G_α partition V to $S = S_\alpha$ and $V \setminus S = \bar{S}_\alpha$. The graph G_α is illustrated in Fig. 1 for an example of a grid graph (V, A) describing the adjacencies. Note however that the algorithm described works for any type of graph, rather than for grid graphs only.

Let the optimal solution to (SD) be $\mathbf{x}^* = (x^*_j)$. The key to the efficient algorithm to the (SD) problem is the threshold theorem:

Theorem 1 (The threshold theorem [15]). *The optimal solution $\mathbf{x}^*$ to (SD) satisfies $x^*_j < \alpha$ for all $j \in S_\alpha$, and $x^*_j \geq \alpha$ for all $j \in \bar{S}_\alpha$.*

The threshold theorem means that for each node we can determine whether the corresponding variable's value in an optimal solution is $< \alpha$ or $\geq \alpha$, depending on whether the respective node belongs to the source or the sink set of the cut. See Fig. 2 for illustration.

By solving for each value of α in the range, the threshold theorem can be used to establish a partition of the nodes in the graph, and the corresponding variables, to sets where in each set all variables get the same value (and same color) in an optimal solution.

Instead of solving for each value of α we find all the *breakpoints* where the cut set is changing. Let S_{λ_q} be the *minimal* source set obtained by solving the minimum cut problem in the graph corresponding to parameter λ_q. Then, for a sequence of monotone increasing values of the parameter, $\lambda_0 < \lambda_1 < \lambda_2 \ldots < \lambda_p$, we get a *nested* collection of source sets of the respective minimum cuts: $\{s\} = S_{\lambda_1} \subset S_{\lambda_2} \subset \ldots \subset S_{\lambda_p} \subset V$. See Fig 3 for illustration. When $\lambda_0 \leq \ell$ then the set of nodes of value $< \lambda_0$ is empty. For $\lambda_p \geq u$ the set of nodes

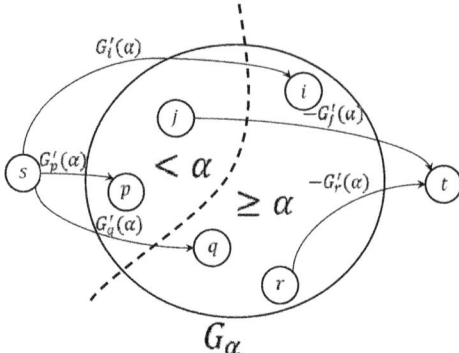

Fig. 2. The threshold theorem: The dashed line represents the arcs of the cut

of value $< \lambda_p$ is V. Therefore, in the optimal corrected image, all pixels in $S_q = S_{\lambda_q} \setminus S_{\lambda_{q-1}}, q = 2, 3, \ldots, p$ have intensity strictly less than λ_q and greater or equal than λ_{q-1}.

Notice that it is sufficient to generate the values of the breakpoints as integers. That is because the values of the variables determined in each set of the partition can take only integer values, so the smallest integer value in the interval $[\lambda_{q-1}, \lambda_q)$ will be the value assigned to all nodes/variables in the set S_q. Hence the values of the breakpoints λ_i do not need to be contained in the set X. However, we will let the set X consist of labels that are integer values.

Since the source set does not change for any $\alpha \in [\lambda_{q-1}, \lambda_q)$, we conclude that for all $j \in S_q$ x_j^* is equal to the smallest value in X that is $\geq \lambda_{q-1}$.

Consider the extension of (SD) to (SD'):

$$\text{(SD') min} \quad \sum_{i \in V} G_i(x_i) + \sum_{i \in V} \sum_{j \in N(i)} F_{ij}(x_i - x_j)$$
$$\text{subject to} \quad x_i \in X \quad \forall i \in V.$$

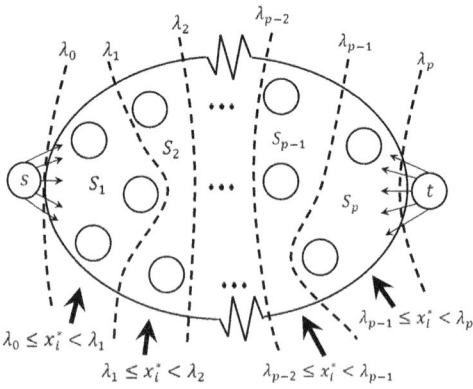

Fig. 3. The parametric cut

Lemma 1. *Given the set of integer breakpoints* $\lambda_0 < \lambda_1 < \lambda_2 \ldots < \lambda_p$, *the optimal solution to (SD') restricted to any set of colors is generated in linear time.*

Proof: The proof is constructive. Let $X = \{i_1, \ldots, i_k\}$. Let V be the set of all the pixels/nodes.

For i_1 let λ_{ℓ_1} be the largest breakpoint smaller or equal to i_1, $\lambda_{\ell_1} = \arg\max \lambda_{\ell_j} \leq i_1$. Assign to all variables with nodes in $S_1 \cup \ldots \cup S_{\ell_1}$ the value i_1. Update $V \leftarrow V \setminus \{S_1 \cup \ldots \cup S_{\ell_1}\}$. Let i_ℓ be the largest value in X less than λ_{ℓ_1+1}. Update $X \leftarrow X \setminus \{i_1, \ldots, i_\ell\}$.

The following iterative step is repeated until all variables values have been assigned and $V = \emptyset$.

Iterative step:

Let i_q be the first (smallest) value in X. Then $i_q \geq \lambda_{\ell_1+1}$. Let λ_{ℓ_q} be the largest breakpoint smaller or equal to i_q, $\lambda_{\ell_q} = \arg\max \lambda_{\ell_j} \leq i_q$. Assign to all variables with nodes in $S_q \cup \ldots \cup S_{\ell_q}$ the value i_q. Update $V \leftarrow V \setminus \{S_q \cup \ldots \cup S_{\ell_q}\}$. Let i_ℓ be the largest value in X less than λ_{ℓ_q+1}. Update $X \leftarrow X \setminus \{i_q, \ldots, i_\ell\}$.

The correctness of the procedure follows from the threshold theorem. $\qquad\square$

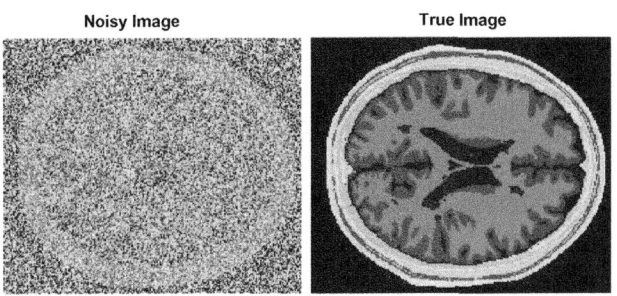

Fig. 4. Brain image 1, true and noisy

4 Experimental Results

4.1 Denoising by Modifying the Ratio between the Separation and Deviation Penalties

The implementation solves the MRF problem with parametric coefficients S and D multiplying the respective terms of separation and deviation. Note that only changes in the ratio $\frac{S}{D}$ have an effect on the optimal solution, rather than the actual values of S and D.

$$(\text{MRF}) \min \; D \sum_{j \in V} G_j(r_j - x_j) + S \sum_{(i,j) \in A} u_{ij}|x_i - x_j|$$
$$\text{subject to} \quad x_i \in X \quad \text{for } i \in V \,.$$

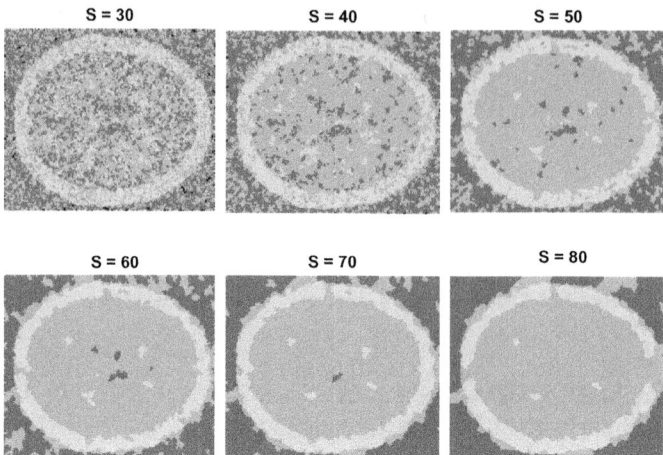

Fig. 5. The output for increasing values of S when applied to noisy brain image 1

The effect of modifying the ratio $\frac{S}{D}$ is illustrated here for two examples of brain images. The first set of true and noisy images are given in Fig. 4. In that image there are four small lesions. We then apply the separation-deviation algorithm with $D = 2$ and for increasing values of S, as shown in Fig. 5. The lesions show very clearly in the high separation (S values of 60 or 70) images in yellow color.

4.2   Increasing Deviation for a Selected Color

The algorithmic tool allows to select a particular color, either by the color code, or by clicking on a pixel that has the desired color. The deviation penalty is then increased for all integer color codes in a small interval around the selected color. For color code q the interval is $[q - 5, q + 5]$. The size of this interval can be adjusted by the user.

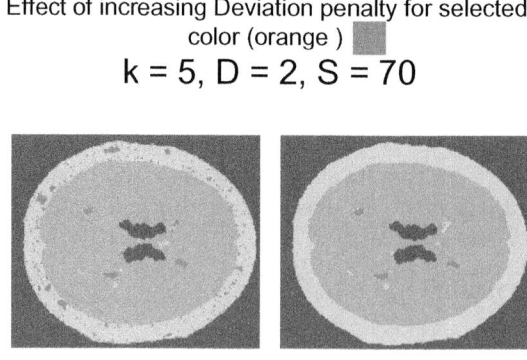

Fig. 6. Increased deviation penalty for a selected color in brain image 1

We show here, for brain image 1, that if the color orange is selected, then it shows as the color of 3 out of the 4 lesions, see Fig. 6 above. When the deviation for that color is increased the lesions become better segmented and more prominent. Of course, the color orange also appears in other areas of the brain shell where it is of no clinical significance. This issue will be addressed in the next prototype of the interactive tool, where the deviation increase will apply only in a user-defined window.

4.3 Comparison of Image Segmentation with Separation-deviation to the Normalized Cut Approach

We now compare our software for image segmentation with the normalized cut approach introduced by Shi and Malik, [25]. This normalized cut approach utilizes the spectral technique in finding the Fielder eigenvalue and the corresponding eigenvector. The method is described and Shi's software implementation is provided in: http://www.cis.upenn.edu/~jshi/software/

The input to that code is the number of desired segments in the output image.

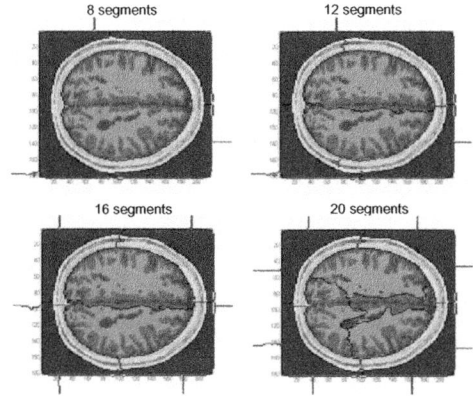

Fig. 7. Normalized cut software segmentation of true brain image 2 for 8, 12, 16 and 20 segments

The code preprocesses the input image, first by converting it to gray scale and then resizing it to 160×160. The algorithm is then applied to the the preprocessed image.

We show here the segmentation of a brain image, brain image 2. This is shown in Fig. 7. Only the 20 segments begins to show the lesion area, but still does not delineate it correctly. This is compared in Fig. 8 to the segmented and traced lesions found with the solution of (SD) applied to the same image. (The software of Shi requires to convert the image first to gray scale, which is why it is not presented in color.)

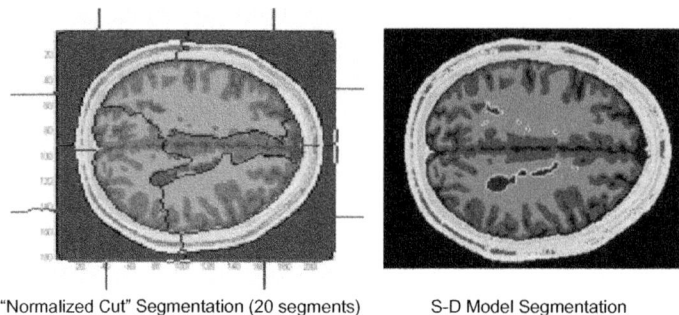

"Normalized Cut" Segmentation (20 segments)   S-D Model Segmentation

Fig. 8. Comparison of the normalized cut software segmentation and the (SD) segmentation of the true brain image 2

5   Conclusions

We demonstrate here that the MRF algorithm is an effective technique for regularization and denoising of images, in theory and in practice. Since the algorithm delivers an optimal solution, and is provably fastest possible, it gives better quality results than any alternative methodology, in terms of minimizing the objective function. The algorithm is shown here to segment successfully the salient features in true images, and to be able to identify hidden important features and de-blur noisy images. These capabilities make the algorithm a useful addition to a segmentation tool box.

References

1. Ahuja, R.K., Hochbaum, D.S., Orlin, J.B.: A cut-based algorithm for the convex dual of the minimum cost network flow problem. Algorithmica 39(3), 189–208 (2004)
2. Ahuja, R.K., Hochbaum, D.S., Orlin, J.B.: Solving the convex cost integer dual network flow problem. Management Science 7, 950–964 (2003)
3. Blake, A., Zisserman, A.: Visual reconstruction. MIT Press, Cambridge (1987)
4. Boykov, Y., Jolly, M.-P.: Interactive graph cuts for optimal boundary & region segmentation of objects in N-D images. In: International Conference on Computer Vision (ICCV), vol. I, pp. 105–112 (2001)
5. Boykov, Y., Veksle, O., Zabih, R.: Markov random fields with efficient approximations. In: Proc IEEE Conference CVPR, Santa Barbara, CA, pp. 648–655 (1998)
6. Boykov, Y., Veksle, O., Zabih, R.: Fast approximate energy minimization via graph cuts. In: Proc 7th IEEE International Conference on Computer Vision, pp. 377–384 (1999)
7. Chan, T.F., Esedoglu, S.: Aspects of total variation regularized l1 function approximation. SIAM J. on Applied Math. 65(5), 1817–1837 (2005)
8. Chandran, B.G., Hochbaum, D.S.: Pseudoflow solver (accessed, January 2007), http://riot.ieor.berkeley.edu/riot/Applications/Pseudoflow/maxflow.html

9. Collins, D.L., Zijdenbos, A.P., Kollokian, V., Sled, J.G., Kabani, N.J., Holmes, C.J., Evans, A.C.: Design and construction of a realistic digital brain phantom. IEEE Transactions on Medical Imaging 17(3), 463–468 (1998)
10. Cox, I.J., Rao, S.B., Zhong, Y.: Ratio regions: A technique for image segmentation. In: Proc. Int. Conf. on Pattern Recognition. B, pp. 557–564 (1996)
11. Geiger, D., Girosi, F.: Parallel and deterministic algorithms for MRFs: surface reconstruction. IEEE Transactions on Pattern Analysis and Machine Interlligence 13, 401–412 (1991)
12. Geman, S., Geman, D.: Stochastic relaxation, Gibbs distributions and the bayesian restoration of images. IEEE Transactions on Pattern Analysis and Machine Intelligence, PAMI 6, 721–741 (1984)
13. Hochbaum D. S., Orlin J. B.: Pseudoflow algorithm in $O(mn \log n^2/m)$ time, UC Berkeley (manuscript) (submitted, 2007)
14. Hochbaum, D.S.: The Pseudoflow algorithm: A new algorithm for the maximum flow problem. Operations Research 4, 992–1009 (2008)
15. Hochbaum, D.S.: An efficient algorithm for image segmentation, Markov random fields and related problems. Journal of the ACM 4, 686–701 (2001)
16. Ishikawa, H., Geiger, D.: Segmentation by grouping junctions. In: IEEE Conference on Computer Vision and Pattern Recognition, CVPR 1998, pp. 125–131 (1998)
17. Li, S.Z., Chan, K.L., Wang, H.: Bayesian image restoration and segmentation by constrained optimization. In: IEEE Computer Society Conference on Computer Vision and Pattern Recognition, CVPR 1996 (1996)
18. Malik, J., Belongie, S., Leung, T., Shi, J.: Contour and texture analysis for image segmentation. Int. J. Comp. Vision 43, 7–27 (2001)
19. Osher, S.J., Fedkiw, R.: Level Set Methods and Dynamic Implicit Surfaces. Springer, New York (2003)
20. Pham, D.L., Xu, C., Prince, J.L.: A survey of current methods in medical image segmentation. Annual Review of Biomedical Engineering 2, 315–337 (2000)
21. Pock, T., Chambolle, A., Cremers, D., Bischof, H.: A convex relaxation approach for computing minimal partitions. In: IEEE Computer Society Conference on Computer Vision and Pattern Recognition, CVPR 2009, pp. 810–817 (2009)
22. Pretorius, P.H., King, M.A., Tsui, B.M.W., LaCroix, K.J., Xia, W.: A mathematical model of motion of the heart for use in generating source and attenuation maps for simulating emission imaging. Med. Phys. 26, 2323–2332 (1999)
23. Rudin, L.I., Osher, S.J., Fatemi, E.: Nonlinear total variation based noise removal algorithms. Phys. D 60, 259–268 (1992)
24. Sarkar, S., Boyer, K.L.: Quantitative measures of change based on feature organization: Eigenvalues and eigenvectors. In: Proc. IEEE Conf. Computer Vision and Pattern Recognition, p. 478 (1996)
25. Shi, J., Malik, J.: Normalized cuts and image segmentation. IEEE Trans. Pattern Anal. Mach. Intell. 22(8), 888–905 (2000)
26. Sharon, E., Galun, M., Sharon, D., Basri, R., Brandt, A.: Hierarchy and adaptivity in segmenting visual scenes. Nature 442, 810–813 (2006)
27. Wang, S., Siskind, J.M.: Image segmentation with ratio cut. IEEE Transactions on Pattern Analysis and Machine Intelligence, PAMI 25(6), 675–690 (2003)
28. Tolliver, D.A., Miller, G.L.: Graph partitioning by spectral rounding: Applications in image segmentation and clustering. In: CVPR 2006, pp. 1053–1060 (2006)

Supervised Scale-Invariant Segmentation (and Detection)

Yan Li, David M.J. Tax, and Marco Loog*

Pattern Recognition Laboratory
Delft University of Technology
The Netherlands
{yan.li,d.m.j.tax,m.loog}@tudelft.nl
http://prlab.tudelft.nl

Abstract. The scale-invariant detection of image structure has been a topic of study within computer vision and image analysis since long. To date, Lindeberg's scale selection method has probably been the most fruitful and successful approach to this problem. It provides a general technique to cope with the detection of structures over scale that can be successfully expressed in terms of Gaussian differential operators. Any detection or segmentation task would potentially benefit from a similar approach to deal with scale. For many of the real-world image structures of interest, however, it will often be impossible to explicitly design or handcraft an operator that is capable of detecting them in a sensitive and specific way. In this paper, we present an approach to the scale-selection problem in which the construction of the detector is driven by supervised learning techniques. The resulting classification method is designed so as to achieve scale-invariance and may be thought of as a supervised version of Lindeberg's classical scheme.

Keywords: Scale selection, scale-invariance, image segmentation, detection, learning and classification.

1   Introduction

Image structures, such as blobs, edges or corners, may appear in images at different scales. To detect them, it is often desired for a detector to select the locally appropriate scales. A well-known scale selection scheme was proposed by Lindeberg [18,17] for image structures which can be detected by differential operators, such as the Laplacian, the Hessian, etc. [27]. The operator under consideration is multiplied with a scale-dependent normalization factor, i.e., it is scale normalized, and applied to an image to get a response at all scales and locations. Subsequently, the scale where the normalized detector attains the maximum response over scales is selected as the local scale of the structure.

* Partly supported by the Innovational Research Incentives Scheme of the Netherlands Research Organization [NWO, VENI Grant 639.021.611]

A.M. Bruckstein et al. (Eds.): SSVM 2011, LNCS 6667, pp. 350–361, 2012.

Scale selection schemes have been at the basis of many successful computer vision and image precessing applications [10,21,22,1,24]. A potential problem, however, is that the schemes are merely applicable to the detection of relatively simple structures. For more complicated or very specific structures, it will often be impossible to explicitly design or handcraft an operator that is capable to detect these. Examples of such structures range from blobs that are textured or have a particular shape to faces, bikes, cars, potted plants, or other image objects.

Next to scale selection, scale-invariance is a desired property in many computer vision and image analysis tasks because an input image can have an arbitrary and unknown inner scale. Informally, the 'inner scale' of a pixel is proportional to the area in the real world that the pixel represents [8]. Employing the proper scale normalization, differential operators in combination with Lindeberg's scale selection are indeed scale-invariant [18]. As with scale selection, many more advanced computer vision techniques rely, at a lower level, on some form of Lindeberg's approach to make the overall scheme scale-invariant as well [21,22,1,24]. Other, more committed, attempts to achieve scale-invariance are to offset scaling with the log-polar and Fourier transform [16,25,15] or to incorporate features from various scales and estimate the local scales of the image under consideration [13,14].

1.1 Work's Novelty and Related Methods

This paper develops a supervised learning approach [4,2] that allows one to construct nontrivial, scale-invariant detection, classification, or segmentation approaches based on available training data in combination with general machine learning and pattern recognition methods. All in all, the approach proposed can be seen as a supervised variation to Lindeberg's classical scheme [18,17].

One critical advantage of our proposal is that learning techniques enable one to develop methods that can potentially handle the more complex structures encountered in real-world segmentation or detection tasks. Like for any supervised learning scheme, in order to apply the technique one needs examples of the task to solve, i.e., a training set. That is, we need to have a collection of raw images, e.g. X-rays, and the desired corresponding output one would like to obtain from them, e.g. an expert segmentation, in order for the learner, e.g. a classifier, to be able to capture the desired relationship. Now, a second advantage is in fact that our approach allows the user to pick the classifier and features of his or her liking. A third critical advantage is that scale selection is made task-dependent by integrating supervision into the selection process. The reason for doing so is that, even when the image data remains the same, different tasks may require different scales to solve them at. Current selection schemes, which are all unsupervised, obviously cannot accommodate this.

Also closely related to our work are face detection schemes that, at test phase, take care of scale and location variations simply by applying the detector to all scales and locations and afterwards finding its maximum responses [12,28]. In a sense, these are supervised approaches that follow Lindeberg's scheme as well. A crucial difference with our approach, however, is in the training of this detector. The face detection techniques need a set of scale and location aligned faces at training phase, which basically takes care of the problem of scale. In many segmentation and detection setting, however, it is difficult to properly align different training instances. Take for instance any medical image segmentation task, how could one identify, even within a single image, the appropriate scale from location to location? Our approach solves the scale selection problem implicitly and does not rely on any a priori knowledge about inner scales in the training or test phase.

The problem covered in the current work has also been discussed in [20], where a supervised method was proposed by viewing classifiers as special types of scale-dependent structure detectors or filters based on which some sort of scale selection could be performed. One of the main shortcomings of this approach, however, is that it is not scale-invariant. A key contributions of this work is to remove this restriction.

1.2 Remark and Outline

To avoid confusion, we use the word structure to mean an image feature, e.g. blobs, edges, or more complicated structures, and the word feature refers to the supervised learning setting where it can mean any kind of measurement that can be made in an image, e.g., Gaussian derivatives, N-jets, differential invariants, texture features, etc.

The remainder of the paper is organized as follows. The next section sets the stage more specifically, it provides some notations used in the paper, and sketches the basics of supervised pixel-based segmentation techniques. Section 3 describes our proposed method. Some illustrative experiments can be found in Section 4. Section 5 concludes the paper.

2 Scale Space Theory and Pixel-Based Segmentation

2.1 Scale Space and Gaussian Derivatives

We will employ linear, or Gaussian, scale space [7,17,27] and limit ourselves to images on $\mathbb{R}^2$, though this limitation is not essential. Given an image $\ell : \mathbb{R}^2 \to \mathbb{R}$, the multi-scale image representation $L : \mathbb{R}^2 \times \mathbb{R}^+ \to \mathbb{R}$ is obtained as a convolution with a Gaussian kernel g_σ for varying scale σ. That is, the scale space representation of ℓ is given by

$$L(x, y; \sigma) = (\ell * g_\sigma)(x, y). \tag{1}$$

The linear scale space representation is mainly used for its Gaussian image derivatives and especially the so-called N-jet [6], which we denote by $J_\sigma^N[\ell]$. The latter is the collection of all Gaussian image derivatives up to order N at a particular scale σ [8,7,27]. N-jets are basic features that are often employed in supervised image analysis techniques to capture the local image structure of interest (see, for instance, [9,19,11]). Also in our experiments, we will use N-jets. The basic theory we present, however, can be used in combination with other features and multi-scale image representations as well as long as scale can properly be dealt with.

2.2 Supervised Pixel Classification

In the test phase, the trained classifier is applied to a new and previously unseen image ℓ_j from which the same feature vectors are extracted. In this way, for every location in ℓ_j, an estimate $\hat{c}_j(x, y)$ of the true class label at (x, y) is obtained by $C[F_j(x, y)]$. Most classifiers can also output an estimated posterior probability $P(c_j(x, y) = k \,|\, F_j(x, y))$ of the true class label $c_j(x, y)$ being equal to k given the feature observed feature vector $F_j(x, y)$ [5] for which we note that

$$C[F_j(x, y)] = \operatorname*{argmax}_{k \in \{1, \ldots, K\}} P(c_j(x, y) = k \,|\, F_j(x, y)). \qquad (2)$$

The posteriors can be viewed as a confidence measure of the classification result and the larger the posterior is, the more confident the classifier is. In this work, we are going to extend the basic pixel-based classification scheme to incorporate scale-invariance by exploiting these posteriors, interpreting them as the output of a complex filter procedure, and apply Lindeberg's idea of maxima selection to it.

(2.3 . . . and Detection)

This work does not explicitly deal with the detection task. We do however want to point out that detection can be formulated in terms of classification (see for example [12,23,26,28]). In our setting, this would, for instance, mean that the desired corresponding outputs that should be provided for the training phase are not necessarily accurate expert segmentations. Instead, for supervised detection it may suffice to label one or a few locations within the structure to be detected with one class label, say *object*, while all other locations are labeled with the label *background*. Strong local maxima among the posterior probabilities $P(c_j(x, y) = object \,|\, F_j(x, y))$ would correspond to a detection of a structure from the *object* class.

3 Supervised Scale-Invariant Segmentation

Our method builds further on standard pixel-based segmentation but is extended so as to take into account scale variations. The idea is to build a classifier that can be applied to all image locations at all feature scales, i.e., instead of considering classification results $C[F_j(x, y)]$, we initially consider its extension to $C[F_j(x, y, \sigma)]$, which provides labels, or for our purpose posteriors $P(c_j(x, y, \sigma) = k \,|\, F_j(x, y, \sigma))$, for the complete scale space of an image ℓ_j.

Ultimately, we are interested in a single overall segmentation and not a segmentation for every scale. Here is where the scale selection comes in. For a particular image location (x, y) in ℓ_j we check over scale which class label receives the highest posterior and assign that label to that location (cf. [20]):

$$\hat{c}_j(x,y) = \operatorname*{argmax}_{k\in\{1,\dots,K\}} \max_{\sigma\in\mathbb{R}^+} P(c_j(x,y,\sigma) = k \mid F_j(x,y,\sigma)). \qquad (3)$$

This approach also solves the scale selection problem in a supervised way. It draws the analogy with Lindeberg's scheme and (implicitly) selects the scale at which the classifier is most confident of its decision, i.e., where classes can be best separated from each other.

The way this classification approach takes into account scale may already be interesting in itself, but we aimed for the segmentation approach to be scale-invariant.

3.1   Additional Remarks

Scale-invariance in the current context means that if we rescale an image ℓ_j with a factor $a > 0$ to an image $\ell_{j'} := \ell_j \circ S_a$ that the corresponding classification result scales in the same way, i.e., $\hat{c}_{j'}(x,y) = \hat{c}_j \circ S_a$. Now this is achieved by relying on scale-invariant features. That is, we generally require that $F_j(x,y,\sigma)$ in the original image ℓ_j equals $F_{j'}(ax, ay, a\sigma)$ in the scaled image $\ell_{j'}$. With this choice of features, corresponding feature vectors are mapped to the same location in feature space and therefore classified in the same way, which results in the desired scale-invariance. In the case of Gaussian derivative features from an N-jet at scale σ, this means for example that every nth order derivative should be normalized by σ^n.

It is copacetic that there are no restrictions on the classification scheme to use. With the choice of scale-invariant features, any choice of classifier results in a scale-invariant segmentation approach and this allows us to employ the full arsenal of machine learning and pattern recognition techniques [4,2].

4   Illustrative Experiments

Our contribution is primarily of a conceptual nature with no need for extensive experimental validation. Nonetheless, we provide some basic, yet nontrivial, illustrations of our scale-invariant segmentation approach as defined through Equation (3). We applied the method to two different tasks. The first one is the segmentation of two simple geometric shapes from the background. The second one comprises a texture segmentation task.

4.1   Classifiers and Features

Before we can apply our segmentation scheme, we need to choose features to describe for every location the relevant local image structure. In basically all of

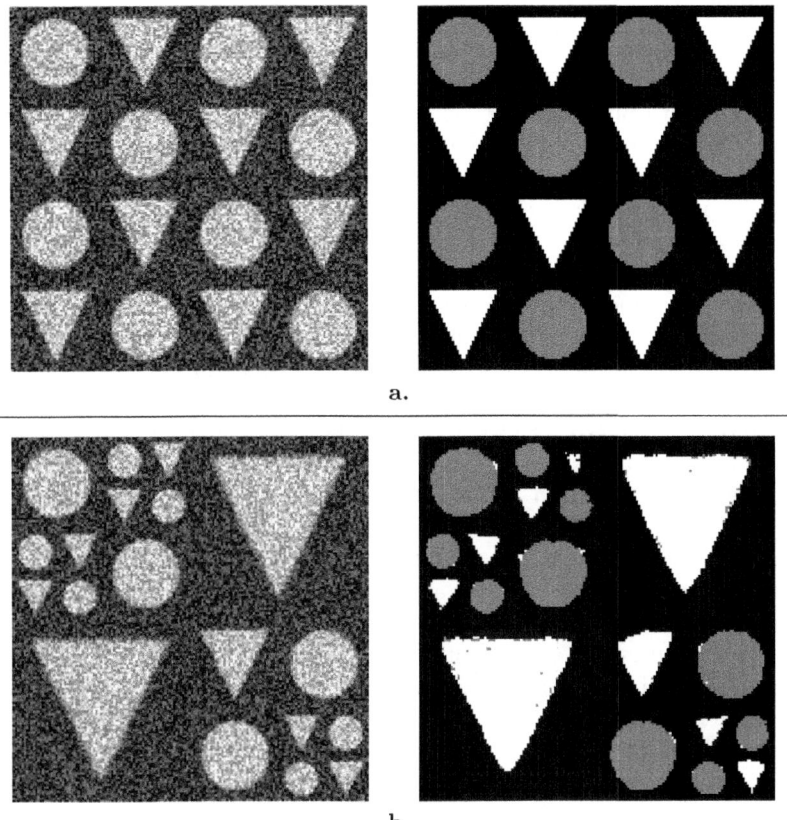

Fig. 1. a. Triangular and circular shapes and their corresponding segmentation used in the training phase. The segmentation is formulated as a three-class pixel classification problem. **b.** An example test image with triangles and circles of different size and the corresponding classification result.

the experiments, we choose the scale-normalized 6-jet, which results in a total of $D = 28$ features for every location. The normalization is depends on the order n of the derivative; every derivative is scaled by σ^n, which makes the features scale-invariant as required.

We also limit ourselves to a relatively straightforward classification technique, namely classical quadratic discriminant analysis (QDA) [4,2]. This classifier makes multivariate normality assumptions about every individual class and based on that constructs a classifier. For every class a feature mean and a class-conditional covariance matrix should be estimated from the training data. This quadratic model is accurate enough for our illustratory purposes and more advanced techniques such as support vector machines, nearest neighbor methods, and boosting approaches provide little extras in the current setting.

4.2 Shapes

Figure 1.a shows on the left examples of noisy triangles and circles, eight each, that should be segmented from the equally noisy background. The right displays the ground truth pixel labeling, which is used as training output. Obviously, a simple blob detector would probably be able to pick out the 16 objects from the input image. It would however be more challenging to design detectors that are more specific and respond merely to one of the two geometric structures. Our scheme therefore also tries to discriminate between the two different shapes and should respond differently to them, i.e., by giving different label outputs. Consequently, we model this problem as a three-class classification problem. The gray-scale in the righthand image of Figure 1.a is of no significance and only indicates that there are indeed three different classes in the image and which pixels belong to which class.

The procedure is tested on the image on the left of Figure 1.b. It also contains scaled versions of the triangular and circular shapes in order to test the scale-invariance of our approach. After extraction the 6-jets from the training data from a range of scales, a QDA is trained and applied to the test image. The resulting segmentation can be found in Figure 1.b on the righthand side. It shows that our procedure is fairly accurate and that the majority of the pixels has been labeled correctly in spite of the relatively straightforward classifier and scale space features. The most notable mistakes seem to be on the small scale triangles. Some of these segmentations are deformed and the one at the top even has been missed almost entirely. The main reasons for these glitches is that the images used are discrete and the 'size' of the added noise does not scale with the shape scale. As a result, scale-invariance will only hold approximately and over a restricted range of scales, which is reflected in somewhat deteriorated performance on the small scale structures.

4.3 Textures

Experiments similar to those in the previous subsection have been performed on two times two Brodatz textures [3]. The two pairs of textures can be found in Figures 2 and 4. For both pairs, we use the two images as the training set and assume them to be from different classes. The corresponding test images, which include both textures from the training set, are displayed in Figures 3.a and 5.a, respectively. All four images are constructed from scaled versions of the original training textures, one of which is in a circular area in the center while the other fills the remainder of the image. The aim is to segment the one texture from the other.

All texture intensities have been normalized to mean zero and unit standard deviation. As a result, a generic blob detector is unable to localize the texturized blobs in the middle of the test images. We really need to employ more rich features that are capable of capturing the relevant higher-order structure and combine these in order to perform the detection or segmentation successfully. This is what QDA, the classifier, does. Figures 3.b and 5.b give the segmentations

Fig. 2. Two training images taken from the Brodatz collection of textures [3]. On the left is D53, the right shows D55.

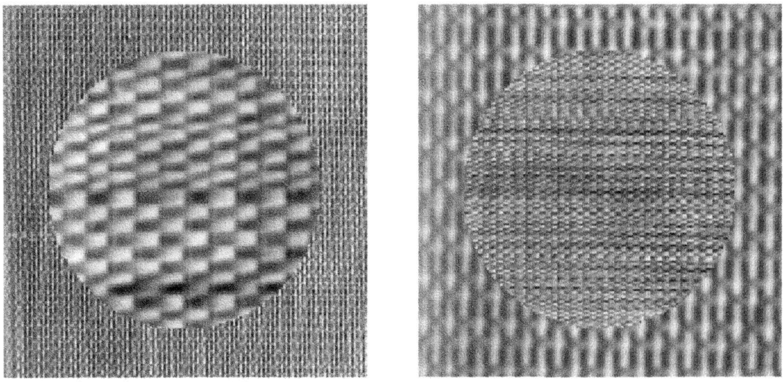

a.

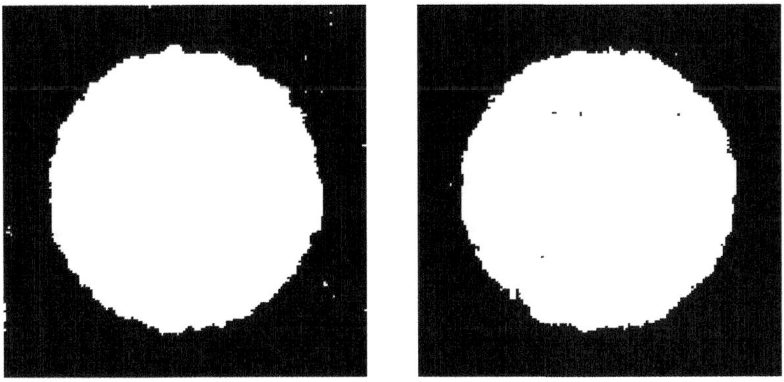

b.

Fig. 3. a. Two example test images in which the texture scales are varied and set differently from those in the training set in Figure 2. Both images contain both textures. **b.** Segmentation results obtained with 6-jets in combination with QDA.

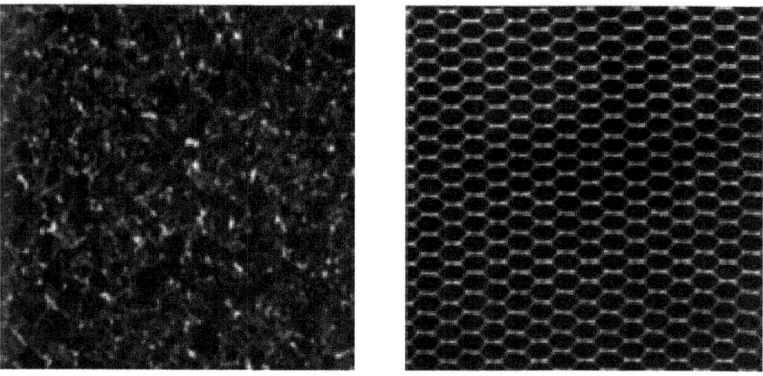

Fig. 4. Compare to Figure 2. Two training images taken from the Brodatz textures collection [3]. On the left is D33, the right displays D34.

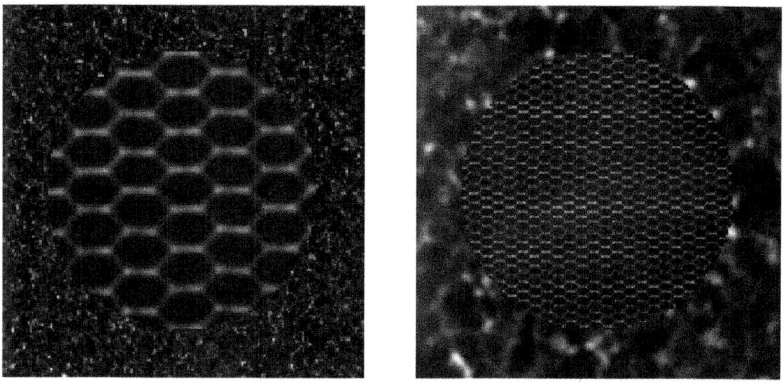

a.

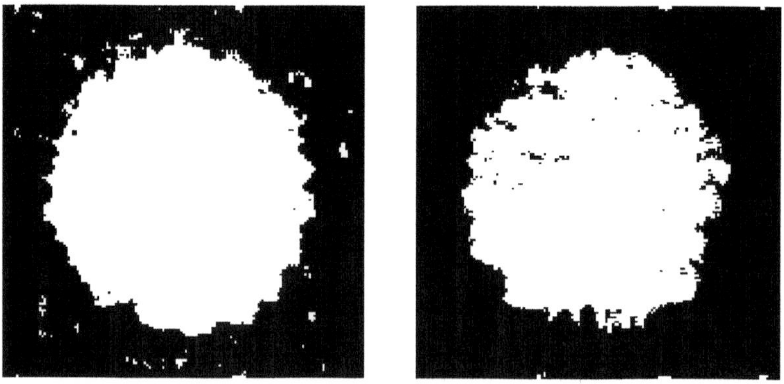

b.

Fig. 5. Compare to Figure 3. **a.** Two example test images in which the texture scales are varied and set differently from those in the training set in Figure 4. Both images contain both training textures. **b.** Segmentation results obtained with 6-jets in combination with QDA.

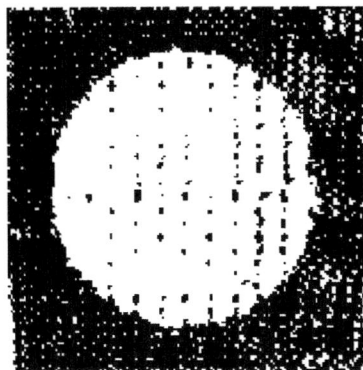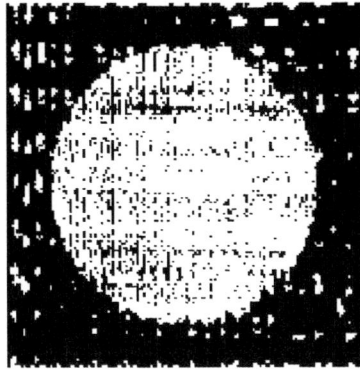

Fig. 6. Segmentation results obtained with 2-jets in combination with QDA. Compare to Figure 3.b.

for the test images in Figures 3.a and 5.a, respectively. As for the results in the previous subsection, similar comments can be made about the reasons for misclassification in these experiments. In the case of these textures, however, there may be two additional reasons at play. First of all, textures are generally more difficult to segment than a shape consisting of a homogenous intensities even though the latter may be noisy. Secondly, the training set does not contain any examples of the two textures bordering, which causes unreliable classification results at such boundaries in the test images. It is indeed at these locations where the segmentation seems most inaccurate.

The first texture segmentation task is probably simpler than the second one. There is a strong difference in orientation between the two textures, which basically sets them apart and one might suspect that a descriptor based on simple second-order, or even first-order, derivatives should be able to capture this difference. Figure 6 shows what happens to the segmentations corresponding to the test images in Figure 3.a if we replace the 6-jets with 2-jets in our procedure. Indeed, to quite a large extent the segmentation is still successful, but the results cannot match the accuracy from those in Figure 3.b, which shows the importance of including higher-order derivatives. Results using the 1-jet are worse even.

5   Discussion and Conclusion

A scale-invariant supervised approach to image segmentation has been presented that draws inspiration from Lindeberg's classical scale selection approach. There are two major advantages compared to other supervised scale-invariant segmentation techniques. Firstly, we are not necessarily committed to specific features that have been designed to achieve invariance in a rather intricate way, as for example in [15]. Our scheme allows the inclusion of any scale-invariant feature set, allowing for very problem specific choices. More important might be the second point, which is the fact that we stay close to the general pixel classification framework and can exploit the full arsenal of powerful pattern recognition and machine learning techniques. In our experiments, we only scratched the surface of possible techniques. They nonetheless show the potential of the approach.

Possibly the most restricting feature of our method is that it is supervised, so we do need training data in order for our approach to work. As a general rule, we may expect to need more complex features, more complex classifiers, and a larger number of examples, with an increasingly complex segmentation problem that we want to tackle. The interplay of these aspects of learning are at the core of general pattern recognition and machine learning research. It is however interesting to study these aspects within the more confined context of image segmentation and detection as this may lead to stronger, more generally applicable guidelines to come to the selection of the right classifier, the right features, etc.

One specific topic for further research we want to mention here concerns Equation (3) and in particular the maximum operator over all scales, which basically picks out a single scale and allows for a close link with Lindeberg's scale-selection scheme. The question remains however if we can do better. An answer might be found in the analysis of the deep structure of the probabilistic posterior scale space (cf. [7,17,27]).

References

1. Bay, H., Tuytelaars, T., Van Gool, L.: SURF: Speeded up robust features. In: Leonardis, A., Bischof, H., Pinz, A. (eds.) ECCV 2006. LNCS, vol. 3951, pp. 404–417. Springer, Heidelberg (2006)
2. Bishop, C.: Pattern recognition and machine learning. Springer, Heidelberg (2006)
3. Brodatz, P.: Textures: A Photographic Album for Artists & Designers. Dover, New York (1966)
4. Duda, R., Hart, P., Stork, D.: Pattern classification, vol. 2. Wiley, Chichester (2001)
5. Duin, R., Tax, D.: Classifier conditional posterior probabilities. In: Advances in Pattern Recognition, pp. 611–619 (1998)
6. Florack, L., Ter Haar Romeny, B., Viergever, M., Koenderink, J.: The Gaussian scale-space paradigm and the multiscale local jet. International Journal of Computer Vision 18(1), 61–75 (1996)
7. Florack, L.: Image Structure. Kluwer Academic Publishers, Dordrecht (1997)
8. Florack, L., ter Haar Romeny, B., Koenderink, J., Viergever, M.: Scale and the differential structure of images. Image and Vision Computing 10(6), 376–388 (1992)
9. Folkesson, J., et al.: Segmenting articular cartilage automatically using a voxel classification approach. IEEE Trans. on Medical Imaging 26(1), 106–115 (2007)
10. Frangi, A.F., Niessen, W.J., Vincken, K.L., Viergever, M.A.: Multiscale vessel enhancement filtering. In: Wells, W.M., Colchester, A.C.F., Delp, S.L. (eds.) MICCAI 1998. LNCS, vol. 1496, pp. 130–137. Springer, Heidelberg (1998)
11. van Ginneken, B., Stegmann, M., Loog, M.: Segmentation of anatomical structures in chest radiographs using supervised methods: a comparative study on a public database. Medical Image Analysis 10(1), 19–40 (2006)
12. Hjelmås, E., Low, B.: Face detection: A survey. Computer Vision and Image Understanding 83(3), 236–274 (2001)

13. Janssen, J., et al.: Scale-invariant segmentation of dynamic contrast-enhanced perfusion MR images with inherent scale selection. J. Visualization and Computer Animation 13(1), 1–19 (2002)
14. Kang, Y., Morooka, K., Nagahashi, H.: Scale invariant texture analysis using multi-scale local autocorrelation features. In: Kimmel, R., Sochen, N.A., Weickert, J. (eds.) Scale-Space 2005. LNCS, vol. 3459, pp. 363–373. Springer, Heidelberg (2005)
15. Kokkinos, I., Yuille, A.: Scale invariance without scale selection. In: IEEE Conf. on Computer Vision and Pattern Recognition, pp. 1–8. IEEE, Los Alamitos (2008)
16. Leung, M., Peterson, A.: Scale and rotation invariant texture classification. In: The 26th Asilomar Conference on Signals, Systems and Computers, pp. 461–465 (1992)
17. Lindeberg, T.: Scale-Space Theory in Computer Vision. Kluwer Academic, Dordrecht (1994)
18. Lindeberg, T.: Feature detection with automatic scale selection. Int. J. of Computer Vision 30(2), 79–116 (1998)
19. Loog, M., Ginneken, B.: Segmentation of the posterior ribs in chest radiographs using iterated contextual pixel classification. IEEE Trans. on Medical Imaging 25(5), 602–611 (2006)
20. Loog, M., Li, Y., Tax, D.M.J.: Maximum Membership Scale Selection. In: Benediktsson, J.A., Kittler, J., Roli, F. (eds.) MCS 2009. LNCS, vol. 5519, pp. 468–477. Springer, Heidelberg (2009)
21. Lowe, D.G.: Distinctive image features from scale-invariant keypoints. Int. J. Comput. Vision 60, 91–110 (2004)
22. Mikolajczyk, K., Schmid, C.: Scale & affine invariant interest point detectors. Int. J. Computer Vision 60(1), 63–86 (2004)
23. Papageorgiou, C., Oren, M., Poggio, T.: A general framework for object detection. In: Sixth International Conference on Computer Vision, pp. 555–562. IEEE, Los Alamitos (2002)
24. Platel, B., Kanters, F., Florack, L., Balmachnova, E.: Using multiscale top points in image matching. In: International Conference on Image Processing, ICIP 2004, vol. 1, pp. 389–392. IEEE, Los Alamitos (2005)
25. Pun, C., Lee, M.: Log-polar wavelet energy signatures for rotation and scale invariant texture classification. IEEE Trans. PAMI, 590–603 (2003)
26. Shotton, J., Blake, A., Cipolla, R.: Contour-based learning for object detection. In: Tenth IEEE International Conference on Computer Vision, ICCV 2005, vol. 1, pp. 503–510. IEEE, Los Alamitos (2005)
27. Ter Haar Romeny, B.: Front-End Vision and Multi-Scale Image Analysis. Kluwer Academic, Dordrecht (2002)
28. Yang, M., Kriegman, D., Ahuja, N.: Detecting faces in images: A survey. IEEE Transactions on Pattern Analysis and Machine Intelligence 24(1), 34–58 (2002)

A Geodesic Voting Shape Prior to Constrain the Level Set Evolution for the Segmentation of Tubular Trees*

Youssef Rouchdy and Laurent D. Cohen

CEREMADE, UMR 7534, Université Paris Dauphine,
75775 Paris Cedex 16, France
youssef.rouchdy@gmail.com, cohen@ceremade.dauphine.fr

Abstract. This paper presents a geodesic voting method to segment tree structures, such as retinal or cardiac blood vessels. Many authors have used minimal cost paths, or similarly geodesics relative to a weight potential P, to find a vessel between two end points. Our goal focuses on the use of a set of such geodesic paths for finding a tubular tree structures, using minimal interaction. This work adapts the geodesic voting method that we have introduced for the segmentation of thin tree structures to the segmentation of tubular trees. The original approach of geodesic voting consists in computing geodesics from a set of end points scattered in the image to a given source point. The target structure corresponds to image points with a high geodesic density. Since the potential takes low values on the tree structure, geodesics will locate preferably on this structure and thus the geodesic density should be high. Geodesic voting method gives a good approximation of the localization of the tree branches, but it does not allow to extract the tubular aspect of the tree. Here, we use the geodesic voting method to build a shape prior to constrain the level set evolution in order to segment the boundary of the tubular structure. We show results of the segmentation with this approach on 2D angiogram images and 3D simulated data.

1 Introduction

In this paper we present a novel method for the segmentation of tree structures. These methods are based on minimal paths with a metric designed from the images and can be applied to the segmentation of numerous structures, such as: microglia extensions; neurovascular structures; blood vessel; pulmonary tree. The vascular tree is modeled as a tubular structure. We consider among the methods used to segment the vascular tree three classes of approaches according to the method used to extract the tubular aspect of the tree: surface models; centerline based models; and 4D curve models. The first category extracts directly the surface of the vessel, see [1]. For the second approach, centerlines based models, centerlines are extracted first and a second process is required to segment the

---

* This work was partially supported by ANR grant MESANGE ANR-08-BLAN-0198.

A.M. Bruckstein et al. (Eds.): SSVM 2011, LNCS 6667, pp. 362–373, 2012.

vessel surface, see [2]. The last approach, 4D curve model, consists in segmenting the vessel centerlines and surfaces simultaneously as a path in a (3D+radius) space [3,4]. For a review of these methods, see [5,6].

Minimal paths techniques were extensively used for extraction of tubular tree structures. These approaches are more robust than the region growing methods, particularly in the presence of local perturbations due to the presence of stenosed branches of the tree or imaging artefacts where the image information might be insufficient to guide the growing process. Several minimal path techniques have been proposed to deal with this problem [7,8,9]. These techniques consist in designing a metric from the image in such a way that the tubular structures correspond to geodesic paths according to this metric. Solving the problem from the practical point of view consists of a front propagation from a source point within a vessel which is faster on the branches of the vascular tree. These methods required the definition by the user of a starting point (propagation source) and end points. Each end point allows to extract a branch of the tree as a minimal path from this point to the source point, the points located on the minimal path are very likely located on the vessel of interest. Few works have been devoted to reduce the interaction of the user in the segmentation of tree structure to the initialization of the propagation from a single point. Authors of [10] defined a stopping criteria from a medialness measure, the propagation is stopped when the medialness drops below a given threshold. This method might suffer from the same problem as the region growing, the medialness might drop below the given threshold in the presence of pathology or imaging artefacts. Wink et al. [11] proposed to stop the propagation when the geodesic distance reaches a certain value. However, this method is limited to the segmentation of a single vessel and the definition of the threshold of the geodesic distance is not straightforward. Cohen and Deschamps [12] proposed to stop the propagation following a criterion based on some geometric properties of the region covered by the front. In [9], assuming the the total length of the tree structure to be visited is given, the stopping criteriuon is based on the Euclidean length of the minimal path.

Li et al. [4] proposed a 4D curve model with a key point searching scheme to extract multi-branch tubular structures. The vascular tree is a set of 4D minimal paths, giving 3D centerlines and width. While this method has the advantage to segment vessel centerlines and surfaces simultaneously, it requires the definition of eight parameters. One point inside the tubular structure and the radius are used to initialize the Fast marching propagation, three parameters are used to set the Fast Marching potential and three distance parameters limit the propagation to the inside of the tubular structure to avoid leakage outside the tree. These last three parameters may require an important intervention of the user since they are crucial to extract the whole structure. If these distance parameters are not suitable, parts of the tree structure may be missed during the propagation.

In this paper, we present a method to extract tree structures without using any *a priori* information. Furthermore, the user has to provide only a single point on the tree structure. The method is generic: it can be used to extract any type of tree structure in 2D as well as in 3D. It is based on the geodesic voting method introduced in [13,14]. It consists in computing geodesics from a given source point to a set of end points scattered in the image. The target structure corresponds to image points with a high geodesic density. The geodesic density is defined at each pixel of the image as the number of geodesics that pass over this pixel. Since the potential takes low values on the tree structure, geodesics will locate preferably on this structure and thus the geodesic density should be high on the tree structure. While the original voting method allows to extract tree structures it does not permit to extract the walls of the vessels. Here, we introduce a shape prior constraint constructed from the geodesic voting method to constrain the evolution of a level set active contour in order to extract the walls of the tree. We use a Bayesian approach to introduce this prior into the level set formulation. We end up with a minimization problem of a global energy composed of two terms. The first term corresponds to a deformation energy for a standard region based level set method and the second term introduces the shape prior constraint. In Section 2, we present the tools needed in Section 3 to introduce the new geodesic voting method. In Section 4, we applied our approach to the segmentation of vessels from 2D angiogram images and 3D simulated data.

2   Background

2.1   Minimal Paths

In the context of image segmentation Cohen and Kimmel proposed, in [15], a deformable model to extract contours between two points given by the user. The model is formulated as finding a geodesic for a weighted distance:

$$\min_{y} \int_0^L \big(w + P(y(s))\big)\mathrm{d}s, \tag{1}$$

where s is the arclength, L is the length of the curve and the minimum is considered over all curves $y(s)$ traced on the image domain Ω that link the two end points, that is, $y(0) = x_0$ and $y(L) = x_1$. The constant w imposes regularity on the curve. $P > 0$ is a potential cost function computed from the image, it takes lower values near the edges or the features. For instance $P(y(s)) = I(y(s))$ leads to darker lines while $P(y(s)) = g(||\nabla I||)$ leads to edges, where I is the image and g is a decreasing positive function.

To compute the solution associated to the source x_0 of this problem, [15] proposed a Hamiltonian approach: Find the geodesic weighted distance U that solves the eikonal equation :

$$||\nabla U(x)|| = w + P(x)   \forall x \in \Omega \tag{2}$$

The ray y is subsequently computed by back-propagation from the end point x_1 by solving the Ordinary Differential Equation (ODE): $y'(s) = -\nabla U(y)$. To solve the eikonal equation (2), we use the Fast Marching algorithm introduced in [16]. The idea behind the Fast Marching algorithm is to propagate the wave in only one direction, starting with the smaller values of the action map U and progressing to the larger values using the upwind property of the scheme. Therefore, the Fast Marching method permits to solve the equation (2) in complexity $O(n \log(n))$, where n is the number of grid points, for details see [16,15].

2.2 Geodesic Voting for Segmentation of Tree Structures

We have introduced in [13,14] a new concept to segment a tree structure from only one point given by the user in the tree structure. This method consists in computing the geodesic density from a set of geodesics extracted from the image. Assume you are looking for a tree structure for which a potential cost function has been defined as above and has lower values on this tree structure. First we provide a starting point x_0 roughly at the root of the tree structure and we propagate a front in the whole image with the Fast Marching method, obtaining the minimal action U. Then assume you consider an end point anywhere in the image. Backtracking the minimal path from the end point you will reach the tree structure somewhere and stay on it till the start point is reached. So a part of the minimal path lies on some branches of the tree structure. The idea of this approach is to consider a large number of end points $\{x_k\}_{k=1}^N$ on the image domain, and analyze the set of minimal paths y_k obtained. For this we consider a voting scheme along the minimal paths. When backtracking each path, you add 1 to each pixel you pass over. At the end of this process, pixels on the tree structure will have a high vote since many paths have to pass over it. On the contrary, pixels in the background will generally have a low vote since very few paths will pass over them. The result of this voting scheme is what we can call the geodesic density. This means at each pixel the density of geodesics that pass over this pixel. The tree structure corresponds to the points with high geodesic density. The set of end points for which you consider the geodesics can be defined through different choices. This could be all pixels over the image domain, random points, scattered points according to some criterion, or simply the set of points on the boundary of the image domain, see [14]. We define the voting score or the geodesic density at each pixel p of the image by

$$\mu(p) = \sum_{k=1}^{N} \delta_p(y_k) \tag{3}$$

where the function $\delta_p(y)$ returns 1 if the path y crosses the pixel p, else 0. Once the geodesic voting is made, the tree structure is obtained by a simple thresholding of the geodesic density μ. As shown in Figure 1, the contrast between the background and the tree is large and the threshold can be chosen easily. We used for all experiments the following value

$$Th = \frac{\max(\text{geodesic density})}{100} \tag{4}$$

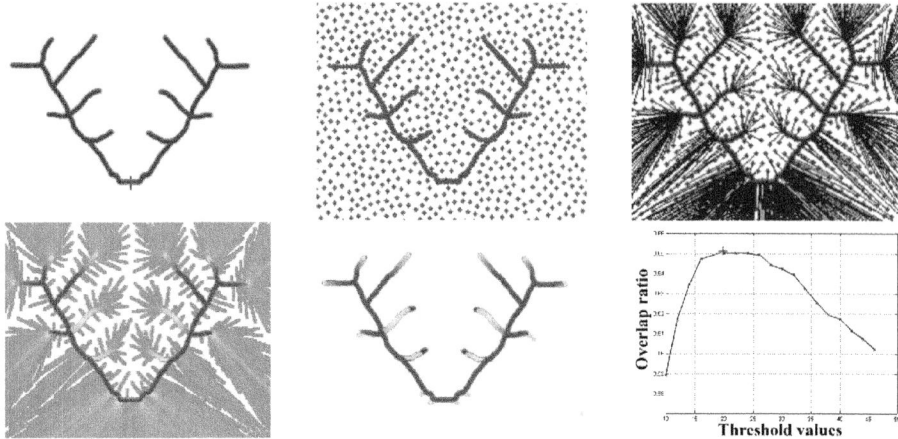

Fig. 1. Geodesic voting. First row: the left panel shows the synthetic tree, the red cross represents the root of the tree; the center panel shows the set of end points (here farthest points, see [14]); the right panel shows in blue the geodesics extracted from the set farthest points to the root. Second row: the left panel shows the geodesic density; the center panel shows the geodesic density after thresholding; the right panel plots the effect of the variation of the threshold on the overlap ratio, the red cross represents the value Th (given by the equation (4)).

as threshold to extract the tree structure using the voting maps. Figure 1 (panel: second row on the right) shows the effect of the threshold on the overlap ratio[1] that measures the similarity between the the manually segmented data A and the segmentation result B. This figure shows that the threshold can be chosen in a large range that contains the threshold Th, given by the equation (4).

2.3   Active Contours without Edges

In this section we describe the level set method that we will use in the next section to introduce our active contour model. The active contour models consist in evolving a curve (2D case) or surface (3D case) constrained by image-based energy toward the target structure. Chan and Vese [17] proposed a region based model adapted to segment an image with poor boundaries (edge information). This model is a piece-wise constant approximation of the Mumford and Shah functional [18]:

$$\mathcal{V}(\phi, c_1, c_2) = \int_{\Omega} \Big(\lambda_1 \big(u_0 - c_1\big)^2 H_\epsilon(\phi) + \lambda_2 (u_0 - c_2)^2 (1 - H_\epsilon(\phi)) + \mu \delta_\epsilon(\phi)|\nabla \phi| + \nu H_\epsilon(\phi)\Big) \mathrm{d}x, \tag{5}$$

---

[1] The overlap ratio is defined by the relation: $O(A, B) = \frac{2|A \cap B|}{|A| + |B|}$, where $|A|$ and $|B|$ are respectively the number of the foreground voxels in the image A and B. $|A \cap B|$ is the number of voxels in the shared regions (intersection of the foreground of the two images)

where ϕ defines the boundary as its zero level set; Ω is the image domain; u_0 is a given image function; λ_1, λ_2, ν, and μ are positive parameters; c_1 and c_2 are two scalar constants used to separate the image into two regions of constant image intensities. The two last terms in the equation introduce regularization constraints, where H_ϵ and δ_ϵ are respectively the regularized Heaviside and Dirac functions, in this work they are approximated by:

$$H_\epsilon(\tau) = \frac{1}{2}\left(1 + \frac{2}{\pi}\text{artang}\left(\frac{\tau}{\epsilon}\right)\right); \quad \delta_\epsilon(\tau) = \frac{1}{\pi}\frac{\epsilon}{\epsilon^2 + \tau^2}. \tag{6}$$

3   From the Voting Tree to the Tubular Tree

While the Chan and Vese energy constraint introduces regularization to smooth the level set funcion ϕ and to deal with noise, it does not introduce a bias towards the target structure. Bayesian models were proposed in the literature to incorporate prior knowledge about the target structure to constrain the evolution of the level set [19]. The first level set method with prior knowledge about shape was introduced by Leventon et al. [19]. Recent improvements of this approach were proposed for example in [20]. The geodesic voting method described in Section 2.2 gives a good approximation of the localization of each branch of the tree.

In this section we introduce a shape prior constraint using a Bayesian framework to segment the walls of the tree structure. The idea is to use the geodesic voting method to construct the shape prior that constrains the evolution of the level set propagation. After thresholding the geodesic density μ defined by the equation (3) we get an approximation of the target tree structure as explained in Section 2.2. However this geodesic density does not allow to extract the tubular aspect of the tree. Indeed the thresholded geodesic density gives only an approximation of the centerlines of the tree structure. Our aim here is to use this rough tree skeleton to build a prior that constrains the evolution of level set active contour in order to extract the boundary of the tree.

From now on we call the voting tree the tree structure obtained after thresholding the geodesic density. To construct the shape prior from the voting tree we use the largest radius of the tubular structure. The largest radius is obtained from the target image. It does not have to be precise: it is sufficient to inspect the target tree visually and to give an approximate value. A uniformly tubular tree containing the target tree structure is obtained by morphological dilation of the voting tree with a radius that corresponds to the largest radius of the tubular tree. The prior that we will use to constrain the level set method corresponds to the signed distance from the boundary $\mathcal{S}$ of the tubular tree obtained after dilatation, which we denote $\tilde{\phi}$. The signed distance $\tilde{\phi}$ is defined by:

$$\tilde{\phi}(x) = \begin{cases} D(x), & \text{if} \quad x \text{ is inside } \mathcal{S}, \\ -D(x), & \text{otherwise}, \end{cases}$$

where D is a distance from S: $D(x) = \inf_{y \in \mathcal{S}} \text{d}(x, \mathcal{S})$ with d a given metric, we use in this work the Euclidean metric. The distance $\tilde{\phi}$ is then used to constrain the level set evolution in the target image. Let $\mathbb{P}(\phi|\tilde{\phi}, u)$ be the posterior probability of the level set ϕ given the image function u and the level set shape prior. The Bayesian formulation of this probability is given by Bayes' theorem:

$$\mathbb{P}(\phi|\tilde{\phi}, u) = \frac{\mathbb{P}(\tilde{\phi}, u|\phi)\,\mathbb{P}(\phi)}{\mathbb{P}(\tilde{\phi}, u)} \propto \mathbb{P}(\tilde{\phi}|\phi)\,\mathbb{P}(u|\phi)\,\mathbb{P}(\phi) \tag{7}$$

where $\mathbb{P}(\tilde{\phi}|\phi)$ is the shape prior term, we suppose that this probability follows a Gaussian distribution and that $\mathbb{P}(u|\phi)\,\mathbb{P}(\phi)$ is derived from the Chan and Vese model, see equation (5). Therefore, the maximum of the posterior probability (7) is equivalent to the lowest energy of the $(-\log)$ functional, and after integration over the image domain we end up with the following Bayesian model:

$$E_b(\phi, c_1, c_2) = \mathcal{V}(\phi, c_1, c_2) + \gamma \int_\Omega \frac{(\phi - \tilde{\phi})^2}{2\sigma^2}\delta_\epsilon(\phi)\mathrm{d}x, \tag{8}$$

the factor term δ_ϵ allows us to restrict the shape prior within the region of interest. For a fixed ϕ, we deduce the values of c_1 and c_2:

$$c_1(\phi) = \frac{\displaystyle\int_\Omega u_0\,H_\epsilon(\phi)\mathrm{d}x}{\displaystyle\int_\Omega H_\epsilon(\phi)\mathrm{d}x}, \quad c_2(\phi) = \frac{\displaystyle\int_\Omega u_0\,(1 - H_\epsilon(\phi))\,\mathrm{d}x}{\displaystyle\int_\Omega (1 - H_\epsilon(\phi))\,\mathrm{d}x} \tag{9}$$

As usual, we use an artificial parameter t in the Euler-Lagrange formulation associated to Equation (8) :

$$\frac{\partial\phi}{\partial t} = \left(\mu\mathrm{div}\left(\frac{\nabla\phi}{|\nabla\phi|}\right) - \nu - \lambda_1(u_0 - c_1)^2 + \lambda_2(u_0 - c_2)^2\right)\delta_\epsilon(\phi) +$$
$$\frac{\gamma}{2\sigma^2}\left(2\left(\phi - \tilde{\phi}\right)\delta_\epsilon(\phi) + \left(\phi - \tilde{\phi}\right)^2\frac{\partial\delta_\epsilon}{\partial\phi}(\phi)\right) = 0 \tag{10}$$
$$\text{in } \Omega \times \mathbb{R}^+; \ \phi(x, 0) = \phi_0(x) \text{ in } \Omega; \quad \frac{\delta_\epsilon(\phi)}{|\nabla\phi|}\frac{\partial\phi}{\partial n} = 0 \text{ on } \partial\Omega$$

The estimation of the solution of the model (8) can be summarized in the following steps:

- initialize $\phi_0 = \tilde{\phi}$, $n = 0$;
- compute $c_1(\phi_n)$ and $c_2(\phi_n)$ by the relations (9);
- compute ϕ_{n+1} by solving the PDE (10) with respect to ϕ;
- update periodically the level set ϕ_n by a signed distance;
- repeat these three steps until convergence (ϕ_n is stationary).

Figure 2 illustrates the segmentation with our approach and shows a comparison with a classical level set method, we will give more detains in the next section.

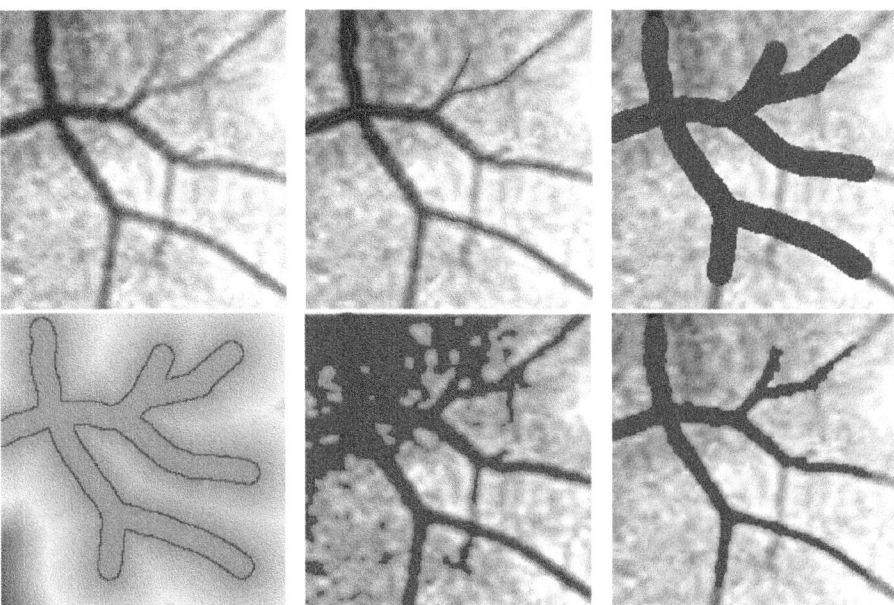

Fig. 2. Segmentation of vessels from a 2D angiogram image. First row: the left panel shows a 2D angiogram image;the center panel shows in red the voting tree; the right panel shows in red the voting tree after morphological dilatation. Second row: the left panel shows the signed distance computed from the dilated voting tree; the center panel shows in red the segmentation results obtained with a Chan and Vese method without shape prior; the right panel shows the segmentation result obtained with our approach.

4   Results and Discussion

We show results obtained with our algorithm on 2D images, see Table 1 and Figure 3. We applied our approach on ten cropped retinal images provided by DRIVE (Digital Retinal Images for Vessel Extraction) [21]. The DRIVE data were acquired using a Canon CR5 non-mydriatic 3CCD camera with a 45 degree field of view (FOV). Each image was captured using 8 bits per color plane at 768 by 584 pixels. The FOV of each image is circular with a diameter of approximately 540 pixels. For this database, the images have been cropped around the FOV. The DRIVE data is composed of 40 images for which manual segmentations are also provided. Considering the complexity of the retinal images and the properties of our algorithm, we have cropped ten different images from the 40 images availabe and evaluated our method on them. In tables 1 and 1, we compare our approach using the three evaluation measures: Dice, Specificity, and Sensitivity. The maximum value of the Dice index is 1, which corresponds to a perfect overlap between the manual and automatic segmentations. It shows that the results obtained with our approach are coherent with the manual segmentation.

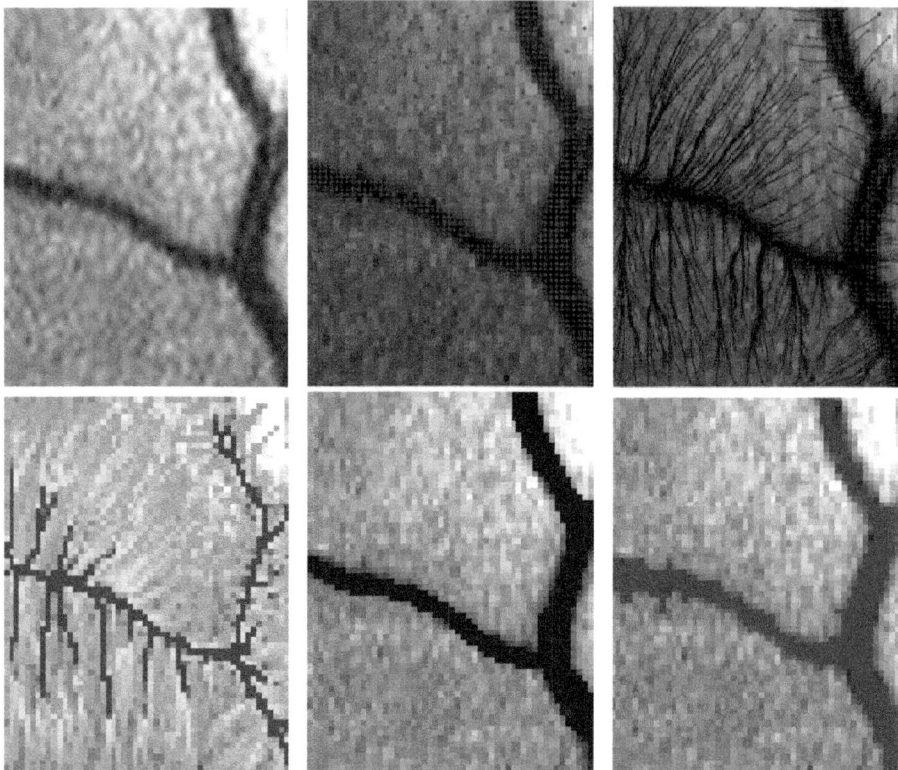

Fig. 3. Segmentation of vessels from one of the ten cropped 2D-retinal images given in table 1. First row: the left panel shows the original image; the center panel shows in red the farthest points detected; the right panel shows in blue the paths extracted from the farthest points to the source point. Second row: the left panel shows the computed geodesic density (green corresponds to a low density and red to a high density); the center panel shows the manual segmentation; the right panel shows the segmentation result obtained with our approach.

For our experiments we have considered the following potential $P(x) = I(x)^3$, where I is the grayscale intensity image of the DRIVE images. Figure 3 shows the segmentation result obtained with our approach. The shape prior allows us to constrain the propagation inside the tubular tree. Figure 2 (second row, center column) shows that the propagation without shape constraints ($\gamma = 0$ in the Equation (8)) can leak outside the tree structure.

We have also applied our approach on 3D simulated data of carotid bifurcation lumen created from the simulated data provided by MICCAI challenge [22], by adding Gaussian noise, see figure 4. The results obtained for these simulated data are better than those obtained for the DRIVE data in terms of the following overlap metrics: Dice, sensitivity, and specificity.

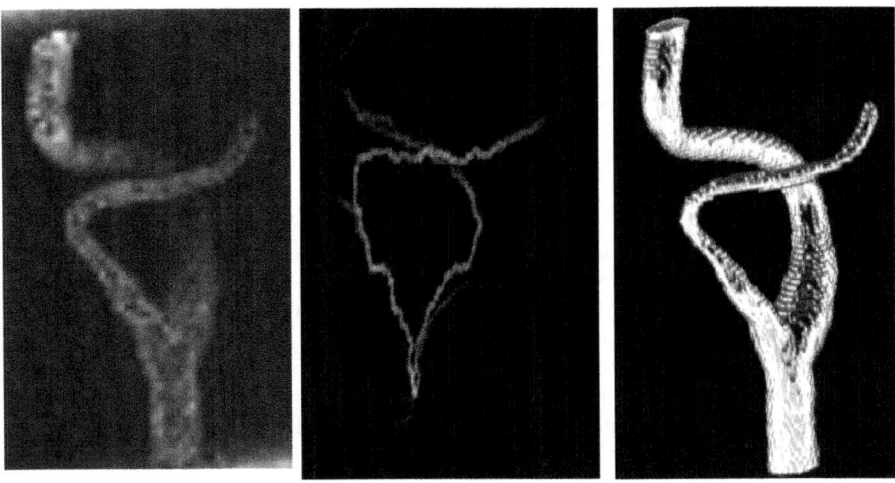

Fig. 4. Lumen segmentation from 3D simulated data (MIP visualization). The left panel shows the original image, the center panel shows the geodesic density, the right panel shows the segmentation result obtained with our approach.

Table 1. Comparison of our segmentations with the manual segmentation, on the ten cropped images from the DRIVE data, in terms of the following statistics: Dice similarity, sensitivity and specificity

Test data	T1	T2	T3	T4	T5	T6	T7	T8	T9	T10
Dice index	0.93	0.73	0.73	0.72	0.67	0.71	0.79	0.78	0.73	0.80
Sensitivity	0.91	0.61	0.64	0.58	0.53	0.60	0.70	0.70	0.70	0.70
Specificity	0.95	0.90	0.83	0.95	0.90	0.95	0.90	0.88	0.93	0.78

Table 2. Mean and standard deviation values of the statistics: Dice measure, sensitivity, and specificity, for all the data test

Statistics	Dice measure	Sensitivity	Specificity
Mean	0.76	0.67	0.90
Standard deviation	0.07	0.10	0.05

5 Conclusion

In this paper we have presented a new method for the segmentation of tree structures. This method is adapted to segment automatically tubular tree structure from a single point given by the user, no a priori information about the tree is required. In contrast, the methods previously described in the literature for the segmentation of tree structures are not fully automatic and require a priori information of the tree to be segmented. We have applied our approach to segment tubular tree structures from 2D retinal images and compared it with the manual segmentation on ten images. The next step is to validate our approach in 3D on a large data set.

References

1. Frangi, A.F., Niessen, W.J., Hoogeveen, R.M., van Walsum, T., Viergever, M.A.: Model-based quantitation of 3D magnetic resonance angiographic images. IEEE Trans. Med. Imaging 18(10), 946–956 (1999)
2. Bouix, S., Siddiqi, K., Tannenbaum, A.: Flux driven automatic centerline extraction. Medical Image Analysis 9(3), 209–221 (2005)
3. Li, H., Yezzi, A.: Vessels as 4D curves: Global minimal 4D paths to extract 3D tubular surfaces and centerlines. IEEE Trans. Med. Imaging 26, 1213–1223 (2007)
4. Li, H., Yezzi, A., Cohen, L.: 3D multi-branch tubular surface and centerline extraction with 4D iterative key points. In: Yang, G.-Z., Hawkes, D., Rueckert, D., Noble, A., Taylor, C. (eds.) MICCAI (1) 2009. LNCS, vol. 5762, pp. 1042–1050. Springer, Heidelberg (2009)
5. Kirbas, C., Quek, F.: A review of vessel extraction techniques and algorithms. ACM Comput. Surv. 36(2), 81–121 (2004)
6. Lesage, D., Angelini, E.D., Bloch, I., Funka-Lea, G.: A review of 3D vessel lumen segmentation techniques: Models, features and extraction schemes. Medical Image Analysis 13(6), 819–845 (2009)
7. Avants, B.B., Williams, J.P.: An adaptive minimal path generation technique for vessel tracking in CTA/CE-MRA volume images. In: Delp, S.L., DiGoia, A.M., Jaramaz, B. (eds.) MICCAI 2000. LNCS, vol. 1935, pp. 707–716. Springer, Heidelberg (2000)
8. Deschamps, T., Cohen, L.D.: Minimal paths in 3D images and application to virtual endoscopy. In: Vernon, D. (ed.) ECCV(2) 2000. LNCS, vol. 1843, pp. 543–557. Springer, Heidelberg (2000)
9. Deschamps, T., Cohen, L.D.: Fast extraction of minimal paths in 3D images and applications to virtual endoscopy. Medical Image Analysis 5(4), 281–299 (2001)
10. Gülsün, M.A., Tek, H.: Robust vessel tree modeling. In: Metaxas, D., Axel, L., Fichtinger, G., Székely, G. (eds.) MICCAI (1) 2008, Part I. LNCS, vol. 5241, pp. 602–611. Springer, Heidelberg (2008)
11. Wink, O., Niessen, W.J., Verdonck, B., Viergever, M.A.: Vessel axis determination using wave front propagation analysis. In: Niessen, W.J., Viergever, M.A. (eds.) MICCAI 2001. LNCS, vol. 2208, pp. 845–853. Springer, Heidelberg (2001)
12. Cohen, L.D., Deschamps, T.: Segmentation of 3D tubular objects with adaptive front propagation and minimal tree extraction for 3D medical imaging. Computer Methods in Biomechanics and Biomedical Engineering 10(4) (2007)
13. Rouchdy, Y., Cohen, L.D.: Image segmentation by geodesic voting. application to the extraction of tree structures from confocal microscope images. In: The 19th International Conference on Pattern Recognition, Tampa, Florida, pp. 1–5 (2008)
14. Rouchdy, Y., Cohen, L.D.: The shading zone problem in geodesic voting and its solutions for the segmentation of tree structures. application to the segmentation of microglia extensions. In: Computer Vision and Pattern Recognition Workshop MMBIA, Miami, Florida, pp. 66–71 (2009)
15. Cohen, L.D., Kimmel, R.: Global minimum for active contour models: A minimal path approach. International Journal of Computer Vision 24(1), 57–78 (1997)

16. Sethian, J.A.: Level set methods and fast marching methods. Cambridge University Press, Cambridge (1999)
17. Chan, T.F., Vese, L.A.: Active contours without edges. IEEE Trans. Med. Imaging 10(2), 266–277 (2001)
18. Mumford, D., Shah, J.: Optimal approximations by piecewise smooth functions and associated variational problems. Communications on Pure and Applied Mathematics 42(5), 577–685 (1989)
19. Leventon, M.E., Faugeras, O.D., Grimson, W.E.L., Wells III, W.E.: Level set based segmentation with intensity and curvature prior. In: MMBIA, pp. 4–11 (2000)
20. Cremers, D., Rousson, M., Deriche, R.: A review of statistical approaches to level set segmentation: Integrating color, texture, motion and shape. International Journal of Computer Vision 72, 215 (2007)
21. Staal, J.J., Abramoff, M.D., Niemeijer, M., Viergever, M.A., van Ginneken, B.: Ridge based vessel segmentation in color images of the retina. IEEE Trans. Med. Imaging 23(4), 501–509 (2004)
22. Hameeteman, K., Freiman, M., Zuluaga, M.A., Joskowicz, L., Rozie, S., van Gils, M.J., van den Borne, L., Sosna, J., Berman, P., Cohen, N., Douek, P., Snchez, I., Aissat, M., van der Lugt, A., Krestin, G.P., Niessen, W.J., van Walsum, T.: Carotid lumen segmentation and stenosis grading challenge. In: MICCAI 2009 (2009)

Amoeba Active Contours

Martin Welk

Institute for Biomedical Image Analysis
University for Health Sciences, Medical Informatics and Technology
Eduard-Wallnöfer-Zentrum 1, 6060 Hall/Tyrol, Austria
martin.welk@umit.at
http://ibia.umit.at

Abstract. We introduce an algorithm for active contour segmentation in which the level set function encoding the contour is processed by median filtering using morphological amoebas. These are adaptive structure elements introduced by Lerallut et al. which can be combined with different morphological operations. Recently it has been proven that iterated amoeba median filtering of an image approximates the well-known self-snakes partial differential equation. Following this approach we prove a partial approximation property of amoeba active contours with respect to geodesic active contours. Experiments prove the viability of the algorithm and confirm the theoretical results.

1   Introduction

The concept of morphological amoebas for structure-adaptive morphological filtering has been introduced by Lerallut et al. [19,20]. In this approach, structure elements adapt flexibly to image structures by taking into account spatial distance of pixels as well as image contrast. By penalising large deviations in image values, amoebas can grow around corners or along anisotropic image features. Once amoeba structure elements are constructed, a great variety of morphological filters can be applied.

One candidate for the filtering step is median filtering which assigns to each pixel the median of all grey-values of the given image within the structure element as its new grey-value. A classic result by Guichard and Morel [13] establishes a relation to partial differential equation (PDE) based image filtering: In its continuous-scale limit median filtering approximates mean curvature motion [2], i.e. the PDE $u_t = |\nabla u| \operatorname{div}(\nabla u/|\nabla u|)$.

Median filtering with amoeba structure elements has been investigated in [19,20]. In [29] it was proven that iterated amoeba median filtering approximates the self-snakes image filter PDE [23,30], i.e. that a space-continuous formulation of amoeba median filtering asymptotically equals a time step of the self-snakes evolution. As in [13] the time step size goes to zero with the square of the radius of the structure element (amoeba).

Self-snakes stand in close relationship to *geodesic active contours* [9,15], a well-established PDE method for image segmentation. In view of the approximation

A.M. Bruckstein et al. (Eds.): SSVM 2011, LNCS 6667, pp. 374–385, 2012.

property between amoeba median filtering and self-snakes it is natural to ask whether a similar amoeba-based process can be designed that performs an active contour segmentation. In this paper, we will demonstrate that this is indeed possible and that the resulting algorithm has similar properties as geodesic active contours. For a special case we will prove an approximation property in the same sense as in [13,29]. While the main contribution of the present paper is of theoretical nature, the new discrete approach to active contours might also turn out useful in applications because nonstandard discretisations of this kind may reduce e.g. numerical dissipation effects that are difficult to circumvent with finite-difference schemes.

Related work. The discrete filters that are in the focus of the present paper take their motivation from two sources: first, the classic median filter as introduced by Tukey [26] which has developed into a standard tool in image processing later on, see e.g. [16]; second, the idea of image-adaptive structure elements [6,7,24,28] which also includes Lerallut et al.'s morphological amoebas [19,20]. The space-continuous description of amoebas resorts to the representation of an image by an image manifold which has been used in the context of the Beltrami framework [14,31] and also underlies the bilateral filter [3,25].

Geodesic active contours were formulated by Caselles et al. [9] and Kichenassamy et al. [15], based on earlier work on active contours [8,21].

The paradigmatic PDE approximation result by Guichard and Morel [13] for the median filter has been followed by results for further discrete filters [3,11,27], for amoeba median filtering see [29].

Structure of the paper. In Section 2 we describe morphological amoebas and develop the amoeba active contour algorithm. By a space-continuous analysis in Section 3 approximation of the geodesic active contour PDE is proven in the radially symmetric case. Experiments demonstrate the viability of the approach, and its similarity to geodesic active contours, see Section 4. Conclusions are presented in Section 5.

2   Amoeba Active Contour Filtering

Let us recall first the principle of amoeba filters as introduced in [19,20].

The first step of any amoeba filter consists in the construction of image-adaptive structure elements, called amoebas, for all pixels in the image. The structure element for pixel p is made up by those pixels which are close to p in some *amoeba metric*. Instead of considering only the spatial distance in the image domain, as for non-adaptive morphological structure elements, the amoeba metric measures the distance of pixels along the *image manifold,* i.e. a surface interpolating the $\mathbb{R}^3$ points $(x, y, \sigma f(x, y))$. Here, (x, y) are point coordinates in the image domain, $f(x, y)$ is the grey-value at (x, y), and the scaling parameter $\sigma > 0$ weights grey-value differences (tonal distances) against spatial distances.

In the second step, some morphological operation is applied to the image with the previously computed structure elements, such as dilation, erosion, opening or closing. As a particularly interesting example, median filtering has been studied in [19,20,29]. Like the non-adaptive median filter, this filter can be iterated, giving rise to *iterated amoeba median filtering*.

Space-discrete and space-continuous amoeba metrics. Concerning the amoeba metric, let us discuss first the space-continuous setting. Natural choices for the Riemannian metric on the image manifold $\{(x, y, \sigma f(x, y))\}$ are those induced by metrics in the embedding space $\mathbb{R}^3$. The simplest case, the Euclidean metric, leads to the amoeba metric $ds^2 = d_2 s^2 = dx^2 + dy^2 + \sigma^2 df^2$. An alternative is an L^1 metric $ds = |dx| + |dy| + \sigma|df|$ which, however, lacks the desirable rotational invariance in space and will not be considered further here. As a compromise, one can choose a combined L^2-L^1 metric that is Euclidean in the two spatial dimensions but L^1 in combining the spatial and tonal distances, $ds = d_1 s = \sqrt{dx^2 + dy^2} + \sigma|df|$. A straightforward generalisation is

$$ds = d_\varphi s = \varphi\left(\sqrt{dx^2 + dy^2}, \sigma\, df\right) \tag{1}$$

where φ is a twice differentiable nonnegative function, homogeneous of degree 1, strictly increasing in both variables, and fulfils the triangle inequality, see [29].

   The distance $d(p, q)$ between two points $p = (x_p, y_p, \sigma f_p)$, $q = (x_q, y_q, \sigma f_q)$ on the image manifold is the minimum of the expression

$$L_\varphi(C) := \int_C d_\varphi s\ , \tag{2}$$

taken over all curves C on the image manifold that connect p and q. Note, however, that for $q \to p$ and smooth u, this distance is asymptotically equal to the corresponding distance of p and q in $\mathbb{R}^3$, i.e.

$$d(p, q) \approx \varphi\left(\sqrt{(x_p - x_q)^2 + (y_p - y_q)^2}, \sigma|f_p - f_q|\right)\ . \tag{3}$$

In a digitised image, a space-discrete formulation of the distance measurement is used. Following [19,20,29] $d(p, q)$ is the minimum of

$$L_\varphi(c) := \sum_{k=0}^{m-1} \varphi\left(\sqrt{(x_k - x_{k+1})^2 + (y_k - y_{k+1})^2}, \sigma|f_k - f_{k+1}|\right) \tag{4}$$

over all discrete curves $(p_0 = p, p_1, \ldots, p_m = q)$, where $p_k = (x_k, y_k, \sigma f_k)$. A discrete curve is a sequence of points in which each pair of subsequent points are neighbours in the image domains. In [19,20], this model is used with $d_\varphi \equiv d_1$ and 4-neighbourhoods, while [29] uses general d_φ and 8-neighbourhoods. We will follow the latter model, notwithstanding that, as [29] mentions, accuracy could be further improved by digital distance transforms [4,5,17,18].

Active contours. In an active contour algorithm [8,21], a contour curve evolves from some initial shape towards a shape that separates the given image into two segments (typically, a foreground object and the background). The initial shape is provided either by user interaction or some automatic method.

The evolution equation for *geodesic active contours* [9,15] is given by

$$c_t = \left(g(|\nabla f|)\kappa - \langle \nabla g(|\nabla f|), \boldsymbol{n} \rangle \right) \boldsymbol{n} \tag{5}$$

where $\boldsymbol{n}$ is the inward normal vector, and κ the curvature of the contour curve c. The nonnegative "edge-stopping function" g depends monotonically decreasing on the local gradient of the input image f. The name *geodesic* active contours indicates that the contour found by this evolution is a local minimum of the arc length, thus, a geodesic, in some image-dependent metric.

The contour c that evolves according to (5) can be represented in different ways, which leads to different implementations of the active contour method. The concept of a contour as parametric curve leads to a representation by sample points. This is on one hand comparably efficient since it represents a curve as a truly one-dimensional object; on the other hand, the evolution of sample points to inter-pixel positions necessitates interpolation. Moreover, due to length changes of the evolving curve over- and undersampling occurs, requiring re-sampling steps in the algorithms. Further difficulties are encountered when segments with multiple connected components cause the need for topology changes in the contour.

Alternatively, level-set methods [22] represent the contour c as zero-level set of a function u over the two-dimensional image domain. For example, a signed distance function of the contour can serve this purpose. The evolution equation (5) is then rewritten into an evolution of $u = u(x, y, t)$ as

$$u_t = |\nabla u|\, \mathrm{div}\left(g(|\nabla f|)\frac{\nabla u}{|\nabla u|} \right) = g(|\nabla f|)u_{\xi\xi} + \langle \nabla g, \nabla u \rangle \tag{6}$$

where ξ denotes a unit vector in level line direction of u, $\xi \perp \nabla u$. Topology changes are implicitly handled in this case, and resampling becomes a non-issue. However, the numerical evaluation in a 2D spatial domain raises the computational cost. This can be mitigated by narrow-band approaches [1] that restrict the computation to the immediate neighbourhood of the actual contour.

In all cases, the contour evolution takes place under the influence of the image being segmented; the image itself is not changed in this process.

Here lies the difference between active contours (snakes) and *self-snakes*. A self-snakes evolution is obtained from an active contour evolution in level-set formulation by identifying the level-set function for the contour with the image, thereby evolving the image itself.

Active contour filtering using morphological amoebas. To design an amoeba-based algorithm for active contours, the identification of input image and evolving image must be removed, leading to the following procedure:

1. Compute amoeba structure elements based on the input image f.
2. Initialise the evolving image u with a level-set function for the initial contour.

3. Evolve the image u by median filtering with the amoebas from Step 1 as structure elements.

In contrast to the iterated amoeba median filtering as described in [29], amoebas depend on the immutable input image and are therefore computed just once for the entire evolution. This saves computational expense and opens the way for further computational optimisations.

Introduction of dilation/erosion terms. Particularly if the initial contour is far from the actual segment boundary, and if the segment boundary is of complex topology, the geodesic active contour evolution (5) or (6) can stop in an undesired local minimum away from the desired contour. For such cases it is recommended in the literature [10,15] to modify (5) by an additional force term $\pm\nu\boldsymbol{n}$. This "balloon force" resembles morphological dilation or erosion and pushes the evolution into a chosen direction, thereby preventing it from stopping prematurely in regions with little contrast.

A similar behaviour can be achieved in the amoeba-based active contour model. To this end, one can bias the median filter: Instead of always selecting the element with index $m/2$ within the ordered sequence $g_0, \ldots, g_m$ of the grey-values in the amoeba, one chooses the element with index αm for some $\alpha \neq 1/2$ (the α-quantile), or the element with index $m/2 + b$ with some fixed offset b. We will use the latter modification in one of our experiments.

3   Space-Continuous Analysis

We turn now to analysing the amoeba active contour filter in a space-continuous setting, and aim at establishing a relationship to a PDE formulation. Analogous to the proceeding in [29], we approximate the input image f and the level-set function u locally by Taylor expansions up to second order, and compute then approximately the amoeba shape, and the median of u within that shape.

For the purposes of the present contribution, we restrict ourselves to the Euclidean amoeba metric $d_\varphi \equiv d_2$. We will not carry out an analysis in full generality but consider the special case in which the input image and initial contour are radially symmetric, which in particular implies that the level lines of the level-set function u and of the input image f always coincide. This special case is motivated by the idea that relevant parts of the segment boundary found by an active contour evolution should be almost aligned with level lines of the input image. Also, analysis of the biased method is beyond the scope of the present paper.

We consider expansions of σf and u within a ϱ-neighbourhood of $(x_0, y_0) = (0, 0)$. Here, $(0, 0)$ is not the centre of radial symmetry; we assume that ∇u and ∇f do not vanish at this point. Without loss of generality, we assume $u(0, 0) = 0$, $f(0, 0) = 0$, and assume that the gradients of u and f are in x direction. The Taylor expansions of u and f then read

$$\sigma f(x, y) = \alpha x + \gamma x^2 + \delta y^2 + \mathcal{O}(\varrho^3) \tag{7}$$

$$u(x, y) = \mu x + \nu x^2 + \lambda y^2 + \mathcal{O}(\varrho^3) . \tag{8}$$

Due to the required radial symmetry the mixed monomial xy does not occur. By the locally invertible coordinate transform $z = \mu x + \nu x^2 + \lambda y^2$ we obtain

$$\sigma f = \frac{\alpha}{\mu} z + \frac{\beta}{\mu^2} z^2 + \mathcal{O}(\varrho^3) \tag{9}$$

$$u = z + \mathcal{O}(\varrho^3) . \tag{10}$$

Note that the curvatures of level lines of σf and u are equal, such that the coordinate transform straightens not only the level lines of u but also those of σf, making the y^2 contribution vanish.

The contour of the amoeba $\mathcal{A}$ with centre $p = (0,0)$ and amoeba radius ϱ is made up by all those points $q = (x, y)$ for which $d_2(p, q) = \varrho^2$, i.e. $x^2 + y^2 + (\sigma f(x, y))^2 - \varrho^2 = \mathcal{O}(\varrho^4)$ or

$$y^2 \left(1 - \frac{2\lambda}{\mu^2} z\right) + \left(\frac{1+\alpha^2}{\mu^2} z^2 + 2\left(\frac{\alpha\beta}{\mu^3} - \frac{\nu}{\mu^4}\right) z^3\right) - \varrho^2 = \mathcal{O}(\varrho^4) . \tag{11}$$

A given level line $u = z$ of u intersects the contour of $\mathcal{A}$ in two points. Their y coordinates are solutions of (11), understood as quadratic equation for y, i.e. $y = \pm Y(z) + \mathcal{O}(\varrho^3)$ with

$$Y(z) = \sqrt{\varrho^2 - \frac{1+\alpha^2}{\mu^2} z^2 \left(1 + \frac{\lambda}{\mu^2} z - \frac{\alpha\beta/\mu^3 - \nu/\mu^4}{\varrho^2 - (1+\alpha^2)z^2/\mu^2} z^3\right)} . \tag{12}$$

Thus, the length of the level line segment within $\mathcal{A}$ is up to $\mathcal{O}(\varrho^3)$ equal to $2Y(z)$. It is nonnegative for $z \in [Z_-, Z_+]$ where

$$Z_\pm = \pm \frac{\varrho\mu}{\sqrt{1+\alpha^2}} + \mathcal{O}(\varrho^2) , \tag{13}$$

and goes to zero with $\mathcal{O}(\sqrt{|z - Z_\pm|})$ when approaching the boundaries.

The part of the amoeba $\mathcal{A}$ in which u takes values $z \in [a, b] \subseteq [Z_-, Z_+]$ has an area approximately given by the integral $2 \int_a^b Y(z)\tau(z)\, \mathrm{d}z$ where $\tau(z) :- \partial x/\partial z = 1/\mu - 2\nu z/\mu^2 + \mathcal{O}(\varrho^2)$ represents the inverse density of level lines. The median M of u within $\mathcal{A}$ therefore satisfies the condition

$$\int_{Z_-}^{M} Y(z)\tau(z)\, \mathrm{d}z = \int_{M}^{Z_+} Y(z)\tau(z)\, \mathrm{d}z + \mathcal{O}(\varrho^4) , \tag{14}$$

which yields, with a loss of accuracy due to the approximation of the integration boundaries via (13),

$$\int_{0}^{\varrho\mu/\sqrt{1+\alpha^2}} \big(Y(z)\tau(z) - Y(-z)\tau(-z)\big)\, \mathrm{d}z = 2\int_{0}^{M} Y(z)\tau(z)\, \mathrm{d}z + \mathcal{O}(\varrho^{7/2}) . \tag{15}$$

Since $M = \mathcal{O}(\varrho^2)$ and $Y(0) = \varrho$, we have $\int_0^M Y(z)\tau(z)\,\mathrm{d}z = \varrho M/\mu + \mathcal{O}(\varrho^4)$, and by the substitution $z = \varrho\mu\zeta/\sqrt{1+\alpha^2}$ we obtain

$$
\begin{aligned}
M &= \frac{(\lambda - 2\nu)\varrho^2}{1 + \alpha^2} \int_0^1 \zeta\sqrt{1 - \zeta^2}\,\mathrm{d}\zeta - \frac{(\alpha\beta\mu + \nu)\varrho^2}{(1+\alpha^2)^2} \int_0^1 \frac{\zeta^3\,\mathrm{d}\zeta}{\sqrt{1-\zeta^2}} + \mathcal{O}(\varrho^{5/2}) \\
&= \frac{\varrho^2}{6}\left(\frac{2\lambda - 4\nu}{1 + \alpha^2} - \frac{4\alpha\beta\mu - 4\nu}{(1+\alpha^2)^2}\right) + \mathcal{O}(\varrho^{5/2}) .
\end{aligned}
\tag{16}
$$

Based on the expansions (9), (10) and the variable substitution for z we can express the coefficients in terms of derivatives of u and f. We have $\mu = u_x$, $\nu = \frac{1}{2}u_{xx}$, $\lambda = \frac{1}{2}u_{yy}$, $\alpha = \sigma f_x$, $\beta = \gamma - \alpha\nu/\mu = \frac{\sigma}{2}(f_{xx} - f_x u_{xx}/u_x)$. Giving up our special choice of coordinates, we replace x and y by unit vectors $\eta \parallel \nabla f$ and $\xi \perp \nabla f$ in gradient and level line direction, respectively. Thus the last equation expresses that, in the radially symmetric case, one step of the amoeba active contour filter asymptotically approximates for $\varrho \to 0$ one time step of size $\varrho^2/6$ of an explicit scheme for the PDE

$$
u_t = \frac{u_{\xi\xi}}{1 + \sigma^2|\nabla f|^2} - \frac{2\sigma^2 f_\eta f_{\eta\eta} u_\eta}{(1 + \sigma^2|\nabla f|^2)^2} = g(|\nabla f|)u_{\xi\xi} + \langle \nabla g(|\nabla f|), \nabla u \rangle , \tag{17}
$$

i.e. (6) with the Perona-Malik-type edge stopping function (compare [9,15])

$$
g(s) := (1 + \sigma^2 s^2)^{-1} . \tag{18}
$$

It is still an open question whether this approximation property holds in exactly the same form for situations other than the radially symmetric case discussed here. Nevertheless, even this partial equivalence result links amoeba active contours to the framework of PDE active contour methods and makes it an interesting candidate for a non-standard discrete realisation of active contours.

4   Experiments

Our first experiment (Figure 1) demonstrates the viability of the amoeba active contour approach and its similarity to geodesic active contours. Starting from an initial contour that generously surrounds almost the entire image area of the test image, Figure 1(a), our amoeba active contour algorithm adapts to the outline of the depicted human head section within 600 iterations with amoeba radius 10, see Figure 1(b, c). By our approximation result (16) the corresponding evolution time for an active contour PDE is $T = 10000$.

Indeed, computation of geodesic active contours (6) up to $T = 10000$ by an explicit finite difference scheme gives a similar result, see Figure 1(d). Slight differences, in particular a stronger rounding of contours, can be attributed to the blurring effect of the central difference approximation of derivatives.

The theoretical link between amoeba active contours and geodesic active contours established in Section 3 is rooted in a space-continuous setting. In fact,

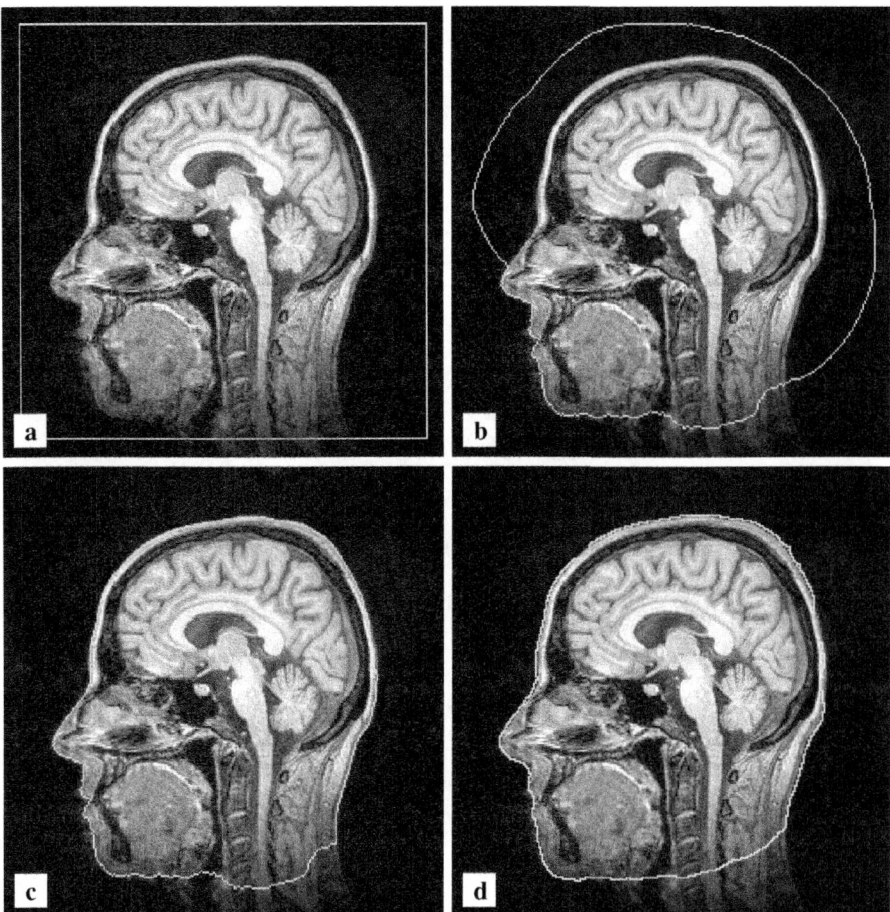

Fig. 1. (a) MR image of a human head with initial contour. (b) Amoeba active contours (unbiased), amoeba radius $\varrho = 10$, $\sigma = 0.1$, 200 iterations. (c) Amoeba active contours, same parameters but 600 iterations. (d) Geodesic active contours (6) with edge-stopping function (18), $\sigma = 0.1$, computed by an explicit time-stepping scheme with time step size $\tau = 0.25$, 40000 iterations.

the application of both filters to digital images reveals some differences in detail which can be attributed to their fundamentally different discrete realisation. The already mentioned numerical dissipation of finite difference discretisations stands in contrast to the very fine adaptivity of amoeba shapes to image structures, which is also reflected in the resulting active contours.

Furthermore, while the disposition to "lock in", i.e. become stationary at image structures with strong gradients, is a feature of both active contour approaches, such a behaviour is more pronounced in the case of amoeba active contours. The reason is that the underlying median filter already in its non-adaptive formulation possesses non-constant steady states, so called root signals [12].

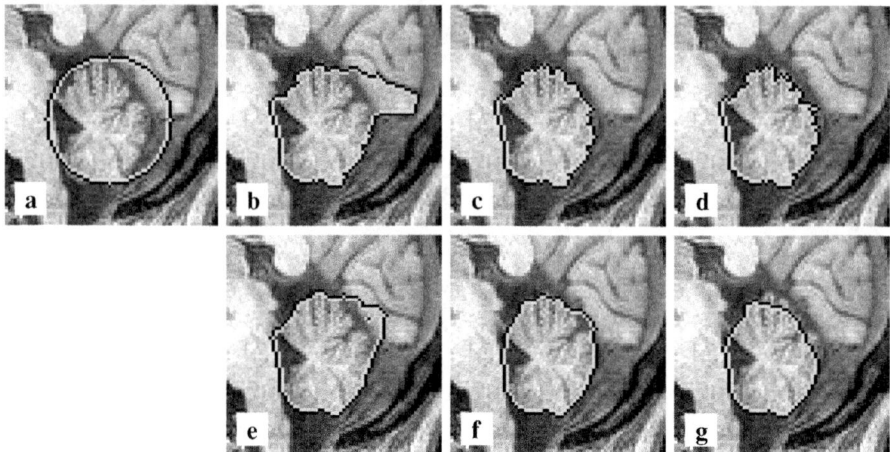

Fig. 2. (a) MR image with initial contour (detail). (b) Amoeba active contours (unbiased), amoeba radius $\varrho = 10$, $\sigma = 0.1$, 20 iterations. (c) Same but $\varrho = 12$, $\sigma = 0.1$, 10 iterations. (d) Same but $\varrho = 12$, $\sigma = 0.1$, 60 iterations. (e) Geodesic active contours (6), $\sigma = 0.1$, $\tau = 0.25$, 960 iterations. (f) Same but 3000 iterations. (g) Same but 57600 iterations.

It is therefore natural also for an amoeba median filter to develop root signals, the more if the amoeba shapes themselves are kept fixed as in our case. This property contributes on one hand to stabilising the segmentation result. On the other hand it means that some minimal amoeba size is needed for reasonable segmentation. Experiments suggest that ϱ should not be smaller than 10.

As speed optimisation has not been in the focus of our work so far, a proper comparison of the two active contour algorithms in terms of runtimes cannot be made at this point. To this end, additional optimisation effort for both algorithms would be required. To state a rough trend we mention that in our present, non-optimised implementations both algorithms are roughly comparable in speed, the amoeba-based algorithm being about 15 % faster than the PDE scheme in the case of Figure 1(c) vs. (d) (but sometimes also a bit slower in other examples).

In our second experiment (Figure 2) we use the same test image as before but aim at segmenting the cerebellum. As our initial contour, Figure 2(a), is not very precise, the amoeba active contour with amoeba radius $\varrho = 10$, the amoeba active contour locks in at some sharp contours outside the desired region (b). With a slightly enlarged amoeba radius $\varrho = 12$ a fairly good segmentation is reached (c). Further evolution of the amoeba contours becomes stationary at a contour that cuts off some small details (d). Running geodesic active contours up to evolution time $T = 240$ (which matches the amoeba evolution of the third frame above) still does not segment the cerebellum well (e); this is achieved only after considerably longer evolution time (f). Continuing geodesic active contour evolution, again a stationary contour is reached, see Figure 2(g).

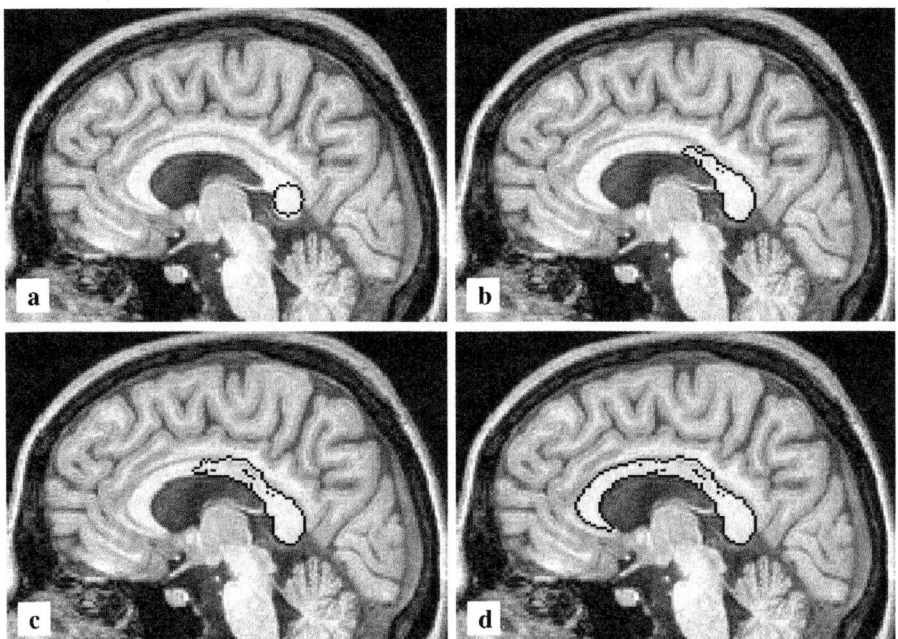

Fig. 3. (a) MR image with initial contour (detail). (b) Amoeba active contours with dilation bias, amoeba radius $\varrho = 20$, $\sigma = 2.0$, bias $b = 10$, 5 iterations. The bias $b = 10$ means that within each amoeba the 10-th greyvalue above the median index was chosen. (c) Same but 15 iterations. (d) Same but 30 iterations.

In our third experiment (Figure 3) we demonstrate the modification of amoeba active contours by a dilation bias $b = 10$ in order to force an expansive evolution of the contour. Thus, within each increasing sequence of grey-values of an amoeba the value 10 positions after the median was selected (the maximum if the amoeba contained less than 20 pixels). Together with amoeba radius $\varrho = 20$ and a comparatively large contrast parameter $\sigma = 2.0$ this allowed to segment the corpus callosum from a small initial contour within the structure.

5   Conclusion

In this paper we have developed a new variant of an active contour algorithm for image segmentation based on iterated amoeba median filtering of a level-set function. We proved that in a radially symmetric setting the continuous-scale limit of our amoeba active contour method coincides with the well-known geodesic active contour equation. Experiments verify that both algorithms behave structurally similar. Due to their entirely different discrete filter strategies, they differ in the representation of contour details.

Ongoing work is directed at extending our theoretical analysis. This will include the study of non-radially symmetric situations as well as different amoeba metrics and the relation between the biased approach and additional force terms. A further goal are algorithmic optimisations.

The revenue of this effort will be, firstly, a deeper theoretical insight into the relations between discrete and continuous image filters will be gained. Secondly, based on the so established approximation properties genuinely discrete filters can be used as unconventional discretisations of PDE filters and improve the practical implementation of the latter.

Acknowledgements. The author thanks Michael Breuß for helpful discussions on the topic. Implementation is partially based on earlier work by Oliver Vogel.

References

1. Adalsteinsson, D., Sethian, J.A.: A fast level set method for propagating interfaces. Journal of Computational Physics 118(2), 269–277 (1995)
2. Alvarez, L., Lions, P.-L., Morel, J.-M.: Image selective smoothing and edge detection by nonlinear diffusion. II. SIAM Journal on Numerical Analysis 29, 845–866 (1992)
3. Barash, D.: Bilateral filtering and anisotropic diffusion: Towards a unified viewpoint. In: Kerckhove, M. (ed.) Scale-Space 2001. LNCS, vol. 2106, pp. 273–280. Springer, Heidelberg (2001)
4. Borgefors, G.: Distance transformations in digital images. Computer Vision, Graphics and Image Processing 34, 344–371 (1986)
5. Borgefors, G.: On digital distance transforms in three dimensions. Computer Vision and Image Understanding 64(3), 368–376 (1996)
6. Braga-Neto, U.M.: Alternating sequential filters by adaptive neighborhood structuring functions. In: Maragos, P., Schafer, R.W., Butt, M.A. (eds.) Mathematical Morphology and its Applications to Image and Signal Processing. Computational Imaging and Vision, vol. 5, pp. 139–146. Kluwer, Dordrecht (1996)
7. Breuß, M., Burgeth, B., Weickert, J.: Anisotropic continuous-scale morphology. In: Martí, J., Benedí, J.M., Mendonça, A.M., Serrat, J. (eds.) IbPRIA 2007. LNCS, vol. 4478, pp. 515–522. Springer, Heidelberg (2007)
8. Caselles, V., Catté, F., Coll, T., Dibos, F.: A geometric model for active contours in image processing. Numerische Mathematik 66, 1–31 (1993)
9. Caselles, V., Kimmel, R., Sapiro, G.: Geodesic active contours. In: Proc. Fifth International Conference on Computer Vision, pp. 694–699. IEEE Computer Society Press, Cambridge (1995)
10. Cohen, L.D.: On active contour models and balloons. Computer Vision, Graphics, and Image Processing: Image Understanding 53(2), 211–218 (1991)
11. Didas, S., Weickert, J.: Combining curvature motion and edge-preserving denoising. In: Sgallari, F., Murli, A., Paragios, N. (eds.) SSVM 2007. LNCS, vol. 4485, pp. 568–579. Springer, Heidelberg (2007)
12. Eckhardt, U.: Root images of median filters. Journal of Mathematical Imaging and Vision 19, 63–70 (2003)
13. Guichard, F., Morel, J.-M.: Partial differential equations and image iterative filtering. In: Duff, I.S., Watson, G.A. (eds.) The State of the Art in Numerical Analysis. IMA Conference Series (New Series), vol. 63, pp. 525–562. Clarendon Press, Oxford (1997)

14. Kimmel, R., Sochen, N., Malladi, R.: Images as embedding maps and minimal surfaces: movies, color, and volumetric medical images. In: Proc. 1997 IEEE Computer Society Conference on Computer Vision and Pattern Recognition, pp. 350–355. IEEE Computer Society Press, San Juan (1997)

15. Kichenassamy, S., Kumar, A., Olver, P., Tannenbaum, A., Yezzi, A.: Gradient flows and geometric active contour models. In: Proc. of Fifth International Conference on Computer Vision, pp. 810–815. IEEE Computer Society Press, Cambridge (1995)

16. Klette, R., Zamperoni, P.: Handbook of Image Processing Operators. Wiley, New York (1996)

17. Ikonen, L.: Priority pixel queue algorithm for geodesic distance transforms. Image and Vision Computing 25(10), 1520–1529 (2007)

18. Ikonen, L., Toivanen, P.: Shortest routes on varying height surfaces using gray-level distance transforms. Image and Vision Computing 23(2), 133–141 (2005)

19. Lerallut, R., Decencière, E., Meyer, F.: Image processing using morphological amoebas. In: Ronse, C., Najman, L., Decencière, E. (eds.) Mathematical Morphology: 40 Years On. Computational Imaging and Vision, vol. 30. Springer, Dordrecht (2005)

20. Lerallut, R., Decencière, E., Meyer, F.: Image filtering using morphological amoebas. Image and Vision Computing 25(4), 395–404 (2007)

21. Malladi, R., Sethian, J.A., Vemuri, B.C.: A topology independent shape modeling scheme. In: Vemuri, B. (ed.) Geometric Methods in Computer Vision. Proceedings of SPIE, vol. 2031, pp. 246–258. SPIE Press, Bellingham (1993)

22. Osher, S., Sethian, J.A.: Fronts propagating with curvature-dependent speed: Algorithms based on Hamilton–Jacobi formulations. Journal of Computational Physics 79, 12–49 (1988)

23. Sapiro, G.: Vector (self) snakes: a geometric framework for color, texture and multiscale image segmentation. In: Proc.1996 IEEE International Conference on Image Processing, Lausanne, Switzerland, vol. 1, pp. 817–820 (September 1996)

24. Shih, F.Y., Cheng, S.: Adaptive mathematical morphology for edge linking. Information Sciences 167(1–4), 9–21 (2004)

25. Tomasi, C., Manduchi, R.: Bilateral filtering for gray and color images. In: Proc. Sixth International Conference on Computer Vision, pp. 839–846. Narosa Publishing House, Bombay (1998)

26. Tukey, J.W.: Exploratory Data Analysis. Addison–Wesley, Menlo Park (1971)

27. van den Boomgaard, R.: Decomposition of the Kuwahara–Nagao operator in terms of linear smoothing and morphological sharpening. In: Talbot, H., Beare, R. (eds.) Mathematical Morphology: Proc. Sixth International Symposium, pp. 283–292. CSIRO Publishing, Sydney (2002)

28. Verly, J.G., Delanoy, R.L.: Adaptive mathematical morphology for range imagery. IEEE Transactions on Image Processing 2(2), 272–275 (1993)

29. Welk, M., Breuß, M., Vogel, O.: Morphological amoebas are self-snakes. Journal of Mathematical Imaging and Vision (in press, 2011)

30. Whitaker, R.T., Xue, X.: Variable-conductance, level-set curvature for image denoising. In: Proc. 2001 IEEE International Conference on Image Processing, Thessaloniki, Greece, pp. 142–145 (October 2001)

31. Yezzi Jr., A.: Modified curvature motion for image smoothing and enhancement. IEEE Transactions on Image Processing 7(3), 345–352 (1998)

A Hybrid Scheme for Contour Detection and Completion Based on Topological Gradient and Fast Marching Algorithms - Application to Inpainting and Segmentation[*]

Y. Ahipo[1], D. Auroux[2], L.D. Cohen[3], and M. Masmoudi[4]

[1] Spring Technologies, Toulouse, France
[2] Laboratoire J. A. Dieudonné, Université de Nice Sophia Antipolis, France
[3] CEREMADE, UMR CNRS 7534, Université Paris Dauphine, France
[4] Institut de Mathématiques de Toulouse, France
cohen@ceremade.dauphine.fr

Abstract. We combine in this paper the topological gradient, which is a powerful method for edge detection in image processing, and a variant of the minimal path method in order to find connected contours. The topological gradient provides a more global analysis of the image than the standard gradient, and identifies the main edges of an image. Several image processing problems (e.g. inpainting and segmentation) require continuous contours. For this purpose, we consider the fast marching algorithm, in order to find minimal paths in the topological gradient image. This coupled algorithm quickly provides accurate and connected contours. We present then two numerical applications, to image inpainting and segmentation, of this hybrid algorithm.

Keywords: topological gradient, fast marching, contour completion.

1 Introduction

Contour detection is a major issue in image processing. For instance, in classification and segmentation, the goal is to split the image into several parts. This problem is strongly related to the detection of the connected contours separating these parts. It is quite easy to detect edges using local image analysis techniques, but the detection of continuous contours is more complicated and needs a global analysis of the image.

Several image processing problems like image inpainting and denoising (or enhancement) are classically solved without detecting edges and contours. The goal of image enhancement is to denoise the image without blurring it. A classical idea is to identify the edges in order to preserve them, and to smooth the image outside them. In this particular case, contour completion is not prerequisite, as the quality of the result is not much related to the completeness of the identified edges. But for most of the other image processing problems (segmentation,

---

[*] This work was partially supported by ANR grant MESANGE ANR-08-BLAN-0198.

A.M. Bruckstein et al. (Eds.): SSVM 2011, LNCS 6667, pp. 386–397, 2012.

inpainting, classification), the detection of connected contours can drastically simplify the resolution and improve the quality of the results. For instance, the image segmentation problem is a very good example, as the goal is to split the image into its characteristic parts. In other words, one has to find connected contours, which define different subsets of the image.

For solving all these problems, various approaches have been considered in the literature. By lack of space, we will only give a general quote on the most commonly used models: the structural approach by region growing, the stochastic approaches and the variational approaches, which are based on various strategies like level set formulations, the Mumford-Shah functional, active contours and geodesic active contours methods or wavelet transforms.

Another approach to define edges as cracks, is based on the topological asymptotic analysis [8, 2]. The goal of topological optimization is to look for an optimal design (i.e. a subset) and its complementary. Finding the optimal subdomain is equivalent to identifying its characteristic function. At first sight, this problem is not differentiable. But the topological asymptotic expansion gives the variation of a cost function $j(\Omega)$ when we switch the characteristic function from one to zero in a small region [9]. More precisely, we consider the perturbation of the main domain Ω by the insertion of a small crack (or hole) σ_ρ: $\Omega_\rho = \Omega \backslash \sigma_\rho$, ρ being the size of the crack. The topological sensitivity theory provides then an asymptotic expansion of the considered cost function when the size of the crack tends to zero. It takes the general form: $j(\Omega_\rho) - j(\Omega) = f(\rho)g(x) + o(f(\rho))$, where $f(\rho)$ is an explicit positive function going to zero with ρ, and $g(x)$ is the topological gradient at point x. Then, in order to minimize the criterion (or at least its first order expansion), one has to insert small cracks at points where the topological gradient is the most negative. An efficient edge detection technique, based on the topological gradient, was introduced in [8]. But the identified edges are usually not connected, and the results can be degraded. This is beyond the scope of this paper to give a more detailed description, which can be found in [8, 2].

In the inpainting problem, we assume that there is a hidden part of the image, and our goal is to recover this part from the known part of the image. Here the interior of the missing part is not empty, it is neither a random set nor a narrow line, we assume that it is a quite *large* part of the image. This problem has been widely studied and some of the most common approaches are: learning approches (neural networks, radial basis functions, ...) [13, 14], minimization of an energy cost function based on a total variation norm [3], morphological component analysis methods separating texture and cartoon [7].

We now consider the crack detection technique, within the framework of the identification of the image edges, either in the hidden part of the image for the inpainting application, or in the whole image for the segmentation application [2]. The topological asymptotic analysis provides very quickly the location of the edges, as they are precisely defined as the most negative points of the topological gradient. The main issue of such a technique is the need for connected contours. This can easily be understood as the hidden part of the image is filled using the Laplace operator in each subdomain of the missing zone, and a discontinuous contour would lead to some blurred reconstruction. Up to now, one had to thresh-old the topological gradient with a not too small value, in order to identify con-nected contours, but this leads to thick identified edges, and also to consider more noisy points as potential edges. In order to overcome this limitation, we consider a minimal path technique for connecting the edges.

Minimal paths have been first introduced for finding the global minimum of active contour models, using the fast marching technique [4]. They have then been used to find contours or tubular structures and also for perceptual grouping using a path or a set of paths minimizing a functional [5, 15, 6, 12, 10]. In our case, the energy to be minimized will be proportional to the topological gradient. As the topological gradient takes its minimal values on the edges of the image, the idea is indeed to find contours for contour completion from the various minima and small values of the topological gradient.

For perceptual grouping, a set of keypoints is considered as starting points and a set of minimal paths connecting some pairs of these keypoints is considered as a contour completion. This approach is extremely satisfactory in 2D problems, with quite few key points. It is also extremely fast. In 3D images, minimal paths find tubular structures, but in order to identify minimal surfaces, this approach is much more difficult to consider. It was dealt in the case of a surface connecting two curves in [1]. We only consider here the 2D case.

The application of the minimal path technique to the topological gradient al-lows us to obtain an automatic identification of the main (missing or not) edges of the image. These edges will be continuous, by construction, and will allow us to simply apply the Laplace operator to fill the image for inpainting appli-cations, or will directly provide the segmented image, with very good results. Another advantage of this technique is to be very fast, as it does not degrade the $\mathcal{O}(n.\log(n))$ complexity of the topological gradient based algorithm intro-duced in [2]. We also refer to these citations for the inpainting and segmentation algorithms by topological asymptotic expansion, and for a detailed presenta-tion of the topological gradient. We assume here that the topological gradient is available.

In section 2, we propose an algorithm based on the minimal path and fast marching techniques in order to identify the valley lines of the topological gra-dient, which correspond to the main edges of the image. Then, we report the results of several numerical experiments in section 3. We also compare this hy-brid scheme with the fast marching algorithm applied to the standard gradient. Two particular image processing problems are considered: segmentation and in-painting. Finally, some conclusions are given in section 4.

2   A 2D Algorithm Based on the Minimal Paths and Fast Marching Methods

2.1   Minimal Paths

In this section, we describe the standard minimal path technique, adapted to our needs. We refer to [4] for more details about the minimal paths method.

In the following, let Ω be the considered image domain. We assume that Ω is a regular subset of $\mathbb{R}^2$. In order to compute some minimal paths, we need to define a potential function, measuring in some sense for any point of Ω the cost for a path to contain this point. As we want to identify paths in the topological gradient image, and considering that this potential function must be positive, we will define a potential function as follows:

$$P(x) = g(x) - \min_{y \in \Omega}\{g(y)\}, \quad \forall x \in \Omega, \tag{1}$$

where g is the topological gradient, defined in all the domain Ω. We can see that the points where the topological gradient g reaches its minimal values are quite costless. We denote by $C(s)$ a path, or curve, in the image, where s represents the arclength. We now define a functional, measuring the cost of such a path:

$$J(C) = \int_C (P(C(s)) + \alpha)\, ds, \tag{2}$$

where α is a positive real coefficient that represents regularization. In our applications, α is usually very small, as the goal is to connect the most negative parts of the topological gradient, whatever the Euclidean distance is.

We now consider a key point $x_0 \in \Omega$ of the image, and x will represent any point of the image. The energy $J(C)$ of a given path C can be seen as a distance between the two endings of C, weighted by the potential function. The goal is to find the minimal energy integrated along the path C. We can now define the weighted distance between key point x_0 and point x by

$$D(x; x_0) = \inf_{C \in A(x,x_0)} J(C) = \inf_{C \in A(x,x_0)} \int_C (P(C(s)) + \alpha)\, ds, \tag{3}$$

where $A(x, x_0)$ is the set of all paths going from point x_0 to point x in the image.

The distance $D(x; x_0)$ introduced in equation (3) is then simply the instant t at which the front, initialized at key point x_0, reaches point x. An efficient way to compute this distance function is the fast marching algorithm, and is justified by the fact that the distance satisfies the following Eikonal equation

$$\|\nabla_x D(x; x_0)\| = P(x) + \alpha, \tag{4}$$

with the initialization $D(x_0; x_0) = 0$. We refer to [4, 11, 1, 15, 10, 12] for more details about the fast marching technique and the justification of equation (4). If n is the size of the image, the complexity of this fast marching method is bounded by $\mathcal{O}(n.\log(n))$, which is also the complexity of the topological gradient algorithm.

2.2   Multiple Minimal Paths

The main issue is now to extend this minimal path technique to more than one keypoint, in order to connect several points. This is exactly what we need, in order to connect the identified edges by the topological gradient, as we have many identified keypoints (for example all negative local minima of the topological gradient) that we want to connect. As explained in [5], the first step of a multiple minimal path algorithm is to reduce the set of keypoints, for computational reasons. Moreover, the selected keypoints should not be too close one to each other. One usually chooses a total number N of keypoints, and the first one. Then the $N - 1$ other keypoints can be chosen for instance as described in [5].

The next step consists of connecting these N points. One has to compute the distance function from each of these key points, and the common minimal paths algorithms provide then the Voronoï diagram of the distance and the corresponding saddle points (minimal distance along the edges of the diagram, maximal distance from the keypoints). The Voronoï diagram defines a partition of the image in as many subsets as the number of keypoints. Each subset is defined by the set of points that are closer to the corresponding keypoint than to all others. The saddle points minimize the distance function on the edges of the diagram: minimal distance on the edge, maximal distance to the keypoints. It is useful to compute these saddle points to save computation time since it reduces the domain of the image where the fast marching computes or updates the weighted distance map..

Finally, the idea is to consider the saddle-points as initial conditions for minimizing the distance function. For each saddle-point as an initial point, a minimization is performed towards each of the two corresponding keypoints (recall that the saddle-points are located at the interface between two subsets of the Voronoï diagram). Each minimization produces a path between the saddle-point (initial condition) and a keypoint (local minimum of the distance function). This step is usually called back-propagation, as it consists of a gradient descent from the saddle-point, back to the linked keypoints. The back-propagation step is straightforward as there is no local minimum of the distance function, except the keypoints. The union of all these paths gives a continuous path, connecting the keypoints together.

The interesting part of the approach introduced in [5] is that each keypoint should not be connected to all the others, but only to at most two others, as we are looking for a set of closed connected path. Thus, the keypoints have to be ordered automatically in a way such that they are only connected to the other keypoints that are closest to them in the energy sense [5]. For this reason, we sort all the saddle-points from smaller to larger distance, and we first try to connect the pairs of keypoints corresponding to the saddle-points of smallest distance. These keypoints are indeed more likely to be connected than distant keypoints, corresponding to saddle-points of large potential. Once the close keypoints are connected, we repeat the process with the new closest pairs of keypoints, provided each point remains connected to at most two other ones. At the end of the process, all the keypoints are connected to at most two other keypoints, and the union of all minimal paths between the keypoints represents one (or several) continuous contour of the image. An interesting feature of this method is that the key points are by construction widely distributed around.

If all the selected keypoints are on the same contour of the image, we are almost sure that at the end, they will all be connected together and we will retrieve the corresponding contour, as the potential function (related to the topological gradient) is very low on this contour. If, on the contrary, one keypoint is not part of the contour, the large values of the topological gradient, and hence of the potential function, will isolate this keypoint from the other ones, and it will not disturb the contour completion process.

2.3   Main Algorithm

The hybrid algorithm we propose is then the following:

Fast marching algorithm applied to the topological gradient:
- Compute the topological gradient of the image.
- Set N the number of keypoints and choose the N keypoints: the main one will be for example the global minimum of the topological gradient, the other ones being the most negative local minima of the topological gradient.
- Compute the distance function (3) with all these keypoints, and the corresponding Voronoï diagram.
- Compute the set of saddle-points: on each edge of the Voronoï diagram, determine the point of minimal distance.
- Sort all these points of minimal distance, from smaller to larger distance.
- For each of these saddle-points, from smaller to larger distance, check if it will not be used to connect two keypoints, one of which is already connected to two other keypoints.
- If this is not the case, perform the back-propagation from this point: use this saddle-point as an initialization for a descent type algorithm in order to connect the two corresponding keypoints.

It is straightforward to see that this algorithm converges, and that at convergence, all the keypoints are connected to at most two other keypoints. This provides one or several continuous contours, containing the keypoints. As the first keypoint is usually the global minimum of the topological gradient, it is on one of the main edges of the image. Consequently, using this algorithm, we can identify this edge. Then, it is possible to restart the algorithm, using other keypoints that are not on this identified edge, by initializing for instance the first keypoint as the minimum of the topological gradient outside the neighborhood of this edge.

3   Numerical Experiments

3.1   Numerical Results for 2D Segmentation

We first consider a two dimensional grey level image, represented in figure 1 on the left. The opposite of the Euclidian norm of its standard gradient is represented on the right. Note that we represent its opposite in order to have comparable images with the topological gradient, which has negative values (see below).

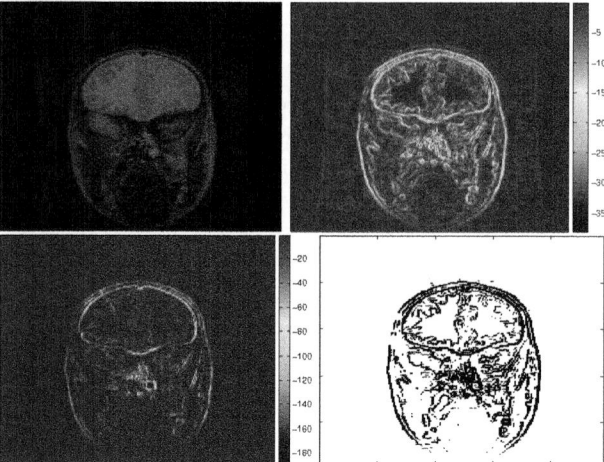

Fig. 1. Top: Original image (left); L^2 norm of its (standard) gradient (right). Bottom: Topological gradient (left); edges by thresholding the topological gradient (right).

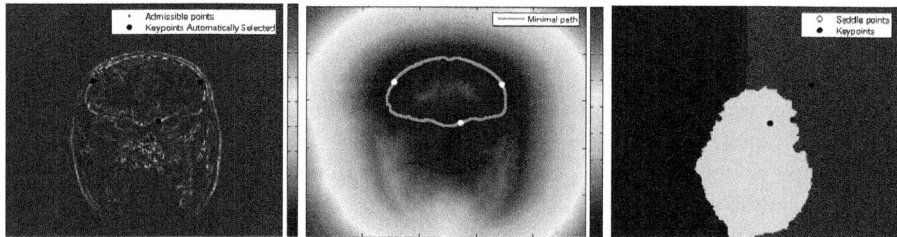

Fig. 2. Admissible set of points in blue, and 3 keypoints automatically selected in black (left); distance function (middle) computed from these 3 keypoints with the fast marching algorithm and identified minimal paths between the keypoints. Corresponding Voronoï diagram, with the 3 keypoints and saddle points ; (right).

The topological gradient is represented on the bottom part of figure 1. As it quantifies in a global way whether a pixel is part of an edge or not, it is much less sensitive to noise and small variations of the image than the standard gradient. For instance, the topological gradient takes much larger absolute values on the edges than outside, contrary to the standard gradient. In the segmentation algorithm presented in [2], the idea until now was to threshold the topological gradient in order to define the edge set. Such a threshold is represented in figure 1 on the right side. One can see that, in order to obtain at least the main connected edge, the threshold coefficient has been set to a low value, leading to add many unwanted points to the edge set, but also to thick edges. And even in this case, the main contour is not totally continuous. Then, the idea is to apply the variant of the fast marching algorithm we proposed in section 2.3.

Using an automatic thresholding for identifying the most negative values of the topological gradient, figure 2 shows on the left the set of points (or admissible keypoints, in blue), in which we will choose the keypoints for the minimal path algorithm. The first keypoint is set to the minimum of the topological gradient. Then, we have set the number of keypoints to $N = 3$. From the first keypoint, we start the minimal path algorithm, and we choose the second keypoint as being the point (in the admissible set) maximizing the weighted distance to the first keypoint. Then, we start again the minimal path algorithm from these two points, and we set the third keypoint in a similar way. These three keypoints are represented by black points in the left side of figure 2. Note that the keypoints can also be (manually) provided by the user, for instance with the aim of identifying a specific edge of the image.

From these keypoints, we run the minimal path algorithm, in order to compute the distance map. Figure 2 shows on the middle this distance function. One can clearly see that the distance does not correspond to the Euclidean metric in the plane, as the distance remains very small on the common edge of the 3 keypoints, whereas it takes much larger values outside. The corresponding Voronoï diagram is represented in figure 2 on the right. The three keypoints are still represented by black points. Each color represents the subset Ω_i of points that are closer to keypoint i than to the others. For any $i \neq j$, we consider the interface $\Gamma_{ij} = \Omega_i \cap \Omega_j$ between two subsets of the Voronoï diagram. Γ_{ij} represents then the set of points equidistant from keypoints i and j. We now minimize the distance function on Γ_{ij} in order to find a saddle-point: same distance to keypoints i and j, minimal distance on Γ_{ij}. These saddle points are represented by blue points on the right side of figure 2.

From these saddle points, the idea is finally to perform a descent-type algorithm in order to minimize the distance function from the saddle points to the keypoints. We consider a saddle-point on an edge Γ_{ij} as an initial condition for two minimizations of the distance function, one towards each of the corresponding keypoints (i and j). Each of these two minimizations provides a continuous path from the saddle-point to one of the two keypoints. The union of these two paths connects the two keypoints. This process is done for all pairs of consecutive keypoints.

The final set of paths is represented in green on the distance function in figure 2 (middle). The three keypoints are also represented (in white). These paths correspond to the contour of the original image that contains the 3 keypoints. By applying again this algorithm, with other keypoints (selected outside the first identified contours), it is possible to detect other contours of the image.

Finally, we illustrate the fact that the topological gradient provides better information about the edges of the image than the standard gradient, as previously observed (see figure 1). Figure 3 shows the original image where we have manually selected 3 keypoints in blue on an edge of the image. From these keypoints, we have run the fast marching algorithm applied to both the standard gradient and the topological gradient (hybrid scheme) and the identified paths are then represented. The topological gradient clearly provides the best identification of

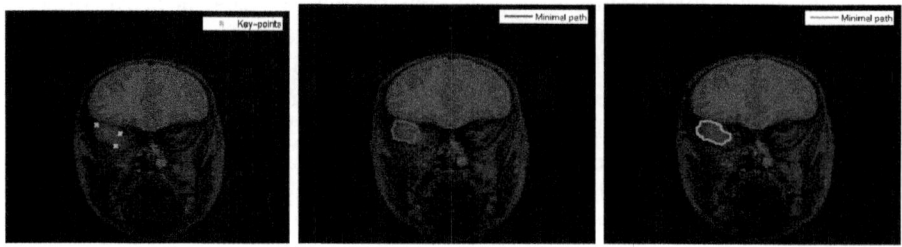

Fig. 3. Original image with 3 selected keypoints (left); Contours identified by the fast marching algorithm applied to the standard gradient (middle) and topological gradient (right)

the edge. This can easily be explained by the bad shape of the standard gradient in this region (see figure 1). On the contrary, the topological gradient is less sensitive to small local variations, and the edge is more visible in figure 1.

3.2   Numerical Results for a New Way of 2D Inpainting

We now consider another application of this technique to image inpainting and improve the results presented in [2]. The idea of the topological gradient algorithm is to identify the missing edges in the occluded part of the image, and then to reconstruct the image from the solution of a Poisson problem with Neumann boundary conditions (see references in [2]). In this application also, it is crucial to have connected contours, otherwise the reconstruction with the Laplacian will not be satisfactory. Figure 4 shows an example of image, in which we added a mask on a quite large part of the image ($\simeq$ 800 pixels). The goal of inpainting is to reconstruct as precisely as possible the original image from the occluded image. We also want the inpainted image to have sharp (unblurred) edges. Figure 4 shows the corresponding topological gradient, provided by the inpainting algorithm (see [2] for more details about this algorithm). In this case, the topological gradient gives some information about the most probable location of the missing edges. In [2], the idea is then to threshold the topological gradient, and define the edge set of the occluded zone as being the set of points below the threshold. The main issue is that the identified missing edges must be connected in order to avoid blurry effects (due to the Laplacian) in the reconstruction. Then, the threshold is sometimes set manually in order to have connected contours. In our example, the identified edge set is represented by white points in figure 4.

Figure 4 shows the corresponding inpainted image. One can see that the reconstruction is not very good, particularly in the top part. This is mainly due to the fact that the identified edges are either connected but thick with a lot of wrong identifications (if the threshold is too small) or discontinuous (otherwise).

The idea of our method is to apply the fast marching algorithm on the topological gradient obtained during the inpainting process, in order to identify connected contours in the hidden part of the image.

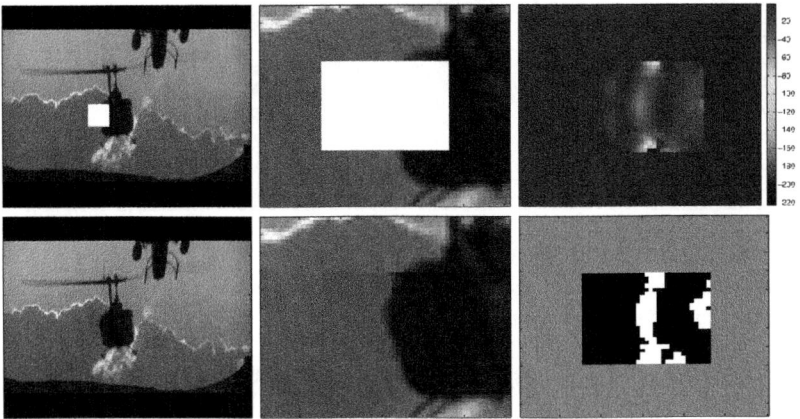

Fig. 4. Top: Occluded image (by a white rectangle) (left); zoom of the occluded zone and topological gradient (right); Bottom: Inpainted image using the edge set from the topological gradient and zoom of the occluded zone, and edge set (right)

After thresholding the topological gradient, several points (identified by blue circles) have been identified and define the admissible set of keypoints, represented in figure 5-left. We choose then the most negative point of the topological gradient as the first keypoint, and then the further admissible point as the second one. The keypoints are represented by a large black point on the same image. They are located on the edge of the domain, as the inpainting topological gradient always takes its minimal values there.

Then the minimal path algorithm is run, and it provides a path between the keypoints, represented in green in figure 5-right. We can see that the path follows very well the valley line of the topological gradient, from one side to the other. By choosing 3 keypoints instead of 2, there will be another keypoint on the bottom edge, near the first one, and it will simply add a small contour located all along on the edge of the domain, and consequently there is absolutely no impact on the reconstruction of the hidden part of the image.

Figure 5 shows on the right the same identified path represented on the occluded image. This allows one to see that the path clearly gives a good approximation of the missing edges, and also that the topological gradient is very powerful for this identification problem. The corresponding identified edge set is represented in figure 6-left. This image should be compared with the thresholded edge set of figure 4-right. And we can conclude that the minimal path algorithm is an excellent tool for extracting the valley lines of the topological gradient.

Finally, using this minimal path as the set of missing edges in the occluded zone, the inpainting topological gradient algorithm produces a much better reconstructed image, shown in figure 6. The quality of the image is very good, as the missing edges used for the reconstruction are connected, and the Laplace operator will not produce any blurring effect due to a discontinuous contour.

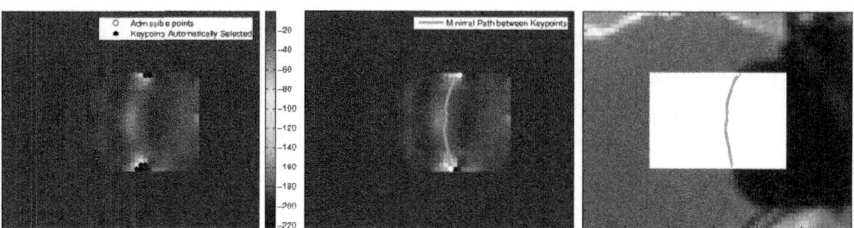

Fig. 5. Admissible set of keypoints (left); selected keypoints on the topological gradient and minimal path between the keypoints represented on the occluded image as corresponding identified missing edge (right)

Fig. 6. Inpainted image using the fast marching algorithm for closing the contours identified by the topological gradient in the hidden part of the image, and corresponding zoom

This example confirms that the quality of all topological gradient applications in image processing can be improved, by replacing a thresholding technique by a minimal path algorithm.

4   Conclusions and Perspectives

We have introduced a hybrid scheme, based on one side on the topological gradient for edge detection, and on the other side on the fast marching and minimal paths methods for contour completion. These approaches allow us to extract connected contours in 2D images, and to solve the main issue of all topological gradient based algorithms for image processing problems (discontinuity of the edges). Moreover, the minimal path algorithm does not degrade the complexity of the topological asymptotic analysis.

We have considered two specific applications in image processing: segmentation and inpainting. In the first one (segmentation), we showed that the topological gradient is more efficient than the standard gradient for edge detection, and the hybrid scheme provides better results than the fast marching method applied to the standard gradient of the image. In the second application (inpainting), we showed that the hybrid scheme particularly improves the quality of the inpainted image, as the contour completion ensures a non-blurred inpainted image, and as it also helps removing the manual thresholding of the topological gradient.

An interesting and natural perspective is to apply this hybrid scheme to 3D images and movies. The topological gradient can very easily be extended to 3D images as well as the minimal path technique.

References

[1] Ardon, R., Cohen, L.D., Yezzi, A.: A new implicit method for surface segmentation by minimal paths in 3D images. Applied Mathematics and Optimization 55(2), 127–144 (2007)

[2] Auroux, D., Masmoudi, M.: Image processing by topological asymptotic expansion. J. Math. Imaging Vision 33(2), 122–134 (2009)

[3] Chan, T., Shen, J.: Non-texture inpainting by curvature-driven diffusions (CDD). Tech. Rep. 00-35, UCLA CAM (September 2000)

[4] Cohen, L.D.: Minimal paths and fast marching methods for image analysis. In: Paragios, N., Chen, Y., Faugeras, O. (eds.) Mathematical Models in Computer Vision: the Handbook. Springer, Heidelberg (2005)

[5] Cohen, L.D.: Multiple contour finding and perceptual grouping using minimal paths. J. Math. Imaging Vision 14(3), 225–236 (2001)

[6] Dicker, J.: Fast marching methods and level set methods: an implementation. Ph.D. thesis, University of British Columbia (2006)

[7] Elad, M., Starck, J.L., Querre, P., Donoho, D.L.: Simultaneous cartoon and texture image inpainting using morphological component analysis (MCA). J. Appl. Comput. Harmonic Anal. 19, 340–358 (2005)

[8] Jaafar-Belaid, L., Jaoua, M., Masmoudi, M., Siala, L.: Image restoration and edge detection by topological asymptotic expansion. C. R. Acad. Sci., Ser. I 342(5), 313–318 (2006)

[9] Masmoudi, M.: The topological asymptotic, Computational Methods for Control Applications. In: Glowinski, R., Karawada, H., Périaux, J. (eds.) GAKUTO Internat. Ser. Math. Sci. Appl., Tokyo, Japan, vol. 16, pp. 53–72 (2001)

[10] Rawlinson, N., Sambridge, M.: The fast marching method: an effective tool for tomographic imaging and tracking multiple phases in complex layered media. Explor. Geophys. 36, 341–350 (2005)

[11] Sethian, J.A.: Level set methods and fast marching methods. Cambridge University Press, Cambridge (1999)

[12] Telea, A., van Wijk, J.: An augmented fast marching method for computing skeletons and centerlines. In: Proc. Eurographics IEEE-TCVG Symposium on Visualization. Springer, Vienna (2002)

[13] Wen, P., Wu, X., Wu, C.: An interactive image inpainting method based on rbf networks. In: 3rd Int. Symposium on Neural Networks, pp. 629–637 (2006)

[14] Zhou, T., Tang, F., Wang, J., Wang, Z., Peng, Q.: Digital image inpainting with radial basis functions. J. Image Graphics 9(10), 1190–1196 (2004) (in Chinese)

[15] Zhu, F., Tian, J.: Modified fast marching methods and level set method for medical image segmentation. J. X-ray Sci. Tech. 11(4), 193–204 (2003)

A Segmentation Quality Measure Based on Rich Descriptors and Classification Methods

David Peles and Michael Lindenbaum

Computer Science Department
Technion, Haifa 32000, Israel
davidpelz@yahoo.com,mic@cs.technion.ac.il

Abstract. Most segmentation methods are based on a relatively simple score, designed to lend itself to relatively efficient optimization. We take the opposite approach and suggest more complex segmentation scores that are based on a mixture of on-line and off-line learning processes and rely on rich descriptors. The score is evaluated by a segmentation process which uses exploration-exploitation to search for good segments in various scales and shapes. We test our algorithm in a foreground-background segmentation task, given a minimal prior which is just a single seed point inside the object of interest. Results on two image databases are presented and compared with earlier approaches.

1 Introduction

Most recent segmentation algorithms are based on optimizing a quality function (a score), which depends on different criteria. Optimizing nontrivial quality measures over a large set of possible segmentations is not an easy task. Therefore, the quality measures designed in this context are compromised for the sake of an elegant and effective optimization process, and while they are intuitively plausible, they are often simplistic. In the spectral approaches, for example, the optimal solution is elegantly specified by a single (or few) eigenvectors, but the quality measure is the simple (normalized) sum of pairwise similarities [1].

This paper proposes the opposite approach: we use a learning based score which relies on a rich characterization describing the segment and the background. For example, we use the distribution of superpixels scores inside and outside the segment as 10 features. This gives a more detailed description than a single scalar representing the distance between these distributions. We found that the richer description yield better results.

We propose two (alternative) scores: One, denoted *direct*, generates the score directly from a discriminative classifier response, and another, denoted *indirect*, generates the score from several such responses. For both scores, the richer description together with training on realistic segment candidates leads to a reliable and accurate segment evaluation.

To test the score utility we construct a simple segmentation algorithm and test it on the task of weakly supervised segmentation: the algorithm gets as

A.M. Bruckstein et al. (Eds.): SSVM 2011, LNCS 6667, pp. 398–410, 2012.

input an image and a point inside the desired object. Using a mixed exploration-exploitation approach, it searches for the best segment in the image that contains this point.

The proposed approach is related to learning edge detection scores (e.g., [2,3]), but is closer to the the work of Ren and Malik [4], who also train a classifier to distinguish between good and bad image segmentations. The proposed approach differs from their work in several important issues: a. a much richer set of characteristics, derived (partially) from on-line training, b. an indirect score shown to performs better than direct scoring, c. a training process which uses near misses (in contrast to the strongly erroneous segmentation used in [4]).

While the focus of this paper is on the segmentation score, the search algorithm, used to test the score, gives an iterative segmentation algorithm. This part of the paper is related, of course, to earlier iterative segmentation approaches which were based either on active contours (e.g., [5]), graph cuts (e.g., [6]) or the composition approach [7]. Unlike these earlier approaches, our algorithm can recover from local score maxima and uses two different scores, a simple one for proposing the segmentation and a more complex and reliable one, relying on online learning and a rich set of descriptor, for evaluating it. Many iterative segmentation approaches use extensive user interaction to get very accurate segmentation [8,9]. We however, aim to achieve fully automatic segmentation. The minimal information provided is used only to identify the desired segment. See section 5 for a further discussion of previous work.

The main contributions of this paper are therefore:

1. A new segmentation score, based on discriminative offline trained classifier. We show that this score performs better than simpler measures.
2. Feature generation using an online learning process capable of adapting to the features that discriminate the segment from those outside it.
3. A mixed exploration-exploitation search procedure which, together with the score, forms a new segmentation algorithm.

We describe the score in section 2 and the search procedure which relies on it in section 3. Experiments and discussion follow in sections 4 and 5.

2   Segment Score

We propose to estimate the quality of a segment using a classifier based approach. Its advantage over intuitively specified scores is that we may use a rich set of characteristics, which, as we found, indeed improves the segmentation score accuracy. The learning process provides a systematic method for combining these features. The input to the off-line training process is a set of images along with manually marked ground truth segments. Additional segments are generated during the training process and used with the ground truth segments to generate positive and negative examples for training the classifiers. In general, we consider segment hypothesis h_i to be better than segment hypothesis h_j if h_i better overlaps the ground truth than h_j. We measure this overlap in terms of F-measure: ($F = \frac{2*Precision*Recall}{Precision+Recall}$).

We considered two (alternative) variations for generating a score using classifiers:
Direct scoring independently estimates the score of every segment hypothesis.
Indirect scoring estimates the score of each segment hypothesis by comparing
it with other available hypotheses.

Both scores rely on classifiers which are trained off-line and use diverse set
of features: the boundary size and the shape, the type of junctions along the
boundary, and so forth. We augment this characterizations with features, generated by a classifier, trained on-line to discriminate between the image parts
inside and outside the segment. This online learning replaces the commonly used
distance between distributions. We next describe the score and its components.

We would like to emphasize that we use classifiers in different parts of the
score estimation procedure. We use offline trained classifiers for estimating the
segment score (sections 2.2, 2.3). We also use online\offline trained classifiers to
generate features (section 2.1).

2.1   Feature Generation

Feature Generation Using a Discriminative Online Classifier. In a good
segmentation the region inside the segment tends to be more homogeneous than
the full image and different from the region outside the segment. This intuitive
observation is often quantified by demanding a low variance within the segment
[5], or a large distance between the distributions inside and outside it [8]. The
first method is clearly too simple for realistic heterogeneous objects. The second
approach is better but requires the construction of the two distributions, which
in turn either requires many samples or uses restrictive parametric assumptions.

We propose, instead, to quantify the difference between the inside and outside
regions of the segment using a classifier, trained online, on examples from these
regions. We start by over-segmenting the image and describing every superpixel
by a vector of features quantifying its size, shape, color, and texture. Let $\mathcal{SP} =
\{\mathbf{sp}_i\}$ be the set of superpixels in the image. The objects of classification are
adjacent pairs of such superpixels, which gave better discrimination than single
superpixels. Let $\mathcal{SPP} = \{\mathbf{spp}_{ij} = (\mathbf{sp}_i, \mathbf{sp}_j)|\mathbf{sp}_i$ is adjacent to $\mathbf{sp}_j\}$ be the set
of adjacent superpixel pairs. Every superpixel pair in $\mathcal{SPP}$ is characterized by a
feature vector composed of the feature vectors of its component superpixels, of
the common boundary length and of the relative position.

The online learning is done in the context of a particular segment hypothesis.
Superpixels pairs for which both superpixels are inside (outside) the segment
hypothesis are used as positive (negative) examples for training the classifier. A
gentle boost classifier trained on $\mathcal{SPP}$ using cross-validation produces a decision
value $\phi'(\mathbf{spp}_{ij})$ for every superpixel pair. The score $\phi(\mathbf{sp}_i)$ assigned to every
single superpixel is the average of the decision values associated with all super-
pixel pairs in which it participates. Denote this score the SP-score. See Figure 2.
The set of scores assigned to all the superpixels inside and outside the segment
hypothesis is described by two 5-bin histograms, which specify 10 components
in the feature vector describing the hypothesis.

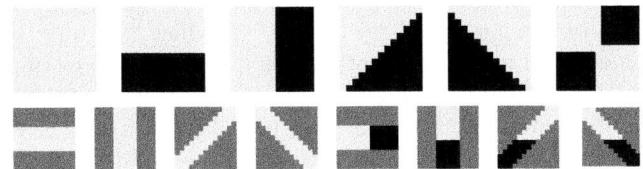

Fig. 1. Masks (13×13) for creating junction feature vectors. Top: masks for the binary segment hypothesis image. Bottom: masks for the edge probability image.

This approach has several advantages over the alternatives described above: First, being based on discriminative learning, it neither makes assumptions regarding the features distribution nor requires a large number of examples. Therefore it matches the limited number of superpixels. Second, it performs feature selection and ignores features which do not distinguish between inner and outer superpixels. Third, it characterizes the similarity between the appearance inside and outside using two histograms of SP-scores. This gives a detailed, informative, description to the classifier. Finally, the score assigned to every superpixel, reflects its affinity with the segment and is useful for guiding and accelerating the segment search process.

Junctions-based features. A junction is a pixel touching at least 3 different superpixels. The properties of junctions that lie on the hypothesized boundary tell about the validity of the hypothesis. It is relatively difficult to parameterize the appearance of junctions. Therefore, we adopt an approximate analysis: Every junction is described by 14 inner products. The first between 8 masks and a window in the edge probability map [2]. The others between 6 other masks and a window in the hypothesis binary representation (1 inside the segment hypothesis and 0 outside it); See Figure 1. A boosting based classifier is trained offline (using cross-validation) to distinguish junctions lying on the boundary of true segments from those which do not. The junction classifier achieved 72.6% accuracy. The distribution of the junctions score (decision values) was represented by a 5-bin histogram, which provides 5 features to the segment classifier.

2.2   Directly Estimated Segment Score

The segment score may be specified directly or indirectly. In the direct approach, a (boosting) segment classifier is used for independently estimating the score of every segmentation hypothesis. The classifier uses a vector of features describing the hypothesized segment and the image. The segment classifier's decision value is used as the score for the hypothesis.

To train this classifier we use sets of good and bad segments as examples, from all the relevant images in the database (using cross-validation). The positive (good) examples are simply manually marked segments. Taking negative examples as "random segments", as suggested by [4], would make classification easy but not reflect the true difficulty of scoring during the search. Therefore, the negative examples are generated (automatically) by simulating the hypothesis search process with a simulated segment score, which is a noisy version of the F-measure associated with the hypothesis. This way, we get examples similar to those created by the real search process. To increase the number of positive examples, we added small variations of the ground truth hypotheses. While these hypotheses introduce some error to the training set, our tests show that this technique benefits the overall performance.

2.3   Enhanced, Indirect Estimation of the Segment Score

Alternatively, we may use a classifier that examines a pair of segmentation hypotheses h_i, h_j and decide which one is better. Many such comparisons are then used to rank all the segmentation hypotheses. Our classifier uses a combined feature vector $(V(h_i), V(h_j), F(h_i, h_j))$ where $V(h)$ is the feature vector associated with h and $F(h_i, h_j)$ is the overlap between h_i and h_j. The training is done over segment pairs from the entire database, where each pair consists of two segment hypotheses from the same image. One segment is considered better if its overlap with the ground truth segment is larger (in terms of F-measure). In the segment evaluation (test) phase, all pairs of all the segments considered so far by the segment search procedure are compared.

For a pair of segment hypotheses (h_i, h_j), this classifier outputs the probability $P(F(h_i) > F(h_j))$ Where $F(\cdot)$ is the F-measure of the segment hypothesis and the ground truth. By assuming independence of these pairwise probabilities, we may estimate the probability that a segment h_i better overlaps the ground truth segment than all other available segment hypotheses in $\mathcal{H}$:

$$IndirectScore(h_i) = P(F(h_i) > \max_{j \neq i} F(h_j)) = \prod_{j \neq i} P(F(h_i) > F(h_j)) \qquad (1)$$

This choice of the indirect score worked better than other methods we tested: a. using FAS-Tournament task [10] to search for hypotheses ordering with minimal number of contradictions and b. ordering hypotheses by the number of wins over other hypotheses in $\mathcal{H}$.

A major advantage of the pairs classifier is the ease of generating positive and negative examples for the training process. In fact, every pair of segment examples which differ in their F-measure provides a positive and a negative example. We found that both direct and indirect scores lead to good results, but that the latter is more accurate and more stable.

3   Segment Search

The algorithm uses the score described above to search for the best segment which contains the input seed. The search process maintains a collection of segment hypotheses $\mathcal{H} = \{h_1, h_2, \ldots, h_{\|\mathcal{H}\|}\}$, and uses them to suggest new hypotheses to be tested. The hypothesis collection is initialized with a fixed number (12)

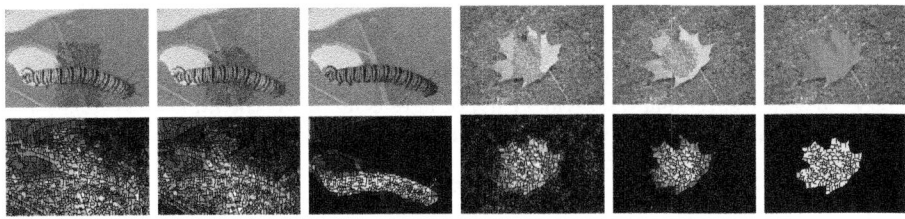

Fig. 2. Examples of segment hypotheses (first row) and their corresponding online superpixel scores (second row). Columns:1,2,4,5 initial hypotheses, Columns:3,6: Final hypotheses reached by the algorithm.

of initial segment hypotheses, which are disks and rectangles of various scales; see Figure 2 for examples. Every search step chooses one "base hypothesis" from the collection $\mathcal{H}$ and generates a few new related hypotheses. The scores of the new hypotheses are evaluated and these hypotheses are added to $\mathcal{H}$.

3.1 Selecting a Base Hypothesis

The base hypothesis is selected so that it is likely to lead to a good new related hypothesis. The criterion is based on an exploration-exploitation mixture.

Exploitation component: The new hypothesis h is close to the base hypothesis $h_i = base(h)$ and is likely to be good if the score of the base segment h_i itself is good. Moreover, h is likely to be better than h_i if $score(h_i) > score(base(h_i))$. A simple score summing these two intuitive considerations is $a(h_i) = score(h_i) + [score(h_i) - score(base(h_i))]$ Denote the relative exploitation grade $EtGrade(h_i)$ as the fraction of hypotheses in $\mathcal{H}$ having an $a(\cdot)$ score lower than h_i's.

Exploration component: Selecting base hypotheses which are not similar to previously examined segment hypotheses is expected to result in a wider search. Let the F-measure $F(h_1, h_2) \in [0, 1]$ be the measure of similarity. Then $b(h_i) = 1 - \max_{h_j \in H \setminus \{h_i\}} F(h_i, h_j)$ is a simple score reflecting the exploration preference. Denote the relative exploration grade $ErGrade(h_i)$ as the fraction of hypotheses in $\mathcal{H}$ having a $b(\cdot)$ score lower than h_i's.

The final score for choosing the base segment hypothesis is

$$pref(h_i) = \alpha ErGrade(h_i) + (1 - \alpha)EtGrade(h_i) \qquad (2)$$

where α is a weight decreasing with iteration count. The hypothesis h_i which maximizes $pref(h_i)$ is selected as base hypothesis for the next iteration. This exploration-exploitation would encourage a wider search in the first search iterations while focusing on the best hypotheses in the last iterations.

Initialization:

- Compute over-segmentation for the input image.
- Generate several initial segment hypotheses centered at the input seed (disks and rectangles of different size) and add them to $\mathcal{H}$.

Iterative search for a good segment:

1. Choose a base segment hypothesis $h \in \mathcal{H}$ using exploration-exploitation strategy.
2. Use h to generate several new segment hypotheses and add them to $\mathcal{H}$.
3. Evaluate the scores of the newly added hypotheses.
4. If using indirect score, re-evaluate the scores of all other hypotheses in $\mathcal{H}$.
5. Repeat steps 1-4 until maximal number of iterations is reached.
6. Output the highest scored segment hypothesis in $\mathcal{H}$.

Fig. 3. Algorithm Summary

3.2 Generating New Hypotheses

To generate a new hypothesis h from $base(h)$, we find a segment minimizing an energy function E, composed of a data term and a boundary term. The data term is specified using the online classifier associated with $base(h)$, which specifies a SP-score $\phi(s_i)$ for every superpixel. The distribution of these scores, inside $base(h)$ and outside it, are approximated by two Gaussians. A posterior probability $P(l_i|\phi(s_i))$ ($l_i \in \{in, out\}$) easily follows. The boundary term penalizes long boundaries without strong edge support.

$$E = -\sum_i \log P(l_i|\phi(s_i)) - \lambda \sum_{i,j} (l_i \neq l_j) \log P(Edge_{s_i,s_j}). \qquad (3)$$

The first sum is over all superpixels and the second is over adjacent superpixel pairs. $P(Edge_{s_i,s_j})$ is the edge probability [2], corresponding to the boundary between s_i and s_j. The hypothesis is found by minimizing the energy E using the min-cut algorithm [11]. Several (3, in our implementation) new hypotheses are generated in each iteration, using different λ values. All of them are evaluated by the classifier based score and inserted into the hypothesis set $\mathcal{H}$. Figure 3 summarizes our algorithm's main steps.

4 Experiments

4.1 Data and Implementation Details

Data. We used two image databases, GrabCut (50 color images) [6], and Weizmann [12] (100 grayscale images). We used the images for which the ground truth segmentation contained a single connected component. (All Grabcut images and 89/100 of the Weizmann images).

Table 1. A superpixel descriptor

#	Description
1	Mean brightness.
2*	Mean chroma components in LAB space (*color images)
3	Mean gradient components and gradient size
2	Superpixel (area, perimeter)
8	Gradient orientations weighted histogram
18	Gabor filter responses [17]
1	Mean log probability of the superpixels border [2]

Table 2. Features for the segment classifier

#	Description
10	SP-scores distribution for (in,out).
2	Mean SP-scores near boundary SP (in,out).
3	Accuracy of the online classifier: (inner, outer, mean) SP.
5	Boundary junctions decision value distribution.
5	Boundary probability distribution[2]
1	Mean of $(positiveCurvature)^2$
1	Mean of $(negativeCurvature)^2$
2	Segment area (including, excluding) holes. divided by image size.
1	Segment area (including holes). divided by $(outerBorderLength)^2$

Initialization. A seed point is specified by eroding the ground truth segment (by 5 pixels) and selecting the point in it closest to the center of mass.

Over-Segmentation. We used the watershed over-segmentation algorithm [13]. To avoid very large superpixels which behave like outliers in the learning processes, small amplitude smoothed noise was added to the image.

Initial Hypothesis. We used 12 initial hypotheses: 4 disks, 4 horizontal and 4 vertical rectangles, with area, roughly 2,4,8 and 20 percent of the image size. All initial hypotheses were centered at the seed coordinate; see Figure 2.

Learning. We used the [14] implementation for the gentle-boost learning [15], with decision trees as weak classifiers. The online classifier uses (5,2) trees (i.e., 5 decision trees with 2 branches each). Both the segment and segment pair classifiers used (40,8) trees. The junction classifier used (20,8) trees. We chose gentle-boost for its robustness to outliers [16], which is important for training the online classifier using erroneous hypotheses containing mislabeled superpixels.

The Online Classifier. Every superpixel pair is described by a feature vector containing the features of each superpixel (Table 1), the difference in the superpixels' center of mass, the common border length, and the edge log probability of this border.

The Segment Classifier. The features used to describe the segment are listed in Table 2. The segment classifier is trained on an equal number of positive and negative example segments.

4.2   Score Validity

To test the scores independently of the search, we used a search algorithm that relies on the simulated score (a noisy version of the F-measure). Of all hypotheses generated from one image, we chose the best one according to various criteria, and recorded its true F-measure. See Table 3.

Table 3. The F-measure of the hypotheses selected by different scores

Algorithm	Mean F	Median F
Best possible hypothesis	0.96	0.97
Best generated hypothesis	0.94	0.96
Indirect score	0.87	0.92
Direct score	0.86	0.92
Direct Score - Minimal Features + SP scores distribution	0.85	0.91
Direct score - Minimal Features	0.83	0.89

The first row in Table 3 corresponds to the best hypothesis specifiable by the oversegmentation (and limited by its inaccuracy). The second row corresponds to the limitation of the search mechanism, due mostly to the inaccuracy of MinCut when the segment of interest has very thin parts. The best results are obtained with the indirect score, followed by the direct score; See Figure 4 (a,b), which also demonstrate the improved stability and monotonic behavior of the indirect score. We experimented with different types of feature vectors. The last line corresponds to a minimal feature vector containing the segment size, the online classifier accuracy, and the boundary log probability. This is roughly the same information used to specify the MinCut. The segmentation achieved by this score is not as good as that achieved when the distribution of superpixels scores is added as 10 more features, which in turn is lower than the score calculated with the full feature set.

For the indirect score, the segment pair classifier is the main procedure "behind the scenes." We tested this classifier separately and found a very good correlation between the decision value produced and the difference in F-measures; see Figure 4 (b,c,d).

4.3 Segmentation Results

The search algorithm (section 3) provides hypotheses and tests them using the score. A typical search progression is described in Figure 4 (e,f), which shows the quality of the hypotheses (F-measures) as the search progresses.

Tables 4, 5 compare the segmentation accuracy achieved by the proposed algorithms, the GrabCut algorithms, [6] (implementation of [18] and optimized smoothness parameter), [19] and the segmentation-by-composition [7] on the databases.

The accuracy of our algorithm (Weizmann database) is similar to that of [7] but is much faster (~3 minutes in Matlab on a standard PC). Note that our initialization is automatic. Also note that while our algorithm is initialized only by a single seed point and [6] is initialized with a full bounding box, our algorithm is as accurate as [6] over the GrabCut database and better than [6] over Weizmann's database. The algorithm of [19] indeed achieves better results, but requires much more informative and user demanding initialization: it gets a tightly specified bounding box in which the segment must be close enough to each side of the box.

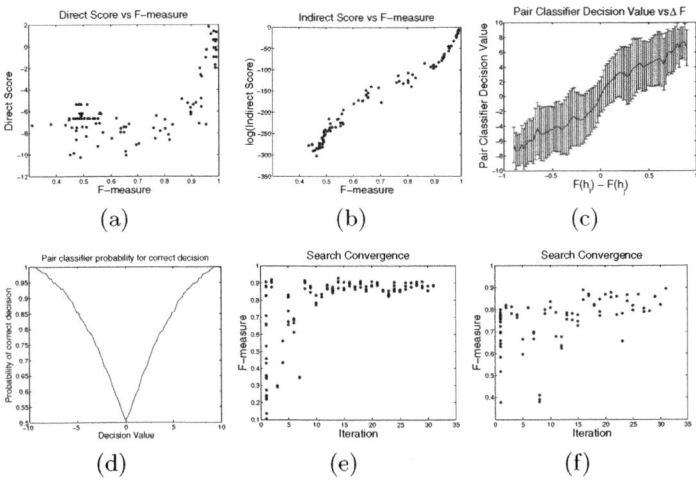

Fig. 4. (a,b): Comparison of direct (a) and indirect (b) scores of hypotheses from a single image. (c,d): Pair classifier statistics: (c) The decision value of the pair classifier is highly correlated with the difference in F-measures (the error bars represent the standard-deviation). (d) A higher absolute value of the pair classifier decision value implies a higher probability for correct decision. (e,f): The F-measures associated with the search in two typical images. Note that the exploration continues and may provide hypotheses with lower scores than that of hypotheses found earlier.

Table 4. Segmentation accuracy (F-measure) on the Weizmann database

Algorithm	Initialization	Mean F	Median F
Direct score	single pixel seed	0.83	0.89
Indirect score	single pixel seed	0.87	0.92
Indirect score (Min Features)	single pixel seed	0.84	0.91
Indirect score (Min Features+Junctions)	single pixel seed	0.845	0.91
Indirect score (Min Features+ SP scores distribution)	single pixel seed	0.86	0.915
GrabCut [6]	bounding box	0.8	0.87
Segmentation By Composition* [7]	single pixel seed**	0.87±0.1	-

Table 5. Segmentation accuracy on the GrabCut color image database

Algorithm	Initialization	Mean Error in the Bounding Box
Indirect score	single pixel seed	8.35%
GrabCut [6]	bounding box	7.2-8.9%
GrabCut-Pinpoint [19]	bounding box	3.7-4.5%

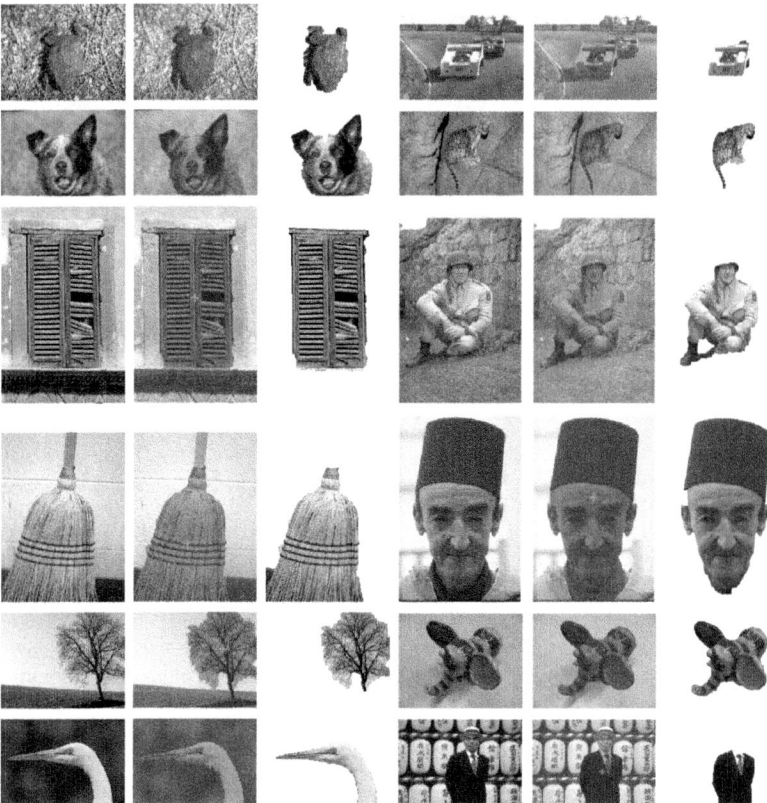

Fig. 5. Segmentation results (Indirect score): Columns 1,4: Original image. Columns 2,5: initial (automatically set) seed (in green) and segmentation results (in red). Columns 3,6: The extracted segment (background removed).

Examples of our results are shown Figure 5. Note that while some segmentation tasks seem easy, some involve heterogeneous descriptors in the segment and/or outside it. Some of the inaccuracies may be attributed to true segmentation ambiguity while others are probably caused by over-fitting of the online classifier or inaccuracy in the initial over-segmentation process.

5   Discussion

This paper proposed a discriminative classifier approach for evaluating image segmentations. Particular insights from this work are:

1. The simple scores used for segmentation hypotheses generation are not the best for their evaluation.
2. Using a rich set of segment descriptors for evaluating the segment quality is practical (when using classification tools) and provides improved, state-of-the-art segment evaluation.

3. Indirect scoring, based on segment comparison, leads to better evaluation results and to better segmentations. Interestingly, humans find it easier to answer the question "which segment is better?" than to answer the question "how good is this segment?".

It is instructive to relate our work to the segmentation by composition approach [7]. In [7] a pixel gets a high score if it is inside a (preferably large) region which matches the segment statistics (e.g., can be constructed from other regions in the segment) and it does not belong to a large region which matches the background statistics. The criterion applied in our approach is, in principle similar but is expressed using the online classifier score. Our algorithm is in a sense simpler than [7] because it does not calculate an explicit construction in various scales. Yet, being discriminative, it is able to select the relevant features automatically and achieves comparable results.

We believe that the proposed scores may be used to improve segmentation algorithms. One practical application for these scores is selecting the best segment out of a group of segment hypotheses produced by different algorithms\parameters. We are now working on a multiscale extension to this algorithm, and on more effective search methods.

References

1. Shi, J., Malik, J.: Normalized cuts and image segmentation. Trans. Pattern Analysis and Machine Intelligence 22, 888–905 (2000)
2. Martin, D.R., Fowlkes, C.C., Malik, J.: Learning to detect natural image boundaries using local brightness, color, and texture cues. IEEE Transactions on Pattern Analysis and Machine Intelligence 26, 530–549 (2004)
3. Dollar, P., Tu, Z., Belongie, S.: Supervised learning of edges and object boundaries. In: 2006 IEEE Computer Society Conference on Computer Vision and Pattern Recognition, vol. 2, pp. 1964–1971 (2006)
4. Ren, X., Malik, J.: Learning a classification model for segmentation. In: Proc. 9th Int'l. Conf. Computer Vision, vol. 1, pp. 10–17 (2003)
5. Chan, T., Vese, L.: Active contours without edges. IEEE Transactions on Image Processing 10, 266–277 (2001)
6. Rother, C., Kolmogorov, V., Blake, A.: Grabcut: Interactive foreground extraction using iterated graph cuts. In: ACM SIGGRAPH 2004, pp. 309–314 (2004)
7. Bagon, S., Boiman, O., Irani, M.: What is a good image segment? a unified approach to segment extraction. In: Forsyth, D., Torr, P., Zisserman, A. (eds.) ECCV 2008, Part IV. LNCS, vol. 5305, pp. 30–44. Springer, Heidelberg (2008)
8. Boykov, Y.Y., Jolly, M.P.: Interactive graph cuts for optimal boundary and region segmentation of objects in n-d images. In: ICCV, vol. 1, pp. 105–112 (2001)
9. Li, Y., Adelson, E., Agarwala, A.: Scribbleboost: Adding classification to edge-aware interpolation of local image and video adjustments. Computer Graphics Forum 27(4), 1255–1264 (2008)
10. Ailon, N., Charikar, M., Newman, A.: Aggregating inconsistent information: ranking and clustering. In: STOC 2005, New York, NY, pp. 684–693 (2005)
11. Boykov, Y., Veksler, O., Zabih, R.: Efficient approximate energy minimization via graph cuts. IEEE Transactions on PAMI 20, 1222–1239 (2001)

12. Alpert, S., Galun, M., Basri, R., Brandt, A.: Image segmentation by probabilistic bottom-up aggregation and cue integration. In: Proceedings of the IEEE Conference on Computer Vision and Pattern Recognition (2007)
13. Meyer, F.: Topographic distance and watershed lines. Signal Processing 38, 113–125 (1994); Mathematical Morphology and Its Applications to Signal Processing
14. Vezhnevets, A.: GML-Adaboost, MSU Graphics and Media Lab, Computer Vision Group, http://graphics.cs.msu.ru
15. Friedman, J., Hastie, T., Tibshirani, R.: Additive logistic regression: a statistical view of boosting. Annals of Statistics 28, 2000 (1998)
16. Lienhart, R., Kuranov, A., Pisarevsky, V.: Empirical analysis of detection cascades of boosted classifiers for rapid object detection. Annals of Statistics 28, 2000 (1998)
17. Sandler, R., Lindenbaum, M.: Optimizing gabor filter design for texture edge detection and classification. Int. J. Comput. Vision 84, 308–324 (2009)
18. Damski, I., Sharfshtein, A.: GrabCut implementation, www1.idc.ac.il/toky/CompPhoto-09/Projects/Stud_projects/IrenaAviad/Web
19. Lempitsky, V., Kohli, P., Rother, C., Sharp, T.: Segmentation with a bounding box prior. In: ICCV 2009 (2009)

Framelet-Based Algorithm for Segmentation of Tubular Structures

Xiaohao Cai[1], Raymond H. Chan[1], Serena Morigi[2], and Fiorella Sgallari[2]

[1] Department of Mathematics,
The Chinese University of Hong Kong, Shatin, N.T., Hong Kong
{xhcai,rchan}@math.cuhk.edu.hk
[2] Department of Mathematics-CIRAM, University of Bologna, Bologna, Italy
{morigi,sgallari}@dm.unibo.it

Abstract. Framelets have been used successfully in various problems in image processing, including inpainting, impulse noise removal, super-resolution image restoration, etc. Segmentation is the process of identifying object outlines within images. There are quite a few efficient algorithms for segmentation that depend on the partial differential equation modeling. In this paper, we apply the framelet-based approach to identify tube-like structures such as blood vessels in medical images. Our method iteratively refines a region that encloses the possible boundary or surface of the vessels. In each iteration, we apply the framelet-based algorithm to denoise and smooth the possible boundary and sharpen the region. Numerical experiments of real 2D/3D images demonstrate that the proposed method is very efficient and outperforms other existing methods.

1 Introduction

In this paper, we consider the segmentation problem of branching tubular objects from 2D and 3D images. This kind of problem arises in several application fields, for example, extracting roads in aerial photography, and anatomical surfaces of tubular structures like blood vessels in Magnetic Resonance Angiography (MRA) images. Because of the necessity to obtain as much fine details as possible in real time, automatic, robust and efficient methods are needed.

There are several vessel segmentation algorithms that are based on deformable models, see [21] for an extended review. Because the explicit deformable model representation is usually impractical, level set techniques to evolve a deformable model have been introduced, and they provide implicit representation of a deformable model. However, the level set segmentation approach is computationally more expensive as it needs to cover the entire domain of interest, which is generally one dimension higher than the original one. Interested readers are referred to recent literature on the level set segmentation strategy for tubular structures [18,20,24].

A new model for active contours to detect objects in a given image based on techniques of curve evolution, Mumford-Shah functional and level sets was

A.M. Bruckstein et al. (Eds.): SSVM 2011, LNCS 6667, pp. 411–422, 2012.

proposed in [10]. A generalization of the active contour without edges model was proposed in [23] for object detection using logic operations. This logic framework suffers from the same limits as in the active contour model and is not suitable for detecting tubular structures.

In [16], a geometric deformable model for the segmentation of tubular-like structures was proposed. The model is characterized mainly by two components: the mean curvature flow and the directionality of the tubular structures. The major advantage of this technique is the ability to segment twisted, convoluted and occluded structures without user interactions; and it can follow the branching of different layers, from thinner to larger structures. The dependence on the grid resolution chosen to solve the discretized partial differential equation (PDE) model is still an open problem. The authors in [16] have also applied a variant of the proposed PDE model to the challenging problem of composed segmentation in [17].

There are some work on texture classification and segmentation using wavelets or wavelet frames [25,1]. Framelet-based approach is a versatile and effective tool for many different applications in image processing, see [4,7,8,3]. Recently, the authors in [13] proposed to combine the framelet-based image restoration model of [5] and the total variation based segmentation model of [10,9,2] to do segmentation. In this paper, we also derive a segmentation algorithm that uses the framelet-based approach. However our method is not based on minimizing any variational model and hence it is different from the method in [13]. In fact, our algorithm gradually updates an interval that contains pixel values of possible boundary pixels. Like the method in [16], our method also has the ability to segment twisted, convoluted and occluded structures. In addition, our method is very effective in denoising and can extract more details from the given image.

The rest of the paper is organized as follows. In Section 2, we recall some basic facts about tight frames and framelet-based algorithms. Our segmentation algorithm is given in Section 3. Section 4 discusses how to find an interval that contains pixel values of possible boundary pixels. In Section 5 we test our algorithm on various real 2D and 3D images. Comparisons with other methods will be also given. Conclusions are given in Section 6.

2   Framelet-Based Algorithm

In this section, we briefly introduce the framelet-based algorithm. For theories of tight frames and framelets, we refer the readers to [11] for more details. In order to apply the framelet-based algorithm, one only needs to know the filters corresponding to the framelets. For the framelets derived from the piecewise linear B-spline, the corresponding filters are:

$$h_0 = \frac{1}{4}[1, \ 2, \ 1], \quad h_1 = \frac{\sqrt{2}}{4}[1, \ 0, \ -1], \quad h_2 = \frac{1}{4}[-1, \ 2, \ -1], \tag{1}$$

see [22]. The framelet coefficients of any given vector $\mathbf{v}$ corresponding to filter h_i can be obtained by convolving h_i with $\mathbf{v}$. In matrix terms, we can construct, for each filter, its corresponding filter matrix which is just the Toeplitz matrix with diagonals given by the filter coefficients, e.g. $H_0 = \frac{1}{4}\text{tridiag}[1,\ 2,\ 1]$. Then the 1D framelet forward transform is given by

$$A = \begin{bmatrix} H_0 \\ H_1 \\ H_2 \end{bmatrix}. \qquad (2)$$

To apply the framelet transform onto $\mathbf{v}$ is equivalent to computing $A\mathbf{v}$, and $H_i\mathbf{v}$ gives the framelet coefficients corresponding to the filter h_i, $i = 1, 2, 3$.

The d-dimensional framelet system is constructed by tensor products from the 1D framelets, see [14]. For example, in 2D, there are nine framelets given by $h_{ij} \equiv h_i^T \otimes h_j$ for $i, j = 1, 2, 3$, where h_i is given in (1). For any 2D image f, the framelet coefficients with respect to h_{ij} are obtained by convolving h_{ij} with f. In 3D, there are twenty-seven filters and the framelet coefficients can also be obtained by convolutions. Let A represent the corresponding framelet forward transform matrix (cf. (2)). In 2D case, A will be a stack of nine block-Toeplitz-Toeplitz-block matrices, see [4]. In this notation, given any image f, the matrix-vector product $A \cdot \text{vec}(f)$ gives all the framelet coefficients. Here $\text{vec}(f)$ denotes the vector obtained by concatenating the columns of f.

All framelet transforms have a very important property, the *"perfect reconstruction property"*: $A^T A = I$, the identity matrix, see [22]. Unlike the wavelets, in general, $AA^T \neq I$. The framelet-based algorithms, as given in [8,4,3], are of the following generic form:

$$\mathbf{f}^{(i+\frac{1}{2})} = \mathcal{U}(\mathbf{f}^{(i)}), \qquad (3)$$

$$\mathbf{f}^{(i+1)} = A^T \mathcal{T}_\lambda(A\mathbf{f}^{(i+\frac{1}{2})}), \quad i = 1, 2, \ldots. \qquad (4)$$

Here $\mathbf{f}^{(i)}$ is an approximate solution, $\mathcal{U}$ is a problem-dependent operator, and $\mathcal{T}_\lambda(\cdot)$ is the soft-thresholding operator defined as follows. Given vectors $\mathbf{v} = [v_1, \cdots, v_n]^T$ and $\lambda = [\lambda_1, \cdots, \lambda_n]^T$, $\mathcal{T}_\lambda(\mathbf{v}) \equiv [t_{\lambda_1}(v_1), \cdots, t_{\lambda_n}(v_n)]^T$, where

$$t_{\lambda_k}(v_k) \equiv \begin{cases} \text{sgn}(v_k)(|v_k| - \lambda_k), & \text{if } |v_k| > \lambda_k, \\ 0, & \text{if } |v_k| \leq \lambda_k. \end{cases} \qquad (5)$$

For how to choose λ_k, see [15].

Algorithm (4) is usually called the *isotropic* framelet-based algorithm. This is because the thresholding operator $\mathcal{T}_\lambda$ is applied on all the framelet coefficients $A\mathbf{f}^{(i+\frac{1}{2})}$ in (4). In [7], the *anisotropic* framelet-based algorithm was proposed. The main idea is that the filter h_1 in (1) is the central-difference apart from a scalar multiple. Hence the corresponding framelet coefficients are related to the gradient ∇f of the image f. One should therefore rotate these coefficients along the tangential and normal direction, and threshold only the components along the tangential direction, see [26]. For the coefficients corresponding to other filters, we threshold as in the isotropic framelet-based algorithm. In [7], it was shown that the anisotropic thresholding scheme can give better restoration than the isotropic one, and can follow edges more closely. Later in the numerical tests, we have tried both thresholding schemes, and found that anisotropic thresholding can give the tubular structures better.

3   Framelet-Based Algorithm for Segmentation

The technology of MRA imaging is based on detection of signals from flowing blood and suppression of signals from other static tissues, so that the blood vessels appear as high intensity regions in the image, see Fig. 1(a). The structures to be segmented are vessels of variable diameters which are close to each other. Partial occlusions and intersections make the segmentation very challenging. Moreover, the real image can be affected by speckle noise. In general in medical images, speckle noise and weak edges make it difficult to identify the structures in the image. Fortunately, the MRA images also contain some properties that can be used to construct our algorithm. From Fig. 1(a), we see that the pixels near the boundary of the vessels are not exactly of one value, but they are in some range, whereas the values of the pixels in other parts are far from this range. Thus the main idea of our algorithm is to approximate this range accurately. We will obtain the range iteratively by a framelet-based algorithm. The main steps are as follows. Suppose in the beginning of the ith iteration, we are given an approximate image $f^{(i)}$, and an approximate range $[\alpha_i, \beta_i]$ for $\alpha_i \leq \beta_i$ which contains the pixel values of all the possible boundary pixels. Then we (i) use the range to threshold the image into three parts—below, inside, and above the range; (ii) denoise and smooth the inside part by the framelet-based algorithm to get a new image $f^{(i+1)}$; and (iii) refine the range to $[\alpha_{i+1}, \beta_{i+1}]$ by using $f^{(i+1)}$. We stop when $f^{(i+1)}$ becomes a binary image. In the followings, we elaborate each of the steps. Without loss of generality, we assume all images have dynamic range in $[0, 1]$.

Step (i): Thresholding the Image into Three Parts. Using the range $[\alpha_i, \beta_i] \subseteq [0, 1]$, we can separate the image $f^{(i)}$ into three parts—below, inside, and above the range, see Fig. 1(b). To emphasize the boundary, we threshold those pixel values that are smaller than α_i to 0, those larger than β_i to 1, and those in between, we stretch them between 0 and 1 using a simple linear contrast stretch, see [19]. More precisely, let Ω be the index set of all the pixels in the image, $f_j^{(i)}$ be the pixel value of pixel j in image $f^{(i)}$, and

$$M_i = \max\{f_j^{(i)} \mid \alpha_i \leq f_j^{(i)} \leq \beta_i, j \in \Omega\},$$
$$m_i = \min\{f_j^{(i)} \mid \alpha_i \leq f_j^{(i)} \leq \beta_i, j \in \Omega\}.$$

Then we define

$$f_j^{(i+\frac{1}{2})} = \begin{cases} 0, & \text{if } f_j^{(i)} \leq \alpha_i, \\ \frac{f_j^{(i)} - m_i}{M_i - m_i}, & \alpha_i \leq f_j^{(i)} \leq \beta_i, \\ 1, & \text{if } \beta_i \leq f_j^{(i)}, \end{cases} \quad \text{for all } j \in \Omega. \tag{6}$$

Fig. 1(c) shows the threshold and stretched image from Fig. 1(b). In the following, we write (6) simply as $f^{(i+\frac{1}{2})} = \mathcal{U}(f^{(i)})$ (cf. (3)), and we denote

$$\Lambda^{(i)} = \{j \mid m_i < f_j^{(i)} < M_i, j \in \Omega\}, \tag{7}$$

the index set for pixels with values inside the range (m_i, M_i), i.e., the index set for pixels of $f^{(i+\frac{1}{2})}$ with values neither 0 nor 1. Our next step is to denoise and smooth the image $f^{(i+\frac{1}{2})}$ on $\Lambda^{(i)}$.

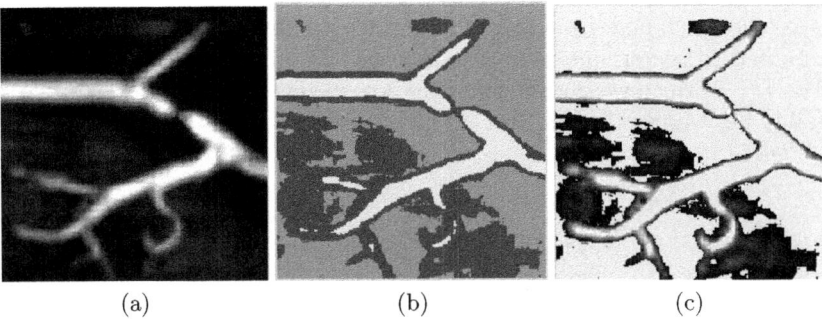

(a)   (b)   (c)

Fig. 1. (a) Given image. (b) Three parts of the given image (green–below, red–in between, and yellow–above). (c) Threshold and stretched image by (6) (yellow pixels are with value 0 or 1).

Step (ii): Framelet-Based Iteration. To denoise and smooth the image $f^{(i+\frac{1}{2})}$ on $\Lambda^{(i)}$, we apply the framelet-based iteration (4) on $\Lambda^{(i)}$. More precisely, if $j \notin \Lambda^{(i)}$, then we set $f_j^{(i+1)} = f_j^{(i+\frac{1}{2})}$; otherwise, we use (4) to get $f_j^{(i+1)}$. To write it out clearly, let $\mathbf{f}^{(i+\frac{1}{2})} = \text{vec}(f^{(i+\frac{1}{2})})$, and $P^{(i)}$ be the diagonal matrix where the diagonal entry is 1 if the corresponding index is in $\Lambda^{(i)}$, and 0 otherwise. Then

$$\mathbf{f}^{(i+1)} \equiv (I - P^{(i)})\mathbf{f}^{(i+\frac{1}{2})} + P^{(i)}A^T \mathcal{T}_\lambda(A\mathbf{f}^{(i+\frac{1}{2})}). \tag{8}$$

By reordering the entries of the vector $\mathbf{f}^{(i+1)}$ into columns, we obtain the image $f^{(i+1)}$. Note that the effect of (8) is to denoise and smooth the image on $\Lambda^{(i)}$, see [4]. Since the pixel values of all pixels outside $\Lambda^{(i)}$ are either 0 or 1, the cost of matrix-vector multiplications in (8), such as $A\mathbf{f}^{(i+\frac{1}{2})}$, can be reduced significantly by taking advantage of this.

Step (iii): Refining the Range. The process of finding the new range $[\alpha_{i+1}, \beta_{i+1}]$ from $f^{(i+1)}$ is very similar to the process of finding the initial interval $[\alpha_0, \beta_0]$ from the given image. We postpone it till the next section. We will see that $[\alpha_{i+1}, \beta_{i+1}] \subsetneq [0,1]$ for all $i \geq 0$. This point guarantees the convergence of our method, see Theorem 1.

Stopping Criterion. We stop the iteration when all the pixels of $f^{(i+\frac{1}{2})}$ are either of value 0 or 1, or equivalently when $|\Lambda^{(i)}| = 0$. For the binary image $f^{(i+\frac{1}{2})}$, all the pixels with value 1 constitute the tubular structures. In the numerical tests, we use the matlab command "`contour`" and "`isosurface`" respectively to obtain the boundary of $f^{(i+\frac{1}{2})}$ in 2D and 3D respectively.

4   Initializing and Refining the Range

In this section, we discuss how to find $[\alpha_i, \beta_i]$ given $f^{(i)}$. When $i = 0$, the initial guess $f^{(0)}$ is chosen to be the given image. Recall that $[\alpha_i, \beta_i]$ is an interval containing the pixel values of the possible boundary pixels. Our idea of finding it is as follows: (i) find the average $\mu^{(i)}$ of the pixel values of the possible boundary pixels; and (ii) determine a suitable interval $[\alpha_i, \beta_i] \subsetneq [0, 1]$ that contains $\mu^{(i)}$.

For $i = 0$, since we do not have any knowledge of where the boundary will possibly be, we use the gradient of $f^{(0)}$ to find it. Define the gradient image g of $f^{(0)}$ as:

$$g_j = \left[\sum_{\ell=1}^{d} (\partial_{x_\ell} f_j^{(0)})^2 \right]^{1/2}, \quad \text{for all } j \in \Omega,$$

where ∂_{x_ℓ} is the forward-difference in the x_ℓ-direction, and $d = 2$ for 2D and $d = 3$ for 3D. Our first approximation of the boundary is composed of those pixels where $g_j > \epsilon$ for a given ϵ. (In our numerical tests, we choose $\epsilon \in [10^{-3}, 10^{-1}]$.) Thus let $\Gamma = \{j \mid g_j > \epsilon, j \in \Omega\}$, the index set of those pixels; and let μ_Γ be the average of $f^{(0)}$ on Γ, i.e.

$$\mu_\Gamma = \frac{1}{|\Gamma|} \sum_{j \in \Gamma} f_j^{(0)},$$

where $|\Gamma|$ is the cardinality of Γ. Obviously the smaller the ϵ is, the larger cardinality of $|\Gamma|$ will be.

Naturally, those pixels in Γ can be separated into two parts by μ_Γ: one part contains the pixels near to the tubulars (yellow part of Fig. 1(b)), and the other part is near to the background (green part of Fig. 1(b)). More precisely, we define

$$\Gamma_+ = \{j \mid f_j > \mu_\Gamma, j \in \Gamma\} \quad \text{and} \quad \Gamma_- = \{j \mid f_j < \mu_\Gamma, j \in \Gamma\}.$$

Let μ_+ and μ_- be the averages of $f^{(0)}$ on Γ_+ and Γ_- respectively. Note that μ_+ (and respectively μ_-) is the average of those possible boundary pixels that are close to the tubulars (and respectively close to the background). We use them to compute $\mu^{(0)}$, the average of the possible boundary pixels. Define

$$\Lambda^{(-1)} = \{j \mid \mu_- < f_j^{(0)} < \mu_+, j \in \Omega\}.$$

Then

$$\mu^{(0)} = \frac{1}{|\Lambda^{(-1)}|} \sum_{j \in \Lambda^{(-1)}} f_j^{(0)}. \tag{9}$$

For $i \geq 0$, we define $\mu^{(i+1)}$ similarly as in (9):

$$\mu^{(i+1)} = \frac{1}{|\Lambda^{(i)}|} \sum_{j \in \Lambda^{(i)}} f_j^{(i+1)}, \tag{10}$$

where $\Lambda^{(i)}$ is given by (7) and $f^{(i+1)}$ is given by (8).

Finally we discuss how to choose the interval $[\alpha_i, \beta_i]$ that contains $\mu^{(i)}$ for $i \geq 0$. Our idea is first to compute coarse interval $[\alpha_i^L, \beta_i^H]$ that contains $\mu^{(i)}$ by

$$\alpha_i^L = \frac{1}{|\{j : f_j^{(i)} \leq \mu^{(i)}\}|} \sum_{\{j : f_j^{(i)} \leq \mu^{(i)}\}} f_j^{(i)}, \quad \beta_i^H = \frac{1}{|\{j : f_j^{(i)} \geq \mu^{(i)}\}|} \sum_{\{j : f_j^{(i)} \geq \mu^{(i)}\}} f_j^{(i)},$$

where $j \in \Lambda^{(i-1)}$ for $i \geq 1$ and $j \in \Omega$ for $i = 0$. From the above formulas, we can see that $[\alpha_i^L, \beta_i^H]$ will never be $[0, 1]$ if the given image is not a binary image. (If $\alpha_i^L = \beta_i^H$, then all remaining pixels have the same pixel value. Hence we set them all to 1 and the image is thus a binary image, and the algorithm stops.)

Next we compute, for all $\alpha \in [\alpha_i^L, \beta_i^H]$ ($\alpha_i^L \neq \beta_i^H$),

$$c(\alpha) \equiv \frac{1}{|\{j : f_j^{(i)} \geq \alpha\}|} \sum_{\{j : f_j^{(i)} \geq \alpha\}} f_j^{(i)} - \frac{1}{|\{j : f_j^{(i)} \leq \alpha\}|} \sum_{\{j : f_j^{(i)} \leq \alpha\}} f_j^{(i)},$$

where $j \in \Omega$. Let the range of $c(\alpha)$ be $[c_m, c_M]$ and $\ell = c_M - c_m$. Then

$$\alpha_i \equiv \min\{\alpha \in [\alpha_i^L, \beta_i^H] \mid c(\alpha) = c(\mu^{(i)}) - \gamma\ell\}, \tag{11}$$

$$\beta_i \equiv \max\{\alpha \in [\alpha_i^L, \beta_i^H] \mid c(\alpha) = c(\mu^{(i)}) + \gamma\ell\}. \tag{12}$$

Here $\gamma \in (0, 1/2)$ is a parameter that controls the length of the interval $[\alpha_i, \beta_i]$. Clearly the larger the interval is, the more pixels are to be considered as possible boundary pixels. So the smaller the γ is, the faster convergence of our method will have. In the numerical tests, we choose $\gamma = 1/5$. We give the full algorithm below and show that it always converges to a binary image.

Algorithm: Framelet-based algorithm for segmentation
1. Initialize: set $f^{(0)} = f$, $\mu^{(0)}$ by (9), and $[\alpha_0, \beta_0]$ by (11) and (12).
2. Do $i = 0, 1, \ldots$, until stopped
   (a) Compute $f^{(i+\frac{1}{2})} = \mathcal{U}(f^{(i)})$ by (6).
   (b) Stop if $f^{(i+\frac{1}{2})}$ is a binary image.
   (c) Update $f^{(i+\frac{1}{2})}$ to $f^{(i+1)}$ by (8).
   (d) Update $\mu^{(i+1)}$ by (10), and then $[\alpha_{i+1}, \beta_{i+1}]$ by (11) and (12).
3. Extract the boundary from the binary image $f^{(i+\frac{1}{2})}$.

Theorem 1. *Our framelet-based algorithm will converge to a binary image.*

Proof. Obviously, we just need to prove that $|\Lambda^{(i)}| = 0$ at some finite step i, see (6) and (7). If the given image $f^{(0)}$ is a binary image, we are done. Without loss of generality, we assume that $f^{(0)}$ is not a binary image. Given $\Lambda^{(i-1)}$ defined by (7) for any $i \geq 1$, note that the pixel values of those pixels not in $\Lambda^{(i-1)}$ will not be changed by (8), i.e., they will stay at either 0 or 1. Then $[\alpha_i, \beta_i]$ will be obtained by (11) and (12), where $[\alpha_i, \beta_i] \subseteq [\alpha_i^L, \beta_i^H] \subsetneq [0, 1]$. Since $[m_i, M_i] \subseteq [\alpha_i, \beta_i]$, we have $m_i \neq 0$ or $M_i \neq 1$. By (6) and (7), the pixels satisfying $f^{(i)} \leq m_i$ or $f^{(i)} \geq M_i$ are set to 0 or 1 respectively. Thus, there will be at least one pixel in $\Lambda^{(i-1)}$ with value neither 0 nor 1 that is set to 0 or 1 by (6). Hence $|\Lambda^{(i)}| < |\Lambda^{(i-1)}|$, where $|\cdot|$ denotes the cardinality of the set. Since $|\Lambda^{(0)}|$ is finite, there must exist some i such that $|\Lambda^{(i)}| = 0$. $\qquad\square$

Finally, let us estimate the computation cost of our method for a given image with n pixels: (i) the cost of computing each $\mu^{(i)}$ and $[\alpha_i, \beta_i]$ is $O(n)$, see (9)–(12) respectively; and (ii) the cost of steps (a) and (c) in the above algorithm is $O(n)$, see (6) and (8). Hence the cost of our method is $O(n)$ per iteration. We remark that our algorithm usually converges within a few iterations, see the numerical results in the next section.

5   Numerical Examples

In this section, we test our proposed framelet-based segmentation method on three 2D/3D real images. All the data are obtained from [16] and [17]. We use the piece-wise linear filters given in (1) with only the first level, i.e. no downsampling. We tried both isotropic and anisotropic thresholding schemes, see the discussion at the end of Section 2. The thresholding parameters λ_k used in (5) are chosen to be $\lambda_k \equiv 2^{-1/2}$ for isotropic thresholding. For anisotropic thresholding, $\lambda_k = 0.1 \times 2^{-1/2}$ for the components along the tangential direction and $\lambda_k = 2^{-1/2}$ for other coefficients.

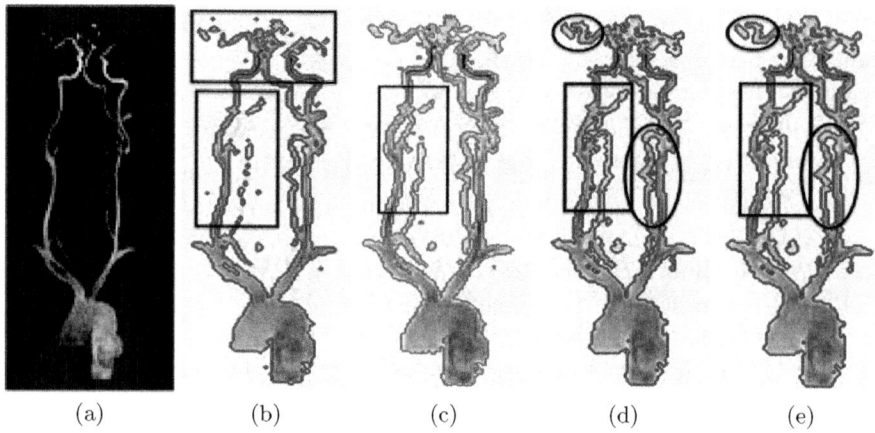

(a)   (b)   (c)   (d)   (e)

Fig. 2. Carotid vascular system segmentation. (a) Given image. (b) and (c) Results by the methods in [10] and [16] respectively. (d) and (e) Results by our method with isotropic and anisotropic thresholding schemes respectively.

Example 1. The test image is a 182×62 MRA image of a carotid vascular system, see Fig. 2(a). The results by our method using isotropic and anisotropic thresholding are given by Fig. 2(d) and (e) respectively. With the parameters $\gamma = 1/5$ and $\epsilon = 1.6 \times 10^{-2}$, our method converges in 6 iterations for both thresholding schemes. The first and second rows of Table 1 give $|\Lambda^{(i)}|$ at each iteration, from which, we can see that only very few pixels (comparing with $|\Omega| = 182 \times 62 = 11,284$) need to be classified after 3 iterations. For the purpose of

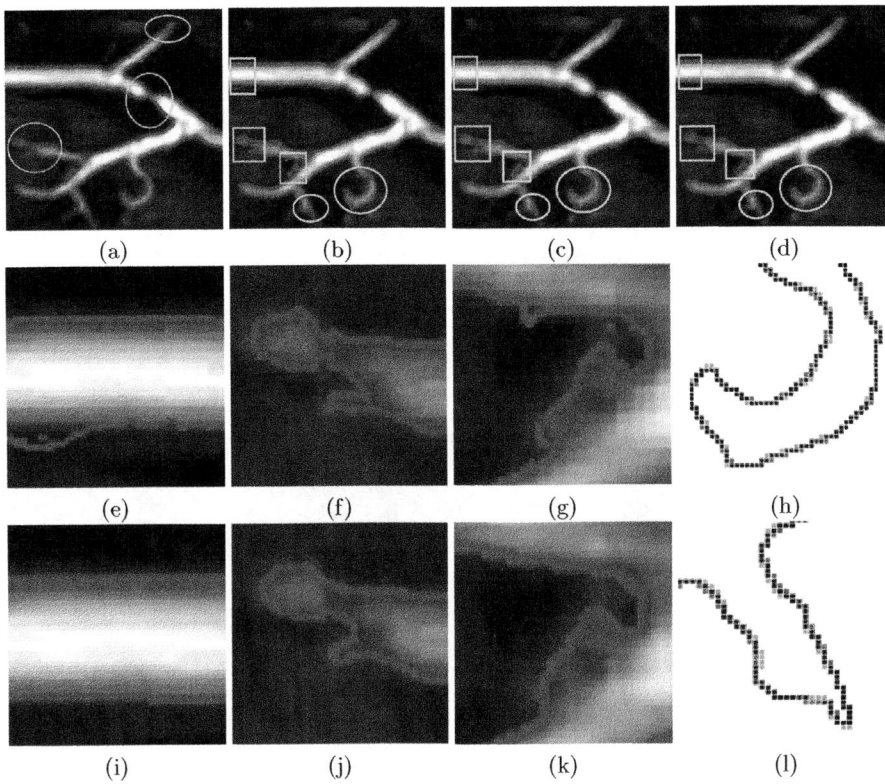

Fig. 3. Kidney vascular system segmentation. (a) and (b) Results by the methods in [10] and [16] respectively. (c) and (d) Results by our method with isotropic and anisotropic thresholding respectively. (e)–(g) and (i)–(k) Zoomed in rectangular parts of (b) and (c) respectively. (h) and (l) Superimposed boundaries inside the ellipses of (c) and (d) (red—in (c), green—in (d), and black–intersection of (c) and (d)).

Table 1. Cardinality of $\Lambda^{(i)}$ at each iteration of the three examples

$n = |\Omega|$	$|\Lambda^{(0)}|$	$|\Lambda^{(1)}|$	$|\Lambda^{(2)}|$	$|\Lambda^{(3)}|$	$|\Lambda^{(4)}|$	$|\Lambda^{(5)}|$	$|\Lambda^{(6)}|$	$|\Lambda^{(7)}|$	$|\Lambda^{(8)}|$	$|\Lambda^{(9)}|$
Fig. 2(d)	11284	2374	307	83	23	7	1	0	-	-	-
Fig. 2(e)	11284	2374	233	48	13	5	1	0	-	-	-
Fig. 3(c)	66049	8314	1834	565	137	29	18	4	0	-	-
Fig. 3(d)	66049	8314	1557	406	95	19	5	1	0	-	-
Fig. 4(d)	8120601	104329	21333	5460	1430	326	70	9	3	1	0
Fig. 4(e)	8120601	104329	20020	4984	1260	299	72	19	6	0	-

comparison, we also give the results by the methods in [10] and [16] respectively, see Fig. 2(b) and (c). Clearly, the result of Fig. 2(b) is not satisfactory since the tubulars obtained are disconnected. By comparing the parts inside the rectangles in Fig. 2 (c) with those in Fig. 2(d) and (e), we see that our method can extract

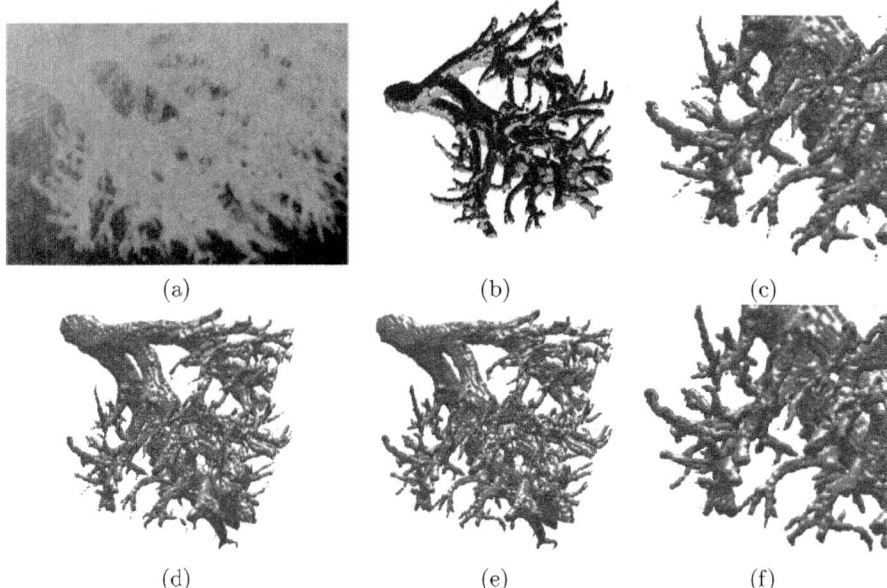

(a)   (b)   (c)

(d)   (e)   (f)

Fig. 4. Segmentation of the kidney volume data set. (a) Given CTA image. (b) Result by the method in [17]. (d) and (e) Results by our method with isotropic and anisotropic thresholding respectively. (c) and (f) Zoomed in the bottom-left corners of (d) and (e) respectively.

more details than the method in [16]. Finally, the parts inside the ellipses of Fig. 2(d) and (e) demonstrate that the anisotropic thresholding keeps the edge better than isotropic thresholding.

Example 2. The test image is a 257×257 MRA image of a kidney vascular system as shown in Fig. 1(a). This example shows the ability of our method to reconstruct structures which present small occlusions along the coherence direction. With the parameters $\gamma = 1/5$ and $\epsilon = 5 \times 10^{-3}$, our method converges in 7 iterations for both thresholding schemes. The third and fourth rows in Table 1 give $|A^{(i)}|$ at each iteration. The result of Fig. 3(a) by the method in [10] is not good since it can not connect the small occlusions along the coherence direction, while this can be done by our method and the method in [16], see Fig. 3(b), (c) and (d). Furthermore, our method is better than the method in [16] by comparing the rectangular parts of Fig. 3(b) with those in Fig. 3(c) and (d), since our method can detect smoother edges. More precisely, see Fig. 3(e)–(g) and (i)–(k), which are the results of zooming in the rectangular parts of Fig. 3(b) and (c) respectively. This also shows that our method is very effective in denoising. In order to compare Fig. 3(c) with (d) explicitly, we superimpose the boundaries of them, see Fig. 3(h) and (l). Clearly, the boundary of the tubulars is tighter and more pixels at the tips are obtained by anisotropic thresholding.

Example 3. This is a 3D example where we extracted a volumetric data set of size $201 \times 201 \times 201$ from a $436 \times 436 \times 540$ CTA (Computed Tomographic Angiography) image of the kidney vasculature system, see Fig. 4(a). With the parameters $\gamma = 1/5$ and $\epsilon = 6 \times 10^{-2}$, our method converges in 9 and 8 iterations for isotropic and anisotropic thresholding respectively. The last two rows in Table 1 show $|\Lambda^{(i)}|$ at each iteration. By comparing our method with the method in [17], the results show that our method can give many more details, see Fig. 4(b), (d) and (e). The zoomed in bottom-left corners of Fig. 4(d) and (e) clearly give the difference of the results of our method by isotropic and anisotropic thresholding, see Fig. 4(c) and (f). We see that more pixels at the tips of the tubular structures are detected and the tubular structures are connected better by the anisotropic thresholding than by the isotropic thresholding.

6 Conclusions and Future Work

In this paper, we introduced a new segmentation method based on the framelet-based approach. The numerical results demonstrate the ability of our method for segmenting tubular structures. The method can be implemented fast and give very accurate, smooth boundaries or surfaces. In addition, since the pixel values of more and more pixels will be set to either 0 or 1 during the iteration, by taking advantage of this, one can construct a sparse data structure to accelerate the method. Moreover, one can use different tight frame systems such as those from contourlets and curvelets [12,6] to better capture the boundary. These are directions we will explore in the future.

References

1. Arivazhagan, S., Ganesan, L.: Texture segmentation using wavelet transform. Pattern Recognition Letters 24, 3197–3203 (2003)
2. Bresson, X., Esedoglu, S., Vandergheynst, P., Thiran, J., Osher, S.: Fast global minimization of the active contour/snake model. J. Math. Imaging Vision 28, 151–167 (2007)
3. Cai, J.F., Chan, R.H., Shen, L.X., Shen, Z.W.: Simultaneously inpainting in image and transformed domains. Numer. Math. 112, 509–533 (2009)
4. Cai, J.F., Chan, R.H., Shen, Z.W.: A framelet-based image inpainting algorithm. Appl. Comput. Harmon. Anal. 24, 131–149 (2008)
5. Cai, J.F., Osher, S., Shen, Z.W.: Split Bregman methods and frame based image restoration. Multiscale Modeling and Simulation 8, 337–369 (2009)
6. Candès, E., Demanet, L., Donoho, D., Ying, L.: Fast discrete curvelet transforms. Multiscale Modeling and Simulation 5, 861–899 (2006)
7. Chan, R.H., Setzer, S., Steidl, G.: Inpainting by flexible Haar-wavelet shrinkage. SIAM J. Imaging Sci. 1, 273–293 (2008)
8. Chan, R.H., Chan, T.F., Shen, L.X., Shen, Z.W.: Wavelet algorithms for high-resolution image reconstruction. SIAM J. Sci. Comput. 24, 1408–1432 (2003)
9. Chan, T.F., Esedoglu, S., Nikolova, M.: Algorithms for finding global minimizers of image segmentation and denoising models. Technical Report 54, UCLA (2004)

10. Chan, T.F., Vese, L.A.: Active contours without edges. IEEE Trans. Image Process. 10, 266–277 (2001)
11. Daubechies, I.: Ten lectures on wavelets. Lecture Notes, vol. CBMS-NSF(61). SIAM, Philadelphia (1992)
12. Do, M.N., Vetterli, M.: The contourlet transform: an efficient directional multiresolution image representation. IEEE Trans. Image Process. 14, 2091–2106 (2004)
13. Dong, B., Chien, A., Shen, Z.W.: Frame based segmentation for medical images. Technical Report 22, UCLA (2010)
14. Dong, B., Shen, Z.W.: MRA based wavelet frames and applications. IAS Lecture Notes Series, Summer Program on The Mathematics of Image Processing, Park City Mathematics Institute (2010)
15. Donoho, D.L.: De-noising by soft-thresholding. IEEE Trans. Inform. Theory 41, 613–627 (1995)
16. Franchini, E., Morigi, S., Sgallari, F.: Segmentation of 3D tubular structures by a PDE-based anisotropic diffusion model. In: Dæhlen, M., Floater, M., Lyche, T., Merrien, J.-L., Mørken, K., Schumaker, L.L. (eds.) MMCS 2008. LNCS, vol. 5862, pp. 224–241. Springer, Heidelberg (2010)
17. Franchini, E., Morigi, S., Sgallari, F.: Composed segmentation of tubular structures by an anisotropic PDE model. In: Tai, X.-C., Mørken, K., Lysaker, M., Lie, K.-A. (eds.) SSVM 2009. LNCS, vol. 5567, pp. 75–86. Springer, Heidelberg (2009)
18. Gooya, A., Liao, H., et al.: A variational method for geometric regularization of vascular segmentation in medical images. IEEE Trans. Image Process. 17, 1295–1312 (2008)
19. Gonzales, R.C., Woods, R.E.: Digital Image Processing, 3rd edn. Prentice Hall, Englewood Cliffs (2008)
20. Hassan, H., Farag, A.A.: Cerebrovascular segmentation for MRA data using levels set. International Congress Series, vol. 1256, pp. 246–252 (2003)
21. Kirbas, C., Quek, F.: A review of vessel extraction techniques and algorithms. ACM Computing Surveys 36, 81–121 (2004)
22. Ron, A., Shen, Z.W.: Affine Systems in L2(Rd): The Analysis of the Analysis Operator. J. Funct. Anal. 148, 408–447 (1997)
23. Sandberg, B., Chan, T.F.: A logic framework for active contours on multi-channel images. J. Vis. Commun. Image R. 16, 333–358 (2005)
24. Scherl, H., et al.: Semi automatic level set segmentation and stenosis quantification of internal carotid artery in 3D CTA data sets. Medical Image Analysis 11, 21–34 (2007)
25. Unser, M.: Texture classification and segmentation using wavelet frames. IEEE Trans. Image Process. 4, 1549–1560 (1995)
26. Weickert, J.: Anisotropic Diffusion in Image Processing. Teubner, Stuttgart (1998)

Weakly Convex Coupling
Continuous Cuts and Shape Priors

Bernhard Schmitzer and Christoph Schnörr

University of Heidelberg

Abstract. We introduce a novel approach to variational image segmentation with shape priors. Key properties are convexity of the joint energy functional and weak coupling of convex models from different domains by mapping corresponding solutions to a common space. Specifically, we combine total variation based continuous cuts for image segmentation and convex relaxations of Markov Random Field based shape priors learned from shape databases. A convergent algorithm amenable to large-scale convex programming is presented. Numerical experiments demonstrate promising synergistic performance of convex continuous cuts and convex variational shape priors under image distortions related to noise, occlusions and clutter.

1   Introduction

1.1   Overview, Related Work

Various continuous variational approaches to image labeling and segmentation have been presented in the recent literature [4,14,13] based on tight convex relaxations of the underlying combinatorial problem. While the relaxation of the binary two-class case can be shown to compute the global combinatorial optimum after thresholding [4], the non-binary case of multiple labels [14,13] also returns high-quality combinatorial solutions in practice, as numerical experiments based on primal-dual iterations show. Unlike algorithms for fully discrete graph-cut approaches [2] that may get stuck in a poor local minimum in the nonbinary case, their continuous convex counterparts do not suffer from such problems. Moreover, a broad range of robust first-order minimization algorithms from sparse convex programming are available for efficiently solving such large-scale problems [9,3].

Variational approaches comprise a data term and a regularization term. In connection with image labeling, the data term is a linear form that does not impose any restriction on the type of image features to be processed. Concerning the regularization term, a large class of alternatives to the standard Potts prior has been suggested in [12], all of which do not compromise convexity of the variational approach.

While features and regularization terms can be handled quite flexibly within a convex variational framework, this is not the case for another major clue to reliable segmentations: shape. Substantial research work has been done on

A.M. Bruckstein et al. (Eds.): SSVM 2011, LNCS 6667, pp. 423–434, 2012.

variational representations of shape statistics as additional penalty terms, ranging from sampled contours and kernel techniques from machine learning to sophisticated manifolds of invariant shape representations [6,18,16]. Analogous ideas have been applied to embedding functions in connection with level set based approaches to shape representation and segmentation [5,7,11]. In this connection, we point out two properties of prior work that motivated the work presented in this paper:

- shape representations may not conform to the representation of image segmentations (contour spaces vs. regions or set of pixels after discretization);
- shape penalty functionals are nonconvex (except for the less attractive case of Gaussian shape statistics based on sampled contours);
- nonconvex level set based functionals have been used for image segmentation;
- adding a shape penalty functional compromises convexity of the overall approach.

1.2   Contribution, Organization

The approach introduced in this paper comprises

(i) *separate convex modeling* of variational approaches to segmentation and shape priors, respectively, and

(ii) *weak convex coupling* of these models in terms of a convex, but possibly indefinite, quadratic form

$$\|Ax - B\mu\|^2, \tag{1}$$

that measures similarity of segmentations x and shapes μ, respectively, *mapped to a common space* by linear mappings A and B.

As a consequence, our corresponding approach avoids all deficiencies of previous approaches, from the viewpoint of optimization.

Concerning (i) and segmentation, we use convex models of continuous cuts as introduced in [4,14,13]. Concerning (i) and shape priors, we apply strengthened local polytope convex relaxations of binary Markov Random Fields (MRFs) [17,8,15] whose structure and parameters are learned offline from a shape database using large-scale convex optimization in a preprocessing step [10]. Concerning (ii), we adopt the framework presented in [1] for coupling two proper convex and lower semicontinuous functionals defined on *different* spaces.

The main objective of this paper is to *introduce the general framework*. Therefore, we make no attempt to present a complete list of fully-fledged model components, but rather confine ourselves to demonstrating the key aspect – coupling of convex models – using preliminary versions of individual models. Specifically, concerning segmentation, we merely employ *binary* continuous cuts [4] and deliberately *do not elaborate the issue of feature extraction*, having in mind that, as discussed in the previous section, any image features computed in a preprocessing step could be used. We apply two different MRF models as shape priors in order to indicate the potential of this research direction: a naive MRF directly defined on the pixel grid, and a hierarchically defined MRF that compares favourably from the viewpoint of learning and automatically extracts a part-based probabilistic representation of object shapes taken from different viewpoints. Under strong noise levels simulating feature imperfections, we numerically demonstrate promising synergistic performance of convex continuous cuts and convex variational shape priors.

(a)   (b)   (c)   (d)

Fig. 1. Binary image segmentation: (a) noisy input, (b) only continuous cuts with low regularization, (c) only continuous cuts with strong regularization, (d) convex coupling of continuous cuts with low regularization and shape prior

Our approach is general and can be extended in various directions. We indicate this below as we go along.

2   Variational Models

2.1   Segmentation by Continuous Cuts

We adopt the approach [4] for globally optimal foreground-background separation by convex optimization.

Let Ω denote the uniform pixel grid of size $|\Omega| = N$ corresponding to the domain $[0,1]^2 \subset \mathbb{R}^2$, and $G \in \mathbb{R}^{2N \times N}$ a discrete gradient matrix corresponding to functions $x \colon \Omega \to \mathcal{C}, \mathcal{C} = [0,1]^N$. Given any similarity values $s \colon \Omega \to \mathbb{R}^N$ extracted from image data beforehand, that locally indicate fore- or background in terms of its components $s_i, i = 1, \ldots, N$, we look for a minimizer $x_{\min} \in \mathcal{C}$ of the functional

$$E_{\mathrm{TV}}(x) = \alpha \,\mathrm{TV}(x) + \langle x, s \rangle, \tag{2}$$

where $\alpha > 0$ is a regularization parameter, and the (discretized) total variation measure as regularizer is given by

$$\mathrm{TV}(x) = \sum_{i=1}^{N} \|(Gx)_i\| = \sum_{i=1}^{N} \left((Gx)_{i,1}^2 + (Gx)_{i,2}^2 \right)^{1/2}$$

$$= \sigma_{\mathcal{D}}(Gx) = \sup_{z} \left\{ \langle Gx, z \rangle - \delta_{\mathcal{D}}(z) \; : \; z \in \mathbb{R}^{2N} \right\},$$

$$\mathcal{D} = \{ z \in \mathbb{R}^{2N} : (z_{i,1})^2 + (z_{i,2})^2 \leq 1, \, 1 \leq i \leq N \}.$$

In [4] it is pointed out how these minimizers are related to finding an optimal solution to the two-level Mumford-Shah energy functional by thresholding.

2.2   MRF Based Shape Priors

General Variational Formulation. Shape priors consist of feature vectors y of binary variables $y_i, 1 \leq i \leq M$ and are defined statistically in terms of a joint distribution

$$P(y) = \frac{1}{Z(\theta)} \exp\left(\langle \theta, \phi(y) \rangle\right) \tag{3}$$

$$Z(\theta) = \sum_{y \in \{0,1\}^M} \exp\left(\langle \theta, \phi(y) \rangle\right),$$

where plausible choices corresponding to familiar shapes are assigned a higher probability. The form of $\phi(y)$ is given by an associated undirected graph G with vertices $V = \{1, \ldots, M\}$ and edges $E \subseteq \{(i,j): 1 \leq i < j \leq M\}$. Then $\phi(y) = (y_1, y_2, \ldots, y_M, \ldots, y_i y_j, \ldots)$, $(ij) \in E$.

This corresponds to a minimally represented binary graphical model [17] with strictly convex and essentially smooth log-partition function $Z(\theta)$, for any choice of the model parameters $\theta \in \mathbb{R}^{(|V|+|E|)}$. These parameters are learned from shape databases, as described in Section 3.

The most probable configuration y_{max} is given as solution to the problem

$$\underset{y}{\mathrm{argmax}} \left\{ \langle \theta, \phi(y) \rangle \ : \ y \in \{0,1\}^M \right\}.$$

This discrete combinatorial problem can be reformulated as linear problem on the marginal polytope $\mathcal{M}(G)$ of the graph G[17]:

$$\underset{y}{\max} \left\{ \langle \theta, \phi(y) \rangle \ : \ y \in \{0,1\}^M \right\} = \underset{\mu}{\sup} \left\{ \langle \theta, \mu \rangle \ : \ \mu \in \mathcal{M}(G) \right\}.$$

As the number of constraints that define $\mathcal{M}(G)$ grows exponentially with the size of the graph, this problem is in general unfeasible. Thus one is forced to relax the optimization set to the local polytope $\mathcal{L}(G)$ and to check carefully the global optimum of this convex relaxation. Methods have been proposed to tighten the standard local polytope relaxation by identifying and reintroducing violated constraints of the original optimization set, see for example [15].

Whatever set of constraints one chooses, they can be written as N_{con} affine inequalities leading to the variational problem formulation

$$\underset{\mu}{\inf} \left\{ E_{\mathrm{MRF}}(\mu) : K\mu \leq k, \ K \in \mathbb{R}^{N_{\mathrm{con}} \times (|V|+|E|)}, \ k \in \mathbb{R}^{N_{\mathrm{con}}} \right\} \tag{4}$$

with

$$E_{\mathrm{MRF}}(\mu) = -\langle \theta, \mu \rangle.$$

3   Variational Shape Priors

We describe two specific instances of the general framework (3).

3.1 Ising Shape Prior

As a baseline, we investigate a direct application of the two dimensional Ising Model: Every pixel is treated as binary feature, hence $M = N$ in (3).

The set E of edges defining the model is determined by the model parameters θ. Vanishing parameter values indicate missing edges. Parameter values are determined using an approach proposed by Hoefling and Tibshirani [10]. The corresponding algorithm maximizes the pseudo-likelihood of a set of training samples $\mathbf{T}$ as a function of the parameter vector θ, constrained by a ℓ_1-norm penalty $\rho\|\theta\|_1$. This penalty enforces sparse connectivity of the graph which is desirable for various reasons: Learning dense graphs tends to overfit the training data; dense graphs lead to a larger linear problem, and the local polytope relaxation tends to become weak. In our simulations about 3% of all possible edges were actually existent (see Fig. 2a).

In the training data some pixels were either always black or always white. As they do not contain any correlation information about other pixels they have been removed from the learning procedure and their unary θ-weight has been set to $\pm\infty$ by hand. The corresponding vertices have no connections to other vertices.

3.2 Hierarchical Part Based Shape Prior

Besides the "flat" Ising Model discussed above, we also study a simple hierarchical model that can demonstrate the benefits of articulated object descriptions and support nonlocal interactions without compromising sparsity. Shapes will be decomposed into a torso and successive limbs. Limbs may depend on another and on the torso: E.g. a lower arm is only expected if the upper arm is present which, in turn, can only be present if the torso is there. Our model learns such statistical relationships from data and penalizes implausible constellations as follows.

First the set of training samples is divided into groups of characteristic views, i.e. viewpoints of the object that yield similar shapes. Then to each group the following procedure is applied: pixels in the image are combined into families depending on the sample subsets in which they are active (=1) or not (=0). For example, usually there will be a "torso" family of pixels that are active in every sample of a specific view, and a "background" family of pixels that are always black in that group.

Correspondingly, a hierarchical dependency graph of families is constructed for every group: family i of pixels is considered dependend on family j if family j is always active if i is active.

This relation is transitive, and the dependency graph only represents the transitive reduction of this relation to keep the representation minimal. As to be expected, in all these graphs the torso-families will be the root vertices.

Using the groups of characteristic views and their dependency graphs a joint prior-graph with parameter vector θ is constructed in the following way: The core of the graph is constituted by the torso-families of all characteristic view

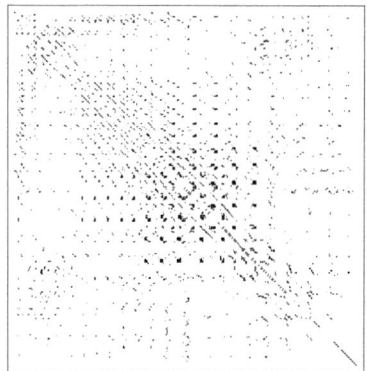

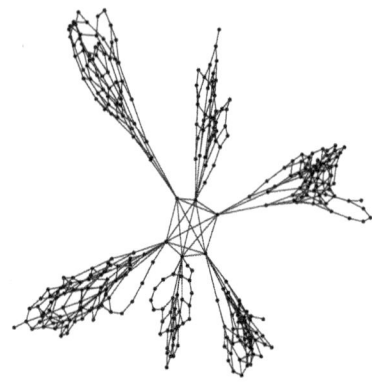

(a) Plot of the adjacency matrix of the graph G of the Ising prior learned from a set of training samples. The nodes corresponding to pixels of constant value in all training samples are not shown. Density: approx. 3%.

(b) Illustration of the graph structure of the hierarchical part based prior: The nodes of the fully connected subgraph in the center represent the different possible torsos.

Fig. 2. Illustrating the two MRF based priors

groups. They are fully connected amongst each other. Then to each torso node the dependency graph of the associated characteristic view is appended (compare Fig. 2b). The parameter vector is initialized with $\theta_i = 0, 1 \leq i \leq M, \theta_{ij} = 0, (ij) \in E$. Then all pairs of torso nodes are considered: Plausibility dictates that at most one torso can be present at a time. Thus the simultaneous presence of two torsos has to be penalized. Therefore all θ_{ij} where i and j are two different torso nodes, are set to $-p$ where p is the positive penalty parameter. Then all dependency graphs of the torso nodes are processed: Whenever a node i is dependend on a node j then θ_i will be decreased by p and θ_{ij} will be increased by p. Thus the unlikely case where y_i is 1 while y_j is 0 will be penalized by $-p$ relative to the other possible combinations.

We are aware that this preliminary version of our learning procedure is somewhat heuristic, but fully data driven and controlled just by a single user parameter p.

The grouping into pixel families and the introduction of a partial ordering relation on the vertices capture some of the most striking features of the sample data. A learning procedure that maximizes the likelihood would deduce similar relations. Thus this step is only an approximation where the explicit forming of families each corresponding to one vertex helps reducing the problem size and simplifies the graph structure.

The clustering into characteristic views subsequently allows multiple vertices to be associated with the same pixel. This simplifies differentiating "for what reason" (i.e. which characteristic view) a pixel is white. Hence, no further edges between different views are required which drastically reduces the number of cycles in the graph and thus supports the applied polytope relaxations.

4   Coupling Convex Models

In this section an approach by Attouch et al.[1] is briefly introduced. We will subsequently use its potential for combining TV-based segmentation with shape prior knowledge in a variational framework.

4.1   Variational Approach

The approach [1] is based on the following functional jointly defined on two real Hilbert spaces $\mathcal{U}$ and $\mathcal{V}$:

$$E(x, \mu) = f(x) + g(\mu) + \frac{\lambda}{2} Q(x, \mu), \qquad x \in \mathcal{U}, \mu \in \mathcal{V}, \qquad (5)$$

where

$$\mathcal{U}, \mathcal{V} : \text{real Hilbert spaces,}$$

$$\begin{aligned} f : \mathcal{U} &\to \mathbb{R} \cup \{+\infty\} \\ g : \mathcal{V} &\to \mathbb{R} \cup \{+\infty\} \end{aligned} : \text{closed convex proper functions,}$$

$$Q : (x, \mu) \in \mathcal{U} \times \mathcal{V} \to \mathbb{R}^+ : \text{nonnegative, quadratic form.}$$

A minimizing pair of vectors (x, μ) is given by the limit of the series generated by the following alternating proximal update steps:

$$\begin{aligned} x_{n+1} &= \underset{\xi}{\operatorname{argmin}} \left\{ f(\xi) + \tfrac{\lambda}{2} Q(\xi, \mu_n) + \tfrac{\beta_f}{2} \|\xi - x_n\|^2 : \xi \in \mathcal{U} \right\}, \\ \mu_{n+1} &= \underset{\eta}{\operatorname{argmin}} \left\{ g(\eta) + \tfrac{\lambda}{2} Q(x_{n+1}, \eta) + \tfrac{\beta_g}{2} \|\eta - \mu_n\|^2 : \eta \in \mathcal{V} \right\}. \end{aligned} \qquad (6)$$

Here $\lambda \geq 0$ is a coupling constant that regulates the strength of the interaction. $\beta_f > 0$ and $\beta_g > 0$ are damping constants that affect the convergence rate of the algorithm but not convergence itself. Indeed, as we work in finite-dimensional Euclidean spaces, [1, Thm. 2.1] immediately yields

Theorem 1. *The sequence (x_n, μ_n) generated by algorithm (6) converges to a minimum (x_∞, μ_∞) of the functional E in (5). Moreover, $f(x_n) \to f(x_\infty)$, $g(\mu_n) \to g(\mu_\infty)$, $Q(x_n, \mu_n) \to Q(x_\infty, \mu_\infty)$, $\|x_{n+1} - x_n\| \to 0$ and $\|\mu_{n+1} - \mu_n\| \to 0$ as $n \to \infty$.*

We point out that this result is non-trivial because functionals f and g are required to be convex and closed, but may be *non-smooth*, as in our applications. As a consequence, heuristic alternating minimization without complementing proper proximal point mappings, as is sometimes done in the field of computer vision, might not yield a minimizing sequence.

From the viewpoint of modeling, the result is appealing as well, because the quadratic form $Q(\cdot, \cdot)$ enables to couple and to compare points from "two different worlds" in various mathematically sound ways in a common third space.

4.2 Shape Constrained Cuts

We present a preliminary application of the idea of coupling two convex functionals to the image segmentation problem. We set $\mathcal{U} = \mathbb{R}^N, \mathcal{V} = \mathbb{R}^{|V|+|E|}$, and investigate the following joint energy:

$$E(x,\mu) = \lambda_1 E_{\mathrm{TV}}(x) + \lambda_2 E_{\mathrm{MRF}}(\mu) + \frac{\lambda}{2}\|x - P\mu\|^2$$

$$+\delta_{\mathcal{C}}(x) + \sup_{\omega}\left\{\langle\omega, K\mu\rangle - \left(\delta_{\mathbb{R}^{N_{\mathrm{con}}}_+}(\omega) + \langle\omega, k\rangle\right)\right\}, \tag{7}$$

where P is a linear mapping to be specified. Note that the terms in the second line constrain solutions $x \in \mathcal{U}, \mu \in \mathcal{V}$, to the optimization problem to the convex sets $x \in \mathcal{C}$ and $\{\mu\colon K\mu \leq k\}$, respectively.

Comparison to Eq. (5) yields

$$f(x) = \lambda_1 E_{\mathrm{TV}}(x) + \delta_{\mathcal{C}}(x) \tag{8}$$

$$g(\mu) = \lambda_2 E_{\mathrm{MRF}}(\mu) + \sup_{\omega}\left\{\langle\omega, K\mu\rangle - \left(\delta_{\mathbb{R}^{N_{\mathrm{con}}}_+}(\omega) + \langle\omega, k\rangle\right)\right\} \tag{9}$$

$$Q(x,\mu) = \|x - P\mu\|^2. \tag{10}$$

It is easy to verify that these choices satisfy the criteria that are required for the sequence generated by (6) to converge.

In our example we chose the coupling space to which x and μ are mapped to be identical with the space x lives in. Thus the matrix A in Eq. (1) is the identity on $\mathbb{R}^N$. The matrix P in the coupling term maps the initially abstract feature vector μ to the pixel space. The coupling then encourages the segmentation proposal x to be close to the image represented by the features μ under this map. In our setup μ contains not only rows for unary marginals of variables y_i but also pairwise marginals of y_iy_j corresponding to edges $(ij) \subset E$. These pairwise marginals are not considered by the map.

In the Ising-Model prior P takes the marginals of the pixel features y_i and maps them to the appropriate pixel positions i in the image space. It is given by

$$\left(\mathbb{I}_{|V|\times|V|}\ 0_{|V|\times|E|}\right)$$

where $\mathbb{I}_{n\times n}$ denotes the n-dimensional identity matrix and $0_{m\times n}$ a $m \times n$ matrix with all entries being 0.

For the hierarchical part based prior, in each column of P that corresponds to a unary marginal of a pixel family the entries of the corresponding pixels are 1. The subspace of pairwise marginals is again ingored. Despite the unusual generation of the parameter vector θ this model is still part of the family given by Eq. (3).

5 Experiments and Discussion

5.1 Setup

Training Data. Shape priors are learned from 2D views of 3D object models. The training samples are views of the model from equidistant angles rotated around its vertical axis. We chose a fixed window size that in applications may cover different region sizes of a given image, corresponding to different levels of a multiscale representation of the image.

In the first scenario the training data consist of 25×25 pixel views of a bunny (see Fig. 3). To demonstrate the generality of our approach a second scenario with 50×50 pixel views of a horse was also studied.

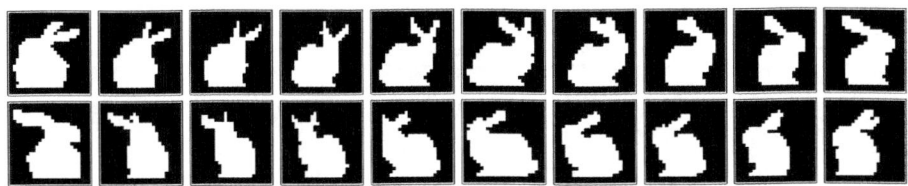

Fig. 3. Some of the training samples of scenario 1: 25×25 pixel b/w views of a 3D-model of a bunny

In scenario 1 the Ising prior was trained with 50 equidistant views, the hierarchical part based model with 100 equidistant views.

Input Data. As input data *new views with random angles* between the training views were taken and distorted to simulate *noise, occlusion and clutter*.

Gaussian noise of mean $\mu_{\text{noise}} = 0$ and some standard deviation σ_{noise} followed by projecting each pixel value onto the interval $[0, 1]$ simulates feature imperfections. Occlusion by another object in the foreground was simulated by drawing black circles on the input image. Clutter could be generated by other objects in the image with texture similar to the object in question and was simulated by drawing white circles on the input image. In the given setup the similarity vector s was obtained by $s = (1/2)_N - u$ where $(1/2)_N$ denotes the vector of length N with each entry being $1/2$ and $u \colon \Omega \to \mathcal{C}$ is the distorted input image.

Parameter Values. Unless otherwise noted in the simulations the following parameter settings have been used: TV regularization parameter $\alpha = 0.5$ (eqn. (2)), part based MRF implausibility penalty $p = 100$ (cf. Section 3.2), coupling constants $\lambda_1 = \lambda_2 = 1, \lambda = 10$ (joint energy (7)), proximal damping parameters $\beta_f = \beta_g = 0.01$ (algorithm (6)). By default the noise level was set to $\sigma_{\text{noise}} = 0.75$.

The part based MRF penalty p is chosen to be on a higher energy scale than the continuous cuts segmentation to favour segmentations of familiar shapes. The coupling parameter λ was set to a high value to make the difference of x and $P\mu$ negligible. In plots that illustrate segmentation results only x is shown but $P\mu$ would not look much different.

5.2 Results

Figure 4 shows (a) several input images with distortions and each time the segmentations given by (b) the uncoupled continuous cuts approach, (c) the

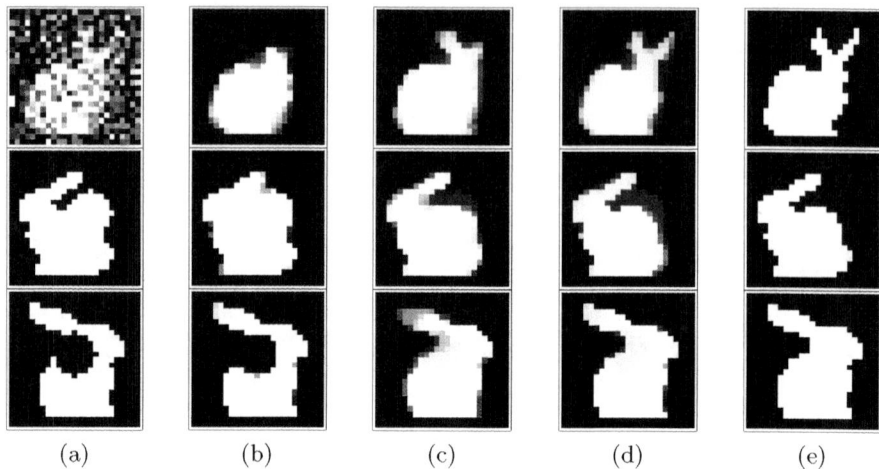

Fig. 4. Modelling different kinds of distortion: Row 1: noise $\sigma_{\mathrm{noise}} = 0.50$, row 2: clutter, row 3: occlusion. Segmentation results: (a) input, (b) only continuous cuts, (c) Ising prior, (d) part based prior, (e) ground truth.

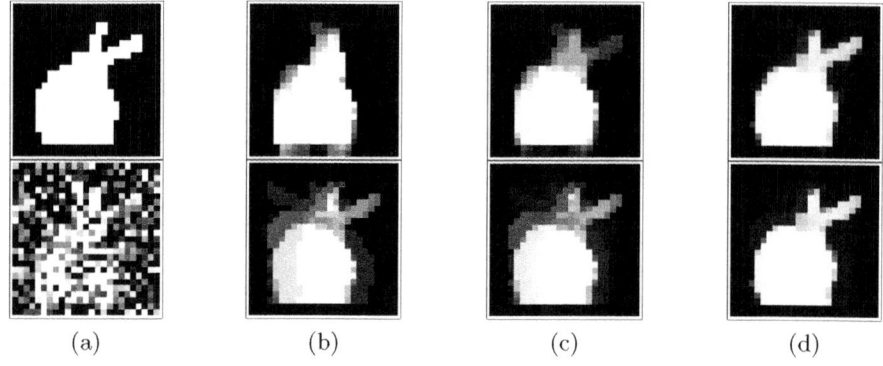

Fig. 5. Segmentation results of the coupled part based prior for various coupling strengths λ: (a) ground truth and noise-distorted input. The upper row shows x, the lower row $P\mu$: (b) $\lambda = 0.1$, (c) $\lambda = 1$, (d) $\lambda = 10$.

Fig. 6. Four pairs of noisy input, and part based prior supported segmentations of different views. All views were processed with the same prior.

coupled Ising prior and (d) the coupled simple part based prior. Furthermore, distortions due to noise, occlusion and clutter was examined, as described in the previous section.

(a)   (b)   (c)   (d)

Fig. 7. Binary image segmentation of a horse: (a) noisy input, (b) only continuous cuts with high scale resolution $\alpha = 0.1$, (c) only continuous cuts with low scale resolution $\alpha = 0.3$, (d) coupled shape prior with high scale resolution $\alpha = 0.1$. Without prior a high resolution (b) can not filter out noise, with low resolution (c) details (legs) are lost. The prior can help to distinguish noise and details.

Clearly the coupling to a shape prior enhances the segmentation results: In row 1 the ears of the bunny are lost to regular segmentation while they are still restored by the coupled models. Similarly dealing with clutter and occlusion works much better with prior.

In Figure 5 it is shown how increased coupling strength forces the variables x and μ to correspond to the same image. In our simulations it turned out that most of the times a large coupling constant yields best results.

6 Conclusions and Further Work

We introduced a novel approach to variational image segmentation with shape priors. Key properties are convexity of the joint energy functional and weak coupling of convex models from different domains by mapping corresponding solutions to a common space. Specifically, we combined state-of-the-art variational approaches for TV-based image segmentation and for MRF-based learning of image classes from examples. A convergent algorithm amenable to large-scale convex programming was presented. Numerical experiments demonstrated promising synergistic performance of convex continuous cuts and convex variational shape priors under (simulated) noise, occlusion and clutter.

Our further work will further explore components of the coupling quadratic form Q in (5). For example, linear mappings that represent collections of linear shape features could be used to compare shapes after projecting them onto a low-dimensional feature space. Another line of research concerns elaborating the part-based variational shape prior so as to cope with several objects simultaneously and efficiently.

Acknowledgement. This work was supported by the DFG, grant GRK 1653.

References

1. Attouch, H., Bolte, J., Redont, P., Soubeyran, A.: Alternating proximal algorithms for weakly coupled convex minimization problems. Applications to dynamical games and PDE's. Journal of Convex Analysis 15(3), 485–506 (2008)
2. Boykov, Y., Veksler, O., Zabih, R.: Fast approximate energy minimization via graph cuts. IEEE Trans. Patt. Anal. Mach. Intell. 23(11), 1222–1239 (2001)

3. Chambolle, A., Pock, T.: A first-order primal-dual algorithm for convex problems with applications to imaging. Journal of Mathematical Imaging and Vision 40, 120–145 (2011)

4. Chan, T.F., Esedoglu, S., Nikolova, M.: Algorithms for finding global minimizers of image segmentation and denoising models. SIAM J. Appl. Math. 66(5), 1632–1648 (2006)

5. Charpiat, G., Faugeras, O., Keriven, R.: Approximations of shape metrics and application to shape warping and empirical shape statistics. Found. Comp. Math. 5(1), 1–58 (2005)

6. Cremers, D., Kohlberger, T., Schnörr, C.: Shape statistics in kernel space for variational image segmentation. Patt. Recognition 36(9), 1929–1943 (2003)

7. Dambreville, S., Rathi, Y., Tannenbaum, A.: A framework for image segmentation using shape models and kernel space shape priors. IEEE Trans. Patt. Anal. Mach. Intell. 30(8), 1385–1399 (2008)

8. Deza, M.M., Laurent, M.: Geometry of Cuts and Metrics. Springer, Heidelberg (1997)

9. Esser, E., Zhang, X., Chan, T.: A general framework for a class of first order primal-dual algorithms for convex optimization in imaging science. SIAM J. Imag. Sci. 3(4), 1015–1046 (2010)

10. Hoefling, H., Tibshirani, R.: Estimation of sparse binary pairwise markow networks using pseudo-likelihoods. Journal of Machine Learning Research 10, 883–906 (2009)

11. Lecumberry, F., Pardo, A., Sapiro, G.: Simultaneous object classification and segmentation with high-order multiple shape models. IEEE Trans. Image Proc. 19(3), 625–635 (2010)

12. Lellmann, J., Becker, F., Schnörr, C.: Convex optimization for multi-class image labeling with a novel family of total variation based regularizers. In: ICCV (2009)

13. Lellmann, J., Kappes, J., Yuan, J., Becker, F., Schnörr, C.: Convex multi-class image labeling by simplex-constrained total variation. In: Tai, X.-C., Mørken, K., Lysaker, M., Lie, K.-A. (eds.) SSVM 2009. LNCS, vol. 5567, pp. 150–162. Springer, Heidelberg (2009)

14. Pock, T., Chambolle, A., Cremers, D., Bischof, H.: A convex relaxation approach for computing minimal partitions. In: CVPR, pp. 810–817 (2009)

15. Sontag, D.A.: Approximate Inference in Graphical Models using LP Relaxations. Ph.D. thesis, Massachusetts Institute of Technology (2010)

16. Sundaramoorthi, G., Mennucci, A., Soatto, S., Yezzi, A.: A new geometric metric in the space of curves, and applications to tracking deforming objects by prediction and filtering. SIAM Journal on Imaging Sciences 4(1), 109–145 (2011)

17. Wainwright, M.J., Jordan, M.I.: Graphical models, exponential families, and variational inference. Foundations and Trends in Machine Learning 1(1-2), 1–305 (2008)

18. Younes, L., Michor, P., Shah, J., Mumford, D.: A metric on shape space with explicit geodesics. Rend. Lincei Mat. Appl. 9, 25–57 (2008)

Wasserstein Barycenter
and Its Application to Texture Mixing

Julien Rabin[1,*], Gabriel Peyré[2], Julie Delon[3], and Marc Bernot[4]

[1] CMLA, ENS de Cachan
rabin@cmla.ens-cachan.fr
[2] Ceremade, Univ. Paris-Dauphine
peyre@ceremade.dauphine.fr
[3] LTCI, Telecom ParisTech
delon@enst.fr
[4] Thales Alenia Space
marc.bernot@thalesaleniaspace.com

Abstract. This paper proposes a new definition of the averaging of discrete probability distributions as a barycenter over the Monge-Kantorovich optimal transport space. To overcome the time complexity involved by the numerical solving of such problem, the original Wasserstein metric is replaced by a sliced approximation over 1D distributions. This enables us to introduce a new fast gradient descent algorithm to compute Wasserstein barycenters of point clouds.

This new notion of barycenter of probabilities is likely to find applications in computer vision where one wants to average features defined as distributions. We show an application to texture synthesis and mixing, where a texture is characterized by the distribution of the response to a multi-scale oriented filter bank. This leads to a simple way to navigate over a convex domain of color textures.

Keywords: Optimal transport, texture synthesis and mixing, barycenter of distributions.

1 Introduction

This paper considers the use of the Monge-Kantorovich optimal transport theory [1] in image synthesis. Optimal transport cost has been widely used as a metric to compare histogram features, often referred to as the "Earth Mover's Distance", since the seminal work of Rubner *et al.* [2]. Another interesting aspect of transportation approaches which is investigated here is the transportation mapping itself. Indeed, it allows various image modifications, such as *e.g.* color transfer [3], texture mapping [4], or contrast equalization of video [5] (see [6] for other applications).

* Research partially financed by the Centre National d'tudes Spatiales (MISS project), the Office of Naval research (grant N00014-97-1-0839), by the European Research Council (advanced grant "Twelve labours"), and the French National Research Agency (grant NatImages ANR-08-EMER-009).

A.M. Bruckstein et al. (Eds.): SSVM 2011, LNCS 6667, pp. 435–446, 2012.

1.1   Previous Work on Texture Synthesis and Mixing

Texture synthesis is a popular problem in computer graphics, which consists in synthesizing a new image f visually similar to a given exemplar f^0.

Texture synthesis by recopy. Synthesis with high fidelity to the exemplar is performed by copying pixels with some coherence constraints on small patches [7,8]. The quality of the synthesis is improved by copying patches or more general sets of pixels [9,10,11,12,13].

Texture synthesis by statistical modeling. While copy-based methods probably yield the best synthesis quality, they often copy large blocks from the original input, and offer little or indirect control about the synthesis process.

   Procedural methods use parametric models of textures, for instance built on top of a Gaussian noise [14]. They are popular in image synthesis because of their ease of use and their low computational cost.

   Texture modeling considers sets of statistical constraints learned from the exemplar, and use the stationarity of the texture for the estimation. Popular approaches use Markov random fields [15,16] or Gibbs distributions built on top of multiscale filters [17].

   The wavelet decomposition is often use to build statistical models, with first order histograms [18,19] or higher order constraints [20].

Texture mixing. The texture mixing problem consists in synthesizing a new texture from a collection $\{f^j\}_{j \in J}$ of exemplars. The mixing should integrate in a meaningful way the colors and texture attributes of the exemplars.

   Heeger and Bergen [18] and Bar-Joseph *et al.* [21] perform the mixing by combining multiscale wavelet coefficients. Averaging statistics of grouplet coefficients enables the mixing of geometrical turbulent textures [22].

   Patch-based methods perform mixing using patches from the set of exemplars, which creates non-homogeneous textures, see for instance [23,9,13]. These methods tend to produce clusters of features and offer little understanding about the mixing process and how to control it.

   Texture metamorphosis approaches [24,25,26] perform the mixing by finding correspondences between elementary features (or textons) between the textures, and progressively morphing between the shapes of the features. These methods are extended by Matusik et al. [27] to perform convex combination of textures, by warping patches and averaging 1D histograms.

1.2   Contributions

The main theoretical contribution of this work is the introduction of a novel definition for the barycenter of statistical distributions in the Monge-Kantorovich space. We also propose an approximate definition more amenable for numerical computations which rely on 1-D projections. We then introduce a Newton gradient descent algorithm to compute the corresponding Wasserstein barycenter. The last contribution of the paper is a general framework for the statistical synthesis of color textures, that encompasses several existing texture models as particular cases. This general framework, together with the Wasserstein barycenter computation, allows to perform color texture mixing.

2   Wasserstein Distance and Its Approximation

This paper considers discrete density distributions in $\mathbb{R}^d$ that are represented as point clouds $X = \{X_i\}_{i \in I} \subset \mathbb{R}^d$, where $I = \{1, \ldots, N\}$ is the list of point indexes. Since any permutation of X corresponds to the same distribution, one considers metrics taking into account

$$[X] = \left\{ (X_{\sigma(j)})_{j=0}^{N-1} \setminus \sigma \in \Sigma_N \right\}, \tag{1}$$

where Σ_N is the set of all permutations of N elements. The methods developed in the paper can be extended to weighted point clouds and density defined continuously over $\mathbb{R}^d$.

2.1   Wasserstein Distance

The quadratic Wasserstein distance $W_2(X, Y)$ between two point clouds of same size $|I| = N$ is defined as

$$W_2(X, Y)^2 = \min_{\sigma \in \Sigma_N} W_\sigma(X, Y) \quad \text{where} \quad W_\sigma(X, Y) = \sum_{i \in I} \|X_i - Y_{\sigma(i)}\|^2. \tag{2}$$

We note that the methods developed in this paper extend to arbitrary strictly convex distances such as $\|X_i - Y_{\sigma(i)}\|^p$ for $p > 1$. One can prove that W_p defines a metric on the set of discrete distributions $[X]$.

Linear program formulation. Computing this distance requires to compute the optimal assignment $i \mapsto \sigma^*(i)$ that minimizes $W_\sigma(X, Y)$ in (2). It is possible to recast this problem as a linear programming one

$$W_2(X, Y)^2 = \min_{P \in \mathcal{P}_N} \sum_{i,j \in I^2} P_{i,j} \|X_i - Y_j\|^2 \tag{3}$$

where $\mathcal{P}_N$ is the set of bistochastic matrices, *i.e.* nonnegative matrices which rows and columns sum to 1. The problem (3) can be solved with standard linear programming algorithms and more dedicated methods in $O(N^{2.5} \log(N))$ operations (see [28]).

1D case. The case $d = 1$ has some special structure that allows for a much faster solution. Indeed, if one denotes by σ_X and σ_Y the permutations that order the points

$$\forall 0 \leqslant i < N - 1, \quad X_{\sigma_X(i)} \leqslant X_{\sigma_X(i+1)} \quad \text{and} \quad Y_{\sigma_Y(i)} \leqslant Y_{\sigma_Y(i+1)} \tag{4}$$

the optimal permutation $\sigma^\star$ that minimizes (2) is

$$\sigma^\star = \sigma_Y \circ \sigma_X^{-1}, \tag{5}$$

so that point $X_{\sigma_X(i)}$ is assigned to the point $Y_{\sigma_Y(i)}$. The Wasserstein distance together with the optimal assignment can thus be computed in $O(N \log(N))$ operations using a fast sorting algorithm.

2.2   Sliced Wasserstein Distance

The computation of the Wasserstein distance W_2 is however computationally too demanding for the application to image processing we have in mind, where N can be quite large. Moreover, W_2 is too difficult to handle in problems requiring the optimization of point clouds with functional involving the Wasserstein distance.

For these reasons, we now consider an alternative metric between distributions, which is based on transport costs between 1-D projections. We denote by SW_2 the *Sliced Wasserstein Distance* defined as the sum of 1-D quadratic Wasserstein distances between projected point clouds

$$\mathrm{SW}_2(X,Y)^2 = \int_{\theta \in \Omega} W_2(X_\theta, Y_\theta)^2 \, d\theta \quad \text{where} \quad X_\theta = \{\langle X_i, \theta \rangle\}_{i \in I} \subset \mathbb{R}$$

$$= \int_{\theta \in \Omega} \min_{\sigma_\theta \in \Sigma_N} \sum_{i \in I} |\langle X_i - Y_{\sigma_\theta(i)}, \theta \rangle|^2 \, d\theta, \tag{6}$$

where $\Omega = \{\theta \in \mathbb{R}^d \setminus \|\theta\| = 1\}$ is the unit sphere. The Sliced Wasserstein distance allows us to use the special case of the 1-D assignment, that can be solved in closed form easily using (5).

Remark: This metric is used in [29] to perform shape retrieval.

3   Barycenter in Wasserstein Space

3.1   Wasserstein Barycenter

Given a family $\mathcal{Y} = \{Y^j\}_{j \in J}$ of point clouds, we are interested in computing a weighted average point cloud $X^\star$, that is defined, by analogy to the Euclidean setting as the minimizer of Problem $(\mathcal{B})$

$$(\mathcal{B}) \qquad \mathrm{Bar}(\rho_j, Y^j)_{j \in J} \in \underset{X}{\mathrm{argmin}} \, \mathrm{E}_\mathcal{Y}(X) = \sum_{j \in J} \rho_j W_2(X, Y^j)^2, \tag{7}$$

where $\{\rho_j \geqslant 0\}_{j \in J}$ is a set of weights, that is constrained to satisfy $\sum_j \rho_j = 1$.

Except in the special case of normal distributions [30], there is no known closed form solution to the problem (7). Independently to our work, Agueh and Carlier have performed a mathematical analysis of this problem [31]. They show the existence of a solution and a dual formulation for continuous distributions.

1D case. In the 1D case, the Wasserstein barycenter can be computed in $O(N \log(N))$ operations using the permutations σ_{Y^j} that order the sets of values $Y^j \subset \mathbb{R}$ as in (4). The barycenter then reads

$$\forall 0 \leqslant i < N, \quad \left(\mathrm{Bar}(\rho_j, Y^j)_{j \in J}\right)_i = \sum_{j \in J} \rho_j Y^j_{\sigma_{Y^j}(i)}.$$

This barycenter was used for texture mixing applications in [27]. This paper generalizes this approach to point clouds in arbitrary dimensions.

Sliced Wasserstein barycenter. If one considers the problem $(\mathcal{B})$ over arbitrary densities (not necessary clouds of a fixed number N of points, and not even necessary discrete), Agueh and Carlier demonstrate in [31] that the barycenter can be found by solving a linear problem[1]. It can be shown that if the densities of $\mathcal{Y}$ are supported on N discrete points, then the barycenter is supported on at most $N^{|J|}$ points. The difficulty with this approach is that the computation time is polynomial in $N^{|J|}$, which is prohibitive for imaging applications, and that there is a combinatorial explosion of the cardinality of the barycenter solution. Moreover, the restriction of problem $(\mathcal{B})$ to point clouds (Equation (7)) boils down to a multi-assignment problem, which unfortunately turns out to be NP-hard [28].

To solve simultaneously these two issues, we propose to define a "Sliced Wasserstein Barycenter" that approximate the original one

$$(\mathcal{S}\text{-}\mathcal{B}) \qquad \text{S-Bar}(\rho_j, Y^j)_{j \in J} \in \operatorname*{argmin}_{X} \mathrm{SE}_{\mathcal{Y}}(X) = \sum_{j \in J} \rho_j \mathrm{SW}_2(X, Y^j)^2 . \qquad (8)$$

One should note at this point that the minimizer of $\mathrm{SE}_{\mathcal{Y}}$, which is easier to manipulate, has also the desired property of being restricted to point clouds with the same dimension as the measure set $\mathcal{Y} = \{Y^j\}_{j \in J}$. In practice, as detailed in Section 3.2, a fast computation of an approximate barycenter is obtained by performing a gradient descent of this energy, which leads to a local minimizer.

3.2   Gradient Descent Algorithm

Finding the barycenter by minimizing (8) corresponds to the minimization of a non-convex functional. This energy is discretized using a finite set ω of directions, with $|\omega| > d$. In the numerical experiment, we use random directions drawn from the uniform distribution on the unit sphere of $\mathbb{R}^d$.

$$\mathrm{SE}_{\mathcal{Y}}(X) = \sum_{\theta \in \omega} \mathrm{SE}_{\mathcal{Y}_\theta}(X) \quad \text{where} \quad \mathrm{SE}_{\mathcal{Y}_\theta}(X) = \frac{1}{2} \sum_{j \in J} \rho_j W_2(X_\theta, Y^j_\theta)^2 . \qquad (9)$$

It is possible to find a stationary point of this functional with a gradient descent algorithm.

---

[1] Note that the proof given in [31] for the continuous case can be extended to our discrete setting.

Newton gradient descent. The algorithm starts from some initial point cloud $X^{(0)} \subset \mathbb{R}^d$, that can be for instance chosen to be any of the clouds Y^j, and sets $k = 0$. At each iteration k, the point cloud $X^{(k)}$ is updated with the following Newton descent step:

$$X^{(k+1)} = X^{(k)} - \eta \, H^{-1} \sum_{\theta \in \omega} \nabla \mathrm{SE}_{\mathcal{Y}_\theta}(X^{(k)}) \tag{10}$$

where $\nabla \mathrm{SE}_{\mathcal{Y}_\theta}(X^{(k)})$ is the gradient of $\mathrm{SE}_{\mathcal{Y}_\theta}$ at point $X^{(k)}$, $H \in \mathbb{R}^{d \times d}$ is the Hessian matrix and $\eta > 0$ the fixed step size. H^{-1} refers to as the inverse of the Hessian matrix, which is invertible since we choose ω s.t. $|\omega| > d$.

Gradient computation. For each $\theta \in \omega$, computing the gradient $\nabla \mathrm{SE}_{\mathcal{Y}_\theta}(X^{(k)})$ requires, for each $j \in J$, to compute the optimal 1-D assignment σ_θ^j that minimizes

$$\min_{\sigma_\theta^j \in \Sigma_N} \sum_{i \in I} |(X_\theta^{(k)})_i - (Y_\theta^j)_{\sigma_\theta^j(i)}|^2.$$

This is computed as detailed in (5) by sorting the values of $X_\theta^{(k)}$ and Y_θ^j, which requires $O(N \log(N))$ operations.

Each element i of the gradient may then be expressed as

$$\sum_{\theta \in \omega} \nabla \mathrm{SE}_{\mathcal{Y}_\theta}(X^{(k)})_i = H X_i^{(k)} - \sum_{\theta \in \omega, j \in J} \rho_j \langle Y_{\sigma_\theta^j(i)}^j, \theta \rangle \theta$$

where the Hessian matrix reads

$$H = \nabla^2 \sum_{\theta \in \omega} \mathrm{SE}_{\mathcal{Y}_\theta}(X^{(k)}) = \sum_{\theta \in \omega} \theta \theta^{\mathrm{T}} = \left(\sum_{\theta \in \omega} \theta_i \theta_j \right)_{0 \leqslant i,j < d}.$$

Observe that the Hessian matrix is independent of point X and thus can be precomputed.

Convergence of the algorithm. The convergence of the proposed algorithm can be proven[2] for any $\eta \in]0, 2[$ (the case where $\eta = 1$ corresponding to the classical Newton descent step, which has been used in all numerical experiments).

Numerical examples. Figures 1 and 2 show 2-D illustrations of point clouds obtained by the proposed Sliced Wasserstein barycenter algorithm, using $|\omega| = 20$ directions (for theses examples with $N = 5.10^3$ points, 100 iterations are required in average). Note that the obtained barycenter actually depends on the initialization $X^{(0)}$ of the algorithm.

---

[2] The proof is omitted here due to the lack of space.

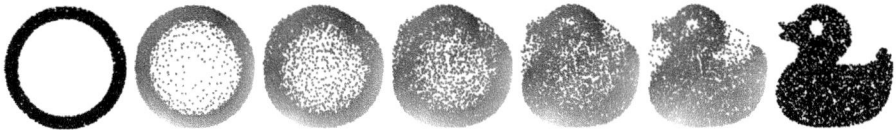

Fig. 1. Sliced Wasserstein barycenters S-Bar$(1 - \rho, Y^1, \rho, Y^2)$ for an increasing values of $\rho \in \{0.1, 0.3, 0.5, 0.7, 0.9\}$. Color of points depends on their index in Y^1 (*i.e.* the ring)

Fig. 2. Two examples (left and right panels) of barycenters S-Bar$(\rho_j, Y^j)_{j=1,2,3}$. The top, bottom left, and bottom right corners of each triangle display respectively the distribution Y^1, Y^2, Y^3. The color of the obtained point clouds corresponds to the weighted average of colors of distributions.

3.3   Computing the Projection on a Distribution

In many applications, one is not only interested in computing the Wasserstein distance $W_2(X, Y)$, but also in the optimal permutation $\sigma^\star$ that minimizes $W_\sigma(X, Y)$ in (2). This optimal permutation allows one to compute the orthogonal projection

$$\text{Proj}_{[Y]}(X)_i = X_{\sigma^\star(i)} \tag{11}$$

where the set $[Y]$ of all the point clouds that represent the same statistical distribution as Y is defined in (1).

Sliced projection. Observe also that in the special case of two distributions $(|J| = 2)$ with initialization $X^{(0)} := Y^1$ and with weights $\{\rho_1 = 0, \rho_2 = 1\}$, our algorithm is similar to the original algorithm proposed in [29] to compute an approximate projection of point cloud Y^1 onto Y^2 (*i.e.* approximate optimal assignment). The only difference here is that, as suggested in [29], we make use of a *stochastic* gradient descent (*i.e.* using random set of orientations $\omega_k \in \Omega$ at each iteration k). It should be mentioned that such a gradient descent strategy is not stable when considering our barycentric energy.

We thus define the Sliced Projection of as the point clouds $X^{(\infty)}$ where $X^{(k)}$ is converging

$$\text{S-Proj}_{[Y]}(X) = X^{(\infty)}. \tag{12}$$

Figure 3 shows on a 2-D example (with $N = 10^3$ and $|\omega_k| = 10$) that the sliced projection is in practice very close to the orthogonal projection $\text{Proj}_{[Y]}$.

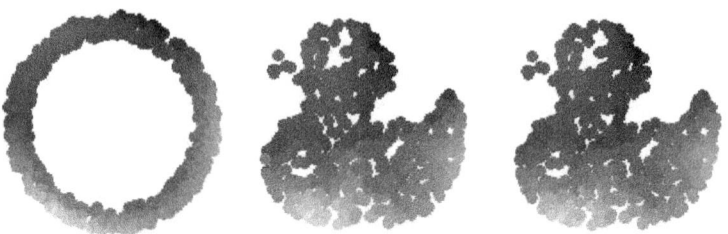

Fig. 3. Left: initial distribution $X \subset \mathbb{R}^2$. Middle: sliced projection $\text{S-Proj}_{[Y]}(X) = X^{(\infty)}$ defined in (12). Right: Wasserstein projection $\text{Proj}_{[Y]}(X)$ defined in (11), computed by linear programming. Color of points only depends on their index in Y^1 (*i.e.* the ring).

4   Texture Synthesis and Mixing

This section applies the sliced Wasserstein barycenter (8) and the sliced Wasserstein projection (12) to perform texture mixing.

4.1   Multiscale Oriented Decompositions

We consider color textures exemplars $f^j \in \mathbb{R}^{P \times 3}$ of P pixels, where each pixel value is a 3D vector $f^j(x) \in \mathbb{R}^3$. A general framework for texture modeling makes use of the projection of the image on a set of atoms $\{\psi_{\ell,n}\}_{\ell \in L, n}$. All the atoms $\psi_{\ell,n} \in \mathbb{R}^P$ for a given $\ell \in L$ typically share a common scale and orientation, while n indexes a position.

Similarly to several previous works [18,19,20], we use a steerable wavelet tight frame [32]. In this case, the atoms $\psi_{\ell,n}$, where $\ell = (s, \theta)$, are parametrized by a dyadic scale 2^s (that indicates the size of the atoms), an orientation $\theta \in [0, \pi)$ and a position $2^s n \in [0, 1]^2$.

For the numerical experiments, we considered 4 dyadic scales and 4 orientations, together with a coarse scale frame (low pass residual) and a high frequency frame (details). The total number of frames in our experiments is thus $|L| = 4 \times 4 + 2 = 18$.

4.2   First Order Statistical Mixing

First order texture model. Following the work of Heeger and Bergen [18], the simplest texture model considers the first order statistics of the projection on the frames atoms. The model thus retains the distributions

$$\forall \ell \in L, \ \forall j \in J, \quad Y^{\ell,j} = \{\langle f^j, \psi_{\ell,n}\rangle\}_n.$$

All the models are computed in $O(|J|P\log(P))$ operations with the fast steerable pyramid transform [32]. Note that each coefficient $\langle f^j, \psi_{\ell,n}\rangle \in \mathbb{R}^3$ is a 3D vector obtained by projecting each channel of f^j onto the atom $\psi_{\ell,n}$.

First order texture mixing. Given a set $\{\rho_j\}_{j\in J}$ of weights, the first step of the mixing algorithm computes the barycentric model from the exemplar distributions, for both all scale and orientation $\ell \in L$ and for the pixel values

$$\forall \ell \in L, \quad Y^\ell = \text{S-Bar}(\rho_j, Y^{\ell,j})_{j\in J} \subset \mathbb{R}^3 \quad \text{and} \quad Y = \text{S-Bar}(\rho_j, f^j)_{j\in J} \subset \mathbb{R}^3$$

using the gradient descent algorithm detailed in Section 3.2 for $|J|$ distributions in $\mathbb{R}^3$.

Following [18], the synthesis of the mixed texture is then obtained by iteratively enforcing the statistical distribution using projections. The algorithm starts from a random white noise color image $f^{(0)} \in \mathbb{R}^{P\times 3}$. At iteration k, the algorithm computes the set of coefficients $\langle f^{(k)}, \psi_{\ell,n}\rangle$, and enforces the statistical constraints using the sliced projections

$$\forall \ell \in L, \quad \{c_{\ell,n}^{(k)}\}_n = \text{S-Proj}_{[Y^\ell]}(\{\langle f^{(k)}, \psi_{\ell,n}\rangle\}_n) \tag{13}$$

using the stochastic gradient descent for only one distribution in $\mathbb{R}^3$. Since the steerable pyramid is a tight frame, an intermediate image is reconstructed in $O(P\log(P))$ operations as

$$\tilde{f}^{(k)} = \sum_{\ell \in L, n} c_{\ell,n}^{(k)} \psi_{\ell,n}. \tag{14}$$

The color pixel values distribution is then enforced as

$$\{f^{(k+1)}(x)\}_x = \text{Proj}_{[Y]}(\{\tilde{f}^{(k)}(x)\}_x)$$

using once again the gradient descent method for a single distribution in $\mathbb{R}^3$.

Figure 4 shows the mixing of two and three textures using this first order model. Observe that this first order method generalizes the original method of Heeger and Bergen [18] by taking into account color distributions (whereas the original framework considers only 1D distributions obtained by a PCA change of color representation) and also by mixing several distributions.

4.3   Higher Order Statistical Mixing

Joint distribution model. The main limitation of the Heeger-Bergen method [18] and our extension is that it does not take into account the spatial correlation between wavelet coefficients. As a result, it yields poor synthesis results in the case of structured textures, such as a text or a brick-wall. Portilla and Simoncelli show in [20] that imposing self- and cross-correlation constraints between the scale and orientation of the steerable frames enables to preserve higher order statistical features in the texture synthesis process. The main limitations of this approach are firstly that the definition of the constraints are not fully generic, so

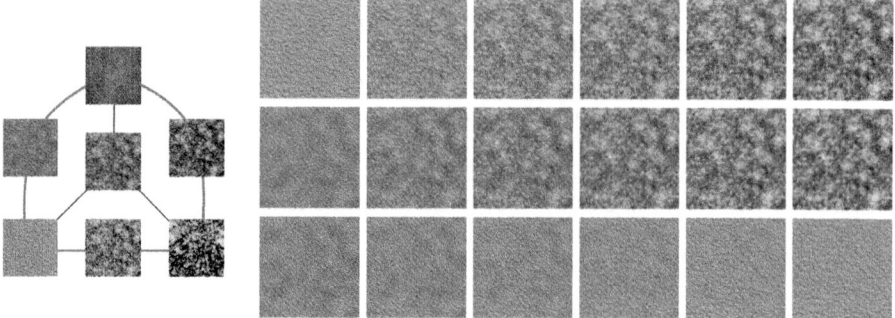

Fig. 4. Texture Interpolation of three textures using the first order statistical model. Left: The three original textures are given at the vertices of a triangle. Right: Details of interpolations on the edges of the triangle.

that the projectors have to be designed by hand; secondly, the authors showed that it yields poor results for texture mixing.

This section extends the first order model described in Section 4.2 to handle higher order statistical features. Following the methodology introduced by Portilla and Simoncelli, we consider for each index ℓ – in addition to aforementioned the 3D-distributions $Y^{\ell,j}$ – the joint distributions of wavelet coefficients located in different spatial positions, $i.e.$

$$\forall \ell \in L, \ \forall j \in J, \quad C_{\mathcal{N}}^{\ell,j} = \{Y^{\ell,j}\{m\}, m \in \mathcal{N}(n)\}_n \subset \mathbb{R}^{3|\mathcal{N}|}.$$

where $\mathcal{N}(n) = n + \mathcal{N}$ is a given neighborhood pattern around each n. The joint distribution $C_{\mathcal{N}}^{\ell,j}$ thus has $3 \times |\mathcal{N}|$ dimensions. In the following experiments, $\mathcal{N}$ is defined as a square neighborhood of size 4×4, but more sophisticated pattern could be implemented.

Higher order statistical texture mixing. The texture mixing framework introduced in section 4.2 is extended to take into account the joint distributions $\{C_{\mathcal{N}}^{\ell,j}\}_{j \in J}$. In a preliminary stage, the sliced Wasserstein barycenters $C_{\mathcal{N}}^{\ell} = $ S-Bar$(\rho_j, C_{\mathcal{N}}^{\ell,j})_{j \in J} \subset \mathbb{R}^{3|\mathcal{N}|}$ are computed using the stochastic gradient descent algorithm. Then, during the iterated synthesis algorithm, for each $\ell \in L$, the coefficients $\{c_{\ell,n}^{(k)}\}_n$ defined in (13) are clustered into blocks and projected into the statistical constraints

$$\forall \ell \in L, \quad \{\hat{c}_{\ell,n}^{(k)}\}_n = \text{S-Proj}_{[C_{\mathcal{N}}^{\ell}]}(\{c_{\ell,m}^{(k)}, m \in \mathcal{N}(n)\}_n)$$

The images $f^{(k+1)}$ are then reconstructed from these coefficients $\{\hat{c}_{\ell,n}^{(k)}\}_n$ using (14).

An example is proposed in Figure 5 to illustrate this high-order texture mixing framework. The proposed algorithm is therefore run on point clouds with at most $N = 2^{16}$ points in $\mathbb{R}^{16}$ with $|\omega| = 10$ directions. It should be noticed that the authors of [20] already introduced a color model, but that was not published. However, our method is the first extension of such joint probability constraints to texture mixing.

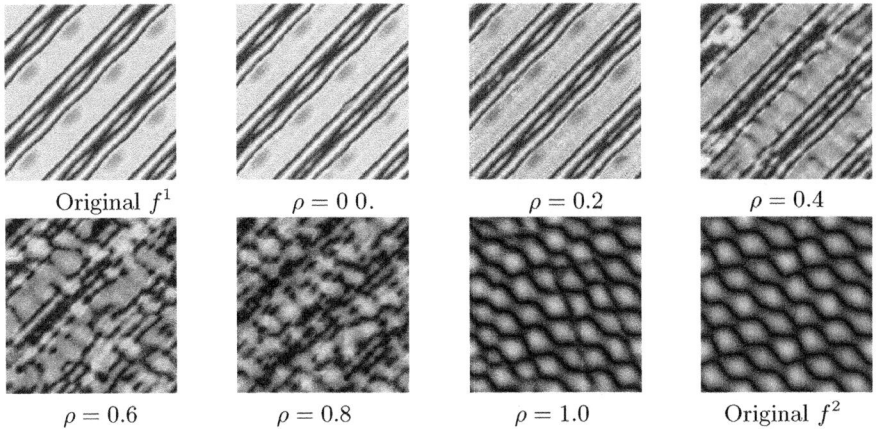

Original f^1	$\rho = 0\ 0.$	$\rho = 0.2$	$\rho = 0.4$
$\rho = 0.6$	$\rho = 0.8$	$\rho = 1.0$	Original f^2

Fig. 5. High Order Texture Interpolation between two textures

5   Conclusion

This paper has tackled the problem of defining average of histograms and distributions features. This shows that optimal distance metrics are useful beyond pairwise comparison of distributions.

The second point made by this paper is that the optimal transport in itself encompasses important information and enables to enforce complicated, high dimensional, statistical features. This paper presented an illustrative example in the field of texture synthesis.

Some extensions of this work are foreseen. First, more complex statistical information could be modeled using adaptive neighborhood $\mathcal{N}$, requiring a statistical learning step before the texture synthesis. Moreover, the extension of the proposed framework for any Wasserstein metric W_p (with $p \geqslant 1$) is currently studied to obtain a more general definition of barycenter, including a "Wasserstein median" when $p = 1$.

References

1. Villani, C.: Topics in Optimal Transportation. American Math. Society (2003)
2. Rubner, Y., Tomasi, C., Guibas, L.J.: The Earth Mover's Distance as a Metric for Image Retrieval. International Journal of Computer Vision 40, 99–121 (2000)
3. Pitié, F., Kokaram, A.: The Linear Monge-Kantorovitch Colour Mapping for Example-Based Colour Transfer. In: Proc. of CVMP 2006 (2006)
4. Dominitz, A., Tannenbaum, A.: Texture mapping via optimal mass transport. IEEE Transactions on Visualization and Computer Graphics 16, 419–433 (2009)
5. Delon, J.: Movie and video scale-time equalization application to flicker reduction. IEEE Trans. Image Proc. 15, 241–248 (2006)
6. Ambrosio, L., Caffarelli, L.A., Brenier, Y., Buttazzo, G., Villani, C.: Optimal Transportation and Applications. Mathematics and statistics edn. Lecture Notes in Mathematics, vol. 1813. Springer, Heidelberg (2003)
7. Efros, A.A., Leung, T.K.: Texture synthesis by non-parametric sampling. In: Proc. of ICCV 1999, p. 1033 (1999)

8. Wei, L.Y., Levoy, M.: Fast texture synthesis using tree-structured vector quantization. In: Proc. Siggraph 2000, pp. 479–488 (2000)
9. Efros, A., Freeman, W.: Image quilting for texture synthesis and transfer. ACM Trans. on Graphics, 341–346 (2001)
10. Ashikhmin, M.: Synthesizing natural textures. In: SI3D 2001: Proceedings of the 2001 Symposium on Interactive 3D Graphics, pp. 217–226 (2001)
11. Lefebvre, S., Hoppe, H.: Parallel controllable texture synthesis. ACM Trans. on Graphics 24, 777–786 (2005)
12. Kwatra, V., Essa, I., Bobick, A., Kwatra, N.: Texture optimization for example-based synthesis. ACM Trans. on Graphics 24, 795–802 (2005)
13. Kwatra, V., Schdl, A., Essa, I., Turk, G., Bobick, A.: Graphcut textures: Image and video synthesis using graph cuts. ACM Trans. on Graphics 22, 277–286 (2003)
14. Perlin, K.: An image synthesizer. In: Proc. Siggraph 1985, pp. 287–296. ACM Press, New York (1985)
15. Bonet, J.S.D.: Multiresolution sampling procedure for analysis and synthesis of texture images. In: Proc. Siggraph 1997, pp. 361–368. ACM Press, New York (1997)
16. Paget, R., Longstaff, I.D.: Texture synthesis via a noncausal nonparametric multiscale markov random field. IEEE Trans. Image Proc. 7, 925–931 (1998)
17. Mumford, D., Gidas, B.: Stochastic models for generic images. Q. Appl. Math. LIV, 85–111 (2001)
18. Heeger, D.J., Bergen, J.R.: Pyramid-Based texture analysis/synthesis. In: Proc. Siggraph 1995. Annual Conference Series, ACM SIGGRAPH, pp. 229–238 (1995)
19. Cook, R., DeRose, T.: Wavelet noise. ACM Trans. on Graphics 24, 803–811 (2005)
20. Portilla, J., Simoncelli, E.P.: A parametric texture model based on joint statistics of complex wavelet coefficients. Int. Journal of Computer Vision 40, 49–70 (2000)
21. Bar-Joseph, Z., El-Yaniv, R., Lischinski, D., Werman, M.: Texture mixing and texture movie synthesis using statistical learning. IEEE Transactions on Visualization and Computer Graphics 7, 120–135 (2001)
22. Peyré, G.: Texture synthesis with grouplets. IEEE Trans. Patt. Anal. and Mach. Intell. 32, 733–746 (2010)
23. Hertzmann, A., Jacobs, C.E., Oliver, N., Curless, B., Salesin, D.H.: Image analogies. In: ACM (ed.) Proc. Siggraph 2001, pp. 327–340. ACM Press, New York (2001)
24. Liu, Z., Liu, C., Shum, H.Y., Yu, Y.: Pattern-based texture metamorphosis. In: Proc. Pacific Graphics 2002, pp. 184–193. IEEE Computer Society, Los Alamitos (2002)
25. Tonietto, L., Walter, M.: Texture metamorphosis driven by texton masks. Computers and Graphics 29, 697–703 (2005)
26. Tal, A., Elber, G.: Image morphing with feature preserving texture. Comput. Graph. Forum 18, 339–348 (1999)
27. Matusik, W., Zwicker, M., Durand, F.: Texture design using a simplicial complex of morphable textures. ACM Trans. on Graphics 24, 787–794 (2005)
28. Burkard, R., Dell'Amico, M., Martello, S.: Assignment Problems. SIAM, Philadelphia (2009)
29. Rabin, J., Peyré, G., Cohen, L.D.: Geodesic shape retrieval via optimal mass transport. In: Daniilidis, K., Maragos, P., Paragios, N. (eds.) ECCV 2010. LNCS, vol. 6315, pp. 771–784. Springer, Heidelberg (2010)
30. Dowson, D.C., Landau, B.V.: The Fréchet distance between multivariate normal distributions. Journal of Multivariate Analysis 12, 450–455 (1982)
31. Agueh, M., Carlier, G.: Barycenters in the Wasserstein space. To appear in SIAM Journal on Mathematical Analysis (2011)
32. Simoncelli, E.P., Freeman, W.T., Adelson, E.H., Heeger, D.J.: Shiftable multiscale transforms. IEEE Trans. Info. Theory 38, 587–607 (1992)

Theoretical Foundations of Gaussian Convolution by Extended Box Filtering

Pascal Gwosdek[1], Sven Grewenig[1], Andrés Bruhn[2], and Joachim Weickert[1]

[1] Mathematical Image Analysis Group, Dept. of Mathematics and Computer Science,
Campus E1.1, Saarland University, 66041 Saarbrücken, Germany
{gwosdek,grewenig,weickert}@mia.uni-saarland.de
[2] Vision and Image Processing Group,
Cluster of Excellence Multimodal Computing and Interaction,
Saarland University, Campus E 1.1, 66041 Saarbrücken, Germany
bruhn@mmci.uni-saarland.de

Abstract. Gaussian convolution is of fundamental importance in linear scale-space theory and in numerous applications. We introduce iterated extended box filtering as an efficient and highly accurate way to compute Gaussian convolution. Extended box filtering approximates a continuous box filter of arbitrary non-integer standard deviation. It provides a much better approximation to Gaussian convolution than conventional iterated box filtering. Moreover, it retains the efficiency benefits of iterated box filtering where the runtime is a linear function of the image size and does not depend on the standard deviation of the Gaussian. In a detailed mathematical analysis, we establish the fundamental properties of our approach and deduce its error bounds. An experimental evaluation shows the advantages of our method over classical implementations of Gaussian convolution in the spatial and the Fourier domain.

Keywords: Gaussian scale-space, box filter, image processing, computer vision.

1 Introduction

Convolution with a Gaussian is one of the most widely used linear filter operations in signal and image processing. It forms the backbone of Gaussian scale-space theory [4,9,13] which has been introduced in Japanese and English papers of Iijima [8] long before it became popular in the western world by Witkin's work [16]. The strong regularisation properties of Gaussian convolution render the filtered signal infinitely times differentiable and stabilise the numerical evaluation of higher order derivatives. Gaussian convolution is inevitable for the detection of edges [2,11] and interest points [6,10] that play a central role in computer vision. The rapid decay properties of the Gaussian both in the spatial and the Fourier domain and the fact that it is the only filter that is rotationally invariant and separable under convolution make Gaussian convolution a perfect low-pass filter in linear systems theory.

A.M. Bruckstein et al. (Eds.): SSVM 2011, LNCS 6667, pp. 447–458, 2012.

Many applications require an accurate and efficient implementation of Gaussian convolution in order to ensure the high quality of the results, to meet runtime requirements, or even to guarantee convergence. However, this can be challenging: It comes down to a convolution of the input signal with a kernel function with infinite support. The m-dimensional Gaussian kernel

$$K_\sigma(\boldsymbol{x}) = \frac{1}{(2\pi\sigma^2)^{\frac{m}{2}}} \exp\left(-\frac{|\boldsymbol{x}|^2}{2\cdot\sigma^2}\right) \tag{1}$$

of standard deviation σ has a characteristic 'bell curve' shape which drops off rapidly towards $\pm\infty$. This is why in practice one often applies a discrete convolution with a sampled and renormalised kernel that is cut off at $n \cdot \sigma$. However, this method becomes inefficient for large σ, as the number of operations grows linearly in the number of samples of both the signal and the kernel. A more efficient alternative for those cases is the computation as a point-wise multiplication in the frequency domain [1]. To this end, a Fourier transform is applied to both the kernel and the signal, the multiplication is performed, and the result is transformed back into the spatial domain. Since the Gaussian kernel in the frequency domain can immediately be evaluated, this method reduces to two fast Fourier transforms, and one point-wise multiplication.

Although these spatial and Fourier-based implementations are the most popular algorithms for Gaussian convolution, and their trade-offs are well investigated [5], there are also further alternatives: Approximations with recursive filters [3,17] offer a runtime behaviour that scales linearly in the number of pixels. However, these filters require a special boundary treatment and a higher implementational effort than other methods which poses additional challenges [14]. Since Gaussian scale-space is equivalent to evolving the image under a homogeneous diffusion problem, one can also implement Gaussian convolution with efficient numerical methods for partial differential equations, e.g. with implicit finite difference schemes [7]. Unfortunately, this requires the fast solution of linear systems of equations which is also a nontrivial task. Gaussian convolution can also be approximated by discrete convolution with binomial kernels. They have a finite support and offer some interesting properties from an implementational viewpoint, but do not allow to approximate Gaussians with arbitrary standard deviations. This can constitute a drawback in scale-space applications which aim at representations at arbitrary scales.

A simple but extremely fast discrete approximation of Gaussian smoothing can be achieved by convolution with iterated box filters [15]. A box filter uses a normalised kernel with identical coefficients within its finite support. By the central limit theorem, a sufficiently high number of iterations with a box filter approximates a Gaussian arbitrarily well. However, this has the same drawback as convolution with binomial kernels: It introduces a quantisation to the range of standard deviations that can be approximated.

In our paper we address this problem. We advocate a modification of the box filter that is based on a new discretisation of the continuous box kernel. In particular, we concentrate on establishing important properties of the resulting *extended box filter*: It combines the simplicity and algorithmic efficiency of the conventional box filter with a good approximation of theoretic properties of Gaussian filtering. In an experimental evaluation, we show that the extended box filter approximates the Gaussian filter significantly better than a classical box filter and offers advantages over spatial and Fourier-based approximations of Gaussian convolution. Moreover, our method introduces only marginal runtime overheads over classical box filtering.

Our paper is structured as follows. In Section 2, we first recapitulate the basic notations and definitions of conventional box filtering. Thereafter, we present our new method in Section 3. After an experimental evaluation in Section 4, we conclude with a summary in Section 5.

2   Conventional Box Filtering

Box filters are usually defined in a purely discrete context. However, in order to derive a new discretisation in this paper, we start with a short review of a continuous definition:

Definition 1. *A continuous box filter B_Λ with a real-valued length $\Lambda \in \mathbb{R}^+ := \{a \in \mathbb{R} \ : \ a > 0\}$ is a convolution*

$$(B_\Lambda * f)(x) := \int_{-\infty}^{\infty} B_\Lambda(x-y) \cdot f(y)\, dy \qquad (2)$$

of a signal f with a box kernel

$$B_\Lambda(x) := \begin{cases} \frac{1}{\Lambda}, & x \in (-\lambda, \lambda) \\ 0, & else \end{cases} \qquad (3)$$

for $x \in \mathbb{R}$ and $\Lambda = 2\lambda$.

In the literature, one usually finds the continuous length Λ being rounded to the closest odd integer L [15]:

Definition 2. *A discrete box filter B_L of length $L = h(2l + 1)$, $l \in \mathbb{N}_0$, and sampled at an equidistant grid of spacing $h > 0$ is a convolution*

$$(B_L * f)(hk) := \sum_{m \in \mathbb{Z}} B_L(h(k-m)) \cdot f(hm) \qquad (4)$$

of a signal f with a discrete box kernel

$$B_L(hk) := \begin{cases} \frac{h}{L}, & -l \leqslant k \leqslant l \\ 0, & else \end{cases} \qquad (5)$$

for $k \in \mathbb{Z}$.

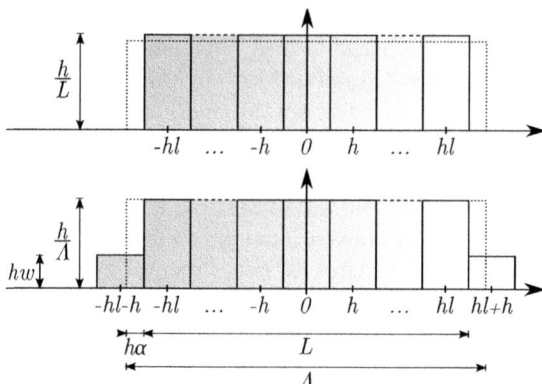

Fig. 1. Visualisation of box kernels. **Top**: Continuous box kernel B_A (**dotted**) and its conventional discrete approximation B_L. **Bottom**: Corresponding discrete extended box kernel E_A.

An illustration of this construction is depicted in Figure 1. Note that we introduce an arbitrary grid spacing h, and couple the length L to a multiple of this distance. For $h \to 0$, B_L thus approaches B_A (cf. Definition 1). If we set $h = 1$, we obtain the formulation in [15].

On discrete data, it can be implemented very efficiently in an iterative 'sliding window' manner, i.e.

$$(B_L * f)_i = (B_L * f)_{i-1} + \frac{h}{L}(f_{i+l} - f_{i-l-1}), \qquad (6)$$

with $(\cdot)_k$ or f_k denoting the discrete value at sampling point hk. After the initialisation of the first sample, the method needs one multiplication and two additions per pixel and dimension, independent of the size of the kernel. Thus, it enjoys a linear complexity in time.

A d-fold convolution of the kernel with the signal approximates a Gaussian convolution. This removes artefacts that arise from the piecewise linearity of the box kernel, as well as from the lack of a rotational invariance property in the multi-dimensional case. The resulting operation is equivalent to the convolution with a C^{d-1}-continuous kernel B_L^d of variance $\sigma^2(B_L^d)$ [15]:

$$\sigma^2(B_L^d) = d\,\frac{L^2 - 1}{12}. \qquad (7)$$

Note that this formula only allows a *discrete* set of standard deviations to be chosen. In the literature, it is suggested to handle this problem by a series of box filters of different length [15]. Unfortunately, this idea does not solve the problem: By practical considerations, d is typically chosen from the range $\{3, 4, 5\}$, such that the distance between admissible σ cannot be reduced arbitrarily. Moreover, the kernel resulting from a convolution of box kernels with different lengths does not fulfil the continuity properties mentioned above.

In contrast to this suggestion, we are now going to derive a better discretisation of the continuous formulation which does not have this problem by construction. Still, it possesses all advantages of the discrete box filter.

3   Extended Box Filter

Our goal is now to find a better discretisation E_Λ of the continuous box filter B_Λ than is given by the conventional discrete approximation B_L. In doing so, we focus in particular on the following criteria:

1. E_Λ must be continuous over Λ to allow kernels with arbitrary variance.
2. For $\Lambda = L \in \mathbb{N}_{\mathrm{odd}}$, it must equal the discrete box filter B_L of length L.
3. For $h \to 0$, it must approach the continuous case, i.e. $\lim_{h\to 0} \sigma^2(E_\Lambda) = \sigma^2(B_\Lambda)$.

To this end, we decompose Λ into an integer part and a real-valued remainder:

$$\Lambda = h(2l + 1 + 2\alpha) = L + 2h\alpha \tag{8}$$

such that $0 \leqslant \alpha < 1$ and $l \in \mathbb{N}_0$. With this formalism, we are now able to set up an 'extended' variant of the discrete box filter:

Definition 3. *An extended box filter E_Λ with a real-valued length $\Lambda \in \mathbb{R}^+$ and discretised on a uniform grid of spacing $h > 0$ is a convolution*

$$(E_\Lambda * f)(hk) := \sum_{m\in\mathbb{Z}} E_\Lambda(h(k-m)) \cdot f(hm) \tag{9}$$

of a signal f with an extended box kernel

$$E_\Lambda(hk) := \begin{cases} \frac{h}{\Lambda}, & -l \leqslant k \leqslant l \\ hw, & k \in \{-(l+1), l+1\} \\ 0, & else \end{cases} \tag{10}$$

with

$$l := \left\lfloor \frac{\Lambda}{2h} - \frac{1}{2} \right\rfloor, \qquad w := \frac{1}{2}\left(\frac{1}{h} - \frac{2l+1}{\Lambda}\right), \tag{11}$$

and $k \in \mathbb{Z}$. $\lfloor x \rfloor$ denotes the so-called floor function, which computes the largest integer not greater than $x \in \mathbb{R}$.

Both constraints in (11) are necessary in order to ensure that all weights sum up to 1. A visualisation of the extended box kernel (10) is depicted in Figure 1. It is immediately clear that our new filter preserves many advantages of the original box filter. It is separable in space, and an efficient 'sliding-window' implementation is still possible and beneficial: Apart from the first value in a row, only four additions and two multiplications are needed per pixel and dimension (since both weighting factors are constants):

$$(E_\Lambda * f)_i = (E_\Lambda * f)_{i-1} + \left(\frac{h}{\Lambda} - hw\right)(f_{i+l} - f_{i-l-1})$$

$$+ hw\ (f_{i+l+1} - f_{i-l-2}). \qquad (12)$$

This means, the computational complexity of a box filtering step is in $\mathcal{O}(n)$ in the number of pixels, and is thus in particular independent of the length of the chosen box kernel. Let us now discuss some mathematical properties of our construction. First of all, we immediately see that w depends proportionally on α (cf. Figure 1):

$$w = \frac{1}{2h}\left(1 - \frac{(2l+1)h}{\Lambda}\right) = \frac{1}{2h}\left(1 - \frac{\Lambda - 2\alpha h}{\Lambda}\right) = \frac{1}{2h} \cdot \frac{2\alpha h}{\Lambda} = \frac{\alpha}{\Lambda}. \qquad (13)$$

Using this equivalence, we can formulate the variance of E_Λ by considering the components of Λ only. Like for the conventional box filter, we regard the more general case for a convolution kernel that corresponds to a d-fold application of a single extended box kernel E_Λ:

Theorem 1. *The variance* $\sigma^2(E_\Lambda^d)$ *of a d-fold iterated extended box kernel is given by*

$$\sigma^2(E_\Lambda^d) = \frac{dh^3}{3\Lambda}\left(2l^3 + 3l^2 + l + 6\alpha(l+1)^2\right). \qquad (14)$$

Proof. By symmetry considerations, we see that the expectation value of E_Λ is zero. For the variance $\sigma^2(E_\Lambda)$ of one (non-iterated) box kernel, it follows

$$\sigma^2(E_\Lambda) = \sum_{k=-(l+1)}^{l+1} E_\Lambda(hk) \cdot (hk - 0)^2$$

$$= \sum_{k=-l}^{l} \frac{h}{\Lambda}(hk)^2 + hw\left(-(hl+h)\right)^2 + hw(hl+h)^2$$

$$= \frac{2h^3}{\Lambda}\sum_{k=1}^{l} k^2 + 2h^3 w(l+1)^2$$

$$\overset{(13)}{=} \frac{h^3}{3\Lambda}\left(2l^3 + 3l^2 + l + 6\alpha(l+1)^2\right).$$

From probability theory, we obtain the variance $\sigma^2(E_\Lambda^d)$ for the iterated extended box kernel as the sum of single variances. This concludes the proof. □

For $h = 1$ and $\Lambda = 2l + 1 \in \mathbb{N}_{\text{odd}}$, i.e. $\alpha = 0$, this is just a generalisation of Equation (7). This means that the extended box filter falls back to the notion of the conventional box filter in these cases:

Theorem 2. *The extended box kernel E_Λ constitutes a generalisation of the discrete box kernel B_L for the case $h = 1$, i.e. $E_L = B_L$ for $L \in \mathbb{N}_{odd}$ and*

$$\forall L \in \mathbb{N}_{odd} : \lim_{\Lambda \to L^+} \sigma^2(E_\Lambda^d) = \sigma^2(B_L^d) \quad and \quad \lim_{\Lambda \to (L+2)^-} \sigma^2(E_\Lambda^d) = \sigma^2(B_{L+2}^d). \quad (15)$$

Proof. It is clear that $E_L = B_L$ for $L \in \mathbb{N}_{odd}$, because in this case we get $\alpha = 0$ and $w = 0$. Thus, it immediately follows that $\lim_{\Lambda \to L^+} \sigma^2(E_\Lambda^d) = \sigma^2(B_L^d)$. So, it remains to show the case $\Lambda \to (L+2)^-$, for which we first consider a single extended box kernel E_Λ:

$$
\begin{aligned}
\lim_{\Lambda \to (L+2)^-} \sigma^2(E_\Lambda) &= \lim_{\Lambda \to (L+2)^-} \frac{1}{3\Lambda} \left(2l^3 + 3l^2 + l + 3(\Lambda - L)(l+1)^2 \right) \\
&= \frac{1}{3(L+2)} (2l^3 + 9l^2 + 13l + 6) \\
&= \frac{1}{3(2l+3)} (2l+3)(l^2 + 3l + 2) \\
&= \frac{(L+2)^2 - 1}{12} ,
\end{aligned}
$$

where we have used that $\alpha = \frac{\Lambda - L}{2}$ and $L = 2l + 1$. It follows immediately that

$$\lim_{\Lambda \to (L+2)^-} \sigma^2(E_\Lambda^d) = \sigma^2(B_\Lambda^d).$$

This shows that E_Λ^d is a consistent generalisation of B_L^d with respect to Λ. $\qquad \square$

Now that we have shown that the extended box filter extends the previous discrete definition, we want to show that it is a good discretisation of the continuous box filter we are about to approximate:

Theorem 3. *The extended box kernel E_Λ is a suitable discretisation of a box kernel B_Λ in the continuous domain, i.e. for d-fold application,*

1. its variance approximates the continuous analogue arbitrarily well:

$$\lim_{h \to 0} \sigma^2(E_\Lambda^d) = \sigma^2(B_\Lambda^d), \quad and \quad (16)$$

2. the order of consistency is $\mathcal{O}(h^2)$.

Proof. We can deduce an approximation of the continuous setting by computing the limit of $\sigma^2(E_\Lambda^d)$ for the grid spacing $h \to 0$. Since we are interested in the order of consistency, we must consider the variance in (14) and rewrite it:

$$
\begin{aligned}
\sigma^2(E_\Lambda^d) &= \frac{(2hl)^3}{12\Lambda} + dh\frac{(2hl)^2}{4\Lambda} + dh^2\frac{2hl}{6\Lambda} + 2dh^3\frac{\alpha}{\Lambda}(l^2 + 2l + 1) \\
&= d\frac{(2hl)^3}{12\Lambda} + dh(1 + 2\alpha)\frac{(2hl)^2}{4\Lambda} + dh^2(1 + 12\alpha)\frac{2hl}{6\Lambda} + dh^3\frac{2\alpha}{\Lambda}.
\end{aligned}
$$

Input: Signal u^0, standard deviation σ, iterations d.
Output: Signal $u^d := E_\Lambda^d * u^0$

$l \;\leftarrow\;$ largest integer such that $\sigma^2(B_L^d) \le \sigma^2$ (by (7))

$\alpha \;\leftarrow\; (2l+1)\dfrac{l(l+1) - \frac{3\sigma^2}{d}}{6(\frac{\sigma^2}{d} - (l+1)^2)}$ (by $\sigma^2(E_\Lambda^d) \le \sigma^2$, (8), and (14))

$w \;\leftarrow\; \dfrac{\alpha}{2l+1+2\alpha}$, $\hat{w} \;\leftarrow\; \dfrac{1-\alpha}{2l+1+2\alpha}$ (by (8) and (13))

For all $j \in \{1,\dots,d\}$
  Compute u^j[0] (for the first pixel)
  For all i > 0
    Compute u^j[i] $\leftarrow$ u^j[i-1] $+$ w $\cdot$ (u^{j-1}[i+l+1] $-$ u^{j-1}[i-l-2])
              $+$ $\hat{w}$ $\cdot$ (u^{j-1}[i+l   ] $-$ u^{j-1}[i-l-1])

Fig. 2. Algorithm for 1-D extended box filtering. Boundaries can be handled on-the-fly.

Now we replace $2hl$ by $\Lambda - (1+2\alpha)h$ and get for the first three terms:

$$d\,\frac{(\Lambda - (1+2\alpha)h)^3}{12\Lambda} \;=\; \frac{d\Lambda^2}{12} - \frac{dh}{4}(1+2\alpha)\Lambda + \frac{dh^2}{4}(1+2\alpha)^2 + \mathcal{O}(h^3),$$

$$dh(1+2\alpha)\frac{(\Lambda - (1+2\alpha)h)^2}{4\Lambda} \;=\; \frac{dh}{4}(1+2\alpha)\Lambda - \frac{dh^2}{2}(1+2\alpha)^2 + \mathcal{O}(h^3),$$

$$dh^2(1+12\alpha)\frac{\Lambda - (1+2\alpha)h}{6\Lambda} \;=\; \frac{dh^2}{6}(1+12\alpha) + \mathcal{O}(h^3).$$

Finally, this yields

$$\sigma^2(E_\Lambda^d) \;=\; \frac{d\Lambda^2}{12} - h^2 \cdot \frac{d}{12}\left(12\alpha^2 - 12\alpha + 1\right) + \mathcal{O}(h^3).$$

Thus, the consistency order is $\mathcal{O}(h^2)$ and we can state that

$$\lim_{h \to 0} \sigma^2(E_\Lambda^d) \;=\; d\frac{\Lambda^2}{12} \;=\; d\int_{-\frac{\Lambda}{2}}^{\frac{\Lambda}{2}} \frac{1}{\Lambda} \cdot x^2 \mathrm{d}x \;=\; d\int_{-\infty}^{\infty} B_\Lambda(x)x^2 \mathrm{d}x \;=\; \sigma^2(B_\Lambda^d). \qquad \square$$

4   Experiments

In order to investigate the properties of the extended box filter on real data, we have implemented the algorithm for application on images. Technically, this means we are dealing with 2-D images $f \in \Omega \subset \mathbb{R}^2$, and assume reflecting boundary conditions to preserve the average grey value. Using the separability of the kernel, we apply a 'sliding window' technique in both directions (cf. Figure 2). This operation is highly parallel, and can thus be significantly accelerated by the streaming SIMD extension mechanism (SSE) of modern desktop processors, by use of all CPU cores, and by graphics processors.

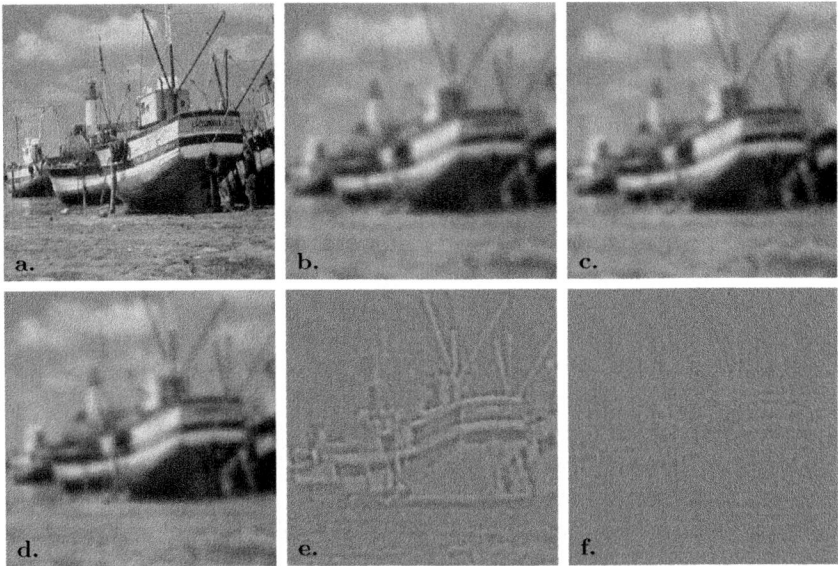

Fig. 3. Visual quality for *Boat*, 512 × 512 pixels. **a:** Original, **b:** conventional and **c:** extended box filtering with $d = 3$, $\sigma = 5.0$, **d:** Discrete Gaussian filtering with $\sigma = 5.0$ (truncated at 10σ), and **e,f:** differences of b and c to d, respectively, scaled by a factor 10 to increase visibility. 50% grey means the error is zero.

4.1 Qualitative Gain

Our aim in designing the extended box filter is to propose a fast but accurate way to perform Gaussian convolution for arbitrary standard deviations. Consequently, we are interested in the accuracy of the proposed method.

To evaluate the accuracy, we use the well-known *Boat* test image from the USC SIPI database (cf. Figure 3a), and convolve it with discrete box kernels. These results are then compared to a ground truth obtained by a convolution with a discretised Gaussian kernel that has been truncated at 10σ and renormalised. Please note that this ground truth is also subject to discretisation artefacts and may not exactly reflect the desired solution. A more complicated alternative to this implementation has been proposed in [12], but this variant also suffers from similar truncation problems. Finally, let us note that we chose the *Boat* test image as a good representative for many real-world examples, since it contains many different frequencies and both homogeneous and textured regions.

In the first part of our experiment, we use conventional box kernels and a varying number of iterations d, and compare these results to the reference solution. Instead of focussing on one specific standard deviation σ, we evaluate many different values. As an error measure for equivalence, we use the mean square error (MSE) given by

$$MSE(a, b) = \frac{1}{N} \sum_{i=1}^{N} (a_i - b_i)^2, \qquad (17)$$

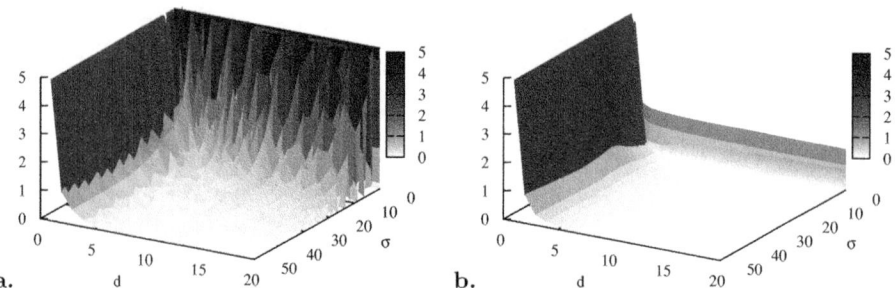

Fig. 4. Plot of the mean square error to discrete Gaussian convolution on *Boat*, 512 ×
512 pixels, depending on σ and d. **a:** Conventional box filter. **b:** Extended box filter.

where N describes the number of pixels. The results for this experiment are
given in Figure 4a. For large σ, we see that a box filter of order $d = 5$ is already
sufficient to approximate the Gaussian very well. However, one also realises that
small standard deviations cannot be represented well at all. This effect is caused
by the integer length of the box kernel, and re-occurs for larger d and larger σ
for similar reasons.

In the second part, we repeated the same experiment with the proposed kernel. This is shown in Figure 4b. Compared to the conventional box filter, the
novel approach attenuates errors much stronger. For any σ, an order of $d = 5$
yields almost identical results to Gaussian filtering. This justifies our model as
a qualitatively equal alternative.

To conclude this experiment, we compare the visual quality of both approaches. Figure 3 depicts a sample output of both methods under a standard
deviation $\sigma = 5.0$, and further shows the desired result as given by Gaussian
convolution. Albeit the visual quality differences are relatively low, the difference images show that our extended box filter performs much better than the
conventional box filtering.

4.2   Runtime

In the last experiment, we are interested in the tradeoff between the accuracy
and the runtime of extended box filtering compared to other techniques. To this
end, we convolve the *Boat* test image using a discrete Gaussian truncated at 3σ,
an FFT-based approach, a conventional, and an extended box filter (both with
$d = 5$). Runtimes were acquired on a single-core 3.2 GHz Pentium 4 with 2MB
L2 cache and 2 GB RAM.

Table 1 shows the results of this experiment. The truncated Gaussian con-
vinces for small standard deviations, but scales linearly in σ such that this
method becomes infeasible for large σ. Although the runtime of all remaining
methods is independent from the standard deviation, box filters have a clear
advantage if we consider larger images: While they only scale linearly in the
number of pixels n, the FFT-based methods have a complexity of $\mathcal{O}(n \log(n))$.
In return, the FFT-based approach offers a much better approximation quality

Table 1. CPU runtime t in milliseconds vs. mean square error (MSE) between the result and the ground truth for different techniques on *Boat* (512×512 pixels)

	$\sigma = 0.5$ MSE	t	$\sigma = 5.0$ MSE	t	$\sigma = 25.0$ MSE	t
Truncated Gaussian	0.000	8	0.001	45	0.007	148
FFT-based	1.032	148	0.000	148	0.000	148
Conventional box	9.580	0	1.400	26	0.154	27
Extended box	0.030	41	0.051	43	0.098	43

for large σ. In this context, the extended box filter is a good tradeoff between classical box filtering and the FFT-based approaches: It provides a convincing approximation quality for all standard deviations at a slightly higher runtime than a classical box filter.

5   Summary

In view of the omnipresence of Gaussian convolution in scale-space theory and its numerous applications in image processing and computer vision, it is surprising that one can still come up with novel algorithms that are extremely simple and offer a number of advantages. In our paper we have shown that a small modification of classical box filtering leads to an extended box filter which can be iterated in order to approximate Gaussian convolution with high accuracy and high efficiency. In contrast to classical box filtering, it does not suffer from the restriction that only a distinct set of standard deviations of the Gaussian are allowed. Although the main focus of our paper is on establishing the essential mathematical properties of extended box filtering, we have also presented experiments that illustrate the advantages over spatial and Fourier-based implementations of Gaussian convolution. In our ongoing research we will perform a more extensive evaluation with a large number of alternative implementations, taking also into account the potential of modern parallel hardware such as GPUs.

Acknowledgements. Our research was partly funded by the Cluster of Excellence "Multimodal Computing and Interaction", and by the Deutsche Forschungsgemeinschaft under project We2602/7-1. This is gratefully acknowledged.

References

1. Bracewell, R.N.: The Fourier transform and its applications, 3rd edn. McGraw-Hill, New York (1999)
2. Canny, J.: A computational approach to edge detection. IEEE Transactions on Pattern Analysis and Machine Intelligence 8, 679–698 (1986)
3. Deriche, R.: Fast algorithms for low-level vision. IEEE Transactions on Pattern Analysis and Machine Intelligence 12, 78–87 (1990)

4. Florack, L.: Image Structure, Computational Imaging and Vision, vol. 10. Kluwer, Dordrecht (1997)
5. Florack, L.: A spatio-frequency trade-off scale for scale-space filtering. IEEE Transactions on Pattern Analysis and Machine Intelligence 22(9), 1050–1055 (2000)
6. Förstner, W., Gülch, E.: A fast operator for detection and precise location of distinct points, corners and centres of circular features. In: Proc. ISPRS Intercommission Conference on Fast Processing of Photogrammetric Data, Interlaken, Switzerland, pp. 281–305 (June 1987)
7. Gourlay, A.R.: Implicit convolution. Image and Vision Computing 3, 15–23 (1985)
8. Iijima, T.: Theory of pattern recognition. Electronics and Communications in Japan pp. 123–134 (November 1963) (in English)
9. Lindeberg, T.: Scale-Space Theory in Computer Vision. Kluwer, Boston (1994)
10. Lowe, D.G.: Distinctive image features from scale-invariant keypoints. International Journal of Computer Vision 60(2), 91–110 (2004)
11. Marr, D., Hildreth, E.: Theory of edge detection. Proceedings of the Royal Society of London, Series B 207, 187–217 (1980)
12. Norman, E.: A discrete analogue of the Weierstrass transform. Proceedings of the American Mathematical Society 11(596-604) (1960)
13. Sporring, J., Nielsen, M., Florack, L., Johansen, P. (eds.): Gaussian Scale-Space Theory, Computational Imaging and Vision, vol. 8. Kluwer, Dordrecht (1997)
14. Triggs, B., Sdika, M.: Boundary conditions for Young - van Vliet recursive filtering. IEEE Transactions on Signal Processing 54(5), 1–2 (2006)
15. Wells, W.M.: Efficient synthesis of Gaussian filters by cascaded uniform filters. IEEE Transactions on Pattern Analysis and Machine Intelligence 8(2), 234–239 (1986)
16. Witkin, A.P.: Scale-space filtering. In: Proc. Eighth International Joint Conference on Artificial Intelligence, vol. 2, pp. 945–951. Karlsruhe, West Germany (1983)
17. Young, I.T., van Vliet, L.J.: Recursive implementation of the Gaussian filter. Signal Processing 44, 139–151 (1995)

From High Definition Image
to Low Space Optimization

Micha Feigin[1], Dan Feldman[2], and Nir Sochen[1]

[1] Deparment of Mathematics, Tel-Aviv University Tel-Aviv, 69978, Israel
[2] Center for Mathematics of Information
California Institute of Technology
Pasadena, CA 91125, USA

Abstract. Signal and image processing have seen in the last few years
an explosion of interest in a new form of signal/image characterization
via the concept of sparsity with respect to a dictionary. An active field
of research is dictionary learning: Given a large amount of example sig-
nals/images one would like to learn a dictionary with much fewer atoms
than examples on one hand, and much more atoms than pixels on the
other hand. The dictionary is constructed such that the examples are
sparse on that dictionary i.e each image is a linear combination of small
number of atoms.

This paper suggests a new computational approach to the problem of
dictionary learning. We show that smart non-uniform sampling, via the
recently introduced method of *coresets*, achieves excellent results, with
controlled deviation from the optimal dictionary. We represent dictionary
learning for sparse representation of images as a geometric problem, and
illustrate the coreset technique by using it together with the K−SVD
method. Our simulations demonstrate gain factor of up to 60 in computational time with the same, and even better, performance. We also
demonstrate our ability to perform computations on larger patches and
high-definition images, where the traditional approach breaks down.

Keywords: Sparsity, dictionary learning, K−SVD, coresets.

1 Introduction

One of the major problems in image processing is image characterization. By
image characterization we mean a system that gets a two-dimensional function
or, in the discrete case, a matrix with non-negative entries, as a query and
provides an answer whether or not this function/matrix is an image. Other
option is that the system provides a probability measure on the space of all
two-dimensional such functions/matrices. We are still far from achieving this
ultimate goal, yet few breakthroughs where recorded since the inception of image
processing as a branch of scientific research. Many characterization, in the past,
used the decay rate of the coefficients of certain transformations. That led to a
characterization in a linear space of functions. In the last decade a new approach
that involves redundant representations and sparsity seems promising. In this

A.M. Bruckstein et al. (Eds.): SSVM 2011, LNCS 6667, pp. 459–470, 2012.
© Springer-Verlag Berlin Heidelberg 2012

framework, a signal is represented again as a superposition of signals. But unlike the representation with a basis of a linear space, the number of basic signals (a.k.a. atoms) in this new approach exceeds the dimension of the signal such that a given signal may have many different representations. Uniqueness is achieved only for a subset of signals which can be represented with a limited number of atoms, called sparse signals. For this class of signals the sparsest representation is unique. This approach shifts the focus of attention from the general law of decay of coefficients to the outliers of such behavior, namely the large coefficients of such an expansion. The class of sparse signals does not form a linear space which reflects the non-linearity of the set of images. At the same time, we still use linear techniques which helps a lot in practice.

The sparsity approach has appealing features for image processing, but it suffers from few problems. First it is clear that sparsity is a notion which is attached to a given dictionary. Clearly, there is no one universal dictionary that can represent any image in a sparse way. This calls upon the need to construct dictionary for each class of images or for each application. Constructing a dictionary for a large number of images from the same class/application goes under the name dictionary learning and is an active field of research. This is the main topic of this paper, and we demonstrate our ideas on the K−SVD method [3]. Second is the very extensive use of computational time and in memory space. Because of the prohibitive computational time and the numerical instabilities in computing with large size matrices, sparsity techniques are applied to small images only. In fact, 8×8 to 16×16 is the most common sizes in image processing. It means that these are patches of images rather than images themselves. Moreover, one may wish to construct dictionaries for the same class of images. Using, implicitly, the approximate self-similarity nature of images it is costumed to use the patches of an image as a class of similar patches and to construct a dictionary per image. Here, again, the curse of limited space and time interfere and high definition images (1024×1024 say) have a huge number of patches of such a small size which makes the dictionary learning task computationally prohibitive. This paper brings the spell of coresets to cure the curse of space and time limitations. Informally, a coreset C for Y is a compressed representation of Y that well approximates the original data in some problem-dependent sense. The coreset C is used to give approximate answers for queries about Y.

We show that the optimization problem can be solved on the coreset, which is much smaller, without sacrificing too much accuracy. Coresets techniques were first introduced in the computational geometry field, and in the recent years used to solve some well known open problems in computer science and machine learning. The subject became more mature theoretically in the last few years. Coresets present a new approach to optimization in general and have huge success especially in tasks which use prohibitively large computation time and/or memory space. In particular, coresets suggest ways to use existing serial algorithms for distributed (parallel) computing, and provide solutions under the streaming model, where the space (memory) for solving the problem at hand is significantly smaller than its input size.

2 Coresets for Dictionaries

Approximation algorithms in computational geometry often make use of random sampling, feature extraction , and ϵ-samples. Coresets can be viewed as a general concept that includes all of the above, and more. See a comprehensive (but not so updated) survey on this topic in [1]. Coresets have been the subject of many recent papers (see references in [1]) and several surveys [2, 4].

In our context, the input is an $d \times n$ matrix Y that represents n points in $\mathbb{R}^d$, and we consider a $d \times k$ matrix D, whose columns represent k points in $\mathbb{R}^d$. The matrix D is called a *dictionary*. Typically n is much larger than k, and k is much larger than d. Let $\text{cost}(\cdot, \cdot)$ be some function of Y and D. We interpret $\text{cost}(Y, D)$ as the result of a query D on a matrix Y.

2.1 k-Dictionary Queries

Let D be a dictionary. The column vectors of the matrix D are called the *points of D*. We write $z \in D$ if z is one of the columns of D. Let y be a point (vector) in $\mathbb{R}^d$, and let $\text{err}(y, D)$ be a non-negative real function that represents the error of approximating y by D.

Through the rest of this section we won't assume anything further about the function $\text{err}(\cdot, \cdot)$. Following some possible definitions of err that are relevant for this paper, when D is a $d \times k$ matrix:

1. $\text{err}(y, D) = \min_{x \in \mathbb{R}^k} \|Dx - y\|_2$ is the Euclidean distance between the point y and the subspace that is spanned by the columns of D.
2. $\text{err}(y, D) = \min_{x \in \{e_i\}} \|Dx - y\|_2$ is the distance between y and its closest point of D. Here, $\{e_i\} = \{e_i\}_1^k$ denotes the standard base of $\mathbb{R}^k$.
3. $\text{err}(y, D) = \min_{x \in \mathbb{R}^k, \|x\|_0 = 1} \|Dx - y\|_2$ is the distance between y and the closest line that intersects both the origin of $\mathbb{R}^d$ and a point in D. Here $\|x\|_0$ denotes the number of non-zeros entries of x.
4. For an integer $j \geq 0$, $\text{err}_j(y, D) = \min_{x \in \mathbb{R}^k, \|x\|_0 \leq j} \|Dx - y\|_2$ is the distance between y and its closest subspace over the set of $O\left(\binom{k}{j}\right)$ subspaces that are spanned by at most j points of D.

More generally, we can replace $\|Dx - y\|_2$ by $\|Dx - y\|_p^q$ for $p, q \geq 0$ in the above examples.

Problem 1 (k-Dictionary Query). The input to this problem is an integer $k \geq 1$ and a $d \times n$ matrix Y. For a given (query) $d \times k$ dictionary D, the desired output is $\text{cost}(Y, D) = \sum_{y \in Y} \text{err}(y, D)$.

Suppose that, given $y \in Y$ and a $d \times k$ dictionary D, the error $\text{err}(y, D)$ can be computed in time T. Then $\text{cost}(Y, D)$ can be computed in $O(Tn)$ time and using $O(dn)$ space. Here, "space" means memory, or number of non-zeros entries. In this paper we wish to pre-process the input Y in such a way that an ε-approximation of the output $\text{cost}(Y, D)$ could be computed in time $O(Tc)$ and space $O(dc)$, where c is sub-linear (actually, independent) in n. To this end, we introduce the concept of ε-coreset for the k-dictionary problem.

2.2   Coreset for a Single k-Dictionary Query

A matrix C is called a *weighted matrix* if every column $y \in C$ is associated with a weight $w(y) \geq 0$. For a weighted matrix C and a dictionary D, we define

$$\text{cost}(C, D) = \sum_{y \in C} w(y) \text{err}(y, c).$$

A (non-weighted) matrix Y is considered a weighted matrix with $w(y) = 1$ for every $y \in Y$.

Definition 1 (ε-coreset). *Let Y be an $d \times n$ matrix, and $\mathcal{D}$ be a set of $d \times k$ dictionaries. Let C be a weighted $d \times c$ matrix. We say that C is an ε-coreset for Y if, for every $D \in \mathcal{D}$, we have*

$$(1 - \varepsilon)\text{cost}(Y, D) \leq \text{cost}(C, D)| \leq (1 + \varepsilon)\text{cost}(Y, D). \tag{1}$$

For example, it is easy to verify that Y is an ε-coreset of itself. However, an ε-coreset C is efficient if $c << n$. In this case, $\text{cost}(C, D)$ can be computed in $Tc << Tn$ time using only $cd << nd$ space.

Algorithm $\texttt{Coreset}(Y, D_0, c)$.
Input: a $d \times n$ matrix Y, an integer $c \geq 1$, and a matrix D_0 (of arbitrary size).
Output: a weighted $c \times d$ matrix C that satisfies Theorem 1.

Pick a non-uniform random sample $S = \{s_1, s_2, \cdots, s_c\}$ of c i.i.d. columns from Y, where $y \in Y$ is chosen with probability proportional to $\text{err}(y, D_0)$. That is, for every $s \in S$ and $y \in Y$, the probability that $s = y$ is

$$\text{pr}(y) = \frac{\text{err}(y, D_0)}{\sum_{y \in Y} \text{err}(y, D_0)} = \frac{\text{err}(y, D_0)}{\text{cost}(Y, D_0)}.$$

Return the weighted matrix C whose columns are the vectors of S (in some arbitrary order), where each $y \in C$ is weighted by

$$w(y) = \frac{1}{c \cdot \text{pr}(y)}. \tag{2}$$

The following lemma can be easily proved using Chernoff-Hoeffding's inequality.

Lemma 1. *Let Y be a $d \times n$ matrix, and $\delta, \varepsilon > 0$ be. Let C be the output of the algorithm* $\texttt{Coreset}$ *with input parameters Y, D_0 and*

$$c \geq \frac{10 \ln(1/\delta)}{\varepsilon^2}.$$

Let D be a fixed $d \times k$ dictionary. Then, with probability at least $1 - \delta$,

$$|\text{cost}(Y, D) - \text{cost}(C, D)| \leq \varepsilon \text{cost}(Y, D_0) \cdot \max_{y \in Y} \frac{\text{err}(y, D)}{\text{err}(y, D_0)}.$$

Corollary 1. *Let $b \geq 1$ be an integer and D_0 be a dictionary, such that for every $D \in \mathcal{D}$:*

(i) $\mathrm{cost}(Y, D_0) \leq b \cdot \mathrm{cost}(Y, D)$.
(ii) $\mathrm{err}(y, D) \leq \mathrm{err}(y, D_0)$ *for every $y \in Y$.*

Put $\varepsilon, \delta > 0$. Let C be the weighted matrix that is returned by the algorithm Coreset *with input parameters $c \geq 10b^2 \ln(1/\delta)/\varepsilon^2$ and D_0. Then, for a fixed dictionary $D \in \mathcal{D}$ (which is independent of C), we have*

$$(1 - \varepsilon)\mathrm{cost}(Y, D) \leq \mathrm{cost}(C, D) \leq (1 + \varepsilon)\mathrm{cost}(Y, D),$$

with probability at least $1 - \delta$.

Proof. We have $\mathrm{cost}(Y, D_0) \leq b\mathrm{cost}(Y, D)$ by property (i). Replacing ε with ε/b in Lemma 1 yields

$$|\mathrm{cost}(Y, D) - \mathrm{cost}(C, D)| \leq (\varepsilon/b)\mathrm{cost}(Y, D_0) \max_{y \in Y} \frac{\mathrm{err}(y, D)}{\mathrm{err}(y, D_0)}$$

$$\leq \varepsilon\mathrm{cost}(Y, D) \max_{y \in Y} \frac{\mathrm{err}(y, D)}{\mathrm{err}(y, D_0)} \leq \varepsilon\mathrm{cost}(Y, D),$$

where the last inequality follows from property (ii).

2.3   Coreset for all k-Dictionary Queries

In order to have an ε-coreset for a set $\mathcal{D}$ of more than one dictionary, there are still two problems that remain to be solved. Firstly, we need to compute D_0 that satisfies Properties (i) and (ii) of Corollary 1 with sufficiently small b. This will be handle in the next section for our specific applications. Secondly, Definition 1 of ε-coreset demands that C will approximate $\mathrm{cost}(C, D)$ *simultaneously* for *every* $D \in \mathcal{D}$. However, Corollary 1 holds, with probability at least $1 - \delta$, only for a fixed dictionary $D \in \mathcal{D}$, i.e, a single query. If the size of $\mathcal{D}$ is finite, we can replace δ with $\delta/|\mathcal{D}|$ in Corllary 1 and use the union bound to obtain an ε-coreset for Y of size

$$c = O\left(\frac{\ln(|D|)b^2 \ln(1/\delta)}{\varepsilon^2}\right). \tag{3}$$

However, in the applications of this paper the size of $\mathcal{D}$ is infinite. In this case, we use the result of [7] that is based on PAC-learning theory. Roughly speaking, the result states that to obtain an ε-coreset, it suffices to replace the term $\ln(|D|)$ in (3) by some dimension v that represents the complexity of the set $\mathcal{D}$. This dimension is similar to the classic notion of VC-dimension that is used in machine learning [14]. Usually v is proportional to the number of parameters that are needed to represent a dictionary D of $\mathcal{D}$, which is, in the general case, the number dk of entries in the matrix D.

Theorem 1. *Let Y be a $d \times n$ matrix, $\varepsilon, \delta, b > 0$ and $v \leq O(dk)$ be the dimension that corresponds to a set $\mathcal{D}$ of $d \times k$ matrices. Let D_0 be a matrix as defined in Corollary 1. Let C be the weighted matrix that is returned by the algorithm* Coreset *with input parameters $c \geq v10b^2 \ln(1/\delta)/\varepsilon^2$ and D_0. Then, with probability at least $1 - \delta$, C is an ε-coreset of Y.*

3   Example Application: Approximating the Optimal Dictionary

3.1   The k-Dictionary Problem

In this section we consider the following problem. The input is a parameter $k \geq 1$, a $d \times n$ matrix Y, and an error function $\mathrm{err}(\cdot, \cdot)$. The output is a $d \times k$ dictionary D^* that minimizes $\mathrm{cost}(Y, D) = \sum_{y \in Y} \mathrm{err}(y, D)$ over a given set $D \in \mathcal{D}$. For $\varepsilon > 0$, a $(1 + \varepsilon)$-approximation for this problem is a matrix D such that $\mathrm{cost}(Y, D) \leq (1 + \varepsilon)\mathrm{cost}(Y, D^*)$.

For the error function 1, 2 and 3 in the beginning of Section 2.1 above, provable $(1+\varepsilon)$-approximation algorithms are provided in [7–9], using coresets techniques.

For the more general and popular case in the context of sparse dictionaries where $j > 2$ (the fourth error function in Section 2.1) D^* minimizes the distances to a set of $\binom{k}{j}$ subspaces under constraints. Very little is known about minimizing distances of points to multiple affine subspaces (that are neither points or lines). For example, the problem of computing a pair of planes in three dimensional space that minimize the sum (or sum of squared) distances from every point to its closest plane is an open problem. This holds even for a corresponding constant factor approximation.

In fact, it is not clear how to compute the closest distance to a set of subspaces efficiently even for a *given* dictionary D and a point $y \in y$ in time that is polynomial in j. This is equivalent to answer Problem 1 for a matrix Y with a single column y. Nevertheless, a lot of heuristics have been suggested over the years for approximating distance to subspaces (called pursuit algorithms, see a detailed description of these methods in [3]), for approximating points by subspaces in general (see [6] and references therein), and for the k-dictionary problem in particular (see references in [3]).

3.2   Coreset for the k-Dictionary Problem

We prove that, under natural assumptions on the input matrix Y and the set of candidate dictionaries $\mathcal{D}$, we can choose input parameters c and D_0 such that the algorithm $\texttt{Coreset}(Y, D_0, c)$ from Section 2.2 returns a small ε-coreset C of Y.

There are two issues that need to be resolved: Firstly, it is not clear how to compute $b \geq 1$ and the input D_0 from Theorem 1 that will satisfy the two properties of Corollary 1. Note that the size c of the output coreset depends on b, so we would like to choose D_0 such that b is small. Secondly, we need to compute the dimension v of the set $\mathcal{D}$ of all possible $d \times k$ dictionaries. Recall that by Theorem 1, v is required in order to compute c.

In order to satisfy property (i), we use the fact that in our experiments on both the synthetic and the real data, the variance of entries in every column vector y of the input matrix Y is generally small. This is because y usually represents a small block of pixels in an image. Letting D_0 be the $d \times 1$ matrix (vector) of ones guarantees that $\mathrm{err}(y, D_0)$ is not larger than the variance of y.

Hence $\mathrm{cost}(Y, D_0) = \sum_{y \in Y} \mathrm{err}(y, D_0)$ is the sum of these variances which is usually not much larger than $\mathrm{cost}(Y, D)$ for every $D \in \mathcal{D}$. This satisfies property (i). Adding this vector D_0 for every possible output dictionary $D \in \mathcal{D}$ satisfies property (ii) since in this case $\mathrm{err}(y, D) \leq \mathrm{err}(y, D_0)$ for every $y \in Y$.

Another option that will satisfy the above two properties is relevant when running a heuristic that uses several iterations in order to compute D^*. In this case, we may run the heuristic on the original input data Y only for a single iteration, and choose D_0 to be the returned dictionary. Property (i) will hold under the assumption that the ratio b between the initial dictionary and the rest of the dictionaries that will be computed by the heuristic is not very large. Property (ii) will be satisfied under the assumption that $\mathrm{err}(y, \cdot)$ is highest on the first iteration of the heuristic, for every column y of Y.

We chose the first option (where $D_0 = (1, \cdots, 1)$), for practical reasons (due the simplicity of implementation), and for theoretical reasons (since the second option assumes that the heuristic is based on iterations). Interestingly enough, we found out that existing heuristics (such as the K–SVD algorithm in [3]) already add such a constant vector D_0 for every dictionary that they output, for different reasons. We summarize our decision and its justification in the following theorem.

Theorem 2. *Let $D_0 = (1, \ldots, 1)^T$ be the d-dimensional vector of ones. Let $\mathcal{D}$ be a set of $d \times k$ dictionaries, such that each $D \in \mathcal{D}$ contains D_0, and suppose that $OPT = \min_{D \in \mathcal{D}} \mathrm{cost}(Y, D) > 0$. Let v denote the dimension of $\mathcal{D}$.*

Let Y be a $d \times n$ matrix, and define $b = \mathrm{cost}(Y, D_0)/OPT$. Let $c \geq 10vb^2 \ln(1/\delta)/\varepsilon^2$ for some $\varepsilon, \delta > 0$. Then $\mathtt{Coreset}(Y, D_0, c)$ returns, with probability at least $1 - \delta$, an ε-coreset of Y.

Determining c. Although the set of dictionaries that a heuristic tests (queries) during its running time is finite, we still cannot use the union bound with Corollary 1, since these dictionaries depend on the input coreset C. However, it is reasonable to assume that, for a given Y, not all the possible $d \times k$ matrices D have a positive probability (over the randomness of C) to be queried by the algorithm. That is, we believe that the dimension v of the candidate set $\mathcal{D}$ is significantly smaller than dk, but a more involved theoretical analysis of the corresponding heuristic is needed. Also, it is not clear how to compute b in Theorem 2. Practically, we will simply apply the algorithm $\mathtt{Coreset}$ with a value of c that is determined by the available memory and time at hand. Hence, the parameters δ, ε, b and v that are defined in Theorem 2 will be used only for the theoretical analysis. Nevertheless, they guarantee that the required sample size c is generally "small" for a reasonable values of δ and ε. The theoretical bound on c also teaches us about the relation between the input parameters. For example, the fact that c in Theorem 1 is independent of n implies that the ratio between the computation time of $\mathrm{cost}(Y, D)$ and $\mathrm{cost}(C, D)$ converges to infinity for asymptotically large n, while the error that is introduced by the coresets remains approximately the same. Indeed, we observe these two phenomenas in our experiments; see Fig. 1(a) and 2(a).

4   Experimental Results

Hardware. We ran the experiments on a standard personal modern Laptop. Namely, IBM Lenovo W500, as provided by the manufacturer, without additional hardware. In particular, we used the CPU "Intel Core 2 Duo processor T9600 (2.80 GHz)" with 2GB memory. See manufacturer's website (http://www-307.ibm.com/pc/support/site.wss/document.do?lndocid=MIGR-71785) for exact hardware details.

Software. The operation system that we used is "Windows Vista Business" and the Matlab version is 2010b. For the K−SVD and OMP algorithms, we used the implementation of Rubinstien that was generously uploaded on the Internet [13]. This implementation was used as a "black box" without changing a line of code. The time and space improvements are therefore only due to the replacing of the original input matrix Y with its coreset.

4.1   Synthetic Data

As in previously reported works [3, 10, 11], we first try to construct coresets of synthetic data. In [3] it was shown how the K−SVD algorithm approximated the original dictionary D^* that generated a synthetic data matrix Y. In the following experiments we replace Y by its (usually much smaller) coreset C, and compare the results of applying K−SVD on C instead of Y. The construction of C is done using algorithm Coreset with D_0 and different values of c, as defined in Theorem 2. The construction of the generative dictionary D^* and the input matrix Y was based on the suggested experiments in [3]. *Generating the dictionary D^* and the matrix Y.* A random (dictionary) matrix D^* of size $d \times k = 20 \times 50$ was generated with i.i.d. uniformly distributed entries. Each column was normalized to a unit norm. Then, a $20 \times n$ matrix Y was produced for different values of n. Each column y of Y was created using a linear combination D^*x of $\|x\|_0 = j = 3$ random and independent different columns of D^*, with uniformly distributed i.i.d. coefficients. White Gaussian noise with varying signal-to-noise ratio (SNR) $\sigma = 20$ was added to the resulting vector D^*x. That is, $Y = D^*X + N$ where N is a matrix that represents the Gaussian noise in each entry, and every column x of X corresponds to a column vector y in Y as defined above. We run the experiment with 11 different assignments for n, that were approximately doubled in every experiment: from $n = 585$ to $n = 500,000$. For every such value of n, 50 trials were conducted, when in every trial new dictionary D^* and matrices Y and X were constructed. *Applying K−SVD on Y.* We run the K−SVD implementation of [13], where the maximum number of iterations was set to 40. The rest of parameters were the defaults of the implementation in [13]. We denote the output dictionary by D_Y. *Generating the coreset C.* We implemented and run the algorithm Coreset(c, D_0) from Section 2.2 on the input matrix Y where the size of the coreset was set to $c = 5000$. The parameter D_0 was always set to be the column vector of d ones. See Section 3.2 for more details.

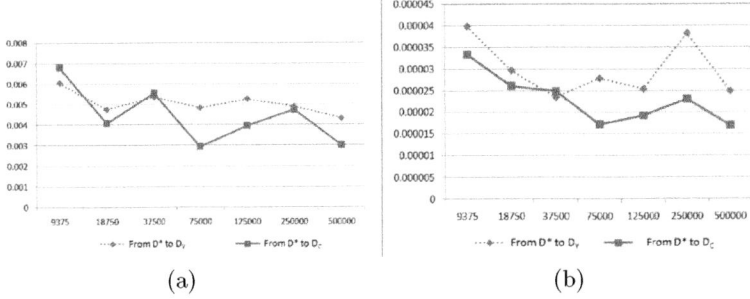

(a)   (b)

Fig. 1. Comparison of the differences between the dictionaries D_Y D_C and D^* over the number n of rows in the matrix Y. The dictionaries D_Y, D_C are respectively the dictionaries that were constructed using the original matrix Y, and its coreset C. The original generator dictionary of Y is denoted by D^*.

Applying K–SVD *on* C. We called to the K–SVD algorithm using the same parameters as the above call for Y, except for the maximum number of iterations. After setting the number of iterations to 40 for the input C (as in the runs on Y), we got results that are only slightly worse than on Y, but significantly faster (up to 100 times). We therefore decided to sacrifice time in order to get better results, and used 120 iterations on the K–SVD with the input C. We denote the output dictionary by D_C.

Approximating the sparse coefficients matrix. In order to approximate the entries of the matrix X, we used the OMP heuristic as defined in [12] and implemented in [13]. The objective of OMP is to minimize $\|Y - D_Y X_Y\|_F$ for the given dictionary D_Y and the input matrix Y, over every matrix X_Y whose columns are sparse ($\|x\|_0 = j = 3$ for every column $x \in X_Y$). This is done by minimizing $\|y - D_Y x\|_F$ for every column $y \in Y$ (one by one) over the set of j-sparse vectors x. Similarly, we computed X_C that suppose to minimize $\|Y - D_C X_C\|$ using the OMP heuristic, as done for Y and D_Y.

Measurement. To measure how close D_Y is to D^*, compared to the difference between D_C and D^*, we used the same error measurement $\text{Distance}(D, D^*)$ that was used in the original K–SVD paper [3], and implemented in [13]. The computation of $\text{Distance}(D, D^*)$ for two dictionaries D and D^* is done by sweeping through the columns of D^* and finding the closest column (in distance) in the computed dictionary D, measuring the distance via $1 - |d_i^T \tilde{d}_i|$, where d_i is a column in D^* and $\tilde{d}_i$ is its corresponding element in the recovered dictionary D. The average distance is denoted by $\text{Distance}(D, D^*)$. That is, $\text{Distance}(D, D^*)$ is the sum of distances over every i, $1 \le i \le k$, divided by k. *The Results.* In Fig. 2(a) we compare the difference (the y-axis) between the dictionaries (the two lines) for different values of n (the x-axis). For example, the dotted line show the average value, for every assignment of n, of $\text{Distance}(D_C, D^*)$ over the 50 trials , between the generation dictionary D^* and the dictionary that returned when running K–SVD with the input matrix Y. The variance over the sets of 50 experiments that corresponds to the average in Fig 2(a) is shown in Fig. 2(b).

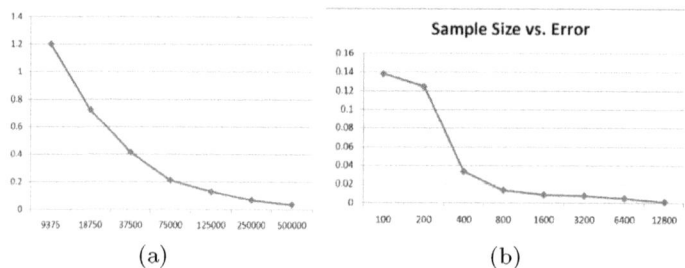

(a)   (b)

Fig. 2. (a) Ratio between running times of K−SVD over coreset C and original input Y. **(b)** The distance between the approximated dictionary D_C and the generating dictionary D for different sizes of coreset C.

The comparison between the running times appears in Fig 2. The x-axis shows the values of n as in Fig. 2, while the y-axis is the ratio between the running time of constructing D_Y, the dictionary of Y, and the running time of constructing D_C, the dictionary of C. The construction time for D_C is the sum of the time it took to construct the coreset C from Y, and the time for constructing D_C from C.

Discussion. In Fig. 2(a) we see that the coreset is usually good at least as the original set for reconstructing the generating dictionary D^*. By Theorem 2, the quality of the coreset C depends on its size c, but not of n. Indeed, the error in Fig. 2 seems to be independent of the value of n. In Fig. 2(b) we see that the results are also more stable on the coreset runs.

Since the size of the coreset is the same ($c = 5000$), the value of n is getting larger, and the running time of the K−SVD algorithm is linear in the rows of the input matrix (c or n), it is not surprising that the ratio between running times grows linearly with the value of n; see Fig. 2(a). For $n = 500K$ in Fig 2(a), the ratio between the running time is approximately 1:30 (0.032). For $n = 1M$ this ratio is approximately 1:60. However due to time and memory constraints we didn't repeat the experiment for $n = 1M$ 50 times.

The role of the sample size c. By Theorem 2, the size c of the coreset C is polynomial in $1/\varepsilon$, where ε represents the desired quality of the coreset. In Fig. 2(b) we show results for additional set of experiments for a constant $n = 500K$ and different values of the coreset size c. The number of iterations is still 120, and the rest of the parameters remain the same as in the previous experiments. The y-axis is the log of the distance between the dictionaries (base 10) over 50 trials. Indeed, it seems that the error is reduced roughly linearly with the size of c.

4.2   Coresets for High-Definition Images

In [5] it is explained how to apply image denoising using the algorihtm K−SVD. Fortunately, source code was also provided by Rubinstein [13]. We downloaded high-definition images from the movie "Inception"' that was recently released by

(a)   (b)

Fig. 3. (a) Noisy Image with SNR= 50. The resulting $PSNR$ is 14.15dB. **(b)** Denoised image using [5] on the small coreset. The resulting $PSNR$ is ~ 30.9.

Warner Bros; see web page "http://collider.com/new-inception-images-posters-christopher-nolan/34058/". We used only one of the images, whose size is $4752 \times 3168 = 15,054,336$ pixels; see Fig. 3. We added a Gaussian noise of $SNR = 50$ which yields a noisy image of $PSNR = 14.15$. Then, we partition the noisy image into 8×8 blocks as explained in [5], and convert the blocks into a matrix Y of approximately $n = 12M$ vectors of dimension $d = 8 \times 8 = 64$. We then hoped to apply the K−SVD as explained in [5] using the default parameters in [13]. However, we got "out of memory" error from Matlab already in the construction of Y.

So, instead, we constructed a coreset C of Y in the streaming model using one pass over Y. In this model, coresets are constructed (using our algorithm `Coreset`) from subsets of columns of Y that are loaded one by one and deleted from memory. When there are too many coresets in memory, a coreset for the union of coresets is constructed and the original coresets are deleted. See details in [8]. After constructing such a coreset C of size $c = 10000$ for *all* the columns of Y, we apply the K−SVD on the coreset using sparsity $j = 10$, and $k = 256$ atoms, and 40 iterations. The $PSNR$ was increased, on average of 10 experiments, from 14.15 to 30.9, with variance of ~ 0.002, while the average time for constructing the dictionary was 69 seconds with variance of ~ 7.2

5   Conclusions and Further Work

We tried to repeat our experiments on real data where Y is partitioned into larger blocks of size $d = 50 \times 50 = 2500$ and an overcomplete dictionary of $k > 3000$. Although the construction of C was fast, the running time of the OMP algorithm (that is used by K−SVD and the denoising procedure for applying the dictionary) is extremely slow when d and k are so large. Besides running time problems, it is noted in [3, 5, 13] that K−SVD does not scale for high dimensional spaces (i.e, large blocks size). We believe that this problem can be solved using recent and more involved coresets techniques of clustering data in high dimensional space, and coresets for the OMP algorithm. We leave this for future papers.

References

1. Agarwal, P.K., Har-Peled, S., Varadarajan, K.R.: Approximating extent measures of points. Journal of the ACM 51(4), 606–635 (2004)
2. Agarwal, P.K., Har-Peled, S., Varadarajan, K.R.: Geometric approximations via coresets. Combinatorial and Computational Geometry - MSRI Publications 52, 1–30 (2005)
3. Aharon, M., Elad, M., Bruckstein, A.: K-SVD: An Algorithm for Designing Over-complete Dictionaries for Sparse Representation. IEEE Transactions on Signal Processing 54(11), 4311–4322 (2006)
4. Czumaj, A., Sohler, C.: Sublinear-time approximation algorithms for clustering via random sampling. Random Struct. Algorithms (RSA) 30(1-2), 226–256 (2007)
5. Elad, M., Aharon, M.: Image denoising via sparse and redundant representations over learned dictionaries. IEEE Trans. Image Processing 15(12), 3736–3745 (2006)
6. Feldman, D., Fiat, A., Segev, D., Sharir, M.: Bi-criteria linear-time approximations for generalized k-mean/median/center. In: Proc. 23rd ACM Symp. on Computational Geometry (SOCG), pp. 19–26 (2007)
7. Feldman, D., Langberg, M.: A unified framework for approximating and clustering data (submitted, 2010) manuscript
8. Feldman, D., Monemizadeh, M., Sohler, C.: A PTAS for k-means clustering based on weak coresets. In: Proc. 23rd ACM Symp. on Computational Geometry (SoCG), pp. 11–18 (2007)
9. Har-Peled, S.: Low rank matrix approximation in linear time (2006) manuscript
10. Kreutz-Delgado, K., Murray, J.F., Rao, B.D., Engan, K., Lee, T.W., Sejnowski, T.J.: Dictionary learning algorithms for sparse representation. Neural Computation 15(2), 349–396 (2003)
11. Lesage, S., Gribonval, R., Bimbot, F., Benaroya, L.: Learning unions of orthonormal bases with thresholded singular value decomposition. In: IEEE International Conference on Acoustics, Speech, and Signal Processing, ICASSP 2005, vol. 5, IEEE, Los Alamitos (2005)
12. Pati, Y.C., Rezaiifar, R., Krishnaprasad, P.S.: Orthogonal matching pursuit: Recursive function approximation with applications to wavelet decomposition. In: 1993 Conference Record of The Twenty-Seventh Asilomar Conference on Signals, Systems and Computers, pp. 40–44. IEEE, Los Alamitos (2002)
13. Rubinstein, R.: Technical report,
http://www.cs.technion.ac.il/~ronrubin/software/ksvdbox13.zip
14. Vapnik, V.N., Chervonenkis, A.Y.: On the uniform convergence of relative frequencies of events to their probabilities. Theory of Probability and its Applications 16(2), 264–280 (1971)

Measuring Geodesic Distances
via the Uniformization Theorem

Yonathan Aflalo[1] and Ron Kimmel[2]

[1] Faculty of Electrical Engineering
[2] Faculty of Computer Science
Technion University, Haifa 3200, Israel

Abstract. According to the Uniformization Theorem any surface can
be conformally mapped into a flat domain, that is, a domain with zero
Gaussian curvature. The *conformal factor* indicates the local scaling introduced by such a mapping. This process could be used to compute
geometric quantities in a simplified flat domain. For example, the computation of geodesic distances on a curved surface can be mapped into
solving an eikonal equation in a plane weighted by the conformal factor. Solving an eikonal equation on the weighted plane can then be done
with regular sampling of the domain using, for example, the *fast marching method*. The connection between the conformal factor on the plane
and the surface geometry can be justified analytically. Still, in order to
construct consistent numerical solvers that exploit this relation one needs
to prove that the conformal factor is bounded.

In this paper we provide theoretical bounds over the conformal factor and introduce optimization formulations that control its behavior. It
is demonstrated that without such a control the numerical results are
unboundedly inaccurate. Putting all ingredients in the right order, we
introduce a method for computing geodesic distances on a two dimensional manifold by using the fast marching algorithm on a weighed flat
domain.

1 Introduction

Consistent and efficient distance computation on various domains is a key component in many important applications. Several papers tackle the problem of
geodesic distance computation on triangulated surfaces. The celebrated *fast
marching method* [7,9] enabled the solution in isotropic inhomogeneous domains
that are regularly sampled. It was later generalized [3] through a geometric
interpretation of the numerical update step, that enabled consistent and efficient computation of distances in anisotropic domains. So far, the fast marching
method was implemented on manifolds given as either a triangulated mesh, a
parametrized surface [10,8], or implicitly defined in a narrow band numerically
sampled with a regular grid [5]. Traditionally, the *fast marching method* is executed on the manifold itself where some parametrization is provided. In these
cases, usually there is some processing involved in order to overcome the iregularity of the numerical sampling. This is the case for the unfolding initialization

A.M. Bruckstein et al. (Eds.): SSVM 2011, LNCS 6667, pp. 471–482, 2012.

step in [3]. Here, in order to avoid this procedure, we use a conformal mapping of a given surface and compute distances in a simplified domain. In other words, we conformally map the original curved surface into a flat plane in which we run the fast marching using the conformal factor as a local weight.

1.1   Introduction to Conformal Mapping

Let us consider a two dimensional parametrized manifold $\mathcal{X} \in \mathbb{R}^3$. It can be defined by the functions $x, y, z : \mathbb{R}^2 \rightarrow \mathbb{R}$, such that $(\alpha, \beta) \in \mathbb{R}^2$ defines a coordinate in $\mathcal{X}$ given by $\mathcal{X} = (x(\alpha, \beta), y(\alpha, \beta), z(\alpha, \beta))$. Such a parametrization induces a metric G, a scalar product $\langle u, v \rangle_G = u^T G v$, a gradient $\nabla_G \cdot = G^{-1} \nabla \cdot$ where $\nabla \cdot$ is the usual gradient with respect to α and β, and a Laplace Beltrami operator $\Delta_G \cdot = \dfrac{1}{\sqrt{g}} \nabla^T \left(\sqrt{g} G^{-1} \nabla \cdot \right)$ where $g = \det(G)$. We would like to map the surface $\mathcal{X}$ defined by this manifold into $D \in \mathbb{R}^2$, preserving the angles of intersections of corresponding curves. That is, given any two curves in $\mathcal{X}$, their images in D have to intersect at the same angle as in $\mathcal{X}$. A conformal mapping is a mapping function that has this property at each and every point, and can be introduced by two functions $(u(\alpha, \beta), v(\alpha, \beta))$ that map our manifold in D and obey the following condition $\nabla u = \dfrac{GR}{\sqrt{g}} \nabla v$, where $R = \begin{pmatrix} 0 & 1 \\ -1 & 0 \end{pmatrix}$. This restriction over (u, v) implies four properties

1. $\Delta_G u = 0$.
2. $\Delta_G v = 0$.
3. $\langle \nabla_G u, \nabla_G v \rangle_G = 0$.
4. $\langle \nabla_G u, \nabla_G u \rangle_G = \langle \nabla_G v, \nabla_G v \rangle_G$.

This is equivalent to the Cauchy-Riemann condition if we take the metric $G = I$.

Denoting by J the Jacobian of the mapping $(\alpha, \beta) \rightarrow (u, v)$, the previous conditions can be written as

$$\begin{pmatrix} \|\nabla_G u\|_G^2 & \langle \nabla_G u, \nabla_G v \rangle_G \\ \langle \nabla_G u, \nabla_G v \rangle_G & \|\nabla_G v\|_G^2 \end{pmatrix} = \|\nabla_G u\|_G^2 I \Leftrightarrow (\nabla_G u, \nabla_G v)^T G (\nabla_G u, \nabla_G v) = \|\nabla_G u\|_G^2 I$$

$$\Leftrightarrow (\nabla u, \nabla v)^T G^{-1} (\nabla u, \nabla v) = \|\nabla_G u\|_G^2 I$$
$$\Leftrightarrow J G^{-1} J^T = \|\nabla_G u\|_G^2 I$$
$$\Leftrightarrow G^{-1} = \|\nabla_G u\|_G^2 J^{-1} J^{-T}$$
$$\Leftrightarrow J^T J = G \|\nabla_G u\|_G^2.$$

Hence, any mapping is conformal with respect to a metric G if and only if there exists a scalar function μ, refered to as the *conformal factor*, such that its jacobian J satisfies $J^T J = \mu^2 G$. We also note that

$$\left\| \begin{pmatrix} du \\ dv \end{pmatrix} \right\|^2 = \begin{pmatrix} d\alpha \\ d\beta \end{pmatrix}^T J^T J \begin{pmatrix} d\alpha \\ d\beta \end{pmatrix} = \begin{pmatrix} d\alpha \\ d\beta \end{pmatrix}^T \mu^2 G \begin{pmatrix} d\alpha \\ d\beta \end{pmatrix} = \mu^2 \left\| \begin{pmatrix} d\alpha \\ d\beta \end{pmatrix} \right\|_G^2.$$

It follows that $\left\| \begin{pmatrix} d\alpha \\ d\beta \end{pmatrix} \right\|_G = \dfrac{1}{\mu} \left\| \begin{pmatrix} du \\ dv \end{pmatrix} \right\|.$

Such a mapping would allow us to compute distances on any metric space with a generalized metric G using the computation of distance in an inhomogeneous isotropic flat manifold.

2 Construction of a Discrete Harmonic Map

We start with a theorem that would be useful for our conformal map construction.

Theorem 1. *Given a metric G defined on a regular domain D, and a function f defined on ∂D, the solution f of the following problem*

$$\underset{\substack{f \in C^2(D) \\ f(x)=f_0(x) \ \forall x \in \partial D}}{\operatorname{argmin}} \left\{ \int_D \|\nabla_G f\|_G^2 \right\}$$

satisfies $\Delta_G f = 0$ and $f(x) = f_0(x) \ \forall x \in \partial D$.

The main idea when constructing a discrete conformal map according to Polthier [6] is to find a triangulation $\mathfrak{T} = \{T_1, \ldots, T_{N_T}\}$ (where T_i is a triangle, and N_T is the number of triangles) of our map with N_V vertices, and search for a continuous function u minimizing the Dirichlet energy. For example, we could find u given by

$$u(\gamma) = u_0(\gamma) + \sum_{i=1}^{N_V} u_i \phi_i(\gamma), \text{ where } u_i \text{ are some coefficients, and } \phi_i \text{ are functions}$$

satisfying

1. $\phi_i \in C^0(M)$
2. $\phi_i(V_j) = \delta_{ij} \ \forall i, j \in \{1, \ldots, N_V\}$
3. ϕ_i is linear in each triangle.

V_j designating the jth vertex of $\mathfrak{T}$. After introducing these prerequisites, one can construct the function u, denoted as the discrete harmonic map, using the minimization problem expression of the harmonic function. It can be shown [6] that the discrete Laplace Beltrami operator applied to u at a vertex V_i can be expressed as

$$\Delta u(V_i) = \sum_{\text{edges } (i,j) \text{ at } i} (\cot(\theta_{ij}) + \cot(\psi_{ij}))(u_i - u_j),$$

where $u_j = u(V_j)$ and θ_{ij} and ψ_{ij} represent the angles supporting the edge $V_i V_j$, where V_j is a neighbor of V_i , and $u_i = u(V_i)$. We then have to solve the following system of equations to find an harmonic function u

$$\sum_{\text{edges } (i,j) \text{ at } i} (\cot(\theta_{ij}) + \cot(\psi_{ij}))(u_i - u_j) = 0, \quad \forall i. \tag{1}$$

After u has been computed, we have to find another conjugate discrete harmonic function v, such that $\nabla v = \dfrac{GR}{\sqrt{g}} (\nabla u)$. Next, we have to compute the gradient of u and perform a rotation by $\dfrac{\pi}{2}$. For that goal, Polthier [6] proposed to define a mid-edge grid. For each edge (V_i, V_j), define a vertex at the mid-edge as $V_s^* = \dfrac{V_i + V_j}{2}$. This way, each triangle (V_1, V_2, V_3) is associated with a new triangle (V_1^*, V_2^*, V_3^*). If we define Ψ_r, the function associated to the vertex V_r^* in the mid-edge grid (or, equivalently to the edge (V_i, V_j) in the regular grid) we can show that

$$\begin{pmatrix} v_3 - v_1 \\ v_3 - v_2 \end{pmatrix} = \frac{1}{2} \begin{pmatrix} (u_2 - u_1)\cot(\theta_{21}) + (u_2 - u_3)\cot(\theta_{23}) \\ (u_2 - u_1)\cot(\theta_{21}) + (u_3 - u_1)\cot(\theta_{31}) \end{pmatrix},$$

where v_r, v_s are the values of v on the mid-edge vertices V_r^*, V_s^* located along the edges $(V_i, V_j), (V_j, V_k)$ (respectively), and θ_{jk} is the oriented angle supporting the edge (j, k).

We end up with an algorithm, summarized for example in [4,6], that computes the mid-edge conformal flattening.

Algorithm 1. Mid-Edge discrete conformal map

Require: $\mathfrak{T}$ triangulation of the space Ω

Choose a face to cut, $C = \{V_{i_c}, V_{j_c}, V_{k_c}\} \in \mathfrak{T}$, and solve:

$$\sum_{j \in \mathcal{N}(i)} (u_i - u_j)\,(\cot(\theta_{ij}) + \cot(\psi_{ij})) = 0 \quad \forall i \notin \{i_c, j_c, k_c\}$$

Set arbitrary value for u on C and solve :

$$\begin{pmatrix} v_j - v_k \\ v_j - v_l \end{pmatrix} = \frac{1}{2} \begin{pmatrix} (u_l - u_k)\cot(\theta_{lk}) + (u_l - u_j)\cot(\theta_{lj}) \\ (u_l - u_k)\cot(\theta_{lk}) + (u_j - u_k)\cot(\theta_{jk}) \end{pmatrix}$$

For the mid-edge vertex $V_r^* = \dfrac{V_p + V_q}{2}$, set the value of the conformal map on the midedge grid

$$u_r^* = \frac{u_p + u_q}{2}, \quad v_r^* = v_r$$

We also have the value of the conformal factor for each triangle $T_k = (V_p, V_q, V_r)$

$$\mu(T_k) = \|\nabla u(x_q)\|$$
$$= \left(\frac{1}{2 \text{ area } T_q} \left((u_r - u_q)^2 \cot(\theta_p) + (u_p - u_q)^2 \cot(\theta_r) + (u_r - u_p)^2 \cot(\theta_q) \right) \right)^{\frac{1}{2}}.$$

3   Fast Marching on the Conformal Map

In the following experiments, we conformally mapped several functions into $\mathbb{R}^2$ and run the fast marching algorithm on the conformal map using the conformal factor as a local scaling of a uniform isotropic metric tensor. That is, we numerically solve the eikonal equation $\|\nabla f(x,y)\| = \mu(x,y)$. When mapping a surface, we have to take care of the boundary conditions. The way we define the boundary of our target map is important, and can help us control the conformal factor and thereby the numerical accuracy of our scheme. Without controlling the boundary, all the points of the surface boundary could be mapped to a line. While uniforming the metric and solving one problem, we encounter a new one, that is, a non-uniform conformal factor. The conformal factor observes the curvature of the surface on one hand, but, yields a challenging highly non-uniformly sampled numerical domain to operate on the other.

In our first example, Figure 1, we map the surface $z = f(x,y) = \exp(-0.2x^2 - 0.5y^2)$ without controlling the boundary.

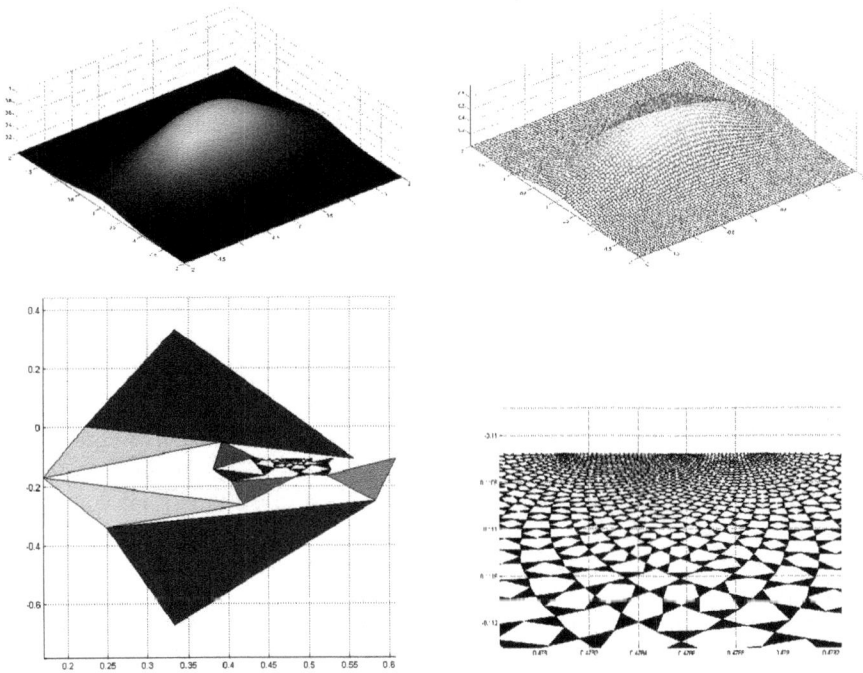

Fig. 1. Left to right, top to bottom: Original surface, midedges surface, conformal map, and zoom in

If we zoom in the area with the smallest triangles we observe that there are three points around which small triangles are concentrated. These points correspond to the corners of the original surface. When we compute the geodesic distances from the corner point $(-2, -2)$ to the rest of the surface points, the result presented in Figure 2 demonstrates numerical inaccuracies caused by the lack of control over the conformal factor.

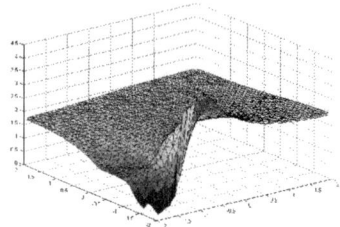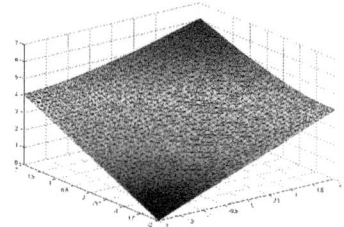

Fig. 2. Geodesic distance from the point (-2,-2) computed with FMM on the conformal map (Left) and with the FMM on the triangulated domain (Rigth)

Our next challenge would be to bound the ratio between the smallest conformal factor and the largest one on the map. Actually, in the above example, the areas ratio is in the order of 10^{-13} and the conformal factor ratio is 10^{-7}. Therefore, it is not trivial to numerically approximate geodesic distances using the FMM on the uniform grid obtained by sampling an arbitrary conformal map. Next, we try to overcome this problem by manipulating the boundary points of the conformal map.

3.1   Controlling the Conformal Factor

We would like to bound the minimal conformal factor. For that goal, we start by studying the computational aspect of the problem. We could try to manipulate the boundary conditions. In Polthier's algorithm, the scheme involves in finding u and v. We find u by solving the system of equations (1). More precisely, this system of equations is defined for each vertex i that does not belong to the boundary of our domain. Define A to be the matrix of cotangent weights, such that the previous equations can be written as $Au = 0$. Let us define $\tilde{A}$ to be the matrix obtained by removing from A the rows and columns that correspond to boundary points. As an example, if the point n belongs to the boundary of our domain, we remove from A the nth row and the nth column. We introduce also P the matrix whose rows are the rows of A corresponding to the removed points from A, and $\tilde{u}$ a vector representing the values of u along the boundary in a lexicographic order. $\tilde{u}$ is filled with the u_i where $i \in \mathcal{B}$, $\mathcal{B}$ being the set of indices of the points along the boundary.

Then, it can be shown that there exists a matrix M whose columns are taken from the identity matrix and from the matrix $\tilde{A}^{-1}P$ such that $u = M\tilde{u}$. It can be also shown that there exist matrices K_i such that $\mu(x_i)^2 = u^T K_i u = \tilde{u}'(M'K_iM)\tilde{u}$.

We would like to control the ratio between the smallest conformal factor and the largest one. We do so by maximizing the following expression

$$\max_{u_j} \frac{\min_i \mu(x_i)^2}{\max_i \mu(x_i)^2},$$
$$\text{s.t.}$$
$$u_j \in [0,1], \forall j \in \mathcal{B}.$$

Actually, the conformal map we get contains some irregularities as some regions of our map are associated with high conformal factor, that are numerically realized as large triangles while some other regions to small conformal factors that correspond to small triangles. Then, when using the conformal factor, we should work with fine grid determined by the smallest triangle to preserve the numerical accuracy captured by the triangulated mesh.

The above problem can be reformulated as

$$\max_{\tilde{u}_j} \left[\frac{\min_i \tilde{u}' \left(M'K_iM \right) \tilde{u}}{\max_i \tilde{u}' \left(M'K_iM \right) \tilde{u}} \right],$$
$$\text{s.t.}$$
$$\tilde{u}_j \in [0,1], \forall j.$$

Since $\tilde{u}$ represents the first coordinate of the boundary points, to avoid foldovers, we have to make sure that its coordinates are increasing and decreasing at most once. The coordinates of $\tilde{u}$ have to grow up to an index from which they decrease. This constraint can be written as

$$Au \leqslant 0, \ A = \begin{pmatrix} 1 & -1 & 0 & \ldots & \ldots & \ldots \\ 0 & 1 & -1 & 0 & \ldots & \ldots \\ \vdots & \ddots & \ddots & \ddots & \ldots & \vdots \\ 0 & \ldots & -1 & 1 & 0 & \ldots \\ \vdots & \ddots & \ddots & \ddots & \ldots & \vdots \\ 0 & \ldots & \ldots & \ldots & -1 & 1 \end{pmatrix}.$$

Actually, without the previous constraint, we could get a conformal map with foldovers as shown in Fig. 3.

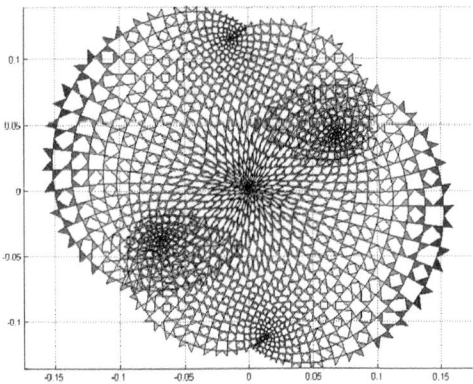

Fig. 3. Unconstrained optimal conformal map

Since the conformal factor can be normalized by restricting $\tilde{u}_j \in [0, 1], \forall j$, we can rewrite our problem and its dual.

$$
\begin{cases}
\max_{\tilde{u}} \left[\min_i \tilde{u}' \left(M' K_i M \right) \tilde{u} \right], \\
\text{s.t.} \\
\tilde{u}_j \in [0, 1], \forall j \\
A\tilde{u} \leqslant 0.
\end{cases}
\Rightarrow
\begin{cases}
\min_i \left[\max_{\tilde{u}} \tilde{u}' \left(M' K_i M \right) \tilde{u} \right]. \\
\text{s.t.} \\
\tilde{u}_j \in [0, 1], \forall j \\
A\tilde{u} \leqslant 0.
\end{cases}
$$

This leads us to the solution of the non-convex optimization problem

$$
\begin{aligned}
&\max_{\tilde{u}} \tilde{u}' K \tilde{u} \\
&\text{s.t.} \\
&B\tilde{u} \leqslant b.
\end{aligned}
\tag{2}
$$

Solving Problem (2) by manipulating the values of u along the boundary, the areas ratio in our example can be increased to 0.34 and the conformal factor ratio becomes 0.59. We can then obtain accurate results, see Fig. 4 and can compare the error between consistent geodesic distances (computed with the Tosca toolbox[1]), and the geodesic distances computed with FMM on a flat regularly sampled domain. We notice that in this case, the error is of the same order as that of the FMM.

We repeat the experiment for another surface given by the peaks function of Matlab with the same boundary condition, see Fig. 4.

So far, we demonstrated the difficulties of working with conformal mapping and showed that manipulating the boundary conditions can lead to a consistent scheme. Next provide more motivation for maximizing the conformal factor.

4   Bounding the Conformal Factor

Let us consider S, a smooth surface embedded in $\mathbb{R}^3$, and G its induced metric. If $u : S \to \mathbb{R}$ is a function defined on the surface, we can define another metric $\bar{G} = e^{2u} G$, that is *conformal* to the original metric, since the two metrics are proportional. The Gaussian curvature $\bar{k}$ of the new metric changes by [2]

$$
\bar{k} = e^{-2u} (k - \Delta_G u)
$$

where k is the original Gaussian curvature, and Δ_G the Laplace-Beltrami operator. In the case of a conformal mapping to the plane, the target curvature of the new metric is zero. Then, the above relation becomes

$$
\Delta_G u = k.
$$

Let us introduce a fundamental property of the Laplace-Beltrami operator:

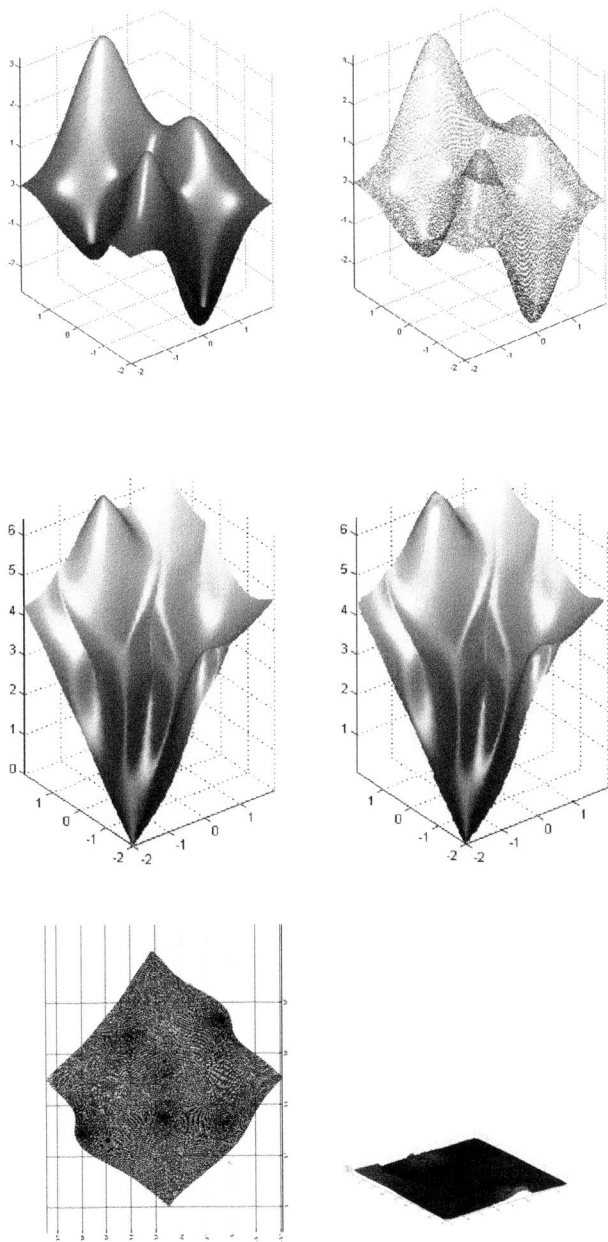

Fig. 4. Left to right, top to bottom: Original surface, midedges surface, geodesic distance with FMM on the surface, geodesic distance with FMM on the conformal map, conformal map optimized for max $\dfrac{\mu_{\min}}{\mu_{\max}}$, the difference between the geodesic distances

Definition 1. *A linear differential operator L of order n on a domain Ω in $\mathbb{R}^d$ given by*

$$Lf = \sum_{\|\alpha\| \leqslant n} a_\alpha(x) \partial^\alpha f$$

is called elliptic if for every x in Ω and every non-zero ξ in $\mathbb{R}^d$,

$$\sum_{\|\alpha\|=n} a_\alpha(x)\xi^\alpha \neq 0.$$

Lemma 1. *The Laplace-Beltrami operator is an elliptic operator.*

Proof. We have $\Delta_G f = \text{trace}\left(G^{-1}\nabla^2 f\right) + v^t \nabla f$ where $v_j = \dfrac{1}{\sqrt{g}}\sum_i \partial_i \left(\sqrt{g}g^{ij}\right)$.
Then, the Δ_G highest order derivative terms are given by $\text{trace}\left(G^{-1}\nabla^2 f\right)$. Taking a vector $\xi \neq 0 \in \mathbb{R}^2$, we have, with the notation of Lemma 1, $\displaystyle\sum_{\|\alpha\|=2} a_\alpha(x)\xi^\alpha = \text{trace}\left(\xi^T G^{-1}\xi\right) \neq 0$ since G^{-1} is a positive definite matrix. This proves that the Laplace-Beltrami operator is elliptic.

The following lemma gives us an upper bound over the conformal factor when the target domain is bounded.

Lemma 2. *Given a C^∞ domain $C \in \mathbb{R}^2$, with a metric G, there exists a function b such that for any function $f : C :\to \mathbb{R}$ s.t. $\forall p \in \partial C : f(p) = 0$, and a positive real number k such that $\|\Delta_G f\| \leqslant k$, we have*

$$\sup_{x \in C} \{\|f(x)\|\} \leqslant b(k).$$

Proof. According to the elliptic regularity theorem, for any $q \in]1, \infty[$, if C is regular, if Δ_G is an elliptic operator, and if $\Delta_G f \in L^q(C)$, then $f \in W^{2,q}(C)$ where $W^{2,q}(C)$ is the $(2,q)$-Sobolev space of C, and there exists a function $g_C^G(q)$ that depends only on C, G and q such that

$$\|f\|_{W^{2,q}} \leqslant g_C^G(q)\|\Delta_G f\|_{L^q}.$$

Moreover, the Sobolev injection theorem states that if $q > 2$, then there exists a function $h_C^G(q)$ that depends only on C and q such that

$$\|f\|_{C^2(C)} \leqslant h_C^G(q)\|f\|_{W^{2,q}}$$

where $\|f\|_{C^1(C)} = \sup_{x \in C}\{\|f(x)\|\}$. We can then conclude that

$$\sup_{x \in C}\{\|f(x)\|\} \leqslant h_C^G(q)g_C^G(q)\mu(C)k = b(k).$$

Using the relation $u = \log \mu$, we can choose the conformal factor such that $\mu = 1$ on ∂C. The previous lemma states that $\log \mu$ is upper bounded, which proves that μ is lower and upper bounded, and that

$$\frac{\sup |\mu|}{\inf |\mu|} \leqslant e^{2b(k)}.$$

This bound justifies using the conformal map for numerically computing geometric measures like geodesic distances. We can then conclude that since it is possible to find a boundary condition for the conformal factor that leads to a global upper bound over the ratio, our optimization programming on the conformal factor is justified. The computation of geometric quantities in the conformal mapping in this case is thereby consistent.

5 Conclusions

Conformal mapping a surface to a plain is a powerful as analysis procedure. Still, in order to justify its usage as a computational tool one needs to control the numerical behavior of this mapping. We proved that a lower bound over the ratio between the minimal and the maximal conformal factor exists. We demonstrated that this theoretical bound does not help much in practice. Next, we formulized optimization problems that maximize this ratio. It allowed us to efficiently and accurately compute geodesic distances using regular sampling of the plain.

Acknowledgment. This research was supported by European Community's FP7- ERC program, grant agreement no. 267414.

References

1. Bronstein, A., Bronstein, M., Kimmel, R.: Numerical Geometry of Non-Rigid Shapes, 1st edn. Springer, Heidelberg (2008)
2. Gu, X., Wang, S., Kim, J., Zeng, Y., Wang, Y., Qin, H., Samaras, D.: Ricci flow for 3D shape analysis. In: Proceedings of ICCV 2007, pp. 1–8 (2007)
3. Kimmel, R., Sethian, J.A.: Fast marching methods on triangulated domains. Proc. National Academy of Science 95, 8341–8435 (1998)
4. Lipman, Y., Funkhouser, T.: Möbius voting for surface correspondence. In: SIGGRAPH 2009: ACM SIGGRAPH 2009 Papers, pp. 1–12 (2009)
5. Mémoli, F., Sapiro, G.: Fast computation of weighted distance functions and geodesics on implicit hyper-surfaces. Journal of Computational Physics 173(2), 730–764 (2001)
6. Polthier, K.: Conjugate harmonic maps and mimimal surfaces. Experimental Mathematics 2, 15–36 (1993)
7. Sethian, J.A.: A fast marching level set method for monotonically advancing fronts. Proc. National Academy of Science 93, 1591–1595 (1996)

8. Spira, A., Kimmel, R.: An efficient solution to the eikonal equation on parametric manifolds. Interfaces and Free Boundaries 6(3), 315–327 (2004)

9. Tsitsiklis, J.N.: Efficient algorithms for globally optimal trajectories. IEEE Transactions on Automatic Control 40, 1528–1538 (1995)

10. Weber, O., Devir, Y.S., Bronstein, A., Bronstein, M., Kimmel, R.: Parallel algorithms for approximation of distance maps on parametric surfaces. ACM Trans. Graph. 104, 104:1–104:16 (2008)

11. Wolansky Incompressible, G.: quasi-isometric deformations of 2-dimensional domains. SIAM J. Imaging Sciences 2(4), 1031–1048 (2009)

Polyakov Action on (ρ, G)-Equivariant Functions Application to Color Image Regularization

Thomas Batard and Nir Sochen

Department of Applied Mathematics, Tel Aviv University,
Ramat-Aviv, Tel Aviv 69978, Israel
{batard,sochen}@post.tau.ac.il

Abstract. We propose a new mathematical model for color images taking into account that color pixels change under transformation of the light source. For this, we deal with (ρ, G)-equivariant functions on principal bundles, where ρ is a representation of a Lie group G on the color space RGB. We present an application to image regularization, by minimization of the Polyakov action associated to the graph of such functions. We test the groups $\mathbb{R}^{+*}$, DC(3) of contractions and dilatations of $\mathbb{R}^3$ and SO(3) with their natural matrix representations, as well as $\mathbb{R}^{+*}$ with its trivial representation. We show that the regularization has denoising properties if the representation is unitary and segmentation properties otherwise.

Keywords: Differential geometry-Fiber bundle-Polyakov functional-Color image regularization.

1 Introduction

Over the last years, mathematical models of images are more and more sophisticated and abstract in order to incorporate more information about images [1],[3],[6],[7]. In this paper, we consider a new framework for color images, dealing with fiber bundles, that generalizes the framework of manifolds [9]. Differential geometry of manifolds has been widely investigated in image processing/analysis [5]. In particular, the notion of Riemannian metric has been introduced in order to provide a measure of variations of images [8]. Lately, vector bundles have been introduced in order to take also into account the vector aspect of the color space RGB [2]. Inspired by geometric models in physics [4], we propose a new mathematical representation for a color image, as a (ρ, G)-equivariant function on a principal bundle of the form $\Omega \times G$, where $\Omega \subset \mathbb{R}^2$ is the domain of the image and (ρ, G) is a Lie group representation. This model allows to take into account that the image of an observed scene is dependent of the light source, and that modifications of the light source modify the image. By the (ρ, G)-equivariance, we assume that the modifications of the light source can be represented by the action of a Lie group G, and that the pixels of the color image change following the representation ρ on the vector space $\mathbb{R}^3$ embedding the color space RGB. In this paper, we propose an application of this new model to image regularization.

A.M. Bruckstein et al. (Eds.): SSVM 2011, LNCS 6667, pp. 483–494, 2012.

The graph φ of such a function realizes the embedding of the manifold $\Omega \times G$ into the manifold $\Omega \times G \times \mathbb{R}^3$. Then, we consider the corresponding Polyakov action $S(\varphi, h, Q)$ where Q is a Riemannian metric on $\Omega \times G \times \mathbb{R}^3$ and h is the induced metric on $\Omega \times G$. Minimizing the functional with respect to the embedding φ provides the evolution equations of the regularization process. We test the group $\mathbb{R}^{+*}$, $DC(3)$ of contractions and dilatations of $\mathbb{R}^3$ and $SO(3)$ with their natural matrix representations, as well as $\mathbb{R}^{+*}$ with its trivial representation.

2   Color Images as (ρ, G)-Equivariant Functions on Principal Bundles

2.1   The General Construction

Definition 1 (Principal bundle). *A smooth locally trivial principal bundle is a quadruplet (P, π, X, G) such that:*
-X and P are two C^∞ manifolds, G is a Lie group, $\pi \colon P \longrightarrow X$ is a surjective map such that the preimage $\pi^{-1}(x)$ of $x \in X$ is diffeomorphic to G, and there is a transitive and free action of G on $\pi^{-1}(x)$.
-for each $x \in X$, there exist a open set $U \supset x$ and a diffeomorphism $\Phi \colon U \times G \longrightarrow \pi^{-1}(U)$ such that $\pi \circ \Phi(y, g) = y \; \forall y \in U, \forall g \in G$.

The principal bundle is trivial if there exists a diffeomorphism $\Phi \colon U \times G \longrightarrow P$ such that $\pi \circ \Phi(y, g) = y \; \forall y \in X, \forall g \in G$.

Definition 2 (Group representation). *Let G be a Lie group, and V be a vector space endowed with a topology. A representation ρ of G on V is a continuous group morphism from G to $GL(V)$. Assuming that V is endowed with a scalar product $<,>$, the representation ρ is said to be unitary if it satisfies*

$$< \rho(g)(v), \rho(g)(w) > = < v, w >, \qquad \forall v, w, \in V, g \in G$$

Definition 3 ((ρ, G)-equivariant function on principal bundle). *Let (ρ, G) be a group representation on $\mathbb{R}^n$. A function $f \in C^\infty(P, \mathbb{R}^n)$ is (ρ, G)-equivariant if it satisfies*

$$f(p \cdot g) = \rho(g)f(p)$$

where $\cdot$ denotes the action of G on the fibers.

Let $I \colon \Omega \subset \mathbb{R}^2 \longrightarrow \mathbb{R}^3$ be a color image given with its coordinates (I^1, I^2, I^3) in the RGB color space. In this paper, we embed RGB into the manifold $\mathbb{R}^3$ in the cartesian coordinates system (z_1, z_2, z_3). Endowing the manifold $\mathbb{R}^3$ of a vector space structure, we have

$$I(x_1, x_2) = I^1(x_1, x_2)e_1 + I^2(x_1, x_2)e_2 + I^3(x_1, x_2)e_3$$

in the corresponding basis (e_1, e_2, e_3).

Let (ρ, G) be a Lie group representation on $\mathbb{R}^3$. Under its action on $\mathbb{R}^3$, the group G acts on the basis (e_1, e_2, e_3). Let $\mathcal{P}$ be the set of basis obtained by the transformations of (e_1, e_2, e_3) under the action of the group G, denoted by $\cdot$. The action of G on $\mathcal{P}$ is transitive and free. Denoting by π the projection of $\Omega \times \mathcal{P}$ on Ω such that

$$\pi(x_1, x_2, g \cdot (e_1, e_2, e_3)) = (x_1, x_2),$$

the quadruplet $(\Omega \times \mathcal{P}, \pi, \Omega, G)$ forms a trivial principal bundle. The global diffeomorphism $\Phi \colon \Omega \times G \longrightarrow \Omega \times \mathcal{P}$ is given by

$$\Phi(x_1, x_2, g) = (x_1, x_2, g \cdot (e_1, e_2, e_3))$$

From the function I, we construct a (ρ, G)-equivariant function J on the principal bundle $(\Omega \times \mathcal{P}, \pi, \Omega, G)$ defined by

$$J(x_1, x_2, g \cdot (e_1, e_2, e_3)) = \rho(g) I(x_1, x_2) \tag{1}$$

In particular, we have

$$J(x_1, x_2, (e_1, e_2, e_3)) = I(x_1, x_2) \tag{2}$$

since $\rho(e) = Id$ by definition of a group representation ρ.

2.2   Interpretation of the (ρ, G)-Equivariance for Color Images

By the function J we construct, we take into account that some transformations of the light source induce color changes on the image. Indeed, by (2) we assign the basis (e_1, e_2, e_3) to the light source of the original image I, that we assume to be composed of red, green and blue lights. Then, we assimilate a basis change given by the action $g \longmapsto g \cdot (e_1, e_2, e_3)$ to a modification of the light source. By (1), the representation ρ tells how the pixels of the color image change under this basis change, and consequently under the corresponding modification of the light source. By the use of the fiber bundle context, we allow that the transformation of the light source change with respect to the points of Ω.

We consider the groups SO(3), and DC(3) of dilatations and contractions of $\mathbb{R}^3$, with their natural representations on $\mathbb{R}^3$, as well as $(\mathbb{R}^{+*}, \times)$ with both its natural and trivial representations on $\mathbb{R}^3$. Let $v \in \mathbb{R}^3$ of coordinates (v_1, v_2, v_3) in the basis (e_1, e_2, e_3).

The Group $(\mathbb{R}^{+*}, \times)$. The action of $exp(a) \in \mathbb{R}^{+*}$ on the basis (e_1, e_2, e_3) gives the basis

$$(exp(a)e_1, exp(a)e_2, exp(a)e_3) \tag{3}$$

This basis change may be interpreted as a homogeneous modification of the intensities of the red, green and blue lights that compose the original light source.

The natural representation of $(\mathbb{R}^{+*}, \times)$ on $\mathbb{R}^3$ is the map defined by

$$\rho(exp(a)) \colon (v_1, v_2, v_3) \longmapsto (exp(a)v_1, exp(a)v_2, exp(a)v_3)$$

This representation makes the colors change in the same way as the light source does in (3).

The trivial representation of $(\mathbb{R}^{+*}, \times)$ on $\mathbb{R}^3$ is the map defined by

$$\rho(exp(a)) \colon (v_1, v_2, v_3) \longmapsto (v_1, v_2, v_3)$$

This representation makes the colors be invariant with respect to a modification of the light source of the form (3).

The Group DC(3). We define the group DC(3) of dilatations and contractions of $\mathbb{R}^3$ as the group of linear transformations represented by the matrices of the form $diag(exp(a), exp(b), exp(c))$ in the basis (e_1, e_2, e_3). The action of such a transformation on the basis (e_1, e_2, e_3) gives the basis

$$(exp(a)e_1, exp(b)e_2, exp(c)e_3) \tag{4}$$

This basis change may be interpreted as an inhomogeneous modification of the intensities of the red, green and blue lights that compose the original light source.

The natural representation of DC(3) on $\mathbb{R}^3$ is the map defined by $\rho(g) \colon v \longmapsto (exp(a)v_1, exp(b)v_2, exp(c)v_3)$ where the matrix representation of g is given by $diag(exp(a), exp(b), exp(c))$ in the basis (e_1, e_2, e_3).

This representation makes the colors change as the light source does in (4).

The Group SO(3). Let A be the matrix representing the rotation R in the basis (e_1, e_2, e_3). The action of R on the basis (e_1, e_2, e_3) gives a new basis represented by the matrix A in the basis (e_1, e_2, e_3).

This basis change may be interpreted as a modification of the three lights composing the original light source in both intensity, saturation and hue.

The natural representation of SO(3) on $\mathbb{R}^3$ is the map defined by

$$\rho(R) \colon v \longmapsto A \begin{pmatrix} v_1 \\ v_2 \\ v_3 \end{pmatrix}$$

This representation makes the colors change as the light source does.

We have seen that the natural representations of the groups $(\mathbb{R}^{+*}, \times)$, DC(3) and SO(3) on $\mathbb{R}^3$ make the colors change in the same way as the basis do. Hence, colors are treated as covectors and not as vectors.

3   Minimization of the Polyakov Action Related to the Graph of (ρ, G)-Equivariant Functions: A Case Study

Following the notations of Section 2.1, the graph of the function $J \circ \Phi$ given by

$$\varphi \colon (x_1, x_2, g) \longmapsto (x_1, x_2, g, \rho(g)I(x_1, x_2))$$

realizes an embedding of the manifold $P = \Omega \times G$ into $P \times \mathbb{R}^3$. Endowing the manifolds P and $P \times \mathbb{R}^3$ of Riemannian metrics h and Q, we can consider the corresponding Polyakov action $S(\varphi, h, Q)$ given by

$$S(\varphi, h, Q) = \int_P h^{\mu\nu} \frac{\partial \varphi^i}{\partial x^\mu} \frac{\partial \varphi^j}{\partial x^\nu} Q_{ij} \sqrt{h} \, dP \tag{5}$$

In this section, we minimize the functionnal (5) with respect to the embedding φ, where h is the metric on P induced by the metric Q on $P \times \mathbb{R}^3$. We test for the different group representations studied in Section 2.2, from which we propose an application to color image regularization.

The Euler-Lagrange equations with respect to the embedding φ are

$$-\frac{1}{2\sqrt{h}} Q^{il} \frac{\partial S}{\partial \varphi^l} = \Delta_h \varphi^i + \Gamma^i_{jk} \partial_\mu \varphi^j \partial_\nu \varphi^k h^{\mu\nu} \tag{6}$$

where Δ_h is the Laplace-Beltrami operator on the Riemannian manifold (P, h), and Γ^i_{jk} are the Levi-Cevita coefficients of the Riemannian manifold $(P \times \mathbb{R}^3, Q)$ with respect to the frame induced by a coordinates system on $P \times \mathbb{R}^3$.

We deduce that the problem of finding the embedding φ minimizing the Polyakov action is dependent of the metric Q we construct on the embedding manifold $\Omega \times G \times \mathbb{R}^3$. In this paper, we construct Riemannian metrics Q of the form $Q = Q_1 \oplus Q_2 \oplus Q_3$ where Q_1, resp. Q_2, resp. Q_3 is a Riemannian metric on Ω, resp. G , resp. $\mathbb{R}^3$. In particular, we construct bi-invariant metrics on G.

3.1 The Case (ρ, G) Is the Natural Representation of $(\mathbb{R}^{+*}, \times)$ on $\mathbb{R}^3$

Riemannian Geometry of $(\mathbb{R}^{+*}, \times)$. The map

$$\psi \colon a \longmapsto exp(a)$$

is a global chart of $(\mathbb{R}^{+*}, \times)$, and makes a be a coordinates system of $(\mathbb{R}^{+*}, \times)$. The metric given by γ, for some constant γ, in the frame $(\partial/\partial a)$ is bi-invariant.

The Induced Metric h. The map $J \circ \Phi \colon \Omega \times \mathbb{R}^{+*} \longrightarrow \mathbb{R}^3$ is defined by

$$J \circ \Phi(x_1, x_2, exp(a)) = (exp(a)I^1(x_1, x_2), exp(a)I^2(x_1, x_2), exp(a)I^3(x_1, x_2))$$

Then, let us consider the embedding

$$\varphi \colon (x_1, x_2, exp(a)) \longmapsto (x_1, x_2, exp(a), J \circ \Phi(x_1, x_2, exp(a)))$$

of $\Omega \times \mathbb{R}^{+*}$ into $\Omega \times \mathbb{R}^{+*} \times \mathbb{R}^3$. We equipp $\Omega \times \mathbb{R}^{+*} \times \mathbb{R}^3$ with the Riemannian metric Q given by

$$Q = diag(1, 1, \gamma, \lambda, \lambda, \lambda)$$

in the frame $(\partial/\partial x_1, \partial/\partial x_2, \partial/\partial a, \partial/\partial z_1, \partial/\partial z_2, \partial/\partial z_3)$, for strictly positive constants λ and γ. Then, the induced metric h on $\Omega \times \mathbb{R}^{+*}$ has a symmetric matrix representation, given by

$h_{11}(x_1, x_2, exp(a)) = 1 + \lambda \, exp(2a)(\sum_{k=1}^{3} I^k_{x_1}(x_1, x_2)^2)$

$h_{12}(x_1, x_2, exp(a)) = \lambda \, exp(2a)(\sum_{k=1}^{3} I^k_{x_1}(x_1, x_2) I^k_{x_2}(x_1, x_2))$

$h_{13}(x_1, x_2, exp(a)) = \lambda \, exp(2a)(\sum_{k=1}^{3} I^k_{x_1}(x_1, x_2) I^k(x_1, x_2))$

$h_{22}(x_1, x_2, exp(a)) = 1 + \lambda \, exp(2a)(\sum_{k=1}^{3} I^k_{x_2}(x_1, x_2)^2)$

$h_{23}(x_1, x_2, exp(a)) = \lambda \, exp(2a)(\sum_{k=1}^{3} I^i_{x_2}(x_1, x_2) I^k(x_1, x_2))$

$h_{33}(x_1, x_2, exp(a)) = \gamma + \lambda \, exp(2a)(\sum_{k=1}^{3} I^k(x_1, x_2)^2)$

in the frame $(\partial/\partial x_1, \partial/\partial x_2, \partial/\partial a)$.

Minimization with Respect to the Embedding φ. All the coefficients of the Levi-Cevita connection of the Riemannian manifold $(P \times \mathbb{R}^3, Q)$ equal zero in the frame $(\partial/\partial x_1, \partial/\partial x_2, \partial/\partial a, \partial/\partial z_1, \partial/\partial z_2, \partial/\partial z_3)$. Then, from the minimization of the Polyakov action with respect to the embedding, we obtain the following evolution equations for $i = 4, 5, 6$ in (6)

$$\frac{\partial J^{i-3}}{\partial t} = \Delta_h J^{i-3} \tag{7}$$

3.2   The Case (ρ, G) Is the Trivial Representation of $(\mathbb{R}^{+*}, \times)$ on $\mathbb{R}^3$

The Induced Metric h. From Section 2, the map $J \circ \Phi \colon \Omega \times \mathbb{R}^{+*} \longrightarrow \mathbb{R}^3$ is

$$J \circ \Phi(x_1, x_2, exp(a)) = (I^1(x_1, x_2), I^2(x_1, x_2), I^3(x_1, x_2))$$

Then, let us consider the embedding

$$\varphi \colon (x_1, x_2, exp(a)) \longmapsto (x_1, x_2, exp(a), I^1(x_1, x_2), I^2(x_1, x_2), I^3(x_1, x_2))$$

of $\Omega \times \mathbb{R}^{+*}$ into $\Omega \times \mathbb{R}^{+*} \times \mathbb{R}^3$. We equipp $\Omega \times \mathbb{R}^{+*} \times \mathbb{R}^3$ with the Riemannian metric Q given by

$$Q = diag(1, 1, \gamma, \lambda, \lambda, \lambda)$$

in the frame $(\partial/\partial x_1, \partial/\partial x_2, \partial/\partial a, \partial/\partial z_1, \partial/\partial z_2, \partial/\partial z_3)$, for strictly positive constants λ, γ. Then, the induced metric h on $\Omega \times \mathbb{R}^{+*}$ is given by $h(x_1, x_2, exp(a)) =$

$$\begin{pmatrix} 1 + \lambda(\sum_{k=1}^3 I_{x_1}^k(x_1, x_2)^2) & \lambda(\sum_{k=1}^3 I_{x_1}^k(x_1, x_2) I_{x_2}^k(x_1, x_2)) & 0 \\ \lambda(\sum_{k=1}^3 I_{x_1}^k(x_1, x_2) I_{x_2}^k(x_1, x_2)) & 1 + \lambda(\sum_{k=1}^3 I_{x_2}^k(x_1, x_2)^2) & 0 \\ 0 & 0 & \gamma \end{pmatrix}$$

in the frame $(\partial/\partial x_1, \partial/\partial x_2, \partial/\partial a)$.

Minimization with Respect to the Embedding φ. By properties of the metric h and the function J, equations (7) may be written $\partial I^{i-3}/\partial t = \Delta_g I^{i-3}$, where g is the Riemannian metric on Ω of matrix representation given by $g_{ij}(x_1, x_2) = \delta_{ij} + \lambda(\sum_{k=1}^3 I_{x_i}^k(x_1, x_2) I_{x_j}^k(x_1, x_2)))$ in the frame $(\partial/\partial x_1, \partial/\partial x_2)$.

3.3   The Case (ρ, G) Is the Natural Representation of DC(3) on $\mathbb{R}^3$

Riemannian Geometry of DC(3). The map

$$\psi \colon (a_1, a_2, a_3) \longmapsto diag(exp(a_1), exp(a_2), exp(a_3))$$

is a global chart of DC(3), and makes (a_1, a_2, a_3) be a coordinates system of DC(3).

The metric on DC(3) given by the matrix representation $diag(\gamma, \gamma, \gamma)$ in the frame $(\partial/\partial a_1, \partial/\partial a_2, \partial/\partial a_3)$, for some constant γ is bi-invariant.

The Induced Metric h. The map $J \circ \Phi \colon \Omega \times DC(3) \longrightarrow \mathbb{R}^3$ is defined by
$J \circ \Phi(x_1, x_2, exp(a_1), exp(a_2), exp(a_3)) =$

$$(exp(a_1)I^1(x_1, x_2), exp(a_2)I^2(x_1, x_2), exp(a_3)I^3(x_1, x_2))$$

Then, let us consider the embedding $\varphi \colon (x_1, x_2, exp(a_1), exp(a_2), exp(a_3)) \longmapsto$

$$(x_1, x_2, exp(a_1), exp(a_2), exp(a_3), J \circ \Phi(x_1, x_2, exp(a_1), exp(a_2), exp(a_3))$$

of $\Omega \times DC(3)$ into $\Omega \times DC(3) \times \mathbb{R}^3$ given by the graph of the function $J \circ \Phi$. We equipp $\Omega \times DC(3) \times \mathbb{R}^3$ with the Riemannian metric Q given by

$$Q = diag(1, 1, \gamma, \gamma, \gamma, \lambda, \lambda, \lambda)$$

in the frame $(\partial/\partial x_1, \partial/\partial x_2, \partial/\partial a_1, \partial/\partial a_2, \partial/\partial a_3, \partial/\partial z_1, \partial/\partial z_2, \partial/\partial z_3)$, for strictly positive constants λ and γ. Hence, the induced metric h on $\Omega \times DC(3)$ has a symmetric matrix representation, given by $h_{ij}(x_1, x_2, exp(a_1), exp(a_2), exp(a_3)) =$

$$
\begin{cases}
\delta_{ij} + \lambda\left(\sum_{k=1}^{3} exp(2a_k)I_{x^i}^k(x_1, x_2)I_{x^j}^k(x_1, x_2)\right) & \text{if } i, j \leq 2 \\[2mm]
\lambda \, exp(2a_{j-2}) \, I_{x^i}^{j-2}(x_1, x_2)I^{j-2}(x_1, x_2) & \text{if } i \leq 2 \text{ and } j \geq 3 \\[2mm]
\delta_{ij}(\gamma + \lambda \, exp(2a_{j-2})I^{j-2}(x_1, x_2)^2) & \text{if } i, j \geq 3
\end{cases}
$$

in the frame $(\partial/\partial x_1, \partial/\partial x_2, \partial/\partial a_1, \partial/\partial a_2, \partial a_3)$.

Minimization with Respect to the Embedding φ. All the coefficients of the Levi-Cevita connection of the Riemannian manifold $(P \times \mathbb{R}^3, Q)$ equal zero in the frame $(\partial/\partial x_1, \partial/\partial x_2, \partial/\partial a_1, \partial/\partial a_2, \partial/\partial a_3, \partial/\partial z_1, \partial/\partial z_2, \partial/\partial z_3)$. Then, from the minimization of the Polyakov action with respect to the embedding, we obtain the following evolution equations for $i = 6, 7, 8$ in (6)

$$\frac{\partial J^{i-5}}{\partial t} = \Delta_h J^{i-5}$$

3.4   The Case (ρ, G) Is the Natural Representation of SO(3) on $\mathbb{R}^3$

Riemannian Geometry of SO(3). The Euler angles $(\theta_1, \theta_2, \theta_3)$ determine a chart ψ of SO(3) given by

$$\psi(\theta_1, \theta_2, \theta_3) = \begin{pmatrix} 1 & 0 & 0 \\ 0 & \cos\theta_1 & \sin\theta_1 \\ 0 & -\sin\theta_1 & \cos\theta_1 \end{pmatrix} \begin{pmatrix} \cos\theta_2 & 0 & \sin\theta_2 \\ 0 & 1 & 0 \\ -\sin\theta_2 & 0 & \cos\theta_2 \end{pmatrix} \begin{pmatrix} \cos\theta_3 & \sin\theta_3 & 0 \\ -\sin\theta_3 & \cos\theta_3 & 0 \\ 0 & 0 & 1 \end{pmatrix}$$

and make $(\theta_1, \theta_2, \theta_3)$ be a coordinates system of SO(3).

The Riemannian metric given by the matrix representation

$$B = \begin{pmatrix} \gamma & 0 & \gamma\sin\theta_2 \\ 0 & \gamma & 0 \\ \gamma\sin\theta_2 & 0 & \gamma \end{pmatrix}$$

in the frame $(\partial/\partial\theta_1, \partial/\partial\theta_2, \partial/\partial\theta_3)$, for some constant γ, is bi-invariant.

The Induced Metric h. The map $J \circ \Phi \colon \Omega \times SO(3) \longrightarrow \mathbb{R}^3$ is defined by

$$J(x_1, x_2, \psi(\theta_1, \theta_2, \theta_3)) = \psi(\theta_1, \theta_2, \theta_3) \begin{pmatrix} I^1(x_1, x_2) \\ I^2(x_1, x_2) \\ I^3(x_1, x_2) \end{pmatrix}$$

Then, let us consider the embedding

$$\varphi \colon (x_1, x_2, \psi(\theta_1, \theta_2, \theta_3)) \longmapsto (x_1, x_2, \psi(\theta_1, \theta_2, \theta_3), J(x_1, x_2, \psi(\theta_1, \theta_2, \theta_3))$$

of $\Omega \times SO(3)$ into $\Omega \times SO(3) \times \mathbb{R}^3$. We equipp $\Omega \times SO(3) \times \mathbb{R}^3$ with the Riemannian metric Q given by

$$Q = diag(1,1) \oplus B \oplus diag(\lambda, \lambda, \lambda)$$

in the frame $(\partial/\partial x_1, \partial/\partial x_2, \partial/\partial \theta_1, \partial/\partial \theta_2, \partial/\partial \theta_3, \partial/\partial z_1, \partial/\partial z_2, \partial/\partial z_3)$, for strictly positive constant λ and γ. The induced metric h on $\Omega \times SO(3)$ has a symmetric matrix representation given by $h_{ij}(x_1, x_2, \theta_1, \theta_2, \theta_3) =$

$$
\begin{cases}
\delta_{ij} + \lambda[\sum_{k=1}^{3} I_{x^i}^{j}(x_1,x_2) I_{x^j}^{j}(x_1,x_2)] \quad \text{if } i,j \leq 2 \\[2mm]
\lambda[\sin\theta_2(I_{x^i}^1(x_1,x_2)I^2(x_1,x_2) - I_{x^i}^2(x_1,x_2)I^1(x_1,x_2)) + \cos\theta_2\sin\theta_3(I_{x^i}^3(x_1,x_2)I^1(x_1,x_2) \\
-I_{x^i}^1(x_1,x_2)I_3(x_1,x_2)) + \cos\theta_2\cos\theta_3(I_{x^i}^2(x_1,x_2)I^3(x_1,x_2) - I_{x^i}^3(x_1,x_2)I^2(x_1,x_2))] \\
\text{if } i \leq 2 \text{ and } j = 3 \\[2mm]
\lambda[\cos\theta_3(I_{x^i}^1(x_1,x_2)I^3(x_1,x_2) - I_{x^i}^3(x_1,x_2)I^1(x_1,x_2)) + \sin\theta_3(I_{x^i}^2(x_1,x_2)I^3(x_1,x_2) \\
-I_{x^i}^3(x_1,x_2)I^2(x_1,x_2))] \quad \text{if } i \leq 2 \text{ and } j = 4 \\[2mm]
\lambda[I_{x^i}^1(x_1,x_2)I^2(x_1,x_2) - I_{x^i}^2(x_1,x_2)I^1(x_1,x_2)] \quad \text{if } i \leq 2 \text{ and } j = 5 \\[2mm]
\gamma + \lambda[(\sin^2\theta_3 + \cos^2\theta_3\sin^2\theta_2)I^1(x_1,x_2)^2 + (\cos^2\theta_3 + \sin^2\theta_2\sin^2\theta_3)I^2(x_1,x_2)^2 \\
+\cos^2\theta_2 I^3(x_1,x_2)^2 + 2(-\sin\theta_3\cos\theta_3 + \sin\theta_3\cos\theta_3\sin^2\theta_2)I^1(x_1,x_2)I^2(x_1,x_2) \\
-2(\sin\theta_2\cos\theta_2\cos\theta_3)I^1(x^1,x^2)I^3(x^1,x^2) - 2(\sin\theta_2 cos\theta_2\sin\theta_3)I^2(x^1,x^2)I^3(x^1,x^2)] \\
\text{if } i = j = 3 \\[2mm]
\lambda[-(\cos\theta_2\cos\theta_3\sin\theta_3)I^1(x_1,x_2)^2 + (\cos\theta_2\cos\theta_3\sin\theta_3)I^2(x_1,x_2)^2 + (\cos\theta_2\cos^2\theta_3 \\
-\cos\theta_2\sin^2\theta_3)I^1(x_1,x_2)I^2(x_1,x_2) - (\sin\theta_2\sin\theta_3)I^1(x_1,x_2)I^3(x_1,x_2) + \\
(\sin\theta_2\cos\theta_3)I^2(x_1,x_2)I^3(x_1,x_2)] \quad \text{if } i = 3 \text{ and } j = 4 \\[2mm]
\gamma\sin\theta_2 + \lambda[\sin\theta_2(I^1(x_1,x_2)^2 + I^2(x_1,x_2)^2) - \cos\theta_2\cos\theta_3 I^1(x_1,x_2)I^3(x_1,x_2) \\
-\cos\theta_2\sin\theta_3 I^2(x_1,x_2)I^3(x_1,x_2)] \quad \text{if } i = 3 \text{ and } j = 5 \\[2mm]
\gamma + \lambda[(\cos^2\theta_3)I^1(x_1,x_2)^2 + (\sin^2\theta_3)I^2(x_1,x_2)^2 + I^3(x_1,x_2)^2 \\
+2(\cos\theta_3\sin\theta_3)I^1(x_1,x_2)I^2(x_1,x_2)] \quad \text{if } i = j = 4 \\[2mm]
\lambda[\cos\theta_3 I^2(x_1,x_2)I^3(x_1,x_2) - \sin\theta_3 I^1(x_1,x_2)I^3(x_1,x_2)] \quad \text{if } i = 4 \text{ and } j = 5 \\[2mm]
\gamma + \lambda[I^1(x_1,x_2)^2 + I^2(x_1,x_2)^2] \quad \text{if } i = j = 5
\end{cases}
$$

in the frame $(\partial/\partial x_1, \partial/\partial x_2, \partial/\partial \theta_1, \partial/\partial \theta_2, \partial/\partial \theta_3)$.

Minimization with Respect to the Embedding φ. On the Riemannian manifold $(P \times \mathbb{R}^3, Q)$, the non-zero coefficients of the subsequent Levi-Cevita connection in the frame $(\partial/\partial x_1, \partial/\partial x_2, \partial/\partial\theta_1, \partial/\partial\theta_2, \partial/\partial\theta_3, \partial/\partial z_1, \partial/\partial z_2, \partial/\partial z_3)$ are

$$\Gamma_{54}^3 = \Gamma_{34}^5 = 1/(2\lambda^3 \cos\theta_2) \qquad \Gamma_{54}^5 = \Gamma_{34}^3 = -\tan\theta_2/(2\lambda^3) \qquad \Gamma_{35}^4 = -1/(2\lambda^3)$$

Then, from the minimization of the Polyakov action with respect to the embedding, we obtain the following evolution equations for $i = 6, 7, 8$ in (6)

$$\frac{\partial J^{i-5}}{\partial t} = \Delta_h J^{i-5}$$

4   Experiments

The minimization of the Polyakov action (5) with respect to the embedding leads to the PDEs

$$\frac{\partial J^k}{\partial t} = \Delta_h J^k, \quad k = 1, 2, 3 \tag{8}$$

for all the group representations presented in this paper, i.e. we obtain heat equations associated to the Laplace-Beltrami operator Δ_h on manifolds of the form $\Omega \times G$, for different Riemannian metrics h and Lie groups G.

For the purpose of application to image regularization, we compute the discrete approximation of the solution of (8) at the points (x_1, x_2, e), by

$$J_{t+dt}^k(x_1, x_2, e) = J_t^k(x_1, x_2, e) + dt\ \Delta_h J_t^k(x_1, x_2, e) \tag{9}$$

of initial condition $J_0^k(x_1, x_2, e) = I^k(x_1, x_2)$. Fig. 1 shows results of the regularization process (9) applied to images on Fig. 1(a) and Fig. 1(b), for the different group representations we have considered. These representations may be classified into two categories: the unitary and the non unitary representations with respect to the Euclidean scalar product on $\mathbb{R}^3$. The natural representations of $\mathbb{R}^{+*}$ and DC(3) are the non unitary representations, whereas the trivial representation of $\mathbb{R}^{+*}$ and the natural representation of SO(3) are the unitary representations.

Then, we observe that the properties of the different regularizations are related to the properties of the corresponding group representations of being unitary or not.

Indeed, the regularization induced by the natural representation of the group DC(3) provides an increase of the contrast (Fig. 1(e)), and the regularization induced by the natural representation of the group $\mathbb{R}^{+*}$ emphasizes edges (Fig. 1(f)). Hence, these two representations appear to be relevant for applications to image segmentation. Conversely, the regularizations induced by the trivial representation of $\mathbb{R}^{+*}$ (Fig. 1(c) and Fig. 1(g)) and the natural representation of SO(3) (Fig. 1(d) and Fig. 1(h)) appear to be relevant for applications to image denoising.

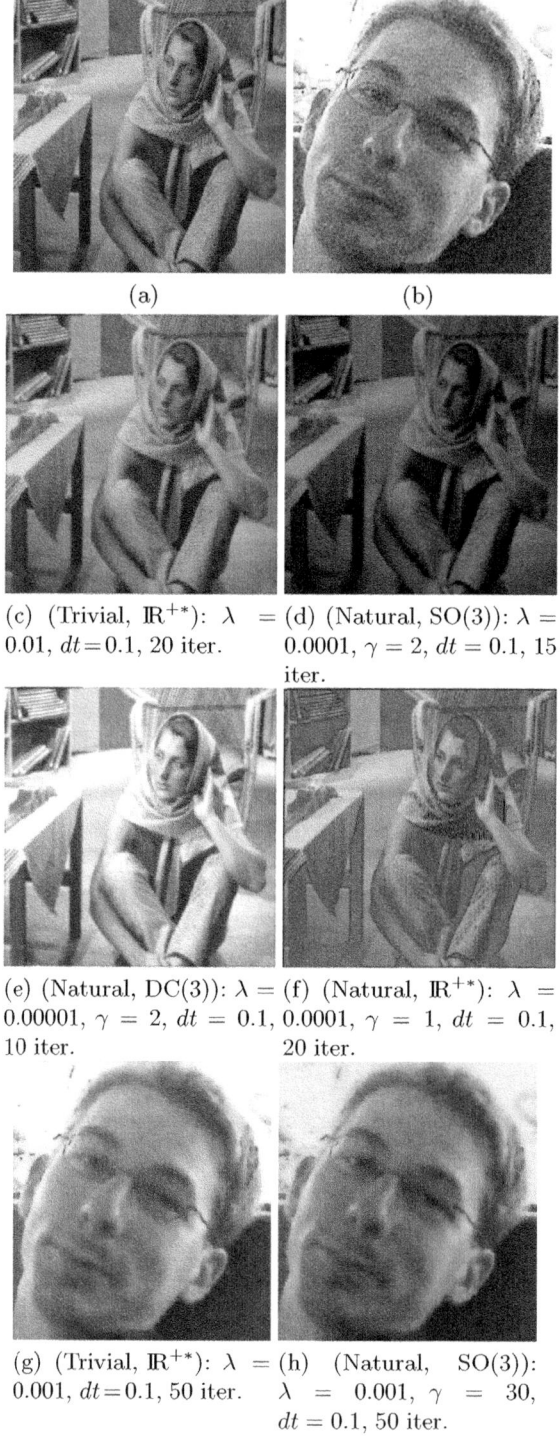

(a) (b)

(c) (Trivial, $\mathbb{R}^{+*}$): $\lambda = 0.01$, $dt = 0.1$, 20 iter. (d) (Natural, $SO(3)$): $\lambda = 0.0001$, $\gamma = 2$, $dt = 0.1$, 15 iter.

(e) (Natural, $DC(3)$): $\lambda = 0.00001$, $\gamma = 2$, $dt = 0.1$, 10 iter. (f) (Natural, $\mathbb{R}^{+*}$): $\lambda = 0.0001$, $\gamma = 1$, $dt = 0.1$, 20 iter.

(g) (Trivial, $\mathbb{R}^{+*}$): $\lambda = 0.001$, $dt = 0.1$, 50 iter. (h) (Natural, $SO(3)$): $\lambda = 0.001$, $\gamma = 30$, $dt = 0.1$, 50 iter.

Fig. 1. Images regularizations for different group representations (ρ, G)

In Section 3.2, we have shown that the regularization induced by the trivial representation of $\mathbb{R}^{+*}$ corresponds to the Beltrami flow [8], whose denoising properties are well-known. Besides a denoising property, the regularization induced by the natural representation of $SO(3)$ has an extra parameter γ, that seems to control the luminance of the image, as it can be seen on Fig. 1(d) and Fig. 1(h). At last, let us mention that the parameters of the experiments (λ, γ, dt and the number of iterations) have been chosen in order to emphasize the properties of the regularizations.

5 Conclusion

In this paper, we have proposed a new framework for color images, dealing with (ρ, G)-equivariant functions on principal bundles. In this context, we have related the action of the Lie group G on the fibers with transformations of the light source of the original image. We have proposed an application to color image regularization by the use of the Polyakov action on graphs of (ρ, G)-equivariant functions on principal bundles, for some particular group representations (ρ, G). We have shown that the regularization has denoising properties if the representation is unitary and segmentation properties otherwise. By studying more closely the physical properties of colors, we expect to determine more relevant Lie group representations in the context of color image processing and analysis. By the metrics on the embedding manifolds we constructed, the minimization of Polyakov action with respect to the embedding yielded heat equations associated to the Laplace-Beltrami operators on the embedded manifolds. In this paper, the metrics on the embedded manifolds were chosen as the metrics induced by the metrics of the embedding manifolds. Further work will be devoted to determine the metrics minimizing the corresponding Polyakov action. Indeed, whereas for embedded manifolds of dimension 2, the metric minimizing the functional corresponds to the metric induced by the metric of the embedding manifold, it does not hold for manifolds of greater dimension anymore.

References

1. Aubert, G., Kornprobst, P.: Mathematical Problems in Image Processing: Partial Differential Equations and the Calculus of Variations, 2nd edn. Springer, Heidelberg (2006)
2. Batard, T., Saint-Jean, C., Berthier, M.: A metric approach of nD images segmentation with Clifford Algebras. Journal of Mathematical Imaging and Vision 33(3), 296–312 (2009)
3. Chan, T., Shen, J.: Image Processing and Analysis: Variational, PDE, Wavelet, and Stochastic Methods. Society for Industrial and Applied Mathematics (2005)
4. Frankel, T.: The Geometry of Physics, 2nd edn. Cambridge University Press, Cambridge (2003)
5. Kimmel, R.: Numerical Geometry of Images: Theory, Algorithms, and Applications. Springer, New York (2003)

6. Paragios, N., Chen, Y., Faugeras, O.: Handbook of Mathematical Models in Computer Vision. Springer, New York (2005)
7. Sapiro, G.: Geometric Partial Differential Equations and Image Analysis. Cambridge University Press, Cambridge (2006)
8. Sochen, N., Kimmel, R., Malladi, R.: A general framework for low level vision. IEEE Transactions on Image Processing 7(3), 310–318 (1998)
9. Spivak, M.: A Comprehensive Introduction to Differential Geometry, 2nd edn. Publish or Perish, Houston (1990)

Curvature Minimization for Surface Reconstruction with Features

Juan Shi[1], Min Wan[1], Xue-Cheng Tai[1,2], and Desheng Wang[1]

[1] Division of Mathematical Sciences, School of Physical and Mathematical Sciences,
Nanyang Technological University, Singapore 637371
{shij0004,wanm0003}@e.ntu.edu.sg, {xctai,desheng}@ntu.edu.sg
[2] Department of Mathematics, University of Bergen,
Johannes Brunsgate 12, 5007 Bergen, Norway
tai@math.uib.no

Abstract. A new surface reconstruction method is proposed based on graph cuts and local swap. We novelly integrate a curvature based variational model and Delaunay based tetrahedral mesh framework. The minimization task is performed by graph cuts and local swap sequentially. The proposed method could reconstruct surfaces with important features such as sharp edges and corners. Various numerical examples indicate the robustness and effectiveness of the method.

1 Introduction

Reconstructing a surface from an unorganized point data set is a significant and challenging problem in the field of computer graphics. The development of scanner techniques and their wide applications in the areas such as animation industry, medical imaging, and archeology have boosted the demand of a good reconstruction method. Extensive research has been conducted and tremendous advances have been made. Therefore, a robust reconstruction method which could recover the surface with the sharp features motivates this study.

Most surface reconstruction methods could be classified into two groups, explicit methods and implicit methods. Explicit methods are local geometric approaches based on Delaunay triangulation and dual Voronoi diagram [3,2,4,13]. One advantage of these methods is their theoretical guarantee that there exists a sub-complex of Delaunay triangulation of the data set, which is homeomorphic to the ground truth surface given a sufficient sampling. However, due to the insufficient sampling density at the sharp features, the explicit approaches could not reconstruct the desired features. Sharp features are high frequency portion in the signal processing language, which means the normal data acquisition resolution could not fulfill the sufficient requirement.

In the last two decades, some researchers turned to the implicit methods to gain flexibility of representation and mathematical facilities [16,35,34,15,26,27,5,22]. The success of the weighted minimal surface model in [35] and its variants prove the effectiveness of this methodology. The most popular regularization term added in the variational model is based on the area, which is designed

A.M. Bruckstein et al. (Eds.): SSVM 2011, LNCS 6667, pp. 495–507, 2012.
© Springer-Verlag Berlin Heidelberg 2012

for noise removal but not for feature preservation. The application of Euler's elastica model in image processing inspires the graphic community and some works oriented to curvature have been proposed, see [15]. However, most of implicit reconstruction methods utilize the regular grid to discretize the energy functional. The consequence of this framework is the staircasing observed in the reconstructed surface. Some smoothing post-processing is needed more or less, but the procedure weakens the feature sharpness.

Graph cuts techniques from combinational optimization have been used in vision problems to find the global minimum of energy functionals for a long time [7,9,8]. Recently, it is also widely used in the field of solving of higher order models [19,6] and surface reconstruction problem [17,18,20,21,25,28,32]. It is a useful tool that can minimize energy functions over implicitly defined surfaces. Compared with the iterative ways such as gradient descent, the main advantages of graph cuts are the efficiency and ability to find global minima.

In this study, we propose a novel method for surface reconstruction. The weighted minimal surface model in [35] has been added with a curvature term. The variational model is first discretized on the tetrahedral mesh. A graph is constructed dual to the mesh and graph cuts are applied. The high order curvature term as well as the closeness term are assigned to the graph edge weights. The energy is calculated based on the last iteration result and graph cuts are performed iteratively. Local swap will be applied on each element of the explicit surface, which is regarded as the mesh partition. The curvature based energy functional is then calculated on 2-manifolds and the change of the energy will be recorded.

Our method integrates Delaunay-based tetrahedral mesh and curvature based variational model. It also takes the advantages of both. The Delaunay triangulation guarantees the existence of reliable recovered surface to the ground truth given sufficient sampling. The curvature based model helps to preserve the features. More important is that the tetrahedral mesh guarantees the better capability of representing piecewise smooth surfaces with sharp corners and cusps. The earlier works based on grid intrinsically could not obtain the sharp features. The input data points are represented by grid data at first place. The precise information is coarsened. Consequently the ground truth are difficult to be reconstructed exactly. In our method, ground truths could be reconstructed exactly, which will be seen in our examples.

The rest of this paper is organized as follows. In Section 2, we review some related works and give an overview of our proposed method. In Section 3, we propose a graph cuts based method as the first stage, global minimization. In Section 4, the local swap method is proposed to recover the remaining features. In Section 5, various numerical experiments are conducted and the results are shown.

2 An Overview of the Proposed Method

We proposed a new method based on the surface model by Zhao et al. in [35], which is solved by a gradient descent method. In [15], Franchini et al. also solved this model. The signed distance function $d(x)$ and the curvature $\kappa(x)$ are calculated on the regular mesh grid. The level set was computed by local RBF reconstruction. In the mean time, models that minimize curvature based functionals have been demonstrated to perform particularly well to avoid the staircasing effect. The Euler's elastica model is of central importance such curvature based model, which was first introduced in image processing in [24, 11, 23] and later in [6, 29].

Inspired by the performance of Euler's elastica model in imaging, we introduce the curvature term into the model for surface reconstruction. We also calculate curvature on tetrahedral mesh to avoid staircasing.

Zhao et al. proposed the weighted minimal surface model as follows:

$$E_{Zhao}(\Gamma) = \left[\int_\Gamma d^p(x)ds \right]^{\frac{1}{p}} . 1 \leq p \leq \infty, \tag{1}$$

where Γ is an arbitrary surface and ds is the surface area. $d(x) = d(x, P)$ here is the distance function from the point x to the nearest point of data set P.

The Euler's elastica of a curve C is given by the energy

$$E_{EL}(C) = \int_C (a + b \cdot |k|^\beta(x))dl, \tag{2}$$

where a and b are two parameters and k is the curvature of C at position x. By setting $b = 0$, $E_{EL}(C)$ measures the total length of the curve. If $a = 0$, $E_{EL}(C)$ measures the total curvature of the curve. For solving this kind of curvature based model, traditionally, the Euler-Lagrange or gradient descent equations are derived. In [6], in order to accelerate the convergence of solution, based on the general formulation of energy functional, we can solve the problem via graph cuts by the connection between minimization problems and binary MRFs.

Our method has been motivated by these methods which adopt the weighted surface area and the curvature function. We firstly introduce our model which can recover not only the smooth parts but also the features such as sharp edges and corners as follows:

$$E(\Gamma) = \int_\Gamma (d(x) + \lambda|\kappa(x)|)ds . \tag{3}$$

Here distance function for each point $d(x)$ is the fidelity term and $\lambda|\kappa(x)|$ is the regularization term which replaced area term in the Zhao et al.'s model, $\kappa(x)$ is the mean curvature at position x. Given the input data set P, we add the non-geometric background points Q; generate mesh in a Delaunay way in order to have the reasonable Delaunay triangulations $P \cup Q$. Then we proposed a two stage strategy:

1. We use graph cuts to minimize the energy functional based on the primal mesh and dual graph. Assign the graph weight according to the energy functional to some extent to get the surface initialization which is for the curvature based evolution;

2. Based on the explicit surface obtained by the first stage, we use local swap here to recover the features without oscillation.

The flowchart of our method is as shown in Fig. 1, and we will describe the details in the following two sections.

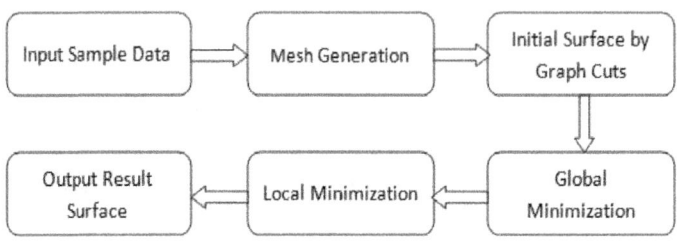

Fig. 1. Flowchart of the proposed method

3   Global Minimization for Surface Reconstruction via Graph Cuts

In this section, a curvature based variational model will be proposed for surface reconstruction and solved by graph cuts. This new energy functional is a generalization from that of the weighted minimal surface model, which is also related to the geodesic active contours approaches [10, 12]. This functional is minimized on an unstructured tetrahedral mesh framework. The method can handle many reconstruction difficulties such as noise, undersampling and non-uniformity.

In this method, the unstructured tetrahedral mesh $\mathcal{T}_h$ is used instead of structured grids, which provides more flexibility and effectiveness. Normally, we use $\{K_i\}_{i=1}^N$ to denote all N tetrahedra in $\mathcal{T}_h$. In such mesh framework, the surface Γ can be approximated by Γ_h, a sub-complex of $\mathcal{T}_h$. Therefore, our energy functional (3) can be approximated by:

$$E(\Gamma_h) = \int_{\Gamma_h} (d(x) + \lambda|\kappa(x)|)ds\,.$$

For convenience reason, we do not distinct Γ and Γ_h in the rest of this paper. The surface triangulation Γ_h can be thought of as the union of the triangular faces shared by tetrahedra in different partitions. In this section, we only discuss two phase problems, in which the ground truth surface $\mathcal{S}$ simply separates the embedding domain $X \subset R^3$ into two connected regions, inside and outside.

We define the level set function:

$$\phi_{\Gamma_h}(K_i) = \begin{cases} c_1 & \text{if } K_i \text{ inside } \Gamma_h, \\ c_2 & \text{if } K_i \text{ outside } \Gamma_h. \end{cases}$$

If we denote $\Gamma_{i,j} = K_i \cap K_j$, which means the shared face of the two neighboring tetrahedrons K_i and K_j, then we have $\Gamma_h = \bigcup \Gamma_{i,j}$.

We define indication function $1_{\{T\}}$ as:

$$1_{\{T\}} = \begin{cases} 1 & \text{if the statement } T \text{ is true,} \\ 0 & \text{if the statement } T \text{ is false.} \end{cases}$$

Hence the energy formulation can be discretized as follows:

$$E(\Gamma_h) = \sum_{i,j} (d_{i,j} + \lambda|\kappa_{i,j}|)S_{i,j}1_{\{\phi_{\Gamma_h}(K_i) \neq \phi_{\Gamma_h}(K_j)\}} , \tag{4}$$

where

$$d_{i,j} = \frac{\int_{\Gamma_{i,j}} d(x)ds}{\int_{\Gamma_{i,j}} ds} , \; S_{i,j} = \int_{\Gamma_{i,j}} ds , \kappa_{i,j} = \frac{\int_{\Gamma_{i,j}} \kappa(x)ds}{\int_{\Gamma_{i,j}} ds}. \tag{5}$$

In level set formulation, the curvature can be calculated by signed distance function as follows:

$$\kappa(x_i) = \nabla \cdot \left(\frac{\nabla d(x_i, \Gamma)}{|\nabla d(x_i, \Gamma)|} \right).$$

In [30], Tong et al. give the corresponding discretization of curvature in details. In order to focus on the steady state solution and not the evolution sequence itself, we first initialize:

$$\kappa^0(x_i) = 0,$$

and

$$\kappa^n(x_i) = \nabla \cdot \left(\frac{\nabla d(x_i, \Gamma^n)}{|\nabla d(x_i, \Gamma^n)|} \right).$$

The energy functional in each iteration is:

$$E(\Gamma^{n+1}) = \int_{\Gamma^{n+1}} (d(x) + \lambda|\kappa^n(x)|)ds = \sum_{i,j} (d_{i,j} + \lambda|\kappa_{i,j}^n|)S_{i,j}^n 1_{\{\phi_{\Gamma_h}(K_i) \neq \phi_{\Gamma_h}(K_j)\}} \cdot$$
$$\tag{6}$$

Therefore, the energy functional can be minimized efficiently by graph cuts, since it is graph representable. A graph dual to the whole mesh is built according to the energy functional and applied with max-flow/min-cut algorithms as in Fig. 2. The two neighboring tetrahedron K_i, K_j can be expressed as two neighboring nodes in the graph i, j respectively. We use triangulation to express the element and small circle is the corresponding graph node for graph cuts computation. In Fig. 3, the weight on the edge (i, j) now is set to $d_{i,j} + \lambda|\kappa_{i,j}^n|$, which can be calculated from (5).

By graph cuts, the proposed energy functional could be minimized globally. However, the iteration result is not satisfactory. The global minimization technique, i.e. graph cuts, is not the main reason for the undesirable result. The reason is the inaccuracy of the curvature calculation. The tetrahedral mesh is intrinsically a much sparser representation compared with the grid representation. The calculation based on such a sparse framework could not obtain a desirable result. From the results of this stage, we can observe that some elements have been recovered. However it is far away from the ground truth. Hence, the local swap based on more precise calculation would be applied sequentially.

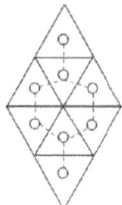

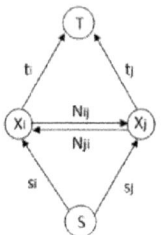

Fig. 2. Primal mesh and dual graph   **Fig. 3.** Graph edge weight assignment

4   Feature Sensitive Local Minimization

Once the tetrahedral mesh is established, finding the embedded surface is equivalent to finding the labeling for all tetrahedra to partition the whole mesh. For each surface Γ, there is one corresponding labeling L. The labeling L is a local minimum with respect to the energy functional in (3) if $E(L) \leq E(L')$ for any L' "near to" L. In the environment of discrete tetrehedal mesh, the labelings near to L are those within the swap of a single tetrahedron. This move is usually referred to by standard moves in computer vision. One good example of the standard moves is simulated annealing [31]. In this section, the object of the swap operation is only changed from image pixels to volumetric tetrahedra.

When the explicit surface expression is obtained, we will adopt the method of [23]. The operator $\mathbf{K}$ maps a point x_i on the surface to the vector:

$$\mathbf{K}(x_i) = 2\kappa(x_i)\mathbf{n}(x_i),$$

where $\mathbf{n}(x_i)$ is the normal vector. The mean curvature normal operator $\mathbf{K}$, known as the Laplace-Beltrami operator for the surface S, is a generalization of Laplacian from flat spaces to manifolds [14]. By using the Gauss' theorem, the integral of the Laplace-Beltrami operator reduces to the following form:

$$\int\int_{\mathcal{A}_M} \mathbf{K}(x)dA = \frac{1}{2}\sum_{j\in N_1(i)}(cot\alpha_{i,j} + cot\beta_{i,j})(x_i - x_j), \tag{7}$$

where $\mathcal{A}_M$ is the 1-ring neighborhood surface area around the point x_i, $\alpha_{i,j}$ and $\beta_{i,j}$ are the two angles opposite to the edge in the two triangles sharing the edge (x_i, x_j) as in Fig. 4, and $N_1(i)$ is the set of 1-ring neighbor vertices of vertex i. The mean curvature normal operator is:

$$\mathbf{K}(x_i) = \frac{1}{2\mathcal{A}_M}\sum_{j\in N_1(i)}(cot\alpha_{i,j} + cot\beta_{i,j})(x_i - x_j). \tag{8}$$

Therefore, the discretization of energy functional (3) can be written as we mentioned:

$$E(\Gamma_h) = \sum_i (d_i + \lambda|\kappa_i|)\Gamma_i, \tag{9}$$

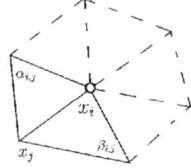

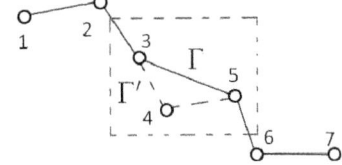

Fig. 4. 1-ring neighbors and angles opposite to an edge

Fig. 5. The change of energy for each local swap

where

$$\Gamma_h = \bigcup \Gamma_i, \ d_i = \frac{1}{3} \sum_{j=1}^{3} d(v_{ij}), \ \kappa_i = \frac{1}{3} \sum_{j=1}^{3} \kappa(v_{ij}). \tag{10}$$

$v_{ij}, j = 1, 2, 3$ are three vertices of Γ_i.

For each known surface, the energy functional could be calculated explicitly by (9). Hence for the labeling swap of a single tetrahedral, the change of energy could also be calculated locally. This swap and comparing procedure is illustrated in Fig. 5. This local swap of a single tetrahedral has the counterpart in image processing field, i.e. stimulated annealing. What is worth mentioning is that stimulated annealing has been questioned of sensitive to the initial labeling. But in our cases, this local swap seldom encounters such problem since the initial labeling is determined by the global minimization stage. This good initial surface makes the local stage work less likely to stuck in a local minimum far away from the global one.

Algorithm 1. Local Swap Procedure

- Step1: Start with the initial surface Γ;
- Step2: For each element, swap it to the other partition and obtain the new surface Γ';
- Step3: Re-calculate the alter energy of (9):$E(\Gamma')$, and compare $E(\Gamma')$ with $E(\Gamma)$,
  Step3.1: If $E(\Gamma') < E(\Gamma)$, confirm this swap;
  Step3.2: If $E(\Gamma') \geq E(\Gamma)$, undo this swap.

---

5 Numerical Experiments

In this section, various examples are presented to illustrate the effectiveness, efficiency and robustness of the proposed method. All experiments were conducted on a PC with Intel Pentium 4 CPU of 3.2GHz and 4GB memory and all examples were synthesized by ourselves. In the mesh generation stage, we adopted the incremental insert algorithm implemented by CGAL [1]. All surfaces are rendered by MeshLab. Only points locations were utilized in the algorithm.

We start by giving illustrative reconstruction examples in Fig. 6 which clearly show the advantage of using curvature information over total variation (TV). As is shown, our algorithm perfectly recovers the sharp edges of cubes. Total variation on the other hand, just recovers the smooth faces. Some more identical examples were also approached in [33,15], readers could compare the performance and find we have recovered the most features.

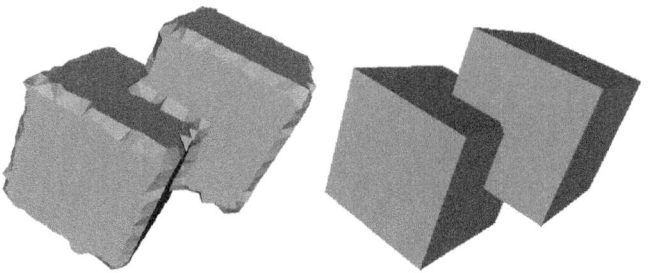

(a) TV reconstruction result   (b) Our reconstruction result

Fig. 6. The comparison of TV result and our result

For all these experiments, we set the value of λ to 0.1. Table 1 gives the sizes of data sets of four surface examples and corresponding CPU time counted in seconds. The first column gives the examples' names. The second column contains the numbers of data points P. The third column is the mesh generation time. The fourth column is the number of generated tetrahedra, the fifth the graph cuts iteration cycles, the sixth the graph cuts time, the seventh the local swap iteration cycles, and the eighth the local swap time.

Table 1. Statistics of four examples

Example	Data Set	Mesh Generation Time	Tetrahedra Number	Global Iteration Cycle	Total Graph Cuts Time	Local Iteration Cycle	Total Swap Time
two cubes	2472	24.1	175851	3	0.33	5	5.57
two spheres	2653	34.2	218119	3	0.42	4	0.31
female symbol	4530	41.8	307968	3	0.63	7	11.57
bolt	2357	34.9	223346	3	0.41	6	2.57

The following figures include data points sets in the first row and the reconstructed surfaces in the second row. In Fig. 7 from left to right, two geometries from basic boolean operation are shown: the unions of two cubes and two spheres. In Fig. 8, surface reconstruction results of four little complicated geometries are shown. In Fig. 9, two platonic solids, i.e. a dodecahedron and an icosahedron, a bucky ball model and a brilliant cut diamond are shown. In Fig. 10, four interesting CAD models are shown, all of which have sharp edges or corners.

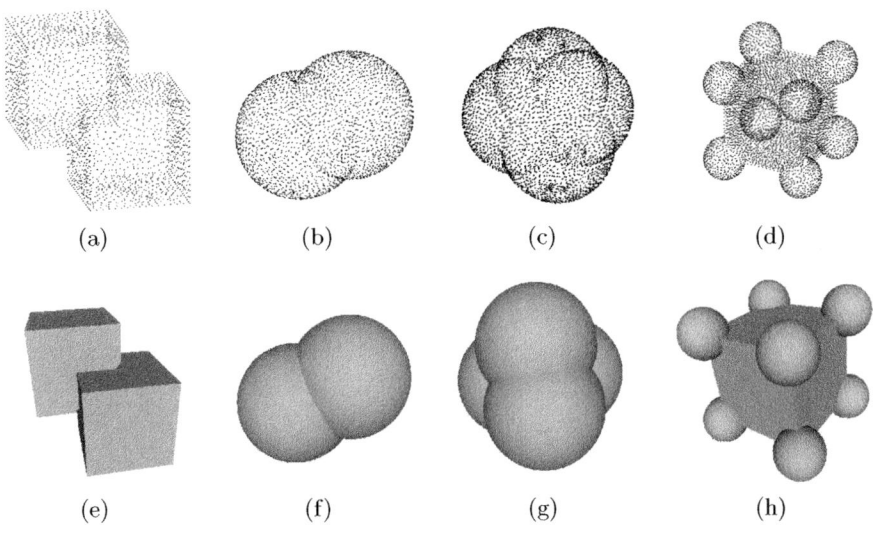

Fig. 7. The unions of geometries

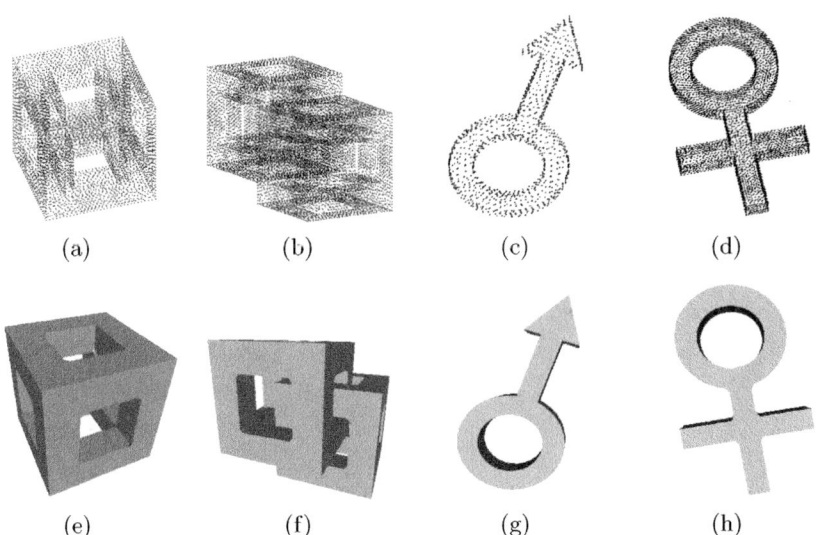

Fig. 8. A perforated cube, two tangling ones, male and female symbol models

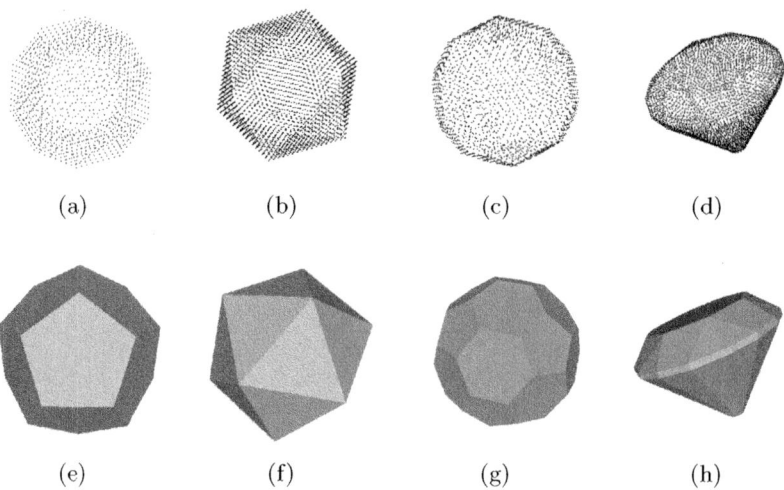

Fig. 9. Two platonic solids: dodecahedron and icosahedron, a buckyball and a brilliant cut diamond

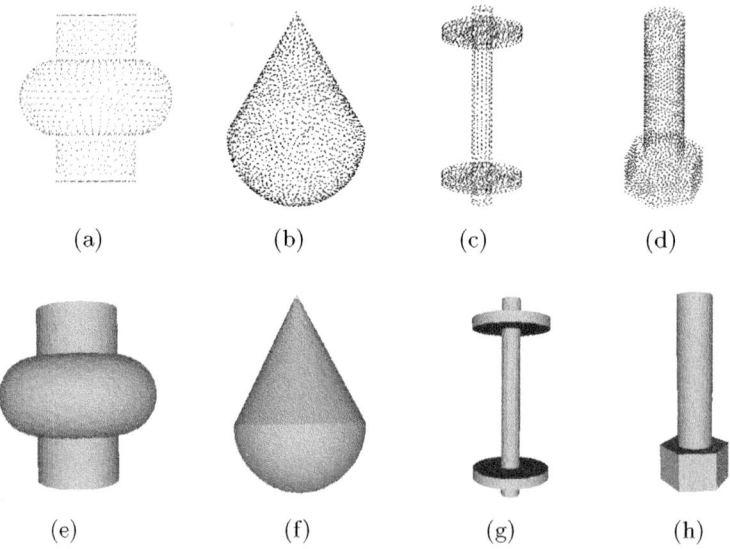

Fig. 10. Four CAD models from left to right are: power wheel, tear drop, dumb bell and bolt

From all our reconstructed examples, most features are recovered especially the sharp edges. The reconstructed surfaces are almost the ground truth surfaces except some place "over-enhancement". The accuracy of our feature-preserving operation will be improved and better performance could be expected in our future works.

Acknowledgement. The research has been supported by MOE (Ministry of Education) Tier II project T207N2202, MOE ARC 29/07 T207B2202, MOE RG 59/08 M52110092 and IDM project NRF2007IDM-IDM002-010 and the Norwegian Research Council (eVita project 166075). The authors would like to thank Egil Bae for valuable discussion and his suggestions on this work.

References

1. Cgal, Computational Geometry Algorithms Library, http://www.cgal.org
2. Amenta, N., Bern, M.: Surface reconstruction by Voronoi filtering. Discrete and Computational Geometry 22(4), 481–504 (1999)
3. Amenta, N., Bern, M., Kamvysselis, M.: A new Voronoi-based surface reconstruction algorithm. In: Proceedings of the 25th Annual Conference on Computer Graphics and Interactive Techniques, pp. 415–421. ACM, New York (1998)
4. Amenta, N., Choi, S., Dey, T.K., Leekha, N.: A simple algorithm for homeomorphic surface reconstruction. In: Proceedings of the Sixteenth Annual Symposium on Computational Geometry, pp. 213–222. ACM, New York (2000)
5. Bae, E., Weickert, J.: Partial differential equations for interpolation and compression of surfaces. Mathematical Methods for Curves and Surfaces, 1–14 (2010)
6. Bae, E., Shi, J., Tai, X.-C.: Graph cuts for curvature based image denoising. IEEE Transactions on Image Processing, November 1 (2010) (preprint) doi:10.1109/TIP.2010.2090533
7. Boykov, V.: Computing geodesics and minimal surfaces via graph cuts. In: Proceedings of Ninth IEEE International Conference on Computer Vision (2003)
8. Boykov, Y., Kolmogorov, V.: An experimental comparison of min-cut/max-flow algorithms for energy minimization in vision. IEEE Transactions on Pattern Analysis and Machine Intelligence, 1124 1137 (2004)
9. Boykov, Y., Veksler, O., Zabih, R.: Fast approximate energy minimization via graph cuts. IEEE Transactions on Pattern Analysis and Machine Intelligence 23(11), 1222–1239 (2001)
10. Caselles, V., Kimmel, R., Sapiro, G.: Geodesic active contours. International Journal of Computer Vision 22(1), 61–79 (1997)
11. Chan, T.F., Kang, S.H., Shen, J.: Euler's elastica and curvature-based inpainting. SIAM Journal on Applied Mathematics, 564–592 (2002)
12. Cohen, R.: Global minimum for active contour models: A minimal path approach. International Journal of Computer Vision 24(1) (1997)
13. Dey, T.K., Goswami, S.: Tight cocone: a water-tight surface reconstructor. In: SM 2003: Proceedings of the Eighth ACM Symposium on Solid Modeling and Applications, pp. 127–134 (2003)

14. Dierkes, U., Hildebrandt, S., Küster, A., Wohlrab, O.: Minimal Surfaces I. Grundlehren der mathematischen Wissenschaften, vol. 295 (1992)

15. Franchini, E., Morigi, S., Sgallari, F.: Implicit shape reconstruction of unorganized points using PDE-based deformable 3D manifolds. Numerical Mathematics: Theory, Methods and Applications (2010)

16. Hoppe, H., DeRose, T., Duchamp, T., McDonald, J., Stuetzle, W.: Surface reconstruction from unorganized points. In: SIGGRAPH 1992: Proceedings of the 19th Annual Conference on Computer Graphics and Interactive Techniques, pp. 71–78 (1992)

17. Hornung, A., Kobbelt, L.: Hierarchical volumetric multi-view stereo reconstruction of manifold surfaces based on dual graph embedding. In: 2006 IEEE Computer Society Conference on Computer Vision and Pattern Recognition, vol. 1 (2006)

18. Hornung, A., Kobbelt, L.: Robust reconstruction of watertight 3d models from non-uniformly sampled point clouds without normal information. In: Geometry Processing 2006: Fourth Eurographics Symposium on Geometry Processing, Cagliari, Sardinia, Italy, 2006, June 26-28, page 41, Eurographics (2006)

19. Ishikawa, H.: Higher-order clique reduction in binary graph cut. In: CVPR, pp. 2993–3000 (2009)

20. Kolmogorov, V., Zabih, R., Gortler, S.: Generalized multi-camera scene reconstruction using graph cuts. In: Rangarajan, A., Figueiredo, M., Zerubia, J. (eds.) EMMCVPR 2003. LNCS, vol. 2683, pp. 501–516. Springer, Heidelberg (2003)

21. Lempitsky, V.S., Boykov, Y.: Global optimization for shape fitting. In: CVPR. IEEE Computer Society, Los Alamitos (2007)

22. Leung, S., Zhao, H.: A grid based particle method for evolution of open curves and surfaces. Journal of Computational Physics 228(20), 7706–7728 (2009)

23. Meyer, M., Desbrun, M., Schröder, P., Barr, A.H.: Discrete differential-geometry operators for triangulated 2-manifolds. Visualization and mathematics 3(7), 34–57 (2002)

24. Mumford, D.: Elastica and computer vision. In: Algebraic Geometry and its Applications (1994)

25. Paris, S., Sillion, F.X., Quan, L.: A surface reconstruction method using global graph cut optimization. International Journal of Computer Vision 66(2), 141–161 (2006)

26. Solem, J.E., Kahl, F.: Surface reconstruction from the projection of points, curves and contours. In: 2nd Int. Symposium on 3D Data Processing, Visualization and Transmission, Thessaloniki, Greece (2004)

27. Solem, J.E., Kahl, F.: Surface reconstruction using learned shape models. Advances in Neural Information Processing Systems 17, 1 (2005)

28. Sormann, M., Zach, C., Bauer, J., Karner, K., Bishof, H.: Watertight multi-view reconstruction based on volumetric graph-cuts. In: Ersbøll, B.K., Pedersen, K.S. (eds.) SCIA 2007. LNCS, vol. 4522, pp. 393–402. Springer, Heidelberg (2007)

29. Tai, X.C., Hahn, J., Chung, G.J.: A Fast Algorithm For Euler's Elastica Model Using Augmented Lagrangian Method. UCLA CAM Report 10-47 (2010)

30. Tong, Y., Lombeyda, S., Hirani, A.N., Desbrun, M.: Discrete multiscale vector field decomposition. ACM Transactions on Graphics 22(3), 445–452 (2003)

31. Van Laarhoven, P.J.M., Aarts, E.H.L.: Simulated annealing: theory and applications. D.Reidel(1988)

32. Vogiatzis, G., Torr, P.H.S., Cipolla, R.: Multi-view stereo via volumetric graph-cuts. In: IEEE Computer Society Conference on Computer Vision and Pattern Recognition, CVPR 2005, vol. 2 (2005)

33. Wan, M., Wang, Y., Wang, D.: Variational surface reconstruction based on delau-
    nay triangulation and graph cut. International Journal for Numerical Methods in
    Engineering 85, 206–229 (2011)
34. Ye, J., Bresson, X., Goldstein, T., Osher, S.: A Fast Variational Method for Surface
    Reconstruction from Sets of Scattered Points. UCLA CAM Report (2010)
35. Zhao, H.K., Osher, S., Fedkiw, R.: Fast surface reconstruction using the level set
    method. In: Proceedings of the IEEE Workshop on Variational and Level Set Meth-
    ods (VLSM 2001), p. 194. IEEE Computer Society Press, Washington, DC (2001)

Should We Search for a Global Minimizer of Least Squares Regularized with an ℓ_0 Penalty to Get the Exact Solution of an under Determined Linear System?

Mila Nikolova

CMLA CNRS ENS Cachan UniverSud
61 av. du Prsident Wilson, 94235 Cachan Cedex, France
nikolova@cmla.ens-cachan.fr
http://www.cmla.ens-cachan.fr/~nikolova/

Abstract. We study objectives $\mathcal{F}_d$ combining a quadratic data-fidelity and an ℓ_0 regularization. Data d are generated using a full-rank $M \times N$ matrix A with $N > M$. Our main results are listed below.

Minimizers $\widehat{u}$ of $\mathcal{F}_d$ are strict if and only if length(support($\widehat{u}$)) $\leqslant M$ and the submatrix of A whose columns are indexed by support($\widehat{u}$) is full rank. Their continuity in data is derived. Global minimizers are always strict.

We adopt a weak assumption on A and show that it holds with probability one. Data read $d = A\ddot{u}$ where length(support($\ddot{u}$)) $\leqslant M - 1$ and the submatrix whose columns are indexed by support($\ddot{u}$) is full rank. Among all strict (local) minimizers of $\mathcal{F}_d$ with support shorter than $M - 1$, the exact solution $\widehat{u} = \ddot{u}$ is the unique vector that cancels the residual. The claim is independent of the regularization parameter. This $\widehat{u} = \ddot{u}$ is usually a strict local minimizer where $\mathcal{F}_d$ does not reach its global minimum. Global minimization of $\mathcal{F}_d$ can then prevent the recovery of $\ddot{u}$.

A numerical example (A is 5×10) illustrates our main results.

Keywords: under-determined systems of linear equations, variational methods, ℓ_0 regularization, linear programming, sparse representation, solution analysis, exact recovery, global minimizers.

1 Introduction

Let $A \in \mathbb{R}^{M \times N}$ be a matrix such that rank $A = M < N$. Given a data vector $d \in \mathbb{R}^M$, we consider an objective function $\mathcal{F}_d : \mathbb{R}^N \mapsto \mathbb{R}$ of the form

$$\mathcal{F}_d(u) = \|Au - d\|^2 + \beta \sum_{i \in \mathbb{I}_N} \phi(u[i]), \quad \beta > 0, \tag{1}$$

$$\phi(t) = \begin{cases} 0 \text{ if } t = 0 \\ 1 \text{ if } t \neq 0 \end{cases} \quad t \in \mathbb{R}, \tag{2}$$

$$\mathbb{I}_N = \{1, \cdots, N\}, \tag{3}$$

A.M. Bruckstein et al. (Eds.): SSVM 2011, LNCS 6667, pp. 508–519, 2012.

where $\| \cdot \|$ is the ℓ_2-norm, β is a regularization parameter and $u[i]$ is the ith entry of the vector u. By an abuse of language, the penalty in (1)-(2) is called the ℓ_0 norm: $\|u\|_0 = \sum_{i \in \mathbb{I}_N} \phi(u[i]) = \sharp \left\{ i \in \mathbb{I}_N : u[i] \neq 0 \right\}$ where $\sharp$ stands for cardinality. The columns of A are denoted by a_i, $i \in \mathbb{I}_N$. It is assumed that $a_i \neq 0$, $\forall i \in \mathbb{I}_N$.

We focus on the (local) minimizers $\widehat{u}$ of an objective $\mathcal{F}_d$ of the form (1)-(2):

$$\widehat{u} \in \mathbb{R}^N \quad \text{such that} \quad \mathcal{F}_d(\widehat{u}) = \min_{u \in \mathcal{O}} \mathcal{F}_d(u) , \tag{4}$$

where $\mathcal{O}$ is a neighborhood of $\widehat{u}$.

Finding the minimum of $\mathcal{F}_d$ in (1)-(2) is an NP-hard numerical problem [5].

The function ϕ in (2) served as a regularizer for a long time. In the context of Markov random fields it was used by Geman and Geman in 1984 [9] and Besag in 1986 [1] as a prior in MAP energies to restore labeled images. The MAP objective reads $\mathcal{F}(u) = \left\| Au - d \right\|^2 + \beta \sum_k \phi(D_k u)$, where D_k is a finite differences operator and ϕ is given by (2). This label-designed form is known as the Potts prior model, or the multi-level logistic model [2], [12]. Leclerc [11] proposed in 1989 the same prior to restore piecewise constant images. Various stochastic and continuation-based algorithms were proposed. Recently, this MAP objective was successfully applied to reconstruct 3D tomographic images using stochastic continuation [16] and stochastic relaxation [17] by Robini and Magnin.

Problems involving the minimization of $\mathcal{F}_d$ in (1)-(2) arise in image processing, morphologic component analysis, compression, dictionary building, inverse problems, compressive sensing, machine learning, among many other fields. The hard-thresholding method proposed by Donoho and Johnstone [7] amounts to minimize (1)-(2) where $u[i]$ are the coefficients of a signal or an image expanded in a wavelet basis ($M = N$). When $M < N$, various (strong) conditions on $\|u\|_0$ (often $\|u\|_0$ is replaced by $\|u\|_p$ for $0 < p \leqslant 1$) and on the relationship between the columns of A (typically, pseudo-orthogonality) are needed to conceive numerical schemes approximating a minimizer of $\mathcal{F}_d$, to establish local convergence and derive the asymptotic of the obtained solution. Haupt and Nowak [10] investigate the statistical performances of the global minimizer of (1)-(2) and propose an iterative bound-optimization procedure. In [18] the authors reformulate the problem so that an approximate solution can be found using difference of convex functions programming. Blumensath and Davies [3] propose an iterative thresholding scheme to find an approximate solution and prove local convergence. Several other references can be evoked, e.g. [15], [8].

Our goal is to analyze the (local) minimizers $\widehat{u}$ (4) of objectives $\mathcal{F}_d$ of the form (1)-(2). We provide a new understanding of the minimization problem. We clarify the possibility to recover exactly an original $\ddot{u}$ from data $d = A\ddot{u}$ as a (local) minimizer of $\mathcal{F}_d$.

The minimization of (1)-(2) might seem close to its constraint variants:

$$\text{given } \varepsilon \geqslant 0, \quad \text{minimize } \|u\|_0 \quad \text{subject to } \|Au - d\|^2 \leqslant \varepsilon , \tag{5}$$

$$\text{given } K \in \mathbb{I}_M, \quad \text{minimize } \|Au - d\|^2 \text{ subject to } \|u\|_0 \leqslant K . \tag{6}$$

The latter problems are abundantly studied in the context of sparse recovery in different fields. An excellent account (involving an exhaustive description of the state of the art) is given in [4], see also the book [13]. We emphasize that in general, in a non asymptotic framework, *there is no equivalence between the problems stated in* (5) *and* (6), *and the minimization of* $\mathcal{F}_d$ *in* (1)-(2). The reason is that all these problems are nonconvex.

1.1   Notations and Definitions

A local minimizer $\widehat{u}$ is strict if that there is a neighborhood $\mathcal{O} \subset \mathbb{R}^N$ containing $\widehat{u}$, such that $\mathcal{F}_d(\widehat{u}) < \mathcal{F}_d(v)$ for any $v \in \mathcal{O}$; such a minimizer is isolated. The identity matrix of size $K \times K$ is denoted I_K. We will often use the notation introduced in (3), namely $\mathbb{I}_K = \{1, \cdots, K\}$.

Definition 1. *For any* $u \in \mathbb{R}^K$, *the support* σ_u *of* u *is defined as*

$$\sigma_u \overset{\text{def}}{=} \left\{ i \subset \mathbb{I}_K \ : \ u[i] \neq 0 \right\}, \quad \sigma_u[1] < \cdots < \sigma_u[\sharp\, \sigma_u] \ ; \tag{7}$$

note that the elements of σ_u *are sorted in a strictly increasing order.*

For any subset $\omega \subset \mathbb{I}_K$, we denote $\omega^c \overset{\text{def}}{=} \mathbb{I}_K \setminus \omega$. Given a matrix $A \in \mathbb{R}^{M \times N}$ and a vector $u \in \mathbb{R}^N$, with any strictly increasing subsequence $\omega \subset \mathbb{I}_N$, say

$$\omega = \big(\omega[1], \cdots, \omega[r]\big) \quad \text{where} \quad r = \sharp\, \omega \ ,$$

we associate the following submatrix A_ω and subvector u_ω

$$A_\omega \overset{\text{def}}{=} \big(a_{\omega[1]}, \cdots, a_{\omega[r]}\big) \in \mathbb{R}^{M \times r} \ , \tag{8}$$

$$u_\omega \overset{\text{def}}{=} \Big(u\big[\omega[1]\big], \cdots, u\big[\omega[r]\big]\Big) \in \mathbb{R}^r \ , \tag{9}$$

as well as the zero-padding operator $Z_\omega \ : \ \mathbb{R}^r \to \mathbb{R}^N$ that inverts (9),

$$u = Z_\omega(u_\omega), \quad u[i] = \begin{cases} 0 & \text{if } i \notin \omega \ , \\ u_\omega[k] & \text{for the unique } k \text{ such that } \omega[k] = i \ . \end{cases} \tag{10}$$

For σ_u given by (7), we have the identity $Au = A_{\sigma_u} u_{\sigma_u} = A Z_{\sigma_u}(u_{\sigma_u}), \forall u \in \mathbb{R}^N$. We denote by A_ω^T the transposed of A_ω.

In what follows, it is systematically assumed that subsequences $\omega \subset \mathbb{I}_N$ are sorted in a strictly increasing order.

Definition 2. *A subset* $\mathcal{S} \subset \mathbb{R}^K$ *is said to be* negligible *(in* $\mathbb{R}^K$) *if* $\mathcal{S}$ *is closed in* $\mathbb{R}^K$ *and its Lebesgue measure in* $\mathbb{R}^K$ *is null,* $\mathbb{L}^K(\mathcal{S}) = 0$. *A property is said to hold for* $\overline{a.e.}$ $v \in \mathbb{R}^K$ *if it holds for all* $v \in \mathbb{R}^K \setminus \mathcal{S}$.

Thus, if a property holds for $\overline{a.e.}$ $v \in \mathbb{R}^K$, then it holds on a subset of $\mathbb{R}^K$ which is open and dense in $\mathbb{R}^K$. By a slight abuse of language, we can say that it holds true with *probability one*.

1.2   Content of the Paper

Section 2 deals with local minimizers in general. Section 3 is devoted to strict minimizers of $\mathcal{F}_d$ and section 4—to global minimizers. Necessary and sufficient conditions to recover exactly an original $\ddot{u}$ from data $d = A\ddot{u}$ as a (local) minimizer of $\mathcal{F}_d$, are presented in section 5 under a weak assumption on A shown to hold for $\overline{a.e.}$ A. A numerical toy example ($M = 5$, $N = 10$) in section 6 illustrates the main theoretical results. The proofs of all statements are outlined in [14].

2   Local Minimizers

The theory starts with a seemingly warning result.

Lemma 1. *For any $d \in \mathbb{R}^M$ and for any $\beta > 0$, $\mathcal{F}_d$ in (1)-(2) has a (local) minimum at $\widehat{u} = 0 \in \mathbb{R}^N$ and the latter is strict.*

Initialization with zero of a surrogate algorithm can be a bad choice.

2.1   Minimizers of $\mathcal{F}_d$ Solve Linear Programming Problems

It is shown below that finding a local minimizer of $\mathcal{F}_d$ is equivalent to solving a linear programming problem.

Theorem 1. *Given $d \in \mathbb{R}^M$ and $\omega \subset \mathbb{I}_N$, consider the problem $(\mathcal{P}_\omega)$ below*

$$(\mathcal{P}_\omega) \quad \left\{ \begin{array}{l} \min_{v \in \mathbb{R}^N} \|Av - d\|^2 \ , \\[2mm] \text{subject to} \quad v[i] = 0, \quad \forall i \in \omega^c = \mathbb{I}_N \setminus \omega \ . \end{array} \right. \tag{11}$$

Let $\widehat{u}$ solve problem $(\mathcal{P}_\omega)$. Then $\widehat{u}$ is a (local) minimizer of $\mathcal{F}_d$ in (1)-(2) and

$$\sigma_{\widehat{u}} \subseteq \omega \ ,$$

where $\sigma_{\widehat{u}}$ is defined according to (7).

The reciprocal statement is quite obvious.

Lemma 2. *Let $\mathcal{F}_d$ have a (local) minimum at $\widehat{u}$. Set $\widehat{\sigma} \stackrel{\text{def}}{=} \sigma_{\widehat{u}}$. Then $\widehat{u}$ solves problem $(\mathcal{P}_\omega)$ for $\omega = \widehat{\sigma}$.*

Remark 1. Problem $(\mathcal{P}_\omega)$ is equivalent to finding a solution $\widehat{u}_\omega$ of the problem below

$$\min_{v \in \mathbb{R}^{\sharp \omega}} \|A_\omega v - d\|^2 \tag{12}$$

and setting $\widehat{u} = Z_\omega(\widehat{u}_\omega)$, where Z_ω is given in (10). The problem in (12) always admits a solution. Using Theorem 1, $\widehat{u}$ is a minimizer $\mathcal{F}_d$ and hence it reads

$$\widehat{u} = Z_\omega(\widehat{u}_\omega) \quad \text{where} \quad \widehat{u}_\omega \text{ satisfies} \quad (A_\omega^T A_\omega)\widehat{u}_\omega = A_\omega^T d \ . \tag{13}$$

One notes that $\widehat{u}_\omega$ is the least squares solution with respect to the submatrix A_ω. It is well known that such a solution cannot deal with noisy data [19], [6].

Remark 1 shows that in general, a minimizer $\widehat{u}$ of $\mathcal{F}_d$ cannot properly reduce any noise corrupting the data d since A_ω is typically far from being unitary.

Since (13) is a linear problem equivalent to $(\mathcal{P}_\omega)$, we can say that $(\mathcal{P}_\omega)$ is *linear* as well.

Most of the results presented in what follows are based on Theorem 1 and Lemma 2. Some direct consequences of these statements are evoked next.

Corollary 1. *For any $d \in \mathbb{R}^M$, the function $\mathcal{F}_d$ has a global minimizer.*

By Lemma 2, if a minimizer $\widehat{u}$ satisfies $\mathcal{F}_d(\widehat{u}) > \beta M$, we are guaranteed that $\widehat{u}$ is a non strict minimizer.

Corollary 2. *Given $d \in \mathbb{R}^M$, consider $\mathcal{F}_d$. Then for any $\omega \subset \mathbb{I}_N$, $\mathcal{F}_d$ has a (local) minimizer $\widehat{u}$ such that $\sigma_{\widehat{u}} \subseteq \omega$.*

3   Strict Minimizers

3.1   Necessary and Sufficient Conditions

Theorem 2. *For a minimizer $\widehat{u}$ of $\mathcal{F}_d$, denote $\widehat{\sigma} = \sigma_{\widehat{u}}$. The minimizer $\widehat{u}$ is strict if and only if*

$$\mathrm{rank} A_{\widehat{\sigma}} = \sharp\widehat{\sigma} \leqslant M . \tag{14}$$

Furthermore, $\widehat{u} = Z_{\widehat{\sigma}}(\widehat{u}_{\widehat{\sigma}})$ and $\widehat{u}_{\widehat{\sigma}}$ reads

$$\widehat{u}_{\widehat{\sigma}} = \left(A_{\widehat{\sigma}}^T A_{\widehat{\sigma}}\right)^{-1} A_{\widehat{\sigma}}^T d . \tag{15}$$

This statement is quite intuitive in the light of Lemma 2 and Remark 1. Note that (14) is a necessary and sufficient condition for (12) to have a unique solution.

The result in (14) allows us to verify if an algorithm minimizing $\mathcal{F}_d$ has converged to a strict (local) minimum, or not.

The next statement is constructive in the sense that it indicates how to escape from a nonstrict local minimum.

Proposition 1. *Let $\bar{u}$ be a local minimizer of $\mathcal{F}_d$. Define $\bar{\sigma} \stackrel{\mathrm{def}}{=} \sigma_{\bar{u}}$ according to (7) and $A_{\bar{\sigma}}$ as in (8). Assume that*

$$\mathrm{rank} A_{\bar{\sigma}} < \min\{M, \sharp\bar{\sigma}\} .$$

Then $\mathcal{F}_d$ has a minimizer $\widehat{u}$ such that $\widehat{\sigma} \stackrel{\mathrm{def}}{=} \sigma_{\widehat{u}} \subsetneq \bar{\sigma}$ and hence

$$\mathcal{F}_d(\widehat{u}) \leqslant \mathcal{F}_d(\bar{u}) - \beta .$$

More precisely,

$$\widehat{u} \in \bar{u} - \mathrm{L}_{\bar{\sigma}} \quad for \quad \mathrm{L}_{\bar{\sigma}} = \left\{ Z_{\bar{\sigma}}(v) \in \mathbb{R}^N \ : \ v \in \ker A_{\bar{\sigma}} \setminus \{0\} \subset \mathbb{R}^{\sharp\bar{\sigma}} \right\} ,$$

where $Z_{\bar{\sigma}}$ is defined according to (10).

Let $\bar{u}$ be a local minimizer of $\mathcal{F}_d$ such that $\text{rank} A_{\hat{\sigma}} < \sharp \sigma_{\bar{u}}$. By Proposition 1, $\bar{u}$ is a nonstrict local minimizer. If a minimization scheme finds such an $\bar{u}$, we will find a strictly deeper minimizer in any direction belonging to $L_{\hat{\sigma}}$.

Let us take stock of the available facts.

- Originals $\ddot{u}$ such that $\sharp \sigma_{\ddot{u}} > M$ cannot be recovered. To recover originals with $\sharp \sigma_{\ddot{u}} \leqslant M$, we do not need to consider minimizers $\hat{u}$ with $\sharp \sigma_{\hat{u}} > M$.
- A minimizer $\hat{u}$ of $\mathcal{F}_d$ is strict if and only if $\text{rank} A_{\sigma_{\hat{u}}} = \sharp \sigma_{\hat{u}} \leqslant M$.
- Noise in data d cannot be properly reduced at a (local) minimizer $\hat{u}$ of $\mathcal{F}_d$.
- Strict minimizers enable exact unambiguous recovery.

These facts motivate the definition below.

Definition 3. *For any $r \in \mathbb{I}_M$, define Ω_r as a subset of r-length supports corresponding to full column rank $M \times r$ submatrices of A by the following 3 properties:*

$$\begin{cases} (1) \; \Omega_r \subseteq \{\omega \subset \mathbb{I}_N \; : \; \omega[1] < \cdots < \omega[r], \; \sharp \omega = r = \text{rank} A_\omega\} \; ; \\ (2) \; \text{if} \; (\varpi, \omega) \in \Omega_r \times \Omega_r, \; \varpi \neq \omega \; \Rightarrow \; A_\varpi \neq A_\omega \; ; \\ (3) \; \Omega_r \; \text{is maximal} \; : \; \text{if} \; \varpi \subset \mathbb{I}_N \; \text{satisfies} \; \sharp \varpi = r = \text{rank} A_\varpi \\ \qquad\qquad\qquad\qquad \Rightarrow \; \exists \omega \in \Omega_r \; : \; A_\varpi = A_\omega \; . \end{cases} \qquad (16)$$

Define as well

$$\Omega \overset{\text{def}}{=} \bigcup_{t=1}^{M-1} \Omega_t \;\; \text{and} \;\; \overline{\Omega} \overset{\text{def}}{=} \Omega \cup \Omega_M \; . \qquad (17)$$

The components of each $\omega \in \Omega_r$ are arranged in increasing order. This prevents the situation when two submatrices corresponding to two different supports are composed out of the same columns but placed in a different order. The cardinality of Ω_r is upper bounded:

$$\sharp \Omega_r \leqslant \frac{N!}{r!(N-r)!} \; .$$

All strict (local) minimizers of $\mathcal{F}_d$ have their supports in $\overline{\Omega}$. If there is an $\omega \in \mathbb{I}_N \setminus \overline{\Omega}$ with $\sharp \omega = \text{rank} A_\omega$, then there is $\varpi \in \overline{\Omega}$ such that $A_\omega = A_\varpi$. We will see that all global minimizers of $\mathcal{F}_d$ are strict (Theorem 4) so they are listed in $\overline{\Omega}$ as well.

3.2   Stability of Strict (local) Minimizers

Here we explore the behavior of the strict local minimizers of $\mathcal{F}_d$ with respect to variations of d.

Corollary 3. *Let $\hat{u} \neq 0$ be a (local) minimizer of $\mathcal{F}_d$ satisfying $\hat{\sigma} \overset{\text{def}}{=} \sigma_{\hat{u}} \in \Omega$ where Ω is given in (17). Define*

$$N_{\hat{\sigma}} \overset{\text{def}}{=} \text{span}\left(\ker A_{\hat{\sigma}}^T\right) \subset \mathbb{R}^M \; .$$

Then $\dim N_{\hat{\sigma}} = M - \sharp \hat{\sigma} \geqslant 1$. For any $d' \in N_{\hat{\sigma}}$, the relevant $\mathcal{F}_{d+d'}$ reaches a strict (local) minimum at the same point $\hat{u}$.

All data living in the $(M - \sharp\widehat{\sigma})$-dimensional vector subspace $N_{\widehat{\sigma}}$ yield the same strict (local) minimizer $\widehat{u}$.

This result shows that minimizing $\mathcal{F}_d$ compresses the data.

Even though $\mathcal{F}_d$ can admit numerous strict (local) minimizers, we show that each one of them is continuous in d.

Definition 4. *Let* $\mathcal{F}_d : \mathbb{R}^M \to \mathbb{R}$ *and* $\mathcal{O} \subseteq \mathbb{R}^M$ *be an open domain. We say that* $\mathcal{U} : \mathcal{O} \to \mathbb{R}^N$ *is a local minimizer function for the family* $\mathcal{F}_{\mathcal{O}} \overset{def}{=} \{\mathcal{F}_d \ : \ d \in \mathcal{O}\}$ *if for any* $d \in \mathcal{O}$, *the function* $\mathcal{F}_d$ *reaches a strict local minimum at* $\mathcal{U}(d)$.

Theorem 3. *Let* $\widehat{u}$ *be a (local) minimizer of* $\mathcal{F}_d$ *satisfying* $\widehat{\sigma} = \sigma_{\widehat{u}} \in \overline{\Omega}$. *Then:*

(i) There exists a local minimizer function $\mathcal{U} : \mathbb{R}^M \to \mathbb{R}^N$ *such that* $\forall d' \in \mathbb{R}^M$,

$$\widehat{u}' \overset{def}{=} \mathcal{U}(d') = Z_{\widehat{\sigma}}(\widehat{u}'_{\widehat{\sigma}}) \quad for \quad \widehat{u}'_{\widehat{\sigma}} = \left(A_{\widehat{\sigma}}^T A_{\widehat{\sigma}}\right)^{-1} A_{\widehat{\sigma}}^T d'$$

is a strict (local) minimizer of $\mathcal{F}_{d'}$ *satisfying* $\sigma_{\widehat{u}'} \subseteq \widehat{\sigma}$ *and* $\mathcal{U}(d) = \widehat{u}$.
(ii) There is a closed subset $D_{\widehat{\sigma}} \subset \mathbb{R}^M$ *with* $\mathbb{L}^M(D_{\widehat{\sigma}}) = 0$ *such that*

$$d' \in \mathbb{R}^M \setminus D_{\widehat{\sigma}} \quad \Rightarrow \quad \sigma_{\widehat{u}'} = \widehat{\sigma} \ ,$$

for $\widehat{u}' = \mathcal{U}(d')$. *Moreover,* $d' \mapsto \mathcal{F}_{d'}(\mathcal{U}(d'))$ *is* $\mathcal{C}^{\infty}$ *on* $\mathbb{R}^M \setminus D_{\widehat{\sigma}}$ *which is an open and dense subset of* $\mathbb{R}^M$.

Obviously, $\mathcal{U}$ is linear with respect to d' and (ii) holds true for $\overline{a.e.}$ $d' \in \mathbb{R}^M$. Note that $d' \mapsto \mathcal{F}_{d'}(\mathcal{U}(d'))$ is discontinuous on $D_{\widehat{\sigma}}$. This explains why global minimizers can correspond to various $\widehat{\sigma} \in \overline{\Omega}$ when data d' ranges over $\mathbb{R}^M$.

4   Global Minimizers

It is useful to state the following quite intuitive result:

Lemma 3. *Given* $d \in \mathbb{R}^M$, *if* $\widehat{u}$ *is a global minimizer of* $\mathcal{F}_d$, *then*

$$\mathcal{F}_d(\widehat{u}) \leqslant \beta M \quad and \quad \sharp\sigma_{\widehat{u}} \leqslant M \ .$$

A minimizer $\widehat{u}$ of $\mathcal{F}_d$ such that $\sigma_{\widehat{u}} > M$ cannot be global for any $\beta > 0$.
    We have a (strong) sufficient condition for the obtention of a null global minimizer.

Lemma 4. *Consider that* $\beta \geqslant \|d\|^2$. *Then* $\mathcal{F}_d$ *has a global minimum at* $\widehat{u} = 0$.

There are also good news on global minimizers, as seen below.

Theorem 4. *If* $\hat{u}$ *is a global minimizer of* $\mathcal{F}_d$, *then* $\mathrm{rank}A_{\widehat{\sigma}} = \sharp\widehat{\sigma} \leqslant M$, *where* $\widehat{\sigma} \overset{def}{=} \sigma_{\hat{u}}$. *All global minimizers of* $\mathcal{F}_d$ *are strict, for any* $d \in \mathbb{R}^M$ *and any* $\beta > 0$.

In [14], abstract conditions on β that ensure exact recovery of an original $\ddot{u}$ with $\sharp \sigma_{\ddot{u}} \leqslant M - 1$ as a global minimizer of $\mathcal{F}_d$ are derived. Actually, they seem hard to exploit in practice.

The statement below provides a simple and *strong necessary condition for a global minimizer* of $\mathcal{F}_d$.

Theorem 5. *Let $\mathcal{F}_d$ reach a global minimum at $\hat{u}$. Then*

$$either \quad \hat{u}[i] = 0 \quad or \quad |\hat{u}[i]| \geqslant \frac{\sqrt{\beta}}{\|a_i\|_2}, \quad \forall i \in \mathbb{I}_N . \tag{18}$$

Observe that the bound does not depend on d.

5 Exact Recovery

5.1 Originals with an M-Length Support

Remind that all matrices belonging to Ω_M—see (16)—are invertible. Hence for any $\omega \in \Omega_M$, the solution given in (15) in Theorem 2 reads $\hat{u}_\omega = A_\omega^{-1} d$.

Proposition 2. *For $\ddot{u} \in \mathbb{R}^N$ with $\ddot{\omega} = \sigma_{\ddot{u}} \in \Omega_M$, let $d = A\ddot{u} \in \mathbb{R}^M$. Define the set*

$$\mathrm{U}_M \overset{\mathrm{def}}{=} \left\{ \hat{u} \in \mathbb{R}^N \; : \; \sigma_{\hat{u}} \in \Omega_M, \; \|A\hat{u} - d\|^2 = 0 \right\} .$$

There is a negligible *subset $\mathrm{Q}_M \subset \mathbb{R}^M$ such that if $\ddot{u}_{\ddot{\omega}} \in \mathbb{R}^M \setminus \mathrm{Q}_M$, we have*

$$\sharp \mathrm{U}_M = \sharp \Omega_M .$$

Each $\hat{u} \in \mathrm{U}_M$ is a strict (local) minimizer of $\mathcal{F}_d$ and $\mathcal{F}_d(\hat{u}) = \beta M, \; \forall \hat{u} \in \mathrm{U}_M$.

This proposition shows that an original $\ddot{u}$ with $\sharp \sigma_{\ddot{u}} = M$ cannot (with probability one) be recovered exactly by minimizing $\mathcal{F}_d$.

5.2 An Assumption on A That Holds for $\overline{\text{a.e.}}$ A

We ask the following question: what conditions ensure that

$$A_\omega^T A_\omega \neq A_\omega^T A_\varpi \left(A_\varpi^T A_\varpi \right)^{-1} A_\varpi^T A_\omega, \quad (\omega, \varpi) \subset \Omega_r \times \Omega_t, \quad t \in \mathbb{I}_M, \; r \in \mathbb{I}_t ? \tag{19}$$

The issue of this question is crucial for exact recovery. The next lemma shows a situation when (19) systematically fails.

Lemma 5. *Let $(\varpi, \omega) \in \Omega_t \times \Omega_r$ for $t \in \mathbb{I}_M$ and $r \in \mathbb{I}_t$. Then*

$$\varpi \subseteq \omega \quad \Rightarrow \quad A_\varpi^T A_\varpi = A_\varpi^T A_\omega (A_\omega^T A_\omega)^{-1} A_\omega^T A_\varpi .$$

It turns out that this (un)property is good for exact recovery.

Proposition 3. *Let* $\ddot{u} \in \mathbb{R}^N$ *be such that* $\ddot{\varpi} \overset{def}{=} \sigma_{\ddot{u}} \in \Omega$ *and* $d = A\ddot{u}$. *Then a (local) minimizer* $\hat{u}$ *of* $\mathcal{F}_d$ *satisfies* $\hat{u} = \ddot{u}$ *if and only if*

$$\omega \in \{\varpi \in \overline{\Omega} \ : \ \ddot{\varpi} \subseteq \varpi\} \text{ and } \hat{u} \text{ solves } (\mathcal{P}_\omega) .$$

This claim is independent of the value of β.

We can also note that $A_\omega (A_\omega^T A_\omega)^{-1} A_\omega^T = I_M$, $\forall \omega \in \Omega_M$, in which case

$$A_\varpi^T \, A_\omega (A_\omega^T A_\omega)^{-1} A_\omega^T \, A_\varpi = A_\varpi^T A_\varpi, \quad \forall \varpi \in \Omega_r, \ \forall r \in \mathbb{I}_M.$$

This result, combined with Proposition 2, is the reason why we restrict our attention only to Ω_t and Ω_r for $(t, r) \in \mathbb{I}_{M-1} \times \mathbb{I}_{M-1}$.

Proposition 4. *For* $t \in \mathbb{I}_{M-1}$ *and* $r \in \mathbb{I}_t$, *define the subsets of matrices:*

$$\mathcal{H}_t = \bigcup_{\varpi \in \Omega_t} \left\{ A_\omega \ : \ \omega \in \Omega_t, \ \varpi \neq \omega, \ A_\omega (A_\omega^T A_\omega)^{-1} A_\omega^T = A_\varpi (A_\varpi^T A_\varpi)^{-1} A_\varpi^T \right\},$$

$$\mathcal{A}_r(t) = \bigcup_{\omega \in \Omega_t} \left\{ A_\varpi \ : \ \varpi \in \Omega_r, \ \varpi \not\subseteq \omega \ \ A_\varpi^T A_\varpi = A_\varpi^T \, A_\omega (A_\omega^T A_\omega)^{-1} A_\omega^T \, A_\varpi \right\}.$$

Then:

(i) *For any* $t \in \mathbb{I}_{M-1}$, *the set* $\mathcal{H}_t$ *is included in a finite union of subspaces of dimension* $M - t \times t$ *in the space of all* $M \times t$ *matrices, hence its Lebesgue measure in the latter space is null.*

(ii) *For any* $t \in \mathbb{I}_{M-1}$ *and* $r \in \mathbb{I}_t$ *each set* $\mathcal{A}_r(t)$ *is included in a finite union of subspaces of dimension* $M - t \times r$ *in the space of all* $M \times r$ *matrices, hence its Lebesgue measure in the latter space is null.*

Example 1. Here we give some elements belonging to the negligible subsets exhibited in Proposition 4. Let $B \in \mathbb{R}^{M \times t}$ satisfies $\operatorname{rank}(B) = t \leqslant M - 1$. Define

$$H = B(B^T B)^{-1} B^T . \tag{20}$$

For any real invertible $t \times t$ matrix W consider $C = BW \ \in \mathbb{R}^{M \times t}$. Then

$$\begin{aligned} C(C^T C)^{-1} C^T &= BW(W^T B^T BW)^{-1} W^T B^T \\ &= B(B^T B)^{-1} B^T = H . \end{aligned}$$

Given two subsets $(\omega, \varpi) \in (\Omega_t \times \Omega_t)$ with $\omega \neq \varpi$, finding a matrix W such that $A_\omega W = A_\varpi$ is seldom possible: W contains t^2 unknowns while there are $Mt > t^2$ equations to be satisfied and W must be invertible.

Let H read as in (20) and W be a $t \times r$ matrix with $\operatorname{rank}(W) = r \leqslant t$. Consider that $C = BW \in \mathbb{R}^{M \times r}$. Then

$$\begin{aligned} C^T H C &= C^T \, B(B^T B)^{-1} B^T \, C \\ &= W^T B^T B(B^T B)^{-1} B^T BW = (BW)^T \, BW = C^T C . \end{aligned}$$

Given $(\omega, \varpi) \in (\Omega_t \times \Omega_r)$ for $1 \leqslant r \leqslant t \leqslant M - 1$, $\varpi \not\subseteq \omega$, finding a W such that $A_\omega W = A_\varpi$ is yet again seldom possible: W has rt unknowns that must satisfy $Mr > tr$ equations.

Proposition 4 tells us that *the assumption on A formulated in H1 below fails to hold only for a negligible subset of matrices A:*

H 1 *The matrix $A \in \mathbb{R}^{M \times N}$, $N > M$, is such that for any $t \in \mathbb{I}_{M-1}$ and $r \in \mathbb{I}_t$*

$$(\varpi, \omega) \in (\Omega_r \times \Omega_t), \quad \varpi \not\subseteq \omega \quad \Rightarrow \quad A_\varpi^T A_\varpi \neq A_\varpi^T A_\omega \left(A_\omega^T A_\omega\right)^{-1} A_\omega^T A_\varpi .$$

5.3 Necessary and Sufficient Conditions

For $r \in \mathbb{I}_{M-1}$ put

$$\Theta_{\varpi, \omega} \overset{\text{def}}{=} \left\{ v \in \mathbb{R}^r \; : \; v^T A_\varpi^T \left(I_M - A_\omega \left(A_\omega^T A_\omega\right)^{-1} A_\omega^T \right) A_\varpi v = 0, \right.$$

$$\left. (\varpi, \omega) \in (\Omega_r \times \Omega_t) \text{ and } \varpi \not\subseteq \omega \right\},$$

$$\Theta_r = \bigcup_{\varpi \in \Omega_r} \bigcup_{t=r}^{M-1} \bigcup_{\omega \in \Omega_t} \Theta_{\varpi, \omega} . \tag{21}$$

We would not like that originals $\ddot{u}$ such that $r = \sharp \sigma_{\ddot{u}}$ have their non zero part living in Θ_r. For such originals we cannot catch the exact solution as a (local) minimizer of $\mathcal{F}_d$. We evaluate the chance that our wish is satisfied.

Lemma 6. *Let H1 holds. For any $r \in \mathbb{I}_{M-1}$, the set Θ_r in (21) is a finite union of (closed) vector subspaces of dimension at most equal to $r - 1$ so $\mathbb{L}^r(\Theta_r) = 0$.*

Since Θ_r in (21) is *negligible*, an original $\ddot{u}$ with $\sharp \sigma_{\ddot{u}} = r \leqslant M - 1$ satisfies $\ddot{u}_{\sigma_{\ddot{u}}} \in \mathbb{R}^r \setminus \Theta_r$ with probability one. Next we give necessary and sufficient conditions for the exact recovery of such an original $\ddot{u}$ as a (local) minimizer of $\mathcal{F}_d$.

Theorem 6. *Let A satisfy H1. Let an original $\ddot{u}$ with $\ddot{\varpi} \overset{\text{def}}{=} \sigma_{\ddot{u}} \in \Omega$ satisfy $\ddot{u}_{\ddot{\varpi}} \in \mathbb{R}^r \setminus \Theta_r$, where Θ_r is negligible in $\mathbb{R}^r$ (Lemma 6). Consider that*

$$d = A\ddot{u} .$$

Then a (local) minimizer $\widehat{u}$ of $\mathcal{F}_d$ satisfies $\widehat{u} = \ddot{u}$ if and only if $\widehat{u}$ solves the problem below

$$\min_{\omega \in \Omega} \left\{ \|A\bar{u} - d\|^2 \; : \; \bar{u} \text{ solves } (\mathcal{P}_\omega) \right\}, \tag{22}$$

where $(\mathcal{P}_\omega)$ is formulated in (11). This claim holds true for any $r \in \mathbb{I}_{M-1}$. It is independent of the value of β.

Let A meet H1. For $\overline{\text{a.e.}}$ original $\ddot{u}$ with $\ddot{\varpi} \overset{\text{def}}{=} \sigma_{\ddot{u}} \in \Omega$ and data $d = A\ddot{u}$, finding the exact solution $\widehat{u} = \ddot{u}$ as a (local) minimizer of $\mathcal{F}_d$ amounts to solve the nonlinear problem in (22).

> In words, among all strict (local) minimizers of $\mathcal{F}_d$ with support shorter than $M - 1$, the exact solution $\widehat{u} = \ddot{u}$ is the unique vector yielding a null residual $\|A\widehat{u} - d\|^2 = 0$.
> Theorem 6 provides a comfortable tool enabling to check if an algorithm minimizing $\mathcal{F}_d$ has found the exact solution, or not.

Next we learn that the exact solution can be "just" a local minimizer of $\mathcal{F}_d$.

Corollary 4. *Let A satisfy H1. Let an original $\ddot{u}$ with $\ddot{\omega} \stackrel{\text{def}}{=} \sigma_{\ddot{u}}$ satisfy $\ddot{\omega} \in \Omega$ and $\ddot{u}_{\ddot{\omega}} \in \mathbb{R}^r \setminus \Theta_r$ where Θ_r is negligible in $\mathbb{R}^r$ (Lemma 6). Suppose that*

$$\exists\, i \in \ddot{\omega} \quad \text{such that} \quad \left|\, \ddot{u}[i]\,\right| < \frac{\beta}{\|a_i\|_2} \ ,$$

where a_i is the ith column of A. Consider that $d = A\ddot{u}$. Than $\mathcal{F}_d$ has a strict minimizer such that $\widehat{u} = \ddot{u}$ and the latter is different from any global minimizer:

$$\mathcal{F}_d(\widehat{u}) > \min_{v \in \mathbb{R}^N} \mathcal{F}_d(v) \ .$$

This statement follows from Theorems 5 and 6. It tells us that the global minimizer of $\mathcal{F}_d$ cannot provide exact recovery for large classes of originals $\ddot{u}$.

Theorem 6 and Corollary 4 clearly show that striving after global minimization of $\mathcal{F}_d$ can prevent the recovery of the exact solution.

6   Numerical Toy-Example

Consider the objective $\mathcal{F}_d$ in (1)-(2) for A and d as given below.

$$A = \begin{bmatrix} 7 & 6 & 5 & 2 & 8 & 7 & 6 & 1 & 2 & 7 \\ 2 & 8 & 1 & 8 & 2 & 3 & 7 & 1 & 5 & 6 \\ 5 & 9 & 1 & 2 & 9 & 6 & 2 & 4 & 8 & 1 \\ 1 & 7 & 0 & 2 & 0 & 4 & 6 & 9 & 9 & 2 \\ 3 & 7 & 9 & 1 & 4 & 4 & 1 & 3 & 8 & 1 \end{bmatrix} \quad \begin{aligned} & \ddot{u} = [0\ 4\ 5\ 0\ 9\ 8\ 0\ 0\ 0\ 0]^T, \\[6pt] & \\[2pt] & d = A\ddot{u} \ . \end{aligned} \tag{23}$$

We chose only integers in (23) for better readability. The arbitrary matrix A has $M = 5$ rows and $N = 10$ columns and it satisfies H1—the test was done using exhaustive search. For $\ddot{\omega} \stackrel{\text{def}}{=} \sigma_{\ddot{u}}$ we have $\sharp\ddot{\omega} = M - 1$, hence $\ddot{\omega} \in \Omega$ (see Definition 3). For any $\beta > 0$, one finds that $\widehat{u} = \ddot{u}$ is the unique strict (local) minimizer of $\mathcal{F}_d$ having a support shorter than $M-1$ and yielding $\|A\widehat{u}-d\|^2 = 0$. This corroborates Theorem 6. For $\beta = 1$, $\widehat{u} = \ddot{u}$ is the global minimizer of $\mathcal{F}_d$. For $\beta = 100$ and $\beta = 10^5$, $\widehat{u} = \ddot{u}$ is no longer a global minimizer. This confirms Corollary 4. For $\beta = 10^5 > \|d\|^2$, the global minimizer is $\bar{u} = 0$, as stated in Lemma 4. All minimizers are calculated using exhaustive combinatorial search.

7   Concluding Remarks

A consequence of Theorem 1 and Lemma 2 is that for any $\beta > 0$, no noise corrupting the data d can be properly reduced at a minimizer of $\mathcal{F}_d$. Equation (14) in Theorem 2 allows us to easily check if an algorithm minimizing $\mathcal{F}_d$ has converged to a strict local minimum, or not.

Assumption H1 holds for $\overline{\text{a.e.}}$ A. Under this assumption, Theorem 6 provides an easy rigorous tool to verify if an algorithm has found, or not, an exact solution whose support is strictly shorter than M. For large classes of data, this exact solution is a local minimizer of $\mathcal{F}_d$ which is different from any global minimizer (Corollary 4). Then global minimization of $\mathcal{F}_d$ can prevent exact recovery.

References

1. Besag, J.E.: On the statistical analysis of dirty pictures (with discussion). Journal of the Royal Statistical Society B 48(3), 259–302 (1986)
2. Besag, J.E.: Digital image processing: Towards Bayesian image analysis. Journal of Applied Statistics 16(3), 395–407 (1989)
3. Blumensath, T., Davies, M.: Iterative thresholding for sparse approximations. The Journal of Fourier Analysis and Applications 14(5) (2008)
4. Bruckstein, A.M., Donoho, D.L., Elad, M.: From sparse solutions of systems of equations to sparse modeling of signals and images. SIAM Review 51(1), 34–81 (2009)
5. Davis, G., Mallat, S., Avellaneda, M.: Adaptive greedy approximations. Constructive Approximation 13(1), 57–98 (1997)
6. Demoment, G.: Image reconstruction and restoration: Overview of common estimation structure and problems. IEEE Transactions on Acoustics Speech and Signal Processing ASSP-37, 2024–2036 (1989)
7. Donoho, D.L., Johnstone, I.M.: Ideal spatial adaptation by wavelet shrinkage. Biometrika 81(3), 425–455 (1994)
8. Gasso, G., Rakotomamonjy, A., Canu, S.: Recovering sparse signals with a certain family of non-convex penalties and DC programming. IEEE Transactions on Signal Processing 57(12), 4686–4698 (2009)
9. Geman, S., Geman, D.: Stochastic relaxation, Gibbs distributions, and the Bayesian restoration of images. IEEE Transactions on Pattern Analysis and Machine Intelligence PAMI-6, 721–741 (1984)
10. Haupt, J., Nowak, R.: Signal reconstruction from noisy random projections. IEEE Transactions on Information Theory 52(9), 4036–4048 (2006)
11. Leclerc, Y.G.: Constructing simple stable description for image partitioning. International Journal of Computer Vision 3, 73–102 (1989)
12. Li, S.Z.: Markov Random Field Modeling in Computer Vision, 1st edn. Springer, New York (1995)
13. Mallat, S.: A Wavelet Tour of Signal Processing (The sparse way), 3rd edn. Academic Press, London (2008)
14. Nikolova, M.: On the minimizers of least squares regularized with an ℓ_0 norm, Technical report (2011)
15. Neumann, J., Schörr, C., Steidl, G.: Combined SVM-based Feature Selection and classification. Machine Learning 61, 129–150 (2005)
16. Robini, M.C., Lachal, A., Magnin, I.E.: A stochastic continuation approach to piecewise constant reconstruction. IEEE Transactions on Image Processing 16(10), 2576–2589 (2007)
17. Robini, M.C., Magnin, I.E.: Optimization by stochastic continuation. SIAM Journal on Imaging Sciences 3(4), 1096–1121 (2010)
18. Thiao, M., Dinh, T.P., Thi, A.L.: DC Programming Approach for a Class of Nonconvex Programs Involving l0 Norm. CCIS, pp. 348–357. Springer, Heidelberg (2008)
19. Tikhonov, A., Arsenin, V.: Solutions of Ill-Posed Problems, Winston, Washington DC (1977)

Weak Statistical Constraints for Variational Stereo Imaging of Oceanic Waves

Guillermo Gallego[1], Anthony Yezzi[1],
Francesco Fedele[2,*], and Alvise Benetazzo[3]

[1] School of Electrical Engineering, Georgia Institute of Technology, Atlanta, USA
[2] School of Civil & Environmental Engineering,
Georgia Institute of Technology, Atlanta, USA
[3] CNR-ISMAR, Venice, Italy

Abstract. We develop an observational technique for the stereoscopic reconstruction of the wave form of oceanic sea states via a variational stereo method. In the context of active surfaces, the shape and radiance of the wave surface are obtained as minimizers of an energy functional that combines image observations and smoothness priors. To obey the quasi Gaussianity of oceanic waves observed in nature, a given statistical wave law is enforced in the stereo variational framework as a weak constraint. Multigrid methods are then used to solve the partial differential equations derived from the optimality conditions of the augmented energy functional. An application of the developed method to two sets of experimental stereo data is finally presented.

1 Introduction

In recent years there has been a growing interest in vision-based remote-sensing observational technology for the measurement of oceanic sea states [7,2,17,5]. This topic is a major concern in ocean engineering because it has a broad impact: the understanding of space-time dynamics of ocean waves enables better forecasting of extreme events, improved design of off-shore structures, validation of theoretical models, etc. Vision systems are non-intrusive, have economical advantages over traditional instrumentation (wave gauges and ultrasonic instruments or buoys) and provide spatio-temporal data whose statistical content is richer than that of previous monitoring methods, but they require more processing power to extract information from the observed video data. The application of vision tools, such as stereography, to oceanography dates back to the first experiments with stereo cameras mounted on a ship by Schumacher [12] in 1939. Stereography gained popularity in studying the dynamics of oceanographic phenomena during the 1980s due to advances in hardware. For example, Shemdin et al. [14] applied stereography for the directional measurement of short ocean waves. Recently, Benetazzo [2] successfully incorporated epipolar techniques in

* Research supported by ONR grant BAA 09-012: "Ocean Wave Dissipation and energy Balance (WAVE-DB):toward reliable spectra and first breaking statistics".

A.M. Bruckstein et al. (Eds.): SSVM 2011, LNCS 6667, pp. 520–531, 2012.

the Wave Acquisition Stereo System (WASS) and showed that the accuracy of WASS is comparable to the accuracy obtained from traditional instrumentation. An alternative trinocular imaging system (ATSIS) for measuring the temporal evolution of 3-D surface waves was proposed in [17].

The three-dimensional reconstruction of an object's surface from multiple images is a classical problem in computer vision [10,6,13], and it is still an extremely active research area. There are many 3-D reconstruction algorithms available in the literature and they are designed under different assumptions that provide a variety of trade-offs between speed, accuracy and viability. Traditional *image-based* stereo methods typically consist of two steps: first, image points are detected and matched across images to establish local correspondences; then depth is inferred by back-projection of correspondences. This is the strategy used in recent observational systems [2,17], and it has the advantages of being simple and fast. However, it also has some major disadvantages that motivated the research on improved stereo reconstruction methods [4,18,8] based on variational theory. Firstly, correspondences rely on strong textures and image matching. They can be poorly estimated if the objects in the scene have a smooth radiance, and can also suffer from the presence of noise and local minima. Furthermore, each space point is reconstructed independently. Therefore, the recovered surface of an object is obtained as a collection of scattered 3-D points. Thus, the hypothesis of the continuity of the surface is not exploited in the reconstruction process. The breakdown of traditional stereo methods in these situations is evidenced by "holes" in the reconstructed surface, which correspond to unmatched image regions [10,2]. This phenomenon may be dominant in the case of the ocean surface, which, by nature, is generally continuous and contains little texture.

Modern *object-based* computer vision methods that rely on Calculus of Variations and Partial Differential Equations (PDE), are able to overcome the disadvantages of traditional stereo [4,18,1,8]. For instance, unmatched regions are avoided by building an explicit model of the smooth surface to be estimated rather than representing it as a collection of scattered 3-D points. Thus, variational methods provide dense and coherent surface reconstructions. Surface points are reconstructed by exploiting the continuity (coherence) hypothesis in the full two-dimensional domain of the surface.

Variational stereo methods combine correspondence establishment and shape reconstruction into one single step and they are less sensitive to matching problems of local correspondences. The reconstructed surface is obtained by minimization of an energy functional designed for the stereo problem. The solution is obtained in the context of active surfaces by deforming an initial surface via a gradient descent PDE derived from the necessary optimality conditions of the energy functional, the so-called Euler-Lagrange (EL) equations. In parallel to the advances in vision tools, the oceanographic community has developed statistical and spectral models for the characterization of oceanic sea states [9,19,15,16] that clearly indicate that oceanic waves are quasi-Gaussian in nature.

Up to date, both traditional and variational stereo techniques do not include in the reconstruction process the prior information of Gaussianity of waves,

which is usually verified a posteriori [2,5]. In this paper, we present a novel
variational framework in which a statistical distribution is enforced as a prior
into the stereo reconstruction of water waves via a weak constraint. Motivated
by the characteristics of the target object in the scene, i.e., the ocean surface,
we first introduce the graph surface representation in the formulation of the
reconstruction problem. Then, we cast the problem as a variational optimization
problem and show how a priori knowledge of statistical wave height models can
be weakly enforced in the variational framework to aid the recovery of the surface
shape. The performance of the algorithm is validated on experimental data and
the statistics of the reconstructed surface are also analyzed. Concluding remarks
are finally presented.

2   The Variational Framework

2.1   Multi-Image Setup and Graph Surface Representation

Let S be a smooth surface in $\mathbb{R}^3$ with generic local coordinates $(u, v) \in \mathbb{R}^2$. Let
$\{I_i\}_{i=1}^{N_c}$ be a set of images of a static scene acquired by cameras with known
calibration parameters $\{\mathbf{P}^i\}_{i=1}^{N_c}$. Space points are mapped into image points according to the pinhole camera model [6]. A surface point (or, in general a 3-D
point) $\mathbf{X} = (X, Y, Z)^\top$ with homogeneous coordinates $\bar{\mathbf{X}} = (X, Y, Z, 1)^\top$ is
mapped to point $\mathbf{x}_i = (x_i, y_i)^\top$ in the i-th image with homogeneous coordi-
nates $\bar{\mathbf{x}}_i = (x_i, y_i, 1)^\top \sim \mathbf{P}^i \bar{\mathbf{X}}$, where the symbol $\sim$ means equality up to a
nonzero scale factor and $\mathbf{P}^i = \mathbf{K}^i[\mathbf{R}^i \mid \mathbf{t}^i]$ is the 3×4 projection matrix with
the intrinsic ($\mathbf{K}^i$) and extrinsic ($\mathbf{R}^i, \mathbf{t}^i$) calibration parameters of the i-th camera.
Point $\mathbf{C}_i = (C_i^1, C_i^2, C_i^3)^\top$ satisfying $\mathbf{P}^i \bar{\mathbf{C}}_i = \mathbf{0}$ is the optical center of the i-th
camera. Let $\pi_i : \mathbb{R}^3 \to \mathbb{R}^2$ note the projection maps, $\mathbf{x}_i = \pi_i(\mathbf{X})$, and $I_i(\mathbf{x}_i)$ be
the image intensity at $\mathbf{x}_i$.

In the variational context of active surfaces, we present a different approach to
the reconstruction problem presented in [18,4] (level set approach) by exploiting
the hypothesis that the surface of the water can be represented in the form of a
graph or elevation map:

$$Z = Z(X, Y), \qquad (1)$$

where Z is the height of the surface with respect to a domain plane that is parameterized by coordinates X and Y. Indeed, slow varying, non-breaking waves
admit this simple representation with respect to a plane orthogonal to gravity direction. The graph representation of the water surface presents some clear
advantages over the more general level set representation of [4,8,18,5]. Surface
evolution is simpler to implement since the surface is not represented in terms
of an auxiliary higher dimensional function (the level set function). The surface
is evolved directly via the height function (1) discretized over a fixed 2-D grid
defined on the $X - Y$ plane. The latter also implies that for the same amount
of physical memory, higher spatial resolution (finer details) can be achieved in
the graph representation than with the level set. The $X - Y$ plane becomes
the natural common domain to parameterize the geometrical and photometric

properties of surfaces. This simple identification does not exist in the level set approach [18]. Finally, the graph representation allows for fast numerical solvers besides gradient descent, like Fast Poisson Solvers, Cyclic Reduction, Multigrid Methods, Finite-Element Methods (FEM), etc. In the level set framework, the range of solvers is not as diverse.

However, there are also some minor disadvantages. A world frame properly oriented with the gravity direction must be defined in advance to represent the surface as a graph with respect to this plane. This is not trivial *a priori* and might pose a problem if only the information from the stereo images is used [2]. Surface evolution is constrained to be in the form of a graph and this may differ from the evolution obtained for an unconstrained surface. As a result, more iterations may be required to evolve the active surface to reach convergence.

2.2   Proposed Energy Functional

Consider the 3-D reconstruction problem from a collection of $N_c \geq 2$ images (we will exemplify with $N_c = 2$). We investigate a generative model of the images that allows for the joint estimation of the shape of the surface S and the radiance function on the surface f as minimizers of an energy functional. Let the energy functional be the sum of a data fidelity term E_{data} and two regularizing terms: a geometry smoothing term E_{geom} and a radiance smoothing term E_{rad},

$$E(S, f) = E_{\mathrm{data}}(S, f) + \alpha E_{\mathrm{geom}}(S) + \beta E_{\mathrm{rad}}(f), \tag{2}$$

where $\alpha, \beta \in \mathbb{R}^+$. The data fidelity term measures the photo-consistency of the model: the discrepancy between the observed images I_i and the radiance model f,

$$E_{\mathrm{data}} = \sum_{i=1}^{N_c} E_i, \qquad E_i = \int_{\Omega_i} \phi_i \, \mathrm{d}\mathbf{x}_i, \tag{3}$$

where a possible photometric matching criterion is

$$\phi_i = \tfrac{1}{2}\big(I_i(\mathbf{x}_i) - f(\mathbf{x}_i)\big)^2. \tag{4}$$

The region of the image domain where the scene is projected is denoted by Ω_i. Assuming that the surface of the scene is represented as a graph $Z = Z(u, v)$, a point on the surface has coordinates

$$\mathbf{X}(u, v) = \big(u, v, Z(u, v)\big)^\top. \tag{5}$$

The chain of operations to obtain the intensity $I_i(\mathbf{x}_i)$ given a surface point with world coordinates $\mathbf{X}(\mathbf{u}) \equiv S(\mathbf{u})$, $\mathbf{u} = (u, v)^\top$, is

$$\mathbf{X}(\mathbf{u}) \mapsto \tilde{\mathbf{X}}_i = \mathsf{M}^i \mathbf{X} + \mathbf{p}_4^i \mapsto \mathbf{x}_i \mapsto I_i(\mathbf{x}_i), \tag{6}$$

where $\tilde{\mathbf{X}}^i = (\tilde{X}_i, \tilde{Y}_i, \tilde{Z}_i)^\top$ are related to the coordinates of $\mathbf{X}$ in the i-th camera frame, $\mathbf{x}_i = (\tilde{X}_i/\tilde{Z}_i, \tilde{Y}_i/\tilde{Z}_i)^\top$ is the projection of $\mathbf{X}$ in the i-th image plane and $\mathsf{P}^i = [\mathsf{M}^i \,|\, \mathbf{p}_4^i]$, with $\mathsf{M}^i = \mathsf{K}^i \mathsf{R}^i \equiv (\mathbf{n}_1^i, \mathbf{n}_2^i, \mathbf{n}_3^i)^\top$ and $\mathbf{p}_4^i = \mathsf{K}^i \mathbf{t}^i$. Also, $|\mathsf{M}^i| = \det(\mathsf{M}^i)$.

The radiance model f is specified by a function $\hat{f}$ defined on the surface S. Then, f in (4) is naturally defined by $f(\mathbf{x}_i) = \hat{f}(\pi_i^{-1}(\mathbf{X}))$, where π_i^{-1} denotes the back-projection operation from a point in the i-th image to the closest surface point with respect to the camera. By abusing notation, let us use f to denote the parameterized radiance $f(\mathbf{u})$, understanding that $f(\mathbf{x}_i)$ in (4) reads the back-projected value in $\hat{f}(\mathbf{X}(\mathbf{u})) = f(\mathbf{u})$.

Motivated by the common parameterizing domain of the shape and radiance of the surface and to obtain the simplest diffusive terms in the necessary optimality conditions of the energy (2), let the regularizers be

$$E_{\text{geom}} = \int_U \tfrac{1}{2}\|\nabla Z(\mathbf{u})\|^2 \, d\mathbf{u}, \qquad E_{\text{rad}} = \int_U \tfrac{1}{2}\|\nabla f(\mathbf{u})\|^2 \, d\mathbf{u}, \tag{7}$$

where $\nabla Z(\mathbf{u}) = (Z_u, Z_v)^\top$, $\nabla f(\mathbf{u}) = (f_u, f_v)^\top$ and subscripts indicate the derivative with respect to that variable.

The definition of the data fidelity term as an integral over the image domain (rather than over the parameter space U) has two advantages: (i) the data term is independent of the choice of domain for the graph, and (ii) the resulting optimality conditions for the minimization of (2) lack image derivatives, which are transferred to the radiance model and can be controlled by the regularizer E_{rad}. This desirable property is inherited from the modeling and mathematical principles that we follow from [18]. The resulting algorithm is less sensitive to image noise than other variational approaches for stereo 3-D reconstruction.

Once all terms in (2) have been specified, they are expressed over a common domain: the parameter space. The Jacobian of the change of variables between integration domains for the data term is, by applying the chain rule to (6),

$$J_i = \left| \frac{d\mathbf{x}_i}{d\mathbf{u}} \right| = -|\mathsf{M}^i|\tilde{Z}_i^{-3}(\mathbf{X} - \mathbf{C}_i) \cdot (\mathbf{X}_u \times \mathbf{X}_v), \tag{8}$$

where $\mathbf{X}_u \times \mathbf{X}_v$ is proportional to the outward unit normal $\mathbf{N}$ to the surface at $\mathbf{X}(u, v)$, and $\tilde{Z}_i = \mathbf{n}_3^i \cdot (\mathbf{X} - \mathbf{C}_i) > 0$ is the depth of the point $\mathbf{X}$ with respect to the i-th camera (located at $\mathbf{C}_i$). With this change, energy (3) becomes

$$E_i = \int_{\Omega_i} \phi_i \, d\mathbf{x}_i = \int_U \phi_i J_i \, d\mathbf{u}, \tag{9}$$

where the last integral is over U: the part of the parameter space whose surface projects on Ω_i in the i-th image. Observe that the Jacobian weights the photometric error ϕ_i proportionally to the cosine of the angle between the unit normal to the surface at $\mathbf{X}$ and the *projection ray* (the ray joining the optical center of the camera and $\mathbf{X}$): $(\mathbf{X} - \mathbf{C}_i) \cdot (\mathbf{X}_u \times \mathbf{X}_v)$. After collecting terms (7) and (9), and noting that the shape $\mathbf{X}$ of the surface solely depends on the height (Eqn. (5)), energy (2) becomes

$$E(Z, f) = \int_U L(Z, \nabla Z, f, \nabla f, u, v) \, d\mathbf{u}. \tag{10}$$

where subscripts indicate the derivative with respect to that variable, and the integrand is the so-called *Lagrangian* L.

2.3   Energy Minimization. Optimality Condition

The energy (10) depends on two functions: the shape Z and the radiance f of the surface. To find a minimizer of such a functional, we derive the necessary optimality condition by setting to zero the first variation of the functional, yielding a coupled system of PDEs (EL equations) along with boundary conditions:

$$g(Z, f) - \alpha \Delta Z = 0 \quad \text{in } U, \tag{11}$$

$$b(Z, f) + \alpha \frac{\partial Z}{\partial \boldsymbol{\nu}} = 0 \quad \text{on } \partial U, \tag{12}$$

$$-\sum_{i=1}^{N_c} (I_i - f) \mathrm{J}_i(Z) - \beta \Delta f = 0 \quad \text{in } U, \tag{13}$$

$$\beta \frac{\partial f}{\partial \boldsymbol{\nu}} = 0 \quad \text{on } \partial U, \tag{14}$$

where the non-linear terms due to the data fidelity energy are

$$g(Z, f) = \nabla f \cdot \sum_{i=1}^{N_c} |\mathrm{M}^i| \tilde{Z}_i^{-3} (I_i - f)(u - C_i^1, v - C_i^2), \tag{15}$$

$$b(Z, f) = \sum_{i=1}^{N_c} \phi_i |\mathrm{M}^i| \tilde{Z}_i^{-3} \big((u - C_i^1) \nu^u + (v - C_i^2) \nu^v\big).$$

The Laplacians ΔZ and Δf arise from the regularizing terms (7), and $\partial * / \partial \boldsymbol{\nu}$ is the the the directional derivative along $\boldsymbol{\nu} = (\nu^u, \nu^v)^\top$, the normal to the integration domain U in the parameter space. A simple classification of the PDEs can be done as follows. For a fixed shape, (13) and (14) form a linear elliptic PDE (of the inhomogeneous Helmholtz type) with Neumann boundary conditions. On the other hand, for a fixed radiance, (11) and (12) lead to a nonlinear elliptic equation in the height Z with nonstandard boundary conditions.

Difficult EL equations, such as (11)-(14), are commonly solved by the steady-state of gradient descent PDEs that evolve the unknown functions in artificial time t. This is the context of the so-called active surfaces. Due to the asymmetry in the complexity of the PDEs, a minimization strategy consisting of a nested iterative scheme is proposed: an outer loop performs a gradient descent in the height, and an inner loop implements a direct optimization for the radiance. Starting from an initial approximate solution, there are two phases within each iteration: (1) compute the optimal radiance for a fixed shape, and (2) evolve the shape, leaving the radiance fixed. To simplify the equations, we approximate the boundary condition (12) by a simpler, homogeneous Neumann boundary condition. This can be interpreted as if the data fidelity term vanished close to the boundary and it is a reasonable assumption since the major contribution to the energy is given by the terms in U, not at the boundary.

Numerical Solution. The optimality PDEs are discretized on a rectangular 2-D grid in the parameter space and then solved numerically using finite-difference methods. Direct optimization of the radiance is achieved by using stationary iterative methods (Jacobi or Gauss-Seidel). Forward differences in time and central differences in space approximate the derivatives in the gradient descent PDE for the height, yielding an explicit updating scheme. The von Neumann stability analysis of the linearized PDE yields a time step $\Delta t \leq 1/(\frac{4\alpha}{h^2} + \frac{1}{2}\max|\dot{g}(Z)|)$, where $\dot{g}(Z)$ is the derivative of (15) and the maximum is taken over the 2-D discretized grid at current time t. The time step may change at every iteration.

Both updating schemes (stationary methods for f and the time-stepping method for Z) are used as relaxation procedures inside a multigrid method [3] that approximately solves the EL equations. Multigrid methods are the most efficient numerical tools for solving elliptic boundary value problems.

3 Weak Enforcement of Wave Height Distributions

The flexibility of the variational framework allows us to incorporate properties of the *physics* of the waves in the model that would be otherwise difficult to take into account in image-based stereo methods. For example, we may include global statistical properties in the form of a weak constraint by considering an extra energy term that penalizes the deviation of the statistics of the reconstructed surface with respect to some target statistics derived from a physical model. In particular, we may penalize the deviation of the height distribution of the water surface with respect to a physically-justified Gaussian model and drive the surface evolution toward (weakly) satisfying such a global property.

If $Z(u, v) = Z(\mathbf{u})$ is the height of the surface (wave) and it is interpreted as a random variable, then its cumulative distribution function (CDF) is

$$\text{cdf}^Z(Z_0) = P(Z \leq Z_0) = \frac{1}{A}\int_U H(Z_0 - Z(\mathbf{u}))d\mathbf{u},$$

where $H(\cdot)$ is the Heaviside function and $A = \int_U d\mathbf{u}$ is the area of the (fixed) domain of integration. Suppose (2) is augmented with an extra energy term $\gamma E_{\text{cdf}}(S)$, $\gamma > 0$, that measures the discrepancy between a target height CDF that we wish to enforce, $G(Z)$, and the experimental CDF of the height:

$$E_{\text{cdf}}(Z) = \int_{-\infty}^{\infty} \frac{1}{2}\big(G(\hat{z}) - \text{cdf}^Z(\hat{z})\big)^2 d\hat{z}. \tag{16}$$

To compute the first variation of (16), we can directly use the definition of the Gâteaux derivative or augment Z with an artificial time variable, $Z = Z(\mathbf{u}, t)$, so that the energy depends on t, differentiate with respect to this variable and exploit the relationship between both derivatives. Carrying out operations in the distributional sense,

$$\frac{\mathrm{d}}{\mathrm{d}t}E_{\text{cdf}} = \frac{\mathrm{d}}{\mathrm{d}t}\int_{-\infty}^{\infty}\frac{1}{2}\Big(G(\hat{z}) - \frac{1}{A}\int_U H(\hat{z} - Z(\mathbf{u}))d\mathbf{u}\Big)^2 d\hat{z}$$

$$= \int_{-\infty}^{\infty}\big(G(\hat{z}) - \text{cdf}^Z(\hat{z})\big)\Big(\frac{1}{A}\int_U \delta(\hat{z} - Z(\mathbf{u}))Z_t d\mathbf{u}\Big)d\hat{z}$$

$$= \int_U \frac{1}{A}\int_{-\infty}^{\infty}\big(G(\hat{z}) - \text{cdf}^Z(\hat{z})\big)\,\delta(\hat{z} - Z(\mathbf{u}))d\hat{z}\,Z_t d\mathbf{u}$$

$$= \int_U \nabla_Z E_{\text{cdf}}\,Z_t d\mathbf{u},$$

where δ is the Dirac delta function and the gradient of (16) with respect to Z is

$$\nabla_Z E_{\text{cdf}}(Z(\mathbf{u})) = \frac{1}{A}\big(G(Z(\mathbf{u})) - \text{cdf}^Z(Z(\mathbf{u}))\big). \qquad (17)$$

As a result of the statistical penalty, a new non-linear term of the form (17) appears in the EL equation (11), while the boundary condition remains unchanged. It is as if the nonlinear term (15) in the PDEs (11) was replaced by $g(Z) \leftarrow g(Z) + \gamma\nabla_Z E_{\text{cdf}}(Z)$. Multigrid methods are still suitable to efficiently solve the new non-linear PDE. However, the time-stepping smoother requires an additional constraint on the time step: the maximum height increment must be of the order of the bin size used to estimate the experimental CDF so that each iteration does not drastically change the CDF of the surface height.

Another reasonable energy to measure the statistical discrepancy between the empirical distribution of the wave field and the one dictated by the physical model is the L^2 difference between probability density functions (PDFs):

$$E_{\text{pdf}}(Z) = \int_{-\infty}^{\infty} \tfrac{1}{2}\big(\dot{G}(\hat{z}) - \text{pdf}^Z(\hat{z})\big)^2 \mathrm{d}\hat{z}, \qquad (18)$$

where $\dot{G}(Z)$ is the target PDF that we wish to enforce. Following similar steps as before, the EL equation (11) would have instead an extra term of the form

$$\nabla_Z E_{\text{pdf}}(Z(\mathbf{u})) = -\frac{1}{A}\frac{\mathrm{d}}{\mathrm{d}Z}\big(\dot{G}(Z(\mathbf{u})) - \text{pdf}^Z(Z(\mathbf{u}))\big). \qquad (19)$$

Enforcing the statistical constraint via the L^2 difference of characteristic functions (i.e. the Fourier transform of the PDFs) is, by Parseval's theorem, equivalent to the above PDF approach.

Theoretical probabilistic models that can be used as target physical wave height distributions are presented in [15,16]. These models are quasi-Gaussian distributions that capture the asymmetry present in real life water waves, which have steep crests and shallow troughs.

4   Applications

Experiment 1. Images of "Canale della Giudecca" in Venice (Italy). Figs. 1 and 2 show an example of a reconstructed water surface from images of the Venice Canal. Cropped images in Fig. 1 are of size 600×450 pixels and show the region of interest to be reconstructed. Fig. 1 also displays the modeled images created by the generative model within our variational method. The data fidelity term compares the intensities of the original and modeled images in the highlighted region. As observed, the modeled image is a good match of the original image. Fig. 2 shows the converged values of the unknowns of the problem: the height and the radiance of the surface, as well as the 3-D representation of the reconstructed surface obtained by combining both 2-D functions. In this experiment, the values of the weights of the regularizers were empirically determined: $\alpha = 0.035$ and

Fig. 1. Left: projection on image 1 of the boundary of the estimated graph, which has been discretized by a grid of 129×513 points. Center: modeled image (computed form surface height and radiance) superimposed on original image 1. Right: modeled image 2 superimposed on original image 2.

Fig. 2. Form left to right: (1) estimated height function $Z(u, v)$ (shape of the water surface) in pseudo-color; (2) height represented by greyscale intensities, from dark (low) to white (high); (3) estimated radiance function $f(u, v)$ (texture on the surface); (4) perspective, three-dimensional wire-frame representation of the estimated surface shape (height) according to grid points; (5) texture-mapped surface obtained by incorporating the radiance function in the wire-frame model. In (4) and (5) the vertical axis has been magnified by a factor of 5 with respect to the horizontal axes for visualization purpose.

$\beta = 0.01$. At the finest of the 5-level multigrid [3] algorithm, the gradient descent PDEs are discretized on a 2-D grid with 129×513 points. The distance between grid points is $h = 5$ cm. Therefore, the grid covers an area of $6.45 \times 25.65 \, m^2$. An example of a surface discretized at the finest grid level is also shown in Fig. 2. The high density of the surface representation is typical of variational methods. The step size h must be chosen so that it approximately matches the resolution in the images: a displacement of 1 pixel is observable at the finest grid level in the multigrid framework and it corresponds to a physical displacement of at least h. Due to perspective, the maximum value of h is determined by the grid points closest to the cameras.

Experiment 2. We apply our variational method, with and without statistical regularizer, to a pair of stereo images acquired at an off-shore platform in the Black Sea. Two cameras mounted 12 m above the mean sea level and with a

Fig. 3. Original image (left), modeled image superimposed on original image (center), error image (right).

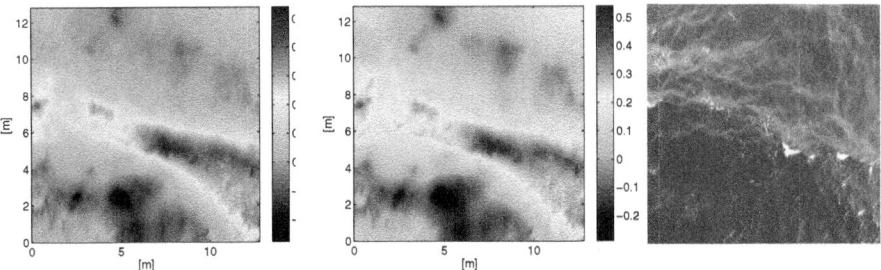

Fig. 4. Pseudo-colored $Z(u,v)$ without statistical regularizer (left). $Z(u,v)$ (center) and $f(u,v)$ (right) with statistical regularizer.

baseline of 2.5 meters acquire images of size 1624×1236 pixels. Fig. 3 (left) shows a sample image from one of the cameras. A grid with 513×513 points and resolution $h = 2.5$ cm, covering an area of $13 \times 13\,m^2$, is used to discretize the graph of the surface. Roughly, 1 image pixel corresponds to a physical displacement of 1.06 cm (1.88 cm) for grid points near (resp. far from) the cameras. Both displacements are of the same order as h. A 6-level full multigrid method [3] with 400 iterations per level, 2 V-cycles per iteration, and 1 pre- and post-relaxation sweeps per cycle, is performed on the linearized optimality PDEs to reach a local solution. The weights of the regularizers used are: $\alpha = 0.1$ and $\beta = 0.025$. Fig. 4 (left) shows the converged height function of the reconstructed surface without imposing a weak statistical constraint, i.e., $\gamma = 0$. Fig. 5 shows the corresponding observed PDF using normalized height $\xi = (Z - \mu_Z)/\sigma_Z$ (zero mean and unit variance). Note the deviations from Gaussianity with large kurtosis. Further, the associated omni-directional spectrum $S(k)$ is also shown in Fig. 5 (dashed line). In a polar-reference frame, $S(k)$ is computed from the two-dimensional power spectrum Ψ of the wave surface Z as $S(k) = \int_0^{2\pi} \Psi(k,\theta)k\,d\theta$, where k is the wavenumber and θ is the angle. According to the wave turbulence theory of Zakharov [19], the spectrum tail initially decays as $k^{-2.5}$ as a result of an energy cascade from large to small scales up to ~ 10 rad/m and then switch to a k^{-3} equilibrium range [11].

Next, 200 V-cycles of multigrid are carried out using the energy augmented by (16), γE_{cdf} with $\gamma/A = 10^{-2}$, to drive the surface toward the target

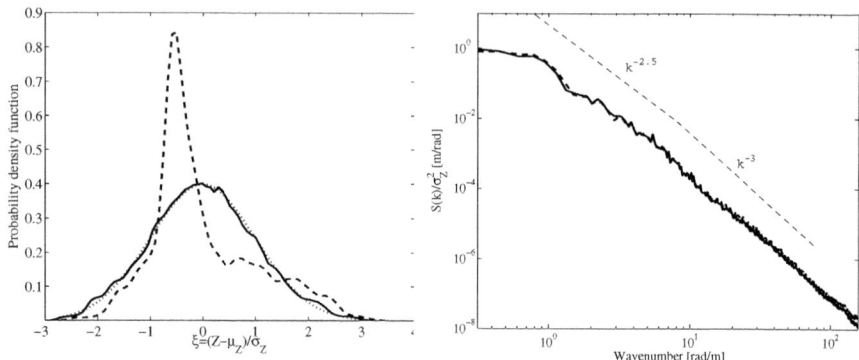

Fig. 5. Left: Observed PDF of the reconstructed wave surface Z with (solid line) and without (dash line) statistical regularization. The Gaussian distribution is plotted for comparison (dotted line). Right: Observed omni-directional spectrum $S(k)$ of the reconstructed surface Z with (solid line) and without (dash line) statistical regularizer.

distribution: Gaussian for simplicity, although other distributions could have been used [15,16]. The converged height and radiance functions are shown in Fig. 4. Both, Z and f generate the modeled image in Fig. 3. The absolute error image with respect to the input image is also displayed. There are subtle differences between height functions with and without the statistical constraint. Both solutions correctly capture the (almost breaking) wave front moving toward the camera. Now, two non-linear terms (photometric fidelity and statistical constraint) compete to evolve the surface. The regions that change the most due to the statistical regularizer are those with smooth texture, corresponding to small photometric error. The statistical regularizer leaves the photometric error and omni-directional spectrum (Fig. 5, right) almost unchanged while significantly modifying the PDF of the height map. The new reconstructed surface is quasi-Gaussian as clearly shown in Fig. 5.

5   Conclusion

Variational stereo is more powerful, flexible, and rigorous, albeit computationally expensive, than earlier traditional, image-based stereo methods. Therefore, we follow this research path by developing a variational stereo method for the case of smooth surfaces representable in the form of a graph supporting a smooth radiance function. Moreover, we show how global properties of ocean waves, such as statistical distributions, can be incorporated in the variational stereo reconstruction framework via a weak constraint. We successfully apply this method in two experiments to reconstruct a small region of the surface of the ocean. The variational stereo method developed can be naturally extended in several ways to process sequences of stereo images to generate a coherent space-time reconstruction of ocean waves. In future research we plan to investigate new energy terms to incorporate more global and/or local properties of the dynamics of ocean waves such as the wave equation, etc.

References

1. Alvarez, L., Deriche, R., Sánchez, J., Weickert, J.: Dense disparity map estimation respecting image discontinuities: A pde and scale-space based approach. Journal of Visual Communication and Image Representation 13, 3–21 (2002)
2. Benetazzo, A.: Measurements of short water waves using stereo matched image sequences. Coastal Engineering 53(12), 1013–1032 (2006)
3. Briggs, W.L., Henson, V.E., McCormick, S.F.: A Multigrid Tutorial, 2nd edn. SIAM, Philadelphia (2000)
4. Faugeras, O.D., Keriven, R.: Variational principles, surface evolution, pdes, level set methods, and the stereo problem. IEEE Trans. Image Proc. 7(3), 336–344 (1998)
5. Gallego, G., Benetazzo, A., Yezzi, A., Fedele, F.: Wave statistics and spectra via a variational wave acquisition stereo system. In: OMAE (2008)
6. Hartley, R.I., Zisserman, A.: Multiple View Geometry in Computer Vision, 2nd edn. Cambridge University Press, Cambridge (2004)
7. Holland, K., Holman, R., Lippmann, T., Stanley, J., Plant, N.: Practical use of video imagery in nearshore oceanographic field studies. IEEE Journal of Oceanic Engineering 22(1), 81–92 (1997)
8. Jin, H.: Variational methods for shape reconstruction in computer vision. Ph.D. thesis, Washington University, St. Louis, MO, USA (2003), director: Soatto, S
9. Longuet-Higgins, M.S.: The effect of non-linearities on statistical distributions in the theory of sea waves. Journal of Fluid Mechanics 17, 459–480 (1963)
10. Ma, Y., Soatto, S., Kosecka, J., Sastry, S.: An Invitation to 3D Vision: From Images to Geometric Models. Springer, Heidelberg (2003)
11. Phillips, O.: The equilibrium range in the spectrum of wind-generated waves. Journal of Fluid Mechanics 4(4), 426–434 (1958)
12. Schumacher, A.: Stereophotogrammetrische wellenaufnahmen. wiss ergeb. dtsch. atlant. exped. forschungs vermessung. meteor 1925-1927. Ozeanographische Sonderuntersuchungen, Erste Lieferung (1939)
13. Seitz, S.M., Curless, B., Diebel, J., Scharstein, D., Szeliski, R.: A comparison and evaluation of multi-view stereo reconstruction algorithms. In: CVPR, vol. 1, pp. 519–528 (2006)
14. Shemdin, O., Tran, H.: Measuring short surface waves with stereography. Photogrammetric Engineering and Remote Sensing 58(311-316) (1992)
15. Socquet-Juglard, H., Dysthe, K., Trulsen, K., Krogstad, H.E., Liu, J.: Probability distributions of surface gravity waves during spectral changes. Journal of Fluid Mechanics 542, 195–216 (2005)
16. Tayfun, A., Fedele, F.: Wave height distributions and nonlinear effects. Ocean Engineering 34(11-12), 1631–1649 (2007)
17. Wanek, J.M., Wu, C.H.: Automated trinocular stereo imaging system for three- dimensional surface wave measurements. Ocean Engineering 33(5-6), 723–747 (2006)
18. Yezzi, A., Soatto, S.: Stereoscopic segmentation. International Journal of Computer Vision 53(1), 31–43 (2003)
19. Zakharov, V.E.: Statistical theory of gravity and capillary waves on the surface of a finite-depth fluid. Eur. J. Mech. B - Fluids 18(3), 327–344 (1999)

Novel Schemes for Hyperbolic PDEs Using Osmosis Filters from Visual Computing

Kai Hagenburg, Michael Breuß, Joachim Weickert, and Oliver Vogel

Mathematical Image Analysis Group,
Faculty of Mathematics and Computer Science,
Building E1.1, Saarland University, 66041 Saarbrücken, Germany
{hagenburg,breuss,weickert,vogel}@mia.uni-saarland.de
http://www.mia.uni-saarland.de

Abstract. Recently a new class of generalised diffusion filters called osmosis filters has been proposed. Osmosis models are useful for a variety of tasks in visual computing. In this paper, we show that these filters are also beneficial outside image processing and computer graphics: We exploit their use for the construction of better numerical schemes for hyperbolic partial differential equations that model physical transport phenomena.

Our novel osmosis-based algorithm is constructed as a two-step, predictor-corrector method. The predictor scheme is given by a Markov chain model of osmosis that captures the hyperbolic transport in its advection term. By design, it also incorporates a discrete diffusion process. The corresponding terms can easily be identified within the osmosis model. In the corrector step, we subtract a stabilised version of this discrete diffusion. We show that the resulting osmosis-based method gives correct, highly accurate resolutions of shock wave fronts in both linear and nonlinear test cases. Our work is an example for the usefulness of visual computing ideas in numerical analysis.

Keywords: diffusion filtering, osmosis, diffusion-advection, drift-diffusion, hyperbolic conservation laws, finite difference methods, predictor-corrector schemes, stabilised inverse diffusion.

1 Introduction

Hyperbolic differential equations (HDEs) model physical wave propagation and transport processes. An important feature of solutions to such partial differential equations (PDEs) is the formation of discontinuities, also called shocks. In image processing shocks correspond to edges. Therefore, it seems natural that concepts from the numerical approximation of HDEs can be useful for constructing discrete filters that deal with the sharpening or evolution of edges. Rudin and Osher [1, 2] have exploited this idea to define edge-enhancing processes. They use the same mechanism as in HDEs to model so-called shock filters. When dealing with noisy images, one often aims at preserving or enhancing edges, while in homogeneous image regions a smoothing should take place. Corresponding to this idea, combinations of shock filters with mean curvature motion [3] or with nonlinear diffusion [4] have been developed. Also, the concept

A.M. Bruckstein et al. (Eds.): SSVM 2011, LNCS 6667, pp. 532–543, 2012.

of stabilised inverse diffusion (SID) has inspired interesting developments, both in a linear [5, 6] and a nonlinear setting [7–9]. In particular, concepts from the numerics of HDEs such as suitable combinations of one-sided differences have been applied to stabilise discretisations of inverse diffusion [5, 9]. Similar ideas from the numerics of HDEs dealing with an improved shock resolution have also been used for optical flow computations [10].

While the influence of ideas from the numerics of HDEs on the field of image processing is undeniable, up to now there are not many works that use techniques from image analysis for improving numerical methods for HDEs. In [11–13] higher order discretisations of HDEs that give a sharp shock resolution but suffer from oscillations are combined with anisotropic diffusion filtering. There, anisotropic diffusion is used to smooth oscillations without destroying the shocks. As an alternative procedure, one may employ a classic first-order scheme featuring diffusive errors to capture the hyperbolic transport. Then, in a second step, the artificial blurring can be removed by linear or nonlinear SID. This methodology is actually older than the SID-approach in image processing, and it is called flux-corrected transport (FCT) [14]. Modern variations of it have been developed for applications in image processing [15–17] and the numerics of HDEs [18].

Our Contribution. The discussion above shows that so far only diffusion or inverse diffusion processes have been used to correct numerical errors in schemes for HDEs. The goal of the present paper is to propose a novel construction of predictor-corrector schemes for HDEs that introduces a different mechanism. To this end, we make use of the recently introduced class of osmosis filters for visual computing problems [19]. They can be regarded as nonsymmetric generalisation of diffusion filters that involve a hyperbolic advection term which allows numerous applications beyond classic diffusion filtering. In contrast to all previous works, we do not correct the numerical errors of a classic HDE scheme by a diffusion filter, but we employ the hyperbolic term of the osmosis process for predicting the hyperbolic transport in the HDE. The Markov chain model corresponding to osmosis filters also includes a diffusion component. In the context of HDEs, this is a reasonable feature, since it is well-known that numerical schemes must incorporate a diffusive mechanism to approximate nonlinear shocks at the correct position, cf. [20]. However, since this diffusion also blurs shocks, we supplement in a corrector step SID to counter this undesired diffusion. As a benefit of the osmosis model, we can do this in a straight forward fashion on a completely discrete basis; see [16] for a similar use of this technique. In linear and nonlinear test cases, we compare our method to a classic second-order MUSCL-Hancock scheme [21, 22] which gives typical results for solvers in the field of HDEs. However, while the MUSCL-Hancock scheme has a similar predictor-corrector format as our proposed method, our approach is substantially easier to implement and much more efficient. We confirm that our osmosis-based algorithm is not only competitive in quality to the MUSCL-Hancock scheme, it even gives much sharper approximations at shocks.

Paper Organisation. In Section 2, we briefly review diffusion filtering and its generalisation to osmosis filters. Then we show in Section 3 how to use osmosis models to design novel predictor-corrector schemes for a fundamental class of HDEs, namely hyperbolic conservation laws. In Section 4, we present numerical experiments. The paper is finished with a conclusion in Section 5.

2 Diffusion Filters and Osmosis

Diffusion filters. Let a continuous-scale 1-D signal $u(x, t)$ be given where we associate x and t with space and time. The diffusion PDE with positive diffusivity function $g(x, t)$ reads in 1D as

$$\partial_t u = \partial_x (g \, \partial_x u). \qquad (1)$$

It has to be supplemented with an initial condition $u(x, 0) := f(x)$, and in case of a bounded domain also with boundary conditions.

In a discrete setting, we use a spatial mesh width h and define the pixel location x_i by $x_i := (i - 1/2)h$ for $i \in \{1, \dots, N\}$. Analogously, we introduce a time discretisation $t_k = k\tau$, so that we obtain a discrete signal $u_i^k \approx u(x_i, t_k)$. Then a standard finite difference discretisation of (1) is given by the explicit scheme

$$\frac{u_i^{k+1} - u_i^k}{\tau} = \frac{1}{h} \left(g_{i+1/2}^k \frac{u_{i+1}^k - u_i^k}{h} - g_{i-1/2}^k \frac{u_i^k - u_{i-1}^k}{h} \right) \qquad (2)$$

where $g_{i+1/2}^k$ denotes the diffusivity between the computational cells i and $i + 1$. Using the mesh ratio $r := \frac{\tau}{h^2}$, our scheme can be rewritten as

$$u_i^{k+1} = u_i^k - rg_{i+1/2}^k u_i^k - rg_{i-1/2}^k u_i^k + rg_{i+1/2}^k u_{i+1}^k + rg_{i-1/2}^k u_{i-1}^k. \qquad (3)$$

It is convenient to express this as a matrix-vector multiplication of the form $\boldsymbol{u}^{k+1} = \boldsymbol{Q}^k \boldsymbol{u}^k$, where $\boldsymbol{Q}^k$ is an $(N \times N)$-matrix with entries

$$q_{i,j}^k := \begin{cases} 1 - rg_{i-1/2}^k - rg_{i+1/2}^k & (j = i) \\ rg_{i-1/2}^k & (j = i - 1) \\ rg_{i+1/2}^k & (j = i + 1) \\ 0 & (\text{else}). \end{cases} \qquad (4)$$

Let us briefly review some important properties of the matrix $\boldsymbol{Q}^k$; cf. [23]. Obviously, the matrix is symmetric. Stability of the iterative scheme (3) can be shown if the entries of $\boldsymbol{Q}^k$ are nonnegative. Since the diffusivity is positive, all off-diagonals contain nonnegative entries, leaving only the diagonal entries without proper clarification. Therefore, for all diagonal entries it must hold that

$$q_{i,i}^k = 1 - rg_{i-1/2}^k - rg_{i+1/2}^k \geq 0. \qquad (5)$$

This implies a stability condition on the time step size τ.

In order to implement homogeneous Neumann boundary conditions $\partial_x u = 0$, we modify the entries for $q_{1,1}^k$ and $q_{N,N}^k$ such that

$$q_{1,1}^k := 1 - rg_{3/2}^k \quad \text{and} \quad q_{N,N}^k := 1 - rg_{N-1/2}^k. \qquad (6)$$

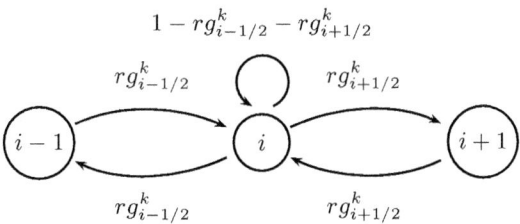

$$1 - rg^k_{i-1/2} - rg^k_{i+1/2}$$

$$rg^k_{i-1/2} \qquad rg^k_{i+1/2}$$

$$i-1 \qquad i \qquad i+1$$

$$rg^k_{i-1/2} \qquad rg^k_{i+1/2}$$

Fig. 1. Diffusion process visualised in terms of a Markov chain model

This can be interpreted as setting the missing terms $g^k_{1/2}$ and $g^k_{N+1/2}$ to 0. It should be mentioned that it is also possible to implement Dirichlet boundary conditions or periodic boundary conditions.

Furthermore, it holds that the sums over all entries in each column of Q^k equal 1. By the symmetry of Q^k this also holds for the row sums. Both properties have an effect on the evolution of the process: The unit column sums imply the preservation of the average grey value. With the unit row sums it is possible to prove a discrete maximum-minimum principle. Moreover, in [23] it is shown that the evolution converges to a constant steady state that is identical to the average grey value of the initial signal. Let us stress that the properties of the discrete minimum-maximum-principle and the trivial steady state solution are consequences of the symmetry of Q^k which implies that unit column sums are equivalent to unit row sums.

We can also express diffusion using Markov chains. Markov chains are described in terms of stochastic matrices that incorporate transition probabilities [24]. A stochastic matrix is a matrix with only nonnegative entries and unit column sums. By taking into account the positivity of the diffusivity and choosing a mesh ratio r such that (5) is satisfied for all i, we can ensure that the matrix Q^k contains only nonnegative entries. Moreover, all column sums are 1. Thus, Q^k is a stochastic matrix, and the entries $q^k_{i,j} \geq 0$ can be interpreted as transition probabilities. In the Markov chain setting it is convenient to use a graph-based representation of the diffusion model. It is given in Figure 1.

Osmosis as a Generalisation of Diffusion Filters. Following [19] let us now consider a nonsymmetric extension of diffusion that is called osmosis. To this end, we assume that we have semi-permeable membranes between adjacent pixels. An osmosis process permits selective transport of particles such that the transition probabilities may be different, depending on the orientation. For example, the transition probability from pixel i to pixel $i + 1$ may differ from the transition probability from pixel $i + 1$ to pixel i. In the Markov model, this leads to the loss of the symmetry in the graph in Figure 1. This is achieved by allowing different diffusivities in different orientation. Such oriented diffusivities are called *osmoticities*. The forward osmoticity from pixel i to $i + 1$ at time level k is denoted by $g^{+,k}_{i+1/2}$, while $g^{-,k}_{i+1/2}$ is the backward osmoticity from pixel $i + 1$ to i. We choose these osmoticities such that the normalisation condition

$$g^{+,k}_{i+1/2} + g^{-,k}_{i+1/2} = 2 \qquad (7)$$

is fulfilled for all i; cf. [19]. Since osmoticities are also supposed to be nonnegative, we conclude that in this case their range is in $[0, 2]$.

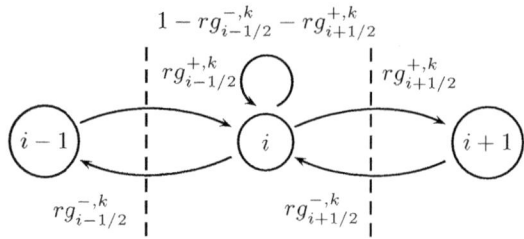

Fig. 2. Osmosis process visualised in terms of a Markov chain model

In Figure 2 we see a graph-based representation of osmosis. This new process is expressed by the scheme

$$u_i^{k+1} = u_i^k \underbrace{- rg_{i+1/2}^{+,k} u_i^k - rg_{i-1/2}^{-,k} u_i^k}_{\text{"outflow"}} \underbrace{+ rg_{i+1/2}^{-,k} u_{i+1}^k + rg_{i-1/2}^{+,k} u_{i-1}^k}_{\text{"inflow"}} \quad (8)$$

This can be rewritten in matrix-vector notation $u^{k+1} = P^k u^k$ with a matrix $P^k :=$ $(p_{i,j}^k)$ with

$$p_{i,j}^k := \begin{cases} 1 - rg_{i-1/2}^{-,k} - rg_{i+1/2}^{+,k} & (j = i) \\ rg_{i-1/2}^{+,k} & (j = i - 1) \\ rg_{i+1/2}^{-,k} & (j = i + 1) \\ 0 & (\text{else}). \end{cases} \quad (9)$$

Homogeneous Neumann boundary conditions are implemented by setting the osmoticities in the boundary locations $x_{1/2}$ and $x_{N+1/2}$ to 0.

Let us comment on the structure of P^k. As in the case with Q^k, the system matrix (9) is a stochastic matrix if r is chosen such that the diagonal entries of P^k are nonnegative. Since P^k has unit column sums, it follows that osmosis preserves the average grey value:

$$\frac{1}{N} \sum_{i=1}^N u_i^{k+1} = \frac{1}{N} \sum_{i=1}^N \sum_{j=1}^N p_{i,j}^k u_j^k = \frac{1}{N} \sum_{j=1}^N \underbrace{\left(\sum_{i=1}^N p_{i,j}^k \right)}_{=1} u_j^k = \frac{1}{N} \sum_{j=1}^N u_j^k. \quad (10)$$

However, P^k is not symmetric. Thus, unit row sums cannot be guaranteed. As a consequence, a discrete maximum-minimum principle does not hold, but the nonnegativity of P^k still implies that a nonnegative initial signal remains nonnegative after filtering. More importantly, the lack of symmetry allows that osmosis can lead to nontrivial steady states. This interesting property is analysed in detail in [19], where it is also exploited for many applications.

As proven in [19], the scheme (8) with normalisation condition (7) approximates on a fixed, given mesh of size h the *1-D osmosis PDE*

$$\partial_t u + \partial_x \left(\frac{g^+ - g^-}{h} u \right) = \partial_{xx} u \quad (11)$$

Fig. 3. Seamless image cloning with osmosis (with permission from [19]). From left to right: **(a)** Original painting of Euler. **(b)** Original drawing of Lagrange (with to-be-cloned face selected). **(c)** Direct cloning on top of Euler's head. **(d)** Cloning with osmosis image editing. See [19] for more details.

where g^+ and g^- are continuous-scale representations of the osmoticities. PDEs of this type are called *advection-diffusion equations* or *drift-diffusion equations*.

It is straight forward to extend osmosis to higher dimensions and colour images; see [19] for details. In [19] it is also shown that osmosis constitutes a versatile framework for many visual computing problems such as clustering, data integration, focus fusion, exposure blending, image editing, shadow removal, and compact image representation. Fig. 3 illustrates this. Let us now explore a new application field for osmosis that goes beyond visual computing tasks: the construction of better numerical schemes for hyperbolic conservation laws.

3   Osmosis Schemes for HDEs

Hyperbolic conservation laws. We aim at constructing numerical approximations of HDEs that can be written as

$$\partial_t u \,+\, \partial_x(\phi(u)) \,=\, 0. \tag{12}$$

Such equations are called *hyperbolic conservation laws (HCLs)*. This is a fundamental class of PDEs with many applications in science and engineering [25]. The design of numerical schemes for HCLs can easily be transferred to other specific HDEs. The function ϕ in (12) is called *flux function*. Its properties, like e.g. linearity or convexity, are important for the features one can expect from solutions of such PDEs. We will write ϕ in the format of a velocity times the underlying density function, i.e. $\phi(u) = au$, where $a := a(u)$ may be nonlinear. This is a very basic choice in the field of HCLs, naturally arising in many settings [25].

Comparing the differential formula for osmosis (11) with the general form of HCLs (12), one can immediately identify the flux $\phi(u)$ and the corresponding flux within the osmosis advection term

$$\phi(u) \,=\, \frac{g^+ - g^-}{h}\, u. \tag{13}$$

In addition, there is the diffusion term $\partial_{xx} u$. The general idea we pursue in the following is to determine useful expressions for g^+ and g^-, so that we can capture the hyperbolic transport by the osmosis model.

Selection of the Osmoticities. For the general construction of osmosis-based algorithms, we stick for simplicity to the 1-D situation. The methodology can be extended to the 2-D case in a straight forward fashion.

In order to approximate the flux $\phi(u) = a(u)u$ of the hyperbolic transport contained in (11), we choose as osmoticities

$$g_{i+1/2}^{+,k} := 1 + \frac{h\, a_{i+1/2}^k}{2} \qquad \text{and} \qquad g_{i+1/2}^{-,k} := 1 - \frac{h\, a_{i+1/2}^k}{2} \tag{14}$$

with velocities $a_{i+1/2}^k$ defined at pixel borders. This setting makes the osmotic transport identical to the desired format $a(u)u$. Let us discuss two examples.

- *Example 1: Osmoticities for linear advection.*
  The linear advection equation

$$\partial_t u + \alpha\, \partial_x u = 0 \tag{15}$$

is a standard example of HDEs, defined via $\phi(u) := \alpha u$ with $\alpha \in \mathbb{R}$. In order to approximate (15), we set all velocities $a_{i+1/2}^k$ to the same value α.

- *Example 2: Osmoticities for Burgers' equation.*
  Burgers' equation is a classic test case for nonlinear HDEs:

$$\partial_t u + \partial_x \left(\frac{1}{2} u^2 \right) = 0, \qquad \text{i.e. } \phi(u) = \frac{1}{2} u^2. \tag{16}$$

Rewriting the flux in the format $\phi(u) = a(u)\, u$ leads to the discrete expression

$$a_{i+1/2}^k = a(u_i^k, u_{i+1}^k) := \frac{1}{2} \frac{u_i^k + u_{i+1}^k}{2} \tag{17}$$

after approximating the density $u_{i+1/2}^k$ at the border between pixels i and $i + 1$ by averaging.

Subtracting the Diffusion. Our osmosis scheme contains the diffusive term $\partial_{xx} u$ which leads to an additional smoothing of the signal. In order to compensate for this effect, we apply a method similar to the fully discrete SID step in [17].

If we use our definitions of $g_{i\pm1/2}^{\pm}$ from (14) within the osmosis filter (8) and carry out further computations, we obtain

$$\tilde{u}_i^k = u_i^k - \underbrace{\frac{\tau}{h} \left(a_{i+1/2}^k \frac{u_{i+1}^k + u_i^k}{2} - a_{i-1/2}^k \frac{u_i^k + u_{i-1}^k}{2} \right)}_{(A)}$$

$$+ \underbrace{r \left(u_{i+1}^k - 2u_i^k + u_{i-1}^k \right)}_{(B)}. \tag{18}$$

The term (A) corresponds to the update formula of an explicit scheme for discretising the hyperbolic transport, while (B) is a discretisation of a time step performed with linear diffusion. It should be noted that (18) varies from the standard Lax-Friedrichs scheme by controlling the diffusive part (B) with the same time step size τ as the transport term (A), see also [25–27].

Let us now subtract the effect of the latter by performing a SID step in the same style as in [16, 17]. This gives the total, corrected result

$$u_i^{k+1} := \tilde{u}_i^k - c_{i+1/2}^k + c_{i-1/2}^k \tag{19}$$

where $c_{i\pm1/2}^k$ denote the fluxes of the stabilised inverse diffusion:

$$c_{i+1/2}^k := \text{minmod}\left(\tilde{u}_i^k - \tilde{u}_{i-1}^k, \eta_{i+1/2}^k\left(\tilde{u}_{i+1}^k - \tilde{u}_i^k\right), \tilde{u}_{i+2}^k - \tilde{u}_{i+1}^k\right) \tag{20}$$

with the minmod function

$$\text{minmod}(a,b,c) := \begin{cases} \max(a,b,c) & \text{if } a > 0 \text{ and } b > 0 \text{ and } c > 0 \\ \min(a,b,c) & \text{if } a < 0 \text{ and } b < 0 \text{ and } c < 0 \\ 0 & \text{else.} \end{cases} \tag{21}$$

Thereby, $\eta_{i+1/2}^k := r$ is the antidiffusion coefficient, as identified in (B). The other arguments of the minmod function serve as stabilisers.

The Complete Algorithm. Now we can summarise our method in a nutshell.

Osmosis-based Method for Approximating $\partial_t u + \partial_x(\phi(u)) = 0$.
Step 1: Determine the velocity function a for a given flux function
   $\phi(u) = a(u)u$.
Step 2: Compute the osmoticities according to (14).
Step 3: Perform one predictor step by applying the osmosis scheme (18).
Step 4: Perform the corrector step (19).
Step 5: Repeat steps 2 to 4 until the stopping time is reached.

---

4   Numerical Experiments

We illustrate the quality of our osmosis-based algorithm with several standard examples from the field of HDEs. Thereby, we focus our attention on the shocks that are the most interesting features of hyperbolic PDEs.

For comparison with standard methods for HDEs, we employ a second-order high-resolution MUSCL-Hancock method [21, 22]. This classic method gives typical results for high-resolution solvers in this field.

Linear Advection in 1D. In our first experiment we consider the linear advection equation (15) with $\alpha = 1$ and periodic boundary conditions. We apply it for transporting a box-like initial signal

$$f(x) := \begin{cases} 1 & (10 \leq x < 30) \\ 0 & (\text{else}). \end{cases} \tag{22}$$

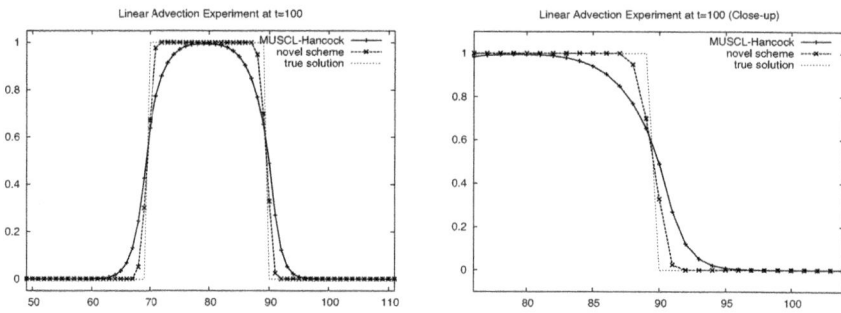

Fig. 4. Linear advection experiment. (a) **Left:** Results at $t = 60$. (b) **Right:** Close-up on the right edge of the signal.

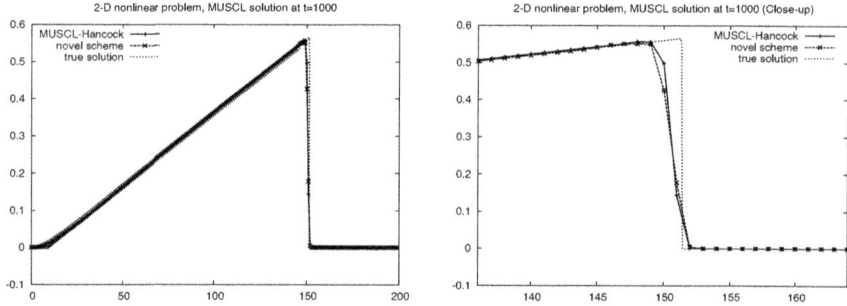

Fig. 5. Burgers' Equation. (a) **Left:** Results at $t = 250$. (b) **Right:** Close-up on the right edge of the signal.

As numerical parameters we choose $N := 200$, $h := 1$, and $\tau := 0.25$. In Figure 4 we show a snapshot taken after 240 time steps of numerical solutions computed by our new scheme and the reference method, together with the exact solution. We observe that our osmosis method gives much sharper discontinuities than the MUSCL-Hancock scheme and comes closer to the exact solution.

Nonlinear Burgers' Equation in 1D. Now we consider the Burgers' equation (16) under the same parameter settings and the same initial condition as in the first test. By the nonlinear evolution, the box signal is shifted to the right. The discontinuity at the right hand side of the box travels as a shock while the rest of the signal is gradually shifted, transforming the box into a ramp. Figure 5 shows the numerical solutions at $t = 250$ for our osmosis-based scheme as well as for the MUSCL-Hancock implementation, together with the exact solution. Both methods give reasonable approximations in this test case.

Nonlinear 2D Experiment. As already mentioned, extending osmosis to 2D is straight forward: One only has to define osmoticities as proposed in (14) for x- and y-direction. Note that our resulting scheme is rotationally invariant w.r.t. the diffusion part, since this is given in 2D by the isotropic Laplace operator [19]. The 2-D MUSCL-Hancock

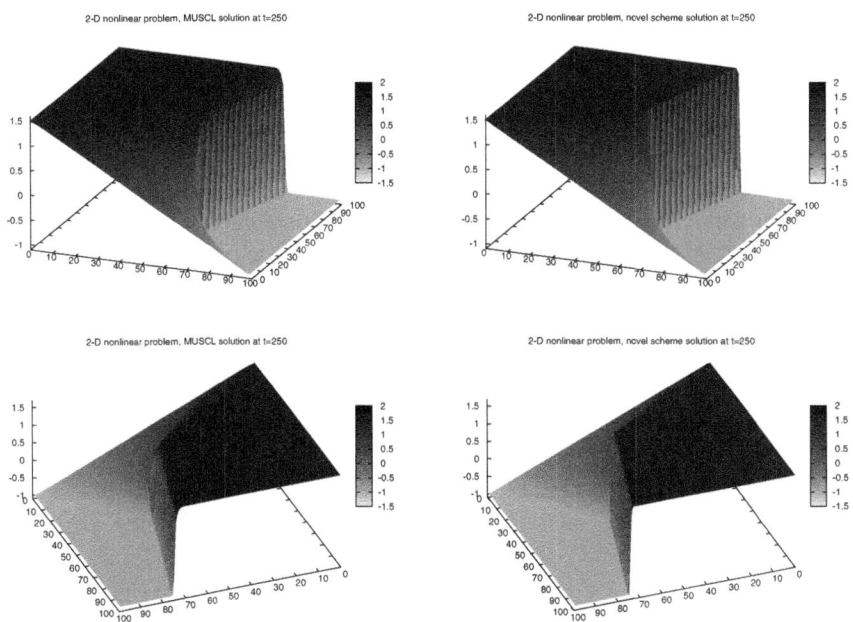

Fig. 6. Steady-state result for the 2-D test. **Left:** MUSCL-Hancock scheme. **Right:** Osmosis scheme. **Top row:** Top-down view. **Bottom row:** Different angle, showing the shock region in detail.

scheme is presumably comparable in this respect, as it uses information from a diamond-shaped stencil of 13 nodes [21, 22].

For our 2-D experiment we consider the nonlinear problem from [11] where the steady state is sought. It combines Burgers' equation with linear advection by choosing the flux function $\phi(u) = \frac{1}{2}u^2$ in x-direction, and $\psi(u) = u$ in y-direction. As initial state on our domain $[0, 100] \times [0, 100]$ we take

$$
f(x, y) := \begin{cases} 1.5 & (x = 1) \\ -2.5x + 1.5 & (y = 1) \\ -1 & (x = 100) \\ 0 & (\text{else}). \end{cases} \tag{23}
$$

These values also define non-zero Dirichlet boundary conditions on three borders of our domain. On the remaining border (at $y = 100$) we impose homogeneous Neumann boundary conditions. We implement the process in a straight forward way using the osmoticities for Burgers' equation and linear advection in x- and y-direction, respectively. The problem is discretised on a grid of size 100×100, and the numerical steady state obtained at $t = 250$ is depicted in Fig. 6. In the smooth regions, our method performs comparable to the MUSCL-Hancock scheme, but we obtain a much sharper shock resolution.

5   Conclusion

We have developed a novel class of schemes for approximating HCLs. They combine recently developed osmosis filters for resolving transport with a stabilised inverse diffusion step. We have shown the strength of our approach for resolving solutions with shocks, which are important features in the fields of hyperbolic differential equations.

Quite frequently, new results in visual computing benefit from the use of modern techniques from numerical analysis. Our work is an example for a fertilisation in the inverse direction. Note that the key for obtaining the results in our paper is the use of a very recent technique from visual computing. However, we do not only propose a novel construction of numerical schemes for HDEs, we also introduce a new application of osmosis filters. Therefore, this paper is an example for the useful interaction of visual computing ideas and numerical analysis. In our future work we will investigate if also other modern PDE-based methods from image analysis can be used with benefit in numerical analysis.

Acknowledgments. The authors gratefully acknowledge the funding given by the *Deutsche Forschungsgemeinschaft* (DFG), grant We2602/8-1.

References

1. Rudin, L.I.: Images, Numerical Analysis of Singularities and Shock Filters. PhD thesis, California Institute of Technology, Pasadena, CA (1987)
2. Osher, S., Rudin, L.I.: Feature-oriented image enhancement using shock filters. SIAM Journal on Numerical Analysis 27, 919–940 (1990)
3. Alvarez, L., Mazorra, L.: Signal and image restoration using shock filters and anisotropic diffusion. SIAM Journal on Numerical Analysis 31, 590–605 (1994)
4. Kornprobst, P., Deriche, R., Aubert, G.: Image coupling, restoration and enhancement via PDEs. In: Proc.1997 IEEE International Conference on Image Processing, Washington, DC, vol. 4, pp. 458–461 (October 1997)
5. Osher, S., Rudin, L.: Shocks and other nonlinear filtering applied to image processing. In: Tescher, A.G. (ed.) Applications of Digital Image Processing XIV. Proceedings of SPIE, vol. 1567, pp. 414–431. SPIE Press, Bellingham (1991)
6. Breuß, M., Welk, M.: Analysis of staircasing in semidiscrete stabilised inverse linear diffusion algorithms. Journal of Computational and Applied Mathematics 206, 520–533 (2007)
7. Pollak, I., Willsky, A.S., Krim, H.: Image segmentation and edge enhancement with stabilized inverse diffusion equations. IEEE Transactions on Image Processing 9(2), 256–266 (2000)
8. Gilboa, G., Sochen, N.A., Zeevi, Y.Y.: Forward-and-backward diffusion processes for adaptive image enhancement and denoising. IEEE Transactions on Image Processing 11(7), 689–703 (2002)
9. Welk, M., Gilboa, G., Weickert, J.: Theoretical foundations for discrete forward-and-backward diffusion filtering. In: Tai, X.-C., Mørken, K., Lysaker, M., Lie, K.-A. (eds.) SSVM 2009. LNCS, vol. 5567, pp. 527–538. Springer, Heidelberg (2009)
10. Breuß, M., Zimmer, H., Weickert, J.: Can variational models for correspondence problems benefit from upwind discretisations? Journal of Mathematical Imaging and Vision 39(5), 230–244 (2011)

11. Grahs, T., Meister, A., Sonar, T.: Image processing for numerical approximations of conservation laws: nonlinear anisotropic artificial dissipation. SIAM Journal on Scientific Computing 23(5), 1439–1455 (2002)
12. Grahs, T., Sonar, T.: Entropy-controlled artificial anisotropic diffusion for the numerical solution of conservation laws based on algorithms from image processing. Journal of Visual Communication and Image Representation 13(1/2), 176–194 (2002)
13. Wei, G.: Shock capturing by anisotropic diffusion oscillation reduction. Computer Physics Communications 144, 317–342 (2002)
14. Boris, J.P., Book, D.L.: Flux corrected transport. I. SHASTA, a fluid transport algorithm that works. Journal of Computational Physics 11(1), 38–69 (1973)
15. Burgeth, B., Pizarro, L., Breuß, M., Weickert, J.: Adaptive continuous-scale morphology for matrix fields. International Journal of Computer Vision 92(2), 146–161 (2011)
16. Breuß, M., Weickert, J.: A shock-capturing algorithm for the differential equations of dilation and erosion. Journal of Mathematical Imaging and Vision 25, 187–201 (2006)
17. Breuß, M., Weickert, J.: Highly accurate schemes for PDE-based morphology with general structuring elements. International Journal of Computer Vision 92(2), 132–145 (2011)
18. Breuß, M., Brox, T., Sonar, T., Weickert, J.: Stabilised nonlinear inverse diffusion for approximating hyperbolic PDEs. In: Kimmel, R., Sochen, N., Weickert, J. (eds.) Scale-Space 2005. LNCS, vol. 3459, pp. 536–547. Springer, Heidelberg (2005)
19. Weickert, J., Hagenburg, K., Vogel, O., Breuß, M., Ochs, P.: Osmosis models for visual computing. Technical report, Department of Mathematics, Saarland University, Saarbrücken, Germany (2011)
20. LeVeque, R.J.: Numerical Methods for Conservation Laws. Birkhäuser, Basel (1992)
21. van Leer, B.: Towards the ultimate conservative difference scheme, V. A second order sequel to Godunov's method. Journal of Computational Physics 32(1), 101–136 (1979)
22. Toro, E.F.: Riemann Solvers and Numerical Methods for Fluid Dynamics - A Practical Introduction, 2nd edn. Springer, Berlin (1999)
23. Weickert, J.: Anisotropic Diffusion in Image Processing. Teubner, Stuttgart (1998)
24. Seneta, E.: Non-negative Matrices and Markov Chains. Series in Statistics. Springer, Berlin (1980)
25. LeVeque, R.J.: Finite Volume Methods for Hyperbolic Problems. Cambridge University Press, Cambridge (2002)
26. Breuß, M.: The correct use of the Lax-Friedrichs method. ESAIM: Mathematical Modeling and Numerical Analysis 38(3), 519–540 (2004)
27. Breuß, M.: An analysis of the influence of data extrema on some first and second order central approximations of hyperbolic conservation laws. ESAIM: Mathematical Modeling and Numerical Analysis 39(5), 965–993 (2005)

Fast PDE-Based Image Analysis in Your Pocket

Andreas Luxenburger, Henning Zimmer, Pascal Gwosdek, and Joachim Weickert

Mathematical Image Analysis Group, Faculty of Mathematics and Computer Science,
Building E1 1, Saarland University, 66041 Saarbrücken, Germany
{luxenburger,zimmer,gwosdek,weickert}@mia.uni-saarland.de

Abstract. The increasing computing power of modern smartphones opens the door for interesting mobile image analysis applications. In this paper, we explore the arising possibilities but also discuss remaining challenges by implementing linear and nonlinear diffusion filters as well as basic variational optic flow approaches on a modern Android smartphone. To achieve low runtimes, we present a fast method for acquiring images from the built-in camera and focus on efficient solution strategies for the arising partial differential equations (PDEs): Linear diffusion is realised by approximating a Gaussian convolution by means of an iterated box filter. For nonlinear diffusion and optic flow estimation we use the recent *fast explicit diffusion (FED)* solver. Our experiments on a recent smartphone show that linear/nonlinear diffusion filters can be applied in realtime/near-realtime to images of size 176×144. Computing optic flow fields of a similar resolution requires some seconds, while achieving a reasonable quality.

1 Introduction

The prevailing problems in image analysis – such as solving partial differential equations (PDEs)– have widely been considered to be a challenging and computationally intensive task. If favourable results had to be computed in a reasonable time, researchers were forced to port their algorithms from desktop architectures to super-computers, which are difficult to work on, not to mention their immense costs.

In the near future, however, this trend could go in the completely opposite direction: On the algorithmic side, researchers spend more and more efforts on simple, yet efficient solution strategies. On the hardware side, the computing power of modern embedded systems such as smartphones is steadily increasing. Furthermore, powerful application development frameworks for standard programming languages ease the implementation on such platforms. Joining the ongoing work from the two mentioned research directions could thus allow to perform challenging image analysis tasks on small handheld devices that are already today in almost everybody's trouser pocket.

In order to prove the basic feasibility of this ambitious goal, the present paper shows some prototypical examples by implementing linear and nonlinear diffusion filters [1], as well as two variational optic flow approaches [2,3] on a recent smartphone (HTC Desire, 1 GHz) running the Android operating system. As users expect mobile image analysis applications to achieve interactive runtimes, we put an emphasis on efficient, but still simple to implement solvers for the occurring partial differential equations (PDEs). In the linear diffusion case, we analytically solve the PDE. This comes down to

A.M. Bruckstein et al. (Eds.): SSVM 2011, LNCS 6667, pp. 544–555, 2012.

a Gaussian convolution of the image, which is approximated by an iterated box filter [4] that achieves a realtime performance for camera images of size 240×160 pixels. In the nonlinear case where an analytical solution is not possible, we opt for an explicit solver that is speeded up by the recently proposed *fast explicit diffusion scheme* [5], resulting in a near-realtime performance. A similar explicit solver, however operating in a coarse-to-fine manner is used for optic flow estimation. Here, our implementation allows to compute flow fields on standard test sequences in the order of some seconds, while achieving a reasonable quality in terms of error measures.

Related Work. Several earlier works applied image analysis on smartphones for tasks like image enhancement and image-based applications. To our surprise there is no diffusion framework with interactive runtimes on smartphones, yet.

A general image processing framework for the Android platform including basic operations such as a box filter has been proposed by Wells [6]. A closed-source nonlinear diffusion filter is available for the iPhone [7], but it does not allow to tune any parameters and is rather slow (20 seconds for an image with 320×320 pixels). Another algorithm that shares principle properties with diffusion schemes is the coherence enhancing shock filter for the iPhone [8] that only needs about 3 seconds on a similar resolution. Recently, the OpenCV framework has been ported to Android devices [9]. It includes many filters and also computer vision methods, e.g. linear diffusion as well as a pyramidal Lucas and Kanade [10] and a Horn and Schunck optic flow algorithm [3]. Based on this framework, there is also an implementation of the combined local-global (CLG) optic flow method [2,11]. However, these approaches cannot provide the optimal runtime possible on mobile devices since OpenCV introduces an additional abstraction layer which was not optimised for particular platforms like Android. Furthermore, standard numerical solvers are used that additionally decrease the performance. Also sparse feature matching approaches based on SIFT or SURF features [12] have been considered on mobile platforms. While SIFT is too slow for interactive applications, SURF only takes about 3 ms per match [13], but does not give dense matches.

Image analysis techniques are also used in more complex applications. Before mobile phones were equipped with gyroscopic sensors, simple optic flow approaches allowed to detect the ego-motion of the phone [14], turning them into wireless pointing devices. Today, similar techniques are still of interest when a highly accurate estimation of the velocity or viewing direction is needed: Recently, a sparse feature matching algorithm was used to create freehand cylindrical panoramas on an Android phone [15]. Related techniques are also used in augmented reality applications [16] where a predefined pattern is recognised and tracked in a live stream from the camera. However, all these techniques are strongly restricted by the computational power of the device. Advanced algorithms are thus usually computed in the cloud, i.e. on remote servers [13].

Paper Organisation. In Sec. 2 we present the models and the solvers for diffusion filtering and optic flow estimation. Apart from basic Android development concepts, Sec. 3 discusses the image acquisition and further optimisations. Screenshots and a performance analysis of our application are presented in Sec. 4. We conclude in Sec. 5 by a summary and an outlook on future work.

2   Models and Solvers

2.1   Diffusion Filtering

We assume to be given a greyscale image $f(x, y) : \Omega \to \mathbb{R}$, where $(x, y)^\top \in \Omega$ denotes the location within the rectangular image domain $\Omega \subset \mathbb{R}^2$. Our goal is then to compute a gradually smoothed result $u(x, y, t) : \Omega \times [0, T] \to \mathbb{R}$, where $t \in [0, T]$ represents the evolution time of the filter, i.e. a larger evolution time leads to a stronger smoothing, and $u(x, y, 0) = f(x, y)$.

Homogeneous Diffusion. The most basic diffusion filter is a linear, homogeneous diffusion process [17] that computes the unknown u as the solution of the parabolic partial differential equation (PDE)

$$u_t = \text{div} \left(\nabla u \right) = \triangle u := u_{xx} + u_{yy} \; , \tag{1}$$

with reflecting boundary conditions. Colour images are treated channel-wise.

It is well-known that an analytical solution to (1) can be computed as $u(x, y, t) = (K_{\sqrt{2t}} * f)(x, y)$, where K_σ denotes a Gaussian of standard deviation σ, and $*$ is the convolution operator. Homogeneous diffusion filtering thus comes down to a Gaussian convolution of the given image. A straightforward way to implement this for discrete, digital images is to perform a discrete convolution with a sampled and truncated Gaussian. However, there are more efficient implementations, e.g. by a d-fold iterated box filter (IBF). This filter approximates the 2-D Gaussian kernel K_σ by a convolution

$$K_\sigma = \underbrace{B_L * B_L * \ldots * B_L}_{d\text{-times}} \quad \text{with} \quad B_L(x, y) := \begin{cases} \frac{1}{L^2}, & x, y \in [-\frac{L}{2}, \frac{L}{2}] \\ 0, & \text{else} \end{cases} , \tag{2}$$

with $L = 2\,l + 1$, and $l \in \mathbb{N}$ [4]. Each B_L can be applied consecutively and is separable in space. Moreover, its implementation requires only two additions per pixel and direction using a sliding-window algorithm: The solution $\tilde{v}_k$ at position k of a 1-D signal v is given by $\tilde{v}_k = \tilde{v}_{k-1} - v_{k-l-1} + v_{k+l}$. Thus, box filters are independent from the standard deviation of the kernel. However, their runtime depends on the number of iterations d, resulting in a trade-off between the approximation error and the runtime.

Nonlinear Isotropic Diffusion. The major problem of homogeneous diffusion is the blurring of semantically important image edges as the filter performs the same smoothing at each location. To overcome this problem, Perona and Malik [1] introduced a nonlinear diffusion process

$$u_t = \text{div} \left(g \left(|\nabla u|^2 \right) \nabla u \right) = \partial_x \left(g \left(|\nabla u|^2 \right) u_x \right) + \partial_y \left(g \left(|\nabla u|^2 \right) u_y \right) \; , \tag{3}$$

where the decreasing diffusivity function $g(|\nabla u|^2) := (1 + |\nabla u|^2/\lambda^2)^{-1}$ reduces the smoothing at evolving edges that are indicated by large values of $|\nabla u|^2$. The parameter λ serves as a contrast parameter. Catté *et al.* [18] later proposed a regularised version where the result u (that occurs in the argument of g) is presmoothed by a Gaussian convolution with standard deviation σ_p. This has several advantages, like reducing the staircasing artefacts and the sensitivity to noise. To process colour images, we apply the filter to each channel, but use a joint diffusivity function where we sum up the gradient magnitudes of each channel in the argument of g.

In the nonlinear case, no analytical solution exists which leaves us with computing an approximate solution by discretising the PDE. The simplest solution scheme is given by an explicit finite difference discretisation of (3) that reads as

$$
\frac{u_{i,j}^{k+1} - u_{i,j}^{k}}{\tau} = \frac{1}{h_x} \left(\frac{g_{i+1,j} + g_{i,j}}{2} \frac{u_{i+1,j}^{k} - u_{i,j}^{k}}{h_x} - \frac{g_{i,j} + g_{i-1,j}}{2} \frac{u_{i,j}^{k} - u_{i-1,j}^{k}}{h_x} \right)
$$

$$
+ \frac{1}{h_y} \left(\frac{g_{i,j+1} + g_{i,j}}{2} \frac{u_{i,j+1}^{k} - u_{i,j}^{k}}{h_y} - \frac{g_{i,j} + g_{i,j-1}}{2} \frac{u_{i,j}^{k} - u_{i,j-1}^{k}}{h_y} \right), (4)
$$

where $u_{i,j}^{k} \approx u(i\,h_x, j\,h_y, k\,\tau)$ with h_x and h_y denoting the grid size in x- and y-direction, τ is the time step size and $g_{i,j}$ approximates the value of the diffusivity at grid point (i, j). Occurring derivatives have been discretised using standard finite difference approximations. Solving (4) for the update $u_{i,j}^{k+1}$ then gives the actual iterative scheme.

Explicit schemes are simple to implement, but stability can only be guaranteed for small time step sizes $((\tau/h_x^2 + \tau/h_y^2) \leq 0.5)$. Thus, a lot of iterations are needed to reach a reasonably large evolution time. This restriction has recently been eased in the fast explicit diffusion (FED) scheme [5]. Here, some extremely large (unstable) time steps are used in combination with some small (stable) steps. As could be shown, the combination of variable step sizes within one cycle guarantees unconditional stability of the whole scheme. This allows FED to advance faster than any other explicit scheme: While classical explicit schemes with n fixed time steps achieve a stopping time in $\mathcal{O}(n)$, FED lifts this to $\mathcal{O}(n^2)$. However, as the result is only guaranteed to be stable after a whole cycle, it is important to update the diffusivities g after the completion of a cycle, and not in between. For computing the varying FED step sizes τ_k, we use the available open source library[1].

Note that we do not discuss anisotropic diffusion filters [19] in this paper. However, they can be implemented in a similar way as the presented nonlinear isotropic filters.

2.2 Variational Optic Flow

For optic flow estimation we are given an image sequence $f(x, y, z)$, where z denotes the temporal dimension of the sequence. We further assume that the sequence has been presmoothed by a Gaussian convolution of standard deviation σ_f. We then aim at computing the flow field $\boldsymbol{w} := (u, v, 1)^{\top}$ that describes the displacements from time z to $z{+}1$. Using a variational approach, the flow field is found by minimising an energy functional consisting of a data term that models constancy assumptions on image features, and a smoothness term that enforces the flow field to be smoothly varying in space.

---

[1] available at http://www.mia.uni-saarland.de/Research/SC_FED.shtml

The energy proposed in the seminal variational optic flow approach of Horn and Schunck [3] can be written as

$$E(u, v) = \int_{\Omega} \left(w^{\top} J \, w + \alpha \left(|\nabla u|^2 + |\nabla v|^2 \right) \right) \mathrm{d}x \, \mathrm{d}y \ , \tag{5}$$

using the motion tensor notation $J := (f_x, f_y, f_z)^{\top} (f_x, f_y, f_z)$ and where the parameter α steers the influence of the smoothness term. To minimise the above energy, we solve the corresponding elliptic Euler-Lagrange equations, which give a necessary condition for a minimiser. For the energy (5), the Euler-Lagrange equations are given by

$$0 = J_{11}u + J_{12}v + J_{13} - \alpha \triangle u \ , \tag{6}$$
$$0 = J_{12}u + J_{22}v + J_{23} - \alpha \triangle v \ , \tag{7}$$

where J_{mn} denotes the entry in row m and column n of the matrix J.

Similar to the solution of the diffusion PDEs, we solve the Euler-Lagrange equations by a stabilised, explicit gradient descent scheme which reads as

$$\frac{u_{i,j}^{k+1} - u_{i,j}^k}{\tau} = \triangle u_{i,j}^k - \frac{1}{\alpha} \left([J_{11}]_{i,j}^k \, u_{i,j}^{k+1} + [J_{12}]_{i,j}^k \, v_{i,j}^k + [J_{13}]_{i,j}^k \right) \ , \tag{8}$$

$$\frac{v_{i,j}^{k+1} - v_{i,j}^k}{\tau} = \triangle v_{i,j}^k - \frac{1}{\alpha} \left([J_{12}]_{i,j}^k \, u_{i,j}^k + [J_{22}]_{i,j}^k \, v_{i,j}^{k+1} + [J_{23}]_{i,j}^k \right) \ , \tag{9}$$

where the stabilisation is achieved by using the new value at time level $k + 1$ for u and v in the first and second equation, respectively. The derivatives occurring in the expressions $[J_{mn}]_{i,j}^k$ are discretised by central finite differences. Solving the equations (8) and (9) for the unknown increments $u_{i,j}^{k+1}$ and $v_{i,j}^{k+1}$ then gives the iterative solution scheme, which we again speed up by using variable FED time step sizes. Additionally, we embed the solution in a multiscale coarse-to-fine strategy (CFED) that computes solutions on small image resolutions and uses them (after upsampling) as initialisation on the next finer scale. This results in a further speed-up as the number of required solver iterations at each level is reduced and because iterations on coarse scales are fast to compute.

If we are given colour image sequences, we sum up the data terms for all channels. This comes down to summing up the motion tensors of each channel in a joint tensor. Furthermore, to increase the robustness of our approach, we follow Bruhn *et al.* [2] and convolve the entries of the motion tensor J with a Gaussian of standard deviation ρ. This results in a combined local-global (CLG) optic flow approach.

3   Implementation on an Android Phone

3.1   Android Basics

Android is an open software platform, i.e. an operating system and software stack for different sorts of devices which has been presented by Google in 2007. It is based on a Linux Kernel and allows to build own applications through the *Android Application Framework* by using large parts of Java SE and Android-specific classes.

The building blocks of applications are `Activity` objects. They are attached to a certain `View` representing the graphical user interface (GUI). With the help of an `Intent` one can specify a certain task and trigger an activity that implements the task. In this way, one can for example capture an image from the camera within a few lines of code. The visual content of an application is organized in a hierarchy of views. Examples for views are different layouts, but also widgets such as scroll bars or check boxes. A view hierarchy is specified in an external XML layout file that assigns an ID to each view. By referring to the ID, a view can be loaded within the application as a programmable object that can be configured and attached to event listeners for user input, e.g. gestures on the touch screen.

For implementing performance-critical application parts like our diffusion filters, we use the *Native Development Kit (NDK)*. It allows to implement routines in native languages such as C/C++, resulting in a significant speedup compared to Java implementations; see our experiments in Sec. 4. The NDK also supports a set of commonly used system headers for native APIs like the `math` library. The incorporation of native code in the application uses a provided build system that lists the native source files and integrates the shared libraries into the application project. These can then be easily accessed in terms of the *Java Native Interface (JNI)* [20]. Finally, the NDK allows for optimisations to the underlying hardware by targeting specific instruction sets for the ARM platform, such as the instruction-level parallelism (NEON).

For more information on Android application development, we refer the interested reader to the excellent textbook of Meier [21].

3.2 Image Processing with Android

Instead of capturing images from the camera via a predefined intent, we use a realtime camera stream by directly accessing the camera hardware. This is achieved by using the `Camera` class and applying three steps: *(i) Image Acquisition:* Obtain the camera frames as raw byte data in YUV format and convert them into the more convenient RGB format for further processing. *(ii) Image Processing:* Apply algorithms to the RGB image (or to a greyscale version). *(iii) Visualisation:* Build a bitmap structure from the processed data and set it as content of an image overlay.

Retrieval of Camera Data. Fortunately, accessing the camera hardware in Android is rather simple and can be achieved within a few lines of code; see [21], pp. 377–381. One starts by adding a `CAMERA` permission to the application manifest, which enables to retrieve a `Camera` object by calling `Camera.open()`. A `SurfaceView` providing a dedicated drawing `Surface` can then be attached to the camera. Internal camera parameters such as the preview size and the frame rate can be retrieved and modified by a `Camera.Parameters` object. On creation and destruction of the underlying surface, the camera starts and stops its preview, respectively. Implementing the `PreviewCallback` interface, raw frame data can be obtained by calling `PreviewCallback.onPreviewFrame()`.

YUV Conversion. Each camera frame is represented in a raw byte array which stores the image row-wise and encodes color information using the YCbCr colour space. Specifically, a planar YUV format (YUV 4:2:0 (NV21)) is used where a plane of 8 bit luma (Y) samples is followed by an interleaved V/U plane containing 8 bit of 2×2 subsampled chroma samples.

For image processing applications, it is most convenient to represent the images in a planar RGB format where all colour channels are orthogonal and represented in the same resolution. Thus, the obtained raw YUV data first has to be converted before applying the respective filters. Unfortunately, the Android framework still lacks such a conversion functionality which leaves us with writing our own conversion routine. Using ARM's built-in *Vector Floating Point Architecture (VFP)* that provides hardware support for half-, single- and double-precision floating point arithmetics, we extract in our conversion routine separated, two dimensional float channels, which are aggregated into a packed 32-bit integer format (ARGB) after processing. Whereas the luminance channel alone provides data for a greyscale format, the conversion into RGB components requires computationally more expensive steps, as can be seen in the the following YUV to RGB conversion formula:

$$\begin{pmatrix} R \\ G \\ B \end{pmatrix} = \begin{bmatrix} 1.164 & 0.000 & 1.596 \\ 1.164 & -0.391 & -0.813 \\ 1.164 & 2.018 & 0.000 \end{bmatrix} \left(\begin{pmatrix} Y \\ U \\ V \end{pmatrix} - \begin{pmatrix} 16 \\ 128 \\ 128 \end{pmatrix} \right) . \tag{10}$$

For an efficient implementation of the above conversion, we follow [22] and use fixpoint arithmetics and process four pixels simultaneously by moving pointers on two scan lines. However, instead of using *SIMD (Single Instruction, Multiple Data)* instructions for an efficient access to precomputed lookup tables, we directly optimise the RGB calculations. Details on these optimisations will be presented in Sec. 3.4.

3.3   Software Design

To save as much system resources as possible, we only created one Activity that incorporates our two main classes: A `FilterCamera` and a `Filter` base class. A `FilterCamera` object specifies a `FrameLayout` that is set as the main content view of the application. It subsumes all components necessary for a camera preview that are discussed in Sec. 3.2. The preview can then be overlaid by at most one `Filter` object.

We designed the diffusion filters as "cartridges" that can be plugged into the camera object and unplugged again. The base class represents a general filter object that provides a filtered image overlay, a user interface that may also be hidden, as well as a resizing option. The filter functionality itself comes with the implementation of the `PreviewCallback` interface and is specified by subclasses which represent objects for the different filters. The actual filtering routines are externalised and encapsulated within a global code library that uses native C libraries via the JNI.

3.4   Optimisations

Since the Android platform is based on a Linux kernel, it constitutes a good basis for image processing applications. Many concepts known from traditional desktop architectures carry over immediately. Moreover, the JNI allows to execute time-critical algorithms as low-level operations. To this end, algorithms can be developed device-independently in C or C++, where the compilers provide strategies for automatic code optimisation. However, interactions with special devices such as the camera, display, or user interface still require special care and must be optimised manually.

An example for such a critical point is the visualisation of images on the display. Android supports many pixel formats including 32-bit RGBA, but it is very time consuming to transfer buffers encoded in this format to the GPU. Furthermore, our algorithms must rely on floating point accuracy to ensure the best possible approximation of the continuous model. On the other hand, the display of our smartphone can only visualise a rather small range of colour values. We thus encode and compress our results in the RGB565 format. Because this format uses only 16 bits per pixel it can be uploaded to the GPU much faster, without sacrificing visual quality. Additionally, we tried to optimise our diffusion and optic flow algorithms that take the major part of the total runtime. Common strategies like loop unrolling or reduction of reads and writes to RAM accelerated the process significantly. However, experiments indicated that no improvement can be achieved by exploiting hardware-specific extensions such as the NEON SIMD unit, probably because many operations are based on stencil operations which cause offset memory fetches. We were surprised that even purely data-parallel operations such as the YUV to RGB conversion could not be accelerated. This might be caused by the high costs for memory reads, or because the compiler by default auto-vectorises such operations already in the scalar case.

4 Experiments

4.1 Our Interactive Camera Applications

We first show screenshots of our interactive diffusion filtering application in Fig. 1. As test device we used an HTC Desire smartphone, with a Snapdragon ARMv7 CPU (1 GHz), 576 MB RAM running Android 2.2 (Froyo). The top left picture shows the application in its idle state. Besides a live preview, the GUI shows two buttons for toggling the selected filter as well as a *filter user interface (FUI)*. The latter shows several statistics like the frame rate and is displayed in the upper left corner of the GUI; see the upcoming screenshots. Furthermore, the user can save an image to the phones image gallery and can toggle the autofocus of the camera. The two following pictures show the context menus for filter and resolution selection. The second row of Fig. 1 shows filter results with linear diffusion on RGB colour images. As one can see, the filter interface offers a slider for tuning the standard deviation of the Gaussian kernel. In the settings dialog, the user can change the type of approximation by means of an iterated box filter or a discrete Gaussian convolution. The *accuracy factor* determines the number of iterations or length of the sampling interval as a multiple of the standard deviation, respectively. Moreover, a split mode can be activated, which leaves the left half of the preview unfiltered. Analogously, the third row shows results for nonlinear isotropic diffusion filtering. Here, the user can tune the contrast parameter λ and noise scale σ_p and the stopping time T. The settings dialog allows to choose among various diffusivity functions and the number of outer FED cycles M.

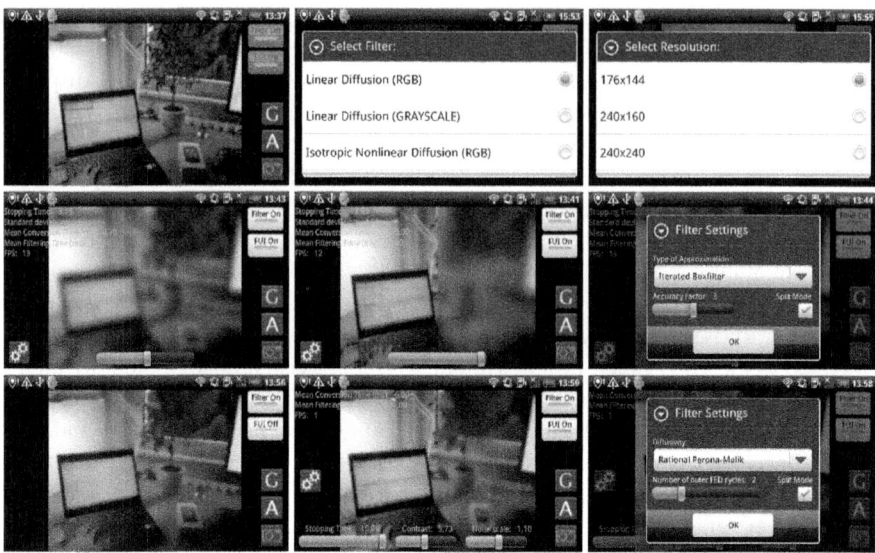

Fig. 1. Our interactive camera application. *First row:* (**a**) GUI with camera preview and controls (image gallery (G), autofocus (A) and image capture). (**b**) Filter selection. (**c**) Resolution selection. *Second row:* (**d**) Linear diffusion for RGB colour images. (**e**) Same in split-screen mode (left half remains unfiltered). (**f**) Settings dialog. *Third row:* (**g**)–(**i**) Same as above, but for nonlinear isotropic diffusion with the Perona Malik diffusivity [1].

Despite the efficient algorithms and the powerful computing platform, we could not achieve an optic flow estimation of reasonable quality in realtime. We thus implemented another simple application where the optic flow computation is performed in an offline process; see Fig. 2. Here, the user can pick two subsequent images of the same size from the gallery, specify the parametric settings and invoke the flow calculation. Besides the seminal approach of Horn and Schunck [3], our application also encompasses the CLG method of Bruhn *et al.* [2] where it achieves runtimes of about 10 seconds for images of size 316×252 pixels.

4.2   Performance Analysis

We now turn to a detailed performance analysis of the linear and nonlinear diffusion routines. To this end, we used simple time stamps that are placed before and after the invocation of a procedure. Resulting runtimes have been averaged over 100 frames to reduce the influence of distorting factors like background processes.

Linear Diffusion. For linear diffusion, we compare two solution strategies: *(i)* a discrete convolution with a sampled Gaussian and *(ii)* an approximation via an iterated box filter (IBF). For both strategies, we exploit separability and symmetry of the two dimensional filter masks. Additionally, we compare implementations of the two strategies in Java and native C (using the JNI). Considering the achieved runtimes shown in Table 1,

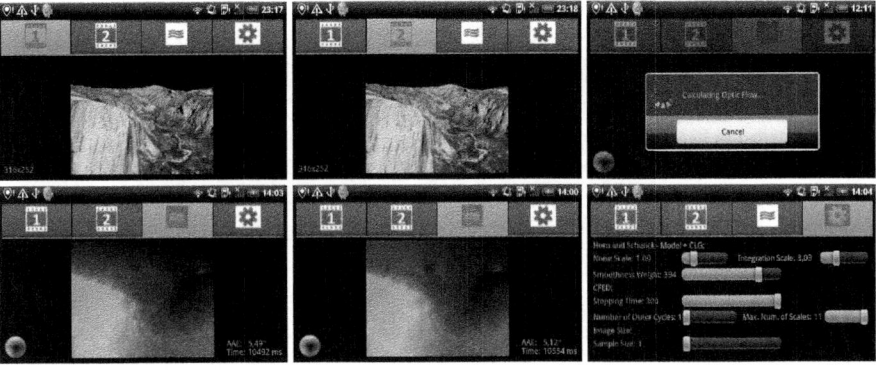

Fig. 2. Offline optic flow application. *First row:* (**a**) First image of the *Yosemite without clouds* sequence. (**b**) Second image. (**c**) Progress dialog. *Second row:* (**d**) Result with the Horn and Schunck model (CFED solver) ($\alpha = 394, \sigma_f = 1.0, T = 300, M = 1 \Rightarrow$ AAE = 5.49°). Flow field visualised by colour code shown in bottom left corner. (**e**) Result with the CLG model ($\rho = 3.03$, other parameters as before $\Rightarrow$ AAE = 5.12°). (**f**) Parameter adjustment dialog.

two observations can be made: The IBF solver is significantly faster than the Gaussian solver, and the native C implementation outperforms its Java counterpart. Furthermore, using greyscale instead of RGB images will give a speedup of a factor 3.

Table 1. Benchmark of linear diffusion on RGB images data with varying resolution and fixed stopping time ($T = 10$). We compare implementations based on a discrete Gaussian convolution with $\sigma = \sqrt{2T} \approx 4.47$ (Gauss) and an approximation via an iterated box filter (IBF). Additionally, we compare a Java to a native C implementation (using the JNI). We wish to note that using the full camera resolution of 5 MP seems infeasible for interactive/realtime applications.

	Java		C	
Resolution [px]	Gauss	IBF	Gauss	IBF
176×144	370 ms	91 ms	361 ms	49 ms
240×160	571 ms	168 ms	564 ms	88 ms
320×240	1174 ms	444 ms	1141 ms	241 ms

Nonlinear Isotropic Diffusion. In the nonlinear diffusion case, we spent some efforts to analyse the runtime fractions of the different processing steps; see Fig. 3. As one can see, the major part of the computation time (98%) is spent for the FED scheme. This is good news as it shows that the runtime of the pre- and post-processing steps (image acquisition, YUV conversion, visualisation, etc.) can be neglected and no further optimisations are required in this respect. It is thus more interesting to further analyse the runtime fractions in the FED scheme itself. Interestingly, the update of the nonlinear operator which comes down to computing the diffusivities after each FED cycle consumes 18% of the time. The rest is spent for the actual iteration steps. As the nonlinear update consumes a considerable amount of time, we also analysed its building blocks, which are a presmoothing that takes 68% of the time and the computation of the diffusivity. Note that the presmoothing is already efficiently realised by two IBF iterations. With a naive implementation, the fraction of the presmoothing would be even higher.

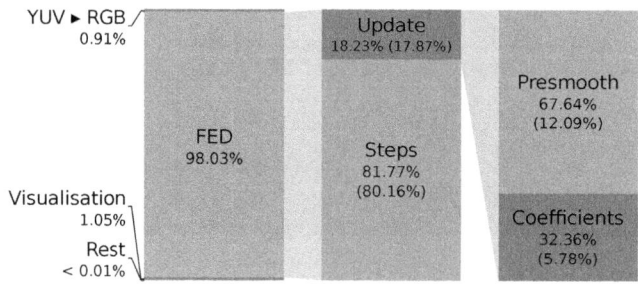

Fig. 3. Runtime analysis for nonlinear Perona-Malik diffusion on RGB images (176×144 pixels, $M = 2, \lambda = 4, \sigma_p = 1.0, T = 10$, resulting in 2 inner FED steps). Brackets give fraction w.r.t. overall running time of 588.1 ms.

5   Conclusions and Outlook

Our paper showed that recent smartphones offer a promising platform for the development of challenging mobile image analysis applications. As a *proof-of-concept*, we presented efficient implementations of linear and nonlinear diffusion as well as basic variational optic flow methods on an Android smartphone (HTC Desire). A main observation is that a careful choice of the solver for the arising PDEs is a key to good performance. For linear diffusion, we used a classical iterated box filter, whereas we opted for the recent FED solver and its coarse-to-fine variant in the context of nonlinear diffusion and optic flow, respectively. These solvers allow for small runtimes and are simple enough to be easily implemented as well as optimised on a mobile platform.

We hope that our work sparks the development of further interesting image analysis applications on mobile devices like smartphones. Here, efficient denoising methods are interesting because the small image sensors and the simple camera optics are prone to noise, especially in low light conditions. Moreover, using optic flow algorithms allows to port challenging computational photography applications like panorama stitching or high dynamic range imaging to mobile platforms.

Acknowledgements. The authors gratefully acknowledge partial funding by the cluster of excellence 'Multimodal Computing and Interaction' (MMCI).

References

1. Perona, P., Malik, J.: Scale space and edge detection using anisotropic diffusion. IEEE Transactions on Pattern Analysis and Machine Intelligence 12, 629–639 (1990)
2. Bruhn, A., Weickert, J., Schnörr, C.: Lucas/Kanade meets Horn/Schunck: Combining local and global optic flow methods. International Journal of Computer Vision 61, 211–231 (2005)

3. Horn, B., Schunck, B.: Determining optical flow. Artificial Intelligence 17, 185–203 (1981)

4. Wells, W.M.: Efficient synthesis of Gaussian filters by cascaded uniform filters. IEEE Transactions on Pattern Analysis and Machine Intelligence 8, 234–239 (1986)

5. Grewenig, S., Weickert, J., Bruhn, A.: From box filtering to fast explicit diffusion. In: Goesele, M., Roth, S., Kuijper, A., Schiele, B., Schindler, K. (eds.) DAGM 2010. LNCS, vol. 6376, pp. 533–542. Springer, Heidelberg (2010)

6. Wells, M.T.: Mobile image processing on the Google phone with the Android operating system (2009),
   `http://www.3programmers.com/mwells/documents/pdf/Final` (retrieved 2011-01-06)

7. GMA3: Moon filter (2010),
   `http://itunes.apple.com/en/app/moon-filter/id387317833`, (retrieved 2011-01-06)

8. Gogolok, R., Steinel, A.: Shockmypic (2009),
   `http://www.shockmypic.com/iphone/` (retrieved 2011-01-06)

9. Bradski, G., Kaehler, A.: Learning OpenCV: computer vision with the OpenCV library. O'Reilly, Sebastopol (2008)

10. Bouguet, J.Y.: Pyramidal implementation of the Lucas Kanade feature tracker – description of the algorithm (2000), `http://trac.assembla.com/dilz_mgr/export/272/doc/ktl-tracking/algo_tracking.pdf`(retrieved 2011-01-06)

11. Harmat, A.: Variational optic flow (2010),
    `http://sourceforge.net/projects/varflow/` (retrieved 2011-01-06)

12. Bay, H., Tuytelaars, T., Van Gool, L.: SURF: Speeded up robust features. In: Bischof, H., Leonardis, A., Pinz, A. (eds.) ECCV 2006, Part I. LNCS, vol. 3951, pp. 404–417. Springer, Heidelberg (2006)

13. Olsson, S., Åkesson, P.: Distributed mobile computer vision and applications on the Android platform. Master's thesis, Faculty of Engineering, Lund University, Sweden (2009)

14. Ballagas, R., Rohs, M., Sheridan, J.G.: Mobile phones as pointing devices. In: Rukzio, E., Hakkila, J., Spasojevic, M., Mäntyjärvi, J. (eds.) Proc. 2005 Pervasive Mobile Interaction Devices, Munich, Germany, vol. 6, pp. 1–4 (2005)

15. Wagner, D., Mulloni, A., Langlotz, T., Schmalstieg, D.: Real-time panoramic mapping and tracking on mobile phones. In: Lok, B., Klinker, G., Nakatsu, R. (eds.) Proc. IEEE Virtual Reality Conference 2010, Waltham, MA, pp. 211–218 (2010)

16. Wagner, D., Schmalstieg, D., Bischof, H.: Multiple target detection and tracking with guaranteed framerates on mobile phones. In: Proc. of IEEE Int. Symposium on Mixed and Augmented Reality 2009, Orlando, FL (2009)

17. Iijima, T.: Basic theory of pattern observation. In: Papers of Technical Group on Automata and Automatic Control. IECE, Japan (1959) (in Japanese)

18. Catté, F., Lions, P.L., Morel, J.M., Coll, T.: Image selective smoothing and edge detection by nonlinear diffusion. SIAM Journal on Numerical Analysis 32, 1895–1909 (1992)

19. Weickert, J.: Anisotropic Diffusion in Image Processing. Teubner, Stuttgart (1998)

20. Liang, S.: Java Native Interface: Programmer's Guide and Reference. Addison–Wesley, Boston (1999)

21. Meier, R.: Professional Android 2 Application Development. Wrox Press Ltd., Birmingham (2010)

22. Dupuis, E.: Optimizing YUV–RGB color space conversion using Intel's SIMD technology (2003), `http://lestourtereaux.free.fr/papers/data/yuvrgb.pdf`, (retrieved 2011-01-07)

A Sampling Theorem for a 2D Surface

Deokwoo Lee and Hamid Krim

Department of Electrical and Computer Engineering
North Carolina State University
Raleigh NC 27606, USA
{dlee4,ahk}@ncsu.edu
http://www.vissta.ncsu.edu/

Abstract. The sampling rate for signal reconstruction has been and remains an important and central criterion in numerous applications. We
propose, in this paper, a new approach to determining an optimal sampling rate for a 2D-surface reconstruction using the so-called Two-Thirds
Power Law. This paper first introduces an algorithm of a 2D surface reconstruction from a 2D image of circular light patterns projected on the
surface. Upon defining the Two-Thirds Power Law we show how the extracted spectral information helps define an optimal sampling rate of the
surface, reflected in the number of projected circular patterns required
for its reconstruction. This result is of interest in a number of applications such as 3D face recognition and development of new efficient 3D
cameras. Substantive examples are provided.

Keywords: Sampling rate, Reconstruction, The *Two-Thirds Power Law*,
Structured light patterns.

1 Introduction

Acquisition of 3D images using an active light source has garnered a lot of interest, and has recently been an important topic in vision and image processing.
The basis of this active 3D imaging technique is in establishing a geometric relationship between a 3D target and the 2D image of structured light patterns
projected on it. In the reconstruction, we assume that the position of the camera
and of the light source are known. We also assume that the camera satisfies a pinhole model and the projected light patterns are parallel [1]. The deformation of
the circular patterns projected on a 3D object provides sufficient information of
the latter's geometrical properties, such as 3D coordinates. In [2], [3], [4] and [5],
various algorithms were proposed to improve an accuracy of reconstruction results using structured light patterns. Our approach to the reconstruction is based
on exploiting the deformed circular patterns projected on a 3D object [12]. Once
3D coordinates are extracted, *the required minimum number of patterns* to be
projected for an efficient reconstruction and minimal computational complexity,
is considered. This is tantamount to determining the required sampling rate on
the surface for its best reconstruction. Akin to determining the *Shannon-Nyquist
Sampling Rate* [6] for a 1D signal, our reconstruction of a 3D signal will seek for

A.M. Bruckstein et al. (Eds.): SSVM 2011, LNCS 6667, pp. 556–567, 2012.

the nontrivial maximal frequency component. Although there have been many contributions ([7], [8]) made for developing a sampling theorem, in this paper, we use the *Shannon-Nyquist Sampling Rate* for a surface reconstruction. The surface of interest has at each point of a projected pattern in $\mathbb{R}^3$ two curvatures, and we use the so-called *Two-Thirds Power Law* [11] to establish a relationship (nonlinear) between a tangential velocity of a curve and these curvatures. Using the equation $V = r\omega$, where V is a tangential velocity, r is the distance from the reference point to an arbitrary point, and $\omega = 2\pi f$ is an angular velocity, we may retrieve the maximum spatial frequency component of the patterns lying on the object (directly related to curvature). The minimum number of patterns is subsequently obtained from the maximum frequency component, which we refer to as a *'curveling rate'* in this paper.

In the following section, we describe an algorithm to extract 3D coordinates using the geometry of the problem. In Section 3, we estimate the frequency components using the *Two-Thirds Power Law* and determine a corresponding sampling rate using the *Shannon-Nyquist Sampling Theorem*. We also substantiate our results by way of experiments in Section 4, followed by some concluding remarks.

2   Geometric Recovery of Surface Coordinates

The substance of this part has appeared in [12] and we hence briefly summarize it here for completeness.

2.1   Geometrical Representation

Let $S \subset \mathbb{R}^3$ be a domain of a 3D object of interest, then a point $P_w \in S$ is represented as

$$P_w = \{(x_w, y_w, z_w) \in \mathbb{R}^3\}, \tag{1}$$

where an index w is used to denote real world coordinates. Let $L \subset \mathbb{R}^3$ be a domain of a circular structured light source, with the origin defined as a center of a pattern (or a curve), then a point $P_L \in L$ is represented as (see Fig. 1)

$$P_L = \{(x_{Lij}, y_{Lij}, z_{Lij}) \in \mathbb{R}^3 \mid x_{Lij}^2 + y_{Lij}^2 = R_j^2, \; z_{Lij} = 0\}, \tag{2}$$

with $i = 1, 2, \ldots, M$, $j = 1, 2, \ldots, N$ respectively indexing the points on the patterns, and the patterns themselves. Let $S_3 \subset \mathbb{R}^3$ be the domain of projected circular patterns on a 3D object, then any point $P_3 \in S_3$ is represented as

$$P_3 = \{(x_{wij}, y_{wij}, z_{wij}) \in \mathbb{R}^3\}, \tag{3}$$

After the patterns projected, P_3 and P_w defined in the intersection of S and S_3 are identical, upon projecting the circular light patterns, P_3 and P_w are to coincide as the intersection of S and S_3,

$$P_3 = \{P_w \mid P_w \in S \cap S_3\} \; or \; P_w = \{P_3 \mid P_3 \in S \cap S_3\}. \tag{4}$$

Let $S_2 \subset \mathbb{R}^2$ be a domain of a 2D image plane of a camera, then any point $P_2 \in S_2$ is represented as

$$P_2 = \{(u_{ij}, v_{ij}) \in \mathbb{R}^2\}. \tag{5}$$

The 3D reconstruction problem consists of establishing a relationship between $P_3 \in S_3$, $P_L \in L$ and $P_2 \in S_2$ (Fig. 1). Let $f : L \to S_3$ be a map of a light

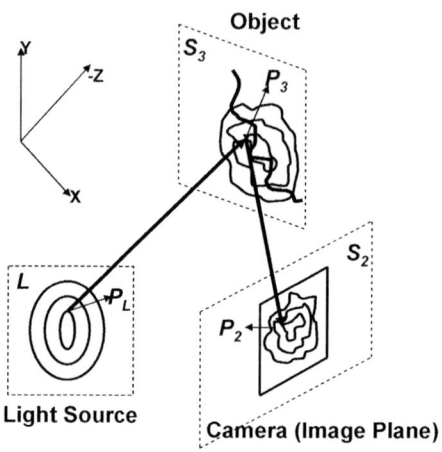

Fig. 1. Geometrical representation of the experimental setup

projection and $g : S_3 \to S_2$ be a map of reflection respectively, then the relevant relationships for surface reconstruction, are

$$f(P_L) = P_3, \quad g(P_3) = P_2. \tag{6}$$

Recall that we assume parallel light projection which preserves (x_{Lij}, y_{Lij}) (i.e. near field projection) and hence the preservation after the pattern projection onto a 3D object so that we have

$$I : (x_{Lij}, y_{Lij}) \to (x_{wij}, y_{wij}), \forall i, j, \tag{7}$$

where I is an identity function. and as discussed previously, under the assumption of parallel projection, (x_{Lij}, y_{Lij}) and (x_{wij}, y_{wij}) obey the following constraints :

$$x_{Lij}^2 + y_{Lij}^2 = R_i^2, \quad x_{wij}^2 + y_{wij}^2 = R_i^2, \tag{8}$$

where i denotes the ith positioned pattern. While coordinates (x_{Lij}, y_{Lij}) are preserved, we note that the depth(z_{wij}) varies and depends on the surface shape. We refer to these variation of depth, z_{wij}, a *deformation factor*. This, in effect, summarizes the reconstruction problem as one of analyzing deformed circular patterns and of depth recovery.

2.2   Mathematical Model

This section details the reconstruction technique of real world 3D coordinates of an object from a planar image of its circularly pattern-lighted surface. The geometrical structure, describing the physical measurement setup, is defined in 3D space and the reference plane is chosen prior to the reconstruction. To solve a reconstruction problem, we opt for two distinct coordinate systems, (X, Z) and (Y, Z) domains(Fig. 2). From Fig. 2, along with associated attributes, we can solve the 3D reconstruction problem. Assuming again that the structured light

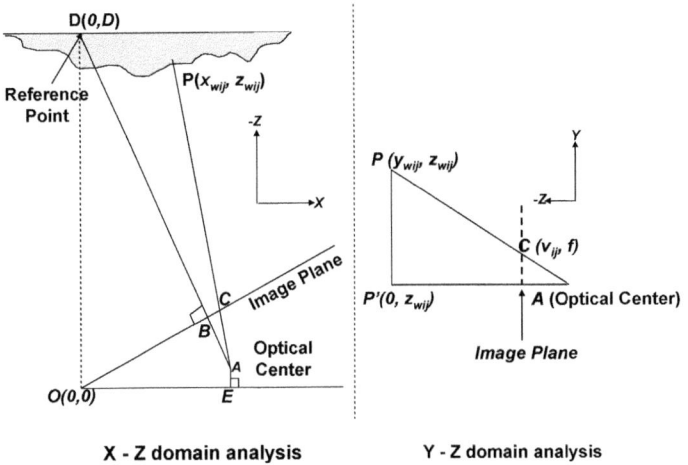

Fig. 2. $(X - Z)$ and $(Y - Z)$ domain analysis

patterns remain parallel, the camera is calibrated to a pinhole model, and its locations together with a chosen reference plane of an object and light source (shown in Fig. 2), we can write the following,

$$\overline{AO} = d, \ \overline{AB} = f, \ \overline{BO} = \sqrt{d^2 - f^2} = d_1,$$
$$d \cos(\angle AOB) = d \cos \theta_2 = \sqrt{d^2 - f^2},$$
$$\overline{OC} = |\overrightarrow{OB} + \overrightarrow{BC}|. \tag{9}$$

Note that the point $C(u_{ij}, v_{ij})$ defined in the 2D image plane, is the result of the reflection of point P, A is the optical center and B is the origin point in the 2D image plane. Since the coordinate system of a 3D object and that of a 2D image plane are different, and upon denoting the $\angle(AOE)$ by θ_1, we transform the domain S_2 to S_3 associated with (X, Z) domain, we can write

$$\theta_1 + \theta_2 = \theta,$$
$$A : (-d \cos \theta_1, d \sin \theta_1),$$

$$B : (-d_1 \cos \theta, d_1 \sin \theta),$$
$$C : (-d_2 \cos \theta, d_2 \sin \theta), \tag{10}$$

where $\theta_1, \theta_2, \theta, d, d_1$ and d_2 are known calibrated quantities. Using the intersection point A of lines $\overline{PC}$ and $\overline{DB}$ (see Fig. 2), we can write a relationship between x_w and z_w. To completely reconstruct the 3D coordinates (x_w, y_w, z_w), we can show the (Y, Z) domain analysis (Fig. 2). Using the above relationships, we can determine 3D coordinates of the deformed curves on a surface,

$$F(x_{wij}) = z_{wij}, \quad H(x_{wij}) = z_{wij}. \tag{11}$$

In [12], we detailed the above relationships and illustrated the approach.

3   Sampling Rate Determination

Using the 2D images of the projected patterns on a surface, and the development described above, we proceed to the surface reconstruction. Each image consists of all curves resulting of the structured circular light patterns. Upon the recovery of 3D real coordinates, the required minimum number of circular patterns for 3D reconstruction is considered. In this section, we develop the minimum sampling rate, to in turn, specify the necessary number of circular patterns required for reconstruction. Recall that the required minimum number of circular pattern is referred to as a *curveling rate*, and preceding its determination, a maximal frequency component of an object should be retrieved. To estimate the frequencies, we apply the *Two-Thirds Power Law* [10] which unveils a relationship between the motion of a shape/curve and its characteristics. Specifically, the radius of curvature(R) of on an osculating circle around a closed curve, and its tangential velocity(V) satisfy the Eq. (12), where K is a constant depending on duration of the motion [10], and α and β are parameters to be estimated [9]. Note that the parameter β is very close to two thirds $(2/3)$, as has been shown by [13]. Projected circular patterns on a 3D object (Fig. 3) have two tangential vectors. Hence, each 3D point of the surface of interest, has two curvature components $(\kappa_{1ij}$ and $\kappa_{2ij})$, the first derivatives of tangential vectors, T_{1ij} and T_{2ij} with respect to the arc length (see Fig. 4). The two-thirds power law can be written as

$$V = r\omega = K \cdot \left(\frac{R}{1 + \alpha R}\right)^{1-\beta}, \tag{12}$$

where r is a distance between the reference point(i.e. nosetip) and the arbitrary point of the object (Fig. 4). According to Eq. (12), we can acquire two frequency components corresponding to κ_{1ij} and κ_{2ij}, respectively.

$$\omega_{1ij} = 2\pi f_{1ij} = \frac{1}{r_{1ij}} \cdot \left(\frac{R_{1ij}}{1 + \alpha R_{1ij}}\right)^{1-\beta}, \tag{13}$$

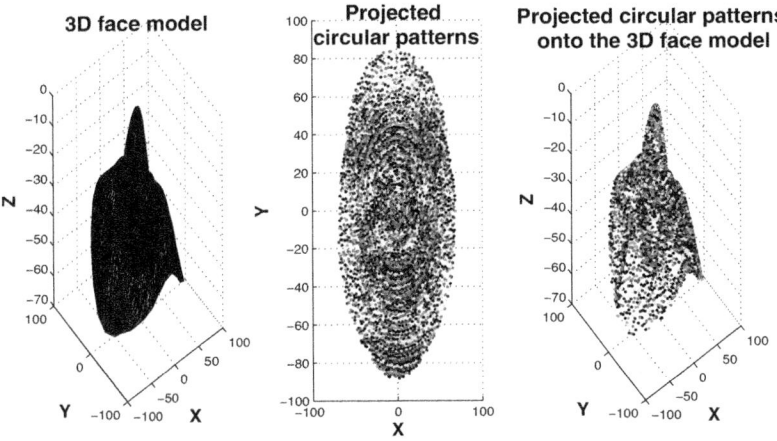

Fig. 3. Simulation of projection of circular patterns, (a). 3D face model, (b). Circular patterns, (c). Overlaying patterns.

$$\omega_{2ij} = 2\pi f_{2ij} = \frac{1}{r_{2ij}} \cdot \left(\frac{R_{2ij}}{1 + \alpha R_{2ij}}\right)^{1-\beta}, \tag{14}$$

$$r_{1ij} = r_{2ij} = r_{ij}, \ i = 1, 2, \ldots, M, \ j = 1, 2, \ldots, N, \tag{15}$$

where R_{1ij} and R_{2ij} are $\frac{1}{\kappa_{1ij}}$ and $\frac{1}{\kappa_{2ij}}$, respectively, and M is the number of points of each curve and N is a number of curves(patterns) on the surface. To determine the minimum sampling rate($2 \times \max(f_{ij})$) which is determined by the *Nyquist Rate*, the maximum frequency component, $\max(f_{ij})$ is required, and we define $\max(f_{ij})$ as

$$\max(f_{ij}) = \max[sup(f_{1ij}), sup(f_{2ij})]. \tag{16}$$

Using a relationship between a frequency component, f_{ij} and the corresponding r_{ij}, the maximum frequency is calculated. Prior to measuring the maximum f_{ij}, $sup(f_{1ij})$ and $sup(f_{2ij})$ should be acquired, and each of which satisfies the following,

$$sup(f_{kij}) \leq \frac{1}{2\pi} sup\left(\frac{1}{r_{kij}} \cdot \left(\frac{R_{kij}}{1 + \alpha R_{kij}}\right)^{1-\beta}\right), \ k = 1, 2. \tag{17}$$

Prior to measurement of curvatures and r_{ij}'s of all the points of the deformed circular patterns, a normalization of data points is carried out. The normalization yields the determination of the intrinsic characteristics of each curve projection on the surface.

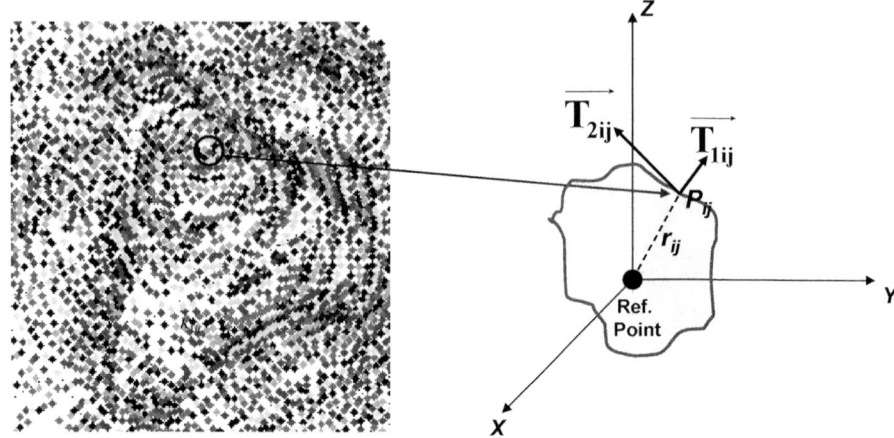

Fig. 4. Two tangential vectors of a point on a 3D object. The first derivative of a tangential vector is a curvature.

4   Experimental Results

To substantiate the measurement of frequency components steps, some simulated examples are shown in this section. Initial number of projected patterns are different from each other and related to the characteristics of objects. The maximum frequency component of the jth curve, $\max[(f_{ij})]_{j=1}^{N}$, measured through all the points of a curve is shown in Fig. 5. From the simulated result in Fig. 5, the maximum frequency component is 0.1501 and the minimum sampling rate is $0.1501 \times 2 = 0.3002$. To substantiate the determination of the above surface sampling rate, we propose some numerical examples. The initial number (N_{ini}) of the projected patterns are determined by the characteristics of the surface. The first (Fig. 6)and the second (Fig. 9) example surfaces consists of 89 and 110 circular patterns, respectively. These numbers correspond to infinite sampling rate in a continuous domain. *Curveling rate*(N_s), the minimum number of patterns defined as

$$N_s = \lceil 2 \times \max[f_{ij}]_{i=1,j=1}^{M,\ N} \times N_{ini} \rceil, \tag{18}$$

and estimated N_{ini}, $\max[f_{ij}]$ and N_s are provided in Table.1. To evaluate the accuracy of a reconstruction, the L_2-norm distance between the original($S_O \subset \mathbb{R}^3$) and the reconstructed surfaces($S_R \subset \mathbb{R}^3$) is computed through all the pixels.

$$d_2(S_O, S_R) = ||S_O - S_R||_2 = \left(\sum_{p_O, p_R \in \mathbb{R}^3} [p_O - p_R]^2 \right)^{1/2}, \tag{19}$$

where $d_2(S_O, S_R)$ is an L_2-norm distance(geometric error), and $p_O \in S_O$ and $p_R \in S_R$ represent 3D Euclidean coordinates of the original and the

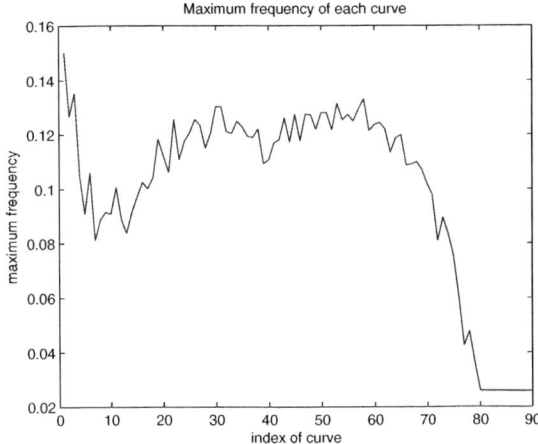

Fig. 5. Maximum frequency components of curves(from the $j = 1$th to Nth curve) on the 3D face model(Fig. 6). Frequency is defined in a unit space(arc length) and quantities are relative each other. In this example, $\max[f_{ij}]$ is 0.1501 and the sampling rate is 0.3002.

Table 1. Estimated *curveling rate* of two face models

	N_{ini}	$\max[f_{ij}]$	N_s
Face 1	89	0.1501	27
Face 2	110	0.1984	44

reconstructed surfaces, respectively. Simulated examples of S_R and $d_2(S_O, S_R)$ are shown in Fig.s. 7, 8, 10 and 11.

5   Conclusion and Future Works

In this paper we have presented an algorithm to determine the sampling rate of a surface (or defining the minimum number of light patterns to be projected on a surface whose maximal curvatures may be known) subjected to an active light source probing. Such a rate, in turn plays a key role in the efficient representation of a surface and its subsequent reconstruction from these patterns. While our primary application of interest lies in the area of biometrics and face modeling, the two-thirds-based sampling criterion may be exploited in many different settings where surface representation and sampling are of interest (e.g. surface archiving). We have also shown some illustrative examples. Although our sampling rate does not recover the surface perfectly as the *Shannon-Nyquist Sampling Rate* does for 1D signals, the sampling criterion we proposed does not show a considerable information loss to be recognized. In the future, there are

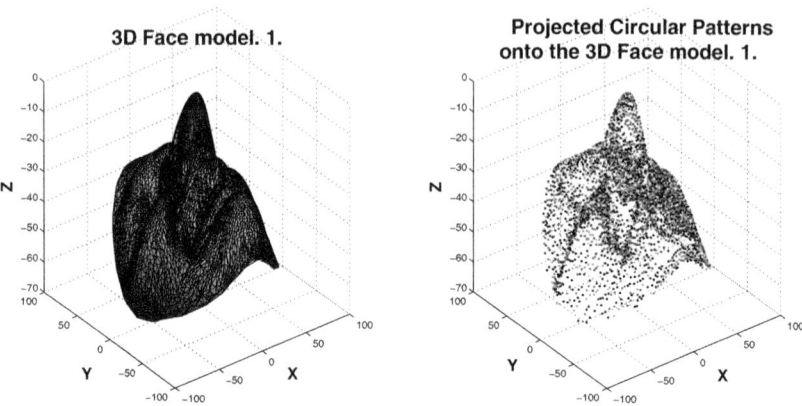

Fig. 6. Original 3D *Face1* model(left), Projected circular patterns onto the 3D face model(right)

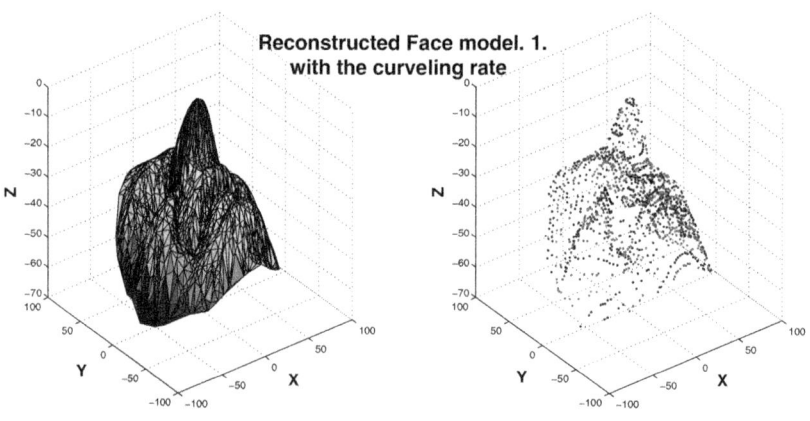

Fig. 7. Reconstructed 3D *Face1* model from the curveling rate 27

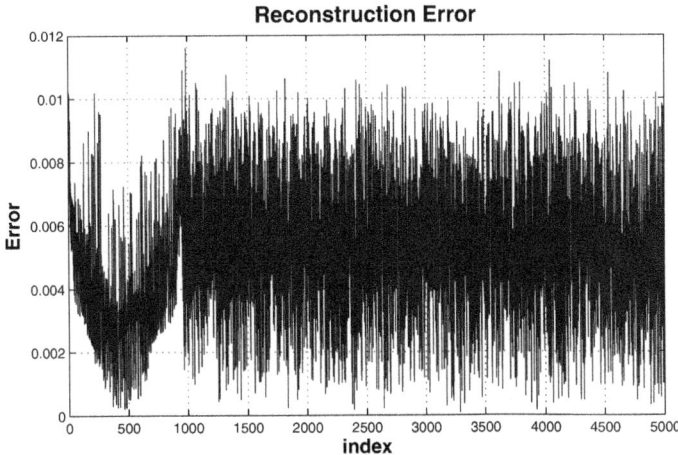

Fig. 8. The L_2-norm distance between the original and the reconstructed *Face1* model (Fig.s 6 and 7)

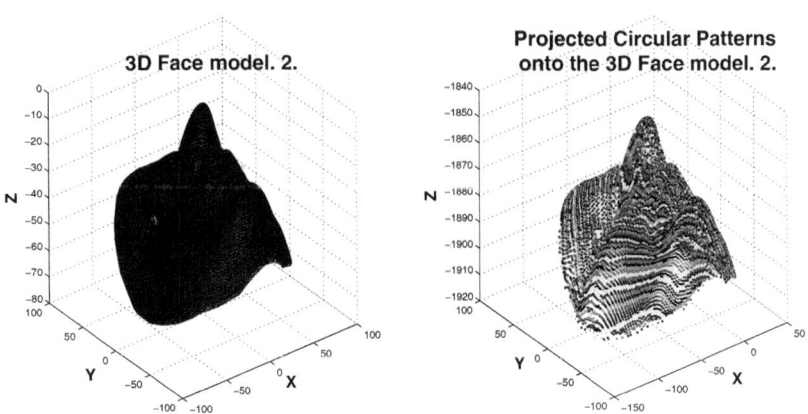

Fig. 9. Original 3D *Face2* model(left), Projected circular patterns onto the 3D face model(right)

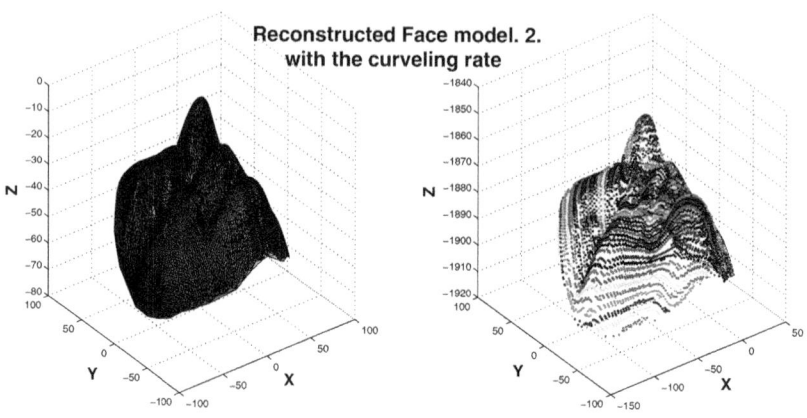

Fig. 10. Reconstructed 3D *Face2* model from the curveling rate 44

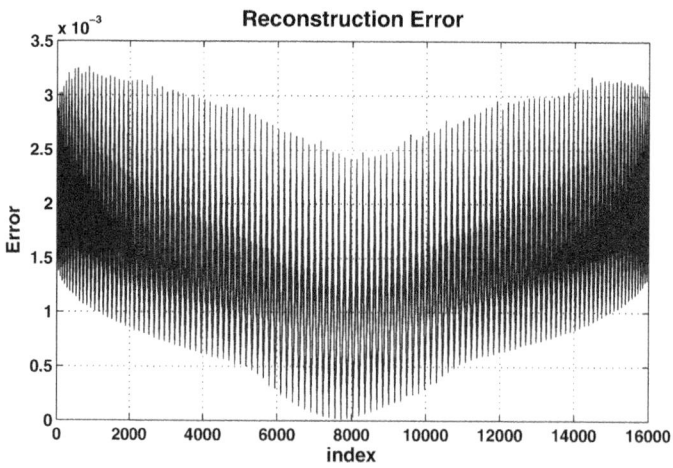

Fig. 11. The L_2-norm distance between the original and the reconstructed *Face2* model (Fig.s 9 and 10)

some technical issues to be considered - quantifying the algorithm efficiency (i.e. computational complexity) and the reconstruction accuracy compared to the previous methods is needed and being in progress.

References

1. Faugeras, O., Luong, Q.-T.: The Geometry of Multiple Images (2001)
2. Wei, Z., Zhou, F., Zhang, G.: 3D coordinates measurement based on structured light sensor. Sensors and Actuators A: Physical 120, 527–535 (2005)
3. Asada, M., Ichikawa, H., Tsuji, S.: Determining of surface properties by projecting a stripe pattern. In: Proc. Int. Conf. on Pattern Recognition, pp. 1162–1164 (1986)
4. Dipanda, A., Woo, S.: Towards a real-time 3D shape reconstruction using a structured light system. Pattern Recognition 38, 1632–1650 (2005)
5. Batlle, J., Mouaddib, E., Salvi, J.: Recent Progress in Coded Structured Light as a Technique to solve the Correspondence Problem: A Survey. Pattern Recognition 31, 963–982 (1998)
6. Papoulis, A.: Signal Analysis. McGraw-Hill, New York (1977)
7. Eldar, Y.C.: Compressed Sensing of Analog Signals in Shift-Invariant Spaces. IEEE Transactions on Signal Processing 57(8) (August 2009)
8. Jerri, A.J.: The Shannon Sampling Theorem - Its Various Extensions and Applications: A Tutorial Review. Proceedings of The IEEE 65(11) (November 1977)
9. Maoz, U., Portugaly, E., Flash, T., Weiss, Y.: Noise and the two-thirds power law
10. de' Sperati, C., Viviani, P.: The Relationship between Curvature and Velocity in Two-Dimensional Smooth Pursuit Eye Movements. The Journal of Neuroscience 17, 3932–3945 (1997)
11. Schaal, S., Sternad, D.: Origins and violations of the 2/3 power law in rhythmic 3D movements. Experimental. Brain Research,, 60–72 (2001)
12. Lee, D., Krim, H.: 3D surface reconstruction using structured circular light patterns. In: Blanc-Talon, J., Bone, D., Philips, W., Popescu, D., Scheunders, P. (eds.) ACIVS 2010, Part I. LNCS, vol. 6474, pp. 279–289. Springer, Heidelberg (2010)
13. Lacquaniti, F., Terzuolo, C., Viviani, P.: The law relating the kinematic and figural aspects of drawing movements. Acta Psychologica, 115–130 (1983)
14. Pollefeys, M., Koch, R., Van Gool, L.: Self-Calibration and Metric Reconstruction Inspite of Varying and Known Intrinsic Camera Parameters. International Journal of Computer Vision, 7–25 (1999)
15. Armangue, X., Salvi, J., Batlle, J.: A Comparative Review Of Camera Calibrating Methods with Accuracy Evaluation. Pattern Recognition 35, 1617–1635 (2000)
16. Sturm, P.: On Focal Length Calibration from Two Views. In: IEEE International Conference on Computer Vision and Pattern Recognition, pp. 145–150 (2001)

Quadrature Nodes Meet Stippling Dots

Manuel Gräf[1], Daniel Potts[1], and Gabriele Steidl[2]

[1] Faculty of Mathematics, Chemnitz University of Technology, Germany
{m.graef,potts}@mathematik.tu-chemnitz.de
http://www.tu-chemnitz.de/~{grman,potts}/
[2] Department of Mathematics, University of Kaiserslautern, Germany
steidl@mathematik.uni-kaiserslautern.de
http://www.mathematik.uni-kl.de/~steidl/

Abstract. The stippling technique places black dots such that their density gives the impression of tone. This is the first paper that relates the distribution of stippling dots to the classical mathematical question of finding 'optimal' nodes for quadrature rules. More precisely, we consider quadrature error functionals on reproducing kernel Hilbert spaces (RKHSs) with respect to the quadrature nodes and suggest to use optimal distributions of these nodes as stippling dot positions. Interestingly, in special cases, our quadrature errors coincide with discrepancy functionals and with recently proposed attraction-repulsion functionals. Our framework enables us to consider point distributions not only in $\mathbb{R}^2$ but also on the torus $\mathbb{T}^2$ and the sphere $\mathbb{S}^2$. For a large number of dots the computation of their distribution is a serious challenge and requires fast algorithms. To this end, we work in RKHSs of bandlimited functions, where the quadrature error can be replaced by a least squares functional. We apply a nonlinear conjugate gradient (CG) method on manifolds to compute a minimizer of this functional and show that each step can be efficiently realized by nonequispaced fast Fourier transforms. We present numerical stippling results on $\mathbb{S}^2$.

1 Introduction

The traditionally artistic stippling technique places black dots to approximate different tones. For an illustration see Fig. 1. Stippling is closely related to dithering, where the dots have to lie on the image grid, see, e.g. [17] and the references therein. A popular stippling method proposed in [18] is based on weighted centroidal Voronoi tessellations and Lloyd's iterative algorithm [13,5]. A capacity-constrained variant of Lloyd's algorithm was introduced in [3]. Recently, a novel stippling approach was proposed in [21]: Consider a gray-value image $u : \mathcal{G} \to [0,1]$ on a grid $\mathcal{G} := \frac{1}{n}(\mathbb{Z}/n\mathbb{Z}) \times \frac{1}{m}(\mathbb{Z}/m\mathbb{Z})$. Since 'black' is 0 and 'white' 1, we use the weight $w := 1 - u$. Now one intends to find the positions $p_i \in [0,1]^2$, $i = 1, \dots, M$, of M black dots by minimizing the attraction-repulsion functional

$$E(\boldsymbol{p}) := \sum_{i=1}^{M} \sum_{x \in \mathcal{G}} w(x)\|p_i - x\|_2 - \frac{\lambda}{2} \sum_{i,j=1}^{M} \|p_i - p_j\|_2, \quad \boldsymbol{p} := \left(p_i\right)_{i=1}^{M} \in \mathbb{R}^{2M}. \quad (1)$$

A.M. Bruckstein et al. (Eds.): SSVM 2011, LNCS 6667, pp. 568–579, 2012.

Here $\lambda := \frac{1}{M} \sum_{x \in \mathcal{G}} w(x)$ is an equilibration parameter between the first sum which describes attracting forces caused by the image gray values and the second one which enforces repulsion between dots. The original idea for considering

Fig. 1. Left: Original image. Right: Stippling result on $\mathbb{T}^2$ with $M = 20000$ dots.

minimizers of this functional as 'good' dot positions comes from electrostatic principles used in [17]. This paper is related to the continuous version of the above attraction-repulsion functional with more general functions $\varphi : [0, \infty) \to \mathbb{R}$:

$$E_\varphi(\boldsymbol{p}) := \frac{\lambda}{2} \sum_{i,j=1}^{M} \varphi(\|p_i - p_j\|_2) - \sum_{i=1}^{M} \int_{[0,1]^2} w(x)\varphi(\|p_i - x\|_2)\, \mathrm{d}x. \qquad (2)$$

where $w : [0,1]^2 \to [0,1]$ and $\lambda := \frac{1}{M} \int_{[0,1]^2} w(x)\, \mathrm{d}x$. The function $\varphi(r) = -r$ was used in (1) and $\varphi(r) = -\log(r)$ in [17] . In [21] the authors mentioned $\varphi(r) = -r^\tau$, $0 < \tau < 2$ and $\varphi(r) = r^{-\tau}$, $\tau > 0$ for $r \neq 0$.

Contribution. In this paper we relate stippling processes with the classical mathematical question of finding best nodes for quadrature rules. We provide theoretical results on the connection between seemingly different concepts, namely quadrature rules, attraction-repulsion functionals, L_2–discrepancies and least squares functionals. For the later approach we provide numerical minimization algorithms.

In the theoretical part, we start with worst case quadrature errors on RKHSs in dependence on the quadrature nodes. While in the literature, this was mainly done for constant weights $w \equiv 1$, see [10,15], we incorporate a weight function related to the image into the quadrature functional. The corresponding quadrature error $\mathrm{err}_K(\boldsymbol{p})$ which depends on the reproducing kernel K can be defined for RKHSs on $\mathcal{X} \in \{\mathbb{R}^2, [0,1]^2\}$ as well as for RKHSs on compact manifolds like $\mathcal{X} \in \{\mathbb{T}^2, \mathbb{S}^2\}$. We aim to minimize this quadrature error in order to obtain optimal quadrature nodes $\boldsymbol{p}$. It turns out that for special kernels K (on special spaces $\mathcal{X}$) this quadrature error (or at least its minimizers) covers the following approaches:

1. Attraction-Repulsion Functionals

   An interesting case of RKHSs appears for radial kernels $K(x, y) = \varphi(\|x - y\|_2)$ depending only on the distance of the points. We will show that in this case the quadrature error $\mathrm{err}_K(\boldsymbol{p})$ can be considered as a generalization of (2) which works not only on $[0, 1]^2$ but also to compact manifolds. Hence our approach goes far beyond the setting in [21] or [17]. To get the special functional (1) from our general quadrature error we must stress conditionally positive definite, radial kernels of order 1.

2. L_2–Discrepancies

   We prove that for $\mathcal{X} \in \{[0, 1]^2, \mathbb{T}^2, \mathbb{S}^2\}$ and discrepancy kernels K, the quadrature errors on RKHSs defined by these kernels coincide with L_2–discrepancy functionals. For various applications of L_2–discrepancy functionals, see [15] and the references therein. Interestingly, this is also related to 'capacity constraints' in stippling techniques, see [2,3]. Note that a relation between the distance kernels $K(x, y) = \|x - y\|_2$ on $\mathbb{T}^2$ and $\mathbb{S}^2$ and the corresponding discrepancy kernels was shown numerically in [9].

3. Least Squares Functionals

   Finally, we consider RKHSs of bandlimited functions with bandlimited kernels on $\mathcal{X} \in \{\mathbb{T}^2, \mathbb{S}^2\}$. The reason for addressing these spaces is that we want to approximate functions on $\mathcal{X}$ by bandlimited functions in order to apply fast Fourier techniques. We prove that for these RKHSs the quadrature error can be rewritten as a least squares functional.

In the numerical part we approximate functions and kernels on $\mathcal{X} \in \{\mathbb{T}^2, \mathbb{S}^2\}$ by their bandlimited versions and minimize the corresponding quadrature error which takes in this case the form of a least squares functional. Due to the page limitation we restrict our attention to the sphere $\mathbb{S}^2$. We are not aware of any results on $\mathbb{S}^2$–stippling in the literature. Note that a stippling example on the torus $\mathbb{T}^2$ was given in Fig. 1. We propose a nonlinear CG method on manifolds to compute a minimizer of the least squares functional on $\mathbb{S}^2$. This method was also successfully used for the approximation of spherical designs, i.e., for $w \sim 1$ in [8] and is generalized in this paper. In particular, each CG step can be realized in an efficient way by the *nonequispaced fast spherical Fourier transform* (NFSFT). This reduces the asymptotic complexity of the proposed algorithm drastically, e.g., from $\mathcal{O}(MN^2)$ to $\mathcal{O}(N^2 \log^2 N + M \log^2(1/\epsilon))$ arithmetic operations per iteration step, where ϵ is the described accuracy and N corresponds to the bandwidth. In other words, only by the help of the NFSFT the computation becomes possible in a reasonable time.

Organization of the paper. In Sect. 2 we introduce our quadrature framework in RKHSs and show the relation to the attraction-repulsion functional. Sect. 3 relates this approach to discrepancy functionals. Sect. 4 deals with band-limited functions on $\mathbb{S}^1, \mathbb{T}^2$ and $\mathbb{S}^2$. We show that in this case the quadrature error can be written as a least squares functional. Moreover, we address the topic that bandlimited functions can be evaluated at a point set in a fast way by using the nonequispaced fast (spherical) Fourier transform. The same holds true for the evaluation of the functional itself. In Sect. 5, we propose a minimizing procedure on $\mathbb{S}^2$ by the CG method, where each iteration step can be efficiently computed by the NFSFT. Finally, Sect. 6 shows stippling results on $\mathbb{S}^2$ and Sect. 7 concludes the paper. Due to the page limitation proofs of certain theorems of this paper are given in the preprint [9].

2   Quadrature Errors in RKHSs

Let $\mathcal{X} \in \{\mathbb{R}^2, [0,1]^2, \mathbb{T}^2, \mathbb{S}^2\}$. For notational reasons, we restrict our attention to two dimensions although the results are also true for arbitrary dimensions. A symmetric function $K : \mathcal{X} \times \mathcal{X} \to \mathbb{R}$ is said to be *positive semi-definite* if for any $M \in \mathbb{N}$ points $x_1, \ldots, x_M \in \mathcal{X}$ and any $a = (a_1, \ldots, a_M)^\mathsf{T} \neq 0$ the relation $a^\mathsf{T} (K(x_i, x_j))_{i,j=1}^M a \geq 0$ holds true and *positive definite* if we have strict inequality. A (real) *reproducing kernel Hilbert space* (RKHS) is a Hilbert space having a *reproducing kernel*, i.e., a function $K : \mathcal{X} \times \mathcal{X} \to \mathbb{R}$ which fulfills

$$K_x := K(\cdot, x) \in H_K \quad \text{and} \quad f(x) = \langle f, K(\cdot, x) \rangle_{H_K}, \quad \forall x \in \mathcal{X}, \forall f \in H_K.$$

To every RKHS there corresponds a unique positive semi-definite kernel and conversely given a positive semi-definite function K there exists a unique RKHS of real-valued functions having K as its reproducing kernel. By $\|\cdot\|_{H_K}$ we denote the norm of H_K. For more information on RKHSs we refer to [1].

In the following, let $w : \mathcal{X} \to \mathbb{R}_{\geq 0}$ be a nontrivial, continuous function which fulfills $h_w(x) := \int_{\mathcal{X}} w(y) K(x, y) \, \mathrm{d}y \in H_K$. We are interested in approximating

$$I_w(f) := \int_{\mathcal{X}} f(x) w(x) \, \mathrm{d}x \quad \text{for } f \in H_K$$

by a quadrature rule

$$Q(f, \boldsymbol{p}) := \lambda \sum_{i=1}^M f(p_i), \qquad \lambda := \frac{1}{M} \int_{\mathcal{X}} w(x) \, \mathrm{d}x$$

for appropriately chosen points $p_j \in \mathcal{X}$. In the literature mainly the case $w \equiv 1$ was considered, see [10,15] and the references therein. In this paper, we have incorporated an image related weight w into the functional. The *worst case quadrature error* is given by

$$\mathrm{err}_K(\boldsymbol{p}) := \sup_{\substack{f \in H_K \\ \|f\|_{H_K} \leq 1}} |I_w(f) - Q(f, \boldsymbol{p})| = \|I_w - Q(\cdot, \boldsymbol{p})\|_{H_K^*}. \tag{3}$$

In the following we will see that this quadrature error covers various known functionals if we choose the kernel in an appropriate way. To start with, the following theorem shows a relation between the quadrature error functional and the attraction-repulsion functional (2). For a proof we refer to [9].

Theorem 1. (Quadrature Error and Attraction-Repulsion Functional)
*Let K be a positive semi-definite function and H_K the associated RKHS. Then
the relation*

$$\mathrm{err}_K(\boldsymbol{p})^2 = 2\lambda E_K(\boldsymbol{p}) + \|h_w\|_{H_K}^2$$

holds true, where

$$E_K(\boldsymbol{p}) := \frac{\lambda}{2}\sum_{i,j=1}^{M} K(p_i, p_j) - \sum_{i=1}^{M}\int_{\mathcal{X}} w(x)K(p_i, x)\,\mathrm{d}x. \tag{4}$$

In particular, the minimizers of err_K and E_K coincide.

We see that for *radial kernels* $K(x, y) := \varphi(\|x - y\|_2)$ with some function $\varphi : [0, \infty) \to \mathbb{R}$, the functional (4) has the form of an attraction-repulsion functional, where the first sum steers the repulsion of the dots and the second one the attraction. However, we have to ensure that the kernel is positive semi-definite. Positive definite, radial kernels on $\mathbb{R}^2$ are for example the inverse multiquadrics $K(x, y) := (\varepsilon^2 + \|x - y\|_2^2)^{-\tau}$, $\varepsilon > 0, \tau > 1$ related to $\varphi(r) := r^{-\tau}$ in (2). These kernels and other positive semi-definite kernels do not lead to the functional (1).

Nevertheless, in the rest of this section, we will see how the attraction-repulsion functional in (1) fits into our quadrature setting. Of course choosing $\mathcal{X} := \mathbb{R}^2$ and the radial kernel $K(x, y) := -\|x-y\|_2$ yields exactly (1). Unfortunately, this kernel is not positive semi-definite. However, it is conditionally positive definite of order 1. Recall that a radial function $\Phi(x) := \varphi(\|x\|_2)$ is *conditionally positive definite of order 1* if for any $M \in \mathbb{N}$ points $x_1, \ldots, x_M \in \mathbb{R}^2$ the relation

$$a^{\mathsf{T}}\left(\Phi(x_i - x_j)\right)_{i,j=1}^{M} a > 0 \quad \forall a = (a_1, \ldots, a_M)^{\mathsf{T}} \not\equiv 0 \quad \text{with} \quad \sum_{i=1}^{M} a_i = 0.$$

holds true. Although these kernels are in general not positive semi-definite, the following slight modification of such kernels given by

$$\tilde{K}_\Phi(x, y) := \Phi(x - y) - \Phi(y) - \Phi(x) + \Phi(0) + 1 \tag{5}$$

defines again a positive semi-definite kernel $\tilde{K}_\Phi(x, y)$ which gives rise to a RKHS $H_{\tilde{K}_\Phi}$. The spaces $H_{\tilde{K}_\Phi}$ can be characterized as in [23]. Now it is not hard to show that the modification $\tilde{K}$ of a kernel $K(x, y) = \Phi(x - y)$ by (5) does not change the minimizer of the functional, i.e., then the minimizers of E_K and $E_{\tilde{K}}$ coincide. In particular, we have for $K(x, y) := -\|x - y\|_2$ that E_K and $E_{\tilde{K}}$ have the same minimizers, so that we can work with the original kernel K.

Finally, we mention with respect to various choices of φ in (2) that other examples of conditionally positive definite, radial functions of order 1 are $\Phi(x) := -\|x\|_2^\tau$, $0 < \tau < 2$ and the multiquadrics $\Phi(x) := -(\varepsilon^2 + \|x\|_2^2)^\tau$, $0 < \tau < 1$.

3   Discrepancies

The quadrature errors considered in the previous section are closely related to discrepancies which adds another interesting point of view. In the following we consider $\mathcal{X} \in \{[0,1]^2, \mathbb{S}^1, \mathbb{T}^2, \mathbb{S}^2\}$ as metric spaces with measure $\mu_{\mathcal{X}}$ and metric $d_{\mathcal{X}}$. Let $D := \mathcal{X} \times [0, R]$ and let $\mathcal{B}(c, r) := \{x \in \mathcal{X} : d_{\mathcal{X}}(c, x) \leq r\}$ be the ball centered at $c \in \mathcal{X}$ with radius $0 \leq r \leq R$. By $1_{\mathcal{B}(c,r)}$ we denote the characteristic function of $\mathcal{B}(c, r)$. Then the kernel defined by

$$K_{\mathcal{B}}(x, y) := \int_0^R \int_{\mathcal{X}} 1_{\mathcal{B}(c,r)}(x) 1_{\mathcal{B}(c,r)}(y) \, d\mu_{\mathcal{X}}(c) \, dr = \int_0^R \mu_{\mathcal{X}}(\mathcal{B}(x, r) \cap \mathcal{B}(y, r)) \, dr \tag{6}$$

is positive semi-definite. Consider for example the sphere $\mathbb{S}^2 := \{x \in \mathbb{R}^3 : \|x\|_2 = 1\}$ with the parameterization in spherical coordinates $x = x(\theta, \varphi) := (\sin\theta\cos\varphi, \sin\theta\sin\varphi, \cos\theta)^\mathsf{T}$, $(\varphi, \theta) \in [0, 2\pi) \times [0, \pi]$. The geodesic distance on $\mathbb{S}^2$ reads $d_{\mathbb{S}^2}(x, y) = \arccos(x \cdot y)$ and the surface measure is given by $\mu_{\mathbb{S}^2}(x) = \sin\theta d\theta d\varphi$. The balls are spherical caps and the area of the intersection of two caps $\mathcal{B}(c, r)$, $0 \leq r < \pi$ with center distance d is

$$a(r, d) = \begin{cases} 0, & 0 \leq r \leq d/2, \\ 4\left[\arccos\left(\sin\frac{d}{2}/\sin r\right) - \cos r \arccos\left(\tan\frac{d}{2}\cot r\right)\right], & \frac{d}{2} < r < \frac{\pi}{2}, \\ 4r - 2d, & r = \frac{\pi}{2}, \\ 4\left[\arccos\left(\sin\frac{d}{2}/\sin r\right) - \cos r \arccos\left(\tan\frac{d}{2}\cot r\right)\right], & \frac{\pi}{2} < r < \pi - \frac{d}{2}, \\ -4\pi\cos r, & \pi - \frac{d}{2} \leq r < \pi. \end{cases}$$

The corresponding discrepancy kernel is $K_{\mathcal{B}}(d) = \int_0^\pi a(r, d) \, dr$. For examples of discrepancy kernels on $[0, 1]^d$, $\mathbb{S}^1$ and $\mathbb{T}^2$ and their relations to distance kernels $\Phi(x, y) = -\|x - y\|_2$, we refer to [15,9].

Integration on the RKHSs $H_{K_{\mathcal{B}}}$ is related to the notation of discrepancy. Set $t := (c, r) \in D$ and $dt := d\mu_{\mathcal{X}}(c) \, dr$. We define the L_2-discrepancy as

$$\mathrm{disc}_2^{\mathcal{B}}(p) := \left(\int_D \left(\int_{\mathcal{X}} w(x) 1_{\mathcal{B}(t)}(x) \, dx - \lambda \sum_{i=1}^M 1_{\mathcal{B}(t)}(p_i) \right)^2 dt \right)^{\frac{1}{2}}. \tag{7}$$

The expression in the inner brackets relates the integral of w on $\mathcal{B}(c, r)$ with the number of points contained in $\mathcal{B}(c, r)$ for fixed $(c, r) \in D$. The discrepancy is then the squared error of their differences taken over all $t \in D$. This point of view is closely related to *capacity-constrained methods* used in [2,3]. The relation between the discrepancy and the quadrature error is given by the next theorem.

Theorem 2. (Quadrature Error and L_2-Discrepancy)
Let $K_{\mathcal{B}}$ be defined by (6) and let $H_{K_{\mathcal{B}}}$ be the associated RKHS of functions on $\mathcal{X}$. Then $\mathrm{err}_{K_{\mathcal{B}}}$ given by (3) and $\mathrm{disc}_2^{\mathcal{B}}$ determined by (7) coincide

$$\mathrm{err}_{K_{\mathcal{B}}}(p) = \mathrm{disc}_2^{\mathcal{B}}(p).$$

4   Least Squares Functionals for Bandlimited Functions

Let $\mathcal{X} \in \{\mathbb{T}^2, \mathbb{S}^2\}$ and let $\{\psi_l : l \in \mathbb{N}\}$ be an orthonormal basis of $L_2(\mathcal{X})$. Then any real-valued function $w \in L_2(\mathcal{X})$ can be written in the form

$$w(x) = \sum_{l=1}^{\infty} \hat{w}_l \psi_l(x), \quad \hat{w}_l = \langle f, \psi_l \rangle_{L_2} = \int_{\mathcal{X}} w(x)\overline{\psi_l(x)} \, dx.$$

In order to develop fast algorithms for the efficient computation of minimizers $\hat{p}$ of functionals E_K we will work in spaces of bandlimited functions $\Pi_N(\mathcal{X}) :=$ span$\{\psi_l : l = 1, \ldots, d_N\}$ of dimension $d_N := \dim \Pi_N(\mathcal{X})$. More precisely, we will use the spaces $\Pi_N(\mathbb{S}^1) := \operatorname{span}\{e^{-2\pi i n(\cdot)} : n = -N/2, \ldots, N/2\}$, $\Pi_N(\mathbb{T}^2) :=$ span$\{e^{-2\pi i n(\cdot)} : n = (n_1, n_2), n_j = -N/2, \ldots, N/2, j = 1, 2\}$ with even N and $\Pi_N(\mathbb{S}^2) := \operatorname{span}\{Y_n^k : n = 0, \ldots, N; k = -n, \ldots, n\}$. Here Y_n^k denote the *spherical harmonics* of degree n and order k, see [14]. We will apply bandlimited kernels of the form

$$K_N(x, y) := \sum_{l=1}^{d_N} \lambda_l \psi_l(x)\overline{\psi_l(y)} \tag{8}$$

with $\lambda_l > 0$. These kernels are reproducing kernels for the RKHSs $H_{K_N} := \Pi_N(\mathcal{X})$ with the inner product $\langle f, g \rangle_{H_{K_N}} = \sum_{l=1}^{d_N} \hat{f}_l \overline{\hat{g}_l}/\lambda_l$. For the efficient computation of minimizers of E_{K_N} it is useful to rewrite the functional in weighted least squares form.

Theorem 3. (Quadrature Error and Least Squares Functional)
Let the kernel K_N be given by (8). Then the relation $\operatorname{err}_{K_N}(\boldsymbol{p})^2 = \mathcal{E}_N(\boldsymbol{p})$ holds true, where

$$\mathcal{E}_N(\boldsymbol{p}) := \sum_{l=1}^{d_N} \lambda_l \left| \lambda \sum_{i=1}^{M} \overline{\psi_l(p_i)} - \hat{w}_l \right|^2 = \|\boldsymbol{\Lambda}^{\frac{1}{2}} F(\boldsymbol{p})\|_2^2$$

with $\boldsymbol{\Lambda} := \operatorname{diag}(\lambda_l)_{l=1}^{d_N}$ and $F(\boldsymbol{p}) = (F_l(\boldsymbol{p}))_{l=1}^{d_N}$, $F_l(\boldsymbol{p}) := \lambda \sum_{i=1}^{M} \overline{\psi_l(p_i)} - \hat{w}_l$. In particular, the functionals E_{K_N} and $\mathcal{E}_N$ have the same minimizers.

Note that for $\mathcal{X} = \mathbb{S}^2$ and $w \equiv 1$ there is a close relation between the minimizers $\mathcal{E}_N(\boldsymbol{p})$ and *spherical designs*, see [19,8,9].

The evaluation of bandlimited functions

$$f(p_i) = \sum_{l=1}^{d_N} \hat{f}_l \psi_l(p_i), \quad i = 1, \ldots, M$$

on $\mathcal{X} \in \{\mathbb{S}^1, \mathbb{T}^2, \mathbb{S}^2\}$ can be written in matrix-vector form as $\boldsymbol{f} = \boldsymbol{A}_N \hat{\boldsymbol{f}}$, where $\boldsymbol{f} := (f(p_i))_{i=1}^{M}$, $\hat{\boldsymbol{f}} := \left(\hat{f}_l\right)_{l=1}^{d_N}$ appropriately ordered and

$$\boldsymbol{A}_N := \begin{cases} \boldsymbol{F}_N = \left(e^{-2\pi i n p_i}\right)_{i=1,\ldots,M; n=-N/2,\ldots,N/2} & \text{for } \mathbb{S}^1, \\ \boldsymbol{F}_{2,N} = \left(e^{-2\pi i (n_1, n_2)^{\mathsf{T}} \cdot p_i}\right)_{i=1,\ldots,M; n_i=-N/2,\ldots,N/2, i=1,2} & \text{for } \mathbb{T}^2, \\ \boldsymbol{Y}_N = (Y_k^n(p_i))_{i=1,\ldots,M; n=0,\ldots,N, |k|\leq n} & \text{for } \mathbb{S}^2. \end{cases}$$

The main reason for working in spaces of bandlimited function is the existence of fast algorithms for the matrix-vector multiplication with $\boldsymbol{A}_N$ and $\overline{\boldsymbol{A}}_N^{\mathsf{T}}$.

Theorem 4. (Fast Evaluation of Bandlimited Functions)
The nonequispaced fast Fourier transform (NFFT) and the nonequispaced fast spherical Fourier transform (NFSFT) realize the multiplication of a vector with the matrix $\boldsymbol{A}_N$, resp. $\overline{\boldsymbol{A}}_N^T$, with the following number of arithmetic operations: $\mathcal{O}(N \log N + M \log(1/\epsilon))$ for $\mathbb{S}^1$, $\mathcal{O}(N^2 \log N + M \log^2(1/\epsilon))$ for $\mathbb{T}^2$ and $\mathcal{O}(N^2 \log^2 N + M \log^2(1/\epsilon))$ for $\mathbb{S}^2$, where ϵ is the prescribed accuracy.

For the NFFT we refer to [6,4,16,12] and for the NFSFT to [11,12]. It can be shown that using these algorithms the same complexity is required for the evaluation of $\mathcal{E}_N(\boldsymbol{p})$.

5   Efficient Minimization Algorithm on $\mathbb{S}^2$

In this section, we describe the computation of a local minimizers of $\mathcal{E}_N$ in an efficient way for given $\hat{w}_l$, $l = 1, \ldots, d_N$. We restrict our attention to the case $\mathcal{X} = \mathbb{S}^2$, where we will only work with kernels of the form

$$K_N(x,y) := \sum_{n=0}^{N} \sum_{k=-n}^{n} \lambda_n Y_n^k(x) \overline{Y_n^k(y)}. \tag{9}$$

such that

$$\mathcal{E}_N(\boldsymbol{p}) = \sum_{n=0}^{N} \sum_{k=-n}^{n} \lambda_n |\lambda \sum_{i=1}^{m} \overline{Y_n^k(p_i)} - \hat{w}_n^k|^2.$$

Using the considerations of the previous section similar algorithms can be deduced for $\mathcal{X} \in \{\mathbb{S}^1, \mathbb{T}^2\}$.

Due to the good experiences in connection with spherical designs in [8] we apply the nonlinear CG method on the manifold $\mathcal{M} := (\mathbb{S}^2)^M$, cf. [20].

Algorithm: (CG algorithm on Riemannian manifolds)
Initialization: $\boldsymbol{p}^{(0)}$, $\boldsymbol{h}^{(0)} := \nabla \mathcal{E}_N(\boldsymbol{p}^{(0)})$, $\boldsymbol{d}^{(0)} = -\boldsymbol{h}^{(0)}$
For $r = 0, 1, \ldots$ repeat until a convergence criterion is reached

1. $\alpha_r := -\langle \boldsymbol{d}^{(r)}, \boldsymbol{h}^{(r)} \rangle / \langle \boldsymbol{d}^{(r)}, \mathrm{H}\mathcal{E}_N(\boldsymbol{p}^{(r)}) \boldsymbol{d}^{(r)} \rangle$
2. $\boldsymbol{p}^{(r+1)} := \exp_{\boldsymbol{p}^{(r)}} \left(\alpha_r \boldsymbol{d}^{(r)} \right)$
3. $\boldsymbol{h}^{(r+1)} := \nabla \mathcal{E}_N(\boldsymbol{p}^{(r+1)})$
4. Compute β_r by $\beta_r := \dfrac{\langle \boldsymbol{h}^{(r+1)}, \mathrm{H}\mathcal{E}_N(\boldsymbol{p}^{(r+1)}) \tilde{\boldsymbol{d}}^{(r)} \rangle}{\langle \tilde{\boldsymbol{d}}^{(r)}, \mathrm{H}\mathcal{E}_N(\boldsymbol{p}^{(r+1)}) \tilde{\boldsymbol{d}}^{(r)} \rangle}$, $\tilde{\boldsymbol{d}}^{(r)} := \boldsymbol{P}_{g(\alpha_r)}(\boldsymbol{d}^{(r)})$.
5. $\boldsymbol{d}^{(r+1)} := -\boldsymbol{h}^{(r+1)} + \beta_r \tilde{\boldsymbol{d}}^{(r)}$

Here $\exp_{\boldsymbol{p}} : \mathrm{T}_{\boldsymbol{p}}\mathcal{M} \to \mathcal{M}$ denotes the *exponential map* from the tangent space $\mathrm{T}_{\boldsymbol{p}}\mathcal{M}$ to the manifold and $\boldsymbol{P}_{g(\alpha_r)}(\boldsymbol{d}^{(r)})$ the *parallel transport* of $\boldsymbol{d}^{(r)} \in \mathrm{T}_{\boldsymbol{p}^{(r)}}\mathcal{M}$ along the geodesics $\boldsymbol{g}$. For the manifold notation including the concept of parallel transport we refer to [22].

Each CG step requires the evaluation of the gradient of $\mathcal{E}_N$ and the multiplication of the Hessian of $\mathcal{E}_N$ with a vector. By the following corollary both computations can be done in an efficient way.

Theorem 5. (Fast Evaluation of $\nabla \mathcal{E}_N$ and Multiplication with $\mathrm{H}\mathcal{E}_N$)
For a given point $\boldsymbol{p} \in (\mathbb{S}^2)^M$ and given $\hat{w}_n^k$, the gradient $\nabla \mathcal{E}_N(\boldsymbol{p})$ can be evaluated with the arithmetic complexity $\mathcal{O}(N^2 \log^2 N + M \log^2(1/\epsilon))$. The multiplication of a vector with the Hessian $\mathrm{H}\mathcal{E}_N(\boldsymbol{p})$ can be computed with the same complexity.

For the proof we refer to the accompanying paper [9].

6   Numerical Results on $\mathbb{S}^2$

In this section, we present some stippling results on $\mathbb{S}^2$. The proposed algorithms were implemented in Matlab R2010a, where the mex-interface to the NFFT library [12] was used. The internal library parameters were set as follows: cutoff parameter $m = 9$, threshold parameter $\kappa = 1000$, flags PRE_PSI and PRE_PHI_HUT. From the sampling points $\boldsymbol{x} := (x_i)_{i=1}^L \in (\mathbb{S}^2)^L$ of the function w we obtain approximate Fourier coefficients

$$\hat{w}_n^k := \sum_{i=1}^L \omega_i w(x_i) \overline{Y_n^k(x_i)}, \qquad l = 1, \dots, d_N, \tag{10}$$

where the quadrature weights ω_i are chosen such that $\int_{\mathbb{S}^2} f(x)\mathrm{d}x = \sum_{i=1}^L \omega_i f(x_i)$ for all $f \in \Pi_N(\mathbb{S}^2)$. Note that the above sums can be evaluated in an efficient way by the NFSFT. As kernel we use the bandlimited version of the distance kernel $\Phi(x-y) = -\|x-y\|_2 = -2\sin(\mathrm{d}_{\mathbb{S}^2}(x,y)/2)$, $x, y \in \mathbb{S}^2$, where the coefficients in (9) are explicitly given by

$$\lambda_n = \frac{16\pi}{(2n+3)(2n+1)(2n-1)}, \quad n \in \mathbb{N}_0.$$

We apply the CG algorithm for randomly distributed initial points $\boldsymbol{p}^{(0)}$.

The **first example** uses the topography map of the earth from Matlab. This map consists of the earth's elevation data. Since the values ranging from -7473 to 5731 we have scaled them to $[0, 1]$, in order to avoid negative values. The data is sampled on the grid $\boldsymbol{x} := (x (\pi i/180, \pi j/180))_{i=1,j=1}^{180,360}$. For this grid we have computed nonnegative quadrature weights $\omega_{i,j}$ for a polynomial degree $N = 179$ by the simple CG algorithm proposed in [7]. After applying the quadrature rule (10) we obtain a polynomial approximation $w = \sum_{n=0}^{179} \sum_{k=-n}^n \hat{w}_n^k Y_n^k$ of the earth's topography, see the left-hand side of Figure 2. For $M = 200000$ and a kernel of bandwidth $N = 1000$, our algorithm obtained after $r = 3600$ iterations the right image in Figure 2, where one iteration takes about 1.5 minutes.

We remark that a naive evaluation of the attraction-repulsion functional (1) requires at least $\mathcal{O}(M^2)$ arithmetic operations. If we approximate the kernel $K(x,y) = \|x-y\|_2$ by bandlimited kernels K_N of the form (8) with $N^2 \sim M$, then every step in the proposed nonlinear CG method needs $\mathcal{O}(M \log^2 M + M \log^2(1/\epsilon))$ arithmetic operations. For a crude illustration of the performance gain in our implementation to a naive one we run our algorithm with the slow NDFST and fast NFSFT. Under the above assumption one iteration step with the NDFST needs also $\mathcal{O}(M^2)$ operations. In our examples, the algorithm takes per iteration about 1.5 minutes versus 3 hours with the NDFST. This reveals the importance of fast algorithms for huge numbers of points.

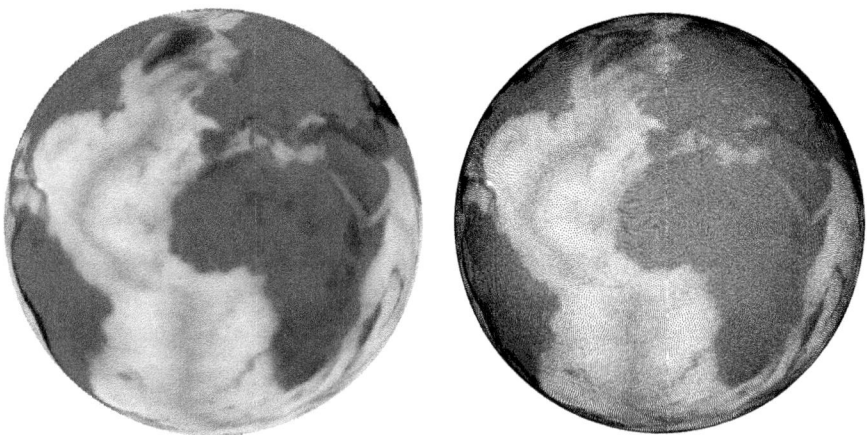

Fig. 2. Left: Original image. Right: Stippling result with $M = 200000$ points.

In the **second example** we map a section of the left bitmap of Figure 1 on the sphere by the same grid as in the first example. The stippling result after $r = 500$ iterations is presented in Figure 3, where we used $M = 100000$ points and a bandwidth $N = 1000$.

7   Conclusions

In this paper, we had a look at the stippling problem from different points of view. Our framework arises primarily from approximation theory but touches many different areas in mathematics as well. The proposed setting is quite general and enables us to consider in some sense 'optimal' point distributions not only in $\mathbb{R}^2$ but also on the torus $\mathbb{T}^2$ and the sphere $\mathbb{S}^2$. Note that even in the seemingly easiest case $w \equiv 1$ the search for optimal point configurations in more than one dimension is a very tough problem which originated many publications. In this case for translationally invariant kernels on $\mathbb{T}^2, \mathbb{S}^2$ the attraction term in (4) is constant and can be omitted.

We have clarified the relation of our quadrature error functionals to recently applied attraction-repulsion functionals and to L_2-discrepancy functionals. For bandlimited functions on $\mathbb{S}^1, \mathbb{T}^2$ and $\mathbb{S}^2$ we suggested to rewrite the quadrature error functional in a least squares form. This is summarized in the figure below. Then the nonlinear CG algorithm can be applied in conjunction with efficient NF(S)FTs for stippling on $\mathbb{S}^2$.

Fig. 3. Left: Original image. Right: Stippling result with $M = 100000$ points.

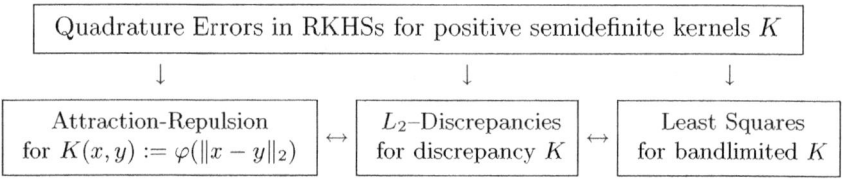

References

1. Aronszajn, N.: Theory of reproducing kernels. Trans. Amer. Math. Soc. 68, 337–404 (1950)
2. Aurenhammer, F., Hoffmann, F., Aronov, B.: Minkowski-type theorems and least-squares clustering. Algorithmica 20, 61–76 (1998)
3. Balzer, M., Schlömer, T., Deussen, O.: Capacity-constrained point distributions: A variant of Lloyd's method. ACM Transactions on Graphics 28(3), Article 86 (2009)
4. Beylkin, G.: On the fast Fourier transform of functions with singularities. Appl. Comput. Harmon. Anal. 2, 363–381 (1995)
5. Du, Q., Faber, V., Gunzburger, M.: Centroidal Voronoi tessellations: Applications and algorithms. SIAM Review 41, 637–676 (1999)
6. Dutt, A., Rokhlin, V.: Fast Fourier transforms for nonequispaced data. SIAM J. Sci. Stat. Comput. 14, 1368–1393 (1993)
7. Gräf, M., Kunis, S., Potts, D.: On the computation of nonnegative quadrature weights on the sphere. Appl. Comput. Harmon. Anal. 27, 124–132 (2009)
8. Gräf, M., Potts, D.: On the computation of spherical designs by a new optimization approach based on fast spherical Fourier transforms. Numer. Math. (2011), doi:10.1007/s00211-011-0399-7
9. Gräf, M., Potts, D., Steidl, G.: Quadrature errors, discrepancies and their relations to halftoning on the torus and the sphere. TU Chemnitz, Fakultät für Mathematik, Preprint 5 (2011)

10. Graf, S., Luschgy, H.: Foundations of Quantization for Probability Distributions. LNM, vol. 1730. Springer, Berlin (2000)
11. Healy, D.M., Kostelec, P.J., Rockmore, D.: Towards Safe and Effective High-Order Legendre Transforms with Applications to FFTs for the 2-sphere. Adv. Comput. Math. 21, 59–105 (2004)
12. Keiner, J., Kunis, S., Potts, D.: Using NFFT3 - a software library for various nonequispaced fast Fourier transforms. ACM Trans. Math. Software 36, Article 19, 1–30 (2009)
13. Lloyd, S.P.: Least square quantization in PCM. IEEE Transactions on Information Theory 28, 129–137 (1982)
14. Müller, C.: Spherical Harmonics. Springer, Aachen (1966)
15. Novak, E., Woźniakowski, H.: Tractability of Multivariate Problems Volume II: Standard Information for Functionals. Eur. Math. Society, EMS Tracts in Mathematics 12 (2010)
16. Potts, D., Steidl, G., Tasche, M.: Fast Fourier transforms for nonequispaced data: A tutorial. In: Benedetto, J.J., Ferreira, P.J. (eds.) Modern Sampling Theory: Mathematics and Applications, pp. 247–270. Birkhäuser, Boston (2001)
17. Schmaltz, C., Gwosdek, P., Bruhn, A., Weickert, J.: Electrostatic halftoning. Computer Graphics Forum 29, 2313–2327 (2010)
18. Secord, A.: Weighted Voronoi stippling. In: Proceedings of the 2nd International Symposium on Non-Photorealistic Animation and Rendering, pp. 37–43. ACM Press, New York (2002)
19. Sloan, I.H., Womersley, R.S.: A variational characterisation of spherical designs. J. Approx. Theory 159, 308–318 (2009)
20. Smith, S.T.: Optimization techniques on Riemannian manifolds. In: Hamiltonian and gradient flows, algorithms and control. Fields Inst. Commun., vol. 3, pp. 113–136. Amer. Math. Soc., Providence (1994)
21. Teuber, T., Steidl, G., Gwosdek, P., Schmaltz, C., Weickert, J.: Dithering by differences of convex functions. SIAM J. Imaging Sci. 4, 79–108 (2011)
22. Udrişte, C.: Convex functions and optimization methods on Riemannian manifolds. Mathematics and its Applications, vol. 297. Kluwer Academic Publishers Group, Dordrecht (1994)
23. Wendland, H.: Scattered Data Approximation. Cambridge University Press, Cambridge (2005)

Discrete Minimum Distortion Correspondence Problems for Non-rigid Shape Matching

Chaohui Wang[1,2], Michael M. Bronstein[3],
Alexander M. Bronstein[4], and Nikos Paragios[1,2]

[1] Laboratoire MAS, Ecole Centrale de Paris, Châtenay-Malabry, France
[2] Equipe GALEN, INRIA Saclay - Île de France, Orsay, France
[3] Institute of Computational Science, Faculty of Informatics
Università della Svizzera Italiana, Lugano, Switzerland
[4] Department of Electrical Engineering, Tel Aviv University, Israel

Abstract. Similarity and correspondence are two fundamental archetype problems in shape analysis, encountered in numerous application in computer vision and pattern recognition. Many methods for shape similarity and correspondence boil down to the minimum-distortion correspondence problem, in which two shapes are endowed with certain structure, and one attempts to find the matching with smallest structure distortion between them. Defining structures invariant to some class of shape transformations results in an invariant minimum-distortion correspondence or similarity. In this paper, we model shapes using local and global structures, formulate the invariant correspondence problem as binary graph labeling, and show how different choice of structure results in invariance under various classes of deformations.

1 Introduction

Recent works in computer vision and shape analysis [1–4] have shown that different approaches to shape similarity and correspondence can be considered as instances of the *minimum distortion correspondence problem*, in which two shapes are endowed with certain structure, and one attempts to find the best (least distorting) matching between these structures. Examples of such structures include multiscale heat kernel signatures [5–7], local photometric properties [8, 9], eigenfunctions of the Laplace-Beltrami operator [10–13], triplets of points [14, 15], and geodesic [2, 3, 16], diffusion [17], and commute time [10, 18] distances. By defining a structure invariant under certain class of transformations (e.g. non-rigid deformations), one obtains correspondence invariant under that class (in the above example, deformation invariant matching). The Gromov-Hausdorff distance [19] is an important particular case of the minimum distortion correspondence problem, in which the matched structures are metric spaces, invariant to isometries of the metric structures.

Some settings of the minimum distortion correspondence problem can be reformulated as labeling problems [20], such that the objective function can be

A.M. Bruckstein et al. (Eds.): SSVM 2011, LNCS 6667, pp. 580–591, 2012.

optimized efficiently using the recently developed discrete optimization algorithms. For example, the dual-decomposition strategy [21], introduced by [22] to perform pairwise Markov random field (MRF) inferences, provides a powerful technique to solve such labeling problems. Based on such a strategy, Torresani *et al.* proposed a pairwise graph matching algorithm [20] to compute correspondence between images using a criterion combining local features and Euclidean distances between nearby features. Such an approach showed better performance than feature-only based methods in deformable 2D object tracking. The increased performance attributed to the use of inter-feature distances as a geometric consistency constraint. However, Euclidean distances are not deformation invariant and can be applied only locally, thus limiting the usefulness of such a constraint.

Main Contribution. In this paper, we study the minimum distortion correspondence problem in the context of non-rigid shape analysis. We formulate *invariant correspondence* as a minimizer of a distortion criterion based on structures invariant to some classes of transformations. In particular, we use local and global structures invariant to important classes of transformations such as non-rigid deformations, changes in topology, and scaling. By such an axiomatic construction of invariant structures, we obtain invariant correspondence. In particular, we show scale invariant shape matching using only singleton and pairwise interactions without higher-order terms. Compared to Torresani *et al.* [20], our use of global structures in non-rigid shapes provides a better regularization to the problem and is better motivated geometrically. Yet, it also increases the computational complexity of the optimization. To address this problem, we use hierarchical matching, in which candidate correspondences are restricted to neighborhoods of matching points from coarser levels.

While the described axiomatic approach is suitable for modeling geometric shape transformations such as bendings, it is not applicable to intra-class shape variations (e.g. different appearances of a human shape). To cope with this case, we show a probabilistic extension of our framework, in which local and global structures are replaced with respective multidimensional distributions, accounting for shape variability.

Related Work. Feature-based shape matching methods for non-rigid shapes were used in numerous recent works [5, 8, 12]. Tree-based [23] and branch-and-bound techniques [24] were used to find the matches between the feature points. Elad and Kimmel [16] used multidimensional scaling (MDS) to represent shapes in a low-dimensional Euclidean space and compare them as rigid objects. The use of an intermediate embedding space was eliminated in [2] using the Gromov-Hausdorff formalism [19]. Bronstein *et al.* [3] proposed an MDS-like algorithm referred to as generalized MDS (GMDS) for the computation of the Gromov-Hausdorff distance and deformation invariant correspondence between shapes. This framework was extended in [17] using diffusion geometry instead of the geodesic one. In [25], Mémoli extended [2, 3] by modeling shapes as metric-measure spaces. He introduced the Gromov-Wasserstein distance based on measure coupling between two metric-measure spaces, and formulated it as a quadratic assignment problem (QAP). Thorstensen and Keriven [9] extended the GMDS framework to textured shapes introducing *photometric stress* as a local matching term in addition to geodesic distance distortion. Dubrovina and Kimmel [13] generalized this approach for the matching of textureless shapes using Laplace-Beltrami eigenfunctions as local geometric descriptors. Mateus *et al.* [11] showed a non-rigid shape correspondence approach with inexact graph matching based on spectral embedding. In the image domain, Torresani *et al.* [20] used graph labeling problem to match 2D images.

2 Problem Formulation

Our shape model is an extension of the *metric model* used in [2, 3, 16]. We assume that the shapes are endowed with *local* and *global* structure, and try to find such a correspondence between the shapes that best preserves these structures. The structures are defined having in mind certain invariance properties required in the particular problem, as discussed in Section 3. Given a shape X, modeled as a connected surface (possibly with boundary) embedded into $\mathbb{R}^3$ (or $\mathbb{R}^2$ in case of planar shapes), its local structure is modeled by a vector field $\mathbf{f}_X : X \to \mathbb{R}^m$ referred to as a *local descriptor*. The global structure of the shape is modeled as a *metric* $d_X : X \times X \to \mathbb{R}$, defined as a positive-definite subadditive function between pairs of points on X.

Given two shapes X and Y with the local descriptors $\mathbf{f}_X$ and $\mathbf{f}_Y$ and metrics d_X and d_Y, respectively, we define a *bijective correspondence* between X and Y as $\mathcal{C} \subset X \times Y$ satisfying $\forall x \in X \, \exists! y \in Y$ such that $(x, y) \in \mathcal{C}$ and $\forall y \in Y \, \exists! x \in X$ such that $(x, y) \in \mathcal{C}$. A good correspondence should match similar descriptors between corresponding points and similar metrics between corresponding pairs of points. This can be quantified using *first-* and *second-order distortion* terms, $\mathrm{dis}(\mathcal{C}) = \|\mathbf{f}(\mathcal{C})\|$ and $\mathrm{dis}(\mathcal{C} \times \mathcal{C}) = \|\mathbf{d}(\mathcal{C} \times \mathcal{C})\|$, measuring the quality of correspondence of local and global structures, respectively. (here, $\mathbf{f}(\mathcal{C})$ is a $|\mathcal{C}| \times 1$ vector with elements $\|\mathbf{f}_X(x) - \mathbf{f}_Y(y)\|$ for all $(x, y) \in \mathcal{C}$; $\mathbf{d}(\mathcal{C} \times \mathcal{C})$ is a $|\mathcal{C}|^2 \times 1$ vector with elements $|d_X(x, x') - d_Y(y, y')|$ for all $(x, y), (x', y') \in \mathcal{C}$; and $\|\cdot\|$ is some norm). In particular,

$$\mathrm{dis}_2(\mathcal{C}) = \sum_{(x,y) \in \mathcal{C}} \|\mathbf{f}_X(x) - \mathbf{f}_Y(y)\|^2;$$

$$\mathrm{dis}_2(\mathcal{C} \times \mathcal{C}) = \sum_{(x,y),(x',y') \in \mathcal{C}} (d_X(x, x') - d_Y(y, y'))^2.$$

The optimal correspondence is found by minimizing a combination of first- and second-order distortion terms,

$$\min_{\mathcal{C}} \mathrm{dis}(\mathcal{C}) + \beta \mathrm{dis}(\mathcal{C} \times \mathcal{C}), \quad \beta \geq 0. \tag{1}$$

The minimizer of problem (1) is the *minimum distortion correspondence* between X and Y. The minimum of problem (1) can be interpreted as the similarity of

X and Y.[1] A particular theoretically important case is a minimum-distortion correspondence with an L_∞ second-order distortion term, referred to as the *Gromov-Hausdorff distance* [19]:

$$d_{\mathrm{GH}}(X, Y) = \frac{1}{2} \min_{\mathcal{C}} \max_{(x,y),(x',y')\in\mathcal{C}} |d_X(x, x') - d_Y(y, y')|.$$

3 Invariance

The choice of the local and global structures ($\mathbf{f}_X, \mathbf{f}_Y$ and d_X, d_Y) defines the invariance properties of the correspondence. Assume that the shape $Y = \tau(X)$ is obtained from X by means of some transformation τ from a class $\mathcal{T}$. If $\mathbf{f}_X \circ \tau = \mathbf{f}_Y$ and $d_X \circ (\tau \times \tau) = d_Y$ for all $\tau \in \mathcal{T}$, our structures are *invariant* under the class of transformations $\mathcal{T}$. As a result, correspondence obtained by the solution of problem (1) is also invariant under $\mathcal{T}$. Important invariance classes can be addressed by appropriate definition of the descriptors and the metric. In particular, we are interested in *inelastic deformations* (bendings), changing the embedding of the shape without changing its intrinsic structure; *topological transformations*, resulting in local changes in the connectivity of the shape, appearing as holes or "gluing" two points on the surface; and *scaling*.

3.1 Choice of the Metric

Geodesic Metric. One of the most straightforward definitions of a metric on a surface is the *geodesic metric*, measuring the length of a shortest path between points x and x',

$$d_X(x, x') = \min_{\gamma\in\Gamma(x,x')} \ell(\gamma),$$

where $\Gamma(x, x')$ denotes the set of all admissible paths between x and x', γ is some admissible path, and $\ell(\gamma)$ is its length. The geodesic metric is intrinsic, dependent only on local distance structure of the shape, and is thus invariant to inelastic deformations [2, 3, 16]. A notable drawback of the geodesic distance is its sensitivity to topological transformations. Connectivity changes alter the admissible paths Γ (e.g., gluing the fingers of the hand creates new paths that have not existed before), and, since the geodesic distance takes the minimum over all path lengths, sometimes the change in the geodesic metric can be very significant.

Diffusion Metric. A more robust definition of an intrinsic metric based on heat diffusion properties has been recently popularized by Lafon *et al.* [27]. Heat

---

[1] The minimizer of problem (1) is not necessarily unique, i.e., there may be two different correspondences $\mathcal{C} \neq \mathcal{C}'$ with $\mathrm{dis}(\mathcal{C}) = \mathrm{dis}(\mathcal{C}')$. Such situations are typical when the shapes have *intrinsic symmetries*. Intrinsic symmetry is manifested by the existence of a self-isometry of X with respect to the metric d_X, i.e., an automorphism $g : X \to X$ satisfying $d_X = d_X \circ (g \times g)$ [24, 26].

diffusion on manifolds is governed by the *heat equation* $\left(\Delta_X + \frac{\partial}{\partial t}\right) u = 0$, where u is the heat distribution and Δ_X is the positive semi-definite *Laplace-Beltrami operator* (LBO), which can be roughly thought of as a generalization of the Laplacian to non-Euclidean domains. The *heat kernel* $h_{X,t}(x, z)$ is the solution of the heat equation with a point heat source at point x at time $t = 0$. For compact manifolds, the Laplace-Beltrami operator has discrete eigendecomposition of the form $\Delta_X \phi_i = \lambda_i \phi_i$, where $\lambda_0 = 0, \lambda_1, \dots \geq 0$ are eigenvalues and $\phi_0, \phi_1, \dots$ are eigenfunctions. Using the eigenbasis of Δ_X, the heat kernel can be presented as

$$h_{X,t}(x, z) = \sum_{i=0}^{\infty} e^{-\lambda_i t} \phi_i(x) \phi_i(z). \tag{2}$$

A family of metrics

$$d_{X,t}(x, y) = \|h_{X,t}(x, \cdot) - h_{X,t}(y, \cdot)\|_{L_2(X)} = \sum_{i=1}^{\infty} e^{-2\lambda_i t} (\phi_i(x) - \phi_i(y))^2, \tag{3}$$

parameterized by the time scale t, is referred to as *diffusion metrics*. Diffusion metric is inversely related to the connectivity of points x and y by paths of length t. Unlike the geodesic distance which measures the length of the shortest path, the diffusion metric has an averaging effect over all paths connecting two points. As a result, diffusion metric is less sensitive to topology and connectivity changes [17]. With an appropriate selection of the time scale t, the effect of topological noise can be reduced.

Commute-Time Metric.   At the same time, the need to select the scale parameter is a disadvantage, as it depends on the shape scale. Moreover, the diffusion metric is not scale invariant, since scale change affects the eigenvalues λ_i and eigenfunctions ϕ_i. A different metric,

$$\delta_X(x, y) = \sum_{i=1}^{\infty} \frac{1}{\lambda_i} (\phi_i(x) - \phi_i(y))^2, \tag{4}$$

called the *commute time* (or *resistance* [28]) *distance*, is similar in its spirit to the diffusion metric, while being scale-invariant. The commute time metric measures the connectivity of points by paths of any length and is related to the expected time it takes a random walk initiating at point x go through point y and return to x.

3.2   Choice of the Descriptor

Similarly to our motivation in the selection of the metric, the choice of the local descriptor is also dictated by the desired invariance properties. Due to their locality, many types of descriptors are usually less susceptible to changes as a result of non-rigid deformations. However, some descriptors have explicit invariance properties by construction.

Table 1. Invariance properties of local (top rows) and global (bottom rows) structures

Structure	Bending	Topology	Scale
Local histogram [24]	Yes	No	No
LBO eigenfunctions [12, 13]	Yes	No	Yes
HKS [5]	Yes	Approx	No
SI-HKS [7]	Yes	Approx	Yes
Geodesic metric [2, 3]	Yes	No	No
Diffusion metric [17]	Yes	Approx	No
Commute-time metric [18]	Yes	Approx	Yes

Heat Kernel Signature. Sun *et al.* [5] introduced intrinsic descriptors based on multi-scale heat kernels, referred to as *heat kernel signatures* (HKS). The HKS is constructed at every point of the shape by considering the values of the heat kernel diagonal at multiple time scales, $\mathbf{f}_X(x) = (h_{X,t_1}(x,x), \ldots, h_{X,t_n}(x,x))$, where $t_1, \ldots, t_n$ are some time scale. The HKS is invariant to inelastic deformations and was also shown to be insensitive to topological transformations [6].

Scale-Invariant Heat Kernel Signature. The disadvantage of HKS is the lack of scale invariance. In a follow-up work, Bronstein and Kokkinos [7] introduced a scale-invariant modification of HKS, referred to as SI-HKS. The main idea is to sample the time scales logarithmically $(t = \alpha^\tau)$ such that shape scaling corresponds to a scale-space shift. Such a shift is then undone by taking the magnitude of the Fourier transform w.r.t. τ. The SI-HKS enjoys the invariance properties of HKS, while in addition also being scale-invariant.

4   Correspondence as a Graph Labeling Problem

Our minimum-distortion correspondence problem can be formulated as a *binary labeling* problem with uniqueness constraints [20] in a graph with vertices defined as pairs of points and edges defined as quadruplets. More formally, let $\mathcal{V} = \{(x,y) : x \in X, y \in Y\} = X \times Y$ be the set of pairs of points from X and Y, and let $\mathcal{E} = \{((x,y),(x',y')) \in \mathcal{V} \times \mathcal{V}$ and $(x,y) \neq (x',y')\}$. Let $\mathcal{L} = \{0,1\}$ further denote the set of binary labels. We can represent a correspondence $\mathcal{C} \subseteq \mathcal{V}$ as binary labeling $u \in \mathcal{L}^{\mathcal{V}}$ of the graph $(\mathcal{V}, \mathcal{E})$, as follows: $u(x,y) = 1$ iff $(x,y) \in \mathcal{C}$ and 0 otherwise. When using L_2 distortions, the correspondence problem (1) can be reformulated as:

$$\min_{u \in \mathcal{L}^{\mathcal{V}}} \sum_{(x,y) \in \mathcal{V}} u_{x,y}(\|\mathbf{f}_X(x) - \mathbf{f}_Y(y)\| - \gamma) +$$

$$\beta \sum_{((x,y),(x',y')) \in \mathcal{E}} u_{x,y} u_{x',y'} |d_X(x,x') - d_Y(y,y')|^2$$

$$\text{s.t.} \sum_y u_{x,y} \leq 1 \ \forall x \in X; \quad \sum_x u_{x,y} \leq 1 \ \forall y \in Y. \tag{5}$$

where $\gamma > 0$ is an occlusion term [20] to penalize unmatched points. We can choose a sufficiently large γ to ensure the bijective correspondence and the equivalence of the two problems.

In general, optimization of this energy is NP-hard [29]. Here, we adopt the graph matching algorithm [20] based on dual-decomposition to perform the optimization of (5). The key idea of this approach is, instead of minimizing directly the energy (5) of the original problem, to maximize a lower bound on it by solving the dual to the linear programming (LP) relaxation of (5). Such approaches demonstrate good global convergence behavior [22]. We first decompose the original problem, which is too complex to solve directly, into a series of subproblems, each of which is smaller and solvable. After getting the solution of the sub-problems, we combine them using a projected-subgradient scheme to get the solution of the original problem. In the numerical experiments, following [20], we decomposed problem (5) into a linear subproblem, a maxflow subproblem and a set of local subproblems.

4.1   Hierarchical Matching

Assuming for simplicity $|X| = |Y| = N$, the number of vertices in the graph is $|\mathcal{V}| = N^2$ and the number of edges, assuming full connectivity, is $\mathcal{O}(N^4)$. The complexity of problem (5) is $\mathcal{O}(|\mathcal{V}|^2|\mathcal{E}|)$ multiplied by the number of iterations, i.e., $\mathcal{O}(N^8)$. This complexity can be reduced by adopting a hierarchical matching strategy: after finding a coarse correspondence between a small number of points, correspondence between nearby points only is looked for. This allows to significantly reduce the graph size.

Let $x_1, x_2, \ldots$ denote a progressive sampling of the shape X, such that $X_n = \{x_1, \ldots, x_n\}$ constitutes an r_n-covering of X (i.e., $d_X(X, X_n) \leq r_n$, where d_X is some metric on X). Such a sequence of points can be found using e.g. farthest point sampling (FPS) strategy [30], in which x_1 is selected arbitrarily and the next point is selected as $x_{k+1} = \arg\max_{x \in X} \min_{i=1,\ldots,k} d_X(x, x_i)$. Same way, $Y_n = \{y_1, \ldots, y_n\}$ will denote an r'_n-covering of Y.

At the first stage of hierarchical matching, correspondence is found between X_{N_1} and Y_{N_1}, where N_1 is some small number (in our experiments, it varied between 4 and 10), solving the labeling problem (5) on the full graph ($\mathcal{V}_1 = X_{N_1} \times Y_{N_1}$, $\mathcal{E}_1 = \{((x, y), (x', y')) \in \mathcal{V}_1 \times \mathcal{V}_1$ and $(x, y) \neq (x', y')\}$. The solution provides a coarse correspondence $\mathcal{C}_1 \subset X_{N_1} \times Y_{N_1}$.

At the $(k + 1)$st level, correspondence is found between $X_{N_{k+1}}$ and $Y_{N_{k+1}}$ (the number of points is increased by a factor typically $2 \leq q = N_{k+1}/N_k \leq 4$), restricting the correspondence candidates for points within a certain radius around x to points within a certain radius around y, where $(x, y) \in \mathcal{C}_k$. This way, the $(k + 1)$st level labeling problem is solved on the graph with vertices

$$\mathcal{V}_{k+1} = \{(x_i, y_i) \in X_{N_{k+1}} \times Y_{N_{k+1}} : \exists (x, y) \in \mathcal{C}_k \text{ s.t. } d_X(x, x_i) < \rho r_k, d_Y(y, y_i) < \rho r'_k\},$$

where $\rho > 1$, and $\mathcal{E}_{k+1} = \{((x, y), (x', y')) \in \mathcal{V}_{k+1} \times \mathcal{V}_{k+1}$ and $(x, y) \neq (x', y')\}$. For $\rho \approx 1$, the size of the ρr_k-neighborhood in $X_{N_{k+1}}$ of a point from X_{N_k} contains $\mathcal{O}(q)$ points. Thus, $|\mathcal{V}_{k+1}| = \mathcal{O}(q^2 N_k)$, and $|\mathcal{E}_{k+1}| = \mathcal{O}(q^4 N_k^2)$ points, a significant reduction compared to $\mathcal{O}(N_{k+1}^2)$ vertices and $\mathcal{O}(N_{k+1}^4)$ edges in a full graph. As a result, the complexity of the optimization becomes $\mathcal{O}(N^4)$.

5   Probabilistic Matching and Shape Prototypes

While invariance to geometric transformations such as bending can be accounted by the selection of local (descriptor) and global (metric) structures, many types of shape variability cannot be accounted for in this way. For example, variability within the shape class (e.g. fat or thin man) results in different local and global structures that cannot be modeled geometrically. At the same time, such a variability can be modeled statistically. Instead of $\mathbf{f}_X$ and d_X, we now have *distributions* $\mathbf{f}_X \sim \mathcal{F}_X$ and $d_X \sim \mathcal{D}_X$, e.g., *Gaussian mixture model* for distances,

$$\mathrm{p}_{xx'}(d) = \sum_{k=1}^{K} \pi_{xx'k} \frac{1}{\sqrt{2\pi}\sigma_{xx'k}} \exp\left\{-\frac{(d-\mu_{xx'k})^2}{2\sigma_{xx'k}^2}\right\}, \sum_{k=1}^{K} \pi_{xx'k} = 1; \ \forall x \neq x' \in X,$$

and descriptors,

$$\mathrm{p}_x(\mathbf{f}) = \sum_{k=1}^{K} \pi_{xk} \frac{\exp\left\{-\frac{1}{2}(\mathbf{f}-\boldsymbol{\mu}_{xk})^{\mathrm{T}}\boldsymbol{\Sigma}_{xk}^{-1}(\mathbf{f}-\boldsymbol{\mu}_{xk})\right\}}{(2\pi)^{m/2}(\det \boldsymbol{\Sigma}_{xk})^{1/2}}, \sum_{i=1}^{K} \pi_{xk} = 1; \ \forall x \in X,$$

where p denotes probability density. The distance distribution between points x and x' is parameterized by $\mathcal{D}_{xx'} = \{\mu_{xx'k}, \sigma_{xx'k}^2, \pi_{xx'k}\}_{k=1}^{K}$; the distribution of descriptors at each points x is parameterized by $\mathcal{F}_x = \{\boldsymbol{\mu}_{xk}, \boldsymbol{\Sigma}_{xk}, \pi_{xk}\}_{i=1}^{K}$, where $\boldsymbol{\mu}_{xk}$ are $m \times 1$ vectors and $\boldsymbol{\Sigma}_{xk}$ are $m \times m$ matrices. We call $\mathcal{X} = ((\mathcal{F}_x)_{x \in X}, (\mathcal{D}_{xx'})_{x \neq x' \in X})$ a *shape prototype*.

In this probabilistic setting, given a shape Y, we determine the correspondence between Y and the prototype $\mathcal{X}$ by solving a problem similar to (5), with the distortion terms replaced by negative log-likelihood functions,

$$\min_{u \in \mathcal{L}^{\mathcal{V}}} - \sum_{(x,y) \in \mathcal{V}} u_{x,y}(\log \mathrm{p}_x(\mathbf{f}_Y(y)) + \gamma) -$$
$$\beta \sum_{((x,y),(x',y')) \in \mathcal{E}} u_{x,y} u_{x',y'} \log \mathrm{p}_{xx'}(d_Y(y,y'))$$
$$\text{s.t.} \sum_y u_{x,y} \leq 1 \ \forall x \in X; \quad \sum_x u_{x,y} \leq 1 \ \forall y \in Y. \tag{6}$$

6   Results

To assess the performance of the presented approach, we performed multiple experiments of shape correspondence and similarity computation under a variety of transformations. Shapes from the TOSCA [31] and Princeton [32] datasets were used in our experiments. Textured shapes acquired with a multicamera system were taken from the INRIA Grenoble dataset [8]. The shapes were represented as triangular meshes with 2000-10000 vertices. Textures were given as RGB values for each vertex. Geodesic distances were computed using fast marching [33]. Diffusion and commute time metrics were computed using the spectral formulae (3) and (4) taking the first 100 eigenvalues. The Laplace-Beltrami operator was approximated using cotangent weights [34]. The heat kernel was approximated using formula (2). Hierarchical matching was implemented in MATLAB with discrete optimization module in C++. Typical running times for pairwise shape matching in the following experiments were about $10 - 20$ sec.

Invariance and the Choice of the Metric/Descriptor. In the first experiment, matching was performed between eight points with equal weight given to the local and global distortion terms in the optimization problem. Three combinations of first- and second-order structures were used: geodesic metric/HKS descriptor, diffusion metric/HKS descriptor, and commute time metric/SI-HKS descriptor. Figure 1 shows the result of correspondence computation between shapes with different transformations for different choice of metric/descriptor. All three methods are invariant to bendings (first row; note that correspondence is defined up to an intrinsic symmetry). The combination geodesic metric/HKS descriptor is sensitive to topology (a human with hands glued to legs, second row) and scale. The combination diffusion metric/HKS descriptor is insensitive to topology but sensitive to scale. Finally, commute time metric with SI-HKS descriptor are invariant to all of the above.

Shape Prototypes. In the second experiment, a shape prototype was created based on 64 examples of a human shape, in which the length of the hands and legs and the size of the head varied. Distance and descriptor distributions were represented using Gaussian mixtures with 5 components. Figure 2 shows a comparison of deterministic and probabilistic matching. Using deterministic matching, the shape of a humanoid alien from the Princeton database is matched to the human shape from TOSCA dataset incorrectly (second column from left), because of different proportions of the head, legs, and hands. On the other hand, matching to the human shape prototype using probabilistic matching produced correct symmetric correspondence (third column). Figure 2 (columns four and five) shows additional examples of shape prototype matching. These results show that the probabilistic matching framework allows to address shape variability that cannot be simply accommodated into the metric model by choosing the metric.

7   Conclusions

We presented a generic framework for invariant matching between shapes, in which matching is performed by minimizing the distortion of local and global geometric structures under the correspondence. Using structures invariant to pre-defined classes of transformations (or, using their statistical distributions if such transformations cannot be modeled explicitly) allows obtaining invariant matching between shapes. Our approach generalizes many previous works in the field, in particular, methods based on metric distortion minimization [2, 3, 17] and global and local features [9, 13, 20], allowing incorporating many existing geometries and local descriptors [5, 7, 8]. In particular, it extends the Gromov-Hausdorff framework [2, 3, 19]. Formulating the problem as graph labeling, we

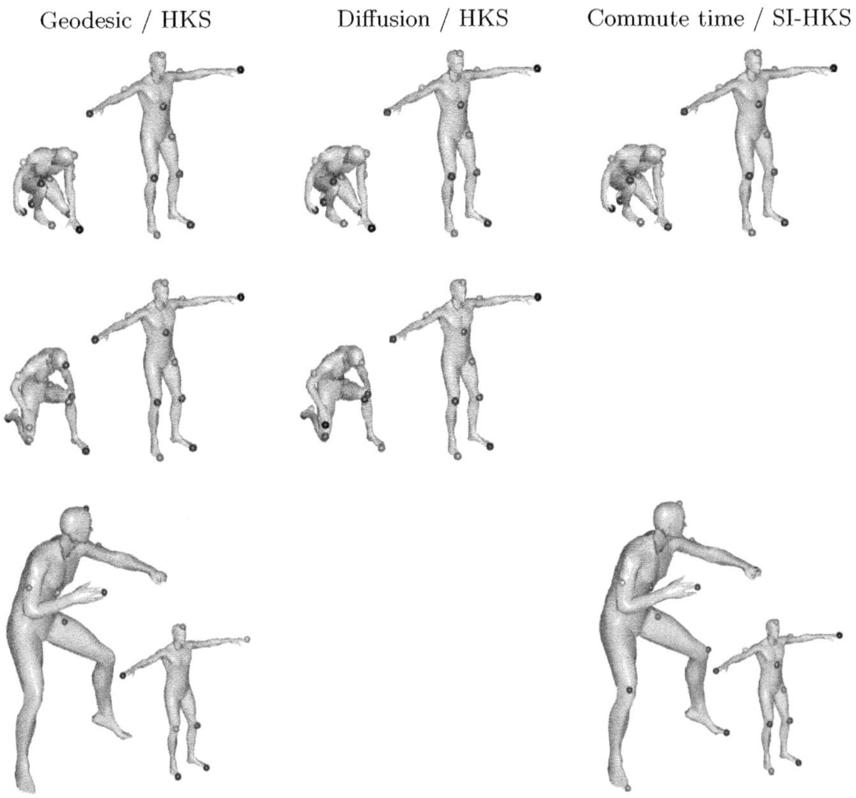

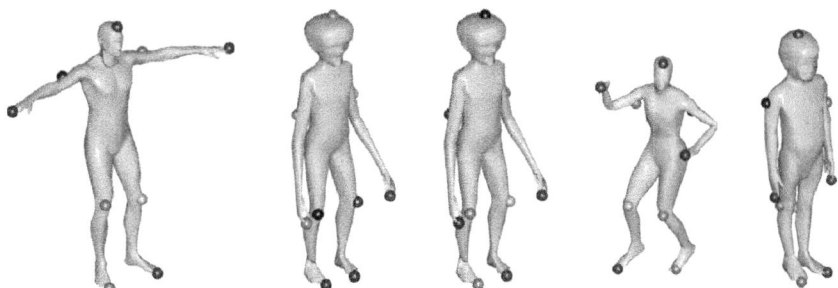

Fig. 1. Invariance to different types of transformations and the choice of the metric/descriptor. Shown is matching between isometric deformations (first row), shapes with different topology (second row), and shapes with different scale (third row), using geodesic metric and HKS descriptors (left), diffusion metric and HKS descriptors (middle), and commute time metric and scale-invariant HKS descriptors (right).

Fig. 2. Matching of an alien shape to the human shape (first column from left) using deterministic (second column) and probabilistic (third column) approaches. Columns four and five: additional probabilistic matching examples.

use powerful optimization method recently developed for this class of problems which are known to have favorable convergence properties. Our approach is especially appropriate for the challenging problems of finding similarity and correspondence between non-rigid shapes.

Limitations and Extensions. The problem of symmetric correspondences, inherent to all approaches based on intrinsic structures, cannot be resolved without resorting to some side information. There are a few potential cures to this problem. First, providing some initial correspondence between the shapes could be used to restrict the vertex set, ruling out symmetric correspondences. Second, exploiting shape orientation could be used to find orientation-consistent matches. Finally, higher-order distortions (in particular, third-order between triplets of points) can be combined to resolve the symmetry problem [15].

References

1. Berg, A., Berg, T., Malik, J.: Shape matching and object recognition using low distortion correspondences. In: Proc. CVPR, vol. 1 (2005)
2. Mémoli, F., Sapiro, G.: A theoretical and computational framework for isometry invariant recognition of point cloud data. Foundations of Computational Mathematics 5, 313–346 (2005)
3. Bronstein, A.M., Bronstein, M.M., Kimmel, R.: Generalized multidimensional scaling: a framework for isometry-invariant partial surface matching. PNAS 103, 1168–1172 (2006)
4. Bronstein, A.M., Bronstein, M.M., Kimmel, R.: Calculus of non-rigid surfaces for geometry and texture manipulation. Trans. Visualization and Computer Graphics 13, 902–913 (2007)
5. Sun, J., Ovsjanikov, M., Guibas, L.J.: A concise and provably informative multi-scale signature based on heat diffusion. Computer Graphics Forum 28, 1383–1392 (2009)
6. Ovsjanikov, M., Bronstein, A.M., Bronstein, M.M., Guibas, L.J.: Shape Google: a computer vision approach to invariant shape retrieval. In: Proc. NORDIA (2009)
7. Bronstein, M.M., Kokkinos, I.: Scale-invariant heat kernel signatures for shape recognition. INRIA Technical Report 7161 (2009)
8. Zaharescu, A., Boyer, E., Varanasi, K., Horaud, R.: Surface Feature Detection and Description with Applications to Mesh Matching. In: Proc. CVPR (2009)
9. Thorstensen, N., Keriven, R.: Non-rigid Shape matching using Geometry and Photometry. In: Proc. CVPR (2009)
10. Rustamov, R.M.: Laplace-Beltrami eigenfunctions for deformation invariant shape representation. In: Proc. SGP, pp. 225–233 (2007)
11. Mateus, D., Horaud, R.P., Knossow, D., Cuzzolin, F., Boyer, E.: Articulated shape matching using laplacian eigenfunctions and unsupervised point registration. In: Proc. CVPR (2008)
12. Hu, J., Hua, J.: Salient spectral geometric features for shape matching and retrieval. Visual Computer 25, 667–675 (2009)
13. Dubrovina, A., Kimmel, R.: Matching shapes by eigendecomposition of the Laplace-Beltrami operator. In: Proc. 3DPVT (2010)
14. Lipman, Y., Funkhouser, T.: Möbius voting for surface correspondence. TOG 28 (2009)

15. Zeng, Y., Wang, C., Wang, Y., Gu, X., Samaras, D., Paragios, N.: Dense non-rigid surface registration using high-order graph matching. In: Proc. CVPR (2010)
16. Elad, A., Kimmel, R.: Bending invariant representations for surfaces. In: Proc. Computer Vision and Pattern Recognition (CVPR), pp. 168–174 (2001)
17. Bronstein, A.M., Bronstein, M.M., Kimmel, R., Mahmoudi, M., Sapiro, G.: A Gromov-Hausdorff framework with diffusion geometry for topologically-robust non-rigid shape matching. IJCV, 1–21 (2010)
18. Qiu, H., Hancock, E.R.: Clustering and embedding using commute times. Trans. PAMI 29, 1873–1890 (2007)
19. Gromov, M.: Structures Métriques Pour les Variétés Riemanniennes. Textes Mathématiques (1) (1981)
20. Torresani, L., Kolmogorov, V., Rother, C.: Feature correspondence via graph matching: Models and global optimization. In: Forsyth, D., Torr, P., Zisserman, A. (eds.) ECCV 2008, Part II. LNCS, vol. 5303, pp. 596–609. Springer, Heidelberg (2008)
21. Bertsekas, D.P.: Nonlinear Programming. Athena Scientific (1999)
22. Komodakis, N., Paragios, N., Tziritas, G.: MRF optimization via dual decomposition: Message-passing revisited. In: Proc. ICCV (2007)
23. Zhang, H., Sheffer, A., Cohen-Or, D., Zhou, Q., van Kaick, O., Tagliasacchi, A.: Deformation-driven shape correspondence. Computer Graphics Forum 27, 1431–1439 (2008)
24. Raviv, D., Bronstein, A., Bronstein, M., Kimmel, R.: Symmetries of non-rigid shapes. In: Proc. NRTL (2007)
25. Memoli, F. On the use of gromov-hausdorff distances for shape comparison
26. Ovsjanikov, M., Sun, J., Guibas, L.J.: Global intrinsic symmetries of shapes. Computer Graphics Forum 27, 1341–1348 (2008)
27. Coifman, R.R., Lafon, S., Lee, A.B., Maggioni, M., Nadler, B., Warner, F., Zucker, S.W.: Geometric diffusions as a tool for harmonic analysis and structure definition of data: Diffusion maps. PNAS 102, 7426–7431 (2005)
28. Grinstead, C.M., Snell, L.J.: Introduction to Probability. AMS (1998)
29. Gold, S., Rangarajan, A.: A graduated assignment algorithm for graph matching. Trans. PAMI 18, 377–388 (1996)
30. Hochbaum, D., Shmoys, D.: A best possible heuristic for the k-center problem. Mathematics of Operations Research 10(2), 180–184 (1985)
31. Bronstein, A.M., Bronstein, M.M., Kimmel, R.: Numerical geometry of non-rigid shapes. Springer, Heidelberg (2008)
32. Shilane, P., Min, P., Kazhdan, M., Funkhouser, T.: The Princeton shape benchmark. In: Proc. SMI, pp. 167–178 (2004)
33. Kimmel, R., Sethian, J.A.: Computing geodesic paths on manifolds. PNAS 95, 8431–8435 (1998)
34. Meyer, M., Desbrun, M., Schroder, P., Barr, A.H.: Discrete differential-geometry operators for triangulated 2-manifolds. Visualization and Mathematics III, 35–57 (2003)

A Correspondence-Less Approach to Matching of Deformable Shapes

Jonathan Pokrass[1], Alexander M. Bronstein[1], and Michael M. Bronstein[2]

[1] Dept. of Electrical Engineering, Tel Aviv University, Israel
evgenyfo@post.tau.ac.il, bron@eng.tau.ac.il
[2] Inst. of Computational Science, Faculty of Informatics,
Università della Svizzera Italiana, Lugano, Switzerland
michael.bronstein@usi.ch

Abstract. Finding a match between partially available deformable shapes is a challenging problem with numerous applications. The problem is usually approached by computing local descriptors on a pair of shapes and then establishing a point-wise correspondence between the two. In this paper, we introduce an alternative correspondence-less approach to matching fragments to an entire shape undergoing a non-rigid deformation. We use diffusion geometric descriptors and optimize over the integration domains on which the integral descriptors of the two parts match. The problem is regularized using the Mumford-Shah functional. We show an efficient discretization based on the Ambrosio-Tortorelli approximation generalized to triangular meshes. Experiments demonstrating the success of the proposed method are presented.

Keywords: deformable shapes, partial matching, partial correspondence, partial similarity, diffusion geometry, Laplace-Beltrami operator, shape descriptors, heat kernel signature, Mumford-Shah regularization.

1 Introduction

In many real-world settings of the shape recognition problem, the data are degraded by acquisition imperfections and noise, resulting in the need to find *partial* similarity of objects. Such cases are common, for example, in face recognition, where the facial surface may be partially occluded by hair. In other applications, such as shape retrieval, correct semantic similarity of two objects is based on partial similarity – for example, a centaur is partially similar to a human because they share the human-like upper body [15].

In rigid shape analysis, modifications of the popular iterative closest point (ICP) algorithm are able to deal with partial shape alignment by rejecting points with bad correspondences (e.g., by thresholding the product of local normal vectors). However, it is impossible to guarantee how large and regular the resulting corresponding parts will be.

Bronstein *et al.* [4] formulated non-rigid partial similarity as a multi-criterion optimization problem, in which one tries to find the corresponding parts in two

A.M. Bruckstein et al. (Eds.): SSVM 2011, LNCS 6667, pp. 592–603, 2012.

shapes by simultaneously maximizing significance and similarity criteria (in [4], metric distortion [14, 19, 5] was used as a criterion of similarity, and part area as significance). The problem requires the knowledge of correspondence between the shapes, and in the absence of a given correspondence, can be solved by alternating between weighted correspondence finding and maximization of part area. In [4], a different significance criterion based on statistical occurrence of local shape descriptors was used.

One of the drawbacks of the above method is its tendency in some cases to find a large number of disconnected components, which have the same area as a larger single component. The same authors addressed this problem using a Mumford-Shah [22, 10]-like regularization for rigid [3] and non-rigid [2] shapes.

Recent works on local shape descriptors (see, e.g., descriptors [16, 33, 24, 11, 18, 21, 32, 29, 27, 9]) have led to the adoption of bags of features [28] approach popular in image analysis for the description of 3D shapes [21, 23, 30]. Bags of features allows to some extent finding partial similarity, if the overlap between the parts is sufficiently large.

In this paper, we present an approach for correspondence-less partial matching of non-rigid 3D shapes. Our work is inspired by the recent work on partial matching of images [12]. The main idea of this approach, adopted here, is to find similar parts by comparing part-wise distributions of local descriptors. This removes the need of correspondence knowledge and greatly simplifies the problem.

The rest of the paper is organized is as follows. In Section 2, we review the mathematical background of diffusion geometry, which is used for the construction of local descriptors. Section 3 deals with the partial matching problem and Section 4 addresses its discretization. Section 5 presents experimental results. Finally, Section 6 concludes the paper.

2   Background

Diffusion Geometry. Diffusion geometry is an umbrella term referring to geometric analysis of diffusion or random walk processes. We models a shape as a compact two-dimensional Riemannian manifold X. In it simplest setting, a diffusion process on X is described by the partial differential equation

$$\left(\frac{\partial}{\partial t} + \Delta\right) f(t, x) = 0, \tag{1}$$

called the *heat equation*, where Δ denotes the positive-semidefinite Laplace-Beltrami operator associated with the Riemannian metric of X. The heat equation describes the propagation of heat on the surface and its solution $f(t, x)$ is the heat distribution at a point x in time t. The initial condition of the equation is some initial heat distribution $f(0, x)$; if X has a boundary, appropriate boundary conditions must be added.

The solution of (1) corresponding to a point initial condition $f(0, x) = \delta(x, y)$, is called the *heat kernel* and represents the amount of heat transferred from x to y in time t due to the diffusion process. The value of the heat kernel $h_t(x, y)$

can also be interpreted as the transition probability density of a random walk of length t from the point x to the point y.

Using spectral decomposition, the heat kernel can be represented as

$$h_t(x,y) = \sum_{i \geq 0} e^{-\lambda_i t} \phi_i(x) \phi_i(y). \tag{2}$$

Here, ϕ_i and λ_i denote, respectively, the eigenfunctions and eigenvalues of the Laplace-Beltrami operator satisfying $\Delta \phi_i = \lambda_i \phi_i$ (without loss of generality, we assume λ_i to be sorted in increasing order starting with $\lambda_0 = 0$). Since the Laplace-Beltrami operator is an *intrinsic* geometric quantity, i.e., it can be expressed solely in terms of the metric of X, its eigenfunctions and eigenvalues as well as the heat kernel are invariant under isometric transformations (bending) of the shape.

Heat Kernel Signatures. By setting $y = x$, the heat kernel $h_t(x,x)$ expresses the probability density of remaining at a point x after time t. The value $h_t(x,x)$, sometimes referred to as the *auto-diffusivity function*, is related to the Gaussian curvature $K(x)$ through

$$h_t(x,x) \approx \frac{1}{4\pi t}\left(1 + \frac{1}{6}K(x)t + \mathcal{O}(t^2)\right). \tag{3}$$

This relation coincides with the well-known fact that heat tends to diffuse slower at points with positive curvature, and faster at points with negative curvature. Under mild technical conditions, the set $\{h_t(x,x)\}_{t>0}$ is fully informative in the sense that it allows to reconstruct the Riemannian metric of the manifold [29].

Sun *et al.* [29] proposed constructing point-wise descriptors referred to as *heat kernel signatures* (HKS) by taking the values of the discrete auto-diffusivity function at point x at multiple times, $\mathbf{p}(x) = c(x)(h_{t_1}(x,x), ..., h_{t_d}(x,x))$, where $t_1, ..., t_d$ are some fixed time values and $c(x)$ is chosen so that $||p(x)||_2 = 1$. Such a descriptor is a vector of dimensionality d at each point. Since the heat kernel is an intrinsic quantity, the HKS is invariant to isometric transformations of the shape.

A scale-invariant version of the HKS descriptor (SI-HKS) was proposed in [9]. First, the heat kernel is sampled logarithmically in time. Next, the logarithm and a derivative with respect to time of the heat kernel values are taken to undo the multiplicative constant. Finally, taking the magnitude of the Fourier transform allows to undo the scaling of the time variable.

Bags of Features. Ovsjanikov *et al.* [23] and Toldo *et al.* [30] proposed constructing global shape descriptors from local descriptors using the *bag of features* paradigm [28]. In this approach, a fixed "geometric vocabulary" is computed by means of an off-line clustering of the descriptor space. Next, each point descriptor is represented in the vocabulary using vector quantization. The bag of features global shape descriptor is then computed as the histogram of quantized descriptors over the entire shape.

3   Partial Matching

In what follows, we assume to be given two shapes X and Y with corresponding point-wise descriptor fields $\mathbf{p}$ and $\mathbf{q}$ defined on them (here we adopt HKS descriptors, though their quantized variants or any other intrinsic point-wise descriptors can be used as well). Assuming that Y is a part of an unknown shape that is intrinsically similar to X, we aim at finding a part $X' \subseteq X$ having the same area A of Y such that the integral shape descriptors computed on X' and Y coincide as closely as possible. In order to prevent the parts from being fragmented and irregular, we penalize for their boundary length. The entire problem can be expressed as minimization of the following energy functional

$$E(X') = \left\| \int_{X'} \mathbf{p} da - \overline{\mathbf{q}} \right\|^2 + \lambda_r L(\partial X') \tag{4}$$

under the constraint $A(X') = A$, where A denotes area and $\overline{\mathbf{q}} = \int_Y \mathbf{q} da$. The first term of the functional constitutes the data term while the second one is the regularity term whose influence is controlled by the parameter λ_r.

Discretization of the above minimization problem with a crisp set X' results in combinatorial complexity. To circumvent this difficulty, in [2, 3] it was proposed to relax the problem by replacing the crisp part X' by a fuzzy membership function u on X, replacing the functional E by a generalization of the Mumford-Shah functional [22] to surfaces. Here, we adopt this relaxation as well as the approximation of the Mumford-Shah functional proposed by Ambrosio and Tortorelli [1]. This yields the problem of the form

$$\min_{u, \rho, \sigma} D(u) + \lambda_r R(u; \rho)$$

$$\text{s.t.} \int_X u da = A, \tag{5}$$

with the data term

$$D(u) = \left\| \int_X \mathbf{p} u da - \overline{\mathbf{q}} \right\|^2 \tag{6}$$

and the Ambrosio-Tortorelli regularity term

$$R(u; \rho) = \frac{\lambda_s}{2} \int_X \rho^2 \|\nabla u\|^2 da + \lambda_b \epsilon \int_X \|\nabla \rho\|^2 da + \frac{\lambda_b}{4\epsilon} \int_X (1 - \rho)^2 da, \tag{7}$$

where ρ is the so-called phase field indicating the discontinuities of u, and $\epsilon > 0$ is a parameter.

The first term of R above imposes piece-wise smoothness of the fuzzy part u governed by the parameter λ_s. By setting a sufficiently large λ_s, the parts become approximately piece-wise constant as desired in the original crisp formulation (4). The second term of R is analogous to the boundary length term in (4) and converges to the latter as $\epsilon \to 0$.

We minimize (5) using alternating minimization comprising the following two iteratively repeated steps:

Step 1: fix ρ and solve for u

$$\min_u \left\| \int_X \mathbf{p} u \, da - \bar{\mathbf{q}} \right\|^2 + \lambda_r \frac{\lambda_s}{2} \int_X \rho^2 \|\nabla u\|^2 da \quad \text{s.t.} \quad \int_X u \, da = A. \tag{8}$$

Step 2: fix the part u and solve for ρ

$$\min_\rho \frac{\lambda_s}{2} \int_X \rho^2 \|\nabla u\|^2 da + \lambda_b \epsilon \int_X \|\nabla \rho\|^2 da + \frac{\lambda_b}{4\epsilon} \int_X (1-\rho)^2 da. \tag{9}$$

4   Discretization and Numerical Aspects

We represent the surface X as triangular mesh with n faces constructed upon the samples $\{\mathbf{x}_1, \ldots, \mathbf{x}_m\}$ and denote by $\mathbf{a} = (a_1, \ldots, a_m)^{\mathrm{T}}$ the corresponding area elements at each vertex. $\mathbf{A} = \mathrm{diag}\{\mathbf{a}\}$ denote the diagonal $m \times m$ matrix created out of $\mathbf{a}$. The membership function u is sampled at each vertex and represented as the vector $\mathbf{u} = (u_1, \ldots, u_m)^{\mathrm{T}}$. Similarly, the phase field is represented as the vector $\boldsymbol{\rho} = (\rho_1, \ldots, \rho_m)^{\mathrm{T}}$.

Descriptors. The computation of the discrete heat kernel $h_t(\mathbf{x}_1, \mathbf{x}_2)$ requires computing discrete eigenvalues and eigenfunctions of the discrete Laplace-Beltrami operator. The latter can be computed directly using the finite elements method (FEM) [26], of by discretization of the Laplace operator on the mesh followed by its eigendecomposition. Here, we adopt the second approach according to which the discrete Laplace-Beltrami operator is expressed in the following generic form,

$$(\Delta f)_i = \frac{1}{a_i} \sum_j w_{ij}(f_i - f_j), \tag{10}$$

where $f_i = f(\mathbf{x}_i)$ is a scalar function defined on the mesh, w_{ij} are weights, and a_i are normalization coefficients. In matrix notation, (10) can be written as $\Delta \mathbf{f} = \mathbf{A}^{-1} \mathbf{W} \mathbf{f}$, where $\mathbf{f}$ is an $m \times 1$ vector and $\mathbf{W} = \mathrm{diag}\left\{ \sum_{l \neq i} w_{il} \right\} - w_{ij}$.

The discrete eigenfunctions and eigenvalues are found by solving the *generalized eigendecomposition* [17] $\mathbf{W}\boldsymbol{\Phi} = \mathbf{A}\boldsymbol{\Phi}\boldsymbol{\Lambda}$, where $\boldsymbol{\Lambda} = \mathrm{diag}\{\lambda_l\}$ is a diagonal matrix of eigenvalues and $\boldsymbol{\Phi} = (\phi_l(x_i))$ is the matrix of the corresponding eigenvectors.

Different choices of $\mathbf{W}$ have been studied, depending on which continuous properties of the Laplace-Beltrami operator one wishes to preserve [13, 31]. For triangular meshes, a popular choice adopted in this paper is the *cotangent weight scheme* [25, 20], in which

$$w_{ij} = \begin{cases} (\cot \beta_{ij} + \cot \gamma_{ij})/2 : & \mathbf{x}_j \in \mathcal{N}(\mathbf{x}_j); \\ 0 & : \quad else, \end{cases} \tag{11}$$

where β_{ij} and γ_{ij} are the two angles opposite to the edge between vertices $\mathbf{x}_i$ and $\mathbf{x}_j$ in the two triangles sharing the edge.

Data Term. Denoting by $\mathbf{P}$ $d \times m$ the matrix of point-wise descriptors on X (stored in columns), we have

$$\int \mathbf{p}u\,da \approx \mathbf{P}\mathrm{diag}\{\mathbf{A}\}\mathbf{u} \tag{12}$$

This yields the following discretization of the data term (6)

$$D(\mathbf{u}) = \|\mathbf{P}\mathbf{A}\mathbf{u} - \bar{\mathbf{q}}\|^2 = \mathbf{u}^\mathrm{T}\mathbf{A}^\mathrm{T}\mathbf{P}^\mathrm{T}\mathbf{P}\mathbf{A}\mathbf{u} - 2\bar{\mathbf{q}}^\mathrm{T}\mathbf{P}\mathbf{A}\mathbf{u} + \bar{\mathbf{q}}^\mathrm{T}\bar{\mathbf{q}}. \tag{13}$$

Gradient Norm. We start by deriving the discretization of a single term $\rho^2\|\nabla u\|^2 da$ in some triangle t of the mesh. Let us denote by $\mathbf{x}_i, \mathbf{x}_j$ and $\mathbf{x}_k$ the vertices of the triangle and let $\mathbf{X}_t = (\mathbf{x}_j - \mathbf{x}_i, \mathbf{x}_k - \mathbf{x}_i)$ be the 3×2 matrix whose columns are the vectors forming the triangle, and by $\alpha_t = \frac{1}{2}\sqrt{\det(\mathbf{X}_t^\mathrm{T}\mathbf{X}_t)}$ its area. Let also $\mathbf{D}_t$ be the sparse $2 \times m$ matrix with $+1$ at indices $(1, j)$ and $(2, k)$, and -1 at $(1, i)$ and $(2, i)$. $\mathbf{D}_t$ is constructed in such a way to give the differences of values of u on the vertices of the triangle with respect to the values at the central vertex, $\mathbf{D}_t\mathbf{u} = (u_j - u_i, u_k - u_i)^\mathrm{T}$. The gradient of the function u is constant on the triangle and can be expressed in these terms by $\mathbf{g}_t = (\mathbf{X}_t^\mathrm{T}\mathbf{X}_t)^{-1/2}\mathbf{D}_t\mathbf{u} = \mathbf{G}_t\mathbf{u}$.

In order to introduce the weighting by ρ^2, let $\mathbf{S}$ be an $n \times m$ sparse matrix with the elements $s_{ti} = \frac{1}{3}$ for every vertex i belonging to the triangle t and zero otherwise. In this notation, $\mathbf{S}\rho$ is a per-triangle field whose elements are the average values of ρ on each of the mesh triangles. We use the Kroenecker product of $\mathbf{S}$ with $\mathbf{1} = (1,1)^\mathrm{T}$ to define the $2n \times n$ matrix $\mathbf{S} \otimes \mathbf{1}$ formed by replicating twice each of the rows of $\mathbf{S}$. This yields

$$\int \rho^2\|\nabla u\|^2 da \approx \sum_t \|(\mathbf{S}\rho)_t \mathbf{G}_t\mathbf{u}\|^2\alpha_t =$$

$$= \left\| \mathrm{diag}\{(\mathbf{S} \otimes \mathbf{1})\rho\} \begin{pmatrix} \sqrt{\alpha_1}\mathbf{G}_1 \\ \vdots \\ \sqrt{\alpha_n}\mathbf{G}_n \end{pmatrix} \right\|^2 = \mathbf{u}^\mathrm{T}\mathbf{G}^\mathrm{T}\mathbf{S}(\rho)\mathbf{G}\mathbf{u}, \tag{14}$$

where $\mathbf{G}$ is the matrix containing $\sqrt{\alpha_t}\mathbf{G}_t$ stacked as rows, and $\mathbf{S}(\rho) = \mathrm{diag}\{(\mathbf{S} \otimes \mathbf{1})\rho\}^2$.

Discretized Alternating Minimization. We plug in the results obtained so far into the two steps of the alternating minimization problem (8)–(9). For fixed ρ, the discretized minimization problem (8) w.r.t. $\mathbf{u}$ can be written as

$$\min_{\mathbf{u}} \mathbf{u}^\mathrm{T}\left(\mathbf{A}^\mathrm{T}\mathbf{P}^\mathrm{T}\mathbf{P}\mathbf{A} + \lambda_\mathrm{r}\frac{\lambda_\mathrm{s}}{2}\mathbf{G}^\mathrm{T}\mathbf{S}(\rho)\mathbf{G}\right)\mathbf{u} - 2\bar{\mathbf{q}}^\mathrm{T}\mathbf{P}\mathbf{A}\mathbf{u} \quad \text{s.t.} \quad \mathbf{a}^\mathrm{T}\mathbf{u} = A \tag{15}$$

Let us now fix $\mathbf{u}$. In a triangle t, we denote $g_t^2 = \|\mathbf{G}_t\mathbf{u}\|^2$ and let $\mathbf{R}(\mathbf{u}) = \mathrm{diag}\{\alpha_1 g_1^2, \ldots, \alpha_n g_n^2\}$. Using this notation, we obtain the following discretization of the integrals in the regularization term (7)

$$\int_X \rho^2\|\nabla u\|^2 da \approx \sum_t (\mathbf{S}\rho)_t^2 g_t^2\alpha_t = \rho^\mathrm{T}\mathbf{S}^\mathrm{T}\mathbf{R}(\mathbf{u})\mathbf{S}\rho. \tag{16}$$

Similar to the derivation of (14),

$$\int \|\nabla \rho\|^2 da \approx \boldsymbol{\rho}^{\mathrm{T}} \mathbf{S} \mathbf{G}^{\mathrm{T}} \mathbf{G} \mathbf{S} \boldsymbol{\rho}, \tag{17}$$

and

$$\int_X (1 - \rho)^2 da \approx \boldsymbol{\rho}^{\mathrm{T}} \mathbf{A} \boldsymbol{\rho} - 2 \mathbf{a}^{\mathrm{T}} \boldsymbol{\rho} + \mathbf{1}^{\mathrm{T}} \mathbf{a}. \tag{18}$$

The discretized minimization problem (9) w.r.t. $\boldsymbol{\rho}$ becomes

$$\min_{\boldsymbol{\rho}} \boldsymbol{\rho}^{\mathrm{T}} \left(2\epsilon \lambda_{\mathrm{s}} \mathbf{S}^{\mathrm{T}} \mathbf{R}(\mathbf{u}) \mathbf{S} + 4\epsilon^2 \lambda_{\mathrm{b}} \mathbf{S} \mathbf{G}^{\mathrm{T}} \mathbf{G} \mathbf{S} + \lambda_{\mathrm{b}} \mathbf{A} \right) \boldsymbol{\rho} - 2\lambda_{\mathrm{b}} \mathbf{a}^{\mathrm{T}} \boldsymbol{\rho}. \tag{19}$$

Since the above is an unconstrained quadratic problem, it has the following closed-form solution

$$\boldsymbol{\rho} = \left(\frac{2\epsilon \lambda_{\mathrm{s}}}{\lambda_{\mathrm{b}}} \mathbf{S}^{\mathrm{T}} \mathbf{R}(\mathbf{u}) \mathbf{S} + 4\epsilon^2 \mathbf{S} \mathbf{G}^{\mathrm{T}} \mathbf{G} \mathbf{S} + \mathbf{A} \right)^{-1} \mathbf{a}. \tag{20}$$

5   Results

In order to test our approach, we performed several partial matching experiments on data from the SHREC 2010 benchmark [8,7] and the TOSCA dataset [6].[1] The datasets contained high-resolution (10K-50K vertices) triangular meshes of human and animal shapes with simulated transformations, with known groundtruth correspondence between the transformed shapes. In our experiments, all the shapes were downsampled to approximately 2500 vertices. Parts were cut by taking a geodesic circle of random radius around a random center point.

For each part, the normalized HKS descriptor was calculated at each vertex belonging to the part. To avoid boundary effects (see Figure 1), descriptors close to the boundary were ignored when calculating $\overline{\mathbf{q}}$ in (4). The distance from the boundary was selected in accordance to the time scales of the HKS. We used ten linearly spread samples in range $[65, 90]$ for the descriptors and the according distance taken from edge was set to 15. Two to three iterations of the alternating minimization procedure were used, exhibiting fast convergence (Figure 2). After three iterations the member function $\mathbf{u}$ typically ceased changing significantly. The phase map ρ assumed the values close to 1 in places of low gradient of the membership function $\mathbf{u}$, and less than 1 in high gradient areas (Figure 2). The importance of the regularization step is is evident observing the change in $\mathbf{u}$ in Figure 2. Figure 3 shows the influence of the parameter λ_{r}, controlling the impact of the regularization. For too small values of λ_{r}, two equally weighted matches are obtained due to symmetry (left). The phenomenon decreases with the increase of the influence of the regularization penalty. However, increasing λ_{r} more causes incorrect matching (second and third columns from the right) due

---

[1] Both datasets are available online at http://tosca.cs.technion.ac.il

to low data term influence. Increasing it even more starts smoothing the result (rightmost column) until eventually making the membership function uniform over the entire shape. The resulting membership function **u** was thresholded in such a way that the outcome area will be as close as possible to the query area. Figures 4–6 show examples of matching results after thresholding. Notice that in Figure 4 the matching result sometimes contain the symmetric counter part of the result due to invariance of the HKS descriptor to intrinsic symmetry. (in this figure, the threshold was adjusted to the value of 0.35 when the membership functions weights are split between two symmetric parts as in Figure 3). The method is robust to shape deformations and geometric and topological noise as depicted in Figure 6. Note that the figures show part-to-whole shape matching, but because of the low scale HKS descriptors the same procedure works for matching to other parts as well.

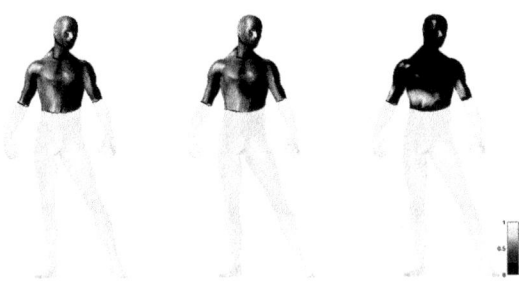

Fig. 1. An RGB visualization of the first three component of HKS descriptors computed on the full shape (left) and on a part of the shape (center). The L_2 difference between the two fields is depicted in the rightmost figure. Note that the difference is maximal on the boundary decaying fast away from it; the error decay speed depends on the scale choice in the HKS descriptor.

Table 1 summarizes quantitative evaluation that was performed on a subset of the SHREC database, for which groundtruth correspondence and its bilaterally symmetric counterpart were available. This subset included a male, a dog and a horse shape classes with different geometric, topological and noise deformations (98 shapes in total). The query set was generated by selecting a part from a deformed shape (1000 queries in each deformation category) and matched to the null shape with parameters and thresholds as described above.

Complexity. The code was implemented in Matalb with some parts written in C with MEX interface. The quadratic programming problem (8) in Step 1 was solved using QPC[2] implementation of a dual active set method. The experiments were run on 2.3GHz Intel Core2 Quad CPU, 2GB RAM in Win7 32bit environment. The running time (including re-meshing and descriptor calculation) per part was $40 - 50$ sec.

---

[2] available online at http://sigpromu.org/quadprog

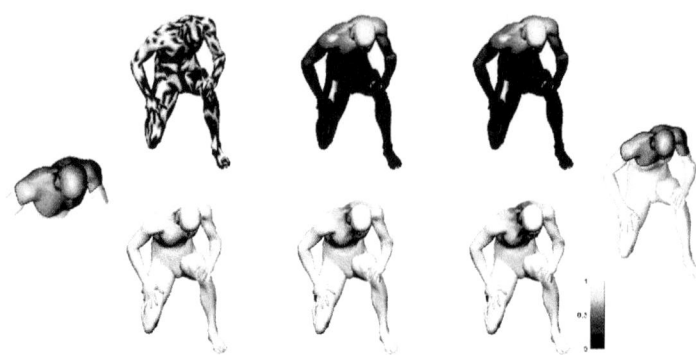

Fig. 2. Convergence of the alternating minimization procedure. Depicted are the membership function **u** (top row) and the phase field ρ (bottom row) at the first three iterations.

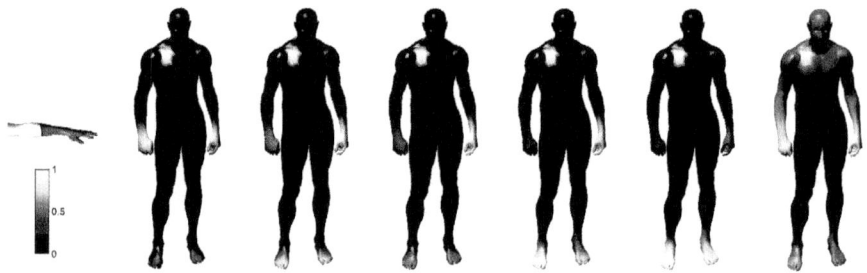

Fig. 3. The influence of the parameter λ_r, controlling the impact of regularization. The leftmost figure depicts a query part; figures on the right are the membership function u for different values of λ_r.

Table 1. Part matching performance on transformed shapes from the SHREC benchmark. At each query a random part (location and size) was selected from a deformed shape and matched to the null shape. Overlap is reported compared to the groundtruth correspondence between the shapes (in parentheses taking into account the intrinsic bilateral symmetry).

Transformation	Queries	Avg. overlap
Isometry	1000	75% (85%)
Isometry + Shotnoise	1000	75% (85%)
Isometry + Noise	1000	71% (82%)
Isometry + Microholes	1000	68% (82%)
Isometry + Holes	1000	66% (76%)
All	5000	71% (82%)

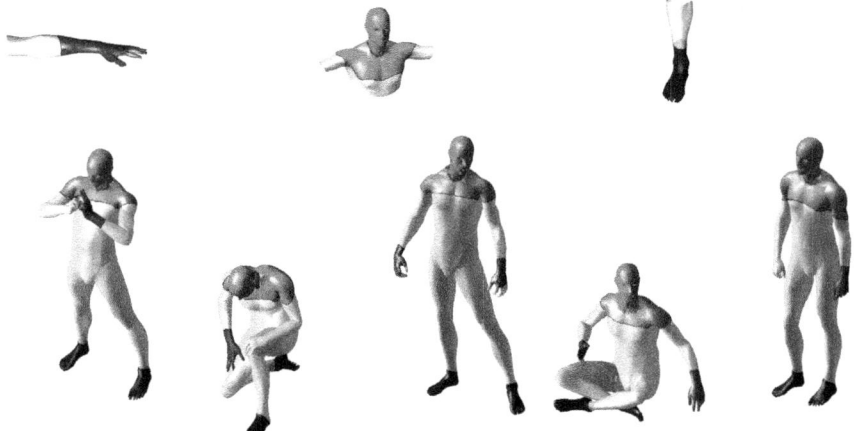

Fig. 4. Examples of matching of random parts of shapes (first row) to approximately isometric deformations of the shapes (second row). Color code indicates different parts.

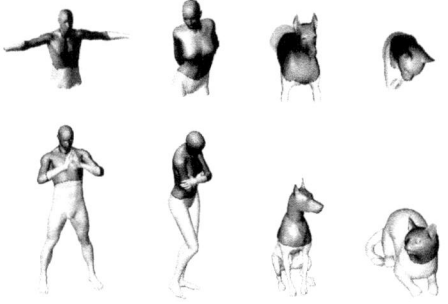

Fig. 5. Examples of matching of random parts of shapes (first row) to to approximately isometric deformations of the shapes (second row).

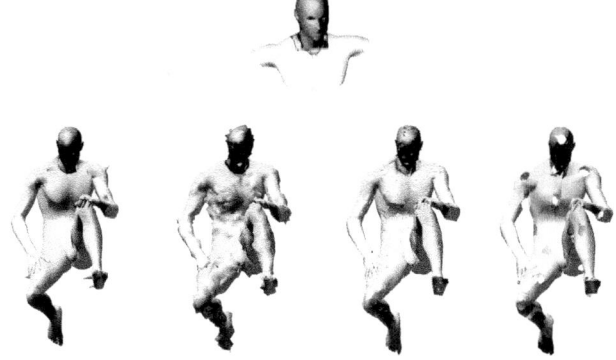

Fig. 6. Results of matching a part of shape (first row) to shapes distorted by different transformations (second row). Shown left-to-right are: shot noise, noise, micro holes and holes.

6   Conclusions

We presented a framework for finding partial similarity between shapes which does not rely on explicit correspondence. The method is based on regularized matching of region-wise local descriptors, and can be efficiently implemented. Experimental results show that our approach performs well in challenging matching scenarios, such as the presence of geometric and topological noise. In the future work, we will extend the method to the setting of two partially-similar full shapes, in which two similar parts have to be found in each shape, and then consider a multi-part matching (puzzle) scenario.

References

1. Ambrosio, L., Tortorelli, V.M.: Approximation of functionals depending on jumps by elliptic functionals via-convergence. Comm. Pure Appl. Math. 43(8), 999–1036 (1990)
2. Bronstein, A.M., Bronstein, M.: Not only size matters: regularized partial matching of nonrigid shapes. In: Prof. NORDIA (2008)
3. Bronstein, A.M., Bronstein, M.M.: Regularized partial matching of rigid shapes. In: Forsyth, D., Torr, P., Zisserman, A. (eds.) ECCV 2008, Part II. LNCS, vol. 5303, pp. 143–154. Springer, Heidelberg (2008)
4. Bronstein, A.M., Bronstein, M.M., Bruckstein, A.M., Kimmel, R.: Partial similarity of objects, or how to compare a centaur to a horse. IJCV 84(2), 163–183 (2009)
5. Bronstein, A.M., Bronstein, M.M., Kimmel, R.: Generalized multidimensional scaling: a framework for isometry-invariant partial surface matching. Proc. National Academy of Science (PNAS) 103(5), 1168–1172 (2006)
6. Bronstein, A.M., Bronstein, M.M., Kimmel, R.: Numerical geometry of non-rigid shapes. Springer-Verlag New York Inc., Secaucus (2008)
7. Bronstein, A.M., Bronstein, M.M., Bustos, B., Castellani, U., Crisani, M., Falcidieno, B., Guibas, L.J., Kokkinos, I., Murino, V., Ovsjanikov, M., et al.: SHREC 2010: robust feature detection and description benchmark. In: Proc. 3DOR (2010)
8. Bronstein, A.M., Bronstein, M.M., Castellani, U., Dubrovina, A., Guibas, L.J., Horaud, R.P., Kimmel, R., Knossow, D., von Lavante, E., Mateus, D., et al.: SHREC 2010: robust correspondence benchmark. In: Eurographics Workshop on 3D Object Retrieval (2010)
9. Bronstein, M.M., Kokkinos, I.: Scale-invariant heat kernel signatures for non-rigid shape recognition. In: Proc. CVPR (2010)
10. Chan, T.F., Vese, L.A.: Active contours without edges. IEEE Trans. Image Processing 10(2), 266–277 (2001)
11. Clarenz, U., Rumpf, M., Telea, A.: Robust feature detection and local classification for surfaces based on moment analysis. Trans. Visualization and Computer Graphics 10(5), 516–524 (2004)
12. Domokos, C., Kato, Z.: Affine puzzle: Realigning deformed object fragments without correspondences. In: Daniilidis, K., Maragos, P., Paragios, N. (eds.) ECCV 2010. LNCS, vol. 6312, pp. 777–790. Springer, Heidelberg (2010)
13. Floater, M.S., Hormann, K.: Surface parameterization: a tutorial and survey. In: Advances in Multiresolution for Geometric Modelling, vol. 1 (2005)
14. Gromov, M.: Structures Métriques Pour les Variétés Riemanniennes. Textes Mathématiques, vol. (1) (1981)

15. Jacobs, D., Weinshall, D., Gdalyahu, Y.: Class representation and image retrieval with non-metric distances. Trans. PAMI 22(6), 583–600 (2000)
16. Johnson, A.E., Hebert, M.: Using spin images for efficient object recognition in cluttered 3D scenes. Trans. PAMI 21(5), 433–449 (1999)
17. Lévy, B.: Laplace-Beltrami eigenfunctions towards an algorithm that "understands" geometry. In: Proc. Shape Modeling and Applications (2006)
18. Manay, S., Hong, B.W., Yezzi, A.J., Soatto, S.: Integral invariant signatures. LNCS, pp. 87–99 (2004)
19. Mémoli, F., Sapiro, G.: A theoretical and computational framework for isometry invariant recognition of point cloud data. Foundations of Computational Mathematics 5, 313–346 (2005)
20. Meyer, M., Desbrun, M., Schroder, P., Barr, A.H.: Discrete differential-geometry operators for triangulated 2-manifolds. Visualization and Mathematics III, 35–57 (2003)
21. Mitra, N.J., Guibas, L.J., Giesen, J., Pauly, M.: Probabilistic fingerprints for shapes. In: Proc. SGP (2006)
22. Mumford, D., Shah, J.: Optimal approximations by piecewise smooth functions and associated variational problems. Communications on pure and applied mathematics 42(5), 577–685 (1989)
23. Ovsjanikov, M., Bronstein, A.M., Guibas, L.J., Bronstein, M.M.: Shape Google: a computer vision approach to invariant shape retrieval. In: Proc. NORDIA, Citeseer (2009)
24. Pauly, M., Keiser, R., Gross, M.: Multi-scale feature extraction on point-sampled surfaces. In: Computer Graphics Forum, vol. 22, pp. 281–289 (2003)
25. Pinkall, U., Polthier, K.: Computing discrete minimal surfaces and their conjugates. Experimental Mathematics 2(1), 15–36 (1993)
26. Reuter, M., Wolter, F.-E., Peinecke, N.: Laplace-spectra as fingerprints for shape matching. In: Proc. ACM Symp. Solid and Physical Modeling, pp. 101–106 (2005)
27. Sipiran, I., Bustos, B.: A robust 3D interest points detector based on Harris operator. In: Proc. 3DOR, pp. 7–14. Eurographics (2010)
28. Sivic, J., Zisserman, A.: Video Google: a text retrieval approach to object matching in videos. In: Proc. CVPR (2003)
29. Sun, J., Ovsjanikov, M., Guibas, L.: A Concise and Provably Informative Multi-Scale Signature Based on Heat Diffusion. In: Computer Graphics Forum, vol. 28, pp. 1383–1392 (2009)
30. Toldo, R., Castellani, U., Fusiello, A.: Visual vocabulary signature for 3D object retrieval and partial matching. In: Proc. 3DOR (2009)
31. Wardetzky, M., Mathur, S., Kälberer, F., Grinspun, E.: Discrete Laplace operators: no free lunch. In: Conf. Computer Graphics and Interactive Techniques (2008)
32. Zaharescu, A., Boyer, E., Varanasi, K., Horaud, R.: Surface feature detection and description with applications to mesh matching. In: Proc. CVPR (2009)
33. Zhang, C., Chen, T.: Efficient feature extraction for 2D/3D objects in mesh representation. In: Proc. ICIP, vol. 3 (2001)

Hierarchical Matching of Non-rigid Shapes

Dan Raviv[*], Anastasia Dubrovina[*], and Ron Kimmel

Technion - Israel Institute of Technology

Abstract. Detecting similarity between non-rigid shapes is one of the fundamental problems in computer vision. While rigid alignment can be parameterized using a small number of unknowns representing rotations, reflections and translations, non-rigid alignment does not have this advantage. The majority of the methods addressing this problem boil down to a minimization of a distortion measure. The complexity of a matching process is exponential by nature, but it can be heuristically reduced to a quadratic or even linear for shapes which are smooth two-manifolds. Here we model shapes using both local and global structures, and provide a hierarchical framework for the quadratic matching problem.

Keywords: Shape correspondence, Laplace-Beltrami, diffusion geometry, local signatures.

1 Introduction

The paper addresses the problem of finding point-correspondences between non-rigid almost isometric shapes. The correspondence is required for various applications such as shape retrieval, registration, deformation, shape morphing, symmetry, self-similarity detection, to name a few.

A common approach to detect correspondence between shapes differing by a certain class of transformations consists of employing invariant properties under those transformations to formulate a measure of dissimilarity between the shapes, and minimize it in order to find the correct matching. Here we use a matching scheme based on local and global surface properties, namely, local surface descriptors and global metric structures. The proposed method is demonstrated with two different types of metrics - geodesic and diffusion, and different surface descriptors, that include histograms of geodesic and diffusion distances, heat kernel signatures [33], and related descriptors based on the Laplace-Beltrami operator [12].

The main issue we address is the matching complexity. Given two shapes represented by triangular meshes, direct comparison of their pointwise surface descriptors and metric structures is combinatorial in nature (see [24] for the metric comparison problem). Our main contribution is a multi-resolution matching algorithm that can handle a large number of points, and still produces a correspondence consistent in terms of both pointwise and pairwise surface properties.

---

[*] Equal contributors.

A.M. Bruckstein et al. (Eds.): SSVM 2011, LNCS 6667, pp. 604–615, 2012.

According to the proposed scheme, at the lowest resolution we solve the exact correspondence problem, up to the approximation introduced by the optimization algorithm. We then propagate this information to higher resolutions thus refining the solution.

The rest of the paper is organized as follows: a brief review of some previous efforts is presented in the next section. Section 2 presents the correspondence problem formulation, followed by Section 3, reviewing relevant mathematical background. Section 4 presents the suggested multi-resolution algorithm. Section 5 contains numerical results, and comparison to the state-of-art algorithms, followed by Section 6 that concludes the paper.

1.1 Non-Rigid Correspondence in a Brief

Zigelman et al. [39], and Elad and Kimmel [13] suggested a method for matching isometric shapes by embedding them into a Euclidian space using multidimensional scaling (MDS), thus obtaining isometry invariant representations, followed by rigid shape matching in that space. Since it is generally impossible to embed a non-flat 2D manifold into a flat Euclidean domain without introducing some errors, the inherited embedding error affects the matching accuracy of all methods of this type. For that end, Jain et al. [17] and Mateus et al. [21] suggested alternative isometry-invariant shape representations, obtained by using eigendecomposition of discrete Laplace operators. The Global Point Signature (GPS) suggested by Rustamov [31] for shape comparison employs the discrete Laplace-Beltrami operator, which, at least theoretically, captures the shape's geometry more faithfully. The Laplace-Beltrami operator was later employed by Sun et al. [33], and Ovsjanikov et al. [25], to construct their Heat Kernel Signature (HKS) and Heat Kernel Maps, respectively. Zaharescu et al. [37] suggested an extension of 2D descriptors for surfaces, and used them to perform the matching. While linear methods, such as [37,25] produce good results, once distortions start to appear, ambiguity increases, and alternative formulations should be thought of. Adding the proposed approach as a first step in one of the above linear dense matching algorithms can improve the final results. Hu and Hua [16] used the Laplace-Beltrami operator for matching using prominent features, and Dubrovina and Kimmel [12] suggested employing surface descriptors based on its eigendecomposition, combined with geodesic distances, in a quadratic optimization formulation of the matching problem. The above methods, incorporating pairwise constraints, tend to be slow due to high computational complexity. Wang et al. [36] used a similar problem formulation, casted as a graph labeling problem, and experimented with different surface descriptors and metrics.

Memoli and Sapiro [24], Bronstein et al. [6], and Memoli [22,23] compared shapes using different approximations of the Gromov-Hausdorff distance [14]. Bronstein et al. [7] used the approach suggested in [6] with diffusion geometry, in order to match shapes with topological noise, and Thorstensen and Keriven [35] extended it to handle surfaces with textures. The methods in [24,22,23] were intended for surface comparison rather than matching, and as such they do not produce correspondence between shapes. At the other end, the GMDS

algorithm [7] results in a non-convex optimization problem, therefore it requires
good initializations in order to obtain meaningful solutions, and can be used as a
refinement step for most other shape matching algorithms. Other algorithms employing geodesic distances to perform the matching were suggested by Anguelov
et al. [1], who optimized a joint probabilistic model over the set of all possible
correspondences to obtain a sparse set of corresponding points, and by Tevs *et
al.* [34] who proposed a randomized algorithm for matching feature points based
on geodesic distances between them. Zhang *et al.* [38] performed the matching
using extremal curvature feature points and a combinatorial tree traversal algorithm, but its high complexity allowed them to match only a small number of
points.Lipman and Funkhouser [20] used the fact that isometric transformation
between two shapes is equivalent to a Möbius transformation between their conformal mappings, and obtained this transformation by comparing the respective
conformal factors. However, there is no guarantee that this result minimizes the
difference between pairwise geodesic distances of matched points.

Self-similarity and symmetry detection are particular cases of the correspondence detection problem. Instead of detecting the non-rigid mapping between
two shapes, [28,26,18] search for a mapping from the shape to itself, and thus
are able to detect intrinsic symmetries.

2   Problem Formulation

The problem formulation we use is based on comparison of local and global surface properties that remain approximately invariant under non-rigid ϵ-isometric
transformations. Given a shape X, we assume that it is endowed with a *metric*
$d_X : X \times X \to \mathbb{R}_+ \cup \{0\}$, measuring distances on X, and pointwise structure
$f_X : X \to \mathbb{R}^d$, which is represented by a set of d-dimensional descriptors.

Given two shapes X and Y, endowed with metrics d_X, d_Y and descriptors
f_X, f_Y, we would like to find correspondence that best preserves these properties.
We denote the correspondence between X and Y by a mapping $\mathcal{C} : X \times Y \to \{0, 1\}$ such that

$$\mathcal{C}(x, y) = \begin{cases} 1, & x \in X \text{ corresponds to } y \in Y, \\ 0, & \text{otherwise.} \end{cases} \tag{1}$$

In order to measure how well the correspondence $\mathcal{C}$ preserves the geometric
structures of the shapes we use the following dissimilarity function based on
global and local shape properties,

$$\text{dis}(\mathcal{C}) = \text{dis}_{lin}(\mathcal{C}) + \lambda \cdot \text{dis}_{quad}(\mathcal{C}). \tag{2}$$

The first term, $\text{dis}_{lin}(\mathcal{C})$, measures the dissimilarity between the descriptors of
the two shapes

$$\text{dis}_{lin}(\mathcal{C}) = \sum_{x \in X, y \in Y} d_F(f_X(x), f_Y(y)) \mathcal{C}(x, y), \tag{3}$$

where d_F is some metric in the descriptor space. $\mathrm{dis}_{lin}(\mathcal{C})$ is a linear function of the correspondence $\mathcal{C}$. The second term, $\mathrm{dis}_{quad}(\mathcal{C})$, measures the dissimilarity between the metric structures of the two shapes

$$\mathrm{dis}_{quad}(\mathcal{C}) = \sum_{\substack{x,\tilde{x}\in X \\ y,\tilde{y}\in Y}} (d_X(x,\tilde{x}) - d_Y(y,\tilde{y}))^2 \mathcal{C}(x,y)\mathcal{C}(\tilde{x},\tilde{y}), \tag{4}$$

and it is a quadratic function of $\mathcal{C}$. The parameter $\lambda \geq 0$ (Eq. (2)) determines the relative weight of the linear and the quadratic terms in the total dissimilarity measure. The optimal matching, denoted here by $\mathcal{C}^*$, is obtained by minimizing the dissimilarity measure $\mathrm{dis}(\mathcal{C})$. In order to avoid a trivial solution $\mathcal{C}^*(x,y) = 0, \forall x, y$, we introduce constraints defined by the type of the correspondence we are looking for. For example, when a bijective mapping from X to Y is required, the appropriate constraints on $\mathcal{C}$ are

$$\sum_{x\in X} \mathcal{C}(x,y) = 1, \forall y \in Y, \ \sum_{y\in Y} \mathcal{C}(x,y) = 1, \forall x \in X \tag{5}$$

The resulting optimization problem can be written as

$$\min_{\mathcal{C}} \ \{\mathrm{dis}_{lin}(\mathcal{C}) + \lambda \cdot \mathrm{dis}_{quad}(\mathcal{C})\} \quad s.t. \ (5) \tag{6}$$

Note that the dissimilarity measure $\mathrm{dis}(\mathcal{C})$ is a quadratic function of the correspondence $\mathcal{C}$. In [12], it was shown how to formulate (6) as a quadratic programming problem with binary variables $\mathcal{C}(x,y)$. The optimization problem described above belongs to the class of Integer Quadratic Programming (IQP) problems, also referred to as Quadratic Assignment Problems (QAP), when used with (5). In general, IQP and QAP problems are NP-Hard. Therefore, in order to minimize $\mathrm{dis}(\mathcal{C})$, one has to resort to either some relaxation technique or a heuristic approach (see e.g. [27]). While matching points using local structures alone (by setting $\lambda = 0$, for instance) is a linear problem, and thus can be solved efficiently, it can not guarantee global invariance in the presence of noise and symmetries. A better solution can be found by considering global structures. Unfortunately, solving the quadratic assignment problem for a large number of variables is almost infeasible, even after relaxation. In Section 4 we apply a hierarchical approach for calculating an approximate solution of the above optimization problem.

3   Mathematical Background

3.1   Choice of Metric

Differential geometry: Smooth surfaces, also known as *Riemannian manifolds*, are *differential manifolds* equipped with an inner product in the tangent space, which provides geometric notions such as angels, lengths, areas and curvatures without resorting to the ambient space, and are referred to as *intrinsic* measures.

The simplest example of an intrinsic metric is the *geodesic metric*, defined by the length of the shortest path on the surface of a shape,

$$d_X(x, x') = \inf_{\gamma \in \Gamma(x,x')} \ell(\gamma), \tag{7}$$

where $\Gamma(x, x')$ is the set of all admissible paths between the points x and x' on the surface X, and $\ell(\gamma)$ is the length of the path γ. There exist several numerical methods to evaluate (7) [19,32,5]. We use *fast marching method*, that simulates a wavefront propagation on a triangular mesh, associating the time of arrival of the front with the distance it traveled.

Diffusion geometry: Heat diffusion on the surface X is described by the heat equation,

$$\left(\Delta_X + \frac{\partial}{\partial t} \right) u(t, x) = 0, \tag{8}$$

where a scalar field $u : X \times [0, \infty) \to \mathbb{R}$ is the heat profile at location x and time t, and Δ_X is the *Laplace-Beltrami operator*.

For compact manifolds, the Laplace-Beltrami operator has a discrete eigen-decomposition of the form

$$\Delta_X \phi_i = \lambda_i \phi_i, \tag{9}$$

where $\lambda_0, \lambda_1, ...$ are eigenvalues and $\phi_0, \phi_1, ...$ are the corresponding eigenfunctions, which construct the heat kernel

$$h_t(x, z) = \sum_{i=0}^{\infty} e^{-\lambda_i t} \phi_i(x) \phi_i(z). \tag{10}$$

The diffusion distance is defined as a cross-talk between two heat kernels [9]

$$d_{X,t}^2(x, y) = \| h_t(x, \cdot) - h_t(y, \cdot) \|_{L_2(X)}^2 = \int_X |h_t(x, z) - h_t(y, z)|^2 dz$$

$$= \sum_{i=0}^{\infty} e^{-2\lambda_i t} \left(\phi_i(x) - \phi_i(y) \right)^2. \tag{11}$$

Since diffusion distances are derived from the Laplace Beltrami operator, they are also intrinsic properties, and, according to [3,11,10], also fulfill the metric axioms.

3.2   Choice of Descriptors

Distance histograms: Given two surfaces X and Y and their metrics d_X and d_Y respectfully, we can evaluate the distances between any two points on each one of the shapes using either choices of metrics. For isometries, a good candidate that matches point $x \in X$ to $y \in Y$ will have similar distances to all other corresponding points. Assuming the surface is well sampled, the distance histograms of corresponding points $x \in X$ and $y \in Y$ have to be similar. Comparison of histograms is a well studied operation. While straight forward bin-to-bin comparison may work, we refer the reader to more robust algorithms such as the *earth moving distances* (EMD) [30].

Heat kernel signatures: Another local descriptor based on the *heat equation*, was presented by Sun *et al.* [33]. It was employed by Bronstein *et al.* [8] for shape retrieval, and was recently adapted to volumes by Raviv *et al.* [29]. Sun *et al.* [33] proposed using the diagonal of the heat kernel $k_t(x,x)$ (10) at multiple scales as a local descriptor, and referred to it as *heat kernel signatures* (HKS). The HKS remains invariant under isometric deformations of X, and it is insensitive to topological noise at small scales. It is also informative in the sense that under certain assumptions one could reconstruct the surface (up to an isometry) from it. Furthermore, the HKS descriptor can be efficiently computed from the eigenfunctions and eigenvalues of the Laplace-Beltrami operator.

Intrinsic symmetry-aware descriptors: Another possible choice for a surface descriptor is one based on the eigendecomposition of the Laplace-Beltrami operator, suggested in [12]. In [12], the focus was on matching intrinsically symmetric non-rigid shapes, and on the fact that in this case there exist more than one possible matching of the two shapes, that preserves their global and local surface properties. The solution proposed in [12] consists of defining distinct sets of descriptors for several possible correspondences, and minimizing the distortion dis($\mathcal{C}$) separately for each of them, to obtain distinct matchings. Thus, when using these descriptors within an hierarchical framework, we can also find more than a single matching of the two shapes, while obtaining denser correspondence.

3.3   Integer Quadratic Programming

A *quadratic program* (QP) is an optimization problem with quadratic objective function and affine constraint functions

$$\min \ \tfrac{1}{2}x^T E x + q^T x + r \ \ s.t. \ \ Gx \preceq h, \ Ax = b. \tag{12}$$

The above problem is called *convex* when the matrix E is positive semi-definite. The *Integer Quadratic Programming* (IQP) has similar form, with the additional constraint on the variables x: $x_i \in \{0,1\}$ (binary variables). While convex QP has one global minimum and can be solved efficiently, IQP is an NP-Hard problem. Two common methods are used to solve an QAP problem [4]. The first is a heuristic approach based on a search procedure. For example, [12] used a branch-and-bound procedure to solve the optimization problem in Eq. (6). This approach usually provides good results assuming the local structures are both robust and unique, and there is no intrinsic symmetry. The second approach is based on relaxation. It is a three step solution, consisting of relaxing the integer constraints, solving a continuous optimization problem and projecting the solution back into integers. As expected, this procedure is highly influenced by the initial conditions. As for complexity, the relaxed IQP problem remains NP-Hard. We use branch-and-bound for initial alignment, and then refine it using a continuous optimization technique.

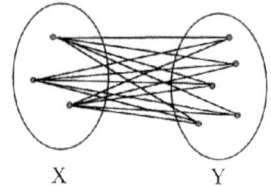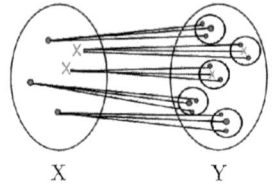

X   Y   X   Y

Fig. 1. In the first step (left) we construct a quadratic correspomdence matrix from all points in X into all points in Y. In each iteration (right) we search for possible matches between points in X from the previous iteration (blue circle) and new sampled points in X (green Xs) and their corresponding neighborhoods (black circles) in Y.

4   Hierarchical Formulation

Solving (6) reveals the main drawback of the quadratic problem formulation. As noted in [12], the dimensionality of the problem allows us to handle up to several dozens of points. Let us assume that X and Y have N and M vertices, respectively. The number of possible correspondences between X and Y is therefore NM, and thus, the dimension of the matrix E in the quadratic problem (12) is $NM \times NM$. Even for a small number of points, e.g. 30, the problem becomes almost infeasible.

Since the problem is not strictly combinatorial by nature, but rather derived from a smooth geometric measure, there should be a way to reduce the complexity. We suggest reducing the high dimensionality of the problem using an iterative scheme. At the first step we follow [12] and solve (6) using a branch-and-bound procedure [2]. Each point $x \in X$ is now matched to a point $c(x) \in Y$ by the mapping c. We denote $y = c(x)$ if $C(x, y) = 1$. In each iteration we search for the best correspondence between x and $c(x)$ neighborhood, instead of all points $y \in Y$, in a manner similar to [36]. Between iterations we add points $x \in X$ and $y \in Y$ using the 2-optimal *Farthest Point Sampling* (FPS) strategy [15], evaluate the neighborhood in Y of the new points, reevaluate the neighborhood of the old points, and continue until convergence. In Figure 1 we show a diagram of the process.

We solve the relaxed version of (6), using quasi-Newton optimization, and project the solution to integers between iterations. Convergence is guaranteed, but only to a local minimum, as for all QAP problems. The solver can now handle up to several hundred of points. Let us further analyze the complexity. We consider the first step to be $\mathbb{O}(N+M)$ as we use a constant (usually around 20) points from each mesh, and only FPS is required, which can be evaluated in linear time. Assuming that each neighborhood in Y consists of K vertices, and a linear growth in each iteration of the matched points from X, then, for the j'th iteration, the quadratic correlation matrix has $jK \times jK$ members which has a complexity of $\mathbb{O}(j^2K^2)$, and the entire iterative framework takes $\mathbb{O}(\Sigma_{j=1}^{N} j^2K^2) = \mathbb{O}(N^3K^2)$. Since each iteration requires a correlation matrix of size j^2K^2, the number of matched points can be significantly higher than the results shown in [12].

5   Results

In this section we provide several matching results obtained using our hierarchical procedure. Figure 2 shows the matching obtained with the proposed framework, combined with different descriptors and metrics, at several hierarchies. The matching was performed using 10 points at the coarse scale, and 30 - 64 points at the finest scale. Figure 2(a) shows the result of matching two cat shapes using geodesic distance histogram descriptors and geodesic distance metric. Figure 2(b) shows the matching result obtained using diffusion distances instead of geodesic ones, and Figure 2(c) - the result obtained using Heat Kernel Signatures [33] and diffusion distances. Note that the last two matchings are in fact reflected ones (follows from the intrinsically symmetric shape matching ambiguity described in [12]). When using the proposed algorithm with Laplace-Beltrami operator-based descriptors [12] and geodesic distances, we were able to obtain both possible correspondences between two cat shapes - the true correspondence and the reflected one. The results are shown in Figure 2(d). As can be seen, all setups provide good results, and we can conclude that the proposed hierarchical framework is independent of the choice of descriptors.

We compared the hierarchical method to [12]'s quadratic matching and [6]'s GMDS framework. Both are based on global structures. Since we followed [12] formulation as our first step, our initial matchings are the same. But, since the complexity of [12] rises rapidly, it can not be used to match more then a few dozen points. In addition, even for a low number of points we have a major quality advantage over [12], since the matched points on the second mesh can move, and are not restricted to the initial sampling. In Figure 3 we see that the ear and the nose of the cat were matched using 10 points, and relocated after several iterations. We also compared the hierarchical matching and the quadratic matching calculation times. The result are shown in Figure 4, for different number of matched points. The quadratic matching succeeded to match only up to 22 points in a reasonable time - less than 4 minutes, while the proposed hierarchical method was able to find 60 matches in shorter time.

Bronstein *et al.* [6] proposed to minimize the Gromov-Hausdorf distance between shapes, which in theory provides the best correspondence between approximate isometries. Since their framework is based on non-convex optimization, the first alignment is critical. We evaluated GMDS results using its own initializer and our quadratic first step, which provided better results. We repeated the experiments shown in 2(a) and measured the geodesic distances between the corresponding points versus the ground truth correspondence. We improved the L_∞ error by 26% and the mean error by 6.25%. It is not surprising, since usually the best correspondence can not be originated from a global structure alone. One can think, for example, on a trivial experiment where only the head rotates. The best correspondence will suffer a distortion in the neck alone, but GMDS will suffer from a distortion in all points.

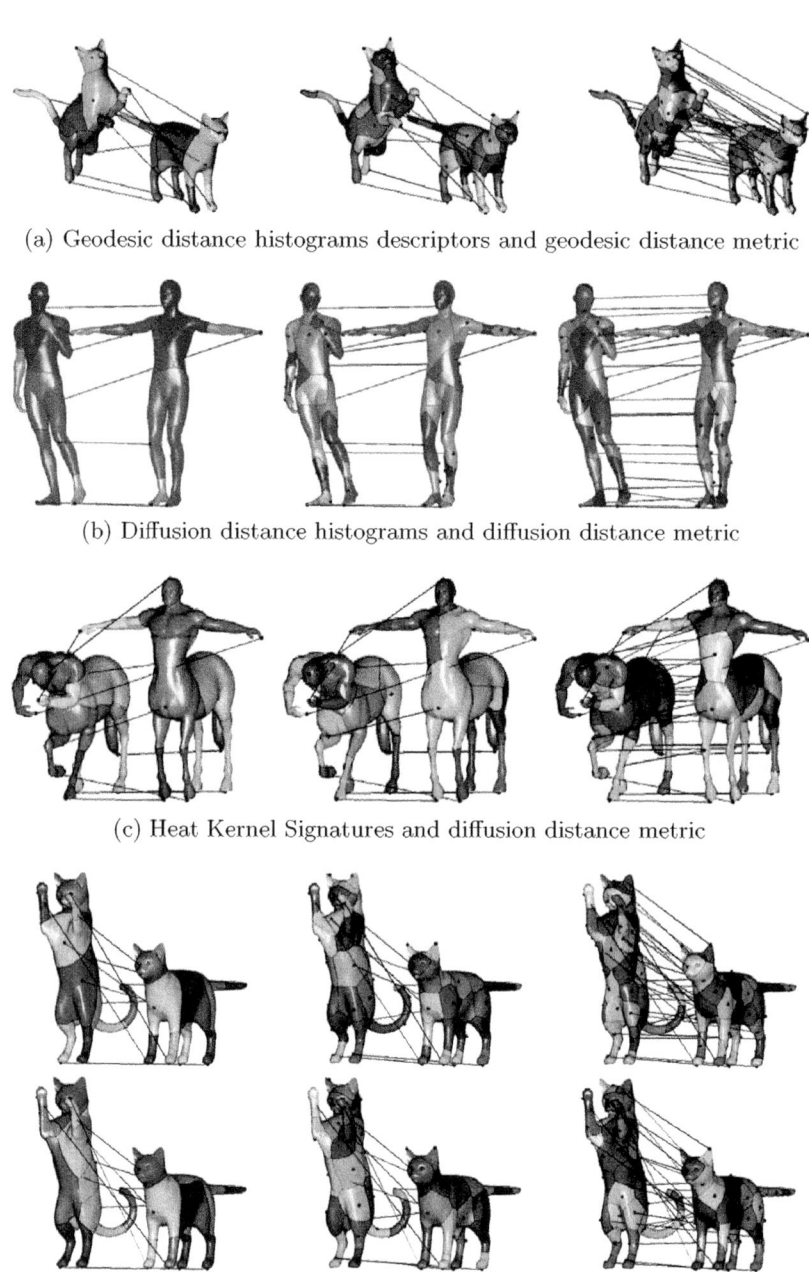

(a) Geodesic distance histograms descriptors and geodesic distance metric

(b) Diffusion distance histograms and diffusion distance metric

(c) Heat Kernel Signatures and diffusion distance metric

(d) The Laplace-Beltrami operator-based descriptors and geodesic distance metric;
the upper row - same orientation correspondence, the lower row - the reflected one.

Fig. 2. Matching results obtained with the proposed framework combined with different descriptors and metrics, at several hierarchies. The hierarchical framework works well with all setups, and it performs equally well with all types of descriptors.

Fig. 3. Using geodesic distances as a global structure, and geodesic based histograms as a local one, the wrong ear-to-nose match gets closer to the correct one during subsequent iterations.

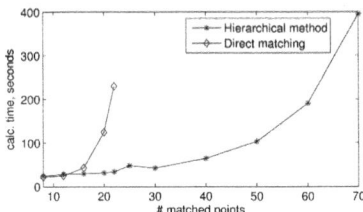

Fig. 4. Graph of calculation time as a function of number of matched points, showing results of the proposed hierarchical method, alongside the quadratic matching algorithm.

6   Conclusions

We presented a hierarchical framework, based on quadratic programming, that solves non-rigid matchings between shapes. While being NP-Hard in general, we solve the assignment problem by taking into account the smooth structure of our shapes using an iterative scheme. We provided numerical results, and compared it to state of art methods.

Acknowledgment. This research was supported by European Community's FP7-ERC program, grant agreement no. 267414.

References

1. Anguelov, D., Srinivasan, P., Pang, H.-C., Koller, D., Thrun, S.: The correlated correspondence algorithm for unsupervised registration of nonrigid surfaces. In: Proc. of the Neural Information Processing Systems (NIPS) Conference, pp. 33–40 (2004)
2. Bemporad, A.: Hybrid Toolbox - User's Guide (2004), http://www.dii.unisi.it/hybrid/toolbox
3. Bérard, P., Besson, G., Gallot, S.: Embedding riemannian manifolds by their heat kernel. Geometric and Functional Analysis 4(4), 373 398 (1994)
4. Boyd, S., Vandenberghe, L.: Convex Optimization. Cambridge University Press, Cambridge (2006)

5. Bronstein, A.M., Bronstein, M.M., Devir, Y.S., Kimmel, R., Weber, O.: Parallel algorithms for approximation of distance maps on parametric surfaces. In: Proc. International Conference and Exhibition on Computer Graphics and Interactive Techniques, SIGGRAPH (2007) (submitted)
6. Bronstein, A.M., Bronstein, M.M., Kimmel, R.: Generalized multidimensional scaling: a framework for isometry-invariant partial surface matching. Proc. of National Academy of Sciences (PNAS), 1168–1172 (2006)
7. Bronstein, A.M., Bronstein, M.M., Kimmel, R., Mahmoudi, M., Sapiro, G.: A gromov-hausdorff framework with diffusion geometry for topologically-robust non-rigid shape matching. International Journal of Computer Vision, IJCV (2009)
8. Bronstein, A.M., Bronstein, M.M., Ovsjanikov, M., Guibas, L.J.: Shape google: a computer vision approach to invariant shape retrieval. In: Proc. NORDIA (2009)
9. Bronstein, M.M., Bronstein, A.M.: Shape recognition with spectral distances. IEEE Transactions on Pattern Analysis and Machine Intelligence 99 (2010) (preprints)
10. Coifman, R.R., Lafon, S.: Diffusion maps. Applied and Computational Harmonic Analysis 21, 5–30 (2006)
11. Coifman, R.R., Lafon, S., Lee, A.B., Maggioni, M., Nadler, B., Warner, F., Zucker, S.W.: Geometric diffusions as a tool for harmonic analysis and structure definition of data: Diffusion maps. Proc. National Academy of Sciences 102(21), 7426–7431 (2005)
12. Dubrovina, A., Kimmel, R.: Matching shapes by eigendecomposition of the laplace-beltrami operator (2010)
13. Elad, A., Kimmel, R.: On bending invariant signatures for surfaces. IEEE Trans. Pattern Analysis and Machine Intelligence (PAMI) 25(10), 1285–1295 (2003)
14. Gromov, M.: Structures metriques pour les varietes riemanniennes. Textes Mathematiques (1) (1981)
15. Hochbaum, D.S., Shmoys, D.B.: A best possible heuristic for the k-center problem. Mathematics of Operations Research 10(2), 180–184 (1985)
16. Hu, J., Hua, J.: Salient spectral geometric features for shape matching and retrieval. Vis. Comput. 25(5-7), 667–675 (2009)
17. Jain, V., Zhang, H., Van Kaick, O.: Non-rigid spectral correspondence of triangle meshes. International Journal on Shape Modeling 13, 101–124 (2007)
18. Kim, V., Lipman, Y., Chen, X., Funkhouser, T.: Mobius transformations for global intrinsic symmetry analysis. In: Proc. Eurographics Symposium on Geometry Processing, SGP (2010)
19. Kimmel, R., Sethian, J.A.: Computing geodesic paths on manifolds. Proceedings of the National Academy of Sciences (PNAS) 95(15), 8431–8435 (1998)
20. Lipman, Y., Funkhouser, T.: Mobius voting for surface correspondence. ACM Transactions on Graphics (Proc. SIGGRAPH) 28(3) (August 2009)
21. Mateus, D., Horaud, R.P., Knossow, D., Cuzzolin, F., Boyer, E.: Articulated shape matching using laplacian eigenfunctions and unsupervised point registration. In: Proc. of the IEEE Conference on Computer Vision and Pattern Recognition, CVPR (2008)
22. Mémoli, F.: On the use of gromov-hausdorff distances for shape comparison. In: Point Based Graphics 2007, pp. 81–90 (September 2007)
23. Mémoli, F.: Spectral gromov-wasserstein distances for shape matching. In: Workshop on Non-Rigid Shape Analysis and Deformable Image Alignment (ICCV Workshop, NORDIA 2009) (2009)
24. Mémoli, F., Sapiro, G.G.: A theoretical and computational framework for isometry invariant recognition of point cloud data. Found. Comput. Math. 5(3), 313–347 (2005)

25. Ovsjanikov, M., Mérigot, Q., Mémoli, F., Guibas, L.: One point isometric matching with the heat kernel. In: Eurographics Symposium on Geometry Processing, SGP (2010)
26. Ovsjanikov, M., Sun, J., Guibas, L.: Global intrinsic symmetries of shapes. Computer Graphics Forum 27(5), 1341–1348 (2008)
27. Pardalos, P.M., Rendl, F., Wolkowicz, H.: The quadratic assignment problem: A survey and recent developments. In: Proc. of the DIMACS Workshop on Quadratic Assignment Problems. DIMACS Series in Discrete Mathematics and Theoretical Computer Science, vol. 16, pp. 1–42 (1994)
28. Raviv, D., Bronstein, A.M., Bronstein, M.M., Kimmel, R.: Full and partial symmetries of non-rigid shapes. International Journal of Computer Vision, IJCV (2010)
29. Raviv, D., Bronstein, A.M., Bronstein, M.M., Kimmel, R.: Volumetric heat kernel signatures. In: Proc. 3D Object recognition (3DOR), Part of ACM Multimedia (2010)
30. Rubner, Y., Tomasi, C., Guibas, L.J.: The earth mover's distance as a metric for image retrieval. International Journal of Computer Vision 40(2), 99–121 (2000)
31. Rustamov, R.M.: Laplace-beltrami eigenfunctions for deformation inavriant shape representation. In: Proc. of SGP, pp. 225–233 (2007)
32. Spira, A., Kimmel, R.: An efficient solution to the eikonal equation on parametric manifolds. Interfaces and Free Boundaries 6(4), 315–327 (2004)
33. Sun, J., Ovsjanikov, M., Guibas, L.: A concise and provably informative multi-scale signature based on heat diffusion. In: Proc. Eurographics Symposium on Geometry Processing, SGP (2009)
34. Tevs, A., Bokeloh, M., Wand, M., Schilling, A., Seidel, H.-P.: Isometric registration of ambiguous and partial data. In: Proc. of the IEEE Conference on Computer Vision and Pattern Recognition, CVPR (2009)
35. Thorstensen, N., Keriven, R.: Non-rigid shape matching using geometry and photometry. In: Zha, H., Taniguchi, R.-i., Maybank, S. (eds.) ACCV 2009. LNCS, vol. 5996, pp. 644–654. Springer, Heidelberg (2010)
36. Wang, C., Bronstein, M.M., Paragios, N.: Discrete minimum distortion correspondence problems for non-rigid shape matching. Technical report, INRIA Research Report 7333, Mathématiques Appliquées aux Systèmes, École Centrale Paris (2010)
37. Zaharescu, A., Boyer, E., Varanasi, K., Horaud, R.P.: Surface feature detection and description with applications to mesh matching. In: Proc. of the IEEE Conference on Computer Vision and Pattern Recognition, CVPR (2009)
38. Zhang, H., Sheffer, A., Cohen-Or, D., Zhou, Q., van Kaick, O., Taghasacchi, A.: Deformation-driven shape correspondence. Computer Graphics Forum (Proc. of SGP) 27(5), 1431–1439 (2008)
39. Zigelman, G., Kimmel, R., Kiryati, N.: Texture mapping using surface flattening via multi-dimensional scaling. IEEE Trans. on Visualization and Computer Graphics 8(2), 198–207 (2002)

Photometric Heat Kernel Signatures

Artiom Kovnatsky[1], Michael M. Bronstein[3],
Alexander M. Bronstein[4], and Ron Kimmel[2]

[1]Department of Applied Mathematics,
art@tx.technion.ac.il
[2]Department of Computer Science,
ron@cs.technion.ac.il
Technion, Israel Institute of Technology, Haifa, Israel

[3]Inst. of Computational Science, Faculty of Informatics,
Università della Svizzera Italiana, Lugano, Switzerland
michael.bronstein@usi.ch

[4]Dept. of Electrical Engineering, Tel Aviv University, Israel
bron@eng.tau.ac.il

Abstract. In this paper, we explore the use of the diffusion geometry framework for the fusion of geometric and photometric information in local heat kernel signature shape descriptors. Our construction is based on the definition of a diffusion process on the shape manifold embedded into a high-dimensional space where the embedding coordinates represent the photometric information. Experimental results show that such data fusion is useful in coping with different challenges of shape analysis where pure geometric and pure photometric methods fail.

1 Introduction

In last decade, the amount of geometric data available in the public domain, such as Google 3D Warehouse, has grown dramatically and created the demand for shape search and retrieval algorithms capable of finding similar shapes in the same way a search engine responds to text queries. However, while text search methods are sufficiently developed to be ubiquitously used, the search and retrieval of 3D shapes remains a challenging problem. Shape retrieval based on text metadata, like annotations and tags added by the users, is often incapable of providing relevance level required for a reasonable user experience.

Content-based shape retrieval using the shape itself as a query and based on the comparison of geometric and topological properties of shapes is complicated by the fact that many 3D objects manifest rich variability, and shape retrieval must often be *invariant* under different classes of transformations. A particularly challenging setting is the case of non-rigid shapes, including a wide range of transformations such as bending and articulated motion, rotation and translation, scaling, non-rigid deformation, and topological changes. The main challenge in shape retrieval algorithms is computing a *shape descriptor*, that would be unique for each shape, simple to compute and store, and invariant under

A.M. Bruckstein et al. (Eds.): SSVM 2011, LNCS 6667, pp. 616–627, 2012.

different type of transformations. Shape similarity is determined by comparing the shape descriptors.

Broadly, shape descriptors can be divided into *global* and *local*. The former consider global geometric or topological shape characteristics such as distance distributions [21,24,19], geometric moments [14,30], or spectra [23], whereas the latter describe the local behavior of the shape in a small patch. Popular examples of local descriptors include spin images [3], shape contexts [1], integral volume descriptors [12] and radius-normal histograms [22]. Using the bag of features paradigm common in image analysis [25,10], a global shape descriptor counting the occurrence of local descriptors in some vocabulary can be computed [7].

Recently, there has been an increased interest in the use of *diffusion geometry* [11,16] for constructing invariant shape descriptors. Diffusion geometry is closely related to heat propagation properties of shapes and allows obtaining global descriptors, such as distance distributions [24,19,8] and Laplace-Beltrami spectral signatures [23], as well local descriptors such as heat kernel signatures [26,9,7]. One limitation of these methods is that, so far, only *geometric* information has been considered. However, the abundance of textured models in computer graphics and modeling applications, as well as the advance in 3D shape acquisition [35,36] allowing to obtain textured 3D shapes of even moving objects, bring forth the need for descriptors also taking into consideration *photometric* information. Photometric information plays an important role in a variety of shape analysis applications, such as shape matching and correspondence [28,33]. Considering 2D views of the 3D shape [32,20], standard feature detectors and descriptors used in image analysis such as SIFT [18] can be employed. More recently, Zaharescu *et al.* [37] proposed a geometric SIFT-like descriptor for textured shapes, defined directly on the surface.

In this paper, we extend the diffusion geometry framework to include photometric information in addition to its geometric counterpart. The main idea is to define a diffusion process that takes into consideration not only the geometry but also the texture of the shape. This is achieved by considering the shape as a manifold in a higher dimensional combined geometric-photometric embedding space, similarly to methods in image processing applications [15,17]. As a result, we are able to construct geometric and photometric local descriptors (color heat kernel signatures or cHKS).

The rest of this paper is organized as follows. In Section 2, we review the mathematical formalism of diffusion processes and their use in shape analysis. In Section 3, we introduce our approach and in Section 4 its numerical implementation details. Section 5 presents experimental results. Finally, Section 6 concludes the paper.

2   Background

Throughout the paper, we assume the shape to be modeled as a two-dimensional compact Riemannian manifold X (possibly with a boundary) equipped with a metric tensor g. Fixing a system of local coordinates on X, the latter can be expressed as a 2×2 matrix $g_{\mu\nu}$, also known as the first fundamental form. The metric tensor allows to express the length of a vector v in the tangent space $T_x X$ at a point x as $g_{\mu\nu} v^\mu v^\nu$, where repeated indices $\mu, \nu = 1, 2$ are summed over following Einstein's convention.

Given a smooth scalar field $f : X \to \mathbb{R}$ on the manifold, its *gradient* is defined as the vector field ∇f satisfying $f(x + dx) = f(x) + g_x(\nabla f(x), dx)$ for every point x and every infinitesimal tangent vector $dx \in T_x X$. The metric tensor g defines the *Laplace-Beltrami operator* Δ_g that satisfies

$$\int f \Delta_g h \, da = - \int g_x(\nabla f, \nabla h) da \tag{1}$$

for any pair of smooth scalar fields $f, h : X \to \mathbb{R}$; here da denotes integration with respect to the standard area measure on X. Such an integral definition is usually known as the Stokes identity. The Laplace-Beltrami operator is positive semi-definite and self-adjoint. Furthermore, it is an *intrinsic* property of X, i.e., it is expressible solely in terms of g. In the case when the metric g is Euclidean, Δ_g becomes the standard Laplacian.

The Laplace-Beltrami operator gives rise to the *heat equation*,

$$\left(\Delta_g + \frac{\partial}{\partial t} \right) u = 0, \tag{2}$$

which describes diffusion processes and heat propagation on the manifold (note that we use a positive-semidefinite Laplace-Beltrami operator, hence the plus sign in the heat equation). Here, $u(x, t)$ denotes the distribution of heat at time t at point x. The initial condition to the equation is some heat distribution $u(x, 0)$, and if the manifold has a boundary, appropriate boundary conditions (e.g. Neumann or Dirichlet) must be specified. The solution of (2) with a point initial heat distribution $u_0(x) = \delta(x, x')$, where $\delta(x', x') = 1$ o.w. 0, is called the *heat kernel* and denoted here by $K_t(x, x')$. Using a signal processing analogy, K_t can be thought of as the "impulse response" of the heat equation.

By the spectral decomposition theorem, the heat kernel can be represented as [13]

$$K_t(x, x') = \sum_{i \geq 0} e^{-\lambda_i t} \phi_i(x) \phi_i(x'), \tag{3}$$

where $0 = \lambda_0 \leq \lambda_1 \leq \ldots$ are the eigenvalues and $\phi_0, \phi_1, \ldots$ the corresponding eigenfunctions of the Laplace-Beltrami operator (i.e., solutions to $\Delta_g \phi_i = \lambda_i \phi_i$). We will collectively refer to quantities expressed in terms of the heat kernel as to *diffusion geometry*. Since the Laplace-Beltrami operator is intrinsic, the diffusion geometry it induces is invariant under isometric deformations of X (incongruent embeddings of g into $\mathbb{R}^3$).

Sun et al. [26] proposed using the heat propagation properties as a local descriptor of the manifold. The diagonal of the heat kernel, $K_t(x, x')$, referred to as the *heat kernel signature* (HKS), captures the local properties of X at point x and scale t. The descriptor is computed at each point as a vector of the values $p(x) = (K_{t_1}(x, x), \ldots, K_{t_n}(x, x))$, where $t_1, \ldots, t_n$ are some time values. Such a descriptor is deformation-invariant, easy to compute, and provably informative [26].

Ovsjanikov et al. [7] employed the HKS local descriptor for large-scale shape retrieval using the *bags of features* paradigm [25]. In this approach, the shape is considered as a collection of "geometric words" from a fixed "vocabulary" and is described by the distribution of such words, also referred to as a *bag of features* or BoF. The vocabulary is constructed offline by clustering the HKS descriptor space. Then, for each point on the shape, the HKS is replaced by the nearest vocabulary word by means of vector quantization. Counting the frequency of each word, a BoF is constructed. The similarity of two shapes X and Y is then computed as the distance between the corresponding BoFs, $d(X, Y) = \|\mathrm{BoF}_X - \mathrm{BoF}_Y\|$.

3   Photometric Heat Kernel Signatures

Let us further assume that the Riemannian manifold X is a submanifold of some manifold $\mathcal{E}$ ($\dim(\mathcal{E}) = m > 2$) with the Riemannian metric tensor h, embedded by means of a diffeomorphism $\xi : X \to \xi(X) \subseteq \mathcal{E}$. A Riemannian metric tensor on X induced by the embedding is the *pullback metric* $(\xi^* h)(r, s) = h(d\xi(r), d\xi(s))$ for $r, s \in T_x X$, where $d\xi : T_x X \to T_{\xi(x)}\mathcal{E}$ is the differential of ξ. In coordinate notation, the pullback metric is expressed as $(\xi^* h)_{\mu\nu} = h_{ij}\partial_\mu \xi^i \partial_\nu \xi^j$, where the indices $i, j = 1, \ldots, m$ denote the embedding coordinates.

Here, we use the structure of $\mathcal{E}$ to model joint geometric and photometric information. Such an approach has been successfully used in image processing [15]. When considering shapes as geometric object only, we define $\mathcal{E} = \mathbb{R}^3$ and h to be the Euclidean metric. In this case, ξ acts as a *parametrization* of X and the pullback metric becomes simply $(\xi^* h)_{\mu\nu} = \partial_\mu \xi^1 \partial_\nu \xi^1 + \ldots + \partial_\mu \xi^3 \partial_\nu \xi^3 = \langle \partial_\mu \xi, \partial_\nu \xi \rangle_{\mathbb{R}^3}$. In the case considered in this paper, the shape is endowed with photometric information given in the form of a field $\alpha : X \to \mathcal{C}$, where $\mathcal{C}$ denotes some colorspace (e.g., RGB or Lab). This photometric information can be modeled by defining $\mathcal{E} = \mathbb{R}^3 \times \mathcal{C}$ and an embedding $\xi = (\xi_g, \xi_p)$. The embedding coordinates corresponding to geometric information $\xi_g = (\xi^1, \ldots, \xi^3)$ are as previously and the embedding coordinate corresponding to photometric information are given by $\xi_p = (\xi^4, \ldots, \xi^6) = \eta(\alpha^1, \ldots, \alpha^3)$, where $\eta \geq 0$ is a scaling constant. Simplifying further, we assume $\mathcal{C}$ to have a Euclidean structure (for example, the Lab colorspace has a natural Euclidean metric). The metric in this case boils down to $(\xi^* h)_{\mu\nu} = \langle \partial_\mu \xi_g, \partial_\nu \xi_g \rangle_{\mathbb{R}^3} + \eta^2 \langle \partial_\mu \xi_p, \partial_\nu \xi_p \rangle_{\mathbb{R}^3}$, which hereinafter we shall denote by $\hat{g}_{\mu\nu}$.[1]

---

[1] The joint metric tensor $\hat{g}$ has inherent ambiguities. The diffusion geometry induced by $\hat{g}$ is invariant the joint isometry group $\mathrm{Iso}_{\hat{g}} = \mathrm{Iso}((\xi_g^* h)_{\mu\nu} + \eta^2 (\xi_p^* h)_{\mu\nu})$. Ideally, we would like $\mathrm{Iso}_{\hat{g}} = \mathrm{Iso}_g = \mathrm{Iso}((\xi_g^* h)_{\mu\nu}) \times \mathrm{Iso}_p = \mathrm{Iso}((\xi_p^* h)_{\mu\nu})$ to hold. In practice, $\mathrm{Iso}_{\hat{g}}$ is bigger: while every composition of a geometric isometry with a photometric isometry is a joint isometry, there exist some joint isometries which cannot be obtained as a composition of geometric and photometric isometries. Experimental results show that no realistic geometric and photometric transformations lie in $\mathrm{Iso}_{\hat{g}} \setminus (\mathrm{Iso}_g \times \mathrm{Iso}_p)$.

Fig. 1. Textured shape (left); values of the heat kernel (x placed on the foot, $t = 1024$) arising from regular purely geometric (middle) and mixed photometric-geometric (right) diffusion process.

The Laplace-Beltrami operator $\Delta_{\hat{g}}$ associated with such a metric gives rise to diffusion geometry that combines photometric and geometric information (Figure 1). We define the *photometric* or *color heat kernel signature* (cHKS) as the diagonal of the heat kernel associated with the joint geometric-photometric diffusion induced by $\Delta_{\hat{g}}$. The cHKS fuses local geometric and photometric information of the shape.

4   Numerical Implementation

Let $\{x_1, \ldots, x_N\} \subseteq X$ denote the discrete samples of the shape, and $\xi(x_1), \ldots, \xi(x_N)$ be the corresponding embedding coordinates (three-dimensional in the case we consider only geometry, or six-dimensional in the case of geometry-photometry fusion). We further assume to be given a *triangulation* (simplicial complex), consisting of *edges* (i, j) and *faces* (i, j, k) where each $(i, j), (j, k)$, and (i, k) is an edge (here $i, j, k = 1, \ldots, N$).

A function f on the discretized manifold is represented as an N-dimensional vector $(f(x_1), \ldots, f(x_N))$. The discrete Laplace-Beltrami operator can be written in the generic form

$$(\hat{\Delta} f)(x_i) = \frac{1}{a_i} \sum_{j \in \mathcal{N}_i} w_{ij}(f(x_i) - f(x_j)), \tag{4}$$

where w_{ij} are weights, a_i are normalization coefficients, and $\mathcal{N}_i$ denotes a local neighborhood of point i. Different discretizations of the Laplace-Beltrami operator can be cast into this form by appropriate definition of the above constants. For shapes represented as triangular meshes, a widely-used method is the *cotangent scheme*, which preserves many important properties of the continuous Laplace-Beltrami operator, such as positive semi-definiteness, symmetry, and locality [31]. Yet, in general, the cotangent scheme does not converge to the continuous Laplace-Beltrami operator, in the sense that the solution of the discrete eigenproblem does not converge to the continuous one (pointwise convergence exists if the triangulation and sampling satisfy certain conditions [34]).

Belkin *et al.* [5] proposed a discretization which is convergent without the restrictions on "good" triangulation required by the cotangent scheme. In this scheme, $\mathcal{N}_i$ is chosen to be the entire sampling $\{x_1, \ldots, x_N\}$, $a_i = \frac{1}{4\pi\rho^2}$, and $w_{ij} = S_j e^{-\|\xi(x_i) - \xi(x_j)\|^2/4\rho}$, where ρ is a parameter, S_j denotes area of all triangles sharing the vertex j. In the case of a Euclidean colorspace, w_{ij} can be written explicitly as

$$w_{ij} = S_j \exp\left\{-\frac{\|\xi_g(x_i) - \xi_g(x_j)\|^2}{4\rho} - \frac{\|\xi_p(x_i) - \xi_p(x_j)\|^2}{4\sigma}\right\} \tag{5}$$

where $\sigma = \rho/\eta^2$, which resembles the weights used in the *bilateral filter* [29]. Experimental results also show that this operator produces accurate approximation of the Laplace-Beltrami operator under various conditions, such as noisy data input and different sampling [27,5].

In matrix notation, equation (4) can be written as $\hat{\Delta}f = A^{-1}Wf$, where $A = \mathrm{diag}(a_i)$ and $W = \mathrm{diag}\left(\sum_{l \neq i} w_{il}\right) - (w_{ij})$. The eigenvalue problem $\hat{\Delta}\Phi = \Lambda\Phi$ is equivalent to the generalized symmetric eigenvalue problem $W\Phi = \Lambda A\Phi$, where $\Lambda = \mathrm{diag}(\lambda_0, \ldots, \lambda_K)$ is the diagonal matrix of the first K eigenvalues, and $\Phi = (\phi_0, \ldots, \phi_K)$ is the matrix of the eigenvectors stacked as columns. Since typically W is sparse, this problem can be efficiently solved numerically. Heat kernels can be approximated by taking the first largest eigenvalues and the corresponding eigenfunctions in (3). Since the coefficients in the expansion of h_t decay as $\mathcal{O}(e^{-t})$, typically a few eigenvalues (K in the range of 10 to 100) are required.

5   Results

In order to evaluate the proposed method, we used the SHREC 2010 robust large-scale shape retrieval benchmark methodology [6]. The query set consisted of 270 real-world human shapes from 5 classes acquired by a 3D scanner with real geometric transformations and simulated photometric transformations of different types and strengths, totalling in 54 instances per shape (Figure 2). Geometric transformations were divided into *isometry+topology* (real articulations and topological changes due to acquisition imperfections), and *partiality* (occlusions and addition of clutter such as the red ball in Figure 2). Photometric transformations included *contrast* (increase and decrease by scaling of the L channel), *brightness* (brighten and darken by shift of the L channel), *hue* (shift in the a channel), *saturation* (saturation and desaturation by scaling of the a, b channels), and *color noise* (additive Gaussian noise in all channels). *Mixed* transformations included isometry+topology transformations in combination with two randomly selected photometric transformations. In each class, the transformation appeared in five different versions numbered 1–5 corresponding to the transformation strength levels. One shape of each of the five classes was added to the queried corpus in addition to other 75 shapes used as clutter (Figure 3).

Retrieval was performed by matching 270 transformed queries to the 75 null shapes. Each query had exactly one correct corresponding null shape in the dataset. Performance was evaluated using the precision-recall characteristic. *Precision* $P(r)$ is defined as the percentage of relevant shapes in the first r top-ranked retrieved shapes. *Mean average precision* (mAP), defined as $mAP = \sum_r P(r) \cdot rel(r)$, where $rel(r)$ is the relevance of a given rank, was used as a single measure of performance. Intuitively, mAP is interpreted as the area below the precision-recall curve. Ideal retrieval performance results in first relevant match with mAP=100%. Performance results were broken down according to transformation class and strength.

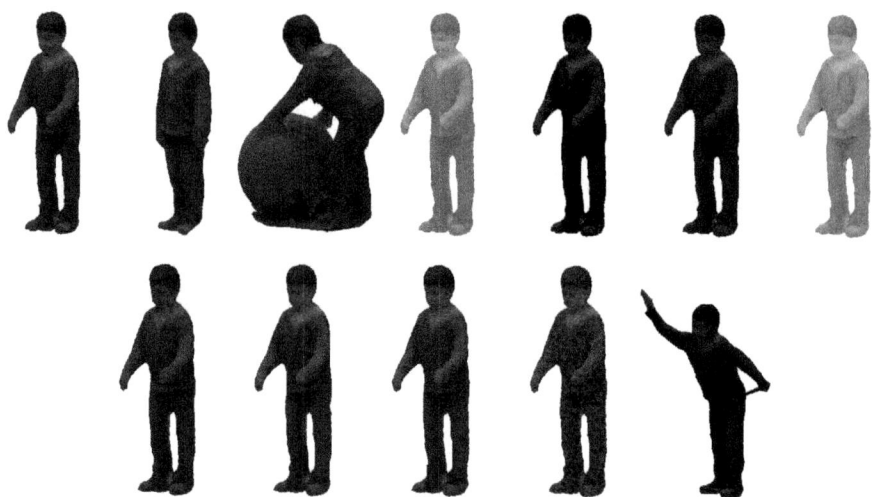

Fig. 2. Examples of geometric and photometric shape transformations used as queries (shown at strength 5). First row, left to right: null, isometry+topology, partiality, two brightness transformations (brighten and darken), two contrast transformations (increase and decrease contrast). Second row, left to right: two saturation transformations (saturate and desaturate), hue, color noise, mixed.

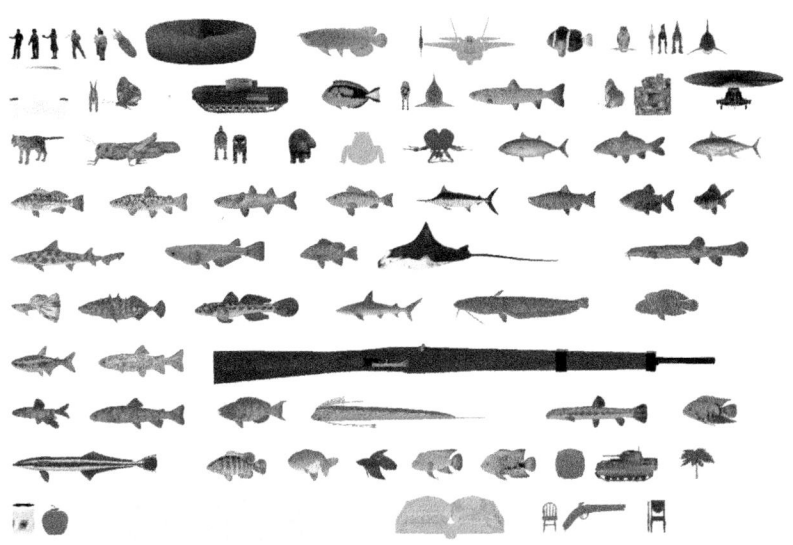

Fig. 3. Null shapes in the dataset (shown at arbitrary scale for visualization purposes)

In additional to the proposed approach, we compared purely geometric, purely photometric, and joint photometric-geometric descriptors. As a purely geometric descriptor, we used bags of features based on HKS according to [7]; purely photometric shape descriptor was a color histogram. As joint photometric-geometric descriptors, we used bags of features computed with the MeshHOG [37] and the proposed color HKS (cHKS).

For the computation of the bag of features descriptors, we used the Shape Google framework with most of the settings as proposed in [7]. More specifically, HKS were computed at six scales ($t = 1024, 1351.2, 1782.9, 2352.5$, and 4096). Soft vector quantization was applied with variance taken as twice the median of all distances between cluster centers. Approximate nearest neighbor method [2] was used for vector quantization. The Laplace-Beltrami operator discretization was computed using the Mesh-Laplace scheme [4] with scale parameter $\rho = 2$. Heat kernels were approximated using the first 200 eigenpairs of the discrete Laplacian. The MeshHOG descriptor was computed at prominent feature points (typically 100-2000 per shape), detected using the MeshDOG detector [37]. The vocabulary size in all the cases was set to 48.

In cHKS, in order to avoid the choice of an arbitrary value η, we used a set of three different weights ($\eta = 0, 0.05, 0.1$) to compute the cHKS and the corresponding BoFs. The distance between two shapes was computed as the sum of the distances between the corresponding BoFs for each η, weighted by η, and 1 in case of $\eta = 0$, $d(X, Y) = \|\mathrm{BoF}_X^0 - \mathrm{BoF}_Y^0\|_1^2 + \sum_\eta \eta \|\mathrm{BoF}_X^\eta - \mathrm{BoF}_Y^\eta\|_1^2$.

Tables 1–4 summarize the results of our experiments. Geometry only descriptor (HKS) [7] is invariant to photometric transformations, but is somewhat sensitive to topological noise and missing parts (Table 1). On the other hand, the

Table 1. Performance (mAP in %) of ShapeGoogle using BoFs with HKS descriptors

Transform.	Strength				
	1	≤ 2	≤ 3	≤ 4	≤ 5
Isom+Topo	100.00	100.00	96.67	95.00	90.00
Partial	66.67	60.42	63.89	63.28	63.63
Contrast	100.00	100.00	100.00	100.00	100.00
Brightness	100.00	100.00	100.00	100.00	100.00
Hue	100.00	100.00	100.00	100.00	100.00
Saturation	100.00	100.00	100.00	100.00	100.00
Noise	100.00	100.00	100.00	100.00	100.00
Mixed	90.00	95.00	93.33	95.00	96.00

Table 2. Performance (mAP in %) of color histograms

Transform.	Strength				
	1	≤ 2	≤ 3	≤ 4	≤ 5
Isom+Topo	100.00	100.00	100.00	100.00	100.00
Partial	100.00	100.00	100.00	100.00	100.00
Contrast	100.00	90.83	80.30	71.88	63.95
Brightness	88.33	80.56	65.56	53.21	44.81
Hue	11.35	8.38	6.81	6.05	5.49
Saturation	17.47	14.57	12.18	10.67	9.74
Noise	100.00	100.00	93.33	85.00	74.70
Mixed	28.07	25.99	20.31	17.62	15.38

Table 3. Performance (mAP in %) of BoFs using MeshHOG descriptors

Transform.	Strength				
	1	≤ 2	≤ 3	≤ 4	≤ 5
Isom+Topo	100.00	95.00	96.67	94.17	95.33
Partial	75.00	61.15	69.93	68.28	68.79
Contrast	100.00	100.00	100.00	98.33	94.17
Brightness	100.00	100.00	100.00	100.00	99.00
Hue	100.00	100.00	100.00	100.00	100.00
Saturation	100.00	100.00	100.00	98.75	99.00
Noise	100.00	100.00	88.89	83.33	78.33
Mixed	100.00	100.00	100.00	93.33	83.40

color-only descriptor works well only for geometric transformations that do not change the shape color. Photometric transformations, however, make such a descriptor almost useless (Table 2). MeshHOG is almost invariant to photometric transformations being based on texture gradients, but is sensitive to color noise (Table 3). The fusion of the geometric and photometric data using our approach (Table 4) achieves nearly perfect retrieval for mixed and photometric transformations and outperforms other approaches. Figure 4 visualizes a few examples of the retrieved shapes ordered by relevance, which is inversely proportional to the distance from the query shape.

Table 4. Performance (mAP in %) of ShapeGoogle using w-multi-scale BoFs with cHKS descriptors

			Strength		
Transform.	1	≤2	≤3	≤4	≤5
Isom+Topo	100.00	100.00	96.67	97.50	94.00
Partial	68.75	68.13	69.03	67.40	67.13
Contrast	100.00	100.00	100.00	100.00	100.00
Brightness	100.00	100.00	100.00	100.00	100.00
Hue	100.00	100.00	100.00	100.00	100.00
Saturation	100.00	100.00	100.00	100.00	100.00
Noise	100.00	100.00	100.00	100.00	100.00
Mixed	100.00	100.00	96.67	97.50	98.00

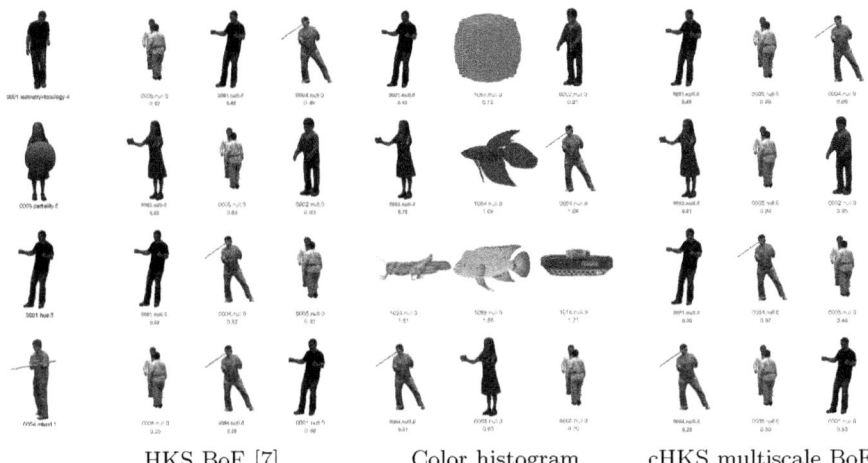

HKS BoF [7]   Color histogram   cHKS multiscale BoF

Fig. 4. Retrieval results using different methods. First column: query shapes, second column: first three matches obtained with HKS-based BoF [7], third column: first three matches obtained using color histograms, fourth column: first three matches obtained with the proposed method (cHKS-based multiscale BoF). Shape annotation follows the convention *shapeid.transformation.strength*; numbers below show distance from query. Only a single correct match exists in the database (marked in green), and ideally, it should be the first one.

6   Conclusions

In this paper, we explored a way to fuse geometric and photometric information in the construction of shape descriptors. Our approach is based on heat propagation on a manifold embedded into a combined geometry-color space. Such diffusion processes capture both geometric and photometric information and give rise to local and global diffusion geometry (heat kernels and diffusion distances), which can be used as informative shape descriptors. We showed experimentally that the proposed descriptors outperform other geometry-only and photometry-only descriptors, as well as state-of-the-art joint geometric-photometric descriptors. In the future, it would be important to formally characterize the isometry group induced by the joint metric in order to understand the invariant properties of the proposed diffusion geometry, and possibly design application-specific invariant descriptors.

Acknowledgements. This research was supported by Israel Science Foundation (ISF) grant number 623/08, by the USA Office of Naval Research (ONR); European Community's FP7- ERC program, grant agreement no. 267414 and the Swiss High-Performance and High-Productivity Computing (HP2C).

References

1. Amores, J., Sebe, N., Radeva, P.: Context-based object-class recognition and retrieval by generalized correlograms. Trans. PAMI 29(10), 1818–1833 (2007)
2. Arya, S., Mount, D.M., Netanyahu, N.S., Silverman, R., Wu, A.Y.: An optimal algorithm for approximate nearest neighbor searching. J. ACM 45, 891–923 (1998)
3. Assfalg, J., Bertini, M., Bimbo, A.D., Pala, P.: Content-based retrieval of 3-d objects using spin image signatures. IEEE Transactions on Multimedia 9(3), 589–599 (2007)
4. Belkin, M., Sun, J., Wang, Y.: Constructing Laplace operator from point clouds in Rd. In: Proc. Symp. Discrete Algorithms, pp. 1031–1040 (2009)
5. Belkin, M., Sun, J., Wang, Y.: Discrete Laplace operator on meshed surfaces. In: Proc. Symp. Computational Geometry, pp. 278–287 (2009)
6. Bronstein, A.M., Bronstein, M.M., Castellani, U., Falcidieno, B., Fusiello, A., Godil, A., Guibas, L.J., Kokkinos, I., Lian, Z., Ovsjanikov, M., Patané, G., Spagnuolo, M., Toldo, R.: Shrec 2010: robust large-scale shape retrieval benchmark. In: Proc. 3DOR (2010)
7. Bronstein, A.M., Bronstein, M.M., Ovsjanikov, M., Guibas, L.J.: Shape google: a computer vision approach to invariant shape retrieval. In: Proc. NORDIA (2009)
8. Bronstein, M.M., Bronstein, A.M.: Shape recognition with spectral distances. Trans. PAMI (2010) (to appear)
9. Bronstein, M.M., Kokkinos, I.: Scale-invariant heat kernel signatures for non-rigid shape recognition. In: Proc. CVPR (2010)
10. Chum, O., Philbin, J., Sivic, J., Isard, M., Zisserman, A.: Total recall: Automatic query expansion with a generative feature model for object retrieval. In: Proc. ICCV (2007)
11. Coifman, R.R., Lafon, S.: Diffusion maps. Applied and Computational Harmonic Analysis 21, 5–30 (2006)
12. Gelfand, N., Mitra, N.J., Guibas, L.J., Pottmann, H.: Robust global registration. In: Proc. SGP (2005)
13. Jones, P.W., Maggioni, M., Schul, R.: Manifold parametrizations by eigenfunctions of the Laplacian and heat kernels. PNAS 105(6), 1803 (2008)
14. Kazhdan, M., Funkhouser, T., Rusinkiewicz, S.: Rotation invariant spherical harmonic representation of 3D shape descriptors. In: Proc. SGP, pp. 156–164 (2003)
15. Kimmel, R., Malladi, R., Sochen, N.: Images as embedded maps and minimal surfaces: movies, color, texture, and volumetric medical images. IJCV 39(2), 111–129 (2000)

16. Lévy, B.: Laplace-Beltrami eigenfunctions towards an algorithm that understands geometry. In: Proc. Shape Modeling and Applications (2006)
17. Ling, H., Jacobs, D.W.: Deformation invariant image matching. In: ICCV, pp. 1466–1473 (2005)
18. Lowe, D.: Distinctive image features from scale-invariant keypoint. IJCV (2004)
19. Mahmoudi, M., Sapiro, G.: Three-dimensional point cloud recognition via distributions of geometric distances. Graphical Models 71(1), 22–31 (2009)
20. Ohbuchi, R., Osada, K., Furuya, T., Banno, T.: Salient local visual features for shape-based 3d model retrieval, pp. 93–102 (June 2008)
21. Osada, R., Funkhouser, T., Chazelle, B., Dobkin, D.: Shape distributions. TOG 21(4), 807–832 (2002)
22. Pan, X., Zhang, Y., Zhang, S., Ye, X.: Radius-normal histogram and hybrid strategy for 3d shape retrieval, pp. 372–377 (June 2005)
23. Reuter, M., Wolter, F.-E., Peinecke, N.: Laplace-spectra as fingerprints for shape matching. In: Proc. ACM Symp. Solid and Physical Modeling, pp. 101–106 (2005)
24. Rustamov, R.M.: Laplace-Beltrami eigenfunctions for deformation invariant shape representation. In: Proc. SGP, pp. 225–233 (2007)
25. Sivic, J., Zisserman, A.: Video google: A text retrieval approach to object matching in videos. In: Proc. CVPR (2003)
26. Sun, J., Ovsjanikov, M., Guibas, L.J.: A concise and provably informative multi-scale signature based on heat diffusion. In: Proc. SGP (2009)
27. Thangudu, K.: Practicality of Laplace operator (2009)
28. Thorstensen, N., Keriven, R.: Non-rigid shape matching using geometry and photometry. In: Proc. CVPR (2009)
29. Tomasi, C., Manduchi, R.: Bilateral fitering for gray and color images. In: Proc. ICCV, pp. 839–846 (1998)
30. Vranic, D.V., Saupe, D., Richter, J.: Tools for 3D-object retrieval: Karhunen-Loeve transform and spherical harmonics. In: Proc. Workshop Multimedia Signal Processing, pp. 293–298 (2001)
31. Wardetzky, M., Mathur, S., Kälberer, F., Grinspun, E.: Discrete Laplace operators: no free lunch. In: Conf. Computer Graphics and Interactive Techniques (2008)
32. Wu, C., Clipp, B., Li, X., Frahm, J.-M., Pollefeys, M.: 3d model matching with viewpoint-invariant patches (vip), pp. 1–8 (June 2008)
33. Wyngaerd, J.V.: Combining texture and shape for automatic crude patch registration, pp. 179–186 (October 2003)
34. Xu, G.: Convergence of discrete Laplace-Beltrami operators over surfaces. Technical report, Institute of Computational Mathematics and Scientific/Engineering Computing, China (2004)
35. Yoon, K.-J., Prados, E., Sturm, P.: Joint estimation of shape and reflectance using multiple images with known illumination conditions (2010)
36. Zaharescu, A., Boyer, E., Horaud, R.P.: Transformesh: a topology-adaptive mesh-based approach to surface evolution (November 2007)
37. Zaharescu, A., Boyer, E., Varanasi, K.: R Horaud. Surface feature detection and description with applications to mesh matching. In: Proc. CVPR (2009)

Human Activity Modeling as Brownian Motion on Shape Manifold

Sheng Yi, Hamid Krim, and Larry K. Norris

North Carolina Sate University, USA

Abstract. In this paper we propose a stochastic modeling of human activity on a shape manifold. From a video sequence, human activity is extracted as a sequence of shape. Such a sequence is considered as one realization of a random process on shape manifold. Then Different activities are modeled by manifold valued random processes with different distributions. To solve the problem of stochastic modeling on a manifold, we first regress a manifold values process to a Euclidean process. The resulted process then could be modeled by linear models such as a stationary incremental process and a piecewise stationary incremental process. The mapping from manifold to Euclidean space is known as a stochastic development. The idea is to parallelly transport the tangent along curve on manifold to a single tangent space. The advantage of such technique is the one to one correspondence between the process in Euclidean space and the one on manifold. The proposed algorithm is tested on database [5] and compared with the related work in [5]. The result demonstrate the high accuracy of our modeling in characterizing different activities.

1 Introduction

Human activity recognition is of great interest in a wide range of applications, spanning areas such as security surveillance, person identification and content-based image retrieval. In addition to security applications, the explosively increasing daily usage of video cameras has and continues to motivate a increasing interest in motion analysis and understanding in video for diverse applications. Recent progress on human activity analysis from video data has been well documented in [9] [8].

Of the various possible representations of human activity [12], we choose to view any given activity of interest as a shape sequence [5] [11]. Different shape representations will lead to different shape manifolds. In Kendall's shape theory [13], a shape is considered to be a set of land marks on the boundary of an object. Due to the simple geoemtry, the Kendall pre-shape space is the popular platform for different modelings, which is invariant to translation and scaling, and geometrically is a hyper sphere. For example, an AR/ARMA model of human activities was proposed in [11] by projecting the shape sequences onto the tangent space of Kendall's preshape space. To overcome the problem of systematically picking consistent landmarks of shapes, we consider a shape as a simple

A.M. Bruckstein et al. (Eds.): SSVM 2011, LNCS 6667, pp. 628–639, 2012.

and closed planar curve. Such a shape formulation was proposed in [1] with a numerically efficient computation for tangent space of each point on manifold and the geodesic path between any two shapes.

With a similar goal of classifying the shape sequences as in [11], our goal in this paper is to build stochastic model for the process on shape manifold and then in return use the estimated model parameter to classify different activities. The idea is to develop a stochastic model of a process on the shape manifold (representing a shape sequence trajectory on a manifold) on a non-linear space by regressing the problem onto a linear space. In [11] a shape on a preshape sphere is projected onto a tangent space at the mean shape. Adopting such a tangent approximation is, however, valid in only a sufficiently small neighborhood. The invertibility of such a projection on pre-shape sphere only holds when the shape sequence does not cross the "north or south poles" of the hypersphere. Generally on a smooth manifold, the condition of such an orthogonal projection is restricted to a local area of the manifold. In contrast, our proposed regression is intrinsically constructed by a curve development [3] as a 1-1 mapping of an evolution curve on any smooth manifold to a curve in a flat space.

We exploit the afore-described approach to develop in this paper, an intrinsic stochastic model with a goal to classify activities. Assuming a proper human silhouette segmentation[1] of each frame in a video sequence of interest, a specific activity may be summarized by a sequence of individual closed curves/shapes in form of an evolution curve on the underlying shape manifold. Any reasonable modeling for a activity, for instance like "running", is expected to describe the different data samples of "running". In this paper, the set of the different representative curves of "running" are viewed as the realizations of the "running process", which more precisely is a random process on the shape manifold. As a result, any activity process of interest may hence be modeled as a manifold valued random process.

In the balance of this paper, we first provide a brief (but sufficient for this development) introduction to manifold geometry and to stochastic analysis on manifolds. The preprocessing on shapes is introduced in Section 2, to make the shape manifold finite dimensional. In Section 3, we introduce the stochastic curve development for a human activity process as a mapping from a manifold to a flat space. In Sections 4 and 5, we construct a connection on the shape manifold, and derive the corresponding curve development result for a given human activity.

2   Background

In this section we first provide a brief review of the required background in differential geometry and stochastic differential equation to allow us to define a shape manifold as a working space. We also describe the required tools of parallel transportation and curve development on a shape manifold.

---

[1] Note that errors in segmentation clearly imply errors down the processing stream, and this investigation is a research topic in and of itself, and is left for future work.

2.1   A Infinite Dimensional Shape Manifold

According to [1], a planar shape is a simple and closed curve $\alpha(s)$ in $\mathbb{R}^2$,

$$\alpha(s) : I \rightarrow \mathbb{R}^2, \tag{1}$$

where an arc-length parameterization is adopted. A shape is represented by a *direction index function* $\theta(t)$. With such a parameterization, $\theta(s)$ may be associated to the shape by

$$\frac{\partial \alpha}{\partial s} = e^{j\theta(s)}. \tag{2}$$

Due to the fact that the rotation index for simple closed curve is restricted to 1 in [1], the ambient space of the manifold of θ is an affine space based on $\mathbb{L}^2$,

$$\theta \in A(\mathbb{L}^2). \tag{3}$$

Further more, according to the restriction for a planar curve to be a closed curve, and invariant over rigid Euclidean transformations. The shape manifold M is defined as a level set of function $\phi : A(\mathbb{L}^2) \rightarrow R^3$,

$$\phi(\theta) = (\int_0^{2\pi} \theta ds, \int_0^{2\pi} \cos(\theta) ds, \int_0^{2\pi} \sin(\theta) ds) \tag{4}$$

Using the function ϕ defined above, in [1] the shape manifold M is defined as following

$$M = \phi^{-1}(\pi, 0, 0) \tag{5}$$

One of the most important properties of M is that the tangent space TM is well defined. Such a property not only simplifies the analysis, but also makes possible the numerical computation,

$$T_\theta M = \{f \in \mathbb{L}^2 | f \perp span\{1, \cos(\theta), \sin(\theta)\}\} \tag{6}$$

2.2   Connection on Manifold

To study a random process on manifold, we need to overcome the difficulty resulted from the curvature of the manifold. The Riemannian structure of the manifold can be defined by the connection in the principle bundle.In this paper the connection is defined in the frame bundle $\mathbb{F}(M)$, which is a special case of principle bundle.

Definition 1 (Principal Fiber Bundle). *A principal fiber bundle is a set (P, G, M), where P, M are C^∞ manifolds, and G is a Lie group such that*
(1) G acts freely on the right of P, $P \times G \rightarrow P$. For $g \in G$, we shall also write R_g for the map $g : P \rightarrow P$
(2) M is the quotient space of P by an equivalence relation under G (any shape subjected to a $g \in G$ is equivalent to itself), and the projection $\pi : P \rightarrow M$ is C^∞, so for $m \in M$, G is simply transitive on $\pi^{-1}(m)$
(3) P is locally trivial. Thus for any open set $U \subset M$, $\pi^{-1}(U) \sim U \times G$

A point u in $\mathbb{F}(M)$ can be written as, $u = (x, b)$, where $m \in M$ and $b = e_1, e_2, ..., e_n$ is a orthogonal basis of the associated tangent space $T_m M$. The group G acting on a fibre is $SO(n)$). Referring the definition of principle bundle, the equivalent class for each point $m \in M$ is all the orthogonal basis for tangent space $T_m M$. The rotation matrix can be utilize to transform one basis to another.

Definition 2 (Connection). *A connection on the principal bundle (P, G, M) is a n-dimensional distribution H on P, where $n = dim(M)$, such that*
(1) $H \in C^\infty$
(2) for every $p \in P$, $H_p + V_p = T_p P$, where V_p is a vertical space and H_p is a horizontal space of $T_p P$. A vector $Y \in T_p P$ is vertical if $\pi_*(Y) = 0$
(3) for every $p \in P$, $g \in G$, $(R_g)_*(H_p) = H_{pg}$.

With the definition of a connection on a manifold in hand, we can achieve a horizontal lift from a manifold to a linear frame bundle. A more expanded and detailed discussion of connections may be found in [3] [4].

Definition 3 (Horizontal Lift). *Let γ be a piecewise C^∞ curve in M, $\gamma : [0, 1] \rightarrow M$. Let $p \in \pi^{-1}(\gamma(0))$. Then there exists a unique lift $\tilde{\gamma}$ of γ such that $\tilde{\gamma}_*(t) \in H_{\tilde{\gamma}(t)}$ and $\tilde{\gamma}(0) = p$. We say that $\tilde{\gamma}$ is the horizontal lift of γ that starts at $p \in P$*

3   Dynamics of Human Activity on a Shape Manifold

Hsu in [2] proposes *an efficient analysis framework* to construct an invertible mapping from a manifold-valued random process to a Euclidean-valued random process. The essence of the mapping is to compute a Euclidean process that can drive a stochastic differential equation (SDE) to generate a manifold-valued random process. In a Euclidean space the random process have been extensively studied and there are many tools available for modeling. In contrast to the orthogonal projection method onto a tangent space around a mean, Hsu's theory provides a *one to one correspondence* between a process on a manifold and one on a Euclidean space. This improved accuracy of representation is primarily due to the so-called *"rolling without sliding" property* of a parallel transport.

As in [2], any random process on the shape manifold, may then be written as a solution to some $SDE(X_0, V, Z)$. Generally we have,

$$X_t = X_0 + \int_0^t \sum_i V_i(X_s) \circ dZ_s^i, \tag{7}$$

where, X_0 is the initial condition, V_i is a smooth vector field defined on M and Z_s^i is a Euclidean valued random process driving Equation (7). The stochastic integration here is the Stratonovich integration. More intuitively Equation (7) can be understood as $dX_t = \sum_i V_i(X_s) \circ dZ_s^i$. Thus the dynamic described by Equation (7) is characterized by both a vector field V and a driving process Z_t.

However, the form of the Euclidean process Z is varies with different choices of V. In contrast to our goal to construct a $1-1$ mapping from manifold process to Euclidean process, there is no one one correspondence $X \leftrightarrow Z$ without proper knowledge about V. In [2] this problem is solved by setting V equal to the horizontal lift of X_t, which in unique for a given connection. Provided the uniqueness of horizontal lift, the resulting driving process Z_t will have the one one correspondence to X_t. Let the vector field U_t be the horizontal lift of X_t in $F(M)$, Equation (7) may be rewritten as

$$X_t = X_0 + \int_0^t \sum_i U^i(X_s) \circ dW_s^i. \tag{8}$$

According to the definition of the orthogonal frame bundle and stochastic horizontal lift in Section 2 we know the horizontal vector field $U(t)$ can be written as,

$$U_t = \{e_1, e_2, \cdots, e_i, \cdots, e_n\} \tag{9}$$

where e_i is the basis of $T_{X_t}M$. In Equation 8 the differential dX_t is represented in a selected basis U_t with corresponding driving process W_t. For an orthogonal basis one can write , Consequently, the stochastic development of X_t is,

$$dW_t = U_t^{-1} \circ dX_t \tag{10}$$

$\forall i = 1, 2, 3, \cdots$ Equation 10 can be represented in vector as

$$dW_t^i = < e_i, dX_t > \tag{11}$$

Such rewriting of Equation (7) provides a representation of the random process X_t on a manifold with the Euclidean random process W_t, which generate the original process X_t by acting on vector field U_t as in Equation (8). In the above discussion, we provide a $1-1$ mapping from $X_t \in M$ to $W_t \in R^{dim(M)}$. The critical point for implementing such a mapping is the specific form of the connection H which we discuss in Section 4.

4   Flat Connection on a Shape Manifold

The construction of connection H is critical to the implementation of the curve development. Theoretically there may exist many different choices of H for a given manifold. Once the exact form of H is determined, the geometry of a manifold is specified accordingly. Among different kind of connections, we adopt the flat connection H for the efficiency of calculation. Flat connection do not always exist. However if the frame bundle $F(M)$ of a manifold M has a global section then the flat connections are easy to define. In the following, we provide a constructive proof of the existence of the global section. Thus the implementation of the flat connection proceeds by constructing a smooth global section $\sigma : M \rightarrow \mathbb{F}(M)$ of the linear frame bundle $\mathbb{F}(M)$. Then for each $m \in M$ we define $H_{\sigma(m)}$

to be the tangent space of the submanifold $\sigma(M)$ at $\sigma(m) \in F(M)$. Let u be any point of fiber over $m \in M$. Then there is a unique $g \in GL(n)$ such that,

$$u = R_g(\sigma(m)) \tag{12}$$

The horizontal subspace H_u is then defined as

$$H_u = R_{g*}(H_{\sigma(m)}) \tag{13}$$

In the construction of the smooth section σ, we smoothly assign to each point $\theta \in M$ a basis $\{E_k\}_{k=1,2,3...}$ for the tangent space $T_\theta M$. From Section 2, we know that the tangent space of M can be written as

$$T_\theta \hat{M} = \{v \in S | v \perp span\{1, \cos(\theta), \sin(\theta)\}\} \tag{14}$$

To construct smoothly distributed basis $\{E_k\}_{k=1,2,3...}$ on manifold, we first construct a Fourier-like global section $\tilde{\sigma}$ in the ambient space $A(L^2[0, 2\pi])$,

$$\tilde{\sigma} : \theta \rightarrow \{1, cos\theta, sin\theta, ..., cosi\theta, sini\theta, ...\} \tag{15}$$

Then $\tilde{\sigma}$ is properly projected to the tangent space TM as $\sigma : \theta \rightarrow \{E_k\}_{k=1,2,3...}$ following the geometry defined by Equation (14). The details of this procedure implementation are as as follows, Firstly, one can easily show that the following set of continuous functions is a linearly independent set,

$$\{1, cos\theta, sin\theta, ..., cosi\theta, sini\theta, ...\} \tag{16}$$

Let B_i be the result of a Gram Schmidt orthogonalization of the above basis in ambient space.

$$\{v_1, v_2, v_3, B_{i=1,2,3,\cdots}\} =$$
$$ON\{1, cos\theta, sin\theta, ..., cosi\theta, sini\theta, ...\}$$

where $\{v_1, v_2, v_3\}$ are the first three basis vectors from the Gram Schmidt procedure which correspond to the normal space of the tangent space,

$$span\{v_1, v_2, v_3\} = span\{1, \cos(\theta), \sin(\theta)\} \tag{17}$$

These basis vectors are excluded because they are orthogonal to the tangent space of M. Then B_i is the ambient representation of the basis of $T_\theta M$. The orthogonal projection from L^2 onto S can be written as a Fourier approximation of B_i and denoted by $\hat{B}_i$. Letting ϕ_j denote the Fourier basis functions, it follows that

$$\hat{B}_i = \sum_{j=1}^{N} < B_i, \phi_j > \phi_j^* \tag{18}$$

where $< B_i, \phi_j >$ is the inner product defined in L^2 In such setting, we would smooth assign a basis for $T_m \tilde{M}$ by a Gram Schmidt procedure applied to $\hat{B}_{i(k)=k}$.

$$E_k = ON\{\hat{B}_1, \hat{B}_2, \hat{B}_3, \cdots \hat{B}_N\} \tag{19}$$

Thus the resulted global section used to define the connection H is,

$$\sigma : \theta \in M \to E_k \tag{20}$$

In the shape manifold M equipped with the flat connection H as defined in Equation (21), the horizontal lift U_t of X_t with initial condition U_0 is computed as following.

$$U_t = R_g \circ \sigma(X_t) \tag{21}$$

where $g \circ \sigma(X_0) = U_0$.

In the ambient space $A(S)$, $\forall t, U_t, \sigma(X_t)$ can be represented by $N \times N$ invertible matrix. For example, $U_t = [e_1, e_2, \cdots, e_N]$, where $e_i \in R^N$ span the tangent space $T_m M$. In such setting U_t can be calculated as a matrix multiplication in the ambient space.

$$(U_t)_{ij} = \sum_k g_{ik}\dot{\sigma}(X_t)_{kj} \tag{22}$$

where $g \cdot \sigma(X_0) = U_0$. Consequently the development W_t of X_t can be written as,

$$W_t = \int_0^t (U_t)^{-1} dX_t \tag{23}$$

In Figure (1), a few numerical results are demonstrated for W_t for three activities: walking, running and bending.

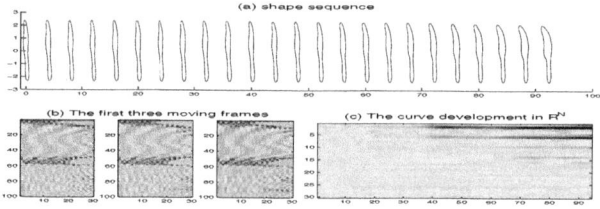

Fig. 1. curve development of $X_t \in M$ in $R^{N=30}$: (a) the original shape sequence represented by angle functions X_t (b) the horizontal lift U_1, U_2, U_3. (c) the development $(U_t)^{-1} dX_t$

5   Stochastic Analysis in a Euclidean Space

As discussed in the Section 3, the random process X_t on the shape manifold is now mapped to a Euclidean random process W_t. The recognition for human activity is thus reduced to comparing different processes in the flat space. In this section, we show that the resulted W_t exhibits a strong non-stationarity trend. A stationarity test is performed on W_t with the "double windows" method as proposed in [6]. The evolutionary spectrum shows that $\|W_t\|^2$ is non-stationary for most of the activity in the motion data base in [5]. The evolution spectrum Y_t is estimated by a double sliding window method.

Figure 2 shows several results of the evolutionary spectrum, $T_1 = 11, T_2 = 51$.

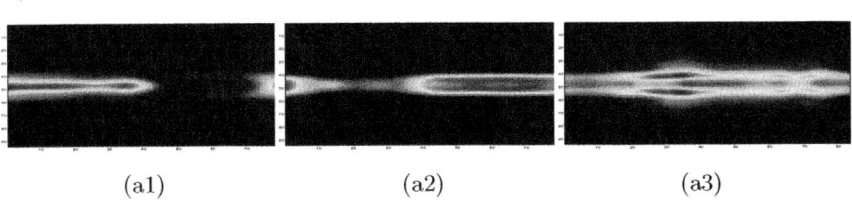

(a1)   (a2)   (a3)

Fig. 2. The evolutionary spectrum of $\|W_t\|$: a1, a2, a3 are the EPSD corresponding to the original shape sequence (a), (b), (c) in figure 1;

Given the non-stationary Euclidean process W_t, which is a stochastic development of X_t, we first analyze it as a Brownian motion. As in Section 5.1, a self-covariance matrix K of the increment dW_t of W_t is estimated from observations. We subsequently proceed to discuss activity classification by introducing a metric for K. Computing the increments of W_t to achieve stationarity also carries a potential of increasing the noise, particularly when the process is non-homogeneous.

According to the comparison in 5.1, the performance of the Brownian Motion Model is sufficient for the classification of human activities. However, from the view of model fitting, the Brownian motion model still assume the first order incremental dW_t to be stationary, which is still not necessarily truth for all the data. Instead of imposing the strong assumption of higher order stationarity, in Subsection 5.2, we further introduce random process segmentation according to the local stationarity. While additional computational cost is incurred to segment the process W_t, we develop the piece-wise Brownian model to further relax the assumption of global stationarity and the resulted modeling can be better fitted to different data.

In the experiments, we test both of the two model on the activity classification database in [5]. The experiment result is compared with [5] and [11] for each of the database. The activity data in [5] includes 10 different activities. Each one has 9 video sequence for 9 different actors. We perform level set segmentation to extract the contour of shape as in [14].

5.1   Human Activity as a Brownian Motion on Manifold

In this section we model W_t as a Euclidean Brownian motion. Consequently, the model for X_t is a Brownain motion on the shape manifold M, which can be written as the following stochastic differential equation on M,

$$X_t = X_0 + \int_0^t \sum_i U^i(X_s) \circ dW_s^i. \tag{24}$$

where W_t is a high dimensional Euclidean Brownian motion and $U(X)$ is the horizontal field calculated in Section 4.

Then by the distribution of dW_t we can characterize different activities. Since dW_t is IID and Gaussian, the time sampling is used as the sampling for random variable dW. We next proceed to estimate the covariance matrix $K(dW^i, dW^j)$ as the feature of choice for the underlying distribution.

$$K(dW^i, dW^j) = E((dW^i - E(dW^i))(dW^j - E(dW^j))) \qquad (25)$$

where dW^i is the ith element of the vector dW. The distance between two different covariance matrices is defined by the Frobenius norm as,

$$D(K1, K2) = \|K1 - K2\|_F. \qquad (26)$$

The results of $D(K1, K2)$ for the data base in [5] is shown in Figure 3. Using the distance matrix D, we may carry the recognition/classification task by using, for example, the leave-one-out algorithm. The nearest neighborhood algorithm is used for classification. If we let N_B be the total number of realizations of B. and $N(B, A)$ the number of realizations of B classified as A activity, we have

$$P(A|B) = \frac{N(B, A)}{N(B)} \qquad (27)$$

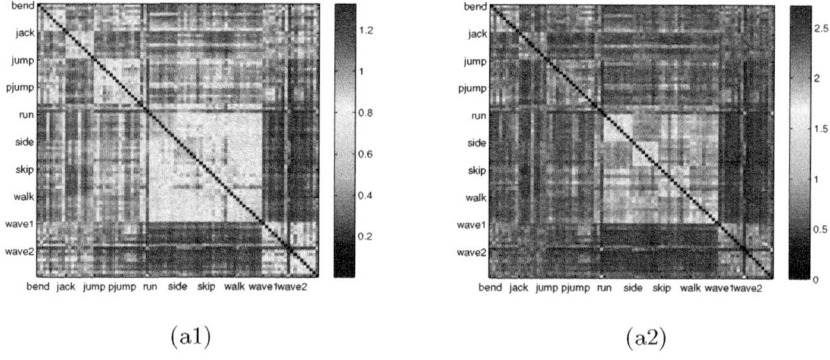

(a1)   (a2)

Fig. 3. (a1) Distance matrix for database in [5] with Brownian Motion (a2) Distance matrix for database in [5] with Piece-wise Brownian Motion

To perform a consistent comparison of results published in [5], we need to change our experiment to the same setting. In [5] the number of activity observations is increased by segmenting any video sequence for a given activity into many overlapped chunks. The segments are assumed independent and the classification is carried out. The performance of our proposed method is summarized in the following tables.

Table 1. Table of recognition rate. the number in () is the result in [5].

$P(act1|act2)$	bend	jack	jump	pjump	run	side	skip	walk	wave1	wave2
bend	1	0	0	0	0	0	0	0	0	0
jack	0	1 (0.98)	0 (0.02)	0	0	0	0	0	0	0
jump	0	0(0.02)	1 (0.971)	0	0	0	0	0	0	0
pjump	0.0556(0)	0	0	0.944(1)	0	0	0	0	0	0
run	0	0	0	0(0.108)	0.944(0.892)	0.0556 (0)	0	0	0	0
side	0	0	0	0	0	1	0	0	0	0
skip	0	0	0	0	0	0	1	0	0	0
walk	0	0	0	0(0.09)	0	0(0.09)	0	1(0.948)	0(0.35)	0
wave1	0	0	0	0(0.09)	0	0	0	0(0.019)	1(0.972)	0
wave2	0	0	0	0	0	0	0	0	0(0.09)	1(0.991)

From the above comparison, our manifold valued Brownian motion model achieve better performance for almost all the activities except slightly lower for case "pJump".

5.2   Human Activity as a Piecewise Brownian Motion on Manifold

Alternatively to assuming dW_t to be stationary, we directly address the non-stationarity by segmenting the W_t into several local stationary segments. For each segment we carry out the Brownian motion modeling. Such a topic has been extensively investigated in time series analysis [6]. We apply a sliding window computation of an evolution spectrum $Y_{t,\omega}$ of dW_t as in [6] and detect transient points.

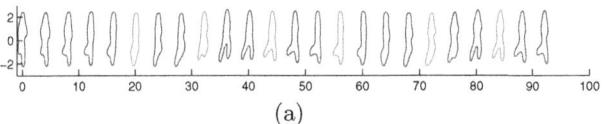

(a)

Fig. 4. Nonstationary time series segmentation: (a),(b),(c) is the segmentation results of activity bending, running, skipping

For the ith segment of dW_t, we estimate K^i according to Equation 25. We next define the distance between two sequence $dW1_t$ and $dW2_t$ is defined by

$$D_{seg}(K1, K2) = median(min_i(D_{cov}(K1^i, K2^j))) +$$
$$median(min_j(D_{cov}(K1^i, K2^j))).$$

The distance matrix is calculated for both the databases according to the above equation. The result is illustrated as in figure 3.

In Figure 3, we show that the distance matrix D_{cov} calculated from the same data set as in [5]. By then using the conditional probability of recognition, as a performance metric as in Equation (27), we demonstrate the classification performance in table 5.2. However here we provide no comparison with the result

Table 2. Table of recognition rate for data base in [5]

$P(act1|act2)$	bend	jack	jump	pjump	run	side	skip	walk	wave1	wave2
bend	0.7778	0	0	0.1111	0	0	0	0	0.1111	0
jack	0	0.7778	0	0	0.1111	0	0	0	0	0.1111
jump	0	0	0.5556	0.2222	0	0.1111	0	0.1111	0	0
pjump	0.2222	0	0.1111	0.3333	0	0.1111	0.1111	0	0.1111	0
run	0	0	0	0	1	0	0	0	0	0
side	0	0	0	0	0	1	0	0	0	0
skip	0	0	0	0	0	0	1	0	0	0
walk	0	0	0	0	0	0	0	1	0	0
wave1	0	0	0	0.2222	0	0	0	0	0.5556	0.2222
wave2	0	0	0	0	0	0	0	0	0.4444	0.5556

in [5]. Because as mentioned in the previous section, in [5] the data sample is increase by segment the original sequence into small overlapping chunks. Such setting is making our stationary segmentation trivial. If following the same way as in [5], then each small chunk would be a signal stationary segment. Therefore as for the modified database, the piecewise brownian model is the same as the global brownian model. So here we only provide our result on the original database.

6   Conclusion

In this paper, we provide a systematic framework for the stochastic modeling of human activity on shape manifold. In theory, such framework is one one mapping from random process on manifold to the random process in Euclidean space. In the resulted flat space, the representative random process of activity is modeled as both global and local Brownian Motion process. The experiment well demonstrate the performance of the proposed modelings of activities.

Acknowledgment. The authors would like to thank Dr. Huiling Le for her insightful discussion and help on this work. Also we would like to thank Dr. Anuj Srivastava for providing the original code for their work in [1].

References

1. Klassen, E., Srivastava, A., Mio, W., Joshi, S.H.: Analysis of planar shapes using geodesic paths on shape spaces. IEEE Trans. Pattern Analysis and Machine Intelligence (2004)
2. Hsu, E.P.: Stochastic analysis on manifold. Graduate Studies in Mathematics, 38
3. Bishop, R.L., Crittenden, R.J.: Geometry of Manifold. Academic Press, New York (1964)
4. Kobayashi, S., Nomizu, k.: Foundations of differential geometry, vol. 1. John Wiley & Sons, West Sussex (1996)

5. Blank, M., Gorelick, L., Shechtman, E., Basri, M.I.R.: Action as Space-Time Shapes. In: IEEE ICCV (2005)
6. Priestley, M.B., Subba Rao, T.: A Test for Non stationarity of Time-Series. Journal of the Royal Statistical Society. Series B 31(1), 140–149 (1969)
7. Keogh, E., Chu, S., Hart, D., Pazzani, M.: Segmenting time series: A survey and novel approach, Data Mining in Time Series Databases. World Scientific, Singapore (2004)
8. Aggarwal, J.K., Cai, Q.: Human Motion Analysis: A Reivew. Computer Vision and Image Understanding 73, 428–440 (1999)
9. Turaga, P., Chellappa, R., Subrahmanian, V., Udrea, O.: Machine recognition of human activities: A survey. IEEE Transactions on Circuits and Systems for Video Technology 18, 1473–1488 (2008)
10. Elgammal, A.M., Lee, C.-S.: Inferring 3D body pose from silhouettes using activity manifold learning. In: Proceedings of the Conference on Computer Vision and Pattern Recognition (CVPR 2004), Washington, DC, vol. 2, pp. 681–688 (June 2004)
11. Veeraraghavan, A., Roy-Chowdhury, A.K., Chellappa, R.: Matching Shape Sequences in Video with Applications in Human Movement Analysis. IEEE Trans. Pattern Analysis and Machine Intelligence 27(12) (December 2005)
12. Veeraraghavan, A., Chowdhury, A.R., Chellappa, R.: Role of shape and kinematics in human movement analysis. In: IEEE Computer Society Conference on Computer Vision and Pattern Recognition (2004)
13. Kendall, D.: Shape Manifolds, Procrustean Metrics and Complex Projective Spaces. Bull. London Math. Soc. 16, 81–121 (1984)
14. Chen, P., Steen, R., Yezzi, A., Krim, H.: Joint brain parametric T1-Map segmentation and RF inhomogeneity calibration. International Journal of Biomedical Imaging (269525), 14 p. (2009)

3D Curve Evolution Algorithm with Tangential Redistribution for a Fully Automatic Finding of an Ideal Camera Path in Virtual Colonoscopy

Karol Mikula and Jozef Urbán

Slovak University of Technology,
Department of Mathematics and Descriptive Geometry
Radlinského 11, 813 68 Bratislava, Slovakia
mikula@math.sk,jozo.urban@gmail.com
http://www.math.sk

Abstract. In this paper we develop new method, based on 3D evolving curves, for finding the optimal trajectory of the camera in the virtual colonoscopy - the medical technology dealing with colon diagnoses by computer. The proposed method consists of three steps: 3D segmentation of the colon from CT images, finding an initial trajectory guess inside the segmented 3D subvolumes, and driving the initial 3D curve to its optimal position. To that goal, the new fast and stable 3D curve evolution algorithm is developed in which the initial curve is driven by the velocity field in the plane normal to the evolving curve, the evolution is regularized by curvature and accompanied by the suitable choice of tangential velocity. Thanks to the asymtotically uniform tangential redistribution of grid points, originally introduced in this paper for 3D evolving curves, and to the fast and stable semi-implicit scheme for solving our proposed intrinsic advection-diffusion PDE, we end up in fast and robust way with the smooth uniformly discretized 3D curve representing the ideal path of the camera in virtual colonoscopy.

Keywords: Virtual colonoscopy, evolving 3D curves, tangential velocity, asymptotically uniform redistribution, distance function, segmentation.

1 Introduction

According to the official evidence from 2007, the colon cancer is the third most spread cancer desease (after the breast and lung cancers) in the countries included in the World Health Organization.

A classical optical colonoscopy is an examination of the colon (large intestine) which can successfully detect colon polyps and colorectal tumours. The examination takes 15-60 minutes and it is performed by a colonoscope which is a flexible tube with a miniature camera and which may also provide a tool for removing a tissue. The colonoscope is introduced into the colon through the rectum, it moves along the colon and a physician can see the situation in the colon on the screen. Because this examination is uncomfortable and painful,

A.M. Bruckstein et al. (Eds.): SSVM 2011, LNCS 6667, pp. 640–652, 2012.

the patient receives medication absorbing the pain or it is done under the general anesthesia. On the other hand, the virtual colonoscopy, introduced in [5], is an examination performed by using the computed tomography (CT). The colon is inflated (with air or CO_2) and then the patient is scanned in two positions (on the abdomen and on the back) by CT. The colon can be viewed similarly to the classical optical colonoscopy, but the physician controls the so-called virtual camera by using its computed "ideal" trajectory. The computation of the ideal path is the important part of the process and new method for its finding is proposed in this paper. It is worth to note that the results achieved by the virtual colonoscopy are comparable to the classical approach [8]. Moreover, the images can be viewed at any time, it provides the option to view panoramas of the colon surface, to make its unfolding etc. In addition to these benefits, the virtual colonoscopy allows to examine colon parts impassable for colonoscope and it avoids risk of a perforation of the colon. A disadvantage is the radiation during CT examination and the fact that if the physician has found a polyp or tumour, it cannot be removed by the virtual approach.

The goal of this paper is to develop new fast and robust method for the fully automatic extraction of the ideal path of the camera in virtual colonoscopy. Here, we understand the ideal path as the 3D curve which passes along the centerline of the colon, which is smooth and which discrete point representation is uniformly distributed. Our proposed method consists of three basic steps. First, the 3D segmentation of the colon from CT images is performed. Due to the quality of CT data, the classical approaches like the thresholding and the region growing are used, see e.g. [1]. As the result we get all simply-connected parts of the large and small intestines filled with the gas. The next step consists in finding an initial guess for the camera trajectory in every simply-connected segmented subvolume of the intestine. Such initial guess is obtained by using the Dijkstra algorithm [3] for computing approximate distance from point sources inside the segmented subvolumes followed by the backtracking in steepest descent direction [13]. The third step is the core of our approach. It consists in driving the initial guess to its optimal position in smooth and stable way. To that goal we use a vector field given by the gradient of distance function to the segmented intestine borders which is computed by a 3D generalization of the approach from [2] based on the numerical solution of the time relaxed eikonal equation. Then, new 3D curve evolution algorithm is developed in which the initial curve is driven by the velocity given by the projection of the computed vector field to the plane normal to the evolving curve, the evolution is regularized by curvature, which makes it smooth, and it is accompanied by the suitable choice of the tangential velocity which makes the curve uniformly discretized during the evolution. In this paper, we develop new asymptotically uniform tangential grid point redistribution method for 3D evolving curves in parametric representation. Our new method is based on ideas from [7], where the authors used special $\kappa_1 - \kappa_2 - \omega - L$ 3D curve evolution formulation. The new method can be also understood as non-trivial generalization of 2D approaches from [9,10,11]. Our final 3D curve evolution model, in the form of an intrinsic advection-diffusion partial differential equation with a driving force, is solved by the fast and stable semi-implicit scheme and we end up with the smooth uniformly discretized 3D curve representing the ideal trajectory of the camera in virtual colonoscopy. It is worth to note that the overall CPU for all steps in our approach is about 8 seconds on standard PC. Thus the method is highly competitive and it is being implemented into the medical software of the TatraMed spol s r.o., Bratislava company.

2 The Colon Segmentation and the Initial Trajectory Guess

The 3D colon image data sets obtained by CT are given by a sequence of 2D slices (512×512 pixels) with a typical slice thickness about 0.75 mm. For our further goals, it is sufficient to subsample data and work with 3D images of the typical size $256 \times 256 \times 400$ voxels. First, we use thresholding corresponding to the air (about -1000 HU) in CT scans and detect all subvolumes filled by the gas. All voxels of these subvolumes get value 1, the others were set to 0. Next, we apply the region growing method in order to find all simply connected parts of the large and small intestines. The first seed is put to a corner of the 3D image and the region growing algorithm finds all voxels outside the body, their value is put to 0, so this subvolume is ignored. Next, we go subsequently through the whole 3D image and the seed for the next region growing is the next voxel found with the value 1. This seed and all voxels found by the region growing get number 2 which is set also as the number of this first inner body subvolume. We continue such procedure until all seeds for the next region growings are found. During the current region growing all detected voxel values are set to the number of the currently segmented subvolume which is given by the increment of the previously detected subvolume number. We also count the number of detected voxels in each subvolume which gives us the approximate size of the segmented structures. The last segmentation step consists in removing all spuriously detected subvolumes inside the body. By the checking of the size, we remove small inner structures filled by the gas (detected e.g. in lungs). Then we compute the distance function of all inner voxels to the border of the segmented subvolume (by the method from section 3) and if the global maximum of the distance function (maximal thickness of the structure) is less then a prescribed threshold we ignore such subvolume (representing e.g. the gas between the body and the CT desk). In such way we end up with one (rarely) or several simply connected subregions of the colon (and also of the small intestine) for which we find then the optimal virtual camera trajectories. The visualization of our segmentation result is presented in Figure 1 left. In order to illustrate our final results, on the right we show the camera trajectories inside segmented subvolumes of both large and small intestines found by our method described in the sequel.

The initial trajectory guess in any colon subvolume is constructed by computing a distance from a point source by the Dijkstra algorithm (in which the graph edges connecting neighbouring voxels have value 1) followed by the backtracking. First, we take any point of the subvolume and fix the distance at this point to

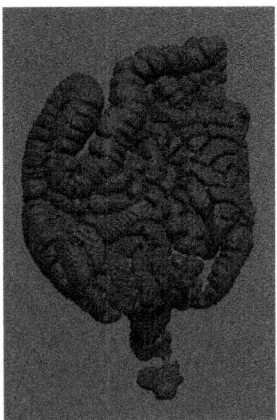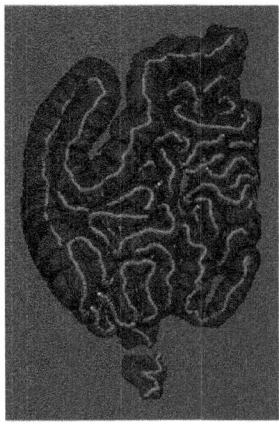

Fig. 1. The segmentation of the large and small intestine (left) and the virtual camera paths visualization for all segmented subvolumes (right)

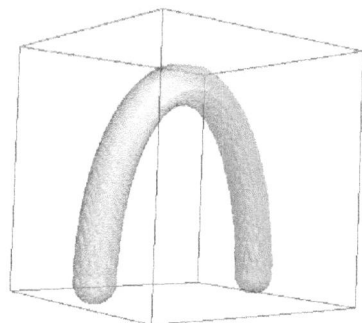

Fig. 2. The 2D and 3D image data sets used for testing the proposed method

0. Then we compute distance in the sense of the Dijkstra algorithm to this fixed point for all voxels inside the subvolume. Since the colon is a long organ we take a point with the maximal distance as the first endpoint of the segmented subvolume. Now, we fix the zero value at this point and use the Dijkstra algorithm again, a point with the maximal distance will represent the second endpoint of the subvolume. From this second endpoint we start the backtracking of computed distances in the steepest descent direction, we end at the first endpoint of the subvolume. The voxel coordinates of such path represent the parametric 3D curve, the initial guess of the trajectory inside the subvolume. It is worth to note that, for our approach, the only important issue is that the extracted path is a parametric 3D curve inside subvolume which connects two endpoints, its exact localization or smoothness is not important, because it will be improved later by the suitable 3D curve evolution. Also, the very precise localization of the two endpoints does not play any crucial role, because the virtual camera does not really touch the subvolume border in the first and last point. In order to illustrate and test the particular steps of our method, we constructed 2D and

3D artificial data shown in Figure 2, the connected circles on the left mimic the typically alternating very thin and thick colon parts (mostly problematic for the algorithms) and U-like volume on the right mimics an overal colon shape. As one can see in Figures 3-4, the initial trajectory guess is nor smooth nor centered, it touches very often the boundary of segmented volume. On the other hand, it gives the first parametric representation of the 3D curve which can be evolved to the optimal position by the approach discussed in the next section.

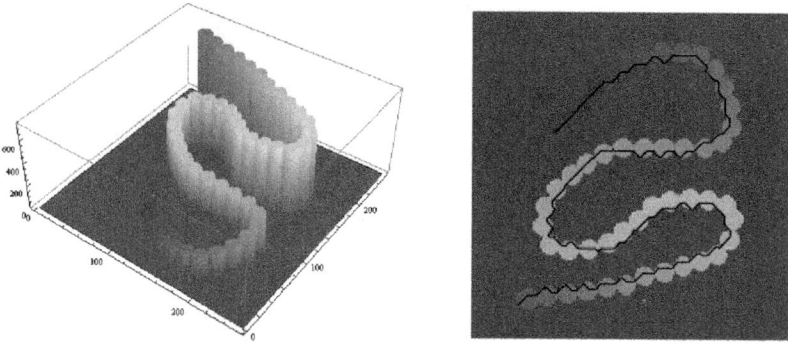

Fig. 3. The graph of the distances (left) and the initial trajectory guess (right)

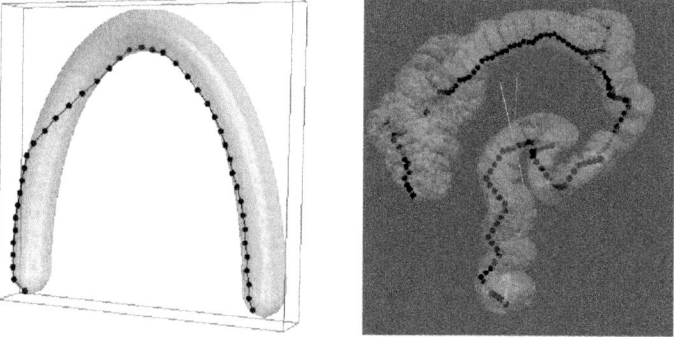

Fig. 4. Initial 3D curve in the test data (left) and in the real segmented colon (right)

3   Finding the Optimal Camera Trajectory

In this section we discuss important issues leading to suitable 3D curve evolution model which will drive the initial curve to its optimal position. Our model will be based on carefull construction of the velocity in normal direction, on the regularization of the motion by curvature and on the suitable tangential velocity yielding the uniform discretization of the evolving curve. We show that all these issues are necessary ingredients in order to get smooth and correctly centered virtual camera trajectory in real colon data of complicated shape.

In order to get the velocity field by which the 3D curve will be moving to its optimal position we solve the eikonal equation with the zero fixed values at the boundary voxels of the segmented subvolume. Its solution is a distance function which has a ridge along the centerline of the segmented subvolume and the gradient of such distance function points towards the ridge. The initial 3D curve should be driven in a smooth way into that ridge position. Our method for finding the distance function, which gives the above mentioned vector field, is based on the numerical solution of the time relaxed eikonal equation $d_t + |\nabla d| = 1$ by the so-called Roy-Tourin scheme [12,2]. Let us denote by d_{ijk}^n approximate solution d in time step n in the middle of voxel with spatial coordinates (i, j, k), τ_D the length of the time step and h_D the size of the voxel. Let us define expressions M_{ijk}^{pqr}, where $p, q, r \in \{-1, 0, 1\}, |p| + |q| + |r| = 1$, by

$$M_{ijk}^{pqr} = (\min(d_{i+p,j+q,k+r}^n - d_{ijk}^n, 0))^2. \tag{1}$$

Then the scheme is given by

$$d_{ijk}^{n+1} = d_{ijk}^n + \tau_D - \tag{2}$$
$$\frac{\tau_D}{h_D} \sqrt{\max(M_{ijk}^{-1,0,0}, M_{ijk}^{1,0,0}) + \max(M_{ijk}^{0,-1,0}, M_{ijk}^{0,1,0}) + \max(M_{ijk}^{0,0,-1}, M_{ijk}^{0,0,1})}.$$

The values at specified points to which the distance is computed numerically are fixed to zero. In all other points the numerical values are increasing monotonically and if they become changeless we can fix them on the fly [2]. Since the colon is an ablong organ, the method (2) is sufficiently fast and easily implementable and applicable to any complicated shape. After computing the distance function we compute the vector field $\mathbf{v} = \nabla d$ by using the central finite difference approximation of the partial derivatives. In Figures 5-6 we show visualization of the computed 2D distance function and the associated vector field.

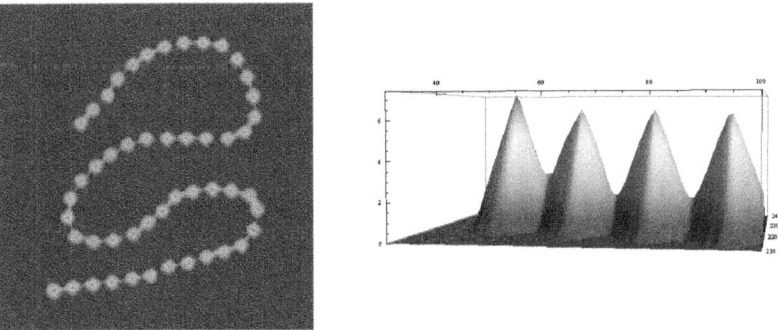

Fig. 5. The distance function to the boundary of 2D testing shape (left) and its detailed graph (right)

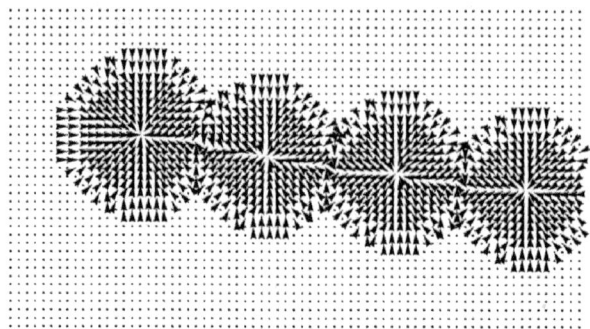

Fig. 6. The detail of the vector field given by the gradient of the distance function

In our parametric approach, the evolving 3D curve is represented by discrete points $\mathbf{r}_i^n = (x_i^n, y_i^n, z_i^n)$, where $i = 0, \ldots, m$, denotes the grid point number and n represents the discrete time stepping. We consider that the endpoints of the curve (i.e. the 0-th and the m-th point) are fixed. The simplest model for the motion of the curve in the vector field $\mathbf{v}$ is given by $\partial_t \mathbf{r} = \mathbf{v}(\mathbf{r})$, numerical discretization of which can be written as $\mathbf{r}_i^{n+1} = \mathbf{r}_i^n + \tau \mathbf{v}(\mathbf{r}_i^n)$, where τ being a discrete time step. The results achieved by this approach can be seen in Figure 7 where all the grid points were moved into the ridge position, but due to the specific direction and length of the velocity field (which is nonzero also on the ridge), they are packed together and thus it is difficult to get smooth virtual camera path by using such final curve state. A group of points may contain curve self-intersections (due to numerics) and their distances are irregular so there is no guarantee that the curve would not cross the edge of the colon on the way between the far-distant points. Such situation is not rare in the real data where the colon has complicated structure similar to our connected circles testing example. We note that many standard approaches to virtual colonoscopy, see e.g. [6,14], uses a combination (e.g. a weighted sum) of two distance functions, the one constructed in the previous section (distance to one fixed endpoint) and the one computed here (distance to the colon borders), followed by a minimization procedure. As we can see, it can lead to a serious troubles in trajectory representation or to a stacking in local minima, which are then solved by some heuristic and/or semi-automatic approaches.

The main difficulty of the above simple approach is given by the fact that the grid points just moved independently on each other by numerical discretization of ODE in direction of the basic velocity field $\mathbf{v}$. There is no mechanism by which the neighbouring points influence each other and thus move smoothly without degeneracy of their distances. All these problems will be solved, without any heuristic, by our new approach described below.

We know that the motion of the curve can be decomposed into the movement in tangential and normal directions and that the overall shape of the evolving curve with the fixed endpoints is determined only by the normal component of the velocity. The tangential velocity influences the redistribution of points along

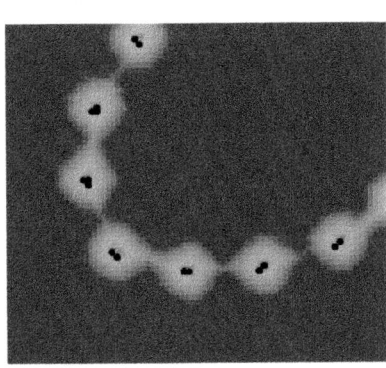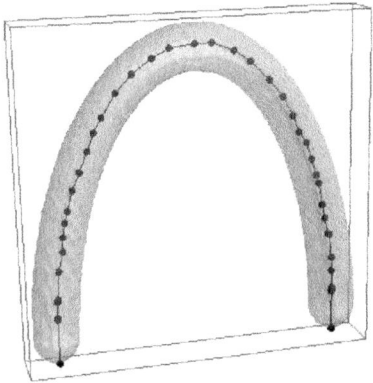

Fig. 7. The results obtained using the velocity field given by the gradient of the distance to the boundary of segmented object in 2D (left) and 3D (right) test data

the curve, thus if it is not controlled, it can cause accumulation of grid points as in the above mentioned examples. As the first modification of the vector field $\mathbf{v}$ we shall consider its projection to the evolving curve normal plane. This makes the model nonlinear (because the curve normal plane depends on the current curve shape) but it greatly improves the result. Moreover, if we want that the evolving curve points are tied together we have to move from the ordinary to a partial differential equation. A natural intrinsic PDE arising in this case is the one obtained by adding the curvature regularization to the motion by using the curvature vector $k\mathbf{N}$ which is again in the curve normal plane. Let $\mathbf{T}$ be the unit tangent vector to the curve, the projection of vector field $\mathbf{v}$ to the curve normal plane is then defined by $\mathbf{N_v} = \mathbf{v} - (\mathbf{T}.\mathbf{v})\mathbf{T}$ and the regularized motion of the curve in the normal plane is given by

$$\partial_t \mathbf{r} = \mu \mathbf{N_v} + \epsilon k\mathbf{N} , \tag{3}$$

where μ and ϵ are the model parameters. Its explicit numerical disretization is

$$\frac{\mathbf{r}_i^{n+1} - \mathbf{r}_i^n}{\tau} = \mu \left(\mathbf{N_v}\right)_i^n + \epsilon \frac{2}{h_{i+1}^n + h_i^n} \left(\frac{\mathbf{r}_{i+1}^n - \mathbf{r}_i^n}{h_{i+1}^n} - \frac{\mathbf{r}_i^n - \mathbf{r}_{i-1}^n}{h_i^n} \right) \tag{4}$$

$i = 1, \ldots, m - 1$, where the second term on the right hand side represents the discretization of the curvature vector $k\mathbf{N}$, see e.g. [4], and the first term is the approximation of the vector $\mathbf{N_v}$ at the i-th curve grid point, both at the previous time step n. The distances between the grid points are given by the expressions

$$h_i^n = \sqrt{(x_i^n - x_{i-1}^n)^2 + (y_i^n - y_{i-1}^n)^2 + (z_i^n - z_{i-1}^n)^2} . \tag{5}$$

Since we removed the improper tangential component of the velocity and used the curvature regularization, the ridge in the testing data is found in much more regular way, see Figure 8. The only problem which is still remaining is the nonuniform distribution of the grid points at the final state and also during the subsequent curve evolution which may cause problems during the motion inside complicated shapes. The uniform curve representation would guarantee that the properties of the projected vector field are taking into account uniformly and thus the motion of the curve is done in the most correct way. The uniform curve discretization will be controlled by adding a suitable tangential velocity into the mathematical model and its numerical discretization.

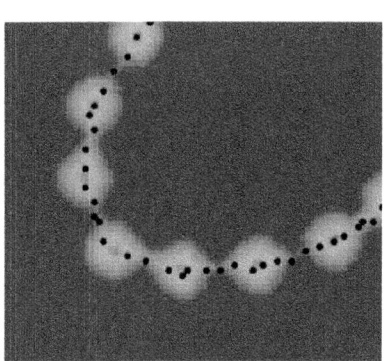 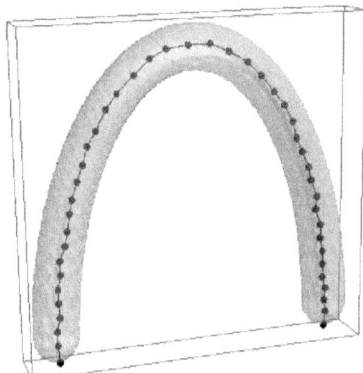

Fig. 8. The results obtained using the projection of the original vector field into the normal plane to the evolving 3D curve accompanied by the curvature regularization

In order to determine suitable tangential velocity we introduce the local orthogonal basis smoothly varying along the 3D curve, cf. [7]. It will consist of $\mathbf{T}$ and two orthogonal vectors in the normal plane defined as $\mathbf{N_1} = \frac{\mathbf{N_v}}{|\mathbf{N_v}|}$ and $\mathbf{N_2} = \mathbf{N_1} \times \mathbf{T}$ (if $|\mathbf{N_v}| = 0$ we redefine $\mathbf{N_1}$ due to the smoothness requirement, e.g. in discrete settings by the averaged value from the neighboring grid points). Let us define $k_1 = k\mathbf{N}.\mathbf{N_1}$ and $k_2 = k\mathbf{N}.\mathbf{N_2}$, the projections of the curvature vector onto $\mathbf{N_1}$ and $\mathbf{N_2}$. Then the curvature vector satisfies $k\mathbf{N} = k_1\mathbf{N_1} + k_2\mathbf{N_2}$ and the evolution equation (3) can be written as

$$\partial_t \mathbf{r} = U\mathbf{N_1} + V\mathbf{N_2} + \alpha\mathbf{T}, \tag{6}$$

with free parameter α representing the tangential component of the velocity, and with the normal components given by

$$U = \epsilon k_1 + \mu|\mathbf{N_v}|, \quad V = \epsilon k_2. \tag{7}$$

Let us consider the curve Γ with fixed endpoints parametrized by the position vector $\mathbf{r}$, and define its local lenght parameter $g = \left|\frac{\partial \mathbf{r}}{\partial u}\right|$, $u \in [0, 1]$. In discrete settings it is approximated by $g \approx \frac{|\mathbf{r}_i - \mathbf{r}_{i-1}|}{h}$ with $h = \frac{1}{m}$. One can prove, cf. [7], that the time change of the local length of the evolving 3D curve is given by

$$\partial_t g = g\partial_s \alpha - g(Uk_1 + Vk_2) \tag{8}$$

where s is the arclength parameter. This relation gives us also the time evolution for the total length L

$$\frac{dL}{dt} = - < U k_1 + V k_2 >_\Gamma L \tag{9}$$

where $< U k_1 + V k_2 >_\Gamma$ denotes the averaged quantity along the curve. In the discrete settings it is computed by

$$< U k_1 + V k_2 >_\Gamma^n = \frac{1}{L^n} \sum_{l=1}^m h_l^n (U_l^n k_{1l}^n + V_l^n k_{2l}^n), \quad L^n = \sum_{l=1}^m h_l^n \tag{10}$$

where $U_l^n, k_{1l}^n, V_l^n, k_{2l}^n$ are approximations of the corresponding quantities on discrete curve segments, cf. (4). In order to define the tangential velocity leading to the asymptotically uniform grid points redistribution, it is worth to study the fraction $\frac{g}{L} \approx \frac{|\mathbf{r}_i - \mathbf{r}_{i-1}|}{Lh} = \frac{|\mathbf{r}_i - \mathbf{r}_{i-1}|}{\left(\frac{L}{m}\right)} = \frac{h_i}{\left(\frac{L}{m}\right)}$ representing in discrete settings the ratio of the actual and averaged lengths of the curve segments. The goal is to desing a model in which this ratio tends to 1 or such that the quantity $\theta = \ln\left(\frac{g}{L}\right)$ converges to 0. Using (8) and (9) we get for its time evolution the relation $\partial_t \theta = \partial_s \alpha - (U k_1 + V k_2) + < U k_1 + V k_2 >_\Gamma$. On the other hand, if we set $\partial_t \theta = (e^{-\theta} - 1)\omega_r$, where ω_r is a speed of redistribution process, we get that $\theta \to 0$ as $t \to \infty$ and we obtain equation for the tangential velocity α guarateeing the asymptotically uniform redistribution of 3D curve grid points

$$\partial_s \alpha = U k_1 + V k_2 - < U k_1 + V k_2 >_\Gamma + \left(\frac{L}{g} - 1\right)\omega_r. \tag{11}$$

Since $\mathbf{T} = \partial_s \mathbf{r}$ and $k\mathbf{N} = \partial_{ss}\mathbf{r}$ we get our final 3D curve evolution model in the form of the following intrinsic advection-diffusion PDE with driving force

$$\partial_t \mathbf{r} = \mu \mathbf{N}_\mathbf{v} + \epsilon \partial_{ss}\mathbf{r} + \alpha \partial_s \mathbf{r} \tag{12}$$

with α given by (11) and accompanied by the Dirichlet boundary conditions (fixed endpoints of the curve). Our final step is the numerical discretization of (11)-(12). We proceed similarly to [9,11] and get the discrete tangential velocity

$$\alpha_i^n = \alpha_{i-1}^n + h_i^n (U_i^n k_{1i}^n + V_i^n k_{2i}^n) - h_i^n < U k_1 + V k_2 >_\Gamma^n + \left(\frac{L^n}{m} - h_i^n\right)\omega_r, \tag{13}$$

for $i = 1, \ldots, m - 1$, setting $\alpha_0^n = 0$ and getting $\alpha_m^n = 0$. The discretization of the equation (12) is performed by using the semi-implicit scheme and we get

$$\frac{h_{i+1}^n + h_i^n}{2} \frac{\mathbf{r}_i^{n+1} - \mathbf{r}_i^n}{\tau} = \mu \frac{h_{i+1}^n + h_i^n}{2} (\mathbf{N}_\mathbf{v})_i^n + \tag{14}$$

$$\epsilon \left(\frac{\mathbf{r}_{i+1}^{n+1} - \mathbf{r}_i^{n+1}}{h_{i+1}^n} - \frac{\mathbf{r}_i^{n+1} - \mathbf{r}_{i-1}^{n+1}}{h_i^n}\right) + \frac{\alpha_i^n}{2}(\mathbf{r}_{i+1}^{n+1} - \mathbf{r}_{i-1}^{n+1}),$$

for $i = 1, \ldots, m-1$, with $\mathbf{r}_0^{n+1}$ and $\mathbf{r}_m^{n+1}$ prescribed. The scheme (14) represents three linear tri-diagonal systems for the x, y, z coordinates of the grid points representing new curve position. They can be written as

$$\mathcal{A}_i^n \mathbf{r}_{i-1}^{n+1} + \mathcal{B}_i^n \mathbf{r}_i^{n+1} + \mathcal{C}_i^n \mathbf{r}_{i+1}^{n+1} = \mathcal{F}_i^n \qquad (15)$$

with coefficients given by

$$\mathcal{A}_i^n = -\frac{\epsilon}{h_i^n} + \frac{\alpha_i^n}{2}, \quad \mathcal{C}_i^n = -\frac{\epsilon}{h_{i+1}^n} - \frac{\alpha_i^n}{2}, \quad \mathcal{B}_i^n = \frac{h_i^n + h_{i+1}^n}{2\tau} - \mathcal{A}_i^n - \mathcal{C}_i^n,$$

$$\mathcal{F}_i^n = \frac{h_i^n + h_{i+1}^n}{2\tau} \mathbf{r}_i^n + \mu \left(\mathbf{N_v}\right)_i^n \frac{h_i^n + h_{i+1}^n}{2}.$$

The tri-diagonal systems are solved by the Thomas algorithm which is the fast procedure, numerically stable provided that the system matrix is strictly diagonally dominant. Such property can be simply fulfilled by chosing appropriately the time step, in practice (thanks to the close to uniform discretization) it can be proportional to the spatial discretization step and thus the numerical evolution to the steady state is realized in the fast and stable way. The results for our testing data as well as for the real virtual colonoscopy data are presented in Figures 9 and 11. In presented computations we used parameters $\mu = \epsilon = \tau = \omega_r = 1$ and the method is robust with respect to their choice. The illustrative 2D experiments were performed by the method from [11]. Figure 10 shows differences in grid point distances for our 3D test data (Figures 7-9 right), the red curve for the basic velocity field, the blue for the projected vector field plus curvature regularization and the violet curve for the final model (11)-(12). In the final model the grid point distances are uniform, the final curve is smoothly centered and can be definitely used for the virtual voyage inside the colon (which will be presented in the form of movie at the conference).

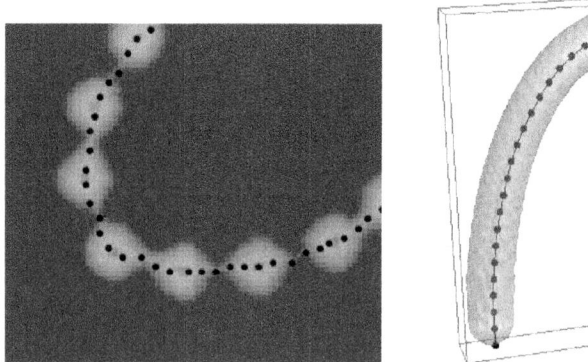
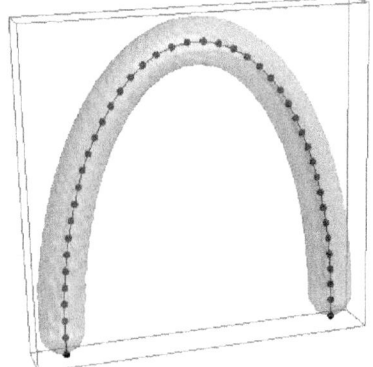

Fig. 9. The results for the test data obtained using tangential redistribution

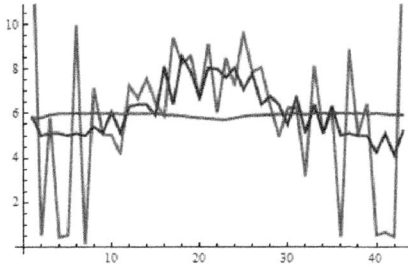

Fig. 10. Comparison of the grid point distances: the basic vector field (red), the projected vector field plus curvature regularization (blue), the final model (11)-(12) (violet)

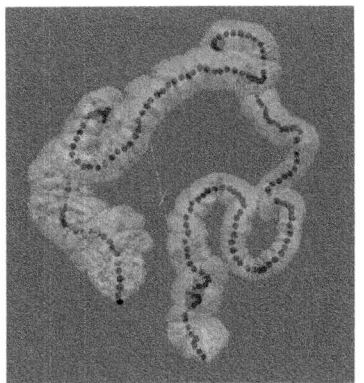
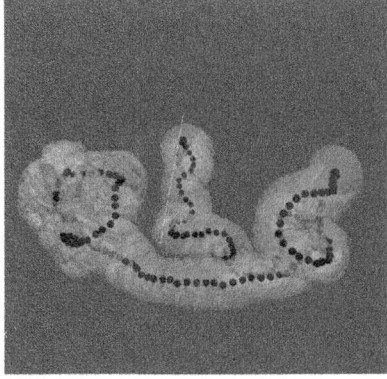

Fig. 11. The results for the real data obtained using the final model (11)-(12)

Acknowledgement. This work has been performed in cooperation with the company TatraMed spol s r.o., Bratislava and supported by the grants APVV-0351-07 and VEGA 1/0733/10.

References

1. Bovik, A.: Handbook of Image and Video Processing. Academic Press, London (2000)
2. Bourgine, P., Frolkovič, P., Mikula, K., Peyriéras, N., Remešíková, M.: Extraction of the intercellular skeleton from 2D images of embryogenesis using eikonal equation and advective subjective surface method. In: Tai, X.-C., Mørken, K., Lysaker, M., Lie, K.-A. (eds.) SSVM 2009. LNCS, vol. 5567, pp. 38–49. Springer, Heidelberg (2009)
3. Dijkstra, E.W.: A note on two problems in connection with graphs. Numerische Mathematik 1, 269–271 (1959)
4. Dziuk, G.: Convergence of a semi discrete scheme for the curve shortening flow. Mathematical Models and Methods in Applied Sciences 4, 589–606 (1994)
5. Hong, L., Kaufman, A.E., Wei, Y., Viswambharan, A., Wax, M., Liang, Z.: 3D Virtual Coloscopy. In: IEEE Symposium on Biomedical Visualization, pp. 26–32 (1995)

6. Hong, L., Muraki, S., Kaufman, A.E., Bartz, D., He, T.: Virtual voyage: Interactive navigation in the human colon. In: ACM SIGGRAPH 1997, pp. 27–34 (1997)
7. Hou, T.Y., Klapper, I., Si, H.: Removing the stiffness of curvature in computing 3-D filaments. J. Comput. Physics 143, 628–664 (1998)
8. Lefere, P., Gryspeerdt, S. (eds.): Virtual colonoscopy: A practical guide. Springer, Heidelberg (2010) ISBN 978-3-540-79879-8
9. Mikula, K., Ševčovič, D.: Evolution of plane curves driven by a nonlinear function of curvature and anisotropy. SIAM J. Appl. Math. 61, 1473–1501 (2001)
10. Mikula, K., Ševčovič, D.: A direct method for solving an anisotropic mean curvature flow of planar curve with an external force. Mathematical Methods in Applied Sciences 27, 1545–1565 (2004)
11. Mikula, K., Ševčovič, D., Balažovjech, M.: A simple, fast and stabilized flowing finite volume method for solving general curve evolution equations. Comm. Comp. Physics 7(1), 195–211 (2010)
12. Rouy, E., Tourin, A.: Viscosity solutions approach to shape-from-shading. SIAM Journal on Numerical Analysis 29(3), 867–884 (1992)
13. Truyen, R., Deschamps, T., Cohen, L.D.: Clinical evaluation of an automatic path tracker for virtual colonoscopy. In: Niessen, W.J., Viergever, M.A. (eds.) MICCAI 2001. LNCS, vol. 2208, p. 169. Springer, Heidelberg (2001)
14. Wan, M., Dachille, F., Kaufman, A.E.: Distance-field based skeletons for virtual navigation. In: Proceedings of IEEE Visualization Conference 2001, pp. 239–245 (2001)

Distance Images and Intermediate-Level Vision

Pavel Dimitrov, Matthew Lawlor, and Steven W. Zucker

Computer Science and Applied Mathematics
Yale University, New Haven, CT, USA

Abstract. Early vision is dominated by image patches or features derived from them; high-level vision is dominated by shape representation and recognition. However there is almost no work between these two levels, which creates a problem when trying to recognize complex categories such as "airports" for which natural feature clusters are ineffective. We argue that an intermediate-level representation is necessary and that it should incorporate certain high-level notions of distance and geometric arrangement into a form derivable from images. We propose an algorithm based on a reaction-diffusion equation that meets these criteria; we prove that it reveals (global) aspects of the distance map locally; and illustrate its performance on airport and other imagery, including visual illusions.

1 Introduction

Consider the problem of finding complex man-made structures, such as airports or medical or industrial complexes, within urban, suburban, and even rural enviroments in satellite imagery. Or, at a finer scale, consider the problem of finding certain crystalline structures among others in microscopic imagery. Such problems are different from the object recognition tasks normally addressed in computer vision. Even though there is significant variation among people or chairs, this variation seems small in comparison with the variation among the complex structures listed above. People have arms and legs and heads; airports have runways and buildings and access roads. Arms and legs have bilateral symmetry; airports do not. However humans can readily detect airports, which suggests that there is an additional level of structure to be found at which such objects can be described. We introduce, in this paper, one such structure: distance from arrangement information about edge elements. It captures the notion that airports consist of elongated structures that are separated from other, possibly more dense structure. A partial differential equation is derived, several of its relevant properties are proven, and its usefulness is demonstrated on the airport problem. At the heart of the matter is the computation of distance-like measures, a notion that is now entering the recognition literature in the form of shape properties.

Object recognition systems must confront the tradeoff between within-class or category variation relative to between-class/category variation. While scale-invariant features (e.g. [1]) and interest detectors can limit some of the within-class variation, an important trend is revealing that edge and shape features working together can improve performance; see e.g. [2,3]. These may involve

A.M. Bruckstein et al. (Eds.): SSVM 2011, LNCS 6667, pp. 653–664, 2012.

not only organizing edge fragments into object boundary parts, but also their relative arrangement as captured by the centroid [4,5]. Such techniques follow a "bottom up" strategy, by which increasingly more context is involved in the recognition ([6].

Centroids suggest involving higher-level shape features directly. E.g., skeleton-based representations are derived from the distance map, or the map of shortest distance to the boundary from every location interior to a shape. Skeleton points are the extrema of the distance map and the centroid is related to shock-based formulations [7]. Computing such skeletons and centroid approximations requires a relatively closed boundary, which is plausible for shape categories such as cups and airplanes. But man-made structures, such as airports and sports complexes are much less structured: although runways are straight, there are huge variations in the buildings, parking facilities and supply roads that flank them. Within-class variation among airports exceeds the between-class variation with highways. Attempts to build templates for them failed, and researchers resorted early to rule-based systems ([8]). But the variation among such complex features precludes such systems: the rules for defining airports in urban areas are quite similar to the rules for defining freeway exchanges; and the rules for defining airports in developing countries are significantly different. Moreover, the image measurements in support of the rules have been elusive, with the consequence that few such models are now in existence. Similar statements can be made about other socially-developed structures, such as medical complexes and sports arenas, and organically developing biological compounds.

We explore the position that new, intermediate-level representations exist and can be useful for recognizing organic, relatively freely-developing structures, such as airports and viral complexes in scanning microscopic imagery.

Mathematically the isoperimetric inequality, (perimeter)^2/area, has something of the flavor we seek, because it integrates a boundary property with a shape property. Although this can be a useful feature, operationally defining the perimeter and the area can be difficult. The problem is illustrated in Fig. 1: edge maps are too local, too broken, and too rigid. High-level features, such as the skeleton, are too global, too susceptible to boundary detail, and too sensitive to closure and interior features. We seek something in between, that extends naturally the unification of top-down shape with bottom-up features [9], and that is reflective of the better parts of both.

2   Global Distance Information Signaled Locally

The key idea behind this paper is to represent locally certain aspects of the distance map; that is, certain global aspects of shape, so that they can be used in an intermediate-level manner. This provides a middle-ground between abstract high-level representations such as skeletons and templates and the lower-levels of layered images. The representation is derived from a partial differential equation, and leads to a non-linear scale space for distances, estimated over increasingly larger domains. We note that there are many applications of pde's in scale space

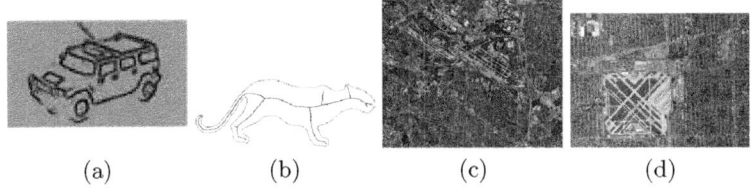

(a)   (b)   (c)   (d)

Fig. 1. The quest for intermediate-level vision is to find useful representational structures between edges (a), reflecting local, bottom-up processing, and global shape features such as the medial axis (b). While informative, edges are incomplete; ideally,
skeletons need perfect boundaries. We seek a representation that captures aspects of
both. For organically- and industrially-developed structures, such as airports (c, d), the
relevant structure is captured by an abstract combination of edge and distance effects,
rather than only local image properties.

analysis (e.g., [10]) but none, to our knowledge, that relate the solutions to
properties of the distance map.

As a warm-up, we note that such problems are widespread in developmental
biology, and we borrow heavily from a plant example: A young leaf in a developing plant consists of a network of veins that form cycles surrounding domains
of cells. A problem arises when these cells grow enough to exceed the nutrient
delivery capability of the existing vasculature: how do the cells in the center of
the domain signal the need to form new veins? What is the nature of the signal,
how is it generated, and what is the value that can be "read out" as a new vein
instruction.

A theoretical solution to this problem has been developed in [11], and we take
their model as a starting point for this paper; the intuition is shown in Fig. 2.
Their key idea is that cells in the developing leaf all produce a hormone called
auxin at the same rate. This hormone then diffuses from cell to cell and is cleared
away at the existing veins. The result is a differential equation (stated in the
next Section), the equilibria of which carry information about the distance from
the veins to the furthest cell. Two properties are salient: the concentration of the
hormone peaks at the furthest cells; and the magnitude of the gradient peaks
at the existing vasculature. It is this gradient peak that provides the signal for
plant development.

We interpret the hormone concentration in [11] as a kind of *distance image*;
that is, an image function whose value corresponds to properties of the distance
map. But in this form it is not at all clear how to apply it to vision problems.

Another clue comes from considering a second motivating example, this one
from visual psychophysics. Although in computer vision we take edge locations to
be calibrated projections of certain positions in space, the human visual system is
not so veridical. We know that arrangements of edge elements can effect apparent
global shape properties, as in the famous Muller-Lyer illusion (Fig. 3). It is
known that optical blur, as first hypothesized by Helmholtz, cannot explain all
of the illusion, and that cognitive effects, such as Gregory, are also only partial.

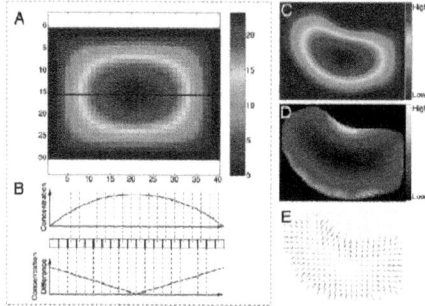

Fig. 2. How do young plants determine where the next vein shoot should go? Consider a rectangular portion of a leaf, surrounded by existing veins. If each cell (or pixel in the rectangle) produces a hormone at a constant rate, the hormone diffuses to neighboring cells and is cleared away by the existing vasculature (boundary condition = 0), the equilibrium distribution shown in (A) results. Taking a cross section through it, the peak in hormone concentration is at the center (B) and the peak in the gradient of concentration at equilibrium is at the existing veins (boundary); this last peak informs the developing leaf about where to start a new vein fragment and in which direction. (C) Concentration and (D,E) gradient of concentration in a real portion of a developing leaf. Figure after [11].)

We interpret the Muller-Lyer by observing that the "wings" at the ends of the horizontal lines effectively define an area context, and this area context is somehow larger when the wings point out than when they point in; it is within this context that the lines appear to be different lengths. So we conclude that line and edge arrangements can effect certain aspects of global shape, such as distance, at least perceptually. Returning to the airport example, we notice an analogy: the arrangement of boundaries, and the spaces between them, are the common thread through the different attempts to define them. Runways are straight and not too close to the buildings around them.

Fig. 3. (left) The Muller-Lyer Illusion: are the horizontal lines equal length? Notice how the outward "wings" provide a context in which the line appears longer than for the inward "wings," even though they are equal in length. (middle) The area enclosed by the wings, here shown in black, is enlarged by the outward "wings." (right) The Muller-Lyer illustion is predicted by Theorem 2; the gradient of concentration values are as shown.

Our goal in this paper is to combine edge representations with information about their arrangement. We call these enhanced edge maps *distance images*, and develop them formally in the next section. Second, we use dimensionality-reduction techniques to distill the major components from these distance-related images. Together they reveal a very curious property of airport definitions: that the distribution in orientation, arrangement, and density of edge elements can be key to defining classifiers. This captures our intuition directly from the Introduction in a manner that is completely novel.

3   Mathematical Formulation

We begin with the formalization of the model for plants, even though it is unrealistic for images, to introduce the type of result we seek. For concreteness, consider the example of the Muller-Lyer illusion in Fig. 3. Imagine that there exists a substance to report distance information, and that it is produced by all of the black pixels at the constant rate K. The set of black pixels, Ω, is a shape and $c : \Omega \rightarrow \Re$ denotes the concentration of the distance substance. Since it diffuses from pixel to pixel, it obeys:

$$c_t = D\nabla^2 c + K \tag{1}$$

where c_t is the derivative of concentration, D is the diffusion constant, and K is the constant production. The *Euclidean distance function* on Ω, denoted $\mathcal{E}_\Omega$, is $\mathcal{E}_\Omega(P) = \inf_{Q \in \partial\Omega} ||P - Q||_2$. The *boundary support* of P, denoted bsupp$(P; \Omega)$, is bsupp$(P; \Omega) = \{Q \in \partial\Omega : ||P - Q|| = \mathcal{E}_\Omega(P)\}$.

At equilibrium we have:

Theorem 1. *Let Ω be a shape and $c : \Omega \rightarrow \mathbb{R}$ the unique function satisfying $c(x, y) = 0$ on $(x, y) \in \partial\Omega$ and $\nabla^2 c = -\frac{K}{D}$.*

Suppose $P \in \Omega$ is such that $\mathcal{E}_\Omega(P) = L = \sup_\Omega \mathcal{E}_\Omega$ and $Q \in$ bsupp$(P; \partial\Omega)$. Suppose the smallest concave curvature radius is pL with $p > 0$. Then,

(a) $c(P) \subset \Theta(L^2)$,
(b) $\frac{K}{2D}L \leq |\nabla c| \leq \frac{K}{D}L\frac{2p+1}{p}$,
(c) $\sup_{\partial\Omega} |\nabla c| = \sup_{\Omega - \partial\Omega} |\nabla c|$

That is, (a) the peak in concentration at P is proportional to the distance squared between the closest boundary point Q and P; (b) the gradient of the concentration reports the (approximate) length between P and Q; and (c) the largest gradient value is on the boundary.

4   Edge Producing Model

We are now ready to develop the model for computer vision applications. Instead of having all of the ground cells (pixels) produce the substance, and having the veins clear it away, we posit a complementary model. Conceptually, for images, let the network of veins be replaced by edges, and let these edge pixels each produce a unit of the substance per unit of time; the other pixels produce no substance. Suppose the substance diffuses anisotropically between pixels, with the diffusion constant faster between edge pixels than between other pixels. To ensure a finite equilibrium concentration, suppose the substance is destroyed (metabolized) everywhere proportional to concentration. This results in a reaction diffusion equation, $c_t = D\nabla^2 c + \rho - \alpha c$, with three terms: the change in concentration at a pixel depends on the amount that enters by diffusion, with the amount produced there ($\rho : \Omega \to \mathbb{R}$ is the production rate) and with the amount destroyed there ($\alpha > 0$ is the destruction constant). Formally, we then have:

Proposition 1. *Consider the dynamical system*

$$\frac{\partial c}{\partial t} = D\nabla^2 c + \rho_\Omega - \alpha c. \tag{2}$$

Suppose that it acts over a domain Ω which a shape as in Theorem 1 and on which we impose a zero-flux boundary condition (Neumann). Let $\rho_\Omega : \Omega \to \mathbb{R}$. Then the following holds.

(a) *If $\alpha > 0$, then $\lim_{t\to\infty} c = c_\alpha$ for a unique steady-state c_α.*
(b) *Let $\alpha = 0$ and $R = \int \rho_\Omega d\Omega / \int d\Omega$ be the average production. Then $\lim_{t\to\infty} c_t = R$ and c converges to $c_\alpha + \mathrm{cst}$. whenever $R = 0$. Further, ∇c_α is unique even when $R \neq 0$.*
(c) *If $A, B \in \mathbb{R}$, then the transformation $\rho_\Omega \mapsto A\rho_\Omega + \alpha B$ induces a unique transformation of the steady state $c_\alpha \mapsto Ac_\alpha + B$ and vice versa. It follows that the gradient of c_α is only affected if $A \neq 1$: $\nabla c_\alpha \mapsto A\nabla c_\alpha$.*

Remark 1. In part (c), if the destruction term is not linear, e.g. $\alpha c + \beta c^2$, then the gradient might be affected by B as well.

Proof. Parts (a) and (b). To show existence we prove that the dynamical system achieves $c_t = 0$. Consider the dynamical system $c_{tt} = D\nabla^2 c_t - \alpha c_t$. The boundary conditions are inherited: since no flux goes through the boundary, there must be no change of concentration in time, i.e. $\nabla c_t \cdot \mathbf{n} = 0$ on $\partial\Omega$. The unique solution of this system is $c_t = 0$.

To prove uniqueness, suppose u_1 and u_2 both satisfy the equation given the boundary conditions and $c_t = 0$. Thus $D\nabla^2 u_1 + \rho_\Omega - \alpha u_1 = D\nabla^2 u_2 + \rho_\Omega - \alpha u_2$ which gives rise to $D\nabla^2 v - \alpha v = 0$ where $v = u_1 - u_2$ and $\nabla v \cdot \mathbf{n} = 0$ where $\mathbf{n}$ is the normal to the boundary. Since v is elliptic and $\alpha > 0$, v vanishes everywhere and uniqueness follows (see [12, p. 329 and 321]). The same reference shows that if $\alpha = 0$, then this uniqueness is up to an additive constant $u = u_1 + \mathrm{cst}$; that is, only ∇u is unique.

Now to show the convergence in (b) whenever $R = 0$, note that $c_{tt} = D\nabla^2 c_t$ assuming $\alpha = 0$. This has a steady-state s.t. $c_t = \mathrm{cst}$. everywhere. Also, $\int c_t = \int \rho_\Omega d\Omega$ which shows that $c_t = R$.

Part (c). Let c_α satisfy equation 2 for $c_t = 0$ and a production function $\rho_\Omega^{(\alpha)}$. Then, $D\nabla^2 c_\alpha - \alpha c_\alpha = -\rho_\Omega^{(\alpha)}$. Suppose $c = Ac_\alpha + B$ satisfies the equation for some ρ_Ω. Since this c is unique, the following verification proves the claim.

$$DV^2 c - \alpha c = -\rho_\Omega$$
$$\therefore DV^2(Ac_\alpha + B) - \alpha(Ac_\alpha + B) = -\rho_\Omega$$
$$\therefore ADV^2 c_\alpha - A\alpha c_\alpha - \alpha B = -\rho_\Omega$$
$$\therefore A(DV^2 c_\alpha - \alpha c_\alpha) = -\rho_\Omega + \alpha B$$
$$\therefore A(-\rho_\Omega^{(\alpha)}) = -\rho_\Omega + \alpha B$$
$$\therefore \rho_\Omega = A(\rho_\Omega^{(\alpha)}) + \alpha B$$

The other direction is derived similarly and the result follows.

Proposition 2. *Let Ω be a shape with two components $\Omega = \Omega_0 \cup \Omega_1$ such that $\Omega_0 \cap \Omega_1 = \partial\Omega_0$. Let D_0 and D_1 be the diffusion coefficients inside Ω_0 and Ω_1 respectively. If $\int_{\Omega_0} \rho_\Omega dv + \int_{\Omega_1} \rho_\Omega dv = 0$ and $\rho_\Omega(\Omega_0) = K \int_{\Omega_0} dv > 0$, then*

$$\lim_{D_0/D_1 \to 0} c_\alpha = c_K$$

where c_K satisfies Theorem 1 for the shape Ω_0 by setting $c_K(\partial\Omega_0) = 0$.

Proof. The convergence of the system derives from Prop. 1(b). As $D_0/D_1 \to 0$ the relative speed of diffusion in Ω_1 increases to infinity. Thus, the concentration over Ω_1 will tend to a constant and, consequently, so will $c(\partial\Omega_0) = c(\Omega_0 \cap \Omega_1)$. The conditions of Theorem 1 are therefore satisfied and the claim follows.

Theorem 2. *Suppose that Ω is a region in an image and that ρ_Ω takes a value of 1 at edge pixels and 0 everywhere else. Let the perimeter P be the number of edge pixels and the area A be the total number of pixels in Ω, i.e. $\int_\Omega d\Omega = A$. Denote by $c_\infty = \lim_{\alpha \to 0} c_\alpha$ and assume that the diffusion coefficient between non-edge pixels $D = 1$ and that the difusion coefficient between edge pixels is much larger than D. Then, for each pixel Q that is not an edge pixel*

$$|\nabla c_\infty(Q)| = \frac{P}{A}L \quad and \quad |\nabla^2 c_\infty(Q)| = \frac{P}{A}$$

Proof. The derivatives of c_∞ are well defined and unique as Prop. 1 shows. They are approximated by c_α to any precision provided that a sufficiently small α is chosen. Thus, given an arbitrary but fixed precision, suppose that α satisfies that requirement. According to Prop. 1(c), we may transform the production function by writing: $\rho_{new} = -\rho_\Omega + \alpha B$ where $\alpha B = \frac{P}{A}$. Thus, $\int_\Omega \rho_{new} d\Omega = -\int_\Omega \rho_\Omega \, d\Omega + \int_\Omega \frac{P}{A} \, d\Omega = -P + P = 0$. Hence, according to Prop. 2, this transormed setup is equivalent to c_K where $K = \frac{P}{A}$ and the claims are true for c_K due to Theorem 1. The result for c_∞ follows from Prop. 1(c) by observing that $\nabla c_\infty = -\nabla c_K$.

The gradient of concentration that emerges in this last result, and that scales with L, is precisely what was shown in Fig. 3.

4.1   Density Scale Space

By varying the destruction constant, α, a scale space for edge density is created. Several examples are shown in Fig. 4. Although this bears some resemblence to image scale spaces [13], there are fundamental differences. In particular, the interpretation of the gradient and the Laplacian of concentration in isoperimetric terms is completely novel.

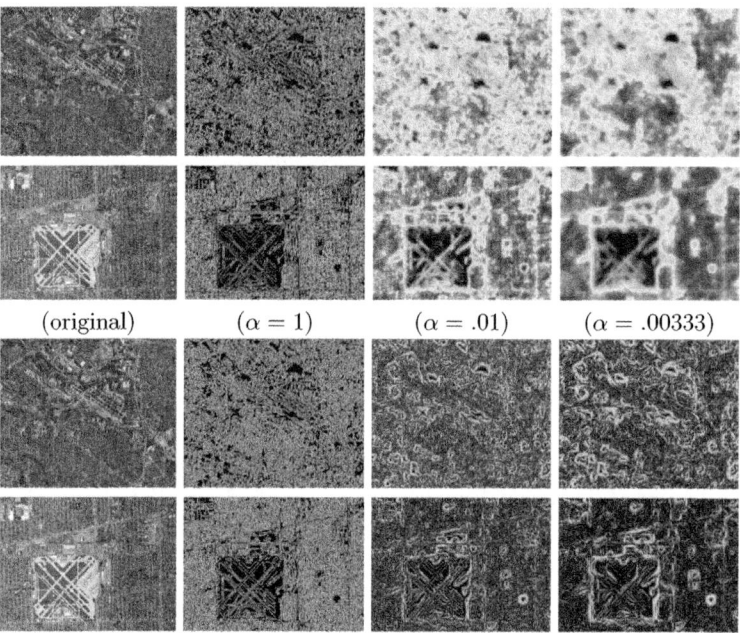

Fig. 4. (top)A concentration scale space for edge density, computed according to Eq. 3. Note how decreasing α permits the "substance" to live longer and hence allow integration of information over a larger area. When α is large, in the limit the result is formally the edge map convolved against a small Gaussian. (bottom) The gradient of concentration. Notice how this concentrates "signals" about edge density very close to the edge locations.

To demonstrate that the Laplacian of concentration across scale can be useful for classification, we build a vector of four values of α as input to a standard linear classifier. The result for this toy experiment is in Fig. 5.

While many more experiments remain to be done on recognition of standard object databases, our goal here is not to focus on classical recognition, but rather to demonstrate that distance images are relevant for airport and other complex features.

Fig. 5. Classification of the truck image with pose, orientation, and scale differences using the P/A measure, which is related to the Laplacian of concentration

5   Diffusion Map Embedding

We now turn to the use of the distance images just computed. A standard approach in machine learning is to use Laplacian eigenmaps [14] to reveal structure in datasets; here we use datasets of distance images from airports and show how they can be useful in recognition. (To keep the paper focussed on distance images, we ignore other sources of information, such as edge geometry.)

Laplacian eigenmaps are applied as follows. Let a data point be a vectorized patch of regular or distance images at different scales (values of α.) (For this paper we use three scales and (17 x 17) patches.) Let $X = \{x_1, x_2, ..., x_N\}$ be the set of data points (typically N=10,000), with each $x_i \in \mathbb{R}^{n=867}$. We seek to find a projection of these data into much lower dimension, under the assumption that they are not randomly distributed throughout $\mathbb{R}^n$ but rather that they lie on (or near) a lower-dimensional manifold embedded in $\mathbb{R}^n$.

The structure of the data are revealed via a symmetric, positivity-preserving, and positive semi-definite *kernel* $k(x, y)$, which provides a measure of similarity between data points. (We use a Gaussian kernel.) The result is a graph, with edges between nearby (according to the similarity kernel) data points. (The similarity value can be truncated to 0 for all but very similiar points.) This codifies the intuition that the natural structure among these "distance images" can be revealed by examining their low-dimensional embedding in significant eigenfunction coordinates. Then nearby (in the Euclidean metric) points can be clustered to reveal airport structure.

The diffusion map is obtained by the following algorithm: Given a set of n input image vectors $x_i \in \mathbb{R}^d$, **Step 1:** $K_0(i,j) \leftarrow e^{-\frac{\|x_i - x_j\|^2}{\sigma^2}}$; **Step 2:** $p(i) \leftarrow \sum_{j=1}^n K_0(i,j)$ approximates the density at x_i; **Step 3:** $\widetilde{K}(i,j) \leftarrow \frac{K_0(i,j)}{p(i)p(j)}$; **Step 4:** $d(i) \leftarrow \sum_{j=1}^n \widetilde{K}(i,j)$; **Step 5:** $K(i,j) \leftarrow \frac{\widetilde{K}(i,j)}{\sqrt{d(i)}\sqrt{d(j)}}$; **Step 6:** $USU^T = K$ (by SVD of K); Steps 2 and 3 normalize for the density of sampling from the manifold, whereas steps 4 and 5 perform the graph laplacian normalization; see [15].

The result of applying this algorithm to the distance map images is illustrated in Fig. 6. Notice how a boomerang-shaped "manifold" is revealed, the basic coordinates of which are edge density (going along it) and edge orientation (going around it). These two dimensions codify our intuition that airports are defined by a certain collected of oriented edges (runways, etc.) arranged in a particular fashion relative to surrounding context.

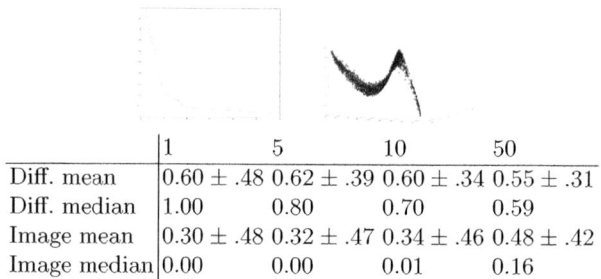

	1	5	10	50
Diff. mean	0.60 ± .48	0.62 ± .39	0.60 ± .34	0.55 ± .31
Diff. median	1.00	0.80	0.70	0.59
Image mean	0.30 ± .48	0.32 ± .47	0.34 ± .46	0.48 ± .42
Image median	0.00	0.00	0.01	0.16

Fig. 6. (top) Illustration of the data patches projected into diffusion coordinates (the first three significant eigenfunctions). (left) Eigenvalue spectrum shows that the first few eigenfuctions capture most of the structure. (right) The dominant dimension captures density; the orthogonal dimensions information about edge distribution. Red points correspond to patches from airport training images; notice how they cluster around the sparse end of the "manifold." This coloring of points can be interpreted as a function on the embedded patch data that defines "airport." (bottom) Table showing performance of distance vs image information for detecting airports. Columns are the number of patches tested and entries show the fraction from airports that were classified as airport by the Nystrom extension.

To test the usefulness of the distance-images, we collected a set of 20 airport images from Google images by randomly spinning the globe, half of which were for training and half for testing. Our goal is to compare distance images against standard image patches for airport detection. Since normal image blur also collects information from a neighborhood around a point, our data points for the embedding consisted of three components: a standard image patch plus either the distance image at two scales or the standard image blurred at two scales.

For training, we roughly outlined airports in the 10 training images and tagged those patches in the embeddings; these are the red points in Fig. 6 for the distance images. (The traditional blurred images are not shown.) To use this training information operationally, we built a characteristic function in embedded patch coordinates that defined "airport."

To test the quality of these patches for finding airports in test images, we used 10 new images. The airport characteristic function was then Nystrom extended onto the embedding of the new patch and scored according to whether or not it was in the airport lobe. The results over all patches (5,000 patches/image; training set = 10,000 patches) are shown in the table (Fig. 6(bottom)).

Counting the mean and the median number of patches that were correctly scored as airport shows that the distance images significantly outperformed the intensity information on this task.

The results on distance images and regular blurred images are shown in Fig. 7. To test the relevance of the distance information for this task, we purposely used square patches and no direct application of edge geometry.

Results: Diffusion Image Patches

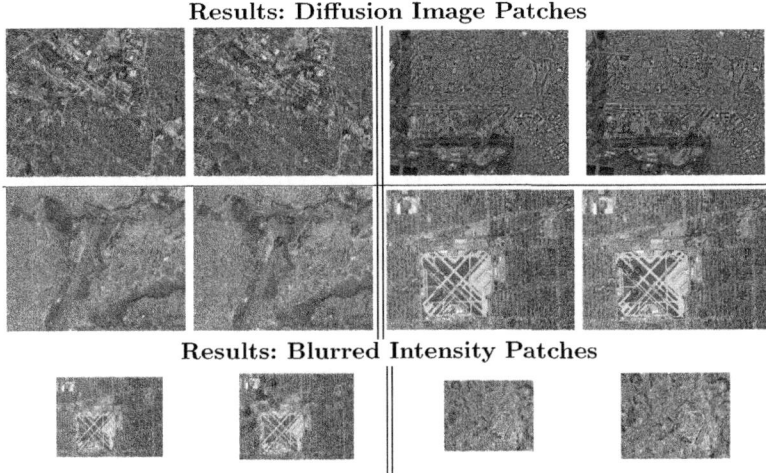

Results: Blurred Intensity Patches

Fig. 7. Results shown as patch boxes superimposed on the original images, with the first few patches on the left and the next few on the right for each pair. (top) Distance image results. (bottom) Blurred intensity image results. Note how the distance image boxes fall on/near the airports, but the intensity image boxes do not.

5.1   Summary and a Psychophysical View

We developed a structural coupling between local edge information and more global arrangement information by postulating a pde whose equilibria provided local signals about global properties of the distance map. An application of these ideas to locating airports demonstrated that these properties are useful. Much more remains to be done with the recognition of such complex structures, and we hope that the distance measures proposed here will find application in more traditional recognition systems. E.g., the experiments did not explicitly incorporate the geometry of boundaries in airports, which could increase performance further.

Finally, while the contribution of this paper was largely theoretical, and the applications were largely from computer vision, we close by returning to a biological perspective. The Muller-Lyer illusion (Fig. 3) was used informally as motivation but it remains unexplained in neurobiological terms. It is curious that the algorithm developed in this paper predicts it and other angular illusions. Since such illusions are represented by activity in the early visual cortex

[16], also the locus of biological edge responses, it is tempting to predict that computations such as these may have a neurobiological equivalent.

Acknowledgments. Research supported by Army Research Organization, Air Force Office of Scientific Research, and the National Science Foundation.

References

1. Lowe, D.: Distinctive image features from scale-invariant keypoints. International Journal of Computer Vision 60, 91–110 (2004)
2. Fidler, S., Leonardis, A.: Towards scalable representations of object categories: Learning a hierarchy of parts. In: CVPR. IEEE Computer Society (2007)
3. Ferrari, V., Jurie, F., Schmid, C.: From images to shape models for object detection. International Journal of Computer Vision 87, 284–303 (2010)
4. Opelt, A., Pinz, A., Zisserman, A.: A boundary-fragment-model for object detection. In: Leonardis, A., Bischof, H., Pinz, A. (eds.) ECCV 2006, Part II. LNCS, vol. 3952, pp. 575–588. Springer, Heidelberg (2006)
5. Opelt, A., Pinz, A., Zisserman, A.: Learning an alphabet of shape and appearance for multi-class object detection. International Journal of Computer Vision 80, 16–44 (2008)
6. Ullman, S., Epshtein, B.: Visual classification by a hierarchy of extended fragments. In: Ponce, J., Hebert, M., Schmid, C., Zisserman, A. (eds.) Toward Category-Level Object Recognition. LNCS, vol. 4170, pp. 321–344. Springer, Heidelberg (2006)
7. Siddiqi, K., Shokoufandeh, A., Dickinson, S.J., Zucker, S.W.: Shock graphs and shape matching. International Journal of Computer Vision 35, 13–32 (1999)
8. McKeown, D.M., Harvey, W.A., McDermott, J.: Rule-based interpretation of aerial imagery. IEEE Trans. Pattern Anal. Mach. Intell. 7, 570–585 (1985)
9. Borenstein, E., Ullman, S.: Combined top-down/bottom-up segmentation. IEEE Trans. Pattern Anal. Mach. Intell. 30, 2109–2125 (2008)
10. Florack, L., ter Haar Romeny, B., Viergever, M., Koenderink, J.: The gaussian scale-space paradigm and the multiscale local jet. 18, 61–75 (1996)
11. Dimitrov, P., Zucker, S.W.: A constant production hypothesis that predicts the dynamics of leaf venation patterning. Proc. Nat. Acad. Sci (USA) 13, 9363–9368 (2006)
12. Courant, R., Hilbert, D.: Methods of mathematical physics, vol. 2. Interscience (1962)
13. Koenderink, J.: The structure of images. Biological Cybernetics 50, 363–370 (1984)
14. Belkin, M., Niyogi, P.: Laplacian eigenmaps for dimensionality reduction and data representation. Neural Computation 6, 1373–1396 (2003)
15. Coifman, R., Lafon, S., Lee, A., Maggioni, M., Nadler, B., Warner, F., Zucker, S.: Geometric diffusions as a tool for harmonic analysis and structure definition of data: Diffusion maps. Proc. Nat. Acad. Sci. (USA) 102, 7426–7431 (2005)
16. Murray, S.O., Boyaci, H., Kersten, D.: The representation of perceived angular size in human primary visual cortex. Nature Neurosci. 9, 429–434 (2006)

Shape Palindromes: Analysis of Intrinsic Symmetries in 2D Articulated Shapes

Amit Hooda[1], Michael M. Bronstein[2],
Alexander M. Bronstein[3], and Radu P. Horaud[4]

[1] Indian Institute of Technology Delhi, India
[2] Inst. of Computational Science, Faculty of Informatics,
Università della Svizzera Italiana, Lugano, Switzerland
[3] Dept. of Electrical Engineering, Tel Aviv University, Israel
[4] INRIA Grenoble, Rhône-Alpes, France

Abstract. Analysis of intrinsic symmetries of non-rigid and articulated shapes is an important problem in pattern recognition with numerous applications ranging from medicine to computational aesthetics. Considering articulated planar shapes as closed curves, we show how to represent their extrinsic and intrinsic symmetries as self-similarities of local descriptor sequences, which in turn have simple interpretation in the frequency domain. The problem of symmetry detection and analysis thus boils down to analysis of descriptor sequence patterns. For that purpose, we show two efficient computational methods: one based on Fourier analysis, and another on dynamic programming.

1 Introduction

Symmetry and self-similarity are frequently encountered in natural and man-made objects at all scales from macro to nano [31]. Because of the relation of symmetry to redundancy of geometric data, the knowledge of the symmetries a shape possesses can be instrumental for its compression, completion, and super-resolution [17]. Many objects that are normally symmetric manifest symmetry breaking as a testimony of some anomaly or abnormal behavior. Therefore, detection of symmetry and asymmetry arises in many practical problems, including applications in medicine, aesthetics, and crystallography. Knowledge of shape symmetries as a prior has been also exploited in shape reconstruction [26], segmentation [25], face detection, recognition, and feature extraction [20].

In pattern recognition and computer vision literature, the problem of symmetry detection was studied mainly in images [16], two-dimensional [32,2,1] and three-dimensional shapes [28,11,18]. A wide spectrum of methods employed for this purpose includes approaches based on dual spaces [8], genetic algorithms [10], moments [6], pair matching [14,7], and local shape descriptors [34]. For an up-to-date overview, the reader is referred to the survey article of Liu *et al.* [13]. Traditionally, symmetries are considered as *extrinsic* geometric properties of shapes that are related to the way the shape is represented in the Euclidean

A.M. Bruckstein et al. (Eds.): SSVM 2011, LNCS 6667, pp. 665–676, 2012.

space. From such a perspective, symmetry is synonymous to invariance to a certain set of global isometric (distance-preserving) transformation of the Euclidean space (a composition of rotation, reflection, and translation). Though adequate for rigid shapes, such a point of view is inappropriate for non-rigid or deformable ones. Due to the deformations such shapes can undergo, the extrinsic symmetries may be lost, while *intrinsically* the shape still remains symmetric. In [23], Raviv *et al.* introduced the notion of *intrinsic symmetry*, defined as a self-isometry of the shape with respect to some intrinsic (e.g., geodesic or diffusion) metric. Such a definition does not make any use of the embedding Euclidean space in which the shape resides and is reduced to the standard notion of an extrinsic symmetry if the Euclidean metric is used. Computationally, such distance-preserving automorphisms can be found using the Gromov-Hausdorff framework, in which an initial set of candidate symmetries is detected using a branch-and-bound algorithm [23,24].

Ovsjanikov *et al.* [21] observed that simple eigenfunctions of the Laplace-Beltrami operator of a shape are invariant, up to a sign change, to intrinsic reflection symmetries with respect to a certain class of diffusion metrics. Consequently, reflection symmetries can be represented as sign sequences under which the corresponding eigenfunctions remain invariant, which provides for a simple algorithm for reflection symmetry detection. Another parametrization of intrinsic symmetries was proposed in [27], who noted that the self-isometry group of shapes with simple (disk- or sphere-like) topology is contained in the low-dimensional Möbius group. Detection of local self-similarity was considered in [19]; the related problem of partial symmetry detection was addressed in [18,33,24]. Finally, detection of symmetries can be considered as a particular case of the more general self-similar structure detection. Recent works focused on detecting repeating grid-like structures [22] or Euclidean structural redundancy in 3D data [3], as well as intrinsic self-similarity [17].

Following [5], we represent 2D shape symmetries as one-dimensional structures by using invariant local descriptors. This way, the symmetry of the shape (both extrinsic and intrinsic) is manifested in specific self-similarity of the associated descriptor sequence. In the case of articulated shapes, deformations of non-rigid joints are manifested as insertions/deletions in the descriptor sequence. Such self-similarity is preserved under non-rigid deformations and can be efficiently detected using Fourier analysis or dynamic programming gapped sequence alignment algorithms used in the field of bioinformatics [30]. Our approach allows detection and classification of both extrinsic and intrinsic symmetries of connected 2D shapes, and is inspired by shape matching by means of dynamic programming [9] and the papers of Bruckstein and Shaked [5,4].

2   Model

Let us be given a simply connected shape modeled as a closed simple planar curve S. The curve is parametrized as $S : [0, L] \to \mathbb{R}^2$. The length of the curve is given by $\ell(S) = \int_0^{l} \|\frac{d}{dt}S(t)\| dt$. In the following, we assume *arclength parametrization*, such that $\|\frac{d}{dt}S(t)\| = 1$ and thus $L = \ell(S)$.

Articulations. The shape is called *articulated* [12] if it can be represented as a collection of rigid *parts* $S_1, \ldots, S_p$ connected by non-rigid *joints* $J_1, \ldots, J_q$, such that $S = \bigcup_{i=1}^{p} S_i \cup \bigcup_{k=1}^{q} J_k$. We further assume that the rigid parts S_i are parametrized over $T_{R,i} \subset [0, L]$ (i.e., $S_i = S(T_{R,i})$) and the joints J_k are parametrized over $T_{J,k} \subset [0, L]$ (i.e., $J_k = S(T_{J,k})$). Let us denote $T_R = \bigcup_i T_{R,i} \subseteq [0, L]$ and $T_J = \bigcup_k T_{J,k} \subseteq [0, L]$. An *articulation* $S' = \mathbf{A}S$ is obtained by applying planar rigid transformations $\mathbf{R}_i$ (rotations and translations) to the rigid parts, and non-rigid transformations $\mathbf{T}_k$ to the joints, $\mathbf{A}S = \bigcup_{i=1}^{p} \mathbf{R}_i S_i \cup \bigcup_{k=1}^{q} \mathbf{T}_k J_k$. Thus, the articulation can be represented as $\mathbf{A} = \{\mathbf{R}_i, \mathbf{T}_k\}$. Obviously, rigid transformations do not change the length of the parts, hence, $\ell(\mathbf{R}_i S_i) = \ell(S_i)$ for $i = 1, \ldots, p$.

Extrinsic Symmetry. Planar transformations preserving Euclidean distances are called *isometries* and include rotations, translations, reflections, and their compositions. The set $\mathrm{Iso}(\mathbb{R}^2)$ of Euclidean isometries together with function composition operator forms the *isometry group* of $\mathbb{R}^2$. The subgroup $\mathrm{Sym}(S) \subset \mathrm{Iso}(\mathbb{R}^2)$ Euclidean isometries to which the shape S is invariant (i.e. $\mathbf{R}S = S$ for all $\mathbf{R} \in \mathrm{Sym}(S)$) is called the *extrinsic symmetry group* of S. Elements of the group from which the entire group can be produced are called the group *generators*.

The structure of the symmetry group tells us "in which way" the object is symmetric. The trivial case is the $C_1 = \{\mathrm{id}\}$ group, containing only the identity transformation (such shapes are usually called *asymmetric*). *Rotation symmetry* is described by a *cyclic group* C_n, generated by the rotations transformation around a fixed center by the angle $2\pi/n$. *Bilateral symmetry* is described by the *dihedral group* D_1, consisting of an identity and a single reflection around a symmetry axis. More general *dihedral symmetry of order* n (described by the semidirect product group $D_n = C_n \times C_2$) is generated by a rotation around a fixed center by the angle $2\pi/n$ and a reflection around an axis passing through the center.

Intrinsic Symmetry. An articulation may break the extrinsic symmetry, such that the resulting shape is no more symmetric in the above sense. Yet, considering the intrinsic geometry of the shape and the group of isometries preserving this geometry, one can define a broader notion of *intrinsic symmetry*, which will hold in this case [23]. In our formulation, the shape S is said to be *intrinsically symmetric* if there exists an articulation $\mathbf{A}$ such that $\mathbf{A}S$ is extrinsically symmetric (i.e., has a non-trivial extrinsic symmetry group). In other words, S can be "symmetrized" by means of an articulation $\mathbf{A} = \{\mathbf{R}_i, \mathbf{T}_k\}$, from which it follows that $\mathbf{R} \bigcup_{i=1}^{p} \mathbf{R}_i S_i = \bigcup_{i=1}^{p} \mathbf{R}_i S_i$, where $\mathbf{R} \in \mathrm{Iso}(\mathbb{R}^2)$.

3   Symmetry Analysis

An important problem in shape analysis is, given a shape S, to automatically determine its symmetry group. This problem is often referred to as *symmetry analysis, classification,* or *detection*. In this section, we present a method for symmetry classification based on the analysis of corresponding shape descriptors. We first show a representation of extrinsic symmetry groups in the descriptor sequence and its Fourier transform domain. Next, we extend our analysis to the intrinsic case using the articulation model.

At each point on the shape contour $S(t)$, we define a scalar or vector descriptor $a_S(t)$, which is local and invariant to shape isometries. A simple example of such a descriptor is the curvature $\kappa(t)$ Since the curvature involves second-order derivatives, it is sensitive to noise (more generally, all *differential invariants* tend to be sensitive to noise). Alternatively, we can use as a_S the *integral invariant* proposed in [15], defined as $I(t) = \int_S \|S(t) - S(t')\| q(t, t') dt'$, where $\|\cdot\|$ denotes the Euclidean distance, and q is a local kernel decreasing with the distance used to localize the descriptor. Such a descriptor is also invariant to Euclidean isometries. Using any local descriptor representation $a_S(t)$, the curve can be considered as a continuous sequence over $\mathbb{R} \bmod \ell(S)$. Assuming that the descriptor is invariant under rigid transformations, the descriptor sequences at the corresponding points of an articulated shape coincide.

Shaked and Bruckstein [5] identified the planar shape transformations with transformations of the descriptor sequences. In other words, given a symmetric shape satisfying $\mathbf{R}S = S$, the shape invariance under the symmetry transformation $\mathbf{R}$ can be related to descriptor sequence invariance, $a_S(t) = (a_S \circ \tau)(t)$, under a re-parametrization transformation τ. We extend this approach to articulated symmetries.

3.1   Extrinsic Symmetry Characterization

Rotation Symmetry. The action of an element of the group C_n (rotation by $2\pi k/n$) is manifested as $a_S(t) = a_S(kL/n + t \bmod L)$. Thus, the descriptor sequence of a C_n-symmetric shape is L/n-*periodic*. Looking at the Fourier transform $\hat{a}_S(\omega) = \int_0^L a_S(t) e^{-2\pi j \omega t} dt$ of the descriptor sequence, the periodicity of $a_S(t)$ is manifested in $\hat{a}_S(\omega)$ being discrete with step $2\pi n/L$.

Dihedral Symmetry. In the simplest case of bilateral (D_1) symmetry, the action of a reflection transformation is manifested as $a_S(t_0 + t \bmod L) = a_S(t_0 - t \bmod L)$, where $S(t_0)$ is a point on the symmetry axis that is mapped to itself. We shall refer to this point as *center of reflection symmetry* (note that the point $t_0 + L/2$ is also a center; we shall call such pairs *conjugate centers*). Thus, the descriptor sequence of a bilaterally-symmetric shape is an *even* function about the point t_0. Consequently, $a_S(t_0 + t \bmod L)$ is an even function about origin, which means that its Fourier transform is real and even. By translation property of Fourier transform, we get $\hat{a}_S(\omega) = \mathcal{F}\{a_S(t_0 + t \bmod L)\} e^{-2\pi\iota t_0\omega}$, which is the polar representation of $\hat{a}_S$. The phase encodes the position of the reflection point t_0; it varies linearly with ω and t_0 is the slope of the line.

More generally, a dihedral group D_n consists of rotation and reflection transformations, and is thus manifested in the descriptor domain as a combination of rotation and reflection symmetries, $a_S(t_i + t) = a_S(t_i + kL/n \pm t \bmod L)$ where $i = 0, \ldots, n-1$ and t_i's represent the n centers of reflection symmetry (note that we do not consider conjugate centers). It can be easily shown that $t_i = t_0 + \frac{L}{2n}i$. Thus, given one of the centers of reflection symmetry and the order of the symmetry group, we can directly find all other centers.

Table 1. Representation of shape symmetries

	Descriptor domain $a_S(t)$	Fourier domain $\hat{a}_S(\omega)$
C_n	L/n periodic	discrete with step $2\pi n/L$
D_1	even	linear phase
D_n	L/n periodic + period even	discrete with step $2\pi n/L$ + linear phase

3.2 Intrinsic Symmetry Characterization

Since we assume the descriptors to be local and invariant to rigid transformations, the descriptor sequences of two articulated shapes coincide on the rigid parts. We call this property *articulation invariance*. Formally, this can be expressed as follows: given an articulation $\mathbf{A} = \{\mathbf{R}_i, \mathbf{T}_k\}$ and the intervals $T'_{R,i} \subset [0, L']$ parametrizing the rigid parts $\mathbf{R}_i S \subset \mathbf{A}S$ (here $L' = \ell(\mathbf{A}S)$), we have $a_S(T_{R,i}) = a_{\mathbf{A}S}(T'_{R,i})$. Explicitly, $T'_{R,i}$ are related to $T_{R,i}$ by $T'_{R,i} = T_{R,i} + t_i \bmod L'$ for $t \in T_{R,i}$, where $t_1, \ldots, t_p \in [0, L']$ are some offsets.

Combining the articulation invariance relation with extrinsic symmetry characterization, we can characterize intrinsic symmetries. Reversing our notation, assume that $\mathbf{A}S$ is intrinsically symmetric: it is related by an articulation transformation to a shape S, which is extrinsically symmetric. Because of articulation invariance, we can consider S instead of $\mathbf{A}S$.

Thus, the intrinsic symmetry case is similar to the extrinsic one, up to insertions into the descriptor sequence at points in which the non-rigid joints are extended and deletions from the descriptor sequence at points in which the non-rigid joints are contracted. If such insertions/deletions are insignificant (i.e., the joints are small compared to the parts sizes), the Fourier domain properties would approximately hold. If the joints are large, we need to explicitly account for insertions/deletions, as described in Sect.4.

4   Numerical Implementation

Our analysis so far assumed a continuous curve, but in practice, the shape is sampled at a finite number of points. We assume that our curve S is sampled at N equidistant points (arclength sampling) $S(t_0), \ldots, S(t_{N-1})$. The corresponding descriptor is also a discrete sequence $a_S(t_0), \ldots, a_S(t_{N-1})$, denoted here by $(a_0, a_1, \ldots, a_{N-1})$. To simplify the notation, we assume all indices hereinafter modulo N.

The detection and classification of shape symmetry is done by analyzing the discrete descriptor sequence $(a_0, \ldots, a_{N-1})$. First, we find the symmetry group generators by attempting to detect rotation and reflection symmetry. Next, we classify the symmetry group according to the detected generators. Rotation and reflection symmetry detection is done using one of the following two methods. *Fourier analysis*, applicable to the extrinsic case and the case of small joints, remains similar to the continuous case. In the case of large joints, we can make use of *dynamic programming* algorithms employed in text sequence alignment to find the self-similar parts.

4.1 Fourier Analysis

In the discrete case, the continuous Fourier transform is replaced by the discrete Fourier transform (DFT), denoted here by $A_k = \sum_{m=0}^{N-1} a_m e^{-2\pi \iota k m / N}$, for $k = 0, \ldots, N-1$.

Rotation Symmetry. If the shape is C_n-symmetric, then the discrete descriptor sequence is periodic with period N/n, i.e., $a_m = a_{m-N/n}$ for $m = 0, \ldots, N-1$, and the corresponding DFTs are equal. Since by the shift property the DFT of $a_{m-N/n}$ is $A_k e^{-2\pi \iota k / n}$, we have $A_k e^{-2\pi \iota k / n} = A_k$. This leads to the result that $k = mn$ for some integer m for all k where $A_k \neq 0$. In other words, the DFT is discrete with step n. This gives an easy way to find the symmetry group of the shape by finding the step size of the DFT of its descriptor sequence. To account for the noise, we only look at a neighborhood of the A_k with maximum the absolute value.

Reflection Symmetry. In the discrete case, the relation between the center of reflectional symmetry and the phase of the kth element of the DFT becomes $\theta_k = \frac{-2\pi k}{N} t_i + 2\pi m_k \in [-\pi, \pi]$, for $k = 0, \ldots, N-1$. Here, the term $2\pi m_k$ is used to adjust for the *phase wrapping* and ensures that θ_k lies between $-\pi$ and π. Using the fact that $t_i \leq \frac{N}{2n}(1+i)$, we get $m_k \geq -\frac{1}{2} + \frac{(1+i)k}{2n}$. In particular, for $k = 1$, no wrapping is required (i.e., $m_1 = 0$) and we can evaluate the reflection point as $t_0 = t_i \bmod \frac{N}{2n}$.

Dihedral Symmetry. For the general case of D_n, the DFT will be discrete as dihedral symmetry also implies C_n rotational symmetry. In that case we check only non-zero values for the phase.

4.2 Dynamic Programming

Finding symmetries in the Fourier domain is possible in the extrinsic case, or in the intrinsic case when the size of the joints is small. For the case of large joints, we need to account for the joints deformations explicitly. The problem of finding common subsequences in two discrete sequences is very common in text analysis and bioinformatics, where dynamic programming algorithms such as the Smith-Waterman (SWAT) algorithm [30] are used for local sequence alignment. Our case is a particular setting of this problem when we compare a sequence to itself.

For each pair of points a_i, a_j in the sequence, we define a similarity $f(a_i, a_j)$. A *gap penalty* $g(a_i)$ or $g(a_j)$ is defined if one of the points is not matched anywhere and a gap is introduced. We construct an $(N+1) \times (N+1)$ matrix H

$$H_{ij} = \max \{ H_{i-1,j-1} + f(a_i, a_j), H_{i-1,j} + g(a_i), H_{i,j-1} + g(a_j), 0 \}$$

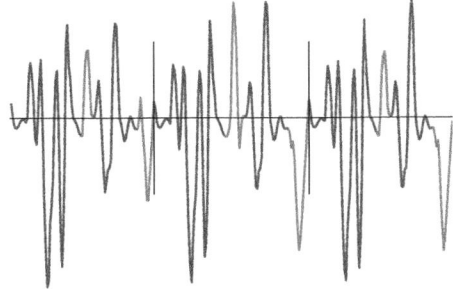

Fig. 1. Rotation symmetry detection using dynamic programming. Left: A C_3 intrinsically symmetric shape which is extrinsically asymmetric. The size of the joint at the knee is substantial. Right: the corresponding descriptor sequence. Similar parts (blue) correspond to rigid parts; gaps (red) correspond to non-rigid joints.

and $H_{i0} = H_{0j} = 0$. Interpreting $H_{i,j}$ as the cost of aligning a_i with a_j, the pair of segments with best similarity is found by going to the maximum element of H. The Smith-Waterman algorithm tries to find an alignment with the best cost recursively.

We define the cost function as $f(a_i, a_j) = c - F(N(a_i, r), N(a_j, r))$ where c is a constant, $N(a_i, r)$ is an r neighborhood of the the point and F is the fraction of points where the two sequences differ by more than a given threshold. Given the length l of the current gap, the gap penalty g is defined as

$$g_l(a) = \begin{cases} 2m & \text{if } l = 0 \\ m & \text{if } l \leq L \\ M & \text{if } l > L \end{cases}$$

where $M \gg m$ are constants and L is the maximum anticipated gap length.

Rotation symmetry. To check for rotational symmetry, we match a part of the shape of length N/n with the rest of the shape, for values of n in $2, 3, \ldots N/2$. If we get a match for some n, we validate it by dividing the shape into n parts, $S_1, S_2, \ldots, S_n$ of length L/n each, and matching consecutive parts. If validated, we find the *principal period* of the sequence by looking for periodicity in these parts. It can be easily shown that the alignment corresponding to the principal period serves as a generator of the C_n symmetry group of the shape.

Dihedral symmetry. The mappings for rotational symmetry can be evaluated as above. For reflectional symmetry, instead of finding all the mappings, we find one mapping and compute the others by composition. For computing the one required mapping, we align the sequence with its reflection. One thing to note here is that the descriptor sequence is defined modulo N, which requires to perform circular matching. Following [29], we duplicate one of the sequences for matching.

4.3   Computational Complexity

We analyze the complexity of our methods in terms of the number of points N sampled on the shape and the order n of the symmetry group.

Fourier Analysis. Descriptor sequence FFT takes $O(N \log(N))$ time. After that, for reflectional symmetry; taking the arguments(angles) and its unwrapping can be done in linear time. Also, reading the step size can also be done in linear time. So the overall time complexity is $O(N \log(N))$. The time can be further reduced by observing that the complete spectrum is not need for either type of symmetry. The slope of the phase vs frequency curve as well as the step size can be inferred from some of the Fourier coefficients. If we take some constant times n coefficients, the complexity can be reduced to $O(nN)$.

Dynamic Programming. The time complexity of SWAT alignment algorithm is $O(NM)$ for sequences of length N and M respectively. For rotational symmetry, we compare subsequences of length N/K with the complete sequence for values of K in $2, 3, \ldots, n$ where n is the order of the group. This gives the time complexity as $O(\log(n)N^2)$.

5   Results

We tested our approach on a set of hand-drawn articulated binary shapes. The shape boundaries were discretized at 300 points. For the local descriptor at point $S(t)$, we used the integral invariant $I(t) = \int_S \|S(t) - S(t')\| q(t,t') dt'$ with $q(t,t') = 1$ if $|t - t'| \leq 5$ and zero otherwise i.e. we take 10 closest points to the point $S(t)$. Dynamic programming algorithm described in Sect. 4 with parameters $c = 0.3, m = 0.5, M = 1.5$ and $L = 7$ was used to find symmetries. Figures 2–4 show the obtained results. We visualize symmetries by colorings of the shape contour; each coloring denotes the corresponding parts. Thin yellow line denotes the gaps. The figures are best viewed in color print. One can observe that intrinsic symmetry is correctly detected even in the presence of very strong articulations and realistic view artifacts. For comparison, in Fig. 5 we show an example of symmetry detection using the voting method of Loy and Ecklundh [14]. One can observe that strong articulation tamper with this algorithm, bringing in some cases to a failure to detect all the symmetries or an incorrect result.

The proposed approach compares favorably to other methods and has multiple advantages. First, unlike Loy-Eklundh [14] and similar approaches, our method can handle intrinsic symmetries. Second, it allows detecting both reflection and rotation symmetries (unlike Ovsjanikov *et al.* [21], limited to reflections only). Third, our approach guarantees the explicit recovery of all group generators (unlike [23,21] which detect approximate self-isometries that can also be compositions of the generators). Fourth, our dynamic programming (SWAT) approach is capable of handling partial symmetries (as opposed, e.g., to [21]). Fifth, our

Fig. 2. Detection of bilateral D_1 symmetry in human figure silhouettes. Each coloring denotes the corresponding patrs. Thin lines mark the gaps.

Fig. 3. Detection of C_3 rotation symmetry in synthetic articulated shapes.

approach has guaranteed low computational complexity (N for Fourier analysis or N^2 for SWAT, where N is the number of samples), compared to the theoretically worst-case $N!$ of [23] (though in practice much lower). Finally, our approach is straightforwardly generalizable to other classes of invariance (e.g. affine) by appropriate choicc of the descriptor. The comparison is summarized in Table 2.

Fig. 4. Detection of D_5 dihedral symmetry in synthetic articulated shapes (first row: extrinsic, second through fourth rows: intrinsic).

Fig. 5. Performance of the Loy-Eklundh [14] symmetry detection algorithm on our shapes. Left to right: correctly detected rotational C_3 symmetry and bilateral D_1 symmetry, partial C_3 symmetry detection (only one rotation), and incorrect bilateral symmetry detection.

6   Conclusions

We presented an approach for detection of symmetries in articulated 2D shapes, based on representing the shape contours using invariant local descriptors and characterizing shape symmetries as patterns in the associated descriptor sequences. Such patterns are preserved under non-rigid deformations and can be efficiently detected using Fourier analysis or dynamic programming sequence alignment algorithms. The main limitation of our approach is the assumption of simple closed curves, which implies that the underlying shape has no disconnected components and has simple topology.

Table 2. Comparison of methods for symmetry detection. Note: [a]Extended to 3D shapes case in [22]. [b]Extension to 2D is straightforward. [c]Extended to intrinsic symmetries in plane-like 3D shapes in [17]. [d]Partial symmetries addressed in [24]. [e]Theoretical worst case that guarantees global optimality. [f]Using fast approximate nearest neighbor search; there is additional factor d^3 depending on the number of eigenfunctions used, d. [g]Computing only the first few frequencies as explained in Sect. 4. [h]Invariance depends on the descriptor. [i]Invariance depends on the metric.

	[14]	[23]	[21]	FA	SWAT
2D/3D	Both[a]	Both[b]	Both[b]	2D	2D
Extrinsic/Intrinsic	Ext[c]	Both	Both	Both	Both
Reflection/Rotation	Both	Both	Ref	Both	Both
Topological noise	+	+	+	-	-
Partial symmetry	+	+[d]	-	-	+
Complexity	N^2	$N!^{(e)}$	$N\log N^{(f)}$	$N^{(g)}$	N^2
General invariance	+[h]	+[i]	-	+[h]	+[h]

References

1. Alt, H., Mehlhorn, K., Wagener, H., Welzl, E.: Congruence, similarity, and symmetries of geometric objects. Discrete Comput. Geom. 3, 237–256 (1988)
2. Atallah, M.J.: On symmetry detection. IEEE Trans. Computers c-34(7) (July 1985)
3. Bokeloh, M., Berner, A., Wand, M., Seidel, H., Schilling, A.: Symmetry detection using line features. Computer Graphics Forum (Special Issue of Eurographics) 28, 697–706 (2009)
4. Bruckstein, A., Shaked, D.: Crazy Cuts: Dissecting Planar Shapes into Two Identical Parts. Mathematics of Surfaces XIII, 75–89 (2009)
5. Bruckstein, A., Shaked, D.: Skew symmetry detection via invariant signatures. Pattern Recognition 31(2), 181–192 (1998)
6. Cheung, K., Ip, H.: Symmetry detection using complex moments. In: Proc. International Conference on Pattern Recognition (ICPR), vol. 2, pp. 1473–1475 (1998)
7. Cornelius, H., Loy, G.: Detecting rotational symmetry under affine projection. In: Proc. International Conference on Pattern Recognition (ICPR), vol. 2, pp. 292–295 (2006)
8. Derrode, S., Ghorbel, F.: Shape analysis and symmetry detection in gray-level objects using the analytical fourier-mellin representation. Signal Processing 84(1), 25–39 (2004)
9. Frenkel, M., Basri, R.: Curve matching using the fast marching method. In: Rangarajan, A., Figueiredo, M.A.T., Zerubia, J. (eds.) EMMCVPR 2003. LNCS, vol. 2683, pp. 35–51. Springer, Heidelberg (2003)
10. Gofman, Y., Kiryati, N.: Detecting symmetry in grey level images: The global optimization approach. In: Proc. International Conference on Pattern Recognition (ICPR), pp. 951–956 (1996)
11. Kazhdan, M., Chazelle, B., Dobkin, D., Funkhouser, T., Rusinkiewicz, S.: A reflective symmetry descriptor for 3D models. Algorithmica 38(1), 201–225 (2003)
12. Ling, H., Jacobs, D.: Shape classification using the inner-distance. Trans. PAMI 29(2), 286–299 (2007)
13. Liu, Y., Hel-Or, H., Kaplan, C.S., van Gool, L.: Computational symmetry in computer vision and computer graphics. Foundations and Trends in Computer Graphics and Vision 5(1-2), 1–195 (2010)

14. Loy, G., Eklundth, J.: Detecting symmetry and symmetric constellations of features. In: Proc. CVPR., vol. 2, pp. 508–521 (2006)
15. Manay, S., Hong, B.-W., Yezzi, A.J., Soatto, S.: Integral invariant signatures. In: Pajdla, T., Matas, J(G.) (eds.) ECCV 2004. LNCS, vol. 3024, pp. 87–99. Springer, Heidelberg (2004)
16. Marola, G.: On the detection of axes of symmetry of symmetric and almost symmetric planner images. Trans. PAMI 11(1) (January 1989)
17. Mitra, N.J., Bronstein, A.M., Bronstein, M.M.: Intrinsic Regularity Detection in 3D Geometry. In: Daniilidis, K., Maragos, P., Paragios, N. (eds.) ECCV 2010. LNCS, vol. 6313, pp. 398–410. Springer, Heidelberg (2010)
18. Mitra, N.J., Guibas, L.J., Pauly, M.: Partial and approximate symmetry detection for 3D geometry. ACM Trans. Graphics (TOG) 25(3), 568 (2006)
19. Mitra, N.J., Guibas, L.J., Pauly, M.: Symmetrization. In: Proc. SIGGRAPH (2007)
20. Natale, F.G.B.D., Giusto, D.D., Maccioni, F.: A symmetry-based approach to facial features extraction. In: Proc. International Conference on Digital Signal Processing Proceedings (ICDSP), vol. 2, pp. 521–525 (1997)
21. Ovsjanikov, M., Sun, J., Guibas, L.J.: Global intrinsic symmetries of shapes. In: Proc. SGP., vol. 27 (2008)
22. Pauly, M., Mitra, N.J., Wallner, J., Pottmann, H., Guibas, L.J.: Discovering structural regularity in 3D geometry. ACM Trans. Graphics (TOG) 27(3), 43 (2008)
23. Raviv, D., Bronstein, A.M., Bronstein, M.M., Kimmel, R.: Symmetries of non-rigid shapes. In: Proc. Non-rigid Registration and Tracking, NRTL (2007)
24. Raviv, D., Bronstein, A.M., Bronstein, M.M., Kimmel, R.: Full and partial symmetries of non-rigid shapes. IJCV 89(1), 18–39 (2010)
25. Riklin-Raviv, T., Kiryati, N., Sochen, N.: Segmentation by level sets and symmetry. In: Proc. CVPR (2006)
26. Shimshoni, I., Moses, Y., Lindernbaum, M.: Shape reconstruction of 3D bilaterally symmetric surfaces. IJCV 39(2), 97–110 (2000)
27. Sorkine, O., Lévy, B., Kim, V.G., Lipman, Y., Chen, X., Funkhouser, T.: Möbius Transformations For Global Intrinsic Symmetry Analysis (2010)
28. Sun, C., Sherrah, J.: 3D symmetry detection using the extended gaussian image. Trans. PAMI 19(2), 164–168 (1997)
29. Uliel, S., Fliess, A., Amiry, A.: A simple algorithm for detecting circular permutations in proteins. Bioinformatics 15, 11–15 (1999)
30. Waterman, M.S., Smith, T.F.: Identification of common molecular subsequences. J. Mol. Biol. 147, 195–197 (1981)
31. Weyl, H.: Symmetry. Princeton University Press (1983)
32. Wolter, J.D., Woo, T.C., Volz, R.A.: Optimal algorithms for symmetry detection in two and three dimensions. The Visual Computer 1, 37–48 (1985)
33. Xu, K., Zhang, H., Tagliasacchi, A., Liu, L., Li, G., Meng, M., Xiong, Y.: Partial intrinsic reflectional symmetry of 3D shapes. ACM Trans. Graphics (TOG) 28(5), 3 (2009)
34. Zabrodsky, H., Peleg, S., Avnir, D.: Symmetry as a continuous feature. Trans. PAMI 17(12), 1154–1166 (1995)

Kernel Bundle EPDiff: Evolution Equations for Multi-scale Diffeomorphic Image Registration

Stefan Sommer[1], François Lauze[1], Mads Nielsen[1,2], and Xavier Pennec[3]

[1] Dept. of Computer Science, Univ. of Copenhagen, Denmark
sommer@diku.dk
[2] Synarc Imaging Technologies, Rødovre, Denmark
[3] Asclepios Project-Team, INRIA Sophia-Antipolis, France

Abstract. In the LDDMM framework, optimal warps for image registration are found as end-points of critical paths for an energy functional, and the EPDiff equations describe the evolution along such paths. The Large Deformation Diffeomorphic Kernel Bundle Mapping (LDDKBM) extension of LDDMM allows scale space information to be automatically incorporated in registrations and promises to improve the standard framework in several aspects. We present the mathematical foundations of LDDKBM and derive the KB-EPDiff evolution equations, which provide optimal warps in this new framework. To illustrate the resulting diffeomorphism paths, we give examples showing the decoupled evolution across scales and how the method automatically incorporates deformation at appropriate scales.

Keywords: LDDKBM, LDDMM, diffeomorphic registration, scale space, computational anatomy, kernels, momentum.

1 Introduction

The Large Deformation Diffeomorphic Metric Mapping (LDDMM) framework plays an increasingly important role in image registration for medical image analysis as it provides good registration results along with a solid mathematical foundation allowing meaningful statistics to be computed on the registration results. It has its foundations in the seminal work of Grenander [3] and Christensen et al. [1] together with the theoretical contributions of Dupuis et al. and Trouvé [2,7]. The theory in its present state is well described in the paper of Younes et al. [9] and the monograph of Younes [8]. The purpose of this paper is to discuss the mathematical foundation behind a multi scale extension of LDDMM, the Large Deformation Diffeomorphic Kernel Bundle Mapping (LDDKBM), and develop the resulting evolution equations for the registration diffeomorphisms.

The LDDMM construction is based on the concept of kernels which encode the scale of the registration. Coarse to fine approaches, such as used in [4] for non-parametric image registration, can be used as tools to guide the search for the optimal registration but a scale mechanism for LDDMM which is truly consistent with the framework must be linked to the kernels. The role of the kernel and

A.M. Bruckstein et al. (Eds.): SSVM 2011, LNCS 6667, pp. 677–688, 2012.

deformation at different scales in LDDMM have been addressed by Risser et al. in [5] where the authors propose a multi-kernel approach which constructs new kernel shapes by adding Gaussian kernels. The method effectively changes only the shape of the kernel and does not allow decoupled momentum across scales. To improve the ability of the registration to adapt to scale information, we developed in [6] the LDDKBM extension of LDDMM which allows decoupling of the energy and momentum at each scale, and it therefore enables the algorithm to select the appropriate deformation at each scale individually. An example of an LDDKBM registration is given in Figure 1.

1.1   Content and Outline

In the next section, we will summarize the LDDMM framework before providing a detailed account of the mathematical foundation behind the LDDKBM extension. We then progress to developing the KB-EPDiff equations describing the evolution of critical paths in the framework and extending the fundamental EPDiff equations in LDDMM. We will present experiments in Section 5 and conclude in Section 6. The paper thus contributes by

(1) providing a detailed account of the theoretical foundation of the LDDKBM framework for multi scale diffeomorphic registration,
(2) deriving the KB-EPDiff equations which are fundamental for the theoretical understanding of LDDKBM and necessary for practical implementations,
(3) and through examples showing the evolution of diffeomorphism paths governed by the KB-EPDiff equations and how the evolution is decoupled across scales.

2   The LDDMM Framework

In the sequel, Ω will denote a hold-all domain of $\mathbb{R}^d$ ($d = 2, 3$ in applications) and V will denote a Hilbert space of vector fields $v : \Omega \to \mathbb{R}^d$ such that V with associated norm $\| \cdot \|_V$ is included in $L^2(\Omega, \mathbb{R}^d)$ and admissible as defined in [8, Chap. 9]. Given a time-dependent vector field $t \mapsto v_t$ with

$$\int_0^1 \|v_t\|_V^2 \, dt < \infty \tag{1}$$

the associated differential equation $\partial_t \varphi_t = v_t \circ \varphi_t$ has with initial condition $\varphi_s = \varphi$ a diffeomorphism φ_{st}^v as unique solution. The set G_V of diffeomorphisms built from V by such differential equations is a Lie group, and V is its tangent space at each point. The inner product on V associated to the norm $\| \cdot \|_V$ makes G_V a Riemannian manifold with right-invariant metric. Setting $\varphi_{00}^v = Id_\Omega$, the map $t \mapsto \varphi_{0t}^v$ is a path from Id_Ω to φ with energy given by (1). A critical path for the energy is a geodesic on G_V.

   In the LDDMM framework, registration is performed through the action of diffeomorphisms in G_V on geometric objects. This approach is very general and

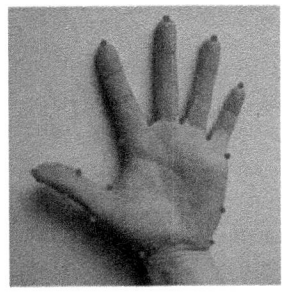
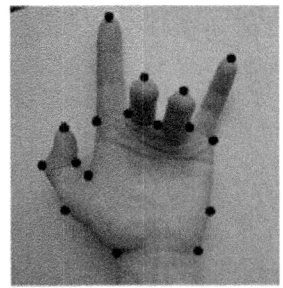

(a) Moving hand and landmarks (red)   (b) Fixed hand and landmarks (black)

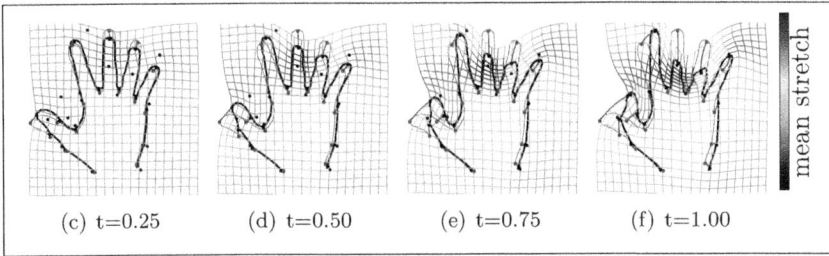

(c) t=0.25   (d) t=0.50   (e) t=0.75   (f) t=1.00

Fig. 1. Matching landmarks of hand (a) to landmarks of hand (b) with LLDKBM multiscale registration with Gaussian kernels of four scales. The critical path of diffeomorphisms determined by the KB-EPDiff equations derived in this paper is shown for four time steps (c)-(f) along with the outline of hand (a) (red line) and deformation of an initially square grid. Black curve shows the transported outline, and the grids are colored with the trace of Cauchy-Green strain tensor (log-scale). As we will see in Section 5, multiple scales are necessary to properly match the hands and movement occur decoupled across scales for the critical path shown.

allows the framework to be applied to both landmarks, curves, surfaces, images, and tensors. In the case of landmarks, the action of a diffeomorphism φ takes the form $\varphi.x = \varphi(x)$, and given landmarks $x_1, \ldots, x_N$ and $y_1, \ldots, y_N$, the registration amounts to a search for φ such that $\varphi.x_i \sim y_i$ for all $i = 1, \ldots, N$. In exact matching, we wish $\varphi.x_i$ be exactly equal to y_i but, more frequently, we allow some amount of inexactness to account for noise and give smoother diffeomorphisms. This is done by defining a quality of match measure U and a regularization measure E_1 to give a combined energy

$$E(\varphi) = E_1(\varphi) + \lambda U(\varphi) .$$

Here λ is a positive real representing the trade-off between regularity and goodness of fit and U is often the L^2-error which in the landmark case takes the form $U(\varphi) = \sum_{i=1}^{N} \|\varphi(x_i) - y_i\|^2$. The regularization term E_1 is defined as

$$E_1(\varphi) = \min_{v_t \in V, \varphi_{01}^v = \varphi} \int_0^1 \|v_s\|_V^2 \, ds . \tag{2}$$

It penalizes highly varying paths and, therefore, a low value of $E_1(\varphi)$ implies that φ is regular.

The regularity is ultimately controlled by the norm on V and this norm is associated to a *reproducing kernel* $K : \Omega \times \Omega \to \mathbb{R}^{d \times d}$. The kernel is often chosen to ensure rotational and translational invariance [8] and the Gaussian kernel $K(x, y) = \exp(\frac{\|x-y\|^2}{\sigma^2})\mathrm{Id}_d$ is a convenient and often used choice. The scaling factor σ is not limited to Gaussian kernels and allows for many kernels to vary the amount of regularization. Larger scales lead in general to higher regularization and smoother diffeomorphisms, whereas smaller kernels penalize higher frequencies less and often gives better matches. This phenomenon is in particular apparent for objects with sparse information and images with e.g. areas of constant intensity.

3   Kernel, Momentum and LDDKBM

The Large Deformation Diffeomorphic Kernel Bundle Mapping (LDDKBM) framework extends LDDMM by equipping diffeomorphism manifolds G_V in LD-DMM with vector bundles allowing deformation to be described at different scales. We start this section by discussing the relation between kernels and momentum in LDDMM before giving details on the mathematical foundation of LDDKBM.

3.1   Kernel and Momentum

The admissibility of V implies that for any $x \in \Omega$, the evaluation $\delta_x : v \mapsto v(x) \in \mathbb{R}^d$ is well-defined and continuous. Thus, for any $a \in \mathbb{R}^d$ the map $a \otimes \delta_x : v \mapsto a^T v(x)$ belongs to the topological dual V^* of V implying the existence of the kernel $K : \Omega \times \Omega \to \mathbb{R}^{d \times d}$ so that, for any constant vector $a \in \mathbb{R}^d$, the vector field $K(\cdot, x)a \in V$ represents $a \otimes \delta_x$ and $\langle K(\cdot, x)a, K(\cdot, y)b \rangle_V = a^T K(x, y)b$ for all points $x, y \in \Omega$ and all vectors $a, b \in \mathbb{R}^d$. This latter property is denoted the reproducing property and gives V the structure of a reproducing kernel Hilbert space (RKHS). Tightly connected to the norm and kernels is the notion of *momentum* given by the linear momentum operator $L : V \to L^2(\Omega, \mathbb{R}^d)$ which satisfies

$$\langle Lv, w \rangle_{L^2(\Omega, \mathbb{R}^d)} = \int_\Omega \left(Lv(x)\right)^T w(x)dx =: \langle v, w \rangle_V$$

for all $v, w \in V$. The momentum operator connects the inner product on V with the inner product in $L^2(\Omega, \mathbb{R}^d)$, and the image Lv of an element $v \in V$ is denoted the momentum of v. The reader can consult [8] for a thorough introduction to reproducing kernels, especially with a view towards the LDDMM framework.

3.2   The Kernel Bundle and LDDKBM

The LDDMM framework is limited to the choice of only one kernel shape and scale but deformation on different scales are often needed for good registration.

To address this problem, we constructed in [6] a multi scale extension of LDDMM resulting in the LDDKBM framework.

In order to use more kernels, we consider a parameter set I_W and subspaces $V_r, r \in I_W$ of the tangent space V where each V_r is equipped with a norm $\|\cdot\|_r$, corresponding kernel K_r, and momentum operator L_r. Typically, I_W will be a discrete set or a closed and bounded interval of $\mathbb{R}^+$ representing different scales. We then let W be the space of functions $w : I_W \to V$, $w_r \in V_r$ such that

$$\int_{I_W} \|w_r\|_r^2 \, dr < \infty \quad \text{and} \quad \int_{I_W} \|w_r\|_r \, dr < \infty .$$

The vector space structures on V_r induce a vector space structure on W, and it can be shown that under reasonable assumptions, the inner product

$$\langle v, w \rangle_W = \int_{I_W} \langle v_r, w_r \rangle_r \, dr, \ v, w \in W$$

turns W into a Hilbert space. Moreover the integral $\Psi(w) = \int_I w_r \, dr$ is well defined for $w \in W$ and allows us to pass from W to V. With this construction, we obtain a vector bundle $G_V \times W$, the *kernel bundle*, allowing kernels of different sizes and shapes, and a map $G_V \times W \to TG_V = G_V \times V$ that provides an extension of TG_V to multiple scales.

Using Ψ we can connect time varying paths $w_t = \{w_{t,r}\}_r$ in W and paths on the manifold G_V by

$$w_t \mapsto \varphi_{0t}^{\Psi(w)} . \tag{3}$$

The path energy is in LDDKBM measured using the norm on W, i.e. we define the energy

$$E_1^W(w_t) = \int_0^1 \|w_s\|_W^2 ds .$$

which induces a regularization measure on diffeomorphisms

$$E_1^W(\varphi) = \min_{w_t \in W, \varphi_{01}^{\Psi(w)} = \varphi} \int_0^1 \|w_s\|_W^2 \, ds . \tag{4}$$

Together with a quality of match measure $U(\varphi)$, this allows a reformulation of the registration problem as the search for a diffeomorphism minimizing

$$E^W(\varphi) = E_1^W(\varphi) + \lambda U(\varphi) . \tag{5}$$

The above formulation should be compared with the standard LDDMM formulation using the regularization (2), and it is immediately clear that the standard LDDMM method is the special case with only one scale and hence $W = V$.

It is interesting to note that W possesses a structure very similar to a RKHS. On V we have for each $x \in \Omega$ and $a \in \mathbb{R}^d$ the evaluation functional $a \otimes \delta_x(v) = a^T v(x)$. Using the integral map Ψ defined above, we define the linear maps on W

$$a \otimes \delta_x^\Psi(w) := \int_{I_W} a \otimes \delta_x(w_r) dr = \int_{I_W} a^T w_r(x) dr = a \otimes \delta_x(\Psi(w)) .$$

As seen from the equation, the maps evaluate w_r at each scale and integrate the results using Ψ. These maps are continuous and hence in the dual W^*. For the elements $K(\cdot, x)a = \{K(\cdot, x)_r a\}_r \in W$, we have

$$\langle K(\cdot, x)a, K(\cdot, y)b \rangle_W = \int_{I_W} \langle K(\cdot, x)_r a, K(\cdot, y)_r b \rangle_r \, dr = \int_{I_W} a^T K_r(x, y) b \, dr$$

$$= a^T \int_{I_W} K_r(x, y) b \, dr = a \otimes \delta_x^\Psi (K(\cdot, y)b) = a^T \Psi (K(x, y)b)$$

which is similar to the reproducing property for LDDMM except for the integration performed by Ψ on the right-hand side of the equation. Also, close to the RKHS situation, we see that

$$\langle K(\cdot, x)a, w \rangle_W = \int_{I_W} \langle K(\cdot, x)_r a, w_r \rangle_r \, dr = \int_{I_W} a^T w_r(x) \, dr = a \otimes \delta_x^\Psi (w) \,, \quad w \in V$$

again with the integration of w occuring in $a \otimes \delta_x^\Psi (w)$.

4   EPDiff and KB-EPDiff

The EPDiff equations in LDDMM describes the evolution of optimal paths for the registration problem. They are most often formulated in the following form: let $a_t = Lv_t$ denote the momentum at time t and assume that φ_t is a path minimizing $E_1(\varphi)$ with $\varphi_1 = \varphi$ minimizing $E(\varphi)$ and v_t is the derivative of φ_t. Then v_t satisfies the system

$$v_t = \int_\Omega K(\cdot, x)a_t(x)dx \,, \quad \frac{d}{dt}a_t = -Da_t v_t - a_t \nabla \cdot v_t - (Dv_t)^T a_t \,.$$

The first equation connects the momentum a_t with the velocity v_t, and the second describes the evolution of the momentum. The EPDiff equations can be interpreted as geodesic equations on the manifold G_V and are important for implementations since we can limit the search for optimal paths to paths satisfying the system.

As we will show in this section, there exists similar equations for LDDKBM: if $\Psi(w_t)$ is the derivative of the path of diffeomorphisms φ_t minimizing (4) with $\varphi = \varphi_1$ minimizing (5) then

$$w_{r,t} = \int_\Omega K_r(\cdot, x)a_{r,t}(x)dx \,,$$

$$\frac{d}{dt}a_{r,t} = \int_{I_W} -Da_{r,t}w_{s,t} - a_{r,t}\nabla \cdot w_{s,t} - (Dw_{s,t})^T a_{r,t} \, ds \,. \tag{6}$$

with $a_{r,t}$ being the momentum for the part $w_{r,t}$ of w_t. In essence, the standard EPDiff equations are integrated over the parameter space I_W to obtain the evolution of the momentum at each scale, and, in particular, the result will imply that the momentum conservation property of LDDMM also holds in LDDKBM. We will derive the KB-EPDiff equations in a more general form which implies the above formulation, and, for doing this, we will follow the strategy in [8] for the LDDMM case.

4.1   Euler-Lagrange Equations

For any time varying path w_t in W, we denote by $\varphi_{t_1 t_2}^{\Psi(w)}$ the diffeomorphism obtained by integrating $\Psi(w_t)$ from time t_1 to time t_2. The end of the integrated path $\varphi_{01}^{\Psi(w)}$ is the diffeomorphism used for the registration. For the energy $E^W(w_t) = E^W(\varphi_{01}^{\Psi(w)})$, we consider a variation $h_t \in W$ and calculate

$$\frac{d}{d\epsilon} E(w_t + \epsilon h_t) = 2 \int_0^1 \langle w_t, h_t \rangle_W \, dt + \frac{d}{d\epsilon} U(\varphi_{01}^{\Psi(w) + \epsilon \Psi(h)}) \, . \tag{7}$$

Following [8], we define $\mathrm{Ad}_\varphi v(x) = (D\varphi \, v) \circ \varphi^{-1}(x)$ for $v \in V$ and get a functional Ad_φ^* on the dual V^* of V by $(\mathrm{Ad}_\varphi^* \rho | v) = (\rho | \mathrm{Ad}_\varphi(v))$. It is shown in [8] that a variation $\tilde{h}_t$ in V of the match functional satisfies

$$\frac{d}{d\epsilon} U(\varphi_{01}^{v + \epsilon \tilde{h}}) = \int_0^1 \left(\mathrm{Ad}_{\varphi_{t1}^v}^* \bar{\partial} U(\varphi_{01}^v) \big| \tilde{h}_t \right) dt$$

with $\bar{\partial} U$ denoting the Eulerian differential of U (see [8, Chap. 10]). Inserting into (7) gives

$$\frac{d}{d\epsilon} E(w_t + \epsilon h_t) = 2 \int_0^1 \langle w_t, h_t \rangle_W \, dt + \int_0^1 \left(\mathrm{Ad}_{\varphi_{t1}^{\Psi(w)}}^* \bar{\partial} U(\varphi_{01}^{\Psi(w)}) \big| \Psi(h_t) \right) dt \, . \tag{8}$$

For each r, we define the operator $\mathrm{Ad}_\varphi^{T,r} v = K_r(\mathrm{Ad}_\varphi^*(L_r v))$ which then satisfies $\left\langle \mathrm{Ad}_\varphi^{T,r} v, w \right\rangle_r = (\mathrm{Ad}_\varphi^*(L_r v) | w)$, and we can now derive the fundamental results [8, Prop. 11.6/Cor. 11.7] in the LDDKBM case:

Proposition 1. *If w_t is an optimal path for E^W then for almost every $r \in I_W$,*

$$w_{t,r} = \mathrm{Ad}_{\varphi_{t1}^{\Psi(w)}}^{T,r} w_{1,r}$$

with $w_{1,r} = -\frac{1}{2} \nabla^{V_r} U(\varphi_{01}^{\Psi(w)})$.

Proof. Assume instead that there exists a time varying h_t in W and $t \in [0,1]$ such that

$$0 < \int_{I_W} \left\langle w_{t,r} - \mathrm{Ad}_{\varphi_{t1}^{\Psi(w)}}^{T,r} w_{1,r}, h_{t,r} \right\rangle_r dr = \int_{I_W} \langle w_{t,r}, h_{t,r} \rangle_r \, dr - \int_{I_W} \left\langle \mathrm{Ad}_{\varphi_{t1}^{\Psi(w)}}^{T,r} w_{1,r}, h_{t,r} \right\rangle_r dr$$

$$= \langle w_t, h_t \rangle + \frac{1}{2} \int_{I_W} (\mathrm{Ad}_{\varphi_{t1}^{\Psi(w)}}^* \bar{\partial} U(\varphi_{01}^{\Psi(w)}) | h_{t,r}) dr$$

$$= \langle w_t, h_t \rangle + \frac{1}{2} (\mathrm{Ad}_{\varphi_{t1}^{\Psi(w)}}^* \bar{\partial} U(\varphi_{01}^{\Psi(w)}) | \Psi(h_t)) \, .$$

But the right hand side vanishes for all t and all h_t by (8) and the fact that w_t is optimal for E^W, a contradiction.

Corollary 1. *Under the same conditions, for almost every $r \in I_W$,*

$$w_{t,r} = \mathrm{Ad}_{\varphi_{t0}^{\Psi(w)}}^{T,r} w_{0,r} \, . \tag{9}$$

The proof of the corollary is identical to the proof of [8, Cor. 11.7].

4.2   Scale Conservation and KB-EPDiff

In LDDKBM, the momentum of a path in general differ across scales. For a path w_t in W, we let a_t be the bundle momentum defined by $a_{t,r} = L_r(w_{t,r})$ recalling that L_r is the momentum operator at scale r. For each t, we can consider a_t to be in the dual W^* by $(a_t|\tilde{w}) = \int_{I_W}(a_{t,r}|\tilde{w}_r)dr$ which is continuous since

$$\left|(a_t|\tilde{w})\right| \le \left|\int_{I_W}(a_{t,r}|\tilde{w}_r)dr\right| = \left|\int_{I_W}\langle w_{t,r}, \tilde{w}_r\rangle_r dr\right| \le \|w_t\|\|\tilde{w}\| \ .$$

Suppose now w_t satisfies the transport equation (9) for almost every $r \in I_W$. Then for all $\tilde{w} \in W$,

$$
\begin{aligned}
(a_t|\tilde{w}) &= \int_{I_W}\langle w_{t,r}, \tilde{w}_r\rangle_r \, dr = \int_{I_W}\left\langle \mathrm{Ad}^{T,r}_{\varphi^{\Psi(w)}_{t0}}w_0, \tilde{w}_r\right\rangle_r dr \\
&= \int_{I_W}\left\langle w_{0,r}, \mathrm{Ad}_{\varphi^{\Psi(w)}_{t0}}\tilde{w}_r\right\rangle_r dr = \left(a_0\Big|\mathrm{Ad}_{\varphi^{\Psi(w)}_{t0}}\tilde{w}\right)
\end{aligned}
\tag{10}
$$

where $\mathrm{Ad}_{\varphi^{\Psi(w)}_{t0}}\tilde{w}$ is the element of W obtained by applying $\mathrm{Ad}_{\varphi^{\Psi(w)}_{t0}}$ to each $\tilde{w}_r$. The above equation shows that the momentum at time t is completely specified by the momentum at time 0 and thus reproduces the momentum conservation property for LDDMM. Note that since $\tilde{w}$ can be chosen arbitraly in (10), the momentum is conserved for each scale separately. By differentiating $\mathrm{Ad}_{\varphi^{\Psi(w)}_{t0}}\tilde{w}$, the momentum conservation property directly implies the equation

$$\partial_t(a_t|\tilde{w}) = -\left(a_t|D\Psi(w_t)\,\tilde{w} - D\tilde{w}\,\Psi(w_t)\right) \tag{11}$$

or, equivalently,

$$\partial_t a_t + \mathrm{ad}^*_{\Psi(w_t)}a_t = 0$$

with $\left(\mathrm{ad}^*_{\Psi(w_t)}a_t|\tilde{w}\right) = \left(a_t|D\Psi(w_t)\,\tilde{w} - D\tilde{w}\,\Psi(w_t)\right)$. Both equations imply the system (6) and extend the EPDiff equations for LDDMM. We denote them KB-EPDiff.

4.3   KB-EPDiff for Landmarks: An Example

To give a concrete application of the KB-EPDiff equations, we redo the calculation for LDDMM landmark matching with scalars kernels to arrive at the corresponding system for LDDKBM. The initial momentum $a_{0,r}$ will in this case be supported at the N landmarks x_i, $i = 1\ldots,N$, i.e. $a_{0,r} = \sum_{i=1}^{N} a_{0,r,i} \otimes \delta_{x_i}$ with vectors $a_{0,r,i} \in \mathbb{R}^d$. We let $x_{t,i}$ denote the trajectory of the ith landmark so that $x_{t,i} = \varphi^{\Psi(w)}_{0t}(x_{0,i})$.

Letting $a_{t,r,i} = (D\varphi^{\Psi(w)}_{t0})^T a_{0,r,i}$, we get from (10)

$$
\begin{aligned}
(a_{t,r}|\tilde{w}) &= \left(\mathrm{Ad}^*_{\varphi^{\Psi(w)}_{t0}}\left(\sum_{i=1}^{N} a_{0,r,i} \otimes \delta_{x_{0,i}}\right)\Big|\tilde{w}\right) = \left(\sum_{i=1}^{N} a_{0,r,i} \otimes \delta_{x_{0,i}}\Big|\mathrm{Ad}_{\varphi^{\Psi(w)}_{t0}}(\tilde{w})\right) \\
&= \sum_{i=1}^{N} u^T_{0,r,i}(D\varphi^{\Psi(w)}_{t0}\,\tilde{w}) \circ \varphi^{\Psi(w)}_{0t}(x_{0,i}) = \left(\sum_{i=1}^{N} a_{t,r,i} \otimes \delta_{x_{t,i}}\Big|\tilde{w}\right) \ .
\end{aligned}
$$

Since $\frac{d}{dt}(D_{x_{t,i}}\varphi_{t0}^{\Psi(w)})^T = -D_{x_{t,i}}\Psi(w_t)^T(D_{x_{0,i}}\varphi_{t0}^{\Psi(w)})^T$, the derivative of the momentum satisfies

$$\frac{d}{dt}a_{t,r,i} = \frac{d}{dt}\left((D\varphi_{t0}^{\Psi(w)})^T a_{0,r,i}\right) = -D_{x_{t,i}}\Psi(w_t)^T a_{t,r,i} \ .$$

We therefore have the trajectory of the landmarks and momentum evolution completely described by the system

$$\Psi(w_t) = \int_{I_W} \sum_{l=1}^N K_r(\cdot, x_{t,l})a_{t,r,l}dr$$

$$\frac{d}{dt}a_{t,r,i} = -\left(\int_{I_W} \sum_{l=1}^N D_1\left(K_s(x_{t,i}, x_{t,l})a_{t,s,l}\right)^T ds\right) a_{t,r,i} \qquad (12)$$

$$x_{t,i} = \varphi_{0t}^{\Psi(w)}(x_{0,i}) \ .$$

Note that the system is finite if I_W is finite.

5 Experiments

We perform two experiments showing the progressing deformation as we move along the critical path of the LDDKBM energy functional specified by the KB-EPDiff equations and showing the different deformation across scale. The first experiment is performed on landmarks from images of hands and the second on a simpler and artificial example to better visualize the scale differences.

5.1 Hand Outlines

We first consider the hand outlines shown in Figure 1 and Figure 2. Using the landmarks (red dots) on the moving hand image, we wish to compute the LDD-KBM match against the landmarks on the fixed image (black dots). The match is computed with three scales of 8, 4, and 2 units of the grid overlayed the figures. After optimizing for the optimal registration, we show in Figure 1 the progression of the deformation as we move along the critical path. The final deformation occurs rightmost for $t = 1$. The initially square grid is seen to progressively deform as time increases and the outline is moved to match the outline of the fixed image.

Figure 2 shows the results of computing the same match with standard LD-DMM with each of the three scales as well as the final match from LDDKBM repeated for comparison. For LDDMM with the largest scale, the match is poor and the sharp bend of the thumb is especially badly modelled. The situation improves for the middle scale though the bend of the thumb is still not sufficiently sharp and the match is bad for the middle fingers. For the smallest scales, the thumb is correctly matched but now the smaller scale is not able to model the even movement of the index finger. The LDDKBM method is by including all scales able to correctly register all the critical areas, and, at the same time, it gives the best match of the landmarks.

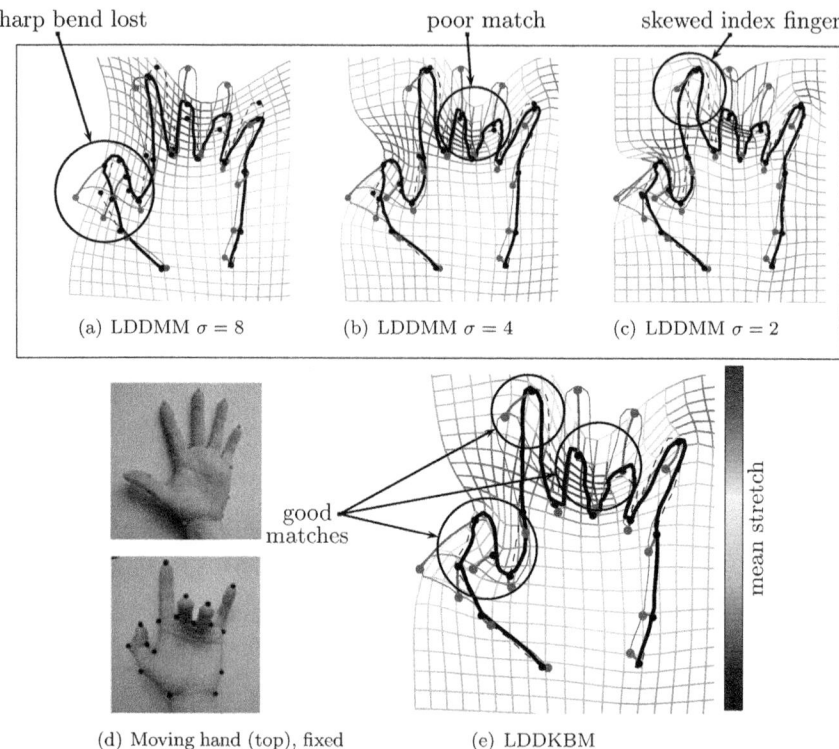

Fig. 2. Matching the hands of Figure 1 and shown in (d) for three scales of LDDMM and LDDKBM. The red landmarks of the moving hand are matched against the black landmarks of the fixed hand with the green crossed points showing the result of the match. The outline of the moving hand (red line) is transported to the black outline and should be compared with the outline of the fixed image (black dashed). The LDDKBM method is able to correctly match all the critical areas on which LDDMM fails, see text.

5.2   KB-EPDiff across Scales

To show how LDDKBM decouples deformation across scale, we extend the experiment presented in [6] where four points (red) are matched to four points (black) with results (green crossed) using LDDKBM with three scales. In Figure 3, the result of the registration is visible in the top right subfigure and the evolution of the critical path generated by the KB-EPDiff equations is shown with time increasing across columns. For the lower rows, the deformation at each scale is here shown independently. We see how most of the transport occurs at the largest scale with the middle scale participating to some degree and starting the acceleration of the two points having to move the farthest. The lowest scale perform almost no horizontal movement but takes care of the fine adjustment allowing the LDDKBM method to achieve an arguably superior registration compared to the corresponding LDDMM registrations which can be found in [6].

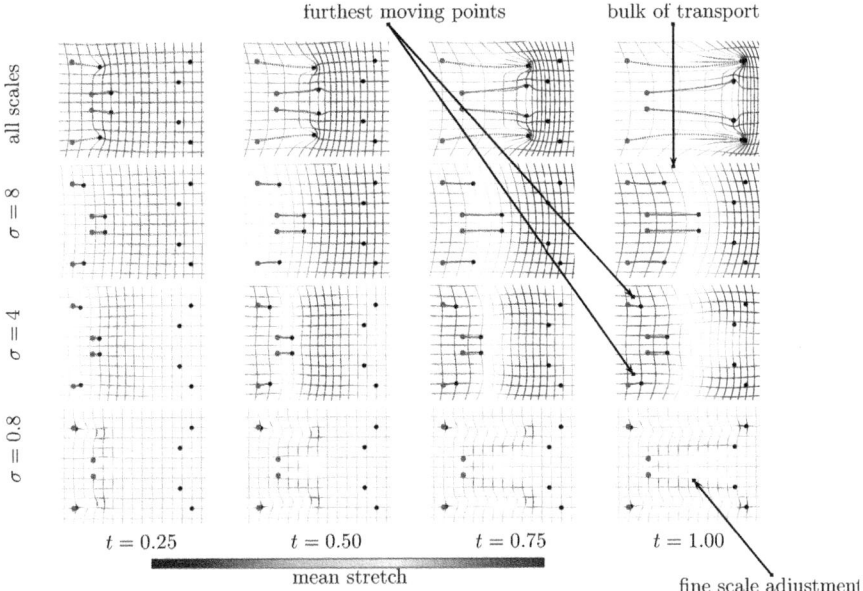

Fig. 3. LDDKBM match of four landmarks (red) to four landmarks (black) with results (green, crossed) for Gaussian kernels of three scales. Top row: critical path determined by KB-EPDiff equations, row 2-4: individual contribution of each of the three scales (scale σ in grid units). The columns shows four time points of the critical path with the rightmost being the final deformation. Initially square grids are shown deformed by the diffeomorphism, and the grids are colored with the trace of Cauchy-Green strain tensor indicative of the mean stretch (log-scale for each row individually). The largest scales contribute to most of the transport movement with smooth deformations while the smallest scale performs fine adjustment of the trajectories to obtain a good match.

6   Conclusion

We have detailed the mathematical foundation behind the LDDKBM framework for registration which extends LDDMM to include deformation at multiple scales. This includes deriving the KB-EPDiff equations describing the evolution of critical paths in the framework, and the resulting differential systems give insight into the geometry behind the framework in addition to being essential for algorithms for computing the improved registrations. We have provided examples showing the evolution governed by the KB-EPDiff equations and how the deformation differ across scales as well as showing the superior registration quality of the LDDKBM method on real images.

A further understanding of the structures behind LDDKBM may allow improved discretization and computational schemes to be developed. Therefore, we expect to look more into the geometry behind the vector bundle construction of LDDKBM and relate the energy to geometric notions generalizing e.g. Riemannian metrics to vector bundles on manifolds.

References

1. Christensen, G., Rabbitt, R., Miller, M.: Deformable templates using large deformation kinematics. IEEE Transactions on Image Processing 5(10), 1435–1447 (2002)
2. Dupuis, P., Grenander, U., Miller, M.I.: Variational problems on flows of diffeomorphisms for image matching (1998)
3. Grenander, U.: General Pattern Theory: A Mathematical Study of Regular Structures. Oxford University Press, USA (1994)
4. Modersitzki, J., Haber, E.: Cofir: Coarse and fine image registration. Real-Time PDE-Constrained Optimization, p. 30322 (2004)
5. Risser, L., Vialard, F.-X., Wolz, R., Holm, D.D., Rueckert, D.: Simultaneous fine and coarse diffeomorphic registration: Application to atrophy measurement in alzheimer's disease. In: Jiang, T., Navab, N., Pluim, J.P.W., Viergever, M.A. (eds.) MICCAI 2010. LNCS, vol. 6362, pp. 610–617. Springer, Heidelberg (2010), http://www.ncbi.nlm.nih.gov/pubmed/20879366
6. Sommer, S., Nielsen, M., Lauze, F., Pennec, X.: A multi-scale kernel bundle for LD-DMM: Towards sparse deformation description across space and scales. In: Székely, G., Hahn, H.K. (eds.) IPMI 2011. LNCS, vol. 6801, pp. 624–635. Springer, Heidelberg (2011)
7. Trouv, A.: An infinite dimensional group approach for physics based models in patterns recognition (1995)
8. Younes, L.: Shapes and Diffeomorphisms. Springer, Heidelberg (2010)
9. Younes, L., Arrate, F., Miller, M.I.: Evolutions equations in computational anatomy. NeuroImage 45(1, suppl. 1), S40–S50 (2009)

Deformable Shape Retrieval by Learning Diffusion Kernels

Yonathan Aflalo[1], Alexander M. Bronstein[2],
Michael M. Bronstein[3], and Ron Kimmel[1]

[1] Technion, Israel Institute of Technology, Haifa, Israel
{yaflalo,ron}@cs.technion.ac.il
[2] Dept. of Electrical Engineering, Tel Aviv University, Israel
bron@eng.tau.ac.il
[3] Inst. of Computational Science, Faculty of Informatics,
Università della Svizzera Italiana, Lugano, Switzerland
michael.bronstein@usi.ch

Abstract. In classical signal processing, it is common to analyze and
process signals in the frequency domain, by representing the signal in
the Fourier basis, and filtering it by applying a transfer function on
the Fourier coefficients. In some applications, it is possible to design an
optimal filter. A classical example is the Wiener filter that achieves a
minimum mean squared error estimate for signal denoising. Here, we
adopt similar concepts to construct optimal diffusion geometric shape
descriptors. The analogy of Fourier basis are the eigenfunctions of the
Laplace-Beltrami operator, in which many geometric constructions such
as diffusion metrics, can be represented. By designing a filter of the
Laplace-Beltrami eigenvalues, it is theoretically possible to achieve invariance to different shape transformations, like scaling. Given a set of
shape classes with different transformations, we learn the optimal filter
by minimizing the ratio between knowingly similar and knowingly dis-
similar diffusion distances it induces. The output of the proposed framework is a filter that is optimally tuned to handle transformations that
characterize the training set.

1 Introduction

Recent efforts have shown the importance of *diffusion geometry* in the field of
pattern recognition and shape analysis. Such methods based on geometric analysis of diffusion or random walk processes that were first introduced in theoretical
geometry [1] have matured into practical applications in the fields of manifold
learning [7] and where more recently introduced to shape analysis [9]. In the
shape analysis community, diffusion geometry methods were used to define low-
dimensional representations for manifolds [7,16], build intrinsic distance metrics
and construct shape distribution descriptors [16,10,5], define spectral signatures
[15] (shape-DNA), local descriptors [18,6], and bags of features [4]. Diffusion embeddings were used for finding correspondence between shapes [11] and detecting
intrinsic symmetries [13].

A.M. Bruckstein et al. (Eds.): SSVM 2011, LNCS 6667, pp. 689–700, 2012.

In many settings, the construction of diffusion geometry boils down to the definition of a *diffusion kernel*, whose choice is problem dependent. Ideally, such an operator should possess certain invariance properties desired in a specific application. For example, the commute time kernel is invariant to scaling transformations of the shape.

In this paper, we propose a framework for *supervised learning* of an optimal diffusion kernel on a training set containing multiple shape classes and multiple transformations of each shape. Considering diffusion kernels related to heat diffusion properties and diagonalized in the eigenbasis of the Laplace-Beltrami operator, we can pose the problem as finding an optimal filter on the Laplace-Beltrami eigenvalues. Optimization criterion is the discriminativity between different shape classes and the invariance to within-class transformations.

The rest of the paper is organized as follows. In Section 2, we review the theoretical foundations of diffusion geometry. Section 3 formulates the problem of optimal kernel learning and its discretization. Section 4 presents experimental results. Finally, Section 5 concludes the paper.

2   Background

2.1   Diffusion Geometry

We model a shape as a Riemannian manifold X embedded into $\mathbb{R}^3$. Equipping the manifold with a measure μ (e.g., the standard area measure), we also define an inner product on real functions on X by $\langle f, g \rangle = \int f g d\mu$. A function $k : X \times X \to \mathbb{R}$ is called a diffusion kernel if it satisfies the following conditions

1. *Non-negativity:* $k(x, x) \geqslant 0$.
2. *Symmetry:* $k(x, y) = k(y, x)$.
3. *Positive semidefiniteness:* for every bounded f,
$$\iint k(x, y) f(x) f(y) d(\mu \times \mu) \geqslant 0.$$
4. *Square integrability:* $\iint k^2(x, y) d(\mu \times \mu) < \infty$.
5. *Conservation:* $\int k(\cdot, y) d\mu = \int k(x, \cdot) d\mu = 1$.

A kernel function can also be considered as a linear operator on all the functions defined on X, $(\mathbf{K}f)(y) = \int k(x, y) f(x) d\mu$. We notice that the operator $\mathbf{K}$ is self-adjoint admitting a discrete eigendecomposition $\mathbf{K}\phi_i = \lambda_i \phi_i$, with $0 \leqslant \lambda_i \leqslant 1$ by virtue of the properties of the kernel. Spectral theorem allows us to write

$$k(x, y) = \sum_{i=0}^{\infty} \lambda_i \phi_i(x) \phi_i(y).$$

2.2 Heat Diffusion

There exists a large variety of possibilities to define a diffusion kernel and the related diffusion operator. Here, we restrict our attention to operators describing *heat diffusion*. Heat diffusion on surfaces is governed by the *heat equation*,

$$\left(\Delta_X + \frac{\partial}{\partial t} \right) u(x,t) = 0; \quad u(x,0) = u_0(x), \tag{1}$$

where $u(x,t)$ is the distribution of heat on the surface at point x in time t, u_0 is the initial heat distribution, and Δ_X is the positive-semidefinite *Laplace-Beltrami operator*, a generalization of the second-order Laplacian differential operator Δ to non-Euclidean domains.

On Euclidean domains $(X = \mathbb{R}^m)$, the classical approach to the solution of the heat equation is by representing the solution as a product of temporal and spatial components. The spatial component is expressed in the Fourier domain, based on the observation that the Fourier basis is the eigenbasis of the Laplacian Δ, and the corresponding eigenvalues are the frequencies of the Fourier harmonics. A particular solution for a point initial heat distribution $u_0(x) = \delta(x-y)$ is called the *heat kernel* $h_t(x-y) = \frac{1}{(4\pi t)^{m/2}} e^{-\|x-y\|^2/4t}$, which is shift-invariant in the Euclidean case. A general solution for any initial condition u_0 is given by convolution $\mathbf{H}^t u_0 = \int_{\mathbb{R}^m} h_t(x-y)u_0(y)dy$, where $\mathbf{H}^t$ is referred to as *heat operator*.

In the non-Euclidean case, the eigenfunctions of the Laplace-Beltrami operator $\Delta_X \phi_i = \lambda_i \phi_i$ can be regarded as a "Fourier basis", and the eigenvalues can be interpreted as the "spectrum". The heat kernel is not shift-invariant but can be expressed as an explicit short time kernel [17] $h_t(x,y) = \sum_{i=0}^{\infty} e^{-t\lambda_i} \phi_i(x)\phi_i(y)$.

It can be shown that the heat operator is related to the Laplace-Beltrami operator as $\mathbf{H}^t = e^{-t\Delta}$, and as a result, it has the same eigenfunctions ϕ_i and corresponding eigenvalues $e^{-t\lambda_i}$. It can be thus seen as a particular instance of a more general family of diffusion operators $\mathbf{K}$ diagonalized by the eigenbasis of the Laplace-Beltrami operator, namely $\mathbf{K}$'s as defined in the previous section but restricted to have the eigenfunctions ϕ_i of Δ_X. The corresponding diffusion kernels can be expressed as

$$k(x,y) = \sum_{i=0}^{\infty} K(\lambda_i)\phi_i(x)\phi_i(y), \tag{2}$$

where $K(\lambda)$ is some function (in the case of $\mathbf{H}_t$, $K(\lambda) = e^{-t\lambda}$) that can be thought of as the *transfer function* of a low-pass filter. Using this signal processing analogy, the kernel $k(x,y)$ can be interpreted as the point spread function at a point y, and the action of the diffusion operator $\mathbf{K}f$ on a function f on X can be thought of as the application of the point spread function by means of a shift-variant version of convolution. In what follows, we will freely interchange between $k(x,y)$ and $K(\lambda)$ referring to both as kernels.

2.3 Diffusion Distances

Since a diffusion kernel $k(x, y)$ measures the degree of proximity between x and y, it can be used to define a metric

$$d^2(x, y) = \|k(x, \cdot) - k(y, \cdot)\|_{L^2(X)}^2, \tag{3}$$

on X, dubbed as the *diffusion distance* by Coifman and Lafon [7]. Another way to interpret the latter distance is by considering the embedding $\Psi : x \mapsto L^2(X)$ by which each point x on X is mapped to the function $\Psi(x) = k(x, \cdot)$. The embedding Ψ is an isometry between X equipped with diffusion distance and $L^2(X)$ equipped with the standard L^2 metric, since $d(x, y) = \|\Psi(x) - \Psi(y)\|_{L^2(X)}$. As a consequence of Parseval's theorem, the diffusion distance can also be written as

$$d^2(x, y) = \sum_{i=0}^{\infty} K^2(\lambda_i)(\phi_i(x) - \phi_i(y))^2. \tag{4}$$

Here as well we can define an isometric embedding $\Phi : x \mapsto \ell^2$ with $\Phi(x) = \{K(\lambda_i)\phi_i(x)\}_{i=0}^{\infty}$, termed as the *diffusion map* by Lafon. The diffusion distance can be casted as $d(x, y) = \|\Phi(x) - \Phi(y)\|_{\ell^2}$.

2.4 Invariance

The choice of a diffusion operator, or equivalently, the transfer function $K(\lambda)$, is related to the *invariance* of the corresponding diffusion distance.

For example, consider the case of scaling transformation, in which a shape X is uniformly scaled by a factor of α. Abusing the notations we denote by αX the new shape, whose Laplace-Beltrami operator now satisfies $\Delta_{\alpha X} f = \alpha^{-2} \Delta_X f$. Since the eigenbasis is orthonormal ($\|\phi_i\| = 1$), it follows that if ϕ_i is an eigenfunction of Δ_X associated to the eigenvalue λ_i, then $\frac{1}{\alpha}\phi_i$ is an eigenfunction of $\Delta_{\alpha X}$ associated with the eigenvalue $\lambda_i \alpha^{-2}$.

In order to obtain diffusion distance d^2 invariant to scaling transformations, we have to ensure that $K^2(\lambda_i \alpha^{-2})\alpha^{-2} = K^2(\lambda_i)$, which is achieved for $K(\lambda) = \lambda^{-1/2}$. This kernel is known as the *commute-time kernel*, and the associated diffusion distance

$$d^2(x, y) = \sum_{i=0}^{\infty} \frac{1}{\lambda_i}(\phi_i(x) - \phi_i(y))^2. \tag{5}$$

as the *commute-time distance*.

2.5 Distance Distributions

Though diffusion metrics contain significant amount of information about the geometry of the underlying shape, direct comparison of metrics is problematic since it requires computation of correspondence between shapes. A common

way to circumvent the need of correspondence is by representing a metric by
its distribution, and measuring the similarity of two shapes by comparing the
distributions of the respective metrics.

A metric d on X naturally pushes forward the product measure $\mu \times \mu$ on
$X \times X$ (i.e., the measure defined by $d(\mu \times \mu)(x,y) = d\mu(x)d\mu(y)$) to the measure
$F = d_*(\mu \times \mu)$ on $[0,\infty)$ defined as $F(I) = (\mu \times \mu)(\{(x,y) : d(x,y) \in I\})$ for every
measurable set $I \subset [0,\infty)$. F can be fully described by means of a cumulative
distribution function, denoted by

$$F(\delta) = \int_0^\delta dP = \int \chi_{d(x,y)\leq\delta}d\mu(x)d\mu(y) \tag{6}$$

with some abuse of notation (here χ is the indicator function). $F(\delta)$ defined this
way is the measure of pairs of points the distance between which in no larger
than δ; $F(\infty) = \mu^2(X)$ is the squared area of the surface X. The density function (empirically approximated as a histogram) can be defined as the derivative
$f(\delta) = \frac{d}{d\delta}F(\delta)$. Sometimes, it is convenient to work with normalized distributions, $\hat{F} = F/F(\infty)$ and the corresponding density functions, $\hat{f}$, which can be
interpreted as probabilities.

Using this idea, comparison of two metric measure spaces reduces to the comparison of measures on $[0,\infty)$, or equivalently, comparison of un-normalized or
normalized distributions, which is carried out using one of the standard distribution dissimilarity criteria used in statistics, such as L_p or *normalized L_p,
Kullback-Leibler divergence, Bhattcharyya dissimilarity, χ^2 dissimilarity,* or *earth
mover's distance* (EMD).

3   Optimal Diffusion Kernels

The main idea of this paper lies in designing an optimal task-specific transfer
function $K(\lambda)$ such that the resulting diffusion distance distribution will lead to
best discrimination between shapes of a certain class while being insensitive as
much as possible to a certain class of transformations.

Let us be given a shape X and some deformation τ such that $Y = \tau(X)$ is also
a valid shape. Equipping each of the shapes with its Laplace-Beltrami operator,
we define $\Delta_X\phi_i = \lambda_i\phi_i$ on X and $\Delta_{X'}\phi_i' = \lambda_i'\phi_i'$ on Y. A transfer function
$K(\lambda)$ defines the diffusion kernel $k(x,x') = \sum_{i\geq 0} K^2(\lambda_i)\phi_i(x)\phi_i(x')$ on X, and
$k'(y,y') = \sum_{i\geq 0} K^2(\lambda_i')\phi_i'(y)\phi_i'(y')$ on Y. We aim at selecting K in such a way
that for corresponding pairs of points (x,x') and $(y,y') = (\tau(x),\tau(x'))$ the two
kernels coincide as much as possible, while differing as much as possible for non-
corresponding points. Denoting by $P = \{((x,x'),(\tau(x),\tau(x'))) : x,x' \in X\}$ the
set of all corresponding pairs (positives), and by $N = \{((x,x'),(y,y')) : x,x' \in
X, (y,y') \neq (\tau(x),\tau(x'))\}$ the set of all non-corresponding pairs (negatives), we
minimize

$$\min_{K(\lambda)} \frac{\displaystyle\sum_{((x,x'),(y,y'))\in P} (k(x,x') - k'(y,y'))^2}{\displaystyle\sum_{((x,x'),(y,y'))\in N} (k(x,x') - k'(y,y'))^2}. \tag{7}$$

We remark that while there is a multitude of reasonable alternative objective functions, in what follows we choose to minimize the above ratio because as it will be shown it lends itself to a simple algebraic problem.

The choice of an appropriate function K can lead to invariance of the kernel under some transformations. For example, the *commute time* kernel $K(\lambda) = \frac{1}{\sqrt{\lambda}}$ is invariant under global scaling. On the other hand, optimal K should be discriminative enough to distinguish between shapes not being one a transformation of the other. This spirit is similar to linear discriminant analysis (LDA) and Wiener filtering and, to the best of our knowledge, has never been proposed before to construct optimal diffusion metrics.

3.1   Discretization

We represent the surface X as triangular mesh with n faces constructed upon the samples $\{\mathbf{x}_1, \ldots, \mathbf{x}_n\}$ The computation of discrete diffuison kernels $k(\mathbf{x}_1, \mathbf{x}_2)$ requires computing discrete eigenvalues and eigenfunctions of the discrete Laplace-Beltrami operator. The latter can be computed directly using the finite elements method (FEM) [15], of by discretization of the Laplace operator on the mesh followed by its eigendecomposition. Here, we adopt the second approach according to which the discrete Laplace-Beltrami operator is expressed in the following generic form,

$$(\Delta f)_i = \frac{1}{a_i} \sum_j w_{ij}(f_i - f_j), \tag{8}$$

where $f_i = f(\mathbf{x}_i)$ is a scalar function defined on the mesh, w_{ij} are weights, and a_i are normalization coefficients. In matrix notation, (8) can be written as $\Delta \mathbf{f} = \mathbf{A}^{-1}\mathbf{W}\mathbf{f}$, where $\mathbf{f}$ is an $m \times 1$ vector and $\mathbf{W} = \mathrm{diag}\left\{\sum_{l\neq i} w_{il}\right\} - w_{ij}$.

The discrete eigenfunctions and eigenvalues are found by solving the *generalized eigendecomposition* [9] $\mathbf{W}\boldsymbol{\Phi} = \mathbf{A}\boldsymbol{\Phi}\boldsymbol{\Lambda}$, where $\boldsymbol{\Lambda} = \mathrm{diag}\{\boldsymbol{\lambda}\}$ is a diagonal matrix of eigenvalues $\boldsymbol{\lambda} = (\lambda_1, \ldots, \lambda_n)^{\mathrm{T}}$, and $\boldsymbol{\Phi} = (\phi_l(x_i))$ is the matrix of the corresponding eigenvectors. Similarly, we triangulate the shape Y and get $\mathbf{A}'\boldsymbol{\Phi}' = \mathrm{diag}\{\boldsymbol{\lambda}'\}\mathbf{W}'\boldsymbol{\Phi}'$.

Different choices of $\mathbf{W}$ have been studied, depending on which continuous properties of the Laplace-Beltrami operator one wishes to preserve [8,19]. For triangular meshes, a popular choice adopted in this paper is the *cotangent weight* scheme [14,12], in which

$$w_{ij} = \begin{cases} (\cot \beta_{ij} + \cot \gamma_{ij})/2 : & \mathbf{x}_j \in \mathcal{N}(\mathbf{x}_j); \\ 0 & : else, \end{cases} \tag{9}$$

where β_{ij} and γ_{ij} are the two angles opposite to the edge between vertices $\mathbf{x}_i$ and $\mathbf{x}_j$ in the two triangles sharing the edge.

We denote by $P = \{((i_m, j_m), (i'_m, j'_m))\}$ the collection of corresponding pairs of vertex indices on X and Y (that is, $i_m \leftrightarrow i'_m$ and $j_m \leftrightarrow j'_m$), and by N the collection of non-corresponding pairs. Denoting by $\mathbf{C}_+$ and $\mathbf{C}'_+$ two matrices whose ml-th elements are the products $\phi_l(x_{i_m})\phi_l(x_{j_m})$ and $\phi'_l(y_{i_m})\phi'_l(y_{j'_m})$, respectively, for $((i_m, j_m), (i'_m, j'_m)) \in P$, we have $\mathbf{k}_+ = \mathbf{C}_+ K^2(\boldsymbol{\lambda})$ and $\mathbf{k}'_+ = \mathbf{C}'_+ K^2(\boldsymbol{\lambda}')$, where the m-th elements of $\mathbf{k}_+$ and $\mathbf{k}'_+$ are $k(x_{i_m}, x_{j_m})$ and $k(x_{i'_m}, x_{j'_m})$, respectively, and $K^2(\boldsymbol{\lambda}) = (K^2(\lambda_1), \ldots, K^2(\lambda_n))^{\mathrm{T}}$. Exactly in the same way, the vectors $\mathbf{k}_-$ and $\mathbf{k}'_-$ corresponding to the negative pairs in N are obtained.

In order to make possible the optimization over all functions K, we fix a grid $\boldsymbol{\gamma} = (\gamma_1, \ldots, \gamma_r)$ or r points on which $\mathbf{k} = (K^2(\gamma_1), \ldots, K^2(\gamma_r))^{\mathrm{T}}$ is evaluated. In this notation, our optimization problem becomes with respect to the elements of $\mathbf{k}$. Since the grids $\boldsymbol{\gamma}$, $\boldsymbol{\lambda}$ and $\boldsymbol{\lambda}'$ are incompatible, we define the interpolation operators $\mathbf{I}$ and $\mathbf{I}'$ transfering a function from the grid $\boldsymbol{\gamma}$ to the grids $\boldsymbol{\lambda}$ and $\boldsymbol{\lambda}'$: $K^2(\lambda_i) = \mathbf{I}\mathbf{k}$, and $K^2(\lambda'_i) = \mathbf{I}'\mathbf{k}$. This yields $\mathbf{k}_+ = \mathbf{C}_+\mathbf{I}\mathbf{k}$ and $\mathbf{k}'_+ = \mathbf{C}'_+\mathbf{I}'\mathbf{k}$. Substituing the latter result into (7) gives the following minimization problem:

$$\mathbf{k}^* = \arg\min_{\mathbf{k}\geq 0} \frac{\|\mathbf{k}_+ - \mathbf{k}'_+\|^2}{\|\mathbf{k}_- - \mathbf{k}'_-\|^2} = \arg\min_{\mathbf{k}\geq 0} \frac{\|(\mathbf{C}_+\mathbf{I} - \mathbf{C}'_+\mathbf{I}')\mathbf{k}\|^2}{\|(\mathbf{C}_-\mathbf{I} - \mathbf{C}'_-\mathbf{I}')\mathbf{k}\|^2}$$

$$= \arg\min_{\mathbf{k}\geq 0} \frac{\mathbf{k}^{\mathrm{T}}\mathbf{P}\mathbf{k}}{\mathbf{k}^{\mathrm{T}}\mathbf{N}\mathbf{k}} = \mathbf{N}^{-\frac{1}{2}} \arg\min_{\substack{\bar{\mathbf{k}}\geq 0 \\ \|\bar{\mathbf{k}}\|=1}} \bar{\mathbf{k}}^{\mathrm{T}}\mathbf{N}^{-\frac{\mathrm{T}}{2}}\mathbf{P}\mathbf{N}^{-\frac{1}{2}}\bar{\mathbf{k}}, \qquad (10)$$

where $\mathbf{P} = (\mathbf{C}_+\mathbf{I} - \mathbf{C}'_+\mathbf{I}')^{\mathrm{T}}(\mathbf{C}_+\mathbf{I} - \mathbf{C}'_+\mathbf{I}')$ and $\mathbf{N} = (\mathbf{C}_-\mathbf{I} - \mathbf{C}'_-\mathbf{I}')^{\mathrm{T}}(\mathbf{C}_-\mathbf{I} - \mathbf{C}'_-\mathbf{I}')$. Note that the matrices $\mathbf{P}$ and $\mathbf{N}$ are of fixed size $r \times r$ and can be constructed without directly constructing the potentially huge matrices $\mathbf{C}_+$ and $\mathbf{C}'_+$. This makes the above problem computationally efficient even on very large training sets.

3.2   Interpolation Operators

Among a plethora of methods for designing the interpolation operations $\mathbf{I}$ and $\mathbf{I}'$ on one-dimensional intervals, we found that regularized spline fitting produced best results. For that purpose, let $\{s_i(\lambda)\}$ be a set of q functions defined on the interval $[\lambda_{\min}, \lambda_{\max}]$. We represent the kernel transfer function as the sum

$$K^2(\lambda) = \sum_{i=1}^{q} a_i s_i(\lambda) \qquad (11)$$

and look for the vector of coefficients $\mathbf{a} = (a_1, \ldots, a_q)^{\mathrm{T}}$. Denoting $\mathbf{S} = (s_1(\boldsymbol{\lambda}), \ldots, s_q(\boldsymbol{\lambda}))$ with $s_i(\boldsymbol{\lambda}) = (s_i(\lambda_1), \ldots, s_i(\lambda_n))^{\mathrm{T}}$, we have $\mathbf{k} = \mathbf{S}\mathbf{a}$. Similarly, for $\mathbf{S}' = (s_1(\boldsymbol{\lambda}'), \ldots, s_q(\boldsymbol{\lambda}'))$, we have $\mathbf{k}' = \mathbf{S}'\mathbf{a}$.

To impose the smoothness of the kernel $K(\lambda)$, we add the regularization term

$$R(K) = \int_{\lambda_{\min}}^{\lambda_{\max}} \|\delta K^2(\lambda)\|^2 d\lambda = \int_{\lambda_{\min}}^{\lambda_{\max}} \left(\sum_{i=1}^{q} a_i \nabla s_i(\lambda)\right)^2 d\lambda = \mathbf{a}^{\mathrm{T}}\mathbf{R}\mathbf{a}, \quad (12)$$

where the ij-th elements of $\mathbf{R}$ are given by $(\mathbf{R})_{ij} = \int_{\lambda_{\min}}^{\lambda_{\max}} \nabla s_i(\lambda) s_j(\lambda) d\lambda$.

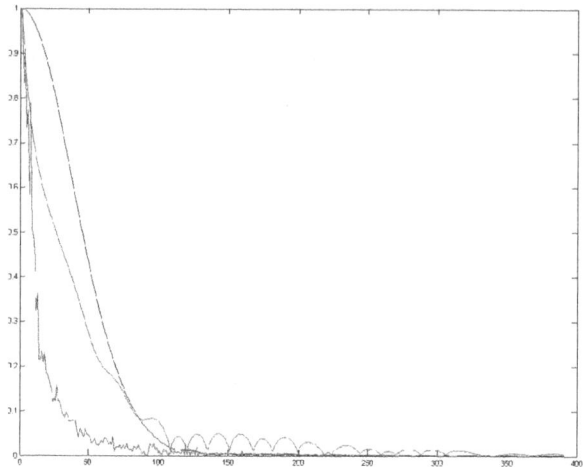

Fig. 1. Optimal kernel designed using straightforward nearest neighbor interpolation (red), splines without smoothness (green), and splines with the smoothness term (blue)

In these terms, the optimization problem (10) becomes

$$
\mathbf{a}^* = \arg\min_{\mathbf{a}} \frac{\mathbf{a}^{\mathrm{T}}(\mathbf{C}_+\mathbf{S} - \mathbf{C}'_+\mathbf{S}')^{\mathrm{T}}(\mathbf{C}_+\mathbf{S} - \mathbf{C}'_+\mathbf{S}')\mathbf{a}}{\mathbf{a}^{\mathrm{T}}(\mathbf{C}_-\mathbf{S} - \mathbf{C}'_-\mathbf{S}')^{\mathrm{T}}(\mathbf{C}_-\mathbf{S} - \mathbf{C}'_-\mathbf{S}')\mathbf{a}} + \eta\mathbf{a}^{\mathrm{T}}\mathbf{R}\mathbf{a}
$$

$$
= \mathbf{N}^{-\frac{1}{2}}\arg\min_{\|\bar{\mathbf{a}}\|=1} \bar{\mathbf{a}}^{\mathrm{T}}\mathbf{N}^{-\frac{\mathrm{T}}{2}}(\mathbf{P} + \eta\mathbf{R})\mathbf{N}^{-\frac{1}{2}}\mathbf{a}, \tag{13}
$$

where now $\mathbf{P} = (\mathbf{C}_+\mathbf{S} - \mathbf{C}'_+\mathbf{S}')^{\mathrm{T}}(\mathbf{C}_+\mathbf{S} - \mathbf{C}'_+\mathbf{S}')$, $\mathbf{N} = (\mathbf{C}_-\mathbf{S} - \mathbf{C}'_-\mathbf{S}')^{\mathrm{T}}(\mathbf{C}_-\mathbf{S} - \mathbf{C}'_-\mathbf{S}')$, and η is a parameter controlling the smoothness of the obtained kernel. The effect of the smoothness term is illustrated in Figure 1.

4　Results

In our experiments, to build the training set, we used the SHREC'10 correspondence benchmark [2]. The dataset contained high-resolution shapes $(10,000 - 30,000$ vertices) organized in seven shape classes with 55 simulated transformations of varying strength in each class (Figure 2) Testing was performed on the SHREC'10 shape retrieval benchmark [3], containing a total of 1184 shapes. Retrieval performance was evaluated using precision/recall characteristic. *Precision* $P(r)$ is defined as the percentage of relevant shapes in the first r top-ranked retrieved shapes (in the used benchmark, transformed shapes were used as queries, while a single relevant null shape existed in the database for each query). *Mean average precision* (mAP), defined as

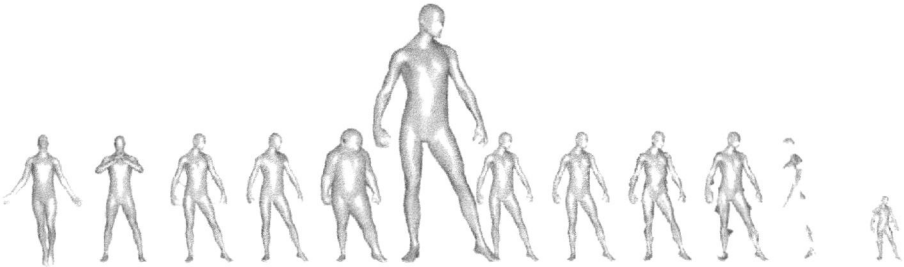

Fig. 2. Transformations of the human shape used as queries (shown in strength 5, left to right): null, isometry, topology, sampling, local scale, scale, holes, micro holes, noise, shot noise, partial, all

$$mAP = \sum_r P(r) \cdot rel(r),$$

where $rel(r)$ is the relevance of a given rank, was used as a single measure of performance. Ideal performance retrieval performance results in first relevant match with mAP=100%. Discretization of the Laplace-Beltrami was based on the cotangent weight formula (9).

In the first experiment, we used our approach to learn a scale-invariant diffusion kernel. We used a training set containing only scaling transformations of the shapes. As can be seen from Figure 3, the learned diffusion kernel is very close to the theoretically-optimal commute-time kernel $K(\lambda) = \lambda^{-1/2}$.

In the second experiment, we extended the training set to include all the shape transformations, resulting in a kernel shown in Figure 4 (red). The learned kernel was used to compute diffusion distance distributions, which were compared to compute the shape similarity, following the spectral distance framework [5]. The performance results with this kernel are summarized in Table 1 (fifth column). For comparison, performance using the commute time kernel is shown (Table 1, sixth column).

In the third experiment, instead of designing a kernel with a discretization of $K(\lambda)$, we used a parametric kernel of the form $K(\lambda) = \exp(-t\lambda)$ and optimized our criterion for the time scale t. The optimal scale was found to be $t^* = 1011$; the performance results with this kernel are summarized in Table 1 (fourth column). For comparison, we show the performance of the same kernel with two other values of the parameter, $t = 700$, and 1700 (Table 1, second and third columns).

In the fourth experiment, we used the diagonal our optimal non-parametric diffusion kernel $k(x,x)$ as a local scalar shape descriptor at each point, similar to the heat kernel signature [18]. A global descriptor was constructed as the histogram of the values of $k(x,x)$ on the entire shapes. We notice that both the local descriptors (Figure 5, top) and the global descriptors (Figure 5, bottom) resulting from our learned diffusion kernel signature computed on two different transformations of a shape are very close one to the other.

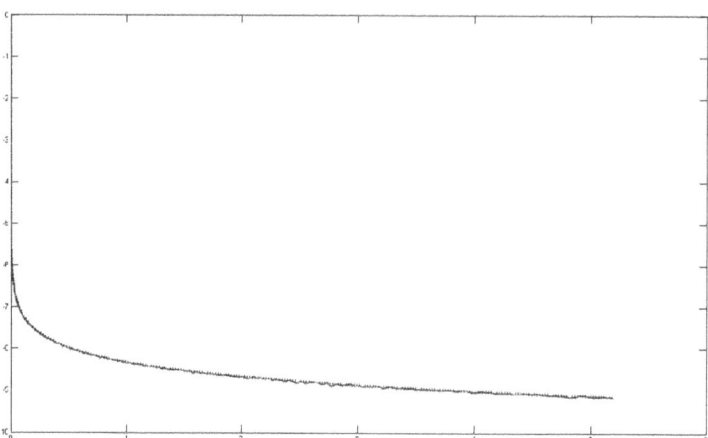

Fig. 3. Theoretical scale invariant (commute time) kernel (blue) and the learned kernel on examples of scaling transformations (red)

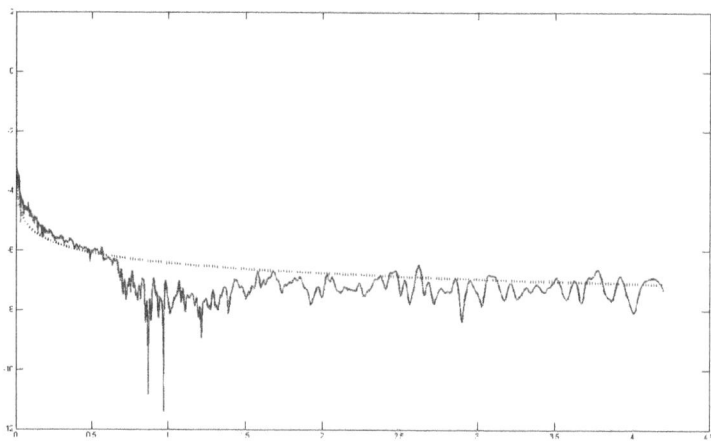

Fig. 4. The lerned kernel using all transformations (red). For comparison, the commute time kernel is shown (blue).

Table 1. Shape retrieval performance (mAP in %) using the spectral distance with different diffusion kernels

Transformation	Heat $(t = 700)$	Heat $(t = 1700)$	Optimal param. $(t^* = 1011)$	Optimal non-param.	Commute time
Isometry	100	99.23	100	98.21	97.95
Topology	91.28	80.36	86.79	77.33	79.16
Holes	82.3	90.33	87.81	72.48	73.94
Micro holes	100	100	100	100	100
Scale	30.97	32.66	32.44	100	100
Local scale	65.64	70.79	70.73	67.92	68.22
Sampling	100	99.23	100	98.21	98.21
Noise	99.23	100	98.46	100	100
Shot noise	99.23	100	99.23	99.23	98.65
Partial	5.54	7.37	6.01	8.06	31.03
All	64.15	64.36	69.56	64.66	64.51

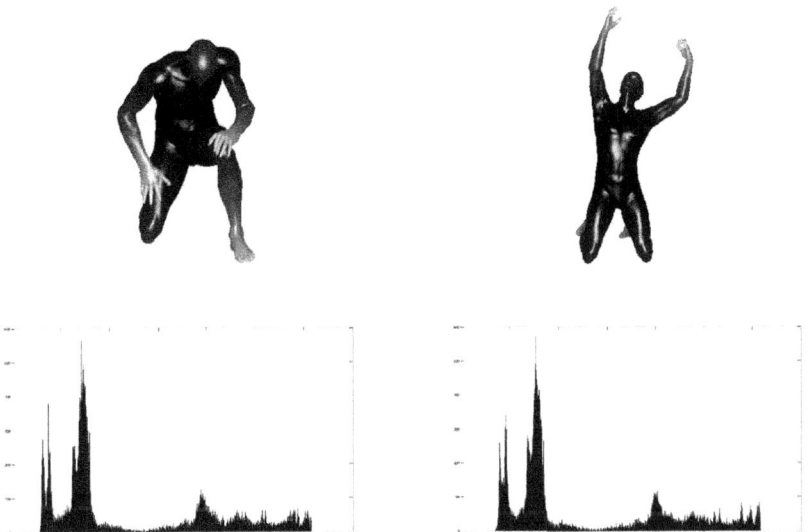

Fig. 5. Top: diagonal of the diffusion kernel $k(x, x)$ used as a local descriptor. Bottom: histogram of the local descriptors.

5   Conclusions

We provided a design framework for kernels that optimize for the ratio between the within class and between classes required for shape recognition under typical type of deformations. So far, our experiments show that the commute time distance is dominating as an optimal filter for the mix of distortions we used. In our future experiments we will investigate the deviation from that type of a filter and try to come up with design framework for specific types of distortions.

References

1. Bérard, P., Besson, G., Gallot, S.: Embedding riemannian manifolds by their heat kernel. Geometric and Functional Analysis 4(4), 373–398 (1994)
2. Bronstein, A.M., Bronstein, M.M., Bustos, B., Castellani, U., Crisani, M., Falcidieno, B., Guibas, L.J., Sipiran, I., Kokkinos, I., Murino, V., Ovsjanikov, M., Patané, G., Spagnuolo, M., Sun, J.: SHREC 2010: robust feature detection and description benchmark. In: Proc. 3DOR (2010)
3. Bronstein, A.M., Bronstein, M.M., Castellani, U., Falcidieno, B., Fusiello, A., Godil, A., Guibas, L.J., Kokkinos, I., Lian, Z., Ovsjanikov, M., Patané, G., Spagnuolo, M., Toldo, R.: Shrec 2010: robust large-scale shape retrieval benchmark. In: Proc. 3DOR (2010)
4. Bronstein, A.M., Bronstein, M.M., Ovsjanikov, M., Guibas, L.J.: Shape google: a computer vision approach to invariant shape retrieval. In: Proc. NORDIA (2009)
5. Bronstein, M.M., Bronstein, A.M.: Shape recognition with spectral distances. Trans. PAMI (2010) (to appear)
6. Bronstein, M.M., Kokkinos, I.: Scale-invariant heat kernel signatures for non-rigid shape recognition. In: Proc. CVPR (2010)
7. Coifman, R.R., Lafon, S.: Diffusion maps. Applied and Computational Harmonic Analysis 21, 5–30 (2006)
8. Floater, M.S., Hormann, K.: Surface parameterization: a tutorial and survey. In: Advances in Multiresolution for Geometric Modelling, vol. 1 (2005)
9. Lévy, B.: Laplace-Beltrami eigenfunctions towards an algorithm that "understands" geometry. In: Proc. Shape Modeling and Applications (2006)
10. Mahmoudi, M., Sapiro, G.: Three-dimensional point cloud recognition via distributions of geometric distances. Graphical Models 71(1), 22–31 (2009)
11. Mateus, D., Horaud, R.P., Knossow, D., Cuzzolin, F., Boyer, E.: Articulated shape matching using laplacian eigenfunctions and unsupervised point registration. In: Proc. CVPR (June 2008)
12. Meyer, M., Desbrun, M., Schroder, P., Barr, A.H.: Discrete differential-geometry operators for triangulated 2-manifolds. In: Visualization and Mathematics III, pp. 35–57 (2003)
13. Ovsjanikov, M., Sun, J., Guibas, L.J.: Global intrinsic symmetries of shapes. Computer Graphics Forum 27, 1341–1348 (2008)
14. Pinkall, U., Polthier, K.: Computing discrete minimal surfaces and their conjugates. Experimental Mathematics 2(1), 15–36 (1993)
15. Reuter, M., Wolter, F.-E., Peinecke N.: Laplace-spectra as fingerprints for shape matching. In: Proc. ACM Symp. Solid and Physical Modeling, pp. 101–106 (2005)
16. Rustamov, R.M.: Laplace-Beltrami eigenfunctions for deformation invariant shape representation. In: Proc. SGP, pp. 225–233 (2007)
17. Spira, A., Sochen, N., Kimmel, R.: Geometric filters, diffusion flows, and kernels in image processing. In: Handbook of Computational Geometry for Pattern Recognition, Computer Vision, Neurocomputing and Robotics. Springer, Heidelberg (2005)
18. Sun, J., Ovsjanikov, M., Guibas, L.J.: A concise and provably informative multi-scale signature based on heat diffusion. In: Proc. SGP (2009)
19. Wardetzky, M., Mathur, S., Kälberer, F., Grinspun, E.: Discrete Laplace operators: no free lunch. In: Conf. Computer Graphics and Interactive Techniques (2008)

Stochastic Models
for Local Optical Flow Estimation

Thomas Corpetti and Etienne Mémin

CNRS/LIAMA, Beijing, China & INRIA/FLUMINANCE, Rennes, France
tcorpetti@gmail.com
etienne.memin@inria.fr

Abstract. In this paper, we present a stochastic interpretation of the motion estimation problem. The usual optical flow constraint equation (assuming that the points keep their brightness along time), embed for instance within a Lucas-Kanade estimator, can indeed be seen as the minimization of a stochastic process under some strong constraints. These constraints can be relaxed by imposing a weaker temporal assumption on the luminance function and also in introducing anisotropic intensity-based uncertainty assumptions. The amplitude of these uncertainties are jointly computed with the unknown velocity at each point of the image grid. We propose different versions depending on the various hypothesis assumed for the luminance function. The substitution of our new observation terms on a simple Lucas-Kanade estimator improves significantly the quality of the results. It also enables to extract an uncertainty connected to quality of the motion field.

Keywords: Optical flow, stochastic formulation, brightness consistency assumption.

1   Introduction

Many computer vision problems are formulated on the basis of the spatial and temporal variations of the image luminance:

$$\frac{df}{dt} = \frac{\partial f}{\partial t} + \boldsymbol{v} \cdot \boldsymbol{\nabla} f = 0, \tag{1}$$

where $\boldsymbol{\nabla}$ is the gradient operator in the x and y directions. When the function f denotes the luminance function this equation is referred in Computer Vision as the Optical Flow Constraint Equation (OFCE) or as the brightness consistency assumption and constitutes the only available information for motion estimation issues. Optical flow estimation has been studied intensively since the seminal work of Horn and Schunck [12] and a huge number of methods based on diverse variations of this constraint have been proposed in the literature [5, 9, 20, 22, 23]. Usually a data model constructed from this constraint is associated with some spatial regularizers that promote motion fields with some spatial (and sometimes temporal) coherency. Many authors have proposed on this basis very efficient

A.M. Bruckstein et al. (Eds.): SSVM 2011, LNCS 6667, pp. 701–712, 2012.

techniques. Readers can refer to [4–6, 13, 15–17, 19, 24–26] for a non exhaustive panel or [11] for a recent review. Comparative performance evaluations of some of these techniques can be found in [1, 2, 10]. Among the developed approaches, the techniques focused first on the design of new regularization terms (able for instance to deal with occlusions, discontinuities or relying on physical grounds [8, 11]) and second on the application of advanced minimization strategies. Surprisingly, apart for some specific applications devoted to some specific types of imagery (fluid, biology, infrared imagery, tomography, IRM, ..., see [21] for a summary), only very few authors have worked on generic alternative data terms to the classical brightness consistency assumption, despite the fact it plays a crucial role in the motion estimation process.

The conventional optical flow constraint relation (1) is in fact defined as the differential of a function known only on spatial and temporal discrete point positions (related to the image sequence spatio-temporal lattice). This is somewhat a strong constraint since in practice, the grid points on which is defined the luminance is transported by a flow itself known only up to the same discrete positions. It results from this discretization process an inherent uncertainty on the points location that can reveal to be of important magnitude when are involved strong motions, large inter frames lapse rate or crude spatial discretization associated for instance to large spatial scales measurements. The idea is therefore to encode such a location uncertainty as a random variable and to incorporate the uncertainty transportation into the brightness consistency assumption. This is done using stochastic rules.

The paper is organized as follows: in section 2 we define a stochastic version of the luminance function, by incorporating isotropic and anisotropic uncertainties. From this formulation, two conservation constraints of the image luminance are derived. If the velocity field is available or if we estimate it simultaneously, we propose in section 3 a way to compute the associated uncertainty. Finally, section 4 presents a local multiscale Lucas and Kanade motion estimator based on the brightness consistency stochastic models.

2   Stochastic Luminance Function and Conservation Constraints

2.1   Notations – Conventions

In this paper we use the following conventions/notations:

- the image luminance is f;
- we represent as a vector $\boldsymbol{X} = (\boldsymbol{X}^1, \ldots, \boldsymbol{X}^m)^T$ a grid of 2D points, $\boldsymbol{X}^s \in \mathbb{R}^2$;
- the "pixel" grid of the images $\boldsymbol{X}_{t-1}$ is represented by the position of a grid $\boldsymbol{X}$ at the initial time, set to $t-1$
- at time $t-1$, this grid is driven by a velocity field $\boldsymbol{v}(\boldsymbol{X}_{t-1}, t-1) : \mathbb{R}^{2m} \times \mathbb{R}^+ \to \mathbb{R}^{2m}$ defined on the initial grid $\boldsymbol{X}_{t-1}$ to generate the new point positions $\boldsymbol{X}_t$ at time t.

2.2   Stochastic Luminance Function

We first write the image luminance as the function of a stochastic process related to the position of image points. If one assumes that the velocity $\boldsymbol{v}$ to estimate transports the grid from $\boldsymbol{X}_{t-1}$ to $\boldsymbol{X}_t$ up to a Brownian motion, we can write:

$$d\boldsymbol{X}_t = \boldsymbol{v}(\boldsymbol{X}_{t-1}, t-1)dt + \boldsymbol{\Sigma}(t, \boldsymbol{X}_t)d\mathbf{B}_t, \tag{2}$$

where $\mathbf{B}_t = (\mathbf{B}_t^1, ..., \mathbf{B}_t^m)^T$ is a multidimensional standard Brownian motion of $\mathbb{R}^{2m}$, $\boldsymbol{\Sigma}$ a $(2m \times 2m)$ covariance matrix and $d\boldsymbol{X}_t = \boldsymbol{X}_t - \boldsymbol{X}_{t-1}$ represents the difference between the grid positions. The luminance function f usually defined on spatial points $\boldsymbol{x} = (x, y)$ at time t is here defined on the grid as a map from $\mathbb{R}^+ \times \mathbb{R}^{2m}$ into $\mathbb{R}^m$ and is assumed to be $C^{1,2}(\mathbb{R}^+, \mathbb{R}^{2m})$. Its differential is obtained following the differentiation rules of stochastic calculus (the so called Îto formulae) that gives the expression of the differential of any continuous function of an Îto diffusion of the form in (2) (see [18] for an introduction to stochastic calculus):

$$df(\boldsymbol{X}_t, t) = \frac{\partial f}{\partial t}dt + \sum_{i=(1,2)} \frac{\partial f(\boldsymbol{X}_t, t)}{\partial x_i}dX_t^i + \frac{1}{2} \sum_{(i,j)=(1,2)\times(1,2)} \frac{\partial^2 f(\boldsymbol{X}_t, t)}{\partial x_i \partial x_j}d < X_t^i, X_t^j >. \tag{3}$$

The term $< X_t^i, X_t^j >$ denotes the joint quadratic variations of X^i and X^j and can be computed according to the following rules: $< B^i, B^j >= \delta_{ij}t$ and $< h(t), h(t) >=< h(t), dB^i >=< B^j, h(t) >= 0$ where $\delta_{ij} = 1$ if $i = j$, $\delta_{ij} = 0$ otherwise, and $h(t)$ is a deterministic function. Compared to classical differential calculus, new terms related to the Brownian random terms have been introduced in this stochastic formulation. A possible way to represent the stochastic part of (2) is to use an isotropic uncertainty variance map $\sigma(X_t, t) : \mathbb{R}^+ \times \mathbb{R}^{2m} \to \mathbb{R}^m$

$$\boldsymbol{\Sigma}(\boldsymbol{X}_t, t)d\mathbf{B}_t = \mathrm{diag}(\sigma(\boldsymbol{X}_t, t)) \otimes \mathbb{I}_2 d\mathbf{B}_t, \tag{4}$$

where $\mathbb{I}_2$ is the (2×2) identity matrix, and $\otimes$ denotes the Kronecker product. Alternatively, one can use anisotropic intensity-based uncertainties along the normal (with a variance σ_η) and the tangent (with a variance σ_τ) of the photometric contour following:

$$\boldsymbol{\Sigma}(\boldsymbol{X}_t, t)d\mathbf{B}_t - \mathrm{diag}(\sigma_\eta(\boldsymbol{X}_t, t)) \otimes \eta dB_t^\eta + \mathrm{diag}(\sigma_\tau(\boldsymbol{X}_t, t)) \otimes \tau dB_t^\tau, \tag{5}$$

where the vectors

$$\boldsymbol{\eta} = \frac{1}{|\boldsymbol{\nabla} f|}\begin{pmatrix} f_x \\ f_y \end{pmatrix}, \tau = \frac{1}{|\boldsymbol{\nabla} f|}\begin{pmatrix} -f_y \\ f_x \end{pmatrix},$$

represent respectively the normal and tangent of the photometric isolines, B^η and B^τ are two scalar independant multidimensional Brownian noises of $\mathbb{R}^m$ and $f_\bullet = \partial f(X_t, t)/\partial \bullet$ for $\bullet = (x, y)$. Let us now express the luminance variations $df(X_t, t)$ under such isotropic or anisotropic uncertainties.

Isotropic Uncertainties. Applying Îto formula (3) to the isotropic uncertainty model yields a luminance variation defined as:

$$df(\boldsymbol{X}_t, t) = \left(\frac{\partial f}{\partial t} + \boldsymbol{\nabla} f \cdot \boldsymbol{v} + \frac{1}{2}\sigma^2 \Delta f \right) dt + \sigma \boldsymbol{\nabla} f \cdot d\boldsymbol{B}_t. \tag{6}$$

Anisotropic Uncertainties. Considering the anisotropic uncertainty model (5) and the mentioned properties regarding the quadratic variations, the term df reads now:

$$df = \left(\frac{\partial f}{\partial t} + \boldsymbol{\nabla} f \cdot \boldsymbol{v} + \frac{\boldsymbol{\nabla} f^T \boldsymbol{\nabla}^2 f \boldsymbol{\nabla} f}{2|\boldsymbol{\nabla} f|^2} (\sigma_\eta^2 - \sigma_\tau^2) + \frac{\sigma_\tau^2 \Delta f}{2} \right) dt + \sigma_\eta \|\boldsymbol{\nabla} f\| dB_t^\eta + \underbrace{\sigma_\tau \boldsymbol{\nabla} f^T \tau dB_t^\tau}_{=0} .$$

$$(7)$$

In this brightness variation model the stochastic term related to the uncertainty along the tangent vanishes (since the projection of the gradient along the level lines is null). It is straightforward to remark that the standard brightness consistency assumption is obtained from (6) or (7) using zero uncertainties ($\sigma = \sigma_\eta = \sigma_\tau = 0$). The proposed stochastic formulation enables thus to use a softer constraint. From this formulation, let us now derive some generic models for the evolution of the image luminance transported by a velocity field with local uncertainties.

2.3   Uncertainty Models for Luminance Conservation

Starting from a known grid $\boldsymbol{X}_{t-1}$ and its corresponding velocity, the conservation of the image luminance can be quite naturally defined as the conditional expectation $E\left(df(\boldsymbol{X}_t, t)|\boldsymbol{X}_{t-1}\right)$ between $t-1$ and t. To compute this term, we exploit the fact (as shown in appendix A) that the expectation of any function $\Psi(\boldsymbol{X}_t, t)$ of a stochastic process $d\boldsymbol{X}_t$ (as in (2)) knowing the grid $\boldsymbol{X}_{t-1}$ reads:

$$E(\Psi(\boldsymbol{X}_t, t)|\boldsymbol{X}_{t-1}) = \Psi(\boldsymbol{X}_{t-1} + \boldsymbol{v}, t) * \mathcal{N}(0, \Sigma), \tag{8}$$

where $\mathcal{N}(0, \Sigma)$ is a multidimensional centered Gaussian. This latter relation indicates that the expectation of a function $\Psi(\boldsymbol{X}_t, t)$ knowing the location $\boldsymbol{X}_{t-1}$ under a Brownian uncertainty of variance Σ is obtained by a convolution of $\Psi(\boldsymbol{X}_{t-1} + \boldsymbol{v}, t)$ with a centered Gaussian kernel of variance Σ.

Assuming Σ known, our new conservation model $\mathcal{H}(f, \boldsymbol{v})$ for the luminance evolution is hence defined as (with g_Σ a gaussian of variance Σ):

$$\mathcal{H}(f, \boldsymbol{v}) = g_\Sigma * (df(\boldsymbol{X}_{t-1} + \boldsymbol{v}, t)) = g_\Sigma * \left(\boldsymbol{\nabla} f \cdot \boldsymbol{v} + \frac{\partial f}{\partial t} + \mathcal{F}(f) \right), \text{with} \tag{9}$$

$$\mathcal{F}(f) = \frac{1}{2}\sigma^2 \Delta f \text{ for isotropic uncertainties or } \mathcal{F}(f) = \frac{\boldsymbol{\nabla} f^T \boldsymbol{\nabla}^2 f \boldsymbol{\nabla} f}{2|\boldsymbol{\nabla} f|^2}(\sigma_\eta^2 - \sigma_\tau^2) + \frac{\sigma_\tau^2 \Delta f}{2} \text{ else.}$$

$$(10)$$

If the brightness conservation constraint strictly holds, one obtains $\sigma = \sigma_\eta = \sigma_\tau = 0$; the Gaussian kernels turn to Dirac distributions and relations (9), (10) correspond to (1). The proposed model provides thus a natural extension of the usual brightness consistency data model. In the next section we propose a way to estimate the uncertainties σ_η and σ_τ.

3   Uncertainty Estimation

Assuming an observed motion field v_{obs} that transports the luminance is available (we will describe in section 5 a local technique for this), it is possible to estimate the uncertainties $\sigma_\eta(x, t)$ and $\sigma_\tau(x, t)$ for each location x at time t.

3.1   Estimation of σ_η

Computing the quadratic variation of the luminance function df between $t - 1$ and t yields, for the isotropic or anisotropic version:

$$d\langle f(X_t, t), f(X_t, t)\rangle = \sigma_\eta^2(X_t, t)\|\nabla f(X_t, t))\|^2, \tag{11}$$

where $\sigma = \sigma_\eta$ in the isotropic formulation. This quadratic variation can also be approximated from the luminance f by
$d\langle f(X_t, t), f(X_t, t)\rangle \approx (f(X_t, t) - f(X_{t-1}, t - 1))^2$. Considering now that the conditional expectation of both previous terms should be identical, one can estimate the variance by:

$$\sigma_\eta(X_t) = \sqrt{\frac{E\left(f(X_t, t) - f(X_{t-1}, t - 1)\right)^2}{E\left(\|\nabla f(X_t, t))\|^2\right)}}. \tag{12}$$

The expectation in the numerator and denominator are then computed at the displaced point $X_{t-1} + v_{obs}(X_{t-1})$ through the convolution of variance $\Sigma(X_{t-1}, t - 1)$. A recursive estimation process is thus emerging from equation (12). For the anisotropic model the uncertainty along the tangent is also needed.

3.2   Estimation of σ_τ

It is not possible to estimate uncertainty along the tangent of the photometric contours in a similar way since, as shown in (7), this quantity does not appear in the noise associated to the luminance variation and therefore is not involved in the corresponding quadratic variations. Writing the Ito diffusion associated to the velocity projected along the tangent yields

$$v_{obs}{}^T \tau = v(X_{t-1}, t - 1)^T \tau dt + \sigma_\tau(t, X_t)dB_t^\tau. \tag{13}$$

This scalar product constitutes a scalar Gaussian random field of mean $\mu = v(X_{t-1}, t - 1)^T \tau$ (assuming $v(x, t)$ is a deterministic function) and covariance $(\mathrm{diag}(\sigma_\tau))$. We assume that the scalar product $v^T \tau$ and the tangent uncertainty $\sigma_\tau(t, x)$ are sufficiently smooth in space and can be respectively well approximated by the local empirical mean and variance over a local spatial neighborhood $N(x)$ of point x:

$$\mu = \frac{1}{|N(x)|} \sum_{x_i \in N(x)} (v_{obs}(x_i, t-1)^T \tau), \sigma_\tau^2 = \frac{1}{|N(x)| - 1} \sum_{x_i \in N(x)} (v_{obs}(x_i, t-1)^T \tau - \mu)^2. \tag{14}$$

The relations in (9-10) provide new models for the variation of the image luminance under isotropic or anisotropic uncertainties. In this section we have presented a technique to estimate such uncertainties from an available velocity field. The next section focuses on application of those extended brightness consistency models for motion estimation.

4   Application of the Proposed Luminance Models

This section aims at defining a simple local motion estimator that embeds the proposed evolution models as an observation term. As the classical Optical Flow Constraint Equation –OFCE– based on (1), an observation model based on a stochastic evolution of the luminance in (9) is subject to the aperture problem. Similarly to the well-known Lucas-Kanade estimator, we cope this difficulty by assuming constant flow within a Gaussian windowing function of variance σ^ℓ. Therefore, the minimum variance estimate v gives:

$$\left(g_{\sigma^\ell} * g_\Sigma * \begin{bmatrix} f_x^2 & f_x f_y \\ f_x f_y & f_y^2 \end{bmatrix} \right) v = -g_{\sigma^\ell} * g_\Sigma * (\mathcal{F}(f) + f_t) \begin{bmatrix} f_x \\ f_y \end{bmatrix}. \tag{15}$$

Let us note that in our model the Gaussian windowing function can be interpreted as the distribution of a new isotropic constant uncertainty term related to the grid resolution and independent of the motion uncertainties that do depend on the image data.

A main advantage of such a formulation of the multiresolution setup is to naturally get rid of the use of a pyramidal image representation. With all these elements, we can define the incremental local motion estimation technique:

Incremental Algorithm

1. Initializations :
   - Fix an initial resolution level $\ell = L$
   - Define $\tilde{f}(X_{t-1}, t) := f(X_{t-1}, t)$; $v = 0$;
2. Estimation for the level ℓ
   (a) Initializations :
       - $n = 1$; $v^0 = 0$;
       - Fix a normal uncertainty σ_η^0
       - Fix a tangent uncertainty σ_τ^0 (if anisotropic formulation)
   (b) Estimate σ_η^n by relation (12)
   (c) Estimate σ_τ^n by measuring the tangential uncertainty of v (relation (14))
   (d) Find v^n by local inversions of the system (15)
   (e) Update motion field : $v := v + v^n$
   (f) Warp the image $f(X_t, t)$: $\tilde{f}(X_{t-1}) = f(X_{t-1} + v, t)$
   (g) $n := n + 1$
   (h) **Loop** to step (b) until convergence ($|v^n| < \epsilon$);
3. Decrease the multiresolution level : $\sigma^\ell = \lambda \sigma^\ell$
4. **Loop** to step 2 until convergence ($\sigma^\ell < \sigma_{min}^\ell$).

The previous framework is a natural and simple implementation of a local motion estimation technique using the proposed models for the evolution of the luminance. A quantitative and qualitative evaluation of such an estimator, with comparisons to the classic OFCE will be presented in the next section.

5   Experimental Results

We present in this section some experimental results of the local motion estimator described in section 4. We show examples on synthetic fluid images and on the Middleburry database[1]. It is important to outline that the estimator defined in section 4 constitutes only a local technique whose aim is only to valid, compare and qualify the observation model based on stochastic uncertainties *vs* the usual ofce in (1). Hence, its performances have to be compared to other local approaches. As a first benchmark we analyze the results obtained on images depicting the evolution of a 2D turbulent fluid flow.

Fluid Images: We used a pair of synthetic images obtained by DNS (Direct Numerical Simulation of Navier-Stokes equations) and representing a 2D turbulent flow. Numerical values of average angular error (AAE) [3] and of the Root Mean Square Error (RMSE) are used as criteria to compare our estimators (isotropic and anisotropic) with some of the state-of-the-art approaches are depicted in table 1. The comparison is done with the following techniques: Horn & Schunck (HS) [12], a commercial software based on correlation (COM, DaVis 7.2 from LaVision GmbH), a pyramidal incremental implementation of the Lucas-Kanade estimator (LK) [14], the proposed framework in section 4 with the OFCE as an observation model (OFCE) (*i.e* with a zero uncertainty), two fluid-dedicated dense motion estimators based on a Div-Curl smoothing with different minimization strategies (DC1–DC2, [8, 27]), a fluid-dedicated dense motion estimator based on a turbulence subgrid model in the data-term (TUR, [7]).

In figure 1, we present an image of the sequence, the estimated flow with the proposed method (anisotropic version) and the error flow field. We have also plotted the velocity spectra of the different techniques and compared them with the ground truth. These spectra are represented in a log-log coordinate and a standard-log coordinate system in order to highlight small and large scales respectively.

On table 1, one can immediately observe that compared to the other local approaches, our method provides very good results since the global accuracy is highly superior than the Lucas-Kanade (LK) and the commercial software (COM). Compared to dense techniques (HS, DC1 and DC2), our numerical results are in the same order of magnitude which is a very relevant point. They are competitive with some dense estimation techniques dedicated to fluid flows analysis with advanced smoothing terms (DC1–DC2, [8, 27]). The comparison between the results OFCE, ISO and ANI is very interesting since it highlights the benefit of the stochastic formulation of the image luminance.

If now one observes the spectra of the velocity we see that the small scales (right part of the graph in fig. 1(g)) are much better recovered by the proposed estimators than by the dense estimators. These latter are generally difficult to estimate and often smoothed out with the spatial regularizers introduced in the dense techniques. Even if the Lucas-Kanade technique seems to exhibit better results on small scales, when observing the figure 1(h), it is obvious to note that large scales are badly estimated with this approach and this yields a very poor overall accuracy (see table 1). As for the large scales the results are comparable

---

[1] http://vision.middlebury.edu/flow/

with the best dense dedicated techniques. We believe hence that our estimator constitutes an appealing alternative to usual local PIV methods. Let us now describe the accuracy of the observation term on some images of the middleburry database.

Table 1. Quantitative comparisons on the DNS sequence

	LK	COM	HS	DC 1	DC 2	TUR	OFCE	ISO	ANISO
AAE	6.07^o	4.58^o	4.27^o	4.35^o	3.04^o	4.49^o	4.53^o	3.59^o	3.12^o
RMSE	0.1699	0.1520	0.1385	0.1340	0.09602	0.1490	0.1243	0.1072	0.0961

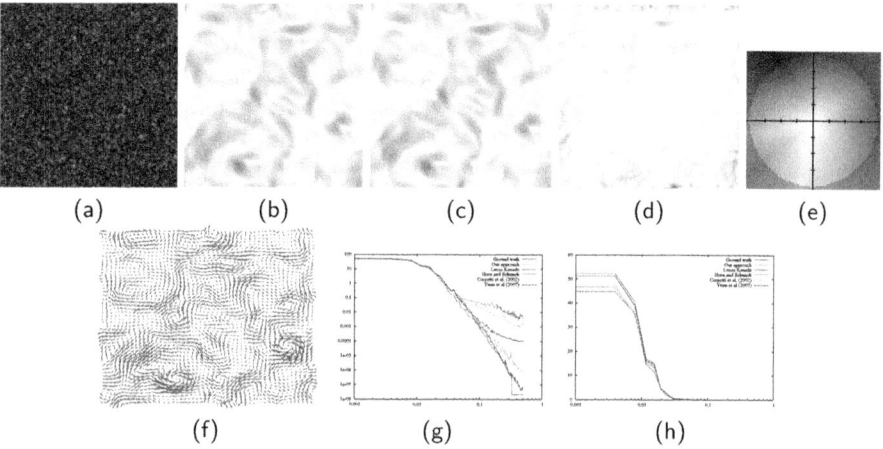

(a)   (b)   (c)   (d)   (e)

(f)   (g)   (h)

Fig. 1. Results on the DNS sequence : Top (a): an image of the sequence; (b): the estimated flow ; (c): the real flow; (d): the difference flow represented with the coding color in (e); **Bottom** (f): visualisation of the estimated flow; (g-h): Spectra of the velocity compared with ground truth and to several method: (g): log-log representation (highlights small scales on the right part) and (h): non log-log representation (highlights large scales on the left part). Color are : Red : ground truth; Green : our approach (anisotropic version); Blue : Lucas-Kanade [14]; Purple : Horn and Schunck [12]; Cyan: Div-Curl smoothing [8] and Black: Div-Curl in mimetic discretization [27]

Middleburry Database: We have tested our approaches on the "Dimetrodon", "Yosemite" and "Venus" sequences. For these sequences ground truths comparisons with others state-of-the-art approaches are available. The dimetrodon sequence is illustrated in figure 2.

The quantitative results are presented in the table 2. When comparing the three first columns that use exactly the same technique but based on the usual OFCE (relation (1)), our luminance model with isotropic (ISO) and anisotropic (ANISO) uncertainties (relation (10)), it immediately points out that the proposed models enable to enhance significantly the quality of the results. This fair comparison of the three observation models onto the same estimator promotes the use of a stochastic formulation under anisotropic uncertainties. In fact this latter version is a softer constraint than the OFCE which, as shown previously, assumes implicitly a perfect measurement without any incertitudes. The estimated motion fields under the anisotropic luminance formulation is represented on figure fig.2 (c) and can be compared with the ground truth in fig.2(d).

Let us in addition remind that the motion estimation technique that has been developed for comparing the models of luminance is quite simple (based on Lucas and Kanade). Therefore, as expected, the errors are mainly localized on discontinuities. However it is very informative to observe that despite the simplicity of this technique, our results in table 2 are very competitive and sometimes outperform advanced dense techniques with a specific process for discontinuities. Apart from regions exhibiting motion discontinuities and where the error can be important, the difference fields of fig.2(d) reveals very good results (white areas) in the other locations. This suggests that the luminance models introduced in this paper is usefull in allowing a global improvement of accuracy. More than the estimated motion fields, such a technique is able to extract the associated uncertainty areas. The norm of the global uncertainty $\sqrt{\sigma_\eta^2 + \sigma_\tau^2}$ map obtained at the end of the process with the best estimator (the anisotropic one) is plotted in fig.2(e). As expected, homogeneous areas where the aperture problem holds correspond to high values whereas small values are linked to photometric contours. Such output of our method is very promising since it it highlights the main structures of the images and gives an indicator of the quality of the estimation. To justify this last point, we have depicted in fig.2(f) the reconstructed errors when we take into account for the evaluation only the points where the incertitude is bellow a given value (blue lines) and the corresponding percentage of points used for the computation (red lines).

We then strongly believe that the stochastic models presented can be exploited in the future to design dense estimators relying on the proposed brightness consistency model.

Table 2. Quantitative results and comparisons on the Dimetrodon, Yosemite and Venus sequence

Results on the Dimetrodon sequence

Met.	OFCE	ISO	ANISO	Bruhn et al.	Black, Ana.	Pyr. LK	Med. Pla.^{TM}	Zitnick et al
Ang. err.	7.95^o	3.95^o	$\mathbf{2.85^o}$	10.99^o	9.26^o	10.27^o	15.82^o	30.10^o

Results on the Yosemite sequence

Met.	OFCE	ISO	ANISO	Bruhn et al.	Black, Ana.	Pyr. LK	Med. Pla.^{TM}	Zitnick et al
Ang. error	4.47^o	3.12^o	2.89^o	$\mathbf{1.69^o}$	2.65^o	5.22^o	11.09^o	18.50^o

Results on the Venus sequence

Met.	OFCE	ISO	ANISO	Bruhn et al.	Black, Ana.	Pyr. LK	Med. Pla.^{TM}	Zitnick et al
Ang. error	12.02^o	10.23^o	8.42^o	8.73^o	$\mathbf{7.64^o}$	14.61^o	15.48^o	11.42^o

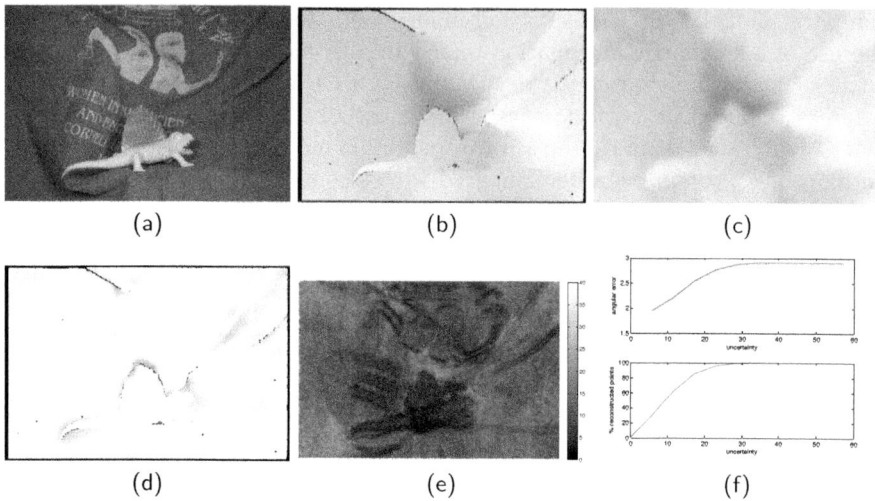

Fig. 2. Dimetrodon sequence (a): an image of the sequence; (b): the ground truth; (c): The estimated motion field with our approach in anisotropic version; (d): difference velocity field and (e): the extracted uncertainty map σ_η and (f): evolution of the error and percentage of correct motion fields when one takes into account only velocity fields with smaller values of $\sqrt{\sigma_\eta^2 + \sigma_\tau^2}$

6 Conclusion

In this paper an observation model for optical flow estimation has been introduced. The new operator is based on a stochastic modeling of the brightness consistency uncertainty. This data model constitutes a natural extension of the usual brightness consistency assumption. Isotropic and anisotropic uncertainty models have been presented. From this new data term, we have designed a simple local motion estimator where the multiresolution is also interpreted in term of a spatial uncertainty.

The performances of this local estimator have been validated on synthetic fluid flows issued from Direct Numerical Simulations and on the middleburry synthetic database. In the first case, the results have exhibited significant performances, especially in the recovery of small scales that are generally smoothed out by spatial regularizers of dense approaches. As for the middleburry database, the simple local implementation of the presented data-term outperforms local approaches. We therefore believe that this stochastic modeling is a very promising alternative to the usual deterministic OFCE for all optical-flow methods.

A Expectation of a Function of a Stochastic Process

The conditional expectation given $\boldsymbol{X}_{t-1}$ of any function $\Psi(\boldsymbol{X}_t, t)$ of a stochastic process defined through Îto diffusion (3) and discretized through an Euler scheme $\boldsymbol{X}_t = \boldsymbol{X}_{t-1} + \boldsymbol{v}(\boldsymbol{X}_{t-1})dt + \boldsymbol{\Sigma}^{1/2}(B_{t+1} - B_t)$ may be written as:

$$E(\Psi(\boldsymbol{X}_t,t)|\boldsymbol{X}_{t-1}) = \int_{\mathbb{R}} \Psi(\boldsymbol{X}_t,t)p(\boldsymbol{X}_t|\boldsymbol{X}_{t-1})d\boldsymbol{X}_t. \tag{16}$$

As the process $\boldsymbol{X}_t$ is known up to the Brownian motion $\boldsymbol{\Sigma}d\boldsymbol{B}_t$, the probability $p(\boldsymbol{X}_t|\boldsymbol{X}_{t-1})$ is a multidimensional Gaussian of variance $\boldsymbol{\Sigma}\sqrt{dt}$ ($dt=1$ here) and we get:

$$E(\Psi(\boldsymbol{X}_t)|\boldsymbol{X}_{t-1}) = \frac{1}{\sqrt{2\pi}\det(\boldsymbol{\Sigma})^{\frac{1}{2}}}\int_{\mathbb{R}} \Psi(\boldsymbol{X}_t,t)\exp\left(-(\boldsymbol{X}_{t-1}+\boldsymbol{v}-\boldsymbol{X}_t)\boldsymbol{\Sigma}^{-1}(\boldsymbol{X}_{t-1}+\boldsymbol{v}-\boldsymbol{X}_t)\right)d\boldsymbol{X}_t.$$
$$\tag{17}$$

By a change of variable $\boldsymbol{Y}_t = \boldsymbol{X}_{t-1} + \boldsymbol{v} - \boldsymbol{X}_t$, this expectation can be written as:

$$E(\Psi(\boldsymbol{X}_t,t)|\boldsymbol{X}_{t-1}) = \frac{1}{\sqrt{2\pi}\det(\boldsymbol{\Sigma})^{1/2}}\int_{\mathbb{R}} \Psi(\boldsymbol{X}_{t-1}+\boldsymbol{v}-\boldsymbol{Y}_t,t)\exp(-\boldsymbol{Y}_t\boldsymbol{\Sigma}^{-1}\boldsymbol{Y}_t)d\boldsymbol{Y}_t$$
$$= \Psi(\boldsymbol{X}_{t-1}+\boldsymbol{v},t) * \mathcal{N}(0,\boldsymbol{\Sigma}). \tag{18}$$

References

1. Baker, S., Scharstein, D., Lewis, J., Roth, S., Black, M., Szeliski, R.: A database and evaluation methodology for optical flow. In: Int. Conf. on Comp. Vis. (2007)
2. Barron, J., Fleet, D., Beauchemin, S.: Performance of optical flow techniques. Int. J. Comput. Vis. 12(1), 43–77 (1994)
3. Barron, J.L., Fleet, D.J., Beauchemin, S.S., Burkitt, T.A.: Performance of optical flow techniques. Int. J. of Comp. Vis. 12(1), 43–77 (1994)
4. Black, M., Anandan, P.: Robust incremental optical flow. In: Eklundh, J.-O. (ed.) ECCV 1994. LNCS, vol. 800, pp. 296–302. Springer, Heidelberg (1994)
5. Brox, T., Bruhn, A., Papenberg, N., Weickert, J.: High accuracy optical flow estimation based on a theory for warping. In: Pajdla, T., Matas, J(G.) (eds.) ECCV 2004. LNCS, vol. 3024, pp. 25–36. Springer, Heidelberg (2004)
6. Bruhn, A., Weickert, J., Kohlberger, T., Schnoerr, C.: A multigrid platform for real-time motion computation with discontinuity-preserving variational methods. Int. J. Com. Vis. 70(3), 257–277 (2006)
7. Cassisa, C., Simoens, S., Prinet, V.: Two-frame optical flow formulation in an unwarping multiresolution scheme. In: Bayro-Corrochano, E., Eklundh, J.-O. (eds.) CIARP 2009. LNCS, vol. 5856, pp. 790–797. Springer, Heidelberg (2009)
8. Corpetti, T., Mémin, E., Pérez, P.: Dense estimation of fluid flows. IEEE Trans. Pattern Anal. Machine Intell. 24(3), 365–380 (2002)
9. Fitzpatrick, J.: The existence of geometrical density-image transformations corresponding to object motion. Com. Vis., Grap., Im. Proc. 44(2), 155–174 (1988)
10. Galvin, B., McCane, B., Novins, K., Mason, D., Mills, S.: Recovering motion fields: an analysis of eight optical flow algorithms. In: Proc. British Mach. Vis. Conf., Southampton (1998)
11. Heitz, D., Mémin, E., Schnoerr, C.: Variational fluid flow measurements from image sequences: synopsis and perspectives. Exp. Fluids 48(3), 369–393 (2010)
12. Horn, B., Schunck, B.: Determining optical flow. Artificial Intelligence 17, 185–203 (1981)
13. Lempitsky, V., Roth, S., Rother, C.: Fusionflow: Discrete-continuous optimization for optical flow estimation. In: IEEE Comp. Vis. Patt. Rec. (2008)

14. Lucas, B., Kanade, T.: An iterative image registration technique with an application to stereovision. In: Int. Joint Conf. on Art. Int., pp. 674–679 (1981)

15. Mémin, E., Pérez, P.: Dense estimation and object-based segmentation of the optical flow with robust techniques. IEEE Trans. Im. Proc. 7(5), 703–719 (1998)

16. Nagel, H.: Extending the oriented smoothness constraint into the temporal domain and the estimation of derivatives of optical flow. In: Faugeras, O. (ed.) ECCV 1990. LNCS, vol. 427, pp. 139–148. Springer, Heidelberg (1990)

17. Nesi, P.: Variational approach to optical flow estimation managing discontinuities. Image and Vision Computing 11(7), 419–439 (1993)

18. Oksendal, B.: Stochastic differential equations. Spinger, Heidelberg (1998)

19. Papenberg, N., Bruhn, A., Brox, T., Didas, S., Weickert, J.: Highly accurate optic flow computation with theoretically justified warping. Int. J. Comput. Vision 67(2), 141–158 (2006)

20. Schunck, B.: The image flow constraint equation. Com. Vis., Grap., Im. Proc. 35, 20–46 (1986)

21. Sun, D., Roth, S., Black, M.: Secrets of optical flow estimation and their principles. In: Proc. IEEE Com. Vis. and Pat. Rec., CVPR 2010, pp. 2432–2439 (2010)

22. Tretiak, O., Pastor, L.: Velocity estimation from image sequences with second order differential operators. In: Proc. 7th Int. Conf. On Pattern Recognition, Montreal, pp. 16–19 (1984)

23. Weber, J., Malik, J.: Robust computation of optical flow in a multi-scale differential framework. Int. J. Comput. Vis. 14(1) (1995)

24. Wedel, A., Pock, T., Braun, J., Franke, U., Cremers, D.: Duality tv-l1 flow with fundamental matrix prior. In: Image Vision and Computing, Auckland, New Zealand (November 2008)

25. Weickert, J., Schnörr, C.: Variational optic-flow computation with a spatio-temporal smoothness constraint. J. Math. Im. and Vis. 14(3), 245–255 (2001)

26. Xu, L., Chen, J., Jia, J.: A segmentation based variational model for accurate optical flow estimation. In: Forsyth, D., Torr, P., Zisserman, A. (eds.) ECCV 2008, Part I. LNCS, vol. 5302, pp. 671–684. Springer, Heidelberg (2008)

27. Yuan, J., Schnörr, C., Mémin, E.: Discrete orthogonal decomposition and variational fluid flow estimation. Journ. of Mathematical Imaging and Vision 28(1), 67–80 (2007)

Optic Flow Scale Space

Oliver Demetz[1], Joachim Weickert[1], Andrés Bruhn[2], and Henning Zimmer[1]

[1] Mathematical Image Analysis Group
Faculty of Mathematics and Computer Science, Building E1.1
Saarland University, 66041 Saarbrücken, Germany
{demetz,weickert,zimmer}@mia.uni-saarland.de
[2] Vision and Image Processing Group
Cluster of Excellence Multimodal Computing and Interaction,
Saarland University, 66041 Saarbrücken, Germany
bruhn@mmci.uni-saarland.de

Abstract. While image scale spaces are well understood, it is undeniable that the regularisation parameter in variational optic flow methods serves a similar role as the scale parameter in scale space evolutions. However, no thorough analysis of this optic flow scale-space exists to date. Our paper closes this gap by interpreting variational optic flow methods as Whittaker-Tikhonov regularisations of the normal flow, evaluated in a constraint-specific norm. The transition from this regularisation framework to an optic flow evolution creates novel vector-valued scale-spaces that are not in divergence form and act in a highly anisotropic way. From a practical viewpoint, the deep structure in optic flow scale space allows the automatic selection of the most accurate scale by means of an optimal prediction principle. Moreover, we show that our general class of optic flow scale-spaces incorporates novel methods that outperform classical variational approaches.

1 Introduction

Starting with Iijima's pioneering work on Gaussian scale-space and its use in optical character recognition many decades ago [12,13], scale-spaces have become versatile tools for analysing and understanding the multiscale structure of images; see e.g. the monographs [8,15,21,23] and the references therein. While partial differential equations (PDEs) of evolution type provide a natural framework for most scale-space concepts [2], it has also been shown that variational regularisation methods create scale-spaces where the regularisation parameter acts as scale [18]. Such variational methods, however, offer much broader application fields than classical data smoothing. In computer vision, for example, they are widely used for solving correspondence problems. For specific applications such as optic flow computation in image sequences, variational methods have become highly sophisticated and constitute the most accurate methods to date [3]. In contrast to classical image regularisation methods, however, these optic flow methods do not regularise image data, but constraint equations. For instance, in the classical method of Horn and Schunck, a grey value constancy assumption replaces the role of a data fidelity term [11]. Therefore, it is unclear in which sense one may interpret variational optic flow methods as scale-spaces. However, in view of the fact that the unreliable normal flow is essentially the only information that can be extracted

A.M. Bruckstein et al. (Eds.): SSVM 2011, LNCS 6667, pp. 713–724, 2012.

directly from the data in an image sequence, it is astonishing that modern variational optic flow methods are capable of achieving results of such a high quality. Thus, there is a clear need to understand their scale-space behaviour.

The goal of our paper is to address this problem. Our contributions are fourfold:

1. We interpret the classical variational methods of Horn and Schunck [11] and of Nagel and Enkelmann [17] as Whittaker-Tikhonov regularisations of the normal flow. We show that this requires to replace the Euclidean norm by a space-variant matrix-induced norm that respects the data constraints.
2. We generalise this framework to a broader class of methods that also allows to come up with new models that have not been considered before. We show that they can offer better performance than the classical variational methods.
3. Going from the regularisation framework to a scale-space representation, we come up with the novel concept of *optic flow scale-spaces*. They are parabolic evolutions of vector-valued data with the regularisation parameter as scale and the normal flow as initial state. However, in contrast to many image scale-spaces they are not of divergence type and hence do not preserve the average value of the initial data. Moreover, they turn out to be highly anisotropic. Both properties are essential for the remarkable performance of optic flow methods.
4. Having a scale-space evolution, it is natural to explore its deep structure. To this end, we show that the optic flow scale-space provides an efficient framework for automatically selecting the best scale that gives the most accurate optic flow field. As a parameter-free scale selection principle we employ the *Optimal Prediction Principle*, which is specifically tailored to the needs of optic flow estimation [25].

This paper is not intended to present a new high accuracy optic flow method. Rather, we want to investigate the scale space behaviour of variational optic flow methods. Thus, we intentionally do not consider algorithmic sophistications such as robust penalisation strategies, constancy assumptions without linearisation, multiscale warping strategies, and nonlocal search methods. Since the focus of our work lies on the fundamental concepts behind optic flow scale space, the discussion of additional side effects would obliterate the core message of the paper.

Related Work. The transformation we apply is related to a proposal by Schnörr [19], who did not pursue this concept further. With respect to the interpretation of variational methods in terms of specific norms, vector spaces, and higher order manifolds, there is a huge amount of literature available; see e.g. Sochen *et al.* [20] and the references therein. Particularly interesting in this context is the work of Ben-Ari and Sochen [5] who derive a class of smoothness terms based on spatially varying norms induced by suitable embeddings of the flow field into higher dimensional vector spaces. Scale selection is a classical issue in Gaussian scale-space theory [15]. More specifically, choosing optimal smoothness parameters is an enduring problem for almost all classes of scale-space and variational methods. In our context, the works by Krajsek *et al.* [14], Mrázek and Navarra [16] and the recent ideas by Zimmer *et al.* [25] are most relevant. While there has been some research on scale spaces for image sequences [7,10], to our knowledge the concept of optic flow scale space has not been considered before.

Organisation of our Paper. In Section 2, we describe the transformation of classical optic flow energies into a regularisation framework for the normal flow in a suitable constraint-induced norm that is further generalised in Section 3. The fourth section introduces our scale space framework for optic flow. An efficient numerical scheme is described in Section 5, while the automatic scale selection is discussed in Section 6. Section 7 presents computational results, and we conclude with a summary and an outlook in Section 8.

2   Variational Optic Flow as Whittaker-Tikhonov Regularisation

As starting point of our derivations of an optic flow scale space we consider the classic method of Horn and Schunck [11]. This variational method estimates the optic flow field $\boldsymbol{w} := (u, v)^\top = (u(x, y, z), v(x, y, z))^\top$ as the minimiser of the energy functional

$$E(\boldsymbol{w}) = \int_\Omega \left((f_x u + f_y v + f_z)^2 + \alpha \left(\|\nabla u\|^2 + \|\nabla v\|^2 \right) \right) \, \mathrm{d}x \, \mathrm{d}y , \qquad (1)$$

where $\Omega \subset \mathbb{R}^2$ represents the spatial image domain, $f : \Omega \times [0, \infty) \to \mathbb{R}$ the image sequence, $\| \cdot \|$ stands for the Euclidean norm, subscripts denote partial derivatives, and $\nabla = (\partial_x, \partial_y)^\top$ is the spatial gradient operator.

The first term of this functional is called data term and models the linearised assumption that corresponding pixels in subsequent frames have similar grey value. Since it depends on two unknown functions u and v that describe the horizontal and vertical displacement field, respectively, its solution is under-determined. Evidently, in order to find a unique solution, additional assumptions on u and v are needed. This is realised by the second term – the so-called smoothness term. It penalises variations of the solution and is weighted by the positive regularisation parameter α.

2.1   Towards Regularisation in a Spatially Varying Norm

In the following, we consider a slightly modified version of the energy (1) with the additional terms $\epsilon^2 (u^2 + v^2) + c$, where ϵ is a small positive constant and $c(x, y, z) = -\epsilon f_z^2/(|\nabla f|^2 + \epsilon^2)$ is a function which does not depend on the unknown and thus does not play a role in the actual minimisation. Later on, these terms will be useful for theoretical reasons. The modified energy then reads as

$$E(\boldsymbol{w}) = \int_\Omega \left((f_x u + f_y v + f_z)^2 + \epsilon^2 (u^2 + v^2) + c + \alpha \left(\|\nabla u\|^2 + \|\nabla v\|^2 \right) \right) \, \mathrm{d}x \, \mathrm{d}y .$$
$$(2)$$

In order to obtain a different much more intuitive understanding of the underlying variational model, we seek to reformulate the latter energy in an image regularisation framework. Since a smoothness term is already present, the main task is now to derive a suitable similarity term. To this end, we make use of the following result (see also [1]): Let $\boldsymbol{w}_n$ be the regularised normal flow, given by

$$\boldsymbol{w}_n = \frac{-f_z \nabla f}{|\nabla f|^2 + \epsilon^2} , \qquad (3)$$

and let $A : \Omega \rightarrow \mathbb{R}^{2 \times 2}$ be a symmetric positive definite matrix in every point of the image domain defined as

$$A^2 = \nabla f \nabla f^\top + \epsilon^2 I \ , \tag{4}$$

where I denotes the unit matrix. Then the following equivalence holds:

$$(f_x u + f_y v + f_z)^2 + \epsilon^2 (u^2 + v^2) + c = (w - w_n)^\top A^2 (w - w_n) \ . \tag{5}$$

This can easily be verified by straightforward calculations. Furthermore, we use the latter quadratic form and the concept of matrix-weighted norms to rewrite the modified functional (2) into an image regularisation-like energy [6]

$$E(w) = \int_\Omega \left(\| w - w_n \|_{A^2}^2 + \alpha \left(\| \nabla u \|^2 + \| \nabla v \|^2 \right) \right) \, \mathrm{d}x \, \mathrm{d}y \ . \tag{6}$$

In this context, for a symmetric positive definite matrix M the corresponding matrix-weighted norm is given by $\| x \|_M^2 := \langle x, x \rangle_M = x^\top M x$. Note that due to the additional term in (2), the matrix A fulfils these requirements by construction.

Having performed the previous rewritings, the following insight becomes explicit: Essentially, the seminal variational optic flow method of Horn and Schunck can be interpreted as Whittaker-Tikhonov regularisation of the normal flow in a matrix-weighted spatially varying norm.

2.2   Analysing the Matrix-Weighted Norm

By analysing the obvious eigenstructure of the constraint matrix A^2, we can gain a deeper understanding of the introduced matrix-weighted norm and thus of the data term. Using the eigendecomposition of A^2 given by

$$A^2 = \left(|\nabla f|^2 + \epsilon^2 \right) \frac{\nabla f}{|\nabla f|} \frac{\nabla f^\top}{|\nabla f|} + \epsilon^2 \frac{\nabla f^\perp}{|\nabla f|} \frac{\nabla f^{\perp \top}}{|\nabla f|} \ , \tag{7}$$

we can express the data term as

$$\| w - w_n \|_{A^2}^2 = \left(|\nabla f|^2 + \epsilon^2 \right) \left\langle w - w_n, \frac{\nabla f}{|\nabla f|} \right\rangle^2 + \epsilon^2 \left\langle w - w_n, \frac{\nabla f^\perp}{|\nabla f|} \right\rangle^2 \ . \tag{8}$$

This shows that the central quantity being under consideration is the normal flow w_n, or rather the difference between the actual solution and the normal flow $w - w_n$. This difference vector is then projected into the local eigensystem of A^2. There, its component perpendicular to the image gradient is basically negligible (since ϵ^2 is small), while its component along the image gradient is the part that actually contributes.

This also confirms the classic explanation of the linearised grey value constancy assumption as a constraint line: The expression $f_x u + f_y v + f_t = 0$ defines a line perpendicular to the image gradient with distance $f_t / |\nabla f|$ from the origin [11]. Also the findings in [25] are in accordance with this interpretation: There, the authors rewrite the assumption into a projection of the difference vector $w - w_n$ onto the image gradient, which exactly comes down to our locally adapted norm for $\epsilon = 0$.

3   Generalisation of the Matrix-Weighted Norm

Up to now, we have mainly reformulated the method of Horn and Schunck and identified the matrix-induced norm $\| \cdot \|_{A^2}$ to play a central role. Consequently, we now propose to generalise this idea to a class of norms with varying anisotropy. To this end, we modify the exponent of the constraint matrix inducing the norm. This yields a data term of the general form

$$\| w - w_n \|_{A^{2-\beta}}^2 \tag{9}$$

with $0 \leq \beta \leq 2$. Larger choices for β do not make sense, since they would invert the anisotropy, i.e. swap the penalisation directions. The parametrisation of the norm has been chosen in such a way that for $\beta = 0$ the original model of Horn and Schunck, and for $\beta = 2$ a pure decoupled vector-valued regularisation of the normal flow in the Euclidean norm is obtained (since the constraint matrix collapses to the identity matrix).

To establish a consistently extended model, we also equip the smoothness term with the same spatially varying norm. This leads to the regulariser

$$\| \nabla u \|_{A^{-\gamma}}^2 + \| \nabla v \|_{A^{-\gamma}}^2 \tag{10}$$

where $\gamma \geq 0$. This generally anisotropic image-driven regulariser allows variations of the flow field across image edges but not along them. In the special case of $\gamma = \beta = 0$ the model corresponds to Whittaker-Tikhonov regularisation [24,22] as used by Horn and Schunck [11]. Another special case of our generalised model is obtained for $\gamma = 2$ and $\beta = 0$: Then, our method resembles the method of Nagel and Enkelmann [17]. Recall that the proposed spatially varying norm naturally arises from the linearised constancy assumption in the data term. In this way our approach differs significantly from the work in [5], which derives such a norm by embedding the flow in a higher dimensional vector space and thus disregards the data term throughout the derivation. Incorporating both generalisations, we finally consider the energy functional

$$E(w) = \int_\Omega \left(\| w - w_n \|_{A^{2-\beta}}^2 + \alpha \left(\| \nabla u \|_{A^{-\gamma}}^2 + \| \nabla v \|_{A^{-\gamma}}^2 \right) \right) \mathrm{d}x \, \mathrm{d}y \,, \tag{11}$$

with α, β and γ as defined before. This energy functional forms the basis for our optic flow scale space introduced in the next section.

4   Optic Flow Scale Space

From the calculus of variations it follows that any minimiser of the functional (11) has to fulfil the associated Euler-Lagrange equation, which reads

$$\frac{w - w_n}{\alpha} = A^{\beta-2} \begin{pmatrix} \operatorname{div}\left(A^{-\gamma} \nabla u\right) \\ \operatorname{div}\left(A^{-\gamma} \nabla v\right) \end{pmatrix} \,, \tag{12}$$

with reflecting Neumann boundary conditions $n^\top A^{-\gamma} \nabla u = 0$ and $n^\top A^{-\gamma} \nabla v = 0$. The reader should keep in mind that the argument of the energy (11) as well as the latter

PDE is vector-valued, since $w = (u, v)^\top$. Equation (12) can be seen as a fully implicit time discretisation of the filter

$$\partial_t w = A^{\beta-2} \begin{pmatrix} \mathrm{div}(A^{-\gamma} \nabla u) \\ \mathrm{div}(A^{-\gamma} \nabla v) \end{pmatrix}, \qquad w(\cdot, 0) = w_n . \tag{13}$$

with a single time step of size α, and the normal flow w_n as initial state at time $t = 0$.

Obviously this temporal evolution constitutes a scale space, whose evolution time t coincides with the regularisation parameter α of the associated energy. Interestingly, the initial state of our *optic flow scale space* is the regularised normal flow, which is the only component of the flow field that can be directly extracted from the image data.

Note that we have transformed the *regularisation-like* energy functional (11) into a *diffusion-like* coupled system of parabolic PDEs (13). In the context of image filtering, relations between such methods have been investigated by Scherzer and Weickert [18].

5   Numerical Realisation

For solving the parabolic problem in (13) we use an explicit scheme. To this end we discretise the two flow components u and v by sampling them on a regular grid and stacking all rows in single vectors $u, v \in \mathbb{R}^N$, where N denotes the number of pixels. Using this single-index notation, we discretise the matrix $A^{\beta-2}$ in pixel i by

$$A_i^{\beta-2} = \begin{pmatrix} a_i & b_i \\ b_i & c_i \end{pmatrix}, \qquad i = 1, \ldots, N . \tag{14}$$

Accordingly, we discretise the diffusive terms using finite differences and obtain a nonadiagonal diffusion matrix $D \in \mathbb{R}^{N \times N}$ (similar to [23]). This leads to the following explicit scheme:

$$\begin{pmatrix} u \\ v \end{pmatrix}^{k+1} = \left(I + \tau \left(\begin{array}{cc|cc} a_1 & & b_1 & \\ & \ddots & & \ddots \\ & & a_N & & b_N \\ \hline b_1 & & c_1 & \\ & \ddots & & \ddots \\ & & b_N & & c_N \end{array} \right) \left(\begin{array}{c|c} D & 0 \\ \hline 0 & D \end{array} \right) \right) \begin{pmatrix} u \\ v \end{pmatrix}^k , \tag{15}$$

where every iteration advances the evolution by the time step size $\tau > 0$. Thus, after k iterations $(u^k, v^k)^\top$ contains the flow field at scale $\alpha = k \cdot \tau$. As a consequence, this explicit scheme inherently samples the whole scale space up to the stopping time α in intervals of size τ. Thereby, the whole iteration matrix remains constant for all iterations, since all terms are exclusively image-driven.

In order to accelerate this explicit scheme, we made use of the recently introduced *fast explicit diffusion (FED)* strategy [9], which performs cycles of explicit iterations with varying time step sizes. In particular, up to 50% of the step sizes can exceed the stability limit significantly, while the overall process remains provably stable. By that, the order of the smoothing time reached in n steps can be increased from $O(n)$ to $O(n^2)$. In our case, this is very beneficial, since the maximal stable time step size of the explicit scheme can decrease drastically with increasing anisotropy or small choices of the parameter β.

6   Scale Selection

In the previous sections, we have set up a novel class of optic flow scale spaces which all evolve in the regularisation parameter α. Evidently, in our optic flow setting there exists one scale within each scale space that provides the flow with the highest accuracy. Since the optic flow on all scales is available by construction, we have access to the deep structure of this scale space and can exploit this information to perform an automatic scale selection. To this end, we adapt the *Optimal Prediction Principle* (OPP) recently developed by Zimmer *et al.* [25]. In short, this principle suggests to rate the quality of an optic flow field between the first and second frame according to its extrapolation quality from the first to the third frame. The underlying assumption is that the velocity of objects (or the camera) remains constant over time. It is shown in [25] that this simple assumption works very well for the automatic estimation of the smoothness weight α.

In our case, for a given flow field $\boldsymbol{w} = (u, v)^{\top}$ between the frames at time z and $z+1$, we assess the extrapolation quality by evaluating the *Average Data Constancy Error* (*ADCE*), which is based on the grey value constancy assumption without linearisation:

$$ADCE_{1,3}(\boldsymbol{w}) = \frac{1}{|\Omega|} \int_{\Omega} \big(f(x + 2u, y + 2v, z + 2) - f(x, y, z) \big)^2 \mathrm{d}x\,\mathrm{d}y\ . \quad (16)$$

It is obvious that if the model assumptions hold, a *good* flow field will lead to *small* values of this error measure. Note that in contrast to the optimisation strategy in [25], we can exploit the following advantageous property of our numerical scheme: It explicitly evolves in the parameter α, hence after each iteration the flow field at cumulated time α is available, and the *ADCE* can be evaluated. This on-the-fly computation of the quality estimate is not possible for most other optic flow methods, because they typically require to solve a new system of equations for each value of α.

Besides the OPP, we also tried other schemes for automatic scale estimation. In particular, we investigated the performance of the decorrelation method by Mrázek and Navarra [16]. However, experiments indicated that the underlying assumptions do not hold for our optic flow scale space.

7   Experiments

In order to investigate the behaviour of our optic flow scale space we perform experiments on several image sequences that are publicly available and for which the ground truth flow field is known. In particular, we use the *Yosemite* sequence with and without clouds [4], the *New Marble*[1] sequence as well as the *Rubberwhale* sequence [3].

---

[1] available from http://i21www.ira.uka.de/image_sequences

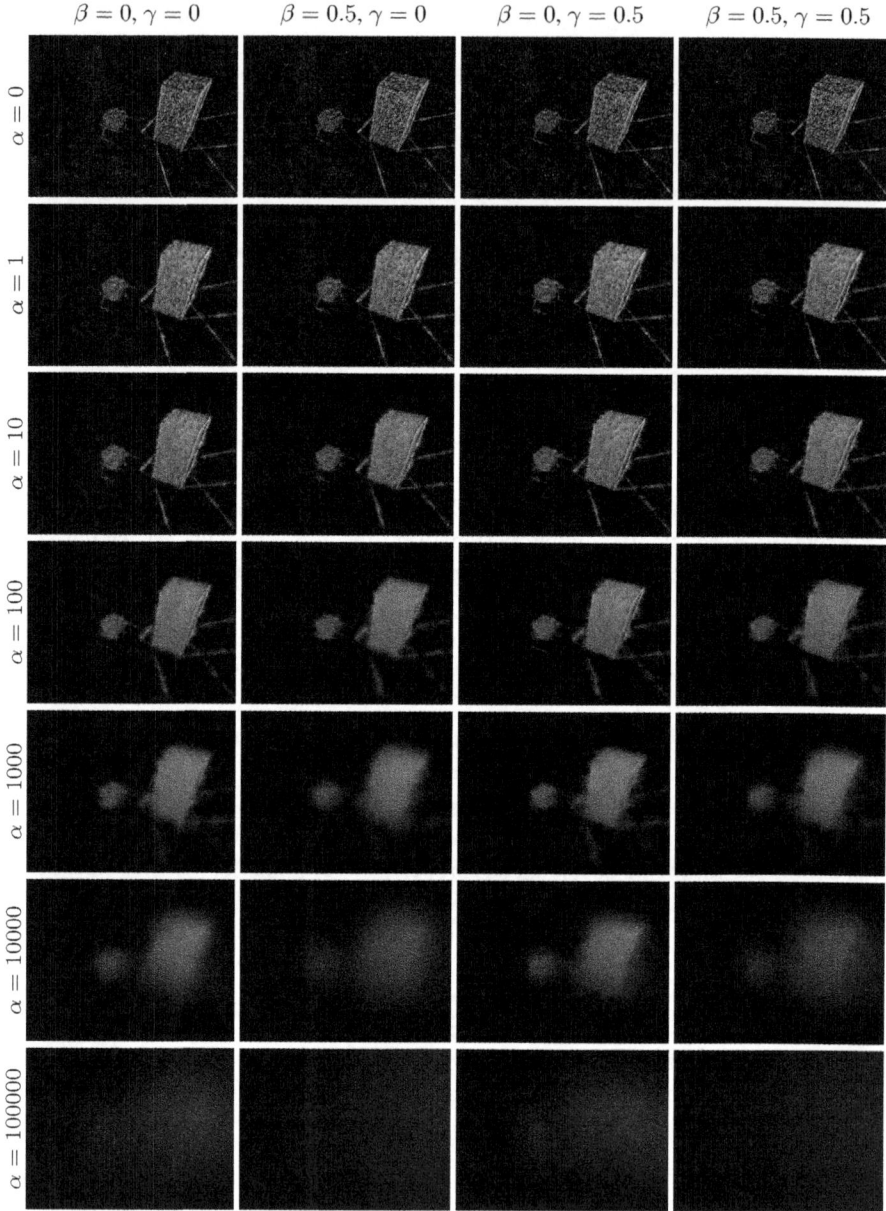

Fig. 1. Scale space at different stopping times for varying β and γ

In our first experiment we compute samples of the scale space for the New Marble sequence at different evolution times and for several choices of β and γ. Figure 1 shows the corresponding flow fields, where colour encodes the direction and brightness indicates the magnitude of the displacements. Here, one can clearly see the scale space behaviour of the proposed diffusion-like optic flow process: Independently of β and γ, the initial state of all these scale spaces ($\alpha = 0$) is given by the noisy normal flow, while for larger values of α the flow fields become successively smoother. In this context, we make two observations: On the one hand, for $\gamma > 0$ discontinuities are preserved for a longer time, since the regulariser is then of image-driven anisotropic nature. On the other hand, results for $\beta > 0$ are slightly less noisy, since the larger eigenvalue of $\boldsymbol{A}^{2-\beta}$ – the one that depends on the magnitude of the image gradient – is now subject to a smaller exponent, cf. Equation (8). This becomes particularly visible in the magnifications shown in Figure 2.

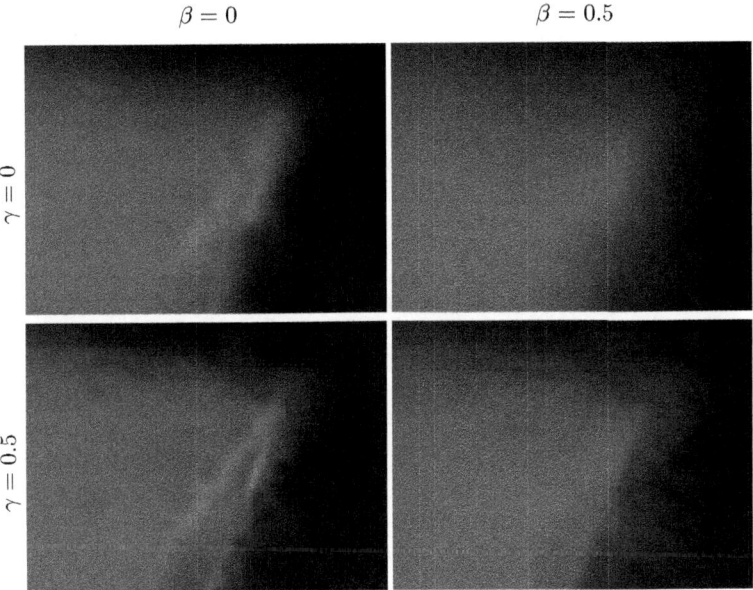

Fig. 2. Zoom into the optic flow fields at scale $\alpha = 1000$ from Figure 1

In a second experiment, we compare the accuracy of the proposed scheme against the two special cases in our framework: Horn and Schunck ($\beta = \gamma = 0$) and Nagel and Enkelmann ($\beta = 0$, $\gamma = 2$). This is done for the aforementioned image sequences by means of the *Average Angular Error* (*AAE*) [4]. Please note that we keep the presmoothing scale fixed at $\sigma = 1$ throughout all experiments, since its impact is not in the focus of our contribution. Table 1 demonstrates that our method consistently leads to improved results. In particular, it shows that the additional degrees of freedom β and γ provided by our general class of scale spaces can be beneficial.

Table 1. Quantitative error measurements in terms of the *AAE* on different image sequences. For our method, β and γ have been optimised. The actual choices are given in brackets.

Image sequence	Horn / Schunck	Nagel / Enkelmann	Our method
New Marble	2.65	2.77	2.53 ($\beta=1.0, \gamma=0.4$)
Rubberwhale	10.58	9.27	9.04 ($\beta=0.5, \gamma=1.0$)
Yosemite	7.51	6.86	6.41 ($\beta=0.4, \gamma=0.9$)
Yosemite no clouds	2.82	3.63	2.76 ($\beta=0.2, \gamma=0.1$)

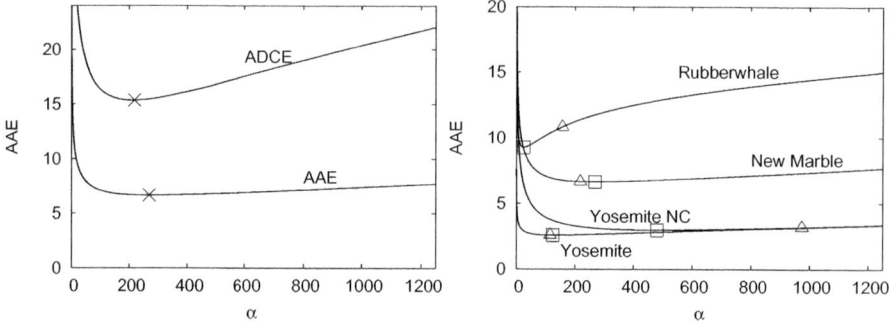

Fig. 3. Automatic scale selection using the Optimal Prediction Principle. **(a) Left:** Graphs of the *ADCE* and *AAE* side-by side. Crosses denote the minimal value of the graph. **(b) Right:** Estimation results of the selected scale α for different image sequences. Triangles indicate the estimated value and squares denote the optimal choice.

In our third experiment, we investigate the automatic selection of the scale parameter α of our model using the OPP. To this end, we first juxtapose the graph of the estimated quality in terms of the *ADCE* with the graph of the measured accuracy given by the *AAE* in Figure 3 (a). This is done for the Yosemite sequence with clouds with $\beta = \gamma = 0.5$. One can see that both graphs have a similar and well aligned shape. In particular, the minima of both curves are attained at almost the same position. Secondly, we compare the estimated values for the regularisation parameter α against those that are optimal with respect to the *AAE*. This is done for all four image sequences with $\beta = \gamma = 0.5$ fixed. As one can see from Figure 3 (b), the OPP works very well in practice: In all cases, the *AAE* at the estimated scale is close to the one of the optimal scale.

In our final experiment we analyse how the two generalisation parameters β and γ influence the accuracy of the estimation. To this end, we have computed the *AAE* for $\beta, \gamma \in [0, 2]$ using the Yosemite sequence with clouds. As in the previous experiments, the selection of the stopping time α within each scale space has been performed automatically using the OPP. The resulting graphs in Figure 4 (a) and (b) show that for both parameters values larger than zero consistently improve the accuracy.

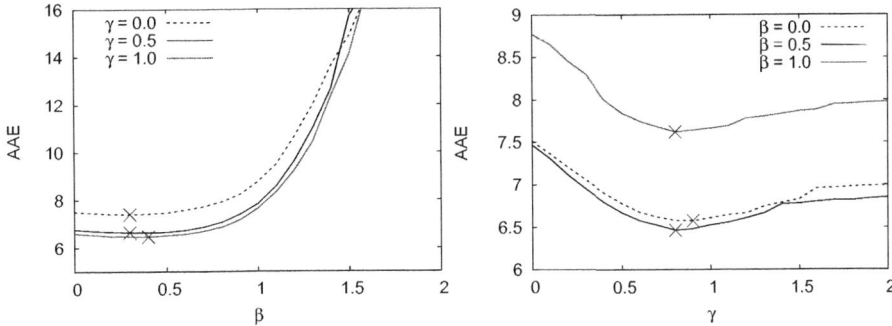

Fig. 4. Influence of the parameters β and γ on the accuracy. **(a) Left:** Behaviour under variations of β for fixed γ. **(b) Right:** Ditto for varying γ and fixed β. In both cases, crosses indicate the minimum of the graphs.

8 Conclusion

It is surprising that in spite of many years of scale space research the scale space character of variational optic flow methods has not been analysed so far. We have shown that such an analysis can be a very worthwhile endeavour: It provides interesting insights in the regularisation behaviour of variational optic flow methods by interpreting them as classical regularisation in a problem-specific norm. Linking this to novel parabolic scale space evolutions that are highly anisotropic and not of diffusion type shows that modern scale space research has only discovered a glimpse of the entire fascinating world of scale space concepts in image processing and computer vision. It would be nice, if our paper serves as a starting point for further research in this direction.

Acknowledgements. Our work has partially been funded by the Cluster of Excellence *Multimodal Computing and Interaction* within the Excellence Initiative of the German Federal Government.

References

1. Abhau, J., Belhachmi, Z., Scherzer, O.: On a decomposition model for optical flow. In: Cremers, D., Boykov, Y., Blake, A., Schmidt, F.R. (eds.) EMMCVPR 2009. LNCS, vol. 5681, pp. 126–139. Springer, Heidelberg (2009)
2. Alvarez, L., Guichard, F., Lions, P.-L., Morel, J.-M.: Axioms and fundamental equations in image processing. Archive for Rational Mechanics and Analysis 123, 199–257 (1993)
3. Baker, S., Scharstein, D., Lewis, J.P., Roth, S., Black, M.J., Szeliski, R.: A database and evaluation methodology for optical flow. Technical Report MSR-TR-2009-179, Microsoft Research, Redmond, WA (December 2009)
4. Barron, J.L., Fleet, D.J., Beauchemin, S.S.: Performance of optical flow techniques. International Journal of Computer Vision 12(1), 43–77 (1994)
5. Ben-Ari, R., Sochen, N.: A geometric framework and a new criterion in optical flow modeling. Journal of Mathematical Imaging and Vision 33, 178–194 (2009)

6. Bertero, M., Poggio, T.A., Torre, V.: Ill-posed problems in early vision. Proceedings of the IEEE 76(8), 869–889 (1988)
7. Fagerström, D.: Spatio-temporal scale-spaces. In: Sgallari, F., Murli, A., Paragios, N. (eds.) SSVM 2007. LNCS, vol. 4485, pp. 326–337. Springer, Heidelberg (2007)
8. Florack, L.: Image Structure. Computational Imaging and Vision, vol. 10. Kluwer, Dordrecht (1997)
9. Grewenig, S., Weickert, J., Bruhn, A.: From box filtering to fast explicit diffusion. In: Goesele, M., Roth, S., Kuijper, A., Schiele, B., Schindler, K. (eds.) DAGM 2010. LNCS, vol. 6376, pp. 533–542. Springer, Heidelberg (2010)
10. Guichard, F.: A morphological, affine, and Galilean invariant scale-space for movies. IEEE Transactions on Image Processing 7(3), 444–456 (1998)
11. Horn, B., Schunck, B.: Determining optical flow. Artificial Intelligence 17, 185–203 (1981)
12. Iijima, T.: Theory of pattern recognition. In: Electronics and Communications in Japan, pp. 123–134 (November 1963) (in English)
13. Iijima, T., Genchi, H., Mori, K.: A theory of character recognition by pattern matching method. In: Proc. First International Joint Conference on Pattern Recognition, Washington, DC, pp. 50–56 (October 1973) (in English)
14. Krajsek, K., Mester, R.: A maximum likelihood estimator for choosing the regularization parameters in global optical flow methods. In: Proc. 2006 IEEE International Conference on Image Processing, Atlanta, GA, pp. 1081–1084 (2006)
15. Lindeberg, T.: Scale-Space Theory in Computer Vision. Kluwer, Boston (1994)
16. Mrázek, P., Navara, M.: Selection of optimal stopping time for nonlinear diffusion filtering. International Journal of Computer Vision 52(2-3), 189–203 (2003)
17. Nagel, H.-H., Enkelmann, W.: An investigation of smoothness constraints for the estimation of displacement vector fields from image sequences. IEEE Transactions on Pattern Analysis and Machine Intelligence 8, 565–593 (1986)
18. Scherzer, O., Weickert, J.: Relations between regularization and diffusion filtering. Journal of Mathematical Imaging and Vision 12(1), 43–63 (2000)
19. Schnörr, C.: On functionals with greyvalue-controlled smoothness terms for determining optical flow. IEEE Transactions on Pattern Analysis and Machine Intelligence 15(10), 1074–1079 (1993)
20. Sochen, N., Kimmel, R., Bruckstein, F.: Diffusions and confusions in signal and image processing. Journal of Mathematical Imaging and Vision 14(3), 195–210 (2001)
21. Sporring, J., Nielsen, M., Florack, L., Johansen, P. (eds.): Gaussian Scale-Space Theory. Computational Imaging and Vision, vol. 8. Kluwer, Dordrecht (1997)
22. Tikhonov, A.N.: Solution of incorrectly formulated problems and the regularization method. Soviet Mathematics Doklady 4, 1035–1038 (1963)
23. Weickert, J.: Anisotropic Diffusion in Image Processing. Teubner, Stuttgart (1998)
24. Whittaker, E.T.: A new method of graduation. Proceedings of the Edinburgh Mathematical Society 41, 65–75 (1923)
25. Zimmer, H., Bruhn, A., Weickert, J.: Optic flow in harmony. International Journal of Computer Vision 93(3), 368–388 (2011)

Group-Valued Regularization Framework for Motion Segmentation of Dynamic Non-rigid Shapes

Guy Rosman[1], Michael M. Bronstein[2], Alexander M. Bronstein[3],
Alon Wolf[4,5], and Ron Kimmel[1,5,*]

[1] Dept. of Computer Science, Technion – Israel Institute of Technology
Haifa, 32000, Israel
{rosman,ron}@cs.technion.ac.il
[2] Institute of Computational Science, Faculty of Informatics
Universit della Svizzera italiana, CH - 6904 Lugano, Switzerland
michael.bronstein@usi.ch
[3] School of Electrical Engineering, Faculty of Engineering
Tel Aviv University, Ramat Aviv 69978, Israel
bron@eng.tau.ac.il
[4] Dept. of Mechanical Engineering, Technion – Israel Institute of Technology
Haifa, 32000, Israel
alonw@tx.technion.ac.il
[5] The Goldstein UAV and Satellite Center, Technion – Israel Institute of Technology
Haifa, 32000, Israel
alonw@tx.technion.ac.il, ron@cs.technion.ac.il

Abstract. Understanding of articulated shape motion plays an important role in many applications in the mechanical engineering, movie industry, graphics, and vision communities. In this paper, we study motion-based segmentation of articulated 3D shapes into rigid parts. We pose the problem as finding a group-valued map between the shapes describing the motion, forcing it to favor piecewise rigid motions. Our computation follows the spirit of the Ambrosio-Tortorelli scheme for Mumford-Shah segmentation, with a diffusion component suited for the group nature of the motion model. Experimental results demonstrate the effectiveness of the proposed method in non-rigid motion segmentation.

Keywords: Motion Segmentation, Lie-groups, Surface Diffusion, Ambrosio-Tortorelli.

1 Introduction

The analysis of articulated motion in three-dimensional space is a key problem in biomechanics [1], mechanical engineering, computer vision [28,31,20,37], and computer graphics [24,29,41,27,44,6,43]. Specific problems of deformation analysis [4] and motion segmentation [5,13] try to infer the articulated motion of an object, given several

* This research was supported in part by The Israel Science Foundation (ISF) grant number 623/08, and by the Goldstein UAV and Satellite Center.

A.M. Bruckstein et al. (Eds.): SSVM 2011, LNCS 6667, pp. 725–736, 2012.
© Springer-Verlag Berlin Heidelberg 2012

instances of the analyzed object in different poses. The desired outcome is the segmentation of the object into rigid parts and motion estimation between the corresponding parts.

Most motion analysis techniques either assume a known prior on the articulated structure of the inspected object (e.g., in the form of a skeleton), or decide on the structure in an *ad hoc* manner, not based on the kinematic model commonly assumed for near-rigid objects [1,4]. Since in many cases such *a priori* assumptions about the data are only approximate, they can lead to errors in the segmentation and motion estimation.

Another common assumption, especially in graphics applications, is that of known correspondence. In computer graphics, the problem is usually referred to as *dynamic mesh segmentation*.

The above assumptions are often too limiting in real-world applications. Instead, we would like to apply the intuition often used when studying real-life near-rigid objects, about the existence of an average rotational motion existing for each body part, but do so without attempting to detect the articulated parts in advance, and without assuming the existence of a clear partition of the surface. In other words, we would like to obtain a "soft" segmentation of the surface, without knowing the number or location of regions in advance, without analyzing the surface features, or having additional priors on the various object parts. In addition, we expect a complete formulation of motion segmentation to incorporate an implicit handling of the correspondence problem, given a reasonable initialization.

Main contribution. In this paper we try to remedy the shortcoming of existing approaches to articulated motion estimation by combining the two tasks of motion estimation and segmentation into a single functional. Unlike existing methods, we propose a principled variational approach, attempting to find a rigid transformation at each surface point, between the instance surfaces, such that the overall transformation is described by a relatively sparse set of such transformations, each matching a rigid part of the object. The functional we propose regularizes the motion between the surfaces, and is guided by the fact that the parameters of the motion transformations (i) should describe the motion at each point with sufficient accuracy; (ii) should vary smoothly within (unknown) rigid parts; (iii) can vary abruptly between rigid parts.

We see the main contribution of this paper in the following three aspects: First, we propose an axiomatic variational framework for articulated motion segmentation. While focusing on the segmentation problem in this paper, our framework is more general and the proposed functionals can be easily incorporated into other applications such as motion estimation, tracking, and surface denoising. Second, we demonstrate that the articulated motion segmentation problem can be solved within the proposed framework by adapting standard tools from variational segmentation to the geometry of the case, and obtain results competitive with domain-specific state-of-the-art tools. Third, we suggest a spatially-coherent algorithm for spatial visualization of group valued data on manifolds, which draws from the same variational principles.

Relation to prior work. The scheme we propose involves diffusing the transformations along the surface, in the spirit of the Ambrosio-Tortorelli scheme [2] for Mumford-Shah segmentation [33]. The diffusion component of our scheme is a diffusion process of Lie group elements, which has recently attracted significant attention in other applications [16,39,18]. In diffusing transformations on the surface, our work is similar to that of Litke et al. [30]. We do not, however, make any assumption on the surface topology; to that end, the proposed method diffuses transformations along the surface, rather than representing the surface in an evenly sampled 2D parametrization plane. When dealing with real-life deformable objects that seldom admit regular global parametrization, such an assumption could be too restrictive.

The idea of combining soft segmentation and motion estimation has been attempted before in the case of optical flow computation (see, e.g., [3]). In optical flows, however, the motion field is merely expected to be piecewise smooth. For truly articulated objects one would expect piecewise-constant flow fields, when expressed in the correct parametrization.

Finally, our work is related, and complementary, to the topic of geometry-based mesh segmentation. While several works from this field can be combined with motion based segmentation techniques, this is not the focus of this work. We point the reader to [7,38,12,26], and references therein, for additional examples of mesh segmentation algorithms.

2   Problem Formulation

Articulation model.   Let us be given a three-dimensional shape, which we model as a two-dimensional manifold X. In the following, we will denote by $\mathbf{x} : X \to \mathbb{R}^3$ the embedding of X into $\mathbb{R}^3$, and use synonymously the notation x and $\mathbf{x}$ referring to a point on the manifold and its Euclidean embedding coordinates, respectively.

We further assume that the shape X is *articulated*, i.e., can be decomposed into *rigid parts* $S_1, \ldots, S_p$ and nonrigid *joints* $J_1, \ldots, J_q$, such that $X = \bigcup_{i=1}^{p} S_i \cup \bigcup_{k=1}^{q} J_k$. An *articulation* $Y = \mathbf{A}X$ is obtained by applying rigid motions $\mathbf{T}_i \in \mathrm{Iso}(\mathbb{R}^3)$ (rotations and translations) to the rigid parts, and non-rigid deformations $\mathbf{Q}_k$ to the joints, such that $\mathbf{A}X = \bigcup_{i=1}^{p} \mathbf{T}_i S_i \cup \bigcup_{k=1}^{q} \mathbf{Q}_k J_k$.

Motion segmentation.   The problem of *motion-based segmentation*, in its simplest setting can be described as follows: given two articulations of the shape, X and Y, extract its rigid parts. Extension to the case of multiple shape poses is straightforward. We therefore consider in the following only a pair of shapes for the sake of simplicity and without loss of generality.

Assuming that the correspondence between the points on two shapes X and Y is known, given two corresponding points $x \in X$ and $y(x) \in Y$, we can find a motion $g \in \mathcal{G}$ such that $g\mathbf{x} = \mathbf{y}$, where $\mathcal{G}$ is some representation of coordinate transformations in $\mathbb{R}^3$, and with some abuse of notation, $g\mathbf{x} \in \mathbb{R}^3$ denotes the action of g on the coordinates of the point x. We can represent the transformation at each point as a field $g : X \to \mathcal{G}$.

Since the articulated parts of the shape move rigidly, if we choose an appropriate motion representation (as detailed below), two points $x, x' \in S_i$ will undergo the same transformation, from which it follows that $g(x)|_{x \in S_i} = \mathrm{const}$. One possibility is to adopt a constrained minimization approach, forcing $g(X) = Y$, where $g(X)$ is a short notation for the set $g(x)\mathbf{x}(x)$ for all $x \in X$. A more convenient possibility is to take an unconstrained formulation,

$$\min_{g:X \to \mathcal{G}} \lambda E_{\text{DATA}}(g) + \rho(g), \tag{1}$$

where ρ denotes some regularization term which is small if g is piecewise constant. $E_{\text{DATA}}(g)$ is our fitting term which penalizes the discrepancy between the transformed template surface $g(X)$ and Y,

$$E_{\text{DATA}}(g) = \int_X \|g(x)\mathbf{x} - \mathbf{y}(x)\|^2 da, \tag{2}$$

where $\mathbf{y}(x)$ denotes the coordinate of the point on Y corresponding to x, $g(x)$ is the transformation at x, and da is a measure on X.

Such a formulation allows us to recover the articulated parts by clustering g into regions of equal value. This formulation (presented in Section 4.4) bears much resemblance to *total variation* regularization common in signal and image processing [35].

One tacit assumption in this problem is that the correspondence between X and Y is known, which is usually not true. We will mention this issue in Section 4.1. Second, it is crucial to observe that the effectiveness of (1) relies on some correct representation $\mathcal{G}$ of the motion. The simplest representation is the *linear motion*, assuming $\mathcal{G} = \mathbb{R}^3$ such that $g\mathbf{x} = \mathbf{x} + \mathbf{t} = \mathbf{y}$ for some $\mathbf{t} \in \mathbb{R}^3$. However, such a simplistic model fails to capture the piecewise constancy of the motion field. It is thus clear that we need a representation of motion that is *redundant* (i.e., an *over-parametrization* using more than three degrees of freedom to describe a transformation) and in which motions of points that move rigidly are described by the same element of $\mathcal{G}$.

One parametrization often used in computer vision and robotics [42,32,27,18] is the representation of rigid motions by the Lie group $SE(3)$ and the corresponding Lie algebra $se(3)$, respectively. Lie groups are topological groups with a smooth manifold structure such that the group action $\mathcal{G} \times \mathcal{G} \mapsto \mathcal{G}$ and the group inverse are differentiable maps. Each Lie group has a Lie algebra associated with it, which can be mapped via the *exponential map* onto the tangent space at the identity operator. The Lie algebra $se(3)$ allows us to represent rotations and rigid motions locally in a consistent linear space using the logarithm and exponential maps at each point. We refer the reader to standard literature on the subject (e.g., [19]) for more information.

The Lie algebra of $SE(3)$ is the group of 4×4 matrices of the form

$$se(3) = \begin{pmatrix} \mathbf{A} & \mathbf{t} \\ \mathbf{0} & 1 \end{pmatrix}, \mathbf{A} \in so(3), \mathbf{t} \in \mathbb{R}^3, \tag{3}$$

where $so(3)$ is the set of 3×3 skew-symmetric matrices.

Under the assumption of $\mathcal{G} = SE(3)$, we have our desired property that the transformation of points undergoing a rigid motion are described by the same group element. Solving problem (1) thus requires a regularization term ρ that favors piecewise constancy of group elements on the shape. We discuss such a regularization in Section 3. We also note that due to the non-Euclidean structure of the group, special care should be taken when parameterizing such a representation [32,18,39,27], as discussed in Section 4.2.

3   Diffusion-Based Regularization

Thinking of the Lie group $\mathcal{G}$ as a Riemannian manifold, we look for a functional defined on maps between manifolds of the form $g : X \rightarrow \mathcal{G}$. Such maps can be regularized by the well-known *Dirichlet energy* [25],

$$\rho_{\text{DIR}}(g) = \frac{1}{2} \int_X \langle \nabla g, \nabla g \rangle_{g(x)} da, \tag{4}$$

where ∇g denotes the *intrinsic gradient* of g on X, $\langle \cdot, \cdot \rangle_{g(x)}$ is the Riemannian metric on $\mathcal{G}$ at a point $g(x)$, and da is the area element of X. The minimizer of the Dirichlet energy is called a *harmonic map*, and it is the solution of a diffusion equation. In the case where both X and $\mathcal{G}$ are Euclidean, ρ_{DIR} reduces to the standard Tikhonov regularization.

Ambrosio-Tortorelli scheme.   Unfortunately, the Dirichlet energy does not favor piecewise-constancy of g, as is desired. We therefore adopt the Ambrosio-Tortorelli scheme [2] for Mumford-Shah regularization [33], in which the Dirichlet energy term is modulated by a diffusivity function $v : X \rightarrow [0, 1]$,

$$\rho_{\text{AT}}(g) = \int_X \left(\frac{1}{2} v^2 \langle \nabla g, \nabla g \rangle_g + \epsilon \langle \nabla v, \nabla v \rangle + \frac{(1-v)^2}{4\epsilon} \right) da, \tag{5}$$

where ϵ is a small positive constant. This allows us to extend our outlook in several ways. The Mumford-Shah functional replaces the notion of a set of regions with closed simple boundary curves with general discontinuity sets. It furthermore generalizes our notion of constant value regions with that of favored smoothness inside the areas defined by these discontinuity curves. This is in order to handle objects which deviate from articulated motion, for example in flexible regions or joints.

Furthermore, the generalized Ambrosio-Tortorelli scheme allows us to explicitly reason about the places in the flow where the nonlinear nature of the data manifold manifests itself. Suppose we have a solution (g^*, v^*) satisfying our piecewise-constancy assumptions of g, and a diffusivity function with 0 at region boundaries and 1 elsewhere. At such a solution, we expect two neighboring points which belong to different regions to have a very small diffusivity value v connecting them, effectively nullifying the interaction between far-away group elements which is dependent on the mapping used for the logarithm map at each point, and hence can be inaccurate [22,32]. While such a solution (g^*, v^*) may not be a minimizer of the functional, it serves well to explain the intuition motivating the choice of the functional.

Diffusion of Lie group elements. In order to efficiently compute the Euler-Lagrange equation corresponding to the generalized Ambrosio-Tortorelli functional (5), we transform the neighborhood of each point into the corresponding Lie algebra elements before applying the diffusion operator. Using Lie algebra representation of differential operators for rigid motion has been used before in computer vision [39], numerical PDE computations [22], path planning and optimal control theory [32,27].

The Euler-Lagrange equation for the generalized Dirichlet energy measuring the map between two manifolds is given as [25]

$$\Delta_X g^\alpha + \Gamma^\alpha_{\beta\gamma} \left\langle \nabla g^\beta, \nabla g^\gamma \right\rangle_{g(x)} = 0, \tag{6}$$

where α, β, γ enumerate the local coordinates of our group manifold, $se(3)$, and we use Einstein's notation according to which corresponding indices are summed over. $\Gamma^\alpha_{\beta\gamma}$ are the *Christoffel symbols* of $SE(3)$, which express the Riemannian metric's local derivatives. We refer the reader to [15] for an introduction to Riemannian geometry. Finally, Δ_X denotes the Laplace-Beltrami operator on the surface X.

In order to avoid the computation of the Christoffel symbols, we transform the point and its neighbors using the logarithm map at that point in $SE(3)$. The diffusion operation is now affected only by the structure of the surface X. After applying the diffusion operator, we use the exponential map in order to return to the usual representation of the transformation.

4   Numerical Considerations

We now describe the algorithm for articulated motion estimation based on the minimization of the functional

$$E(g, v) = \lambda E_{\text{DATA}}(g) + \rho_{AT}(g, v), \tag{7}$$

where $E_{\text{DATA}}(g)$ is the matching term defined by Equation 2, and $\rho_{AT}(g, v)$ is defined in Equation 5. The main steps of the algorithm are outlined as Algorithm 1. Throughout the algorithm we parameterize $g(x)$ based on the first surface, given as a triangulated mesh, with vertices $\{x_i\}_{i=1}^N$, and an element from $SE(3)$ defined at each vertex. The triangulation is used merely to obtain a more consistent numerical diffusion operator, and is not required otherwise. Special care is made in the choice of coordinates during the optimization as explained in Section 4.2.

4.1   Initial Correspondence Estimation

As in other motion segmentation algorithms, some initialization of the matching between the surfaces must be used. One approach [6] is to use nonrigid surface matching for initialization. Another possibility, in the case of high framerate sequences [44], is to exploit temporal consistency. While we focus on the functional itself, we present a possible initialization scheme which assumes a known sparse correspondence between the surfaces (simulating motion capture markers). We then interpolate this sparse set in order to initialize an *iterative closest point* (ICP) search [8], matching the patch around each point to the target mesh. In Figure 3, we use 30 matched points for initialization. This number of points is within the scope of current motion capture marker systems, or of algorithms for global nonrigid surface matching such as spectral methods [23,34,36], or the *generalized multidimensional scaling* (GMDS) algorithm [9].

We expect better initial registration, possibly using a smoothness assumption, to allow fewer markers to be used.

4.2   Diffusion of Lie Group Elements

Rewriting the optimization over the functional in Equation 7 in a fractional step approach [45], we update each function in a suitable representation.

Using the transformation described in Section 3, the update step with respect to the regularization now becomes [18]

$$g^{k+1/2} = \exp\left(-dt\frac{\delta\rho_{AT}}{\delta\tilde{g}}\right)g^k, \quad v^{k+1} = v^k - dt\frac{\delta\rho_{AT}}{\delta v} \tag{8}$$

where $\exp(A) = I + A + A^2/2! + A^3/3! + \dots$ denotes the matrix exponential, $\tilde{g}$ denotes the logarithm transform of g, and dt denotes the time step. $\frac{\delta\rho_{AT}}{\delta\tilde{g}}$ denotes the variation of the regularization term $\rho_{AT}(g)$ w.r.t. the Lie-algebra local representation of the solution, describing the Euler-Lagrange descent direction. $g(x)$ and the neighboring transformations are parameterized by a basis for matrices in $se(3)$, after applying the logarithm map at $g(x)$. The descent directions are given by

$$\frac{\delta\rho_{AT}}{\delta\tilde{g}_i} = v^2\Delta_X(\tilde{g}_i) + v\langle\nabla v, \nabla\tilde{g}_i\rangle \tag{9}$$

$$\frac{\delta\rho_{AT}}{\delta v} = \langle\nabla g, \nabla g\rangle_{g(x)}v + 2\epsilon\Delta_X(v) + \frac{(v-1)}{2\epsilon},$$

where $\tilde{g}_i$ denote the components of the logarithmic representation of g. The discretization we use for Δ_X is a cotangent one suggested by [14], which has been shown to be convergent for relatively smooth and well-parameterized surfaces. It is expressed as

$$\Delta_X(u) \approx \frac{3}{\mathcal{A}_i}\sum_{j\in\mathcal{N}_1(i)}\frac{\cot\alpha_{ij} + \cot\beta_{ij}}{2}[u_j - u_i], \tag{10}$$

for a given function u on the surface X, where $\mathcal{N}_1(i)$ denotes the mesh neighbors of point i, and α_{ij}, β_{ij} are the angles opposing the edge ij in its neighboring faces. $\mathcal{A}_i$ denotes the area of the 1-ring around i in the mesh. After a gradient descent step w.r.t. the diffusion term, we take a step w.r.t. the data term.

$$g^{k+1} = P_{SE(3)}\left(g^{k+1/2} - dt\frac{\delta E_{DATA}}{\delta g}\right), \tag{11}$$

where $P_{SE(3)}(\cdot)$ denotes a projection onto the group $SE(3)$ obtained by correcting the singular values of the rotation matrix. We compute the gradient w.r.t. a basis for small rotation and translation matrices comprised of the regular basis for translation and the skew-matrix approximation of small rotations. We then reproject the update onto the manifold. This keeps the inaccuracies associated with the projecting manifold-constrained data [32,18] at a reasonable level.

Finally, we note that we may not know in advance the points $y(x)$ which match X in Y. The correspondence can be updated based on the current transformations in an efficient manner similarly to the ICP algorithm [8], using a KD-tree.

4.3   A Patch-Based Data Term

The data term we use works to fit the 3 output functions of the transformations defined on the surface. As in the case of the aperture problem in optical flow computation, it is the regularization term that helps us obtain a complete view of the transformations field. However, as in optical flow computation [11], extending the surface matching to a small patch around each point gave us a more robust estimation of the transformations. The revised data term reads

$$E_{\mathrm{DATA}}(g) = \int_X \int_{\mathcal{N}(x)} \|g(x)\mathbf{x} - \mathbf{y}(x)\|^2 da \times da, \tag{12}$$

where $\mathcal{N}(x)$ denotes a small neighborhood around the point x.

Algorithm 1. Articulated Surface Segmentation and Matching

1: Given an initial correspondence.
2: **for** $k = 1, 2, \ldots,$ until convergence **do**
3:     Update $g^{k+1/2}, v^{k+1}$ w.r.t. the diffusion term, according to Equation 8.
4:     Obtain g^{k+1} according to the data term, using Equation 11.
5:     Update $y^{k+1}(x)$, the current estimated correspondence of the deformed surface.
6: **end for**

4.4   Visualizing Lie Group Clustering on Surfaces

Finally, we need to mention the approach taken to visualize the transformations as the latter belong to a six-dimensional non-Euclidean manifold. Motivated by the widespread use of vector quantization in such visualizations, we use a clustering algorithm with spatial regularization. Instead of minimizing the Max-Lloyd cost function, we minimize the function

$$E_{VIS}(g_i, R_i) = \sum_i \int_{R_i} \|g - g_i\|^2 da + \int_{\partial R_i} v^2(s) ds, \tag{13}$$

where ∂R_i denotes the set of boundaries between partition regions $\{R_i\}_{i=1}^N$, g_i are the group representatives for each region, and $v^2(s)$ denotes the diffusivity term along the region boundary. Several (about 50) initializations are performed, as is often customary in clustering, with the lowest cost hypothesis kept.

While this visualization algorithm coupled with a good initialization at each point can be considered as a segmentation algorithm in its own right, it is less general as it assumes a strict separation between the parts. We further note, however, that the diffusion process lowered the score obtained in Equation 13 in the experiments we conducted, indicating a consistency between the two algorithms in objects with well-defined rigid parts.

5   Results

We now demonstrate the results obtained by our method, in terms of the obtained transformation field and the diffusivity function. In Figure 1 we demonstrate matching between two human body poses taken from the TOSCA dataset [10]. As can be seen, the diffusivity function hints at the location of boundaries between parts. In addition, we visualize the transformations obtained using the clustering algorithm described in subsection 4.4.

Figure 1 also demonstrates the results of comparing four poses of the same surface, this time with the patch-based data term described by (12). In our experiments the patch-based term gave a cleaner estimation of the motion as is observed in the diffusivity function.

We also demonstrate our algorithm on a horse taken from [40] as a set of 6 poses in Figure 2. In this figure we compare our results to those of [43], obtained on a similar set

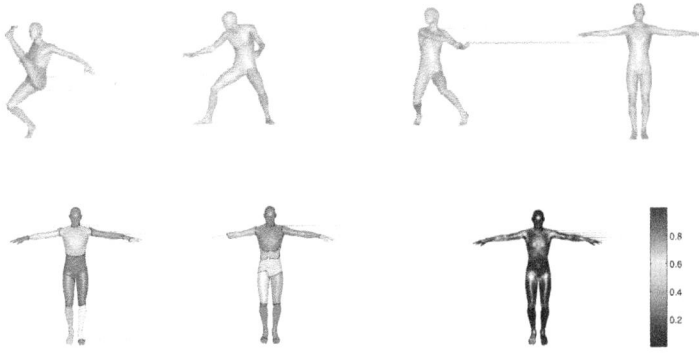

Fig. 1. Segmenting a human figure. Top row: the set of poses used. Bottom row, left-to-right: the transformations obtained from the two left most poses, the transformations obtained from all four poses using Equation 12 as a data term, and the Ambrosio-Tortorelli diffusivity function based on four poses.

Fig. 2. Segmenting a horse dynamic surface motion based on six different poses. Top row: the poses used. Bottom row, left to right: a visualization of the transformations of the surface obtained by our method, and the segmentation results obtained by [43], and the diffusivity function v.

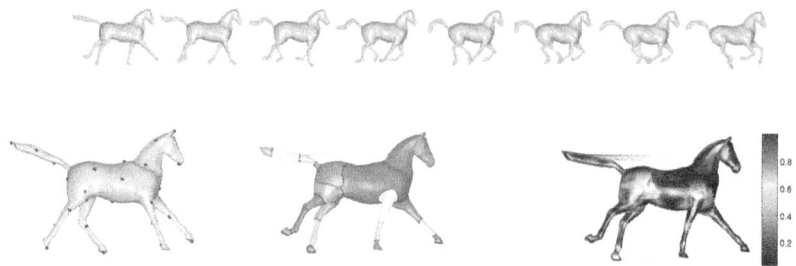

Fig. 3. Segmenting a horse dynamic surface motion with a given sparse initial correspondences. Top row: the eight random poses used. Bottom row, left to right: the set of points used for initializing the convergence, and a visualization of the transformations obtained, and the diffusivity function v.

of poses with 10 frames. In Figure 3 we demonstrate our algorithm while initializing it from a set of 30 points where displacement is known. The relatively monotonous motion range available in the dynamic mesh sequence leads to a less complete, but still quite meaningful, segmentation of the horse, using an initialization which can be obtained by a feasible setup. We also note the relatively low number of poses required for segmentation – in both Figure 2 and Figure 3 we obtain good results despite the fact we use only a few poses, six and eight respectively. This contrasts with 10 poses used in [43], the results of which are shown in Figure 2 for comparison.

Finally, in Figure 3 we demonstrate convergence based on a sparse initialization, with 30 known correspondence points, arbitrarily placed using farthest point sampling [17,21]. This demonstrates a possibility of initializing the algorithm using motion capture markers, coupled with a 3D reconstruction pipeline, for object part analysis. While the examples shown in this paper are synthetic, this experiment demonstrates the algorithm can be incorporated into a real-world system.

6   Conclusion

In this paper we have presented a method for simultaneous segmentation and motion estimation in articulated objects, based on a variational formulation. Several results shown demonstrate the method's effectiveness, and merit its application to specific problems where it can be contrasted and combined with domain-specific algorithms for articulated object analysis. In future work we intend to adapt the proposed algorithm to real data from range scanners, and explore initialization methods as well as use the proposed framework in other applications such as articulated surfaces tracking and denoising.

References

1. Alexander, E.J., Bregler, C., Andriacchi, T.P.: Non-rigid modeling of body segments for improved skeletal motion estimation. Computational Modeling Engineering Science 4, 351–364 (2003)
2. Ambrosio, L., Tortorelli, V.M.: Approximation of functional depending on jumps by elliptic functional via Γ-convergence. Comm. on Pure and Appl. Math. 43(8), 999–1036 (1990)

3. Amiaz, T., Kiryati, N.: Piecewise-smooth dense optical flow via level sets. Int. J. of Comp. Vision 68(2), 111–124 (2006)
4. Andersen, M.S., Benoit, D.L., Damsgaarda, M., Ramsey, D.K., Rasmussen, J.: Do kinematic models reduce the effects of soft tissue artefacts in skin marker-based motion analysis an in vivo study of knee kinematics. Journal of Biomechanics 43, 268–273 (2010)
5. Anguelov, D., Koller, D., Pang, H.-C., Srinivasan, P., Thrun, S.: Recovering articulated object models from 3d range data. In: Proc. Conf. on Uncertainty in Artificial Intelligence, pp. 18–26. AUAI Press, Arlington (2004)
6. Arcila, R., Buddha, S.K., Hétroy, F., Denis, F., Dupont, F.: A framework for motion-based mesh sequence segmentation. In: Int. Conf. on Comp. Graphics, Visual. and Comp. Vision, Plzeň, Czech Republic (2010)
7. Attene, M., Katz, S., Mortara, M., Patane, G., Spagnuolo, M., Tal, A.: Mesh segmentation - a comparative study. In: Proc. IEEE Int. Conf. on Shape Modeling and Applications, pp. 7–18. IEEE Computer Society, Washington, DC (2006)
8. Besl, P.J., McKay, N.D.: A method for registration of 3D shapes. Trans. PAMI 14(2), 239–256 (1992)
9. Bronstein, A.M., Bronstein, M.M., Kimmel, R.: Generalized multidimensional scaling: a framework for isometry-invariant partial surface matching. Proc. National Academy of Science (PNAS) 103(5), 1168–1172 (2006)
10. Bronstein, A.M., Bronstein, M.M., Kimmel, R.: Numerical geometry of non-rigid shapes. Springer-Verlag New York Inc. (2008)
11. Bruhn, A., Weickert, J., Schnörr, C.: Lucas/Kanade meets Horn/Schunck: combining local and global optic flow methods. Int. J. of Comp. Vision 61(3), 211–231 (2005)
12. Chen, X., Golovinskiy, A., Funkhouser, T.: A benchmark for 3D mesh segmentation. ACM Trans. Graphics 28(3) (August 2009)
13. Cremers, D., Soatto, S.: Motion competition: A variational framework for piecewise parametric motion segmentation. Int. J. of Comp. Vision 62(3), 249–265 (2005)
14. Desbrun, M., Meyer, M., Schroder, P., Barr, A.H.: Implicit fairing of irregular meshes using diffusion and curvature flow. In: Proc. SIGGRAPH, pp. 317–324 (1999)
15. DoCarmo, M.P.: Riemannian Geometry. Birkhäuser, Boston (1992)
16. Duits, R., Burgeth, B.: Scale spaces on lie groups. In: Sgallari, F., Murli, A., Paragios, N. (eds.) SSVM 2007. LNCS, vol. 4485, pp. 300–312. Springer, Heidelberg (2007)
17. Gonzalez, T.F.: Clustering to minimize the maximum intercluster distance. Theor. Comput. Sci. 38, 293–306 (1985)
18. Gur, Y., Sochen, N.A.: Regularizing flows over lie groups. J. of Math. in Imag. and Vis. 33(2), 195–208 (2009)
19. Hall, B.C.: Lie Groups, Lie Algebras,and Representations, An Elementary Introduction. Springer, Heidelberg (2004)
20. Hauberg, S., Sommer, S., Pedersen, K.S.: Gaussian-like spatial priors for articulated tracking. In: Proc. CVPR, pp. 425–437. Springer, Heidelberg (2010)
21. Hochbaum, D., Shmoys, D.: A best possible approximation for the k-center problem. Mathematics of Operations Research 10(2), 180–184 (1985)
22. Iserles, A., Munthe-kaas, H.Z., Nrsett, S.P., Zanna, A.: Lie group methods. Acta Numerica, 215–365 (2000)
23. Jain, V., Zhang, H.: Robust 3D shape correspondence in the spectral domain. In: Proc. of Shape Modeling International, pp. 118–129 (2006)
24. James, D.L., Twigg, C.D.: Skinning mesh animations. In: Proc. SIGGRAPH, vol. 24(3), pp. 399–407 (2005)
25. Eells Jr., J., Sampson, J.H.: Harmonic mappings of Riemannian manifolds. American J. of Math. 86(1), 106–160 (1964)

26. Kalogerakis, E., Hertzmann, A., Singh, K.: Learning 3D Mesh Segmentation and Labeling. ACM Trans. Graphics 29(3) (2010)
27. Kobilarov, M., Crane, K., Desbrun, M.: Lie group integrators for animation and control of vehicles. ACM Trans. Graphics 28(2), 1–14 (2009)
28. Kompatsiaris, I., Tzovaras, D., Strintzis, M.G.: Object articulation based on local 3D motion estimation. In: Leopold, H., García, N. (eds.) ECMAST 1999. LNCS, vol. 1629, pp. 378–391. Springer, Heidelberg (1999)
29. Lee, T.-Y., Wang, Y.-S., Chen, T.-G.: Segmenting a deforming mesh into near-rigid components. Vis. Comput. 22(9), 729–739 (2006)
30. Litke, N., Droske, M., Rumpf, M., Schröder, P.: An image processing approach to surface matching. In: Proc. SGP, pp. 207–216. Eurographics Association, Aire-la-Ville (2005)
31. Mateus, D., Horaud, R., Knossow, D., Cuzzolin, F., Boyer, E.: Articulated shape matching using Laplacian eigenfunctions and unsupervised point registration. In: Proc. CVPR, pp. 1–8 (2008)
32. Müller, A., Terze, Z.: Differential-geometric modelling and dynamic simulation of multibody systems. Strojarstvo: Journal for Theory and Application in Mechanical Engineering 51(6), 597–612 (2009)
33. Mumford, D., Shah, J.: Optimal approximations by piecewise smooth functions and associated variational problems. Communications on Pure and Applied Mathematics 42(5), 577–685 (1989)
34. Raviv, D., Dubrovina, A., Kimmel, R.: Hierarchical shape matching. In: Bruckstein, A.M., et al. (eds.) SSVM 2011. LNCS, vol. 6667, pp. 604–615. Springer, Heidelberg (accepted, 2011)
35. Rudin, L.I., Osher, S., Fatemi, E.: Nonlinear total variation based noise removal algorithms. Physica D Letters 60, 259–268 (1992)
36. Ruggeri, M.R., Patanè, G., Spagnuolo, M., Saupe, D.: Spectral-driven isometry-invariant matching of 3D shapes. Int. J. of Comp. Vision 89(2-3), 248–265 (2010)
37. Sapp, B., Toshev, A., Taskar, B.: Cascaded models for articulated pose estimation. In: Proc. CVPR, pp. 406–420. Springer, Heidelberg (2010)
38. Shamir, A.: A survey on mesh segmentation techniques. Computer Graphics Forum 27(6), 1539–1556 (2008)
39. Subbarao, R., Meer, P.: Nonlinear mean shift over riemannian manifolds. Int. J. of Comp. Vision 84(1), 1–20 (2009)
40. Sumner, R.W., Popović, J.: Deformation transfer for triangle meshes. In: Proc. SIGGRAPH, pp. 399–405. ACM, New York (2004)
41. Tierny, J., Vandeborre, J.-P., Daoudi, M.: Fast and precise kinematic skeleton extraction of 3D dynamic meshes. In: Proc. Int. Conf. Pattern Recognition, pp. 1–4 (2008)
42. Tuzel, O., Porikli, F., Meer, P.: Learning on lie groups for invariant detection and tracking. In: Proc. CVPR, pp. 1–8 (2008)
43. Wuhrer, S., Brunton, A.: Segmenting animated objects into near-rigid components. The Visual Computer 26, 147–155 (2010)
44. Yamasaki, T., Aizawa, K.: Motion segmentation for time-varying mesh sequences based on spherical registration. EURASIP Journal on Applied Signal Processing (2009)
45. Yanenko, N.N.: The method of fractional steps: solution of problems of mathematical physics in several variables. Springer, Heidelberg (1971) (translated from Russian)

Wavelet-Based Fluid Motion Estimation

Pierre Dérian, Patrick Héas, Cédric Herzet, and Étienne Mémin

INRIA Rennes-Bretagne Atlantique,
Campus universitaire de Beaulieu, 35042 Rennes Cedex, France
{Pierre.Derian,Patrick.Heas,Cedric.Herzet,Etienne.Memin}@inria.fr
http://irisa.fr/fluminance

Abstract. Based on a wavelet expansion of the velocity field, we present a novel optical flow algorithm dedicated to the estimation of continuous motion fields such as fluid flows. This scale-space representation, associated to a simple gradient-based optimization algorithm, naturally sets up a well-defined multi-resolution analysis framework for the optical flow estimation problem, thus avoiding the common drawbacks of standard multi-resolution schemes. Moreover, wavelet properties enable the design of simple yet efficient high-order regularizers or polynomial approximations associated to a low computational complexity. Accuracy of proposed methods is assessed on challenging sequences of turbulent fluids flows.

Keywords: Optical flow, continuous fluid motion, wavelet multi-resolution analysis, high-order regularization, polynomial approximation.

1 Introduction

Recent years have seen significant progress in signal processing techniques for fluid motion estimation. The wider availability of image-like data, whether coming from experimental facilities (e.g. particle image velocimetry) or from larger-scale geophysical study systems such as lidars or meteorological and oceanographical satellites, strongly motivates the development of computer-vision methods dedicated to their analysis. Correlation-based and variational methods have proven to be efficient in this context. However, the specific nature of fluid motion highly complicates the process. Indeed, one has to deal with continuous fields showing complex structures evolving at high velocities. This is particularly problematic with optical flow methods, where the problem non-linearity requires to resort to an ad-hoc multi-resolution strategy. Although leading to good empirical results, this technique is known to have a number of drawbacks. Moreover, the underdetermined nature of the optical flow estimation problem imposes to add some prior information about the sought motion field. In many contributions dealing with rigid-motion estimation, first-order regularization is considered with success. However, when tackling more challenging problems such as motion estimation of turbulent fluids, this simple prior turns out to be inadequate. Second-order regularizers allowing to enforce physically-sound properties of the flow are considered [2,4,10,11], but their implementation raises up several issues.

A.M. Bruckstein et al. (Eds.): SSVM 2011, LNCS 6667, pp. 737–748, 2012.

In this paper, we propose an optical-flow estimation procedure based on a wavelet expansion of the velocity field. This approach turns out to offer a nice mathematical framework for multi-resolution estimation algorithms, which avoids some of the drawbacks of the usual approach. Note that algorithms based on wavelet expansion of the data [1] or the velocity field [9] have been previously proposed. However, unlike the algorithm presented hereafter, the computational complexity of the later seriously limits its application to small images and/or the estimation of the coarsest motion scales. Moreover, we consider the effective implementation of high-order regularization schemes, based upon very simple constraints on the wavelet coefficients at small scales. We finally assess the relevance of proposed methods on challenging image sequences of turbulent fluid motions. Simulation results prove that the proposed approach outperforms the most effective state-of-the-art algorithms.

2   Optical Flow Background

Optical flow estimation is an ill-posed inverse problem. It consists in estimating the apparent motion of a 3D scene through image brightness $I(\boldsymbol{x}, t)$ variations in space $\boldsymbol{x} = (x_1, x_2)^T \in \Omega \subset \mathbb{R}^2$ and time $t \in \mathbb{R}$. Optical flow, identified by a 2D velocity field $\boldsymbol{v}(\boldsymbol{x}, t) : \Omega \times \mathbb{R}^+ \mapsto \mathbb{R}^2$ is the projection on the image plane of the 3D scene velocity. Its estimation involves two main aspects: a *data model* that links image data to the velocity field and a *regularization* scheme to overcome the ill-posedness.

2.1   Non-linear Data Model

Data models are commonly built upon assumptions about the temporal variations of the image brightness. The integration of a conservation assumption leads to the well-known *Displaced Frame Difference* (DFD) equation, which is studied in the following. However, the approach remains valid for any other integrated data model. Let us denote by $I_0(\boldsymbol{x})$ and $I_1(\boldsymbol{x})$ two consecutive image samples of the continuous sequence $I(\boldsymbol{x}, t)$ which has been discretized in time with a unit interval. Under rigid motion and stable lighting conditions, $\boldsymbol{v} = (v_1, v_2)^T$ satisfies the standard DFD equation, which reads:

$$\forall \boldsymbol{x} \in \Omega, \ f_{\mathrm{DFD}}(I, \boldsymbol{v}) = I_1(\boldsymbol{x} + \boldsymbol{v}(\boldsymbol{x})) - I_0(\boldsymbol{x}) = 0 \ . \tag{1}$$

The estimated motion field $\hat{\boldsymbol{v}}$ is obtained by minimizing a cost function, which we chose quadratic in the following to clarify the presentation:

$$\hat{\boldsymbol{v}} = \arg\min_{\boldsymbol{v}} J_{\mathrm{DFD}}(I, \boldsymbol{v}) \ , \ \text{with} \ J_{\mathrm{DFD}}(I, \boldsymbol{v}) = \frac{1}{2} \int_\Omega |f_{\mathrm{DFD}}(I, \boldsymbol{v})|^2 d\boldsymbol{x} \ . \tag{2}$$

The data model being non-linear w.r.t. the velocity field $\boldsymbol{v}$, estimation of optical flow therefore requires a specific optimization approach.

2.2 Classical Multi-resolution Strategy

Indeed, the equations for the inversion are only valid if the solution remains in the linearity region of the image intensity function. A standard approach for tackling non-linearity is to rely on an incremental multi-resolution strategy. This approach consists in choosing some sufficiently coarse low-pass-filtered version of the images at which the linearity assumption is valid, and to estimate a first displacement field assumed to correspond to a coarse representation of the motion. Then, a so-called Gauss-Newton strategy is used by applying successive linearizations around the current estimate and warping accordingly a representation of the images of increasing resolution. More explicitly, let us introduce the following incremental decomposition of the displacement field at resolution[1] 2^j:

$$\boldsymbol{v}_j = \tilde{\boldsymbol{v}}_j + \boldsymbol{v}'_j \tag{3}$$

where $\boldsymbol{v}'_j$ represents the unknown incremental displacement field at resolution 2^j and $\tilde{\boldsymbol{v}}_j \triangleq \sum_{i<j} \mathcal{P}_j(\boldsymbol{v}'_i)$ is a coarse motion estimate computed at the previous scales; $\mathcal{P}_j(\boldsymbol{v}'_i)$ denotes a projection operator which projects $\boldsymbol{v}'_i$ onto the grid considered at resolution 2^j. In order to respect the Shannon sampling theorem, the coarse scale data term is derived by a low-pass filtering of the original images with a kernel[2] $\mathcal{G}_j$, followed by a subsampling at period 2^j. Using (3), at coarse scale, image $I_j(\boldsymbol{x})$ and the *motion-compensated* image $\tilde{I}_j(\boldsymbol{x})$ are then defined as:

$$\begin{cases} I_j(\boldsymbol{x}) = \downarrow_{2^j} \circ (\mathcal{G}_j \star I_0(\boldsymbol{x})) \\ \tilde{I}_j(\boldsymbol{x}) = \downarrow_{2^j} \circ (\mathcal{G}_j \star I_1(\boldsymbol{x} + \tilde{\boldsymbol{v}}_j(\boldsymbol{x}))) , \end{cases} \tag{4}$$

where $\downarrow_{2^j}$ denotes a 2^j-periodic subsampling operator. It yields a functional J_{OBS}^j defined as a linearized version of (1) around $\tilde{\boldsymbol{v}}_j(\boldsymbol{x})$:

$$J_{\text{OBS}}^j(I_j, \boldsymbol{v}'_j) = \frac{1}{2} \int_{\Omega_j} \left[\tilde{I}_j(\boldsymbol{x}) - I_j(\boldsymbol{x}) + \boldsymbol{v}'_j(\boldsymbol{x}) \cdot \nabla \tilde{I}_j(\boldsymbol{x}) \right]^2 d\boldsymbol{x} . \tag{5}$$

Finally, the sought motion estimate $\hat{\boldsymbol{v}}$ is given by solving a system of coupled equations associated to resolutions increasing from 2^C to 2^F:

$$\begin{cases} \hat{\boldsymbol{v}} = \boldsymbol{v}'_F + \tilde{\boldsymbol{v}}_F = \boldsymbol{v}'_F + \sum_{i=C}^{F-1} \mathcal{P}_F(\boldsymbol{v}'_i) , \\ \boldsymbol{v}'_j = \arg\min_{\boldsymbol{v}'} J_{\text{OBS}}^j(I_j, \boldsymbol{v}') , \forall j \in \{C, \cdots, F\} . \end{cases} \tag{6}$$

where the finest scale $s = 2^{-F}$ corresponds to the pixel whereas the coarsest scale is noted $s = 2^{-C}$. In practice, equations in (6) are usually solved independently, starting from the coarsest to the finest scale. This coarse-to-fine approach has the

---

[1] In this paper, we shall use the following convention: indices $j \geq 0$ represent the *resolution* 2^j —contrary to [7]. Corresponding *scale* is 2^{-j}.

[2] A Gaussian kernel of variance proportional to 2^j is commonly used.

drawback of freezing (i.e. leaving unchanged), at a given scale, all the previous coarser estimates. Moreover, the major weakness of this strategy is the arbitrary approximation of the original functional in (2) by a set of coarse scale data terms (5), which are defined at different resolutions by a modification of original input images with (4) and by a linearization of model (1) around the previous motion estimate. In the next section, we will see that this multi-resolution strategy has a mathematically-sound formulation within the framework of wavelet representations.

2.3   The Aperture Problem and Usual Regularization Schemes

Previously introduced data models remain under-constrained, as they provide for each time t a single equation for two unknowns (v_1, v_2) at each spatial location $\boldsymbol{x} = (x_1, x_2)^T$. To deal with this under-constrained estimation problem, the most common setting consists in enforcing some spatial coherence to the solution.

Implicit Regularization. The motion field is constrained to be of the form $\boldsymbol{v} = \Phi(\Theta)$, where Φ is a function parametrized by Θ (piece-wise polynomial functions are often used). Implicit regularization schemes penalize discrepancies from model (1) by minimizing J_{DFD} with respect to Θ, i.e.

$$\hat{\boldsymbol{v}} = \Phi\left(\arg \min_{\Theta} J_{\text{DFD}}(I, \Phi(\Theta)) \right) . \tag{7}$$

Associated to a low-order parametric representation, this simple approach reduces drastically the dimension of the problem, hence addressing its under-constrained nature. However, when spatiotemporal gradients of the images vanish, it is impossible to guarantee the existence of an unique solution: this is the *aperture problem*.

Explicit Regularization. Global regularization schemes in their simplest form define the estimation problem through the minimization of a functional composed of two terms balanced by a regularization coefficient $\mu > 0$:

$$J(I, \boldsymbol{v}, \mu) = J_{\text{DFD}}(I, \boldsymbol{v}) + \mu J_{\text{reg}}(\boldsymbol{v}) . \tag{8}$$

Thus, motion estimate $\hat{\boldsymbol{v}}$ satisfies $\hat{\boldsymbol{v}} = \arg \min_{\boldsymbol{v}} J(\boldsymbol{v}, I, \mu)$. The data term J_{DFD} is still defined by (2) . The second term, J_{reg} (the "regularization term"), encourages the solution to follow some prior smoothness model formalized with function f_{reg}:

$$J_{\text{reg}}(\boldsymbol{v}) = \frac{1}{2} \int_{\Omega} f_{\text{reg}}(\boldsymbol{v}, \boldsymbol{x}) d\boldsymbol{x} . \tag{9}$$

An n-order regularization writes in its simplest form:

$$f_{\text{reg}}(\boldsymbol{v}, \boldsymbol{x}) = \sum_{i=1,2} \sum_{j=1,2} \left| \frac{\partial^n v_i}{\partial x_j^n}(\boldsymbol{x}) \right|^2 . \tag{10}$$

A first-order regularizer (i.e. $n = 1$) enforcing weak spatial gradients of the two components v_1 and v_2 of the velocity field $\boldsymbol{v}$ is very often used [6]. Second-order regularizers (i.e. $n > 1$) have been proposed in the literature in the case of fluid flows [2,10,11]. However, since motion variables are considered on the pixel grid, an approximation of continuous spatial derivatives by discrete operators is required. For regular pixel grids, it is usually done using finite difference schemes. Nevertheless, it is well known that ensuring stability of the discretization schemes of high-order regularizer may constitute a difficult problem.

3   Wavelet Formulation

As shown in Sect. 2, the common optical flow estimation approach suffers from two main drawbacks: the necessary "empirical" multi-resolution approach and the implementation of efficient regularizations terms. The use of wavelet bases is a simple answer to both problems. Moreover, it has been shown that a wavelet expansion is appropriate for representing turbulent flows [3].

3.1   Wavelet Decomposition

In order to avoid the limitations of the classical multi-resolution strategy, we consider the projection of each scalar component v_1, v_2 of the velocity field $\boldsymbol{v}$ onto *multi-resolution approximation spaces* exhibited by the wavelet formalism. Let us introduce briefly this context for real 1D scalar signals. We consider a multi-resolution approximation of $\mathbf{L}^2(\mathbb{R})$ as a sequence $\{V_j\}_{j \in \mathbb{Z}}$ of closed subspaces, so-called *approximation spaces*, notably verifying[3]

$$V_j \subset V_{j+1} \; ; \; \lim_{j \to -\infty} V_j = \bigcap_{j=-\infty}^{+\infty} V_j = \{0\} \; ; \; \lim_{j \to +\infty} V_j = \text{Closure}\left(\bigcup_{j=-\infty}^{+\infty} V_j\right) = \mathbf{L}^2(\mathbb{R}) \; .$$

Since approximation spaces are sequentially included within each other, they can be decomposed: $V_{j+1} = V_j \oplus W_j$. Those W_j are the orthogonal complements of approximation spaces, they are called *detail spaces*

Practically, scalar 1D signals being finite, they belong to a given approximation space according to their resolution, i.e. number of samples. Let w be a 1D signal of 2^{F+1} samples, then $w \in V_{F+1} = V_C \oplus W_C \oplus W_{C+1} \oplus \cdots \oplus W_F \subset \mathbf{L}^2([0,1])$, where $0 \leq C \leq F$. The projection of w on this multiscale basis writes:

$$w(x) = \sum_{k=0}^{2^C-1} \langle w, \phi_{C,k} \rangle_{\mathbf{L}^2} \, \phi_{C,k}(x) + \sum_{j=C}^{F} \sum_{k=0}^{2^j-1} \langle w, \psi_{j,k} \rangle_{\mathbf{L}^2} \, \psi_{j,k}(k) \; . \tag{11}$$

Here, $\{\phi_{C,k}\}_k$ and $\{\psi_{j,k}\}_k$ are orthonormal bases of V_C and W_j, respectively. They are defined by *dilatations* and *translations*[4] of the so-called *scale function*

---

[3] See [7] for a complete presentation of wavelet bases.
[4] Written in a general form $f_{j,k}(x) = 2^{j/2} f(2^j x - k)$.

ϕ and its associated *wavelet function* ψ. Functions ϕ and ψ verify the following two-scale relations:

$$\phi(x) = \sqrt{2} \sum_{k \in \mathbb{Z}} h[k]\phi(2x - k) \; ; \; \psi(x) = \sqrt{2} \sum_{k \in \mathbb{Z}} g[k]\phi(2x - k) , \qquad (12)$$

where sequences $h[k] = \langle \phi(x), \sqrt{2}\phi(2x - k) \rangle$ and $g[k] = \langle \psi(x), \sqrt{2}\phi(2x - k) \rangle$ are called *conjugate mirror filters*. Those filters play an important role in the fast implementation with *filter banks* of forward and inverse wavelet transform, i.e. projection on the wavelet basis and reconstruction, from (11) [7]. Finally, the representation of a signal projected onto the multiscale wavelet basis is given by the set of coefficients appearing in (11): $a_{C,k} \triangleq \langle w, \phi_{C,k} \rangle_{\mathbf{L}^2}$ and $d_{j,k} \triangleq \langle w, \psi_{j,k} \rangle_{\mathbf{L}^2}$ are approximation and detail coefficients, respectively. Those results are extended to the case of 2D signals, in order to obtain *separable multiscale orthonormal bases* of $\mathbf{L}^2([0,1]^2)$.

3.2   Wavelet Data Term

In this work, the representation of the velocity field $\boldsymbol{v}$ is obtained by the wavelet decomposition (11) of each component. We denote by Θ_1 and Θ_2 the sets of coefficients respectively associated to v_1 and v_2; $\Theta = (\Theta_1, \Theta_2)^T$ is the set of all coefficients. Denoting the linear reconstruction operator by Φ for convenience, we may write

$$\forall \boldsymbol{x} \in \Omega, \; \boldsymbol{v}(\boldsymbol{x}) = \Phi(\boldsymbol{x})\Theta . \qquad (13)$$

Here the constant coefficients vector Θ is the unknown of our optical flow estimation problem. Replacing $\boldsymbol{v}(\boldsymbol{x})$ by (13) in DFD data term (1), we obtain

$$J_{\mathrm{DFD}}(\Theta) = \frac{1}{2} \int_{\Omega} [I_1(\boldsymbol{x} + \Phi(\boldsymbol{x})\Theta) - I_0(\boldsymbol{x})]^2 \, d\boldsymbol{x} \qquad (14)$$

and the estimation problem becomes

$$\hat{\boldsymbol{v}} = \Phi\hat{\Theta} \in V_{F+1} , \; \text{where} \; \hat{\Theta} = \arg\min_{\Theta} J_{\mathrm{DFD}}(\Theta) . \qquad (15)$$

3.3   Multiscale Estimation

Unknown coefficients are estimated sequentially from coarsest scale C to a chosen finest one L (with $C \leq L \leq F$) using a gradient-descent algorithm. At each scale j, all coefficients from scales C to j are estimated. Coefficients previously estimated at coarser approximation spaces are used to initialize the gradient descent; this strategy enables the update of the latter coarser coefficients while estimating "new" details at current scale j. In other words, solution $\hat{\boldsymbol{v}}$ is sequentially sought within higher resolution spaces: $V_C \subset V_{C+1} \subset \cdots \subset V_L$. This way, the projection of the current solution $\hat{\boldsymbol{v}} \in V_j$ onto every coarser space V_p with $C \leq p < j$ is constantly updated, contrary to the standard incremental approach (Sect. 2.2). The use of wavelet bases thus leads to a "natural" and well-defined multi-resolution framework. At each refinement level, minimization of functional

J_{DFD} is efficiently achieved with a gradient-based quasi-Newton algorithm (L-BFGS) [8], to seek the optimum $\hat{\Theta}$. For any coefficient $\theta_{i,p} \in \Theta_i \subset \Theta$,

$$\frac{\partial J_{\mathrm{DFD}}}{\partial \theta_{i,p}}(\Theta) = \left\langle \frac{\partial I_1}{\partial x_i}(\cdot + \Phi(\cdot)\Theta)\left[I_1(\cdot + \Phi(\cdot)\Theta) - I_0(\cdot)\right], \Phi_p \right\rangle_{\mathbf{L}^2([0,1]^2)} \tag{16}$$

where Φ_p is the wavelet basis atom related to $\theta_{i,p}$. As a consequence, components of the spatial gradient of the data-term functional (14) are simply given by the coefficients of the wavelet decomposition of the two terms

$$\frac{\partial I_1}{\partial x_i}(\boldsymbol{x} + \Phi(\boldsymbol{x})\Theta)\left[I_1(\boldsymbol{x} + \Phi(\boldsymbol{x})\Theta) - I_0(\boldsymbol{x})\right], \quad i = 1, 2,$$

on the considered wavelet basis. It is easy to see that the proposed coarse-to-fine estimation strategy enables to capture large displacements: at large scales, the decomposition of (3.3) is obtained by convolutions with the atoms of the wavelet basis having the largest support. Note that conversely to the algorithm proposed in [9], the low-complexity of gradient computation via fast wavelet transform does not restrict motion estimation to large scales and/or images of small size.

4   Regularizations

4.1   Wavelet Properties

Wavelet-based regularizers which are described in the following are based upon wavelet properties such as polynomial reproduction, differentiation and interpolation. Those aspects are linked to the notion of *vanishing moments* (VM). A wavelet $\psi(x) \in \mathbf{L}^2(\mathbb{R})$ has n VM if :

$$\int_{\mathbb{R}} x^\ell \psi(x) dx = 0, \text{ for } 0 \le \ell < n . \tag{17}$$

Wavelets as polynomial approximations. From (17), a wavelet with n VM is hence orthogonal to any polynomial of degree $n-1$. Consequently, piece-wise[5] polynomials of degree $n-1$ belonging to V_{F+1} are exactly described in V_F, since the atoms of the basis that belong to its orthogonal complement W_F have vanishing coefficients.

Wavelets as Differential Operators. Given a signal $w \in \mathcal{C}^n$, the behavior of its small scales coefficients resulting from an n-VM wavelet decomposition can be related to its nth derivative [7]:

$$\lim_{j \to \infty} \frac{\langle w(x), \psi_{j,k}(x)\rangle}{2^{-j(n+\frac{1}{2})}} \propto \frac{\partial^n w(x)}{\partial x^n} . \tag{18}$$

This result can be extended to the case of 2D signals.

---
[5] On the support of $\{\psi_{F,k}\}$.

Wavelet-Based Interpolation. A multiscale interpolation is the orthogonal projection of a signal w estimated at a given resolution 2^j onto the next finer approximation space V_{j+1}:

$$P_{V_{j+1}} w \left(2(p + \frac{1}{2}) \right) = \sum_{k=-\infty}^{+\infty} w(2k)\varphi_{j+1}(p - k + 1/2) . \tag{19}$$

The interpolation function φ is defined as the autocorrelation of the scaling function: $\varphi = \phi \star \check{\phi}$, where $\star$ and $\check{\cdot}$ denote respectively convolution and time-reverse[6] operators. It can be shown that φ interpolates exactly polynomials of order n if and only if the wavelet associated to scaling function ϕ has $n+1$ VM [7]. This linear interpolation operator $P_{V_{j+1}}$ is also implemented with filter banks using filter h_i, where $h_i[n] = \left(h \star \check{h} \right) [2n + 1]$ and h is defined in (12).

4.2   Polynomial Approximation on a Truncated Basis

As seen in Sect. 2.3, a first way to overcome the under-constrained nature of the optical flow estimation problem consists in reducing the number of unknowns through a parametric formulation of the velocity field. Using the proposed wavelet formulation (15), this can be easily achieved by estimating the velocity field on a truncated wavelet basis. This means that the solution $\hat{v}$ belongs to a lower-resolution space $V_L \subset V_{F+1}$ and therefore is a piece-wise polynomial of order $n-1$ in V_{L+1}. Details coefficients associated to non-estimated small details scales (W_j with $L \leq j \leq F$) are thus not estimated, but set to zero.

$$\hat{v} = \Phi\hat{\Theta} \in V_L , \ L < F + 1 , \ \text{where} \ \hat{\Theta} = \underset{\Theta}{\arg\min} \, J_{\mathrm{DFD}}(\Theta) . \tag{20}$$

Since the basis truncation reduces the number of unknowns, it is theoretically possible to estimate detail coefficients up to penultimate scale $F - 1$, i.e. $\hat{v} \in V_F$. Practically, it is impossible due to the aperture problem.

4.3   High-Order Regularization

It has been previously mentioned that smallest scale coefficients might be interpreted as the signal's n^{th} derivative (Sect. 4.1), with n number of VM of the considered wavelet. The penalization of small scale coefficients' amplitude thus enables to control the amplitude of the derivative of the estimated signal. However, due to the dyadic structure of the discrete wavelet decomposition, only a "piecewise" control is possible. In order to control the derivative at junctions of those dyadic blocks, *interpolated signal* $\tilde{v}$ of the velocity field v on a shifted 2D grid is considered. Small scale coefficients $\{\tilde{\Theta}_F\}$ and $\{\Theta_F\}$ of both $\tilde{v}$ and v are penalized. "Interpolated coefficients" $\tilde{\Theta}$ are expressed as a linear combination of Θ through wavelet inverse and forward transformations (Φ, $\Phi^{-1} = \Phi^T$, resp.) and interpolation: $\tilde{\Theta} = \left(\Phi^T \circ P_{V_{F+1}} \circ \Phi \right) \Theta$. We finally get the regularization term

---

[6] More explicitly, $\check{f} : t \mapsto \check{f}(t) = f(-t)$.

$$J_{\text{reg}}(\Theta) = \frac{1}{2}\|\Theta_F\|^2 + \frac{1}{2}\|\tilde{\Theta}_F\|^2 \text{ and } \nabla J_{\text{reg}}(\Theta) = \Theta_F + \left(\Phi^T \circ P_{V_{F+1}}^T \circ \Phi\right)\tilde{\Theta}_F \quad (21)$$

The gradient in (21) is a linear form which can be efficiently computed using the recursive filter banks presented in Sect. 3.1 and 4.1. The addition of the regularization term (21) therefore does not increase significantly the computational burden. Supplementing (15), the estimation problem becomes:

$$\hat{v} = \Phi\hat{\Theta} \in V_{F+1}, \text{ where } \hat{\Theta} = \arg\min_{\Theta} J_{\text{DFD}}(\Theta) + \mu J_{\text{reg}}(\Theta). \quad (22)$$

5   Results

Daubechies wavelets have been chosen since they have a minimum support size for a given number of VM [7]. Daubechies wavelet with n VM will be referred to as D_n hereafter. Wavelet transform is implemented with periodic boundary conditions.

5.1   Synthetic PIV Sequence

The first data set used for evaluation is a synthetic sequence of Particle Imagery Velocimetry (PIV) images of size 256×256 pixels, representing small particles (of radius below 4 pixels) advected by a 2D periodic forced turbulent flow. The dynamic of the fluid flow is given by numerical simulation of 2D Navier-Stokes

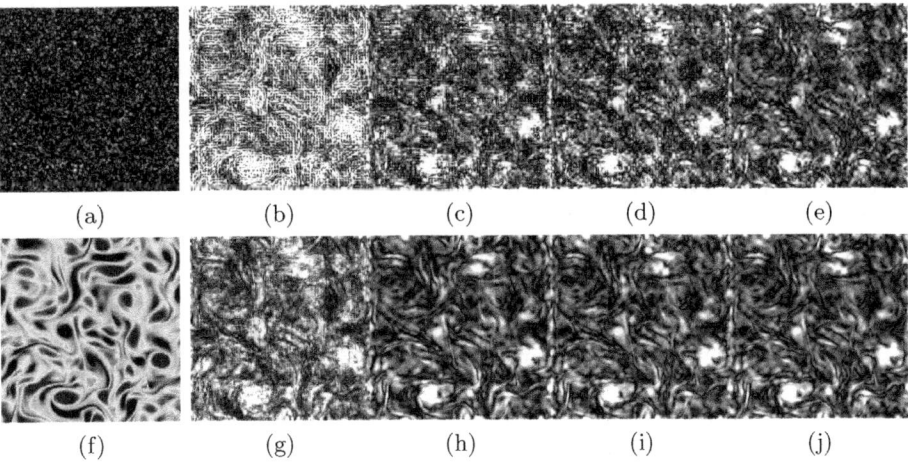

Fig. 1. Sample synthetic *PIV image* (1a) with below the *vorticity* of the underlying reference velocity field (1f). End-point error maps on velocity field estimations for a polynomial approximation (*upper row*) and high-order regularization (*lower row*) with D_n wavelets are presented, i.e. polynomial (resp. derivative) order of $n-1$ (resp. n), for D_1 (1b, 1g), D_2 (1c, 1h), D_3 (1d, 1i) and D_{10} (1e, 1j).

equations at $Re = 3000$, using the vorticity conservation equation and the Lagrangian equation for non-heavy particles transported by the flow (simulation details can be found in [5]). This simulated flow has a null-divergence by construction. An image of the PIV sequence is displayed in Fig. 1a together with its associated ground truth motion vorticity (1f). Estimated velocity field evaluation is based on the Root Mean Square end-point Error (RMSE)[7].

When the true velocity field is decomposed on a Daubechies wavelet with a number of VM higher than 3, the velocity field reconstructed with the $p = 6$ coarser scales (out of 8) carries out more than 99,95% of the total kinetic energy. Those 6 scales are represented with only 12.5% of atoms from the full wavelet basis. Moreover, when n is chosen high enough (> 7), the reconstruction error stabilizes around $0,013$.

Motion Estimation. From the previous analysis of the true velocity fields, it seems that 6 detail scales out of 8 should give an accurate representation of the motion in terms of kinetic energy, as long as the chosen number of VM is high enough. Figure 2a (black curve) shows RMS errors computed on an estimated velocity field with truncated Daubechies wavelet bases (Sect. 4.2) having different VM n, i.e. with a polynomial approximation of order $n - 1$. As expected, RMSE converges rapidly towards a median value of 0.0613 when n increases. Figure 1 (upper row) shows corresponding end-point error maps for motion estimated with wavelets bases D_1, D_2, D_3 and D_{10}. Although errors effectively lower when higher VM wavelets are employed, artifacts due to high-amplitude errors on 6th scale coefficients (small white "dots") and on coarse coefficients (white straight "lines") remain clearly visible. With the proposed high-order regularization scheme (Sect. 4.3), all scales are estimated, which should highly improve results with $n \leq 3$ VM, i.e. for penalization of derivatives of order lower than 3. This is confirmed on Fig. 2a (red dashed curve), with a reduction of 35% and 30% of the RMS obtained with D_1 and D_2 wavelets bases, respectively, whereas the diminution observed using D_{10} wavelet basis is of 10% at best. At the same time, derivative penalization eliminates most of the artifacts observed on estimates with truncated bases, which is displayed on the lower row of Fig. 1. Note also that there are less differences between estimations with different VM, in comparison to the previous case.

Comparison with State-of-the-Art Estimators. Figure 2 is a comparison of RMS errors obtained on the synthetic PIV sequence with the proposed high-order regularizer and various state-of-the-art estimators, after a null-divergence projection. Our wavelet-based estimator clearly outperforms other methods.

5.2   Real PIV Sequence

This data set consists in 128 PIV pictures of a transversal view of a planar concomitant jet flow, of size 1024×1024 pixels. The flow has a "top-hat" velocity

---

[7] Ground-truth velocity fields being given on a shifted grid (by 1/2 pixel) by the numerical simulation, they have been interpolated in order to compute accurate RMSE on the pixel grid.

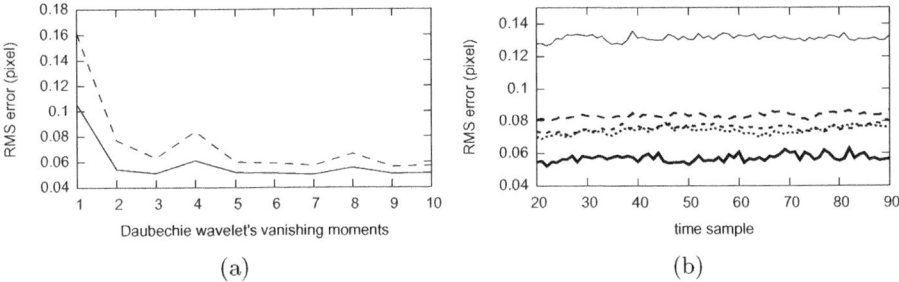

Fig. 2. *Left*: Comparison of RMS errors on velocity fields estimated from a pair of the synthetic PIV sequence, with the proposed methods and Daubechies wavelets with 1 to 10 VM, i.e. for polynomial approximation (resp. derivative regularization) of order 0 to 9 (resp. 1 to 10). RMSE obtained with the polynomial approximation (6 scales out of 8, *dashed line*) and with derivatives regularization (best case, *solid line*). *Right*: Comparison of RMS errors on a sequence of velocity fields estimated with proposed regularization (*thick solid*) and with state-of-the-art methods: correlation (*thin solid*), first order regularization [6] (*long-dashed*), div-curl regularization [10] (*dashed*), self-similar regularization [4] (*dotted*).

profile and is poorly turbulent, but shows two high-shear regions featuring development of Kelvin-Helmholtz instabilities. Motion is estimated with proposed wavelet-based estimator (22), using the following settings: 2 VM and derivatives penalization with factor $\mu = 10^7$. Figure 3 presents a PIV image of the sequence and streamlines of an estimated velocity field along with two consecutive vorticity maps. A qualitative evaluation of the presented motion field shows a remarkably good agreement with the physics of concomitant jets. A very good temporal coherence is also observed, although no prior dynamic model is considered (i.e. successive pairs of images are processed independently).

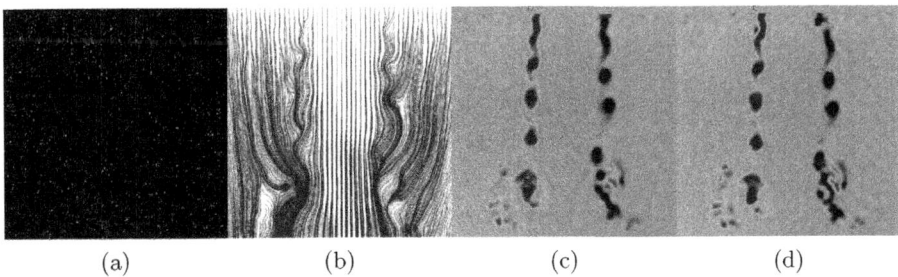

(a)   (b)   (c)   (d)

Fig. 3. Sample estimated motion fields from 2D planar jet PIV dataset: detail of input *PIV image* (3a), *streamlines* (3b) and *vorticity* (3c). Figure 3d is the vorticity field corresponding to motion estimated at the next time step. Three different areas are visible: at the output of the jet (top of the field), shear regions begin to oscillate slowly. The middle region clearly shows the development of vortices characteristic of the Kelvin-Helmholtz instability. Finally, in the lower part of the field, structure of vortices collapse due to their tri-dimensionalization.

6   Conclusion

An optical flow estimation algorithm dedicated to continuous motion has been introduced. The choice of the wavelet formalism sets-up a well-defined multi-resolution framework that avoids most drawbacks of such usual approaches. Being associated to a gradient-based quasi-Newton optimization method, its low complexity makes possible the estimation of the full range of scales composing the motion. Moreover, high numbers of vanishing moments enable to truncate the wavelet basis without increasing the error of the polynomial reconstruction, thus significantly reducing the number of unknowns and the problem complexity. A high-order regularization scheme, involving small scale coefficients penalization, highly enhances estimation results and generally helps reducing errors by removing noise of the solution, as emphasized by experiments on a synthetic PIV sequence. Application to a real PIV sequence shows the capability of the estimation method to reconstruct accurately vortices of large amplitude.

Acknowledgments. The authors acknowledge the support of the French Agence Nationale de la Recherche (ANR), under grant MSDAG (ANR-08-SYSC-014) "Multiscale Data Assimilation for Geophysics".

References

1. Bernard, C.: Wavelets and ill posed problems: optic flow and scattered data interpolation. Ph.D. thesis, École Polytechnique (1999)
2. Corpetti, T., Mémin, E., Pérez, P.: Dense estimation of fluid flows. Pattern Anal. Mach. Intel. 24(3), 365–380 (2002)
3. Farge, M.: Wavelet transforms and their applications to turbulence. Annual Review of Fluid Mechanics 24(1), 395–458 (1992)
4. Heas, P., Memin, E., Heitz, D., Mininni, P.: Bayesian selection of scaling laws for motion modeling in images. In: International Conference on Computer Vision (ICCV 2009), Kyoto, Japan (October 2009)
5. Heitz, D., Carlier, J., Arroyo, G.: Final report on the evaluation of the tasks of the workpackage 2, FLUID project deliverable 5.4. Tech. rep., INRIA - Cemagref (2007)
6. Horn, B., Schunck, B.: Determining optical flow. Artificial Intelligence 17, 185–203 (1981)
7. Mallat, S.: A Wavelet Tour of Signal Processing: The Sparse Way. Academic Press (2008)
8. Nocedal, J., Wright, S.J.: Numerical Optimization. Springer Series in Operations Research. Springer, New York (1999)
9. Wu, Y., Kanade, T., Li, C., Cohn, J.: Image registration using wavelet-based motion model. Int. J. Computer Vision 38(2), 129–152 (2000)
10. Yuan, J., Schnörr, C., Memin, E.: Discrete orthogonal decomposition and variational fluid flow estimation. Journ. of Math. Imaging & Vison 28, 67–80 (2007)
11. Yuan, J., Schnörr, C., Steidl, G.: Simultaneous higher-order optical flow estimation and decomposition. SIAM Journal on Scientific Computing 29(6), 2283–2304 (2007)

Multiscale Weighted Ensemble Kalman Filter for Fluid Flow Estimation

Sai Gorthi, Sébastien Beyou, Thomas Corpetti*, and Etienne Mémin

INRIA / FLUMINANCE, 35042 Rennes Cedex, France
CNRS/LIAMA, PO Box 2728, Beijing 100190, PR China
{sai.gorthi,sebastien.beyou,thomas.corpetti,etienne.memin}@inria.fr
http://www.irisa.fr/fluminance/indexFluminance.html

Abstract. This paper proposes a novel multi-scale fluid flow data assimilation approach, which integrates and complements the advantages of a Bayesian sequential assimilation technique, the Weighted Ensemble Kalman filter (WEnKF) [12], and an improved multiscale stochastic formulation of the Lucas-Kanade (LK) estimator. The proposed scheme enables to enforce a physically plausible dynamical consistency of the estimated motion fields along the image sequence.

1   Introduction

The analysis of geophysical fluid flows is of the utmost importance in domains such as oceanography, hydrology or meteorology for applications of forecasting, studies on climate changes, or for monitoring hazards or events. In all these domains orbital or geostationary satellites provide a huge amount of image data with a still increasing spatial and temporal resolution. Compared to *in situ* measurements supplied by dedicated probes or Lagrangian drifters, satellite images provide a much more denser observation field. They however offer only an indirect access to the physical quantities of interest, and give rise consequently to difficult inverse problems to estimate characteristic features of the flow such as velocity fields or vorticity maps.

Fluid motion estimation techniques differ mainly on the smoothness prior they are handling: first order penalization[13], second order div-curl regularization [2,14], or power law auto-similarity principles [6]. These methods provide accurate instantaneous displacements, however they may exhibit difficulties for mid to small scales measurements due to the smoothing prior used and photometric uniform regions. All these difficulties may thus generate inconsistent measurements along time. For interested readers, a complete overview of fluid motion estimation techniques can be found in [7].

Dynamical consistency of the velocity measurements can be enforced by embedding the estimation problem within an image based assimilation process. Variational assimilations of image information have been recently considered for the estimation of fluid motion fields [1,11]. Those methods, though efficient,

---

* Corresponding author.

A.M. Bruckstein et al. (Eds.): SSVM 2011, LNCS 6667, pp. 749–760, 2012.
© Springer-Verlag Berlin Heidelberg 2012

constitutes batch methods, which requires forward and backward integrations of the dynamical system and the associated tangent linear dynamics respectively. The latter relies implicitly on a linearization of the dynamics and is adapted in practice for short time horizon.

On the other hand, stochastic filters are also well known techniques for data assimilation. Recently, a data assimilation procedure embedding an Ensemble Kalman filter (EnKF) [5] into the particle filter (PF) framework, referred to as Weighted Ensemble Kalman filter (WEnKF), has been proposed [12]. This filter has shown to be efficient on toy examples with synthetic measurements. The objective of this work consists to specify such a procedure from local noisy velocity measurements and their uncertainties.

2   Stochastic Lucas-Kanade Estimator

This section first presents a stochastic formulation of the well known Lucas-Kanade (LK) optical flow approach [9] that will be used to provide local motion measurements in the assimilation method we propose. This technique departs somewhat from the traditional Lucas and Kanade motion estimator. It leads naturally to a continuous multiresolution formulation and enables not only to extract the motion fields at different resolutions but supplies uncertainties of those estimates as well.

In what follows, we represent the image luminance with f, and a grid of 2D points $\mathbf{X} = (X^1, ..., X^n)^T \in R^{2n}$, represents the grid point locations. The image over a regular grid at time $t - 1$, $I = f(\mathbf{X}_{t-1}, t - 1)$, is driven by the velocity field $\mathbf{v}(\mathbf{X}_{t-1}, t - 1)$ to generate new point positions $\mathbf{X}_t$ at time t.

2.1   Luminance Variation with Uncertainties

In a stochastic formulation, if we assume that the 2D grid from $\mathbf{X}_{t-1}$ to $\mathbf{X}_t$ is transported by a velocity field, $\mathbf{v}$, up to a Brownian motion $\mathbf{B}_t = (B_t^1, ..., B_t^n) \in R^{2n}$, we can write: $d\mathbf{X}_t = \mathbf{v}(\mathbf{X}_{t-1}, t - 1)dt + \Sigma(\mathbf{X}_t, t)d\mathbf{B}_t$, here Σ is the covariance matrix and $d\mathbf{X}_t = \mathbf{X}_t - \mathbf{X}_{t-1}$. Assuming uncorrelated uncertainties with local isotropic standard deviation $\sigma(\mathbf{X}_t, t)$, the noise term reads $\Sigma(\mathbf{X}_t, t)d\mathbf{B}_t = $ diag $\sigma(\mathbf{X}_t, t) \otimes \mathcal{I}_2 d\mathbf{B}_t$, $\mathcal{I}_2$ being the 2×2 identity matrix, and $\otimes$ denoting the Kronecker product.

The differential of luminance function f defined for each spatial point at time t is obtained through stochastic calculus differentiation using the celebrated Ito formulae [10] as :

$$df(\mathbf{X}_t, t) = \frac{\partial f(\mathbf{X}_t, t)}{\partial t}dt + \sum_{i=(1,2)} \frac{\partial f(\mathbf{X}_t, t)}{\partial x_i}d\mathbf{X}^i + \frac{1}{2}\sum_{(i,j)} \frac{\partial^2 f(\mathbf{X}_t, t)}{\partial x_i \partial x_j}d\langle \mathbf{X}_t^i, \mathbf{X}_t^j \rangle.$$

The quadratic variation terms $d\langle \mathbf{X}_t^i, \mathbf{X}_t^j \rangle$ are computed based on the properties:

$$d\langle B^i, B^j \rangle = dt; \langle h(t), h(t) \rangle = \langle h(t), B^i \rangle = \langle B^j, h(t) \rangle = 0.$$

With the considered uncertainties this yields a luminance variation:

$$df(\mathbf{X}_t, t) = \left(\frac{\partial f}{\partial t} + \boldsymbol{\nabla} f^T \mathbf{v} + \frac{1}{2}\sigma^2 \Delta f\right) dt + \sigma \boldsymbol{\nabla} f^T d\mathbf{B}_t. \tag{1}$$

The operators $\boldsymbol{\nabla}$ and Δ represents the 2D gradient and Laplacian of the luminance function, respectively. This model obviously comes back to the standard brightness consistency assumption for zero uncertainties ($\sigma = 0$). Interest readers may refer to [3] for a more complete presentation of this estimator.

2.2 Data Model with Uncertainties and Local Estimation

In a minimum least square sense, we define the motion to be estimated $\mathbf{v}(\mathbf{X}_{t-1}, t-1)$, as the minimum conditional variance of the luminance variation. Starting from known grid $\mathbf{X}_{t-1}$, it can estimated by minimizing the expectation $\mathbb{E}(df^2(\mathbf{X}_t, t)/\mathbf{X}_{t-1})$. This conditional expectation given $\mathbf{X}_{t-1}$ of a function of a stochastic processes $\mathbf{X}_t$ driven by an Ito diffusion (1) discretized through an Euler-Maruyama scheme, $\mathbf{X}_t = \mathbf{X}_{t-1} + \mathbf{v}(\mathbf{X}_{t-1}, t-1)dt + \Sigma^{1/2}(\mathbf{B}_{t+1} - \mathbf{B}_t)$, can be expressed as the following convolution: $\mathbb{E}(df^2(\mathbf{X}_t, t)/\mathbf{X}_{t-1}) = df^2(\mathbf{X}_{t-1} + \mathbf{v}, t) \star g_\Sigma$, where $g_\Sigma = N(0, \Sigma)$ is a multidimensional zero mean Gaussian. From the illumination variation equation (1), the cost function to be minimized reads hence:

$$\mathcal{H}(f, \mathbf{v}) = g_\Sigma \star \left(\frac{\delta f}{\delta t} + \boldsymbol{\nabla} f \cdot \mathbf{v} + \frac{1}{2}\sigma^2 \Delta f\right)^2. \tag{2}$$

To alleviate the ill-posed nature of (2)[1] we assume a locally constant flow within a Gaussian window of variance λ^ℓ centered at location (x, y) as in the standard Lucas-Kanade estimator. At point (x, y), the estimate $\mathbf{v}$ should hence minimize:

$$\arg\min_{v} g_{\lambda^\ell} \star \mathcal{H}(f, \mathbf{v}). \tag{3}$$

Differentiating (3) with respect to $\mathbf{v}$ and equating to zero, at any position (x, y) (with f_x, f_y, f_t representing the spatial (x, y) and temporal derivatives of f) yields:

$$\left(g_{\lambda^\ell} \star g_\Sigma \star \begin{bmatrix} f_x^2 & f_x f_y \\ f_x f_y & f_y^2 \end{bmatrix}\right)\mathbf{v} = -g_{\lambda^\ell} \star g_\Sigma \star \left(\frac{1}{2}\sigma^2 \Delta f + f_t\right)\begin{bmatrix} f_x \\ f_y \end{bmatrix}. \tag{4}$$

2.3 Multiresolution Analysis and Uncertainty Estimation

A multi-resolution analysis (formulated within an incremental framework) of this stochastic formulation can be accomplished by a coarse-to-fine decrease of the variance parameter associated to the local smoothing Gaussian window, λ^ℓ, in (4). Furthermore, the quadratic variation of luminance function between $t-1$ and t can be written as

$$d\langle f(\mathbf{X}_t, t), f(\mathbf{X}_t, t)\rangle = \sigma^2 \| \boldsymbol{\nabla} f(\mathbf{X}_{t-1} + \mathbf{v}(\mathbf{X}_{t-1}, t), t) \|^2. \tag{5}$$

---

[1] Single equation, with two unknown components.

In a probabilistic sense the variance parameters, in (5) can be estimated as:

$$\sigma = \sqrt{\frac{\mathbb{E}(f(\mathbf{X}_{t-1} + \mathbf{v}(\mathbf{X}_{t-1}, t), t) - f(\mathbf{X}_{t-1}, t))^2}{\mathbb{E}(\| \nabla f(\mathbf{X}_{t-1} + \mathbf{v}(\mathbf{X}_{t-1}, t), t) \|^2)}} \quad a.s. \tag{6}$$

This provides us a spatial distribution of the motion estimate uncertainties.

3   Monte Carlo Implementation of Stochastic Filtering with the Weighted-Ensemble Kalman Filter

In this section we briefly review the main principles driving the construction of the Weighted Ensemble Kalman filter, proposed in [12], and discuss its advantages and limitations in the context of fluid flow analysis. This technique is a particle implementation of a nonlinear stochastic filtering problem build upon an ensemble Kalman update stage. In the following section we recall briefly the basic elements constituting such filter.

3.1   Stochastic Filtering, Filtering Distribution

Stochastic filters aim at estimating the posterior probability distribution $p(\mathbf{x}_{0:k}|\mathbf{y}_{1:k})$ of a state variable trajectory $\mathbf{x}_{0:k}$ starting from an initial state $\mathbf{x}_0$ up to the state at the current time $\mathbf{x}_k \in \mathcal{R}^n$ given a complete measurements trajectory $\mathbf{y}_{1:k}$. The state variable trajectory is obtained through the integration of a dynamical system:

$$\mathbf{x}_t = M(\mathbf{x}_{t-\delta t}) + \boldsymbol{\eta}_t, \tag{7}$$

where M denotes a deterministic linear/nonlinear dynamical operator, corresponding to a discrete representation (through numerical integration with time step δt) of a physical conservation law describing the state evolution. And $\boldsymbol{\eta}_t$ is usually a white Gaussian noise of covariance $Q_{\delta t}$, that accounts for the uncertainties in the deterministic state model. However, as the true initial state is unknown, observation $\mathbf{y}_k \in \mathcal{R}^m$ of the state occurring at discrete instants are assumed to be available. These observations and the state variable are linked through:

$$\mathbf{y}_k = H(\mathbf{x}_k) + \boldsymbol{\gamma}_k, \tag{8}$$

a measurement equation where $\boldsymbol{\gamma}_k$, the observation noise, is a white Gaussian noise with covariance matrix R, and H stands for the linear/nonlinear mapping from the state variable space to the observation space. We note that the (integration) time step used for the state variable dynamics δt is usually much smaller (about 10-100 times), than the latency δk between two subsequent measurements. A sequence of measurements or observations from time 1 to k will be denoted by a set of vectors of dimension m as: $\mathbf{y}_{1:k} = \{\mathbf{y}_i, i = 1, \ldots, k\}$ where the latency between two successive measurements is arbitrarily set to $\delta k = 1$.

A recursive expression of the filtering distribution $p(\mathbf{x}_{0:k}|\mathbf{z}_{1:k})$, describing the distribution of the hidden Markov process we want to estimate conditioned upon the whole set of past observations $\mathbf{z}_{1:k}$, can be obtained from Bayes' law and the assumption that the measurements depends only on the current state:

$$p(\mathbf{x}_{0:k}|\mathbf{y}_{1:k}) = p(\mathbf{x}_{0:k-1}|\mathbf{y}_{1:k-1})\frac{p(\mathbf{y}_k|\mathbf{x}_k)p(\mathbf{x}_k|\mathbf{x}_{k-1})}{p(\mathbf{y}_k|\mathbf{y}_{1:k-1})}. \tag{9}$$

3.2   Linear Gaussian Models and the Kalman Filter

For a Gaussian initial distribution, linear dynamics and linear measurement operator, denoted by $\mathbf{M}$ and $\mathbf{H}$ respectively, the distribution $p(\mathbf{x}_k|\mathbf{y}_{1:k})$ remains a Gaussian distribution whose first and second moment, $\mathbf{x}_k^a = \mathbb{E}(\mathbf{x}_k/\mathbf{y}_{1:k})$ and $\mathbf{P}_k^a = \mathbb{E}((\mathbf{x}-\mathbf{x}_k^a)(\mathbf{x}-\mathbf{x}_k^a)^T/\mathbf{y}_{1:k})$, can be explicitly computed from the well known recursive Kalman equations [8]:

$$\mathbf{x}_k^f = \mathbf{M}\mathbf{x}_{k-1}^a, \ \ \mathbf{P}_k^f = \mathbf{M}\mathbf{P}_{k-1}^a\mathbf{M}^T + \mathbf{Q}_k, \tag{10}$$

and

$$\mathbf{K}_k = \mathbf{P}_k^f\mathbf{H}^T(\mathbf{H}\mathbf{P}_k^f\mathbf{H}^T + \mathbf{R})^{-1}, \mathbf{x}_k^a = \mathbf{x}_k^f + \mathbf{K}_k(\mathbf{y}_k - \mathbf{H}\mathbf{x}_k^f), \mathbf{P}_k^a = (\mathbb{I} - \mathbf{K}_k\mathbf{H})\mathbf{P}_k^f, \tag{11}$$

here superscripts f and a on state variable and covariance denote the respective quantities before and after analysis (update) at time k, respectively. The prediction or forecast step (10) brings forward the first two moments of the state vector, from its previous time step $k-1$, through the dynamical model parameters, while the analysis or the correction step (11) provides the first two moments of the state characterizing the Gaussian filtering distribution at time k. The matrix $\mathbf{K}_k$ is referred to as the Kalman gain matrix.

3.3   Particle Implementation of the Nonlinear Filtering

For nonlinear dynamics or nonlinear measurement equation a direct sampling from the filtering distribution is impossible since it would require the complete knowledge of the filtering distribution – which is in the general case a non Gaussian multimodal distribution – at a previous time.

Particle filtering techniques introduce a discrete approximation of the sought density as a sum of N weighted Diracs:

$$p(\mathbf{x}_{0:k}|\mathbf{y}_{1:k}) \approx \sum_{i=1}^{N} w_k^{(i)}\delta_{\mathbf{x}_{0:k}}(\mathbf{x}_{0:k}), \tag{12}$$

centered on hypothesized locations of the state space sampled from a proposal distribution $\pi(\mathbf{x}_{0:k}|\mathbf{z}_{1:k})$ (also called the importance distribution) approximating the true filtering distribution. Each sample is then weighted by a weight, $w_k^{(i)}$, accounting for the ratio between the two distributions. Any importance function can be chosen (with the only restriction that its support contains the filtering distribution one). Under weak hypotheses the importance ratio can be recursively defined as:

$$w_k^{(i)} \propto w_{k-1}^{(i)} \frac{p(\mathbf{y}_k|\mathbf{x}_k^{(i)})p(\mathbf{x}_k^{(i)}|\mathbf{x}_{k-1}^{(i)})}{\pi(\mathbf{x}_k^{(i)}|\mathbf{x}_{0:k-1}^{(i)}, \mathbf{y}_{1:k})}. \tag{13}$$

By propagating the particles from time $k-1$ through the proposal density $\pi(\mathbf{x}_k^{(i)}|\mathbf{x}_{0:k-1}^{(i)}, \mathbf{y}_{1:k})$, and by weighting the sampled states with $w_k^{(i)}$, a sampling of the filtering law is obtained. When the proposal distribution is set to the prior, the weights updating rule (13) simplifies to the data likelihood $p(\mathbf{y}_k|\mathbf{x}_k^{(i)})$. This particular instance of the particle filter is called the *Bootstrap filter* and constitutes the most common filtering method based on particle filter. Nevertheless, such an importance function does not take into account the current observation and depends only weakly on the past data through the filtering distribution estimated at the previous instant. High dimensional probability distribution spaces being excruciatingly difficult to sample, it is very important to devise an importance function that enables focusing on the most meaningful areas of the state space. To that end it is essential to consider proposal distributions that take into account more significantly the past and current measurements. Along this idea, the weighted ensemble Kalman filter defines the proposal distribution from the sampling mechanisms of ensemble Kalman filtering techniques.

3.4 Ensemble Kalman Filtering

The Ensemble Kalman filter [4] can be interpreted as a Monte Carlo implementation of the Kalman filter recursion for the propagation of the two first moments. The Ensemble filter relies hence intrinsically on a Gaussian approximation of the filtering distribution.

More precisely, let us assume that we have sampled N members from initial filtering distribution $p(\mathbf{x}_0/\mathbf{y}_0)$, denoted by $\mathbf{x}_0^{(i)}, i = 1, ..., N$. Propagating these samples, iteratively, through the Kalman prediction and correction steps, provides us the Gaussian approximations of the prediction and filtering distributions.

The prediction step consists in propagating the ensemble members $\mathbf{x}_{k-1}^{a,(i)}$ and their associated uncertainties (noise) through the state dynamics in order to obtain a predicted particles or forecast ensemble as:

$$\mathbf{x}_k^{f,(i)} = \sum_{t=k-1}^{k-\delta t} \left(M(\mathbf{x}_t^{f,(i)}) + \boldsymbol{\eta}_{t+\delta t}^{(i)} \right), \quad \mathbf{x}_{k-1}^{f,(i)} = \mathbf{x}_{k-1}^{a,(i)}. \tag{14}$$

From this, the empirical mean, $\overline{\mathbf{x}}_k^f$, of the forecast ensemble and the corresponding empirical forecast covariance matrix $\mathbf{P}_k^{fe}$ are computed. Using this ensemble based forecast covariance, an ensemble based Kalman gain matrix $\mathbf{K}_k^e$ can be computed. With this Kalman gain and the observation model the forecast ensemble members are then corrected towards the current observation.

This correction consists to update the forecast ensemble members $\mathbf{x}_k^{f,(i)}$, through the Kalman update equations, with a set of perturbed observation $\mathbf{y}_k + \boldsymbol{\gamma}_k^{(i)}$ obtained from samples of the observation noise $\{\boldsymbol{\gamma}_k^{(i)}, i = 1, ..., N\}$. This provides an analysis ensemble members $\{\mathbf{x}_k^{a,(i)}, i = 1, ..., N\}$ defined as:

$$\mathbf{x}_k^{a,(i)} = \mathbf{x}_k^{f,(i)} + \mathbf{K}_k^e\left(\mathbf{y}_k + \boldsymbol{\gamma}_k^{(i)} - \mathbf{H}\mathbf{x}_k^{f,(i)}\right). \tag{15}$$

Here, we note that, in the Kalman gain or in the update stage, computation of the high dimensional covariance matrix or inverse of the $n \times n$ covariance term, $(\mathbf{HP}_k^{f_e}\mathbf{H}^T + R)^{-1}$, are never explicitly computed nor stored. Rather, Kalman gain and update are efficiently implemented by defining and employing matrices with ensemble of perturbations. In most of the geophysical applications, the state vector related usually to temperature, pressure or velocity fields is of much higher dimension than the number of samples used in EnKF N. i.e., $n >> N$, thus, handling the perturbation matrices (instead of the actual corresponding covariance matrices) approximately brings down the number of operations from $O(n^2)$ to $O(nN)$. The inverse needed in the Kalman gain can be efficiently computed through the singular value decomposition of a $n \times N$ matrix [5].

3.5   Weighted EnKF

Starting from the descriptions of the previous section, a hybrid filtering procedure that takes advantage of both the particle filter and the EnKF can be devised. We briefly describe the approach proposed in [12].

The importance sampling principle indicates that a wide range of proposal distributions can be considered. We will experimentally show that a proposal distribution defined by the EnKF procedure constitutes an efficient proposal mechanism for particle filter techniques in high dimensional spaces.

Relying on the usual assumption of the EnKF (*i.e.* considering the dynamics as a discrete Gaussian system), the conditional distribution $p(\mathbf{x}_k|\mathbf{x}_k^{(i)}, \mathbf{y}_k^o)$ can be approached by a Gaussian distribution of respective mean and covariance [12]:

$$\overline{\boldsymbol{\mu}}_k^{(i)} = (\mathbb{1} - \mathbf{K}_k^e\mathbf{H})\sum_{t=k-1}^{k-\Delta t} M(\mathbf{x}_t^{f,(i)}) + \mathbf{K}_k^e\mathbf{y}_k^o, \quad \boldsymbol{\Sigma}_k^e - (\mathbb{1} - \mathbf{K}_k^e\mathbf{H})\mathbf{P}_k^{f_e}. \tag{16}$$

This distribution provides us a natural expression for the proposal distribution. In order to make the estimation of the filtering distribution exact (up to the sampling), each member of the ensemble must be weighted at each instant, k, with appropriate weights, $w_k^{(i)}$, defined from (13). With a systematic resampling scheme and for high dimensional systems represented on the basis of a very small number of particles the weights simplify as [12]:

$$w_k^{(i)} \propto p(\mathbf{y}_k^o|\mathbf{x}_k^{(i)}), \text{ and } \sum_{i=1}^N w_k^{(i)} = 1. \tag{17}$$

The Weighted ensemble Kalman filter (WEnKF) procedure can be simply summarized by the algorithm 1.

Algorithm 1. The WEnKF algorithm, one iteration.

Require: Ensemble at instant $k - 1$: $\{\mathbf{x}_{k-1}^{(i)}, i = 1, \ldots, N\}$
observations $\mathbf{y}_k^o$
Ensure: Ensemble at time k: $\{\mathbf{x}_k^{(i)}, i = 1, \ldots, N\}$
 EnKF step: Get $\mathbf{x}_k^{(i)}$ from the assimilation of $\mathbf{y}_k^o$ with an EnKF procedure;
 Compute the weights $w_k^{(i)}$ according to (17);
 Resample: For $j = 1 \ldots N$, sample with replacement index $a(j)$ from discrete probability $\{w_k^{(i)}, i = 1, \ldots, N\}$ over $\{1, \ldots, N\}$ and set $\mathbf{x}_k^{(j)} = \mathbf{x}_k^{a(j)}$;

4   WEnKF Assimilation of SLK Observations

In this Section, we present our WEnKF formulation based on the SLK optical flow estimates. In what follows, we detail the dynamical model, the observation model, and the strategy we adopt to incorporate the uncertainties supplied by the SLK estimator.

Dynamical model: As in this work we considered only 2D incompressible fluid flows, we will rely for the dynamics on the vorticity-velocity formulation of the Navier-Stokes equation with a stochastic forcing function:

$$d\xi = -\nabla\xi \cdot \mathbf{v}dt + \nu\Delta\xi dt + \eta dB, \tag{18}$$

where the state vector $\mathbf{x} = \xi = v_x - u_y$, represents the vorticity of the velocity field $\mathbf{v} = [u, v]^T$, ν is the kinematic viscosity and ηdB is a random forcing term (see following section). The velocity field can be recovered from its vorticity using the Biot-Savart kernel. The numerical simulation of this dynamical model is detailed in [12].

Observation model: The measurements on which we will rely on are set directly as the curl map (*i.e.* vorticity) of the SLK velocity estimates (4). Assuming the observation is a corrupted version of the true vorticity map (state), we define the observation model as:

$$\mathbf{y}_k = \mathbf{x}_k + \boldsymbol{\gamma}_k, \tag{19}$$

where $\boldsymbol{\gamma}_k$ is a Gaussian random field whose variance is fixed to the spatially varying uncertainties associated to the measurements. These uncertainties are provided by our motion estimator from equation (6) where the expectations have been approached with ensemble empirical mean over the displaced image corresponding to each ensemble members.To mitigate the effect of outliers a Gaussian smoothed version of these variances is considered.

Random fields sampling: To simulate the random forcing term dB in the dynamics (18) and the random field of the observation model (19), homogeneous Gaussian fields, correlated in space, but uncorrelated in time are used. Their covariance have a general form given by:

$$Q_{iso}(\mathbf{r}, \tau) = \mathbb{E}[d\mathbf{B}(\mathbf{x}, t)d\mathbf{B}^T(\mathbf{x} + \mathbf{r}, t + \tau)] = \mathbf{g}_\lambda(\mathbf{r})dt\delta(\tau), \tag{20}$$

where $\mathbf{g}_\lambda(\mathbf{r})$ describes the spatial correlation structure with cutoff parameter λ. These random fields are in practice sampled in the Fourier domain.

WEnKF implementation: With this dynamics and observation models the WEnKF can be directly implemented as follow. At $k = 0$, the ensemble of states $\{\mathbf{x}_0^{a,(i)}, i = 1, ..., N\}$ is initialized with noisy versions of the SLK vorticity map obtained from the two first images of the sequence. At the current time, the ensemble obtained at the previous measurement instant is propagated through the stochastic state dynamics (18) to generate the forecast ensemble members $\mathbf{x}_k^{f,(i)}$. The EnKF update is then performed with the new observation in order to sample the proposal distribution. The importance sampling weighting based on the likelihood and a resampling process of the particles with respect to those weights are performed. The empirical mean of the analysis ensemble provides the vorticity estimate at time k. The corresponding velocity field is finally obtained from the Biot-Savart law.

Although this direct WEnKF filtering of the SLK vorticity maps does provide good results as we shall see it, the estimation may fail for long range velocities. To overcome this limitation and to further improve the performance of the WEnKF we propose in the next section a multiscale extension of WEnKF.

5   Multiscale SLK-WEnKF Filtering

The idea of multiscale WEnKF consists to provide an improved proposal distribution from velocity measurements at different scales. The update step operates iteratively in an incremental coarse-to-fine way by introducing motion measurements obtained at different scales through the Gaussian smoothing parameter λ_ℓ in (4). More precisely, at scale $\ell \in [0, \ell_f]$ the proposal ensemble is build from successive analysis steps as follow:

$$\mathbf{x}_k^{a,(i),\ell} = \mathbf{x}_k^{f,(i),\ell} + \mathbf{K}_k^{e,\ell}\left(\mathbf{y}_k^\ell + \gamma_k^{(i),\ell} - \mathbf{H}\mathbf{x}_k^{f,(i),\ell}\right), \tag{21}$$

$$\mathbf{x}_k^{f,(i),\ell} = \mathbf{x}_k^{f,(i),\ell-1} - \overline{\mathbf{x}}_k^{a,\ell-1}, \tag{22}$$

where the measurements $\mathbf{y}_k^\ell$ are supplied by the stochastic Lucas and Kanade motion estimates between the backwarped image $\widetilde{I}_k^\ell = f(\mathbf{X}_{k-1} + \sum_{l=0}^{\ell-1} \overline{\mathbf{x}}_k^{a,l}, k)$ and image $I_{k-1} = f(\mathbf{X}_{k-1}, k-1)$ within the range of scale $[\lambda^{\ell-1}, \lambda^\ell]$. The quantity $\overline{\mathbf{x}}_k^{a,\ell}$ denotes the empirical mean of the analysis ensemble. The initial analysis ensemble is fixed to a null value ($\overline{\mathbf{x}}_k^{a,\ell_c-1} = 0$) and the initial forecast is set to the forecast ensemble computed from the dynamics ($\mathbf{x}_k^{f,(i),0} = \mathbf{x}_k^{f,(i)}$). At each scale, the Gaussian random fields attached to the measurements are drawn with the uncertainties provided by the stochastic Lucas and Kanade formulation computed from the couple of images ($\widetilde{I}^\ell, I_{k-1}$) and the current analysis ensemble (6). Let us note that compared to the previous single scale filtering where the proposal was based on a single ensemble Kalman update, here several updates associated to different Kalman gains are considered. In the experimental section three successive scales will be considered in such a filtering. The final proposal corresponds to the sum of the analysis ensemble obtained at the different scales: $\mathbf{x}_k^{a,(i)} = \sum_L^l \mathbf{x}_k^{a,(i),\ell}$. In the same way as for the previous filter, these ensemble members are then resampled according to the importance weights computed from the likelihood associated to the original couple of images (I_{k-1}, I_k).

6   Experimental Results and Comparisons

In this section, we present the results obtained by the application of the single scale and the multiscale WEnKF denoted as 1L-WEnKF and 3L-WEnKF respectively as the latter has been applied on a set of 3 three successive scale ranges. Those filters have been compared with state-of-the-art fluid motion estimators [6,11,14] on a sequence of 100 simulated PIV images with a known ground truth corresponding to the numerical simulation (DNS) of a forced 2D turbulence at Reynolds 3000 available at http://www.fluid.irisa.fr. Quantitative comparisons with the ground truth in terms of the Root-Mean-Square-Error (RMSE) of vorticity and velocity are both presented in figure 1.

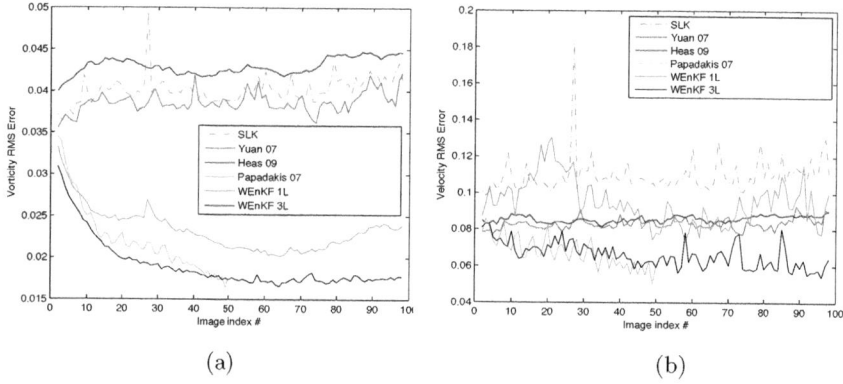

(a)   (b)

Fig. 1. Comparison with State of the Art: RMSE in (a) Vorticity (b) Motion field

As we notice from the fig. 1(a), the RMSE in vorticity of the SLK approach is close to the state of art approaches [14,6] (0.04), though the RMSE values in velocity are higher (fig. 1(b)). The RMSE in vorticity by assimilating the SLK observation through 1L-WEnKF is much lower (0.03), while the error in terms of velocity estimates is close to the approach of Yuan et al. [14]. However, the 3L-WEnKF assimilation shows better results both in terms of vorticity or velocity. These errors are lower than all the fluid motion estimators that have been tested and are at the same level as the errors provided by the batch variationnal assimilation techniques[2] [11] (which corresponds thus to a smoothing filter as opposed to a recursive filter as in our case).

---

[2] The results were unfortunately available only for 50 images.

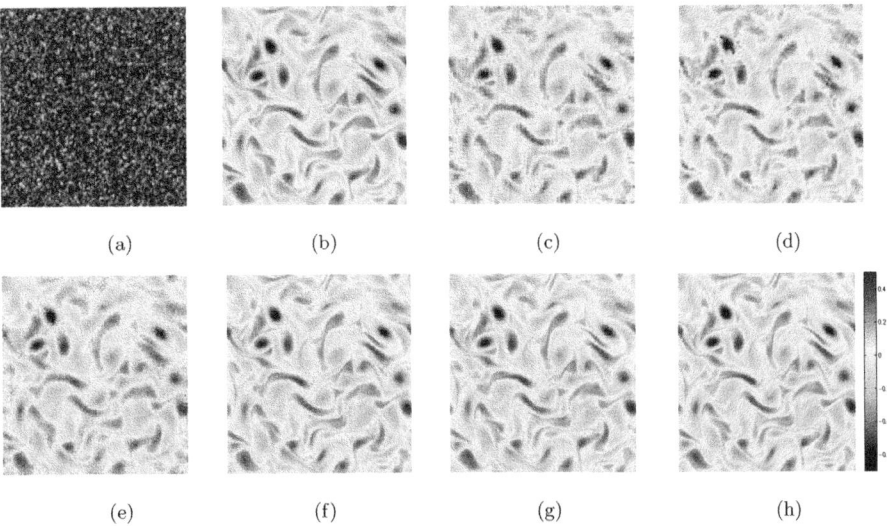

Fig. 2. (a) Particle image 50 (b) True vorticity and the estimates of (c) SLK (d) Yuan et al. [14] (e) Heas et. al. [6] (f) Papadakis et al. [11] (g) WEnKF assimilation and (h) 3L-WEnKF assimilation

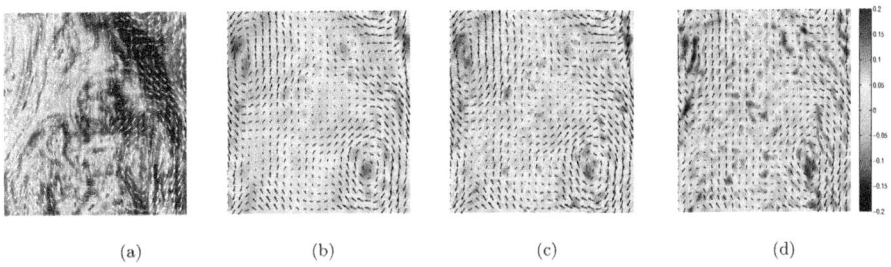

Fig. 3. Real image sequence of a 2D turbulent flow of a soap film (a) an image with the 3L WEnKF velocity field superimposed; Vorticity maps with their corresponding velocity fields (b) SLK (c) 1L- WEnKF and (d) 3L-WEnKF.

For a visual comparison we show in fig. 2, the vorticity maps obtained by the different methods for the 50th images of the sequence. The vorticity estimated by the 3L-WEnKF assimilation (fig. 2 (h)) corresponds to the lowest error.

Our next set of results corresponds to a real world image sequence of a 2D turbulence generated from the wake of a soap film behind a comb. The flow is visualized through a Schlieren technique at a rate of 2500 frames per second. A typical image of the sequence is shown in figure 3(a) in false color. The estimated vorticity maps and velocity fields corresponding to SLK, 1L WEnKF and 3L WEnKF are shown in figs. 3(b), (c) and (d), respectively. We note that though the 1L-WEnKF assimilation of SLK brings out some details at a smaller scale than the SLK measurements, the 3L-WEnKF assimilation recovers even finer details. It is however important to remarks that all those results remains consistent and are close when interpreted at a larger scale.

7   Conclusion

In this paper, we have proposed an efficient multiscale extension of the Weighted Ensemble Kalman filter for fluid flow motion estimation problem. This filter is a particle filter relying on a proposal distribution built from the ensemble Kalman filtering mechanism. The particular instance we considered here incorporates measurements issued from a stochastic extension of the Lucas and Kanade estimator.

Acknowledgements. The authors acknowledge the support of the French Agence Nationale de la Recherche (ANR), under grant PREVASSEMBLE (ANR-08-COSI-012).

References

1. Corpetti, T., Heas, P., Memin, E., Papadakis, N.: Pressure image asimilation for atmospheric motion estimation. Tellus 61A, 160–178 (2009)
2. Corpetti, T., Heitz, D., Arroyo, G., Memin, E., Santa-Cruz, A.: Fluid experimental flow estimation based on an optical-flow scheme. Experiments in Fluids 40, 80–97 (2006)
3. Corpetti, T., Memin, E.: Stochastic Models for Local Optical Flow Estimation. In: Bruckstein, A.M., et al. (eds.) SSVM 2011. LNCS, vol. 6667, pp. 701–712. Springer, Heidelberg (2011)
4. Evensen, G.: Sequential data assimilation with a non linear quasi-geostrophic model using Monte Carlo methods to forecast error statistics. J. Geophys. Res. 99(C5)(10), 143–162 (1994)
5. Evensen, G.: The ensemble Kalman filter, theoretical formulation and practical implementation. Ocean Dynamics 53(4), 343–367 (2003)
6. Heas, P., Memin, E., Heitz, D., Mininni, P.: Bayesian selection of scaling laws for motion modeling in images. In: Proc. Int. Conf. Computer Vision (2009)
7. Heitz, D., Memin, E., Schnoerr, C.: Variational fluid flow measurements from image sequences: synopsis and perspectives. Exp. in Fluids 48(3), 369–393 (2010)
8. Kalman, R.E.: A new approach to linear filtering and prediction problems. Transactions of the ASME - Journal of Basic Engineering 82, 35–45 (1960)
9. Lucas, B., Kanade, T.: An iterative image registration technique with an application to stereovision. In: Int. Joint Conf. on Artificial Intel. (IJCAI), pp. 674–679 (1981)
10. Oksendal, B.: Stochastic differential equations. Spinger, Heidelberg (1998)
11. Papadakis, N., Memin, E.: An optimal control technique for fluid motion estimation. SIAM Journal on Imaging Sciences 1(4), 343–363 (2008)
12. Papadakis, N., Memin, E., Cuzol, A., Gengembre, N.: Data assimilation with the weighted ensemble kalman filter. Tellus-A 62(5), 673–697 (2010)
13. Ruhnau, P., Kohlberger, T., Schnoerr, C., Nobach, H.: Variational optical flow estimation for particle image velocimetry. Exp. in Fluids 38, 21–32 (2005)
14. Yuan, J., Schnoerr, C., Memin, E.: Discrete orthogonal decomposition and variational fluid flow estimation. J. Mathematical Imaging and Vision 28(1), 67–80 (2007)

Over-Parameterized Optical Flow Using a Stereoscopic Constraint

Guy Rosman, Shachar Shem-Tov, David Bitton, Tal Nir,
Gilad Adiv, Ron Kimmel, Arie Feuer, and Alfred M. Bruckstein*

Abstract. The success of variational methods for optical flow computation lies in their ability to regularize the problem at a differential (pixel) level and combine piecewise smoothness of the flow field with the brightness constancy assumptions. However, the piecewise smoothness assumption is often motivated by heuristic or algorithmic considerations. Lately, new priors were proposed to exploit the structural properties of the flow. Yet, most of them still utilize a generic regularization term.

In this paper we consider optical flow estimation in static scenes. We show that introducing a suitable motion model for the optical flow allows us to pose the regularization term as a geometrically meaningful one. The proposed method assumes that the visible surface can be approximated by a piecewise smooth planar manifold. Accordingly, the optical flow between two consecutive frames can be locally regarded as a homography consistent with the epipolar geometry and defined by only three parameters at each pixel. These parameters are directly related to the equation of the scene local tangent plane, so that their spatial variations should be relatively small, except for creases and depth discontinuities. This leads to a regularization term that measures the total variation of the model parameters and can be extended to a Mumford-Shah segmentation of the visible surface. This new technique yields significant improvements over state of the art optical flow computation methods for static scenes.

Keywords: Optical Flow, Epipolar Geometry, Ambrosio-Tortorelli.

1 Introduction

Optical flow is defined as the motion field between consecutive frames in a video sequence. Its computation often relies on the brightness constancy assumption, which states that pixel brightness corresponding to a given scene point is constant throughout the sequence. Optical flow computation is a notoriously ill-posed problem. Hence, additional assumptions on the motion are made in order to regularize the problem. Early methods assumed spatial smoothness of the optical flow [1,2]. Parametric motion models [3,4], and more recently machine learning [5] were introduced in order to take into account the specificity of naturally occurring video sequences. In parallel, the regularization process was made much more robust [6,7,8,9].

In this paper, we focus on optical flow computation in stereoscopic image pairs, given a reliable estimation of the fundamental matrix. This problem has already been

---

* This research was supported by the Israel Science foundation (ISF) grant no. 1551/09.

A.M. Bruckstein et al. (Eds.): SSVM 2011, LNCS 6667, pp. 761–772, 2012.

addressed in [10,11,12,13]. The papers [10,11] expressed the optical flow as a one-dimensional problem. This was done either by working on a rectified image pair [10], or by solving for the displacement along the epipolar lines [11]. A different approach [12,13] merely penalized deviation from the epipolar constraint. In addition, [12] proposed a joint estimation of the stereoscopic optical flow and the fundamental matrix. Finally, in order to treat the problem of occluded areas and object boundaries, Ben-Ari and Sochen [14] suggest to explicitly account for regions of discontinuities.

Yet a third body of works turned to a complete modeling of the scene flow [15,16,17]. While this approach is the most general, we focus in this paper on static scenes, for which a more specific parameterization can be found.

While the reported experimental results in the aforementioned papers are very convincing, their regularization methods still rely on the traditional assumption that optical flow should be piecewise smooth. Here, motivated by the over-parameterization approach presented in [18], the optical flow is obtained by estimation of the space-time dependent parameters of a motion model, the regularization being applied to the model parameters. In [19], we used homogeneous coordinates to express a homography model, which allows to select a geometrically meaningful coordinate systems for this problem. Here we elaborate upon this model by adding an Ambrosio-Tortorelli scheme, which gives a physically meaningful interpretation for the minima obtained in the optimization process.

In the case of a static scene, the optical flow can be factored into a model determined by the camera motion and an over-parameterized representation of the scene. The scene motion is described locally as a homography satisfying the epipolar constraint and parameterized by the equation of a local planar approximation of the scene. Assuming that the scene can be approximated by a piecewise smooth manifold, enforcing piecewise spatial smoothness on the homography parameters becomes an axiomatically justified regularization criterion which favors piecewise smooth planar regions.

2   Background

2.1   The Variational Framework

In the variational framework for optical flow, brightness constancy and smoothness assumptions are integrated in an energy functional. Let $(u(x, y, t), v(x, y, t))$ denote the optical flow at pixel coordinates (x, y) and time t. Brightness constancy determines the data term of the energy functional

$$E_D(u, v) = \int \Psi\left(I_z^2\right),\tag{1}$$

where

$$I_z = I(x + u, y + v, t + 1) - I(x, y, t)\tag{2}$$

and $\Psi(s^2) = \sqrt{s^2 + \varepsilon^2}$ is a convex approximation of the L_1 norm for a small ε.

$\mathcal{M}(a, x, y, t)$ denotes a generic model of the optical flow at pixel (x, y) and time t, where $a = (a_i(x, y, t))_{i \in \{1,...,n\}}$ is a family of functions parameterizing the model, i.e.,

$$\begin{pmatrix} u(x, y, t) \\ v(x, y, t) \end{pmatrix} = \mathcal{M}(a, x, y, t).\tag{3}$$

We begin with the smoothness term proposed by Nir et al. in [18],

$$E_S(\mathbf{a}) = \int \Psi \left(\sum_{i=1}^{n} ||\nabla a_i||^2 \right). \tag{4}$$

In order to refine the discontinuities and obtain a physically meaningful regularization, we extend the smoothness prior using the Ambrosio-Tortorelli scheme [20,21].

$$E_{S,AT}(\mathbf{a}) = \int v_{AT}^2 \Psi \left(\sum_{i=1}^{n} ||\nabla a_i||^2 \right) + \epsilon_1 (1 - v_{AT})^2 + \epsilon_2 ||\nabla v_{AT}||^2, \tag{5}$$

where v_{AT} is a diffusivity function, ideally serving as an indicator of the discontinuities set in the flow field. Choosing $\epsilon_1 = \frac{1}{\epsilon_2}$ and gradually decreasing ϵ_2 towards 0 can be used to approximate the Mumford-Shah [22] model via Γ-convergence process, but we do not pursue this direction in this paper.

While the Ambrosio-Tortorelli scheme has been used in the context of optical flow [23,24,25], in our case this seemingly arbitrary choice of regularization and segmentation has a physical meaning. The regularization of the flow becomes a segmentation process of the *visible surface* in the scene into planar patches, each with his own set of plane parameters. In addition, it helps us obtain accurate edges in the resulting flow.

Furthermore, the generalized Ambrosio-Tortorelli scheme allows us to explicitly reason about the places in the flow where the nonlinear nature of the data manifold manifests itself. Suppose we have a piecewise-planar, static, scene, and an ideal solution (a^*, v_{AT}^*) where a^* is piecewise constant, and the diffusivity function v_{AT}^* is 0 at planar region boundaries and 1 elsewhere. At such a solution, we expect two neighboring points which belong to different regions to have a very small diffusivity value v_{AT} connecting them, effectively nullifying the interaction between different planes' parameters. Furthermore the cost associated with this solution is directly attributed to the discontinuity set measure in the image. The proposed ideal solution therefore becomes a global minimizer of the functional, as determined by the measure of discontinuities in the $2\frac{1}{2}$-*D sketch* [26]. This is directly related to the question raised by Trobin et al. [27] regarding the over-parameterized affine flow model and its global minimizers.

The complete functional now becomes:

$$E(\mathbf{a}) = E_D(\mathcal{M}(\mathbf{a}, x, y, t)) + \alpha E_{S,AT}(\mathbf{a}). \tag{6}$$

In the remainder of this paper, we will propose a motion model enforcing the epipolar constraint and show how to minimize the proposed functional.

2.2 Epipolar Geometry

Let us introduce some background on epipolar geometry, so as to motivate the choice of the motion model. A complete overview can be found in [28,29].

Given two views of a static scene, the optical flow is restricted by the epipolar constraint. Figure 1 shows that a pixel $\mathbf{m}$ in the left image is restricted to a line $\mathbf{l}'$ called an epipolar line in the right image. All the epipolar lines in the left (resp. right) image go through $\mathbf{e}$ (resp. $\mathbf{e}'$), which is called the left (resp. right) epipole.

In projective geometry, image points and lines are often represented by 3D homogeneous coordinates

$$\mathbf{m} = \left\{ \lambda \begin{pmatrix} x \\ y \\ 1 \end{pmatrix} \mid \lambda \in \mathbb{R}^{\star} \right\}. \tag{7}$$

Image points and their corresponding epipolar lines are related by the fundamental matrix $\mathcal{F}$

$$\mathbf{l}' = \mathcal{F}\mathbf{m}. \tag{8}$$

Consider a plane π, visible from both cameras, and the planar homography H_π which corresponds to the composition of the back-projection from the left view to a plane (π) and the projection from (π) to the right view (see Figure 1). The homography H_π gives rise to a useful decomposition of the fundamental matrix

$$\mathcal{F} = [\mathbf{e}']_\times H_\pi, \tag{9}$$

where $[\mathbf{e}']_\times$ is a matrix representation of the cross product with $\mathbf{e}'$.

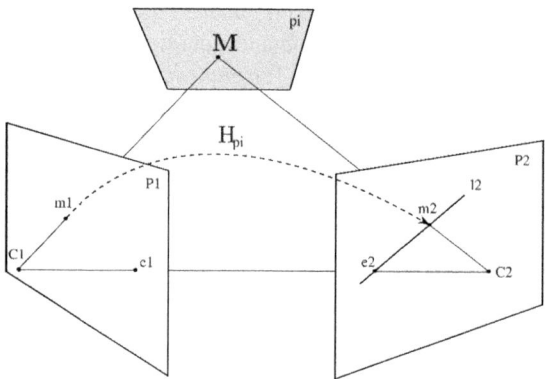

Fig. 1. Epipolar geometry

3   Estimation of the Fundamental Matrix

One of the main challenges in estimating optical flow using the epipolar geometry is to retrieve an accurate and robust estimation of the fundamental matrix. Mainberger et. al. [30] showed that robustness of the fundamental matrix estimation could be achieved by using dense optical flow instead of applying RANSAC or LMedS methods to a sparse set of matches. Hence, we use as initialization the Horn-Schunck with Charbonnier penalty function optical flow implementation provided by Sun et al. [31], modified to use color images. This represents a baseline nonlinear optical flow method, as in [31]. In addition to allowing the computation of the fundamental matrix, this initialization also serves as a starting point for our optical flow computation algorithm.

Many methods aimed at estimating the fundamental matrix can handle large numbers of correspondences. Among those, we choose a robust M-estimation method based on the symmetric epipolar distance, the implementation of which is made very efficient by the use of the Levenberg-Marquardt algorithm, as explained in [32].

4 A Flow Model Based on Local Homographies

We now proceed to develop the model and motivation for the flow equations. Suppose the camera is calibrated, with projection matrices

$$P(t) = P_0 = (I \,|\, \mathbf{0}), \quad P(t+1) = P_1 = (R \,|\, \mathbf{t}). \tag{10}$$

where R is a rotation matrix and $\mathbf{t}$ is a translation vector expressing camera motion between the two consecutive frames at t and $t+1$. We assume that locally, the scene is well approximated by the plane

$$\mathbf{v}^T \mathbf{x} + d = 0 \tag{11}$$

where $(\mathbf{x}^T, d)^T = (x, y, 1, d)^T$ denotes the 3D scene point visible at pixel $\mathbf{x}$ in homogeneous coordinates. The corresponding point of $\mathbf{x}$ at time $t+1$ is

$$\mathbf{x}' = P_1 \begin{pmatrix} \mathbf{x} \\ d \end{pmatrix} = R\mathbf{x} + \mathbf{t}d = (R - \mathbf{t}\mathbf{v}^T)\mathbf{x} \tag{12}$$

in homogeneous coordinates. $\mathbf{v}$ designates the normal of the local planar approximation of the scene, and $-(\mathbf{v}^T\mathbf{x})^{-1}$ is the depth of the scene at time t. The planar homography expressed in (12) gives a geometrically meaningful motion model parameterized by $\mathbf{v}$. From now on, consider $\mathbf{v}$ as a function of the pixel coordinates. Under the assumption that the scene can be approximated by a piecewise smooth manifold, $\mathbf{v}$ must be piecewise smooth.

We now derive the motion parameterization. In general, the camera parameters are not known, but we can re-parameterize the planar homography using $\mathbf{e}'$ and $\mathcal{F}$. In the following derivation we assume a calibrated view for simplicity's sake. Let $H(x, y, t)$ denote the planar homography motion model. We have

$$H \propto R - \mathbf{t}\mathbf{v}^T. \tag{13}$$

For any compatible planar homography H_0 (cf. [29], 13.1.1.1, we will provide a specific choice later on),

$$\exists (\mathbf{v_0}, \mu): \quad H_0 = \mu(R - \mathbf{t}\mathbf{v_0}^T) \tag{14}$$

$$H = H_0 - \mu\mathbf{t}(\mathbf{v} - \mathbf{v_0})^T. \tag{15}$$

As $\mathbf{t}$ and $\mathbf{e}'$ are parallel, we can also write

$$H = H_0 + \mathbf{e}' \frac{-\mu\mathbf{e}'^T\mathbf{t}}{||\mathbf{e}'||^2} (\mathbf{v} - \mathbf{v_0})^T. \tag{16}$$

Hence, $H(x, y, t)$ can be parameterized by the function

$$\mathbf{a}(x, y, t) = \frac{-\mu e'^{\mathrm{T}} \mathbf{t}}{||e'||^2} (\mathbf{v}(x, y, t) - \mathbf{v_0}), \tag{17}$$

so that

$$H(x, y, t) = H_0 + e' \mathbf{a}(x, y, t)^{\mathrm{T}}. \tag{18}$$

The parameterization $\mathbf{a}$ is the unknown field we want to compute in order to model and estimate the optical flow. The piecewise smoothness of $\mathbf{a}$ is a direct consequence of the piecewise smoothness of $\mathbf{v}$, as testified by (17). More precisely, minimization of the Ambrosio-Tortorelli regularization term favors segmentation of the visible surface into planar patches where the data evidence permits it.

When the cameras are not calibrated, the relationship between the parameterization $\mathbf{a}$ and $\mathbf{v}$ is still linear. In fact, the calibration matrices mainly affect the relative weighting of the model parameters smoothness. Our experiments show that even without controlling the relative smoothness of the model parameters, the optical flow can be estimated accurately.

Note that the parameterization $\mathbf{a}$ can also be derived directly from the fundamental matrix decomposition (9).

For H_0, we can choose the special matrix

$$H_0 = S = [e']_\times \mathcal{F}. \tag{19}$$

Each column of S with the corresponding column of $\mathcal{F}$ and e' form an orthogonal basis of $\mathbb{R}^3$ so that (9) is satisfied. S is a degenerate homography which projects points in the left image to points of the line represented by e' in the right image. Next, we use the notations

$$\mathbf{x} = \begin{pmatrix} x_1 \\ x_2 \\ x_3 \end{pmatrix}, \quad e' = \begin{pmatrix} x_{e'} \\ y_{e'} \\ z_{e'} \end{pmatrix}, \quad H_0 = \begin{pmatrix} \mathbf{h_1}^{\mathrm{T}} \\ \mathbf{h_2}^{\mathrm{T}} \\ \mathbf{h_3}^{\mathrm{T}} \end{pmatrix}, \tag{20}$$

to signify the 3D point coordinates, the epipole's 2D homogeneous coordinates, and the homography matrix rows, respectively. The parameterization of H is introduced into the expression of the optical flow

$$\mathcal{M}(\mathbf{a}, x, y, t) = \begin{pmatrix} u \\ v \end{pmatrix} = \lambda \begin{pmatrix} \mathbf{h_1}^{\mathrm{T}} \mathbf{x} + x_{e'} \mathbf{a}^{\mathrm{T}} \mathbf{x} \\ \mathbf{h_2}^{\mathrm{T}} \mathbf{x} + y_{e'} \mathbf{a}^{\mathrm{T}} \mathbf{x} \end{pmatrix} - \begin{pmatrix} x \\ y \end{pmatrix}, \quad \lambda = \frac{1}{\mathbf{h_3}^{\mathrm{T}} \mathbf{x} + z_{e'} \mathbf{a}^{\mathrm{T}} \mathbf{x}}. \tag{21}$$

where $\begin{pmatrix} x \\ y \end{pmatrix}$ are the corresponding pixels in the left image.

4.1 Euler-Lagrange Equations

By interchangeably fixing $a_i, i = 1...n$ and v_{AT}, we obtain the Euler-Lagrange equations which minimize the functional.

Minimization with respect to a_i. Fixing v_{AT}, we obtain

$$\forall i, \quad \nabla_{a_i}(E_D + \alpha v_{AT}^2 E_S) = 0. \tag{22}$$

the variation of the data term with respect to the model parameter function a_i is given by

$$\nabla_{a_i} E_D(u, v) = 2\Psi'\left(I_z^2\right) I_z \nabla_{a_i} I_z, \tag{23}$$

where

$$\nabla_{a_i} I_z = \lambda^2 x_i (x_{e'} \mathbf{h_3}^T \mathbf{x} - z_{e'} \mathbf{h_1}^T \mathbf{x}) I_x^+ + \lambda^2 x_i (y_{e'} \mathbf{h_3}^T \mathbf{x} - z_{e'} \mathbf{h_2}^T \mathbf{x}) I_y^+, \tag{24}$$

and

$$I_x^+ = I_x(x + u, y + v, t + 1) \tag{25}$$
$$I_y^+ = I_y(x + u, y + v, t + 1). \tag{26}$$

For the smoothness term, the Euler-Lagrange equations are

$$\nabla_{a_i} E_s = 2v_{AT}\Psi\left(\sum_{i=1}^n ||\nabla a_i||^2\right) + 2v_{AT}^2 div\left(\Psi'\left(\sum_j ||\nabla a_j||^2\right)\nabla a_i\right) \tag{27}$$

thus, the energy is minimized by solving the nonlinear system of equations

$$\Psi'\left(I_z^2\right) I_z \nabla_{a_i} I_z - \alpha\nabla\left(v_{AT}^2\Psi'\left(\sum_{i=1}^n ||\nabla a_i||^2\right)\right)^T \nabla a_i -$$
$$\alpha v_{AT}^2 div\left(\Psi'\left(\sum_j ||\nabla a_j||^2\right)\nabla a_i\right) = 0. \tag{28}$$

Minimization with respect to v_{AT}. Fixing a_i, we obtain

$$2\alpha v_{AT}\Psi\left(\sum_{i=1}^n ||\nabla a_i||^2\right) + 2\epsilon_1(v_{AT} - 1) - \epsilon_2 \Delta v_{AT} = 0 \tag{29}$$

4.2 Implementation

Minimization with respect to v_{AT} is straightforward, as the equations are linear with respect to v_{AT}, therefore we will only elaborate on the minimization with respect to a_i

The nonlinear Euler-Lagrange equation minimizing a_i, are linearized by adopting three embedded loops, similarly to [18]. First, the warped image gradient (I_x^+, I_y^+) is frozen, and so is λ. At each iteration k, we have

$$(\nabla_{a_i} I_z)^k = x_i d^k \tag{30}$$

where

$$d^k = (\lambda^k)^2 (x_{e'} \mathbf{h_3}^\mathrm{T} \mathbf{x} - z_{e'} \mathbf{h_1}^\mathrm{T} \mathbf{x})(I_x^+)^k$$
$$+ (\lambda^k)^2 (y_{e'} \mathbf{h_3}^\mathrm{T} \mathbf{x} - z_{e'} \mathbf{h_2}^\mathrm{T} \mathbf{x})(I_y^+)^k,$$

and the following approximation is made using first order Taylor expansions

$$I_z^{k+1} \approx I_z^k + d^k \sum_{i=1}^{3} x_i da_i^{\ k} \tag{31}$$

where

$$da^k = a^{k+1} - a^k. \tag{32}$$

The system of equations (28) becomes

$$\Psi'\left((I_z^{k+1})^2\right)\left(I_z^k + d^k \sum_{j=1}^{3} x_j da_j^{\ k}\right) x_i d^k - \alpha \operatorname{div}\left(\Psi'\left(\sum_j ||\nabla a_j^{\ k+1}||^2\right)\nabla a_i^{\ k+1}\right) = 0.$$

A second loop with superscript l is added to cope with the nonlinearity of Ψ'.

$$(\Psi')_{\mathrm{Data}}^{k,l}\left(I_z^k + d^k \sum_{j=1}^{3} x_j da_j^{\ k,l+1}\right) x_i d^k - \alpha \operatorname{div}\left((\Psi')_{\mathrm{Smooth}}^{k,l}\nabla a_i^{\ k,l+1}\right) = 0$$

where

$$(\Psi')_{\mathrm{Data}}^{k,l} = \Psi'\left(\left(I_z^k + d^k \sum_{i=1}^{3} x_i da_i^{\ k,l}\right)^2\right), \quad (\Psi')_{\mathrm{Smooth}}^{k,l} = \Psi'\left(\sum_j ||\nabla a_j^{\ k,l}||^2\right).$$

At this point, the system of equations is linear and sparse in the spatial domain. The solution a, as well as the diffusivity term v_{AT} are obtained through Gauss-Seidel iterations. In the case of the Ambrosio-Tortorelli regularization term, the diffusion term of the equation is modulated by v_{AT}.

5   Experimental Results

We now demonstrate motion estimation results using our algorithm, both visually and in terms of the average angular error (AAE). No post-processing was applied to the optical flow field obtained after energy minimization. The algorithm was tested on image pairs from the Middlebury optical flow test set [33], as well as all images with a static scene and publicly available ground truth optical flow from the training set. Results from the training set are presented in Table 1.

The flow, parameters, and diffusivity field resulting from our method are presented in Figure 3. The optical flow is shown with color encoding and a disparity map.

Results from the test set are shown in Figure 2. A smoothness parameter α of 400 was used in all experiments, and the Ambrosio-Tortorelli coefficients were set to $\epsilon_1 =$

$20, \epsilon_2 = 5 \times 10^{-5}$. The proposed method produced the best results to date on the static Yosemite and Urban scenes. The algorithm is not designed, however, for non-static scenes, where the computed epipolar lines have no meaning. One possible solution to this shortcoming is to resort to a 2D search [13]. Such a combined approach is left for future work.

In the Teddy and Grove test images, the initialization of our algorithm introduced errors in significant parts of the image, which our method could not overcome. This behavior is related to the problem of finding a global minimum for the optical flow, which is known to have several local minima. Improving the global convergence using discrete graph-based techniques, has been the focus of several papers (see [34,35,36], for example), and is beyond the scope of this work. We expect better initialization to improve the accuracy to that of the Yosemite and Urban image pairs.

Our optical flow estimation for the Yosemite and Urban sequences gives the best results to date, achieving an AAE of 1.25 for the Yosemite sequence test pair and 2.38 for the Urban sequence, as shown in Figure 2. When the fundamental matrix estimate was improved (by estimating from the ground truth optical flow), we reduced the AAE to 0.66 for Yosemite!

Table 1. AAE comparison for static scenes of the Middlebury training set and for the Yosemite sequence

	AAE	STD
Grove2	2.41	7.16
Grove3	5.53	15.76
Urban2	2.15	9.22
Urban3	3.84	16.88
Venus	4.29	12.01
Yosemite	0.85	1.24

(a) Middlebury training set

Method	AAE	Method	AAE
Brox et al. [7]	1.59	Roth/Black [5]	1.43
Mémin/Pérez [4]	1.58	Valgaerts et al. [12]	1.17
Bruhn et al. [8]	1.46	Nir et al. [18]	1.15
Amiaz et al. [37]	1.44	Our method	0.85

(b) Yosemite sequence

Fig. 2. Average angular error values of our algorithm, compared on the middlebury test set. The smoothness coefficient was set to $\alpha = 400$ in all experiments. Red marks the row of the suggested algorithm.

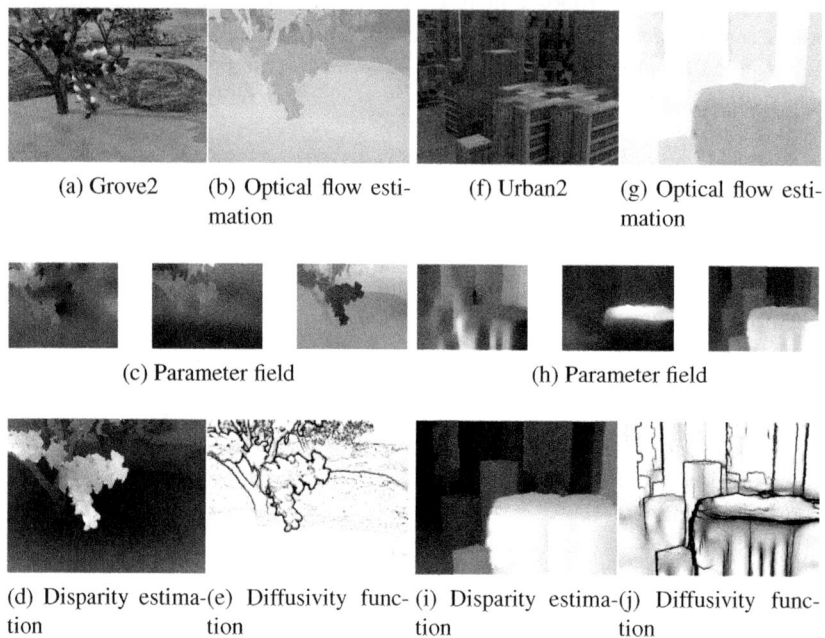

(a) Grove2   (b) Optical flow esti-   (f) Urban2   (g) Optical flow estimation   mation

(c) Parameter field   (h) Parameter field

(d) Disparity estima-(e) Diffusivity func-(i) Disparity estima-(j) Diffusivity function   tion   tion   tion

Fig. 3. Grove2 and Urban2 sequence results

It is interesting to look at the results obtained for scenes with planar regions, such as the Urban2 (Figure 3) image pair. In Urban2, the scene is composed of many planar patches, modeled by constant patches in the model parameters. In both these scenes, as well as others, the resulting diffusivity field clearly marks the contours of planar regions in the image such as the buildings in Urban2 and the tree and soil ridges in Grove2.

6   Conclusions

A new method for optical flow computation was presented, which hinges on a guiding principle that optic flow regularization should have a strong theoretical foundation. The method is applicable to static scenes and retrieves meaningful local motion parameters related to the scene geometry. At each pixel, the parameters provide an estimation of the plane tangent to the scene manifold, up to a fixed shift and scale. To that extent, they can be seen as a higher level output than optical flow in the computer vision hierarchy.

An interesting aspect of our energy functional, which was already mentioned in [18], is that given a carefully selected over-complete parameter field, the different parameters support each other to find a smooth piecewise constant parameter patches, while the incorporated Ambrosio-Tortorelli scheme prevents diffusion across discontinuities. Furthermore, the Ambrosio-Tortorelli scheme allows us to combine regularization and segmentation, resulting in a physically meaningful regularization process, while minimizing the dependency on the relative scaling of the coefficients.

Finally, although the performance demonstrated already goes beyond the latest published results, there is still much gain to be expected from better fundamental matrix estimation and algorithm initialization. In addition, when more than two frames are available and the camera pose is known, augmenting the model with time-smoothness is expected to systematically improve the results.

References

1. Horn, B.K.P., Schunck, B.G.: Determining optical flow. Artificial Intelligence 17, 185–203 (1981)
2. Lucas, B.D., Kanade, T.: An iterative image registration technique with an application to stereo vision. In: International Joint Conference on Artificial Intelligence, pp. 674–679 (1981)
3. Bergen, J., Anandan, P., Hanna, K., Hingorani, R.: Hierarchical model-based motion estimation. In: Sandini, G. (ed.) ECCV 1992. LNCS, vol. 588, pp. 237–252. Springer, Heidelberg (1992)
4. Mémin, E., Pérez, P.: Hierarchical estimation and segmentation of dense motion fields. International Journal of Computer Vision 46, 129–155 (2002)
5. Roth, S., Black, M.J.: On the spatial statistics of optical flow. International Journal of Computer Vision 74, 33–50 (2007)
6. Black, M.J., Anandan, P.: A framework for the robust estimation of optical flow. In: International Conference on Computer Vision, pp. 231–236 (1993)
7. Brox, T., Bruhn, A., Papenberg, N., Weickert, J.: High accuracy optical flow estimation based on a theory for warping. In: Pajdla, T., Matas, J(G.) (eds.) ECCV 2004. LNCS, vol. 3024, pp. 25–36. Springer, Heidelberg (2004)
8. Bruhn, A., Weickert, J., Schnörr, C.: Lucas/Kanade meets Horn/Schunck: Combining local and global optic flow methods. International Journal of Computer Vision 61, 211–231 (2005)
9. Cohen, I.: Nonlinear variational method for optical flow computation. In: Proc. Eighth Scandinavian Conference on Image Analysis, vol. 1, pp. 523–530 (1993)
10. Birchfield, S., Tomasi, C.: Depth discontinuities by pixel-to-pixel stereo. International Journal of Computer Vision 35, 269–293 (1999)
11. Slesareva, N., Bruhn, A., Weickert, J.: Optic flow goes stereo: A variational method for estimating discontinuity-preserving dense disparity maps. In: Kropatsch, W.G., Sablatnig, R., Hanbury, A. (eds.) DAGM 2005. LNCS, vol. 3663, pp. 33–40. Springer, Heidelberg (2005)
12. Valgaerts, L., Bruhn, A., Weickert, J.: A variational model for the joint recovery of the fundamental matrix and the optical flow. In: Rigoll, G. (ed.) DAGM 2008. LNCS, vol. 5096, pp. 314–324. Springer, Heidelberg (2008)
13. Wedel, A., Pock, T., Braun, J., Franke, U., Cremers, D.: Duality TV-L1 flow with fundamental matrix prior. Image Vision and Computing New Zealand, 1–6 (2008)
14. Ben-Ari, R., Sochen, N.: Variational stereo vision with sharp discontinuities and occlusion handling. In: International Conference on Computer Vision, pp. 1–7. IEEE Computer Society (2007)
15. Pons, J.P., Keriven, R., Faugeras, O., Hermosillo, G.: Variational stereovision and 3d scene flow estimation with statistical similarity measures. In: International Conference on Computer Vision, vol. 1, p. 597 (2003)
16. Huguet, F., Devernay, F.: A variational method for scene flow estimation from stereo sequences. In: Computer Vision and Pattern Recognition, pp. 1–7 (2007)
17. Basha, T., Moses, Y., Kiryati, N.: Multi-view scene flow estimation: A view centered variational approach. In: Computer Vision and Pattern Recognition, pp. 1506–1513 (2010)

18. Nir, T., Bruckstein, A.M., Kimmel, R.: Over-parameterized variational optical flow. International Journal of Computer Vision 76, 205–216 (2008)
19. Anonymous: Over-parameterized optical flow using a stereoscopic constraint (Technical report)
20. Ambrosio, L., Tortorelli, V.M.: Approximation of functional depending on jumps by elliptic functional via Γ-convergence. Communications on Pure and Applied Mathematics 43, 999–1036 (1990)
21. Shah, J.: A common framework for curve evolution, segmentation and anisotropic diffusion. In: Proceedings of the 1996 Conference on Computer Vision and Pattern Recognition, CVPR 1996, p. 136. IEEE Computer Society, Washington, DC (1996)
22. Mumford, D., Shah, J.: Optimal approximations by piecewise smooth functions and associated variational problems. Communications on Pure and Applied Mathematics 42, 577–685 (1989)
23. Amiaz, T., Kiryati, N.: Piecewise-smooth dense optical flow via level sets. International Journal of Computer Vision 68, 111–124 (2006)
24. Brune, C., Maurer, H., Wagner, M.: Detection of intensity and motion edges within optical flow via multidimensional control 2, 1190–1210 (2009)
25. Ben-Ari, R., Sochen, N.A.: Stereo matching with mumford-shah regularization and occlusion handling. IEEE Trans. Pattern Anal. Mach. Intell. 32, 2071–2084 (2010)
26. Marr, D.: Vision: a computational investigation into the human representation and processing of visual information. W.H. Freeman, San Francisco (1982)
27. Trobin, W., Pock, T., Cremers, D., Bischof, H.: An unbiased second-order prior for high-accuracy motion estimation. In: Rigoll, G. (ed.) DAGM 2008. LNCS, vol. 5096, pp. 396–405. Springer, Heidelberg (2008)
28. Faugeras, O., Luong, Q.T.: The Geometry of Multiple Images. The MIT Press (2001) ISBN: 0262062208
29. Hartley, R.I., Zisserman, A.: Multiple View Geometry in Computer Vision, 2nd edn. Cambridge University Press (2004) ISBN: 0521540518
30. Mainberger, M., Bruhn, A., Weickert, J.: Is dense optic flow useful to compute the fundamental matrix? In: Campilho, A., Kamel, M.S. (eds.) ICIAR 2008. LNCS, vol. 5112, pp. 630–639. Springer, Heidelberg (2008)
31. Sun, D., Roth, S., Black, M.J.: Secrets of optical flow estimation and their principles. In: CVPR, pp. 2432–2439 (2010)
32. Klappstein, J.: Optical-Flow Based Detection of Moving Objects in Traffic Scenes. PhD thesis, Ruprecht-Karls-Universität, Heidelberg (2008)
33. Baker, S., Roth, S., Scharstein, D., Black, M., Lewis, J., Szeliski, R.: A Database and Evaluation Methodology for Optical Flow. In: International Conference on Computer Vision, pp. 1–8 (2007)
34. Lempitsky, V., Rother, C., Roth, S., Blake, A.: Fusion moves for markov random field optimization. IEEE Trans. Pattern Anal. Mach. Intell. 99 (2009)
35. Liu, Y., Cao, X., Dai, Q., Xu, W.: Continuous depth estimation for multi-view stereo. In: Computer Vision and Pattern Recognition, pp. 2121–2128 (2009)
36. Kim, W., Park, J., Lee, K.: Stereo matching using population-based mcmc. International Journal of Computer Vision 83, 195–209 (2009)
37. Amiaz, T., Lubetzky, E., Kiryati, N.: Coarse to over-fine optical flow estimation. Pattern Recognition 40, 2496–2503 (2007)

Robust Optic-Flow Estimation with Bayesian Inference of Model and Hyper-parameters

P. Héas, C. Herzet, and E. Mémin

Fluminance, INRIA Rennes-Bretagne Atlantique,
Campus universitaire de Beaulieu, 35042 Rennes, France
{patrick.heas,cedric.herzet,etienne.memin}@inria.fr

Abstract. Selecting optimal models and hyper-parameters is crucial for accurate optic-flow estimation. This paper solves the problem in a generic variational Bayesian framework. The method is based on a conditional model linking the image intensity function, the velocity field and the hyper-parameters characterizing the motion model. Inference is performed at three levels by considering maximum a posteriori problem of marginalized probabilities. We assessed the performance of the proposed method on image sequences of fluid flows and of the "Middlebury" database. Experiments prove that applying the proposed inference strategy on very simple models yields better results than manually tuning smoothing parameters or discontinuity preserving cost functions of classical state-of-the-art methods.

Keywords: Motion modeling, marginalized posterior, Bayesian inference, regularization coefficients, robust hyper-parameters, cost-functions.

1 Introduction

Choosing appropriate models and fixing hyper-parameters is a tricky and often hidden process in optic-flow estimation. Most of the motion estimators proposed so far have generally to rely on successive trials and a empirical strategy for fixing the hyper-parameters values and choosing the adequate model. Besides of its computational inefficiency, this strategy may produce catastrophic estimate without any relevant feedback for the end-user, especially when motions are difficult to apprehend as for instance for complex deformations or non-conventional imagery. Imposing hard values to these parameters may also yield poor results when the lighting conditions or the underlying motions differ from those the system has been calibrated with. At the extreme, the estimate may be either too smooth or at the opposite exhibits nonexistant strong motion discontinuities.

However, Bayesian analysis has been intensively studied in the past for hyper-parameter estimation and for model selection [1][2]. In particular, in the context of interpolation of noisy data, a powerful hierarchical Bayesian model has been proposed in the seminal work of [3]. In optic-flow estimation, state-of-the-art inference techniques [4,5] remain limited since they select the weight of different model candidates rather than really selecting one and/or do not consider model deviations from Gaussianity. Such non-Gaussian models are nevertheless very common

A.M. Bruckstein et al. (Eds.): SSVM 2011, LNCS 6667, pp. 773–785, 2012.

in computer vision where we have to cope with with motion discontinuities and observation outliers due to noise or varying lighting conditions. Non-gaussian robust statistics are commonly used to manage such problems. Another problem raised by the use of the robust norms is the choice of their hyper-parameters, since in general they are parametrical models. This choice is crucial and different tuning of these parameters can lead to motion estimates which are drastically different. Finally, although it is crucial for accurate motion measurement, very little emphasis has been devoted in the computer vision literature to the problem of model selection for optic-flow estimation. In particular, except in a particular case [6], no proper Bayesian formulation has been proposed in the literature for the selection of optimal optic-flow data and regularization models.

In the perspective of solving this crucial problem, this work presents a generic Bayesian modeling framework for robust optic-flow estimation. It yields the design of non-parametrical estimation methods, able to reliably decide among several data and regularization models with optimal tuning of regularization coefficients and robust model hyper-parameters. The effectiveness of our approach is illustrated on challenging image sequences of turbulent diffusive flows and computer vision scenes. In particular the proposed method achieves with very simple models better performances than classical state of the art algorithms.

The notational conventions adopted in this paper are as follows. Italic lowercase indicates a scalar quantity, as in a; boldface lowercase indicates a vector quantity, as in $\mathbf{a}$; the kth element of vector $\mathbf{a}$ is denoted $\mathbf{a}(k)$; capital boldface letters indicate matrices, as in $\mathbf{A}$; the element corresponding to the ith row and jth column of $\mathbf{A}$ is denoted as $\mathbf{A}(i,j)$; we will use the notation $\Lambda_{\mathbf{a}}$ to define a diagonal matrix whose elements are those of vector $\mathbf{a}$; calligraphic letters, $e.g.$, $\mathcal{A}$, represent the set of values that a variable or vector can take on; capital normal letters, as A, denote random variables.

2   Short Overview of Optic-Flow Estimation

2.1   Data and Prior Models

Let $I : (\mathbf{s}, t) \in \mathcal{S} \times \mathcal{T} \to I(\mathbf{s}, t) \in \mathbb{R}$ be an image intensity function where $\mathcal{S} \subseteq \mathbb{R}^2$ (resp. $\mathcal{T} \subseteq \mathbb{R}$) is the image spatial (resp. temporal) domain. Moreover, let the optic flow be defined by a function $v : (\mathbf{s}, t) \in \mathcal{S} \times \mathcal{T} \to v(\mathbf{s}, t) \in \mathbb{R}^2$ which associates a two-dimensional motion vector to every spatio-temporal position. Using this formalism, the optic-flow problem can then be restated as the problem of identifying $v(\mathbf{s}, t)$ from the (partial) knowledge of $I(\mathbf{s}, t)$. Note that, in practice, complexity and storage constraints often limit the estimation of the motion field over a $finite$ subset $\mathcal{S}_r \times \mathcal{T}_r$ of $\mathcal{S} \times \mathcal{T}$. We will consider this scenario hereafter and use the notation $\mathbf{v}$ to denote the vector made up of the concatenation of $v(\mathbf{s}, t)$'s $\forall \mathbf{s} \in \mathcal{S}_r, \forall t \in \mathcal{T}_r$. The estimation of the optic flow requires a mathematical characterization of the link between the image intensity and the motion field. One standard way to relate $v(\mathbf{s}, t)$ to $I(\mathbf{s}, t)$ is the (so-called) "Optic Flow Constraint" (OFC),

$$\frac{\partial I}{\partial t}(\mathbf{s},t) + \nabla_\mathbf{s}^T I(\mathbf{s},t)\, v(\mathbf{s},t) = 0, \qquad \forall \mathbf{s} \in \mathcal{S}_r, \forall t \in \mathcal{T}_r, \qquad (1)$$

which is valid under rigid motion and stable lighting hypotheses. For other configurations, many other models have been proposed in the literature to relate the image intensity function to the sought motion fields [7]. All these models obey the same general formulation:

$$\Phi_{I,v}(\mathbf{s},t) = 0 \qquad \forall \mathbf{s} \in \mathcal{S}_r, \forall t \in \mathcal{T}_r, \qquad (2)$$

where $\Phi_{I,v}$ is an operator on I and v. In the sequel, we will refer to $\Phi_{I,v}$ as the *data model* and use the notation $\boldsymbol{\Phi}$ to denote the vector made up of the concatenation of the $\Phi_{I,v}(\mathbf{s},t)$'s $\forall \mathbf{s} \in \mathcal{S}_r, \forall t \in \mathcal{T}_r$. Note that an important family of data models is defined by linear operators, *i.e.*,

$$\boldsymbol{\Phi} = \mathbf{A}_\Phi \mathbf{v} + \mathbf{b}_\Phi, \qquad (3)$$

where $\mathbf{A}_\Phi$ and $\mathbf{b}_\Phi$ are respectively a matrix and a vector characterizing the operator. The system of equations defined in (2) is commonly underdetermined *i.e.*, it does not univocally specify a solution for v. A proper conditioning of the problem requires therefore to include some additional constraints specifying the nature of the sought solution, *e.g.*,

$$\Pi_v(\mathbf{w}) = 0, \qquad \mathbf{w} \in \mathcal{W}, \qquad (4)$$

where Π_v denotes an operator on v which is parameterized by a (possibly multidimensional) index $\mathbf{w}$. In the sequel, we will refer to Π_v as the *prior model* and use the notation $\boldsymbol{\Pi}$ to denote the vector formed by the concatenation of the $\Pi_v(\mathbf{w})$'s $\forall \mathbf{w} \in \mathcal{W}$. The choice of the prior model is commonly made (but does not have to) so that some form of regularity is ensured. For example, a possible choice for Π_v is as follows [8]

$$\Pi_v(\mathbf{s},t) \triangleq \nabla_\mathbf{s} v(\mathbf{s},t), \qquad \forall \mathbf{s} \in \mathcal{S}_r, \forall t \in \mathcal{T}_r \qquad (5)$$

where $\nabla_\mathbf{s} v$ denotes the Jacobian of v and we made the identification $\mathbf{w} \triangleq (\mathbf{s},t)$, $\mathcal{W} \triangleq \mathcal{S}_r \times \mathcal{T}_r$. In other application, Π_v can enforce the solution to satisfy some physical constraints on motion regularity (see *e.g.*, [6]). Among the possible prior operators, we will have a particular emphasis on the family of linear operators, *i.e.*,

$$\boldsymbol{\Pi} = \mathbf{A}_\Pi \mathbf{v} + \mathbf{b}_\Pi, \qquad (6)$$

where $\mathbf{A}_\Pi$ and $\mathbf{b}_\Pi$ are respectively a matrix and a vector characterizing the operator.

In practice, the data and the prior models may not perfectly describe the sought motion field. Looking for v satisfying both (2) and (4) may then lead to aberrant or unstable solutions. Hence, a common approach to avoid such issues consists in minimizing an energy functional composed of two terms balanced by a regularization coefficient γ:

$$L(I, \mathbf{v}, \gamma) = f_d(\Phi_{I,v}) + \gamma f_r(\Pi_v), \qquad (7)$$

where γ is a positive parameter. The *"data term"* $f_d(\Phi_{I,v})$ (resp. *"regularization term"* $f_r(\Pi_v)$) penalizes discrepancies from the considered data model (2) (resp. prior model (4)) whereas γ tunes the tradeoff between the two terms. The choice of the cost functions f_d and f_r is of great importance since it implicitly defines the type of solution we are looking for. One possible choice for f_d is, for example,

$$f_d(\Phi_{I,v}) = \|\Phi\|_2^2 \triangleq \sum_{(\mathbf{s},t) \in \mathcal{S}_r \times \mathcal{T}_r} (\Phi_{I,v}(\mathbf{s},t))^2, \qquad (8)$$

which measures the Euclidean distance of Φ to zero. In the context of strong model deviations, *e.g.*, when dealing with observation outliers, the use of the ℓ_2 norm may be inefficient. In such scenarios, *"robust cost functions"* are commonly considered to penalize model discrepancies:

$$f_d(\Phi_{I,v}) = \rho_d(\Phi, \tau_d), \qquad (9)$$

where $\rho_d : \mathbb{R}^{|\mathcal{S}_r||\mathcal{T}_r|+1} \to \mathbb{R}$ is an even continuously differentiable concave function with some suitable properties and τ_d is a parameter. Robust cost functions are also commonly referred to as *"M-estimators"*. Well-known instances of M-estimators include *Leclerc's* cost function and an approximation of the ℓ_1 norm (see *e.g.*, [9][10]) . Similarly, $f_r(\Pi_v)$ can be defined by either a quadratic norm or, in the context of strong motion spatial discontinuities, a robust cost function $\rho_r(\Pi, \tau_r)$, where τ_r is a parameter.

2.2   Standard Optic-Flow Estimation

Practical estimation of the optic flow requires to find tractable and accurate implementations of the following problem:

$$\hat{\mathbf{v}} = \arg\min_{\mathbf{v}} L(I, \mathbf{v}, \gamma). \qquad (10)$$

Different cases can be distinguished according to whether the data/prior operators are linear or not, the cost functions f_d and f_r implement quadratic norm or robust cost functions. When the operators are linear and the cost functions quadratic, the problem is convex. There is therefore one unique minimum which can be efficiently accessed by numerical procedure such as Conjugate Gradient Squared (CGS) algorithm [11] or multi-grid algorithms [12].

When f_d and f_r are robust cost functions, the direct application of standard optimization algorithms may lead to cumbersome procedures. Instead, one common technique consists in expressing (10) as a sequence of quadratic problems. This approach is based on the concavity of M-estimators and Fenchel-Legendre duality. In particular, it can be shown [9][13] that

$$\min_{\mathbf{v}} L(I, \mathbf{v}, \gamma) = \min_{\mathbf{v}, \mathbf{z}} B(I, \mathbf{v}, \mathbf{z}, \gamma), \qquad (11)$$

where $B(I, \mathbf{v}, \mathbf{z}, \gamma) \triangleq \Phi^T \Lambda_{\mathbf{z}_d} \Phi + \psi_\Phi(\mathbf{z}_d) + \gamma(\Pi^T \Lambda_{\mathbf{z}_r} \Pi + \psi_\Pi(\mathbf{z}_r))$, $\mathbf{z} \triangleq [\mathbf{z}_d^T \mathbf{z}_r^T]^T$ and $\psi_\Phi(\mathbf{z}_d)$ (resp. $\psi_\Pi(\mathbf{z}_r)$) is the Fenchel-Legendre dual function of f_d (resp.

f_r). Minimizing B instead of L often eases the resolution of the optimization problem. Indeed, considering iterative conditional minimization of B, we have

$$\mathbf{v}^{(n)} = \arg \min_{\mathbf{v}} B(I, \mathbf{v}, \mathbf{z}^{(n)}, \gamma), \qquad (12)$$

$$\mathbf{z}^{(n+1)} = \arg \min_{\mathbf{z}} B(I, \mathbf{v}^{(n)}, \mathbf{z}, \gamma). \qquad (13)$$

Now, since B is a quadratic function with respect to $\boldsymbol{\Phi}$ and $\boldsymbol{\Pi}$, (12) can be solved efficiently by applying standard optimization procedures. Moreover, (13) usually possesses an easy analytical solution. In the case of linear model, we have:

$$\mathbf{z}_d^{(n+1)} = \frac{1}{2\tau_d} \Lambda_{\mathbf{A}_{\boldsymbol{\Phi}}\mathbf{v}^{(n)}+\mathbf{b}_{\boldsymbol{\Phi}}}^{-2} \nabla_{\boldsymbol{\Phi}} \rho_d(\mathbf{A}_{\boldsymbol{\Phi}}\mathbf{v}^{(n)} + \mathbf{b}_{\boldsymbol{\Phi}}, \tau_d), \qquad (14)$$

$$\mathbf{z}_r^{(n+1)} = \frac{1}{2\tau_r} \Lambda_{\mathbf{A}_{\boldsymbol{\Pi}}\mathbf{v}^{(n)}+\mathbf{b}_{\boldsymbol{\Pi}}}^{-2} \nabla_{\boldsymbol{\Pi}} \rho_d(\mathbf{A}_{\boldsymbol{\Pi}}\mathbf{v}^{(n)} + \mathbf{b}_{\boldsymbol{\Pi}}, \tau_r). \qquad (15)$$

Hence, the minimization of B via (12)-(13) reduces to solving a sequence of tractable quadratic problems.

3   A Bayesian Framework for Model and Hyper-parameter Selection

In the previous section we emphasized that the optic-flow estimation problem requires to make assumptions about: *i)* the observation and data models, $\boldsymbol{\Phi}_{I,v}$ and $\boldsymbol{\Pi}_v$; *ii)* the costs functions, f_d and f_r; *iii)* the hyper-parameters, γ, τ_d and τ_r. The choice of these quantities often dramatically influences the performance achieved by the estimation algorithms. Quite surprisingly this problem has been mainly overlooked in the current literature.

In this section, we propose a Bayesian inference method to make proper decisions about the models, the cost functions and the value of the hyper-parameters. In a first part, we give a Bayesian reformulation of the standard optic-flow problem (10) which motivates our subsequent derivations. We then devise a Bayesian method for the estimation of the models, cost functions and hyper parameters based on the so-called *"Bayesian evidence framework"* [3].

3.1   Bayesian Formulation of the Optic-Flow Estimation Problem

In this section, we emphasize that the general optic-flow estimation problem

$$(\mathbf{v}^{\star}, \mathbf{z}^{\star}) = \arg \min_{\mathbf{v}, \mathbf{z}} B(I, \mathbf{v}, \mathbf{z}, \alpha, \beta), \qquad (16)$$

can also be expressed as a maximum a posteriori (MAP) problem. For the sake of conciseness, we focus exclusively on the case of linear operators (3), (6) but keep in mind that non-linear operators can be made (locally) linear by a first-order Taylor expansion. We consider the following probabilistic model relating I, $\mathbf{v}$ and $\mathbf{z}$:

$$p(\mathbf{b}_\Phi | \mathbf{z}_d, \mathbf{v}, \beta, \Phi_{I,v}) \triangleq Z_{\mathbf{b}_\Phi}^{-1} \exp\left\{-\frac{(\mathbf{b}_\Phi + \mathbf{A}_\Phi \mathbf{v})^T \beta \Lambda_{\mathbf{z}_d}(\mathbf{b}_\Phi + \mathbf{A}_\Phi \mathbf{v})}{2}\right\}, \quad (17)$$

$$p(\mathbf{z}_d | \beta, \Phi_{I,v}) \triangleq Z_{\mathbf{z}_d}^{-1} \exp\left\{-\frac{\beta \psi_\Phi(\mathbf{z}_d)}{2}\right\} \sqrt{\det(\beta \Lambda_{\mathbf{z}_d})^{-1}}, \quad (18)$$

$$p(\mathbf{v} | \mathbf{z}_r, \alpha, \Pi_v) \triangleq Z_{\mathbf{v}}^{-1} \exp\left\{-\frac{(\mathbf{v} - \mathbf{m}_\Pi)^T \boldsymbol{\Gamma}_\Pi^{-1} (\mathbf{v} - \mathbf{m}_\Pi)}{2}\right\}, \quad (19)$$

$$p(\mathbf{z}_r | \alpha, \Pi_v) \triangleq Z_{\mathbf{z}_r}^{-1} \exp\left\{-\frac{\alpha \psi_\Pi(\mathbf{z}_r)}{2}\right\} \sqrt{\det(\mathbf{A}_\Pi^T \alpha \Lambda_{\mathbf{z}_r} \mathbf{A}_\Pi)^{-1}}, \quad (20)$$

where $Z_{\mathbf{b}_\Phi}$, $Z_{\mathbf{z}_d}$, $Z_{\mathbf{v}}$ and $Z_{\mathbf{z}_r}$ are normalization constants, α, β are two positive parameters and

$$\boldsymbol{\Gamma}_\Pi \triangleq (\alpha \mathbf{A}_\Pi^T \Lambda_{\mathbf{z}_r} \mathbf{A}_\Pi)^{-1}, \qquad \mathbf{m}_\Pi = \boldsymbol{\Gamma}_\Pi \mathbf{A}_\Pi \Lambda_{\mathbf{z}_r} \mathbf{b}_\Pi. \quad (21)$$

Equation (17) defines a family of Gaussian distributions on $\mathbf{b}_\Phi$ parameterized by $\mathbf{z}_d$; the probability of each Gaussian of this family is given in (18) and depends on ψ_Φ, the M-estimator dual function. It is interesting to note that $\mathbf{b}_\Phi$ is a function of the observed image I; $p(\mathbf{b}_\Phi | \mathbf{z}_d, \mathbf{v}, \beta, \Phi_{I,v})$ can therefore be seen as a probabilistic "observation model" relating I to $\mathbf{v}$.

Similarly, (19) defines a family of Gaussians parameterized by $\mathbf{z}_r$; (20) is a prior on the probability of occurrence of each instance of this family. $p(\mathbf{v} | \mathbf{z}_r, \alpha, \Pi_v)$ can therefore be regarded as a probabilistic "prior model" on $\mathbf{v}$.

Based on these definitions, we can now define the following MAP estimation problem:

$$(\mathbf{v}^\star, \mathbf{z}^\star) = \arg\max_{(\mathbf{v}, \mathbf{z})} \{\log p(\mathbf{b}_\Phi, \mathbf{v}, \mathbf{z}, |\alpha, \beta, \Phi_{I,v}, \Pi_v)\} \quad (22)$$

where

$$p(\mathbf{b}_\Phi, \mathbf{v}, \mathbf{z} | \alpha, \beta, \Phi_{I,v}, \Pi_v) = p(\mathbf{b}_\Phi | \mathbf{v}, \mathbf{z}_d, \beta, \Phi_{I,v}) \, p(\mathbf{z}_d | \beta, \Phi_{I,v})$$
$$p(\mathbf{v} | \mathbf{z}_r, \alpha, \Pi_v) \, p(\mathbf{z}_r | \alpha, \Pi_v). \quad (23)$$

It is quite easy to see (by direct substitution) that (22) is equivalent to (16) if we set $\gamma \triangleq \frac{\alpha}{\beta}$. This connection gives a physical interpretation to the assumptions which are implicitly made when considering standard problem (16). For example, the optimization of B with respect to $\mathbf{z}_d$ (resp. $\mathbf{z}_r$) is equivalent to selecting the best probabilistic data (resp. prior) model among a family of Gaussians with different covariance matrices.

3.2   Bayesian Inference for Robust Optic-Flow Estimation

Since standard optic-flow estimation algorithms based on (16) implicitly consider probabilistic model (17)-(20), it is legitimate to wonder whether this model can also be exploited to infer the data/prior models, the cost functions and the hyper-parameters? In this section, we will assume so and propose a Bayesian methodology to estimate these quantities based on model (17)-(20). Note that this Bayesian approach differs from the learning strategies proposed in [14], since it neither requires training data nor ground truth. We will use the notation ω_d to refer to the couple $(\Phi_{I,v}, f_d)$ specifying the data model and cost function. Similarly, ω_r will refer to (Π_v, f_r). Finally, we will use the following short-hand notations: $\boldsymbol{\theta} \triangleq [\alpha, \beta, \tau_d, \tau_r]^T$ and $\omega \triangleq [\omega_d, \omega_r]^T$. We will infer $\mathbf{v}, \mathbf{z}, \boldsymbol{\theta}$ and ω from the following set of problems:

$$(\mathbf{v}^\star, \mathbf{z}^\star) = \arg\max_{(\mathbf{v},\mathbf{z})} \left\{ \log p_{\mathrm{B}_\Phi, \mathrm{V}, \mathrm{Z}|\Theta,\Omega}(\mathbf{b}_\Phi, \mathbf{v}, \mathbf{z}|\boldsymbol{\theta}^\star, \omega^\star) \right\}, \tag{24}$$

$$\boldsymbol{\theta}^\star = \arg\max_{\boldsymbol{\theta}} \left\{ \log p_{\mathrm{B}_\Phi|\mathrm{Z},\Theta,\Omega}(\mathbf{b}_\Phi|\hat{\mathbf{z}}(\boldsymbol{\theta}, \mathbf{v}^\star), \boldsymbol{\theta}, \omega^\star) \right\}, \tag{25}$$

$$\omega^\star = \arg\max_{\omega} \left\{ \log p_{\mathrm{B}_\Phi|\mathrm{Z},\Omega}(\mathbf{b}_\Phi|\hat{\mathbf{z}}(\boldsymbol{\theta}^\star, \mathbf{v}^\star), \omega) \right\}, \tag{26}$$

with $\hat{\mathbf{z}}(\boldsymbol{\theta}) = [\hat{\mathbf{z}}_d^T(\boldsymbol{\theta}, \mathbf{v}) \; \hat{\mathbf{z}}_r^T(\boldsymbol{\theta}, \mathbf{v})]^T$,

$$\hat{\mathbf{z}}_d(\boldsymbol{\theta}, \mathbf{v}) = \frac{1}{2\tau_d} \Lambda_{\mathbf{A}_\Phi \mathbf{v} + \mathbf{b}_\Phi}^{-2} \nabla_\Phi \rho_d(\mathbf{A}_\Phi \mathbf{v} + \mathbf{b}_\Phi, \tau_d), \tag{27}$$

$$\hat{\mathbf{z}}_r(\boldsymbol{\theta}, \mathbf{v}) = \frac{1}{2\tau_r} \Lambda_{\mathbf{A}_\Pi \mathbf{v} + \mathbf{b}_\Pi}^{-2} \nabla_\Pi \rho_r(\mathbf{A}_\Pi \mathbf{v} + \mathbf{b}_\Pi, \tau_r). \tag{28}$$

The system (24)-(26) is inspired from the Bayesian evidence framework proposed in [3]. It defines three levels of inference. In the first level (24), $\mathbf{v}$ and $\mathbf{z}$ are estimated by relying on hyper-parameter and model estimates $(\boldsymbol{\theta}^\star, \omega^\star)$. In the second level (25), the dependence on $\mathbf{v}$ is marginalized out and $\boldsymbol{\theta}$ is inferred by assuming $\omega = \omega^\star$. Finally, in the last level (26), $\omega^\star$ is computed by maximizing a likelihood function in which both the dependence on $\mathbf{v}$ and $\boldsymbol{\theta}$ has been removed.

Note that the set of problems defined in (24)-(26) is slightly different from the one presented in [3] since the dependence on $\mathbf{z}$ is not removed in (25)-(26). Instead, we constraint $\mathbf{z}$ to have a particular structure, namely (27)-(28). As will see in the remainder of this section, this digression from the original Bayesian evidence framework allows a tractable implementation of the inference algorithm. On the other hand, it also forces an interconnection between all level of inference: $\boldsymbol{\theta}^\star$ depends on $\mathbf{v}^\star$ through $\hat{\mathbf{z}}(\boldsymbol{\theta}, \mathbf{v}^\star)$ whereas $\mathbf{v}^\star$ is the maximum of a function depending on $\boldsymbol{\theta}^\star$, etc. We consider the following iterative procedure to find an estimate satisfying all the equations of the system (24)-(26):

$$\dot{\omega} = \arg\max_{\omega} \left\{ \log p_{\mathrm{B}_\Phi|\mathrm{Z},\Omega}(\mathbf{b}_\Phi|\hat{\mathbf{z}}(\boldsymbol{\theta}_\omega^{(\infty)}, \mathbf{v}_\omega^{(\infty)}), \omega) \right\}, \tag{29}$$

where the sequence $\{\boldsymbol{\theta}_\omega^{(n)}, \mathbf{v}_\omega^{(n)}\}_{n=0}^\infty$ is defined as follows

$$(\mathbf{v}_\omega^{(n)}, \mathbf{z}_\omega^{(n)}) = \arg\max_{(\mathbf{v},\mathbf{z})} \left\{ \log p(\mathbf{b}_\Phi, \mathbf{v}, \mathbf{z}|\boldsymbol{\theta}_\omega^{(n-1)}, \omega) \right\}, \tag{30}$$

$$\boldsymbol{\theta}_\omega^{(n)} = \boldsymbol{\theta}_\omega^{(n-1)} + \mu \mathbf{P} \nabla_{\boldsymbol{\theta}} \log p(\mathbf{b}_\Phi|\hat{\mathbf{z}}(\boldsymbol{\theta}_\omega^{(n-1)}, \mathbf{v}_\omega^{(n)}), \boldsymbol{\theta}_\omega^{(n-1)}, \omega), \tag{31}$$

and μ is positive step factor, and $\mathbf{P}$ a positive definite matrix. In practice, convergence is obtained after about 10 iterations in average. Matrix $\mathbf{P}$ is chosen to be a finite difference approximation of the Hessian so that (31) constitutes an iteration of a Quasi-Newton ascent method (see [11]). Step μ is fixed according to the strong Wolf conditions [11]. Clearly, any fixed point of recursions (29)-(31) satisfies (24)-(26). We detail hereafter the strategy we considered to implement each step of the proposed algorithm:

Step (30): The problem to solve is equivalent to the standard optic-flow estimation problem (16) (see Section 3.1) and can be solved by iterative conditional maximizations (12)-(13).

Step (31): The update requires the gradient of $\log p(\mathbf{b}_\Phi|\hat{\mathbf{z}}(\boldsymbol{\theta}, \mathbf{v}_\omega^{(n)}), \boldsymbol{\theta}, \omega)$ which can be efficiently computed by noticing that [15]:

$$\nabla_{\boldsymbol{\theta}} \log p(\mathbf{b}_\Phi|\mathbf{z}, \boldsymbol{\theta}, \omega) = \int p(\mathbf{v}|\mathbf{z}, \boldsymbol{\theta}, \omega) \nabla_{\boldsymbol{\theta}} \log p(\mathbf{b}_\Phi, \mathbf{v}|\mathbf{z}, \boldsymbol{\theta}, \omega) d\mathbf{v}. \tag{32}$$

This leads to

$$\frac{\partial}{\partial \alpha} \log p(\mathbf{b}_\Phi|\mathbf{z}, \boldsymbol{\theta}, \omega) = -\frac{1}{2} \left\langle \alpha^{-1} + (\mathbf{v} - \mathbf{m}_\Pi)^T \mathbf{A}_\Pi^T \Lambda_{\mathbf{z}_r} \mathbf{A}_\Pi (\mathbf{v} - \mathbf{m}_\Pi) \right\rangle,$$

$$\frac{\partial}{\partial \beta} \log p(\mathbf{b}_\Phi|\mathbf{z}, \boldsymbol{\theta}, \omega) = -\frac{1}{2} \left\langle \beta^{-1} + (\mathbf{b}_\Phi - \mathbf{A}_\Phi \mathbf{v})^T \Lambda_{\mathbf{z}_d} (\mathbf{b}_\Phi - \mathbf{A}_\Phi \mathbf{v}) \right\rangle,$$

$$\frac{\partial}{\partial \tau_d} \log p(\mathbf{b}_\Phi|\mathbf{z}, \boldsymbol{\theta}, \omega) = -\frac{1}{2} \left\langle \text{tr}\left(\Lambda_{\mathbf{z}_d}^{-1} \Lambda_{\frac{\partial \mathbf{z}_d}{\partial \tau_d}}\right) + (\mathbf{b}_\Phi - \mathbf{A}_\Phi \mathbf{v})^T \beta \Lambda_{\frac{\partial \mathbf{z}_d}{\partial \tau_d}} (\mathbf{b}_\Phi - \mathbf{A}_\Phi \mathbf{v}) \right\rangle,$$

$$\frac{\partial}{\partial \tau_r} \log p(\mathbf{b}_\Phi|\mathbf{z}, \boldsymbol{\theta}, \omega) = -\frac{1}{2} \left\langle \text{tr}\left(\Lambda_{\mathbf{z}_r}^{-1} \Lambda_{\frac{\partial \mathbf{z}_r}{\partial \tau_r}}\right) + (\mathbf{v} - \mathbf{m}_\Pi)^T \alpha \mathbf{A}_\Pi^T \Lambda_{\frac{\partial \mathbf{z}_r}{\partial \tau_r}} \mathbf{A}_\Pi (\mathbf{v} - \mathbf{m}_\Pi) \right\rangle,$$

where we use the notation $\langle \cdot \rangle$ to denote the expectation with respect to $p(\mathbf{v}|\mathbf{z}, \boldsymbol{\theta}, \omega)$. Note that $\mathbf{v}$ only appears in linear and quadratic forms in the expressions of the partial derivatives defined above. As a consequence, the latter derivatives are only a function of the mean and covariance of $p(\mathbf{v}|\mathbf{z}, \boldsymbol{\theta}, \omega)$. Now, it is easy to see that $p(\mathbf{v}|\mathbf{z}, \boldsymbol{\theta}, \omega)$ is a Gaussian distribution with mean and covariance defined as

$$\mathbf{m}_{\mathbf{v}|\mathbf{z}, \boldsymbol{\theta}, \omega} \triangleq \langle \mathbf{v} \rangle = \boldsymbol{\Gamma}_{\mathbf{v}|\mathbf{z}, \boldsymbol{\theta}, \omega} (\mathbf{A}_\Phi^T \Lambda_{\mathbf{z}_d} \mathbf{b}_\Phi + \mathbf{A}_\Pi^T \Lambda_{\mathbf{z}_r} \mathbf{b}_\Pi), \tag{33}$$

$$\boldsymbol{\Gamma}_{\mathbf{v}|\mathbf{z}, \boldsymbol{\theta}, \omega} \triangleq \langle (\mathbf{v} - \mathbf{m}_{\mathbf{v}|\mathbf{z}, \boldsymbol{\theta}, \omega})(\mathbf{v} - \mathbf{m}_{\mathbf{v}|\mathbf{z}, \boldsymbol{\theta}, \omega})^T \rangle = \left(\alpha \mathbf{A}_\Phi^T \Lambda_{\mathbf{z}_d} \mathbf{A}_\Phi + \beta \mathbf{A}_\Pi^T \Lambda_{\mathbf{z}_r} \mathbf{A}_\Pi\right)^{-1}. \tag{34}$$

Therefore, the computation of the gradient of $p(\mathbf{v}|\mathbf{z}, \boldsymbol{\theta}, \omega)$ only requires tractable linear operations.

Step (29): The decision on the model and the cost function ω is made by maximizing $\log p(\mathbf{b}_\Phi|\hat{\mathbf{z}}(\boldsymbol{\theta}_\omega^{(\infty)}, \mathbf{v}_\omega^{(\infty)}), \omega)$. We assume that ω takes on its values in a *finite* set so that solving (29) only requires to evaluate $\log p(\mathbf{b}_\Phi|\hat{\mathbf{z}}(\boldsymbol{\theta}_\omega^{(\infty)}, \mathbf{v}_\omega^{(\infty)}), \omega)$ for these values. We use the Laplace's method to derive an approximation of $p(\mathbf{b}_\Phi, \omega)$. This method approximates the integral of a function by fitting a Gaussian at its maximum and computing the volume under the Gaussian. For a k dimensional variable $\mathbf{x}$ and a function $f(\mathbf{x})$, the Laplaces' approximation reads:

$$\int f(\mathbf{x}) d\mathbf{x} \simeq f(\hat{\mathbf{x}})(2\pi)^{k/2} [-\det\{\nabla_\mathbf{x}^2 \log f(\hat{\mathbf{x}})\}]^{-1/2}, \tag{35}$$

where $\nabla_\mathbf{x}^2$ represents the Hessian operator and $\hat{\mathbf{x}} = \arg\max_\mathbf{x} f(\mathbf{x})$. Hence, if $p(\boldsymbol{\theta})$ is a flat non-informative prior we get the following approximation:

$$p_{\mathbf{B}_\Phi|Z,\Omega}(\mathbf{b}_\Phi|\hat{\mathbf{z}}(\boldsymbol{\theta}_\omega^{(\infty)}, \mathbf{v}_\omega^{(\infty)}), \omega) = \int p(\mathbf{b}_\Phi|\hat{\mathbf{z}}(\boldsymbol{\theta}_\omega^{(\infty)}, \mathbf{v}_\omega^{(\infty)}), \boldsymbol{\theta}, \omega) \, p(\boldsymbol{\theta}) \, d\boldsymbol{\theta}, \tag{36}$$

$$\propto p(\mathbf{b}_\Phi|\hat{\mathbf{z}}(\boldsymbol{\theta}_\omega^{(\infty)}, \mathbf{v}_\omega^{(\infty)}), \boldsymbol{\theta}_\omega^{(\infty)}, \omega)(-\det \mathbf{H}_{\boldsymbol{\theta}}), \tag{37}$$

where $\mathbf{H}_{\boldsymbol{\theta}} = \nabla_{\boldsymbol{\theta}}^2 \log p(\mathbf{b}_\Phi | \hat{\mathbf{z}}(\boldsymbol{\theta}_\omega^{(\infty)}, \mathbf{v}_\omega^{(\infty)}), \boldsymbol{\theta}_\omega^{(\infty)}, \omega)$. Finally, we obtain:

$$\log p_{\mathbf{B}_\Phi | Z, \Omega}(\mathbf{b}_\Phi | \hat{\mathbf{z}}(\boldsymbol{\theta}_\omega^{(\infty)}, \mathbf{v}_\omega^{(\infty)}), \omega) \simeq \log(-\det \mathbf{H}_{\boldsymbol{\theta}}) + \frac{1}{2} \log \det \boldsymbol{\Gamma}_{\mathbf{v}|\mathbf{z},\boldsymbol{\theta},\omega}$$

$$-\frac{1}{2}\mathbf{m}_{\mathbf{v}|\mathbf{z},\boldsymbol{\theta},\omega}^T \boldsymbol{\Gamma}_{\mathbf{v}|\mathbf{z},\boldsymbol{\theta},\omega}^{-1} \mathbf{m}_{\mathbf{v}|\mathbf{z},\boldsymbol{\theta},\omega} + \frac{1}{2}\mathbf{b}_\Phi^T \Lambda_{\mathbf{z}_d} \mathbf{b}_\Phi + \frac{1}{2}\mathbf{m}_\Pi^T \mathbf{A}_\Pi^T \Lambda_{\mathbf{z}_r} \mathbf{A}_\Pi \mathbf{m}_\Pi^T. \quad (38)$$

where $\mathbf{m}_{\mathbf{v}|\mathbf{z},\boldsymbol{\theta},\omega}$ and $\boldsymbol{\Gamma}_{\mathbf{v}|\mathbf{z},\boldsymbol{\theta},\omega}$ are the a posteriori mean and covariance of $\mathbf{v}$ defined in (34)-(33) and evaluated at $\boldsymbol{\theta} = \boldsymbol{\theta}_\omega^{(\infty)}$ and $\mathbf{z} = \hat{\mathbf{z}}(\boldsymbol{\theta}_\omega^{(\infty)}, \mathbf{v}_\omega^{(\infty)})$.

4 Experiments

In the following, basic experiments have been designed in order to provide a proof of concept on the capabilities of the Bayesian inference technique. In our experiments, hyper-parameters and motion have been estimated at the different level of a standard multi-resolution algorithm. The inference of the models has been performed only on the finest resolution level.

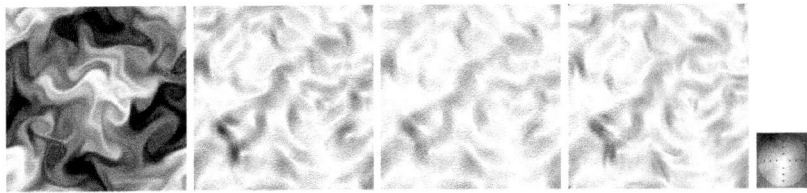

Fig. 1. Generated images $I(t)$, ground truth, our estimate and the one of [16] in color representation. Color and intensity code vector orientations and magnitudes [17].

4.1 Fluid Motion Image Sequence

We first consider a synthetic sequence of scalar images of 256×256 pixels representing the evolution of two-dimensional turbulent flow [18]. The dynamical process was obtained by direct numerical simulation of the Navier-Stokes equations coupled with the advection-diffusion of a passive scalar equation: $\frac{\partial I(\mathbf{s},t)}{\partial t} + \nabla^T I(\mathbf{s},t)v(\mathbf{s},t) = \nu \Delta I(\mathbf{s},t)$, where ν represents a unknown diffusion coefficient.Fig. 1 presents a scalar image of the sequence together with the ground truth motion. In our simulation we defined the set of possible value for ω as follows. Advection-diffusion equation with different values of ν constitutes the set of possible data models. A 1-st order regularizer (5) is used to implement the prior model. Both the ℓ_2 and Leclerc's cost functions are possible choices for the data and prior cost functions. We considered the estimation of hyper-parameters α, β, τ_d and τ_r.

Fig. 1 and Fig 2 displays the *posterior* motion estimate $\hat{\mathbf{v}}$, the Mean End Point (MEP) error and the Mean Barron Angular (MBA) [17] error obtained applying the proposed Bayesian inference framework. Comparing with results of [16] displayed in the same figures, one can notice that the use of Bayesian inference with

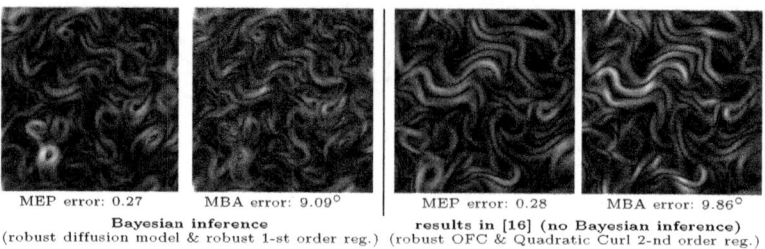

MEP error: 0.27   MBA error: 9.09° │ MEP error: 0.28   MBA error: 9.86°
Bayesian inference   **results in [16] (no Bayesian inference)**
(robust diffusion model & robust 1-st order reg.)   (robust OFC & Quadratic Curl 2-nd order reg.)

Fig. 2. Motion field, MEP and MBA errors [17] corresponding to estimation with (left) or without (right) Bayesian inference

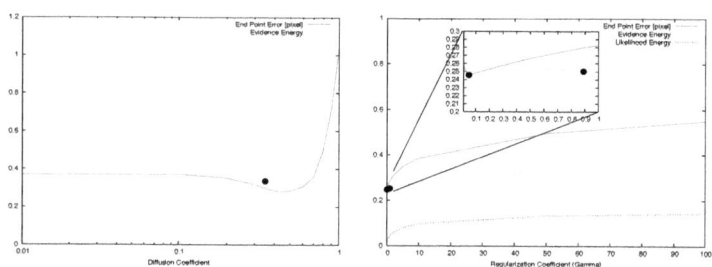

Fig. 3. Left: the energy of model probability w.r.t. coefficient ν (green) is minimum for $\hat{\nu}$ minimizing the MEP error (red). Right: the energy of the regularization coefficient probability (green) reaches its minimum close to the MEP error curve minimum (red).

a simple robust 1-st order regularization outperforms the most accurate state of the art fluid motion estimators [18]. Therefore, fitting an inappropriate regularizers while selecting data models by Bayesian inference yields better results than fine regularizers adjusted by manually tuning hyper-parameters. Fig. 3 shows that the inferred diffusion and regularization coefficients also minimize the MEP error.

4.2   Computer Vision Scenes

In this section, we assess the performance of the proposed Bayesian inference method with image sequences from the Middleburry database [17]. We show the power of the proposed method by emphasizing its effectiveness for very simple observation and prior models: the basic model for the data (monochromatic model, OFC equation (1) without any image gradient preservation) and a basic 1-st order regularizer (5). Obviously, using more sophisticated models would likely improve our results.

"Venus" sequence: We considered ℓ_2, ℓ_1 or Leclerc's function for f_r while we chose for f_d the Leclerc's cost function. Our scheme selects ℓ_1 norm as the best prior cost function estimate. The estimated motion field $\hat{\mathbf{v}}$ obtained with $\hat{\omega}$ and $\hat{\boldsymbol{\theta}}$ is shown in Fig. 4 together with the maps of data outliers and motion spatial

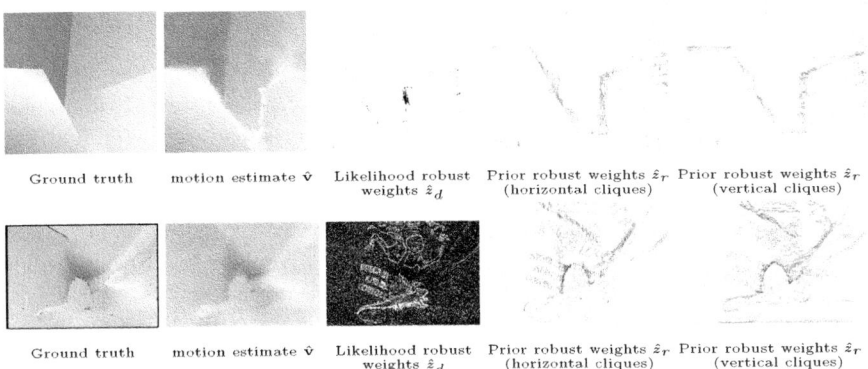

Fig. 4. Ground truth, estimate $\hat{\mathbf{v}}$, robust weights $\hat{z}_d$ and $\hat{z}_r$. Top (resp. bottom) line represents result for the frames 10-11 of the "Venus" (resp. "Dimetrodon") sequence.

discontinuities related to estimate $\hat{\tau}_d$ and $\hat{\tau}_r$. As shown in the table of Fig. 5, errors obtained with Bayesian inference for these very simple observation and prior models are comparable to error of manually tuned hyper-parameters of affine regularization model or specialized data term dedicated to such scenes composed of rigid objects [9,12].

"*Dimetrodon*" *sequence:* We considered ℓ_2, ℓ_1 norms or Leclerc's cost function for f_d and f_r. Our scheme selects the ℓ_1 norm for both f_d and f_r and adjusts hyper-parameters in less than an hour. The left table in Fig. 5 shows that this combination performs the best in terms of MEP and MBA errors. The estimated motion field $\hat{\mathbf{v}}$ and error maps obtained with $\hat{\omega}$ and $\hat{\boldsymbol{\theta}}$ are displayed in Fig. 4 together with maps of data outliers and motion spatial discontinuities related to estimate $\hat{\tau}_d$ and $\hat{\tau}_r$. The right table in Fig. 5 shows that Bayesian inference enables to obtain a higher accuracy, or at least, results comparable to more refined method with manually tuned parameters.

f_d	f_r	$\hat{\alpha}/\hat{\beta}$	$\hat{\tau}_d$	$\hat{\tau}_r$	$-\log p(\mathbf{b}_\Phi|\hat{\mathbf{z}},\hat{\omega})$	End-point error	Barron error
ℓ_2	ℓ_2	12.48	0	0	443013	0.201	3.656
Leclerc	ℓ_2	14.97	0.32	0	418662	0.199	3.542
ℓ_1	ℓ_2	1.85	20.0	0	337990	0.191	3.309
ℓ_2	Leclerc	9.58	0	2.00	437326	0.206	3.760
Leclerc	Leclerc	15.01	0.32	1.34	418602	0.199	3.542
ℓ_1	Leclerc	1.83	20.0	0.39	338097	0.191	3.308
ℓ_2	ℓ_1	3.24	0	10.0	434656	0.258	4.883
Leclerc	ℓ_1	12.28	0.34	10.0	417340	0.204	3.657
ℓ_1	ℓ_1	**1.70**	**20.0**	**10.0**	**335564**	**0.190**	**3.303**

	Venus	Dimetrodon
Bayesian inference	**8.348**	**3.303**
Bruhn&al [12]	8.732	10.993
Black&Anandan [9]	7.641	9.261
Lucas-Kanade [19]	14.614	10.272
Media Player^{TM}	15.485	15.824
Zitnick & al [20]	11.423	30.105

Fig. 5. Left: selection of the most likely cost functions for the data and the regularization terms. Right: comparison with state of the art based on the MBA error criterion

5   Conclusion

A generic and efficient Bayesian inference scheme has been proposed for selecting models and hyper-parameters in robust optic-flow estimation. Motion fields, models together with their hyper-parameters are treated as interdependent random variables. Optic-flow, regularization coefficients, M-estimator parameters, *prior* and *likelihood* motion models are simultaneously inferred in this context by maximizing the *posterior* marginalized distributions. Experiments prove that the proposed Bayesian inference scheme succeeded to select appropriate model and hyper-parameters. In particular, using very simple models, we achieve an accuracy comparable to state of the art results. An intensive evaluation adapting models to the Middleburry optic-flow database could provide a fair judgement of the proposed framework performances.

References

1. Jaynes, E.T.: Bayesian methods: General background (1986)
2. Molina, R., Katsaggelos, A.K., Mateos, J.: Bayesian and regularization methods for hyperparameter estimation in image restoration. IEEE Trans. Image Processing 8, 231–246 (1999)
3. MacKay, D.J.C.: Bayesian interpolation. Neural Computation 4, 415–447 (1992)
4. Krajsek, K., Mester, R.: Bayesian model selection for optical flow estimation. In: Hamprecht, F.A., Schnörr, C., Jähne, B. (eds.) DAGM 2007. LNCS, vol. 4713, pp. 142–151. Springer, Heidelberg (2007)
5. Nir, T., Bruckstein, A.M., Kimmel, R.: Over-parameterized variational optical flow. Int. J. Comput. Vision 76, 205–216 (2008)
6. Heas, P., Memin, E., Heitz, D., Mininni, P.: Bayesian selection of scaling laws for motion modeling in images. In: International Conference on Computer Vision (ICCV 2009), Kyoto, Japan (2009)
7. Liu, T., Shen, L.: Fluid flow and optical flow. Journal of Fluid Mechanics 614, 253 (2008)
8. Horn, B., Schunck, B.: Determining optical flow. Artificial Intelligence 17, 185–203 (1981)
9. Black, M., Anandan, P.: The robust estimation of multiple motions: Parametric and piecewise-smooth flow fields. Computer Vision and Image Understanding 63, 75–104 (1996)
10. Geman, D., Reynolds, G.: Constrained restoration and the recovery of discontinuities. IEEE Trans. Pattern Anal. Mach. Intell. 14, 367–383 (1992)
11. Nocedal, J., Wright, S.J.: Numerical Optimization. Springer Series in Operations Research. Springer, New York (1999)
12. Bruhn, A., Weickert, J., Kohlberger, T., Schnorr, C.: A multigrid platform for real-time motion computation with discontinuity-preserving variational methods. International Journal of Computer Vision 70, 257–277 (2006)
13. Jordan, M.I., Ghahramani, Z., Jaakkola, T.S., Saul, L.: An introduction to variational methods for graphical models. Machine Learning 37, 183–233 (1999)
14. Sun, D., Roth, S., Lewis, J.P., Black, M.J.: Learning optical flow. In: Forsyth, D., Torr, P., Zisserman, A. (eds.) ECCV 2008, Part III. LNCS, vol. 5304, pp. 83–97. Springer, Heidelberg (2008)

15. Noels, N., Steendam, H., Moeneclaey, M.: The true cramer-rao bound for phase-independent carrier frequency estimation from a psk signal. In: IEEE Global Telecommunications Conference 2002, pp. 1137–1141 (2002)
16. Yuan, J., Schnoerr, C., Memin, E.: Discrete orthogonal decomposition and variational fluid flow estimation. Journ. of Math. Imaging & Vison 28, 67–80 (2007)
17. Baker, S., Scharstein, D., Lewis, J., Roth, S., Black, M., Szeliski, R.: A database and evaluation methodology for optical flow. In: Int. Conf. on Comp. Vis., ICCV 2007 (2007)
18. Carlier, J., Wieneke, B.: Report 1 on production and diffusion of fluid mechanics images and data. Fluid project deliverable 1.2 (2005), http://www.fluid.irisa.fr
19. Lucas, B., Kanade, T.: An iterative image registration technique with an application to stereovision. In: Int. Joint Conf. on Artificial Intel. (IJCAI), pp. 674–679 (1981)
20. Zitnick, C., Jojic, N., Kang, S.B.: Consistent segmentation for optical flow estimation. In: Tenth IEEE International Conference on Computer Vision (ICCV 2005), vol. 2, pp. 1308–1315 (2005)

Regularization of Positive Definite Matrix Fields Based on Multiplicative Calculus

Luc Florack

Department of Mathematics & Computer Science and Department of Biomedical
Engineering, Eindhoven University of Technology, The Netherlands
L.M.J.Florack@tue.nl

Abstract. Multiplicative calculus provides a natural framework in
problems involving positive images and positivity preserving operators.
In increasingly important, complex imaging frameworks, such as diffusion tensor imaging, it complements standard calculus in a nontrivial
way. The purpose of this article is to illustrate the basics of multiplicative calculus and its application to the regularization of positive definite
matrix fields.

1 Introduction

Images are typically positive-valued, as they capture some kind of signal energy
or attenuation. However, positivity is rarely adopted as an *a priori* axiom in
image analysis, probably due to the popularity of positivity violating operators
from standard differential calculus. However, already in 1887 Volterra introduced
the so-called *multiplicative calculus* [1], which appears to be the natural framework in the context of positive functions, and admits a positivity preserving
differential calculus. It has not received much attention in the image literature,
although it has been advocated in other application contexts, such as in survival
analysis and Markov processes, cf. Gill and Johansen [2].

In the context of non-commutative matrix algebras, a comprehensive account
does not seem to be available to the best of our knowledge. Yet precisely this
case is becoming increasingly relevant in image analysis. Positive matrix valued
functions and positivity preserving operators are for instance encountered in the
context of diffusion tensors [3,4,5,6], and deformation tensors [7].

We start by considering scalar functions [8,9,10,11,12] and subsequently turn
to matrix valued functions [13,14,15]. The latter are considerably more complicated as a result of the non-commutative nature of the matrix product. We will
give a concrete example in the context of a multiplicative multi-scale representation, or regularization, of a positive definite matrix-valued image.

2 Theory

Loosely speaking, the key to understand multiplicative calculus is a formal substitution, whereby one replaces addition and subtraction by multiplication and

A.M. Bruckstein et al. (Eds.): SSVM 2011, LNCS 6667, pp. 786–796, 2012.

division, respectively. As a corollary one is then led to replace multiplication in standard calculus by exponentiation in the multiplicative case, and (thus) division by exponentiation with the reciprocal exponent.

Until stated otherwise, we will assume commutative multiplication, so that no ambiguity arises with respect to the ordering of product factors or the meaning of division signs. We will consider appropriately chosen, positive functions of a single variable ($n = 1$), furnished with the usual rules $(f\,g)(x) = f(x)\,g(x)$, $(f^{\lambda})(x) = (f(x))^{\lambda}$ for $f, g \in V$, $\lambda \in \mathbb{R}$, $x \in \mathbb{R}^n$, et cetera. We subsequently address the non-commutative case of positive definite matrix fields.

2.1   Multiplicative Differentiation

Quotes ($'$) and asterisks ($*$) will be used to denote standard and multiplicative differentiation, respectively. The multiplicative derivative is defined as follows:

$$f^*(x) = \lim_{h \to 0} \left(\frac{f(x + h)}{f(x)} \right)^{1/h}. \tag{1}$$

Clearly $f^* : \mathbb{R} \to \mathbb{R}^+$ is positive definite iff $f : \mathbb{R} \to \mathbb{R}^+$ is positive definite, and

$$\ln f^*(x) = (\ln f)'(x), \tag{2}$$

whence, more generally, using self-explanatory notation for k-fold differentiation,

$$\ln f^{*\,(k)}(x) = (\ln f)^{(k)}(x), \tag{3}$$

cf. Fig. 1. Eq. (3) tells us that if a (positive) function is differentiable to some order in standard sense, it is also so in multiplicative sense, vice versa.

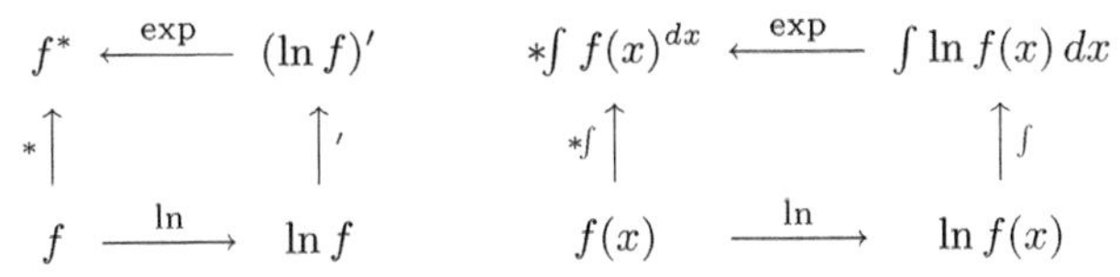

$$
\begin{array}{ccc}
f^* & \xleftarrow{\ \exp\ } & (\ln f)' \\
{\scriptstyle *}\uparrow & & \uparrow{\scriptstyle '} \\
f & \xrightarrow{\ \ln\ } & \ln f
\end{array}
\qquad\qquad
\begin{array}{ccc}
*\!\int f(x)^{dx} & \xleftarrow{\ \exp\ } & \int \ln f(x)\,dx \\
{\scriptstyle *\!\int}\uparrow & & \uparrow{\scriptstyle \int} \\
f(x) & \xrightarrow{\ \ln\ } & \ln f(x)
\end{array}
$$

Fig. 1. Left: Commuting diagram for multiplicative and standard differentiation: $f^*(x) = \exp\left((\ln f)'(x)\right)$. Right: Commuting diagram for multiplicative and standard antiderivation: $*\!\int f(x)^{dx} = \exp\left(\int \ln f(x)\,dx\right)$.

It is clear that multiplicative calculus—*in the commutative case*—is merely a disguise of standard calculus via the commuting diagram, Fig. 1. Still, it may simplify analysis in some cases, which is an advantage by itself. More importantly, however—and this is our main motivation here—its generalization to the non-commutative case has no obvious standard counterpart.

2.2   Multiplicative Integration

Antiderivatives are introduced in multiplicative calculus as follows:

$$*\!\!\int f(x)^{dt} = c\,F(x) \text{ for some constant } c \in \mathbb{R}^+ \quad \text{iff} \quad F^* = f\,. \tag{4}$$

Note that we denote the measure dt as a formal ("infinitesimal") exponent.

Multiplicative definite integrals can be introduced via a limiting procedure akin to the Riemann sum approximation of their standard counterparts:

$$*\!\!\int_a^b f(x)^{dx} = \lim_{\triangle x_i \to 0} \prod_{i=1}^N f(\xi_i)^{\triangle x_i} \quad \text{with } \xi_i \in [x_{i-1}, x_i] \text{ and } x_0 = a,\ x_N = b, \tag{5}$$

in which $\triangle x_i = x_i - x_{i-1}$. The relationship between Eqs. (4) and (5) is formalized by the following fundamental theorem of multiplicative calculus:

$$*\!\!\int_a^b F^*(x)^{dx} = \frac{F(b)}{F(a)}\,. \tag{6}$$

Again, by virtue of commutativity of multiplication there exists a one-to-one mapping between standard and multiplicative antiderivatives, cf. Fig. 1. In the non-commutative case this is no longer self-evident, as we will see in Section 2.7.

2.3   Linear Functions and Linear Mappings

Linear functions can be defined as those functions that have a constant multiplicative derivative (i.e. allowing a constant offset):

$$f^*(x) = a \quad \text{or, equivalently,} \quad f(x) = b\,a^x \quad \text{with } a, b \in \mathbb{R}^+. \tag{7}$$

Thus exponential functions are the linear functions of multiplicative calculus. In general, $*$-linear mappings $A : V \to W$ are defined without offset (b parameter in Eq. (7)), as follows. If $u, v \in V$, $\lambda, \mu \in \mathbb{R}$, then

$$A(u^\lambda v^\mu) = A(u)^\lambda A(v)^\mu\,. \tag{8}$$

Multiplicative derivation and antiderivation are important examples. We have

$$(f^\lambda g^\mu)^* = (f^*)^\lambda (g^*)^\mu \quad \text{respectively} \quad *\!\!\int (f^\lambda g^\mu)^{dx} = \Big(*\!\!\int f^{dx}\Big)^\lambda \Big(*\!\!\int g^{dx}\Big)^\mu. \tag{9}$$

2.4   Taylor Expansions

Analogous to the standard Taylor expansion of an analytic function we have

$$f(x) = \left(\prod_{k=0}^M \big(f^{(*k)}(a)\big)^{\frac{1}{k!}(x-a)^k} \right) \big(f^{(*(M+1))}(\xi)\big)^{\frac{1}{(M+1)!}(x-a)^{M+1}}, \tag{10}$$

for some ξ in-between x and a. The rightmost factor is the multiplicative Lagrange remainder. In particular this leads to the linear approximation

$$f(x) \approx f(a) \, f^*(a)^{x-a} \,. \tag{11}$$

Multiplicative approximations have the advantage of preserving positivity. They are valid up to a multiplicative factor close to unity. As an illustration, consider the sigmoidal function:

$$f(x) = \frac{1}{1 + e^{-x}} \,. \tag{12}$$

Its standard ("s") and multiplicative ("m") 1st order Taylor approximations are

$$f(x) \approx f_s(x) = \frac{1}{2} + \frac{1}{4}x \quad \text{resp.} \quad f(x) \approx f_m(x) = \frac{1}{2}\exp\left(\frac{1}{2}x\right) \,. \tag{13}$$

The former is seen to violate positivity as soon as $x \leq -2$.

As a second example, consider the standard Gaussian function,

$$f(x) = \frac{1}{\sqrt{2\pi}} \exp\left(-\frac{1}{2}x^2\right) \,. \tag{14}$$

Its 2nd order Taylor approximations are given by

$$f(x) \approx f_s(x) = \frac{1}{\sqrt{2\pi}} - \frac{1}{2\sqrt{2\pi}}x^2 \quad \text{resp.} \quad f(x) \approx f_m(x) = \frac{1}{\sqrt{2\pi}}\exp\left(-\frac{1}{2}x^2\right) \,. \tag{15}$$

The latter is in fact exact. In addition to preserving positivity, multiplicative expansions typically provide better approximations for compactly supported positive filters, cf. Fig. 2.

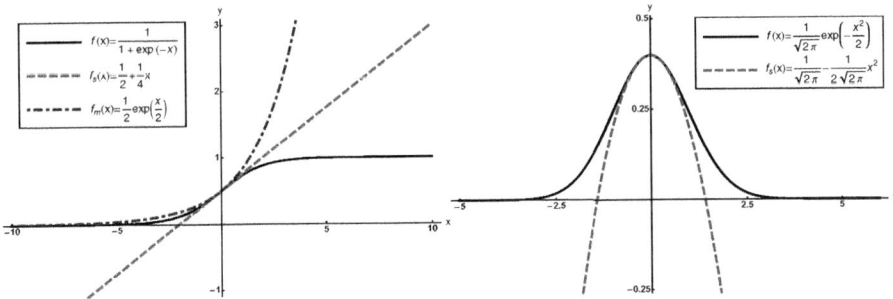

Fig. 2. Left: Sigmoidal function and its 1st order standard and multiplicative Taylor expansions, recall Eq. (13). Right: Gaussian function and its 2nd order standard Taylor expansion, recall Eq. (15). The second order multiplicative Taylor expansion is exact and thus coincides with the original Gaussian function.

2.5   Critical Points

The following claims are easily verified. If $f^*(x) > 1$ then f is strictly increasing at $x \in \mathbb{R}$. If $f^*(x) < 1$ then f is strictly decreasing at $x \in \mathbb{R}$. If $f^*(x) = 1$ then f has a critical point at $x \in \mathbb{R}$, viz. a local minimum if $f^{**}(x) > 1$, a local maximum if $f^{**}(x) < 1$, and an indifferent or degenerate critical point if $f^{**}(x) = 1$. These observations may provide the foundations for a multiplicative variational calculus for multiplicative energy functionals for image optimization problems, and are easily generalized to the multivariate setting.

2.6   Differential Equations

An important class of image processing techniques employs differential equations (ODEs or PDEs). As an illustration consider the following multiplicative initial value problem for $x \in \mathbb{R}$:

$$u^{**} = A \quad \text{with initial conditions } u^*(0) = B \text{ and } u(0) = C, \qquad (16)$$

with $A, B, C > 0$ given constants. A straightforward computation, using Eq. (3), yields the following unique solution (the multiplicative counterpart of a parabola):

$$u(x) = \exp\left(\frac{1}{2}ax^2 + bx + c\right), \qquad (17)$$

in which $a = \ln A, b = \ln B, c = \ln C$. Qualitative behaviour is governed by the convexity parameter A, with $0 < A < 1$ producing bounded (Gaussian) solutions, $A = 1$ unilaterally unbounded exponential solutions, and $A > 1$ bilaterally unbounded solutions. In particular this explains the coincidence of a Gaussian function and it 2nd order multiplicative Taylor expansion.

As another example, consider the multiplicative heat equation for $(x, t) \in \mathbb{R}^n \times \mathbb{R}^+$:

$$u_t^* = \Delta^* u \quad \text{with initial condition } u(x, 0) = f(x), \qquad (18)$$

in which we use multiplicative derivation with respect to both the evolution parameter $t \in \mathbb{R}^+$, i.e. $u_t^* = \partial_t^* u$, as well as with respect to the (Cartesian) coordinates $x \in \mathbb{R}^n$. The multiplicative ($*$-linear) Laplacian is defined here as

$$\Delta^* = \exp \circ \Delta \circ \ln, \qquad (19)$$

cf. Eq. (3). *Only* in the commutative case this implies (v.i.)

$$\Delta^* u = \partial_{x^1 x^1}^* u \ldots \partial_{x^n x^n}^* u. \qquad (20)$$

The solution is straightforward, since in the logarithmic domain the problem reduces to the standard heat equation for $\ln u$ with $\ln f$ as initial condition:

$$u(x, t) = \exp\left((\phi_t * \ln f)(x)\right), \qquad (21)$$

in which

$$\phi_t(x) = \frac{1}{\sqrt{4\pi t}^n} \exp\left(-\frac{\|x\|^2}{4t}\right) .$$ (22)

Any $*$-linear combination of multiplicative derivatives of u solves Eq. (18).

Eq. (18) is a special case of a pseudo-linear scale space [16,17]. Also, the so-called log-Euclidean scale space for diffusion tensor images [3,4,5,6] is governed by a multiplicative system similar to Eq. (18), in which case u and f are positive definite *matrix fields*, and exp and ln the extensions applicable to such matrices [5,14]. However, in this non-commutative case equivalence of Eqs. (19) and (20) does *not* hold, a consequence of the Campbell-Baker-Hausdorff formula:

$$\ln(\exp X \exp Y) = X + Y + \text{commutator terms involving } [X,Y].$$ (23)

2.7   The Non-commutative Case

As anticipated by Eq. (23), extension to non-commutative multiplication is non-trivial, yet highly relevant in modern image analysis practice. For instance, we must account for non-commutative multiplication when handling (positive definite) matrix valued functions, such as diffusion tensor images or strain tensor images. This case has received remarkably little attention. A few results have been provided by Gantmacher [13] and Slavík [15]. Gantmacher's definition differs from ours (Slavík discusses further alternatives). Consistency with our notation for the scalar case suggests the following definition:

$$\boldsymbol{X}^*(x) = \exp\left(\boldsymbol{X}'(x)\,\boldsymbol{X}^{-1}(x)\right) .$$ (24)

One must remain cautious, for $\boldsymbol{X}'\,\boldsymbol{X}^{-1} = \ln'\boldsymbol{X} = \boldsymbol{X}^{-1}\,\boldsymbol{X}'$ generally holds *only* in the commutative case, such as the scalar case ($m = 1$), or the special case whereby $\boldsymbol{X}(x) \propto x$ is linear in standard sense, recall Eq. (23). In other words, Eq. (2), or Fig. 1, neither holds for Eq. (24) nor for its mirror form (with $\boldsymbol{X}'(x)$ and $\boldsymbol{X}^{-1}(x)$ interchanged). Given Eq. (24), Eq. (5) remains applicable, provided we arrange factors on the right hand side as follows:

$$*\int_a^b \boldsymbol{X}(x)^{dx} = \lim_{\triangle x_i \to 0} \boldsymbol{X}(\xi_N)^{\triangle x_N} \dots \boldsymbol{X}(\xi_1)^{\triangle x_1} \quad \text{with } \xi_i \in [x_{i-1}, x_i],$$ (25)

in which $x_0 = a$, $x_N = b$. Eq. (24) entails a definite *choice* with respect to the ordering of the factors $\boldsymbol{X}'$ and $\boldsymbol{X}^{-1}$ in the multiplicative derivative, which affects the corresponding form of the antiderivative, Eq. (25). Thus we have at least three distinct ways to introduce multiplicative differential and integral calculus in the context of matrix functions, viz. (i) Eq. (24) in combination with Eq. (25), (ii) the analogous scheme with reverse ordering of $\boldsymbol{X}'$ and $\boldsymbol{X}^{-1}$, respectively of the $*$-infinitesimal factors as they occur in the defining limiting procedure of the multiplicative integral, and (iii) the matrix equivalent of the ln/exp-formalism of Eq. (2).

As an illustration of the first option (i), Eqs. (24–25), consider the following matrix-valued 1st order ODE, in which L is a non-stationary matrix, cf. Gantmacher [13]:

$$\dot{F} = L\,F\,, \tag{26}$$

subject to an initial condition at $t = t_0$, say. The multiplicative nature of the evolution of $F(t, t_0)$ is apparent from Eq. (26). The simplicity of Eq. (26) is deceptive. The complication arises due to the fact that L is non-stationary, as a result of which $[L(s), L(t)] \neq 0$ for $s \neq t$, causing complications due to Eq. (23). In multiplicative form Eq. (26) simplifies to

$$F^* = \exp(L) \quad \text{with} \quad F(t{=}t_0, t_0) = I\,, \tag{27}$$

immediately yielding the solution via antiderivation, Eq. (25), with $X = \exp(L)$, $a = t_0$ and $b = t$. Only in the stationary case $(L(t) = L_0)$ this solution reduces to $F(t, t_0) = \exp((t - t_0)L_0)$. The reader is referred to the literature for a proof based on standard calculus [13], and for an application in cardiac strain tensor analysis [7].

The multiplicative integral suggests a straightforward numerical approximation, viz. by using the discrete form of Eq. (25) without limiting procedure.

2.8 Multiscale Representation of Positive Definite Matrix Fields

The log-Euclidean paradigm provides an example of a representation that takes positivity into account *a priori* [3,4,6]. Here we consider the paradigm in the context of multiscale representation, or regularization [5].

We denote a positive definite tensor image by $X : \mathbb{R}^n \to \mathbb{S}_n^+$, where $\mathbb{S}_n^+ \subset \mathbb{S}_n \subset \mathbb{M}_n$ denotes the set of $\mathbb{R}$-valued symmetric positive definite $n \times n$ matrices, $\mathbb{S}_n$ the set of $\mathbb{R}$-valued symmetric $n \times n$ matrices, and $\mathbb{M}_n$ the set of all $\mathbb{R}$-valued $n \times n$ matrices. Its pointwise inverse is $X^{\mathrm{inv}} : \mathbb{R}^n \to \mathbb{S}_n^+$, so that $(X^{\mathrm{inv}} X)(x) = (X X^{\mathrm{inv}})(x) = I$, at each point $x \in \mathbb{R}^n$. $C^\omega(\mathbb{R}^n, \mathbb{M}_n)$ denotes the class of analytical functions $X : \mathbb{R}^n \to \mathbb{M}_n$. Self-explanatory definitions hold for $C^\omega(\mathbb{R}^n, \mathbb{S}_n^+) \subset C^\omega(\mathbb{R}^n, \mathbb{S}_n) \subset C^\omega(\mathbb{R}^n, \mathbb{M}_n)$.

The scale space representation of $X \in C^\omega(\mathbb{R}^n, \mathbb{S}_n^+)$ is generated by the regularization operator (detailed below)

$$\mathscr{F} : C^\omega(\mathbb{R}^n, \mathbb{S}_n^+) \times \mathbb{R}^+ \to C^\omega(\mathbb{R}^n, \mathbb{S}_n^+) : (X, t) \mapsto \mathscr{F}(X, t)\,, \tag{28}$$

with $\mathscr{F}(X, 0) = X$ for all $X \in C^\omega(\mathbb{R}^n, \mathbb{S}_n^+)$. We use the shorthand notation $X_t \equiv \mathscr{F}(X, t)$. The isotropic Gaussian scale space kernel in n dimensions is given by Eq. (22). Elsewhere it has been argued that the requirement that regularization and inversion should commute,

$$\mathscr{F}(X, t)^{\mathrm{inv}} = \mathscr{F}(X^{\mathrm{inv}}, t)\,, \tag{29}$$

naturally produces the log-Euclidean paradigm [5].

Recall that the matrix exponential map maps a general matrix to a nonsingular matrix [13,18,19]. For our purpose it suffices to consider elements of $\mathbb{S}_n \subset \mathbb{M}_n$, which are diagonalizable with real eigenvalues, in which case the range of the exponential map equals $\exp(\mathbb{S}_n) = \mathbb{S}_n^+$:

$$\exp : \mathbb{S}_n \to \mathbb{S}_n^+ : \boldsymbol{A} \mapsto \exp \boldsymbol{A} . \tag{30}$$

The logarithmic map, restricted to $\mathbb{S}_n^+$, has prototype

$$\ln : \mathbb{S}_n^+ \to \mathbb{S}_n : \boldsymbol{B} \mapsto \ln \boldsymbol{B} . \tag{31}$$

It is the unique inverse of the exponential map on $\mathbb{S}_n$, with $\ln(\mathbb{S}_n^+) = \mathbb{S}_n$.

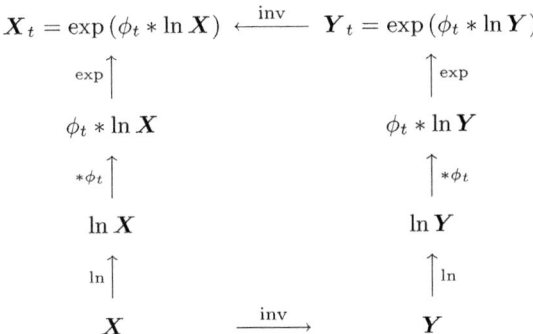

Fig. 3. Commuting diagram for blurring and inversion

Fig. 3 shows the multiscale representation consistent with Eq. (29). Indeed, if $\boldsymbol{X} \in C^\omega(\mathbb{R}^n, \mathbb{S}_n^+)$, then $\boldsymbol{X}_t = \mathscr{F}(\boldsymbol{X}, t)$ constructed according to

$$\mathscr{F}(\boldsymbol{X}, t) = \exp(\phi_t * \ln \boldsymbol{X}) , \tag{32}$$

satisfies the desired commutativity property, Eq. (29). This follows immediately by inspection of Fig. 3 and Eq. (32), using the identities

$$\exp(-\boldsymbol{A}) = (\exp \boldsymbol{A})^{\mathrm{inv}} \quad \text{and} \quad \ln \boldsymbol{B}^{\mathrm{inv}} = -\ln \boldsymbol{B} , \tag{33}$$

for $\boldsymbol{A} \in \mathbb{S}_n$, $\boldsymbol{B} \in \mathbb{S}_n^+$. For an application to DTI, cf. Florack and Astola [5].

Formulae for standard derivatives of Eq. (32) are highly nontrivial, cf. the explicit computations to 1st and 2nd order by Florack and Astola [5]. The log-Euclidean paradigm, however, suggests the following way to introduce multiplicative derivation for the non-commutative case, recall the three options discussed in Section 2.7:

$$\boldsymbol{X}^* \stackrel{\text{def}}{=} \exp(\ln' \boldsymbol{X}) , \tag{34}$$

for the one-dimensional case. This is similar to Eq. (2) for the scalar case, but recall that in Eq. (34) exp and ln are the matrix exponential and logarithm,

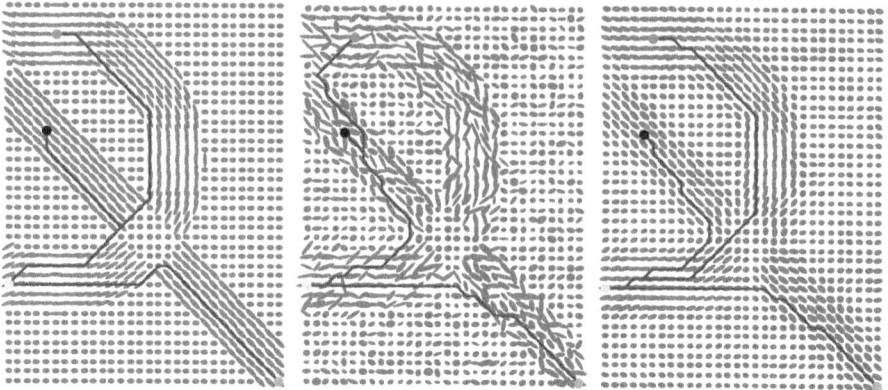

Fig. 4. Two-dimensional synthetic images illustrating a positive symmetric tensor field in terms of ellipsoidal glyphs (principal axes and radii reflect eigendirections and corresponding eigenvalues). Overlayed are some fixed end-point geodesics obtained by applying Dijkstra's shortest path algorithm, in which the tensor field itself is interpreted as the dual Riemannian metric. This complies with the Riemannian rationale for geodesic tractography in diffusion tensor imaging [20,21]. The left image shows the result for the originally synthesized, smooth image. The middle image shows the result of the same algorithm after the image has been perturbed by pixel-uncorrelated noise. The lack of robustness of geodesic tractography is a consequence of the ill-posedness of differentiation. The right image demonstrates the regularizing effect of log-Euclidean regularization, Eq. (32), and its effect on the performance of the algorithm.

respectively. For the multivariate case this leads to the following operational-ization of a multiplicative k^{th} order partial derivative of $\boldsymbol{X}_t = \mathscr{F}(\boldsymbol{X}, t)$, recall Eq. (32):

$$\partial^*_{i_1 \ldots i_k} \boldsymbol{X}_t \stackrel{\text{def}}{=} \exp\left(\partial_{i_1 \ldots i_k} \phi_t * \ln \boldsymbol{X}\right) . \tag{35}$$

This is consistent with the scale space paradigm given by Eqs. (18–19) and Eqs. (21–22), in which diffusion and Gaussian convolution are now applied component-wise to matrix entries via the ln/exp detour. This definition of mul-tiplicative derivation thus seems to fit most naturally with the log-Euclidean paradigm [3,4,5,6]. Adhering to this definition, log-Euclidean blurring can thus be seen as the multiplicative counterpart of a standard diffusion process, i.e. the counterpart of Eqs. (18–19) for positive symmetric matrix-valued functions. See Fig. 4 for an example of multiplicative diffusion for regularizing geodesic tractography.

3   Conclusion and Discussion

Multiplicative calculus and its applications to image analysis raises many im-portant questions not addressed in this short paper. One question pertains to the extension of standard variational techniques for image optimization prob-lems to the multiplicative case. How to set up such a framework rigorously? In image analysis such a framework would have the intrinsic advantage that positivity of solutions would be guaranteed *a priori*. Additional questions arise in the context of (non-commutative) matrix fields. Which of the three proposed options for multiplicative differential calculus (if any) is the most natural one in a given application context, what are their mutual relations, how do they relate to standard differential calculus, and, in the log-Euclidean case, what does the corresponding antiderivative look like?

Despite major open questions it has been argued that multiplicative calculus provides a natural framework for biomedical image analysis, particularly in problems in which positive images or positive definite matrix fields and positivity preserving operators are of interest. We therefore believe that this subject is of broad interest. However, it seems that many fundamental problems have not been addressed in the mathematical literature sofar, especially regarding the non-commutative case. This is an impediment for progress in biomedical image analysis.

Examples have been given in the context of regularization of positive definite matrix fields to illustrate the relevance of multiplicative calculus in image analysis, and to support the recommendation for further investigation into both practical as well as fundamental issues.

Acknowledgments. Shufang Liu has conducted a useful literature survey. Laura Astola has generated the synthetical data and tractography results of Fig. 4.

References

1. Volterra, V.: Sulle equazioni differenziali lineari. Rendiconti Academia dei Lincei (series 4)3, 393–396 (1887)
2. Gill, R.D., Johansen, S.: A survey of product-integration with a view toward application in survival analysis. Annals of Statistics 18, 1501–1555 (1990)
3. Arsigny, V., Fillard, P., Pennec, X., Ayache, N.: Log-Euclidean metrics for fast and simple calculus on diffusion tensors. Magnetic Resonance in Medicine 56(2), 411–421 (2006)
4. Fillard, P., Pennec, X., Arsigny, V., Ayache, N.: Clinical DT-MRI estimation, smoothing, and fiber tracking with log-Euclidean metrics. IEEE Transactions on Medical Imaging 26(11) (November 2007)
5. Florack, L.M.J., Astola, L.J.: A multi-resolution framework for diffusion tensor images. In: Aja Fernández, S., de Luis Garcia, R. (eds.) CVPR Workshop on Tensors in Image Processing and Computer Vision, Anchorage, Alaska, USA, June 24-26. IEEE, Los Alamitos (2008); Digital proceedings
6. Pennec, X., Fillard, P., Ayache, N.: A Riemannian framework for tensor computing. International Journal of Computer Vision 66(1), 41–66 (2006)
7. Florack, L., van Assen, H.: A new methodology for multiscale myocardial deformation and strain analysis based on tagging MRI. International Journal of Biomedical Imaging (2010), Published online: doi:10.1155/2010/341242
8. Bashirov, A.E., Kurpinar, E.M., Özyapici, A.: Multiplicative calculus and its applications. Journal of Mathematical Analysis and Applications 337, 36–48 (2008)
9. Grossman, M., Katz, R.: Non-Newtonian Calculus. Lee Press, Pigeon Cove (1972)

10. Guenther, R.A.: Product integrals and sum integrals. International Journal of Mathematical Education in Science and Technology 14(2), 243–249 (1983)
11. Rybaczuk, M., Kędzia, A., Zieliński, W.: The concept of physical and fractal dimension II. the differential calculus in dimensional spaces. Chaos, Solitons and Fractals 12, 2537–2552 (2001)
12. Stanley, D.: A multiplicative calculus. PRIMUS: Problems, Resources, and Issues in Mathematics Undergraduate Studies IX(4), 310–326 (1999)
13. Gantmacher, F.R.: The Theory of Matrices. American Mathematical Society (2001)
14. Higham, N.J.: Functions of Matrices: Theory and Computation. SIAM (2008)
15. Slavík, A.: Product Integration, its History and Applications. Matfyzpress, Prague (2007)
16. Florack, L.M.J., Maas, R., Niessen, W.J.: Pseudo-linear scale-space theory. International Journal of Computer Vision 31(2-3), 247–259 (1999)
17. Welk, M.: Families of generalised morphological scale spaces. In: Griffin, L.D., Lillholm, M. (eds.) Scale-Space 2003. LNCS, vol. 2695, pp. 770–784. Springer, Heidelberg (2003)
18. Fung, T.C.: Computation of the matrix exponential and its derivatives by scaling and squaring. International Journal for Numerical Methods in Engineering 59, 1273–1286 (2004)
19. Moler, C., Van Loan, C.: Nineteen dubious ways to compute the exponential of a matrix. SIAM Review 20(4), 801–836 (1978)
20. Lenglet, C., Deriche, R., Faugeras, O.: Inferring white matter geometry from diffusion tensor MRI: Application to connectivity mapping. In: Pajdla, T., Matas, J(G.) (eds.) ECCV 2004. LNCS, vol. 3024, pp. 127–140. Springer, Heidelberg (2004)
21. Prados, E., Soatto, S., Lenglet, C., Pons, J.P., Wotawa, N., Deriche, R., Faugeras, O.: Control theory and fast marching techniques for brain connectivity mapping. In: Proceedings of the IEEE Computer Society Conference on Computer Vision and Pattern Recognition, New York, USA, vol. 1, pp. 1076–1083. IEEE Computer Society Press (2006)

Author Index

Adiv, Gilad 761
Aflalo, Yonathan 471, 689
Ahipo, Y. 386
Auroux, D. 386

Bae, Egil 279
Bajaj, Chandrajit 62
Bar, Leah 183
Batard, Thomas 483
Bauer, Sebastian 98
Becker, Florian 206
Benetazzo, Alvise 520
Berkels, Benjamin 98, 326
Bernot, Marc 435
Bertaccini, Daniele 194
Beyou, Sébastien 749
Bischof, Horst 314
Bitton, David 761
Boykov, Yuri 122, 279
Breuß, Michael 532
Bronstein, Alexander M. 580, 592, 616, 665, 689, 725
Bronstein, Michael M. 580, 592, 616, 665, 689, 725
Bruckstein, Alfred M. 761
Bruhn, Andrés 447, 713

Cai, Xiaohao 411
Chan, Raymond H. 86, 194, 411
Chung, Ginmo J. 169
Cohen, Laurent D. 255, 362, 386
Corpetti, Thomas 701, 749
Creusen, Eric J. 1, 14

Dela Haije, Tom C.J. 1, 14
Deledalle, Charles-Alban 231
Delon, Julie 435
Demetz, Oliver 713
Dérian, Pierre 737
Dimitrov, Pavel 653
Doerr, Benjamin 26
Duan, Yuping 144
Dubrovina, Anastasia 604
Duits, Remco 1, 14
Duval, Vincent 231

Fedele, Francesco 520
Feigin, Micha 459
Feldman, Dan 459
Feuer, Arie 761
Florack, Luc 786
Frick, Klaus 74

Gallego, Guillermo 520
Genctav, Murat 267
Ghosh, Arpan 1
Goesele, Michael 62
Gorthi, Sai 749
Gräf, Manuel 568
Grewenig, Sven 447
Gwosdek, Pascal 447, 544

Hagenburg, Kai 532
Hahn, Jooyoung 144, 169
Héas, Patrick 737, 773
Herzet, Cédric 737, 773
Hochbaum, Dorit S. 338
Hoffmann, Sebastian 26
Hong, Byung-Woo 243
Hooda, Amit 665
Horaud, Radu P. 665
Hornegger, Joachim 98
Huan, Zhongdan 218
Huang, Haiyang 218

Imiya, Atsushi 302
Itoh, Hayato 302

Johannsen, Daniel 26
Jung, Miyoun 255

Kalbe, Thomas 62
Kimmel, Ron 134, 471, 604, 616, 689, 725, 761
Klapp, Iftach 157
Kotowski, Marc 326
Kovnatsky, Artiom 616
Krim, Hamid 556, 628
Kuijper, Arjan 62

Lang, Annika 50
Lauze, François 677

Lawlor, Matthew   653
Lee, Deokwoo   556
Lellmann, Jan   206
Lenzen, Frank   206
Li, Yan   350
Lindenbaum, Michael   398
Liu, Jun   218
Loog, Marco   350
Luxenburger, Andreas   544

Mainberger, Markus   26
Marnitz, Philipp   74
Masmoudi, M.   386
Mémin, Étienne   701, 737, 749, 773
Mendlovic, David   157
Mikula, Karol   640
Morigi, Serena   38, 194, 411

Neumann, Frank   26
Ni, Kangyu   243
Nielsen, Mads   677
Nikolova, Mila   86, 110, 508
Nir, Tal   761
Nishiguchi, Haruhiko   302
Norris, Larry K.   628

Paragios, Nikos   580
Peles, David   398
Pennec, Xavier   677
Petra, Stefania   206
Peyré, Gabriel   255, 435
Pock, Thomas   314
Pokrass, Jonathan   592
Potts, Daniel   568

Rabin, Julien   435
Raviv, Dan   604
Rosman, Guy   725, 761
Rouchdy, Youssef   362
Rucci, Marco   38
Rumpf, Martin   98, 326

Sakai, Tomoya   302
Salmon, Joseph   231
Schaller, Carlo   326

Schmitzer, Bernhard   423
Schnörr, Christoph   206, 423
Schwarzkopf, Andreas   62
Sgallari, Fiorella   38, 194, 411
Shem-Tov, Shachar   761
Shi, Juan   122, 495
Shi, Yu-Ying   291
Soatto, Stefano   243
Sochen, Nir   157, 183, 459, 483
Sommer, Stefan   677
Steidl, Gabriele   568

Tai, Xue-Cheng   122, 144, 169, 218, 279,
    291, 495
Tang, Ching Hoo   26
Tari, Sibel   267
Tax, David M.J.   350
ter Haar Romeny, Bart   1
Teuber, Tanja   50

Unger, Markus   314
Urbán, Jozef   640

Vilanova, Anna   1
Vogel, Oliver   532

Wan, Min   495
Wang, Chaohui   580
Wang, Desheng   495
Wang, Li-Lian   291
Wang, Yu   144, 169
Weickert, Joachim   26, 447, 532, 544, 713
Welk, Martin   374
Wen, You-Wei   86
Werlberger, Manuel   314
Wetzler, Aaron   134
Wolf, Alon   725

Yezzi, Anthony   520
Yi, Sheng   628
Yuan, Jing   122, 279

Zimmer, Henning   544, 713
Zucker, Steven W.   653